ANNALS OF THE NEW YORK ACADEMY OF SCIENCES

Volume 394

The New York Academy of Sciences
2 East 63rd Street
New York, New York 10021

THE BRATTLEBORO RAT

ANNALS OF THE NEW YORK ACADEMY OF SCIENCES

VOLUME 394

THE BRATTLEBORO RAT

Edited by Hilda Weyl Sokol and Heinz Valtin

The New York Academy of Sciences
New York, New York
1982

Library of Congress Cataloging in Publication Data

Main entry under title:

The Brattleboro rat.

(Annals of the New York Academy of Sciences; v. 394)

Proceedings of the International Symposium on the Brattleboro Rat, held at Dartmouth Medical School, Sept. 4–7, 1981.

Bibliography: p.

Includes index.

1. Brattleboro rat—Congresses. I. Sokol, Hilda Weyl. II. Valtin, Heinz. III. International Symposium on the Brattleboro Rat (1981 : Dartmouth Medical School) IV. Series. [DNLM: 1. Diabetes insipidus—Congresses. 2. Hypertension, Renal—Congresses. 3. Vasopressins—Congresses. 4. Pituitary hormones, Posterior—Congresses. 5. Rats, Inbred strains—Congresses. W1 AN626YL v. 394 / WK 520 B824 1981]

Q11.N5 vol. 394 [QL737.R666] 500s [599.32'33] 82–18797
ISBN 0–89766–178–8
ISBN 0–89766–179–6 (pbk.)

Subject index prepared by Albert P. Rouslin.

Cover:
Line drawing from photograph by Lewis B. Kinter and Lewis Lainey.

PCP
Printed in the United States of America
ISBN 0–89766–178–8 (cloth)
ISBN 0–89766–179–6 (paper)

ANNALS OF THE NEW YORK ACADEMY OF SCIENCES

VOLUME 394

October 30, 1982

THE BRATTLEBORO RAT *

Editors and Conference Organizers
HILDA WEYL SOKOL AND HEINZ VALTIN

CONTENTS

* This volume is the result of a symposium entitled International Symposium on The Brattleboro Rat, held on September 4–7, 1981, at Dartmouth Medical School in Hanover, New Hampshire.

Short Research Reports

Part III. Electrolyte Balance

Short Research Reports

Part IV. Hemodynamic Mechanisms

Short Research Reports

Part V. Water Balance

Short Research Reports

Part VI. Endocrinopathies Other Than Diabetes Insipidus

Short Research Reports

Part VII. Disorders of Behavior and the Central Nervous System

Short Research Reports

Financial assistance was received from:

- NATIONAL INSTITUTE OF ARTHRITIS, DIABETES, AND DIGESTIVE AND KIDNEY DISEASES
- NATIONAL SCIENCE FOUNDATION
- MARCH OF DIMES
- BLUE SPRUCE FARMS, INC.
- FERRING PHARMACEUTICALS
- BRISTOL-MYERS COMPANY
- SMITH KLINE AND FRENCH LABORATORIES
- BRATTLEBORO RETREAT
- ABBOTT LABORATORIES
- ALZA CORPORATION
- AMERICAN CYANAMID COMPANY
- ARTHUR D. LITTLE FOUNDATION
- BURROUGHS WELLCOME COMPANY
- KROC FOUNDATION
- MERCK SHARP AND DOHME RESEARCH LABORATORIES
- SQUIBB INSTITUTE FOR MEDICAL RESEARCH
- WARNER-LAMBERT COMPANY, INC.

Dedicated to

JOSEF RUDINGER

Emanuel Escher

JOSEF RUDINGER
1924–1975

DEDICATION TO JOSEF RUDINGER

Irving L. Schwartz

Department of Physiology and Biophysics
Mount Sinai Medical School
New York, New York 10029

Introduced by Heinz Valtin

There are several reasons why we chose to dedicate this symposium to Josef Rudinger. He was a chemist who was both interested in and understood biology. He loved to come to Dartmouth, to our group, and to our homes. On his many sojourns, all of us were repeatedly impressed by the breadth of his knowledge and by his ability to track our arguments whether they were physiological, pharmacological, biochemical, or clinical. He was a catalyst par excellence, drawing together diverse interests—as we see them at this symposium—in order to get to the heart of a scientific question. He was an internationalist, lecturing all over the world, and more often than not, lecturing in the language of his hosts. Above all he was a thoughtful, decent, and kind human being. Simply said, it is because we greatly respected and loved him that we wanted to dedicate this symposium to Joe Rudinger.

We have asked a close friend and scientific collaborator of Josef Rudinger, Irving Schwartz, to deliver the dedication. In 1960 Dr. Schwartz and his colleagues offered an eminently reasonable hypothesis for the antidiuretic action of the neurohypophyseal hormones. They proposed that this action might depend on an interchange between the disulfide bridge of the hormone and thiol groups on the receptor. Although this hypothesis was supported by several diverse lines of evidence, critical experiments (which depended on a state-of-the-art level of peptide synthesis) were lacking. Dr. Schwartz (then at the University of Cincinnati) and Dr. Rudinger (then at the Czechoslovak Academy of Science) had set about independently to carry out these critical experiments by synthesizing and evaluating the biological activities of analogues of the neurohypophyseal hormones, which lacked the disulfide bridge: both expected to prove the hypothesis to be correct. Rudinger's and Schwartz's synthetic and biological studies were completed at approximately the same time, in late 1963. Contrary to their expectations, all of their analogues had antidiuretic activity, which meant that the thiol-disulfide hypothesis was disproved; the observed interchange could not be responsible for the intrinsic activity of the hormone, although it might still serve to facilitate its action. With characteristic generosity Joe Rudinger suggested that he and Irving Schwartz pool their new data and publish them together, with Irving Schwartz as senior author. They did so in a report to the U.S. National Academy of Sciences in 1964. To me, this joint communication stands as an example of mature science and of fruitful collaboration among investigators to whom the truth matters more than the credit.

H. V.

I am greatly indebted to Dr. Valtin for giving me the privilege of participating in the dedication of this meeting to Josef Rudinger. Josef Rudinger was a chemist by formal training but in addition, by self education, he was a biologist, a linguist, a philosopher, and perhaps most important, a wide-ranging statesman of science, as Dr. Valtin has already indicated to you. For those of you who are familiar with Joe's contributions, and for those of you who knew him personally, there will be no difficulty in understanding why we are singling him out for remembrance at the outset of this symposium. However, for those of you who have not encountered his work or his personality, I would like to tell you a few things about him, things which in my view render this dedication so very appropriate.

Dating as far back as the latter half of his career as a graduate student, Joe's major scientific interest centered on the neurohypophyseal peptides: their structures, primary, secondary, and tertiary; their structure-activity relationships; their synthesis in the chemical laboratory; their biosynthesis; their release from the neurohypophysis; their "carriage" in the blood stream; their linkage to target tissue; the chain of events in target tissue which follows hormone-receptor interaction leading to ultimate physiological and pharmacological effector events; and finally, the enzymic and other mechanisms by which their hormonal actions are modulated or terminated. In other words, his central scientific interest embraced the biology of neurohypophyseal hormones in the broadest sense; and indeed the breadth of this interest and the range of his talent and motivation enabled him to move freely from the most sophisticated developments in peptide chemistry, to state-of-the-art levels in conformational analysis, to forefront advances in general pharmacology and receptor theory, and ultimately to the Brattleboro rat—its care, feeding, and experimental use.

He was perhaps best known—to physiologists, pharmacologists, and endocrinologists if not to chemists—for having been the first with Honzl and Zaoral to prepare synthetic peptide hormone (oxytocin) analogues, contemporaneously with Boissonnas in Switzerland, and Katsoyannis and du Vigneaud in the United States. He is also well known for having prepared a group of hormone analogues that he called hormonogens. These analogues, one of which is triglycyl lysine-vasopressin, are inactive depot forms of the parent hormones. For example, triglycyl lysine vasopressin (tGLVP) provides a relatively slow release of lysine vasopressin (LVP) so that this active peptide has a regulated delivery to its receptors. The risk of injecting high doses of LVP with potential side effects, such as hypertension and possibly associated strokes and paralysis or coronary artery occlusion, are well known. None of these catastrophic side effects has ever been seen after administration of triglycyl LVP. Therefore, thanks to Josef Rudinger, clinicians are able to take advantage of the therapeutically desirable activities of this hormonogen in a way that was never possible with the natural parent hormone.

Studies of the structure-activity relations of neurohypophyseal peptides by no means constituted his only important contributions. He was responsible for a long succession of elegant advances in the arena of peptide synthesis. These advances included outstanding studies on the azide method of coupling, the use of the tosyl group for the protection of the amino function, and the synthesis of peptides containing glutamine or diamino acids—just to mention a few.

Josef Rudinger was also an inspiring leader and teacher who contributed importantly to the development of an outstanding school of peptide science in Prague. He lectured extensively to scientific audiences in Europe, the Western Hemisphere, and Asia. He was a consummate editor and a diligent and elegant teacher of undergraduates, graduate students, post-doctoral fellows, and peers. In 1958 he organized the first of the now well-known biannual European Peptide Symposia, meetings that later gave birth to the alternating biannual American Peptide Symposia, and which many believe to be the principal international event devoted to the advance of peptide chemistry. In an earlier memorial,[1] Professor J. Fruton of Yale University recalled how Joe Rudinger "enriched and enlivened the proceedings" of the European Peptide Symposia with his "incisive comments always offered with courtesy and good humor." Fruton went on to say: "We were fortunate in having him as a major participant in the Third American Peptide Symposium, and you will remember the pleasure we all derived from his company at the scientific sessions and at the social gatherings. At these Symposia we saw him not only as a great scientist but also as a citizen of the world—for his charm and his integrity and his easy command of languages all helped to build lasting friendships among scientists from countries with widely different political regimes."

Professor Fruton's reference to Josef Rudinger's linguistic ability reminded me of my first encounter with Joe, which took place at the Second International Pharmacology Meeting in Prague in 1963. He had organized a session on neurohypophyseal peptides which he was chairing and in which I was speaking. There were no provisions for simultaneous multilingual translation at this session, but nonetheless there were several scientists in the audience who did not speak English. During my talk, one of my slides caught the attention of a German scientist (who was among those not fluent in English) and in due course he asked me a complex question—in German, which I didn't understand. Josef Rudinger, an ever alert chairman, spotted my problem before the last phrase came off my questioner's tongue, and immediately translated the question into the King's English, extracted my answer (which was given in less than the King's English), and whipped it back to the questioner in fluent High German. At this point a Russian, who understood no English but a little German, got into the act with a question which neither our German colleague nor I had the faintest glimmer of understanding. Again with incredible speed Joe translated the Russian question into German, the German into English, and both the German and English responses back into Russian; and he managed not only to satisfy the Russian but to tweak the curiosity of a Chinese scientist who seemed to understand some of the Russian-German-English interchange and came up with a question of her own—in Chinese! At this point I saw a touch of sweat on Joe's brow, but he valiantly attacked the Chinese question and managed to get it translated into Russian, German, and English, and to deliver the response in Chinese. I was astounded, as was everyone in the room, and I couldn't wait to corner Joe after the session to find out where he was born, lived, traveled, et cetera to get such facility with language. When I finally got to him and asked "where were you born?" the answer that came back was "Jerusalem."

Josef Rudinger was indeed born in Jerusalem, in 1924, but when he was three years of age his family returned to their native Czechoslovakia where he received his primary and part of his secondary school education. In 1939 he moved again, this time to England where he finished his secondary education and, in 1941, entered King's College in the University of Durham (now the University of Newcastle-on-Tyne) where he spent one year before joining the Royal Air Force in 1942, in which service he spent three years. After the end of World War II, in 1945, he returned to his studies, graduating in 1947 with a Bachelor of Science degree and "First Class Honors in Chemistry." He remained at Newcastle University as a graduate student for two more years and then returned to Czechoslovakia in 1949 to join Professor F. Sorm at the Institute of Chemical Technology in Prague, thereby getting established as a member of a group that later became the nucleus of the Institute of Organic Chemistry and Biochemistry of the Czechoslovak Academy of Science. In 1961 he was awarded a doctorate in Chemistry by the Czechoslovak Academy and shortly thereafter he became the head of a laboratory for peptide chemistry at the Institute of Organic Chemistry and Biochemistry, a position which he held formally until his resignation in 1970. He served as a Visiting Professor in the Department of Biochemistry at Yale University in 1965, and in the Laboratory of Molecular Biology of the Swiss Federal Institute of Technology (better known as ETH) in Zurich from 1968 to 1970. From 1970 until his death in April 1975, he was "ordentlicher Professor" at the Institute of Molecular Biology and Biophysics at ETH. During the course of this five-year stay in Zurich he managed to roam about the world serving as a Visiting Lecturer at the Weizmann Institute of Science in Israel, the University de Paris Sud in France, the Mount Sinai School of Medicine in the United States, as well as the Universities of Lausanne and Basel in his host country.

Although his two initial years of laboratory work as a graduate student in England involved studies on alkaloids, he soon turned to the area that was to become his principal scientific interest, the chemistry and biology of peptides, in which field he made outstanding contributions from 1949 to 1975. I will not present his contributions to peptide chemistry and to the understanding of structure-activity relationships of peptide hormones in any more detail than I have already given because that might diminish the overall portrait and flavor of the personality that I have tried to sketch for you. These contributions have been thoroughly documented in the memorial reviews by M. Brenner [2] and V. Pliska,[3] in the *Proceedings of the Fourteenth European Peptide Symposium*; the guide for ultimate detail is Joe Rudinger's complete bibliography.

In closing these dedicatory comments, let me cite one more brief quotation from Professor J. Fruton's remarks at the Fourth American Peptide Symposium: "For many of us the loss of Josef Rudinger deprives us of a cherished friend. For all of us it means that a great chemist and a great scientific statesman has been halted prematurely in a career that could have done so much more for the advancement of our area of science and for the promotion of international good will."

I.L.S.

References

1. Fruton, J. S. 1975. Remembering Josef Rudinger. *In* Peptides: Chemistry, Structure and Biology. R. Walter & J. Meienhofer, Eds.: 689–690. Science Publishers, Inc. Ann Arbor, Mich.

2. BRENNER, M. 1976. In memoriam, Josef Rudinger: Contributions to peptide chemistry. *In* Pcptides 1976. A Loffet, Ed.: 3–32. Bruxelles: Editions de l'Université de Bruxelles. Brussels, Belgium.
3. PLIŠKA, V. 1976. J. Rudinger's contributions to structure-activity relationships of peptide hormones: A biologist's view. *In* Peptides 1976. A. Loffet, Ed.: 33–51. Editions de l'Université de Bruxelles. Brussels, Belgium.

INTRODUCTORY REMARKS

Hilda Weyl Sokol and Heinz Valtin

Department of Physiology
Dartmouth Medical School
Hanover, New Hampshire 03755

Soon after the discovery of the Brattleboro rat in 1961, we realized that we were dealing with an unusual and potentially useful model of diabetes insipidus. But we did not immediately foresee the other abnormalities—all, we still believe, secondary to the absence of vasopressin—that these animals manifest. These additional aberrations became the subjects of intense study by numerous investigators throughout the world, and they encompassed disciplines ranging from molecular biology to clinical investigations and applications. With the recent explosion of research on brain peptides, for which vasopressin and oxytocin are practically prototypes, work with Brattleboro rats entered a new crescendo.

The idea of convening an international symposium to discuss the multiple abnormalities of the Brattleboro rat originated with Drs. Lewis B. Kinter and Reinier Beeuwkes III. It seemed important that anyone working with this animal should be aware of its total biology, especially since the interaction of many systems could influence the interpretation of any given result. Several years of planning culminated in the International Symposium on the Brattleboro Rat, which was held at Dartmouth Medical School from September 4 to 7, 1981. Some one-hundred-and-twenty scientists gathered from four continents for lively presentations and discussions. Although several invited participants were unable to attend, we are glad that in most instances we were able to include their contributions in this published proceedings, which will serve as a handbook on the Brattleboro rat.

This symposium would not have been possible without the generous financial contributions of the following organizations: the National Institute of Arthritis, Diabetes, and Digestive and Kidney Diseases; the National Science Foundation; the March of Dimes; Blue Spruce Farms, Inc.; Ferring Pharmaceuticals; Bristol-Myers Company; Smith Kline and French Laboratories; the Brattleboro Retreat; Abbott Laboratories; Alza Corporation; American Cyanamid Company; Arthur D. Little Foundation; Burroughs Wellcome Company; the Kroc Foundation; Merck Sharp and Dohme Research Laboratories; The Squibb Institute for Medical Research; Warner-Lambert Company, Inc.; and The New York Academy of Sciences, which agreed to publish these proceedings.

We wish to acknowledge the invaluable efforts of our colleagues and staff, without whom the conference could not have succeeded. We were assisted at every step by the local organizing committee, which consisted of the renal/neuroendocrine group at Dartmouth Medical School: Brian R. Edwards, Miklos Gellai, Victoria L. Herzberg, Frederick T. LaRochelle, Jr., William G. North, Judy E. Stern, Paul Stern, and Larry A. Walker. For nearly two years, Mary L. Kenyon and Ethel B. Garrity added the needs of the symposium to their already demanding administrative tasks in the Department of Physiology. Margaret A. Colbeck served as Administrative Director with unbelievable energy, imagination, thoroughness, thoughtfulness, and kindness. She led an always willing and cheerful staff who assured the comfort of our guests and the smooth running of the meeting: Elizabeth Craig-Ykema, Joseph E. Melton, Dirk B. Mendel, Richard H. Drew, and Geraldine M. North. We are most grateful to Bill Boland and Mary K. Brennan of The New York Academy of Sciences, who guided this volume to completion. Finally (in the words of John Morris), we thank the Brattleboro rat, which made it all possible.

THE DISCOVERY OF THE BRATTLEBORO RAT, RECOMMENDED NOMENCLATURE, AND THE QUESTION OF PROPER CONTROLS *

Heinz Valtin

Department of Physiology
Dartmouth Medical School
Hanover, New Hampshire 03755

HISTORY OF THE BRATTLEBORO RAT

The discovery of the Brattleboro rat epitomizes the truism expressed by Louis Pasteur, that ". . . chance only favours the prepared mind";[1] Dr. Henry A. Schroeder surely had such a mind. In 1960, Dr. Schroeder, then Associate Professor of Clinical Physiology at Dartmouth Medical School, was raising a colony of Long-Evans hooded rats in his private laboratories on a remote mountain in West Brattleboro, Vermont (FIGURE 1). (Following his resignation, due to illness, from his positions as Associate Professor of Medicine and Director of the Hypertension Division at Washington University School of Medicine in St. Louis, Dr. Schroeder had "retired" in 1957 to his summer home in West Brattleboro, only to resume his studies on the role of trace elements in cardiovascular diseases and aging[2] two years later.) † Early in 1961, Dr. Schroeder's alert assistant, Mr. Tim Vinton, noticed that the drinking bottles attached to one among hundreds of cages were always nearly empty. This cage contained a mother and her litter of seventeen young, born on February 24, 1961. Together, Mr. Vinton and Dr. Schroeder quickly determined that some of the pups were drinking excessive amounts and that this abnormality could be corrected by giving them vasopressin. It thus appeared that some of the newborn animals had hereditary hypothalamic diabetes insipidus.

Most investigators might have discarded the animals as being of only peripheral interest, and a nuisance to the main purpose of their research. But not Harry Schroeder; he offered four of the six diseased rats from the original litter to the two people at Dartmouth Medical School who had worked with problems of vasopressin: to Dr. Kurt Benirschke, at that time Chairman of the Department of Pathology,[3] and to me.[4] Having determined that the occurrence of this disorder in a common laboratory animal was apparently unique, we quickly decided that our first objective must be to produce more diseased offspring. The first animals were transferred to Hanover, New Hampshire on June 3, 1961, and after some preliminary testing, matings were begun in the middle of July.

In August, 1961 we had the good fortune to have Dr. Hilda Weyl Sokol join our Department. Having rather recently completed a Ph.D. thesis on the cytophysiology of the neurohypophysis,[5] she was not only vitally interested in

* Supported principally by U.S. Public Health Service Research Grant AM-08469.
† Further details about Dr. Schroeder are given in the Introduction to the Henry A. Schroeder Memorial Lecture in this volume.

0077-8923/82/394-001 $1.75/0 © 1982; NYAS

our rats, but also brought to our group expert techniques that proved invaluable towards defining the defect in these animals.[6] Another person who helped us tremendously from the very beginning was Dr. Wilbur H. Sawyer. In his thorough and reliable—yet always quiet and modest—approach, Bill Sawyer did the initial bioassays[7, 8] which, when combined with the anatomical findings,[6] virtually nailed down the defect in the Brattleboro strain.

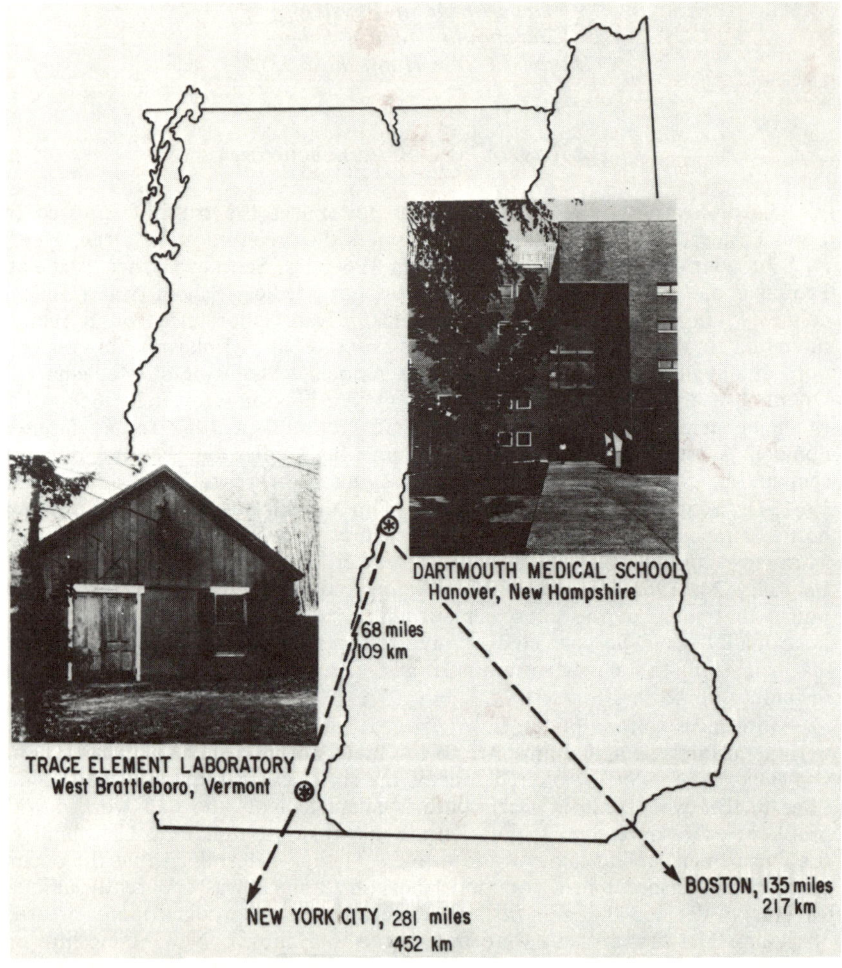

FIGURE 1. Map of the states of New Hampshire (on the right) and Vermont (on the left), showing the location of Brattleboro, Vermont in relation to Dartmouth Medical School. The insert on the right depicts the main building of Dartmouth Medical School; that on the left is a picture of Dr. Henry A. Schroeder's shed—named the "Rathaus"—where the Brattleboro strain was discovered.

For nearly nine months, Hilda Sokol and I coddled and nursed newborn rats in an effort to produce live offspring. Repeatedly, our efforts were frustrated when newly born animals were found dead in the cage or failed to thrive and died within a few days after birth, or when matings proved infertile. Success was finally achieved when the defective gene was transferred into normal stock and the resulting heterozygotes were crossed with one another or with one of the original affected animals. Soon thereafter, our very able assistant, Ethel Garrity, joined us; for years, she supervised the colony and shipped Brattleboro rats to other investigators.

The first oral report on these animals was delivered to the American Physiological Society in August of 1962,[9] and a brief written account appeared shortly thereafter.[10] In the same year, Hilda Sokol "introduced" the rats to the American Society of Zoologists.[11] In the audience was Ernst Scharrer, who might be called the father of neurosecretion;[12] his immediate and prescient reaction, which he voiced during the discussion, was: "You have a gold mine there." The strain was given its name—after the town of its birth—in the first full-length report, published in 1964.[13]

It is fun to recall a few episodes, some of which seemed momentous at the time, when the very survival of the Brattleboro strain hung in balance. One incident illustrates the orientation of our disciplines. When Kurt Benirschke and I met to decide what to do with the first four diseased animals, he, a good pathologist, said: "Let's look at their hypothalami." And while I agreed that that would be a good idea eventually, I thought that we should first try to breed the animals. (In fairness and with a smile, I must quickly add that, even without me, Dr. Benirschke would surely have followed the same sequence; in fact, he is currently Director of Research at the San Diego Zoo, and in this capacity is bending his vast energies to the preservation of endangered species.)

At the same time that we were trying to perpetuate the Brattleboro strain, we had also begun experiments on the original diseased rats. In September, 1961, a well-meaning colleague, Dr. William O. Berndt, observed me clumsily trying to administer an oral water load to one of these precious animals. He suggested a better method, which I promptly tried and thereby delivered 10 ml of tap water into the rat's trachea. The rat gasped; I gasped; and for the next four days I drew on my clinical experience to treat the animal for what I presumed to be pneumonia. Collections were continued during this period and the results were reported as the urinary response of Brattleboro rats to stress.[13]

Early in 1962, when we finally succeeded in producing more diseased rats, we were still inexperienced in handling young, jumpy rats. One day, while we were changing cages, one of the new, diseased animals got away. Anyone who has tried to catch a rat scurrying across the floor, under cabinets, and behind drawers can appreciate the contortions we went through to retrieve this priceless animal.

From the very beginning, we made Brattleboro rats available to other laboratories. This was not simply an altruistic move, for we learned a great deal from the interactions with other scientists. It is one of our great satisfactions that the field has mushroomed and that the rats have been utilized so effectively as to justify the assemblage of many distinguished scientists at this symposium. Among the early recipients of Brattleboro rats was Dr. Josef Rudinger, the Czech scientist to whom this symposium is dedicated. In the spring of 1965, Dr. Rudinger and his wife, Edita, visited Dartmouth. (He was at that time Visiting Professor of Biochemistry at Yale.) As he left Hanover, he took several rats with him to launch them on a journey via Boston and New Haven to New York, and thence to Prague by Air India, which was at that time the only direct flight from the U.S.A. to Czechoslovakia. On May 22, 1965, Joe Rudinger wrote to me from New Haven: "So far, so good. We had a very pleasant trip to Boston, an interesting stay there, and an uneventful journey back [to New Haven], and I trust I can say the same on behalf of the rats. They seem cheerful enough to me, though I am a layman as far as rat psychology is concerned." From this "Prague branch," as Joe Rudinger called it, the Brattleboro strain spread throughout middle and eastern Europe, and as far east as Siberia.

Recommended Nomenclature

Brattleboro rats have been called by many names. Some of the appellations and abbreviations that have been used include the following: Brattleboro homozygotes; homozygous Brattleboro rats; diabetes insipidus rats (DI); Brattleboro rats (BR or BB); Brattleboro diabetes insipidus rats; Brattleboro strain; heterozygous Brattleboro rats (HB or HE-DI or HET-DI or HZ); homozygous diabetes insipidus (Brattleboro) rats (HO-DI or HOM-DI); and homozygous normal (HO-NO). We have previously suggested a uniform nomenclature,[14] and it seems appropriate to publicize it once more in these proceedings.

The Brattleboro strain arose from a stock of Long-Evans hooded rats, and the inability to synthesize vasopressin is inherited, in Brattleboro rats, as an autosomal recessive trait at a single gene locus.[15] As Figure 2 shows, the heterozygote, having one normal and one abnormal allele for producing vasopressin, has both a plasma concentration and a rate of urinary excretion of the hormone that are intermediate between homozygotes and normal rats, and it also has an intermediate ability to concentrate urine. Therefore, the mode of inheritance is semirecessive, and three genotypes and phenotypes can be identified, for which we suggest the following nomenclature and abbreviations (Table 1): (1) Brattleboro homozygotes (DI or di/di); (2) Brattleboro heterozygotes (HZ or di/+); and (3) normal, Long-Evans rats (LE or +/+). We recommend that these designations be used in all reports; in addition, we advocate that, wherever possible, the word "Brattleboro" be included in the title, since that practice would greatly facilitate retrieval.

Proper Controls

A number of controls have been employed in studies with Brattleboro homozygotes: (a) Each homozygote serving as its own control, tested before and after being treated with vasopressin; (b) Brattleboro heterozygotes; (c) Long-Evans rats; and (d) rats of other strains. Cautions can be voiced about each type of control.

Testing a given function before and after treating a homozygote with vasopressin often yields useful results, especially when that function returns to the value seen in "normal" rats. Thus, the fact that high plasma sodium concentration and osmolality seen in untreated homozygotes decline to normal levels during prolonged treatment with vasopressin,[16] seems to justify the conclusion that the hypernatremia and hyperosmolality are the direct consequence of contraction of the body fluid volumes. Simultaneous correction of hypokalemia, on the other hand, may have a more complicated interpretation involving possibly indirect influences.[17] And when a function is not corrected or only partially corrected, the interpretation may be even more complex. A result of this type, especially when it entails the central nervous or endocrine system,[18] will not necessarily mean that the absence of vasopressin is only partly or not at all responsible for the abnormality under investigation; it could mean that the exogenous vasopressin, especially when administered systemically, has not reached a critical site in sufficient concentration to mimic the release of endogenous vasopressin from nerve terminals within the central nervous system or into the hypothalamic portal circulation.

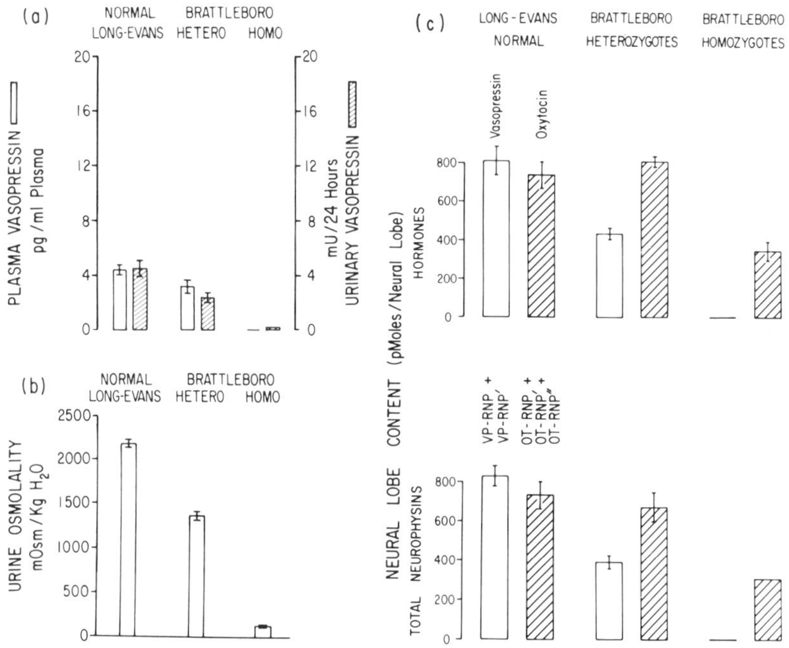

FIGURE 2. (a) Results of radioimmunoassays for vasopressin in plasma (open columns) and urine (striped columns) of Long-Evans (normal) rats, Brattleboro heterozygotes (Hetero), and Brattleboro homozygotes (Homo). The parallelism between results in plasma, obtained by Möhring and Möhring,[25] and those in urine obtained by Miller and Moses,[26] is striking. Each bar with brackets denotes mean ± SEM. From Valtin.[27] With permission from the *American Physiological Society*. (b) Urine osmolalities on the same three types of animals for which results of assays are given in (a), showing that the intermediate level for vasopressin in Brattleboro heterozygotes results in an intermediate deficiency for concentrating urine. (c) The intermediate position of heterozygotes is also apparent in the amount of vasopressin and vasopressin-associated neurophysins (VP-RNP, open columns) that are stored in the posterior pituitary gland. From Valtin *et al.*[19] With permission from Karger.

Brattleboro heterozygotes are permissible controls when one wants to contrast a given state during total lack of vasopressin with that state in the presence of vasopressin. Inasmuch, however, as heterozygotes have an intermediate defect for producing vasopressin (FIGURE 2c) and its associated neurophysins,[19] and since many functions such as water balance are dose-dependent (FIGURE 2b), one must be careful not to equate a heterozygote with a "normal" rat. On

TABLE 1

RECOMMENDED NOMENCLATURE AND ABBREVIATIONS

1. Brattleboro homozygotes	DI or di/di
2. Brattleboro heterozygotes	HZ or di/+
3. Long-Evans rats (normal)	LE or +/+

Use the term "Brattleboro" in all titles.

the other hand, heterozygotes may for certain purposes offer an advantage over Long-Evans rats (see next paragraph) in that they are presumably more similar genetically to their homozygous littermates than to individuals of other strains.

In our opinion, the best controls for most studies are Long-Evans rats, from which the Brattleboro strain arose. Since most colonies are inbred, a Long-Evans rat obtained from one source undoubtedly differs in many respects from a Long-Evans rat raised at another source. For the same reason, it seems unlikely that much advantage can be gained by working with Long-Evans rats derived from the original Rockland Farms colony, since twenty years of at least some inbreeding must have altered that stock as well. We think the best control is a Long-Evans rat from the same colony that is used periodically for outbreeding and the production of "new" heterozygotes. This practice not only avoids excessive inbreeding but also provides controls that are genetically similar to the rats that carry the Brattleboro gene. Strains other than Long-Evans do not seem to constitute proper controls for most purposes.

INTERPRETATIONS

Very frequently, Brattleboro homozygotes are used to test the role of vasopressin in certain functions. If that function turns out to be normal in untreated homozygotes, it would be wrong to conclude simply that vasopressin is not involved; it might be very much involved in the maintenance of that particular function in individuals that, unlike Brattleboro homozygotes, can produce vasopressin. Systemic blood pressure can serve as a case in point. TABLE 2 shows that the mean arterial blood pressure in untreated Brattleboro homozygotes is not significantly different from that in Long-Evans rats. This finding permits us to conclude only that vasopressin is not *essential* to the maintenance

TABLE 2

MEAN ARTERIAL BLOOD PRESSURE

Strain	Number of Rats *	mm Hg (Mean±SEM)
Long-Evans (Normal)	40	109.8 ± 9.33
Brattleboro Homozygotes	40	111.3 ± 7.51

* Includes both sexes.

of blood pressure, not necessarily that it is not involved. In fact, a likely role for vasopressin can be unmasked in Brattleboro homozygotes by taking them through certain maneuvers, such as hemorrhage.[20]

Any truly vital function has a number of controlling factors and when one of these fails, the others are likely to compensate. Homer Smith expressed it this way: "Where multiple controls are superimposed on a function, such as sodium excretion, it is conceived that normal regulatory mechanisms may be obscured by compensatory reactions."[21] There is much evidence that vasopressin may play a role in the maintenance of normal blood pressure[22] or in the causation of hypertension.[23] It is likely that compensatory stimulation of other controlling variables, such as the autonomic or renin-angiotensin systems,[24] is at least partly responsible for the normal systemic pressure in Brattleboro homozygotes (TABLE 2).

CONCLUDING RECOMMENDATIONS

The two abnormal genotypes of the Brattleboro strain should be referred to as Brattleboro homozygotes (DI or di/di) and Brattleboro heterozygotes (HZ or di/+).

Retrieval of published reports on these animals would be greatly aided if the word "Brattleboro" were inserted in each title.

For most purposes, it is probably best to choose normal Long-Evans rats for controls (LE or +/+). Brattleboro heterozygotes have an intermediate defect for producing vasopressin and thus are inappropriate controls if one wishes to contrast a homozygote against a "normal" rat.

When exogenous vasopressin fails to correct a given abnormality in Brattleboro homozygotes, the results must be interpreted cautiously, since the failure may have arisen from insufficient concentration of vasopressin at some critical sites, especially within the central nervous system.

When a vital function, such as systemic arterial blood pressure, is normal in Brattleboro homozygotes, such a finding permits us to conclude that vasopressin is not essential for that function; the finding does not exclude the possibility that vasopressin may play a role in maintaining that function in normal individuals, which have the hormone.

ACKNOWLEDGMENT

I am grateful to Hilda Weyl Sokol for reviewing this manuscript and offering some excellent suggestions.

REFERENCES

1. STRAUSS, M. B., Ed. 1968. Familiar Medical Quotations. p. 108a. Little, Brown. Boston, Mass.
2. SCHROEDER, H. A., W. H. VINTON, JR. & J. J. BALASSA. 1963. Effects of chromium, cadmium and lead on the growth and survival of rats. J. Nutrition **80:** 48–54.
3. DINGMAN, J. F., K. BENIRSCHKE & G. W. THORN. 1957. Studies of neurohypophyseal function in man. Diabetes insipidus and psychogenic polydipsia. Am. J. Med. **23:** 226–238.

4. VALTIN, H., I. D. WILSON & S. M. TENNEY. 1959. CO_2 diuresis, with special reference to role of left atrial stretch receptor mechanism. J. Appl. Physiol. **14:** 844–848.
5. SOKOL, H. W. 1956. Cytophysiological studies on the teleost hypophysis. Ph.D. thesis. Radcliffe College, Cambridge, Mass.
6. SOKOL, H. W. & H. VALTIN. 1965. The morphology of the neurosecretory system in rats homozygous and heterozygous for hypothalamic diabetes insipidus (Brattleboro strain). Endocrinology **77:** 692–700.
7. SAWYER, W. H., H. VALTIN & H. W. SOKOL. 1964. Neurohypophysial principles in rats with familial hypothalamic diabetes insipidus (Brattleboro strain). Endocrinology **74:** 153–155.
8. VALTIN, H., W. H. SAWYER & H. W. SOKOL. 1965. Neurohypophysial principles in rats homozygous and heterozygous for hypothalamic diabetes insipidus (Brattleboro strain). Endocrinology **77:** 701–706.
9. VALTIN, H., H. A. SCHROEDER & K. BENIRSCHKE. 1962. Familial hypothalamic diabetes insipidus in rats. The Physiologist **5:** 225 (Abstract).
10. VALTIN, H., H. A. SCHROEDER, K. BENIRSCHKE & H. W. SOKOL. 1962. Familial hypothalamic diabetes insipidus in rats. Nature **196:** 1109–1110.
11. SOKOL, H. W. & H. VALTIN. 1962. The morphology of the neurosecretory system in a strain of rats with familial hypothalamic diabetes insipidus. Am. Zool. **2:** 560 (Abstract).
12. SCHARRER, E. & B. SCHARRER. 1940. Secretory cells within the hypothalamus. Res. Publs. Assoc. Res. Nerv. Ment. Dis. **20:** 170–194.
13. VALTIN, H. & H. A. SCHROEDER. 1964. Familial hypothalamic diabetes insipidus in rats (Brattleboro strain). Am. J. Physiol. **206:** 425–430.
14. VALTIN, H. 1976. Animal model of human disease: Hereditary hypothalamic diabetes insipidus in the Brattleboro strain of rat. Am. J. Pathol. **83:** 633–636.
15. SAUL, G. B. II, E. B. GARRITY, K. BENIRSCHKE & H. VALTIN. 1968. Inherited hypothalamic diabetes insipidus in the Brattleboro strain of rats. J. Hered. **59:** 113–117.
16. CHENG, W.-T., W. G. NORTH & M. GELLAI. 1982. Replacement therapy with arginine vasopressin in Brattleboro homozygous rats. Ann. N.Y. Acad. Sci. (This volume.)
17. OPAVA-STITZER, S., E. FERNANDEZ-REPOLLET & P. STERN. 1982. Sodium and potassium balance in the Brattleboro rat. Ann. N.Y. Acad. Sci. (This volume.)
18. MCCANN, S. M., J. ANTUNES-RODRIGUES, R. NALLAR & H. VALTIN. 1966. Pituitary-adrenal function in the absence of vasopressin. Endocrinology **79:** 1058–1064.
19. VALTIN, H., W. G. NORTH, F. T. LAROCHELLE, JR., H. W. SOKOL & J. F. MORRIS. 1978. Biochemical and Anatomical Aspects of ADH Production. *In* Proc. 7th Int. Cong. Nephrol. M. Bergeron, Ed.: 313–320. Karger. Basel.
20. LAYCOCK, J. F., W. PENN, D. G. SHIRLEY & S. J. WALTER. 1979. The role of vasopressin in blood pressure regulation immediately following haemorrhage in the rat. J. Physiol. (Lond.) **296:** 267–275.
21. SMITH, H. W. 1957. Salt and water volume receptors. An exercise in physiologic apologetics. Am. J. Med. **23:** 623–652.
22. ANDREWS, C. E. & B. M. BRENNER. 1981. Relative contributions of arginine vasopressin and angiotensin II to maintenance of systemic arterial pressure in the anesthetized water-deprived rat. Circ. Res. **48:** 254–258.
23. CROFTON, J. T., L. SHARE, R. E. SHADE, W. J. LEE-KWON, M. MANNING & W. H. SAWYER. 1979. The importance of vasopressin in the development and maintenance of DOC-salt hypertension in the rat. Hypertension **1:** 31.

24. MÖHRING, J., G. KOHRS, B. MÖHRING, M. PETRI, E. HOMSY & D. HAACK. 1978. Effects of prolonged vasopressin treatment in Brattleboro rats with diabetes insipidus. Am. J. Physiol. **234** (Renal Fluid Electrolyte Physiol. 3):F106–F111.
25. MÖHRING, B. & J. MÖHRING. 1975. Plasma ADH in normal Long-Evans rats and in Long-Evans rats heterozygous and homozygous for hypothalamic diabetes insipidus. Life Sci. **17:** 1307–1314.
26. MILLER, M. & A. M. MOSES. 1971. Radioimmunoassay of urinary antidiuretic hormone with application to study of the Brattleboro rat. Endocrinology **88:** 1389–1396.
27. VALTIN, H. 1977. Genetic models for hypothalamic and nephrogenic diabetes insipidus. *In* Disturbances In Body Fluid Osmolality. T. E. Andreoli, J. J. Grantham & F. C. Rector, Jr., Eds.: 197–215. American Physiological Society. Bethesda, Md.

DISCUSSION OF THE PAPER

K. LEDERIS (*University of Calgary, Calgary, Alberta*): While we are waiting for questions for Dr. Valtin, may I ask him what is the global distribution of the rats now?

H. VALTIN (*Dartmouth Medical School, Hanover, N.H.*): Brattleboro rats are currently being bred and used on at least four continents: Europe (both Western and Eastern), Oceania, Asia, and North America. We have representatives at this conference from all of these areas. I think the animals are also being raised and used in South America; I don't know about Africa; and I am quite certain that there are none in Antarctica. Several national laboratories maintain colonies of Brattleboro rats, among them the National Institutes of Health in the U.S. (being represented on the panel this morning by Dr. Carl T. Hansen), Zeist in the Netherlands (represented by Dr. Gubbels), and Carshalton, in England. The commercial supplier in the U.S. is Blue Spruce Farms, Inc., of Altamont, New York, and Mr. Hans Kappel has come from there to participate in the panel discussion.

LEDERIS: I wish to thank Dr. Valtin for this brief historical review. Before inviting the next speaker, I regret to inform you that my co-chairman, Dr. Dlouhá, regrettably is unable to attend this meeting. Fortunately for us, her colleague Dr. Josef Zicha has kindly agreed to present her paper.

POSTNATAL DEVELOPMENT AND DIABETES INSIPIDUS IN BRATTLEBORO RATS

H. Dlouhá,* J. Křeček, and J. Zicha

Institute of Physiology
Czechoslovak Academy of Sciences
Prague 4, CS 142 20, Czechoslovakia

The syndrome of hereditary diabetes insipidus of hypothalamic origin has been described in adult rats.[1] The task of this paper is to ascertain whether this syndrome is also present in young animals and at what age it begins to be manifest. Inasmuch as this type of diabetes insipidus is caused by the defect of vasopressin synthesis, it is obvious that its symptoms cannot be disclosed before the age at which the presence of vasopressin is detected in normal rats.

Neurons of the vasopressin-synthesizing hypothalamic nuclei (i.e., n. supraopticus and n. paraventricularis) are found between days 13–17 of fetal life in rats.[2,3] Radioimmunoassayable vasopressin was observed in the rat brain at day 14[4] and in the pituitary at day 16 of fetal life.[5] In his pioneer study from 1947 Hans Heller[6] estimated the vasopressin content in the neurohypophysis to be less than 3 mU and more than 1.5 mU/gland in 1-day-old rats. This is in perfect accordance with recent data obtained using radioimmunoassay.[7]

The osmoregulatory role of neurohypophysis is effectuated mainly by its influence on kidney function. It is mediated by the blood transfer of vasopressin. However, no antidiuretic activity was found in the blood plasma of normally hydrated rats until the age of four weeks.[8] Antidiuretic activity corresponding to 4–5 μU/ml was observed in plasma of 23-day-old rats after water deprivation for six hours. Very low but detectable (about 1 μU/ml) activity appeared in plasma of 10-day-old rats loaded by a 3% salt solution.[9] From these data it could be predicted that the first signs of impaired concentrating ability of the kidney, accompanied by polyuria and polydipsia, ought to appear in homozygous Brattleboro rats in the course of the second to fourth postnatal week.

Young Brattleboro rats were exposed to water deprivation for six hours at the ages of 10, 14, 18, and 22 days. Urine was obtained by perineal stimulation. 10- and 14-day-old animals were kept without mother at 30° C, the older ones at 25° C. Urine osmolarity was determined by means of the Knauer osmometer. As shown in FIGURE 1, values of urinary osmolarity in homo- and heterozygotes did not differ in 10- and 14-day-old rats whereas two distinct populations were found in 18- and 22-day-old animals. It was assumed that the group with lower osmolarity consisted of insipidic animals while the group with higher urinary osmolarity were animals without diabetes insipidus. This assumption was confirmed by measurement of water intake carried out at the age of 27 days. Water intake did not exceed 25 $ml \cdot 100\ g^{-1} \cdot 24\ h^{-1}$ in individuals with high urine osmolarity and it exceeded 50% of the body weight in animals with low urinary osmolarity.[10] Hence, the difference in concentrating

* Mailing address: Dr. H. Dlouhá, CSc, Institute of Physiology, Czechoslovak Academy of Sciences, Vídeňská 1083, 142 20 Prague 4, Czechoslovakia.

0077–8923/82/0394–0010 $1.75/0 © 1982, NYAS

ability between homo- and heterozygotes appeared between day 14 and 18 of postnatal life.

The beginning of the third week of postnatal life represents a borderline between two developmental periods (i.e. suckling and weaning periods) in rat.[11] Until the end of the second week young rats consume only mother's milk as the source of calories, water, minerals, and other nutrients. Starting with the onset of the third week, the rat begins to supplement the nutrition by consumption of solid food and water. The process of weaning is accomplished at the age of 28 days.[11, 12]

An attempt was made to compare the milk and water consumption of homo- and heterozygous Brattleboro rats in the course of suckling and weaning periods. In order to keep the nest conditions, namely the permanent contact of pups

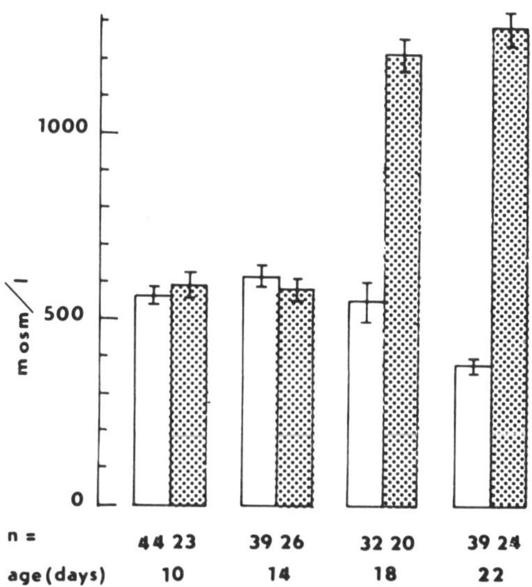

FIGURE 1. Effect of water deprivation for six hours on urinary osmolarity in young Brattleboro rats of different ages. Ordinate: urinary osmolarity in mOsm/L. White bars: mean ± SEM in homozygotes; dark bars: mean ± SEM in heterozygotes.

with their mother, for the course of the whole experiment the radioisotope method described by Babický *et al.*[12, 13] was used. Milk consumption was estimated by means of $^{85}SrCl_2$ (40 μCi) injected s.c. in the mothers in 0.2 ml of saline. 24 and 48 hours later, activity of individual littermates was determined *in vivo* by a whole body counter. The activity was expressed in percent of the dose administered to the mother.[12] Determination of the water consumption was carried out by labeling of the water supply with colloidal ^{198}Au stabilized with a solution of gelatin in a final concentration of 0.1% with about 0.2 μCi/L specificity. An approximate amount of water consumed by the young was calculated from the ratio of radioactivity in the mother and young and the total amount consumed in the course of 24 hours.[13]

Milk consumption was estimated in 27 Brattleboro rats. Nine of them were established to be homozygous and 18 heterozygous for diabetes insipidus by subsequent measurement of water intake carried out at the age of 35–40 days. As shown in FIGURE 2, no distinctly higher milk intake was observed in homozygotes in comparison with heterozygotes. Highest values of milk consumption related to one animal were achieved at the age of 13–18 days, i.e. at the age at which the suckling begins the weaning period. Milk consumption stopped at the age of 27 days independent of whether rats were homo- or heterozygous. This pattern of the age-dependent milk consumption fits well with that found in Wistar rats.[12]

Results of experiments with water intake are presented in FIGURE 3. No water was consumed until day 13 and only negligible intake was registered at the age of 15 days. Starting at day 16, a substantial amount of water was drunk by homozygotes and the consumption increased in subsequent days. From that day a marked difference in water consumption was observed between homo- and heterozygotes. It could be concluded that no signs of polydipsia are present in homozygous Brattleboro rats in the course of the suckling period in which

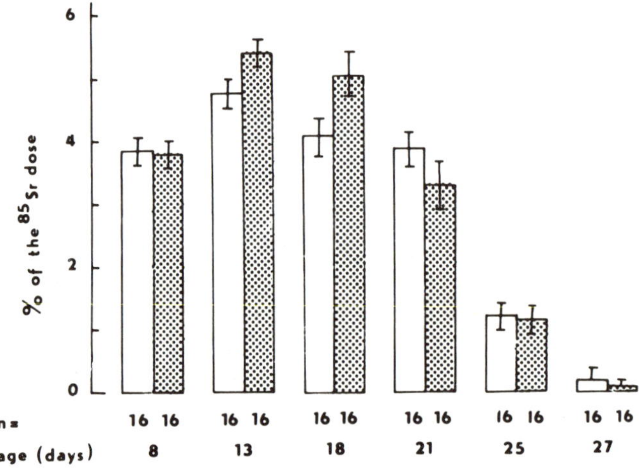

FIGURE 2. Mother's milk consumption in young Brattleboro rats of different ages. Ordinate: consumption in % of $^{85}SrCl_2$ dose applied to the mother. See FIGURE 1 for further details.

only milk is consumed. Polydipsia begins at the age of 15–16 days and from that age water is preferred to milk as drinking fluid.[14]

The syndrome of polydipsia and polyuria thus appears in homozygous Brattleboro rats at the onset of the weaning period. 14-day-old rats are unable to use increased water intake to equilibrate osmotic imbalances of body fluids. In homozygotes, this deficiency is reflected in the osmolarity of the blood plasma, which is higher than in heterozygotes at that age (FIGURE 4). Hence, increased osmolarity of the blood plasma indicates presence of an osmoregulatory imbalance in 14-day-old rats although neither polydipsia nor decreased concentrating ability are observed at that age.

This assumption is supported by data on urinary osmolarity obtained in dehydrated, water-loaded or arginine vasopressin-injected Brattleboro rats aged 10 and 14 days (FIGURE 5). Water diuresis was induced in suckling and

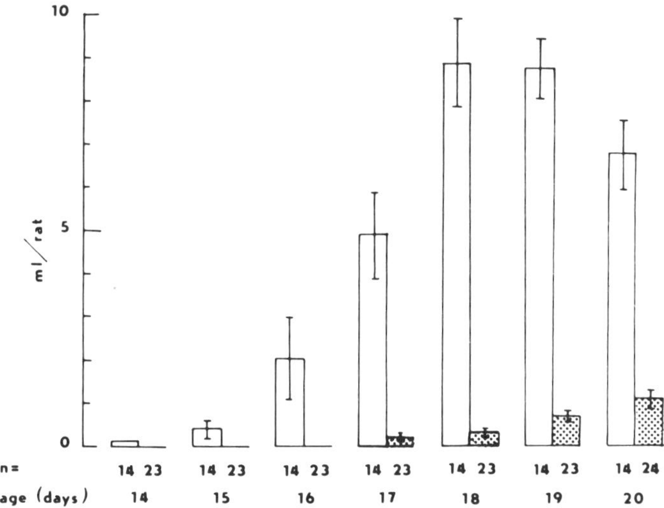

FIGURE 3. Water consumption in young Brattleboro rats of different ages. Ordinate: estimated consumption per day in ml/animal. See FIGURE 1 for further details.

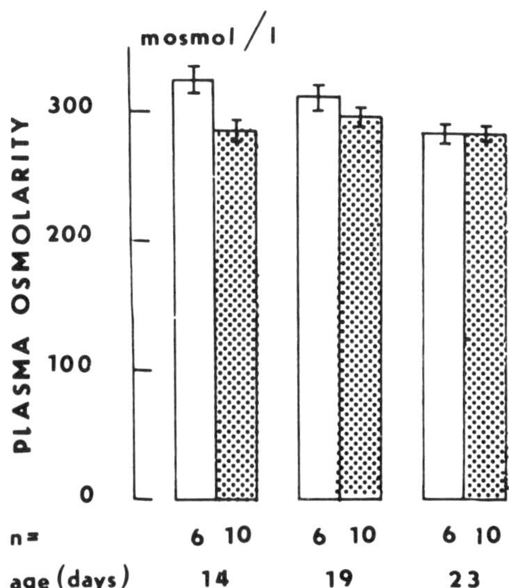

FIGURE 4. Osmolarity of the blood plasma in young Brattleboro rats of different ages. Ordinate: osmolarity in mOsm/L. See FIGURE 1 for further details.

weanling rats by two infusions of water (each 4.5% of body weight) in 30-min intervals through a stomach tube. Urine was sampled 15 min after the second water load. After a further 15 min, arginine vasopressin (AVP, Sandoz Ltd. Basel, 25 mU/100 g b.w.) was injected s.c. and urine collected 60 min later. Urinary osmolarity was measured in non-loaded animals as well. In non-loaded rats 18-days-old, the urinary osmotic pressure was higher in hetero- than in homozygotes but between 10 and 14 days the reverse was true. Water loading decreased urine osmolarity; values were lower in hetero- than in homozygotes both at 14 and 18 days of age. The effect of AVP was to raise the urinary osmotic pressure: in homozygotes age did not affect this response but in

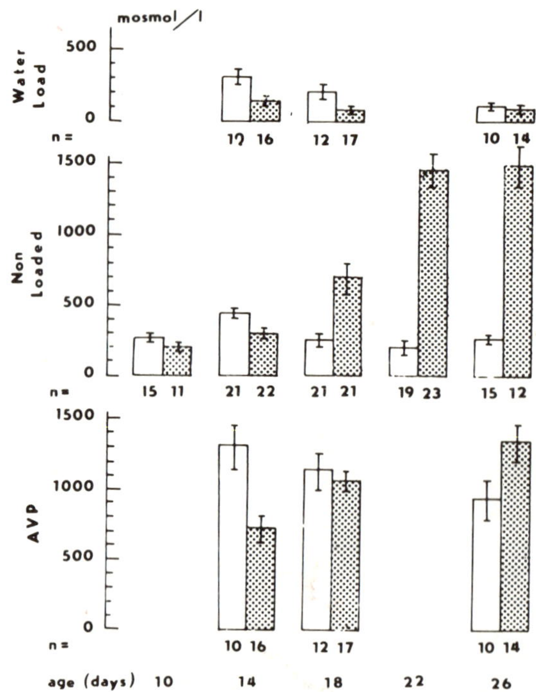

FIGURE 5. Urinary osmolarity in dehydrated, water-loaded, and vasopressin-treated young Brattleboro rats of different ages. Ordinates: osmolarity in mOsm/L. See FIGURE 1 for further details.

heterozygotes the size of the response increased with age. After treatment with AVP, urinary osmotic pressure was higher in heterozygotes than in homozygotes aged 26 days but in 14-day-old rats the reverse was true.[15]

Under all these experimental conditions higher urinary osmotic concentration was observed in homozygous rats before the appearance of the syndrome of diabetes insipidus. Since no antidiuretic activity was found in plasma, even in rats without diabetes insipidus at the age of 10–14 days, it is hardly possible to explain this particular kidney function of suckling homozygous Brattleboro rats merely by the absence of vasopressin. Actually, data are available that indicate that homo- and heterozygous Brattleboro rats differ, apart from osmoregulation, in other properties as well before the age of 14 days.

Valtin *et al.*,[16] and many others, demonstrated that body weight of homozygotes is lower than heterozygotes' body weight soon after birth. Malnutrition can be excluded as the cause of the growth retardation since the milk consumption does not differ between homo- and heterozygous sucklings (FIGURE 2). An attempt to abolish the growth retardation by vasopressin supplementation was unsuccessful.[17] However, supplementation by vasopressin was started only after the fourth week of postnatal life. Growth retardation induced in neonatal rats is difficult to repair in later stages of development.[18, 19] Therefore, it might be too late to try to repair a growth disturbance from the fourth week of age.

Growth hormone (GH) deficiency was supposed by many authors to be responsible for the lower weight of Brattleboro homozygotes.[17, 20, 21] However, there is low correlation between the plasma GH level and growth rate in rats in the first four weeks of postnatal life.[23-24] Moreover, growth of suckling rats does not seem to be dependent on their own GH.[25]

Recently, Boer *et al.*[26] demonstrated that apart from body growth, the proportionality of brain growth is changed in Brattleboro homozygotes. The weight of the cerebellum is lower and the bulbus olfactorius is higher in comparison with heterozygotes. In rats, the brain size is determined in the course of two critical periods, the first of which occurs at the late stage of fetal life.[27] Vasopressin begins to be produced in hypothalamic nuclei even at that period.[28] The system of vasopressinergic fibers penetrating into different parts of the central nervous system is established between day 18–20 of fetal life.[5] The conclusion of the authors of this discovery, that lack of vasopressin in mutant may have its expression in brain development, seems to be substantiated. Supplementation of vasopressin in homozygous Brattleboro rats in the course of late fetal and early postnatal age is, therefore, highly desirable in order to prevent some disturbances observed in these animals later in life. In this way, perhaps, deviations of the kidney function observed in homozygotes before the manifestation of the syndrome of diabetes insipidus could be clarified.

Vasopressin is involved not only in osmoregulation but also in volume regulation.[29] Development of the regulation of blood and extracellular volumes in rats lasts until the age of 40 days.[30] At this age, plasma and blood volumes are stabilized.[31]

Plasma and blood volumes were estimated in Brattleboro rats by means of Evans blue dilution technique. Results are expressed in FIGURE 6. No difference was observed both in the blood and plasma volume between homo- and heterozygotes until the age of 30 days. Significantly higher values were obtained in homozygotes aged 42 and 60 days. At that period sexual maturation occurs in rats.[32] Only at that age is the development of the kidney function accomplished by achievement of the adult pattern of distribution of the single-nephron glomerular filtration (SNGFR).[33] Therefore, SNGFR distribution was estimated in developing Brattleboro rats. Baines[34] modification of Hansen's [^{14}C]ferrocyanide single-injection method was used and the ratios of superficial to juxtamedullary (S/J) and intercortical to juxtamedullary (I/J) nephrons were determined. As shown in FIGURE 7, significantly higher values of both S/J and I/J were found in homozygotes at the age of 60 days, while no significant difference was observed in younger animals.

There is still a controversy concerning the effect of volume expansion on the distribution of SNGFR.[35] From data presented in this paper it could be suggested that vasopressin deficiency is accompanied by both the increased blood volume and the redistribution of SNGFR towards nephrons situated in more superficial layers of the kidney cortex. However, it is not possible to conclude definitely whether there is another relationship between them except for the age-coincidence of their maturation.

The absence of vasopressin seems to be manifested in volume-regulatory processes later than in osmoregulation. It could be hypothesized that the changed volume regulation observed in homozygotes could be considered as a manifestation of adaptation to impaired osmoregulation.

Conclusion

The expression of the genetic disorder producing diabetes insipidus in homozygous Brattleboro rats represents a long-time process, lasting at least from the late fetal period until the end of sexual maturation. Both vasopressin deficiency and the process of natural development of water metabolism are involved in its manifestation.

The syndrome of diabetes insipidus, i.e. polyuria and polydipsia, is absent till the end of the second week of postnatal life and it appears at the onset of weaning period. Increased plasma osmolarity is observed already at the age of 14 days, but polydipsia appears only at the age at which the weanling rat begins to be able to drink water. Polydipsia, of course, is accompanied by polyuria. Other differences between homo- and heterozygous Brattleboro rats also develop

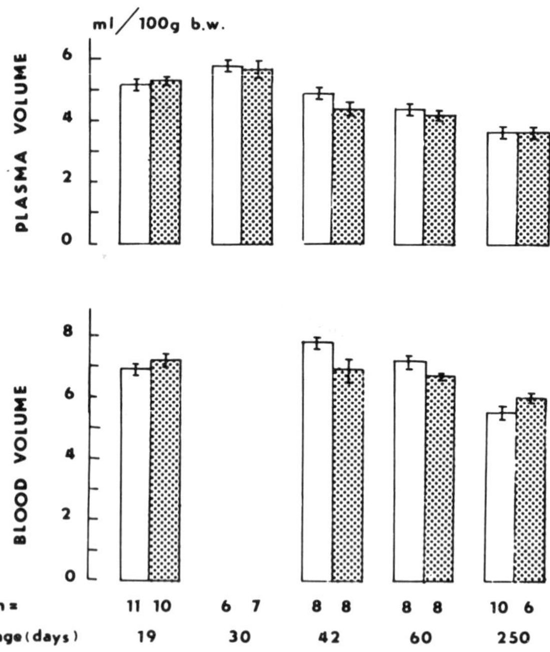

FIGURE 6. Plasma and blood volumes in young Brattleboro rats of different ages. Ordinate: volume in ml/100 g b.w. See FIGURE 1 for further details.

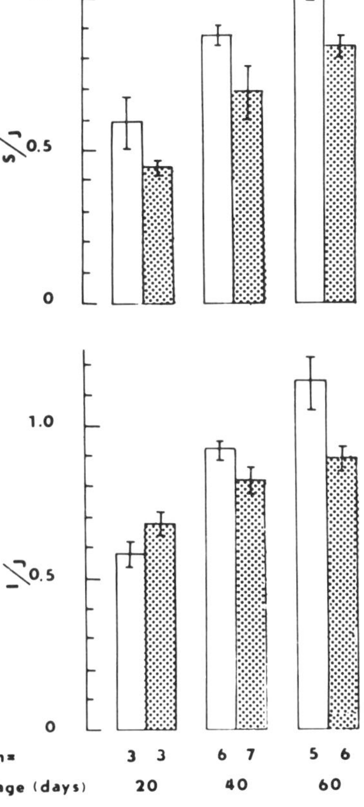

FIGURE 7. SNGFR distribution in kidneys of young Brattleboro rats of different ages. Ordinate: ratio of single nephron filtration rate; S/J: ratio of superficial to juxtamedullary nephron; and I/J: ratio of intercortical to juxtamedullary nephron. See FIGURE 1 for further details.

gradually in the course of ontogeny. Some of them appear before and others after the manifestation of diabetes insipidus. These abnormalities cannot be ascribed to the absence of vasopressin in blood because no vasopressin is detected in blood of normal rats at that early age. Absence of vasopressin in brain and its consequences seem to be the more probable cause of early functional abnormalities. Only at the age of sexual maturation, i.e. after the establishment of diabetes insipidus, SNGFR redistribution accompanied by increased blood volume becomes manifest. These abnormalities could be related to the absence of the volume-regulatory role of vasopressin, since prepuberty and puberty are periods of maturation of volume regulation in rats.

REFERENCES

1. VALTIN, H. & H. A. SCHROEDER. 1964. Familial hypothalamic diabetes insipidus in rats (Brattleboro strain). Am. J. Physiol. **206:** 425–430.
2. IFFT, J. D. 1972. An autoradiographic study of the time of final division of neurons in rat hypothalamic nuclei. J. Comp. Neurol. **144:** 193–204.
3. ANDERSON, C. H. 1978. Time of neuron origin in the anterior hypothalamus of the rat. Brain Res. **154:** 119–122.

4. SINDING, C., A. G. ROBINSON, S. M. SEIF & P. G. SCHMID. 1980. Neurohypophyseal peptides in the developing rat fetus. Brain Res. **195:** 177–186.
5. BOER, G. J., D. F. SWAAB, H. B. M. UYLINGS, K. BOER, R. M. BUIJS & D. N. VELIS. 1980. Neuropeptides in rat brain development. Prog. Brain Res. **53:** 207–227.
6. HELLER, H. 1947. Antidiuretic hormone in pituitary glands in new-born rats. J. Physiol. (London) **106:** 28–32.
7. SINDING, G. & A. G. ROBINSON. 1977. A review of neurophysins. Metabolism **26:** 1355–1368.
8. HELLER, J. 1960. The physiology of the antidiuretic hormone. V. The antidiuretic activity of rat plasma during ontogeny. Physiol. Bohemoslov. **9:** 289–293.
9. KŘEČEK, J., H. DLOUHÁ & J. KŘEČKOVÁ. 1961. The neurohypophysis and osmoregulation of new-born rats in weaning period. *In* Development of Homeostasis. P. Hahn, Ed.: 95–101. Publishing House of Czechoslovak Academy of Science. Prague.
10. DLOUHÁ, H., J. KŘEČEK & J. ZICHA. 1976. The renal concentrating ability of newly born Brattleboro rats (hereditary diabetes insipidus). Experientia **32:** 59–60.
11. KŘEČEK, J., J. KŘEČKOVÁ & H. DLOUHÁ. 1956. On problems of the regulation of water intake in new-born mammals. Physiol. Bohemoslov. 5 (Suppl.): 33–37.
12. BABICKÝ, A., I. OŠŤÁDALOVÁ, J. PAŘÍZEK, J. KOLÁŘ & B. BÍBR. 1970. Use of radioisotope techniques for determining the weaning period in experimental animals. Physiol. Bohemoslov. **19:** 457–467.
13. BABICKÝ, A., L. PAVLÍK, J. PAŘÍZEK, I. OŠŤÁDALOVÁ & J. KOLÁŘ. 1972. Determination of the onset of spontaneous water intake in infant rats. Physiol. Bohemoslov. **21:** 467–471.
14. BABICKÝ, A., DLOUHÁ, J. KŘEČEK & J. ZICHA. 1982. Suckling and weaning periods in Brattleboro rats. (Submitted for publication.)
15. DLOUHÁ, H., J. KŘEČEK & J. ZICHA. 1977. Growth and urinary osmolarity in young Brattleboro rats. J. Endocr. **75:** 329–330.
16. VALTIN, H., W. H. SAWYER & H. W. SOKOL. 1965. Neurohypophyseal principles in rats homozygous and heterozygous for hypothalamic diabetes insipidus (Brattleboro strain). Endocrinology **79:** 1058–1064.
17. SOKOL, H. W. & J. SISE. 1973. The effect of exogenous vasopressin and growth hormone on the growth of rats with hereditary hypothalamic diabetes insipidus. Growth **37:** 127–142.
18. KENNEDY, G. C. 1957. Development with age of hypothalamic restraint upon the appetite of the rat. J. Endocr. **16:** 9–17.
19. MACHO, L. 1979. Development of Thyroid and Adrenal Function During Ontogenesis. Veda Publishers. Bratislava, Czech.
20. ARIMURA, A., S. SAWANO, T. W. REDDING & A. V. SCHALLY. 1968. Studies on retarded growth of rats with hereditary hypothalamic diabetes insipidus. Neuroendocrinology **3:** 187–192.
21. NAIK, D. V. 1972. Growth and growth hormone studies in rats with hereditary hypothalamic diabetes insipidus. *In* Fourth Int. Congr. Endocr. Vol. 256: Abstract 179. Exc. Med. Found. Washington, D.C.
22. EDÉN, S., K. ALBERTSON-WIKLAND & D. ISAKSSON. 1978. Plasma levels of growth hormone in female rats of different ages. Acta Endocr. **88:** 676–690.
23. OJEDA, S. R. & H. E. JAMESON. 1977. Developmental pattern of plasma and pituitary growth hormone (GH) in the female rats. Endocrinology **100:** 881–889.
24. RIEUTORT, G. H. 1974. Pituitary content and plasma levels of growth hormone in fetal and weanling rats. J. Endocr. **60:** 261–268.
25. WALKER, D. G., M. E. SIMPSON, C. W. ASLING & H. M. EVANS. 1960. Growth

and differentiation in the rat following hypophysectomy at 6 day of age. Anat. Rec. **106:** 539–554.
26. Boer, G. J., R. M. Buijs, D. F. Swaab & G. J. De Vries. 1980. Vasopressin and developing rat brain. Peptides **1** (Suppl. 1): 203–209.
27. Winick, M. & A. Noble. 1965. Quantitative changes in DNA, RNA and protein during prenatal and postnatal growth in the rat. Devel. Biol. **12:** 451–466.
28. Buijs, R. M., D. N. Velis & D. F. Swaab. 1980. Immunocytochemical demonstration of vasopressin and oxytocin in the rat central nervous system by light and electron microscopy. Prog. Brain Res. **53:** 159–167.
29. Gauer, O. H., J. P. Henry & C. Behm. 1970. The regulation of extracellular fluid volume. Ann. Rev. Physiol. **32:** 547–595.
30. Bengele, H. H. & S. Solomon. 1974. Development of renal response to blood volume expansion in the rat. Am. J. Physiol. **227:** 364–368.
31. Garcia, J. F. 1957. Changes in blood plasma and red cells volume in the male rat as the function of age. Am. J. Physiol. **190:** 19–24.
32. Ramírez, V. D. 1973. Endocrinology of puberty. Handb. Physiol. Sec. 7, Vol. IV, Part 1: 1–28.
33. Dlouhá, H., B. Bíbr, J. Ježek & J. Zicha. 1976. Single nephron filtration rate ratios of superficial, intercortical and juxtamedullary nephrons in rats during development. Pflügers Arch. **366:** 277–279.
34. Baines, A. D. 1973. Redistribution of nephron function in response to chronic and acute solutes load. Am. J. Physiol. **224:** 237–244.
35. Chabardès, D., P. Poujeol, S. Deiss, J. P. Bonvalet & C. de Rouffignac. 1974. Intrarenal glomerular filtration rate distribution in salt loaded rats. Pflügers Arch. **349:** 191–202.

Discussion of the Paper

C. T. Hansen (*National Institutes of Health, Bethesda, Md.*): If you use a background strain that has a high lactating ability, and contrast that to a strain that has poor ability to lactate, would you not expect the occurrence of the syndrome to be related to the degree of the lactation performance of the mother?

J. Zicha (*Czechoslovak Academy of Science, Prague*): We started to breed Brattleboro rats in 1967. Since that time we have always used litters of heterozygous mothers mated with homozygous males. No attempts to alter the lactation of Brattleboro mothers were ever performed. Nevertheless, it is very unlikely that eventual malnutrition could change the timing of the developmental process. It was demonstrated in Wistar rats (Babický *et al. Physiol. Bohemoslov.* **22:**499 and 557, 1973) that increased or reduced size of the litter did not influence either the onset or the end of the weaning period.

G. J. Boer (*Netherlands Institute of Brain Research, Amsterdam*): We usually breed genetically homogeneous DI or heterozygote litters by mating homozygous DI females with homozygous DI or normal males, respectively, for studies on postnatal brain development. Our studies suggest that the pre- and postnatal conditions in the mother herself may influence brain water content in the developing rats. Also, an increase in diuresis has been seen in homozygous DI rats that were supplemented with vasopressin during the postnatal period.

J. Zicha: Thank you for your comments. As far as I know there was only one experiment in which vasopressin (lysine-VP) was supplemented to homozygous Brattleboro rats during the first month of life, i.e. in the suckling and weaning period (Wright & Kutscher, *Pharmacol. Biochem. Behav.* **6:** 505, 1977). Polyuria and polydipsia were exaggerated later in homozygous rats that were treated with vasopressin during the neonatal period. This indicates that homozygous rats must adapt their water metabolism to the absence of vasopressin during the first month of postnatal life. No data are available to indicate whether postnatal vasopressin treatment could modify the onset or the end of the weaning period.

As far as problems with lactation in homozygous Brattleboro mothers are concerned, they are probably worthy of future investigations using isotopic techniques of Dr. Babický (*Physiol. Bohemoslov.* **19:** 457, 1970 and **21:** 467, 1972).

INCIDENCE AND SOME FUNCTIONAL CHARACTERISTICS OF HYDRONEPHROSIS IN BRATTLEBORO RATS

Ian W. Henderson, J. Ann Oliver, Catherine M. Milne,*
and Richard J. Balment *

Department of Zoology
University of Sheffield
Sheffield S10 2TN, England

* *Department of Zoology*
University of Manchester
Manchester M13 9PL, England

The present report concerns spontaneous hydronephrosis in Brattleboro rats. The usual criteria and definitions regarding this strain of rat are employed: DI—rats homozygous for the trait of diabetes insipidus; HZ—rats heterozygous for the condition.[1] In addition, the defective gene responsible for diabetes insipidus of Brattleboro rats has been eliminated by successive HZ crosses to produce a "wild-type" Brattleboro rat.

Hydronephrosis is a generic term used to define a condition wherein there is a progressive dilatation of the renal pelvis, usually associated with degenerative changes (atrophy, destruction, or compression) of renal medullary tissue. Obstruction of urine flow at any site between pelvis and urethra is the usual cause, although the pathogenesis is far from uniform. The condition can be produced experimentally in a variety of ways, the most common being to partially occlude the ureter. There is, however, strong evidence at least in laboratory rodents, and possibly man, that heritable factors affecting the kidney and/or ureter are involved in the appearance of hydronephrosis.[2-7]

In the Brattleboro rats, the severe hydronephrosis in some cases leads to almost total destruction of the kidney (FIGURE 1). The colony initiated from two DI males and six HZ females (courtesy of Professor J. C. Sloper, Charing Cross Hospital Medical School, London) began to display hydronephrosis after approximately 30 generations. The incidence in subsequent generations is given for a random selection of rats in FIGURE 2. A breeding program is presently underway crossing DI with the Long Evans rat—the strain from which the Brattleboros were originally derived[8]—to eliminate the hydronephrosis. It is however premature to speculate upon the genetics of the disease and its relationship to the inheritance or otherwise of diabetes insipidus. The most salient features of the data presented in FIGURE 2 include: hydronephrosis affects the right kidney more than the left; male rats are more susceptible than females to hydronephrosis—moreover it is apparently absent from "wild-type" females; hydronephrosis is present from an early age, although its severity may increase. Weanling male DI rats display the condition and certainly bilateral hydronephrosis is more often present in older than younger rats; and the pattern of incidence of the condition in general excludes dietary or infectious causes and to date frank ureteral occlusions (spermatic vessels in males, for example) and other anatomical variations have not been detected.

0077-8923/82/0394-0021 $1.75/0 © 1982, NYAS

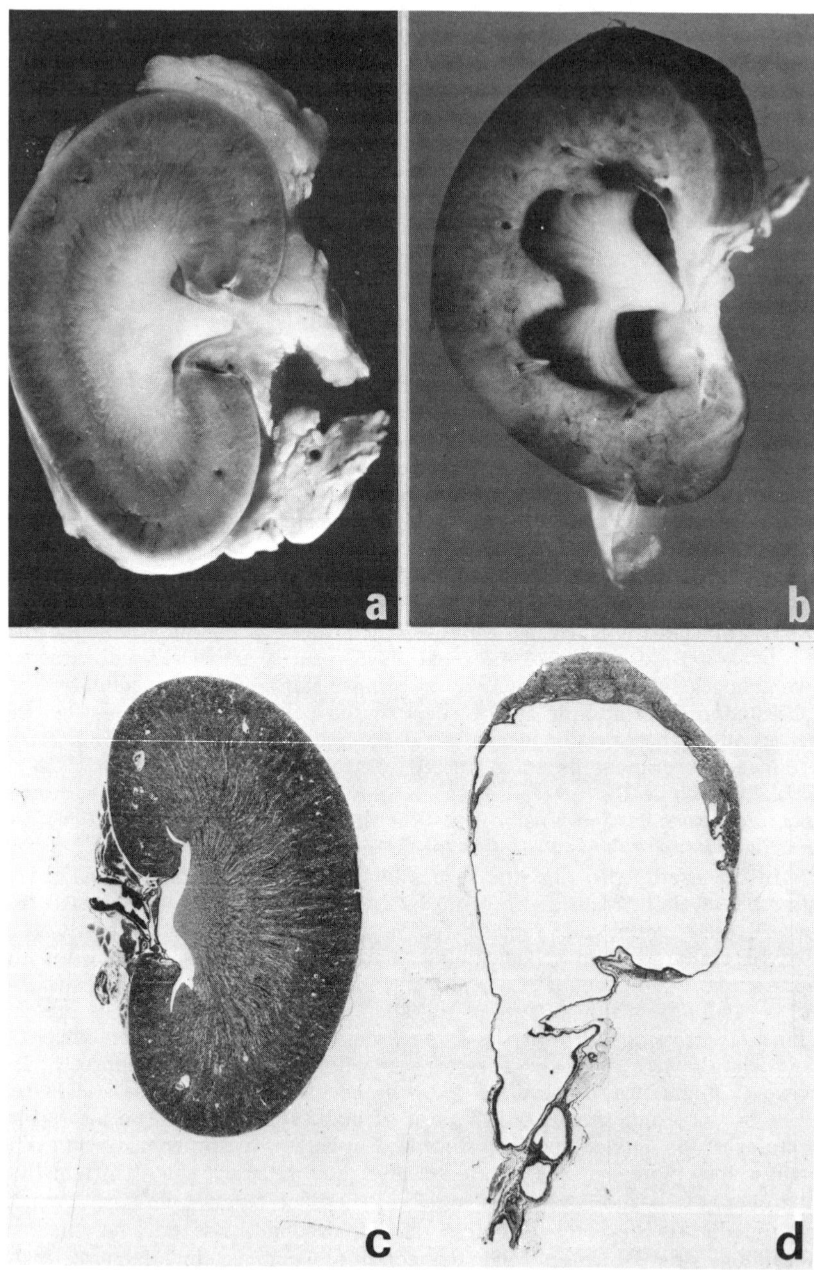

FIGURE 1. Hydronephrotic and normal kidneys of male DI Brattleboro rats: (a) normal kidney; (b) right kidney with moderate hydronephrosis; (c) histological section of normal kidney; and (d) histological section of kidney with severe hydronephrosis and tissue loss. c and d are stained with Haematoxylin/Azan and cut at six microns. (×2.5 throughout).

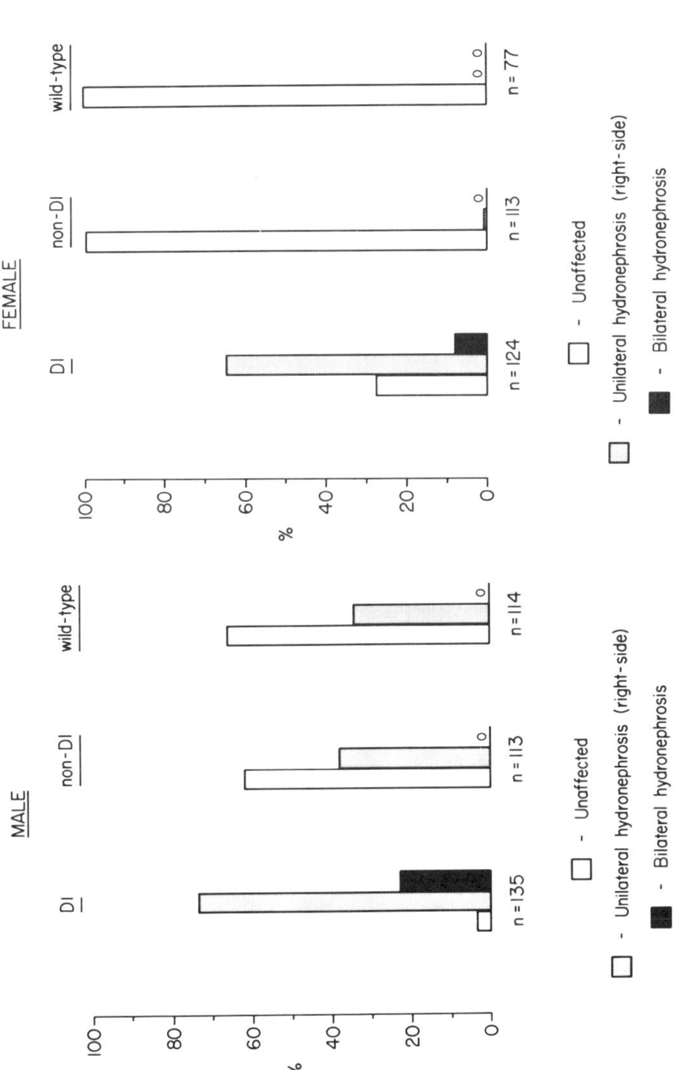

FIGURE 2. Incidence of hydronephrosis in male and female Brattleboro rats of the various genotypes. Non-DI = HZ. See text for details.

TABLE 1

NUMBER OF GLOMERULI, KIDNEY WEIGHTS, AND WATER CONTENT OF HYDRONEPHROTIC AND NORMAL KIDNEY OF MALE DI BRATTLEBORO RATS

	Right Kidney (hydronephrotic)	Left Kidney (normal)
Weight (mg)	915.9 ± 30.4	846.3 ± 35.9 *
Percentage Water	75.2 ± 0.53	73.6 ± 0.61 *
Glomerular Numbers	13,081 ± 774	15,302 ± 850 †

Means ± SE, N = 12 rats.
* $p < 0.001$, using paired *t* test.
† $p < 0.04$, using paired *t* test.

Some general aspects of the anatomy and function of the hydronephrotic kidney have been examined in DI male rats. The data are preliminary and largely of a descriptive rather than diagnostic nature, so that little can be said of the etiology or its detailed effects upon water and electrolyte metabolism and endocrine status. The following descriptions apply to kidneys with moderate hydronephrosis such as is shown in FIGURE 1b, as opposed to the severe condition that is sometimes found (FIGURE 1d). When such moderate hydronephrosis is present, the diseased kidney is visibly dilated and when drained of pelvic urine is actually heavier than normal. This is largely a result of a greater tissue water content (TABLE 1). Glomerular numbers, assessed by the method of Damadian *et al.*,[9] are reduced in the hydronephrotic kidney when compared with the contralateral normal organ: on average the hydronephrotic kidney has 2,220 ± 824 (mean ± SE) fewer glomeruli.

The overall water and electrolyte balances of normal male DI rats have been compared with those possessing right-sided hydronephrosis. Animals were held in metabolism cages [10] and daily intakes and outputs determined. There are no obvious differences between the two groups and it would appear that

TABLE 2

WATER AND SOLUTE TURNOVER IN MALE DI BRATTLEBORO RATS WITH UNILATERAL (RIGHT SIDE) HYDRONEPHROSIS COMPARED WITH THOSE WITH NORMAL KIDNEYS

	Normal Rats	Hydronephrotic Rats
Body weight (g)	309 ± 22.7	339 ± 18.3
Water intake (ml)	80.4 ± 4.5	75.2 ± 5.9
Urine (ml)	72.1 ± 4.7	67.5 ± 5.6
Urinary osmolarity (mOsm/L)	130.0 ± 8.5	127.3 ± 14.4
Osmotically free water excretion (ml)	41.7 ± 4.1	40.2 ± 6.1
Sodium excretion (mEq)	0.224 ± 0.02	0.219 ± 0.01
Sodium balance (mEq)	0.67 ± 0.02	0.59 ± 0.04
Potassium excretion (mEq)	1.06 ± 0.07	1.04 ± 0.05
Potassium balance (mEq)	0.58 ± 0.12	0.45 ± 0.06
Calcium balance (mEq)	0.514 ± 0.02	0.469 ± 0.02
Magnesium balance (mEq)	0.262 ± 0.01	0.229 ± 0.01
Chloride balance (mEq)	0.324 ± 0.01	0.315 ± 0.03

Means ± SE. 5 animals per group. Units (ml, mEq) are expressed per 100 g body weight per 24 hours. Balances are intake less urinary output.

potential imbalances resulting from functional lesions in the hydronephrotic kidney are compensated for by the left kidney (TABLE 2). The elevated water turnover, copious hypo-osmotic urine, as well as electrolyte balances and excretory rates are similar in unilaterally hydronephrotic and normal DI rats, comparing favorably with those published previously.[10] In these animals renal venous renin activity of the normal kidney, at 678 ± 35 ng equivalents angiotensin I/ml/16-h incubation, was significantly greater than that of the venous effluent of the hydronephrotic kidney (495 ± 26 ng equivalents; $p < 0.005$; $N = 15$).

Acute studies separating right from left renal functions have however revealed significant differences between hydronephrotic and normal kidneys. DI male rats were anesthetized with Inactin (100 mg/kg body wt, i.p.) and ureteral, venous, and arterial catheters implanted. Either 0.45% sodium chloride solution or Bretag's solution [11] was infused intravenously and renal clearances of 30 minutes collected. The renal clearance of ^{125}I-labeled sodium

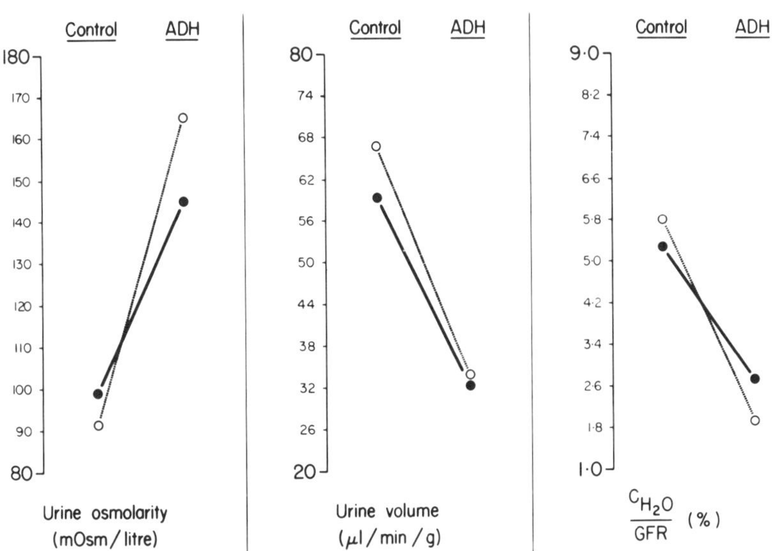

FIGURE 3. The renal response to intravenous infusion of antidiuretic hormone (ADH) in normal and hydronephrotic kidneys of male DI Brattleboro rat. (○—○) = Left kidney, normal and (●—●) = Right kidney, hydronephrotic.

iothalamate was taken as a measure of glomerular filtration rate (GFR). In rats infused with hypo-osmotic saline (150 μl/min), GFRs and urine production rates were similar in hydronephrotic and normal kidneys (GFRs: 730.9 ± 73.1 and 756.9 ± 86.5; urine production rates: 59.5 ± 10.4 and 66.9 ± 6.9 μl/min/g kidney, respectively; $N = 6$). Intravenous vasopressin (5 μU/min for 60 minutes) produced a greater effect in terms of increased urinary osmolarity and decreased fractional osmotically free water excretion in the normal than in the diseased kidney (FIGURE 3).

Rats infused with Bretag's solution (260 μl/min) showed the hydronephrotic kidney to be relatively sodium- and chloride-losing; the average sodium and chloride excretory rates in the hydronephrotic kidney were, respectively, 0.63 and 1.30 μmols/min/g kidney compared with 0.16 and 0.94 in the normal (left) kidney. The hydronephrotic kidney renal-plasma flow, determined with *para*-amino-hippuric acid (PAH), was approximately 60% that of the normal kidney, and hydronephrotic kidney urine had a lower pH (5.56 versus 5.83).

The development of hydronephrosis, previously noted in Brattleboro rats,[12] has been related to age, sodium intake, and the induction of hypertension.[13] In other obstructive nephropathies, both clinical and experimental, the inconsistent hypertension [14-16] probably reflects varying degrees of renal tissue damage and insufficiency with end-state sodium retention.[17] In general, hydronephrotic kidneys in which tissue loss is not massive are sodium-losing [3, 18] and renal medullary damage may account for defective diluting and concentrating capacities in the distal nephron,[19, 20] to give on occasion an apparent state of nephrogenic diabetes insipidus.[6, 21, 22] The associations between reduced renal-blood flows [3, 23-25] and altered activities of the renin-angiotensin system [26-28] and reno-medullary prostaglandins [29, 30] are less clearly defined. The hydronephrosis described in the present report is particularly pertinent to possible interplay between vasopressin, renin, adrenocortical, and even gonadal steroids. DI rats have high plasma renin activities and a reduced adrenocortical activity.[31] In addition there is a different plasma renin response to exogenous vasopressin in males and females and this is dependent on gonadal factors.[32] Thus, together with the varying incidences of hydronephrosis both between the sexes and between DI and HZ rats, it is clear that this model will provide insight into factors that cause and/or exacerbate hydronephrosis.

References

1. Valtin, H. 1967. Hereditary hypothalamic diabetes insipidus in rats (Brattleboro strain). A useful experimental model. Am. J. Med. **42:** 814–827.
2. Lozzio, B. B., A. I. Chernoff, E. R. Machado & C. B. Lozzio. 1967. Hereditary renal disease in a mutant strain of rats. Science **156:** 1742–1744.
3. Freidman, J., J. M. Hoyer, B. McCormick & J. E. Lewy. 1979. Congenital unilateral hydronephrosis in the rat. Kidney Int. **15:** 567–571.
4. Astarabadi, R. & E. T. Bell. 1962. Spontaneous hydronephrosis in albino rats. Nature **195:** 392–393.
5. Silverstein, E. L., L. Sokoloff, O. Mickelson & G. E. Jay. 1961. Primary polydipsia and hydronephrosis in an inbred strain of mice. Am. J. Pathol. **38:** 143–160.
6. Silverstein, E. & L. Tobian. 1961. Pitressin resistant diabetes insipidus with massive hydronephrosis. Am. J. Med. **30:** 819–822.
7. Machado, E. A. & B. B. Lozzio. 1972. Congenital renal structural alterations in a mutant strain of rats. Invest. Urol. **10:** 78–83.
8. Valtin, H., H. A. Schroeder, K. Benirschke & H. Sokol. 1962. Familial hypothalamic diabetes insipidus in rats. Nature **196:** 1109–1110.
9. Damadian, R. V., E. Shawayri & N. S. Bricker. 1965. On existence of non-urine forming nephrons in the diseased kidney of the dog. J. Lab. Clin. Med. **65:** 26–39.
10. Balment, R. J., I. Chester Jones, I. W. Henderson & J. A. Oliver. 1976. Effects of adrenalectomy and hypophysectomy on water and electrolyte metabolism in male and female rats with inherited hypothalamic diabetes insipidus (Brattleboro strain). J. Endocrinol. **71:** 193–217.
11. Bretag, A. H. 1969. Synthetic interstitial fluid for isolated mammalian tissue. Life Sci. **8:** 319–329.

12. Dlouhá, H., J. Křeček & J. Zicha. 1977. Hypertension in rats with hereditary diabetes insipidus—the role of age. Pflüg. Arch. Eur. J. Physiol. **369:** 177–182.

13. Dlouhá, H., J. Křeček & J. Zicha. 1980. Unilateral nephrectomy and intake of 0.6% NaCl solution as an hypertensinogenic stimulus in Brattleboro rats: its age dissociation. Physiol. Bohem. **29:** 97–106.

14. Weidmann, P., C. Beretta-Piccoli, D. Hirsch, F. C. Reubi & S. G. Massry. 1977. Curable hypertension with unilateral hydronephrosis. Studies on the role of circulating renin. Ann. Intern. Med. **87:** 437–440.

15. Sparks, J. C. & D. Susic. 1975. Rapid onset of salt induced hypertension in rats with hereditary hydronephrosis. Res. Commun. Chem. Pathol. Pharmacol. **11:** 425–434.

16. Silk, M. R. 1967. Hypertension secondary to hydronephrosis in adult and young animals. Invest. Urol. **5:** 30–34.

17. Legrain, M., M. Bilker & R. Kuss. 1967. Obstructive nephropathy in adults. Proc. 3rd International Congress of Nephrology. **2:** 336–353. Karger. Basel/New York.

18. Suki, W., G. Eknoyan, F. C. Rector & D. W. Seldin. 1966. Patterns of nephron perfusion in acute and chronic hydronephrosis. J. Clin. Invest. **45:** 122–131.

19. Eknoyan, G., W. N. Suki, M. Martinez-Maldonado & M. A. Anhalt. 1970. Chronic hydronephrosis: observations on the mechanism of the defect in urine concentration. Proc. Soc. Exptl. Biol. Med. **134:** 634–639.

20. Berlyne, G. M. & A. Macken. 1962. On the mechanism of renal inability to produce a concentrated urine in chronic hydronephrosis. Clin. Sci. **22:** 315–324.

21. Berlyne, G. M. 1961. Distal tubular function in chronic hydronephrosis. Quart. J. Med. **30:** 339–356.

22. Shapiro, S. R., S. Woerner, R. D. Adelman & J. M. Palmer. 1978. Diabetes insipidus and hydronephrosis. J. Urol. **119:** 715–719.

23. Idbohrn, H. & A. Muren. 1957. Renal blood flow in experimental hydronephrosis. Acta Physiol. **38:** 200–206.

24. Rao, N. R. & R. H. Heptinstall. 1968. Experimental hydronephrosis. A microangiographic study. Invest. Urol. **6:** 183–204.

25. Hsu, C. H., T. W. Kurtz, J. Rosenzweig & J. M. Weller. 1977. Intrarenal haemodynamics and ureteral pressure during ureteral obstruction. Invest. Urol. **14:** 442–445.

26. Susic, D., J. C. Sparks & D. Kentera. 1975. The renin-angiotensin system in rats with hereditary hydronephrosis. Pflüg. Arch. Eur. J. Physiol. **358:** 265–274.

27. Vaughan, E. D., F. R. Buhler & J. H. Laragh. 1964. Normal renin secretion in hypertensive patients with primary unilateral chronic hydronephrosis. J. Urol. **112:** 153–156.

28. Belman, A. B., K. A. Kropp & N. M. Simon. 1968. Renal pressor hypertension secondary to unilateral hydronephrosis. New Engl. J. Med. **278:** 1133–1136.

29. Susic, D., J. C. Sparks, E. A. Machado & D. Kentera. 1978. The mechanism of renomedullary antihypertensive action: haemodynamic studies in hydronephrotic rats with one-clip kidney renal clip hypertension. Clin. Sci. Mol. Med. **54:** 361–367.

30. Nishikawa, K., A. Morrison & P. Needleman. 1977. Exaggerated prostaglandin biosynthesis and its influence on renal resistance in the isolated hydronephrotic kidney J. Clin. Invest. **59:** 1143–1149.

31. Milne, C. M., R. J. Balment, I. W. Henderson, W. Mosley & I. Chester

JONES. 1982. Adrenocortical function in the absence of vasopressin. (This volume.)

32. OLIVER, J. A., I. W. HENDERSON, A. MCKEEVER & R. J. BALMENT. 1982. The renin-angiotensin system in the absence of antidiuretic hormone. (This volume.)

DISCUSSION OF THE PAPER

J. F. MORRIS (*Oxford University, Oxford, England*): Have you any information about the age at which the hydronephrosis develops? You mentioned just one rat of 200 grams. The other question, with relationship to human hydronephrosis: what is the incidence of inferior accessory renal arteries in these rats? It's quite high in humans and is said to be associated with ureteric obstruction and hydronephrosis.

I. W. HENDERSON: I can't answer your second question. Regarding your first question, the incidence of hydronephrosis is less in weanling rats, although it is apparent at 25 days, and it is aggravated with age.

J. ZICHA (*Czechoslovak Academy of Science, Prague*): Could you compare the functional changes in hydronephrotic kidneys of homozygous Brattleboro rats with the data reported in MRC/H rats? Machado and Lozzio (*Invest. Urol.* **10:** 78, 1972) described a similar sex difference in the incidence of hydronephrosis as well as the greater susceptibility of the right kidney in this strain of rats.

I do not know the extent of the genetic determination of hydronephrosis in Brattleboro rats. Nevertheless, we are sure that the incidence of hydronephrosis in homozygous rats could be modified by postnatal intervention (FIGURE 1). The basal incidence of hydronephrosis among DI homozygous female rats of our Brattleboro colony was about 8%. If homozygous females started to drink 0.6% NaCl solution instead of water in adulthood (NaCl 80), no significant change in the occurrence of hydronephrosis was observed. However, the administration of hypotonic saline as the only drinking fluid prior to sexual maturation (NaCl 30, NaCl 35, NaCl 50) resulted in a much higher incidence of hydronephrosis.

H. W. SOKOL (*Dartmouth Medical School, Hanover, N.H.*): There is a strain of mice with primary polydipsia in which the males also have a much higher incidence of hydronephrosis (Silverstein *et al., Am. J. Pathol.* **38:** 143–160, 1961). Merely restricting water intake to normal levels virtually eliminated hydronephrosis, indicating that hydronephrosis was not a genetic defect, per se, but a consequence of function. With regard to hydronephrosis in the Brattleboro rat, Jan Möhring found (personal communication) a rather high incidence (ca. 10%) in the Long-Evans rat colony he maintained in Heidelberg.

HENDERSON: In reply to Dr. Zicha, it would not be really appropriate to compare these Brattleboro rats with the MRC/H strain since the latter display additional gross renal abnormalities and they are derived from Wistar rats. It is very difficult to compare all the information in the published studies at this point, although, of course, dietary sodium chloride and basal water turnover may affect the development of hydronephrosis. In reply to Dr. Sokol's comment I am not suggesting for one moment, that the syndrome is unique to the Brattleboro rat. Indeed it has been described in a number of laboratory rodent colonies.

M. MILLER (*Veteran's Administration Medical Center, Syracuse, N.Y.*): In regard to Dr. Sokol's comment on the water intake in the polydipsic mouse, the mouse with nephrogenic diabetes insipidus also has an exceedingly high incidence of hydronephrotic change. This is a characteristic of these animals and occurs in association with the development of the diabetes insipidus. Since androgens have a very profound effect on tissue growth, is there any evidence that what you may be seeing is not a sex-linked phenomenon, but rather an effect of the high androgen environment in these male animals? Are you undertaking or do you plan to undertake any studies to eliminate the androgen effect early on in life and look at whether or not the incidence of subsequent hydronephrosis is the same?

HENDERSON: Yes, I think androgens and estrogens are probably relevant, especially since they apparently affect renin release in addition to the actions you allude to. It does of course occur in young animals that presumably have not been exposed to a high concentration of gonadal steroids.

L. BANKIR (*Hôpital Necker, Paris, France*): If I understood correctly you have bred these rats for 10 years and in the beginning you did not observe such a high frequency of hydronephrosis. It occurred progressively and became worse more recently, is that right?

HENDERSON: It has emerged within the last two years to produce the sort of incidence described.

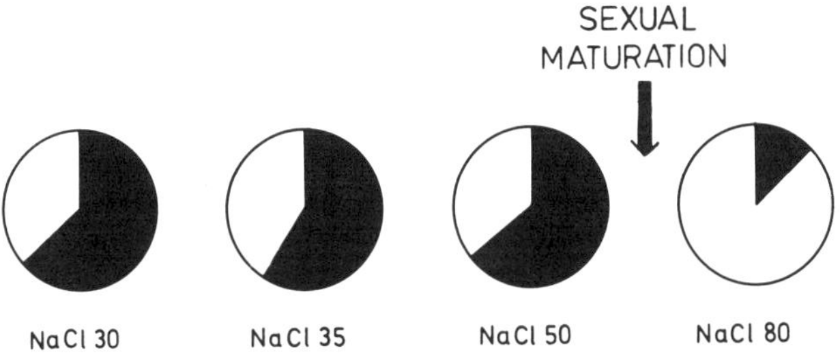

FIGURE 1. Frequency of hydronephrosis in female homozygous Brattleboro rats. The age-dependent effect of saline drinking.

BANKIR: In our colony we have observed cases of hydronephrosis a few times, also on the right side only, and I am now concerned that it might become more frequent.

You also mentioned a possible ureter obstruction to explain the occurrence of hydronephrosis. Couldn't it be, on the contrary, a too loose valve at the opening of the ureter in the bladder, which would enable a chronic urine reflux? DI rats often have a very large and distended bladder, which means that the intrabladder pressure must be high and could favor a reflux.

HENDERSON: Yes, I think that is a very good point, which will obviously have to be tested. However, it is the right kidney that is predominantly affected.

C. I. JOHNSTON (*Prince Henry's Hospital, Melbourne, Australia*): It is common clinically that reflux nephropathy affects one side more severely than the other. I don't think, therefore, that unilateral hydronephrosis rules out hydronephrosis due to reflux.

HENDERSON: Why should it be predominantly right sided?

JOHNSTON: One side becomes more incompetent than the other and therefore it goes up the side with the lowest pressure and resistance. For example, in pregnancy the right ureter always dilates more than the left.

CARE AND BREEDING OF THE BRATTLEBORO RAT: A PANEL DISCUSSION

Panel: Carl T. Hansen, Eduard J. Gubbels, and Hans Kappel

H. KAPPEL (*Blue Spruce Farms, Inc., Altamont, N.Y.*): On behalf of Dr. Hansen and Dr. Gubbels I would like to express our appreciation for the opportunity to participate in this meeting on the Brattleboro rat. I would like to comment on hydronephrosis in the Brattleboro rat. Hydronephrosis is a genetic problem in some rat colonies the severity of which may be increased if proper precautions are not taken. We know of a colony of Sprague-Dawley rats in which 18% of the animals were hydronephrotic. Our Long-Evans rat, as I recall, had a 2% incidence. If one develops a colony from just a few pairs, one risks a high degree of hydronephrosis if it is present in the original breeders. Hydronephrosis can be reduced or eliminated by selecting future breeders from non-hydronephrotic parents.

I'd like to comment briefly on some problems we've had since we began raising DI rats. For example, we have experienced failures to reproduce, small litters, and cannibalism. The first step we took to solve some of these problems was to breed the DI gene back into a good producing Long-Evans stock. This accomplished two very important things: better productivity and a genetically related normal Long-Evans control. Some of the initial breeding problems may also have been caused by stresses such as environmental changes, housing and bedding conditions, and size of colony, just to mention a few. Since the growth of the colony to an appreciable size, most of the problems are no longer evident.

D. M. GASH (*University of Rochester, Rochester, N.Y.*): I understand that it is possible to breed pure homozygotes, is that true?

H. KAPPEL: Yes, although at first we were unsuccessful. We feel this was due to the small population of animals and possibly some stresses. When the size of the colony was increased, these problems seem to have disappeared, and now we are mating homozygous pairs, i.e. DI/DI, with an average litter size of 8.0. In our DI/Heterozygote matings, the average litter size is 8.6.

H. W. SOKOL (*Dartmouth Medical School, Hanover, N.H.*): I have some slides that compare reproductive capacity of homozygous (DI) and heterozygous (HZ) Brattleboro rats. The litter size of DI/DI matings is similar to that reported by Mr. Kappel. In the first study (TABLE 1) 10 HZ and 10 DI littermates (one of each in a cage) was bred to the same male and about 80% of the females became pregnant (apparently two of the males were impotent!). The average litter size was smaller in the DI rats than in the heterozygotes, i.e. 8 versus 11 pups, respectively. Offspring survival rate was also somewhat less for DI rats—perhaps due to trauma at birth. TABLE 2 includes data collected by Jan Möhring when he compared LE/LE with HZ♀/DI♂ matings, indicating similar reproductive capacity. Results of my second study, which included three female littermates (HZ, DI, and DI treated with vasopressin) bred to the same male also appear in TABLE 2. This time all the females became pregnant. The number of pups per litter was the same as before, with 11 for the heterozygotes and eight in both control (peanut oil-injected) and VP-treated DI females. It was discouraging that vasopressin therapy throughout the pregnancy

0077–8923/82/0394–0030 $1.75/0 © 1982, NYAS

TABLE 1

A COMPARISON OF REPRODUCTIVE CAPACITY BETWEEN HETEROZYGOUS (HZ) AND HOMOZYGOUS (DI) BRATTLEBORO RATS

	HZ	DI
Number of rats mated	10	10
Number of pregnant rats	8	7
Number of days between placing DI ♂ in cage and delivery of litter (range)	25.4 (24–27)	29.7 (24–42)
Number of pups in litter (range)	11 (2–14)	8 (4–11)
Proportion of pups surviving at time of weaning	96%	84%

did not improve litter size; it did, however, decrease the average number of days between initial exposure to a male and parturition. Untreated DI females, on the average, conceived one cycle later than heterozygotes. Despite the reduced number of pups per litter in DI/DI matings, I encourage the use of this method of breeding since it obviates measuring fluid and urine osmolality for genotype identification and avoids the danger of an incorrect assignment.

LEDERIS: Are there any comments to Dr. Sokol's slides?

J. F. MORRIS (*Oxford University, Oxford, England*): With respect to the question of breeding homozygotes to homozygotes, for a year or so in Oxford, D. Chapman and I have been doing that and getting eight or nine pups. I know

TABLE 2

THE REPRODUCTIVE CAPACITY OF NORMAL LONG-EVANS (LE) AND HETEROZYGOUS (HZ) AND HOMOZYGOUS (DI) BRATTLEBORO RATS

	Möhring *		Sokol †		
	LE	HZ	HZ	DI	DI(+ VP)
Rats mated	36	34	10	10	10
Pregnant rats	36	34	10	10	10
Days between mating and parturition (range)	29.1 (23–70)	27.4 (22–59)	28.7 (24–43)	32.7 (24–40)	29.7 (24–39)
Pups in litter (range)	9.8 (1–16)	11.0 (1–17)	10.8 (6–13)	7.9 (1–11)	7.8 (3–13)
% of pups weaned	96	94	98.1	92.4	92.3

* LE females were bred to LE males and HZ females were bred to DI males in Heidelberg, Germany. Routine breeding data (1976, 1978–79) kindly supplied by Jan Möhring. Some females were bred several times.

† These rats were bred to DI male rats in Hanover, N.H. (1979). This was a pilot study using female littermates bred to the same male. One DI female in each cage was injected daily with 100 mU vasopressin tannate in oil (VP); the other DI and HZ females received peanut oil vehicle only. All females were bred only once.

that a colleague at Charing Cross Hospital in London has been doing this for some years, with no problems at all; so it is quite feasible. I am concerned about your comment, Mr. Kappel, about the size of the breeding population; you didn't define what that size was. We regularly have something like 14 breeding pairs that stay together and have found really very few unexplained problems. What do you regard as optimum colony size? While I was here at Dartmouth we found that non-dusty bedding was very important in the care of the young very early on. I think small things like this can just tip the balance between survival and non-survival of the homozygote pups and I suspect that some of the pup cannibalism may be due to the fact that the animals are a little sick and the mothers respond to that, rather than to something more intrinsic.

KAPPEL: The size of the population I mentioned is now around 200 breeding pairs. It is quite true that dusty shavings and sickness cause some of the stresses. My observation is that in a large population of animals, some of these mechanical stresses may not play as large a role as we may suspect. We as humans feel much more secure in company than in isolation, and so do rats. I think that psychological aspects may also play a role. We had similar problems with another mutant strain of rats, the "jaundiced" or "Gunn" rat, which is also reproduced by Homozygous/Homozygous and Homozygous/Heterozygous matings.

MORRIS: When you start doing a new experiment it often doesn't work correctly two or three times, and then the fourth and subsequent times, it goes well. Could it not be that gaining experience, rather than colony size, is what counts?

LEDERIS: Dr. Hansen, do you have any comments?

C. T. HANSEN (*National Institutes of Health, Bethesda, Md.*): Well, first of all, I suspect the success in the homozygote/homozygote matings depends on the heterogeneity of the background stock. I am sure that a very heterogeneous stock should have far less problems that a homogeneous stock. I want to touch on that point a little bit later. There is one point that I think that everybody should keep in mind when dealing with double homozygote matings: the gene for diabetes insipidus may not be fully expressed. Although expecting the old classical 3:1 ratio or 50:50 ratio, depending on the type of mating, and after looking at many homozygotes, I suspect there is some significant variation in their expression of diabetes insipidus, i.e. it is not simply an all-or-none characteristic. There can be some master selection in favor of some individuals within a certain range and only these individuals survive and reproduce offspring. In effect, there is a counteracting force of selection towards the more normal end. Thus, I have not chosen to go to this type of mating system.

L. BANKIR (*Hôpital Necker, Paris, France*): I want to draw attention to another point that may be important in DI/DI matings. It is important that the air in the rat rooms be humidified. When we first bred the rats in a normal room without humidification, we had little success. Some females became pregnant but they often ate their pups at birth and the number of pups per litter was very small (2–4). We now have a room with constant humidity (70% saturation) and it increased the success of our breeding with a mean of 8 (4–11) offspring per litter. Nearly all females get pregnant and they don't eat their offspring, so it is very important to have the air properly humidified. Reduced fertility in winter was also corrected by a 12 hour light–12 hour dark cycle, which is now used all year round.

LEDERIS: Are there any comments with regard to optimal relative humidity?

E. J. GUBBELS (*Central Institute for the Breeding of Laboratory Animals,*

Zeist, the Netherlands): We breed our colonies (homozygous males/heterozygous females) with the normal humidity of 60–70% and do not have any problems. We changed our breeding system from permanent mating to intermittent mating, but that didn't seem to influence the selection against the homozygous genotype. We do not do homozygous/homozygous matings because of the large amount of labor needed in that type of breeding system.

BANKIR: In our case, humidity was important only when we tried to mate DI females with DI males. We had good fertility when mating heterozygous females with DI males. It was only when we attempted DI/DI matings that we had poor results. When we humidified the room, fertility improved. In commercial firms like yours, with well-equipped animal rooms that are humidified, those problems do not arise. When investigators just want to breed a small colony of DI rats for their own needs they might not think of this problem and keep their rats in a room that is too dry.

GUBBELS: There might be a more important point than whether you get seven or eight offspring from the DI/DI matings, namely small colonies of Brattleboro rats. Many investigators are breeding DI rats and there seem to be differences arising among the populations that may account for the differences between the results obtained in the different research institutes. We may be comparing apples and pears. In all kinds of research everybody is looking for inexpensive ways of getting animals since animals are the last part of the research that they want to pay for. In my country every university has its own breeding colonies; then the scientists go to international conferences and try to compare their results with results from other laboratories. That they find differences is not so surprising, because every new population is founded with only a few breeding animals; this leads to some kind of genetic bottle-neck and ultimately to differences between subpopulations. One of the most important things that this kind of symposium could accomplish is to establish an international agreement for the standardization of the Brattleboro rat.

HANSEN: I want to describe briefly the program I have under way that deals with this particular point. We have maintained the Brattleboro rat for about 10 years and are one of the primary sources for breeding stock. (At the National Institutes of Health I am responsible for maintaining a variety of unique model systems.) I am as concerned as Dr. Gubbels is that eventually a whole series of subpopulations of DI rats will be scattered around the countryside and in different places in the world. Many investigators go to an enormous amount of effort to standardize their experimental techniques and make very sure that the reagents, the chemicals, etc. are absolutely pure; then they buy a garden variety mouse or rat at the cheapest price and wonder why their results don't agree with those obtained in another laboratory, not realizing that the animal itself is the cause. I am sure that as this meeting progresses we will have more examples of such disagreements. Thus, I think the strategy of Blue Spruce Farms to continue to develop the di gene on the Long-Evans background will eventually lead to a good uniform heterogeneous and vigorous stock.

I would like to describe very briefly an alternative program in which I am in the process of establishing the di gene in a number of inbred rat strains, to yield what is called a congenic rat. The purpose for doing this is not only to stabilize the background, but to develop better controls with which to compare the Brattleboro gene. As pointed out this morning, the value of heterozygous Brattleboro rats as controls is limited. I am introducing this gene into six inbred strains, three of which have been defined on the basis of their behavior and three on blood pressure levels.

The hypertensive strains include the spontaneous hypertensive rats (SHR) in which a systolic pressure of 200 mm Hg is found in males at about 10 weeks of age. (We have recently discovered other strains that are mildly hypertensive, with blood pressures of 150 mm Hg in one strain and 120 mm Hg in another.) The resulting genotypes will be the normal SHR +/+, the heterozygote SHR +/di, and the homozygote SHR di/di. Since the genetic background of these three groups are identical, any differences would then be due to the effect of this di gene. The other three strains exhibit specific behavioral patterns. Roman high avoidance (RHA) rats have been selected for their response to avoid electric shock in a standard environment. The other two strains are the Maudsley reactors and the Maudsley nonreactors, identified by the number of times the rats urinate and defecate in an open field test; the non-reactors constitute the normal group. This breeding scheme transfers the di gene to several different strains, which is very valuable. For example, by comparing SHR +/+ with SHR di/di rats one could assess the role of vasopressin in the development of hypertension. Since it is difficult to extrapolate results from an inbred strain (which represents a unique repeating constellation of genes) to a broader population, the presence of the di gene in several different strains provides researchers with a broader spectrum so that their results have greater credibility. It is possible that compensatory mechanisms found in the homozygous Brattleboro rat, as mentioned by Dr. Valtin in his introduction this morning, may be different in the SHR di/di rat. Although the process of sorting all these things out will be laborious, it has the potential of elucidating the effects of the di gene per se.

I would like to make one final comment on the phenotypic expression of the gene in the various strains. I am able to distinguish Brattleboro homozygotes from heterozygotes by six weeks of age at which time the heterozygotes are larger. In the case of the SHR strain, the homozygote (i.e., SHR di/di) can be unequivocably identified at 2.5 weeks of age when they are runts weighing only about 10 grams. We attribute this poor growth to poor lactation since SHR rats stop lactating around 14 days post partum; thus, the two-week old DI pups are relatively deprived of liquid until they are able to drink water, at which time they exhibit catch-up growth. The RHA di/di rat can also be identified as early as three weeks of age on the basis of size. An interesting defect is observed in the ACI strain bearing the di gene in which 20% of the rats have only one kidney. This may be a good model in which to study the development of hydronephrosis. With these examples I have tried to point out that superimposing the di gene on various genetic backgrounds can provide insights into the functioning of this gene, since different patterns are found in different inbred strains.

T. BERL (*University of Colorado, Denver, Col.*): Given your comments, I wonder about the criteria that are to be used to define the rat as truly vasopressin deficient. How do you determine that this is really a DI rat? Some of these rats may have minimal amounts of vasopressin around or at least some form of vasopressin that would be biologically active. How do you select the animals and designate them as being truly vasopressin deficient, for example?

HANSEN: I was afraid that you were going to ask that question. Very basically, I think that everybody would have to agree that the level of vasopressin in the homozygotes is zero. If there is a variation in the growth of the individual DI rats (this is a particularly important point because most breeders identify the rats by size), one could conceivably, by appropriate selection, produce an animal that looked phenotypically normal.

K. Berecek (*University of Alabama, Birmingham, Ala.*): You define a homozygous DI rat on the basis of lack of vasopressin in the plasma and urine, and large fluid intake and urine output. If I obtain a homozygous rat from you, will it be functionally the same as a homozygous Brattleboro rat from Blue Spruce Farms? The reason I am asking this is based upon problems we've encountered with inbred spontaneously hypertensive rats (SHR). Such rats from different suppliers may have equivalent magnitudes of hypertension; however, these animals respond differently to various stimuli, so they are really not the same. It is difficult to feel confident that the work that I do on one particular colony is at all comparable to the work that somebody else does on another colony. The only characteristic various SHR colonies share appears to be high arterial pressure. They can be very different depending upon the way they are bred or the environmental conditions in which they are raised. Is this the same with the DI rats? As an investigator, should I limit all my studies to one particular colony?

Hansen: The answer to the last question is an emphatic yes. While the gene is fully penetrant, the secondary responses vary from group to group because the genetic background is different. Although every DI rat is going to have excessive water intake and urine output, physiological responses are going to be somewhat different because they are controlled by different genes.

Lederis: I would like to add a comment to this; it pertains to your question as well as to your answer. Recently, we ran low on available experimental animals in our colony and purchased some. My laboratory people observed first that these animals appeared to drink somewhat less and to wet their cages a little less than our local DI/DI rats. Eventually their pituitaries were removed and assayed, and lo and behold appreciable amounts of vasopressin were found. How do you explain that DI rats from a reputable supplier had vasopressin in their pituitaries?

Gubbels: I think that the gene is much less penetrant than Dr. Hansen assumed; there is a variation in expression, so there may be some vasopressin synthesis in some animals.

Lederis: But these rats were obtained as di/di animals from a reputable supplier.

Gubbels: I am sure you could get them from any supplier. We can't guarantee complete lack of vasopressin unless we measure for it.

Lederis: What would these animals be genetically?

Sokol: I would say that any Brattleboro rat with vasopressin is, by definition, a heterozygote. Measurement of plasma AVP by radioimmunoassay in every experimental animal would certainly be very expensive and impractical; but DI breeding pairs could certainly be verified (i.e., no detectable plasma regused from normal. Do we now have to think of a heterozygote which can AVP) in this manner. The safest procedure for producing homozygous (di/di) Brattleboro rats is by DI/DI matings. HZ/DI matings require very careful screening procedures for separating homozygous and heterozygous offspring.

B. T. Pickering (*University of Bristol, Bristol, England*): Until now, we have tended to think, as it were, of three varieties of Brattleboro rats: the homozygote, which has zero vasopressin in its gland; the normal, which has whatever the figure is; and the heterozygote, which has a finite value somewhat reduced from normal. Do we now have to think of a heterozygote which can manifest pituitary vasopressin levels anywhere between just above zero and the normal value? Is it possible for this gene to be expressed at different rates so that we can get different levels in the heterozygote?

HANSEN: As we begin to understand more and more of the structure and function of DNA, it appears that the old classical segregation ratios may not necessarily be the case. For example, the H_2 system in the mouse initially was thought to behave as a single gene entity. It has turned out to be a very complex gene. I suspect that if we make a sufficient number of crosses, we might begin to see some evidence of intragenic recombination. Genes that we now think behave as simple units may in fact be very complex and heterogeneous entities.

LEDERIS: To get back to Professor Pickering's comment, it may be necessary to determine for each batch of purchased animals whether there is vasopressin, that is, whether the animals are or are not di/di.

R. G. ALLEN (*Veterans Administration Hospital, Portland, Ore.*): The word gene is being bandied around a lot, and I am concerned since my interests lie in molecular biology. These rats either have a copy of the DNA or they don't. Are they not expressing the protein for some reason even if they have a copy of the DNA? What you are really talking about is a defective protein expression mechanism so far, not defective genes.

HANSEN: I won't argue with that. But, before we can get down to the molecular level you are describing, we have to deal with the whole phenotype.

P. STERN (*Dartmouth Medical School, Hanover, N.H.*): It may be wrong then, from all that we have discussed today, to call this a recessive gene. A Brattleboro heterozygote expresses something that we can measure relatively easily.

H. VALTIN (*Dartmouth Medical School, Hanover, N.H.*): Some years ago John Stewart, a geneticist who worked with us, said that we should call it a semi-recessive gene for that reason.

GUBBELS: The usual term for it is an incomplete dominance.

HANSEN: Which implies a heterogeneity of function. What it is saying is that we have a far more complex genetic system then we had originally thought.

J. ZICHA (*Czechoslovak Academy of Sciences, Prague, Czech.*): I should like to point out that besides genetic factors, postnatal animal care could also be a source of observed differences among various colonies of Brattleboro rats. For example, premature weaning (prior to the age of 28 days) could alter many aspects that have been studied in Brattleboro rats yielding different results in various laboratories. Data are available in Wistar rats suggesting that endocrine regulation, water, and electrolyte metabolism, behavior, etc., could be influenced by premature weaning (Křeček, In *The Post-Natal Development of Phenotype,* London-Butterworths, Prague-Academia 1970, pp. 33–41; Křeček: In *The Biopsychology of Development,* New York, Academic Press 1971, pp. 233–248).

P. TANKOSIC (*Faculty of Medicine, Nancy, France*): After crossings between male and female DI rats, we have also observed some survival problems with newborn DI rats. Litters were usually found at the end of the cage opposite to the drinking bottle. When we reversed the position of the water bottle, that is just above the litter, the locomotor activity required of the mother to reach the drinking spout was reduced to a great extent. We have shown that the growth of these young DI rats nursed by a DI mother is similar to young rats bred by a suckling Long-Evans rat.

ON THE REPRODUCTIVE AND DEVELOPMENTAL DIFFERENCES WITHIN THE BRATTLEBORO STRAIN

G. J. Boer, K. Boer, and D. F. Swaab

Netherlands Institute for Brain Research
IJdijk 28, 1095 KJ Amsterdam
the Netherlands

Introduction

Reproductive and developmental disturbances have been described from the moment the diabetes insipidus (DI) Brattleboro rat was discovered. They were summarized as semi-sterility, high incidence of fetal death, stillbirths, and small litter size, and the presence of runts, early postnatal deaths and a protracted postnatal body development.[1-4] Since the high frequency of stillbirth, early death of newborns, and runts was present only in some of the early DI pedigrees, the absence of vasopressin in this mutant [1] was not thought to be a causal factor in this respect.[2] The postnatal growth impairment moreover was thought to be due to growth hormone insufficiency.[4, 5] Now that the Brattleboro strain has been distributed, a number of breeding companies have stabilized the mutancy, and new data are present on the possible role of neurohypophyseal hormones in reproduction [6] and development.[7] Re-examination of these processes seems to be in order, especially since vasopressin not only acts as an antidiuretic hormone on the kidney, but also acts as a neurotransmitter [8] distributed throughout the brain by an extensive exohypothalamic network of vasopressinergic fibers [9-12] in a sex-dependent way.[13]

Animals

Cross-breeding of Brattleboro rats at the Dutch Central Institute for Breeding of Laboratory Animals (TNO-CPB Zeist, the Netherlands) has also provided homozygous normal (HOM-N) Brattleboro rats as a control for the anomalies of the deficient homozygous (HOM) DI rat. The animal institute raises the presently used heterozygous (HET) DI (i.e., non-deficient for vasopressin) and HOM-DI animals by mating HET-DI females with HOM-DI males and discriminates between genotypes within the litter on the basis of water metabolism.

All rats received tap water and standard chow *ad libitum;* illumination was provided from 7 a.m. until 7 p.m. At the time of sacrifice the pituitaries were removed in order to establish the genotype radioimmunologically [14] by the absence (HOM-DI) or presence of vasopressin (HET-DI and HOM-N).

Fertility

Daily vaginal smears in 11 HOM-DI and 7 HET-DI females (series A, Table 1) revealed that 8 of the HOM-DI and all of the 7 HET-DI females

0077-8923/82/0394-0037 $1.75/0 © 1982, NYAS

TABLE 1

LITTER SIZE AND PUP WEIGHT AFTER BIRTH IN DIFFERENT SCHEMES OF BRATTLEBORO BREEDING *

Series†	♀ × ♂	Number of Litters	Mean Litter Size (± SEM, range)	Weighing of Pups‡	Mean Pup Weight in g ± SEM (number of pups)		
					All Litters	6/7/8 Litters	9/10/11 Litters
A	HOM × HOM	30	7.6 ± 0.6 (1-12)	at birth	5.90 ± 0.05 (228)	5.99 ± 0.10 (47)	5.75 ± 0.06 (139)
	HET × HET	22	8.9 ± 0.8 (1-13)		5.83 ± 0.03 (194)	5.99 ± 0.09 (26)	5.79 ± 0.06 (48)
	N × N	23	9.9 ± 0.5** (4-13)		6.14 ± 0.02 (225) **	6.35 ± 0.04 (35) **	6.11 ± 0.04 (88) **
B	HOM × HOM	14	8.9 ± 1.0 (4-19)	day 1	6.66 ± 0.07 (125)	6.63 ± 0.12 (22)	6.81 ± 0.14 (56)
	HET × HET	12	9.5 ± 0.5 (7-11)		7.08 ± 0.06 (113) **	6.98 ± 0.14 (30)	7.11 ± 0.07 (83) **
C	HOM × HOM	36	8.3 ± 0.4 (1-12)	day 1	6.29 ± 0.03 (312)	6.41 ± 0.04 (85)	6.24 ± 0.04 (178)
	HOM × N	17	8.0 ± 0.6 (1-13)		6.42 ± 0.03 (136) **	6.42 ± 0.04 (55)	6.48 ± 0.07 (58) **
D	HOM × HOM	4	8.0 ± 1.7 (3-10)	day 1	6.48 ± 0.18 (32)	——	——
	N × N	4	11.1 ± 2.1 (5-14)		6.80 ± 0.11 (45)	——	——
E§	HOM × HOM	40	8.6 ± 0.5 (1-12)	day 1	6.17 ± 0.09 (82)	6.31 ± 0.17 (23)	6.14 ± 0.13 (47)
	HOM × N	33	8.5 ± 0.5 (4-13)		6.63 ± 0.12 (82) **	7.41 ± 0.13 (19) **	6.86 ± 0.13 (38) **
A	HET × HET	HOM pups		at birth	5.89 ± 0.07 (48)	——	——
		non-HOM pups ¶			5.82 ± 0.04 (146)	——	——

* Homozygous diabetes insipidus (HOM), heterozygous diabetes insipidus (HET), and homozygous normal (N) Brattleboro rats were used of matched ages within the various series. Females were mated either until the morning of a sperm-positive vaginal smear and thereafter singly caged throughout pregnancy (B, C and D) or first kept in groups until prenatal day 20 (A), or without daily vaginal smear caged with males for a week and put into single cages after visual inspection of their bellies, i.e. around day 18/19 of pregnancy (E).

† The various series were used in different investigations from which the experimental results were largely published elsewhere (A, ref. 15; B, ref. 22 and 26; C and D, ref. 22; E, forthcoming publications).

‡ Weighing of pups on day 1 was usually carried out between 9.00 and 11.00 a.m.

§ Pup weight data of this series were based on nine litters of both groups.

¶ Non-HOM pups: 66% HET and 33% N young.

** Statistically significantly different as compared with HOM-values at $p < 0.05$ using Student's *t*-test.

mated during their first estrus cycle upon exposure to males, whereas the remaining three HOM-DI females mated during the second cycle. However, while all HET-DI's became pregnant, five of the HOM-DI females failed to do so. The irregular cyclicity in HOM-DI females,[15] probably due to the higher circulating oxytocin levels,[14] might be related to the subfertility observed.[15] A male factor however, might also be involved since in the larger series C (TABLE 1) a relatively low (67%) pregnancy response also was found, whereas upon sperm-positivity of a HOM-N male, 96% of the DI females became pregnant. In a later and similar breeding set-up (series E, TABLE 1), no such differences were found, suggesting that the female contribution to the subfertility is the major one. Preliminary observations indicate no differences in sexual behavior between HOM-DI and HOM-N males.[16]

SIZE AND GENETIC CONSTITUTION OF LITTER

As shown in TABLE 1, average litter size of HOM-DI females was always lower than that of HET-DI and HOM-N rats within a series. The suggestion that this is caused by the diminished viability of the HOM-DI embryos[2] was not supported by the present data, since the 48/194 proportion of HOM-DI young in litters from HET-DI parent of series A is not different from the theoretically expected level of 25%. The litter size is, moreover, independent of the male's genotype, as no difference was found whenever HOM-DI females were mated with HOM-DI or HOM-N males (series C and E). The small litter size might thus be linked to a disturbance of implantation owing to the high oxytocin levels.[17]

GESTATION LENGTH AND COURSE OF PARTURITION

The onset of parturition was studied within series A (TABLE 1). Since there was a negative correlation between litter size and gestation length and since small litters were more common in HOM-DI and HET-DI than in HOM-N rats,[15] only rats having six or more young were used for the data presented in FIGURE 1. For HET-DI and HOM-N females, a peak in deliveries occurred at the transition period from dark to light at day 22 post-coitally, whereas HOM-DI Brattleboro's gave birth during the entire dark period between days 21 and 22, and even in the light period of day 21. Median gestation lengths were similar for both non-DI rat groups. HOM-DI median gestation length was 3.5 h shorter. Within these (larger than six pups) litters still a smaller average litter size was present for HOM-DI mothers (9.0 versus 10.5 and 10.1 for HET-DI and HOM-N). However, normally this would postpone labor[18] and so it does not explain the results. The elevation of plasma oxytocin in DI seems a sufficient explanation for the shortened pregnancy length, since advancement of labor has been obtained by oxytocin infusion (e.g. Reference 19), or stimulated neurohypophyseal hormone release.[20] The disturbed timing of birth for the HOM-DI females within the day/night rhythm might be related to the absence of vasopressin in cells of the suprachiasmatic nucleus, a nucleus correlated with the circadian rhythmicity.[21]

In the same group of mothers as described above, also used for determining the course of labor, the birth intervals throughout (always >5) were indis-

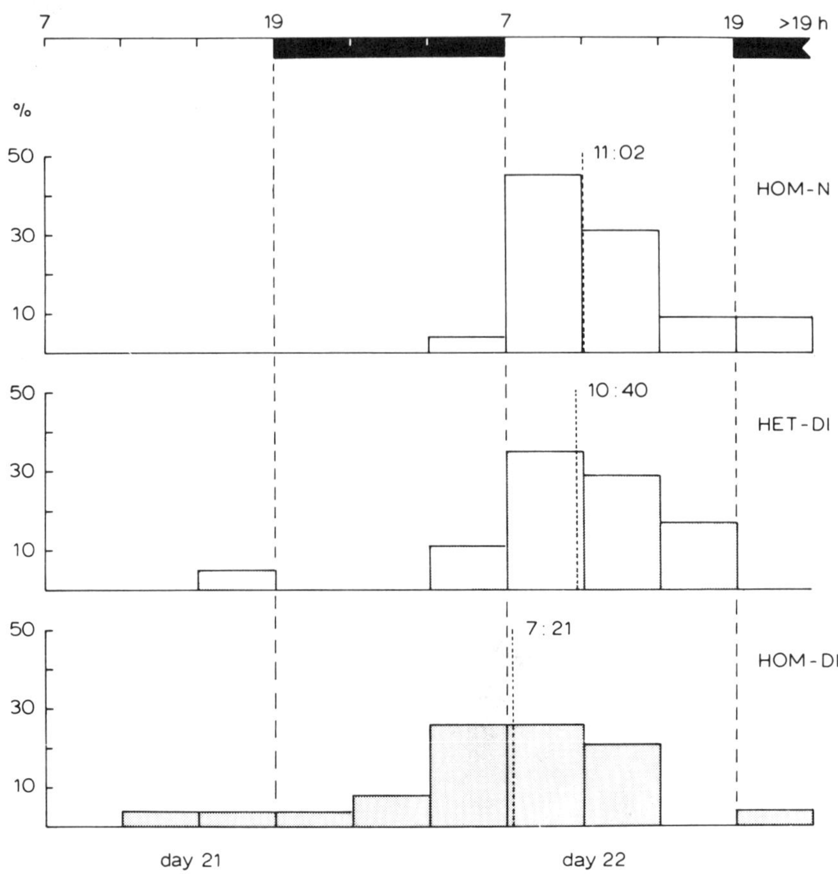

FIGURE 1. Frequency distribution and median gestation length in HOM-N, HET-DI, and HOM-DI Brattleboro rats of series A (cf. TABLE 1), having more than 6 pups at delivery (see text). Median gestation length of HOM-DI animals is shortened by more than 3.5 h. (From Boer *et al.*[15] With permission from the *Journal of Reproduction and Fertility*.)

tinguishable between the HET-DI and HOM-DI genotype mothers, but for both significantly lower than for HOM-N females (FIGURE 2; median overall birth interval differs about 10 min). The increased circulatory oxytocin levels in HOM-DI females might also be responsible for the acceleration of labor. The fact that HET-DI females give birth more rapidly as well, while gestation length was not shortened, is less easy to explain.[15]

PUP WEIGHT AFTER BIRTH

Birth weight is known to be inversely related to litter size. Therefore weight data of pups are also given separately for litters with 6 to 8 and 9 to 11 young (TABLE 1). Generally, however, the weight of HOM-DI newborns remained significantly lower as compared with the non-deficient controls of the respective series.

The significant difference as measured at birth between HOM-DI and HOM-N litters (series A, TABLE 1) is not likely to be influenced by the 3.5 h earlier onset of birth for the HOM-DI female (FIGURE 1). The average speed of growth of the HOM-DI pups in their first day of life is approx. 25 mg/h (from FIGURE 3 data) and would then only account for a 0.09 g lower weight. The weight differential of HOM-DI pups at birth is up to three times that figure (series A, TABLE 1). Moreover, the moment of birth is not of importance for the day 1 weighings in series B to E, since these data are all obtained on the same day 23 post-coitally. For series C and E, the growth impairment was also found when HOM-DI and HET-DI pups were delivered in genetically homogeneous litters by HOM-DI females. These data confirmed and extended the results of Wright and Kutscher[5] who found a similar weight difference in the 50/50 HOM-DI/HET-DI offspring of HET-DI females mated with HOM-DI males.

However, some observations do not fit in with prenatally stunted growth of HOM-DI young. In series A the birth weights of the 50/25/25 HET-DI/HOM-DI/HOM-N offspring of the HET-mothers are not different from those of the homogeneous HOM-DI litters of HOM-DI mothers, whereas in series B a difference is certainly present on the day after birth. Furthermore, within the offspring of the HET-DI females no weight deficit is seen for the HOM-DI newborns as compared to the non-DI littermates (TABLE 1), which in fact is

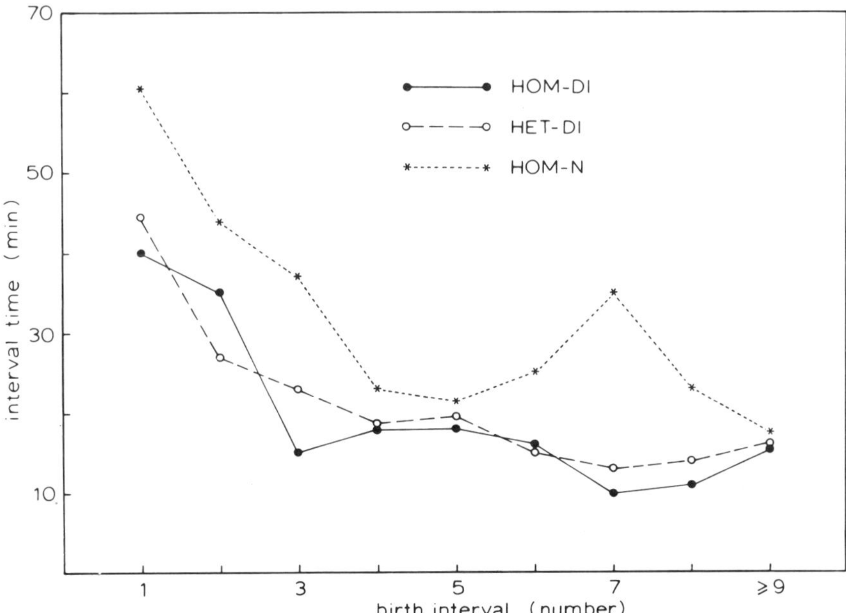

FIGURE 2. Course of parturition of HOM-DI, HET-DI, and HOM-N Brattleboro rats given as time between births (min) for the successive intervals (number). (From Boer *et al.*[15] With permission from the *Journal of Reproduction and Fertility*.)

in contrast to the above-cited results of Wright and Kutscher.[5] No clear explanation can be given.

Early Postnatal Growth

When the growth pattern of the litters of series B was followed at a constant litter size of five, a 30% body weight deficit for the HOM-DI pups emerges during the first ten days of life (Figure 3). Although this supports the presence of a retarded development for DI, the magnitude of the impairment appeared to be highly affected by the nature of the rearing dam. For instance, the homogeneously genotyped HOM-DI and HET-DI litters of series C, brought at seven to eight pups after birth and both nursed by their own HOM-DI mothers, showed a much smaller (11%) weight deficit that was still statistically significant ($p < 0.001$).[22] From Figure 3 it can be seen that this change is largely due to a more disturbed development of the (control) HET-DI's under HOM-DI mother care than under that of the (vasopressin non-deficient) HET-DI mothers. Since the nursing performance of the HOM-DI mother might theoretically also have affected the growth of the HOM-DI pup, a third series D was incorporated (Figure 3): HOM-N and HOM-DI pups (from homologous parents) were matched in weight on their same postnatal day 1, and for both genotypes two six-pup litters were composed and thereafter reared by HOM-N mothers (see for details Reference 22). HOM-N development then appeared to be at control level, while the HOM-DI pups grew even better than did the HET-DI young by a HOM-DI mother (Figure 3). However, even under this condition the growth deficit of HOM-DI Brattleboro rats remains apparent (12%, $0.001 < p < 0.005$).[22]

These considerations illustrate the need for a very carefully chosen breeding scheme whenever an inherited defect is under investigation during development. Prenatal as well as postnatal environmental conditions can best be kept constant both for control and experimental Brattleboro rats by raising them with a HOM-DI mother (series C and E, Table 1). Any intra-litter competition among the group is, in addition, excluded this way.

Early postnatal treatment with vasopressin does not prevent body growth disturbances in the DI young.[5, 23] The supposed involvement of a deficiency of GH production [3-5] can be discarded, since GH is normally expressing its trophic influence only after the tenth postnatal day [7, 22] while a severe growth retardation is seen already by this time in HOM-DI Brattleboro rats. The absence of vasopressin in early prenatal life of the mutant, where normally the hypothalamo-neurohypophyseal and exohypothalamic vasopressin systems of the brain would have developed and probably functioned as well,[24, 25] might be related to the growth impairment. This is possible since anomalies in hormonal and neurotransmitter systems other than that of vasopressin affect development in a permanent way.[7]

Summary

Once they become pregnant, the poorly fertile vasopressin-deficient HOM-DI Brattleboro females reproduce their offspring quite normally. Shortened gestation length and more rapid course of labor as well as the smaller litter size, may be explained by the concomitant higher oxytocin levels in these animals, whereas disturbed timing of birth may be related to the absence of vasopressin.

Although rearing by a HOM-DI female causes a reduced postnatal growth of the rat pup, the mother itself is not the only factor determining the stunted neonatal growth of HOM-DI young as compared to HET-DI and HOM-N controls. The absence of vasopressin in the prenatal stage is perhaps related to this defect.

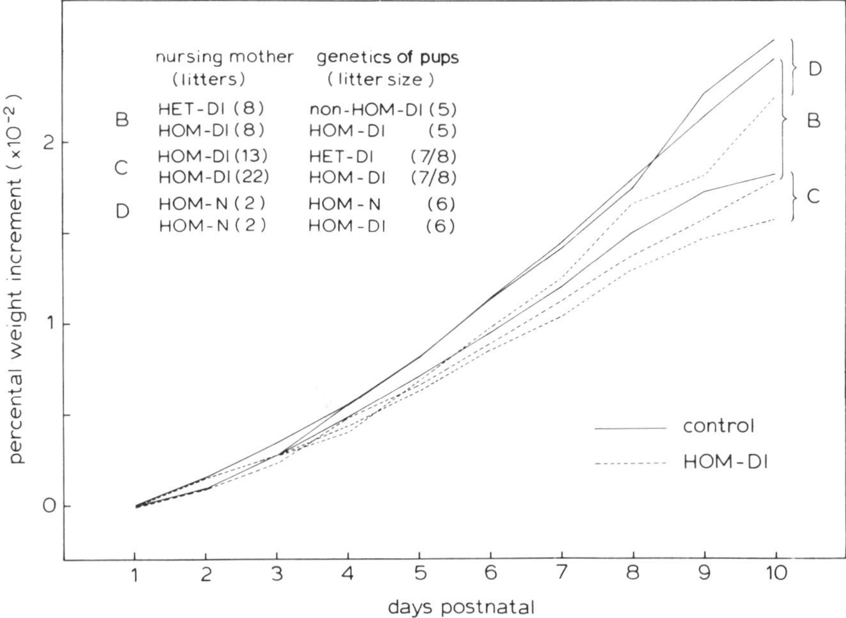

FIGURE 3. Early neonatal growth of normal and HOM-DI Brattleboro rats at different schemes of reproduction. In series B, HOM-DI pups, raised and nursed by HOM-DI mothers, were compared with the non-HOM-DI offspring (66% HET-DI and 33% HOM-N) of HET mothers (litter size brought to 5 pups by day 3). In series C, HOM-DI were compared with HET-DI young, both delivered and nursed by HOM-DI females (litter size 7/8 at day 3). In series D, HOM-DI pups raised by HOM-DI females were reared postnatally (from day 1 onwards with litter size 6) by HOM-N mothers at the same stage of lactation and compared with HOM-N pups from HOM-N females (Adapted from Boer *et al.*[22])

ACKNOWLEDGMENTS

The authors are grateful to Mrs. J. Sels and W. Chen-Pelt for typing the manuscript.

REFERENCES

1. VALTIN, H. & H. A. SCHROEDER. 1964. Familial hypothalamic diabetes insipidus in rats (Brattleboro strain). Am. J. Physiol. **206:** 425–430.

2. SAUL, G. B., E. B. GARRITY, K. BENIRSCHKE & H. VALTIN. 1968. Inherited hypothalamic diabetes insipidus in the Brattleboro strain of rats. J. Hered. **59:** 113–117.
3. ARIMURA, A., S. SAWANO, T. W. REDDING & A. V. SCHALLY. 1968. Studies on retarded growth of rats with hereditary hypothalamic diabetes insipidus. Neuroendocrinology **3:** 187–192.
4. SOKOL, H. W. & J. SISE. 1973. The effect of exogenous vasopressin and growth hormone on the growth of rats with hereditary hypothalamic diabetes insipidus. Growth **37:** 127–142.
5. WRIGHT, W. A. & C. L. KUTSCHER. 1977. Vasopressin administration in the first month of life: effects on growth and water metabolism in hypothalamic diabetes insipidus rats. Pharm. Biochem. Behav. **6:** 505–509.
6. SWAAB, D. F. & K. BOER. 1980. Technical developments in the study of neuroendocrine mechanisms in rat pregnancy and parturition. *In* Animal Models in Fetal Medicine. P. W. Nathanielsz, Ed.: 169–234. Elsevier/North-Holland Biomedical Press. Amsterdam.
7. BOER, G. J., D. F. SWAAB, H. B. M. UYLINGS, K. BOER, R. M. & D. N. VELIS. 1980. Neuropeptides in rat brain development. Prog. Brain Res. **53:** 207–227.
8. BUIJS, R. M. & D. F. SWAAB. 1979. Immunoelectronmicroscopical demonstration of vasopressin and oxytocin synapses in the rat limbic system. Cell Tiss. Res. **204:** 355–365.
9. WEINDL, A., M. V. SOFRONIEW & I. SCHINKO. 1976. Psychotrope Wirkungen hypothalamischer Hormone: Immunohistochemische Identifikation Extrahypophysärer Verbindungen neuroendokriner Neurone. Arzneimittelforsch. **26:** 1191–1194.
10. BROWNFIELD, M. S. & G. P. KOZLOWSKI. 1977. The hypothalamo-choroidal tract. I. Immunohistochemical demonstration of neurophysin pathways to telencephalic plexuses and cerebrospinal fluid. Cell Tiss. Res. **178:** 111–127.
11. BUIJS, R. M. 1978. Intra- and extrahypothalamic vasopressin and oxytocin pathways in the rat. Pathways to the limbic system, medulla oblongata and spinal cord. Cell Tiss. Res. **192:** 423–435.
12. STERBA, G., W. NAUMANN & G. HOHEISEL. 1980. Exohypothalamic axons of the classic neurosecretory systems and their synapses. Prog. Brain Res. **53:** 141–158.
13. DE VRIES, G. J., R. M. BUIJS & D. F. SWAAB. 1981. Ontogeny of the vasopressinergic neurons of the nucleus suprachiasmaticus and their exohypothalamic projections in the rat brain. Presence of a sex-linked difference in the lateral septum. Brain Res. **218:** 67–78.
14. DOGTEROM, J., TJ. B. VAN WIMERSMA GREIDANUS & D. F. SWAAB. 1977. Evidence for the release of vasopressin and oxytocin into cerebrospinal fluid: measurements in plasma and CSF of intact and hypophysectomized rats. Neuroendocrinology **42:** 108–118.
15. BOER, K., G. J. BOER & D. F. SWAAB. 1981. Reproduction in the Brattleboro diabetes insipidus rat. J. Reprod. Fertil. **61:** 273–280.
16. BOER, K. & N. E. VAN DE POLL. Unpublished observations.
17. BUCHANAN, G. D. & M. D. SMITH. 1972. Effects of estrogen and oxytocin on pregnancy in the rat. Anat. Rec. **172:** 280.
18. SCHOFIELD, B. M. 1968. Parturition. Adv. Reprod. Physiol. **3:** 9–32.
19. FUCHS, A.-R. 1972. Prostaglandin effects on rat pregnancy. I. Failure of induction of labor. Fert. Steril. **23:** 410–416.
20. BOER, K., D. W. LINCOLN & D. F. SWAAB. 1975. Effects of electrical stimulation of the neurohypophysis on labour in the rat. J. Endocr. **65:** 163–176.
21. MOORE, R. Y. 1978. Central neural control of circadian rhythms. *In* Frontiers in Neuroendocrinology. W. F. Ganong & L. Martini, Eds. **5:** 185–206. Raven Press. New York, N.Y.

22. BOER, G. J., C. M. H. VAN RHEENEN-VERBERG & H. B. M. UYLINGS. 1982. Impaired brain development of the diabetes insipidus Brattleboro rat. Develop. Brain Res. (In press.)
23. BOER, G. J., H. B. M. UYLINGS, A. J. PATEL, K. BOER & R. KRAGTEN. 1982. The regional impairment of brain development in the Brattleboro diabetes insipidus rat; some vasopressin supplementation studies. (This volume.)
24. BOER, G. J., R. M. BUIJS, D. F. SWAAB & G. J. DE VRIES. 1980. Vasopressin and the developing rat brain. Peptides **1** (Suppl. 1): 203–209.
25. BUIJS, R. M., D. N. VELIS & D. F. SWAAB. 1980. Ontogeny of vasopressin and oxytocin in the fetal rat: early vasopressinergic innervation of the fetal brain. Peptides **1:** 315–324.
26. BOER, G. J., H. B. M. UYLINGS, C. M. F. VAN RHEENEN-VERBERG & B. FISSER. 1978. Postnatal brain development in rats with hereditary diabetes insipidus (Brattleboro strain). *In* Hormones and Brain Development. G. Dörner & M. Kawakami, Eds.: 253–258. Elsevier. Amsterdam.

DECREASED SENSITIVITY TO OXYTOCIN OF UTERI FROM HOMOZYGOUS BRATTLEBORO RATS *

Jaya Haldar, Linda Kupfer, and Hilda Weyl Sokol

Department of Pharmacology
College of Physicians and Surgeons
Columbia University
New York, New York 10032

and

Department of Physiology
Dartmouth Medical School
Hanover, New Hampshire 03755

Although most female homozygous diabetes insipidus (DI) rats of our Brattleboro rat colony can become pregnant, litters are smaller and somewhat less viable. This may be due to three factors: (a) DI rats experience difficulties during delivery, (b) stillborns occur more frequently, and (c) mothers often destroy their pups. Difficult parturition could result from inadequate myometrial stimulation by oxytocin (OT). We have tested the possibility that uteri from DI rats are less sensitive to OT and that lack of vasopressin can influence uterine responses. The results of these experiments are reported here.

Materials and Methods

For determining oxytocic responses, DI rats weighing between 160–205 g, and Long-Evans rats weighing between 200–230 g, were obtained from Blue Spruce Farms (Altamont, N.Y.). Rats were placed in individual metabolic cages and were supplied with food and water *ad libitum.* To determine the severity of diabetes insipidus, urine flow and osmolality were recorded in the DI rats for one week. Oxytocic responses were measured both by *in vitro* and *in vivo* methods in untreated Long-Evans and DI rats and in DI rats that received subcutaneous injections of 1.0 unit of Pitressin tannate in oil each day for 14 days. All rats received 1 mg diethylstilbestrol the evening before the experiment. Oxytocic responses *in vitro* were determined following the method of Holton [1] as modified by Munsick [2] in a Mg^{2+}-free solution. Uterine contractions were recorded isometrically. *In vivo* experiments followed the method of Bisset *et al.*[3] Integrator readings 10 min before and 10 min after the injection were noted and the results expressed as net increase in activity during the later 10 min period. Injections were given intravenously using Syntocinon (Parke Davis) as standard.

Results

Table 1 presents the data on urine flow and osmolality of (a) control Long-Evans, (b) DI, and (c) DI rats treated with Pitressin. The severity of vasopressin deficiency in DI rats used in these experiments is obvious since urine flow for this group was about 23 times higher than that of control rats.

* Supported by National Institutes of Health grants AM 25772, 01940, and 08469.

0077–8923/82/0394–0046 $1.75/0 © 1982, NYAS

TABLE 1

URINE FLOW AND CONCENTRATION IN NORMAL AND DI RATS

Rat Type	Urine Flow (ml/100 g/24 h)	Urine Osmolality (mOsm/kg)
Long-Evans	3.5 ± 0.8	2252 ± 80
DI	80 ± 3.5	150 ± 20
DI + Pitressin	16 ± 2.0	1704 ± 65

As expected, osmolality of DI rat urine was low (6.7% of control). A decrease in urine flow and increase in osmolality was obtained when DI rats were treated with Pitressin. These values in the treated rats were comparable to those in the control animals.

FIGURE 1 illustrates the dose-response curves constructed from *in vitro* oxytocic responses of control, DI, and Pitressin-treated DI groups. It is clear that the uteri from control rats were more sensitive to OT than those of DI rats. The mean ED_{50} (±SE) for control rat uteri was 1.84 ± 0.29 mU ($N = 9$) and for DI uteri was 7.8 ± 1.4 mU ($N = 9$). Uteri from DI rats that were treated with Pitressin showed increased sensitivity to OT when compared with untreated DI uteri. Mean ED_{50} for this group was 3.1 ± 0.9 mU ($N = 6$).

Typical responses of *in vivo* preparations to OT are shown in FIGURE 2. Although in all three groups the frequency and amplitude of contractions increased with the dose, the control rats were most sensitive, followed by the Pitressin-treated group, and the untreated DI rats were least sensitive.

FIGURE 3 shows the dose-response curves for *in vivo* oxytocic responses in these three groups. The ED_{50}'s for these groups followed a pattern similar to

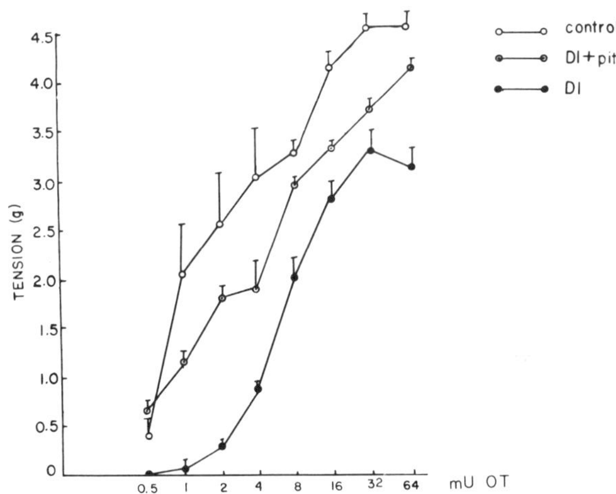

FIGURE 1. Dose-response curves of isolated rat uterus *in vitro*. Control (○—○), DI + Pit (⊙—⊙), and DI (●—●).

that obtained for *in vitro* responses. ED_{50}'s ($\pm$SE) were 15 $\pm$ 2.5, 20 $\pm$ 3.9, and 40 $\pm$ 5 mU for control, Pitressin-treated DI, and untreated DI, respectively.

Discussion

The results demonstrate that DI rat uteri are less sensitive to OT measured both by *in vitro* and *in vivo* methods. The *in vitro* method may not necessarily reveal true oxytocic activity since the bathing fluid does not contain Mg^{2+} which can influence the oxytocic activity of the neurohypophyseal hormones.[2] However, since our *in vivo* data also demonstrate less sensitivity of the DI rat uterus to OT, our *in vitro* data most likely represent a genuine difference. Furthermore, the present findings are in keeping with the reports of Geonzon *et al.*[4] who observed that OT receptors from DI uteri had about 85% less binding capacity than those from control uteri.

Injection of Pitressin for 14 days increased the sensitivity of DI rat uteri to OT, although not to control levels. This is not surprising, since at the end of this treatment period neither urine flow nor osmolality were back to normal. To achieve uterine sensitivity equal to that of control rats, a longer period of Pitressin treatment in DI rats might be required since it has been shown by Harrington and Valtin[5] that correction of the concentrating defect in DI rats requires injections of Pitressin for 28 days.

One reason for the lower sensitivity of DI rat to OT might be high circulating OT in DI rats.[6] Constant exposure to high OT levels might have desensitized receptors involved in eliciting oxytocic responses. It is possible that

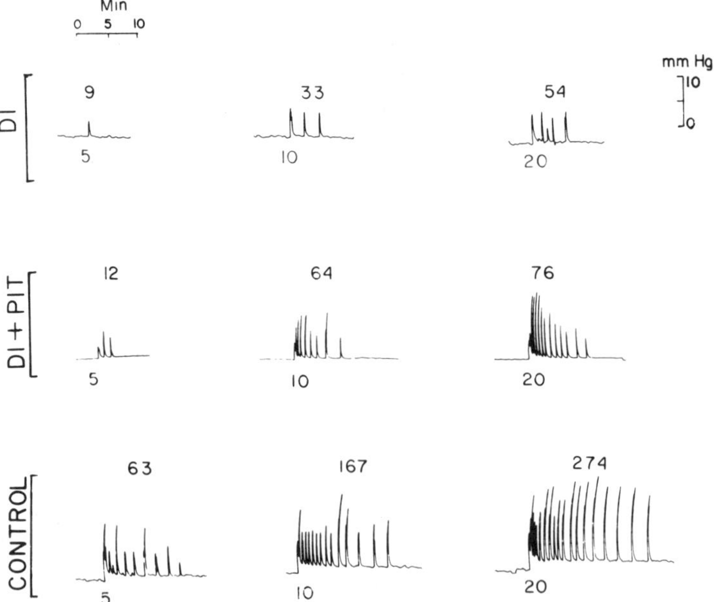

FIGURE 2. Dose-response curves of rat uterus *in vivo*. Oxytocin was injected intravenously in doses of 5, 10, and 20 mU. Numbers above uterine responses represent the net increase in the integrator counts recorded during 10-min period following injections.

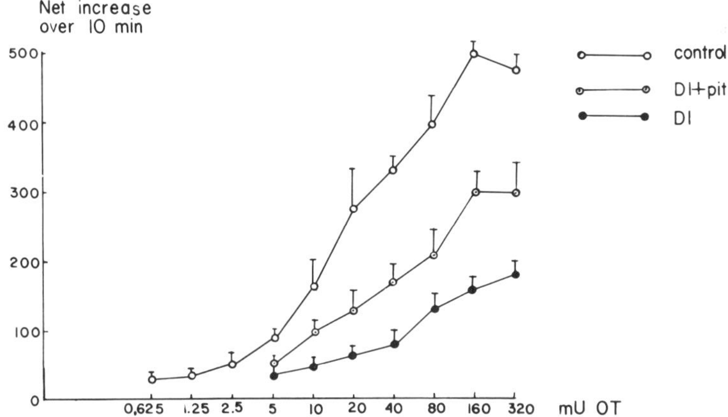

FIGURE 3. Dose-response curves of rat uterus *in vivo*. Control (○—○), DI + Pit (⊙—⊙), and DI (●—●).

Pitressin treatment for 14 days diminished the circulating OT level in DI rats sufficiently so that the sensitivity of the uteri to OT approached normal. It has indeed been shown by Sunde and Sokol[7] that in DI rats the pituitary levels of neurophysin II and III (OT neurophysin) increase to normal following Pitressin treatment. However, to confirm the hypothesis that the elevated circulating level of OT is one of the causes for reduced sensitivity of DI rat uteri to OT, plasma OT levels before and after Pitressin treatment must be determined.

If desensitization indeed occurs in the myometrium during pregnancy in DI rats, this could contribute to the observed problems associated with delivery.

ACKNOWLEDGMENT

Technical assistance of Ita Segal is greatly appreciated.

REFERENCES

1. HOLTON, P. A. 1948. A modification of the method of Dale and Laidlaw for standardisation of posterior pituitary extract. Brit. J. Pharmacol. **3:** 328–334.
2. MUNSICK, R. A. 1960. Effect of magnesium ion on the response of the rat uterus to neurohypophysial hormone and analogues. Endocrinology **66:** 451–457.
3. BISSET, G. W., J. HALDAR & J. LEWIN. 1966. Actions of oxytocin and other biologically active peptides on the rat uterus. Memoirs of the Society for Endocrinology **14:** 185–198.
4. GEONZON, R. M., H. J. GOREN, M. D. HOLLENBERG, K. LEDERIS & D. MORGAN. 1980. Oxytocin binding and responsiveness in the uterus of the Brattleboro rat. J. Physiol **303:** 52P.
5. HARRINGTON, A. R. & H. VALTIN. 1965. Vasopressin effect on urinary concentration in rats with hereditary hypothalamic diabetes insipidus (Brattleboro Strain). Proc. Soc. Exp. Biol. Med. **118:** 448–450.
6. EDWARDS, B. R., F. T. LAROCHELLE & M. GELLAI. 1982. Concentration of urine by dehydration in Brattleboro homozygous diabetes insipidus rats: Is there any role for oxytocin? Ann. N.Y. Acad. Sci. (This volume.)
7. SUNDE, D. A. & H. W. SOKOL. 1975. Quantification of rat neurophysins by polyacrylamide gel electrophoresis (PAGE): Application to the rat with hereditary hypothalamic diabetes insipidus. Ann. N.Y. Acad. Sci. **248:** 345–364.

Henry A. Schroeder
1906–1975

INTRODUCTION TO THE HENRY A. SCHROEDER MEMORIAL LECTURE

Heinz Valtin

Department of Physiology
Dartmouth Medical School
Hanover, New Hampshire 03755

We honor two persons today in the Schroeder Memorial Lecture: One is Dr. Henry A. Schroeder, about whom I spoke earlier today;[1] and the other is Dr. John F. Morris, the third Schroeder Memorial Lecturer.*

For some 17 years—until his death in 1975—Dr. Schroeder was a member of the Department of Physiology at Dartmouth Medical School. He was a man of many interests, many talents, and great courage. I shall tell a little of his background because it illustrates the last attribute, courage, and says a great deal about his unflagging energy, his unbounded intellectual curiosity, and his sense of humor about life and about himself. I relate these facts with the permission of Dr. Schroeder's widow, Janet Gregg Schroeder—herself an accomplished sculptress whose work is represented in the National Portrait Gallery—whose loving devotion and strength sustained Dr. Schroeder through a prolonged and debilitating illness.

In 1957, while a member of the Faculty of Medicine at Washington University in St. Louis and while on a world tour to collect scientific specimens, Dr. Schroeder was told that he had a very malignant tumor and but a few years to live. On the basis of that information, he and his wife decided to retire to their beautiful summer home in West Brattleboro, Vermont. But two years later, Dr. Schroeder appeared to be well and he resumed his research. He founded the Trace Element Laboratory, which was based in part at his home in West Brattleboro and partly at the Brattleboro Retreat. Through the wisdom and imagination of Dr. S. Marsh Tenney, then head of Dartmouth Medical School and chairman of its Department of Physiology, Dr. Schroeder became affiliated with our medical school. Among his important discoveries was that of the association between the hardness or softness of a community's water supply and its incidence of cardiovascular diseases. In recognition of that work, the American Heart Association invited Dr. Schroeder to deliver its prestigious Brown Lecture.

It was, of course, among Dr. Schroeder's extensive colonies of rats and mice that the Brattleboro rat was discovered.[1] It is therefore singularly fitting that the third Henry A. Schroeder Memorial Lecture should be delivered at this symposium. The lecturer is Dr. John F. Morris of the Department of Human Anatomy at the University of Oxford. John Morris is a product of that remarkable "neuroendocrine tradition" at Bristol, England, which was founded by Hans Heller and carried on by—among many others—Barry Cross, Karl Lederis, and Brian Pickering. John graduated with First Class Honors from

* The first was Sir George Pickering in 1976, who delivered a lecture entitled: "Medicine on the Brink: The Dilemma of a Learned Profession";[2] the second was Dr. Michael J. Dunn who, on September 8, 1978 spoke about "Prostaglandins: Their Biochemistry, Physiology, and Clinical Applicability."

Bristol University in 1964, and he received both the M.B.Ch.B. and M.D. degrees from that institution. Since 1977, he has been a Lecturer in Human Anatomy at the University of Oxford and Tutorial Fellow in Medicine at St. Hugh's College. Last year, John Morris was honored by appointment to the endowed Wellcome-Franks Medical Fellowship.

John Morris is one of the world's foremost authorities on the ultrastructure of granulated endocrine cells. One aspect of his work that I admire greatly is its dynamic quality. John ventures far beyond the descriptive; all is treated quantitatively, and it is remarkable what physiological and biochemical information about the processes of hormone synthesis, transport, storage, release, and breakdown he can extract from his otherwise static pictures.

We were fortunate, five years ago, to host John Morris for a sabbatical year. It was during that sojourn that he studied the ultrastructure of the hypothalamo-neurohypophyseal system of the Brattleboro rat.[3] He has since extended that work as a prototype of peptidergic neurons, and that will be the topic of his lecture today.

Before John begins, I would like to relate one more episode, because it exemplifies the benefits that can be derived from international exchange of scientific information. One bit of information that John Morris brought with him to Dartmouth in 1975 was that neurosecretory granules appeared most stable if fixed at an acid pH;[4] to Bill North of our group, and with John's agreement, this finding suggested that the pH within granules might be around 4.5. At the same time, Bill had completed the purification of the rat neurophysins, and a little earlier Burford and Pickering[5] had suggested that rat neurophysin III might be a metabolic product of rat neurophysin II. Bill North and John Morris put their heads together and reasoned that if the suggestion were correct, incubation of purified neurophysin II with hypothalamic extract should yield neurophysin III if run at pH 4.5 but not if run at pH 7. The experiment worked and led to what I think was the first demonstration of enzymatic activity within neurosecretory granules[6]—a requirement if the suggestion of Sachs and Takabatake[7] were to be sustained, namely, that vasopressin and oxytocin are generated from their precursors by enzymatic cleavage within neurosecretory granules.

References

1. Valtin, H. 1982. The discovery of the Brattleboro rat, recommended nomenclature, and the question of proper controls. Ann. N.Y. Acad. Sci. (This volume.)
2. Pickering, G. W. 1978. Medicine on the brink: The dilemma of a learned profession. Perspect. Biol. Med. **21:** 551–560.
3. Morris, J. F., H. W. Sokol & H. Valtin. 1977. One neuron—one hormone? Recent evidence from Brattleboro rats. *In* Neurohypophysis. A. M. Moses & L. Share, Eds.: 58–66. Karger. Basel.
4. Cannata, M. A. & J. F. Morris. 1973. Changes in the appearance of hypothalamo-neurohypophysial neurosecretory granules associated with their maturation. J. Endocrinol. **57:** 531–538.
5. Burford, G. B. & B. T. Pickering. 1973. Intra-axonal transport and turnover of neurophysins in the rat. A proposal for a possible origin of the minor neurophysin component. Biochem. J. **136:** 1047–1052.

6. North, W. G., H. Valtin, J. F. Morris & F. T. LaRochelle, Jr. 1977. Evidence for metabolic conversions of rat neurophysins within neurosecretory granules of the hypothalamo-neurohypophysial system. Endocrinology **101:** 110–118.
7. Sachs, H. & Y. Takabatake. 1964. Evidence for a precursor in vasopressin biosynthesis. Endocrinology **75:** 943–948.

THE BRATTLEBORO MAGNOCELLULAR NEUROSECRETORY SYSTEM: A MODEL FOR THE STUDY OF PEPTIDERGIC NEURONS *

John F. Morris

Department of Human Anatomy
University of Oxford
Oxford OX1 3QX
England

The hypothalamic magnocellular neurosecretory system projecting to the neural lobe (HNS) was the first peptidergic nervous system to be characterized in detail. The parvicellular neurosecretory system projecting to the median eminence was defined next, though the identity of a number of the active principles remains in doubt. The past six years have produced an enormous outpouring of information on new peptidergic neuronal systems in both the central and peripheral nervous systems. The characterization of the neurohemal systems has been followed by the demonstration that both magnocellular and parvicellular peptidergic neurons have axons that terminate on other neurons in a wide variety of areas of the central nervous system and of other peptidergic neuronal systems which have no apparent neurohemal terminations.[1]

The HNS has been the subject of many ultrastructural studies but few of these have used the Brattleboro rat. This mutant, first described by Valtin and his colleagues,[2] which lacks both vasopressin [3-5] and its related neurophysin [6,7] in the homozygous state, has a number of advantages for the elucidation of problems common to the study of peptidergic neurons. The location of the magnocellular neurons is, for the most part, clearly defined; the normal and abnormal neurons can be readily recognized and differentiated on the basis of their granule contents; the neurons have a characteristic architecture in which synthesis, storage, and release appear to be largely separated anatomically; and the nerve endings contain a population of microvesicles that also characterizes peptidergic and other synapses. Finally, two steady-states can be achieved: hyperactivity in the untreated Brattleboro homozygote; and approximately normal secretion in homozygotes whose fluid balance has been normalized by vasopressin administration.[3,8]

Brattleboro Rats and Dale's Principle

If Dale's Principle—that any neuron makes only one transmitter which is released from all its terminals—is applied to the HNS then, at first sight, the question is whether there are separate oxytocinergic and vasopressinergic

* Supported by the National Institute of Arthritis, Metabolism, and Digestive Diseases (U.S. Public Health Research Grant AM–08469–6M), the Agricultural Research Council (AG 43/94), and a Fogarty Fellowship to J.F.M. (IF05TW 2304–01).

0077–8923/82/0394–0054 $1.75/0 © 1982, NYAS

neurons. Physiological evidence for two cell types is compelling. The hormones can be released independently and this correlates with electrical activation of different populations of magnocellular neurons.[9] Immunocytochemical studies[5, 10–12] concur, and both approaches reveal preferential localizations of both oxytocinergic and vasopressinergic cell bodies in the hypothalamus and of their respective fibers in the neural lobe.[5, 10–12] It is not possible to separate oxytocinergic from vasopressinergic cells on general ultrastructural criteria. However, in the Brattleboro homozygote, two populations of neuronal perikarya emerge clearly. One—the presumed oxytocin neurons—contains 160–170 nm neurosecretory granules of the type that characterize magnocellular neurons of normal animals (FIGURE 1). The other lacks these granules but contains instead a population of smaller dense-cored granules 80–110 nm diameter (FIGURE 2). A few small granules of this type can be found in the oxytocinergic neurons of Brattleboro homozygotes and in normal Long-Evans controls, but they are four to eight times less numerous than in the abnormal neurons. In the neural lobe of Brattleboro homozygotes, separate nerve endings and swellings containing the 170 nm and 100 nm granules can also be distinguished (FIGURE 3). Again, a few small dense bodies are found among the 170 nm neurosecretory granules (NSG) in the presumed oxytocinergic endings but they are infrequent. In hypersecreting vehicle-injected Brattleboro homozygotes, many sections of nerve endings are devoid of any dense-cored granules.[9] Clearly one population of neurons does not produce 170 nm NSG, but what is the significance of the 100 nm dense granules? They have been recognized in two other ultrastructural studies of the system[13, 14] both of which assume that, by virtue of their size, they can be identified as aminergic though any verification of this is lacking. Some of the small dense bodies in the perikarya might be primary lysosomes,[15] which would explain their presence in small numbers in oxytocin perikarya and those in the nerve endings could be formed from acid phosphatase-containing axonal reticulum. However, after vasopressin treatment for 30 days, the small dense granules accumulate in the abnormal neurons in much the same way that the 170 nm NSG do in the putative oxytocinergic neurons (TABLE 2) which suggests that they too might be released. The small granules might contain a defective precursor.[16] Numerous studies now make it clear that peptide hormones, neurohormones, and neurotransmitters are formed from higher molecular weight precursors along with other peptides. It might seem then that Dale's principle should be rephrased as 'one neuron–one precursor' for peptidergic neurons until we recall that activity of a secreted peptide is determined by the presence of receptors on the targets and thus different peptides released together could affect different targets.

The formation of immuno- and affinity-identified vasopressin, oxytocin, and their neurophysins from two M_r 20,000 precursors—propressophysin and prooxyphysin—has recently been demonstrated.[17, 18] In Brattleboro rats, which lack propressophysin, a third M_r 20,000 precursor—protein X—is converted by trypsin to a neurophysin-like protein and neurophysin-binding peptides which differ from those formed from prooxyphysin or propressophysin. It is possible that protein X is contained in the 100 nm granules of both normal and abnormal neurons of Brattleboro rats—if so, two similar precursors are packed in very different granules. The abnormal neurons of Brattleboro homozygotes are, however, devoid of neurophysin as judged by immunocytochemistry and, though the immunoreactivity of protein X is not stated, this makes it less likely that it is contained in the abnormal neurons. Alternatively it might be packaged

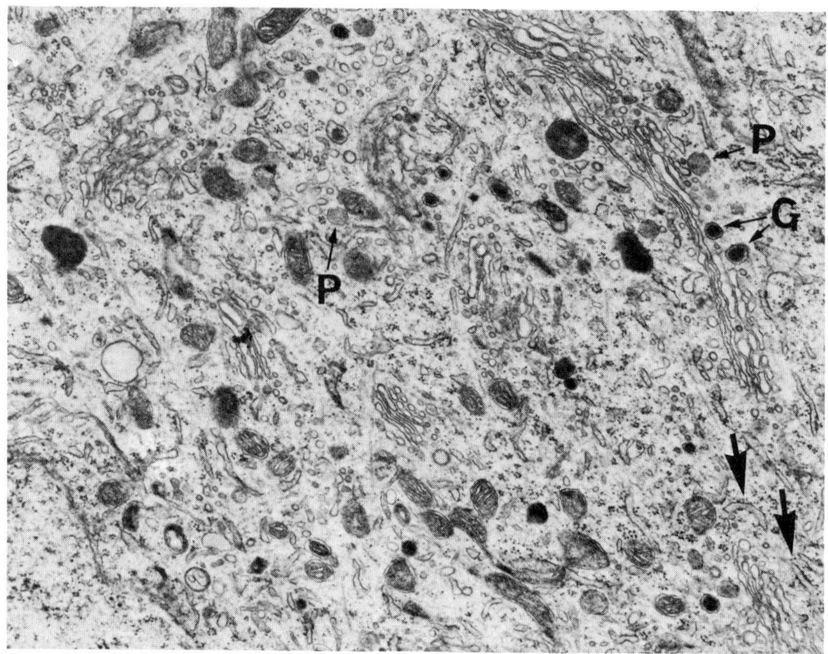

FIGURE 1. Part of a magnocellular supraoptic neuron from a Brattleboro homozygote. The cell contains normal 160 nm neurosecretory granules (G) and is therefore presumed oxytocinergic. Some of the granules are pale (P) indicating greater maturity.[8] The rough endoplasmic reticulum (arrow) is undilated and the Golgi apparatus is present in multiple small units. ×14,600.

in the 160 nm granules of the oxytocin cells. Propressophysin is a glycoprotein, but neither prooxyphysin nor protein X are glycosylated.[18] This explains the histochemical staining of glycoprotein in the granules of only one group of neural lobe axons (the vasopressinergic axons) and the absence of this staining in Brattleboro homozygotes.[19]

During the past two to three years, a number of other peptides—gastrin,[20] somatostatin, LHRH, and, in particular, enkephalins [21–23] have been localized by immune techniques in the neural lobe.† Enkephalins inhibit hormone release from the neural lobe [24] and, like many other peptides, can be associated with granules of about 100-nm diameter.[25] Enkephalins in particular are reported to coexist with vasopressin and oxytocin in the magnocellular neurons [21] so that the small dense granules in the magnocellular neurons could be related to enkephalins. The identification of somatostatin in magnocellular neurons has been shown to be artefactual cross-reaction.[12] Also, in view of the demonstration of nonspecific binding of immunoglobulins to anionic sites [26] and the anionic nature of the neurosecretory material, caution must be exercised before these peptides are accepted as contents of the magnocellular neurons on the

† Also angiotensin II, see Hoffman *et al.*, this volume.

basis of immunocytochemistry. Identification of the contents of the small dense granules of Brattleboro homozygotes must await the isolation of the granules.

Brattleboro Rats and the Intragranular-Extragranular Hormone Controversy

For many years the existence or nonexistence of a physiologically significant pool of extragranular hormones has been argued.[13, 27, 28] The possibility of cytosolic hormone has also been raised for parvicellular neurons, e.g., LHRH[29] on the basis of immunocytochemical staining. Kalimo[13] argues for cytoplasmic hormone on the grounds that in Brattleboro homozygotes "there are very few NSG in the whole HNS of DI-rats, far fewer than in neurons functioning to meet the demands of basal secretion in normal rats." This argument is often used but it neglects the fact that the number of granules present reflects the storage of granules and not their turnover. Furthermore in the Brattleboro neural lobe, whereas many areas are very depleted of granules (FIGURE 4), occasional regions have a large accumulation (FIGURE 5) so that the gland must be sampled systematically. The true test is whether there are sufficient granules to account for the hormones present in Brattleboro homozygotes. Stereological

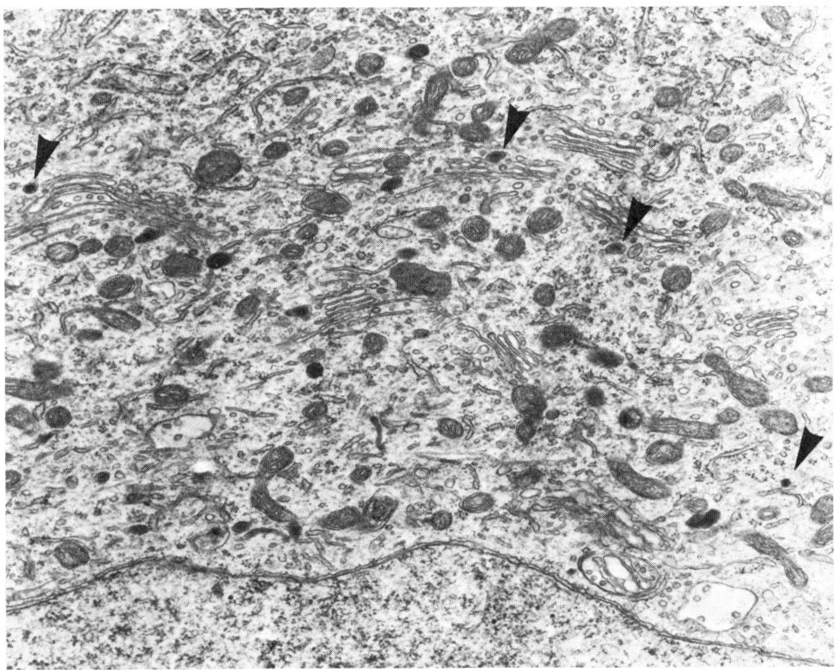

FIGURE 2. Part of a magnocellular supraoptic neuron from a Brattleboro homozygote. This abnormal cell contains no 160 nm granules but does contain a population of small 100 nm dense granules (arrows). Rough endoplasmic reticulum and Golgi apparatus are as in the normal cell (FIGURE 1). ×14,600.

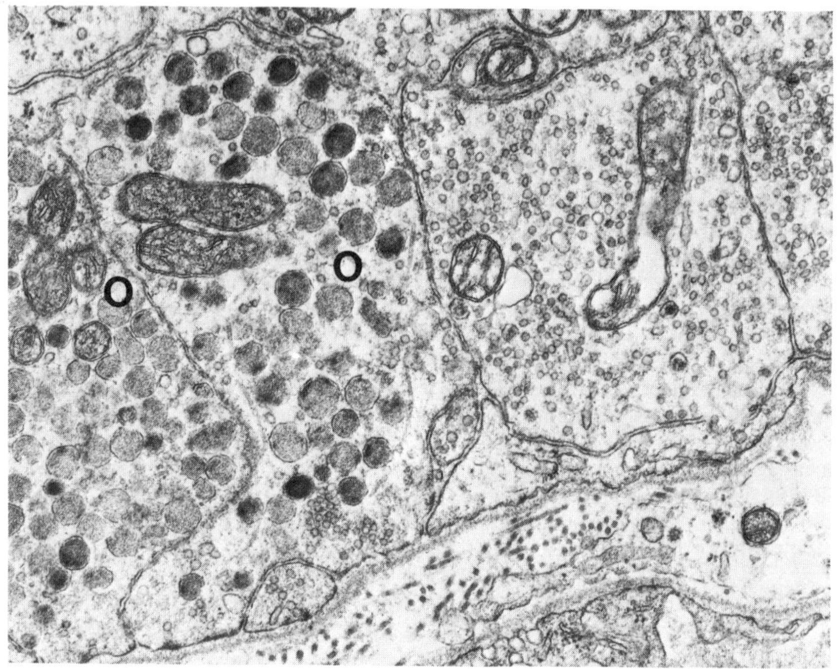

FIGURE 3. Adjacent nerve endings in the neural lobe of a Brattleboro homozygote. Two endings (O) contain 160 nm presumed oxytocinergic granules. The endings on the right contain no 160 nm granules but, like the abnormal perikarya in the supraoptic nucleus, contain 100 nm dense granules. $\times 27,000$.

analysis of the neural lobe of Wistar rats indicates that each granule contains $84 \pm 4 \times 10^3$ molecules of hormone.[30] When the hormone content of Brattleboro homozygotes is calculated from assessments of the number of granules in the neural lobe and its hormone content, the average calculated oxytocin content of a granule in untreated homozygotes is $63 \pm 19 \times 10^3$ mols and in vasopressin-treated homozygotes it is $77 \pm 19 \times 10^3$ mols. These values are not significantly different from those in the normal Wistar animals (Morris, North, La Rochelle & Valtin, unpublished observations). It cannot therefore be argued that the neural lobe of Brattleboro homozygotes contains insufficient granules to account for the hormone present—but merely that many fewer are stored than in normal rats.

The smooth endoplasmic reticulum (SER) in Brattleboro homozygotes is reportedly increased although no measurements have been made.[14] The transport of neurosecretory material to the neural lobe via SER has been suggested by a number of authors, particularly in states in which demand for hormone is high.[31-33] TPPase activity in axonal reticulum[32] could indicate that a continuation of the Golgi apparatus is present throughout the system and this might be responsible for packaging granules in distal parts of the axons. The unequivocal demonstration of hormone and neurophysin transport by the SER will not be easy to achieve—the Brattleboro rat with its hypertrophied SER and lack of vasopressin in one population of axons may be useful in this respect.

The Brattleboro Rat and the Study of Neuronal Architecture

Non-neurohemal Contacts

It is now clear that both the magnocellular and parvicellular neurosecretory systems send fibers to many regions of the brain other than the neurohemal contact sites.[1] Other parts of this symposium deal with the effects of these projections and will not be considered further here. The Brattleboro rat has been useful in verifying claims of immuno-identified vasopressin in such projections.[12]

Architecture of Magnocellular Fibers in the Neural Lobe

The projection to the neural lobe is characterized by the varicose nature of the axons in both the hypothalamus and neural lobe. In the neural lobe the axonal dilatations are separable into nerve endings, which have a significant population of microvesicles in addition to the granules, and nerve swellings, which lack this microvesicle population but are often sites of massive lysosomal activity.[15, 34] It can reasonably be asked whether the axonal swellings are a

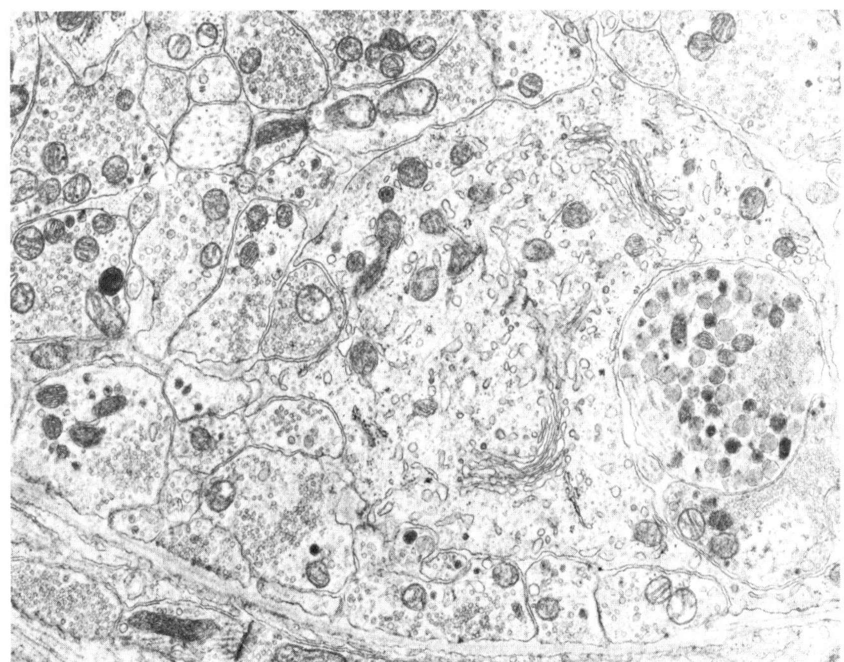

FIGURE 4. Representative area of the neural lobe of a Brattleboro homozygote. All granules are scarce. 160 nm oxytocinergic granules are seen in one nerve ending. Other nerve endings in the field contain the 100 nm dense granules which characterize the abnormal neurons. ×14,100.

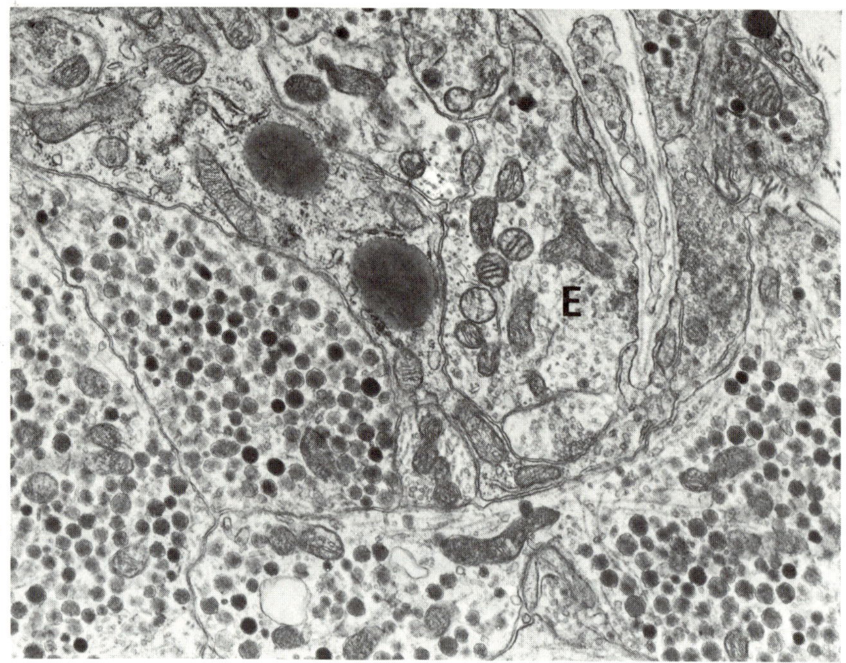

FIGURE 5. Area of neural lobe of a Brattleboro homozygote in which many 160 nm oxytocinergic granules are present, mostly in nerve swellings. A nerve ending containing only a few 100 nm dense granules (E) is also present. ×14,100.

permanent feature of the axons or whether they merely reflect the presence of large accumulations of granules. The neural lobe of the Brattleboro homozygote contains, in some areas, large numbers of nerve endings containing no, or only a few, granules (FIGURE 6). This suggests that swellings do not disappear because they are empty. The incidence of nerve endings and nerve swellings in the neural lobe is similar for both vasopressin- and vehicle-treated Brattleboro homozygotes and for normal Long Evans controls—about 20 profiles of endings/100 μm^2 tissue based on a mid-transverse section of the gland, which is representative for Wistar neural lobes.[34] However the Brattleboro neural lobe is double the weight of the normal neural lobe. This could indicate that the Brattleboro neural lobe contains almost twice the number of nerve endings as the normal Long Evans rat. Is this difference present at birth? Or does it develop with the functional hyperactivity of the system? The Brattleboro rat offers a good system for the study of such function-related changes in neuronal architecture.

THE BRATTLEBORO RAT IN THE STUDY OF HYPERACTIVITY OF PEPTIDERGIC NEURONS

Plasma oxytocin concentration in Brattleboro homozygotes is elevated five-fold above that found in either Brattleboro heterozygotes or normal Wistar rats, reaching 50 pg/ml. This is equivalent to the level found after 48–96 h water deprivation and indicates that Brattleboro homozygotes are subject to severe chronic osmotic stress.[35] ‡

Granule Formation in the Perikarya

The perikarya of magnocellular neurons in Brattleboro homozygotes are considerably hypertrophied in comparison to those of normal Long-Evans rats.[8, 13, 14] Administration of vasopressin that normalizes urine output and water intake causes the oxytocin cells (as identified by their content of 170 nm NSG) to reduce to normal size but the abnormal neurons (with 100 nm granules) remain somewhat hypertrophied (Table 1). There are no reports of dilated rough endoplasmic reticulum (RER) in Brattleboro homozygotes. This suggests that the dilated RER, which characterizes some cells in the early stage of dehydration, is a temporary phenomenon that reflects an imbalance between RER and Golgi during cellular hypertrophy. The Golgi apparatus in Brattleboro homozygotes takes the form of many small complexes dispersed

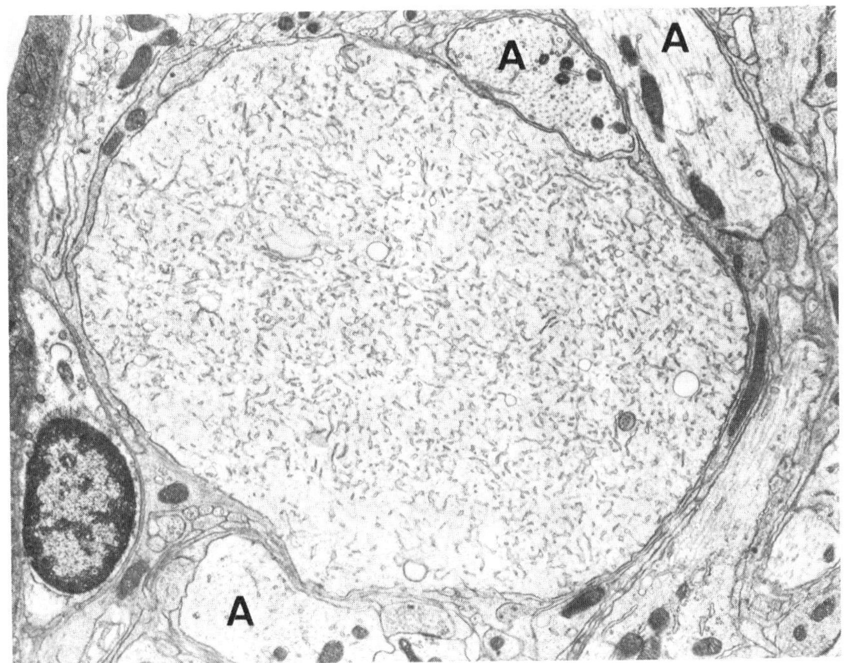

Figure 6. Very large nerve swelling in a Brattleboro homozygote with three large axons (A) around it. The swelling is filled with smooth reticulum membranes. No secretory granules are seen. ×8,200.

‡ See also Edwards *et al.* (this volume) who found lower (ca. 10 pg/ml) basal plasma oxytocin levels in Brattleboro homozygotes representing a two-fold increase over Long-Evans plasma; Balment *et al.;* and North *et al.*, this volume.

TABLE 1

CYTOPLASMIC AREA, GRANULE POPULATIONS, AND LYSOSOME POPULATIONS IN NORMAL AND ABNORMAL MAGNOCELLULAR NEURONS OF THE SUPRAOPTIC NUCLEI OF BRATTLEBORO HOMOZYGOTES * AND IN NORMAL LONG-EVANS RATS †

	DI-VP		DI-OIL		LE-OIL
	Normal	Abnormal	Normal	Abnormal	Normal
Cytoplasmic area (μm^2)	237 ± 37 ‡	315 ± 40	428 ± 36	419 ± 45	219 ± 13
170 nm Granules (number/average cell profile)	60 ± 7	—	60 ± 6	—	102 ± 14
100 nm Dense granules (number/average cell profile)	8 ± 1	26 ± 4	10 ± 2	40 ± 7	—
Immature 170 nm granules (%)	45 ± 3	—	59 ± 3	—	34 ± 3
Large dense lysosomes (number/average cell profile)	59 ± 4	89 ± 7	72 ± 5	115 ± 11	36 ± 2

* Given 1 I.U./day Pitressin tannate in oil (DI-VP) or oil vehicle (DI-OIL).
† Given oil vehicle (LE-OIL).
‡ Mean ± SEM.

throughout the cell (FIGURE 1). This morphological response to hyperactivity could well be investigated further since the functions that underlie the characteristic morphology of the Golgi apparatus are still far from clear.

Many reports stress a paucity of neurosecretory granules in the perikarya of Brattleboro homozygotes but do not quantify it.[13, 14, 36, 37] When the increased cell size is allowed for, oxytocinergic magnocellular perikarya in the supraoptic nucleus can be shown to contain about half the number of NSG found in Long-Evans controls[8] (TABLE 1). Vasopressin administration does not increase the number but does cause an increase in the maturity of the NSG population indicating slower turnover of granules[8] (TABLE 1). In the abnormal neurons the 100 nm small dense granules are found in somewhat greater numbers in hypersecreting than in vasopressin-treated animals (TABLE 1). Brattleboro homozygote perikarya contain large numbers of dense lysosomes (FIGURE 1, TABLE 1) and in some neurons nematosomes can be found.[14]

NEUROSECRETORY AXONS AND TRANSPORT

The magnocellular axons of Brattleboro homozygotes are enlarged two to five times compared to normal[37-39] and contain increased numbers of microtubules.[37] This increase develops postnatally,[38] a change probably attributable to the increased secretory activity. There are no reports of changes in axons and their microtubules when hypersecretion is halted by vasopressin administration though such information would be useful. There appear to be more myelinated fibers in the hypothalamoneurohypophyseal tract of Brattleboro homozygotes[13] though it is not clear that these are magnocellular axons.

The Neural Lobe and Storage of Secretory Granules

Neurosecretory granules appear to progress from undilated axons to nerve endings and later to larger nerve swellings in the neural lobe of the normal rat.[39, 40, 43] It is the relatively newly formed granules[41, 42] in the nerve endings[43] that are acutely releasable. In the Brattleboro homozygote the neural lobe contains many fewer NSG than in normal rats (Table 2) though their numbers match the hormone content (see above). Vasopressin administration causes a threefold increase in NSG (Table 2). It might have been predicted that the reduced number of granules in the Brattleboro homozygote would be found predominantly in endings, but this does not appear to be the case since about 30–40% of NSG in both untreated and vasopressin-treated Brattleboro homozygotes are found in nerve endings as in normal rats[34, 45] By contrast, 70–80% of the small 100 nm granules in the abnormal neurons are found in nerve endings (Table 3). The reasons for these differences are not clear but once again the Brattleboro rat offers a system for investigating how peptidergic neurons control the location of the granules that they contain.

Neural Lobe Nerve Endings and Hormone Release[46]

The release of granules by exocytosis appears to occur specifically from nerve endings[44] and in Brattleboro homozygotes 67% of nerve ending profiles are devoid of electron dense granules. Treatment with vasopressin for 30 days

Table 2

Size of Nerve Endings, Their Basement Membrane Contact, and the Granules, Microvesicles, Vacuoles, and Mitochondria that They Contain in the Neural Lobes of Brattleboro Homozygotes * and in Normal Long-Evans Rats †

	DI-VP	DI-OIL	LE-OIL
Area (μm^2)	1.6 ± 0.1 ‡	1.7 ± 0.1	1.9 ± 0.1
Contact with basement membrane (μm/ending)	1.15 ± 0.05	1.5 ± 0.1	1.2 ± 0.1
% Endings without granules	25	67	14
Microvesicles (% of ending)	16 ± 1	13 ± 1	13 ± 1
Vacuoles (% of ending)	0.8 ± 0.1	1.8 ± 0.2	1.0 ± 0.1
Mitochondria (% of ending)	8.8 ± 0.5	8.9 ± 0.5	5.5 ± 0.5
160 nm neurosecretory granules (number/μm^2 neural lobe)	1.5 ± 0.4	0.6 ± 0.2	9.5 ± 0.6
100 nm dense granules (number/μm^2 neural lobe)	0.38 ± 0.03	0.15 ± 0.02	—

* Given 1 I.U./day Pitressin tannate in oil (DI-VP) or oil vehicle (DI-OIL).
† Given oil vehicle (LE-OIL) for 30 days.
‡ Mean ± SEM.
The data for the nerve endings is taken from Morris & Nordmann.[46]

TABLE 3

DISTRIBUTION (%) OF 160 NM NEUROSECRETORY GRANULES AND 100 NM DENSE GRANULES AMONG UNDILATED AXONS, NERVE ENDINGS, AND NERVE SWELLINGS IN THE NEURAL LOBES OF BRATTLEBORO HOMOZYGOTES *

		Undilated Axon	Nerve Ending	Nerve Swelling
160 nm Granules	DI-OIL	7	36	57
	DI-VP	5	39	55
100 nm Granules	DI-OIL	12	64	20
	DI-VP	13	76	9

* Given 1 I.U./day Pitressin tannate in oil (DI-VP) or oil vehicle (DI-OIL) for 30 days.

causes an increase in the number of both 170 nm NSG and 100 nm dense granules in their respective endings and fewer endings are devoid of granules (TABLE 2). The size of individual nerve ending profiles that transect the basement membrane is increased in hypersecreting Brattleboro homozygotes and the basement membrane contact zone of the nerve endings is increased. This change is reversed by 30 days of vasopressin administration. The nerve endings of Brattleboro homozygotes also contain more mitochondria than those of normal Long-Evans rats whether or not vasopressin has been given.[46] The increased mitochondria in Brattleboro homozygotes presumably reflect the increased activity—in particular the need for a high capacity calcium sink in the rapidly firing neurons[47]—but it is not clear why their number remains elevated after vasopressin treatment unless their half-life is particularly long.

The rapid exocytosis of granules from the nerve endings leaves behind their membranes. There is controversy concerning the organelle responsible for membrane retrieval and both microvesicles and vacuoles have been implicated.[27, 44, 48, 49] The Brattleboro rat offers an excellent system for the study of this problem. The proportion of area of the nerve ending profiles occupied by microvesicles is not increased in either the normal or abnormal neurons of the hypersecreting animals nor is it greater than in Long-Evans controls. Conversely, the vacuole population is increased in hypersecreting Brattleboro homozygotes given only vehicle and is decreased to control levels by vasopressin treatment. Furthermore the modal size of the vacuoles in the 160 nm NSG and the 110 nm dense-cored granule containing terminals corresponds to that of the dense granules they contain. This suggests that granule membranes are retrieved intact by vacuoles as secretion proceeds.[46]

What then is the role of the microvesicles that characterize the nerve endings? Release of hormone is a calcium-dependent process[48] and the intracellular calcium concentration is tightly controlled. Microvesicles in both normal rats[50] and Brattleboro homozygotes (FIGURE 7) can be shown to contain calcium and a microvesicle-rich fraction of the neural lobe of normal animals will accumulate calcium at physiological concentrations.[51, 52] Microvesicles characterize all synapses and, in many, some or all microvesicles also contain neurotransmitter. There is no evidence that neural lobe microvesicles contain neurohormones so that the Brattleboro neural lobe, in which two classes of dense-cored granules can be distinguished unequivocally from the microvesicles, offers an attractive model for the study of yet another aspect of neural function.

The Hypogonadal (*hpg*) Mouse—A Parvicellular Analogue of the Brattleboro Rat

The hypogonadal (*hpg*) mouse was first recognized in 1977 as a mutant characterized by failure of the postnatal development of gonads and accessory sexual organs. The trait is inherited in an autosomal recessive manner. The cause of the hypogonadism appears to be a failure of production of gonadotrophin hormone releasing hormone (GnRH) since radioimmunoassay detects in *hpg* mice less than 1% of the GnRH found in normal mice [53] and immunocytochemistry detects no GnRH in the *hpg* hypothalamus (Charlton and Parry, personal communication). As a result, the pituitary and plasma contents of gonadotrophins are very low.[53] The pituitary is, however, capable of responding to GnRH [54, 55] both by releasing gonadotrophins [54] acutely and by hypertrophy and increased secretory activity of the gonadotroph cells.[55] If GnRH is given daily for 20 days the FSH content of gonadotrophins increases to normal levels while the LH level increases only two-fold to one tenth of that found in the anterior pituitary of normal mice.[55] The *hpg* mouse is, therefore, very useful for the study of GnRH action on the pituitary and the cascade of effects that

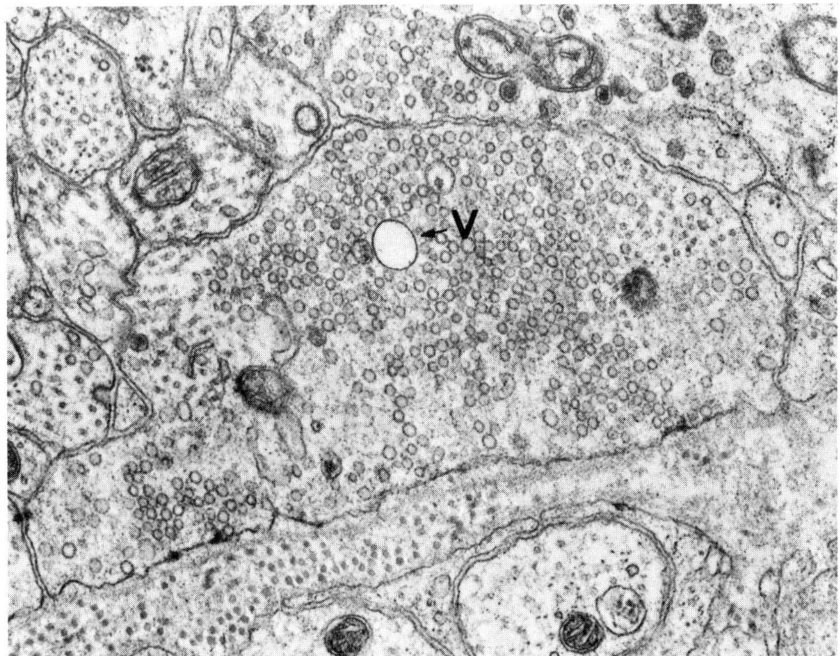

FIGURE 7. Nerve ending in Brattleboro homozygote in which only two 100 nm dense granules can be seen. Microvesicles almost fill the ending and a vacuole (V) is also present. ×30,000.

control reproduction. By comparison with the Brattleboro rat, however, it is much less easy to examine the structure of the neural component since the GnRH neurons are not closely packed within a hypothalamic nucleus nor would it be easy to detect the abnormal terminals in the median eminence.

In conclusion, structural studies on the Brattleboro rat have so far only scratched the surface of the manifold opportunities that the mutant offers for the study of many interesting aspects of 'peptidergic neurology.' It is to be anticipated that any future symposium on this topic will again amply demonstrate its value in this rapidly developing facet of neurobiology.

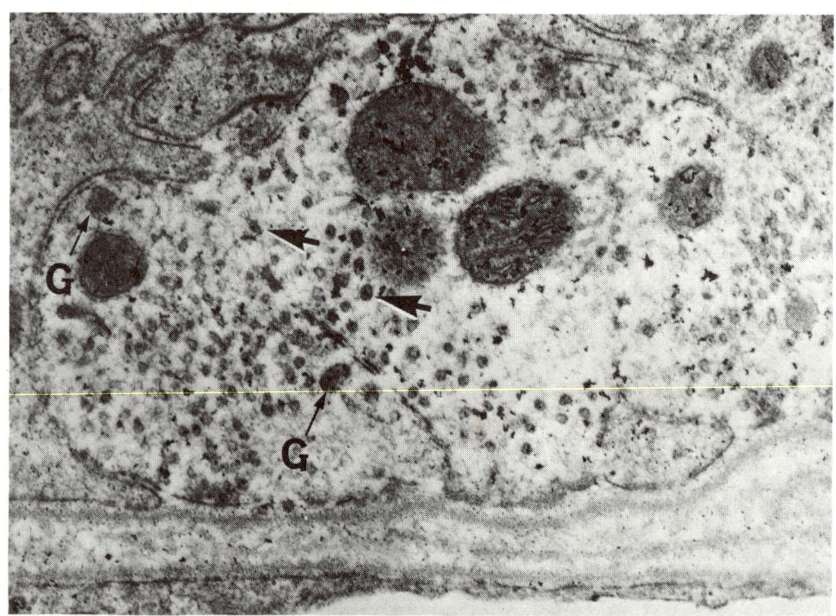

FIGURE 8. Nerve endings in a Brattleboro homozygote prepared by an oxalate-pyroantimonate technique[50] to reveal calcium as electron dense deposits. Such deposits are seen in the mitochondria and in many of the microvesicles (arrows). Two 100 nm granules (G) are unreactive. ×100,000.

ACKNOWLEDGMENTS

It is a pleasure to acknowledge my deep gratitude to Professor H. Valtin and his colleagues at Dartmouth, W. G. North, F. T. La Rochelle, and H. W. Sokol who introduced me to the Brattleboro rat; to B. T. Pickering (Bristol), J. J. Nordmann (Bordeaux), F. D. Shaw (London), and D. Chapman (Oxford) with whom some of these studies have been discussed and carried out. Glenys Davies kindly typed the manuscript.

REFERENCES

1. BURGEN, A., H. W. KOSTERLITZ & L. L. IVERSEN, Eds. 1980. Neuroactive Peptides. The Royal Society. London.
2. VALTIN, H., A. SCHROEDER, K. BENIRSCHKE & H. W. SOKOL. 1962. Familial hypothalamic diabetes insipidus in rats. Nature **196:** 1109–1110.
3. VALTIN, H., W. H. SAWYER & H. W. SOKOL. 1965. Neurohypophysial principles in rats homozygous and heterozygous for hypothalamic diabetes insipidus (Brattleboro strain). Endocrinology **77:** 701–705.
4. VALTIN, H., H. W. SOKOL & D. SUNDE. 1975. Genetic approaches to the study of the regulation and actions of vasopressin. Rec. Prog. Horm. Res. **31:** 447–486.
5. VANDESANDE, F. & K. DIERICKX. 1976. Immuno-cytochemical demonstration of the inability of the homozygous Brattleboro rat to synthesize vasopressin and vasopressin-associated neurophysin. Cell Tiss. Res. **165:** 307–316.
6. BURFORD, G. D., C. W. JONES & B. T. PICKERING. 1971. Tentative identification of a vasopressin-neurophysin and an oxytocin-neurophysin in the Rat. Biochem. J. **124:** 809–813.
7. SUNDE, D. A. & H. W. SOKOL. 1975. Quantification of rat neurophysins by polyacrylamide gel electrophoresis (PAGE): application to the rat with hereditary hypothalamic diabetes insipidus. Ann. N.Y. Acad. Sci. **248:** 345–364.
8. MORRIS, J. F., H. W. SOKOL & H. VALTIN. 1977. One neuron—one hormone? Recent evidence from Brattleboro rats. *In* Neurohypophysis. Int. Conf. Key Biscayne, Fla. 1976. A. M. Moses & L. Share, Eds.: 58–66. Karger, Basel.
9. POULAIN, D. A., J. B. WAKERLEY & R. E. J. DYBALL. 1977. Electrophysiological differentiation of oxytocin and vasopressin-secreting neurones. Proc. Roy. Soc. London Ser. B **196:** 367–384.
10. VANDESANDE, F. & K. DIERICKX. 1975. Identification of the vasopressin producing and of the oxytocin producing neurons in the hypothalamic magnocellular neurosecretory system of the rat. Cell Tiss. Res. **164:** 153–162.
11. SWAAB, D. F., F. NIJVELDT & C. W. POOL. 1975. Distribution of oxytocin and vasopressin in the rat supraoptic and paraventricular nucleus. J. Endocr. **67:** 461–462.
12. VAN LEEUWEN, F. W., C. DE RAAY, D. F. SWAAB & B. FISSER. 1979. The localization of oxytocin, vasopressin, somatostatin and luteinizing hormone releasing hormone in the rat neurohypophysis. Cell Tiss. Res. **202:** 189–210.
13. KALIMO, H. & U. K. RINNE. 1972. Ultrastructural studies on the hypothalamic neurosecretory neurones of the rat. II. The hypothalamo-neurohypophysial system in rats with hereditary hypothalamic diabetes insipidus. Z. Zellforsch. **134:** 205–225.
14. TASSO, F. & S. RUA. 1978. Ultrastructural observations on the hypothalamo-posthypophysial complex of the Brattleboro rat. Cell Tiss. Res. **191:** 267–286.
15. BOUDIER, J.-A., J.-L. BOUDIER, A. MASSACRIER, P. CAU & D. PICARD. 1979. Structural and functional aspects of lysosomes in the neurosecretory neurons. Biol. Cell. **36:** 185–192.
16. VALTIN, H., H. W. SOKOL, W. H. SAWYER & A. R. HARRINGTON. 1965. Possible synthesis of a "defective polypeptide" in a strain of rats with hereditary hypothalamic diabetes insipidus. Proc. 2nd Intl. Congress Endocrinology. p. 1287. Excerpta Medica Foundation. Amsterdam.
17. RUSSELL, J. T., M. J. BROWNSTEIN & H. GAINER. 1979. Trypsin liberates an arginine vasopressin-like peptide and neurophysin from a M_r 20,000 putative common precursor. Proc. Natl. Acad. Sci. USA **76:** 6086–6090.
18. RUSSELL, J. T., M. J. BROWNSTEIN & H. GAINER. 1980. Biosynthesis of vasopressin, oxytocin and neurophysins: isolation and characterization of two

common precursors (propressophysin and prooxyphysin). Endocrinology **107:** 1880–1891.

19. Tasso, F., D. Picard & J. J. Dreifuss. 1976. Ultrastructural identification of granules containing oxytocin and vasopressin. Nature **260:** 621–622.
20. Rehfeld, J. F. 1978. Localisation of gastrins to neuro- and adenohypophysis. Nature **271:** 771–773.
21. Martin, R. & K. H. Voigt. 1981. Enkephalins co-exist with oxytocin and vasopressin in nerve terminals of the rat neurohypophysis. Nature **289:** 502–504.
22. Rossier, J., E. Battenberg, Q. Pittmann, A. Bayon, L. Koda, R. Miller, R. Guilleman & R. Blood. 1979. Hypothalamic enkephalin neurones may regulate the neurohypophysis. Nature **277:** 653–655.
23. Sar, M., W. E. Stumpf, R. Miller, K. J. Chang & P. J. Cuatrecasas. 1978. Immunohistochemical localization of enkephalins in rat brain and spinal cord. J. Comp. Neurol. **182:** 17–38.
24. Clarke, G., P. Wood, L. Merrick & D. W. Lincoln. 1979. Opiate inhibition of peptide release from the neurohumoral terminals of hypothalamic neurones. Nature **282:** 746–748.
25. Pickel, V. M., K. K. Sumal, S. C. Beckley, R. J. Miller & D. J. Reis. 1980. Immunocytochemical localisation of enkephalin in the neostriatum of rat brain: a light and electron microscopic study. J. Comp. Neurol. **189:** 721–740.
26. Grübe, D. 1980. Immunoreactivities of gastrin (G-) cells. II. Non-specific binding of immunoglobulins to G cells by ionic interactions. Histochemistry **66:** 147–167.
27. Morris, J. F., J. J. Nordmann & R. E. J. Dyball. 1978. Structure-function correlation in mammalian neurosecretion. Int. Rev. Exp. Path. **18:** 1–95.
28. Krisch, B. 1979. Indication for a granule-free form of vasopressin in immobilization-stressed rats. Cell Tiss. Res. **197:** 95–104.
29. Kozlowski, G. P., L. Chu, G. Hostetter & B. Kerdelhue. 1980. Cellular characteristics of immunolabeled luteinizing hormone releasing hormone (LHRH) neurons. Peptides **1:** 37–46.
30. Morris, J. F. 1976. Hormone storage in individual neurosecretory granules of the pituitary gland: a quantitative ultrastructural approach to hormone storage in the neural lobe. J. Endocr. **68:** 209–224.
31. Alonso, G., J. Gabrion, N. Lutz-Bucher & I. Assenmacher. 1980. Localization subcellulaire de la vasopressin dans les axones neurosécrétoires du système hypothalamo-neurohypophysaire chez le rat. J. Physiol. (Paris). **76:** 1B.
32. Castel, M. & H.-D. Dellmann. 1980. Thiamine pyrophosphatase activity in axonal smooth endoplasmic reticulum of neurosecretory neurons. Cell Tiss. Res. **210:** 205–221.
33. Dreyfuss, F., A. Burlet, M. Chateau & P. Czernichow. 1980. Localization ultrastructurale par immunocytochemie de la vasopressine non-granulaire dans la neurohypophyse du rat: rôle possible du réticulum endoplasmique lisse dans le transport de l'hormone. Biol. Cell. **35:** 141–146.
34. Morris, J. F. 1976. Distribution of neurosecretory granules among the anatomical compartments of the neurosecretory processes of the pituitary gland: A quantitative ultrastructural approach to hormone storage in the neural lobe. J. Endocr. **68:** 225–234.
35. Dogterom, J., Tj.B. van Wimersma Greidanus & D. F. Swaab. 1977. Evidence for release of vasopressin and oxytocin into cerebrospinal fluid: measurements in plasma and CSF of intact and hypophysectomised rats. Neuroendocrinology **24:** 108–118.
36. Orkand, P. M. & S. L. Palay. 1966. The fine structure of the supraoptic nucleus in normal rats compared with that in rats with hereditary diabetes insipidus. Anat. Rec. **154:** 396.
37. Orkand, P. M. & S. L. Palay. 1967. Effect of treatment with exogenous

vasopressin on the structural alterations in the hypothalamo-neurohypophysial system of rats with hereditary diabetes insipidus. Anat. Rec. **157:** 295.

38. Grainger, F. & J. C. Sloper. 1974. Correlation between microtubular number and transport activity. Cell Tiss. Res. **153:** 101–113.
39. Grainger, F. & J. C. Sloper. 1976. Microtubular number in the tractus hypophyseus of newborn normal rats and newborn rats with congenital diabetes insipidus. Cell Tiss. Res. **169:** 405–414.
40. Heap, P. F., C. W. Jones, J. F. Morris & B. T. Pickering. 1975. Movement of neurosecretory product through the anatomical compartments of the rat neural lobe—an electron microscopic autoradiographic study. Cell Tiss. Res. **156:** 483–497.
41. Haddad, A., S. P. Guaraldo, G. Pelletier, I. L. G. Brasiliero & F. Marchi. 1980. Glycoprotein secretion in the hypothalamo-neurohypophysial system of the rat. Cell Tiss. Res. **209:** 399–422.
42. Sachs, H. & E. W. Haller. 1968. Further studies on the capacity of the neurohypophysis to release vasopressin. Endocrinology **83:** 251–262.
43. Nordmann, J. J. & J. Labouesse. 1981. Neurosecretory granules: evidence for an aging process within the neurohypophysis. Science **211:** 595–597.
44. Morris, J. F. & J. J. Nordmann. 1980. Membrane recapture after hormone release from nerve endings in the neural lobe of the rat pituitary gland. Neuroscience **5:** 639–649.
45. Nordmann, J. J. 1977. Ultrastructural morphometry of the rat neurohypophysis. J. Anat. **123:** 213–218.
46. Morris, J. F. & J. J. Nordmann. 1981. Membrane retrieval by vacuoles after exocytosis in the neural lobe of Brattleboro rats. Neuroscience. (Submitted for publication.)
47. Dyball, R. E. J. 1974. Single unit activity in the hypothalamo-neurohypophysial system of Brattleboro rats. J. Endocr. **60:** 135–143.
48. Douglas, W. W. 1974. Exocytosis and exocytosis-vesiculation sequence: with special reference to neurohypophysis, chromaffin and mast cells, calcium and calcium ionophores. *In* Secretory Mechanisms of Exocrine Glands. N. A. Thorn & O. H. Petersen, Eds.: 116–136. Munksgaard. Copenhagen.
49. Theodosis, D. T., J. J. Dreifuss, M. C. Harris & L. Orci. 1976. Secretion related uptake of horseradish peroxidase in neurohypophysial axons. J. Cell Biol. **70:** 294–303.
50. Shaw, F. D. & J. F. Morris. 1980. Calcium localisation in the rat neurohypophysis. Nature **287:** 56–58.
51. Nordmann, J. J. & J. Chevallier. 1980. On the role of microvesicles in buffering Ca^{2+} in the neurohypophysis. Nature **287:** 54–56.
52. Torp-Pedersen, C., T. Saermark, M. Bungaard & N. A. Thorn. 1980. ATP-dependent Ca^{2+} accumulation by microvesicles isolated from bovine neurohypophyses. J. Neurochem. **35:** 552–557.
53. Cattenach, B. M., C. A. Iddon, H. M. Charlton, S. A. Chiappa & G. Fink. 1977. GnRH deficiency in a mutant mouse with hypogonadism. Nature **269:** 338–340.
54. Iddon, C. A., H. M. Charlton & G. Fink. 1980. Gonadotrophin release in normal and hypogonadal mice after electrical stimulation of the median eminence or injection of LHRH. J. Endocr. **85:** 105–110.
55. McDowell, I. F. W., J. F. Morris & H. M. Charlton. 1981. The gonadotrophs of hypogonadal (*hpg*) mice. J. Anat. **133:** 686–687.

Discussion of the Paper

T. G. Sherman (*State University of New York, Stony Brook, N.Y.*): Do you think that the presence of the extragranular neurophysin plays any sort of role in secretion or intracellular regulation of the granular pool?

J. F. Morris: My belief is that any demonstration of extragranular neurophysin essentially results from the way in which the material is prepared. I don't see how neurophysins could cross biologically intact granule membranes or, if they did so, how they could then leave the cell. It is possible that leakage from the oldest granules *in vivo* could act as a stimulus to granulolysis (Morris & Nordmann. Lysosomal control in neurosecretion. *Biologie Cellulaire* **36:** 193–200, 1979.)

C. I. Johnston (*Prince Henry's Hospital, Melbourne, Australia*): Do you think that the absence of granules in the Brattleboro DI rat is good indirect evidence that it doesn't make a false "peptide transmitter"? Would an abnormal protein be packaged into granules?

Morris: There are other models in which the absence of a functional leader sequence prevents the material being synthesized by the ribosome from crossing the endoplasmic reticulum membrane (Bedouelle *et al.* Mutation which alters the function of the signal sequence of the maltose binding protein. *Nature* **285:**78–81, 1980). If that was the case for the Brattleboro homozygote then one might expect to see neurophysin-immunoreactive material throughout the abnormal cells, but this is not the case. Both types of cell do have some material packaged in granules; we just have no idea what the 110 nm granules of the abnormal cells contain.

J. R. Sladek (*University of Rochester, Rochester, N.Y.*): The last animal you mentioned, the *hpg* mouse, is a very interesting model and I would like to take just a tiny exception to your comment on LHRH distribution in the median eminence wherein, I believe, it is reasonably well circumscribed.

Morris: I quite agree with you if we restrict the problem to the light microscopic level. However, many items of interest related to the organization of the individual terminals require ultrastructural investigation. I don't see any easy way that one could study either abnormal perikarya or terminals ultrastructurally.

J. R. Sladek: Just a note of optimism: with the inception of new techniques for dual localization of peptides within individual neurons, it may well be that in the near future someone will find some specific substance in addition to LHRH that would allow you to localize these putative LHRH-deficient neurons.

C. Sladek (*University of Rochester, Rochester, N.Y.*): I was wondering if you thought that dynorphin might potentially be the peptide contained in the small neurosecretory granules that you showed. Dr. Stan Watson (Michigan) has demonstrated immunocytochemically that dynorphin is contained in supraoptic vasopressin neurons in normal rats and in the non-oxytocin neurons in Brattleboro DI rats.

Morris: I was aware of that. I highlighted the possibility of an enkephalin content of the 110 nm granules because there is also evidence for enkephalin receptors in the neurohypophysis and for an inhibitory effect of enkephalins on neurohypophyseal hormone release. But is it the magnocellular neurons or other neurons that contain the enkephalin? There are two competing sets of evidence. I am worried also by the growing number of peptides that are competing for a space within the system. The granules are virtually full with 84,000 molecules of M_r 20,000 precursor. There would be little room in the granules for substantial amounts of other precursors.

C. SLADEK: I agree with that, but is it possible that the normal vasopressin cell also contains these small dense-core granules, but they are just overlooked because of the massive number of large neurosecretory granules?

MORRIS: They do contain just a very few. But the mere presence of 110 nm granules in the two cell types is no guarantee that they have the same content.

C. SLADEK: That is exactly my point. Possibly other molecules, and dynorphin is just an example, are packaged in a separate population of small dense-cored granules.

MORRIS: It's certainly possible.

E. A. ZIMMERMAN (*Columbia University, New York, N.Y.*): Along that line, what do you think of the recent reports of CCK in oxytocin cells.

MORRIS: Well, five or six peptides have been reported in the magnocellular neurons. I think that we must take seriously the possibility of ionic interactions between proteins and immunoglobulins. Dr. B. Stein, working in my laboratory, found a lot of somatostatin cross-reactivity in gastrin cells in the gut. It appeared to be quite specific for the gastrin cells. When she used Grübe's manipulation of increased salt concentration in the primary antiserum, the reactivity disappeared from the gastrin cells but was, of course, retained in the somatostatin cells. I would like to see those sorts of controls for other peptides claimed in the magnocellular neurons.

BIOCHEMICAL AND FUNCTIONAL ASPECTS OF MAGNOCELLULAR NEURONS AND HYPOTHALAMIC DIABETES INSIPIDUS

Brian T. Pickering * and William G. North †

** Department of Anatomy*
University of Bristol
Bristol BS8 1TD, England

† Department of Physiology
Dartmouth Medical School
Hanover, New Hampshire 03755

The Brattleboro rat has made considerable contributions to our understanding of the biosynthesis of neurohypophyseal hormones and, rather than give a chronological report, we propose here to consider again some of the questions that have been answered, partially answered, or are still begging.

ESTABLISHMENT OF A VASOPRESSIN-NEUROPHYSIN AND AN OXYTOCIN-NEUROPHYSIN

The hormones oxytocin and vasopressin are synthesized in the perikarya of the magnocellular neurons in the hypothalamus, transported to axon terminals in the neural lobe of the pituitary gland, and stored and released from there. Perhaps the most clear-cut contribution of the Brattleboro rat has been the association of the hormones to specific neurophysins (Nps). The neurophysins [1] make up a family of polypeptides having 91–95 amino acid residues and are stored along with the hormones in the neurosecretory granules.[2]

Neurophysin has had a checkered history with regard to the number of different molecules present in a given species. Hope and his colleagues [3] first emphasized that many of the neurophysin molecules of the ox were generated during extraction because the neurohypophysis is a rich source of peptidases. This group was also the first to detect an indication that there might be a specific association of one ox neurophysin with oxytocin and another with vasopressin. They were attempting to separate bovine oxytocin granules from vasopressin granules by centrifugal procedures and obtained differential enrichment across a sucrose density gradient. Bovine NpI was enriched along with oxytocin and bovine NpII with vasopressin.[4] Nevertheless, it was in the rat system that we could first be sure that there was a neurophysin associated with vasopressin and one with oxytocin.[5] Polyacrylamide gel electrophoresis of a simple acid extract of a rat neural lobe separates three neurophysins that run ahead of serum albumin. When the rats received intracisternal injections of [^{35}S]cysteine 24 h before death, most of the radioactivity recovered from the polyacrylamide gel was associated with the two faster moving neurophysins. Moreover, the ratio of radioactivity in these two peaks was the same as the vasopressin:oxytocin ratio determined by biological assay of the gland extracts. Thus, there was an indication of a specific association of one hormone with one neurophysin. But

0077–8923/82/0394–0072 $1.75/0

it was the Brattleboro rat that allowed unequivocal assignment of the faster moving neurophysin to vasopressin and the slower one to oxytocin. The Brattleboro rat, homozygous for diabetes insipidus, is unable to synthesize vasopressin and the putative vasopressin-neurophysin was found to be absent from the glands of such animals and present in reduced amounts in the heterozygote of the strain.[6, 7]

The establishment of a vasopressin-neurophysin and an oxytocin-neurophysin allowed their separate isolation and characterization[6, 8] followed by the generation of specific antisera.[9, 10] Such antisera allowed the confirmation by both immunocytochemical and immunoassay techniques that vasopressin and its associated neurophysin are absent from homozygous Brattleboro rats.[9–13] More importantly, the homozygous Brattleboro provides an essential control for the specificity of these antisera and is invaluable for their use in the localization of the neurohypophyseal hormones in other parts of the brain.

The Significance of the Third Neurophysin

A time-course for the labeling of the neural lobe neurophysins showed that the third neurophysin is labeled with very different kinetics from vasopressin-neurophysin or oxytocin-neurophysin. Whereas they begin to be labeled between one and two hours after the injection of tracer, the third neurophysin does not become labeled until much later.[14] Figure 1 shows this differential labeling for the heterozygous Brattleboro rat and also demonstrates a difference in the rate of decline of radioactivity in the two major components: at early time intervals there is more radioactivity associated with oxytocin-neurophysin (B) than with vasopressin-neurophysin (A), as expected for a heterozygous Brattleboro, but this reverses at later times. The differential fall-off taken together with the late labeling of the third neurophysin (C) prompted a suggestion that it was a metabolic product of oxytocin-neurophysin generated in the granule. The arithmetic was compatible with this: the combined decline of oxytocin-neurophysins and the third component paralleled that of vasopressin-neurophysin.[6] Recent sequence data (Figure 2) and compositional analysis show oxytocin-neurophysin and the third component to be very similar.[15] Since all three neurophysins are sedimented in a granular fraction, it was concluded that this interconversion must occur inside the granule, which must therefore contain the necessary enzyme(s).

The presence of such an enzyme(s) in the granules was then demonstrated.[16] Conversion of a purified oxytocin-neurophysin to the third neurophysin by a preparation of granules was observed to occur at a substrate concentration of 0.1 mM and at a pH of 5. The pH within the granules was believed to be in the vicinity of 5 because they were most stable to fixation at this pH[17] and because they contained high concentrations of neurophysin (pI $\sim$ 5).[18] Recent measurements of the intragranular pH have indicated it to be 5.5.[19]

A Common Biosynthetic Precursor

With the acceptance of the initial synthesis of two neurophysins, one associated with vasopressin and one with oxytocin, strength was given to the proposal of Sachs and his colleagues[20] that the hormones shared common precursors with the neurophysins. Further support came from the observation that the hormones and neurophysins are transported to the neural lobe in parallel.[6] Moreover, as can be seen in FIGURE 3, there is a molar equivalence of the incorporation of radioactive cysteine into hormones and neurophysins; indicating that they may have been labeled from the same pool. The proposal has been fully vindicated in recent years by the demonstration of the precursors themselves.[21, 22] Studies of the incorporation of [^{35}S]cysteine by the supraoptic nuclei have led to the identification of two precursors, one related to oxytocin and its neurophysin, called pro-oxyphysin by the Gainer group,[23] and the other related to vasopressin and its neurophysin. This latter component, propressophysin, is absent from the homozygous Brattleboro rat.[23]

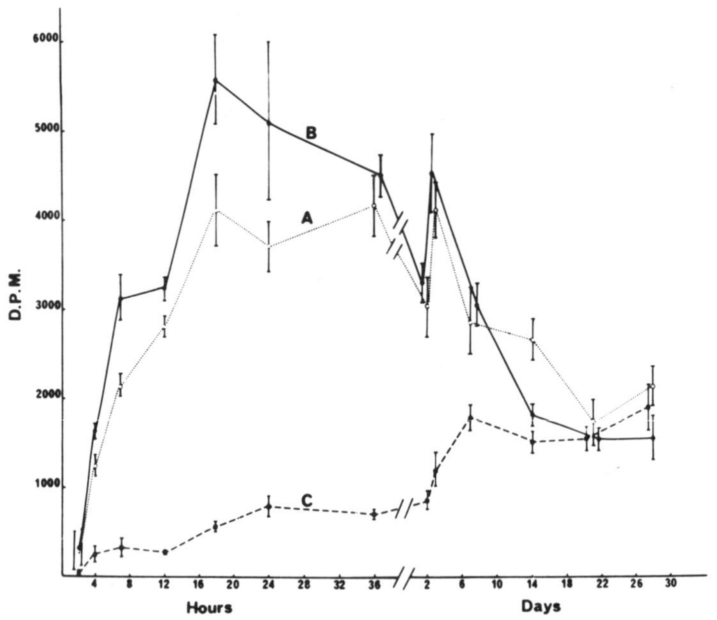

FIGURE 1. Time course for the incorporation of radioactivity into the neurophysins in the neural lobe of the heterozygous Brattleboro rat after an intracisternal injection of 50 μCi [^{35}S]cysteine. (A) vasopressin-neurophysin (rNpI); (B) oxytocin-neurophysin (rNpII); (C) third neurophysin component (rNpIII).

Translation of an RNA species isolated from supraoptic regions of the hypothalami of oxen and rats have also shown the synthesis of similar common precursors.[24–27]

THE SITE OF PROCESSING AND THE ENZYMES INVOLVED

The first indication that processing of the common precursors occurs while the granules are in transit along the axons came from Sachs.[28] He demonstrated that, after infusion of [^{35}S]cysteine, the specific radioactivity of vasopressin

```
                      5                   10                  15
rNp III  Ala.Ala.Leu.Asp.Leu.Asp.Met.Arg.Lys.Cys.Leu.Pro.Cys.Gly.Pro.
rNp II   Ala.Ala.Leu.Asp.Leu.Asp.Met.Arg.Lys.Cys.Leu.Pro.Cys.Gly.Pro.

rNp I    Ala.Thr.Ser.Asp.Met.Glu.Leu.Arg.Gln.Cys.Leu.Pro.Cys.Gly.Pro.

                     20                  25                  30
rNp III  Gly.Gly.Lys.Gly.Arg.Cys.Phe.Gly.Pro.Ser.Ile.Cys.Cys.Ala.Asp.
rNp II   Gly.Gly.Lys.Gly.Arg.Cys.Phe.Gly.Pro.Ser.Ile.Cys.Cys.Ala.Asp.

rNp I    Gly.Gly.Lys.Gly.Arg.Cys.Phe.Gly.Pro.Ser.Ile.Cys.Cys.Ala.Asp.

                     35                  40
rNp III  Glu.Leu.Gly.Cys.Phe.Val.Gly.Thr.Ala.Glu.
rNp II   Glu.Leu.Gly.Cys.Phe.Val.Gly.Thr.Ala.Glu.

rNp I    Glu.Leu.Gly.Cys.Phe.Leu.Gly.Thr.Leu.Val.
```

FIGURE 2. Partial amino acid sequences of the rat neurophysins.

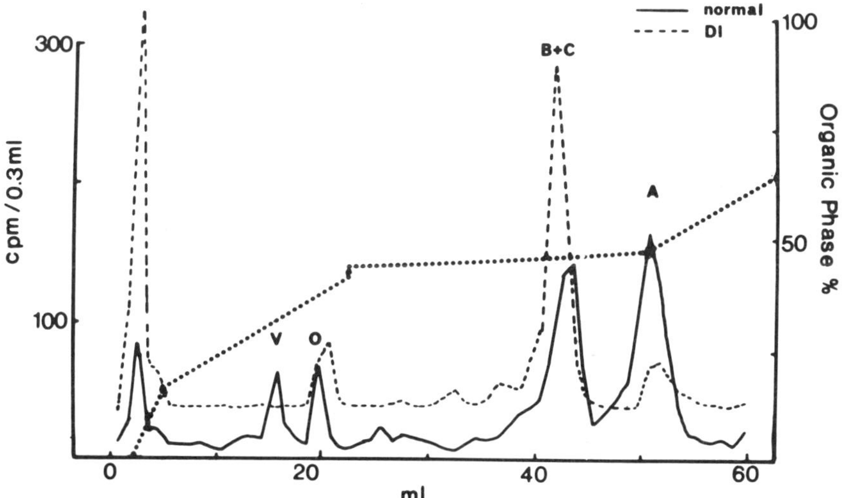

FIGURE 3. High performance liquid chromatography of neural lobe extracts from normal (—) and homozygous Brattleboro (- - - -) rats which had received 50 μCi [^{35}S]cysteine 24 h previously. The column of ODS was eluted with a gradient (· · · · ·) achieved with increasing proportions of 60% acetonitrile in 0.2 M NaH_2PO_4 adjusted to pH 2.1 with H_3PO_4. (A) vasopressin-neurophysin (rNpI); (B) oxytocin-neurophysin (rNpII); (C) third neurophysin component (rNpIII); (V) vasopressin; (O) oxytocin. Note that the radioactivity ratios A/V and B+C/O approximate seven for each extract. (R. W. Swann, unpublished observations).

isolated from hypothalamus was lower than that of neural lobe hormone. The intra-axonal transport velocity of neurophysins and hormones also pointed towards processing during transport.[6] Firm evidence for this came after the identification of the precursors and the demonstration that precursor:product ratios in supraoptic nuclei, median eminence, and neural lobe were in accord with processing en route.[21, 29] Granule movement is not obligatory for processing since it continues normally within granules that have been arrested in the perikarya as a result of treatment with colchicine.[30] Thus it would appear that the granule is a package of substrate (precursor) and enzyme which interact in a time-dependent manner.

That the granules possess at least one proteolytic enzyme was apparent from the formation of the third neurophysin component as described above. The questions that arise are whether this is also the enzyme that converts precursor to neurophysin plus hormone and what are the characteristics of such an enzyme.

The enzyme displays maximal activity between pH 4.0 and 5.0 and has an α-chymotrypsin-like specificity to cleave peptide bonds.[16] In fact conversion of oxytocin-neurophysin to the third neurophysin involves cleavage at a Phe—Ser bond and this also results from α-chymotrypsin digestion at pH 7.6.[9] Thus an acid protease with α-chymotrypsin-like specificity is a component of granules.

Generation of hormone and neurophysin products from precursors by the action of this enzyme would necessitate the existence of susceptible bonds involving aromatic or long-chain aliphatic residues. An alternate proposal for processing of the precursors, advanced by Gainer and coworkers,[23] envisages trypsin-like enzymes cleaving paired basic residues according to the scheme found for conversion of proinsulin to insulin.[31]

In support of this view, they have demonstrated liberation of vasopressin-, oxytocin-, and neurophysin-like fragments from putative precursors after limited treatment with trypsin.[32] Moreover, Fletcher *et al.*[33] have shown that the enzyme of insulin granules with trypsin-like specificity is active at pH 5, although such an enzyme has yet to be demonstrated in neurosecretory granules.

What is the Nature of the Precursor?

Having established that the hormones and neurophysins are derived from common precursors, which are processed inside the granule as it moves from perikaryon to axon terminal, a number of questions still remain.

Each precursor contains the sequences of the hormone and its respective neurophysin, but in what order and what else goes to make up the missing 4,000–10,000 daltons between the sum of hormone and neurophysin (11,000) and the molecular weight of precursor (15,000–23,000)? Injection of radioactive sugars, either intracisternally or into supraoptic nuclei, leads to the arrival of a radioactive component(s) in the neural lobe.[34, 35] However, the Brattleboro homozygote does not have such a glycopeptide and Gainer and his colleagues have concluded that while the vasopressin precursor, propressophysin, is glycosylated the oxytocin one is not.[23] In the bovine system *in vitro*, Schmale and Richter[26] have shown that a vasopressin-related propeptide of 19,000 daltons is subsequently glycosylated to give a 23,000 dalton product. Similarly the Gainer group have shown that propressophysin, which is glycosylated, has an apparent molecular weight of 20,500 and pro-oxyphysin, which is not, has a molecular weight of 18,700.[23] Thus it would appear that the two hormone precursors have similar sized polypeptide chains, some 4,000 daltons or more, bigger than the sum of hormone and neurophysin, but the vasopressin one has some carbohydrate chains attached to it.

What is the role of this carbohydrate? The simple answer is that we don't know! From studies in the ACTH system, Loh and Gainer [36] proposed that the carbohydrate directed enzymatic attack during processing by protecting some otherwise susceptible bonds. One would have to ask why propressophysin needs such protection while pro-oxyphysin does not. Moreover, inhibition of glycosylation by intracisternal injection of tunicamycin did not lead to misprocessing of any product that became packaged, but rather inhibited packaging.[37]

What is the order of the sequences within the precursors? An argument can be made that the hormone is on the amino-terminal side of neurophysin so that oxytocin might arise from transamidation of a Gly—Ala bond, for this would explain the constancy of the NH_2-terminal alanine in all neurophysins.[38] A suggested possibility for the vasopressin precursor has the glycopeptide moiety on the NH_2-terminal side of the hormone which thus becomes sandwiched between this and the neurophysin.[29] Another view places the hormone at the carboxyl end of the molecule.[9, 26]

Biosynthesis in the Brattleboro

It will be interesting to see which of these steps in the biosynthesis of vasopressin is missing in the Brattleboro rat. The oxytocin system seems to be unaffected by the mutation since the low levels of this hormone seen in the homozygote can be restored to normal by removal of the osmotic stress by treatment with vasopressin.[39] Even in the vasopressin system, the mutation seems to be restricted to the synthesis of vasopressin-containing granules. The cells that would normally fulfill this function are present in apparently normal numbers.[40] They are capable of packaging secretory product since they contain a population of smaller granules, as Morris has described.[2] One question that has arisen time and time again is: Do the cells produce, and do these granules represent an altered form of vasopressin and its neurophysin? All one can say is that there is very little evidence to support this suggestion. Russell *et al.*[23] have pointed out the occurrence of 'peptide X', which can be recognized in Brattleboro homozygotes, although it is not clear if this peptide is unique to them. The neurohypophyses of Brattleboro homozygotes as well as normals contain a number of other neuropeptides, e.g. somatostatin, enkephalins, and angiotensin, and there is some evidence that some of these may exist within the oxytocin and vasopressin neurons.[41, 42]

Thus we are in sight of a complete understanding of the biosynthetic pathway to vasopressin and oxytocin and the Brattleboro rat has played a significant part in this achievement. However, as can be seen from FIGURE 4, the neural lobe contains a number of still unrecognized peptides, some of which are absent from the Brattleboro. Many of these questions will be answered when we have the complete amino acid sequence of the hormone precursors.

[**Note added in proof:** Recent analysis using the cDNA technique has allowed Land *et al.*[43] to conclude that the primary precursor in the vasopressin system has the sequence: signal peptide–vasopressin–neurophysin–glycopeptide.]

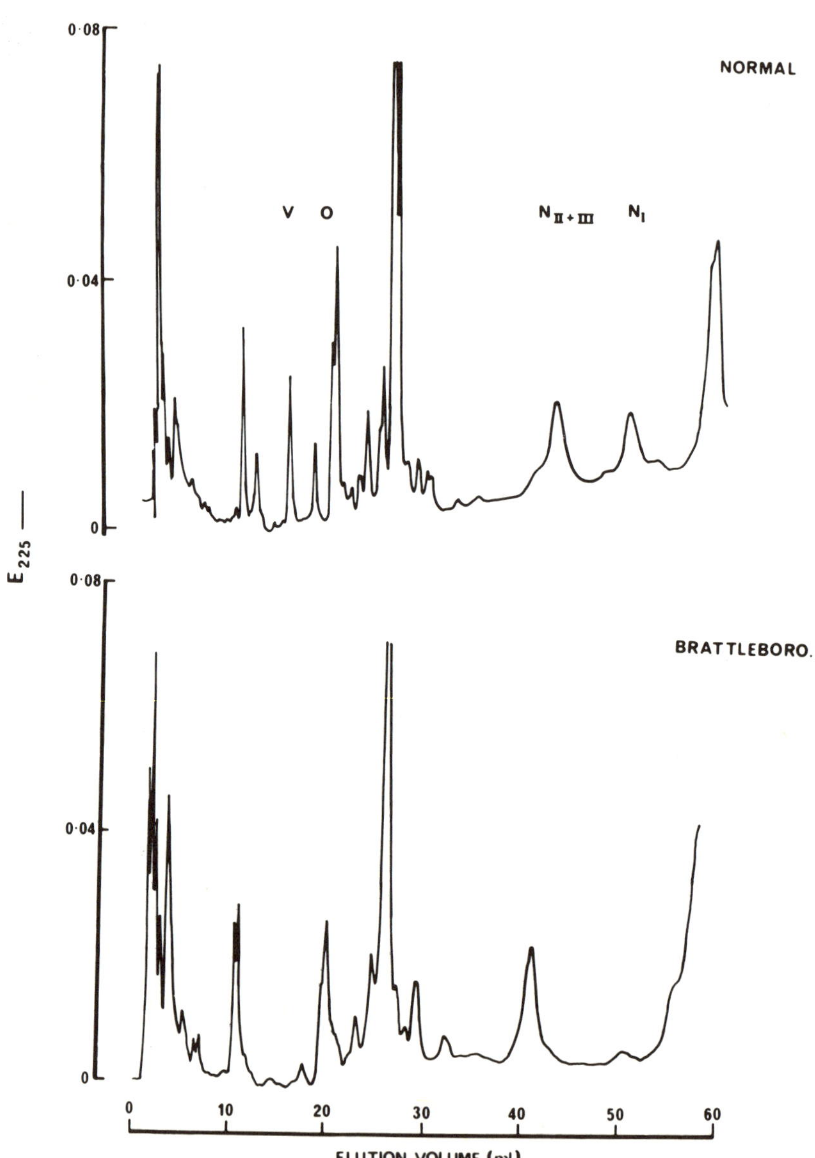

FIGURE 4. High performance liquid chromatography of neural lobe extracts. Details as in FIGURE 3. (R. W. Swann, unpublished observations).

REFERENCES

1. WALTER, R., Ed. 1975. Neurophysins: Carriers of Peptide Hormones. Ann. N.Y. Acad. Sci. **248**.
2. MORRIS, J. F. 1982. The Brattleboro magnocellular neurosecretory system: A model for the study of peptidergic neurons. Ann. N.Y. Acad. Sci. (This volume.)
3. DEAN, C. R. & D. B. HOPE. 1967. The isolation of purified neurosecretory granules from bovine pituitary posterior lobes. Comparison of granule protein constituents with those of neurophysins. Biochem. J. **104:** 1082–1088.
4. DEAN, C. R., D. B. HOPE & T. KAZIC. 1968. The total hormone-binding capacity of the neurophysins and the oxytocin and vasopressin content of the posterior pituitary. Brit. J. Pharmacol. **34:** 192–193P.
5. BURFORD, G. D., C. W. JONES & B. T. PICKERING. 1971. Tentative identification of a vasopressin-neurophysin and an oxytocin-neurophysin in the rat. Biochem. J. **124:** 809–813.
6. PICKERING, B. T., C. W. JONES, G. D. BURFORD, M. MCPHERSON, R. W. SWANN, P. F. HEAP & J. F. MORRIS. 1975. The role of neurophysin proteins: suggestions from the study of their transport and turnover. Ann. N.Y. Acad. Sci. **248:** 15–35.
7. SUNDE, D. A. & H. W. SOKOL. 1975. Quantification of rat neurophysins by polyacrylamide gel electrophoresis (PAGE): Application to the rat with hereditary hypothalamic diabetes insipidus. Ann. N.Y. Acad. Sci. **248:** 345–364.
8. NORTH, W. G. & H. VALTIN. 1977. The purification of rat neurophysins by a method of preparative polyacrylamide gel electrophoresis. Anal. Biochem. **78:** 436–450.
9. NORTH, W. G., F. T. LAROCHELLE, JR., J. F. MORRIS, H. W. SOKOL & H. VALTIN. 1978. Biosynthetic specificity of neurons producing neurohypophysial principles. *In* Current Studies of Hypothalamic Function 1978. K. Lederis & W. L. Veale, Eds. **1:** 62–76. Karger. Basel.
10. MCPHERSON, M. A. & B. T. PICKERING. 1978. Preparation of antisera to three rat neurophysins and their use for radioimmunoassays. J. Endocrinol. **76:** 461–471.
11. VANDESANDE, F. & K. DIERICKX. 1976. Immunocytochemical demonstration of the inability of the homozygous Brattleboro rat to synthesize vasopressin and vasopressin-associated neurophysin. Cell Tiss. Res. **165:** 307–316.
12. FISHER, A. W. F., V. GILL, K. WONG, S. RAGHATRAM, W. NORTH & K. LEDERIS. 1981. Anatomical and developmental aspects of the neurons terminating in the posterior pituitary. *In* Neurosecretion. Molecules, Cells, Systems. D. S. Farner & K. Lederis, Eds. Plenum. New York, N.Y.
13. VALTIN, H., W. G. NORTH, F. T. LAROCHELLE, JR., H. W. SOKOL & J. F. MORRIS. 1978. Biochemical and anatomical aspects of ADH production. *In* Proc. Seventh Intl. Cong. Nephrol. M. Bergeron, Ed.: 313–320. Karger. Basel.
14. BURFORD, G. D. & B. T. PICKERING. 1973. Intra-axonal transport and turnover of neurophysins in the rat. A proposal for a possible origin of the minor neurophysin component. Biochem. J. **136:** 1047–1052.
15. NORTH, W. G. & T. I. MITCHELL. 1981. Evolution of neurophysin proteins: partial amino acid sequences of rat neurophysins. FEBS Lett. **126:** 41–44.
16. NORTH, W. G., J. F. MORRIS, F. T. LAROCHELLE, JR. & H. VALTIN. 1977. Enzymatic interconversions of neurophysins. *In* Neurohypophysis. A. M. Moses & L. Share, Eds.: 43–52. Karger. Basel.
17. MORRIS, J. F. & M. A. CANNATA. 1973. Ultrastructural preservation of the dense core of posterior pituitary neurosecretory granules and its implications for hormone release. J. Endocrinol. **57:** 517–529.

18. Dreifuss, J. J. 1975. A review of neurosecretory granules: their contents and mechanisms of release. Ann. N.Y. Acad. Sci. **248:** 184–201.
19. Russell, J. T. & R. W. Holz. 1981. Measurement of ΔpH and membrane potential in isolated neurosecretory vesicles from bovine neurohypophyses. J. Biol. Chem. **256:** 5950–5953.
20. Sachs, H., P. Fawcett, Y. Takabatake & R. Portanova. 1969. Biosynthesis of vasopressin and neurophysin. Rec. Prog. Horm. Res. **25:** 447–491.
21. Gainer, H., Y. Sarne & M. J. Brownstein. 1977. Biosynthesis and axonal transport of rat neurohypophysial proteins and peptides. J. Cell Biol. **73:** 366–381.
22. Brownstein, M. J., A. G. Robinson & H. Gainer. 1977. Immunological identification of rat neurophysin precursor. Nature **269:** 259–261.
23. Russell, J. T., M. J. Brownstein & H. Gainer. 1980. Biosynthesis of vasopressin, oxytocin and neurophysins: Isolation and characterization of two common precursors (propressophysin and prooxyphysin). Endocrinology **107:** 1880–1891.
24. Giudice, L. C. & J. M. Chaiken. 1979. Immunological and chemical identification of a neurophysin-containing protein coded by messenger RNA from bovine hypothalamus. Proc. Natl. Acad. Sci. USA **76:** 3800–3804.
25. Lin, C., P. Joseph-Bravo, T. Sherman & J. F. McKelvy. 1979. Cell-free synthesis of putative neurophysin precursors from rat and mouse hypothalamic poly (A)-RNA. Biochem. Biophys. Res. Commun. **89:** 943–950.
26. Schmale, H. & D. Richter. 1981. Immunological identification of a common precursor to arginine vasopressin and neurophysin II synthesized by in vitro translation of bovine hypothalamic mRNA. Proc. Natl. Acad. Sci. USA **78:** 766–769.
27. Schmale, H., B. Leipold & D. Richter. 1979. Cell-free translation of bovine hypothalamic mRNA. Synthesis and processing of the preproneurophysin I and II. FEBS Lett. **108:** 311–315.
28. Sachs, H. 1960. Vasopressin biosynthesis I. in vivo studies. J. Neurochem. **5:** 297–303.
29. Brownstein, M. J., J. T. Russell & H. Gainer. 1980. Synthesis, transport, and release of posterior pituitary hormones. Science **207:** 373–378.
30. Parish, D. C., E. M. Rodriguez, S. D. Birkett & B. T. Pickering. 1981. Effects of small doses of colchicine on the components of the hypothalamo-neurohypophysial system of the rat. Cell Tissue Res. **220:** 809–827.
31. Steiner, D. F. 1976. Peptide hormone precursors: biosynthesis, processing and significance. *In* Peptide Hormones. J. A. Parsons, Ed.: 49–66. University Park Press. Baltimore, Md.
32. Russell, J. T., M. J. Brownstein & H. Gainer. 1979. Trypsin liberates an arginine vasopressin-like peptide and neurophysin from a M_r 20,000 putative common precursor. Proc. Natl. Acad. Sci. USA **76:** 6086–6090.
33. Fletcher, D. J., J. Quigley, E. J. Baver & B. D. Noe. 1981. Characterization of proinsulin- and proglucagon-converting activities in isolated islet secretory granules. J. Cell Biol. **90:** 312–322.
34. Jones, C. W. & R. W. Swann. 1974. Glycoproteins in the neural lobe. Studies on the incorporation of radioactive sugars into neural lobe components. J. Endocrinol. **63:** 53–54P.
35. Gainer, H. & M. J. Brownstein. 1978. Electrophoretic analyses of proteins transported to the rat posterior pituitary. J. Neurochem. **30:** 1509–1512.
36. Loh, Y. P. & H. Gainer. 1979. The role of the carbohydrate in the stabilization, processing, and packaging of the glycosylated adrenocorticotropin-endorphin common precursor in toad pituitaries. Endocrinology **105:** 474–487.
37. Gonzales, C. G., R. W. Swann & B. T. Pickering. 1981. Effects of tunicamycin on the hypothalamo-neurohypophysial system of the rat. Cell Tissue Res. **217:** 199–210.

38. PICKERING, B. T. 1976. The molecules of neurosecretion: their formation, transport and release. Progr. Brain Res. **45:** 161–179.
39. VALTIN, H. & H. A. SCHROEDER. 1964. Familial hypothalamic diabetes insipidus in rats (Brattleboro strain). Am. J. Physiol. **206:** 425–430.
40. RHODES, C. H., J. I. MORRELL & D. W. PFAFF. 1981. Immunohistochemical analysis of magnocellular elements in rat hypothalamus: Distribution and number of cells containing neurophysin, oxytocin and vasopressin. J. Comp. Neurol. **198:** 45–64.
41. MARTIN. R. & K. H. VOIGT. 1981. Enkephalins co-exist with oxytocin and vasopressin in nerve terminals of rat neurohypophysis. Nature **289:** 502–504.
42. KILCOYNE, M. M., D. L. HOFFMAN & E. A. ZIMMERMAN. 1980. Immunocytochemical localization of angiotensin II and vasopressin in rat hypothalamus: evidence for production in the same neuron. Clinical Science **59**(6): 57S–60S.
43. LAND, H., G. SCHUTZ, H. SCHMALE & D. RICHTER. 1982. Nucleotide sequence of cloned cDNA encoding bovine arginine vasopressin-neurophysin II precursor. Nature **295:** 299–303.

DISCUSSION OF THE PAPER

R. G. ALLEN (*University of Oregon, Portland, Ore.*): Two questions: (1) Could you describe your gel electrophoresis system? (2) How can you calculate the molar ratio of radioactivity? Do you know you are steady-labeling to be able to do this calculation?

B. T. PICKERING: Data have been expressed as moles of immunoreactive neurophysin or immunoreactive hormone. I agree that if you set out to examine the hypothesis that equimolar amounts of hormone and neurophysin are produced, you have to worry about making such a calculation. However, what I was showing was the evidence that came from the experiment: If you measure the ratio, it comes out to 7. The point is that this is an empirical observation and that there is a 7:1 ratio.

ALLEN: With regard to the molar reactivity in your precipitation, you could use affinity-purified antibody in excess, which would get around the problem of different molar reactivity in the precursor of vasopressin, because you are going to take everything down.

PICKERING: I am not worried (about different molar reactivity) because I am using it to answer a different sort of question.

ALLEN: It's kind of hard to reconcile when I see a little radioactive peak of vasopressin-neurophysin but I never see the peak of vasopressin. In the Brattleboro rat there always is a little peak of vasopressin-neurophysin, but never any vasopressin peak.

PICKERING: Do you mean the historical slide I showed you of the work we did ten years ago where there is a little radioactive peak in the vasopressin-neurophysin region? I don't know what that means. If one looks at the stained gel there is no band to correspond with this little radioactive peak. Maybe one day we will know what this means, maybe it's related to peptide 'X'. But it's a very low percentage of the total reactivity recovered from the gel.

MULTIPLE FORMS OF NEUROPHYSIN PRECURSOR SYNTHESIZED FROM BRATTLEBORO RAT HYPOTHALAMIC RNA

Thomas G. Sherman and Jeffrey F. McKelvy

Department of Neurobiology and Behavior
The State University of New York at Stony Brook
Stony Brook, New York 11794

Introduction

The observed close relationship between the synthesis of the neurohypophyseal hormones and the neurophysin proteins has developed beyond the common precursor hypothesis, as originally proposed by Sachs and Takabatake,[1] to the direct observation of synthesis of high molecular weight forms of neurophysin via translation of hypothalamic poly(A)RNA isolated from mouse,[2] rat,[2] and bovine[3,4] hypothalamus. In addition, extensive work has been performed on the precursors synthesized *in vivo* in mouse,[5] bovine,[6] and rat,[7] resulting, as exemplified by the work of Gainer and coworkers,[7,8] in the observation of two 20,000 dalton proteins of isoelectric points (pI) 5.4 and 6.1, which when subjected to limited trypsinization yield a 10,000 dalton protein, thought to be mature neurophysin.

The homozygous Brattleboro rat has played an important role throughout all these studies as a tool to help identify which neurophysin is biosynthetically related to which nonapeptide hormone,[9,10] or which *in vivo* labeled precursor contains which neurophysin and peptide.[7] This exploitation of a genetic anomaly continues at the *in vitro* level in this study by examination of the neurophysin precursors coded for by Brattleboro hypothalamic poly(A)RNA, with direct comparisons made to similarly obtained precursors from normal and salt-loaded rat RNA translation products. Evidence is presented that illustrates that Brattleboro RNA fails to direct the synthesis of one of the three precursor species observed in normal or salt-loaded rat, indirectly supporting the idea that the Brattleboro's genetic lesion exists at or before the level of transcription.

Materials and Methods

Isolation of Rat Hypothalamic Poly(A)RNA

Freshly dissected hypothalami (70–90 mg) from 250-gram female Sprague-Dawley retired breeders or a homozygous strain of male Brattleboro rat (Blue Spruce Farms, Inc., Altamont, N.Y.) were quickly frozen in liquid nitrogen. Female Sprague-Dawley rats designated as salt-loaded had been provided with 2% (w/vol) saline *ad libitum* for 16 days prior to decapitation. Total cytoplasmic RNA was extracted by a guanidine-HCl procedure as described previously.[2] Poly(A)-containing RNA was purified from total cytoplasmic RNA by oligo(dT)-cellulose (Type 3; Collaborative Research, Waltham, Mass.)

0077–8923/82/0394–0082 $1.75/0 © 1982, NYAS

chromatography essentially as described by Aviv and Leder [11] (see footnote, TABLE 1), except that the poly(A)RNA was converted to the potassium salt via ethanol precipitation from 2.5 M potassium acetate and the pellet lyophilized to remove trace ethanol.

Cell-Free Translations

Wheat germ extract (WGE) and rabbit reticulocyte lysate (RRL) were obtained from Bethesda Research Laboratories, Inc., Rockville, Md. Translations were optimized with respect to potassium, magnesium, and poly(A)RNA concentrations and were performed in the presence of 1.0 mCi/ml L-[^{35}S] cysteine (850–1000 Ci/mmole; New England Nuclear, Boston, Mass.) and 1.0 mM diisopropyl fluorophosphate (DFP; Aldrich Chemical Co., Milwaukee, Wisc.). Total incorporation of [^{35}S]cysteine into trichloroacetic acid (TCA)-precipitable, *S*-carboxymethylated protein was performed as described by Hirs [12] as modified by New England Nuclear, except that the *S*-carboxymethylation reaction was carried out in the dark in 0.5 M Tris-HCl, 6 M guanidine-HCl, 2.0 mM disodium ethylenediamine-tetraacetate (Na_2EDTA), pH 8.1.

Immunoprecipitation of Neurophysin Precursors

Immunoglobulin G (IgG) fractions were purified from normal rabbit serum (NRS) and rabbit anti-rat neurophysin serum (kindly provided by Dr. Alan G. Robinson) by protein A-Sepharose CL-4B (Sigma Chemical Co., St. Louis, Mo.) chromatography by a modification of the procedure of Miller and Stone.[13] Serum was diluted in 50 mM sodium phosphate buffer, 0.15 M NaCl, pH 8.0 (PBS) and passed over protein A-Sepharose. IgG fractions were eluted with and lyophilized from 0.1 N acetic acid, and resuspended in PBS at 5.0 mg/ml. Immunoprecipitation of neurophysin precursors was performed as described,[2] except that the immuno-binding buffer was made 10 mM L-cysteine and 1.0 mM DFP. NRS and anti-rat neurophysin IgG were used at a final concentration of 0.5 mg/ml.

Purification of Individual Rat Neurophysins I, II, and III

Total rat neurophysin (RNp) was purified from lyophilized posterior pituitaries (generous gift of Dr. A. F. Parlow) by 0.1 N HCl extraction and chromatography on a lysine[8]-vasopressin-Sepharose CL-4B affinity column (LVP-Sepharose) [14] as modified by David Sunde (personal communication). The neurophysins were separated on a preparative 10% native acrylamide slab gel (3 mm) in the presence of bromophenol blue (BPB),[15] the protein eluted from the excised bands by extensive dialysis against 0.1 M ammonium acetate, pH 5.8, and the bound BPB removed by dialysis against 0.1 N HCl, 0.5 M NaCl. The soluble contents of the dialysis bags, after re-equilibration with 0.1 M ammonium acetate, pH 5.8, were re-chromatographed on LVP-Sepharose to remove solubilized acrylamide.

Gel Electrophoresis and Isoelectric Focusing

Proteins in the immunoprecipitates were fractionated by discontinuous sodium dodecyl sulfate (SDS) slab gel electrophoresis according to the method of Laemmli.[16] Isoelectric focusing (IEF) was performed in polyacrylamide gels as described by O'Farrell.[17] Focused gels were carefully extruded via syringe as a prelude to: (1) SDS-gel electrophoresis in the second dimension.[17, 18] (2) Radioactivity profile determinations by sectioning the gel into 2-mm slices. Each slice was soaked overnight in 0.5 ml Protosol (New England Nuclear, Boston, Mass.), incubated for 18 hours at 37° C and counted in 10 ml Scint-A (Packard Instrument Co., Downers Grove, Ill.). (3) Isolation of the individual precursor species.

Purification and S-*Carboxymethylation of Individual Precursors*

Radioactive protein bands as determined by the section and count procedure were excised, placed into degassed 0.1 M ammonium bicarbonate buffer, pH 8.1, containing 1.0 mM Na_2EDTA, 5.0 mM dithiothreitol (DTT), and 200 μg carrier rat total neurophysin, and incubated under nitrogen at 25° for 12 hours. Each sample was then made 15 mM in iodoacetic acid (IAA), incubated in the dark under nitrogen for 60 minutes, acidified with 1.0 N formic acid, and dialyzed against 0.1 N formic acid in washed, bovine serum albumin (BSA)-adsorbed Spectropor dialysis tubing (Spectrum Medical Industries, Inc., Los Angeles, Calif.). Soluble contents of dialysis bags were lyophilized. Total and individual rat neurophysins were *S*-carboxymethylated by a modification of the method of Chauvet *et al.*[19] RNp I was carboxymethylated with [^{3}H]IAA (176.2 mCi/mmole; New England Nuclear, Boston, Mass.) by the method of Lockridge *et al.*[20]

Peptide Mapping

Samples were trypsinized for 8 hours at 37° in 0.1 M ammonium bicarbonate buffer, 1.0 mM Na_2EDTA, pH 8.0 with a 40:1 mass ratio of DPCC-trypsin (Sigma Chemical Co., St. Louis, Mo.) to carrier RNp.

Separation of the tryptic fragments was carried out by reverse-phase high performance liquid chromatography (HPLC) on a Waters model HPLC (Waters Associates, Milford, Mass.) equipped with two model 6000A pumps, a model 450 variable wavelength detector, a model 660 solvent programmer, and a model U6K sample injector with a 2.0-ml sample loop. The analyses were performed on a LiChrosorb RP-18 (4.6×250 mm), 10 μm column (Brownlee Labs, Reodyne, Inc., Cotati, Calif.) essentially as described by Mahoney and Hermodson.[21]

Results

Isolation and Translation of Hypothalamic Poly(A)RNA

Table 1 illustrates a greatly abridged purification scheme for rat hypothalamic RNA fractions enriched in the polyadenylated species. Typically, as shown, the yields of both total cellular RNA and poly(A)RNA per gram tissue were greater in rats having completed a 16-day salt-loading regimen with 2% saline as compared to normal or homozygous Brattleboro rats. In part, this increase may be explained by the accumulating evidence that osmotic stimuli which promote the release of vasopressin and oxytocin [22, 23] can similarly enhance the production of the hormones,[24–27] and, conceivably, its mRNA, as has been demonstrated in the regulation of prolactin mRNA levels by thyrotropin-releasing hormone (TRH).[28] Since all cells of the supraoptic nuclei are osmoreceptive,[29, 30] the specificity of the hypertonic stimuli extends to oxytocin neurons as well.[22] Therefore, salt-loading is regarded as a general stimulus for the enhancement of neurohypophyseal hormone-directed RNAs.

Following conversion of the poly(A)RNA from the sodium to the potassium salt and removal of trace ethanol, both of which had proved inhibitory in *in vitro* translation, the RNAs directed the incorporation of [^{35}S]cysteine into TCA-precipitable, *S*-carboxymethylated protein in cell-free translation systems derived from wheat germ and rabbit reticulocytes in a manner linear with RNA concentration (FIGURE 1). RRL was found to be significantly more active than

TABLE 1

ISOLATION OF POLY(A)RNA FROM RAT HYPOTHALAMUS

Strain	Hypothalami Number	Hypothalami grams	Total RNA (mg)	Poly(A) RNA * (mg)	260 nm / 280 nm	260 nm † / 230 nm
Normal	107	9.63	6.01	0.58	2.10	2.20
Salt-Loaded	102	9.05	6.42	0.65	2.12	2.16
Brattleboro	55	3.77	2.33	0.15	2.18	2.25

* Yield after one pass over oligo(dT)-cellulose, Type 3. Normal and salt-loaded poly(A)RNA was routinely passed over oligo(dT)-cellulose twice, with recoveries on second pass of 50–60%.

† Ratio increases with progressive loss of guanidine-HCl.

WGE, as shown in TABLE 2, both in terms of stimulation over endogenous protein synthesis activity and in total [^{35}S]cysteine incorporation into proteins coded for by hypothalamic RNA. The non-poly(A)-containing RNA fractions that did not bind to oligo(dT)-cellulose were much less effective in the WGE and RRL translation assays (data not shown).

Immunoprecipitation and SDS-Polyacrylamide Gel Electrophoresis

Using a protein A-Sepharose affinity purified IgG fraction directed against total rat neurophysin we were able to precipitate neurophysin-like immunoreactive radioactivity to an extent proportional to both added poly(A)RNA and total products (FIGURE 1 and TABLE 2).

Normal and salt-loaded hypothalamic RNA translation products in both WGE and RRL each contained immunoprecipitable material (TABLE 2) which upon electrophoresis in an SDS-polyacrylamide slab gel yielded two bands migrating at 17,000 and 18,500 daltons (FIGURE 2A). Brattleboro RNA

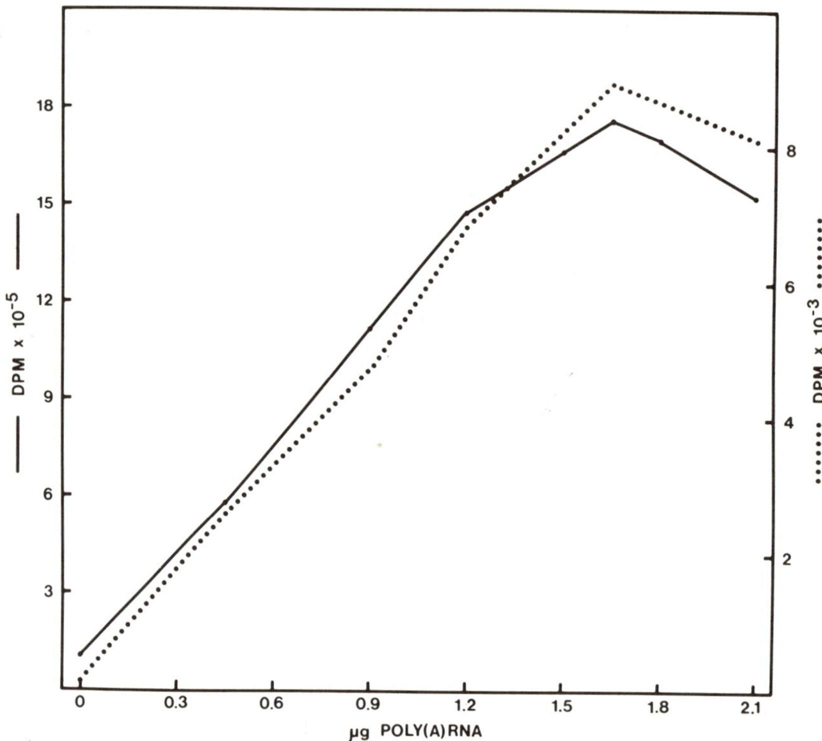

FIGURE 1. Colinearity of salt-loaded poly(A)RNA-dependent incorporation of [^{35}S]cysteine into TCA-precipitable, carboxymethylated protein (—) with neurophysin-like immunoprecipitable radioactivity (······) in RRL. 3 µl aliquots of 30 µl translations were reduced and *S*-carboxymethylated as described in MATERIALS AND METHODS. The remaining 27 µl was diluted 1:1 with IBB and immunoprecipitated with 27 µg anti-rat neurophysin IgG as described in MATERIALS AND METHODS. All dpm's are corrected to 30 µl volumes. ^{35}S counting efficiency was 92%.

similarly directed active protein synthesis *in vitro* with immunoprecipitable radioactivity; yet, when the immunoreactive material was analyzed by one-dimensional SDS-polyacrylamide gel electrophoresis (SDS-PAGE), only one band (17,000 daltons) was observed (FIGURE 2A). In all cases, the observed bands were not detected if the anti-RNp IgG was pre-adsorbed with affinity-purified rat neurophysins (FIGURE 2B) or with NRS IgG (data not shown).

Isoelectric Focusing

FIGURE 3 details the improved resolution obtained when the immunoprecipitates are separated under denaturing conditions by isoelectric focusing. The two bands of normal and salt-loaded samples on SDS-PAGE are resolved into three protein species, with observed isoelectric points of 5.28, 5.60, and 6.15.

Interestingly, the Brattleboro samples contain the two more acidic species of pI 5.28 and 5.60, indicating that the 17,000 dalton band represents two co-migrating proteins of different pI. The Brattleboro's absence of the larger, more neutral species at pI 6.15 is consistent between WGE and RRL translation products (FIGURE 3, A and B).

While the individual precursor species identified qualitatively in terms of molecular weight and pI are accordant between WGE and RRL, the observed quantitative differences—ratios of recovered dpms at pI 5.28:pI 5.60 of 1.30 and 1.04 for normal and salt-loaded RNA, respectively, for WGE; and 0.56 and 0.58, respectively, in RRL—are striking. We feel this does not reflect a difference at the level of immunoprecipitation efficiency or elution recovery from gel slices, but at the level of translation and is a difference not understood at present. The pI 5.28:pI 5.60 ratios for translated Brattleboro RNA of 1.02 and 0.51 for WGE and RRL, respectively, are very similar to those of the salt-loaded rat.

The third protein, migrating at 18,500 daltons of pI 6.15, which is missing in Brattleboro translates, represents a relatively minor immunoprecipitable species, and yet is subject to the greatest percentage change due to long-term salt-loading. Under these conditions this protein is increased two-fold.

FIGURE 3C illustrates the IEF separation of the WGE precursor proteins on a pH gradient similar in composition to that used by Russell *et al.*[7] This has been demonstrated to provide good separation in the pH range 4.5 to 6.5. Comparisons between FIGURE 3 A and C reveal retention of the basic pattern of three distinct species, discounting the argument that our observation of three separate neurophysin-like proteins on IEF is an artifact of our pH gradient.

Two-dimensional gel electrophoresis, when performed by the method of O'Farrell,[17] as shown in FIGURE 4, displays the pI-molecular weight relationships of the three species to best advantage. Frequently, a small spot was detected

TABLE 2

In Vitro TRANSLATABILITY OF HYPOTHALAMIC POLY(A)RNA *

Strain	Poly(A)RNA (μg/ml)	TCA-prec., CM-protein (dpm $\times 10^{-5}$)	Immuno-precipitable protein (dpm $\times 10^{-2}$)	Stimulation †
WGE				
Normal	50	2.39	249	5.8
Salt-Loaded	50	2.67	271	6.5
Brattleboro	60	2.14	279	5.2
RRL				
Normal	55	16.7	716	15.2
Salt-Loaded	55	17.6	936	16.0
Brattleboro	55	15.1	596	13.8

* Total [^{35}S]cysteine incorporation into TCA-precipitable, *S*-carboxymethylated protein, and immunoprecipitable radioactivity corrected for 30 μl volume as in legend to FIGURE 1.

† Fold-stimulation of hypothalamic poly(A)RNA directed protein synthesis over endogenous.

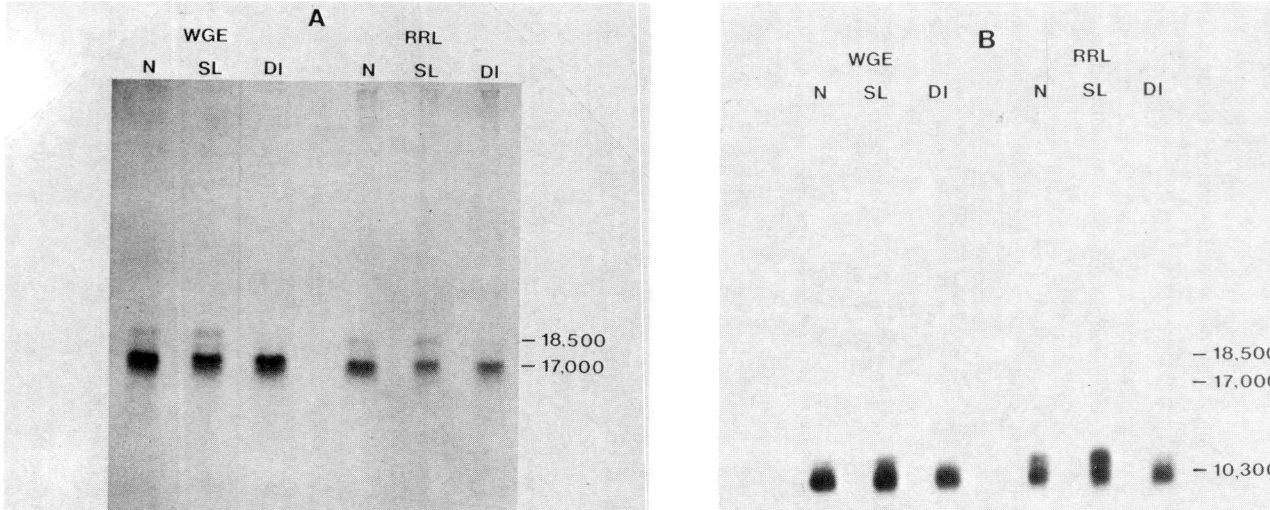

FIGURE 2. 15% SDS-polyacrylamide gel electrophoresis of WGE and RRL immunoprecipitates. Aliquots of the 0.1 N acetic acid eluates from protein A-Sepharose of normal (N), salt-loaded (SL), and Brattleboro (DI) translation immunoprecipitates were lyophilized, dissolved in 20 μl sample buffer and electrophoresed into a 15% (2.7% C_{bis}) acrylamide slab gel as described in MATERIALS AND METHODS at constant current (25 mA) for two additional hours following runoff of the dye front. Gels were prepared for fluorography with EN^{3}HANCE (New England Nuclear, Boston, Mass.), dried, and exposed at −80 degrees for 176 hours (A). In panel (B), six aliquots of affinity-purified IgG are pre-adsorbed with 25 μg LVP-Sepharose purified RNp each and immunoprecipitated as described in MATERIALS AND METHODS, except that to each aliquot from protein A-Sepharose is added 1,000 cpm of [^{3}H]*S*-carboxymethylated RNp prior to lyophilization. Fluorograph is a 282-hour exposure.

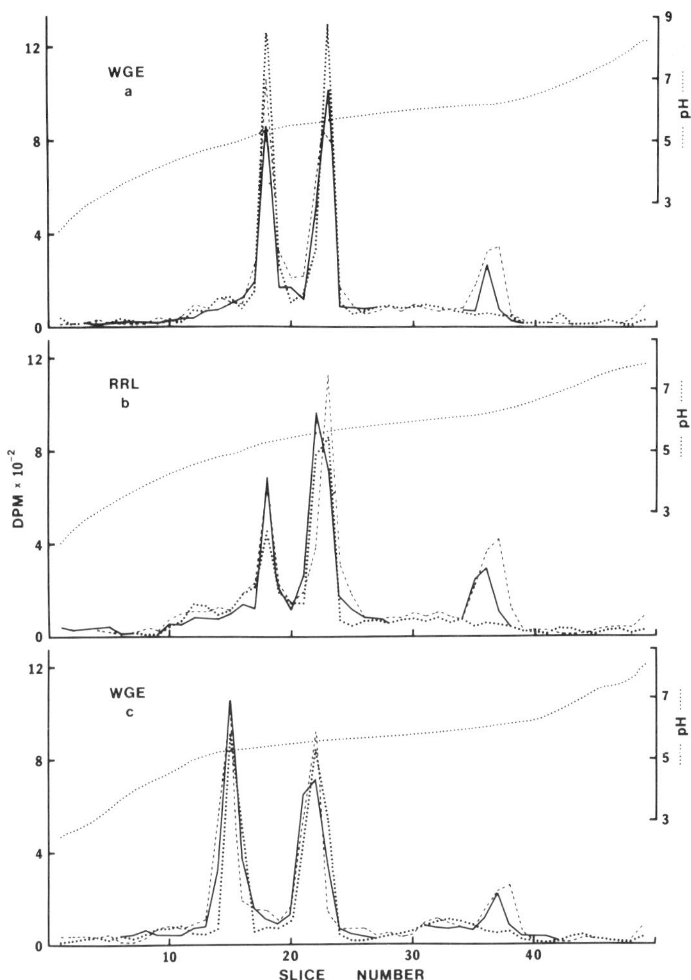

FIGURE 3. Isoelectric focusing profiles of normal (——), salt-loaded (- - - - -), and Brattleboro (·····) rat from WGE and RRL translation immunoprecipitates. IEF was performed by the method of O'Farrell[17] in 10-cm acrylamide gels run in 2 mm I.D. × 12.5 cm gel tubes. Focused gels were sliced into 2-mm sections starting from the acidic end, with each slice placed into a scintillation vial containing 0.5 ml Protosol (New England Nuclear, Boston, MA), incubated for 18 hours at 37 degrees, and counted in a toluene-based scintillation fluid. Panels (a) and (b) are focused in 2% vol/vol ampholyte gradients consisting of the nominal pH range ratios 3–10:4–6; 4:1, vol/vol, (Bio-Rad Laboratories, Richmond, CA). Panel (c) is pH range 3–10:5–7; 2:1, vol/vol.

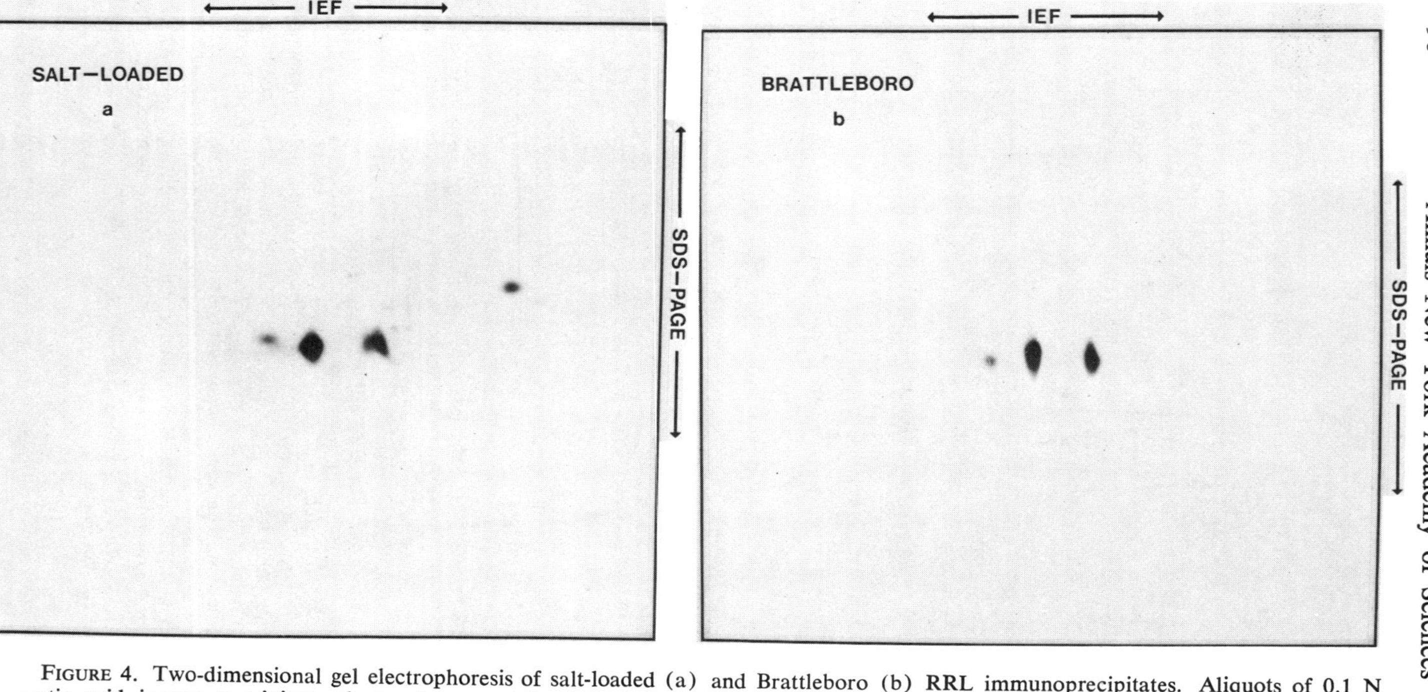

FIGURE 4. Two-dimensional gel electrophoresis of salt-loaded (a) and Brattleboro (b) RRL immunoprecipitates. Aliquots of 0.1 N acetic acid immunoprecipitate eluates from protein A-Sepharose were lyophilized and isoelectric focused as described in the legend to FIGURE 3a. The extruded, focused gels were run in the second dimension into a 12.5% (2.7% C_{bis}) SDS-polyacrylamide slab gel as described by O'Farrell [17, 18] except that the agar used to adhere the IEF gel to the top of the second dimension gel was made 0.125 M Tris-HCl, 0.1% w/vol SDS, pH 6.8. The gels were prepared for fluorography as described in the legend to FIGURE 3. Exposure time was 196 hours. Arrows denote dimension of IEF or SDS-polyacrylamide gel electrophoresis (SDS-PAGE). IEF gradient is acidic to basic, left to right.

on the acidic side of the pI 5.28 precursor on two-dimensional gel analysis, which was not readily apparent in IEF profiles. The presence of the spot appears sensitive to loss when immunoprecipitates are extensively washed, and could be either a tightly adhering nonspecific protein or a relatively weak-affinity neurophysin-like protein.

Peptide Mapping

The individual precursor species were isolated from IEF gels as described in MATERIALS AND METHODS with final recoveries following *S*-carboxymethylation of approximately 45–60%. Following exhaustive trypsinization, the tryptic peptides were chromatographed in reverse phase on a C_{18} column, as described in MATERIALS AND METHODS.

Mature RNp I contains four cysteine-containing tryptic peptides as shown in FIGURE 5D (dotted lines), which is consistent with the predicted close sequence homology of the rat neurophysins to bovine, ovine, and human neurophysins.[19, 31]

The tryptic peptides of either mature neurophysins or precursor species elute at *n*-propanol concentrations considerably less than for the time-zero, undigested forms (FIGURE 5D). With increasing molecular weight, the resolving capabilities of this system are drastically reduced, resulting in comigration of each of the three high molecular weight forms of neurophysin at 44–47% *n*-propanol (FIGURE 5D, solid lines).

As further illustrated in FIGURE 5, each of the three precursors contains a neurophysin sequence plus an additional cysteine-containing peptide. With only one additional cysteine-containing peptide identified in each protein species, it appears that all of the cysteine residues of the precursors are contained exclusively within the neurophysin and nonapeptide hormone sequences.

DISCUSSION

The synthesis and identification of three distinct neurophysin precursors from normal rat hypothalamic poly(A)RNA, one of which is absent in the homozygous Brattleboro animal, suggests that the missing protein species (pI 6.15, 18,500 daltons) is the vasopressin-containing precursor to RNp I. Keeping in mind the questions concerning the specificity of the salt-loading stimulus mentioned earlier, this precursor also exhibited the greatest proportional increase in synthesis under hypertonic conditions as assayed by cell-free translation.

Evidence from Gainer and co-workers[7] on the characteristics of putative *in vivo* labeled precursors points towards a pI 6.10, 20,000 dalton protein as being the pro-form of vasopressin—evidence similarly based on the absence of such a species in the Brattleboro rat. However, direct comparison of our results with theirs is made difficult due to processes operative *in vivo,* such as glycosylation[7] and signal peptide cleavage,[3] which alter the migratory properties of the polypeptide species. However, an observation on one of their results can be offered on the basis of our studies. In the Brattleboro rat, Russell *et al.*[7] observed a peak 'X' radioactivity which exhibited certain properties of the vasopressin precursor (electrophoretic, chromatographic), but did not give rise to the same tryptic peptides, or bind to a lectin affinity resin, as did the putative vasopressin precursor. Although not commented on by these authors, their presentation of data on peak 'X' allows the inference that this species could be an altered form of the vasopressin precursor in the Brattleboro rat. Our data suggest that the existence of an altered form of a vasopressin precursor is unlikely, since an absence of only one polypeptide species was observed to define the difference between normal, salt-loaded, and Brattleboro translation products. It is possible that peak 'X' represents background material in ^{35}S labeling studies, and is not apparent when pro-vasopressin is being synthesized.

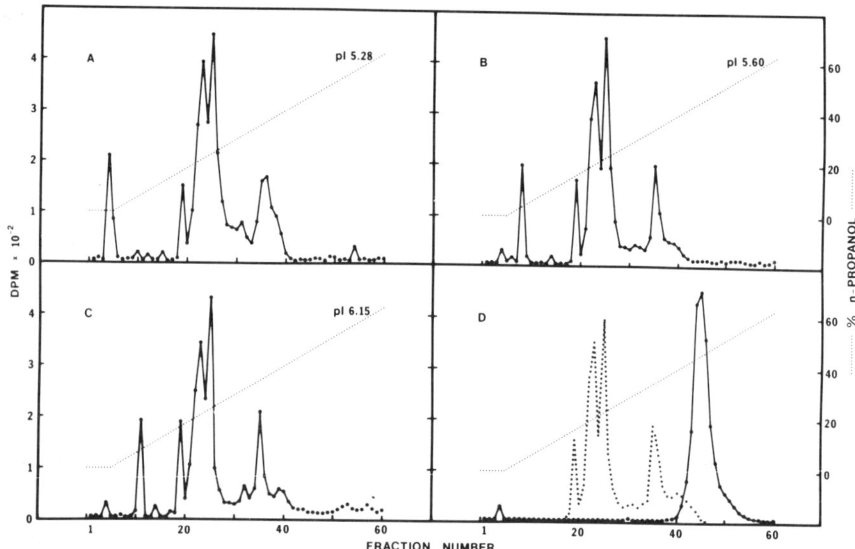

FIGURE 5. One-dimensional reverse-phase peptide mapping of purified individual precursors. Each of the three observed precursors were separated, *S*-carboxymethylated, and trypsinized as described in MATERIALS AND METHODS. Lyophilized tryptic peptides were dissolved in 50 μl 0.1% (vol/vol) trifluoroacetic acid (TFA, 13.5 mM) and chromatographed at 1.0 ml/min on a 10 μm RP–18 reverse-phase column (see MATERIALS AND METHODS) with an 80-minute gradient program. Solvent A: 0.1% TFA. Solvent B: *n*-propanol, 0.1% TFA. Five minutes at 0.1% B, then a linear 75-minute gradient to 80% B. 1.0-ml fractions (1.0 min) were collected and counted. The isoelectric points of the digested precursors are noted in the upper right-hand corner of panels A, B, and C. Panel D: zero-time tryptic digest of pI 5.60 precursor (——), and tryptic digest of [^{3}H]*S*-carboxymethylated RNp I (·····).

The presence of three precursor species in normal rat, each containing a neurophysin sequence, leaves open the question as to which of the remaining two serves as the pro-oxytocin. Work is continuing to define the amino acid sequences of the resolved peptides; and in light of known extrahypothalamic projections to central nervous system areas and spinal cord of oxytocin and vasopressin neurons[32, 33] and the observed microheterogeneity of several pro-hormones,[28, 34] the observation of more than two proneurophysin species is not surprising.

The observation based on our peptide mapping studies, that the neurophysin precursors contain only one additional cysteine-containing peptide, is of particular interest. In a recent paper, Dreth[35] examined the two-domain structure of neurophysin as defined by cysteine sequence homology to other cysteine-rich proteins. The cysteine domain of neurophysin required an additional di-cysteinyl peptide sequence at the C-terminus to maintain homology. This hypothesis may be useful to consideration of the structures of the neurohypophyseal hormone precursors.

References

1. Sachs, H. & Y. Takabatake. 1964. Evidence for a precursor in vasopressin biosynthesis. Endocrinology **75:** 943–948.
2. Lin, C., P. Joseph-Bravo, T. G. Sherman, L. Chan & J. F. McKelvy. 1979. Cell-free synthesis of putative neurophysin precursors from rat and mouse hypothalamic poly(A)RNA. Biochem. Biophys. Res. Commun. **89**(3):943–950.
3. Schmale, H., B. Leipold & D. Richter. 1979. Cell-free translation of bovine hypothalamic mRNA. Synthesis and processing of the prepro-neurophysin I and II. FEBS Lett. **108**(2): 311–316.
4. Giudice, L. C. & I. M. Chaiken. 1979. Immunological and chemical identification of a neurophysin-containing protein coded by messenger RNA from bovine hypothalamus. Proc. Natl. Acad. Sci. USA **76**(8): 3800–3804.
5. Lauber, M., M. Camier & P. Cohen. 1979. Immunological and biochemical characterization of distinct high molecular weight forms of neurophysin and somatostatin in mouse hypothalamus extracts. FEBS Lett. **97**(2): 343–347.
6. Nicolas, P., M. Camier, M. Lauber, Marie-J. O. Masse, J. Mohring & P. Cohen. 1980. Immunological identification of high molecular forms common to bovine neurophysin and vasopressin. Proc. Natl. Acad. Sci. USA **77**(5): 2587–2591.
7. Russell, J. T., M. J. Brownstein & H. Ganier. 1980. Biosynthesis of vasopressin, oxytocin, and neurophysins: isolation and characterization of two common precursors (propressophysin and prooxyphysin). Endocrinology **107**(6): 1880–1891.
8. Gainer, H., Y. Sarne & M. J. Brownstein. 1977. Biosynthesis and axonal transport of rat neurohypophysial proteins and peptides. J. Cell Biol. **73:** 366–381.
9. Valtin, H., W. H. Sawyer & H. W. Sokol. 1965. Neurohypophysial principles in rat homozygous and heterozygous for hypothalamic diabetes insipidus (Brattleboro strain). Endocrinology **77:** 701–705.
10. Burford, G. D., C. W. Jones & B. T. Pickering. 1971. Tentative identification of a vasopressin-neurophysin and an oxytocin-neurophysin in the rat. Biochem. J. **124:** 809–813.
11. Aviv, H. & P. Leder. 1972. Purification of biologically active globin messenger RNA by chromatography on oligothymidylic acid-cellulose. Proc. Natl. Acad. Sci. USA **69**(6): 1408–1412.
12. Hirs, C. H. W. 1967. Reduction and *S*-carboxymethylation of proteins. Methods Enzymol. **11:** 199–203.
13. Miller, T. J. & H. O. Stone. 1978. The rapid isolation of ribonuclease-free immunoglobulin G by protein A-Sepharose affinity chromatography. J. Immunol. Methods **24:** 111–125.
14. Robinson, I. C. A. F., D. H. Edgar & J. M. Walker. 1976. A new method of coupling (8-lysine) vasopressin to agarose; the purification of neurophysin by affinity chromatography. Neurosci. **1:** 35–39.

15. BURFORD, G. D. & B. T. PICKERING. 1972. The number of neurophysins in the rat. Biochem. J. **128:** 941–944.

16. LAEMMLI, U. K. 1970. Cleavage of structural proteins during the assembly of the head of bacteriophage T4. Nature **227:** 680–685.

17. O'FARRELL, P. H. 1975. High resolution two-dimensional electrophoresis of proteins. J. Biol. Chem. **250**(10): 4007–4021.

18. O'FARRELL, P. Z., H. M. GOODMAN & P. H. O'FARRELL. 1977. High resolution two-dimensional electrophoresis of basic as well as acidic protein. Cell **12:** 1133–1142.

19. CHAUVET, M. T., S. CHAUVET & R. ACHER. 1976. The neurohypophysial hormone-binding protein: complete amino-acid sequence of ovine and bovine MSEL-neurophysins. Eur. J. Biochem. **69:** 475–485.

20. LOCKRIDGE, O., H. W. ECKERSON & B. N. LA DU. 1979. Interchain disulfide bonds and subunit organization in human serum cholinesterase. J. Biol. Chem. **254**(17): 8324–8330.

21. MAHONEY, W. C. & M. A. HERMODSON. 1980. Separation of large denatured peptides by reverse phase high performance liquid chromatography. J. Biol. Chem. **255**(23): 11199–11203.

22. DOGTEROM, J., C. M. F. VAN RHEENEN-VERBERG, TJ. B. VAN WIMERSMA GREIDANUS & D. F. SWAAB. 1977. Vasopressin and oxytocin in cerebrospinal fluid of rats. J. Endocr. **72:** 74–75P.

23. JONES, C. W. & B. T. PICKERING. 1969. Comparison of the effects of water deprivation and sodium chloride inhibition on the hormone content of the neurohypophysis of the rat. J. Physiol. (London) **203:** 449–458.

24. ZAMBRANO, D. & E. DEROBERTIS. 1966. The secretory cycle of supraoptic neurons in the rat. Z. Zellforsch. Mikroskop. Anat. **73:** 414–431.

25. GEORGE, J. M. 1972. Functional mapping of the hypothalamus: RNA synthesis in response to hypertonic saline (Abstract). J. Clin. Invest. **51:** 35a.

26. SUNDE, D. A., D. D. ANTHONY, J. OSINCHAK & H. SACHS. 1969. Effect of osmotic stimulation on synthesis of RNA in the rat neurohypophysis. Federation Proc. **28:** 317.

27. TAKABATAKE, Y. & H. SACHS. 1964. Vasopressin biosynthesis. III. In vitro studies. Endocrinology **75:** 934–942.

28. EVANS, G. A., D. N. DAVID & M. G. ROSENFELD. 1978. Regulation of prolactin and somatotropin mRNAs by thyroliberin. Proc. Natl. Acad. Sci. USA **75:** 1294–1298.

29. HASKINS, J. T., J. P. JENNINGS & J. M. ROGERS. 1975. Response of neuroendocrine cell firing pattern type to measured changes in plasma osmolarity. Physiologist **18:** 240.

30. BRIMBLE, M. J. & R. E. J. DYBALL. 1977. Characterization of the responses of oxytocin- and vasopressin-secreting neurons in the supraoptic nucleus to osmotic stimulus. J. Physiol. **271:** 253–271.

31. NORTH, W. G. & T. I. MITCHELL. 1981. Evolution of neurophysin proteins: partial amino acid sequences of rat neurophysin. FEBS Lett. **126**(1): 41–44.

32. BUIJS, R. M. 1978. Intra and extra hypothalamic vasopressin and oxytocin pathways in the rat. Pathways to the limbic system, medulla oblongata and spinal cord. Cell Tissue Res. **192:** 423.

33. SWANSON, L. W. 1977. Immunohistochemical evidence for a neurophysin-containing autonomic pathway arising in the paraventricular nucleus of the hypothalamus. Brain Res. **128:** 346.

34. MAJZOUB, J. A., H. M. KRONENBORG, J. T. POTTS, JR., A. RICH & J. F. HABENER. 1979. Identification and cell-free translation of mRNA coding for a precursor of parathyroid secretory protein. J. Biol. Chem. **254**(15): 7449–7455.

35. DRENTH, J. 1981. The structure of neurophysin. J. Biol. Chem. **256**(6): 2601–2602.

Discussion of the Paper

H. Gainer (*National Institutes of Health, Bethesda, Md.*): I would like to make a comment about the structures shown in Dr. Pickering's paper that bear on Dr. McKelvy's data. Recent cyanogen bromide cleavage and peptide mapping studies from our laboratory have indicated that the order of the constituents in the pro-oxyphysin (precursor) is that oxytocin is on the N-terminal side of neurophysin; and in the pro-pressophysin (precursor) that vasopressin is near the N-terminal, the vasopressin-neurophysin is in the center, and the glycopeptide is at the C-terminal. This order differs from our previous suggestion, only in the glycopeptide assignment for pro-pressophysin (i.e., at the C-terminal and not the N-terminal of the precursor). Hence, if the nonapeptides are at the N-terminals of the prohormones, then the extra cysteine-labeled peptides (other than can be accounted for by neurophysin) in Sherman and McKelvy's peptide maps must be treated cautiously. Since they are peptide-mapping pre-prohormones these extralabeled peptides may be related to the signal sequences which could contain cysteine.

McKelvy: We do not know if the signal sequences contain cysteine. Do you have a complete amino acid analysis and know whether there is more cysteine in the precursor than in the two neurophysins and nonapeptides?

Gainer: To my knowledge, there is no complete sequence yet for the signal peptides of these two prohormones. We have done similar tryptic-peptide maps (in collaboration with Irwin Chaiken's lab at NIH) on the prohormones synthesized *in vivo*. We see extra cysteine-labeled peptide peaks on HPLC as you do. Since these *in vivo* isolated prohormones should not contain the signal sequences, we presume the extra labeled peptides in this case contain the nonapeptides. In fact, they are immunoprecipitable by antibodies to the nonapeptides.

McKelvy: That is good, and as I said we are developing some methods now that should be directly applicable to these kinds of studies with the Brattleboro rats where we can chemically show which hormone sequence is where; they may also shed some light on the question of why there are multiple species, and which one of the things we see is related to which hormone. This is the information that you really need.

Gainer: The *in vitro* translations using rat mRNA described in your paper, compares nicely to the results reported by Schmale and Richter (*Proc. Natl. Acad. Sci. USA* **78:**765, 1981) using bovine mRNA. The use of Brattleboro rats and especially two-dimensional gels in your work is particularly interesting, in that you suggest two pre-prohormones for oxytocin (with apparent pIs of 5.28 and 5.6). In this regard, it is significant that by using two-dimensional gels in *in vivo* experiments, we have also found (*Frontiers Neuroendocrinol.*, in press) two pI forms (pI 5.1 and 5.4) of pro-oxyphysin, and two pI forms (pI 5.6 and 6.1) of pro-pressophysin as well.

THE POSSIBILITY OF TOPOGRAPHICAL ORGANIZATION OF NERVE TERMINALS WITHIN THE NEURAL LOBE OF THE BRATTLEBORO RAT *

A. W. F. Fisher, K. Lederis, and P. Redstone

Departments of Anatomy and Pharmacology
University of Calgary
Calgary, Alberta
Canada T2N 4N1

Introduction

Since the report of Ferrier[1] describing an accurate map of the monkey motor cortex, evidence regarding somatotopic localization in numerous regions of the brain has appeared steadily. With particular reference to the magnocellular neurosecretory system, the recent papers of Swanson and Kuypers[2] and Armstrong *et al.*,[3] clearly suggest the possibility of somatotopic organization within the paraventricular nucleus. The present study was begun in order to check the similarities and differences between the Brattleboro rat hypothalamo-neurohypophyseal system and those of the more commonly used rat species. It was noted that the neural lobe of the Brattleboro rat, in spite of its known deficit of vasopressin, was larger than that of the usual rat species. It was further noted that small injections of horseradish peroxidase (HRP) into the Brattleboro neural lobe produced varying distribution patterns of transported marker within the hypothalamus, strongly suggestive of somatotopic organization within the neural lobe. This possibility was investigated further and the observations support these preliminary findings.

Materials and Methods

Fourteen Brattleboro rats of either sex, were examined. The animals were anesthetized with Nembutal after which horseradish peroxidase (type VI, Sigma Chemical Co. St. Louis, Mo., 20% solution in physiological saline) was injected (0.5 μl over 10 minutes) into the neurointermediate lobe via a glass micropipette of tip diameter 40 μm inserted via the ear bar of a Hoffman-Reiter hypophysectomy instrument.[4] The animals were allowed to survive 18–20 hours, re-anesthetized and perfused through the heart with 50 ml of 0.89% saline followed by 150 ml of 2.5% glutaraldehyde in phosphate buffer (pH 7.2). After fixation, the brain and pituitary were removed, placed in gum acacia/sucrose overnight, and sectioned serially at 50 μm on a Leitz freezing microtome. Alternate sections were reacted with diaminobenzidine (DAB) or tetramethylbenzidine (TMB),[5] dehydrated, and mounted. The serial sections were examined under dark-field illumination and the locations of HRP-filled neurons were drawn with a camera lucida and three-dimensional reconstructions made on acetate sheets. A similar three-dimensional reconstruction of a Brattleboro rat paraventricular nucleus reacted for oxytocin neurophysin was made.[6]

* Supported in part by Medical Research Council (Canada).

0077–8923/82/0394–0096 $1.75/0 © 1982, NYAS

Two attempts were made to remove the neural lobe by suction and replace the lost volume with 5 μl of HRP. In both cases the whole pituitary was removed, with consequent infarction of the infundibular vessels; there was no transport of HRP in either case.

Observations

Horseradish peroxidase injected into the neural and intermediate lobes of the Brattleboro rat pituitary is transported to hypothalamic neurons with an overall distribution pattern similar to that previously reported in the Sprague-Dawley rat.[7, 8] The major differences are that the Brattleboro magnocellular system appears foreshortened, the forniceal and paraventricular nuclei being much closer together (separated by some 50 μm only), and the retrochiasmatic and lateral perivascular components are much smaller.

The pattern of HRP distribution is, however, dependent upon the site of the injection. Although the size of the injection is large, a variable quantity of the injection mass leaks back along the needle and fills the colloid space—the actual injection into the neural lobe is therefore not constant and is frequently small. When the needle track and injection are situated in the antero-superior aspect of the neural lobe, the magnocellular components of the paraventricular (medial, lateral and posterior) nucleus are well seen (FIGURE 1), as are also the forniceal and circularis nuclei. By contrast, the supraoptic (SON) nucleus does not react with DAB sections but faint neurons are seen when reacted with TMB; similarly, the retrochiasmatic nucleus is ill-defined and small. Injections of HRP into the postero-superior aspect of the neural lobe transport to the paraventricular nucleus (medial and lateral) but not to its posterior component. The forniceal nucleus reacts well and scattered neurons in the bed nuclei of the stria terminalis are well seen. The supraoptic nucleus reacts but the middle and posterior react more heavily than the anterior aspects; again, the retrochiasmatic component is ill-defined. In both of these examples of injections into the superior aspects of the neural lobe the needle has entered via the right side, the bulk of the injection is on the left and there is a preponderance of HRP-filled neurons in the left hypothalamus.

Injection of HRP into the colloid space and pars intermedia with no apparent damage to the neural lobe but with obvious diffusion of HRP into it, shows transported marker in the medial paraventricular nucleus, a little faint reaction scattered through the lateral component, and nothing in the posterior. The forniceal nucleus also reacts well, whereas only a few faint neurons are seen in the middle of the supraoptic nucleus. Similar injections into the colloid space, but with obvious damage to the inferior aspects of the neural lobe, show good transport to the medial paraventricular nucleus, scattered faint neurons in the lateral component, and nothing in the posterior, i.e., an identical pattern in the paraventricular nucleus. The forniceal nucleus is well seen as also are the supraoptic and circularis nuclei. Injections posterior and ventral in the neural lobe demonstrate a well stained supraoptic nucleus throughout its length; the forniceal and circularis nuclei are well seen as is most of the paraventricular nucleus but not its posterior component.

Injection of the middle of the neural lobe resulted in good transport to the majority of the nuclear groups; the supraoptic and retrochiasmatic were well seen, the forniceal and paraventricular nuclei were well filled, but the posterior

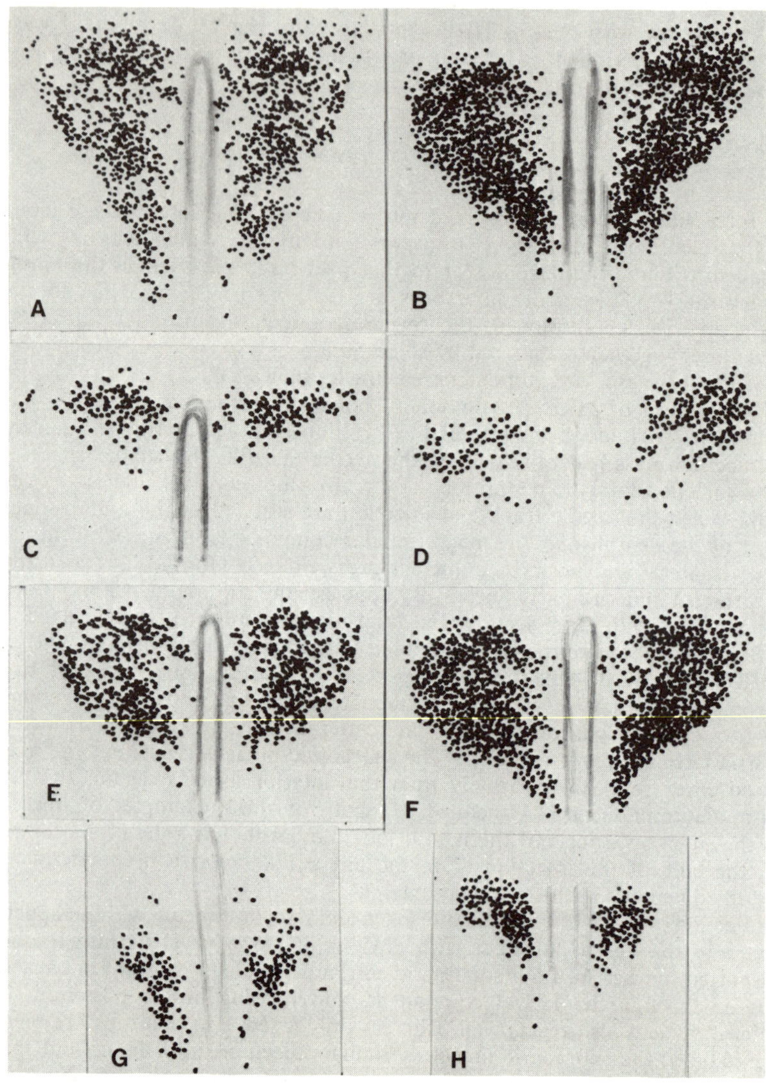

FIGURE 1. Three-dimensional reconstruction of the Brattleboro paraventricular nucleus (PVN) viewed from the front. A, C, E, and G (on the left) were reacted for oxytocin neurophysin. B, D, F, and H (on the right) show HRP transported from the left antero-superior region of the neural lobe. A and B are composite reconstructions of the entire PVN; C and D show the posterior one-third of the PVN, i.e. the posterior nucleus; E and F show the middle one-third of the PVN, i.e. the lateral nucleus; G and H show the anterior one-third of the PVN, i.e. the medial nucleus. Note the preponderance of neurons in the HRP preparations (B, D, and F); also more neurons in the left PVN contain HRP suggesting ipsilateral uptake and transport from the pituitary.

component of the latter and the posterior forniceal group were significantly absent. The circularis and perivascular groups were seen best with injections at this site.

DISCUSSION

The papers of Lavail,[9] Adams,[10] and Herkenham[11] clearly highlight the significance of uptake of HRP by damaged axons as well as by the physiologic method of pinocytosis.[9] The present study not only supports these established observations but also suggests that in certain sites, use can be made of this artefactual process in delineating the topography of neural projections. Moreover, the occurrence of small injections, with little obvious diffusion of peroxidase, in conjunction with visualization of reaction product by DAB instead of the more sensitive TMB, picks up nuances of change otherwise not seen.

Comparison of FIGURE 1a and b shows that considerably more neurons transport HRP from the pituitary than react for oxytocin neurophysin. The inference is that not only do the oxytocin neurophysin cells transport HRP, but also many if not all of those neurons in the various nuclei that are devoid of vasopressin seem to transport the marker. Whether or not these congenitally deficient magnocellular neurons pick up the HRP by physiologic processes, by traumatized terminals and axons, or both remains to be determined. Moreover, the projections of the paraventricular nuclei to the pituitary would appear bilateral but there does seem to be a preference for the ipsilateral side. In all examples where the injection is "pear-shaped" the ipsilateral nucleus is more heavily stained.

The paraventricular and forniceal nuclei seem to project much more heavily to the superior aspects of the neural lobe and, in order to demonstrate the bulk of the paraventricular nucleus, the injection has to be delivered superior and anterior; such injections demonstrate the posterior subgroup as well as the medial and lateral components. In this regard these findings in the Brattleboro agree with the description of Swanson and Kuypers[2] in the "albino" rat in that all aspects of the paraventricular nucleus project to the pituitary, whereas Armstrong *et al.*[3] find that the posterior subgroup does not. It would be interesting to know the site of the injections into the neural lobe in this latter paper. Postero-superior injections also show good filling in the posterior forniceal nucleus and the bed nuclei of the stria terminalis.

The supraoptic nucleus would seem to be distributed to the inferior aspects of the neural lobe, again with an ipsilateral preference, whereas the retrochiasmatic nucleus is well seen only with an injection into the middle of the structure. Similarly, good delineation of the circularis and lateral perivascular groups occurs with injections into the middle of the neural lobe.

Injections of HRP primarily into the colloid space and pars intermedia suggest that the innervation to these regions is from the medial paraventricular and forniceal nuclei, the occasional faint staining of the supraoptic being dependent upon trauma to the adjacent neural lobe.

The results presented could be explained by the individual vagaries of a particular animal. Thus the evidence is inadequate to prove that somatotopy exists in the innervation of the neural lobe. However, there are sufficient repeatable differences to suggest that a crude form of somatotopy may exist. These differences may help to explain some of the discrepant reports in the literature regarding projection of neurons of the paraventricular and supraoptic nuclei. Further studies are in progress to determine if these apparent somatotopic differences exist in other rat species.

References

1. Ferrier, D. 1876. The Functions of the Brain. Smith, Elder & Co. London.
2. Swanson, L. W. & H. G. J. M. Kuypers. 1980. The paraventricular nucleus of the hypothalamus: cytoarchitectonic subdivisions and organization of projections to the pituitary, dorsal vagal complex and spinal cord as demonstrated by retrograde fluorescence double-labelling methods. J. Comp. Neurol. **194:** 555–570.
3. Armstrong, W. E., S. Warach, G. I. Hatton & T. H. McNeill. 1980. Subnuclei in the rat hypothalamic paraventricular nucleus: a cytoarchitectural, horseradish peroxidase and immunocytochemical analysis. Neurosci. **5:** 1931–1958.
4. Price, P., A. W. F. Fisher & P. Redstone. 1977. A simple apparatus for the injection of small (nanolitre) volumes of horseradish peroxidase. Neurosci. Lett. **6:** 21–25.
5. Mesulam, M.-M. 1978. Tetramethyl benzidine for horseradish peroxidase neurochemistry: A noncarcinogenic blue reaction-product with superior sensitivity for visualizing afferents and efferents. J. Histochem. Cytochem. **26:** 106–117.
6. Fisher, A. W. F., K. Wong, W. North & K. Lederis. 1981. (In preparation.)
7. Sherlock, D. A., P. M. Field & G. Raisman. 1975. Retrograde transport of horseradish peroxidase in the magnocellular neurosecretory system of the rat. Brain Res. **88:** 403–414.
8. Fisher, A. W. F., P. G. Price, G. D. Burford & K. Lederis. 1979. A 3-dimensional reconstruction of the hypothalamo-neurohypophysial system of the rat. Cell Tiss. Res. **204:** 343–354.
9. Lavail, J. H. 1975. The retrograde transport method. Fed. Proc. **34:** 1618–1624.
10. Adams, J. C. 1977. Technical considerations on the use of horseradish peroxidase as a neuronal marker. Neuroscience **2:** 141–145.
11. Herkenham, M. & W. J. H. Nauta. 1977. Afferent connections of the habenular nuclei in the rat. A horseradish peroxidase study, with a note on the fiber-of-passage problem. J. Comp. Neurol. **173:** 123–146.

Discussion of the Paper

J. F. Morris (*Oxford University, Oxford, England*): You say that the colloid fissure is innervated by a limited portion of the paraventricular nucleus. At what time after the injection of HRP was the neural lobe fixed, and could there not have been diffusion of HRP from the cleft into the lower part of the neural lobe at an earlier time?

A. W. Fisher: First of all, the animals were killed 18 to 20 hours after the injection. I have two or three examples from Sprague-Dawley rats, as well as the Brattleboro group, where the injection of label has been made into the anterior lobe or into the colloid space. There is obvious diffusion of horseradish into the neural lobe but no obvious damage—the needle stops too short; there is, however, no labeling of hypothalamic neurons. I tend to believe that horseradish peroxidase retrograde labeling works best in the presence of damaged axons and I am not convinced that diffusion of horseradish peroxidase into a site evokes sufficient uptake to allow visualization, particularly when using diaminobenzidine as the visualizing agent.

MORRIS: If only damaged axons take up label why don't the injections into the anterior superior part of the neural lobe fill more cells? Such injections must disrupt the stalk where virtually all axons are entering.

FISHER: The photomicrograph of the antero-superior injection site I showed is but one section of a series. When reconstructed three-dimensionally the needle track and injection site are small and do not damage the bulk of the fibers in the stalk. I believe that the small injection sites we achieved are due to leakage back along the needle track into the colloid space. Uptake and transport of label from damaged axons along the needle track produces specific labeling of the neuronal pools described.

E. A. ZIMMERMAN (*Columbia University, New York, N.Y.*): Dr. Silverman and I will be presenting a paper at the Society of Neuroscience in October (1981) that shows that there is a unilateral projection of PVN to median eminence in the rat, as we have shown before in the monkey. In addition, as we previously reported with Dr. Antunes, there is somatotopy in the PVN-median eminence projection in the monkey. Rostral PVN projects to rostral median eminence, and inferior to inferior, etc. Thus, I am not surprised that you find some somatotopy from PVN to posterior pituitary.

IMMUNOHISTOCHEMICAL ANALYSIS OF VASOPRESSIN NEURONS TRANSPLANTED INTO THE BRATTLEBORO RAT *

J. R. Sladek, Jr.,† J. Schöler, M. D. Notter, and D. M. Gash

Department of Anatomy and Center for Brain Research
University of Rochester School of Medicine
Rochester, New York 14642

Introduction

Neural transplantation has been utilized by several laboratories to study the developmental and growth characteristics of neurons. These studies have yielded important information about the ability of grafted tissue to survive in such areas as the anterior chamber of the eye,[1,2] the cerebral ventricular system,[3,4] the cerebral cortex,[5,6] and elsewhere.[7] Transplanted neurons appear to be capable of reinnervating targets in several brain regions. Insights also have been gained into the ability of transplanted neurons to emulate functions that were attributed to them in their nontransplant state. For example, compelling evidence points to the ability of grafted dopaminergic neurons to restore some degree of motor function in animals which have been induced with neurotoxins to show specific basal ganglion deficits.[8-10] Reinnervation[11] and restoration of function[12] have been reported for other neural loci also.

Additional studies of transplanted endocrine neurons have revealed the ability of vasopressin neurons of the supraoptic nucleus (SON) to reverse the genetic diabetes insipidus of the homozygous Brattleboro rat.[3] Immunohistochemical analyses of these implanted rats revealed a predominance of neurophysin-positive neurons in the donor tissue following 20 or 40 days of growth.[13,14] The present investigation explores this phenomenon further by the analysis of these latter transplants with immunohistochemical staining for vasopressin. In addition, an attempt was made to characterize peptidergic neurons that had been grafted following cell-separation techniques.

Materials and Methods

Surgical procedures described previously were performed to insure optimal placement of fetal hypothalamic tissue into the ventricle of adult, male homozygous Brattleboro rats.[14] A total of nine rats were implanted and monitored for fluid balance for 20 to 40 days at which time they were killed by decapitation and prepared for immunohistochemistry following a new approach that utilizes freeze-dried, paraffin-embedded tissue which simultaneously can be examined

† Address correspondence to: John R. Sladek, Jr., Department of Anatomy, Box 603 University of Rochester School of Medicine, Rochester, New York, 14642

* Supported by U.S. Public Health Service Grants NS 15816 and AG 00847 (J.R.S.) and NS 15109 (D.M.G.). J.S. was supported in part by National Institute of Mental Health training grant MH 14577 and research grant AG 00847.

0077–8923/82/0394–0102 $1.75/0 © 1982, NYAS

for formaldehyde-induced monoamine histofluorescence.[15, 16] Sections were cut at 8 or 10 μm and every tenth section was stained with cresyl violet and luxol fast blue for characterization of the site of implantation. Sections chosen for immunohistochemistry were prepared for the unlabeled antibody demonstration for either rat neurophysin (supplied by A. Robinson), or vasopressin or oxytocin (supplied by G. Nilaver and E. Zimmerman).

Three additional rats received dispersed-cell transplants. Tissue samples from fetal donors were subjected to trypsinization as follows; blocks were rinsed in calcium–magnesium-free buffer and treated with 0.1% trypsin for 10 minutes at 37° C. They were then washed in buffer and placed in Hank's balanced salt solution with 20% calf serum and 0.4 mg/ml deoxyribonuclease. Once dispersed into a single-cell preparation with 90% viability, cells were centrifuged and 500,000 cells in 50 μl of tissue culture medium were implanted into the third ventricle as described above. Following 20 days, the rats were decapitated and prepared for thick-section immunohistochemistry of rat neurophysin or vasopressin according to the method of Grzanna, Molliver, and Coyle.[17]

Results

The majority of transplants were characterized at the light microscopic level by the presence of a slab of donor tissue occupying a large portion of an enlarged third ventricle (Figure 1). This tissue slab extended rostrally from preoptic hypothalamic levels to mamillary and pituitary stalk levels caudally. It extended dorsally into ventricular regions adjacent to the thalamus. It usually appeared isolated laterally from the host by ependymal lining of the ventricle. Ventrally, the transplant appeared contiguous with the underlying median eminence. The dorsal apex appeared either to terminate freely in the third ventricle or to blend with the neuropil of the host, most often in areas where the ventricular lining appeared absent.

Blood vessels continuous with those of the portal vasculature appeared to span the host–transplant interface at the level of the host median eminence and were seen throughout the entire extent of the transplant (Figure 1B, D). In several hosts, large oval areas existed that, in all probability, corresponded to portions of the optic tract or chiasm (Figure 2).

Nissl staining revealed populations of small as well as large, or magnocellular, neurons. They often appeared in clusters in the vicinity of the putative optic chiasm. Although suspected magnocellular elements appeared throughout all portions of the transplant, they occurred in clusters near the interface between the host and transplant in the ventral- and caudal-most regions of the transplant (Figure 3A). Occasionally, suspected magnocellular neurons were found within the pituitary stalk (Figure 4A).

Immunohistochemical analysis with antisera to vasopressin revealed positively stained neuronal perikarya and fibers in both the transplant and host. Two types of perikarya were seen. One corresponded to small, parvicellular neurons of the suprachiasmatic nucleus. They appeared in circumscribed clusters in the vicinity of the optic chiasm and in all instances were seen in association with a dense pattern of vasopressin-positive fibers (Figure 2B). The second group of perikarya were considerably larger and resembled magnocellular neurons. They stained moderately to densely for vasopressin and

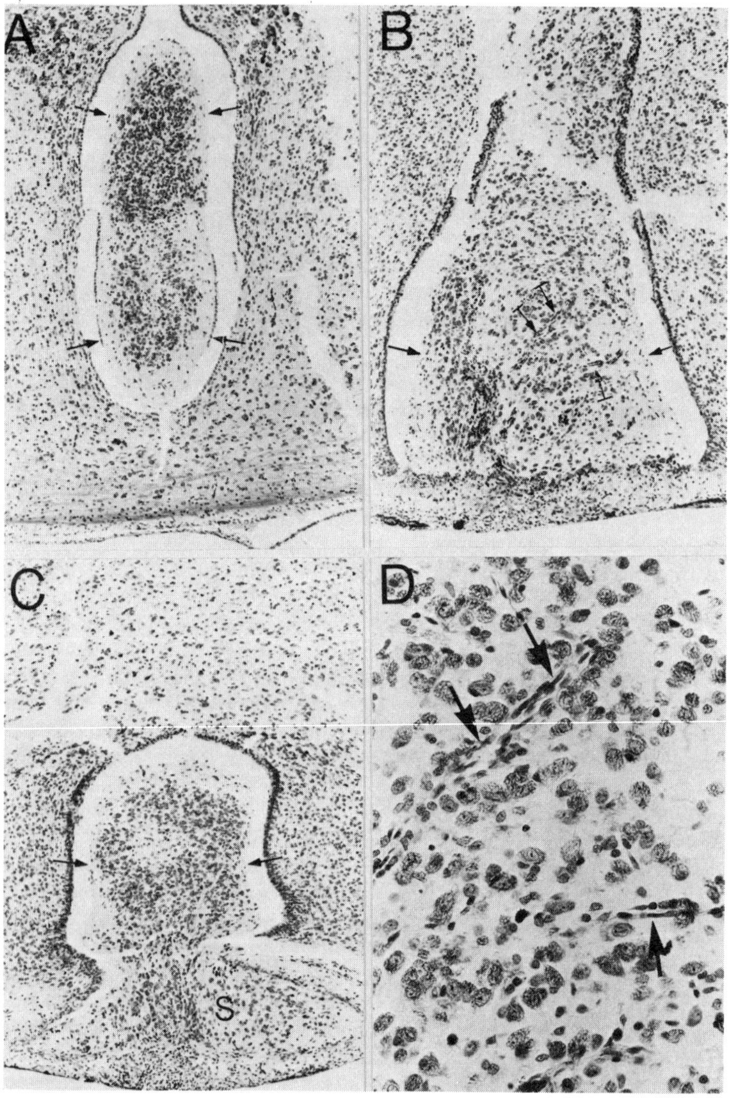

FIGURE 1. These Nissl-stained sections through the frontal plane illustrate the geometric characteristics of a typical transplant (→) proceeding from rostral (A) through caudual (C) levels. The rostral pole of the transplant appears unattached to its ventricular wall surroundings (A). More caudally (B,C) it appears to blend with the lateral walls of the ventricle dorsally (seen to advantage in FIGURE 2A) and with the floor of the ventricle, i.e. the median eminence, ventrally. This latter junction provides a route for blood vessels (→), which appear to branch freely from the underlying, richly vascularized median eminence (D-enlarged view of B). Most caudally (C) the transplant appears continuous with the pituitary stalk(s).[5] A, B, and C × 38. D × 145.

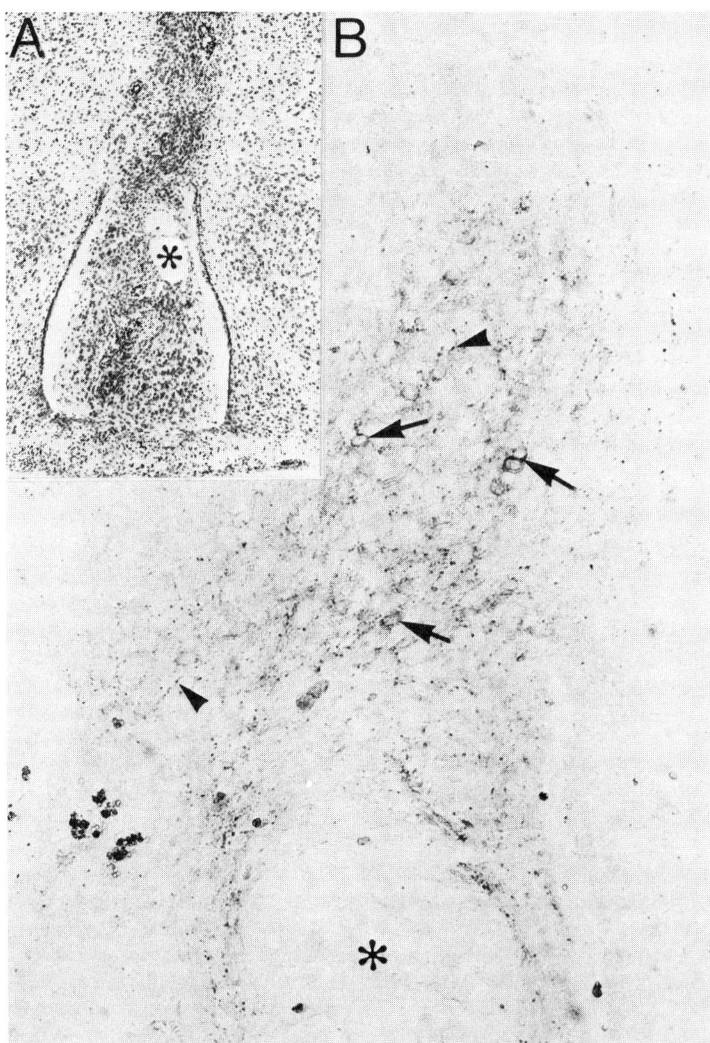

FIGURE 2. A. Nissl stained section is seen (A) at a scanning magnification. The dorsal part of the transplant is characterized by a degenerated optic tract (*) dorsal to which is a region of densely packed small cells, which appear to be part of the suprachiasmatic nucleus as determined from inmmunohistochemical staining for vasopressin (B) on a section near that seen in A. Parvicellular neurons (→) and fibers (►) are indicated; they appear to represent a degree of organization similar to that seen in normal animals. A, × 34. B, × 196.

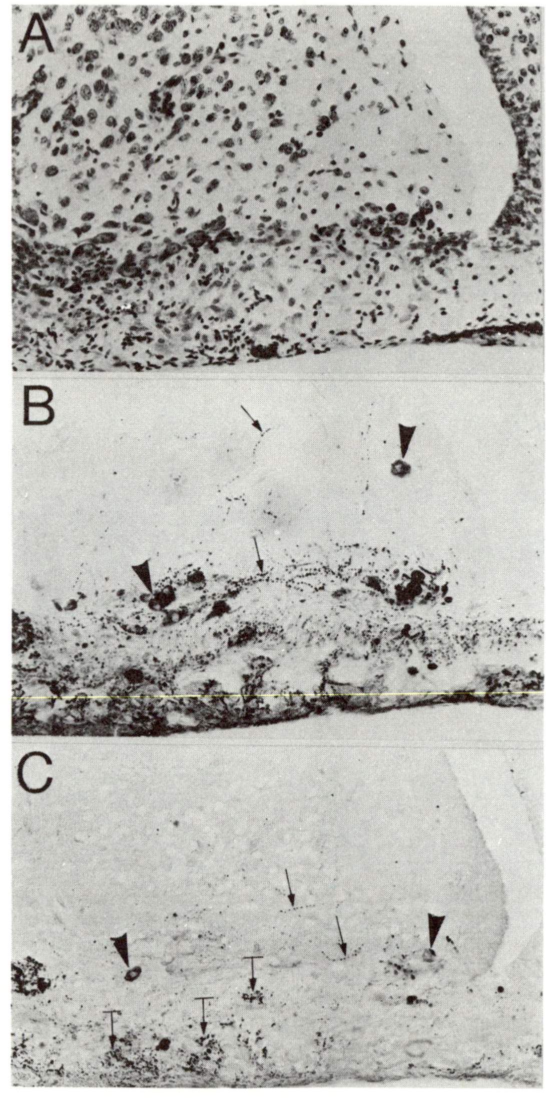

FIGURE 3. The junction between the host and transplant frequently contains an abundance of magnocellular neurons seen with Nissl stain (A), neurophysin (B), and vasopressin (C) immunohistochemistry. Vasopressin- as well as neurophysin-positive fibers course through the transplant (→) and are seen in abundance in the host median eminence (→), which in the unimplanted Brattleboro rat is devoid of vasopressin staining. Positively stained perikarya are evident also (►). A, B, and C × 102.

densely for neurophysin, and frequently occurred in groups or clusters at the interface between the host and transplant with the heaviest concentrations seen within the host median eminence (FIGURE 3).

Fibers that stained positive for vasopressin, oxytocin, and neurophysin were common features of the host median eminence (FIGURES 3, 5 and 6). They were moderate to coarse in size and frequently appeared clustered around blood vessels (FIGURES 5 and 6). They occurred in all layers of the median eminence

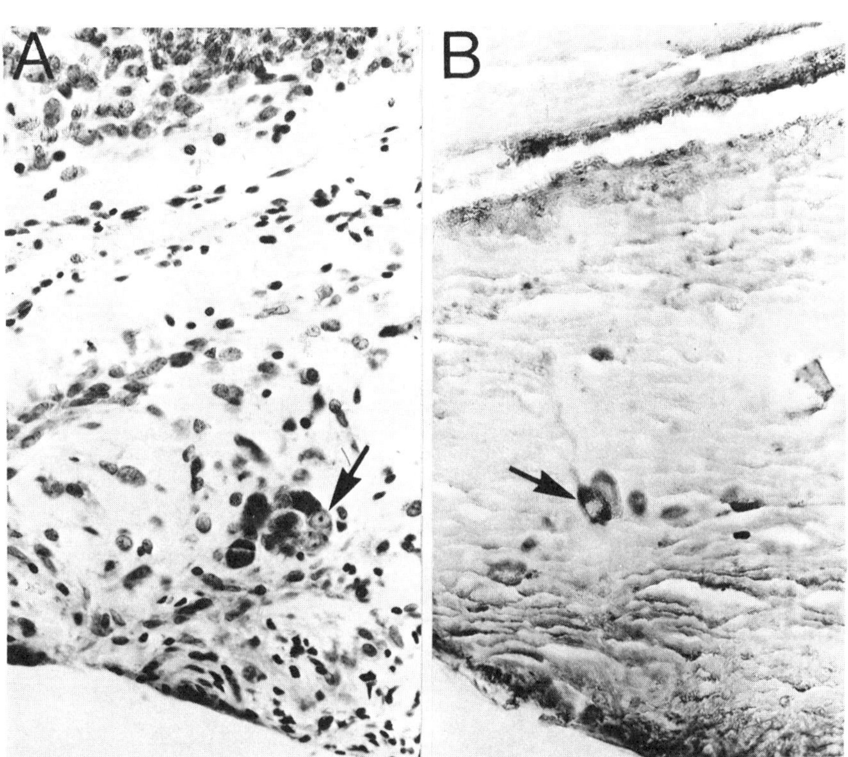

FIGURE 4. The region of the neurohypophyseal stalk is illustrated with Nissl (A) and neurophysin staining (B). Magnocellular neurons (→) appear ectopically located within the stalk, which is suggestive of a migration of transplanted neurons in a direction toward the neural lobes. A and B × 215.

in approximately equal densities. Fibers within the host were of finer diameter, appearing more typically like the varicose processes of magnocellular neurons (FIGURE 3C). Fibers in the median eminence often appeared rather large, somewhat reminiscent of Herring bodies; fine-size fibers also were seen, especially in the external lamina of the host median eminence (FIGURE 5B).

Vasopressin staining was almost completely absent in the host supraoptic nucleus although a rare fiber profile or weakly stained perikaryon was noted (FIGURE 7).

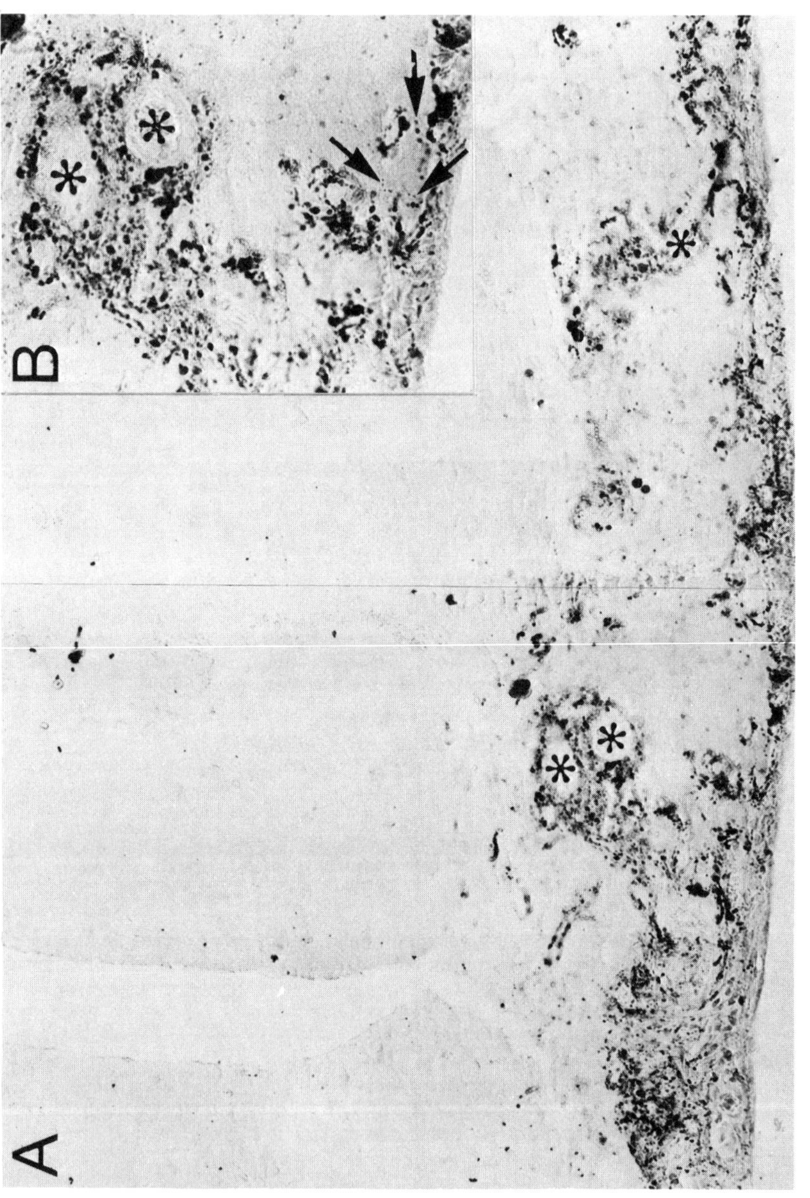

FIGURE 5. A dense accumulation of vasopressin fibers is seen in the host median eminence (A). They occur in abundance in the vicinity of portal blood vessels (*) of the median eminence, which are seen to advantage in B. Fibers are rather coarse in size; however, fine-size profiles also exist especially in the external lamina of the median eminence (→). A, ×170; B, ×300.

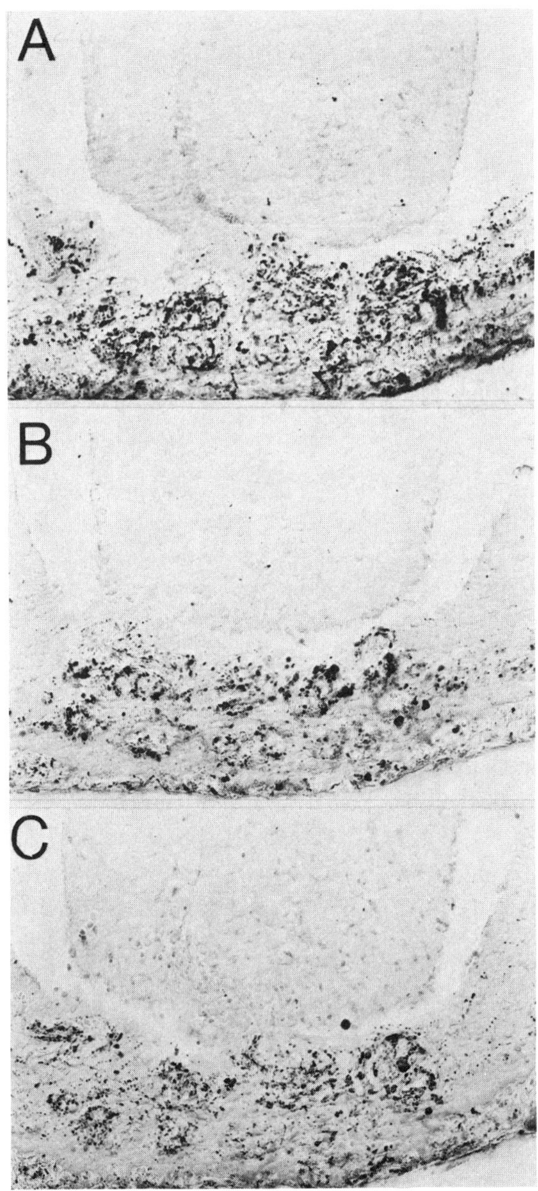

FIGURE 6. Serial sections immunohistochemically stained for neurophysin (A), vasopressin (B), and oxytocin (C) demonstrate dense staining in the host median eminence for each neurohypophyseal principle. Perivascular locations are common. A, B, and C × 108.

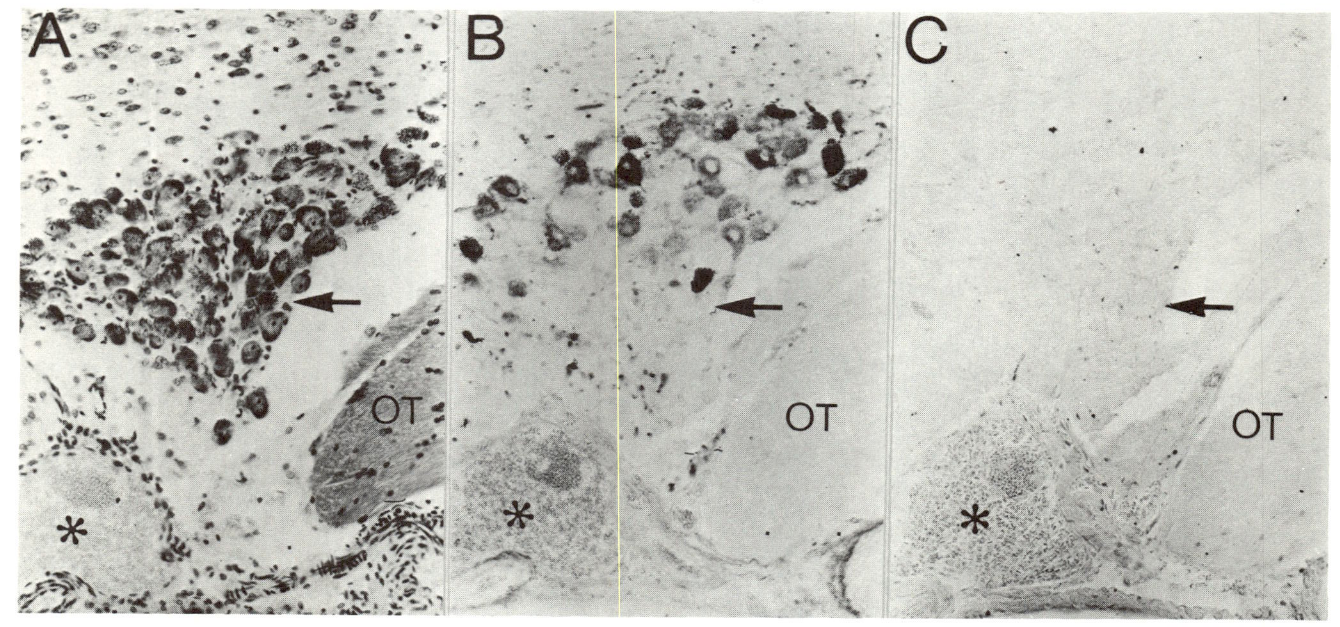

FIGURE 7. The supraoptic nucleus of the host brain is seen following nissl (A), neurophysin (B), and vasopressin staining (C). While magnocellular neurons occupy the entire dorsal to ventral extent of the nucleus it is apparent that the ventrally placed neurons (→) are deficient in neurophysin staining and represent the vasopressin-positive neurons of the normal rat. That vasopressin staining is absent ventrally in C lends support to the belief that the vasopressin antiserum is recognizing the appropriate peptide in the transplanted tissue and in the host median eminence. A blood vessel (*) and the optic tract (OT) are indicated as reference points. A, B, and C $\times 120$.

Dispersed Cell Preparations

Following neurophysin staining, small numbers of neuron-like structures were noted in the third ventricle of the host rats. They usually appeared solitarily as moderately stained structures with prominent neurophysin-filled processes (FIGURE 8). In addition, fibers were noted in their vicinity that could represent axonal terminals or ramifications. These neurons were part of a larger piece of tissue that formed a somewhat spherical mass in juxtaposition to the ventricular wall. No apparent continuity was detected between the masses of transplanted neurons and the host hypothalamus. These same structures were not seen following vasopressin staining.

DISCUSSION

Two types of vasopressin-containing neurons were found within transplanted hypothalamic tissue. Parvicellular, suprachiasmatic-like neurons as well as larger magnocellular neurons were common features. Interestingly, the parvicellular neurons appeared more consistently stained. They usually occupied a position in the dorsal portion of the transplant and appeared in groups resembling the suprachiasmatic nucleus. The magnocellular neurons, however, appeared in several locations; scattered within all portions of the transplant, clustered at the interface with the host, and clustered somewhat ectopically within the median eminence and pituitary stalk. The less consistently vasopressin-stained perikarya of the magnocellular neurons could reflect a higher level of turnover of vasopressin in comparison to the parvicellular neurons. This is an exciting prospect as it might be an indication of two functionally related phenomena, i.e., that the magnocellular, but not parvicellular, neurons are in a position to receive appropriate stimuli for the release of vasopressin and that they release vasopressin into an appropriate channel for delivery to the peripheral circulation. The latter presumably is via hypophyseal portal vessels, which occur in abundance in the transplant as a result of their growth from the underlying, host median eminence. This rich vascularization also might account for the delivery of an appropriate signal for vasopressin release. One such signal could be plasma osmolality monitored by the magnocellular neurons or alternately, by osmosensitive interneurons which, if located in the vicinity of the transplanted magnocellular nuclei, might be expected to survive the grafting procedure with appropriate short connections intact. In either event, behavioral and biochemical studies support the concept of a release of vasopressin by the apparently viable neurons.[14, 18]

The majority of vasopressin-positive magnocellular neurons were located at the interface with the host median eminence. This is in spite of the fact that Nissl staining revealed large numbers of putative magnocellular neurons throughout most levels of the transplant. Moreover, heavily stained neurophysin-bearing perikarya were seen in several locations including dorsal regions of the transplant. While some of the neurophysin-positive neurons dorsally probably contain oxytocin, an interesting possibility is that vasopressin-synthesizing neurons have migrated to the underlying median eminence in an organizational attempt either to form a circumscribed nucleus or to occupy a position closer to the appropriate route of delivery, i.e., in proximity to the neurohypophysis. The occurrence of vasopressin-positive neurons in the pituitary stalk

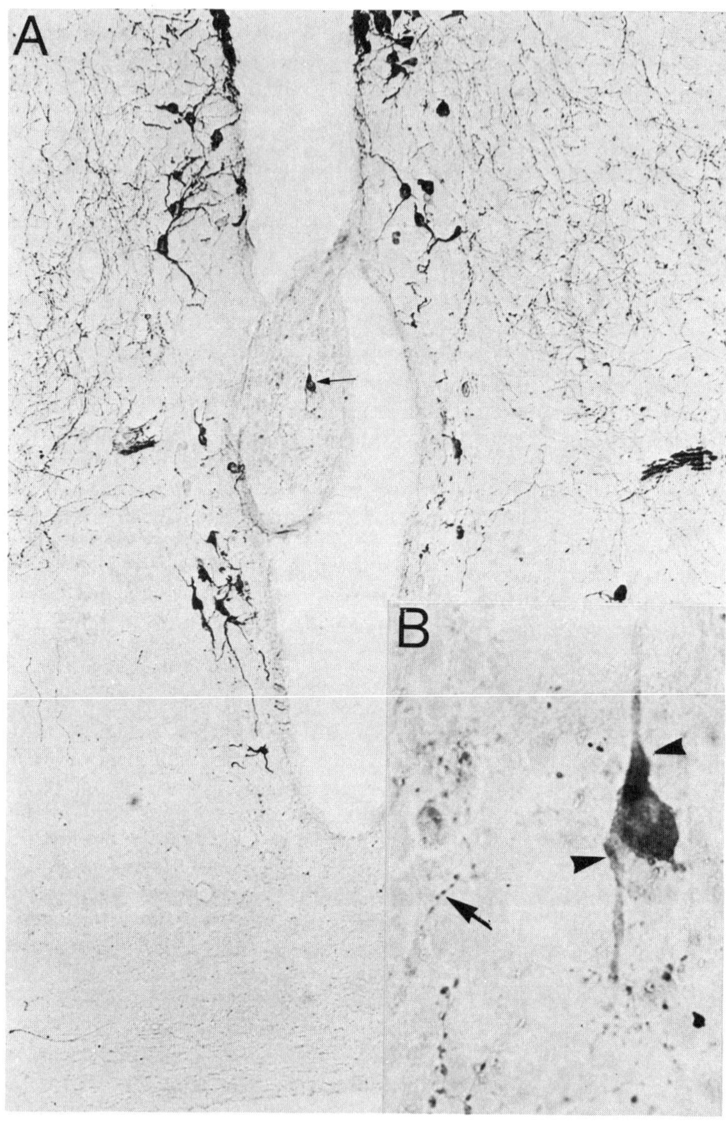

FIGURE 8. A cell dispersion preparation is seen following neurophysin immunohistochemistry. An oval mass of tissue occupies a third ventricular position and may be attached to the walls of the ventricle. A positively stained neuron (→) is seen within this reaggregated mass. In B it is characterized by typically large proximal stumps (►) of processes. Beaded profiles (→) may represent an axonal arborization of this or other neurons within the aggregate. A, × 81; B, × 510.

further suggests a migration. The concept of reorganization appears to be supported by the finding of extraordinarily dense concentrations of vasopressin-containing fibers in the host median eminence. The fibers were found not only in the fibrous and external layers where they occur in control rats, but also more internally, an area normally devoid of vasopressin fibers. Furthermore, the vasopressin fibers were seen in high densities surrounding portal vessels in the host median eminence as well as in the transplant. Many of these fibers were of coarse, large diameters, which is suggestive of an accumulation of hormone prior to release in the distended fibers of the neurohypophysis.

The discovery of viable, chemically identified neurons within the third ventricle of rats that received dispersed-cell implants is consistent with an earlier report of transplanted dopaminergic neurons of the substantia nigra.[19] The presently transplanted neurons were characterized by dense, neurophysin staining within perikarya which extended into neuronal processes. The latter phenomenon is indicative of some degree of growth of these dispersed neurons, which is supported by the occurrence of positively stained fibers at levels of the transplant which were void of similarly stained neuronal perikarya. While a functional responsiveness was not seen in these transplants this procedure offers excellent potential for a more cell-specific transplant that eventually could result in the use of homogeneous neuronal populations.

References

1. Olson, L. & A. Seiger. 1972. Brain tissue transplanted to the anterior chamber of the eye. I. Fluorescence histochemistry of immature catecholamine and 5-hydroxytryptamine neurons reinnervating the rat iris. Zeit. Zellforsch. **135:** 175–194.
2. Olson, L., R. Freeclman, A. Seiger & B. Hoffer. 1977. Electrophysiology and cytology of hippocampal formation transplants in the anterior chamber of the eye. I. Intrinsic organization. Brain Res. **119:** 87–106.
3. Gash, D. M., J. R. Sladek, Jr. & C. D. Sladek. 1980. Functional development of grafted peptidergic neurons. Science **210:** 1367–1369.
4. Stenevi, U., A. Björklund, L. Kromer, C. Paden, J. Gerlach, B. McEwen & A. Silverman. 1980. Differentiation of embryonic hypothalamic transplants cultured on the choroidal pia in brains of adult rats. Cell Tiss. Res. **205:** 217–228.
5. Lund, R. D. & S. D. Hauschka. 1976. Transplanted neural tissue develops connections with host rat brain. Science **193:** 582–584.
6. Jaeger, C. D. & R. D. Lund. 1979. Efferent fibers from transplanted cerebral cortex of rats. Brain Res. **165:** 338–342.
7. Stenevi, U., A. Björklund & N. A. Svengaard. 1976. Transplantation of central and peripheral monoamine neurons to the adult rat brain: Techniques and conditions for survival. Brain Res. **114:** 1–20.
8. Björklund, A. & U. Stenevi. 1979. Reconstruction of the nigrostriatal dopamine pathway by intracerebral nigral transplants. Brain Res. **177:** 555–560.
9. Freed, W. J., J. J. Perlow, F. Karoum, A. Seiger, L. Olson, B. J. Hoffer & R. J. Wyatt. 1979. Restoration of dopaminergic function by grafting of fetal rat substantia nigra to the caudate nucleus: long term behavioral, biochemical, and histochemical studies. Ann. Neurol. **8:** 510–519.
10. Perlow, J. J., W. J. Freed, B. J. Hoffer, A. Seiger, L. Olson & R. J. Wyatt. 1979. Brain grafts reduce motor abnormalities produced by destruction of nigrostriatal dopamine system. Science **204:** 643–647.

11. NYGREN, L. G., L. OLSON & A. SEIGER. 1977. Monoaminergic reinnervation of the transected spinal cord by homologous fetal brain grafts. Brain Res. **129:** 227–235.
12. BJÖRKLUND, A., M. SEGAL & U. STENEVI. 1979. Functional reinnervation of rat hippocampus by locus coeruleus implants. Brain Res. **170:** 409–426.
13. GASH, D. M. & J. R. SLADEK, JR. 1980. Vasopressin neurons grafted into Brattleboro rats: Viability and activity. Peptides **1:** 11–14.
14. GASH, D. M., C. D. SLADEK & J. R. SLADEK, JR. 1980. A model system for analyzing functional development of transplanted peptidergic neurons. Peptides **1** (Suppl. 1): 125–134.
15. SLADEK, J. R., JR., C. D. SLADEK, T. H. MCNEILL & J. G. WOOD. 1978. New sites of monoamine localization in the endocrine hypothalamus as releaved by new methodological approaches. *In* Neural Hormones and Reproduction. Brain-Endocrine Interaction III. D. E. Scott, G. P. Kozlowski & W. Weindl, Eds.: 154–171. Karger. Basel.
16. MCNEILL, T. H. & J. R. SLADEK, JR. 1980. Simultaneous monoamine histofluorescence and neuropeptide immunocytochemistry. V. A detailed methodology. Brain Res. Bull. **5:** 599–608.
17. GRZANNA, R., M. E. MOLLIVER & J. T. COYLE. 1978. Visualization of central noradrenergic neurons in thick sections by the unlabeled antibody method: a transmitter-specific Golgi image. Proc. Natl. Acad. Sci. USA **75:** 2502–2506.
18. GASH, D. M., P. H. WARREN, L. KICK, J. R. SLADEK, JR. & J. R. ISON. 1981. Behavioral modification in Brattleboro rats due to vasopressin administration and neural transplantation. Ann. N.Y. Acad. Sci. This volume.
19. BJÖRKLUND, A, R. H. SCHMIDT & U. STENEVI. 1980. Functional reinnervation of the neostriatum in the adult rat by use of intraparenchymal grafting of dissociated cell suspensions from the substantia nigra. Cell Tiss. Res. **212:** 39–45.

DISCUSSION OF THE PAPER

Q. J. PITTMAN (*University of Calgary, Calgary, Canada*): Do you think that the adrenergic fibers are coming from the brain adrenergic system rather than the sympathetic system?

J. R. SLADEK: Well, both are certainly candidates. I think that one of the isolated blood vessels I showed could well carry postganglionic adrenergic sympathetic fibers; however, until we do some superior cervical ganglionectomies, we won't fully be able to answer that. There is clear evidence for the periventricular noradrenergic fibers contributing under that particular circumstance, when the ventricular wall has been broken; moreover, there is little question that the tuberoinfundibular dopamine fibers are accompanying blood vessels of the median eminence, which grow into the transplant. A third source, which I didn't draw on the schematic slide, is from the deeper lamina of the median eminence, which contains noradrenergic fibers presumably from the ventral norepinephrine system separate from blood vessel innervation. Undoubtedly, they contribute also.

PITTMAN: Does the fluorescence look a little bit different between sympathetic fibers and other fibers? The reason I ask is because I think Dr. Rebecca Loy at the University of California at San Diego arrived at the conclusion that the catecholamine fibers innervating the lesioned hippocampus were sympathetic in origin.

SLADEK: My impression of vascular fiber fluorescence is that it is a much smoother non-varicose fluorescence. The kind of things we see are more typical of centrally sprouted catecholamine fibers. We are using Falck-Hillarp fluorescence, which isn't quite as sensitive as some of the new techniques. Perhaps we could answer that question better with some of the other techniques.

J. J. DREIFUSS (*University of Geneva, Geneva, Switzerland*): John, were you able to immunoassay your animals for vasopressin content and compare it to normal rats? Are you able to offer some evidence that vasopressin is being released and capable of reversing the diabetes insipidus of the transplant recipient?

SLADEK: Jean-Jacques asked a very important question as to whether or not we can detect vasopressin. That is a question best answered by Celia Sladek.

C. SLADEK: We have done radioimmunoassay on a number of transplant recipients and we have been able to measure vasopressin in some of them. We haven't attempted to measure plasma vasopressin in the transplant recipients yet.

H. GAINER (*National Institutes of Health, Bethesda, Md.*): What I am curious about is what is the correlation between the rapid water balance response and the beauty of the pictures we are seeing?

SLADEK: The shortest term we have examined immunohistochemically post-surgery is 20 days. And again, invariably in the 20-day animals we see an abundance of positively stained neurons that are multiprocessed with beaded fibers in cell areas that I have shown you here. We haven't examined shorter survivals, although the behavioral response begins after a period of transient (one or two days) adipsia, which perhaps is surgically related. Thus, important questions are what is the signal, if any, for vasopressin release and what is the route of delivery? Can the blood vessels grow into the graft that quickly? We really need to determine the temporal sequence of this event.

G. J. BOER (*Netherlands Institute for Brain Research, Amsterdam, the Netherlands*): Is there a relation between the innervation by aminergic fibers and the effect on diuresis in the transplant recipient? In other words is it absolutely necessary for the grafted cells to become innervated in order to be functional in water turnover regulation via vasopressin release?

SLADEK: The normal role of norepinephrine on vasopressin release, to the best of my understanding, is that it is inhibitory, and presumably in these animals one phase of inhibition is turned off. In animals that Don Gash has followed for 40 days, at least in some of them, the water balance characteristically starts to fluctuate a bit. And it is in these animals that we see a higher incidence of "reinnervation." My guess is that if the adrenergic fibers are making the final contact, there is some attempt at a regulatory mechanism, albeit not well controlled or finely tuned, at those longer post-survival times.

T. SAITO (*Jichi Medical School, Tochigi, Japan*): Does urinary osmolality of animals bearing hypothalamic transplants increase in response to dehydration or elevation of plasma osmolality?

SLADEK: I don't know the answer to that question. Don have you tested the transplant recipient's response to various challenges?

D. M. GASH (*University of Rochester, Rochester, N.Y.*): We haven't done that yet.

BIOLOGICAL HALF-LIVES AND ORGAN DISTRIBUTION OF TRITIATED 8-LYSINE-VASOPRESSIN AND 1-DEAMINO-8-D-ARGININE-VASOPRESSIN IN BRATTLEBORO RATS

T. Janáky, F. Laczi, and F. A. László

Endocrine Unit and Research Laboratory
First Department of Medicine
University Medical School
Szeged, Hungary

Introduction

Earlier investigations demonstrated that Brattleboro rats with hereditary hypothalamic diabetes insipidus have a considerably disturbed water metabolism.[1-4] Homozygous rats of this strain have polyuria and polydipsia, which result from an apparent inability to produce vasopressin[5] and the associated neurophysin.[6] Besides a partial disturbance of vasopressin formation, AVP release was found also to be impaired in heterozygous Brattleboro rats as compared to R-Amsterdam rats.[4, 7]

Valtin[8] suggested that increased vasopressin elimination might be one pathogenetic factor of diabetes insipidus in Brattleboro rats. An increased hormone metabolism, or enhanced storage of the hormone in certain organs, might lead to accelerated inactivation.

Soon after the synthesis of 1-deamino-8-D-arginine-vasopressin (dDAVP),[9] it emerged that the antidiuretic effect of this compound is higher and lasts for a longer time[10] than in the case of the naturally occurring hormone. We reported earlier that the biological half-life of dDAVP[11] is longer than those of 8-arginine-vasopressin[12] and 8-lysine-vasopressin.[13] It was concluded from this that the prolonged half-life may play a role in the longer duration of the antidiuretic effect. The reason for this is that dDAVP has a greater stability against enzymatic degradation than does the natural vasopressin.[14]

These considerations have led us to investigate the biological half-lives and organ distributions of exogenously administered tritiated LVP and dDAVP in Brattleboro rats.

Materials and Methods

Experiments were performed on 10 homozygous and 10 heterozygous male Brattleboro rats (Central Proefdierenbedrijf TNO, Zeist, the Netherlands), weighing 150–200 g and kept on a standardized diet. Male R-Amsterdam rats (Institute for the Breeding of Laboratory Animals, Gödöllö, Hungary) of the same weight and diet were used as controls. The animals received tap water *ad libitum*.

0077–8923/82/0394–0116 $1.75/0 © 1982, NYAS

Radioactive Materials

The tritiated 8-lysine-vasopressin ([^{3}H]LVP) was prepared by catalytic tritiation from the vasopressin precursor peptide containing dibromotyrosine in the presence of a palladium/aluminium oxide catalyst. Purification was performed by Sephadex G-15 gel-chromatography.[15]

Synthetic dDAVP[16] was iodinated, the resulting 3,5-diiodotyrosine^2-1-deamino-8-D-arginine-vasopressin was tritiated catalytically, and the product was purified by thin-layer chromatography and high-performance liquid chromatography.[17]

In both cases the tritium was incorporated into the molecule on the second amino acid, tyrosine. The specific activities of the [^{3}H]LVP and [^{3}H]dDAVP proved to be 3.5 and 15 Ci/mmole, respectively.

The homogeneity of each of the radioactive hormones was investigated both prior to and after the experiments. Checking of the [^{3}H]LVP was performed by silica-gel G thin-layer chromatography and Amberlite CG-50 ion-exchange chromatography.[18] The radiochemical purity of the [^{3}H]dDAVP was investigated by silica-gel G thin-layer chromatography, Sephadex G-15 gel-chromatography[11] and high-performance liquid chromatography (HPLC).

The coincidence of the radio- and biological activities was checked on the eluates of ion-exchange, Sephadex G-15, and high-performance liquid chromatography for both hormones. During the experimental period the labeled materials were stored in liquid air.

Measurement of Biological Activity

The antidiuretic activities of the [^{3}H]LVP, [^{3}H]dDAVP, and nonlabeled peptides were measured by the procedure of de Wied[19] in female homozygous Brattleboro rats anesthetized with alcohol. The biological activities of the tritiated compounds were compared with those of the corresponding non-labeled peptides.

High-Performance Liquid Chromatography

The high-performance liquid chromatography consisted of a Waters GmbH (Vienna, Austria) Model 6000 A pump and a universal liquid chromatograph injector (Model U6K), coupled to an LKB Uvicord III (Bromma, Sweden) fixed-wavelength (206 nm) UV monitor with an 8 μl flowthrough cell and an LKB Flat Bed recorder. Chromatography was performed using a reversed-phase Nucleosil 5 C_{18} column (25.0 $\times$ 0.46 cm) packed with 5 μm octadecyl-silica, purchased from Chrompack (Middelburg, The Netherlands). Elution was effected isocratically using 0.05 M freshly prepared ammonium acetate buffer (pH 6.5)-methanol (60:40, vol/vol) solution at room temperature. The flow rate was 1.2 ml/min. Fractions of 1.2 ml were collected in counting vials, mixed with 10 ml scintillation cocktail (4 g PPO + 0.05 g POPOP in 1 liter toluene + 500 ml Triton X-100), and counted with a Packard Liquid Scintillation spectrometer. Quench correction was carried out with external standard ratios. The radioactivities of the individual samples were given in dpm.

Determination of Biological Half-Life

In vivo clearances of the tritiated vasopressins were determined with a micromethod. A polyethylene cannula was inserted into the right carotid artery of ether-anesthetized rats, and diluted heparin was injected via this cannula. Then the femoral vein was exposed, and 15 μCi [^{3}H]LVP or [^{3}H] dDAVP in a volume of 200 μl was injected. At various times (20, 40, and 60 sec and 2, 4, 8, 16, 32, and 60 min) after the administration of the labeled hormone, blood samples were taken via the carotid cannula; these samples, in 0.1 ml capillaries, were microcentrifuged to separate the plasma from the formed elements. The radioactivity of 50 μl plasma was measured with a Packard TriCarb liquid scintillation spectrometer in 10 ml liquid scintillation cocktail.

For calculation of the biological half-lives, the logarithms of the radioactivities measured at the different points of time were plotted as a function of time. The regression line fitted to the points of the slow phase was extended to the ordinate, and the values relating to the initial points of the extrapolated line were subtracted from the experimental values. The regression line fitted to the resulting points gives the fast phase. Half-lives were calculated from the slopes of the straight lines relating the natural logarithms of the radioactivities to time [20] (i.e., $t_{½} = \ln 2$ (slope)).

Measurement of Radioactivity in the Organs

One hour after the injection of labeled peptides, the rats were decapitated, the organs were rapidly removed, and blood contamination was washed off the larger organs in cold physiological saline solution. The weighed small organs and 100–200 mg from the larger organs were incubated for 12 hours at 56° C in 1–2.5 ml Soluene 350 (Packard Instruments), depending on the amount of tissue. To decolorize the solutions, 0.5 ml 30% H_2O_2 was added per ml dissolved organ solution, followed by 10 ml liquid scintillation cocktail. The radioactivities were measured three days later with a Packard Liquid Scintillation spectrometer. The results were expressed as a percentage of the total radioactivity added, calculated on 100 mg wet organ weight. The origins of the radioactivities in the plasma and in the homogenized tissue samples were analyzed by HPLC.[12]

Results

The results from the silica-gel G thin-layer chromatography indicated that [^{3}H]LVP and [^{3}H]dDAVP were pure. After scanning, one peak was visible on the scanogram, which coincided with the spot of standard LVP or dDAVP developed with ninhydrin. The R_f values proved to be 0.55 for LVP and 0.50 for dDAVP. The same results were obtained after four weeks, after the completion of the experiments.

The data obtained by ion-exchange chromatography of [^{3}H]LVP on the Amberlite CG 50 column are shown in Figure 1. It may be stated that the radioactivity of the tubes forms one peak (between tubes 20 and 30). The biological activity coincides with this radioactivity peak. The high-performance

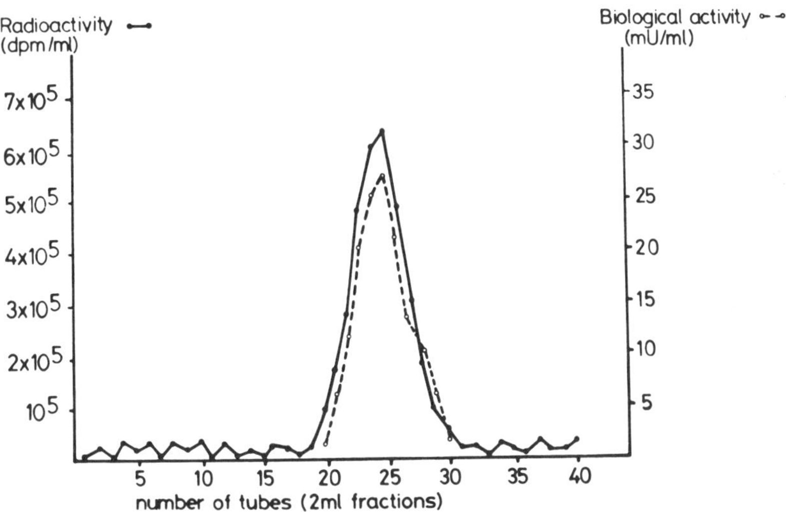

FIGURE 1. Radio- and biological activities of [³H]LVP chromatographed on Amberlite CG 50 column.[18]

liquid radiochromatogram of tritiated dDAVP is shown in FIGURE 2. The retention time of [³H]dDAVP is 18 min, the same as for authentic dDAVP. The biological activity could be demonstrated in the same peak: the radioactivity ran parallel with the biological activity. These results indicated that both compounds were biologically active and homogeneous from a radiochemical aspect.

It may be concluded from TABLE 1 that the biological activities of the

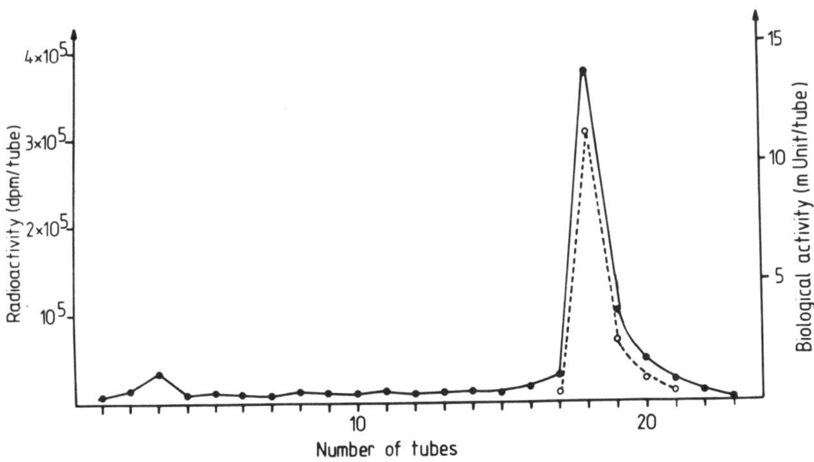

FIGURE 2. High-performance liquid chromatography of [³H]dDAVP. Radioactivity (●—●—●) and biological activity (– ○ – – ○ – – ○ – –).

TABLE 1

BIOLOGICAL ACTIVITIES OF [^{3}H]LVP AND [^{3}H]dDAVP

Compound	Biological Activity (I.U./mg)
Non-labeled LVP	270.0 ± 17.5 *
[^{3}H]LVP (before experiment)	267.5 ± 18.5
[^{3}H]LVP (after experiment)	248.3 ± 21.4
Non-labeled dDAVP	955.0 ± 38.5
[^{3}H]dDAVP (before experiment)	928.0 ± 47.3
[^{3}H]dDAVP (after experiment)	907.0 ± 45.3

* Standard error of mean (SEM).

tritiated peptides are equivalent to those of the non-labeled hormones. After the completion of the experimental series, the biological activity exhibited a moderate decrease in both cases.

TABLE 2 gives the means of the biological half-lives. In the heterozygous Brattleboro and the R-Amsterdam rats, the measured biological half-lives of [^{3}H]LVP barely differ. The value for the homozygous Brattleboro rats is slightly lower, but there is no significant difference between the means found in the three groups. It may be stated that the biological half-lives of [^{3}H]dDAVP in the fast phase for both heterozygous and homozygous Brattleboro rats were shorter than that found for the control R-Amsterdam animals. The half-lives calculated for the slow phase were almost the same for the heterozygous and the R-Amsterdam rats, but the value was somewhat higher for the homozygous than for the heterozygous animals.

The data on the organ distribution of radioactivities one hour after administration of [^{3}H]LVP are shown in TABLE 3. In the R-Amsterdam rats, the radioactivities were found highest in the neurohypophysis, adenohypophysis,

TABLE 2

BIOLOGICAL HALF-LIFE OF [^{3}H]LVP AND [^{3}H]dDAVP IN BRATTLEBORO AND R-AMSTERDAM RATS

	Biological Half-Life of [^{3}H]LVP (min)		Biological Half-Life of [^{3}H]dDAVP (min)	
Group	Fast phase (0.33–4 min)	Slow phase (8–60 min)	Fast phase (0.33–8 min)	Slow phase (16–60 min)
R-Amsterdam ($N = 10$)	1.21 ± 0.07 *	26.55 ± 6.11	2.15 ± 0.22	48.78 ± 8.62
Homozygote Brattleboro ($N = 10$)	0.96 ± 0.08	22.37 ± 5.48	1.82 ± 0.28	54.31 ± 7.83
Heterozygote Brattleboro ($N = 10$)	1.03 ± 0.08	25.10 ± 6.89	1.84 ± 0.42	45.11 ± 4.26

* Standard error of mean (SEM).

and kidney, lower in the small intestine and liver, and very low in muscle, hypothalamus, and cerebral cortex. In the Brattleboro rats the highest radioactivities were similarly found in the neurohypophysis, adenohypophysis, and kidney. The amount of radioactivity enrichment was lower in the neurohypophysis and adenohypophysis of both homozygous and heterozygous Brattleboro rats as compared to the R-Amsterdam rats, but the differences were not significant. The radioactivities of the kidney, cerebral cortex, skeletal muscle, and liver were not substantially different. The small intestine of the Brattleboro rats displayed a radioactivity significantly higher than that of the R-Amsterdam animals.

TABLE 4 gives the mean radioactivities measured in the individual organs one hour after injection of [^{3}H]dDAVP. In the R-Amsterdam rats, the highest radioactivity was to be found in the kidney, followed in turn by the small intestine, the liver, the neurohypophysis, and the adenohypophysis. The radioactivity accumulation sequence was similar in the Brattleboro rats, but there was significantly less ($p < 0.01$) tritiated material in the kidney of the homozygous animals than in that of the control group. The accumulation was significantly lower in the small intestine of both the heterozygous and the homozygous Brattleboro rats ($p < 0.01$ and $p < 0.05$, respectively). The individual groups did not display any significant difference as regards the radioactivities of the liver, muscle, cerebral cortex, adenohypophysis, and neurohypophysis.

TABLE 3

ORGAN-DISTRIBUTION OF THE RADIOACTIVITY AFTER [^{3}H]LVP ADMINISTRATION IN BRATTLEBORO AND R-AMSTERDAM RATS

	% of Total Radioactivity per 100 mg Organ Weight		
		Brattleboro	
Organ	R-Amsterdam ($N = 10$)	Homozygous ($N = 10$)	Heterozygous ($N = 10$)
Kidney	0.45 ± 0.06 *	0.31 ± 0.03	0.44 ± 0.04
Liver	0.14 ± 0.02	0.16 ± 0.04	0.14 ± 0.03
Skeletal muscle	0.06 ± 0.01	0.06 ± 0.01	0.06 ± 0.01
Small intestine	0.16 ± 0.01	0.25 ± 0.04 †	0.23 ± 0.02 †
Neurohypophysis	0.62 ± 0.09	0.45 ± 0.07	0.58 ± 0.08
Adenohypophysis	0.61 ± 0.08	0.47 ± 0.06	0.55 ± 0.07
Hypothalamus	0.06 ± 0.02	0.07 ± 0.02	0.08 ± 0.02
Cerebral cortex	0.04 ± 0.01	0.05 ± 0.01	0.04 ± 0.01

* Standard error of mean (SEM).
† $p < 0.02$.

The HPLC examination permits the finding that one hour after administration of [^{3}H]LVP only 0–10% of the total radioactivity measured in the plasma originates from the intact [^{3}H]LVP; the rest is due to [^{3}H]tyrosine. The radioactivity observed in the kidney, liver, and small intestine stems only from [^{3}H]tyrosine. One hour after the administration of [^{3}H]dDAVP, 25–35% of the radioactivity measured in the plasma, 20–25% of that in the kidney, and 0–5% of that in the liver and in the small intestine comes from the unchanged [^{3}H]dDAVP. The remaining radioactivity originates from [^{3}H]tyrosine eluting together with the solvent front.

Discussion

Previous investigations have shown that after injection, natural vasopressins (AVP and LVP) are cleared rapidly from the circulation via an intensive metabolism by various organs (kidney, liver, small intestine).[12, 20–22]

The half-life of 8-arginine-vasopressin (AVP) was found by Lauson[20] to be 1–4 min, by Gazis and Sawyer[23] to be 2.6 min, and by Janáky[12] to be 2.49 min in the fast phase and 27.9 min in the slow phase; the half-life of

Table 4

Organ Distribution of the Radioactivity After [^{3}H]dDAVP Administration in Brattleboro and R-Amsterdam Rats

	% of Total Radioactivity per 100 mg Organ Weight		
		Brattleboro	
Organ	R-Amsterdam ($N = 10$)	Homozygous ($N = 10$)	Heterozygous ($N = 10$)
Kidney	0.83 ± 0.048 *	0.57 ± 0.047 †	0.75 ± 0.040
Liver	0.20 ± 0.005	0.20 ± 0.012	0.20 ± 0.012
Skeletal muscle	0.07 ± 0.003	0.08 ± 0.005	0.07 ± 0.004
Small intestine	0.57 ± 0.040	0.42 ± 0.018 ‡	0.36 ± 0.028 †
Neurohypophysis	0.20 ± 0.027	0.21 ± 0.013	0.19 ± 0.018
Adenohypophysis	0.14 ± 0.005	0.16 ± 0.016	0.14 ± 0.008
Cerebral cortex	0.08 ± 0.003	0.06 ± 0.005	0.08 ± 0.004

* Standard error of mean (SEM).
† $p < 0.01$.
‡ $p < 0.05$.

8-lysine-vasopressin was found by László[18] to be 2.46 min and by Sjöholm[24] to be 1.9 min. The biological half-life of the synthetic dDAVP proved longer.[11, 25] Our present experiments have revealed that the biological half-lives of [^{3}H]LVP and [^{3}H]dDAVP measured in the fast phase in Brattleboro rats were somewhat shorter than the half-lives for the R-Amsterdam groups, but these differences were not significant. The half-lives calculated on the basis of the slow phase do not differ significantly from that found in R-Amsterdam rats. These observations indicate that vasopressin does not disappear more rapidly from the circulation of the Brattleboro animals; thus, the accelerated elimination of the hormone cannot play a causal role in the development of the water metabolism disturbance.[13]

The different means of determining the biological half-lives, and the fact that not all authors examined both the fast and the slow phases, make it difficult to compare the different results. In our view, the half-lives determined for the fast (diffusion) phase are less informative than those for the slow (elimination) phase.

The half-life of dDAVP is longer than that of LVP because of the different metabolic clearances of the two analogues. This could be an important factor in the greater and more prolonged antidiuretic effect of dDAVP. The metabolism of LVP occurs rapidly;[24] dDAVP is more stable. The desamination is an extremely important structural change, for the removal from the AVP molecule of the amino group on the hemicysteine in position 1 decreases the rate of splitting of the Cys^1-Tyr^2 bond by aminopeptidase.[14, 26] In the plasma after one hour we found 25–35% of the total radioactivity to be due to nonmetabolized [^{3}H]dDAVP.

The kidney of the Brattleboro rat accumulates less radioactivity than that of the control animal in the case of both hormones. It is possible that the metabolism of vasopressin is decreased in the kidney of Brattleboro rat. At the same time we established that, in comparison to [^{3}H]LVP, a larger portion of the radioactivity accumulated in the kidney was [^{3}H]dDAVP; furthermore we found more unmetabolized dDAVP.

After [^{3}H]LVP administration, the small intestine of both the homozygous and the heterozygous Brattleboro rats displayed radioactivities significantly higher than that of the R-Amsterdam rat. This radioactivity comes only from [^{3}H]tyrosine. It is possible that in the Brattleboro rat the small intestine participates to an enhanced extent in the elimination of [^{3}H]LVP from the circulation. In contrast, after administration of [^{3}H]dDAVP the small intestine of the Brattleboro rat accumulates significantly less radioactivity than in the R-Amsterdam animals. However, this lower radioactivity is more than that measured in the case of [^{3}H]LVP. At present we can not give an explanation for this phenomenon.

The neurophysins (NPs) owe their significance to an association with the neurohypophyseal hormones. Three major neurophysins can be isolated from the rat neurohypophysis.[6, 27] One of these three rat neurophysins (RNP-I) has been linked to vasopressin production, and the other two neurophysins to oxytocin production. The complete absence of neural lobe NP-I[6] correlates well with the absence of vasopressin[5] in the homozygous Brattleboro rat. Likewise, there is a similar correspondence in the heterozygous Brattleboro rat, which has 55% of the level of NP-I and 50–60% of the level of vasopressin in normal rats.[5, 7, 27] On this basis it could be expected that less exogenous [^{3}H]LVP would accumulate in the posterior lobe of the pituitary of the Brattleboro rats. Our results do not support this hypothesis. The neurohypophyses of Brattleboro and R-Amsterdam rats do not differ significantly from one another in regard to their abilities to accumulate radioactivity. The binding between neurophysin and vasopressin (oxytocin) results from electrostatic ion-pair interaction between the α-amino group of the first amino acid of the hormone and the free carboxyl group of the protein.[28, 29] The presence of the α-amino group is therefore indispensable for the binding.[30] As a 1-deamino derivative, dDAVP is unable to bind to neurophysins. In all three animal groups the [^{3}H]dDAVP uptake by the neurohypophysis is the same and con-considerably less than that of [^{3}H]LVP. This supports the inability of NP-I to bind to dDAVP.

The accumulation of [^{3}H]dDAVP in the adenohypophysis is similarly essentially lower than that found after the administration of [^{3}H]LVP.

SUMMARY

The biological half-lives and organ distribution of tritiated 8-lysine-vasopressin and 1-deamino-8-D-arginine-vasopressin were determined in R-Amsterdam rats and in homozygous and heterozygous Brattleboro rats with hereditary central diabetes insipidus. It was found that the biological half-lives of [^{3}H]LVP and [^{3}H]dDAVP in the Brattleboro rats did not differ significantly from that found in the control R-Amsterdam rats. The half-life of [^{3}H]dDAVP proved longer than that of [^{3}H]LVP in all three groups of animals. In the case of [^{3}H]LVP the highest radioactivities were observed in the neurohypophyses, adenohypophyses, and kidneys of both the R-Amsterdam and Brattleboro rats. The accumulation of tritiated material was higher in the small intestine of the Brattleboro rats than in that of the R-Amsterdam animals. In all three groups of rats, [^{3}H]dDAVP was accumulated to the greatest extent in the kidney and the small intestine. The kidney and small intestine contained less radioactivity in homozygous Brattleboro rats than in the controls. There was only a slight radioactivity accumulation in the adenohypophysis and neurohypophysis. From the results it was concluded that the decrease in the rate of enzymatic decomposition may play a role in the increased duration of antidiuretic action of dDAVP.

The results have led to the conclusion that the accelerated elimination of vasopressin and its pathologic organ accumulation are probably not involved in the water metabolism disturbance of Brattleboro rats with hereditary diabetes insipidus.

REFERENCES

1. VALTIN, H. & H. A. SCHROEDER. 1964. Familial hypothalamic diabetes insipidus in rats (Brattleboro strain). Am. J. Physiol. **206:** 425–430.
2. SOKOL, H. W. & H. VALTIN. 1965. Morphology of the neurosecretory system in rats homozygous and heterozygous for hypothalamic diabetes insipidus (Brattleboro strain). Endocrinology **77:** 692–700.
3. MILLER, M. & A. M. MOSES. 1971. Radioimmunoassay of urinary antidiuretic hormone with application to study of the Brattleboro rat. Endocrinology **88:** 1389–1396.
4. LACZI, F., É. NAGY & F. A. LÁSZLÓ. 1978. The ADH-reserve capacity in Brattleboro rats. Acta Med. Acad. Sci. Hung. **35:** 173–179.
5. VALTIN, H., W. H. SAWYER & H. W. SOKOL. 1965. Neurohypophysial principles in rats homozygous and heterozygous for hypothalamic diabetes insipidus (Brattleboro strain). Endocrinology **77:** 701–706.
6. BURFORD, G. D., C. W. JONES & B. T. PICKERING. 1971. Tentative identification of a vasopressin-neurophysin and an oxytocin-neurophysin in the rat. Biochem. J. **124:** 809–813.
7. MOSES, A. M. & M. MILLER. 1970. Accumulation and release of pituitary vasopressin in rats heterozygous for hypothalamic diabetes insipidus. Endocrinology **86:** 34–41.
8. VALTIN, H. 1967. Hereditary hypothalamic diabetes insipidus in rats (Brattleboro strain). A useful experimental model. Am. J. Med. **42:** 814–827.
9. ZAORAL, M., J. KOLC & F. SORM. 1967. Amino-acids and peptides. LXX. Synthesis of 1-deamino-8-D-gamma-aminobutyrine vasopressin, 1-deamino-8-D-lysine-vasopressin and 1-deamino-8-D-arginine vasopressin. Coll. Czech. Chem. Comm. **32:** 1250–1257.

10. Vávra, J., A. Machova, V. Holecek, J. H. Cort, M. Zaoral & F. Sorm. 1966. Effect of synthetic analogue of vasopressin in animals and patients with diabetes insipidus. Lancet **1:** 948–952.
11. László, F. A., T. Janáky, L. Balászpiri & J. L. Morgat. 1981. Biological half-life and organ distribution of [^{3}H]-1-deamino-8-D-arginine-vasopressin in the rat. J. Endocrinol. **88:** 181–186.
12. Janáky, T., F. A. László, F. Sirokmán & J. L. Morgat. 1982. Biological half-life and organ distribution of [^{3}H]8-arginine-vasopressin in the rat. J. Endocrinol. **93:** 831–839.
13. Laczi, F., F. A. László, Gy. Kéri & I. Teplán. 1980. Study on the biological half-life and organ distribution of tritiated lysine-vasopressin in Brattleboro rats. Acta Phys. Acad. Sci. Hung. **55:** 129–133.
14. Sawyer, W. H., M. Acosta, L. Balászpiri, J. Judd & M. Manning. 1974. Structural changes in the arginine-vasopressin molecule that enhance antidiuretic activity and specificity. Endocrinology **94:** 1106–1115.
15. Kéri, Gy., I. Teplán & J. Mezõ. 1976. Organic compounds labelled with radioisotopes. COMECON Symposium. Marienske Lazne.
16. Balászpiri, L., G. Tóth & K. Kovács. 1979. Synthetic aspects and highlights of structure-activity relationships in vasopressins. *In* Proceedings of 9th Hungarian Endocrinological Congress. F. A. László, Ed.: 95–102. Hungarian Academy of Sciences. Budapest.
17. Morgat, J. L. 1979. The latest peptide ^{3}H-labelling methods. *In* Proceedings of 9th Hungarian Endocrinological Congress. F. A. László, Ed.: 121–130. Hungarian Academy of Sciences, Budapest.
18. László, F. A., É. Nagy, L. Gáspár, Gy. Kéri & I. Teplán. 1980. Effect of thirsting on the biological half-life and organ distribution of ^{3}H-lysine-vasopressin. Horm. Metab. Res. **9:** 471–475.
19. de Wied, D. 1960. A simple automatic and sensitive method for the assay of antidiuretic hormone with notes on the antidiuretic potency of plasma under different experimental conditions. Acta Physiol. Pharmacol. Neerl. **9:** 69–81.
20. Lauson, H. D. 1974. Metabolism of the neurohypophyseal hormones. Handb. Physiol. Sec. 7, Vol. IV, Part 1: 287–393.
21. Bauman, G. & J. F. Dingman. 1976. Distribution, blood transport, and degradation of antidiuretic hormone in man. J. Clin. Invest. **57:** 1109–1116.
22. Sjöholm, I. & G. Rydén. 1967. Half-life of oxytocin and lysine vasopressin in blood of rat at different hormonal stages. Acta Pharm. Suecica **4:** 23–30.
23. Gazis, D. & W. H. Sawyer. 1978. Estimation of plasma half-lives of vasopressin analogs from vasopressor responses by curve-fitting. Proc. Soc. Exp. Biol. Med. **157:** 584–589.
24. Sjöholm, I. & G. Rydén. 1969. Uptake of oxytocin in tissues after intravenous administration of tritiated oxytocin in rats. Acta Endocrinol. **61:** 432–440.
25. Seif, S. M., T. V. Senser, F. E. Chiarochi, B. B. Davis & A. G. Robinson. 1978. DDAVP (1-desamino-8-D-arginine-vasopressin) treatment of central diabetes insipidus. Mechanism of prolonged antidiuresis. J. Clin. Endocr. Metab. **46:** 381–388.
26. Walter, R. & W. H. Simmons. 1977. Metabolism of neurohypophyseal hormones: considerations from a molecular viewpoint. *In* Neurohypophysis. A. M. Moses & L. Share, Eds.: 167–188. S. Karger. Basel.
27. Sunde, D. A. & H. W. Sokol. 1975. Quantification of rat neurophysins by polyacrylamide gel electrophoresis (PAGE): application to the rat with hereditary hypothalamic diabetes insipidus. Ann. N.Y. Acad. Sci. **248:** 345–364.
28. Ginsburg, M. & M. Ireland. 1964. Binding of vasopressin and oxytocin to

proteins in extracts of bovine and rabbit neurohypophyses. J. Endocrinol. **30:** 131–145.

29. Breslow, E., H. L. Aanning, L. Abrash & M. Schmir. 1971. Physical and chemical properties of the bovine neurophysins. J. Biol. Chem. **246:** 5179–5188.
30. Breslow, E. 1975. On the mechanism of binding of neurohypophyseal hormones and analogs to neurophysins. Ann. N.Y. Acad. Sci. **248:** 423–441.

Discussion of the Paper

A. Moses (*State University Hospital, Syracuse, N.Y.*): With the specific activities you have attained, have you been able to study the binding of vasopressin to cell membrane preparations, such as the kidney?

Janáky: We have not done such binding studies.

J. F. Morris (*Oxford University, Oxford, England*): This is fascinating data, particularly in connection with some of the things that we were discussing this morning. For me, one big interest was to find that the uptake of administered hormone into the adenohypophysis was almost the same as that into the neurohypophysis. In both of these situations you implied binding of hormone to neurophysin. I would like to know how the material could pass through the membranes to reach the neurophysin. The common feature of the adenohypophysis and neurohypophysis is not neurophysin but the fenestrated vasculature, which would allow material to cross the blood-brain barrier and enter the extracellular space. It is interesting that there is no greater uptake into the neurohypophysis, since this would imply that diffusion of hormones (and thus also neurophysins) through the granule membranes and plasma membranes is unlikely to occur.

I. L. Schwartz (*Mt. Sinai School of Medicine, New York, N.Y.*): You account for the prolonged action of dDAVP relative to AVP in terms of the fact that dDAVP, lacking the terminal free amino group of AVP, is not a suitable substrate for aminopeptidase action. This focus on amino peptidase as the sole or major pathway for termination of neurohypophyseal hormone action should be deemphasized because, as shown by Dr. John Glass about 10 years ago, the most important degradative process probably occurs at the C-terminus of neurohypophyseal peptides through the action of carboxyamidopeptidase, an enzyme that finds the D-arginine-containing analogue a less favorable substrate than its L-arginine-containing counterpart, deamino AVP, or AVP itself.

T. Saito (*Jichi Medical School, Tochigi, Japan*): Did you have any chance to check the distribution of tritiated dDAVP or vasopressin in the arterioles? Recently I examined a pressor response to vasopressin and in Brattleboro rats it was significantly enhanced in comparison to the Long-Evans rats. Therefore, I wonder if the vasopressin receptor is increased in arterioles in DI Brattleboro rats.

Janáky: We have measured the distribution of [^{3}H]dDAVP only in the renal arteries. We have found concentrations similar to those in muscle tissue. We have no technique available at this time to measure concentration in smaller arterioles.

A. F. NEGRO-VILAR (*University of Texas, Dallas, Tex.*): I would like to ask Dr. Moses a question concerning the binding of AVP analogues. We have started some studies in which we labeled some vasopressin analogues, including dDAVP, with iodine. We started with such molecules because we thought it would be better to use an analogue that would not be degraded as rapidly as AVP. However, we found that binding to membranes obtained from posterior pituitaries is very low, although it can be displaced by cold arginine vasopressin. So now we are going back to using labeled AVP and measuring binding in membranes obtained from anterior pituitaries. We do find quite a bit of vasopressin in these preparations, and our purpose is to determine more specific receptors.

R. BODNAR (*Queens College, City University of New York, New York, N.Y.*): It appears from your data that the [^{3}H]vasopressin in the cerebral cortex and hypothalamus is crossing the blood-brain barrier. How much of it is crossing the blood-brain barrier?

JANÁKY: From 0.06–0.08%. However, we think this comes mainly from [^{3}H]tyrosine. When we injected [^{3}H]AVP intravenously, we found no detectable radioactivity in the third ventricle. However, when we injected [^{3}H]tyrosine we found a time-dependent concentration curve, reaching a maximum between 5–10 minutes. On the basis of these results we think that AVP does not cross the blood-brain barrier, while [^{3}H]tyrosine does.

I. W. HENDERSON (*University of Sheffield, Sheffield, England*): Is there a difference in the total volume of distribution of the two materials? In other words, does dDAVP, with its longer half-life, have a greater volume of distribution?

JANÁKY: We do not think that the longer half-life of dDAVP would be responsible for a difference in the volume of distribution. We did not measure volume of distribution; however, we made sure that all rats used in these experiments were of the same weight.

Short Research Reports

SPIKE ACTIVITY IN "VASOPRESSIN" NEURONS IN THE BRATTLEBORO RAT

R. E. J. Dyball

Department of Anatomy
King's College
London WC2R 2LS, England

Introduction

The special characteristic of Brattleboro rats is their lack of vasopressin,[1] but it is important to determine what other characteristics of these animals are also abnormal. The neurohypophyseal cells in normal rats show many of the electrical properties of ordinary nerve cells.[2] This investigation was carried out to see whether the abnormal secretory cells in Brattleboro rats also showed normal electrical properties. Single unit spikes with electrical properties similar to those of normal rats could be readily recorded from the supraoptic nucleus of Brattleboro rats. The recordings from Brattleboro rats were interpreted in the light of anatomical and recording studies in normal rats.

Materials and Methods

Electrical Recording Study

Thirteen homozygous DI rats, 14 heterozygous DI rats, and 13 normal Long-Evans rats were prepared for single unit recording from the hypothalamus.[2, 3] The rats were classified as homozygous or heterozygous DI animals by measuring urine volume and by bioassay of neurohypophyseal extracts.[4] Recordings were made from at least 60 antidromically activated units in the supraoptic nuclei of each group of animals.

Golgi Cox Study

Six normal and two homozygous DI rats were stunned by a blow on the head and decapitated. The brain was then rapidly dissected out and immersed in a solution containing 1.14% (w/vol) potassium dichromate, 1.14% mercuric chloride, and 0.45% potassium chromate, for six weeks.[5] Each block was then dehydrated, cleared, embedded in celloidin, sectioned transversely in serial sections at 120 μm, mounted on albuminized slides and counterstained[6] so that the boundaries of the supraoptic nucleus could be determined.

Results

Electrical Recording Study

Clearly recognizable single unit spikes could be recorded from the supraoptic nucleus of Brattleboro rats homozygous for the factor causing diabetes insipidus.

Like those recorded from normal rats the spikes frequently showed an inflection on their rising phase (FIGURE 1) which was often more pronounced when the spike was evoked antidromically by stimulation of the neural stalk. Inflected and uninflected spikes could be recorded both from continuously active cells and from those that fired in bursts (phasic cells: FIGURE 2). As in normal rats, neural stalk stimulation to evoke antidromic spikes appeared to interrupt spontaneous firing leading to a short period of inactivity or inhibition after the stimulus (FIGURE 1).

Recordings from the supraoptic nucleus in Brattleboro rats like those in normal rats fell into three categories: slow irregular, faster continuous, and bursting (phasic: see FIGURE 2). The average firing rate of units in Brattleboro rats was greater than that of normal rats; the distribution in both heterozygotes and homozygotes was significantly different from control ($p < 0.01$, Mann-Whitney test). The proportion of phasic cells was also increased and within the middle range of the distribution (1–8.9 spikes/sec) approached 50% (FIGURE 2). The electrical properties of individual cells in Brattleboro rats were not identical with those in normal rats. They showed a shorter latency of antidromic spike ($p < 0.01$) and spike height was maintained at higher frequencies ($p < 0.02$: TABLE 1).

Golgi-Cox Study

Examination of the Golgi-Cox material showed that two cell types could be distinguished in both normal and DI rats. The majority of cells (putative secretory cells) were oval, about 25 μm in diameter, with one to three ventrally directed dendrites, and a finer process, the axon, directed dorsally and medially. The axon in some cases emerged from a dendrite that was directed dorsally for part of its course (FIGURE 3). No evidence of branching axons was found. A much smaller group (<5% of the total: putative interneurons) consisted of cells that were multipolar (FIGURE 3).

DISCUSSION

Like the spikes recorded from the same nucleus in normal rats, the initial phase of the spikes from the supraoptic nucleus of homozygous Brattleboro rats frequently show an inflection. Such an inflection is not a feature of cells in any particular part of the nucleus nor of those showing a particular firing pattern. It more probably depends upon the exact position of the recording electrode with relation to the cell. Similarly, an apparent post-stimulus inhibition occurs after the antidromic pulses in both normal and Brattleboro rats, emphasizing their similarity.

The present Golgi-Cox study shows that by far the majority of stained cells fall into one anatomical type, probably neurosecretory cells. Thus most of the cells recorded will probably have been of this type. In normal rats three types of firing pattern can be discerned in the supraoptic nucleus, slow irregular, faster continuous, and bursting, and the same is true in Brattleboro rats. On the basis of experiments involving recording through reflex milk ejection and hemorrhage,[7] the majority of cells in the supraoptic nucleus can be classified into one of two types: oxytocin cells that show a short burst of

a

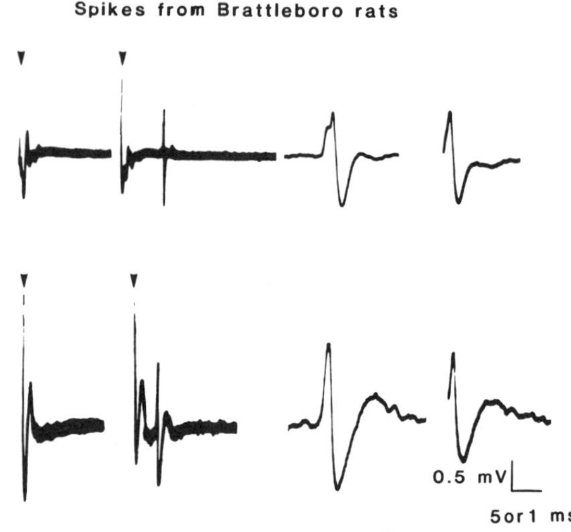

b

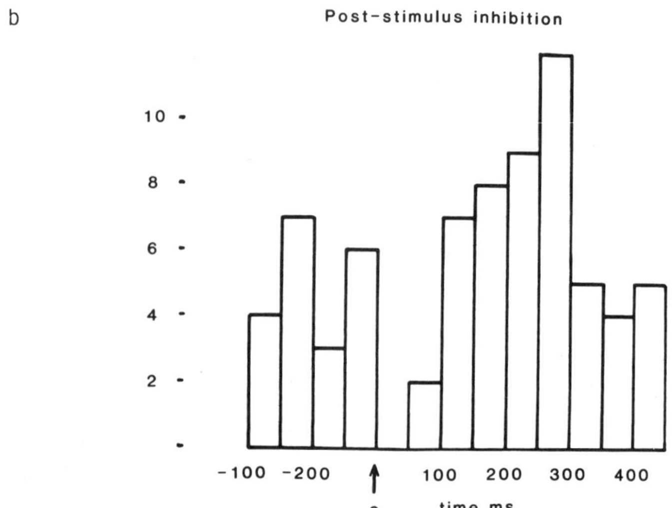

FIGURE 1. (a) Oscilloscope traces from Brattleboro rats. A cell with a marked inflection on the initial phase is shown above one without an inflection. On the left two antidromic stimulus pulses (arrowed) were triggered from a spontaneous spike to collide the first antidromic potential; center: an antidromic spike; and right: a spontaneous spike on an expanded timescale. (b) Histogram to show the spontaneous spikes falling just before and just after application of a series of 100 antidromic stimulus pulses at 3-sec intervals. The stimulus was followed by a short period during which no spikes fell.

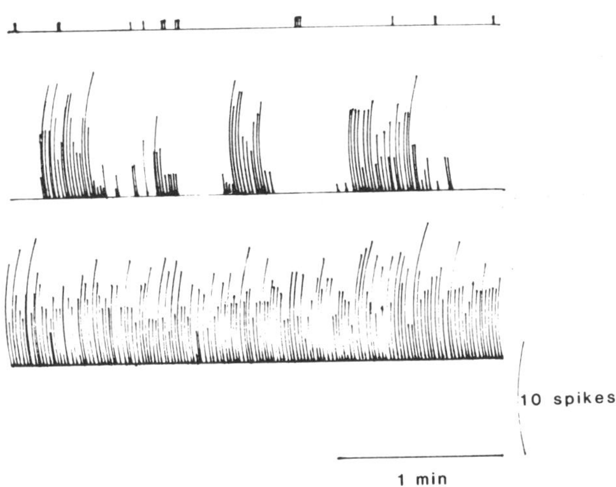

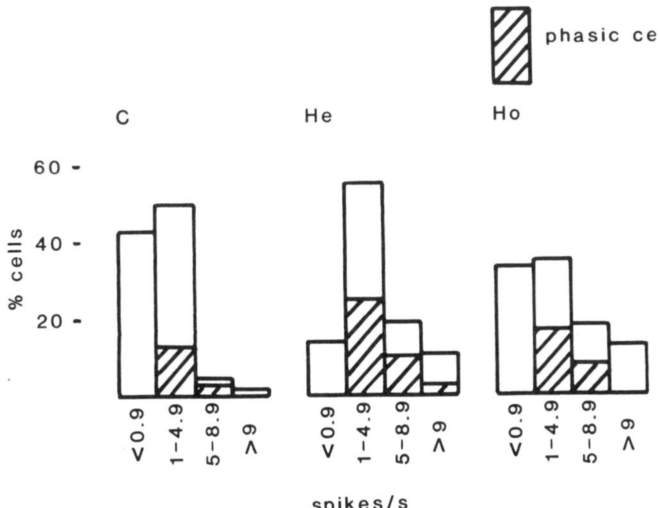

FIGURE 2. (a) Rate meter records to show the three types of firing pattern (slow irregular, bursting, and continuous) encountered in Brattleboro rats. (b) The firing rate distribution in normal rats compared with heterozygous and homozygous Brattleboro rats.

spikes before reflex milk ejection and vasopressin cells that fire phasically when stimulated. It thus seems extremely likely that the bursting (phasic) cell in Brattleboro rats are abnormal ("vasopressin") cells.

The electrical properties of secretory cells in Brattleboro rats are thus similar to those of normal rats. The average firing frequency was greater but Brattleboro rats have a higher serum osmolality than controls [1] and dehydration increases the firing rate of supraoptic cells.[8] For similar reasons an increased number of phasic cells was not unexpected. In the range of firing rate between 1 and 8.9 spikes/sec, the range in which vasopressin cells fire phasically, nearly 50% of the cells in Brattleboro rats show such a pattern. Thus the majority of abnormal "vasopressin" cells are probably electrically active. The shorter than normal latency of antidromic spike in Brattleboro rats may reflect their greater axon diameter.[9] Less expected was the greater frequency of stimulation at which full spike height was maintained, which may reflect changes in electrolyte balance.[1]

TABLE 1

SUMMARY OF THE ELECTRICAL CHARACTERISTICS OF SUPRAOPTIC CELLS FROM BRATTLEBORO RATS

	C	He	Ho
rate	1.8	4.0	3.7
% phasic	16	38	25
% phasic 1–8.9	30	46	46
latency	13.8 ± 0.4	13.9 ± 0.6	10.8 ± 0.5
max Hz	61 ± 5	78 ± 7	84 ± 8

The firing rate (spikes/sec), % of phasic cells, % phasic cells within the range of firing rate 1–8.9 spikes/sec, and latency of antidromic spike (msec) are shown for normal rats and heterozygous and homozygous Brattleboro rats. Also shown is the maximum frequency of stimulation in the different groups at which full spike height was maintained.

CONCLUSIONS

Since the majority of cells in the supraoptic nucleus fall into one anatomical type it is probable that the different types of firing pattern commonly encountered are all displayed by cells in this group. Bursting "vasopressin" cells are encountered in both Brattleboro rats and in normal rats so it is probable that non-secreting "vasopressin" cells are electrically active. The firing patterns of cells in Brattleboro rats are similar to those in dehydrated normal rats and their electrical properties are in many respects similar. Neither the generation of a bursting pattern nor the apparent inhibition that occurs after application of an antidromic stimulus can be dependent upon the presence of vasopressin.

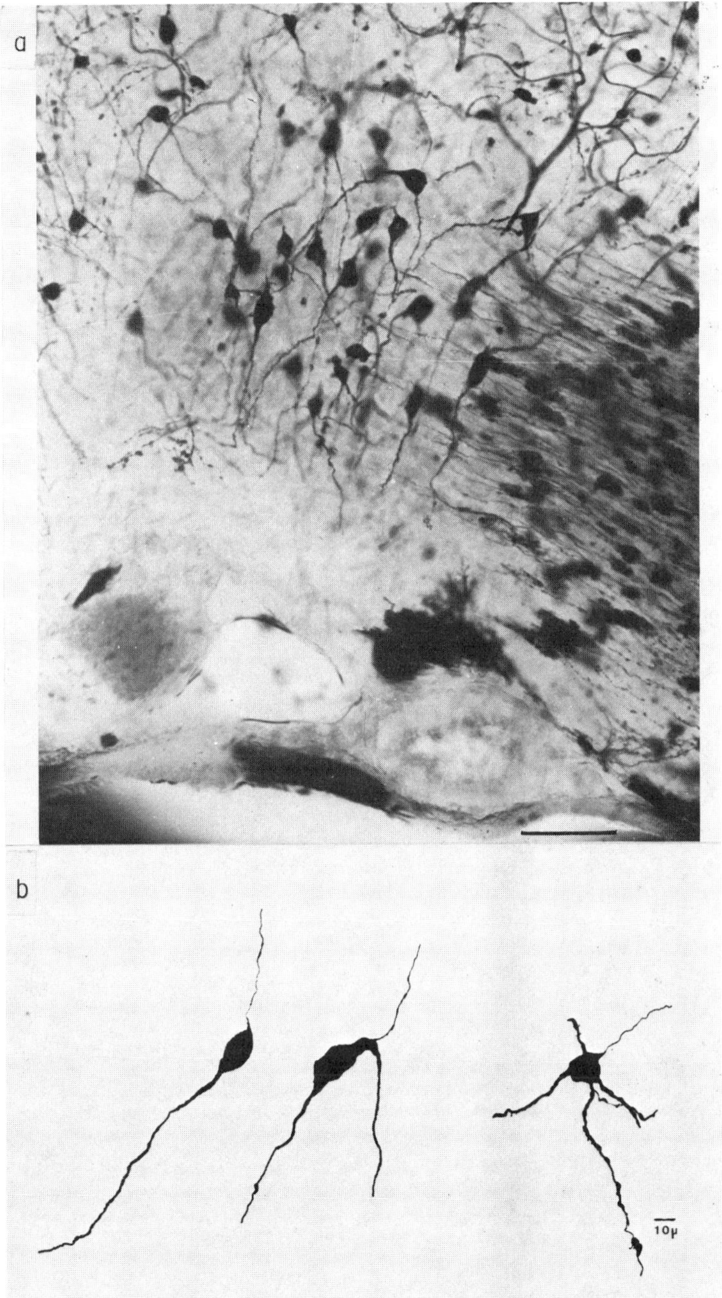

FIGURE 3. Photomicrograph to show part of a 120 μm transverse section of the normal hypothalamus, including the supraoptic nucleus, stained by the Golgi-Cox method (scale mark 100 μm). (b) Tracings of two putative neurosecretory cells (on the left) and one putative interneuron on the right.

References

1. Valtin, H. & H. A. Schroeder. 1964. Familial hypothalamic diabetes insipidus in rats (Brattleboro strain) Am. J. Physiol. **206:** 425–430.
2. Dyball, R. E. J. 1971. Oxytocin and ADH secretion in relation to electrical activity in antidromically identified supraoptic and paraventricular units. J. Physiol. **214:** 245–256.
3. Dyball, R. E. J. 1974. Single unit activity in the hypothalamoneurohypophysial system of Brattleboro rats. J. Endocr. **60:** 135–143.
4. Dekanski, J. 1952. The quantitative assay of vasopressin. Br. J. Pharmac. Chemother. **7:** 567–572.
5. Ramon-Moliner, E. 1970. The Golgi-Cox technique. *In* Contemporary Research Methods in Neuroanatomy. W. J. H. Nauta & S. E. E. Ebbeson, Eds.: 32–55. Springer Verlag. New York, N.Y.
6. Ramon-Moliner, E., M. A. Vane & G. U. Fletcher. 1964. Basic dye counterstaining of sections impregnated by the Golgi-Cox method. Stain Tech. **39:** 65–70.
7. Poulain, D. A., J. B. Wakerley & R. E. J. Dyball. 1977. Electrophysiological differentiation of oxytocin- and vasopressin-secreting neurones. Proc. R. Soc. London Ser. B **196:** 367–384.
8. Wakerley, J. B., D. A. Poulain & D. A. Brown. 1978. Comparison of firing patterns in oxytocin and vasopressin releasing neurones during progressive dehydration. Brain Res. **148:** 425–440.
9. Sokol, H. W. & H. Valtin. 1965. Morphology of the neurosecretory system in rats homozygous or heterozygous for hypothalamic diabetes insipidus (Brattleboro strain). Endocrinology **77:** 692–700.

ANGIOTENSIN IMMUNOREACTIVITY IN VASOPRESSIN CELLS IN RAT HYPOTHALAMUS AND ITS RELATIVE DEFICIENCY IN HOMOZYGOUS BRATTLEBORO RATS *

D. L. Hoffman, L. Krupp, D. Schrag, G. Nilaver, G. Valiquette, M. M. Kilcoyne, and E. A. Zimmerman †

Departments of Neurology and Medicine
College of Physicians & Surgeons
Columbia University
New York, New York 10032

Introduction

It has been twenty years since a role for angiotensin II (AII) in the regulation of blood pressure by the central nervous system was suggested by cross-circulation experiments in dogs.[1] Administration of small doses of AII into the brain revealed additional central actions including thirst, vasopressin (VP) and adrenocorticotropin (ACTH) secretion, sodium appetite, and suppression of renal renin.[2] Although an endogenous brain renin-angiotensin system was proposed ten years ago,[3, 4] its existence is still debated.[2, 5, 6] One issue now apparently resolved concerned the presence of a renin enzyme in brain distinct from cathepsin D.[7] Another has been the inability of most laboratories to extract significant quantities of endogenous AII from brain.[2] Therefore, it is still not known whether brain renin is actually involved in the production of angiotensin II, and whether this might occur within specific neuronal systems. Immunoreactive renin has been demonstrated in neurons in the supraoptic (SON), paraventricular (PVN) and suprachiasmatic (SCN) nuclei of hypothalamus, as well as other brain regions of the mouse.[8] In addition, the possibility that a single cell can contain the entire renin-angiotensin system was recently demonstrated in a neuroblastoma-glioma X cell line in tissue culture.[9]

In recent years AII-like immunoreactivity has been reported in nerve fibers in a number of regions of the brain and spinal cord by immunocytochemical methods.[6, 10–16] The neuronal perikarya most consistently labeled within the cytoplasm were found in the magnocellular nuclei of the hypothalamus: SON and PVN.[13–16] We previously reported that the "AII-like" reactivity in the SON and PVN was localized in vasopressin-containing neurons and behaved like vasopressin: it increased in fibers to the zona externa (ZE) of the median eminence (ME) in response to adrenalectomy, and was deficient in homozygous Brattleboro rats (DI rats), which also lack vasopressin.[15, 16] We also suggested that the antiserum had some preference for 5-valine AII compared to 5-isoleucine AII.[16] In this communication we report the results of further experiments with DI rats, and additional absorption studies on normal rat hypothala-

* Supported by U.S. Public Health Service grants HL 24105, HD 13147, and AM 20337.

† Send all correspondence to E. A. Z.

0077–8923/82/0394–0135 $1.75/0

mus. We also report the use of a more specific antiserum to VP than the one used previously on normal and Brattleboro rats.[17]

Methods

Normal Sprague-Dawley, Long-Evans, and homozygous Brattleboro rats (Blue Spruce Farms, Altamont, New York) were used. Five Brattleboro rats were treated with daily injections of 100 mU VP tannate in oil for two weeks. Both types of normal rats and DI rats were subjected to bilateral adrenalectomy two weeks prior to sacrifice. Others were given 60 μg colchicine into the cerebral ventricles 48 hours prior to sacrifice. Brains were fixed by perfusion with 10% formalin or by immersion in Bouin's solution, blocked and embedded in paraffin. Serial sections 3 or 6 μm, on glass slides were deparaffinized, rehydrated, and incubated with one of four primary rabbit antisera overnight as the first step in the peroxidase antiperoxidase (PAP) technique: [17-19] antiserum to VP (N_1-F, 1:1,000), to oxytocin (OT) (N-3, 1:1,000), to rat neurophysins (NPS) (Robinson #4),[17] 1:4,000), and to AII (Kilcoyne 10 A-6, 1:500). Antisera to synthetic arginine VP (AVP) and synthetic OT were obtained by immunizing rabbits with synthetic peptide (M. Manning) conjugated by the carbodiimide technique to bovine thyroglobulin. Antiserum to AII was prepared by immunizing a rabbit with 5-valine AII amide (Hypertensin®) coupled to rabbit serum albumin by the carbodiimide method.

Specificity of antisera was tested by overnight preincubation of 1 ml of diluted antiserum with synthetic antigen at 4° C prior to use: antiserum to VP or OT with 1 μg AVP or 1 μg OT; antiserum to AII with 1, 10, or 100 μg 5-valine AII or 5-isoleucine AII, or 100 μg AI, AIII (Peninsula Laboratories Inc.), 100 μg AVP and OT, or 600 μg rat or bovine serum albumin.

Results

Analysis of serial 3-μm sections of colchicine-treated normal brains revealed that oxytocin and vasopressin were localized in totally different cells in the SON and PVN, all of which appeared neurophysin positive (antiserum visualizes both rat neurophysins).[17] AII-reactivity was associated with VP- and not at all with OT-reactive perikarya. Although one can not always clearly identify the same cell in serial adjacent sections, AII and VP immunoreactivity was always found in the same cell. There was also an excellent correlation in non-colchicine-treated animals. There was more perikaryal reactivity after colchicine and some reduction in fibers. Dorsomedial cells in the SCN were reactive for AII in the same pattern as for VP and NP, but AII staining was weaker and the cells too small for serial analysis.

Like VP and OT, AII reactivity was found in axonal projections to ME (Figure 1A) and posterior pituitary. Only a rare fiber was found outside the hypothalamus, for example in the amygdala, the nucleus solitarius of the medulla and the spinal cord. In the ME of normal rats some AII and VP fibers projected to the ZE, and after adrenalectomy there was a marked build up of both types of reactivity in ZE (Figure 1B). In fact AII reactivity was stronger than vasopressin after adrenalectomy. DI rat ME did not react with antiserum to AII whether adrenalectomized (Figure 1C) or not.

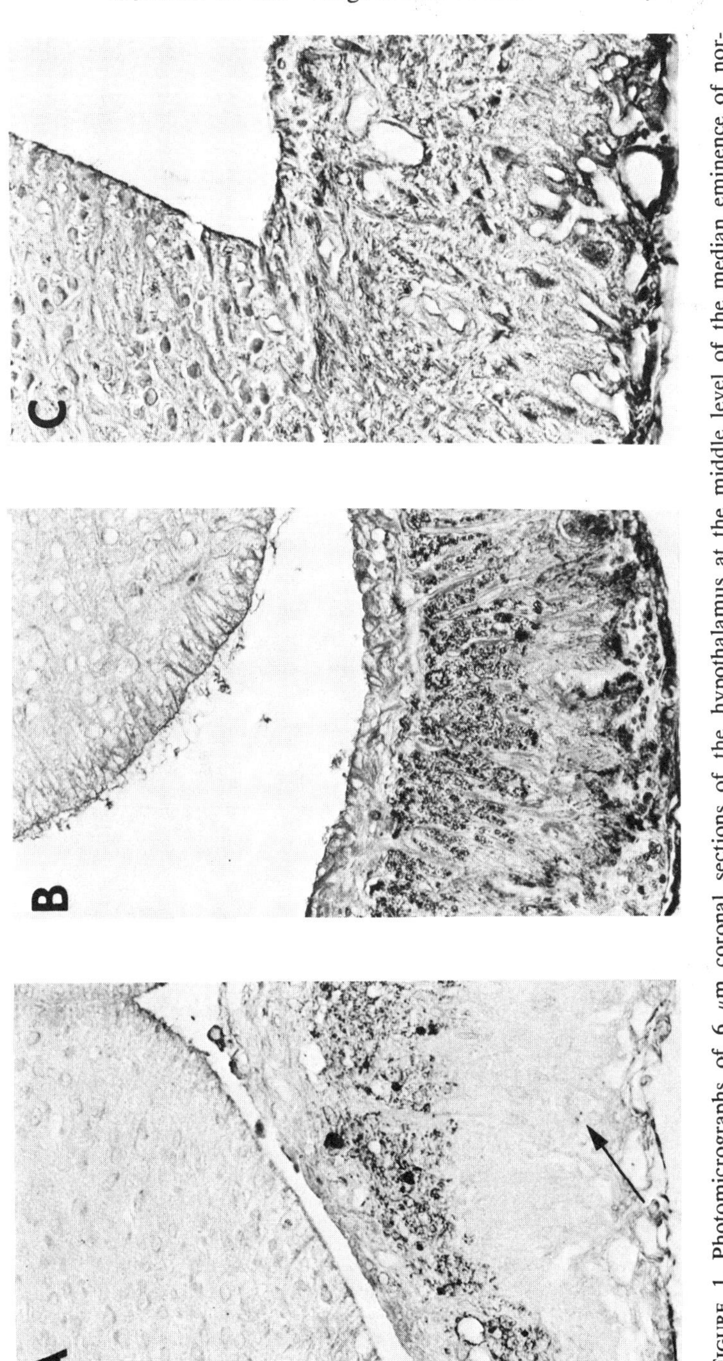

FIGURE 1. Photomicrographs of 6 μm coronal sections of the hypothalamus at the middle level of the median eminence of normal (A) and adrenalectomized Long-Evans (B) rats, and homozygous Brattleboro (C) rats. All reacted with antiserum to angiotensin II and immunoperoxidase technique. Note very few reactive fibers to the zona externa (arrow) of normal rat (A), which increases with adrenalectomy in normal (B) rat, but not in Brattleboro rat (C). ×113.

In addition to producing reaction products in totally different cells, the antisera to OT and VP were shown to be specific by total absorption with homologous and not at all with heterologous antigen. Furthermore antiserum to VP did not react with DI rat hypothalamus. Antiserum to AII absorbed with 100 μg (FIGURE 2A,B) and 10 μg 5-valine or 5-isoleucine AII, and about 50% with 1 μg. AI and AIII absorbed partially, and VP, OT, NP, and albumin not at all.

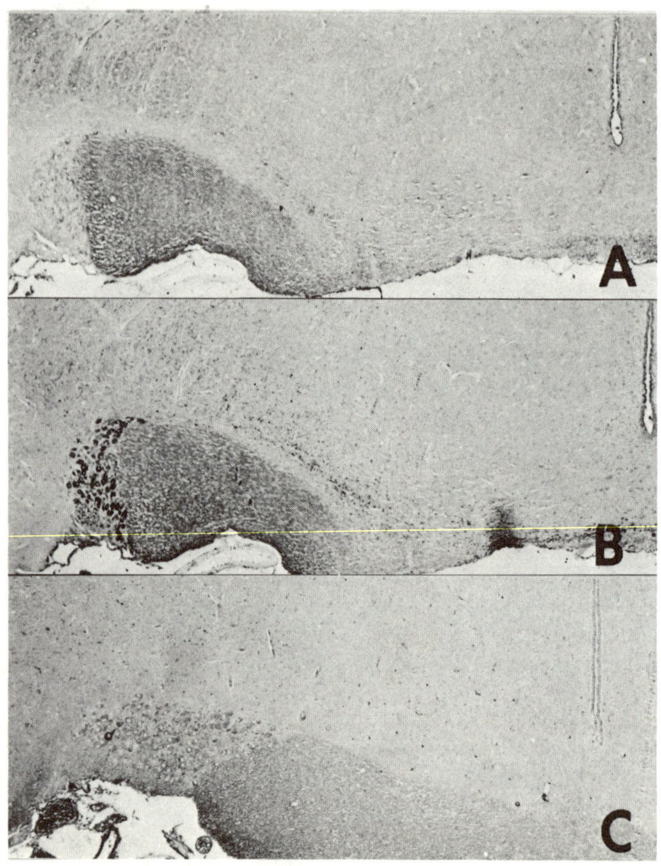

FIGURE 2. Coronal sections of rat brains at the level of the supraoptic nucleus reacted for angiotensin II. (A, B) normal rat, (C) Brattleboro rat. In (A) preabsorption with synthetic angiotensin II eliminated the reactivity in supraoptic neurons seen in (B, unabsorbed). Only a single reactive neuron is found in this homozygous Brattleboro rat section. $\times 80$.

Examination of sections of the hypothalamus of the DI rats with the antiserum to AII revealed that only two of the 17 rats tested had any positive staining. In those two, only one or two reactive perikarya were found in the SON or PVN in some sections (FIGURES 2C, 3). No fibers reacted. We confirmed that these were homozygous Brattleboro rats by lack of reactivity to VP, and the absence of NP staining in half of the SON-PVN cells.[17] The VP-treated and adrenalectomized DI rats were totally unreactive for AII or VP.

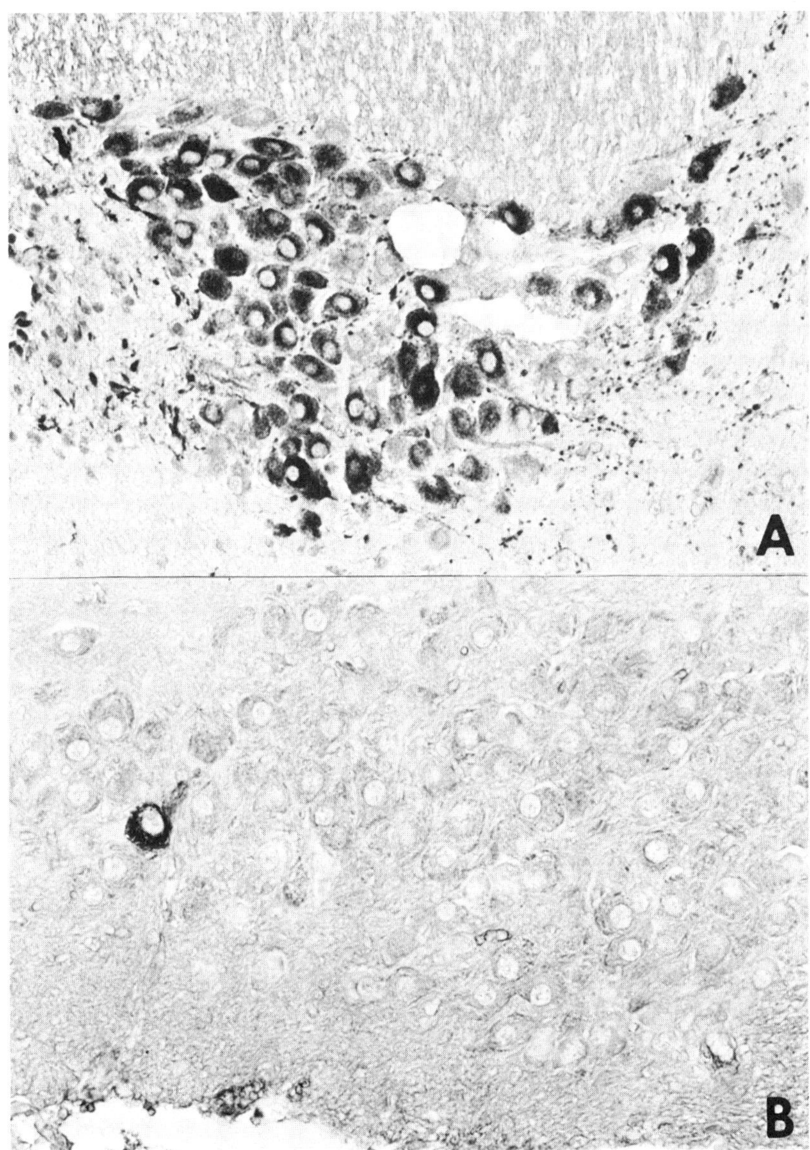

FIGURE 3. Higher magnifications of angiotensin reactivity in normal rat (A) from FIGURE 2B (rotated 45° and reversed), and the reactive cell in Brattleboro rat (B) from FIGURE 2C. ×250.

Summary and Conclusions

These results suggest that vasopressin neurons contain a substance immunologically related to AII. Absorptions controls indicate that its antigenic determinants are more like AII than AI or AIII, and that the antiserum has equal affinity for 5-isoleucine AII and 5-valine AII. The presence of AII material in the perikaryon and its build up after colchicine suggests that it is produced in these cells. Failure to find it in most homozygous DI Brattleboro rats that have high circulating concentrations of AII [20] is additional evidence that it does not come from the general circulation.

The true chemical nature of this AII material cannot be determined by this immunological method. The absorptions suggest that it is not VP, yet it appears to be related. It increases along with VP and VP-NP in fibers to ZE in response to adrenalectomy in normal rats.[19] In addition it is absent, or nearly so, from DI rats that cannot produce VP, VP-NP, or its precursor.[21] Attempts to react radiolabeled precursor (propressophysin) with this antiserum failed (E. A. Zimmerman, H. Gainer, and M. J. Brownstein, unpublished data). Attempts to reduce the possibly rapid turnover of AII reactivity by VP treatment of DI rats were not successful.

At the moment we do not know how to interpret these findings, or those of others who have also used immunocytochemistry. Extrahypothalamic fibers containing AII reactivity could come from PVN like those containing VP,[22] but this remains to be proven. Increases in ZE AII with adrenalectomy support the idea that AII which has effects on corticotropin release [23] may be mediated by the same PVN-ME system as VP.[19] Regarding vasopressin release by AII,[24] one would have expected to see AII in nerve terminals on VP neurons rather than AII in them. It is conceivable that AII-producing neurons lie elsewhere in the brain and the material found here in VP neurons represents cross-reactivity with some other peptide, or alternatively, that there are several AII systems in the brain. It is hoped that these issues may be resolved in future studies by chemical analysis, and by application of monoclonal antibodies to different parts of the renin-angiotensin system.

Acknowledgment

The authors thank J. Haldar for VP-treated DI rat brains.

References

1. Bickerton, R. K. & J. P. Buckley. 1961. Evidence for a central mechanism in angiotensin-induced hypertension. Proc. Soc. Expl. Biol. Med. **106:** 834–856.
2. Ramsay, D. J. 1982. Effects of circulating angiotensin II on the brain. *In* Frontiers in Neuroendocrinology. W. F. Ganong & L. Martini, Eds. Vol. 7. Raven Press. New York, N.Y. (In press.)
3. Ganten, D., A. Marquez-Julio, P. Granger, K. Hayduk, K. P. Karsunky, R. Boucher & J. Genest. 1971. Renin in dog brain. Am. J. Physiol. **221:** 1733–1737.
4. Fischer-Ferraro, C., V. E. Hahmod, D. J. Goldstein & S. Finkielman. 1971. Angiotensin and renin in rat and dog brain. J. Exp. Med. **133:** 353–361.
5. Reid, I. A. 1979. The brain renin angiotensin system: a critical analysis. Fed. Proc. **38:** 2255–2259.
6. Ganten, D., K. Fuxe, M. I. Phillips, J. F. E. Mann & U. Ganten. 1978.

The brain isorenin-angiotensin system: biochemistry, localization and possible role in drinking and blood pressure regulation. *In* Frontiers in Neuroendocrinology. W. F. Ganong & L. Martini, Eds. **5:** 61–99. Raven Press. New York, N.Y.

7. Hirose, S., H. Yokosawa & T. Inagami. 1976. Immunochemical identification of renin in rat brain distinct from acid proteases. Nature (London) **274:** 392–393.
8. Celio, M. R., D. L. Clemens & T. Inagami. 1980. Renin in anterior pituitary, pineal, and neuronal cells of mouse brain: immunohistochemical localization. Biomedical Res. **1:** 427–431.
9. Fishman, M. C., E. A. Zimmerman & E. E. Slater. 1981. Renin and angiotensin: the complete system within the neuroblastoma X glioma cell. Science. (In press.)
10. Nahmod, V. E., S. Finkielman, O. S. deGorodner & D. J. Goldstein. 1977. On the localization and the physiological variations of brain angiotensin. *In* Central Actions of Angiotensin and Related Hormones. J. P. Buckley & C. M. Ferrario, Eds.: 573–579. Pergamon Press. New York, N.Y.
11. Changaris, D. G., W. B. Severs & L. C. Keil. 1978. Localization of angiotensin in rat brain. J. Histochem. Cytochem. **26:** 593–607.
12. Fuxe, K., D. Ganten, T. Hokfelt & P. Bolme. 1976. Immunohistochemical evidence for the existence of angiotensin II containing nerve terminals in the brain and spinal cord in the rat. Neurosci. Lett. **2:** 229–234.
13. Phillips, M. I., J. Weyhenmeyer, D. Felix, D. Ganten & W. E. Hoffman. 1979. Evidence for an endogenous brain renin angiotensin system. Fed. Proc. **38:** 2260–2266.
14. Quinlan, J. T. & M. I. Phillips. 1981. Immunoreactivity for an angiotensin II-like peptide in the human brain. Brain Res. **205:** 212–218.
15. Kilcoyne, M. M., D. L. Hoffman & E. A. Zimmerman. Immunocytochemical localization of angiotensin II and vasopressin in rat hypothalamus: evidence for production in the same neuron. Clinical Sci. **59:** 57s–60s.
16. Zimmerman, E. A., L. Krupp, D. L. Hoffman, E. Matthew & G. Nilaver. 1980. Exploration of peptidergic pathways in brain by immunocytochemistry: a ten year perspective. Peptides **1** (Suppl. 1): 3–10.
17. Sokol, H. W., E. A. Zimmerman, W. H. Sawyer & A. G. Robinson. 1976. The hypothalamic-neurohypophysial system of the rat: localization and quantitation of neurophysin by light microscopic immunocytochemistry in normal rats and in Brattleboro rats deficient in vasopressin and a neurophysin. Endocrinology **98:** 1176–1188.
18. Sternberger, L. A., P. Hardy, J. Cuculis & H. Meyer. 1970. The unlabeled antibody enzyme method of immunohistochemistry. J. Histochem. Cytochem. **18:** 315–323.
19. Stillman, M. A., L. Recht, S. Rosario, S. M. Seif, A. G. Robinson & E. A. Zimmerman. 1977. The effects of adrenalectomy and glucocorticoid replacement in the zona externa of the median eminence of the rat. Endocrinology **101:** 42–49.
20. Gutman, Y. & F. Benzakein. 1974. Antidiuretic hormone and renin in rats with diabetes insipidus. Eur. J. Pharmacol. **28:** 114–118.
21. Brownstein, M. J., J. T. Russell & H. Gainer. 1980. Synthesis, transport, and release of posterior pituitary hormones. Science **207:** 373–378.
22. Swanson, L. W. & P. E. Sawchenko. 1980. Paraventricular nucleus: a site of integration of neuroendocrine and autonomic mechanisms. Neuroendocrinology **31:** 410–417.
23. Maran, J. W. & F. E. Yates. 1977. Cortisol secretion during intrapituitary infusion of angiotensin II in conscious dogs. Am. J. Physiol. **232:** E273–E285.
24. Sladek, C. & R. Joynt. 1980. Role of angiotensin in the osmotic control of vasopressin release by the organ-cultured rat hypothalamo-neurohypophyseal system. Endocrinology **106:** 173–178.

SOMATOSTATIN IS DECREASED IN THE NEUROHYPOPHYSIS OF THE BRATTLEBORO RAT AND MAY PLAY A ROLE IN THE REGULATION OF VASOPRESSIN SECRETION

Hans-Georg Güllner, Elliott C. Kulakowski, and Roger H. Unger

Hypertension-Endocrine Branch
National Heart, Lung, and Blood Institute
National Institutes of Health
Bethesda, Maryland 20205

Veterans Administration Medical Center
Dallas, Texas 75216

It has recently been shown that administration of several neuropeptides, including substance P and the endorphins, causes an increase in plasma vasopressin and that these peptides may play a role in the regulation of vasopressin release from the pituitary gland.[1, 2] The mechanism of this action is not understood. Somatostatin (growth hormone release inhibiting hormone) has been shown to modulate the release of several hormones and to exert a wide spectrum of biological effects.[3] Somatostatin-like immunoreactivity (SLI) is present throughout the central nervous system and in the cells of the diffuse neuroendocrine system.[4–6] The hypothalamus contains the highest concentration of somatostatin. Within the hypothalamus, the concentration of somatostatin is highest in the median eminence and in the paraventricular and arcuate nuclei. It has been suggested that somatostatin reaches neuronal terminals in the median eminence by axoplasmic flow.[7]

Significant amounts of immunoreactive somatostatin are present in extracts of the rat neurohypophysis whereas only negligible amounts of the peptide can be detected in the anterior pituitary. The high somatostatin content in the neurohypophysis led us to further study its possible function. We examined the somatostatin content in the hypothalamo-neurohypophyseal system of the Brattleboro rat and investigated the effect of physiological perturbations that stimulate vasopressin release on somatostatin content in hypothalamus and in the pituitary of rats of the parent Long-Evans strain.

Methods

All animals (male homozygous (di/di) Brattleboro rats, heterozygous control littermates (di/+) and rats of the parent Long-Evans strain) were purchased from Blue Spruce Farms, Altamont, N.Y.

To increase the amount of tissue available for extraction and to improve the accuracy of extraction recovery, tissues from three animals were pooled for each extraction in all experiments. The hypothalamus and pituitary gland were rapidly dissected and the latter separated into the anterior lobe (pars distalis) and the neurointermediate lobe. The isolated tissues were then homogenized and then extracted in 2.0 M acetic acid. Each extract was centrifuged and the supernatant lyophilized. The residue was then reconstituted with assay buffer.

0077–8923/82/0394–0142 $1.75/0 © 1982, NYAS

SLI was measured by radioimmunoassay using an antibody directed against the central portion of the somatostatin molecule.[8] It had previously been established by multiple criteria that the material measured in this assay is immunologically related to somatostatin rather than reflecting interfering substances. Dose-response curves of brain extracts paralleled those of synthetic somatostatin.

Glucagon and insulin were measured with radioimmunoassays that have previously been described.[9,10] The determination of substance P and neurotensin concentrations in tissue extracts by radioimmunoassay[11,12] was kindly performed by Dr. Susan E. Leeman (Department of Physiology, University of Massachusetts, Worcester, Mass.).

To determine whether SLI might be secreted in parallel fashion with vasopressin, the posterior pituitary SLI content and the content of the other peptides were determined in Long-Evans rats given 2% saline as drinking fluid for five days. This procedure has been reported to decrease the vasopressin stores in the neurohypophysis to about 20% of control.[13]

Results

Figure 1 shows the distribution of SLI in the anterior and posterior lobes of Brattleboro rats, of their heterozygous controls, and of Long-Evans rats which had received 2% NaCl as drinking fluid for five days. When compared with unaffected heterozygous littermate controls, the SLI content of the neuro-intermediate lobe of Brattleboro rats was very low (11% of the heterozygous controls: 331 ± 114 pg/mg tissue control, 39 ± 9 pg/mg tissue Brattleboro; $p < 0.01$). Dehydration by substitution of 2% saline for drinking fluid for five days in a group of Long-Evans rats decreased the neurohypophyseal SLI content by about 50% (Figure 1).

To examine whether or not the observed differences in SLI content are specific for somatostatin, we measured the pituitary levels of two other hypothalamic neuropeptides, substance P and neurotensin. In addition, the presence of glucagon and insulin was determined since somatostatin is present in high concentrations in the pancreas[14] and since the tetradecapeptide plays a major role in glucoregulation.[15] There was no difference in substance P and neurotensin levels between the different rat strains (Figure 2). The effect of salt loading was therefore not studied. Glucagon was undetectable in both anterior and posterior pituitary. Insulin was present in small concentrations (1–2 μU/mg tissue) only in the posterior lobe but there was no difference between rat strains and dehydration did not affect the insulin content.

No difference between homozygous Brattleboro rats and heterozygous Brattleboro or Long-Evans rats was observed in the concentrations of these peptides in the hypothalamus (Table 1).

Discussion

The present findings suggest that the content of immunoreactive somatostatin and vasopressin in the neurohypophysis changes in parallel fashion. These changes are peptide specific since the content of other peptides such as substance P and neurotensin or insulin in the Brattleboro rats was not different from that of controls. Since we did not perform immunocytochemical studies or lesioning

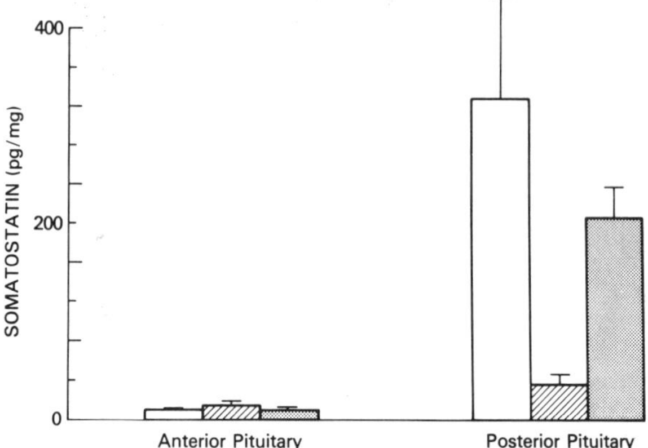

FIGURE 1. Concentration of immunoreactive somatostatin (SLI). The open bars represent heterozygous (di/+) Brattleboro rats, the "hatched" bars represent homozygous (di/di) Brattleboro rats, and the "stippled" bars represent Long-Evans rats which had been given 2% saline as drinking fluid for five days. There was no difference in SLI content between heterozygous Brattleboro rats and Long-Evans rats on an *ad libitum* water intake.

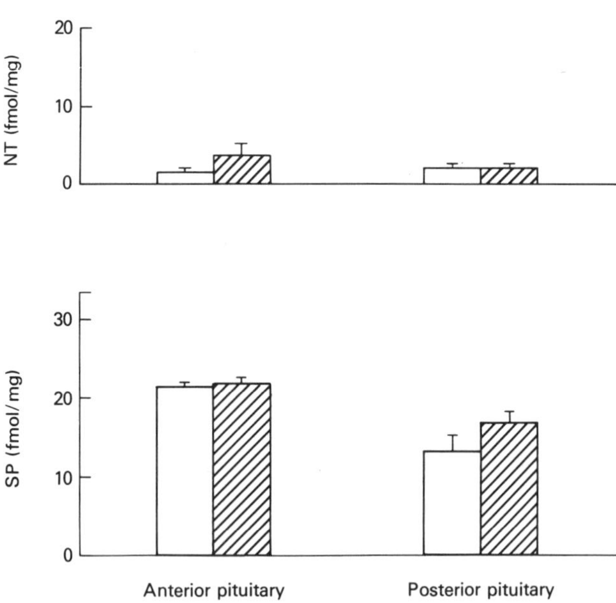

FIGURE 2. Distribution of immunoreactive substance P (SP) and neurotensin (NT) in the pituitary of control Long-Evans rats (open bars) and homozygous (di/di) Brattleboro rats ("hatched" bars).

experiments to trace somatostatinergic neurons in the hypothalamo-neurohypophyseal system we do not know if the somatostatin-containing nerve fibers in the posterior lobe originate in supraoptic and paraventricular nuclei of the hypothalamus or if somatostatin is intrinsic to the posterior pituitary. The latter possibility appears, however, to be very unlikely in view of the parallel changes of somatostatin and vasopressin and in view of the very low background somatostatin concentrations in the anterior pituitary.

The somatostatin content of the posterior pituitary was substantially decreased by salt loading, a physiological perturbation which stimulates the release of vasopressin. This finding and the dramatically reduced SLI content in the posterior pituitary of homozygous Brattleboro rats suggest that somatostatin-containing nerve fibers could be involved in the regulation of neurohypophyseal hormone neurosecretion.

Table 1

Peptide Content of Hypothalamus

Rat Strain	Somatostatin (pg/mg)	Substance P (fmol/mg)	Neuro-tensin (fmol/mg)	Glucagon (pg/mg)	Insulin (μU/mg)
Long-Evans	18092 ± 1052	137 ± 22	40 ± 2	<1	<1
Brattleboro (di/di)	24068 ± 4751	124 ± 4	47 ± 2	<1	<1

References

1. Güllner, H.-G. 1981. Possible participation of substance P in the regulation of body fluid balance. IRCS Med. Sci. **9:** 66–68.
2. Weitzman, R. E., D. A. Fisher, S. Minick, N. Ling & R. Guillemin. 1977. β-endorphin stimulates secretion of arginine vasopressin *in vivo*. Endocrinology **101:** 1643–1646.
3. Reichlin, S., R. Saperstein, I. M. D. Jackson, A. E. Boyd III & Y. C. Patel. 1976. Hypothalamic hormones. Ann. Rev. Physiol. **38:** 389–424.
4. Vale, W., C. Rivier, M. Palkovits, J. M. Saavedra & M. Brownstein. 1974. Ubiquitous brain distribution of inhibitors of adenohypophysial secretion. The Endocrine Society Program of the 56th Annual Meeting, Atlanta. No. 146 (Abstract).
5. Hökfelt, T., S. Efendic, O. Johansson, R. Luft & A. Arimura. 1974. Immunohistochemical localization of somatostatin (growth hormone release-inhibiting factor) in the guinea pig brain. Brain Res. **80:** 165–169.
6. Brownstein, M., A. Arimura, H. Sato, A. V. Schally & J. S. Kizer. 1975. The regional distribution of somatostatin in the rat brain. Endocrinology **96:** 1456–1461.
7. Pelletier, G., F. Labrie, A. Arimura & A. V. Schally. 1974. Electron microscopic immunohistochemical localization of growth hormone-release inhibiting hormone (somatostatin) in the rat median eminence. Am. J. Anat. **140:** 445–450.
8. Harris, V., J. M. Conlon, C. B. Srikant, K. McCorkle, V. Schusdziarra, E. Ipp & R. H. Unger. 1978. Measurements of somatostatin-like immunoreactivity in plasma. Clin. Chim. Acta **87:** 275–283.
9. Harris, V., G. R. Faloona & R. H. Unger. 1979. Glucagon. *In* Methods of

Hormone Radioimmunoassay, 2nd edit. B. M. Jaffe & H. R. Behrman, Eds.: 643–656. Academic Press. New York, N.Y.

10. Yalow, R. S. & S. A. Berson. 1960. Immunoassay of endogenous plasma insulin in man. J. Clin. Invest. **39:** 1157–1175.
11. Mroz, E. & S. E. Leeman. 1979. Substance P. *In* Methods of Hormone Radioimmunoassay. B. M. Jaffe & H. R. Behrman, Eds.: 122–137. Academic Press. New York, N.Y.
12. Carraway, R. E. 1979. Neurotensin and related substances. *In* Methods of Hormone Radioimmunoassay. B. M. Jaffe & H. R. Behrman, Eds.: 139–169. Academic Press. New York, N.Y.
13. Valtin, H., H. W. Sokol & D. Sunde. 1975. Genetic approaches to the study of the regulation and actions of vasopressin. Recent Prog. Horm. Res. **31:** 447–486.
14. Dubois, M. P. 1975. Immunoreactive somatostatin is present in discrete cells of the endocrine pancreas. Proc. Natl. Acad. Sci. USA **72:** 1340–1343.
15. Brown, M., J. Rivier & W. Vale. 1979. Somatostatin: Central nervous system actions on glucoregulation. Endocrinology **104:** 1709–1715.

ANGIOTENSIN STIMULATES OXYTOCIN RELEASE: IMPAIRED RESPONSE IN RATS WITH GENETIC HYPOTHALAMIC DIABETES INSIPIDUS *

Rudolf E. Lang, Wolfgang Rascher, Jürgen Heil
Thomas Unger, and Detlev Ganten

Department of Pharmacology
University of Heidelberg
Im Neuenheimer Feld 366
6900 Heidelberg, Federal Republic of Germany

INTRODUCTION

Angiotensin II (AII) is a strong stimulus for antidiuretic hormone (ADH) secretion.[2, 3, 8, 11, 15, 17, 18] It is still a matter of controversy whether blood-borne AII acts as an ADH releaser, but it has been demonstrated that acute intracerebroventricular (i.c.v.) injection of AII results in a marked increase of plasma ADH in goat, dog, and rat.[2, 8, 11, 14] The physiological significance of AII as a mediator of ADH secretion has been revealed by experiments showing that the increase in ADH during water deprivation was prevented by i.c.v. administration of a competitive AII antagonist.[23]

The specificity of hypovolemia and hyperosmolality, as stimuli for ADH release, however, has been challenged by experiments in which a decrease not only in ADH but also in oxytocin (OT) content of the neurohypophysis following water deprivation or salt loading was observed.[10] Furthermore, hemorrhage or hypertonic saline have been reported to increase concomitantly ADH and OT concentrations in plasma.[22] This simultaneous release of ADH and OT in response to osmotic and hypovolemic stimuli suggested that a common regulatory mechanism for both hormones might exist. To examine this hypothesis, we studied the effects of centrally administered AII on plasma OT levels in normal rats and in rats with hypothalamic diabetes insipidus (DI), which are unable to synthesize ADH. DI rats were included in these experiments to rule out interferences of ADH in the OT measurements and to investigate whether increased levels of plasma AII and altered activity of brain renin[9, 12] may affect OT release in response to exogenous i.c.v. AII.

METHODS

Male Wistar rats (WR), rats of the Brattleboro strain, homozygous (DI) or heterozygous (HZ) for hypothalamic diabetes insipidus and Long-Evans rats (LE), the genetic control for the DI, all weighing 200–250 g, were used. A group of DI rats underwent ADH replacement treatment using 600 ng, s.c., of the ADH analogue 1-desamino-8-D-arginine vasopressin DDAVP (Ferring) daily for 12 days. The response of plasma OT to an osmotic stimulation was tested by measurement of OT 15 min after an infusion of either 3% NaCl or

* Supported in part by Deutsche Forschungsgemeinschaft, SFB 90.

0077-8923/82/0394-0147 $1.75/0 © 1982, NYAS

9.0% NaCl as control (0.4 ml/min per 100 g for 5 min). For injections of AII into the brain ventricles (i.c.v.), the animals were implanted with chronic lateral ventricular cannulae under halothane anesthesia. All i.c.v. injections were carried out 3 to 5 days after cannula implantation in conscious, freely moving rats. A 1.0 μl solution of 0.01 to 10 μg AII (Beckmann) freshly dissolved in 0.9% NaCl or 0.9% NaCl control injections were administered. The specificity of the AII effects was tested by i.c.v. administration of the competitive AII antagonist {Sar^1, Ile^8}-AII at a dose of 1.0 μg alone and in combination with AII. One minute after the i.c.v. injection, at which time the peak pressor response to i.c.v. AII had been determined to occur, the animals were decapitated and trunk blood was collected. OT and ADH were extracted from 2-ml aliquots of plasma by the acetone-ether technique and determined by specific radioimmunoassay (RIA), as developed in our laboratory.[16] The hormones were labeled with ^{125}I by the chloramine-T procedure and purified by chromatography on DEAE Sephadex. The OT and AVP antisera were used in the RIA at a final dilution of 1:600,000 and 1:75,000, respectively. Separation of bound from free hormone was achieved by charcoal. Synthetic ADH showed less than 0.5% cross-reaction in the OT RIA; no cross-reactivity was observed with 1-desamino-8-D-arginine vasopressin up to 1 μg. The ADH antibody cross-reacted with OT by less than 0.1%. The OT assay was sensitive to 1.05 fmol OT, the ADH assay to 0.3 fmol ADH. All results are given as mean ± SEM. Significance of differences was calculated by Student's unpaired *t*-test. The level of confidence is indicated in the tables and figures.

Results

One minute after i.c.v. administration of 10 or 100 ng ANG II, a marked rise of ADH, as well as OT secretion, was observed in WR. The 100 ng dose induced about a tenfold increase of both hormones in the plasma (Table 1). When 100 ng AII were injected in combination with 1,000 ng {Sar^1, Ile^8}-AII, the release of OT and ADH was almost completely blocked. The antagonist alone had no effect on OT release.

Plasma OT was significantly elevated in DI as compared to HZ or LE rats under control conditions. Stimulation by 100 ng AII i.c.v. resulted in a large increase in OT (Figure 1) and ADH (not shown) in plasma of HZ and LE similar to that observed in WR. In contrast, DI showed only a weak response in OT secretion following i.c.v. AII, barely exceeding the control response (Figure 1). Increasing the dose even up to 10 μg, AII produced no further effects on OT release in these animals. To the osmotic stimulus, however, the response of DI and LE rats was the same. OT increased from 26.8 ± 2.4 (control) to 66.3 ± 6.8 (3% NaCl) in DI rats, and from 16.3 ± 4 (control) to 64.2 ± 8.3 (3% NaCl) in LE rats ($N = 10$ each group).

Replacement therapy with DDAVP in DI rats resulted in an increase in body weight (39 ± 5 g vs. 10 ± 5 g) and a fall in plasma sodium 147.6 ± 1.7 mmol/L vs. 154 ± 2.0 mmol/L; $N = 20$) and water intake (30 ± 1 ml vs. 143 ± 7 ml/24 h) as compared to untreated DI controls. Following DDAVP treatment, 100 ng AII i.c.v. caused a rise in plasma OT concentration several fold higher than in unsubstituted DI and similar to the results observed in LE and HZ rats (Figure 1).

TABLE 1

PLASMA OXYTOCIN AND VASOPRESSIN AFTER INTRAVENTRICULAR ADMINISTRATION OF ANGIOTENSIN II AND/OR THE ANGIOTENSIN II ANTAGONIST {SAR[1], ILE[8]}-AII

Angiotensin II Dose	{Sar[1], Ile[8]}-AII (ng)	Plasma OT (fmol/ml plasma)	Plasma ADH (fmol/ml plasma)	*N*
0	0	10.5 ± 1.4	2.3 ± 0.4	10
0	1000	11.3 ± 1.3		9
10	0	88.9 ± 5.1		10
100	0	119.6 ± 11.5	16.4 ± 2.1	10
100	1000	20.6 ± 4.6		9

$p < 0.01$.

DISCUSSION

Our results clearly demonstrate that AII is a potent stimulus not only for ADH but also for OT release. The stimulatory effect of AII on ADH and OT-containing neurons has been suggested by recent electrophysiological studies. It has been reported that i.c.v. administration of AII excited both types of neurons in the paraventricular and supraoptic nuclei of rats.[1] However, these electrophysiological findings have not yet been verified by the direct determination of hormone release into plasma.

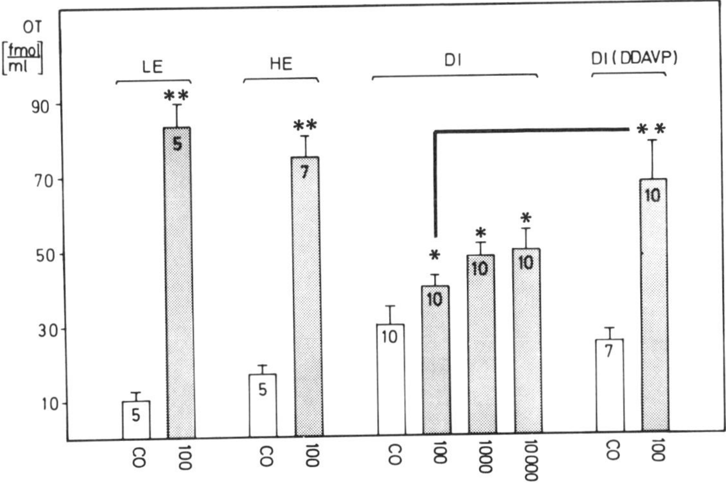

FIGURE 1. Effect of i.c.v. administered AII on plasma oxytocin in Long-Evans rats (LE), rats heterozygous (HE), and homozygous (DI) for hypothalamic diabetes insipidus as well as in DI rats pretreated with the vasopressin analogue DDAVP. The number of animals in each groups is indicated at the top of the columns. The abscissa shows the respective doses of AII given in nanograms per injection. Controls received 0.9% NaCl without AII (CO). * $p < 0.05$; ** $p < 0.01$.

In the present study, the i.c.v. administration of AII was chosen to examine the possible effects of this hormone on OT release since central application of AII reliably produces an increase in ADH secretion[2, 8, 11, 14] whereas the participation of systemically injected AII in ADH control continues to be a subject of debate.[5] The efficacy of the i.c.v. administration on ADH and OT release suggests that AII may reach receptor sites that are situated inside the blood-brain barrier. Centrally administered AII could thereby mimic the action of endogenous AII formed by an intrinsic brain renin-angiotensin system.[7] However, it cannot be ruled out that i.c.v. AII penetrates more easily to receptor sites that are also accessible from the periphery but cannot be reached in sufficient concentration due to the dilution of systemically administered AII. The effects on OT release appear to be mediated by receptors that specifically interact with AII. This is suggested by the observation that the OT release in response to i.c.v. AII was almost completely prevented by the simultaneous administration of the competitive AII antagonist {Sar^1,Ile^8}-AII.

The observation that DI rats produced only a small rise in plasma OT following i.c.v. injection of AII, even in doses thousand-fold higher than those required to induce a several-fold increase of OT in normal rats, was surprising since a normal sensitivity of DI rats to i.c.v. AII with respect to drinking has been reported.[12] The reason for this impaired response of OT to AII appears to be secondary to the defect in ADH synthesis since substitution of DI rats with the ADH analogue DDAVP led to full restoration of the OT response. Several other differences between DI and LE rats may, however, explain the impaired OT release in DI rats.

It has been reported that there is a decreased dopamine turnover in the median eminence of DI rats.[21] Since dopamine seems to be involved in the control of OT secretion,[4] one might assume that the altered neurotransmitter metabolism in DI is associated with the disturbance in OT response. From studies using the OT bioassay, it is known that the OT content in the posterior pituitary gland of HZ and LE is virtually identical, but that pituitary OT is reduced to less than one-third of normal in DI rats.[20] Prolonged treatment with exogenous ADH has been reported to increase pituitary OT content in DI to the levels measured in LE and HZ rats. Our present data, which were obtained by radioimmunoassay, confirm these observations. It is suggested that the stored OT is reduced because of the chronic contraction of body fluid volumes and the consequent stimulation of the hypothalamo-neurohypophyseal system and increased pituitary secretion in DI. The elevated OT levels in plasma of DI as compared to non-diabetic controls may also indicate an increased demand on OT due to the absence of ADH lending further support to this concept. In light of these facts the minimal effect of AII on OT release could be due to an already maximal OT secretion rate from the posterior pituitary in DI rats. Such an interpretation appears unlikely, however, since our results also show that the pituitary of DI rats still has sufficient capacity to respond with a large increase in OT release following an osmotic stimulus.

AII appears to be a mediator of the osmotic ADH control[7] and this could also be true for OT. The attenuated response of OT secretion in response to AII in DI could therefore reflect a decreased sensitivity to AII at the receptor level. Consistent with such a view is the observation that DI rats possess an activated renin-angiotensin system which is indicated by the more than two-fold elevated plasma AII levels.[12] In addition, increased renin activity in the hypothalamus and neurohypophysis of DI has been reported.[9] The chronic elevation of AII may lead to a down regulation of AII receptors. Indeed, a diminished number of AII receptors has been observed in the brain of DI rats.[6] This could explain the blunted release of OT in response to AII. In addition, the reduction of receptors can be taken into consideration as a possible explanation for the attenuated blood pressure responses to AII and the low aldosterone plasma levels in DI. Prolonged substitution of DI with ADH has been shown to reverse elevated AII as well as plasma and brain renin to normal values.[9, 12] Accordingly, in our experiments substitution of DI with the ADH analogue DDAVP, produced a rapid rise in body weight, a decrease in packed cell volume, and a normalization of the OT response to i.c.v. AII.

The actions of angiotensin point to an integral role of this hormone in homeostatic mechanisms serving to regulate blood pressure, salt, and water balance. The physiological significance of OT is still unclear in this context. The antidiuretic potency of OT is negligible as compared with that of ADH. OT has rather been known to control certain processes in the reproductive system of the female; its release is usually evoked by stimulation of the genital tract or by suckling.[13] Whether AII is involved as a mediator in these events remains to be examined.

References

1. Akaishi, T., H. Negoro & S. Kobayasi. 1980. Responses of paraventricular and supraoptic units to angiotensin II, Sar^1-Ile^8-angiotensin II and hypertonic NaCl administered into the cerebral ventricle. Brain Res. **188:** 499–511.
2. Andersson, B. & O. Westbye. 1970. Synergistic action of sodium and angiotensin on brain mechanisms controlling fluid balance. Life Sci. **9:** 601–608.
3. Bonjour, J. P. & R. L. Malvin. 1970. Stimulation of ADH release by the renin-angiotensin system. Am. J. Physiol. **218:** 1555–1559.
4. Bridges, T. E., E. W. Hillhouse & M. T. Jones. 1976. The effect of dopamine on neurohypophysial hormone release in vivo and from the neural lobe and hypothalamus in vitro. J. Physiol. Lond. **260:** 647–666.
5. Claybaugh, J. R., L. Share & K. Shimizu. 1972. The inability of infusions of angiotensin to elevate the plasma vasopressin concentration in the anesthetized dog. Endocrinology **90:** 1647–1652.
6. Cole, F. E., H. L. Blakesley, E. D. Frohlich & A. A. MacPhee. 1981. Angiotensin II receptors are reduced in the CNS of the Brattleboro rat. Endocrine Society 63rd Meeting. Abstr. 880.
7. Ganten, D., K. Fuxe, M. I. Phillips, J. F. E. Mann & U. Ganten. 1978. The brain isorenin-angiotensin system: biochemistry, localization and possible role in drinking and blood pressure regulation. Frontiers Neuroendocrinology **5:** 61–101.
8. Haack, D. & J. Möhring. 1978. Vasopressin-mediated blood pressure response to intraventricular injection of angiotensin II in the rat. Pflügers Arch. **373:** 167–173.
9. Hoffman, W. E., U. Ganten, P. Schelling, M. I. Phillips, P. G. Schmid & D. Ganten. 1978. The renin and isorenin-angiotensin system in rats with hereditary hypothalamic diabetes insipidus. Neuropharmacology **17:** 919–923.
10. Jones, C. W. & B. T. Pickering. 1969. Comparison of the effects of water deprivation and sodium chloride imbibition on the hormone content of the neurohypophysis of the rat. J. Physiol. **203:** 449–458.
11. Keil, L. C., J. Summy-Long & W. B. Severs. 1975. Release of vasopressin by angiotensin II. Endocrinology **96:** 1063–1065.
12. Möhring, J., G. Kohrs, B. Möhring, M. Petri, E. Homsy & D. Haack. 1978.

Effects of prolonged vasopressin treatment in Brattleboro rats with diabetes insipidus. Am. J. Physiol. **234**(2): F106–F111.

13. Moos, F. & P. Richard. 1975. Importance de la liberation d'oxytocine induite par la dilatation vaginale (reflexe de Ferguson) et la stimulation vagale (reflexe vago-pituitaire) chez la ratte. J. Physiol. (Paris) **70:** 307–314.
14. Mouw, D., J. P. Bonjour, R. L. Malvin & A. Vander. 1971. Central action of angiotensin in stimulating ADH release. Am. J. Physiol. **220:** 239–242.
15. Ramsay, D. J., L. C. Keil, M. C. Sharpe & J. Shinsako. 1978. Angiotensin II infusion increases vasopressin, ACTH and 11-hydroxysteroid secretion. Am. J. Physiol. **234**(1): R66–R71.
16. Rascher, W., E. Weidmann & F. Gross. 1981. Vasopressin in the plasma of stroke-prone spontaneously hypertensive rats. Clin. Science **61:** 295–298.
17. Severs, W. B., J. Summy-Long, J. S. Taylor & J. D. Connor. 1970. A central effect of angiotensin: release of pituitary pressor material. J. Pharmacol. Exp. Ther. **174:** 27–34.
18. Sladek, D. C. & R. J. Joynt. 1979. Angiotensin stimulation of vasopressin release from the rat hypothalamo-neurohypophyseal system in organ culture. Endocrinology **104**(1): 148–153.
19. Sladek, C. D. & R. J. Joynt. 1980. Role of angiotensin in the osmotic control of vasopressin release by the organ-cultured rat hypothalamo-neurohypophyseal system. Endocrinology **106**(1): 173–178.
20. Valtin, H. 1967. Hereditary hypothalamic diabetes insipidus in rats (Brattleboro strain). Am. J. Med. **42:** 814–827.
21. Versteeg, D. H. G., M. Tanaka & E. R. De Kloett. 1978. Catecholamine concentration and turnover in discrete regions of the brain of homozygous Brattleboro rat deficient in vasopressin. Endocrinology **103:** 1654–1661.
22. Weitzman, R. E., T. H. Glatz & D. A. Fisher. 1978. The effect of hemorrhage and hypertonic saline upon plasma oxytocin and arginine vasopressin in conscious dogs. Endocrinology **103:** 2154–2160.
23. Yamaguchi, K., H. Hama, T. Sakaguchi, H. Negoro & K. Kamoi. 1980. Effects of intraventricular injection of Sar^1-Ala^8-angiotensin II on plasma vasopressin level increased by angiotensin II and by water deprivation in conscious rats. Acta Endocrinol. **93:** 407–412.

INFLUENCE OF VASOPRESSIN UPON FIRING PATTERNS OF SUPRAOPTIC NEURONS: A COMPARISON OF NORMAL AND BRATTLEBORO RATS

G. Leng and W. T. Mason

Agricultural Research Council Institute of Animal Physiology
Babraham, Cambridge CB2 4AT
England

Introduction

Recent electrophysiological experiments [1] suggest that synaptic interaction between supraoptic neurons may modulate specific aspects of the discharge patterning in vasopressin neurons. In particular, the technique of 'constant collision' stimulation was used to demonstrate that, when the neighbors of a phasic supraoptic neuron are activated antidromically by stimuli applied to the neural stalk, the pattern of phasic discharge in the neuron under study is modified: the burst length is reduced, but the discharge within bursts is unaffected.

It is possible that vasopressin itself, released within the supraoptic nucleus, may influence the activity of neighboring neurons. Several lines of evidence support this hypothesis. First, supraoptic neurons have been reported [2] to have processes that stain densely for neurosecretory material and that terminate near other neurons within the nucleus. Second, vasopressin is able to induce phasic activity in certain invertebrate neurons, and to modify the discharge pattern of certain spontaneously phasic neurons.[3,4] Third, vasopressin applied by iontophoresis inhibits some supraoptic neurons.[5,6] Finally, the effects of constant-collision stimulation upon phasic discharge are absent in Brattleboro rats homozygous for diabetes insipidus (DI rats).[7,8]

In the present paper we report that pre-treatment of DI rats with vasopressin tannate does not restore the effectiveness of constant-collision stimulation. Thus the ineffectiveness of this stimulation in DI rats is not a consequence of the absence of circulating vasopressin, but may be due to the absence of vasopressin available for release. We have also examined the effect of vasopressin upon the discharge of phasic neurons recorded in an *in vitro* brain slice preparation. We have confirmed that vasopressin can inhibit supraoptic phasic neurons, and can modify their discharge pattern by reducing the burst length.

Electrophysiological Evidence for Synaptic Interaction

We recorded from antidromically identified supraoptic neurons in urethane-anesthetized rats. A shock applied to the neural stalk, which activates most supraoptic neurons, will normally activate the particular neuron under study. In these circumstances it is difficult to separate the effects of antidromic invasion of the neuron under study from effects mediated by antidromic invasion of neighboring neurons. We avoided this problem by presenting shocks within 1 ms of recording a spontaneous action potential. Although such shocks will

0077-8923/82/0394-0153 $1.75/0

antidromically invade most supraoptic neurons, the cell under study will not be invaded because the action potential evoked antidromically in that cell will be extinguished by collision with the spontaneous, orthodromic action potential. Low frequency stimulation (<1 Hz) had no detectable effect, but when shocks were presented after every spontaneous action potential, the discharge pattern of the phasic cells was dramatically modified (FIGURE 1.). The mean burst length was consistently and reversibly reduced during stimulation, but other facets of the discharge activity were relatively unaffected: in particular the length of the silent period between bursts was unaffected or slightly increased and the pattern of intraburst activity was unaffected. Inter-spike interval histograms were unchanged during stimulation and mean burst shape was also unaffected.

The original experiments were performed on Wistar rats. We repeated the experiments using DI rats and a control group of Long-Evans rats. Supraoptic neurons from Long-Evans rats behaved in a similar manner to those from Wistar rats. However the effects of constant-collision stimulation upon phasic cells were largely absent in DI rats. The mean burst length was reduced during constant-collision stimulation by 40% or more for 14 of 15 phasic cells in Wistar rats and for 14 of 18 cells in Long-Evans rats, but a similar reduction was seen in only 1 of 23 cells in DI rats.

Effects of Pretreatment with Vasopressin Tannate

We tested the possibility that the difference in the behavior of phasic neurons from DI rats in response to constant collision stimulation is due to the absence of circulating vasopressin by pre-treating DI rats with vasopressin tannate. Five DI rats were given 1 U vasopressin tannate (pitressin tannate in oil, Parke-Davis) by subcutaneous injection daily for 7–10 days prior to experiment. This treatment reduced water intake by at least 50% for each rat, and similarly reduced urine output. Each rat was anesthetized with urethane (ethyl carbamate; 1.3 g/kg i.p.) and the ventral surface of the brain was exposed at the neural stalk and at the optic region.[1] A bipolar stainless steel stimulating electrode was placed upon the neural stalk, and single neurons were recorded from the supraoptic nucleus with 0.15 M NaCl micropipettes. These neurons were antidromically identified as projecting to the neural stalk.

Thirteen phasic cells were recorded from these rats, and were tested with one or more 10–15-min periods of constant-collision stimulation of the neural stalk at a similar range of intensities to those used with Long-Evans and untreated DI rats (0.5–1.5 mA peak-to-peak). As in the untreated DI rats, constant-collision stimulation was significantly less effective than in either Long-Evans or Wistar rats (FIGURE 2). Mean burst length was reduced by 40% or more in 3 of the 13 cells. Overall, no significant difference was observed in the effectiveness of this stimulation between these rats and the untreated DI rats.

In Vitro *Recordings from the Supraoptic Nucleus*

Vasopressin applied by iontophoresis inhibits the activity of some supraoptic neurons.[5,6] To mimic the anticipated slow, local release of vasopressin during

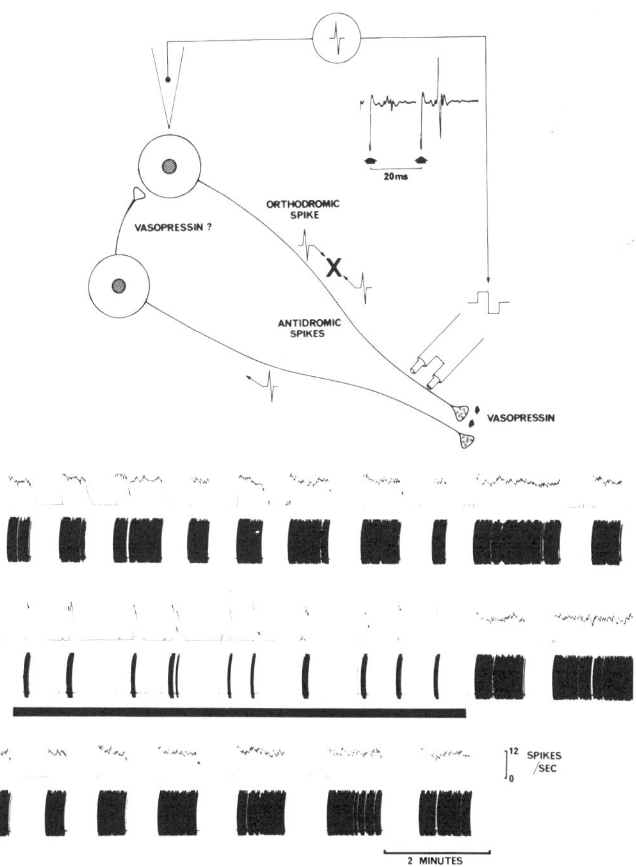

FIGURE 1. Top. Schematic diagram illustrating the technique of constant-collision stimulation. Spontaneous extracellular action potentials are recorded from a supraoptic neuron. Each action potential triggers a shock to the neural stalk, which initiates antidromic action potentials in the axons of the supraoptic neurons. The antidromic action potential evoked in the axon of the recorded neuron is extinguished by collision but other antidromic action potentials persist to invade the cell bodies of neighboring neurons, and presumably activate intranuclear connections. The inset photograph, from the face of a storage oscilloscope, illustrates collision. A spontaneous action potential (not visible) is followed by a pair of shocks to the neural stalk (arrows). Each shock evokes an antidromic action potential in the recorded neuron, but the action potential following the first shock is extinguished by collision whereas that following the second shock invades the cell body.

Bottom. Continuous polygraph record of the activity of a phasic neuron from a Long-Evans rat showing the effects of 1 mA constant-collision stimulation of the neural stalk (black bar). The mean burst length is reversibly reduced during stimulation.

constant-collision stimulation we sought to determine the effects of prolonged exposure to vasopressin rather than the effects of a brief 'bolus.' Accordingly, we recorded from an *in vitro* slice preparation of rat hypothalamus while introducing vasopressin into the bathing medium.

Coronal sections (400 μm) of rat hypothalamus were prepared on a vibratome and were continuously perfused with fresh oxygenated Yamamoto medium at 0.6–1.0 ml/min. The electrical activity of supraoptic neurons was recorded extracellularly with 0.15 M NaCl micropipettes. The neurons were spontaneously active, and some showed phasic activity similar to that seen *in vivo*. Vasopressin (arginine vasopressin; 3×10^{-8}–10^{-7} M) added to the medium produced a marked depression of the discharge activity of each of seven phasic neurons tested. The length of the active phases was reversibly reduced in the presence of vasopressin (FIGURE 3).

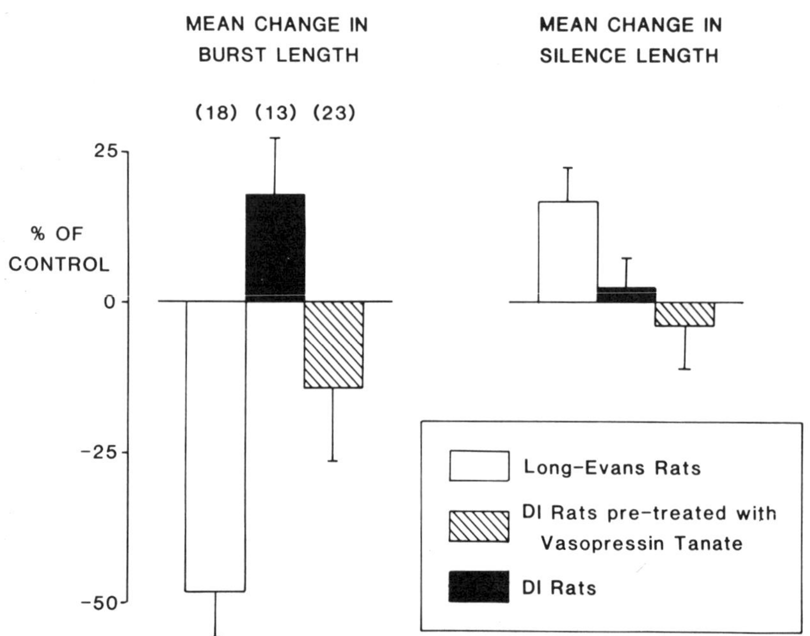

FIGURE 2. Effects of constant-collision stimulation upon phasic cells from Long-Evans rats and from DI rats either untreated or pre-treated with vasopressin tannate. The blocks show the mean percentage change in burst and silence length, and the bars give the standard error of the mean. The number of cells in each group is in parentheses. Stimulation was significantly more effective in Long-Evans rats than in either group of DI rats (Mann-Whitney *U* test, $p < 0.01$). There were no significant differences between pre-treated and untreated DI rats.

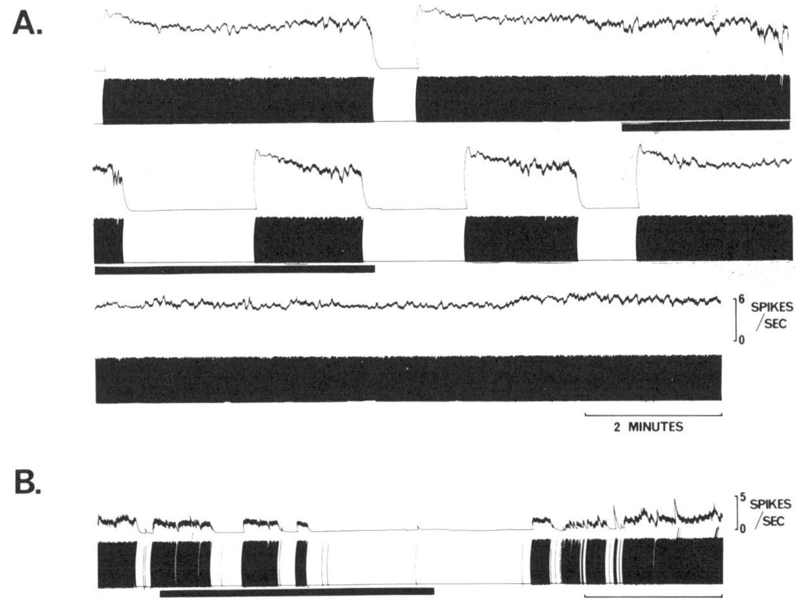

FIGURE 3. (A) Continuous polygraph record of a phasic neuron recorded *in vitro* in a hypothalamic slice preparation. The addition of vasopressin (3×10^{-8} M, black bar) to the perfusion medium reversibly inhibited the activity of the neuron, reducing the burst length. The lag time of about 4 min before effects are seen is due to the lag time of the perfusion system. (B) A second neuron totally but reversibly inhibited by vasopressin.

CONCLUSIONS

Supraoptic neurons respond to systemic osmotic stimulation by releasing more hormone from their nerve endings in the neural lobe. However the response of the vasopressin neurons is not simply to fire more action potentials during osmotic stimulation, but is also to discharge in a 'phasic' pattern that optimizes the efficiency of vasopressin release from the nerve endings. Clearly the presence of vasopressin is not necessary for the induction of phasic firing, since phasic neurons occur in DI rats, but we hypothesize from our experimental findings that vasopressin is released from nerve endings within the nucleus, and acts upon other vasopressin neurons to assist the evolution of relatively efficient discharge patterning.[8, 9]

REFERENCES

1. LENG, G. 1981. The effects of neural stalk stimulation upon firing patterns in rat supraoptic neurones. Exp. Brain Res. **41:** 135–145.
2. SOFRONIEW, M. W. & W. GLASSMAN. 1981. Golgi-like immunoperoxidase staining of hypothalamic magnocellular neurones that contain vasopressin oxytocin or neurophysin in the rat. Neuroscience **6:** 619–643.

3. Barker, J. L. & H. Gainer. 1975. Studies on bursting pacemaker potential activity in molluscan neurones. I. Membrane properties and ionic contributions. Brain Res. **84:** 461–477.
4. Barker, J. L. & T. G. Smith. 1976. Peptide regulation of neuronal membrane properties. Brain Res. **103:** 167–170
5. Moss, R. L., R. E. J. Dyball & B. A. Cross. 1972. Excitation of antidromically identified neurosecretory cells of the paraventricular nucleus by oxytocin applied iontophoretically. Exp. Neurol. **34:** 95–102.
6. Nicoll, R. A. & J. L. Barker. 1971. The pharmacology of recurrent inhibition in the supraoptic neurosecretory system. Brain Res. **35:** 501–511.
7. Leng, G. & J. Wiersma. 1981. Effects of neural stalk stimulation on phasic discharge of supraoptic neurones in Brattleboro rats devoid of vasopressin. J. Endocr. **90:** 211–220.
8. Mason, W. T. 1980. Supraoptic neurones of rat hypothalamus are osmosensitive. Nature **287:** 154–157.
9. Bicknell, R. J. & G. Leng. 1981. Relative efficiency of neural firing patterns for vasopressin release *in vitro*. Neuroendocrinology. **33:** 259–299.

EVIDENCE FOR THE ABSENCE OF ARGININE VASOTOCIN IN MAMMALIAN PINEALS, INCLUDING THE BRATTLEBORO RAT

A. Negro-Vilar, W. K. Samson, and F. Sanchez-Franco

Department of Physiology
University of Texas Health Science Center
Dallas, Texas 75235

In lower species, such as birds, reptiles, and amphibia, arginine vasotocin (AVT) is the antidiuretic principle. In most mammalian species, such a role is subserved by arginine vasopressin (AVP), another peptide whose composition differs by only one amino acid from that of AVT. AVP is widely distributed within hypothalamic and extrahypothalamic regions of the brain [1,2] and many reports suggest that besides its role in regulation of water metabolism, this peptide may participate in other central nervous system functions.[3] It has been suggested that AVT may represent the pineal antigonadotropic principle, and that it may be involved in the secretion of other pituitary hormones, such as ACTH and prolactin.[4,5] Pavel *et al.* have reportedly found AVT in mammalian pineals and in cerebrospinal fluid.[6,7] Their measurements of AVT have been based on complex differential bioassay systems, a technique whose specificity can be questioned. Another group also reported the presence of bioassayable AVT in neural lobes from fetal sheep and seals.[8] However, Pavel [6] acknowledged that chemical analysis is necessary to establish beyond doubt the presence of AVT in pineals, CSF, and other tissues of mammals.

It is obvious that if AVT is postulated to be a pineal hormone involved in different physiological events, a major prerequisite to accept that assumption is the unequivocal demonstration of the presence of the nonapeptide in the pineals and/or other organs of mammals. We will review succinctly the early studies concerning the search for this peptide by either chemical or radioimmunological methods and then will describe our recent studies, as well as those of others, which have afforded conclusive evidence for the absence of AVT in pineals of a variety of mammals.

Chemical and Immunological Determination of AVT in Mammalian Tissues: Early Evidence

Using chemical methods, two early studies [9,10] described a pineal antigonadotrophic principle that, upon isolation and sequencing, was reported to be identical to AVT. However, using exhaustive chemical procedures for the separation and identification of the bovine pineal antigonadotrophic principle, Rosenblum *et al.*[11] concluded that this factor was not vasotocin.

Using a combination of two antibodies to measure AVT indirectly by radioimmunoassay (RIA), Rosenbloom and Fisher [12,13] reported the presence of relatively large amounts of AVT in pineal and subcommissural organs. However, since one of the antisera to AVP cross-reacted with oxytocin (OXY) at a ratio of 4:1 (OXY:AVP) the large amounts of AVT detected may reflect

0077-8923/82/0394-0159 $1.75/0 © 1982, NYAS

cross-reactivity with AVP or OXY, which are known to be present in these areas.[14]

Systematic Search for AVT in Pineals of Different Species by RIA: Comparative and Chronological Studies

In view of the contradictory and sometimes inconclusive evidence discussed above, we decided to initiate studies to obtain definitive proof for the presence, or absence, of AVT in mammalian tissues. We took advantage of the availability of two antisera, one developed by Moore *et al.*[15] which cross-reacts only with AVP (AVP-AS), and the other, developed by one of us (F.S.F.), which has 100% cross-reactivity for both AVP and AVT (AVP-AVT-AS). Neither of the two antisera cross-reacts with OXY or a variety of neurohypophyseal analogues and other peptides. A detailed characterization of the antisera as

Table 1

Arginine Vasopressin (AVP) and Arginine Vasotocin (AVT) in Posterior Pituitaries and Pineal Extracts from Different Species

Species	Tissue	I:AVP *	II:AVP-AVT †	Ratio II/I(AVT)
Toad	Post Pit. (7)‡	0.25 ± 0.07	289 ± 113	1,156
Chicken	Post. Pit. (10)	0.11 ± 0.01	2861 ± 617	26,000
Chicken	Pineal (10)	0.04 ± 0.01	0.180 ± 0.04	4.3
Rat (Adult)	Pineal (5)	0.186 ± 0.01	0.167 ± 0.05	0.9
Rat (Infantile)	Pineal (6)	0.420 ± 0.19	0.480 ± 0.08	1.1
Rabbit (Adult)	Pineal (4)	0.03 ± 0.008	0.03 ± 0.005	1.0
Hamster	Pineal (10)	0.03 ± 0.006	0.03 ± 0.008	1.0
Rat	Pineal (6) + Synthetic AVT	0.650 ± 0.18	29.6 ± 2.7	45.5

* AVP measured with antiserum provided by Moore *et al.*[15]

† AVP-AVT measured with F.S.F. antiserum.

‡ Numbers in parentheses indicate number of samples assayed. Means (in pg. peptide/μg protein) ± SEM.

well as the radioimmunoassay procedures has been published.[16] The amount of AVT in any given tissue was calculated by measuring samples of the tissue with both antisera and then determining the ratio between the value obtained with the AVP-AVT/AVP antisera. Any value significantly higher than one should indicate the presence of AVT in the tissue. Indeed, such is the case as described in Table 1. Posterior pituitaries or pineals obtained from chickens (*Gallus domesticus*) and toads (*Bufo marinus*) contained very large amounts of AVT and negligible levels of AVP in those tissues, yielding ratios much larger than one in every case. On the other hand, pineals obtained from adult rats, rabbits, and hamsters showed low amounts of AVP and ratios of one, indicating that no AVT is present in those glands. Pineals obtained from infantile rats also had detectable amounts of AVP and no AVT. When synthetic AVT was added to rat pineal extracts, the expected increase in ratio (meaning AVT was present) was observed, indicating that the lack of AVT in the tissue extracts was not due to loss after extraction and handling of the tissue. Recovery of the exogenous AVT was >90%.

Other reports appeared almost simultaneously which confirmed our study.[17] Utilizing specific RIA systems for AVT, AVP, and OXY, the authors concluded that no AVT was detectable in rat or bovine pineals, or in rat subcommissural (SCO) organ. AVP, however, was detectable in both pineal and SCO regions.

Shortly after these studies were published, Fernstrom *et al.*[18] reported the presence of radioimmunoassayable AVT in the pineals of several rodents (rat, mouse, hamster, guinea pig, gerbil) as well as cows. The latter glands also contained similar or larger amounts of OXY and AVP, whereas in the rat, no AVP was detected in the pineal, a result at variance with previous reports.[11, 16, 17] The amount of AVT detected in rat pineals was very small (<10 pg/gland) in sharp contrast with the values of 150 pg/gland reported by Rosenbloom *et al.*[13] Therefore, the results raised again the question of whether the displacement measured in the RIA was indeed due to authentic AVT or perhaps to another peptide.

Further Studies Combining Chromatographic and Radioimmunoassay Methods

In view of the above report, we decided to initiate studies in which a large number of pineals were pooled, homogenized, and chromatographed, followed by RIA analysis of the collected fractions to determine the presence of either AVT or AVP. In order to characterize the system, synthetic AVT or AVP (200 ng of either peptide) were run through a Sephadex G-25 (Pharmacia) column (0.9 × 96 cm) and the effluent collected at a rate of 45 drops/tube (≃2.5 ml). Acetic acid (0.2 M) was used to homogenize the tissue and also to equilibrate and wash the column. The effluents collected after running either AVT or AVP through the column are depicted in Figure 1. It can be seen that AVT (left panel) was only recognized by the AVP-AVT antiserum, whereas AVP (right panel) was recognized by both antisera on an equimolar basis. Next, both peptides were dissolved together and run simultaneously through the column. It can be seen (Figure 2) that the AVP-AVT-AS detected the AVT peak earlier than the AVP peak (left panel) at the regular flow rate of 45 drops/tube. If the size of the fractions collected was reduced to 20 drops/tube, a much better separation of the two peaks was observed (right panel) indicating that AVT indeed eluted earlier than AVP under these conditions.

Finally, the homogenates of either 70 pineals from 15-day old male rats or 40 pineals from adult male rats were applied to the column. The results (Figure 3) show that the profile obtained with both antisera is perfectly superimposable, and presented no evidence for the presence of an AVT peak, as would be expected had AVT been present in the pineal extracts.

In further confirmation of the previous findings, a recent study from Fernstrom's laboratory[19] reported the purification of bovine pineal extracts by high performance liquid chromatography followed by RIA for AVP and AVT. Their results, in contrast with their previous findings,[18] indicate that less than 3 pg AVT/gland were found in the cow pineal (vs. 200–400 pg in the early study). In contrast, AVP was present in significant amounts (300 pg/gland). These observations, coupled with those presented in this report, seem to rule out the presence of any significant amounts of AVT in the pineal of different mammalian species.

Search for AVT in Brain Areas Employing the Brattleboro Rat

Having been unable to determine the presence of AVT in mammalian pineals, we then explored the possibility that the nonapeptide could be present in other brain regions, using for that purpose Sprague-Dawley and homozygous Brattleboro rats. Measurements with both antisera were performed on tissue

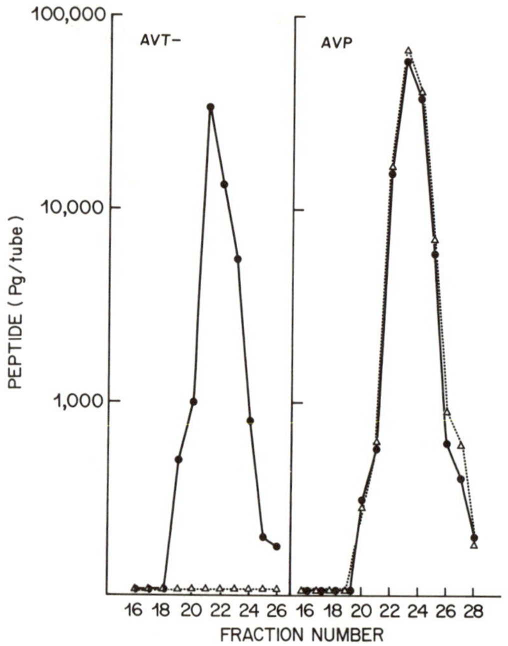

Figure 1. Chromatographic profiles of synthetic arginine vasotocin (AVT, left panel) arginine vasopressin (AVP, right panel) run through a Sephadex column (Pharmacia, G–25). Column size: 0.9 × 96 cm. Acetic acid (0.2 M) was used to equilibrate and wash the column. Effluent was collected in 2.5 ml aliquots and assayed for AVP (△ - - - △), AVP-AVT (● - - - ●). For details of antisera employed see text.

punches of several circumventricular organs, as well as from the neural lobe. AVP values were highest, as expected, in neural lobe followed by median eminence, in normal Sprague-Dawley rats (Table 2). Other circumventricular organs had low but readily detectable levels of AVP. No AVT was detected, as determined by the ratio of the two antisera. Neither antisera detected either AVP or AVT when tissues obtained from Brattleboro rats were analyzed. Similarly negative results were seen when other nuclei (such as supraoptic, paraventricular, arcuate) were measured in Brattleboro rats. These observations

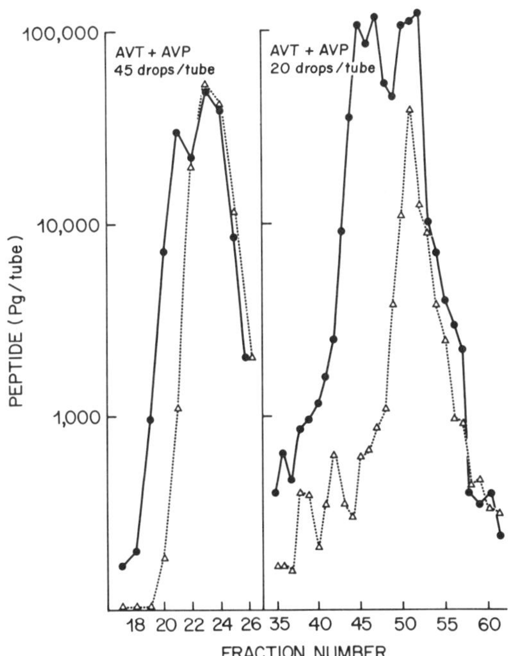

FIGURE 2. Chromatographic profiles of synthetic AVT and AVP run simultaneously in a Sephadex column. Fractions were collected at a rate of 45 drops/tube (left panel) or 20 drops/tube (right panel). (△ --- △), AVP antiserum (● --- ●) AVP-AVT antiserum.

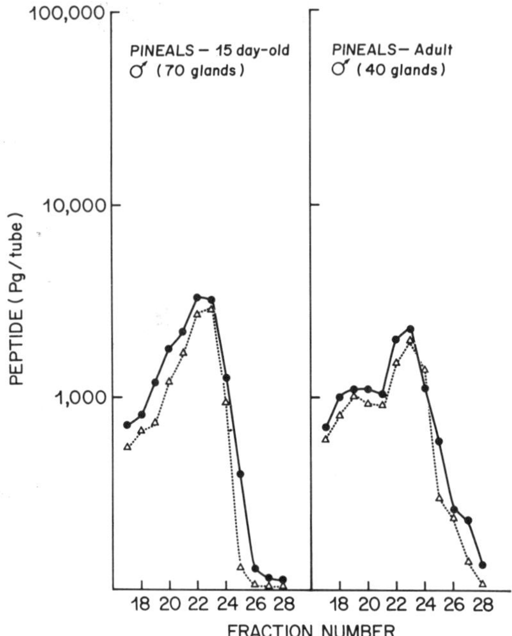

FIGURE 3. Chromatographic profile of pineal extracts from 15-day old (left panel) or adult rats (right panel) run though a Sephadex column. Fractions were assayed for AVP (△ – – – △) or AVP-AVT (● – – – ●).

TABLE 2

LEVELS OF AVP AND AVT IN CIRCUMVENTRICULAR ORGANS, POSTERIOR LOBES, AND PINEALS OF SPRAGUE-DAWLEY AND HOMOZYGOUS BRATTLEBORO RATS

Region	Sprague-Dawley		Brattleboro *	
	AVP †	AVT ‡	AVP	AVT
Median Eminence	807.3 ± 63.5	ND	ND	ND
Organum Vasculosum (OVLT)	1.0 ± 0.13	ND	ND	ND
Subfornical Organ	1.2 ± 0.12	ND	ND	ND
Subcommissural Organ	–	–	ND	ND
Pineal	0.16 ± 0.04	ND	ND	ND
Neural Lobe	$15 \times 10^3 \pm 0.9$	ND	ND	ND

* Homozygous rats. $N = 5$.

† Values are expressed in pg AVP/μg protein. Means ± SEM, $N = 10$.

‡ Measured as the ratio between the AVP-AVT/AVP antisera. Any value not significantly different from 1 was considered as AVT = ND (not detectable, or less than 2 pg/area or gland).

confirm and extend those of Dogterom *et al.*,[17] which reported lack of AVT in both pineal and SCO from Brattleboro rats.

From the data presented in this study as well as from the recent reports in the literature,[16, 17, 19] it can be concluded that AVT is not present in either the pineal or other brain structures. Therefore, although pharmacological effects on hormone secretion may be obtained after injecting AVT by different routes, claims concerning the role of this peptide as a pineal hormone involved in physiological events should acknowledge the lack of evidence for the presence of AVT in mammalian pineals.

References

1. Defendinini, R. & E. A. Zimmerman. 1978. The magnocellular neurosecretory system of the mammalian hypothalamus. *In* The Hypothalamus. S. Reichlin, R. J. Baldessarini & J. B. Martin, Eds.: 137–152. Raven Press. New York, N.Y.
2. Negro-Vilar, A. & J. M. Saavedra. 1980. Changes in brain somatostatin and vasopressin levels after stress in spontaneously hypertensive and Wistar-Kyoto rats. Brain Res. Bull. **5:** 353–358.
3. De Wied, D. & D. H. Versteeg. 1979. Neurohypophyseal principles and memory. Fed. Proc. **38:** 2348–2354.
4. Pavel, S., L. Matrescu & M. Petrescu. 1973. Central corticotropin inhibition by arginine vasotocin in the mouse. Neuroendocrinology **12:** 371–375.
5. Vaughn, M. K. & D. E. Blask. 1978. Arginine vasotocin: A search for its function in mammals. Prog. Reprod. Biol. **4:** 90–115.
6. Pavel, S. 1970. Tentative identification of arginine vasotocin in human cerebrospinal fluid. J. Clin. Endocr. **31:** 369–371.
7. Pavel, S., I. Dimitru, I. Klepsh & M. Dorcescu. 1973/74. A gonadotropin inhibiting principle in the pineal of human fetuses. Evidence for its identity with arginine vasotocin. Neuroendocrinology **13:** 41–46.
8. Vizsolyi, E. & A. M. Perks. 1969. New neurohypophyseal principle in fetal mammals. Nature **223:** 1169–1171.
9. Cheesman, D. W. & B. L. Fariss. 1970. Isolation and characterization of a gonadotrophin-inhibiting substance from the bovine pineal gland. Proc. Soc. Exp. Biol. Med. **133:** 1254–1256.
10. Cheesman, D. W. 1970. Structure elucidation of a gonadotropin inhibiting substance from the bovine pineal gland. Biochim. Biophys. Acta **207:** 247–253.
11. Rosemblum, I. Y., B. Benson & V. J. Hruby. 1976. Chemical differences between bovine pineal antigonadotropin and arginine vasotocin. Life Sci. **18:** 1367–1374.
12. Rosenbloom, A. A. & D. A. Fisher. 1974. Radioimmunoassay of arginine vasotocin. Endocrinology **95:** 1726–1732.
13. Rosenbloom, A. A. & D. A. Fisher. 1975. Radioimmunoassayable AVT and AVP in adult mammalian brain tissue: comparison of normal and Brattleboro rats. Neuroendocrinology **17:** 354–361.
14. Dogterom, J., F. G. M. Snijdewint & R. M. Buijs. 1978. The distribution of vasopressin and oxytocin in the rat brain. Neurosci. Lett. **9:** 341–346.
15. Moore, G., A. Lutterodt, G. Burford & K. Lederis. 1977. A highly specific antiserum for arginine vasopressin. Endocrinology **101:** 1421–1435.
16. Negro-Vilar, A., F. Sanchez-Franco, M. Kwiatkowski & W. K. Samson. 1979. Failure to detect radioimmunoassayable arginine vasotocin in mammalian pineals. Brain Res. Bull. **4:** 789–792.

17. DOGTEROM, J., F. G. M. SNIJDEWINT, P. PEVET & R. M. BUIJS. 1979. On the presence of neuropeptides in the mammalian pineal gland and subcommisural organ. *In* The Pineal Gland of Vertebrates Including Man. J. Ariens-Kappers & P. Pevet, Eds.: 465–470. Elsevier-North Holland. Amsterdam.
18. FERNSTROM, J. D., L. A. FISHER, B. M. CUSACK & M. A. GILLIS. 1980. Radioimmunologic detection and measurement of nonapeptides in the pineal gland. Endocrinology **106:** 243–251.
19. FISHER, L. A., E. R. SPINDEL & J. D. FERNSTROM. 1980. Nonapeptide content of the bovine pineal gland. Program, 62nd Endocrine Soc. Mtg. pp. 103. Washington, D.C.

OXYTOCIN AND OXYTOCIN-ASSOCIATED NEUROPHYSIN EVALUATION BY RIA IN THE BRATTLEBORO RAT: TURNOVER *

William G. North, M. Gellai, and G. Hardy

Department of Physiology
Dartmouth Medical School
Hanover, New Hampshire 03755

INTRODUCTION

Neurophysins are most probably part of the prohormone structures for oxytocin (OT) and vasopressin (VP).[1] In the Brattleboro homozygote (DI) which seems unable to produce VP, the absence of vasopressin-associated neurophysin (VP-RNP) is also notable.[2, 3] A partial deficiency in the synthesis of both substances is seen in rats heterozygous for the condition of hypothalamic diabetes insipidus. This illustrates the close association that exists in magnocellular neurons between the production of neurophysins and the production of hormones, and this association seems to extend to the release of the substances by exocytosis of neurosecretory granules.[4, 5]

An evaluation of the rates of synthesis and release of VP-RNP or oxytocin-associated neurophysin (OT-RNP) should therefore reflect these parameters for the corresponding hormone provided there are not significant differences between them in rate of breakdown during storage or in their possible recapture by neurons. Indeed, if breakdown and recapture by the neurons are not significant contributing factors in turnover, then during steady-state conditions the rate of release of each neurohypophyseal principle should equal its rate of synthesis by the hypothalamo-neurohypophyseal system (HNS). Moreover, during the non-steady-state, the rate of synthesis should equal the rate of release minus the rate of decline in storage levels in the neural lobe. On the further premise that neurohypophyseal principles, once released, are distributed into a single compartment, the rate of release is given by:

$$R = \frac{V_D \cdot \ln\ 2 \cdot [\mathrm{P}]}{T_{1/2}}$$

where R is the rate of release, V_D the volume of distribution, [P] the plasma concentration, and $T_{1/2}$ the half-life of the substance in the plasma.[6]

In this study we measured $T_{1/2}$, V_D, and [P] for immunoactive oxytocin-associated rat neurophysin (OT-RNP) in Long-Evans (LE) rats and in Brattleboro homozygotes. RIAs for OT-RNP and OT were used to measure storage levels of these materials in the neural lobes of LE rats and Brattleboro homozygotes. These data were then coupled to obtain estimates of the turnover of OT-RNP (and OT) by the rat HNS.

* Supported in part by U.S. Public Health Service Grants CA–19613, CA–00552, and AM–08469; and a grant from the New Hampshire Heart Association.

0077–8923/82/0394–0167 $1.75/0 © 1982, NYAS

Methods

LE rats, 3 to 6 months old and weighing from 198 to 300 g, and DI rats, 3 to 12 months old and weighing from 190 to 300 g, were used in this study. LE rats were purchased from Canadian Breeding Laboratories and DI rats were raised at Dartmouth. All rats were placed on food ad libitum. LE rats were given free access to drinking water. Some DI rats were given free access to water, while others had their drinking water withdrawn 24 hours before they were killed. Animals were killed by decapitation and plasma obtained from blood collected from the trunk into chilled tubes. The neural lobe of each animal was extracted for hormones and neurophysins in 2×100 μl of 0.1 M HCl at 0° C.[7]

Radioimmunoassays for OT-RNP and OT in crude extracts of neural lobe, and RIA for OT-RNP in unextracted plasma were performed by previously described methods.[7,8] Radioiodination of OT-RNP and OT employed lactoperoxidase and tracers were isolated by Sephadex chromatography.[7,8]

The conscious, trained rat model[9] was used to determine the half-life and volume of distribution of immunoreactive OT-RNP. A shunt was created between the arterial and venous catheters for the injection of the radioactive material. LE rats received an intravenous bolus (100 μl) of monoiodinated [^{125}I]OT-RNP (0.5 pmole, specific activity 2,500 to 3,000 μCi/nmole) which was flushed into the cannula with 100 μl of saline. Serial arterial blood samples (75 μl) were then collected into heparinized tubes each minute for the first 10 minutes, then each 10 minutes through 60 minutes following the addition of radiotracer. Urine was collected under oil as 5-minute samples for the first hour, and then as 10-minute samples thereafter up to 2 hours post-injection. Aliquots of plasma and urine samples were incubated at 4° C, pH 7.5 for 48 hours in the presence of an excess of antibody to OT-RNP; antibody-bound material was then precipitated with polyethylene glycol, pelleted at $10{,}000 \times g$ and counted.

Results and Discussion

The mean values (±SEM) for OT-RNP and OT found in neural lobes are given in Figure 1. The data show that there are approximately molar equivalents of OT and OT-RNP stored. This is compatible with the view that OT and OT-RNP are generated from a common precursor.[1] The values shown for LE rats are for animals 15 weeks old and the ratio OT-RNP:OT was 1.19 (male) and 1.38 (female). This ratio was 1.06 (male) and 1.37 (female) in the DI rats allowed access to water, and 1.44 (male) and 1.37 (female) in dehydrated animals. The data also indicate that dehydration for 24 hours in DI rats causes storage levels of both substances to fall about 35%. Of particular interest was the much higher storage levels found in female rats, especially in DI rats where there are 2 to 2.5 times those levels in male animals.

Table 1 shows the plasma levels of OT-RNP found by RIA. For LE rats, females had about 1.5 times the plasma levels of males, and the corresponding levels for DI rats given water were 2 times and 1.5 times these values. Plasma OT-RNP rose 2.7 times (males) and 1.9 times (females) as a consequence of dehydration.

The disappearance of immunoreactive radioactivity from the plasma of

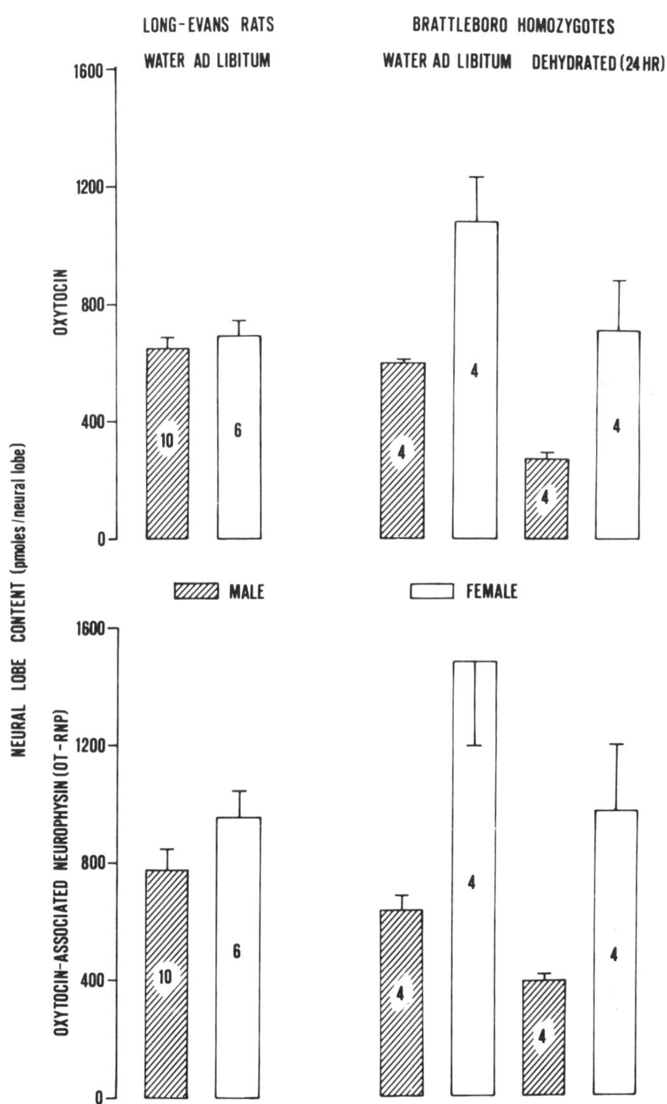

FIGURE 1. Values (means and SEM) for neural lobe content of LE rats and DI rats determined by RIAs for OT and OT-RNP. The values in the histograms represent the number of animals used in each study.

TABLE 1

PLASMA LEVELS OF OT-RNP IN LONG-EVANS AND BRATTLEBORO RATS

	Concentration (fmoles/ml)	
	Water ad libitum	Dehydrated (24 h)
Long-Evans		
males	31 ± 4 ($N = 5$)	
females	46 ± 11 ($N = 6$)	
Brattleboro homozygotes		
males	44 ± 4 ($N = 10$)	120 ± 9 ($N = 9$)
females	96 ± 12 ($N = 8$)	182 ± 26 ($N = 7$)

conscious LE rats showed on analysis a fast and slow component when the natural logarithm of counts/ml plasma was plotted against time. Extrapolation of the slow component to zero time gave a $T_{½}$ of 22.3 minutes, and correction of the fast component for contribution of the slow [10] gave a $T_{½}$ of 1.61 minutes (TABLE 2).

Almost no immunoreactive [^{125}I]OT-RNP was found in the urine, and in fact, radioactivity that did not bind to antibody appeared maximally at times >60 minutes post-injection. Analysis of organs for radioactivity in some animals killed 20 minutes post-injection showed that >85% was concentrated in the kidneys as shown earlier for porcine neurophysins by Johnston and coworkers.[11] Knowing this, one interpretation of the hyperbolic disappearance curves for immunoreactive [^{125}I]OT-RNP is that there is a single V_D and a saturable binding site(s) for uptake and removal of OT-RNP, with an insignificant contribution from net filtration: The fast component of the disappearance curve could represent bound material, the slow component could represent material not bound initially. On this basis the V_D was obtained by summing the intercepts at $t = 0$ of the fast and slow components and dividing the sum into the total counts injected. When V_D for each animal tested was expressed as a percentage of body weight, the value given in TABLE 2 was obtained.

From the above reasoning, it was concluded that the true $T_{½}$ should be that obtained by summing the fractional contributions at $t = 0$ of the fast and

TABLE 2

THE HALF-LIVES AND VOLUME OF DISTRIBUTION OF ^{125}I-LABELED OT-RNP IN THE RAT ($N = 4$)

	Mean ± SEM
Half-life $T_{½}$ (min)	1.61 ± 0.06
	22.30 ± 0.22
Volume of Distribution (% body weight)	7.20 ± 0.39

slow components of each disappearance curve. At $t = 0$, the fast component was 88.1 ± 1.3% of the total counts. Then,

$$T_{1/2\ \mathrm{actual}} = \frac{88.1 \times 1.61}{100} + \frac{11.9 \times 22.3}{100} = 4.07 \text{ minutes.}$$

By use of $T_{1/2\ \mathrm{actual}}$ and the V_D from LE rats, the rates of synthesis and release of OT-RNP were calculated for LE rats and DI rats and these are given in TABLE 3.

Provided the $T_{1/2\ \mathrm{actual}}$ and V_D of OT-RNP are the same in LE and DI rats for the given conditions, then the following conclusions can be drawn: (1) The rates of synthesis and release of OT-RNP in DI rats are normally 1.5 to 2 times that in LE rats, reflecting hypertrophy of magnocellular neurons. (2) Female DI rats have twice the rates of synthesis and release of male DI rats and this finding was similar to that in LE rats. (3) Dehydrated DI rats released 2–3 times as much OT-RNP as DI rats given free access to water. (4) Dehydration causes a smaller increase in rate of synthesis of OT-RNP in female DI rats (1.6 times) than in male DI rats (2.5 times). (5) Because molar storage levels of oxytocin and OT-RNP are similar, then synthesis and release rates of OT-RNP probably reflect synthesis and release rates of oxytocin.

TABLE 3

RATES OF RELEASE AND SYNTHESIS OF OT-RNP IN LONG-EVANS RATS AND BRATTLEBORO RATS

	Release (pmol/day/100 g)		Synthesis (pmol/day/100 g)	
	Water ad libitum	Dehydrated	Water ad libitum	Dehydrated
Long-Evans				
males	55		55	
females	81		81	
Brattleboro homozygotes				
males	78	212	78	190
females	170	321	170	279

REFERENCES

1. PICKERING, B. T. & W. G. NORTH. 1982. Biochemical and functional aspects of magnocellular neurons and hypothalamic diabetes insipidus. Ann. N.Y. Acad. Sci. (This volume.)
2. SUNDE, D. A. & H. W. SOKOL. 1975. Quantification of rat neurophysins by polyacrylamide gel electrophoresis (PAGE): Application to the rat with hereditary hypothalamic diabetes insipidus. Ann. N.Y. Acad. Sci. **248:** 345–364.
3. VALTIN, H., W. H. SAWYER & H. W. SOKOL. 1965. Neurohypophysial principles in rats homozygous and heterozygous for hypothalamic diabetes insipidus (Brattleboro strain). Endocrinology **77:** 701–706.
4. MORRIS, J. F. 1982. The Brattleboro neural lobe: a model for the study of subcellular aspects of section from nerve terminals. Ann. N.Y. Acad. Sci. (This volume.)

5. Dreifuss, J. J. 1975. A review of neurosecretory granules: Their contents and mechanisms of release. Ann. N.Y. Acad. Sci. **248:** 184–201.
6. Valtin, H., J. Stewart & H. W. Sokol. 1974. Genetic control of the production of posterior pituitary principles. Handb. Physiol. Sec. 7, Vol. IV, Part 1: 131–171.
7. North, W. G., F. T. Larochelle, Jr., J. Haldar, W. H. Sawyer & H. Valtin. 1978. Characterization of an antiserum used in a radioimmunoassay for arginine-vasopressin: Implications for reference standards. Endocrinology **103:** 1976–1984.
8. North, W. G., F. T. Larochelle, Jr., J. Melton & R. C. Mills. 1980. Isolation and partial characterization of two human neurophysins: Their use in the development of specific radioimmunoassays. J. Clin. Endocrinol. **51:** 884–891.
9. Gellai, M. & H. Valtin. 1979. Chronic vascular constrictions and measurements of renal function in conscious rats. Kidney Int. **15:** 419–429.
10. Lauson, H. D. 1974. Metabolism of the neurohypophysial hormones. Handb. Physiol. Sec. 7, Vol. IV, Part 1: 287–393.
11. Johnston, C. I., J. S. Hutchinson, B. J. Morris & E. M. Dax. 1975. Release and clearance of neurophysins and posterior pituitary hormones. Ann. N.Y. Acad. Sci. **248:** 272–280.

THE BRATTLEBORO HETEROZYGOTE RAT AS A MODEL FOR NEUROHYPOPHYSEAL AGING: VASOPRESSIN RESPONSE TO DEHYDRATION *

Celia D. Sladek † and Thomas H. McNeill

Departments of Neurology and Anatomy
University of Rochester School of Medicine and Dentistry
Rochester, New York 14642

Chronic water deprivation in aged rats does not result in an elevation of the serum vasopressin (VP) concentration comparable to that observed in young rats.[1] Since the VP stores in the neural lobe relative to either body weight or weight of the neural lobe are significantly reduced in aged rats when compared to young rats,[1] inadequate VP stores may contribute to or be responsible for the failure of old rats to increase serum VP levels following dehydration. This was evaluated by comparing the response of Brattleboro rats, which are heterozygous for the genetic deficit causing diabetes insipidus and have reduced neurohypophyseal VP stores,[2] to the response of aged rats to chronic water deprivation.

Male Fischer 344 rats of 3 or 30 months of age were obtained through the National Institute on Aging (Bethesda, Md.). Male Brattleboro heterozygote (HZ) rats were obtained from the breeding colony at the University of Rochester. This colony is partially derived from Long-Evans (LE) rats, and therefore LE rats (Charles River) were used as controls for the HZ rats. Animals were fed ad libitum, but they were deprived of water for 0, 24, 48, 72, or 96 hours prior to sacrifice. They were sacrificed by decapitation in random order, and trunk blood was collected for determination of serum VP concentration, hematocrit, and plasma osmolality (micro-vapor pressure, Wedco). The neural lobe was weighed, homogenized, and analyzed for VP content. Kidney function was assessed in the aged rats and determined to be adequate.[1] VP was evaluated by radioimmunoassay.[1, 3]

As shown in FIGURE 1, water deprivation resulted in a comparable degree of dehydration in all four groups of rats as judged by the decrease in body weight. Following 72 hours of water deprivation body weight of the aged Fischer and HZ rats was reduced to 85% of their hydrated weight. Plasma osmolality and hematocrit were significantly elevated ($p < .025$ and $p < .001$ respectively) in LE and HZ rats following 96 h of water deprivation (TABLE 1). There was more variability in these parameters in the Fischer 344 rats, but plasma osmolality was significantly increased in the 30 month rats following 72 h of dehydration ($p < .01$) and hematocrit was significantly elevated in the comparably dehydrated young rats ($p < .02$).

Serum VP was significantly elevated following 72 hours of fluid deprivation in the LE, HZ, and 3 month old Fischer rats ($p < .01$, FIGURE 2). However, the serum VP concentration achieved by the HZ rats was consistently lower

* Supported by the National Institute on Aging (grants AG–01456 and AG–05175).

† Recipient of Research Career Development Award NS–00259.

0077–8923/82/0394–0173 $1.75/0 © 1982, NYAS

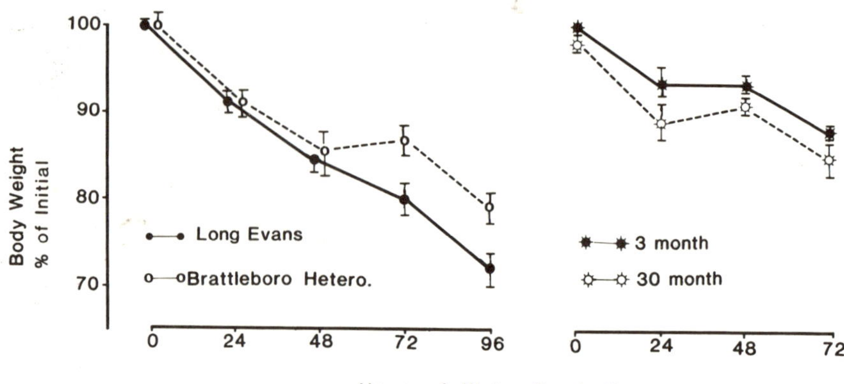

FIGURE 1. The effect of water deprivation on body weight. Mean ± SEM. $N = 10$ per group of LE and HZ rats. $N = 5$ per group of 3 and 30 month old Fischer 344 rats.

than the LE rats and this difference became statistically significant following 96 hours of dehydration ($p < .05$). The aged rats did not exhibit significant elevations in serum VP at any stage of dehydration (FIGURE 2).

As expected, the VP stores in the neural lobe relative to body or posterior pituitary weight in the hydrated HZ rats were only 50% of hydrated LE rat stores (HZ rats—0.55 ± .05 μg VP/mg neural lobe; LE rats—1.03 ± .16 μg/mg; $p < .025$). This difference was comparable to the reduction in neural lobe VP concentration observed in aged rats (30 months—0.88 ± .09 μg/mg; 3 months—1.70 ± 0.12 μg/mg; $p < .025$). During dehydration, the VP concentration in the neural lobe decreased progressively in all four groups of rats. By 72 hours, the neural lobe VP stores of the aged and HZ rats were comparably reduced to 39 and 44%, respectively, of their hydrated controls (FIGURE 3). This reduction was significantly greater than observed in the young Fischer 344 rats ($p < .05$). However, as demonstrated in FIGURE 3 the most rapid rate of depletion of pituitary VP concentration occurred in the

TABLE 1

EFFECT OF DEHYDRATION ON PLASMA OSMOLALITY AND HEMATOCRIT

	Long-Evans	Brattleboro Heterozygote	Fischer 344	
			3 mos.	30 mos.
Plasma Osmolality ($mOsmol/kg\ H_2O$)				
hydrated rats	296 ± 3	299 ± 2	289 ± 2	287 ± 1
dehydrated—72 h	303 ± 3	310 ± 1	310 ± 2	315 ± 10
96 h	310 ± 3	310 ± 5		
Hematocrit (%)				
hydrated rats	41 ± 1	45 ± 2	43 ± 2	42 ± 4
dehydrated rats—72 h	48 ± 2	51 ± 1	52 ± 1	46 ± 3
96 h	51 ± 1	53 ± 1		

LE rats. Thus, after 24 h of dehydration there is no longer significantly more VP in the neural lobe of LE rats compared to HZ rats. This phenomenon has been reported previously by Moses and Miller [4] and may indicate differences in the proportion of hormone stored in a readily releasable form.

The attenuated serum VP response to prolonged dehydration in HZ rats probably reflects the reduced quantity of VP stored by these rats. Since neural lobe VP stores also are reduced in aged rats and decline precipitously during chronic dehydration, the failure of the aged rats to achieve elevated serum VP concentrations may partially reflect this abnormality. However, it appears that additional abnormalities in the mechanisms controlling VP release may occur in aged rats, because the VP response to dehydration was quantitatively different in the aged Fischer rats compared to the HZ rats: The response was attenuated in the HZ rats, but was apparently absent in the old rats. This suggests that, in the aged rats, abnormalities may exist in the mechanisms that normally stimulate VP release. Attenuation of VP release in response to baroreceptor stimulation in some aged humans has been reported previously.[5]

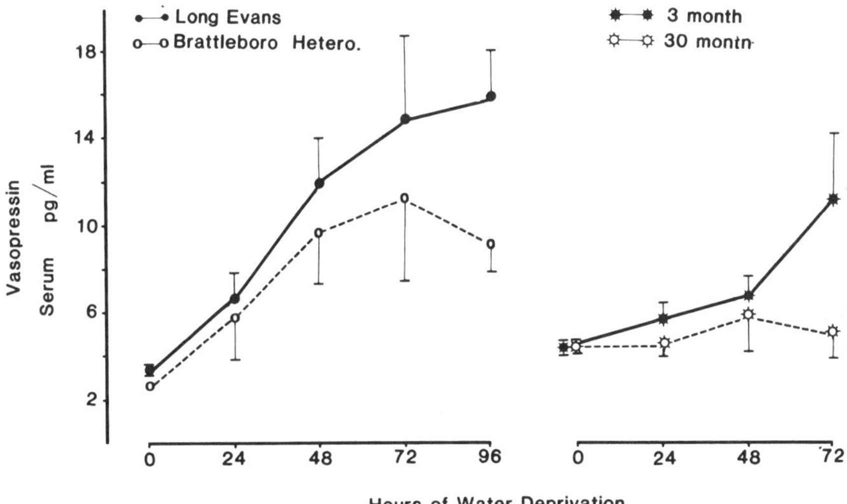

FIGURE 2. The effect of chronic water deprivation on serum VP concentration. Mean ± SEM. *N* same as FIGURE 1.

ACKNOWLEDGMENTS

The authors appreciate the skilled technical assistance of Margaret Mudd, Mark Gallagher and Sue Walker. Also, we thank Dr. Don Gash and Ms. Leslie Dick for providing the HZ rats.

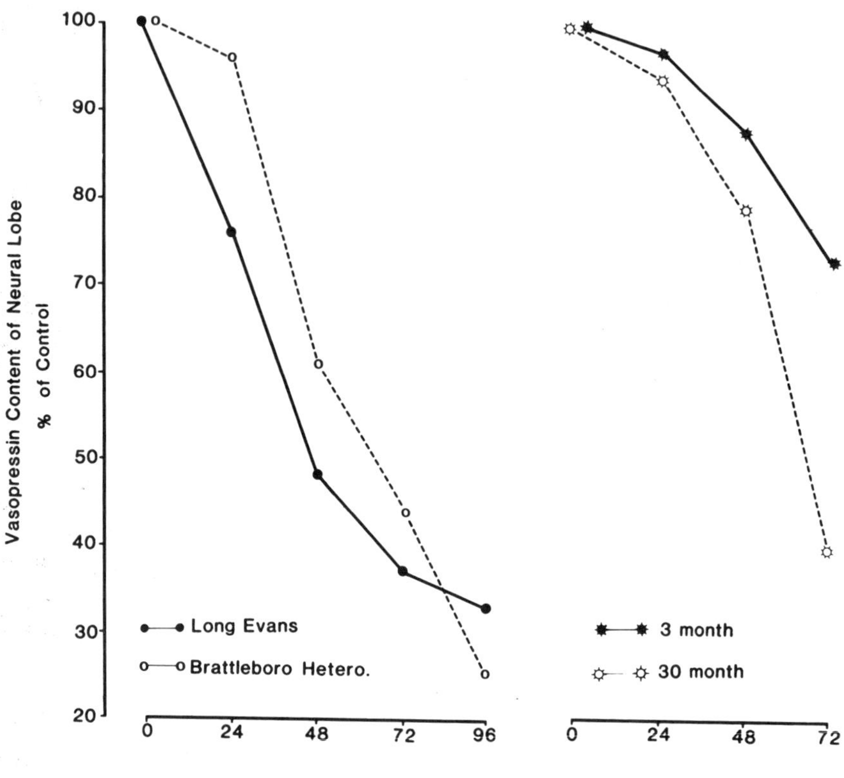

FIGURE 3. Depletion of neural lobe VP content expressed as a percentage of the content in hydrated rats. *N* same as FIGURE 1.

REFERENCES

1. SLADEK, C. D., T. H. MCNEILL, C. M. GREGG, M. L. BLAIR & R. B. BAGGS. 1981. Vasopressin and renin response to dehydration in aged rats. Neurobiol. Aging **2:** 293–302.
2. VALTIN, H., W. H. SAWYER & H. W. SOKOL. 1965. Neurohypophyseal principles in rats homozygous and heterozygous for hypothalamic diabetes insipidus (Brattleboro strain). Endocrinol. **77:** 701–706.
3. SLADEK, C. D. & K. M. KNIGGE. 1977. Cholinergic stimulation of vasopressin release from the rat hypothalamo-neurophypophyseal system in organ culture. Endocrinol. **101:** 411–420.
4. MOSES, A. M. & M. MILLER. 1970. Accumulation and release of pituitary vasopressin in rats heterozygous for hypothalamic diabetes insipidus. Endocrinol. **86:** 34–41.
5. ROBERTSON, G. L. & J. ROWE. 1980. The effect of aging on neurohypophyseal function. Peptides **1** (Suppl 1): 158–162.

THE ONTOGENY OF THE OXYTOCIN/NEUROPHYSIN SYSTEM OF THE BRATTLEBORO RAT

K. L. Wong, A. Fisher, K. Lederis, and W. North *

Departments of Anatomy and Pharmacology
University of Calgary
Calgary, Alberta, Canada T2N 4N1

** Department of Physiology*
Dartmouth Medical College
Hanover, New Hampshire

Introduction

The ontogeny of neuropeptides has interested several laboratories.[1-4] The differentiation of magnocellular neurons in the brains of several strains of rats has been extensively described,[5] such that the process of maturation is thought to commence in the supraoptic area with subsequent migration to other dorsal and lateral locations.[6] However, reports concerning the early development of magnocellular neurons in the diabetes insipidus (Brattleboro) rats are fragmentary and are generally concerned with postnatal development.[2]

The availability of specific and highly purified hormonal peptide and neurophysin antisera has enabled the investigation of the ontogeny of specific neurons in the hypothalamo-neurohypophyseal system. In the present study, immunocytochemical methods and radioimmunoassay were used to examine the ontogeny of oxytocin neurons in the Brattleboro rats and control animals.

Materials and Methods

Male and female adult homozygous diabetes insipidus rats (HOM-DI) were mated. The first appearance of a vaginal plug was designated as day 1 of pregnancy. Fetuses were collected from gestational day 13 through day of delivery (day 22). At the same time fetuses of the Long-Evans strain were collected for comparison. Pregnant females were decapitated and the uterus was exposed. The heads of fetuses were removed and placed either in Bouins fixative for immunocytochemical study or dried in acetone for subsequent radioimmunoassays. Fetal heads fixed in Bouins solution were washed, dehydrated, cleared, and embedded in paraffin. Seven-micron thick serial sections were cut. The dewaxed sections were treated with various antisera.[4] Antisera used in the present studies were: oxytocin neurophysin, oxytocin, vasopressin neurophysin, vasopressin, and vasotocin. Purified oxytocin neurophysin antiserum was obtained by affinity chromatography; the antiserum was first run against vasopressin neurophysin; the eluate was collected and run against oxytocin neurophysin; the column was washed and acidified; and the purified oxytocin neurophysin collected. Concentration of hormones in hypothalamus and pituitary gland was measured by radioimmunoassay as previously described.[7]

0077-8923/82/0394-0177 $1.75/0

Results

16-Day Fetus

Oxytocin neurophysin-containing neurons were visualized both in paraventricular (PVN) (Figure 1a) and supraoptic (SON) (Figure 1b) regions of the HOM-DI fetus. These immunoactive perikarya appeared more abundant in the PVN than in the SON area. By contrast, in the 16-day Long-Evans fetus more immunoactive neurons in the SON than in the PVN region were seen.

17–19 Days

Little change was evident at 17 days, as most of the immunoreactive neurons were still located in the PVN of the HOM-DI fetuses. However, at 18 and 19 days of gestation, increasing numbers of oxytocin neurophysin-positive neurons were found in the SON. At the same time, neurons in the PVN showed axonal processes. In the 18–19-day animals, the intensity of immuno-staining was more pronounced.

20 Days—Newborn

At 20 days the neuronal groupings were seen in their definitive adult position (SON, Figure 2a and PVN, Figure 2b). Retrochiasmatic neurons were consistently seen at 20 days and occasionally visualized in the 19-day fetus. Similarly, forniceal neurons could be clearly identified at 20 days of gestation. Processes of neurons were clearly seen at 20 days, particularly in PVN (Figure 2b). However, these neurons did not reach the fully mature stage at this point. In the 20-day Long-Evans control animals, the immunoreactive neurons resembled the adult locations.

Oxytocin antiserum failed to demonstrate any immunoreactive neurons at all gestational ages examined. Similarly, purified antisera for vasopressin and vasotocin failed to demonstrate immunoreactivity.

The measurement of oxytocin concentration in the hypothalamus and pituitary glands by radioimmunoassay (RIA) is shown in Table 1. From day 16 of gestation to birth (22 days), oxytocin in the pituitary showed no significant change, however, hypothalamic oxytocin levels appeared to increase at day 20 of gestation.

Discussion

Immunoreactive neurons for purified oxytocin neurophysin antiserum have been identified in PVN and SON regions at 16 days of gestation in HOM-DI rats. These findings differ from previous observations on several strains of rats in our and other laboratories.[4, 8, 9] These studies suggested that vasopressin neurophysin is identified earlier in the SON (16 days) than the PVN (18 days), and that oxytocin neurophysin is visualized later (18 days) in both SON and PVN.[4] The early appearance of oxytocin neurophysin neurons in the PVN of the HOM-DI rats may indicate an apparent attempt on the part of the hor-

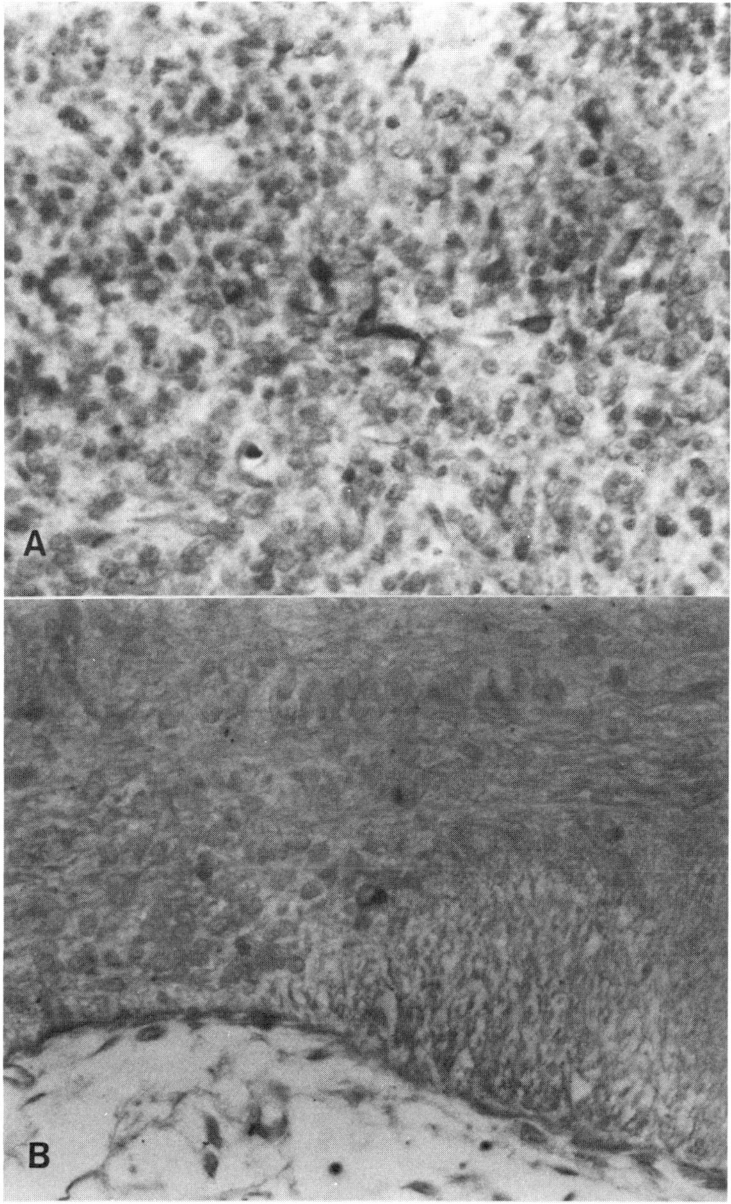

FIGURE 1. Oxytocin neurophysin-positive neurons in the fetal hypothalamus at day 16 of gestation of Brattleboro rat. (A) Immuno-staining neurons in PVN area (B) Immuno-staining neuron as visualized in SON region. Sagittal section. ×225

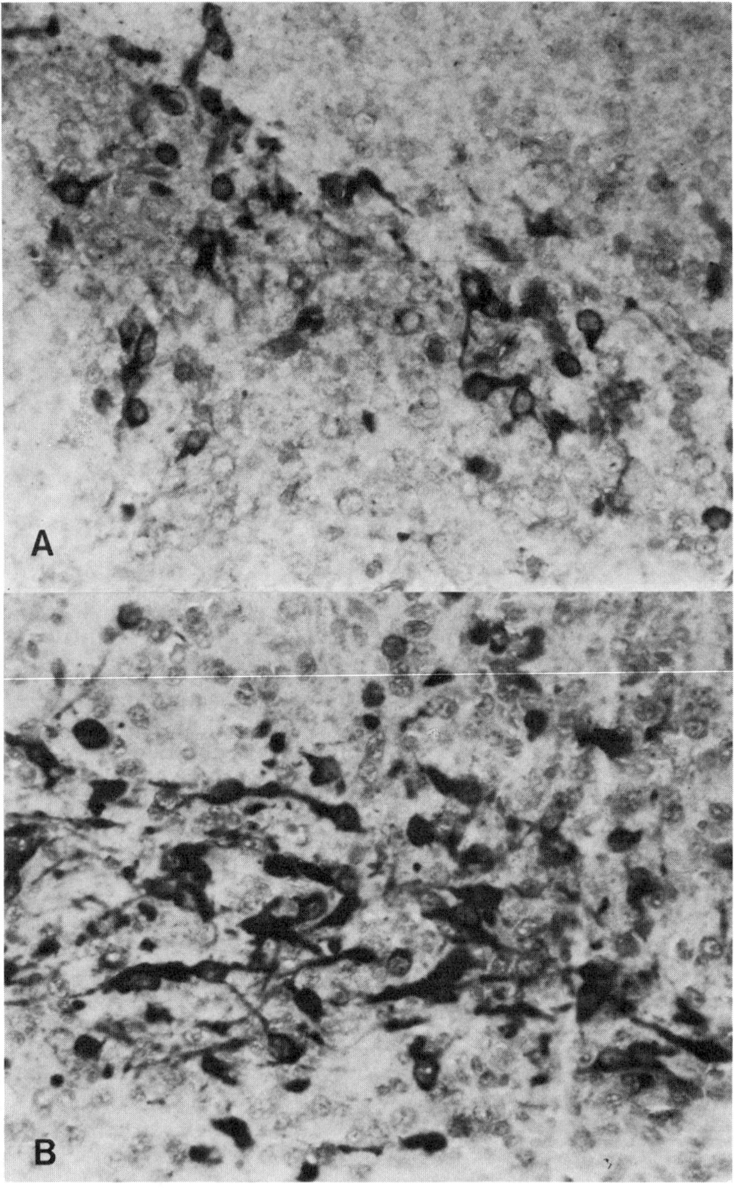

FIGURE 2. Oxytocin neurophysin-positive neurons in the fetal hypothalamus at day 20 of gestation of Brattleboro rat. (A) Immunoreactive neurons located mostly in the dorsal region of SON. (B) More mature-looking neurons found in the PVN area. Sagittal section. ×225

monally deficient brain to produce a similar peptide to vasopressin as early as possible. It is interesting to note that the early presence of immunoreactive oxytocin neurophysin is not correlated with the identification of oxytocin itself. Assayable levels of oxytocin appeared to increase in the hypothalamus at 20 days but still not in sufficient quantity to be identified immunocytochemically. The age at which oxytocin appears in the neural lobe has yet to be ascertained. However, the number of hypothalami and pituitaries assayed at this time does not permit an evaluation of the significance of the observations. It may be speculated that some system is "driving" the early appearance of the neurophysin but fails to influence the subsequent maturation of the actual hormone. Moreover, the regional maturation of neurons in HOM-DI is apparently different from that in the Sprague-Dawley rat.[4] Paraventricular and forniceal groups are formed earlier than supraoptic and retrochiasmatic groups. The lateral perivascular group is seemingly the latest one to develop.

TABLE 1

OXYTOCIN IN THE PITUITARY AND HYPOTHALAMUS OF RAT FETUSES †

Days of Gestation	Homozygous DI		Long-Evans		Sprague-Dawley	
	Hypothalamus	Pituitary	Hypothalamus	Pituitary	Hypothalamus	Pituitary
13 *		6.3				
16	10.0	8.2	8.0	7.2	8.7	6.6
17	11.5	11.0	17.25	13.75	–	–
18	8.2	5.6	7.1	6.75	22.75	32.0
19	9.9	5.5	–	–	9.1	8.7
20	15.5	6.5	15.13	6.6	10.25	18.75
21	16.5	5.8	–	–	8.36	7.14
Newborn	15.5	7.5	–	–	10.8	10.3

* At 13-day gestation, whole head of the fetuses was used for assay.
† pg/animals; means of triplicate assays.

ACKNOWLEDGMENTS

The expert technical assistance of Mrs. W. Ho and Mr. D. Ko is gratefully appreciated.

REFERENCES

1. BUIJS, R. M., D. N. VELIS & D. F. SWAAB. 1980. Ontogeny of vasopressin and oxytocin in the fetal rat: Early vasopressinergic innervation of the fetal brain. Peptides **1:** 315–324.
2. BOER, G. J., D. F. SWAAB, H. B. M. UYLINGS, K. BOER, R. M. BUIJS & D. N. VELIS. 1980. Neuropeptides in rat brain development. Prog. Brain Res. **53:** 207–224.
3. CHOY, V. J. & W. B. WATKINS. 1979. Maturation of the hypothalamus-neurohypophysial system. I. Localization of neurophysin, oxytocin and vasopressin

in the hypothalamus and neural lobe of the developing rat brain. Cell Tiss. Res. **197:** 325–336.

4. Fisher, A. W. F., V. Gill, K. L. Wong, S. Raghavan, W. North & K. Lederis. 1981. Anatomical and developmental aspects of the neurons terminating in the posterior pituitary. *In* Neurosecretion. D. S. Farner & K. Lederis, Eds. Plenum Press. New York, NY.
5. Ifft, J. D. 1972. An autoradiographic study of the time of final division of neurons in rat hypothalamic nuclei. J. Comp. Neurol. **144:** 193–204.
6. Altman, J. & S. A. Bayer. 1978. Development of the diencephalon in the rat II. Correlation of the embryonic development of the hypothalamus with the time of origin of its neurons. J. Comp. Neurol. **182:** 973–994.
7. Moore, G., A. Lutterodt, G. Burford & K. Lederis. 1977. A highly specific antiserum for arginine vasopressin. Endocrinology **101:** 1421–1435.
8. Wolf, G. & B. Trautmann. 1977. Ontogeny of the hypothalamo-neurohypophysial system in rats—an immunohistochemical study. Endokrinologie **69:** 222–226.
9. Khachaturian, H. & J. R. Sladek. 1980. Simultaneous monoamine histofluorescence and neuropeptide immunocytochemistry III. Ontogeny of catecholamine varicosities and neurophysin neurons in the rat supraoptic and paraventricular nuclei. Peptides **1:** 77–95.

BIOSYNTHESIS OF OXYTOCIN AND OXYTOCIN-NEUROPHYSIN IN HOMOZYGOUS BRATTLEBORO RATS *

Joseph C. Rosenior, William G. North,† and Graham J. Moore

Department of Pharmacology and Therapeutics
Faculty of Medicine
The University of Calgary
Calgary, Alberta, Canada T2N 4N1

and

† Department of Physiology
Dartmouth Medical School
Hanover, New Hampshire 03755

INTRODUCTION

The biosynthesis of the neurohypophyseal peptide hormones, [arginine^8] vasopressin (AVP) and oxytocin (OT), and their associated carrier proteins, vasopressin-neurophysin (AVP-NP) and oxytocin-neurophysin (OT-NP) appear to be made via high molecular weight (MW) precursors.[1-4] These studies have also suggested that each peptide hormone and its associated carrier protein may be biosynthesized from a common precursor. In this study an attempt was made to identify high MW precursors of OT and OT-NP in homozygous Brattleboro diabetes insipidus (DI) rats and the results compared to that obtained previously from studies with normal Sprague-Dawley (S-D) rats.[4]

MATERIALS AND METHODS

Animals

Male and female DI rats weighing approximately 180–200 grams were purchased from Blue Spruce Farms (Altamont, N.Y.). All animals were housed in a 12-hour light-dark cycle, fed and watered *ad lib.*

Protein Extraction

Six hypothalami from male and female DI rats containing the supraoptic, paraventricular, and retrochiasmatic nuclei were obtained after decapitation and homogenized in 12 ml acid-ethanol, pH 1.8 and centrifuged. The supernatant was collected and made pH 7.5 with aqueous ammonia, and protein was precipitated overnight at 4° C in 54 ml ether-ethanol (11.2:6.8). The precipitate was recovered by centrifugation and vacuum-dried. The residue was

* Supported by the Medical Research Council (Canada) and U.S. Public Health Service Grant CA–19613.

0077–8923/82/0394–0183 $1.75/0 © 1982, NYAS

weighed and dissolved at a concentration of 4 mg per 0.2 ml in 0.01 M phosphate buffer pH 7.2 containing 2% sodium dodecyl sulphate (SDS) and 8 M urea. Aliquots containing 100 μg of protein (dry weight) were later subjected to gel electrophoresis using tritiated bovine serum albumin ([^{3}H]BSA), ovalbumin ([^{3}H]OVAL) and cytochrome c ([^{3}H]Cyt.c) as internal MW standards.

Gel Electrophoresis and Radioimmunoassay (RIA)

Gel electrophoresis was performed according to the method of Cooper[5] using 10% polyacrylamide gels (12×0.5 cm) containing 0.1% SDS. After electrophoresis, the gels were sliced into 5-mm sections, crushed, and extracted for three days at 4° C with 0.3 ml 0.01 N HCl for OT assay and 0.6 ml 0.01 N HCl for OT-NP assay. Aliquots (0.1 ml–0.2 ml) of the gel extracts were then subjected to RIA for either immunoreactive OT or OT-NP, utilizing a specific antiserum for each peptide.[4] Aliquots of the extracts were also checked for the presence of AVP and AVP-NP by RIA to be certain animals used in these studies were homozygous for diabetes insipidus. Electrophoresis of extracts from male and female DI rats were done on the same set of gels. RIA or gel extracts for OT and OT-NP were done on two different sets of gels. RIA for OT was performed at Calgary, and RIA for OT-NP was performed at Dartmouth.

Results

Gel electrophoresis of hypothalamic extracts in the presence of denaturing agents and RIA or gel extracts for OT gave four characteristic immunoreactive peaks for both male and female DI rats. The peaks corresponded to MW 28K, 18K, 8–10K, and <5K (OT) for male DI rats (Figure 1a) and 32K, 18K, 10K, and <5K (OT) for female DI rats (Figure 1b).

RIA of gel extracts for OT-NP gave four immunoreactive peaks corresponding to MW >70K, 28–32K, 17–18K, and ~10K (OT-NP) for male DI rats (Figure 2a), and >70K, 32K, 18–22K, and ~10–12K (OT-NP) for female DI rats (Figure 2b).

RIA of extracts for AVP and AVP-NP produced no significant amount of immunoreactivity indicating that the animals used in those experiments were homozygous for diabetes insipidus.

Discussion

Several studies have attempted to characterize high MW precursors of the neurohypophyseal hormones in normal rats. However similar studies have been limited in homozygous DI rats which are unable to synthesize AVP and AVP-NP.[6, 7, 9] The results of the present study show that both OT and OT-NP have immunoreactive high MW precursors of approximately 28–32K and 18–22K in both male and female DI rats. There may also be a common immunoreactive precursor of MW >70K, although the amount of immunoreactivity obtained at this MW with the OT antiserum was below the limits of detection. This may also be due to inefficient precipitation of OT from acidified extracts, since peaks of immunoreactivity of the peptide eluting at the migration distance of OT are also much smaller than would be expected (FIGURE 1).

However, the results obtained in this study are in good agreement with results obtained from a previous study with normal male Sprague-Dawley (S-D) rats[4] in which high MW immunoreactive precursors of MW >70K, 28–35K, and 18–19K were obtained for both OT and OT-NP. Hence it is not unreasonable to conclude that OT and OT-NP are biosynthesized in the DI rat via a biosynthetic pathway similar to that observed in S-D rats. On the other hand, the levels of OT-NP, and the OT-NP precursor of MW ~18K, in the hypothalami of male DI rats are considerably lower than those observed in female

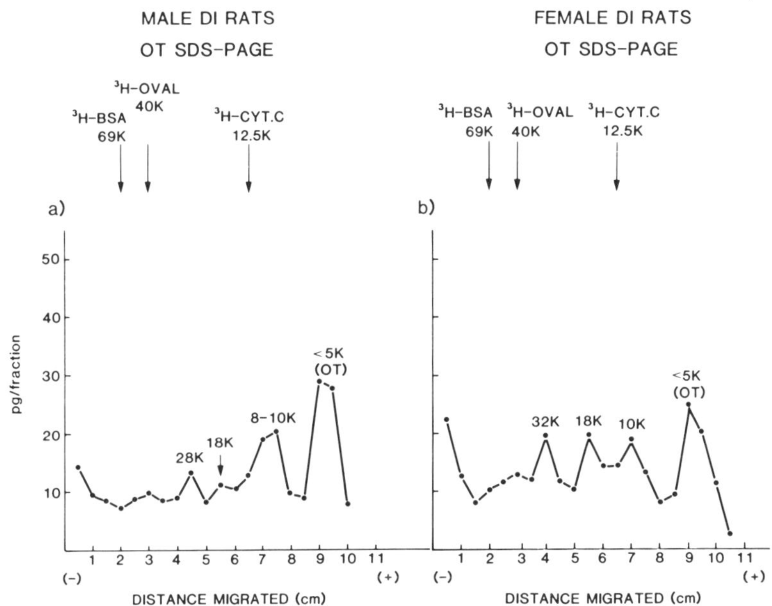

FIGURE 1. Immunoreactive profile for OT of an extract of the hypothalami of male (a) and female (b) homozygous DI rats after SDS-PAGE and RIA of sliced gel extracts.

DI rats and male S-D rats. Lower levels of OT-NP in the neurohypophysis of male DI rats have been reported previously.[7] These findings together with the findings of North and coworkers[8] reported in this symposium, may reflect differences in the rates of synthesis—release of OT and OT-NP between male and female DI rats.

Finally, since no high MW immunoreactive forms were found in the AVP and AVP-NP assays in this study, one would conclude that high MW forms obtained earlier in studies with S-D rats[4] were dependent on AVP being produced in the rat hypothalamus.

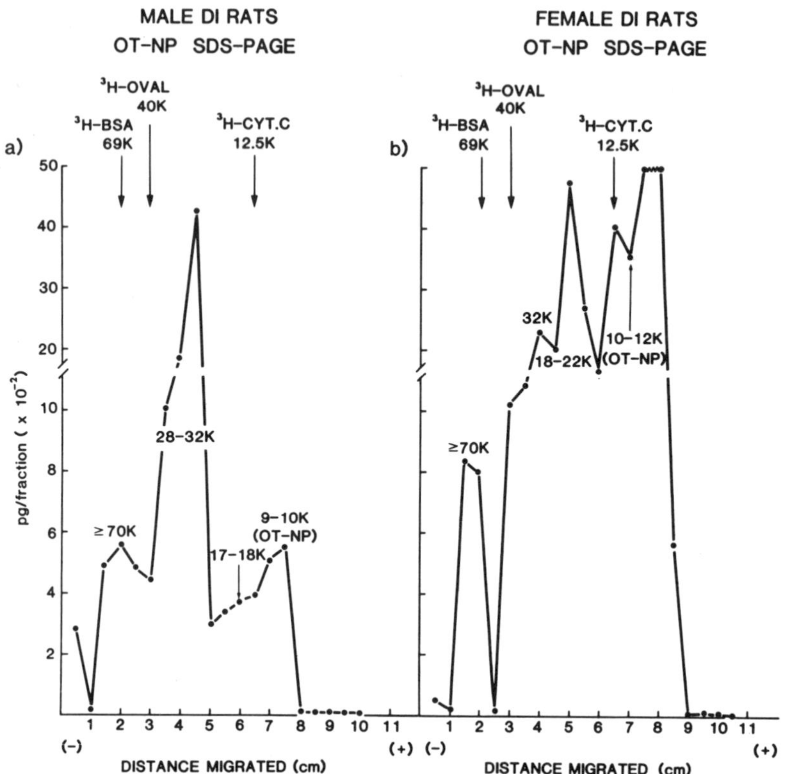

FIGURE 2. Immunoreactive profile for OT-NP of an extract of the hypothalami of male (a) and female (b) homozygous DI rats after SDS-PAGE and RIA of sliced gel extracts.

ACKNOWLEDGMENT

The technical assistance of Gail Hardy is gratefully acknowledged.

REFERENCES

1. MOORE, G. J., R. W. SWANN, A. W. F. FISHER & K. LEDERIS. 1978. A high molecular weight substance in the supraoptic region of the rat hypothalamus which cross-reacts with a specific antiserum for arginine-vasopressin. J. Endocrinol. **79:**23P.
2. BROWNSTEIN, M. J., J. T. RUSSELL & H. GAINER. 1980. Synthesis, transport, and release of posterior pituitary hormones. Science **207:** 373–378.
3. NICHOLAS, P., M. CAMIER, M. LAUBER, M. J. MASSE, J. MÖHRING & P. COHEN. 1980. Immunological identification of high molecular weight forms common to bovine neurophysin and vasopressin. Proc. Natl. Acad. Sci. USA **77:** 2587–2591.

4. Rosenior, J. C., W. G. North & G. J. Moore. 1981. Putative precursors of vasopressin, oxytocin and neurophysins in the rat hypothalamus. Endocrinology **109:** 1067–1072.
5. Cooper, T. G. 1977. The Tools of Biochemistry, 1st edit. Wiley Interscience. New York, p. 219.
6. Norstrom, A. 1974. Biosynthesis of neurohypophysial proteins in rats with hereditary hypothalamic diabetes insipidus (Brattleboro strain). Brain Res. **68:** 309–317.
7. Sunde, D. A. & H. W. Sokol. 1975. Quantification of rat neurophysins by polyacrylamide gel electrophoresis (PAGE): Application to the rat with hereditary hypothalamic diabetes insipidus. Ann. N.Y. Acad. Sci. **248:** 345–361.
8. North, W. G., M. Gellai & G. Hardy. 1981. Oxytocin and oxytocin-associated neurophysin evaluation by RIA in the Brattleboro rat.: Turnover. Ann. N.Y. Acad. Sci. This volume.
9. Russell, J. T., M. J. Brownstein & M. Gainer. 1980. Biosynthesis of vasopressin, oxytocin, and neurophysins: Isolation and characterization of two common precursors (propressophysin and prooxyphysin). Endocrinology **107:** 1880–1891.

SODIUM AND POTASSIUM BALANCE IN THE BRATTLEBORO RAT

Susan Opava-Stitzer, Emma Fernández-Repollet, and Paul Stern *

Department of Physiology
University of Puerto Rico School of Medicine
San Juan, Puerto Rico 00936

and

* *Departments of Maternal and Child Health, and Physiology*
Dartmouth Medical School
Hanover, New Hampshire 03755

Since its discovery in 1961,[1] it has become apparent that the Brattleboro (DI) rat is a useful model for the study of a variety of physiological problems in addition to the obvious one of the role of antidiuretic hormone (vasopressin, ADH) in urine concentration and water balance. In the study of the control of extracellular fluid volume and electrolyte balance, in particular, the DI rat offers a unique opportunity to observe the spontaneous interaction of homeostatic mechanisms when a single disturbance, namely the absence of ADH, has been introduced.

This review will attempt to present a comprehensive description of the state of sodium and potassium balance in the DI rat. Although some data exist on the handling of other electrolytes[51, 54] these have not been studied extensively. Mention will also be made of the multiple factors that may influence electrolyte balance in the Brattleboro rat. New data, obtained by Opava-Stitzer and Fernández-Repollet, will be included when relevant.

Sodium Balance

Among the early observations of Valtin[1, 2] was the finding that, as a result of their inability to conserve water, DI rats were in a fairly continuous state of mild dehydration interrupted by periods of hydration or even overhydration. This was evidenced by an elevated plasma sodium concentration and plasma osmolality compared to normal or heterozygous rats. As shown in Table 1, the plasma sodium concentration of DI rats averaged 152 mEq/L as compared to 145 mEq/L in normal Long-Evans and in heterozygous rats. The plasma osmolality of DI rats was 320 mOsm, significantly higher than the values found in normals and heterozygotes.

In addition, as a result of the absence of ADH, electrolyte concentrations in renal tissue were dramatically affected.[3] The greatest effect, as shown in Table 2, was on interstitial sodium concentration. The concentration of sodium in the papilla of DI rats was significantly reduced compared to Long-Evans rats, a difference that was not eliminated by treatment with one unit of vasopressin per day for three days, despite a significant increase in the papillary sodium concentration. Since there were no differences in the papillary sodium content (mM/g dry solids) of DI and Long-Evans rats, the reduced papillary

0077-8923/82/0394-0188 $1.75/0

TABLE 1

PLASMA SODIUM CONCENTRATION (P_{NA}) AND OSMOLALITY (P_{OSM}) IN NORMAL, HETEROZYGOUS, AND DI RATS

	Long-Evans	Heterozygotes	DI
P_{NA} (mEq/L)	145	145	152 *
P_{OSM} (mOsm/Kg H_2O)	307	301	320 *

* Significantly different from normal, Long-Evans rats. Data from Valtin and Schroeder [1] and Valtin *et al.*[2]

concentration could be attributed to increased water reabsorption by the papillary collecting duct in the absence of ADH, and thus a dilution of papillary interstitial solute. Indeed, Valtin [3] observed a greater papillary water content in untreated DI rats than in Long-Evans controls, and a significant decrease in papillary water content after ADH treatment. It was later shown by Jamison *et al.*[4] in micropuncture studies in the DI rat, that a greater fraction of the glomerular filtrate is reabsorbed by the medullary collecting duct in the absence of ADH than during ADH-induced antidiuresis.

In contrast to the variations noted in the renal papilla, the medullary sodium concentrations of DI and Long-Evans rats were not different and ADH treatment of DI rats resulted in a paradoxical reduction in medullary sodium content reflected here in a reduced sodium concentration in this zone. In cortex, both sodium concentration and content were elevated in untreated DI rats, no doubt reflecting the augmented plasma sodium concentration,[1, 2] and the reduced water reabsorption in the cortex in the absence of ADH.[4]

Valtin's observation that ADH administration to the DI rat did not result in enhanced sequestration of sodium in the medullary interstitium was surprising in view of prior reports,[5, 6] which had shown that the infusion of ADH in dogs undergoing water diuresis significantly raised both the sodium concentration

TABLE 2

RENAL SODIUM CONCENTRATION IN NORMAL (LE), UNTREATED (DI), AND ADH-TREATED (DI-ADH) RATS

	[Na] (mMoles/Kg H_2O)		
	Papilla	Medulla	Cortex
LE	358.4 ± 33.2	223.8 ± 19.0	79.2 ± 2.8
DI	201.5 * ± 31.0	193.5 ± 23.5	89.5 * ± 3.0
DI-ADH	236.8 * ± 36.1	158.0 * ± 21.0	81.6 ± 4.0

Data from Valtin.[3]
* Significantly different from normal rats.

and content of papillary and medullary tissue. In addition, in 1971, Atherton *et al.*[7] demonstrated a dose-dependent increase in sodium content of renal tissue in response to the infusion of small amounts of ADH in water-loaded normal rats. Using a dose of ADH of 100 mU/100 g/day, considerably lower than that used by Valtin, we have consistently observed the accumulation of sodium in the kidneys of DI rats. As shown in TABLE 3, the administration of ADH to DI rats for three weeks resulted in a significant increase in both sodium content and concentration in renal medulla and papilla compared to values in untreated DI rats.

Interestingly, the state of potassium balance had a profound effect on this sodium-sequestering activity of ADH. We studied the effects of administration of ADH for three weeks on the sodium content of kidneys of DI rats maintained for the same period of time on either high potassium or potassium-free diets. As shown in TABLE 4, ADH caused increased tissue sodium deposition in both groups of rats. In the rats on a potassium-free diet, the increase in sodium content was significant only in papilla and was markedly less than that observed in rats on a high potassium diet, or in rats on a normal diet, shown earlier. On a high potassium diet, the increase in sodium content occurred in medulla as well as papilla and the magnitude of the changes was also greater than that seen on a normal diet. It thus appears that during potassium loading the sodium-retaining effects of ADH are exaggerated. This finding may be related to the natriuresis which accompanies potassium loading [8-13] particularly in the presence of aldosterone deficiency.[14] An increased delivery of sodium to the loops of Henle during potassium loading [8, 10, 11] would provide more sodium for the medullary pool and thus make more apparent any effect of ADH on the interstitial deposition of sodium. Furthermore such a mechanism could be exaggerated in the aldosterone-deficient DI rat.[15]

Thus, despite some discrepancies in the literature, there is ample evidence to indicate that ADH can enhance the sequestration of sodium in the renal medulla and papilla, and that the cortico-papillary solute gradient of the DI rat may be further compromised by the lack of this action of ADH as well as its effects on water permeability. Whether or not the sodium-retaining effects of ADH are apparent in any given situation, no doubt depends on a variety of factors which may influence this particular action of ADH. Dosage may be critical. In early studies in particular, in which larger doses of ADH were

TABLE 3

RENAL SODIUM CONTENT AND CONCENTRATION AND PLASMA SODIUM CONCENTRATION IN UNTREATED (DI) AND ADH-TREATED (DI-ADH) DI RATS

		Na_c (mMoles/Kg FFDW)	$[Na^+]$ (mM)
Medulla	DI	634 ± 64 *	152 ± 13 *
	DI-ADH	1271 ± 80	280 ± 19
Papilla	DI	1158 ± 141 *	402 ± 19 *
	DI-ADH	2916 ± 205	640 ± 45
Plasma	DI	—	137 ± 3
	DI-ADH	—	140 ± 1

* Significantly different from ADH-treated DI rats, $p < 0.05$.

TABLE 4

RENAL SODIUM CONTENT (NA_c) AND CONCENTRATION AND PLASMA SODIUM CONCENTRATION IN UNTREATED (DI) AND ADH-TREATED (DI-ADH) DI RATS ON DIFFERENT POTASSIUM DIETS

		Na_c (mMoles/Kg FFDW)		$[Na^+]$ (mM)	
		High K	K-free	High K	K-free
Medulla	DI	716 ± 110 *	610 ± 16	177 ± 32	122 ± 3
	DI-ADH	1087 ± 65	655 ± 21	257 ± 35	136 ± 7
Papilla	DI	1036 ± 134 *	911 ± 60 *	348 ± 41 *	276 ± 24 *
	DI-ADH	3125 ± 94	1360 ± 91	637 ± 51	354 ± 6
Plasma	DI	—	—	136 ± 3	148 ± 1
	DI-ADH	—	—	137 ± 1	142 ± 1 *

* Significantly different from ADH-treated DI rats, $p < 0.05$.

utilized, changes in renal hemodynamics may have occurred, resulting in a reduction in the amount of sodium presented to the long loops of Henle for reabsorption. Such an occurrence would diminish the ability of ADH to increase interstitial sodium content, by a mechanism opposite to that which we have proposed may occur during potassium loading.

As to the mechanism by which ADH exerts its sodium-sequestering action, all explanations are still speculative. It had been proposed as early as 1962 by Levitin *et al.*[5] that ADH might enhance active sodium reabsorption by the loops of Henle or collecting tubules, analogous to the manner in which it increases active sodium transfer across the frog skin[16] and toad bladder.[17] Micropuncture studies,[18-22] however, had until recently consistently failed to show such an action of ADH. In 1980 an *in vitro* effect of ADH on chloride transport of isolated perfused medullary thick ascending limbs of the mouse was demonstrated by Hall and Varney.[23] In the same year Sasaki and Imai[24] reported that ADH stimulated NaCl transport across isolated perfused medullary thick ascending limbs of mouse, rat, and rabbit. The biochemical basis for such an effect had already been shown to exist by Imbert, Morel, and colleagues in a series of publications.[25-31] They have shown the presence of ADH-sensitive adenyl cyclase activity in both thin and thick ascending limbs of rat[28, 29] and rabbit[25, 27] and in the thick ascending limbs of mouse.[31] No such activity could be demonstrated in the human,[30] a fact that may partially explain the inferior concentrating ability of man compared to the rodent and rabbit.

These findings once again allude to the possible importance of a direct effect of ADH on active sodium reabsorption. Even barring such an effect, however, ADH might still affect the medullary sodium gradient according to the model of passive sodium reabsorption in the inner medulla proposed by Kokko and Rector in 1972.[32] By increasing the urea concentration in the interstitium surrounding the loops of Henle of deep nephrons, ADH will also enhance the abstraction of water from descending limbs, resulting in an increased gradient for the passive diffusion of sodium out of thin ascending limbs. A failure to find increased interstitial sodium content after ADH administration would, in fact, challenge the importance of this mode of sodium reabsorption in the inner medulla.

It is also possible that ADH may alter the deposition of sodium in the medullary interstitium indirectly, through alterations in medullary blood flow and/or filtration rate of juxtamedullary nephrons. It has been shown that medullary blood flow is lower in antidiuresis than in water diuresis,[33-36] and thus the countercurrent mechanism of solute deposition would operate more efficiently in the presence of ADH. Measurements of single nephron glomerular filtration rate in rats[37, 38] show that the filtration rate of deep nephrons is higher in antidiuresis than in the absence of ADH. Thus more sodium might be sequestered in the medullary interstitium simply because a greater fraction of filtered sodium is delivered to that region in the presence of ADH.

Finally, it has been reported recently[39] that, compared to normal Long-Evans rats, DI rats have a deficiency of glycosaminoglycans in the papillary interstitium. It has been suggested that these polyanionic molecules may act as ion exchangers, sequestering cations in the interstitium, and may also sterically exclude water from interstitial spaces. If the concentration of papillary glycosaminoglycans can be restored to normal by ADH treatment, an ADH-dependent role for these molecules in the maintenance of the medullary sodium gradient may be established.

In addition to the question of the role of ADH in the maintenance of the interstitial solute gradient, it is also unclear to what extent ADH, or the absence of it, affects urinary sodium excretion ($U_{\mathrm{Na}}V$) and sodium balance. It is generally accepted that antidiuretic hormone has a natriuretic action in dog and rat,[6, 40-49] including the Brattleboro rat,[46, 47] but this is normally seen at supraphysiological doses and may be related to systemic effects of ADH rather than a direct action of ADH on renal sodium reabsorption. Nevertheless, it is relevant to consider the possibility of physiologically important effects of ADH on renal sodium handling and sodium balance, particularly in view of the evidence concerning the effects of ADH on the intrarenal deposition of sodium.

Another of Valtin's and Schroeder's early findings was that urinary sodium excretion was about 50% higher in DI rats than in Long-Evans controls.[1] Friedman and Friedman reported shortly thereafter that food intake was also higher in the DI rat[50] and this could have accounted for the elevated sodium excretion seen by Valtin. Subsequent studies by Balment[51] and Haack,[52] however, showed that the DI rat excretes a greater fraction of its sodium intake than normal rats[51] and may have a food intake either equal to[51] or less than[51, 52] that of normal rats. In 1974, Möhring[53] showed a dose-dependent sodium retention in DI rats in response to single injections of 250 or 500 mU of ADH, but no effect with lower doses. The evidence for sodium retention was a decrease in the fraction of ingested sodium that was excreted. Prolonged administration of low doses of ADH also had no effect on sodium balance.[47] A dose of 500 mU/day for seven days caused sodium retention on the first day only, followed by cycles of positive and negative sodium balance.[47] In contrast, Balment *et al.*[54] reported a decrease in the excreted fraction of ingested sodium in male DI rats from a pretreatment value of 82.5% to 63.3% after one week of treatment with one unit of ADH per day. As shown in Table 5, we have also observed a significant decrease in urinary sodium excretion in DI rats on varying potassium intakes, after three weeks of treatment with 100 mU ADH/100 g body weight/day. Although on a normal diet this reduction in $U_{\mathrm{Na}}V$ was associated with a reduction in sodium intake, on a high potassium diet there was no significant change in sodium intake and on a potassium-free diet sodium intake was actually increased after ADH treatment. In all cases, the fraction of ingested sodium that was excreted in the urine was decreased as a result of ADH administration. The effect of ADH to decrease sodium excretion was greatest on a high potassium diet.

It should be noted that any large differences between sodium intake and urinary sodium excretion must represent either a transient state of sodium retention or increased loss of sodium in the feces.

It seems clear, therefore, that ADH can, in some circumstances, cause sodium retention, an action that is influenced by the state of potassium balance. The same mechanisms previously discussed, by which ADH may enhance medullary sodium trapping, could also explain its effects on sodium excretion.

Möhring[47] measured serum sodium concentration in DI rats during seven to nine days of treatment with 100 mU or 500 mU of ADH per day. Serum sodium concentration was significantly reduced by both doses of ADH, despite differing effects on sodium balance, as discussed earlier. Although plasma volumes were not measured in these experiments, the fall in plasma sodium

TABLE 5

URINARY SODIUM EXCRETION ($U_{Na}V$) AND INTAKE OF UNTREATED (DI) AND ADH-TREATED (DI-ADH) DI RATS WITH DIFFERENT POTASSIUM INTAKES

		$U_{Na}V$ or Na intake (μEq/100 g bw/day)	
		DI	DI-ADH
Normal diet	$U_{Na}V$	1096 ± 59	965 ± 49 *
	Na intake	1510 ± 78	1395 ± 102 *
		(72.5%)	(69.2%)
High K diet	$U_{Na}V$	1340 ± 85	858 ± 61 *
	Na intake	1390 ± 40	1295 ± 75
		(96.4%)	(66.3%)
K-free diet	$U_{Na}V$	798 ± 12	703 ± 58 *
	Na intake	843 ± 37	922 ± 63 *
		(94.7%)	(76.2%)

* Significantly different from untreated DI rats, $p < 0.05$. Numbers in parentheses are ($U_{Na}V$/Na intake) ×100.

concentration was most likely the result of a rise in plasma volume following ADH treatment. In another study,[53] Möhring measured actual changes in plasma volume in DI rats after a single injection of 125 mU of Pitressin. Plasma volume increased significantly (4.14 to 4.74 ml/100 g body weight), accompanied by a significant decrease in hematocrit but, surprisingly, no change in plasma osmolality. Similar findings were associated with chronic ADH treatment.[47]

In our experiments, using doses comparable to Möhring's for a longer period of time, we could demonstrate no significant effect of ADH on plasma sodium concentration (TABLE 4). In view of the concurrent finding of reduced urinary sodium excretion, our failure to see a *fall* in plasma sodium concentration may actually indicate a net increase in the amount of extracellular sodium. It should be emphasized, however, that without precise measurements of plasma or extracellular fluid volume, changes in serum sodium concentration are at best a poor index of whether or not sodium retention has occurred in response to ADH treatment. The values of plasma sodium concentration, which are shown in TABLES 3 and 4, merit further discussion. Plasma sodium concentration varied from 136 ± 3 mEq/L to 148 ± 1 mEq/L in untreated DI rats, depending on potassium intake. We have generally found a reciprocal relationship between plasma sodium and potassium concentration; hence plasma sodium concentration was highest on a K^+-free diet. A reciprocal relationship was also seen in these studies between sodium and potassium content of skeletal muscle. In addition, our values of plasma sodium concentration are lower than those reported by others.[1, 2, 15, 47, 52, 53, 65] We have indeed found at times that plasma sodium concentration of the DI rat was *not* elevated compared to Long-Evans rats. In agreement with our data, Balment *et al.* reported values of plasma sodium concentration of 141.9 ± 1.7 mEq/L in male and 137.6 ± 1.2 mEq/L in female DI rats compared to 144.1 ± 2.0 and 137.7 ± 1.2 mEq/L in male and female heterozygous rats.[51] There are many possible explanations for these differences including biological variation among the different colonies of DI rats studied, as well as differences in environmental and dietary conditions and methods of blood sampling. In addition the state of hydration of the rat at the time of blood sampling may be important (e.g. dehydrated versus overhydrated) and this may be related to the time of day at which the sample is obtained. Thus even a simple measurement such as serum sodium concentration may be complicated by the peculiar physiology of the DI rat. Standardized experimental conditions and appropriate controls are essential.

In summary, studies in the Brattleboro rat indicate that physiological levels of ADH may play a direct role in the maintenance of sodium and water balance through actions on renal sodium handling and renal tissue sodium deposition, effects that are modified by the state of potassium balance. In addition, sodium balance may be affected indirectly by changes in blood pressure, and extracellular fluid volume secondary to the well-known effects of ADH on water balance.

POTASSIUM BALANCE

In his early descriptions of the Brattleboro rat, Valtin[1] noted that, in addition to an elevated urinary sodium excretion, DI rats excreted significantly more potassium in the urine than did Long-Evans rats. The same constraints on interpretation that applied to the finding of increased sodium excretion, however, were also applicable in the case of potassium. In the face of reports of a greater than normal food intake in DI rats,[50] the higher than normal excretion of potassium might simply reflect a greater ingestion of potassium. This topic was not given much attention until 1972 when Möhring[55] first reported that DI rats were hypokalemic. Further studies[15] confirmed this finding revealing, as well, a reduced potassium content of skeletal muscle of DI rats. Balment *et al.*[51] later found that female DI rats had a greater urinary potassium loss than female heterozygous rats. As shown in TABLE 6, when expressed as a fraction of dietary intake, female DI rats excreted more potassium in the urine than did heterozygous females. No such difference was found, however, between DI and heterozygous males. The authors suggested that this difference in potassium handling might be due to sex differences in mineralocorticoid levels.

TABLE 6

URINARY POTASSIUM EXCRETION, AS A FRACTION OF POTASSIUM INTAKE, OF DI AND HETEROZYGOUS BRATTLEBORO RATS

	$U_K V/(\%)$ K Intake	
	DI	Hetero
Female	86.9 ± 1.1 *	78.0 ± 1.8
Male	74.3 ± 1.4	72.4 ± 2.4

* Significantly different from heterozygous rats of the same sex. Data from Balment *et al.*[51]

In an effort to determine whether the hypokalemia of DI rats might be secondary to excessive loss of potassium in the urine we compared food intake and electrolyte excretion of DI and Long-Evans rats housed in metabolism cages for periods of three and nine days.[56] The urinary potassium excretion of DI rats was found to be higher than that of Long-Evans rats on the first four days of study, but not significantly different on the remaining days. The difference in urinary potassium excretion could not be accounted for by differences in food intake between the two groups. As shown in TABLE 7, measurements of plasma potassium concentration and muscle potassium content revealed that the DI rats were potassium deficient at the start of the balance study and after three days, but by nine days of study were potassium repleted, as evidenced by normal plasma potassium concentration and muscle potassium content.

We also studied the alterations in renal potassium content and concentration of DI and Long-Evans rats, which occurred between three and nine days of balance study.[56] After three days in metabolism cages both the potassium content and concentration of the renal medulla and papilla of DI rats were elevated compared to Long-Evans rats. By nine days, however, there were no differences in either of these measurements. The major change between three and nine days had been a decrease in the medullary and papillary potassium concentration of DI rats, primarily due to an increase in the water content of these regions.

Valtin[3] did not note differences in tissue potassium between DI and Long-Evans rats after a brief period in metabolism cages. This, however, may have been due to his method of analysis. Wet and dry weights were determined on

TABLE 7

PLASMA POTASSIUM CONCENTRATION AND MUSCLE POTASSIUM CONTENT OF DI AND LONG-EVANS (LE) RATS PRIOR TO AND AFTER 3 AND 9 DAYS, OF BALANCE STUDY

		Start	3 days	9 days
Plasma $[K^+]$	DI	3.6 ± 0.1 †	3.7 ± 0.2 *†	4.3 ± 0.2
	LE	—	4.3 ± 0.2	4.3 ± 0.1
Muscle K^+ (mMoles/Kg FFDW)	DI	362 ± 6 *†	442 ± 8 *†	468 ± 5
	LE	495 ± 3	465 ± 8	456 ± 5

* Significantly different from Long-Evans rats studied at the same time, $p < 0.05$.
† Significantly different from DI rats at 9 days, $p < 0.05$.

a different group of kidneys than those used for chemical analysis, a procedure that may have obscured small differences in potassium concentration. In our studies, wet and dry weights were obtained on the same tissue used for measurement of electrolytes.

It should be recalled that after nine days in metabolism cages urinary potassium excretion of DI rats was not different from that of normal rats. Thus a prolonged period in metabolism cages had resulted in spontaneous potassium repletion as well as normalization of the rate of potassium excretion in DI rats.

Interestingly, the decrease in potassium excretion after several days in metabolism cages occurred regardless of the potassium intake. As shown in FIGURE 1, despite a wide range of urinary potassium excretion rates, a significant decrease in the $U_K V$ of DI rats was always observed after three to five days in metabolism cages.[56] It should be noted also, that DI rats on a potassium-free diet reduced their urinary potassium to levels even lower than those observed in normal rats on the same diet.[56] This suggested to us that the transiently high urinary potassium excretion observed early in balance studies was not due to a primary defect in renal potassium handling. We postulated, instead, that these alterations in tissue and urinary potassium might be secondary to changes in water balance when DI rats are moved from their regular habitation cages to individual metabolism cages. If this were so, then mild dehydration would be expected to enhance urinary potassium excretion and mimic the observed tissue changes.

When DI rats were acutely dehydrated by deprivation of water for five to eight hours, urinary potassium excretion was increased compared to DI rats drinking *ad lib*. Not only was basal $U_K V$ higher in dehydrated DI rats but also the excretion of potassium in response to a KCl load (200 μEq/100 g body weight, by gavage).[56] In addition, as shown in TABLE 8, five hours of water deprivation significantly decreased the water content of renal papilla and increased the potassium concentration of both medulla and papilla.[56] No changes were observed in gastrocnemius muscle. These results are in agreement with those of Valtin[3] who found an increase in potassium concentration of medulla and papilla of DI rats after 12 hours of dehydration. A more chronic dehydration in the form of water rationing was also found to enhance potassium output.[56] In summary, the potassium deficiency of DI rats originally reported by Möhring[15, 55] was shown to be spontaneously reversible and associated with a transient elevation in urinary potassium excretion. The elevation in potassium excretion may occur when DI rats are forced to compete with other rats for drinking water and thus are more apt to experience periods of dehydration. This may result in increased potassium concentration of renal medulla and papilla through a reduction in water content. The higher potassium concentration, in turn, may augment the gradient for potassium secretion in the loops of Henle and collecting ducts[57-63] and result in inappropriate renal potassium loss and hypokalemia. The simple procedure of allowing DI rats non-competitive access to drinking water reverses this chain of events and results in potassium repletion.

In some of our studies of tissue composition it was noted that alterations in water balance affected not only the concentration of potassium in renal tissue but also the potassium content. Jamison *et al.*[62] have shown that, in the rat, there is medullary recycling of potassium from the collecting tubule to the descending limb of the loop of Henle. The degree of recycling was dependent on the concentration of potassium in the collecting duct fluid.[64] Thus

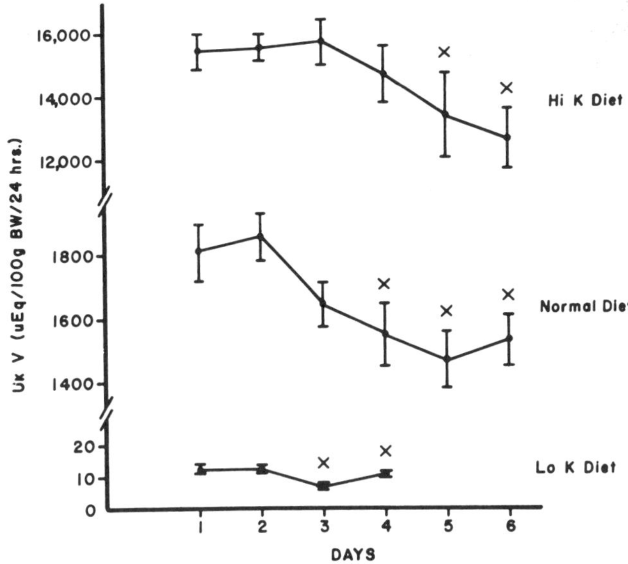

FIGURE 1. Urinary potassium excretion ($U_K V$) in DI rats during six days of balance study on normal ($N = 12$) and high ($N = 6$) potassium diets; and four days of study on a potassium-free diet ($N = 6$). Values are means ± SEM. ×, Significantly different from $U_K V$ on Day 1 in each group, $p < 0.05$. (From Fernandez-Repollet *et al.*[56] With permission of *The Journal of Physiology*.)

during mild dehydration, medullary potassium recycling might be enhanced, thereby increasing the medullary and papillary potassium pool.

Our results also explain a previous discrepancy in the literature between the data of Möhring[15, 55] and an earlier study by Friedman and Friedman[50] in which muscle and plasma potassium levels of DI rats were found to be normal. Friedman's rats, however, had been housed in metabolism cages for more than three weeks prior to study and thus, according to our findings,[56] would already have been potassium repleted.

TABLE 8

RENAL WATER CONTENT AND POTASSIUM CONCENTRATION OF CONTROL (CONT.) AND DEHYDRATED (DEHYD.) DI RATS

		H_2O Content (ml/100 g wet wt.)	[K] (mM)
Medulla	Cont.	80.0 ± 1.5	98.7 ± 2.0
	Dehyd.	78.0 ± 2.5	152.4 ± 30.0 *
Papilla	Cont.	84.0 ± 2.0	73.1 ± 2.1
	Dehyd.	78.4 ± 2.5 *	87.6 ± 2.1 *

* Significantly different from values in control DI rats, $p < 0.05$. Data from Fernandez-Repollet *et al.*[56]

The next logical question is whether ADH administration exerts an effect on potassium balance and whether such an effect is physiological or pharmacological in nature. As was the case with sodium, ADH has also been shown to be kaliuretic.[42, 44, 48, 49] Since a recent report has shown that DDAVP, a nonpressor analogue of arginine vasopressin did not cause either natriuresis or kaliuresis,[48] these may be effects secondary to vasopressor-related actions of ADH.

Möhring[53] studied the acute effects of ADH on potassium balance in Brattleboro rats. Single injections of Pitressin, between 50 and 500 mU, caused a dose-dependent potassium retention, evidenced in a decrease in the fraction of ingested potassium that appeared in the urine. This potassium retention was transient, potassium excretion returning to normal by the third to fourth day after injection. There was no effect of ADH on potassium balance in heterozygous rats. A single injection of 100 mU of vasopressin significantly increased the serum potassium concentration of DI rats (3.32 ± 0.06 to 3.44 ± 0.09 mEq/L).

Möhring also observed that with more prolonged treatment of DI rats with ADH,[47] a dose-dependent retention of potassium occurred on the first day of treatment. Potassium excretion rates then returned to normal despite continued ADH treatment. Kaliuresis occurred after cessation of treatment. Balment[54] could demonstrate no difference in potassium balance of untreated or oil-treated DI rats and DI rats after one week of ADH treatment. This is not really contradictory to Möhring's results, however, since even if potassium had been retained early in treatment, the escape observed by Möhring would have occurred by the time of study.

Möhring[53] initially attributed the effects of ADH on potassium balance to alterations in distal tubular flow rate secondary to effects of ADH on water reabsorption. In support of this he showed a significant inverse correlation between the degree of potassium retention and urine flow rate, during single injections of varying doses of ADH. Such reasoning may be specious, however, and the physiologically significant correlation may, in fact, have been between potassium retention and dose of ADH. This interpretation is supported by Möhring's later findings[47] in which the magnitude of potassium retention was greater but the duration shorter, at a higher dose of ADH. This suggests to us that the degree of potassium retention is dose-dependent but limited to an absolute net potassium gain, which may be determined by the degree of potassium deficiency. It should be recalled that potassium retention did not occur in potassium-replete heterozygous rats.[53] In addition, we have shown[56] that DI rats retain more of an oral potassium load than do normal rats.

Furthermore, we have studied the effects of three weeks of ADH treatment on $U_K V$ of DI rats maintained on normal, high potassium, or potassium-free diets. In DI rats on a potassium-free diet ADH caused a slight but not significant decrease in urinary potassium excretion. On a normal diet, ADH significantly decreased $U_K V$ but potassium intake was also decreased and there was no significant change in the fractional excretion of potassium. In contrast, in DI rats on a high potassium diet ADH significantly *increased* both $U_K V$ and the fraction of ingested potassium that appeared in the urine. Thus ADH had different effects on potassium balance and urinary potassium excretion depending on the state of potassium balance at the time of administration.

The escape from potassium retention observed by Möhring[47] could then be due simply to potassium repletion and restoration of normal serum and tissue potassium levels, although Möhring[47] suggested that this escape may be mediated by the increase in plasma mineralocorticoid concentration, which also occurs in response to ADH treatment.

We have studied the effects of three weeks of ADH treatment on potassium concentration and content of renal tissue in varying states of potassium balance. As shown in TABLE 9, ADH treatment of DI rats on a normal diet caused a significant increase in the potassium content of renal medulla and papilla. This increase, however, was not reflected by changes in potassium concentration, since water content of these zones was also significantly increased by ADH treatment. This increase in water content, though contradictory to Valtin's observations,[3] and paradoxical in view of the finding of reduced water reabsorption into the inner medulla during antidiuresis,[4] was nonetheless consistently observed by us. Whatever the mechanism, this may be a long-term effect of ADH, and thus not seen in acute studies of the transition to antidiuresis[4] or after short-term treatment with ADH.[3, 5]

On high potassium and potassium-free diets (TABLE 9), ADH increased papillary renal potassium content, and in the high potassium group an increase was observed in the medulla as well. As might be anticipated, the increase in potassium content was also much greater on a high potassium intake. In the papilla the change in potassium content was large enough to significantly increase the potassium concentration despite an increase in water content of this zone. No changes were seen in the cortex on any dietary potassium intake as a result of ADH treatment.

In summary, both ADH and the lack of it, affect potassium balance and the deposition of potassium in the renal medulla and papilla. The mechanism for these effects is not known, but it is interesting to speculate that ADH might stimulate the medullary potassium recycling described by Jamison.[62] A role of systemic effects of ADH and/or alterations in renal hemodynamics and filtration rate cannot, however, be ruled out.

OTHER INFLUENCES ON SODIUM AND POTASSIUM BALANCE IN THE DI RAT

Having reviewed the current state of our knowledge of sodium and potassium balance in the Brattleboro rat and of the effects of ADH on the balance of these cations, it would seem of interest to mention those unusual characteristics of the DI rat that have a proven or potential influence on their electrolyte balance. Reduced blood pressure[38, 65, 66] and altered renal hemodynamics may be important, particularly insofar as glomerular filtration rate has been shown to be reduced in the DI rat.[67, 68] Not only gross changes but more subtle differences, such as the lack of nephron heterogeneity recently reported,[38] may affect renal electrolyte handling.

It has been known for some years that the renin-angiotensin system is also markedly altered in the DI rat, either as a direct consequence of the lack of ADH, which is known to inhibit renin secretion in a variety of species[48, 49, 70, 77] or indirectly, as a result of the diabetes insipidus. It is well established that plasma renin activity,[78, 79] plasma renin concentration,[65] and plasma angiotensin II concentration[15, 47, 52] are all increased in the DI rat. Although it has been reported that the increase in renin is confined to the male,[78] this has not been confirmed by others.[65, 79]

ADH treatment of DI rats, either a single injection,[80] or more chronic

TABLE 9

RENAL POTASSIUM CONTENT AND CONCENTRATION OF UNTREATED (DI) AND ADH-TREATED (DI-ADH) DI RATS ON POTASSIUM-FREE, NORMAL, AND HIGH POTASSIUM DIETS

		[K] (mEq/L)			K content (mMoles/Kg FFDW)		
		K-free	Normal	High K	K-free	Normal	High K
Medulla	DI	63.7 ± 2.4	104.8 ± 8.0	110.5 ± 2.5	358 ± 14	437 ± 18	455 ± 17
	DI-ADH	61.3 ± 5.6	109.6 ± 3.5	109.6 ± 3.5	325 ± 11	565 * ± 36	580 * ± 20
Papilla	DI	78.3 ± 1.7	128.5 ± 11.0	150.6 ± 9.0	272 ± 14	361 ± 19	407 ± 18
	DI-ADH	67.6 * ± 3.3	151.0 ± 14.5	235.4 * ± 11.5	313 * ± 12	580 * ± 20	1169 * ± 71

Values are means ± SEM.
* Significantly different from untreated DI rats, $p < 0.05$.

treatment,[47, 52, 79, 80] reduced plasma renin activity,[79] plasma renin concentration,[80] and plasma angiotensin II concentration.[47, 52] A sex difference in the renin response to ADH has also been noted,[79, 81] but not consistently.[80]

Not only is there an increased release of renin and generation of angiotensin II in the DI rat, but an increased storage of renin in the kidneys, as well. Renal renin content was found to be greater in kidneys of DI rats.[80, 82] It is interesting that even when female DI rats had normal plasma renin activity, renal renin content was increased.[80]

Finally, this hyperactivity of the renin angiotensin system is not limited to basal conditions. DI rats also have an exaggerated renin response to common stimuli of renin release, such as dehydration and renal artery clamping.[65]

Despite a hyperactive renin-angiotensin system, Brattleboro rats have been shown to suffer from adrenal insufficiency, specifically, reduced plasma aldosterone[15, 47] and corticosterone,[15, 47, 83] concentrations and adrenal weights.[15, 47, 83] They also had a diminished corticosterone response to mild stress that was not corrected by two to three weeks of ADH treatment.[83] ADH treatment did restore plasma mineralocorticoid levels and adrenal weights to normal despite a concomitant reduction in angiotensin II concentration.[47]

In addition to the possibility that ADH or the lack of it may directly affect the adrenal gland,[84, 85] reduced adrenal function could also be secondary to the mild potassium deficiency[15, 86, 87] and/or the elevated serum sodium concentration[1, 2, 87] found in the DI rat. Pituitary ACTH content was also lower in DI rats[83] and this may play an important role in the adrenal insufficiency.

Thus, although the renin-angiotensin system is highly stimulated in the Brattleboro rat, adrenal function is subnormal. It is not known to what extent these alterations affect electrolyte balance, although Balment[51] did not note any significant changes in sodium or potassium balance of DI rats after adrenalectomy. It is also not known what role the high plasma angiotensin II levels may play in preventing an even greater adrenal insufficiency.

Finally, the Brattleboro rat shows a significant reduction in the urinary excretion of prostaglandins E_2 and $F_{2\alpha}$.[88] The excretion rate of these prostaglandins in the DI rat was only about 20% of the rate observed in Long-Evans rats, and was increased in both DI and normal rats after ADH or DDAVP treatment.[88] Renal prostaglandin synthesis may be diminished in the DI rat due to the absence of a stimulatory effect of ADH on prostaglandin synthesis.[89] Although one might have expected to see an increased prostaglandin production in the DI rat due to hypokalemia,[90, 92] potassium deficiency did not increase prostaglandin excretion in normal rats.[93]

Bankir and coworkers[94] also showed a decreased rate of synthesis of PGE_2 by the papilla, but an increased synthesis by glomeruli, of the DI rat *in vitro*. The finding of a decreased rate of papillary synthesis correlates well with morphological studies,[39, 69] which have shown both gross and fine structural changes in the papillae of DI rats. Notably, interstitial cells were markedly affected even at the subcellular level[39] and were shown to contain fewer lipid droplets,[39] presumably a reflection of decreased prostaglandin synthesis. The known effects of prostaglandins on renal sodium excretion[95] and renin release[95-97] make this an important area for further study in the DI rat.

The many hemodynamic, morphological and hormonal changes listed here are no doubt manifestations of the complex interaction of multiple homeostatic mechanisms. It is to be hoped that a study of these interactions in the DI rat will provide useful information concerning the function of these same mechanisms in the normal rat, and ultimately, man.

REFERENCES

1. VALTIN, H. & H. A. SCHROEDER. 1964. Familial hypothalamic diabetes insipidus in rats (Brattleboro strain). Am. J. Physiol. **206:** 425–430.
2. VALTIN, H., W. H. SAWYER & H. W. SOKOL. 1965. Neurohypophyseal principles in rats homozygous and heterozygous for hypothalamic diabetes insipidus (Brattleboro strain). Endocrinol. **77:** 701–706.
3. VALTIN, H. 1966. Sequestration of urea and non-urea solutes in renal tissues of rats with hereditary hypothalamic diabetes insipidus: effect of vasopressin and dehydration on the countercurrent mechanism. J. Clin. Invest. **45:** 337–345.
4. JAMISON, R. L., J. BUERKERT & F. LACY. 1971. A micropuncture study of collecting tubule function in rats with hereditary diabetes insipidus. J. Clin. Invest. **50:** 2444–2452.
5. LEVITIN, H., A. GOODMAN, G. PIGEON & F. H. EPSTEIN. 1962. Composition of the renal medulla during water diuresis. J. Clin. Invest. **51:** 1145–1151.
6. PERLMUTT, J. H. 1962. Influence of hydration on renal function and medullary sodium during vasopressin infusion. Am. J. Physiol. **202:** 1098–1104.
7. ATHERTON, J. C., R. GREEN & S. THOMAS. 1971. Influence of lysine-vasopressin dosage on the time course of changes in renal tissue and urinary composition in the conscious rat. J. Physiol. **213:** 291–309.
8. KAHN, M. & N. K. BOHRER. 1967. Effects of potassium-induced diuresis on renal concentration and dilution. Am. J. Physiol. **212:** 1365–1375.
9. VANDER, A. J. 1970. Direct effects of potassium on renin secretion and renal function. Am. J. Physiol. **219:** 455–459.
10. BRANDIS, M., J. KEYES & E. E. WINDHAGER. 1972. Potassium-induced inhibition of proximal tubular fluid reabsorption in rats. Am. J. Physiol. **22:** 421–427.
11. SCHNEIDER, E. G., R. E. LYNCH, L. R. WILLIS & F. G. KNOX. 1972. The effect of potassium infusion on proximal sodium reabsorption and renin release in the dog. Kidney Int. **2:** 197–202.
12. SHADE, R. E., J. O. DAVIS, J. A. JOHNSON & R. T. WITTY. 1972. Effects of renal arterial infusion of sodium and potassium on renin secretion in the dog. Circ. Res. **31:** 719–727.
13. FLAMENBAUM, W., T. A. KOTCHEN, R. NAGLE & J. D. MCNEIL. 1973. Effect of potassium on the renin-angiotensin system and $HgCl_2$-induced acute renal failure. Am. J. Physiol. **224:** 305–311.
14. MILLER, P. D., C. WATERHOUSE, R. OWENS & E. COHEN. 1975. The effect of potassium loading on sodium excretion and plasma renin activity in Addisonian man. J. Clin. Invest. **56:** 346–353.
15. MÖHRING, B., J. MÖHRING, G. DAUDA & D. HAACK. 1974. Potassium deficiency in rats with hereditary diabetes insipidus. Am. J. Physiol. **227:** 916–920.
16. USSING, H. H. & K. ZERAHN. 1951. Active transport of sodium as the source of electric current in the short-circuited isolated frog skin. Acta Physiol. Scand. **23:** 110–127.
17. LEAF, A. & E. DEMPSEY. 1960. Some effects of mammalian neurohypophyseal hormones on metabolism and active transport of sodium by the isolated toad bladder. J. Biol. Chem. **235:** 2160–2163.
18. TADORUKO, M., I. HAURAGUCHI, M. TERUOKA & F. SAKAI. 1968. Microperfusion experiments in collecting ducts of the isolated rat renal medulla *in vitro*. Jap. J. Pharmacol. **18:** 272–273.
19. SCHNERMANN, J., H. VALTIN, K. THURAU, W. NAGEL, M. HORSTER, H. FISCHBACK, M. WAHL & G. LIEBAU. 1969. Micropuncture studies on the influence

of antidiuretic hormone on tubular fluid reabsorption in rats with hereditary hypothalamic diabetes insipidus. Pflügers Archiv. **306:** 103–118.

20. Ullrich, K. J., C. Baldamus, E. Uhlich & G. Rumrich. 1969. Einflus von Calciummionen und antidiuretischem Homon auf transtubularen Natrium transport in der Ratteniere. Pflügers Archiv. **310:** 369–376.
21. Helman, S. I., J. Grantham & M. Burg. 1971. Effect of vasopressin on electrical resistance of renal cortical collecting tubules. Am. J. Physiol. **220:** 1825–1832.
22. Antoniou, L. D., T. J. Burke, R. R. Robinson & J. R. Clapp. 1973. Vasopressin-related alterations of sodium reabsorption in the loop of Henle. Kidney Int. **3:** 6–13.
23. Hall, D. A. & D. M. Varney. 1980. Effect of vasopressin on electrical potential difference and chloride transport in mouse medullary thick ascending limb of Henle's loop. J. Clin. Invest. **66:** 792–802.
24. Sasaki, S. & M. Imai. 1980. Effects of vasopressin on water and NaCl transport across the *in vitro* perfused medullary thick ascending limb of Henle's loop of mouse, rat and rabbit kidneys. Pflügers Archiv. **383:** 215–221.
25. Imbert, M., D. Chabardés, M. Montegut, A. Clique & F. Morel. 1975. Vasopressin dependent adenylate cyclase in single segments of rabbit kidney tubule. Pflügers Archiv. **357:** 173–186.
26. Imbert, M., D. Chabardés, M. Montegut, A. Clique & F. Morel. 1975. Présence d'une adenylcyclase stimulée par la vasopressine dans la branche ascendante des anses de néphrons du rein de lapin. C. R. Acad. Sci. Paris **280:** 2129–2132.
27. Morel, F., D. Chabardés & M. Imbert. 1976. Functional segmentation of the rabbit distal tubule by microdetermination of hormone-dependent adenylate cyclase activity. Kidney Int. **9:** 264–277.
28. Imbert-Teboul, M., D. Charbardés, M. Montegut, A. Clique & F. Morel. 1978. Vasopressin-dependent adenylate cyclase activities in the rat kidney medulla: Evidence for two separate sites of action. Endocrinol. **102:** 1254–1261.
29. Morel, F., D. Chabardés & M. Imbert-Teboul. 1980. Distribution of adenylate cyclase activity in the nephron. Curr. Top. Membr. Transport **12:** 415–426. Academic Press. New York, N.Y.
30. Chabardés, D., M. Gagnan-Brunette, M. Imbert-Teboul, O. Gontcharevskaia, M. Montegut, A. Clique & F. Morel. 1980. Adenylate cyclase responsiveness to hormones in various portions of the human nephron. J. Clin. Invest. **65:** 439–448.
31. Morel, F. 1981. Sites of hormone action in the mammalian nephron. Am. J. Physiol. **240:** F159–F164.
32. Kokko, J. P. & F. C. Rector. 1972. Countercurrent multiplication system without active transport in inner medulla. Kidney Int. **2:** 214–223.
33. Fourman, J. & G. C. Kennedy. 1966. An effect of antidiuretic hormone on the flow of blood through the vasa recta of the rat. J. Endocrinol. **35:** 173–176.
34. Fisher, R. D., J. P. Grunfeld & A. C. Barger. 1970. Intrarenal distribution of blood flow in diabetes insipidus: role of ADH. Am. J. Physiol. **219:** 1348–1358.
35. Cross, R. B., J. W. Trace & J. R. Vattuone. 1974. The effect of vasopressin upon the vasculature of the isolated perfused rat kidney. J. Physiol. **239:** 435–442.
36. Banks, R. O. 1976. Influence of antidiuretic hormone on intrarenal blood flow distribution in diabetes insipidus dogs and rats. Proc. Soc. Exptl. Biol. Med. **151:** 547–551.
37. Davis, J. M. & J. Schnermann. 1971. The effect of antidiuretic hormone on

the distribution of nephron filtration rates in rats with hereditary diabetes insipidus. Pflügers Archiv. **330:** 323–334.

38. TRINH-TRANG-TAN, M.-M., M. DIAZ, J.-P. GRUNFELD & L. BANKIR. 1981. ADH-dependent nephron heterogeneity in rats with hereditary hypothalamic diabetes insipidus. Am. J. Physiol. **240:** F372–F380.
39. MCAULIFFE, W. G. 1980. Histochemistry and ultrastructure of the interstitium of the renal papilla in rats with hereditary diabetes insipidus (Brattleboro strain). Am. J. Anat. **157:** 17–26.
40. SAWYER, W. H. 1952. Posterior pituitary extracts and excretion of electrolytes by the rat. Am. J. Physiol. **169:** 583–587.
41. GRINNELL, E. H., J. KRAMAR, W. M. DUFF & T. E. LYDON. 1968. Further studies on the diuretic activity of antidiuretic hormone. Endocrinol. **83:** 199–206.
42. MARTINEZ-MALDONADO, M., G. EKNOYAN & W. N. SUKI. 1971. Natriuretic effects of vasopressin and cyclic AMP: possible site of action in the nephron. Am. J. Physiol. **220:** 2013–2020.
43. KURTZMAN, N. A., P. W. ROGERS, S. BOONJARERN & J. A. L. ARRUDA. 1975. Effect of infusion of pharmacologic amounts of vasopressin on renal electrolyte excretion. Am. J. Physiol. **228:** 890–894.
44. BUCKALEW, V. M., JR. & K. A. DIMOND. 1976. Effect of vasopressin on sodium excretion and plasma antinatriferic activity in the dog. Am. J. Physiol. **231:** 28–33.
45. AKATSUKA, N., W. H. MORAN, M. L. MORGAN & M. F. WILSON. 1977. Effects of steady-state vasopressin levels on the distribution of intrarenal blood flow and electrolyte excretion. J. Physiol. **266:** 567–586.
46. STERN, P. & H. VALTIN. 1978. Lack of clear-cut antidiuretic effect of 8-*p*-chlorophenylthio cyclic AMP. Mineral Elect. Metab. **1:** 330–335.
47. MÖHRING, J., G. KOHRS, B. MÖHRING, M. PETRI, E. HOMSY & D. HAACK. 1978. Effects of prolonged vasopressin treatment in Brattleboro rats with diabetes insipidus. Am. J. Physiol. **234:** F106–F111.
48. JOHNSON, M. D., L. B. KINTER & R. BEEUWKES III. 1979. Effects of AVP and DDAVP on plasma renin activity and electrolyte excretion in conscious dogs. Am. J. Physiol. **236:** F66–F70.
49. SMITH, M. J., JR., A. W. COWLEY, JR., A. C. GUYTON & R. D. MANNING, JR. 1979. Acute and chronic effects of vasopressin on blood pressure, electrolytes, and fluid volumes. Am. J. Physiol. **237:** F232–F240.
50. FRIEDMAN, S. M. & C. L. FRIEDMAN. 1965. Salt and water distribution in hereditary and in induced hypothalamic diabetes insipidus in the rat. Can. J. Physiol. Pharmacol. **43:** 699–705.
51. BALMENT, R. J., I. C. JONES, I. W. HENDERSON & J. A. OLIVER. 1976. Effects of adrenalectomy and hypophysectomy on water and electrolyte metabolism in male and female rats with inherited hypothalamic diabetes insipidus (Brattleboro strain). J. Endocrinol. **71:** 193–217.
52. HAACK, D., E. HOMSY, G. KOHRS, B. MÖHRING, P. OSTER & J. MÖHRING. 1975. Studies on drinking feeding interactions in rats with hereditary hypothalamic diabetes insipidus. *In* Control of Mechanisms of Drinking. G. Peters, J. T. Fitzsimons & L. Peters-Haefeli, Eds.: 41–45. Springer-Verlag. Berlin.
53. MÖHRING, J., B. MÖHRING, A. SCHÖMIG, H. SCHÖMIG-BREKNER & D. HAACK. 1974. Acute effects of vasopressin on potassium and water balance in rats with diabetes insipidus. Am. J. Physiol. **227:** 921–926.
54. BALMENT, R. J., I. W. HENDERSON, I. C. JONES & W. MOSLEY. 1977. Water and electrolyte balance in adrenalectomized rats with diabetes insipidus (Brattleboro strain) given antidiuretic hormone. Gen. Comp. Endocrinol. **33:** 428–433.
55. MÖHRING, J., A. SCHÖMIG, H. BREKNER & B. MÖHRING. 1972. ADH-induced potassium retention in rats with genetic diabetes insipidus. Life Sci. **11:** 65–72.

56. Fernández-Repollet, E., M. Martínez-Maldonado & S. Opava-Stitzer. 1980. Role of water balance in the enhanced potassium excretion and hypokalemia of rats with diabetes insipidus. J. Physiol. **305:** 97–108.
57. Malnic, G., R. M. Klose & G. Giebisch. 1964. Micropuncture study of renal potassium excretion in the rat. Am. J. Physiol. **206:** 674–686.
58. Malnic, G., R. M. Klose & G. Giebisch. 1966. Micropuncture study of distal tubular potassium and sodium transport in rat nephron. Am. J. Physiol. **211:** 529–547.
59. Malnic, G., R. M. Klose & G. Giebisch. 1966. Microperfusion study of distal tubular potassium and sodium transfer in rat kidney. Am. J. Physiol. **211:** 548–559.
60. Finkelstein, F. O. & J. P. Hayslett. 1974. Role of medullary structures in the functional adaptation of renal insufficiency. Kidney Int. **6:** 419–425.
61. Silva, P., B. D. Ross, A. N. Charnet, A. Besarab & F. H. Epstein. 1975. Potassium transport by the isolated perfused kidney. J. Clin. Invest. **56:** 862–869.
62. Jamison, R. L., F. B. Lacy, J. P Pennell & V. M. Sanjana. 1976. Potassium secretion by the descending limb or pars recta of the juxtamedullary nephron *in vivo*. Kidney Int. **9:** 323–332.
63. Bengele, H. H., A. Evan, E. R. McNamara & E. A. Alexander. 1978. Tubular sites of potassium regulation in the normal and uninephrectomized rat. Am. J. Physiol. **234:** F146–F153.
64. Dobyan, D. C., F. B. Lacy & R. L. Jamison. 1979. Suppression of potassium recycling in the renal medulla by short-term potassium deprivation. Kidney Int. **16:** 704–709.
65. Gross, F., G. Dauda, S. Kazda, J. Kynčl, J. Möhring & H. Orth. 1972. Increased fluid turnover and the activity of the renin-angiotensin system under various experimental conditions. Circ. Res. **30–31** (Suppl. II): 173–181.
66. Gellai, M., B. R. Edwards & H. Valtin. 1979. Urinary concentrating ability during dehydration in the absence of vasopressin. Am. J. Physiol. **237:** F100–F104.
67. Martinez-Maldonado, M., A. Stavroulaki-Tsapara & G. Eknoyan. 1974. Renal effects of cyclic AMP in normal and congenital diabetes insipidus rats. Life Sci. **14:** 2025–2030.
68. Gellai, M. & H. Valtin. 1979. Chronic vascular constrictions and measurements of renal function in conscious rats. Kidney Int. **15:** 419–426.
69. Tisher, C. C., R. E. Bulger & H. Valtin. 1971. Morphology of renal medulla in water diuresis and vasopressin-induced antidiuresis. Am. J. Physiol. **220:** 87–94.
70. Buñag, D. D., I. H. Page, J. W. McCubbin. 1967. Inhibition of renin release by vasopressin and angiotensin. Cardiovasc. Res. **1:** 67–73.
71. Vander, A. J. 1968. Inhibition of renin release by vasopressin and vasotocin. Circ. Res. **23:** 605–609.
72. Goodwin, F. J., J. G. G. Ledingham & J. Laragh. 1970. The effects of prolonged administration of vasopressin and oxytocin on renin, aldosterone, and sodium balance in normal man. Clin. Sci. **39:** 641–651.
73. Tagawa, H., A. J. Vander, J. P. Bonjour & R. L. Malvin. 1971. Inhibition of renin secretion by vasopressin in unanesthetized sodium-deprived dogs. Am. J. Physiol. **220:** 949–951.
74. Shade, R. E., J. O. Davis, J. A. Johnson, R. W. Gotshall & W. S. Spielman. 1973. Mechanism of action of angiotensin II and antidiuretic hormone on renin secretion. Am. J. Physiol. **224:** 926–929.
75. Vandongen, R. 1975. Inhibition of renin secretion in the isolated rat kidney by antidiuretic hormone. Clin. Sci. Mol. Med. **49:** 73–76.
76. Hesse, B. & I. Nielsen. 1977. Suppression of plasma renin activity by intravenous infusion of antidiuretic hormone in man. Clin. Sci. Mol. Med. **52:** 357–360.

77. Konrads, A., G. Hofbauer, U. Werner & F. Gross. 1978. Effects of vasopressin and its desamino-D-arginine analog on renin release in the isolated perfused rat kidney. Pflügers Arch. **377:** 81–85.
78. Gutman, Y. & F. Benzakein. 1971. Effect of an increase and lack of antidiuretic hormone on plasma renin activity in the rat. Life Sci. **10:** 1081–1085.
79. Balment, R. J., I. W. Henderson & J. A. Oliver. 1975. The effects of vasopressin on pituitary oxytocin content and plasma renin activity in rats with hypothalamic diabetes insipidus (Brattleboro strain). Gen. Comp. Endocrinol. **26:** 468–477.
80. Gutman, Y. & F. Benzakein. 1974. Antidiuretic hormone and renin in rats with diabetes insipidus. Europ. J. Pharmacol. **28:** 114–118.
81. Oliver, J. A., R. J. Balment & I. W. Henderson. 1976. Sex differences in vasopressin-induced changes in plasma renin activity of Brattleboro and Long-Evans rats. J. Endocrinol. **71:** 79P–80P.
82. Peter, S. & J. Möhring. 1978. The juxtaglomerular apparatus of rats with hereditary hypothalamic diabetes insipidus. Cell. Tiss. Res. **188:** 335–339.
83. McCann, S. M., J. Antunes-Rodrigues, R. Nallar & H. Valtin. 1966. Pituitary-adrenal function in the absence of vasopressin. Endocrinol. **79:** 1058–1064.
84. Jones, I. C. & A. Wright. 1954. Some aspects of zonation and function of the adrenal cortex. IV. The histology of the adrenal in rats with diabetes insipidus. J. Endocrinol. **10:** 266–272.
85. Hilton, J. G. 1960. Adrencorticotropic action of antidiuretic hormone. Circ. **21:** 1038–1046.
86. Campbell, W. B. & J. M. Schmitz. 1978. Effect of alterations in dietary potassium on the pressor and steroidogenic effects of angiotensin II and III. Endocrinol. **103:** 2098–2104.
87. Davis, J. O., U. Urquhart & J. T. Higgins, Jr. 1963. The effects of alterations of plasma sodium and potassium concentration on aldosterone secretion. J. Clin. Invest. **42:** 597–609.
88. Dunn, M. J., H. P. Greely, H. Valtin, L. B. Kinter & R. Beeuwkes III. 1978. Renal excretion of prostaglandins E_2 and $F_{2\alpha}$ in diabetes insipidus rats. Am. J. Physiol. **235:** E624–E627.
89. Walker, L. A., A. R. Whorton, M. Smigel, R. France & J. C. Frohlich. 1978. Antidiuretic hormone increases renal prostaglandin synthesis *in vivo*. Am. J. Physiol. **235:** F180–F185.
90. Galvez, O. G., B. W. Roberts, W. H. Bay & T. F. Ferris. 1976. Studies of the mechanism of polyuria with hypokalemia. Kidney Int. **10:** 583A.
91. Galvez, O. G., B. W. Roberts, W. H. Bay & T. F. Ferris. 1977. Hemodynamic changes with hypokalemia. Clin. Res. **25:** 454A.
92. Zusman, R. M. & H. R. Keiser. 1977. Prostaglandin biosynthesis by rabbit renomedullary interstitial cells in tissue culture. Stimulation by angiotensin II, bradykinin, and arginine vasopressin. J. Clin. Invest. **60:** 215–223.
93. Hood, V. L. & M. J. Dunn. 1978. Urinary excretion of prostaglandin E_2 and prostaglandin $F_{2\alpha}$ in potassium-deficient rats. Prostaglandins **15:** 273–280.
94. Bankir, L., M.-M. Trinh-Trang-Tan, M.-P. Nivez, J. Sraer & R. Ardaillou. 1980. Altered PGE_2 production by glomeruli and papilla of rats with hereditary diabetes insipidus. Prostaglandins **20:** 349–365.
95. Dunn, M. J. & V. L. Hood. 1977. Prostaglandins and the kidney. Am. J. Physiol. **233:** F169–F184.
96. Gerber, J. G., R. D. Olson & A. S. Nies. 1981. Interrelationship between prostaglandins and renin release. Kidney Int. **19:** 816–821.
97. Henrich, W. L. 1981. Role of prostaglandins in renin secretion. Kidney Int. **19:** 822–830.

Discussion of the Paper

T. Berl (*University of Colorado, Denver, Col.*): I am not sure that I understand the mechanism of the high-potassium loss in animals put in competitive drinking situations, because all the factors (decreased water intake, decreased urine flow, low mineralocorticoid) should all add up to decreasing potassium excretion. Have you compared the aldosterone levels in these animals before and after they are put in the metabolic cage?

S. Opava-Stitzer: We did that last week, but we don't have the results yet. We have measured renin, however, and there is no change. But I think the important thing is that the change between a short period and a long period in the metabolism cages is not a gross change in water balance. There are no differences in water intake or urine flow rate or in water content of the muscle, so there are no changes that would alter total body water content. The changes that we see are only in the kidney. Another interesting point is that we are seeing this high potassium content in the kidney, when, in fact, the potassium content of the muscle is greatly reduced. So the kidney does not, in this situation, reflect changes in extracellular fluid or total body potassium.

L. B. Kinter (*Smith Kline and French Laboratories, Philadelphia, Pa.*): I also have some questions about the potassium situation in the DI rat. We have done a lot of work with the hormone analogue dDAVP, administering it chronically for over 30 days. We observe a dose-dependent decrease in urinary potassium excretion although urinary potassium remains a constant function of the total solute excreted. When the highest dose of dDAVP is administered to Brattleboro heterozygotes, no change in potassium excretion is detectable despite a 60% increase in urine osmolality. I conclude that there is a relationship between antidiuretic activity per se (and not pressor activity) and decreased potassium excretion. I agree with you that this effect is probably not due to a direct effect of antidiuresis on renal potassium handling. Rather, I suspect that it is a function of the change in urine osmolality achieved in a progressive and dose-dependent fashion.

Opava-Stitzer: I would agree with you. In fact, I don't think that anything you said conflicts with our data. What we would have to know in order to have a better idea of what's going on in your rats is the state of their potassium balance during the entire period that you were administering dDAVP. Are they hypokalemic when you start the study? Do they become potassium replete at some time during the study as a result of dDAVP or not? The fact that you don't see the decrease in potassium excretion in heterozygous rats would tend to support our hypothesis that vasopressin or dDAVP only has that effect if there is potassium deficiency.

Kinter: We have not done an intake/output balance, which I acknowledge is a deficiency in our experimental protocol. However, our animals always grow in parallel with our controls during the experimental periods. This is consistent with net positive potassium balance in these animals. In contrast to your studies we consistently detect decreased plasma potassium levels in our DI rats. Our animals are always housed individually in metabolic cages and we collect blood samples via chronically implanted catheters. We collect our samples very quickly, in as unstressed an animal preparation as is convenient for our studies.

Furthermore, we observe that in dDAVP-treated DI rats the plasma potassium levels rise to Long-Evans control levels. Since our animals have been kept in metabolic cages for up to 60 days, we don't observe the adaptation of plasma potassium that you have reported. It would be interesting to see if this is related to the different subcolonies used by our two laboratories.

OPAVA-STITZER: I would like to point out that in earlier studies by the Friedmans in which muscle potassium was measured, the DI rat was not found to be hypokalemic. Their rats were in metabolism cages for 22 days before sacrifice and that may be the reason the hypokalemia reported by Möhring could not be demonstrated.

M. D. LINDHEIMER (*University of Chicago, Chicago, Ill.*): That was a nice and detailed account of the many things causing intrarenal sequestration of sodium when you administer vasopressin to DI rats. Would you also add to your theory, based on information in the elegant abstract that Kriz and Bankir submitted to this meeting, that from the moment you give vasopressin, the thick ascending limb may start to hypertrophy?

When you put your rats in metabolic cages and observe the sodium and potassium loss that takes place in the first days, when they may still be acclimating to the cages, is their food intake down? We have all seen the solute loss that occurs when rats are not drinking and eating.

OPAVA-STITZER: I should have clarified this. The rats have already been in the metabolism cages for 60 hours before the day we have designated Day 1, so we miss that transient period of adaptation when they don't drink or eat.

MEMBRANE RESTING POTENTIALS AND EXTRA-CELLULAR POTASSIUM ION CONCENTRATIONS IN THE BRATTLEBORO RAT

John F. Laycock, Stephen J. Walter, Ivor B. Gartside, and Ann M. Jones

Department of Physiology
Charing Cross Hospital Medical School
London W6 8RF, England

The Brattleboro rat with hereditary hypothalamic diabetes insipidus (DI) is hypokalemic compared with rats of the parent Long-Evans strain.[1] Administration of vasopressin (as Pitressin tannate in oil) by subcutaneous injection to the Brattleboro DI rat is associated with dose-dependent potassium retention.[2] Since the resting potentials of cell membranes are associated with the relative distributions of ions, and in particular potassium ions (K^+), across those membranes, a hypokalemia could alter the K^+ equilibrium potentials and therefore the resting potentials. A change in resting potential for a nerve or muscle cell would alter the excitability of that cell. If the reduction in plasma K^+ concentration occurs in conjunction with a similar decrease in other extracellular fluids, such as the cerebrospinal fluid (CSF), then the excitability of cells within the central nervous system (CNS) could be similarly affected. Such an effect, if present, could account for certain behavioral differences that have been observed between Brattleboro DI rats and Long-Evans (or Wistar) rats. The Brattleboro rat, lacking endogenous vasopressin, shows an impaired ability to adopt normal avoidance behavior in response to noxious stimuli. This impairment can be reversed following the administration of vasopressin or vasopressin analogues.[3, 4]

In order to investigate whether the hypokalemia is associated with altered resting potentials and whether the CSF potassium concentration is also reduced in Brattleboro DI rats compared with Long-Evans rats, skeletal muscle fiber resting potentials and cation concentrations in plasma and CSF have been determined in both groups of animals. Some of the work presented here has already been accepted for publication.[5]

Adult male Brattleboro DI and Long-Evans rats weighing between 200–400 g were used in the experiments. Each animal was anesthetized with Inactin (Promonta, Hamburg) administered by intraperitoneal injection at a dose of 0.4 mmoles/kg body weight. Occasionally initial, light anesthesia was induced by a gas mixture of oxygen, nitrous oxide, and halothane. Experiments usually lasted approximately one to two hours, under similar environmental conditions. Statistical comparisons were made using the Student unpaired *t* test unless otherwise stated.

Resting Potential Determinations

In each anesthetized animal, the superficial flexor muscle of the right forelimb was exposed and the connective tissue removed. The surface of the muscle

0077-8923/82/0394-0209 $1.75/0 © 1982, NYAS

was kept moist with saturated paraffin oil. Resting potentials were then determined using microelectrodes filled with 3 M KCl inserted into muscle fibers by means of a micromanipulator, and an Ag/AgCl indifferent electrode placed on the surface of the exposed muscle. The potential differences between the electrodes were amplified by a high input impedence amplifier and displayed on a cathode-ray oscilloscope.[6] The initial reference line on the oscilloscope was displaced when the microelectrode penetrated a muscle fiber and the membrane potential was measured by returning the displayed line back to the reference level using a DC millivolt calibrator placed in series with the indifferent electrode and the amplifier.

For each vertical descent of the microelectrode the first five stable membrane potentials were recorded and assigned to fibers at progressively deeper levels within the muscle (fiber levels 1 to 5). The mean resting potentials at each fiber level for the Brattleboro DI and Long-Evans rats are given in FIGURE 1. There was no significant difference between the mean values for the outermost fibers (fiber level 1) for both groups of animals, and the vast majority of individual values were less than −70 mV. These values were therefore considered to represent the potentials of damaged muscle fibers or fibers not surrounded by true physiological fluid. However the mean values for progressively deeper fibers were significantly more negative (hyperpolarized) for the Brattleboro rats than the corresponding values for the Long-Evans control animals. Within each group of animals the resting potentials had stabilized by the fiber levels 4–5, with approximately 10 mV separating the two groups (−103.6 ± 1.8 and 92.8 ± 1.7 mV at fiber 5 level for Brattleboro DI and Long-Evans rats respectively; mean ± SE; $p < 0.005$.)

Potassium Equilibrium Potential Determinations for Skeletal Muscle

In order to verify the resting potential values, estimates of the K^+ equilibrium potentials for skeletal muscle fibers were calculated using the Nernst equation. Values for intracellular and extracellular potassium concentrations were determined as follows. Sections from the gastrocnemius or anterior tibialis muscles of the Brattleboro and Long-Evans rats were heated to constant weight at 100° C. Total water contents of the muscle sections were calculated from wet and dry weights. Each dried muscle was then ground to a powder and defatted with chloroform. A sample of the fat-free, dried muscle (ffdm) was dissolved with N HNO_3 and the K^+ concentration determined using a dual-channel integrating flame photometer (Evans Electroselenium Ltd). The total wet muscle K^+ concentration was calculated using the percentage water content of each muscle sample (TABLE 1).

The intracellular water content of fast-twitch muscle fibers has been determined as 80% of the total muscle water.[7] Since the muscles used in the present experiments were of the fast-twitch fiber variety, the intracellular K^+ concentration of the muscle fibers was calculated using this percentage water distribution (TABLE 1). There was no significant difference between the mean intracellular K^+ concentrations for muscle fibers from Brattleboro DI and Long-Evans rats.

A blood sample was obtained from each experimental animal and immediately centrifuged. The packed cell volume (PCV) was determined for each blood sample and the plasma Na^+ and K^+ concentrations determined by flame photometry. The plasma cation concentrations estimated from these blood

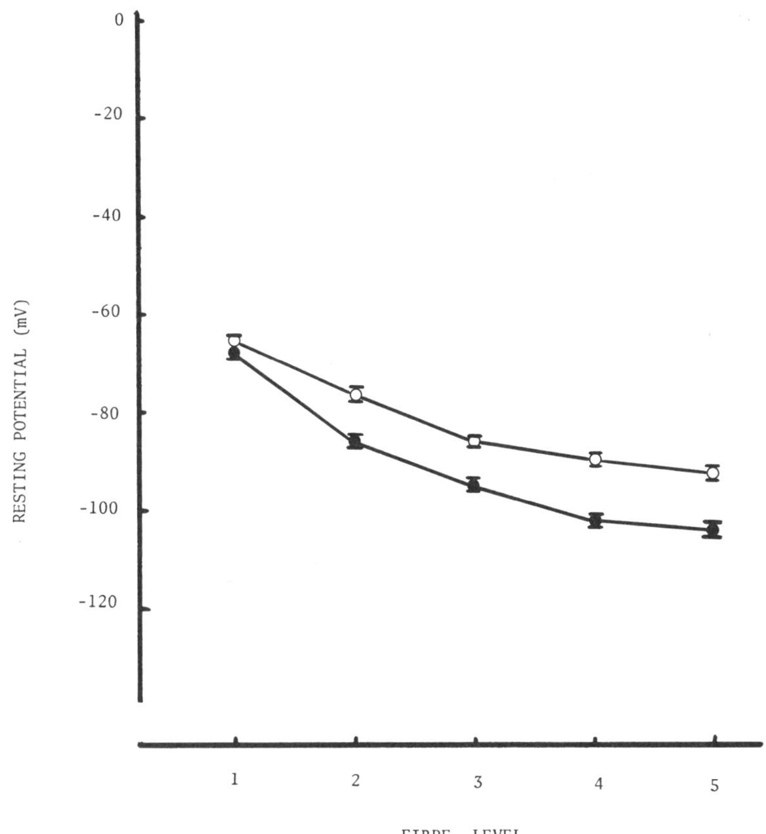

FIGURE 1. The resting potentials (mean ± SEM) of skeletal muscle fibers, at progressively deeper levels (1–5), from Brattleboro DI (●) and Long-Evans (○) rats.[5]

TABLE 1

SKELETAL MUSCLE PERCENTAGE WATER CONTENTS, FAT-FREE DRIED MUSCLE (FFDM) POTASSIUM CONCENTRATIONS, INTRACELLULAR POTASSIUM CONCENTRATIONS, AND POTASSIUM EQUILIBRIUM POTENTIALS IN BRATTLEBORO DI AND LONG-EVANS RATS (MEANS ± SE)

Skeletal Muscle	Brattleboro DI	Long-Evans
Water content (%)	73.6 ± 0.5 (9)	75.3 ± 0.7 (12)
K^+ content (mmol/kg ffdm)	350.5 ± 17.8 (9)	369.3 ± 9.0 (12)
Intracellular K^+ concentration (mmol/L)	157.0 ± 10.6 (9)	149.9 ± 4.6 (12)
K^+ equilibrium potential (mV)	−100.3 ± 2.7 (7)	−90.1 ± 1.6 (11)

Numbers in parentheses represent numbers of animals used.

samples (FIGURE 2) confirmed the finding that the Brattleboro DI rats are significantly hypokalemic compared with Long-Evans rats,[1] while there was no significant difference between the mean plasma Na^+ concentrations. The skeletal muscle K^+ equilibrium potentials were then calculated using the estimated intracellular K^+ concentrations and the measured extracellular (plasma) K^+ concentrations. The mean K^+ equilibrium potential for the Brattleboro DI rats was significantly more negative than the corresponding value for the Long-Evans rats ($p < 0.005$) and the difference was again of the order of 10 mV.

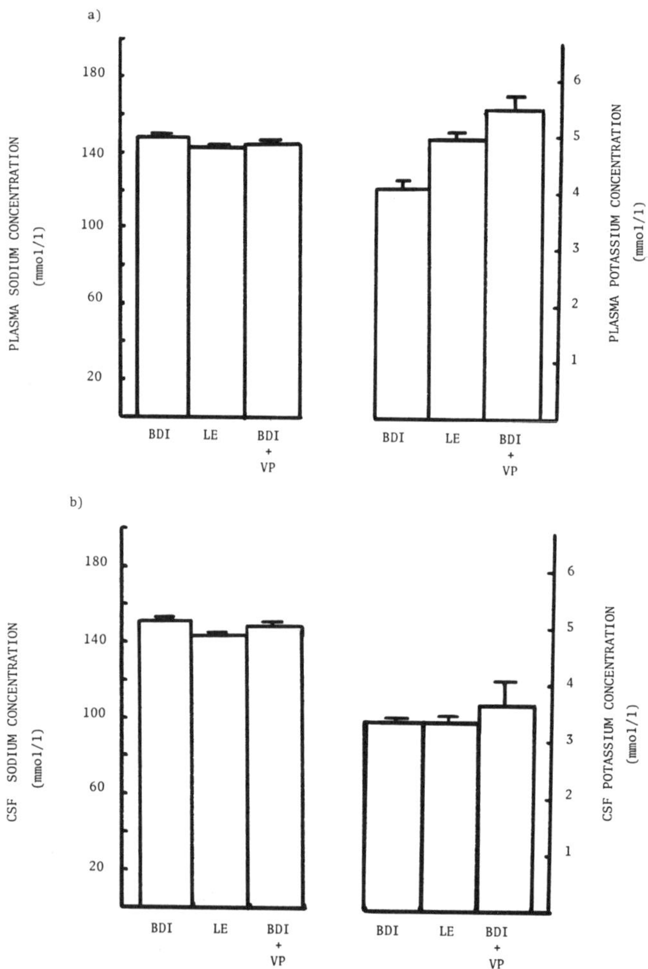

FIGURE 2. The sodium and potassium concentrations (mean ± SEM) of Brattleboro DI rats (BDI), Long-Evans rats (LE), and vasopressin-treated Brattleboro DI rats (BDI + VP) in (a) plasma and (b) CSF.

Potassium Equilibrium Potential Determinations for Erythrocytes

In order to determine whether other, nonexcitable, peripheral cells also have more negative K^+ equilibrium potentials, intracellular K^+ concentrations were determined for erythrocytes. The intracellular K^+ concentrations were estimated by two methods. The first consisted of hemolyzing a known volume of packed red cells from a centrifuged blood sample with a known volume of deionized water. The intracellular K^+ concentration for the erythrocytes could then be determined from the measured K^+ concentration of the sample. This method provided a direct measure of the intracellular K^+ concentration. The second method consisted of hemolyzing a known volume of whole blood with deionized water and then measuring the K^+ concentration of the sample. Using the values for the PCV and the plasma [K^+], the intracellular concentration was then calculated. This second method therefore provided an indirect, calculated measure of the intracellular K^+ concentration. Mean values for the intracellular [K^+] obtained using the two methods described above, for the Brattleboro DI and Long-Evans rats are given in TABLE 2. There was no significant

TABLE 2

PACKED CELL VOLUME (PCV), ERYTHROCYTE INTRACELLULAR POTASSIUM CONCENTRATIONS (DIRECT AND INDIRECT ESTIMATIONS), AND POTASSIUM EQUILIBRIUM POTENTIALS, IN BRATTLEBORO DI AND LONG-EVANS RATS (MEANS ± SE)

Erythrocytes	Brattleboro DI	Long-Evans
PCV	46.5 ± 0.8 (25)	47.7 ± 0.8 (26)
Measured intracellular K^+ concentration (mmol/L)	91.3 ± 1.9 (18)	88.7 ± 2.3 (18)
Calculated intracellular K^+ concentration (mmol/L)	98.1 ± 2.7 (18)	93.6 ± 4.6 (18)
K^+ equilibrium potential (mV)	−86.1 ± 1.7 (18)	−79.1 ± 1.8 (81)

Numbers in parentheses represent numbers of animals used.

difference between the mean values (measured and calculated) within either group of rats. Indeed, there were no significant differences between the corresponding mean values between the two groups of rats.

Using the measured intracellular and extracellular K^+ concentrations in the Nernst equation, the K equilibrium potentials were calculated for the erythrocytes. There was a significant difference ($p < 0.01$) between the mean values for Brattleboro DI rat and Long-Evans rat erythrocytes (TABLE 2) and again the former group of animals had the more negative value.

Cerebrospinal Fluid Electrolytes

Samples of cerebrospinal fluid were obtained from Brattleboro DI and Long-Evans rats by piercing the atlanto-occipital membranes with micropipettes. Those samples noticeably contaminated with blood were discarded and only clear CSF samples used for K and Na determinations using an Aminco Helium glow Photometer. An occasional sample that had a very slight contamination was immediately centrifuged and the clear supernatant kept for analysis. All samples were deposited under water-saturated paraffin oil until they were analyzed. Results from the few centrifuged samples were comparable to those obtained for the clear CSF samples and therefore all values for either Na^+ or K^+ concentrations were considered together. The mean cation concentrations for the Brattleboro DI and Long-Evans rats are shown in FIGURE 2. There was no significant difference between the mean K^+ values for the two groups of rats but the mean Na concentration for the Brattleboro DI rats was significantly higher ($p < 0.005$) then the corresponding value for the Long-Evans rats.

Effects of Vasopressin Administration to Brattleboro DI Rats

Sixteen male adult Brattleboro rats were given vasopressin as Pitressin tannate in oil by subcutaneous injection (500–1000 mU/24 h/rat) daily for three days after an initial 24 h control period. The rats were housed in individual cages with food and water *ad libitum* during the four-day experimental period. Daily vasopressin treatment resulted in an average 75% decrease in water intake so that by the third treatment day the amounts of water drunk by the Brattleboro rats were comparable to the fluid intakes of normal rats of similar body weight. The measurement of water intakes/24 h provided a simple measure of the antidiuretic effect of the vasopressin administration.

Each rat was then used for skeletal muscle fiber resting potential and extracellular (plasma and CSF) Na^+ and K^+ concentration determinations. All resting potential values less than —70 mV were excluded on the basis that they probably represented damaged fibers, as mentioned earlier. A mean value for the resting potential from undamaged 'deep' fibers was therefore obtained for this group of animals, and compared with mean resting potentials from similar fibers in 12 untreated Brattleboro DI rats and 10 Long-Evans rats. The resting potentials were -90.3 ± 1.1, -86.0 ± 1.2, and -82.8 ± 0.9 mV for untreated Brattleboro DI rats, Long-Evans rats, and vasopressin-treated Brattleboro DI rats, respectively. The mean resting potential for the treated Brattleboro rats was significantly less negative than the mean value for the untreated Brattleboro rats ($p < 0.001$) and was even significantly lower than the corresponding value for the Long-Evans rats ($p < 0.05$). The plasma and CSF Na^+ and K^+ concentrations for the vasopressin-treated Brattleboro DI rats are given in FIGURE 2. The mean plasma K^+ concentration was significantly raised in this group of animals compared with the mean for the untreated Brattleboro rats ($p < 0.001$) and was also significantly higher than the mean value for the Long-Evans rats ($p < 0.05$). On the other hand there were no significant differences between the CSF K^+ concentrations of any of the three groups of rats.

DISCUSSION

Brattleboro DI rats are hypokalemic compared with normal Long-Evans rats, confirming earlier observations.[1] In addition, vasopressin administration to the DI rat is associated with a significantly raised plasma potassium concentration, in agreement with the finding that potassium balance is restored to normal in these hormone-treated animals.[2] We have shown that the resting potentials of skeletal muscle fibers appear to be inversely related to the plasma potassium concentrations. Consequently, the relative depolarization of skeletal muscle fiber membranes of vasopressin-treated Brattleboro DI rats compared with Long-Evans rats may be associated with the relative degree of hyperkalemia in the former group of animals. It is then also necessary to relate the dose of vasopressin administered to the degree of potassium balance (and therefore the plasma K^+ concentration), as suggested by other studies.[2] The mean K^+ equilibrium potentials for Brattleboro DI and Long-Evans rats are in general agreement with the mean resting potential values in these two groups of animals, being significantly different to the same order of magnitude and in the same direction. Other peripheral cells also appear to be similarly affected since the mean K^+ equilibrium potential of erythrocytes from Brattleboro DI rats was also significantly more negative than the mean value for Long-Evans rats. It would therefore appear likely that peripheral cells, whether excitable (e.g. muscle) or non-excitable (e.g. erythrocytes), are influenced by the presence—or absence—of vasopressin. While these results support the view that the hormone effect is indirect, and is due to its regulatory influence on plasma K^+ concentration, the possibility that vasopressin acts directly on cell membranes to affect membrane permeabilities or pump mechanisms cannot be discounted.

Since the mean CSF K^+ concentration for untreated and vasopressin-treated Brattleboro DI rats and for the Long-Evans rats are not significantly different from each other, the regulation of this cation in the CSF is presumably vasopressin independent. On the other hand the mean CSF Na^+ concentration was significantly lower in the Long-Evans rat group than in the untreated Brattleboro DI group, while the mean value obtained for the vasopressin-treated DI rats was somewhere between the two and not significantly different from either. These observations are of interest in view of the finding that brain water permeability is apparently increased by intraventricularly administered vasopressin in Rhesus monkeys.[8] Despite the differences in CSF Na^+ concentrations it would seem unlikely that vasopressin influences CNS activity in the same way that it appears to affect peripheral cell membrane potentials.

Vasopressin has been detected in the CSF of rats, dogs, rabbits, and humans,[9–11] but not in the CSF of Brattleboro DI rats.[11] Since vasopressin does not appear to cross the blood–CSF barrier readily,[12] it seems probable that it enters the CSF directly from nerve terminals located on the walls of the third ventricle.[13] Intracerebroventricular administration of antivasopressin-serum to normal Wistar rats impairs active and passive avoidance behavior. Similar impaired avoidance responses to noxious stimuli are observed in Brattleboro DI rats[3, 4] and in hypophysectomized Wistar rats.[15]

These and other observations implicate vasopressin in the acquisition and maintenance of active and passive avoidance behavior, and suggest that the principal pathway by which this neuropeptide affects the CNS is by direct release into the CSF. However, since intravenously administered vasopressin can also influence the memory process,[4] it is likely that other pathways exist. These alternative pathways must involve vasopressin release from the nerve terminals of the posterior pituitary into the general circulation followed by (1) a direct effect across the blood-brain barrier, (2) direct effects via entry into the CSF, or (3) indirect effects via changes in circulating cation concentrations in the extracellular fluid outside the CSF.

References

1. MÖHRING, B., J. MÖHRING, G. DAUDA & D. HAACK. 1974. Potassium deficiency in rats with hereditary diabetes insipidus. Am. J. Physiol. **227:** 916–920.
2. MÖHRING, J., B. MÖHRING, A. SCHÖMIG, H. SCHÖMIG-BRECKNER & D. HAACK. 1974. Acute effects of vasopressin on potassium and water balance in rats with diabetes insipidus. Am. J. Physiol. **227:** 921–926.
3. BOHUS, B., TJ. B. VAN WIMERSMA GREIDANUS & D. DE WIED. 1975. Behavioural and endocrine responses of rats with hereditary hypothalamic diabetes insipidus (Brattleboro strain). Physiol. Behav. **14:** 609–615.
4. DE WIED, D., B. BOHUS & TJ. B. VAN WIMERSMA. 1975. Memory deficit in rats with hereditary diabetes insipidus. Brain Res. **85:** 152–156.
5. GARTSIDE, I. B., A. M. JONES, J. F. LAYCOCK & S. J. WALTER. 1981. The effect of vasopressin on extracellular cation concentrations and muscle resting potentials in the rat. J. Physiol. **318.** (In press.)
6. DE VILLIERS, P. L. G. & P. FATT. 1972. An input probe for microelectrodes suitable for classroom use. J. Physiol. **229:** 24P.
7. HOH, J. F. Y. & B. SALAFSKY. 1971. Effects of nerve cross-union on rat intracellular potassium in fast-twitch and slow-twitch rat muscles. J. Physiol. **216:** 171–179.
8. RAICHLE, M. E. & R. L. GRUBB, JR. 1978. Regulation of brain water permeability by centrally released vasopressin. Brain Res. **143:** 191–194.
9. VORHERR, H., M. W. B. BRADBURY, M. HOGHOUGHI & C. R. KLEEMAN. 1968. Antidiuretic hormone in cerebrospinal fluid during endogenous and exogenous changes in its blood level. Endocrinology **83:** 246–250.
10. DOGTEROM, J., TJ. B. VAN WIMERSMA GREIDANUS & D. F. SWAAB. 1977. Evidence for the release of vasopressin and oxytocin into cerebrospinal fluid: measurements in plasma and CSF of intact and hypophysectomized rats. Neuroendocrinology **24:** 108–118.
11. DOGTEROM, J., TJ. B. VAN WIMERSMA GREIDANUS & D. DE WIED. 1978. Vasopressin in cerebrospinal fluid and plasma of man, dog and rat. Am. J. Physiol. **234:** 463–467.
12. ZAIDI, S. M. A. & H. HELLER. 1974. Can neurohypophysial hormones cross the blood-cerebrospinal fluid barrier? J. Endocr. **60:** 195–196.
13. ZIMMERMAN, E. A. & A. G. ROBINSON. 1976. Hypothalamic neurons secreting vasopressin and neurophysin. Kidney Int. **10:** 12–24.
14. VAN WIMERSMA GREIDANUS, TJ. B., J. DOGTEROM & D. DE WIED. 1975. Intraventricular administration of anti-vasopressin serum inhibits memory consolidation in rats. Life Sciences **16:** 637–644.
15. BOHUS, B., W. H. GISPEN & D. DE WIED. 1973. Effect of lysine vasopressin and $ACTH_{4-10}$ on conditioned avoidance behaviour of hypophysectomized rats. Neuroendocrinology **11:** 137–143.

Discussion of the Paper

T. BENNETT (*University of Nottingham Medical School, Nottingham, England*): What is your definition of hypokalemia?

J. F. LAYCOCK: Our definition of hypokalemia was based on our values for Long-Evans rats. In other experiments we have in fact obtained values which are less (i.e., 3.6 mmol/liter) than those reported here.

BENNETT: In clinical terms, that would not be considered hypokalemia.

LAYCOCK: I accept your point.

P. STERN (*Dartmouth Medical School, Hanover, N.H.*): Perhaps I could make some editorial comments here. It is possible that we are all agreed that the Brattleboro rat has a slight hypokalemia. I think Dr. Opava-Stitzer also said that. I hope that at some point in the symposium someone will discuss appropriate dosages of vasopressin, because the comment was made that 1,000 mU/day was an appropriate dose, and many feel that it may be a high dose. At the same time that potassium levels and resting potentials were changing, sodium concentrations, albeit discretely, were changing also. Sodium concentration may also influence depolarization. Dr. Laycock, could you comment on that?

LAYCOCK: We didn't investigate the effects of sodium on membrane potentials in any way. All we can say is that we didn't get any significant change in the plasma sodium concentration.

W. BAILEY (*Rockefeller University, New York, N.Y.*): Following up on the first question, do you have a feeling for what the "clinical" manifestations of the hypokalemia might be? For example, in humans extreme hypokalemia is associated with muscle weakness and a similar condition may affect the behavior of the Brattleboro rats in experiments requiring strenuous motor performance over a long period of time. The next question I have is what about the effects of adrenocortical stimulation produced by this dose of vasopressin? Using only 500 mU per day, we found rather significant stimulation.

LAYCOCK: In answer to your first question, I have very little knowledge about diabetes insipidus in humans—to my knowledge there is not that much data available. However, I have heard of two patients at Charing Cross over the last year who were slightly hypokalemic, although I must confess it is a very slight hypokalemia.

With respect to your second point, we didn't measure aldosterone in our experiments, but it is quite possible that it may play a role in this situation.

M. D. LINDHEIMER (*University of Chicago, Chicago, Ill.*): I believe Dr. Opava-Stitzer touched on this topic in her talk, but I am still having trouble with the concept of no differences in plasma sodium levels between DI Brattleboro rats and their controls. The homozygous Brattleboro rat is characterized by its hyperosmolar state—in fact, it is this hyperosmolality that's "telling" it to drink. Thus, I am surprised by your data showing that all sodium levels were equal and that potassium was higher in the vasopressin-treated DI Brattleboro rats. This would mean that plasma osmolality of the Brattleboro rat receiving vasopressin was actually higher than in the untreated Brattleboro control. In most laboratories DI rats have a plasma osmolality of 305–310 mOsm, i.e. 8 mOsm above the controls. Thus, I have trouble with your not finding an elevation in sodium since it is mainly the effective osmoles of sodium and its anion that determine plasma tonicity.

LAYCOCK: While we didn't see a *statistically* significant difference in plasma sodium concentration between normal and DI rats their mean values differ by about 4 mM/L.

OPAVA-STITZER: Could I comment on that? There are some strange things going on in the DI rat with respect to serum electrolytes which Möhring also found and had no explanation for. When he treated DI rats with vasopressin, plasma sodium was decreased but plasma osmolality was not and the difference could not be accounted for by urea. There is obviously something else altered in serum.

LINDHEIMER: That's what I was aiming at. We really have to be sure that we are using accurate flame photometers or we'll be looking for idiogenic osmoles, when they may not be there!

STERN: Well, with this adequate reminder about technique, which is always important but especially when dealing with small differences, I think we should move on to the next paper.

LAYCOCK: I have not tried to make too much of the hypokalemic status. All I have tried to show is that hypokalemia is sufficient in the DI rat to affect peripheral cell membrane potentials.

STERN: I think that no one disputes your interesting and well documented observations. It is a more general question that we are all grappling with.

EXAGGERATED NATRIURETIC RESPONSE OF BRATTLEBORO RATS TO EXTRACELLULAR VOLUME EXPANSION

S. Opava-Stitzer, E. Fernández-Repollet, C. Rodriguez-Sargent,*
J. L. Cangiano,* and M. Martínez-Maldonado *

Department of Physiology
University of Puerto Rico School of Medicine
San Juan, Puerto Rico 00936

* *Medical and Research Services*
San Juan Veterans Administration Hospital
San Juan, Puerto Rico 00936

A state of chronic dehydration with reduced plasma volume,[1] decreased blood pressure,[2] and increased plasma renin activity (PRA)[2-4] has been demonstrated in rats with hereditary hypothalamic diabetes insipidus (DI rats). In this situation decreased renal perfusion and glomerular filtration rate[5] might result in sodium retention. On the other hand, the DI rat also suffers from mineralocorticoid deficiency[6-8] which might result in salt wasting. In addition it has recently been shown that in contrast to normal rats, there are no differences between superficial cortical and juxtamedullary nephrons of the DI rat with respect to single nephron filtration rate, glomerular volume, and proximal tubular length.[9] This lack of internephron heterogeneity might also affect renal sodium handling in the DI rat. It thus seems of particular interest to evaluate the natriuretic response of DI rats to volume expansion.

In the present study, the natriuretic responses of DI and normal Long-Evans rats to acute and chronic volume expansion were compared. Other factors involved in sodium handling, namely mineralocorticoids and renal ($Na^+ + K^+$)-ATPase activity, were also studied.

Materials and Methods

Male and female Long-Evans rats of the Brattleboro strain were used. They were either normal (LE) or homozygous for the hypothalamic diabetes insipidus trait (DI rats). Body weight ranged from 170–280 g. Control and experimental groups were matched for weight and sex.

Prior to any experiment, acute or chronic, rats were placed in individual metabolism cages for three to four days.

The effects of chronic volume expansion were studied in nine DI and eight LE rats on a high salt diet. The rats were first maintained on a normal diet during four days of balance study. The balance study consisted of daily measurements of food and water intake, body weight, urine volume and osmolality, and urinary excretion of sodium and potassium. The rats were then fed a high sodium diet (Purina Labchow plus 1% NaCl by weight). Five of the DI and four of the LE rats were injected daily with 0.5 mg/100 g BW of deoxycorticosterone acetate (DOCA, Upjohn). The remaining rats received an equivalent volume of sesame oil.

0077-8923/82/0394-0219 $1.75/0 © 1982, NYAS

The effect of acute volume expansion was studied in six DI and six LE rats following a six-day equilibration period in metabolism cages. On the morning of the seventh day the rats were placed in clean metabolism cages without food but with water *ad libitum*. Two hours later each rat was induced to urinate by touching its nose with an ether-impregnated gauze. This sample was discarded. A control four-hour urine collection was then begun. At the end of this period (and all subsequent periods) urination was induced and this urine added to that which had been spontaneously voided during the collection period. The rats were then weighed and given by gavage a volume of isotonic saline equivalent to 5% of body weight in two doses, ten minutes apart. Urine was collected 30, 60, and 120 min after the second dose of saline. Sodium concentration and volume of all samples were measured.

One week later blood was obtained from the same rats following a four-hour control collection as previously described. The next day the animals were again saline loaded and blood was obtained at 60 minutes after loading, the time when the peak of natriuresis had been found to occur. In this manner, blood sampling did not interfere with the natriuretic response. Hematocrit and serum sodium concentration were measured.

A similar experiment was carried out in 12 DI and 12 LE rats pretreated with either DOCA (6 DI, 6 LE; 2 mg/day) or oil for three days before saline loading. Still another group (6 DI, 6 LE) received free deoxycorticosterone (DOC, Sigma Chemical Co.; 2 mg/day) for three days before the saline-loading study.

Plasma aldosterone concentration and renal $(Na^+ + K^+)$-ATPase activity were measured in another group of DI and Long-Evans rats of both sexes. Thirteen DI and nine Long-Evans rats weighing 201–338 g were housed in metabolism cages for more than one week. A balance study was carried out on the last two days. The conscious rats were then bled by tailcutting for determination of plasma aldosterone concentration (Abbott Labs).[10] Eight of the DI and seven of the LE rats were then anesthetized with ether and the kidneys rapidly excised for determination of renal $(Na^+ + K^+)$-ATPase activity by methods previously described.[11] The microsomal fraction of a whole kidney homogenate was separated by differential centrifugation. Total ATPase activity was determined by incubation of this fraction with a reaction mixture containing a final concentration of 100 mM NaCl, 10 mM KCl, 10 mM imidazole buffer, 6 mM $MgCl_2$, and 6 mM disodium ATP at pH 7.4. Mg^{2+}-ATPase activity was determined by incubation with the same solution, from which sodium and potassium chloride had been omitted, and 2 mM ouabain added. The amount of inorganic phosphate produced in five minutes at 37° C was determined by the method of Fiske and Subbarow[12] and the protein content by the method of Lowry *et al.*[13] $(Na^+ + K^+)$-ATPase activity was expressed as the difference between total and Mg^{2+}-ATPase activity.

Sodium concentration of plasma and urine was determined by flame photometry (Radiometer, Model FLM 2E). Urine osmolality was determined by freezing-point depression (Advanced Instruments osmometer, Model 3W). Data was analyzed using paired-*t* and Student's *t* tests.

Results

As shown in Figure 1, the urinary sodium excretion ($U_{Na}V$) of both oil-treated and DOCA-treated Long-Evans rats was significantly increased on a high sodium diet. Urine flow rate was unchanged but a significant increase in urine osmolality was seen (1827 ± 170 versus 2149 ± 158 mOsm/kg H_2O; $p < 0.005$). Body weight also increased significantly. The only differences between $U_{Na}V$ of DOCA-treated and oil-treated LE rats occurred on days three and five of treatment. Sodium intake was greater than urinary sodium excretion on all days of study.

FIGURE 2 shows the response of DI rats to chronic volume expansion. Sodium excretion was significantly increased by a high salt diet with or without DOCA treatment. No differences were observed between $U_{Na}V$ of DOCA-treated and oil-treated rats on any day of study. Urine flow rate was signifi-

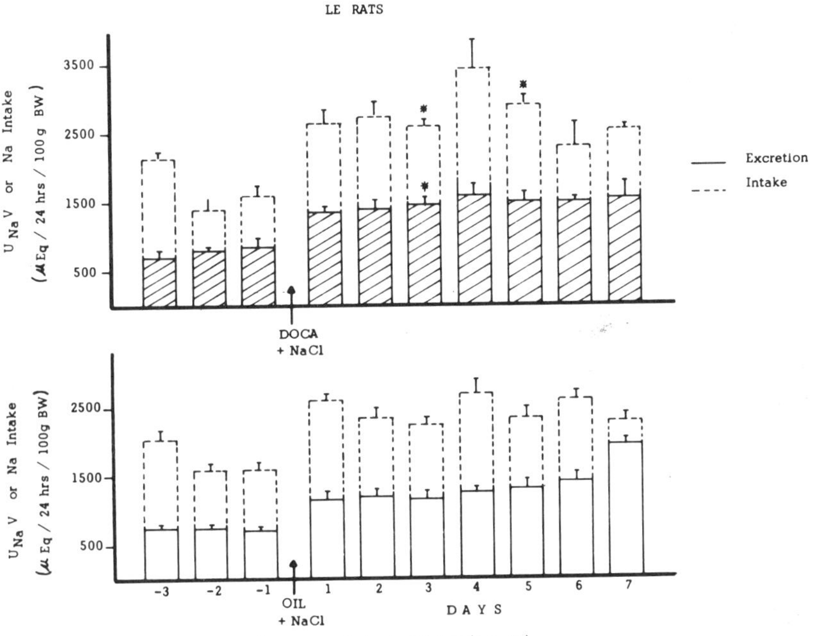

FIGURE 1. Urinary sodium excretion ($U_{Na}V$) and sodium intake of Long-Evans (LE) rats before and during a high salt diet plus DOCA or oil treatment. Values are means ± SEM. * Significantly different from oil-treated rats on the same day, $p < 0.05$.

cantly increased compared to values on a normal diet (95 ± 4 versus 72 ± 2 ml/24 h/100 g BW), while urine osmolality and body weight were unchanged. A negative sodium balance was observed in both oil- and DOCA-treated DI rats on the first day of a high salt diet. Urinary sodium excretion substantially exceeded sodium intake on this day. The urinary sodium excretion of DI and LE rats on the first day of a high salt diet is compared in FIGURE 3. DI rats had a significantly greater natriuresis on this day whether or not DOCA was also administered.

As shown in FIGURE 4, $U_{Na}V$ of DI rats was significantly higher than that of LE rats ($p < 0.01$) during all periods following an oral saline load. Maximal

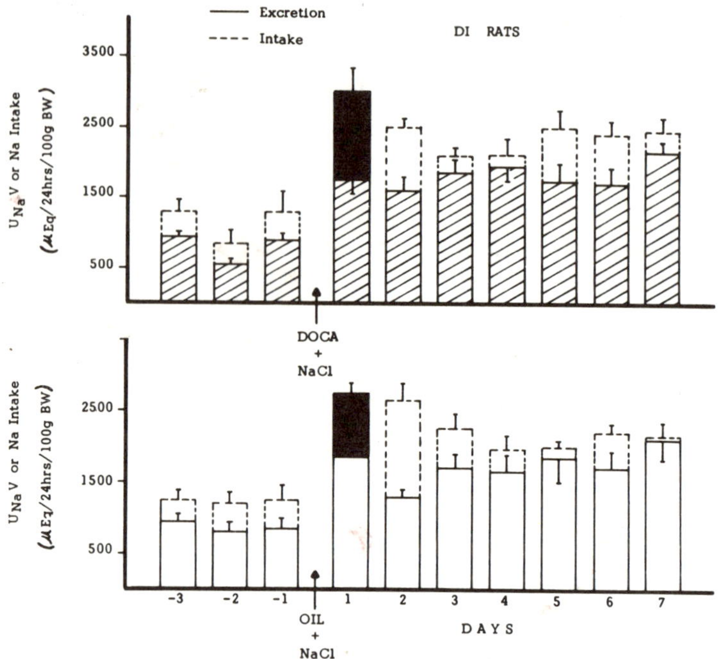

FIGURE 2. Urinary sodium excretion ($U_{Na}V$) and sodium intake of DI rats before and during a high salt diet plus DOCA or oil treatment. Values are means ± SEM. On Day 1 $U_{Na}V$ exceeded Na intake as shown by the shaded area.

$U_{Na}V$ was reached at 60 minutes after loading in both groups. No differences in $U_{Na}V$, hematocrit, or serum sodium concentration of DI and LE rats were observed in the period prior to saline loading. At 60 minutes after loading only serum sodium of DI rats differed from that of LE rats (147 ± 1 and 140 ± 2 mM/L, respectively).

Pretreatment with DOCA (FIGURE 5) significantly enhanced the natriuretic response of LE rats to saline loading but had no effect on the response of DI rats. Due to this enhanced response of LE rats, $U_{Na}V$ was higher in DI rats only at 90 and 120 min after loading. No difference was observed between DOCA-treated DI and LE rats with respect to $U_{Na}V$, hematocrit, and serum sodium concentration, either in the control period prior to loading or at 60 minutes after the saline load.

The effects of DOCA and DOC on the natriuretic response to saline loading are compared in FIGURE 6. These two forms of deoxycorticosterone were equally effective in augmenting the natriuretic response of normal rats, and neither affected the response of DI rats. Oil treatment had no effect in either DI or normal rats.

Plasma aldosterone concentration and renal $(Na^+ + K^+)$-ATPase activity of DI and Long-Evans rats are shown in TABLE 1. Plasma aldosterone concentration was significantly lower while renal $(Na^+ + K^+)$-ATPase activity was significantly higher in DI than in Long-Evans rats.

Discussion

DI rats had an exaggerated natriuretic response to both chronic and acute volume expansion. The natriuretic response of DI rats to a high salt diet was so great as to result in negative sodium balance during the first 24 hours (Figure 2). In contrast, Long-Evans rats, even when DOCA-treated, always excreted less sodium than the amount ingested (Figure 1). The negative sodium balance observed in DI rats was transient, and urinary sodium excretion decreased on subsequent days of the high sodium diet (Figure 2). Nevertheless, throughout the period of high salt intake, the urinary sodium excretion of DI rats represented a greater fraction of sodium intake than did the $U_{Na}V$ of Long-Evans rats.

No significant difference was observed between $U_{Na}V$ of DOCA-treated and oil-treated DI rats on any day of treatment and in Long-Evans rats $U_{Na}V$ was higher in the DOCA-treated group on only two of seven days of treatment. This may be related to the reported resistance to DOCA in this strain of rats.[14-16]

There was also a clear difference in the natriuretic response of DI rats to an acute saline load (Figure 3). Pretreatment with either DOCA or DOC abolished this difference by enhancing the response of Long-Evans rats. Both forms of deoxycorticosterone were used since the development of hypertension in response to free DOC treatment has been reported to be greater than when DOCA is used in the Long-Evans strain.[14]

The increased response of LE rats to a salt load following deoxycorti-

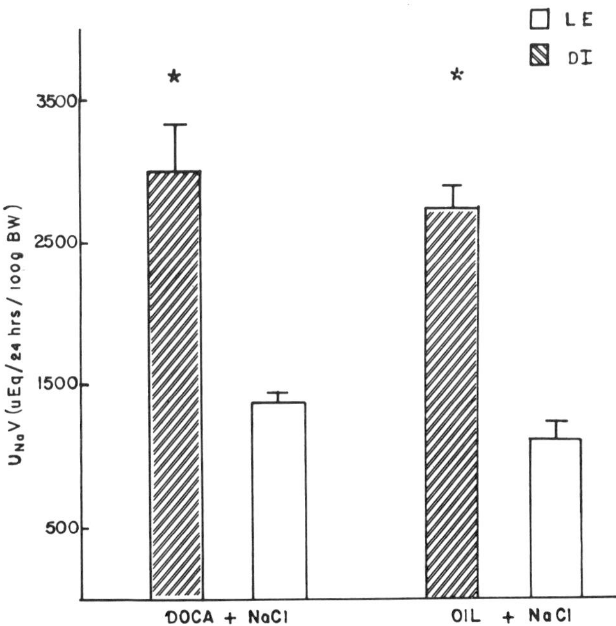

Figure 3. Comparison of $U_{Na}V$ of LE and DI rats on Day 1 of a high salt diet plus DOCA or oil treatment. Values are means ± SEM. * Significantly different from LE rats on the same treatment, $p < 0.05$.

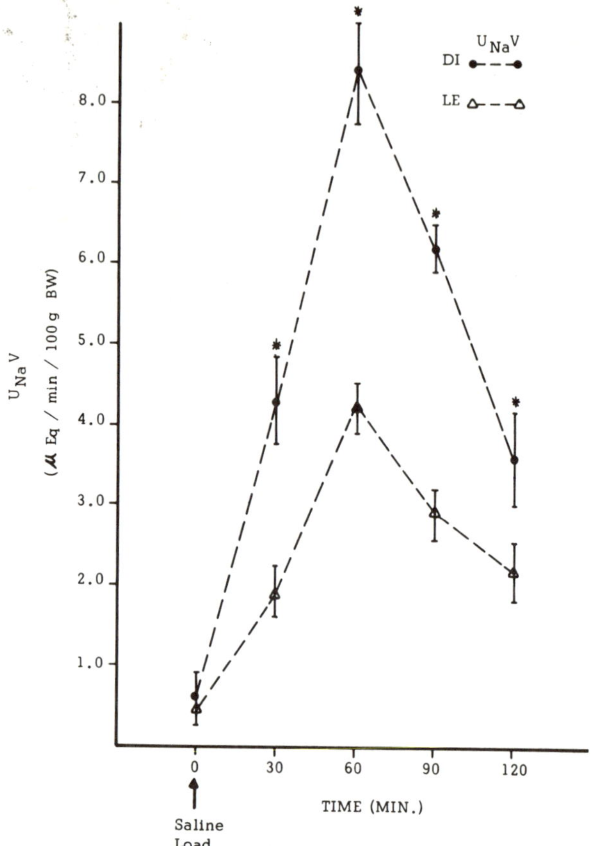

FIGURE 4. Urinary sodium excretion ($U_{Na}V$) of DI and LE rats before and after an acute oral saline load. Values are means ± SEM. * Significantly different from LE rats at the same time, $p < 0.01$.

costerone was no doubt due to volume expansion prior to saline loading. DOCA-treated LE rats had a higher $U_{Na}V$ (0.78 ± 0.15 versus 0.42 ± 0.10 μEq/min/100 g BW) and a lower hematocrit (45 ± 1 versus 47 ± 1) than oil-treated LE rats in the control period. In contrast, the hematocrit and urinary sodium excretion of DOCA-treated DI rats were not different from those of oil-treated DI rats in the control period. This suggests a lack of effect of DOCA in DI rats, possibly due to a reduction in the number and/or decreased affinity of tubular mineralocorticoid receptors, secondary to chronic mineralocorticoid deficiency.[6-8] It also suggests that the natriuretic response of DI rats after DOCA treatment was not due to DOCA-induced volume expansion and therefore different in nature from the similar response in untreated DI rats. Furthermore, since there was no significant difference between the hematocrits of DI (41 ± 1) and LE (41 ± 1) rats after saline loading, the greater natriuresis of DI rats would not seem to be attributable to a greater degree of volume expansion.

It has been shown that the absence of aldosterone is associated with reduced

$(Na^+ + K^+)$-ATPase activity,[17-19] and aldosterone replacement in adrenalectomized rats restores normal levels of enzyme activity.[18, 19] It is not clear, however, whether the hormone has a direct effect on the enzyme or whether it affects activity indirectly by increasing the reabsorption of sodium across the luminal membrane of the tubule, thereby providing more substrate for previously dormant enzyme. Recent studies by Petty *et al.*[20] have shown that restoration of normal enzyme activity in cortical collecting tubules of adrenalectomized rabbits, by aldosterone replacement, was prevented by simultaneous administration of amiloride, which blocks luminal sodium uptake in the collecting duct. These data support the hypothesis that mineralocorticoids affect $(Na^+ + K^+)$-ATPase activity indirectly, through changes in tubular sodium transport. In our studies renal $(Na^+ + K^+)$-ATPase activity was higher in the DI rat than the normal rat (TABLE 1), despite mineralocorticoid deficiency and an apparent resistance to deoxycorticosterone. This finding is consistent with the hypothesis that the level of plasma mineralocorticoids is not the primary

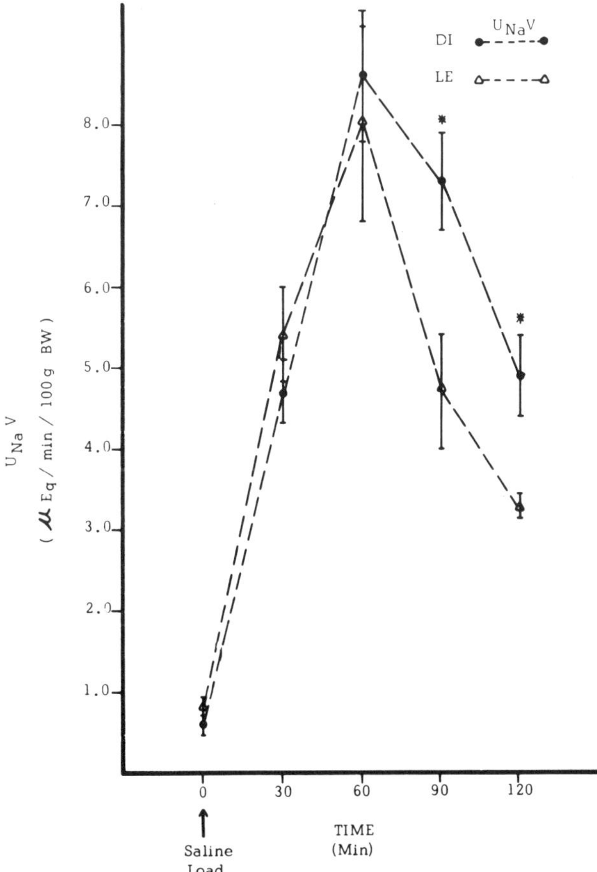

FIGURE 5. Urinary sodium excretion ($U_{Na}V$) of DOCA-treated DI and LE rats before and after an acute oral saline load. Values are means ± SEM. * Significantly different from LE rats at the same time, $p < 0.01$.

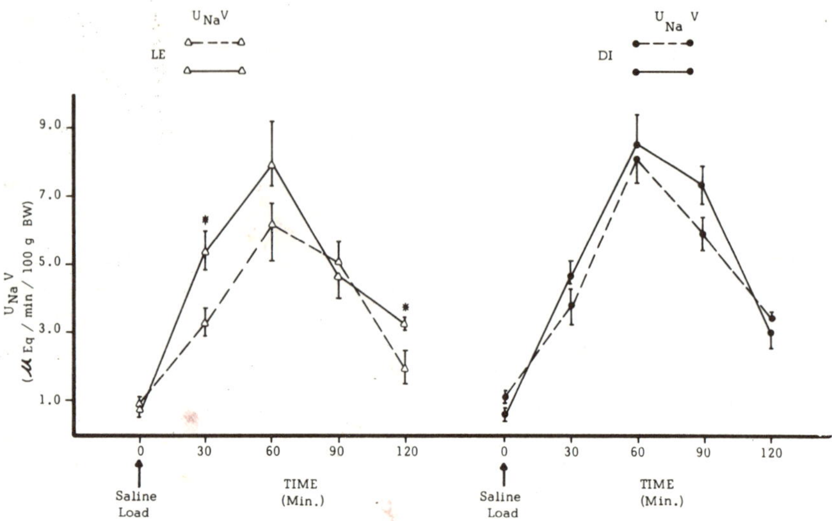

FIGURE 6. Comparison of the effects of DOCA and DOC on the natriuretic response of DI and LE rats to an acute oral saline load. Solid lines represent DOCA-treated rats; broken lines DOC-treated rats. Values are means ± SEM. * Significantly different from DOC-treated rats at the same time, $p < 0.05$.

regulator of renal ($Na^+ + K^+$)-ATPase activity. In addition the exaggerated natriuresis of the DI rat cannot be attributed to reduced renal ($Na^+ + K^+$)-ATPase activity. Even though we measured enzyme activity under basal conditions, the acute volume expansion induced in our experiments (either by oral saline loading or during the first 24 hours of a high salt diet) would not be expected to alter enzyme activity, as shown by others.[21, 22]

An alternate explanation for both the exaggerated natriuretic response and increased ($Na^+ + K^+$)-ATPase activity in the DI rat could be reduced passive sodium reabsorption by the loops of Henle of juxtamedullary nephrons in the absence of a high interstitial urea concentration. According to the Kokko model of sodium reabsorption in the inner medulla[23] reabsorption of sodium

TABLE 1

PLASMA ALDOSTERONE CONCENTRATION AND RENAL ($Na^+ + K^+$)-ATPASE ACTIVITY IN NORMAL LONG-EVANS RATS AND RATS WITH DIABETES INSIPIDUS (DI RATS)

	Aldosterone (ng%)	($Na^+ + K^+$)-ATPase (μmol PO_4/mg protein/h)
Long-Evans	12.4 ±1.6 (N=9)	73 ±6 (N=7)
DI	6.9 * ±1.2 (N=13)	123 † ±7 (N=7)

* Significantly different from normal Long-Evans rats, $p < 0.02$.
† $p < 0.001$.

chloride by thin ascending limbs of juxtamedullary nephrons occurs passively down a concentration gradient established by the abstraction of water in the descending limb. This water abstraction occurs due to the presence of urea in the interstitium and the low permeability of this segment to urea and NaCl. This would not, however, be a significant means of sodium reabsorption in the DI rat with its extremely low inner medullary urea concentration. There might thus be a chronically increased delivery of sodium to the thick ascending limb resulting in an adaptive increase in ($Na^+ + K^+$)-ATPase activity. This increased enzyme activity could then prevent salt wasting under basal conditions and no differences would be observed between DI and normal rats in the absence of salt loading. During volume expansion, however, when proximal tubular reabsorption is also depressed, the sodium not reabsorbed by the thin ascending limbs might appear in the urine, accounting for the augmented natriuretic response of the DI rat.

If this hypothesis is correct then there should be no difference between the sodium excretion of DI and normal rats during conditions of medullary urea washout. In fact during hypotonic mannitol infusion DI rats had CH_2O curves that were not different from those of normal Wistar rats[24] suggesting similar sodium handling under these conditions. In addition, in our studies, Long-Evans rats on a high salt diet increased their $U_{Na}V$ by increasing urinary sodium concentration (and osmolality) rather than urine volume, suggesting that the interstitial sodium concentration was increased. In contrast, urine flow rate, but not osmolality, was increased in the DI rat.

One possible pitfall in our experiments is that we measured whole kidney microsomal enzyme activity. It remains to be seen whether the increased activity is due to a selective change in the medullary enzyme.

An alternative explanation for the enhanced natriuretic response of the DI rat to oral salt loading must also be considered. It has been shown in the rat that intraportal isotonic saline loading results in a greater natriuresis than a similar load given intravenously.[25,26] It has been postulated that either a reflex[27] involving a hepatic sodium receptor, or a hepatic humoral factor,[25] may mediate this enhanced response. Since in our experiments saline was administered by gavage, we cannot rule out the possibility that the absence of ADH in some way facilitated the hepatic–mediated portion of the natriuresis in the DI rat.

In summary, the DI rat has been shown to have an exaggerated natriuretic response to both acute and chronic volume expansion. Despite aldosterone deficiency and an apparent unresponsiveness to mineralocorticoids, renal ($Na^+ + K^+$)-ATPase activity was elevated compared to normal rats. These findings and other data suggest that in the absence of ADH there may be reduced sodium reabsorption by the thin ascending limbs of juxtamedullary nephrons, resulting in both the enhanced natriuretic response and the increased renal ($Na^+ + K^+$)-ATPase activity. In addition, the absolute level of mineralocorticoids in the plasma would not seem to be the primary determinant of renal ($Na^+ + K^+$)-ATPase activity in the rat.

References

1. Möhring, J., B. Möhring, A. Schömig, H. Schömig-Brekner & D. Haack. 1974. Acute effects of vasopressin on potassium and water balance in rats with diabetes insipidus. Am. J. Physiol. **227:** 921–926.
2. Gross, F., G. Dauda, S. Kazda, J. Kynčl, J. Möhring & H. Orth. 1972.

Increased fluid turnover and the activity of the renin-angiotensin system under various experimental conditions. Circ. Res. **30** (Suppl. 2): 173–181.

3. GUTMAN, Y. & F. BENZAKEIN. 1971. Effect of an increase and lack of antidiuretic hormone on plasma renin activity in the rat. Life Sci. **10:** 1081–1085.
4. BALMENT, R. J., I. W. HENDERSON & J. A. OLIVER. 1975. The effects of vasopressin on pituitary oxytocin content and plasma renin activity in rats with hypothalamic diabetes insipidus (Brattleboro strain). Gen. Comp. Endocrinol. **26:** 568–477.
5. GELLAI, M. & H. VALTIN. 1979. Chronic vascular contrictions and measurements of renal function in conscious rats. Kidney Int. **15:** 419–426.
6. MÖHRING, B., J. MÖHRING, G. DAUDA & D. HAACK. 1974. Potassium deficiency in rats with hereditary diabetes insipidus. Am. J. Physiol. **227:** 916–920.
7. MÖHRING, J., G. KOHRS, B. MÖHRING, M. PETRI, E. HOMSY & D. HAACK. 1978. Effects of prolonged vasopressin treatment in Brattleboro rats with diabetes insipidus. Am. J. Physiol. **234:** F106–F111.
8. MCCANN, S. M., J. ANTUNES-RODRIGUEZ, R. NALLAR & H. VALTIN. 1966. Pituitary-adrenal function in the absence of vasopressin. Endocrinol. **79:** 1058–1064.
9. TRINH-TRANG-TAN, M.-M., M. DIAZ, J. P. GRUNFELD & L. BANKIR. 1981. ADH-dependent nephron heterogeneity in rats with hereditary hypothalamic diabetes insipidus. Am. J. Physiol. **240:** F372–F380.
10. OGIHARA, T., K. IINUMA, K. NISHI, Y. ARAKAWA, A. TAGAKI, K. KURATA, K. MIYAI & Y. KUMAHARA. 1977. A non-chromatographic non-extraction radioimmunoassay for serum aldosterone. J. Clin. Endo. Metab. **45:** 726–731.
11. WESTENFELDER, C., G. J. AREVALO, R. L. BARANOWSKI, N. A. KURTZMAN & A. I. KATZ. 1977. Relationship between mineralocorticoids and renal Na^+-K^+-ATPase: Sodium reabsorption. Am. J. Physiol. **233:** F393–599.
12. FISKE, C. H. & Y. SUBBAROW. 1925. The colorimetric determination of phosphorus. J. Biol. Chem. **66:** 375–400.
13. LOWRY, D. H., N. J. ROSEBROUGH, A. L. FARR & R. J. RANDALL. 1951. Protein measurement with the folin phenol reagent. J. Biol. Chem. **193:** 265–275.
14. BROWNIE, A. C., S. GALLANT, P. A. NICKERSON & L. M. JOSEPH. 1978. The occurrence of 11-deoxycorticosterone (DOC)—induced hypertension in the Long-Evans rat. Endocrine Res. Comm. **5:** 71–80.
15. HALL, C. E., S. AYACHI & O. HALL. 1973. Hypertension following adrenal enucleation and its absence during deoxycorticosterone treatment in Long-Evans rats. Endocrinol. **92:** 1175–1180.
16. HOLLAND, O. B., C. GOMEZ-SANCHEZ & T. ZIEGLER. 1979. Hypertension with mineralocorticoid administration to the Long-Evans rat. Clin. Sci. **56:** 109–113.
17. JORGENSEN, P. L. 1968. Regulation of the (Na^+-K^+)-activated ATP hydrolyzing enzyme system in rat kidney. I. The effects of adrenalectomy and the supply of sodium on the enzyme system. Biochim. Biophys. Acta **151:** 212–224.
18. JORGENSEN, P. L. 1969. Regulation of the (Na^+-K^+)-activated ATP hydrolyzing enzyme system in the rat kidney. II. The effect of aldosterone on the activity in kidneys of adrenalectomized rats. Biochim. Biophys. Acta **192:** 326–334.
19. SCHMIDT, U. J. SCHMID, H. SCHMID & U. C. DUBACH. 1975. Sodium and potassium activated ATPase. A possible target of aldosterone. J. Clin. Invest. **55:** 655–660.
20. PETTY, K. J., J. P. KOKKO & D. MARVER. 1981. Specific influence of aldosterone on Na^+-K^+-ATPase activity in the cortical collecting tubule. Clin. Res. **29:** 473A.
21. KATZ, A. I. & M. D. LINDHEIMER. 1975. Relation of Na^+-K^+-ATPase to acute

changes in renal tubular sodium and potassium transport. J. Gen. Physiol. **66:** 209–222.
22. MARTÍNEZ-MALDONADO, M. & A. SCHWARTZ. 1975. Renal ATPase as a receptor for drugs acting on the kidney. Methods Pharmacol. **4:** 199–225.
23. KOKKO, J. P. & F. C. RECTOR. 1972. Countercurrent multiplication system without active transport in inner medulla. Kidney Int. **2:** 214–223.
24. MARTÍNEZ-MALDONADO, M. & S. OPAVA-STITZER. 1978. Free water clearance curves during saline, mannitol, glucose and urea diuresis in the rat. J. Physiol. **280:** 487–497.
25. PERLMUTT, J. H., O. AZIZ & F. J. HABERICH. 1975. A comparison of sodium excretion in response to infusion of isotonic saline into the vena porta and vena cava of conscious rats. Pflugers Arch. **351:** 1–14.
26. VALDIVIESO, A. J. & G. O. PEREZ. 1981. Renal response to isotonic saline infusions into portal and jugular vein in sodium-loaded conscious rats. Proc. Soc. Exptl. Biol. Med. **167:** 261–266.
27. PASSO, S. S., J. R. THORNBOROUGH & A. B. ROTHBALLER. 1973. Hepatic receptors in control of sodium excretion in unanesthetized rats. Am. J. Physiol. **224:** 373–375.

DISCUSSION OF THE PAPER

L. B. KINTER (*Smith Kline and French Laboratories, Philadelphia, Pa.*): I would like to take issue with thc conclusion that the mineralocorticoid state is depressed in Brattleboro rats. We have measured 24-hour urinary excretion of free aldosterone in homozygous and heterozygous Brattleboro rats housed individually in metabolic cages for periods in excess of seven days. We consistently observe a much greater urinary aldosterone excretion rate in the DI animals, although we also find lower plasma aldosterone in the DI rats than in the heterozygous or normal rats, as reported by others. When DI rats are treated more than seven days with dDAVP, urine osmolalities increase, ranging from 1,200–1,500 mOsm and aldosterone excretion decreases. Paradoxically, and in agreement with previous studies, plasma aldosterone is higher, although not significantly, in the dDAVP-treated animal. I have to conclude that the adrenal cortex of DI rats is secreting more aldosterone per unit time than the adrenal of heterozygous Brattleboro rats. This is entirely reasonable given the well-established fact that DI rats are volume-depleted and have high renin plasma levels. The fact that the plasma aldosterone levels do not reflect that difference may be explained by enhanced excretion via the kidneys.

E. FERNANDEZ-REPOLLET: Thank you for this information. The only comment I have to make is that previous studies by Möhring *et al.* have shown that there is a reduction in the weight of the adrenal gland, and in the width of the zona glomerulosa. This correlates much better with low levels of plasma aldosterone than with enhanced secretion of aldosterone.

M. D. LINDHEIMER (*University of Chicago, Chicago, Ill.*): My statement is directed toward that nice data of Dr. Kinter's. When you are measuring aldosterone in plasma versus the urinary excretion of free aldosterone I would be concerned about systematic collection errors in a study involving an animal that excretes 200 ml of urine per day versus one whose output is much smaller.

ADRENOCORTICAL FUNCTION IN THE BRATTLEBORO RAT *

Catherine M. Milne, Richard J. Balment,‡ Ian W. Henderson,† Warwick Mosley,† and Ian Chester Jones †

Department of Zoology
University of Manchester
Manchester M13 9PL, England

and

† *Department of Zoology*
University of Sheffield
Sheffield, England

Introduction

The vasopressin-deficient Brattleboro rat (DI) is a useful model with which to elucidate adrenocortical function in the absence of antidiuretic hormone. The present investigations stem from studies suggesting that hypothalamic diabetes insipidus either primarily or secondarily is associated with hypoadrenocorticism. Thus peripheral plasma corticosterone concentrations may be reduced in the DI rat,[1-3] though concentrations within the normal range have been reported.[4,5] In addition, the adrenocortical response may be impaired although both the type and the severity of the stress imposed are influential.[6] Indeed vasopressin may be necessary for normal corticotrophin releasing factor(CRF)-stimulated ACTH release.[6-8] *In vitro,* DI rat adrenocortical tissue is also less responsive to ACTH.[9]

The reduced peripheral plasma aldosterone concentrations in the DI rat[2] are not readily understood in the light of an apparently vasopressin-dependent potassium wasting[10,11] and a highly activated renin-angiotensin system.[12,13]

The present investigations were aimed to reconcile certain of these perhaps incongruent observations, by determining steroidogenic capacity of the adrenal cortex in the DI rat and relating this to plasma hormone dynamics and adrenocortical morphology.

Impairment of the hypothalamic-pituitary-adrenal axis in the vasopressin-deficient DI animal is also evident, though to a lesser degree, in the heterozygous (HZ) Brattleboro rat.[2,3] This may be related to the partial defect in vasopressin secretion reported in the HZ animal.[14] Thus adrenocortical function of DI and HZ Brattleboro rats has been compared with that in vasopressin-replete animals of the parent Long-Evans strain.

Materials and Methods

Male Long-Evans (LE), and Brattleboro rats with (DI) and without (HZ) inherited hypothalamic diabetes insipidus were bred in departmental colonies.

* This work was supported in part by Science Research Council grant No. GR/A/62415.

‡ Address correspondence to R.J.B.

0077-8923/82/0394-0230 $1.75/0 © 1982, NYAS

Age-matched groups (100–180 days) were maintained on 12 h light/12 h dark regimes and allowed free access to food and tap water. Adrenocortical morphology was examined in rats given Oxoid 86 diet (Na^+: 0.17 mEq/g; K^+: 0.14 mEq/g). Adrenal glands were rapidly excised from ether anesthetized animals, weighed, and fixed in Bouin's fluid. Mean adrenocortical zone width estimates were made on mid-sections of both glands cut at six microns, and stained with hematoxylin and eosin.

Plasma hormone concentrations and corticosteroid secretory dynamics were examined in rats fed Labsure diet (Na^+: 0.12 mEq/g; K^+: 0.22 mEq/g). Abdominal aortic blood was collected into heparinized syringes within 90 sec of exposure to ether. Five minute blood collections were also taken from the left adrenal vein of Inactin (0.11 g/kg body wt i.p.; Byk Gulden)-anesthetized rats by direct needle (0.6 mm diameter) puncture. Plasma was stored at —20° C until analysis.

Experiments were carried out in the laboratory between 9 a.m. and 5 p.m. Peripheral plasma steroid levels were within the same range of those from animals bled by decapitation in the animal house.

Corticosteroid Production and Plasma Clearance Rates

The isotope dilution technique [15] was applied. Under Inactin anesthesia cannulae were placed in the left carotid artery and right jugular vein and a tracheotomy was performed. In preliminary experiments, three LE rats were infused (20 μl/min) with [^{3}H]aldosterone (45 Ci/mmol, Amersham) in 0.9% sodium chloride (0.7 μCi/ml). A steady state of isotope in plasma occurred between 75 and 120 min (Figure 1). A similar pattern of equilibrium occurred in rats infused with [^{3}H]corticosterone (50 Ci/mmol, Amersham).

Rats were thus infused with either [^{3}H]aldosterone or [^{3}H]corticosterone for 90 min. The venous cannula was then clamped and 3 ml of blood rapidly collected from the carotid artery. Separated plasma (100 μl) was counted for ^{3}H and the endogenous concentrations of unlabeled corticosterone and aldosterone measured by radioimmunoassay. The level of pure labeled steroid, separated from labeled metabolites, was then determined by extraction and chromatography of the plasma, following the addition of 0.001 μCi of pure [^{14}C]aldosterone (57 mCi/mmol, New England Nuclear) or [^{14}C]corticosterone (57 mCi/mmol, Amersham) as a recovery marker.

Plasma (0.8 ml) was diluted with 2.1 ml 0.05 M NaOH and allowed to equilibrate for 30 min with the ^{14}C steroid. A 0.1-ml aliquot was assayed for ^{3}H and ^{14}C and the remainder extracted with 2×10 ml dichloromethane and the extract washed with 0.1 M acetic acid and water before drying down under nitrogen. The dried residue was redissolved in 0.2 ml cyclohexane:toluene: methanol (50:50:7.5 vol/vol CTM) and applied to a Sephadex LH20 column. The columns were run with CTM and fractions containing aldosterone or corticosterone collected and counted for ^{3}H and ^{14}C. The purity of steroid obtained by the single Sephadex LH20 chromatography sequence was satisfactory since subsequent acetylation with acetic anhydride and pyridine followed by paper chromatography did not significantly alter the ^{3}H:^{14}C ratio. The plasma concentration of ^{3}H-labeled steroid was then computed by subtraction of the percentage of ^{3}H counts removed by purification from the counts found in the original plasma, using ^{14}C steroid recovery to assess procedural losses.

Plasma clearance rate (PCR) and production rate (R) were calculated as follows:

$$PCR = \frac{r}{x_c} (ml/min)$$

where r is the rate of ^{3}H steroid infusion (cpm/min) and x_c is the steady-state concentration of pure labeled steroid (cpm/ml). R (μg/min) equals PCR (ml/min) times the plasma steroid concentration (μg/ml).

Radioimmunoassays

Corticosterone

A slightly modified assay described by Kime [16] was used. Standards (0.02–1.0 ng) in ethanol and plasma samples extracted with 2×0.5 ml ethanol were

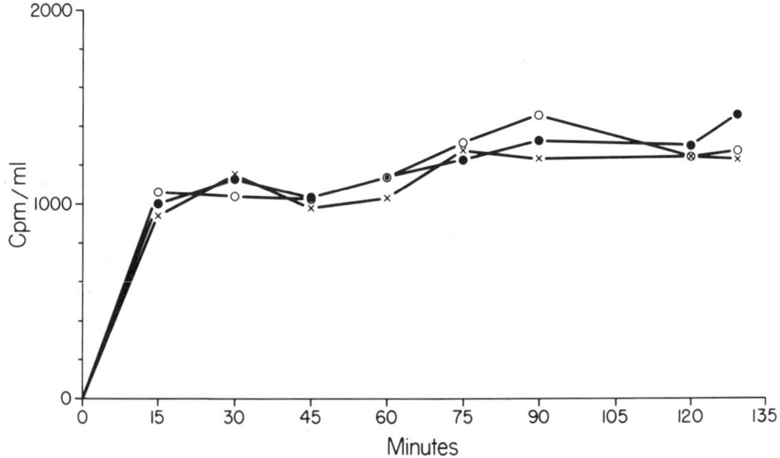

FIGURE 1. Temporal changes in plasma ^{3}H during constant intravenous infusion (0.014 μCi/ml) of [^{3}H]aldosterone. Arterial blood (0.2 ml) was withdrawn at 15-min intervals and replaced by an equal volume of 0.9% NaCl in three separate animals.

evaporated under vacuum. The antiserum against corticosterone-21-hemi-succinate-bovine serum albumin was raised in rabbits. At the separation stage of the assay 100 μl dextran-charcoal was added and, after centrifugation, the whole of the supernatant decanted into counting vials. Intra- and inter-assay variations were 9%.

Aldosterone

Plasma aldosterone was initially separated from other steroids by Sephadex LH20 chromatography. 50 μl aliquots of adrenal venous plasma or 1 ml of peripheral plasma were added to 50 μl of 0.1 μCi/ml [^{3}H]aldosterone (recovery tracer 80 Ci/mmol) and allowed to equilibrate for 30 min. Steroids were extracted with dichloromethane, the solvent dried under nitrogen and the residue redissolved in 3 × 0.2 ml CTM and applied to the column. CTM was run through the column, the aldosterone fraction collected, and the solvent dried under nitrogen. The residue was redissolved in 0.3 ml borate buffer; 50 μl of this were taken for recovery calculation and 2 × 100 μl for assay.

The antiserum was raised in rabbits against aldosterone-3-carboxymethoxime-bovine serum albumin (BSA) in the Department of Chemical Pathology, St. Mary's Hospital Medical School, London. One hundred μl 0.04 μCi/ml [^{3}H]aldosterone (80 Ci/mmol, Amersham) and 100 μl diluted antisera were added to the standards (5–200 pg) and samples then incubated overnight at 4° C. One ml dextran-charcoal (w/vol 0.0025% dextran, 0.25% activated charcoal, 0.05% BSA in borate buffer) was added to all the tubes and left for 10 min at 4° C. After centrifugation (1,200 × *g* for 15 min at 4° C), the supernatant was decanted and counted together with recovery samples. 18-hydroxycorticosterone (11,18,21-trihydroxypregnene-3,20-dione), which was not separated from aldosterone by the chromatography step, showed a cross-reactivity of 6%. The limit of detection of the standard curve was 2 pg. The intra- and inter-assay variations were 13% and 14%, respectively.

Progesterone

Steroids were extracted from plasma samples and standards with ether. The aqueous layer was frozen in an acetone/solid CO_2 bath and the organic layer decanted and then evaporated under vacuum. Two hundred μl [^{3}H]progesterone (0.02 μCi/ml, 80–110 Ci/mmol, Amersham) and antisera (200 μl, obtained from the World Health Organization) were added to the residues. The subsequent procedures including dextran-charcoal separation of free and bound steroid have been described.[17]

The limit of detection of the standard curve was 10 pg. Intra-assay variation was 8.0% and inter-assay variation 14.3%.

Unlabeled steroids and all other chemicals were obtained from Sigma Chemical Co., London. Solvents used were of analytical grade (British Drug Houses, Poole, Dorset) and were used without further purification. The purity of all radioactive steroids was checked by chromatography on paper and Sephadex LH20 columns.

Statistical Analysis

Values are presented as mean ± standard errors and all comparisons employed the Student's *t*-test.

Results

The reduced growth rate of Brattleboro DI compared with LE rats is evident when body weights at the same age are compared (Table 1). The lower rate of growth in the DI animals has been attributed to reduced pituitary growth hormone production[18] and/or to depletion of metabolic energy asso-

TABLE 1

ADRENOCORTICAL CHARACTERISTICS OF MALE DI AND HZ BRATTLEBORO AND LE RATS

		BW	Relative	Adrenal Zonation *			
			Adrenal	Total	Zona	Zona	Zona
			Weight	Cortical	Glomerulosa	Fasciculata	Reticularis
				Width	Width	Width	Width
		(g)	(mg/100 g BW)	(μm)	(μm)	(μm)	(μm)
	LE	(10) 400 ± 9	(10) 7.9 ± 0.2	(8) 929 ± 29	(8) 50 ± 2	(8) 579 ± 25	(8) 300 ± 14
	HZ	358 ± 11	8.6 ± 0.2	1263 ± 27	77 ± 2	748 ± 17	440 ± 25
	DI	261 ± 7	9.9 ± 0.6	1126 ± 29	90 ± 2	660 ± 21	375 ± 17
Statistical comparisons (p)	LE vs. HZ	<0.01	<0.05	<0.001	<0.001	<0.001	<0.01
	LE vs. DI	<0.001	<0.01	<0.01	<0.001	<0.05	<0.02
	HZ vs. DI	<0.001	NS	<0.01	<0.001	<0.01	NS

Values are means ± SE; * 16 measurements per gland.
Numbers of animals given in parenthesis.

ciated with their large water turnover.[19] In this and most other parameters measured in the study, the HZ rats, though more similar to LE, are intermediate between LE and DI animals. The size of the adrenal gland is not reduced in proportion to the lower body weight of the DI rat. Indeed, gland weight and adrenocortical width per 100 g body weight are greater in DI animals than either LE or HZ rats. It was also interesting to find that although the zonae fasciculata and reticularis occupy similar proportions of the adrenal cortex in all groups, the zona glomerulosa was more extensive in the DI rat occupying 8% of the cortex compared with 6% in HZ and 5% in LE rats.

Depressed peripheral plasma corticosterone and aldosterone concentrations have been confirmed in the DI rat (FIGURE 2). Progesterone, which is largely present in the circulation as an intermediate product of steroidogenesis, is also found at reduced levels in the DI rat. Although the concentrations of all three steroids were 100-fold greater in adrenal venous blood the same pattern of depressed hormone titres in the DI rat was evident. The free flow method of adrenal venous blood collection employed in this study resulted in similar blood flow rates (μl/100 g BW/min) in LE (27.2 ± 0.6, $N = 8$), HZ (30.7 ± 2.0, $N = 6$), and DI (30.4 ± 4.7, $N = 6$) rats.

The peripheral plasma corticosterone and aldosterone concentrations in the Inactin anesthetized rats for the clearance studies (TABLE 2) were similar to those obtained previously. The observed low rate of plasma hormone clearance in the DI rat alongside the depressed peripheral plasma corticosterone gave much reduced hormone production. The calculated corticosterone production in DI rats is less than a quarter of that in LE and HZ rats, even after correction for body weight differences. By contrast, the lower peripheral plasma aldosterone of the DI rat apparently reflects an elevated rate of plasma hormone clearance. There is no difference between the groups in the calculated rate of aldosterone production. It is perhaps significant that both Brattleboro rat groups (DI and HZ) show higher rates of plasma clearance of aldosterone compared with LE rats when related to body weight.

DISCUSSION

Previous observations have indicated reduced adrenocortical volumes in the DI rat compared with normal rats of similar body size.[1, 11] However, when animals of the same age are compared, the amount of adrenal tissue per 100 g body weight in the DI rat is greater than, or similar to, that of normal animals. The established influence of ACTH on adrenal blood flow[20] and the abnormal ACTH secretory patterns of the DI rat[3] might have been expected to produce altered adrenal blood flow in these animals. However, in the present study the rates of adrenal venous blood collection, under the free flow conditions employed, were similar in all groups of rats. In view of the relationship between adrenal blood flow and steroid secretion,[21] the steroid concentrations observed in the adrenal venous effluent collected in this way are unlikely to reflect the usual pattern of steroid secretion in these animals. Direct collection of adrenal venous blood thus may not provide a reliable index of adrenal steroid secretion. Accordingly, the isotope dilution technique has been applied, which necessitates far less manipulation of the animals.

Corticosterone and aldosterone levels in the peripheral circulation are much reduced in the vasopressin-deficient DI rat. Progesterone, thought to appear largely as an incidental intermediate of corticosteroid synthesis, is also present in reduced concentration in peripheral and adrenal venous blood of the DI rat. This might result from enhanced conversion of progesterone to corticosterone within the adrenal but is more likely, in view of the associated low titres of corticosterone, to indicate a lower overall rate of adrenal steroidogenesis in the DI animal. This was supported by the calculated depressed production rate for corticosterone, the principal product of the rat adrenal cortex, and is also consistent with the defective hypothalamic-pituitary-adrenal axis of the DI animal.[8]

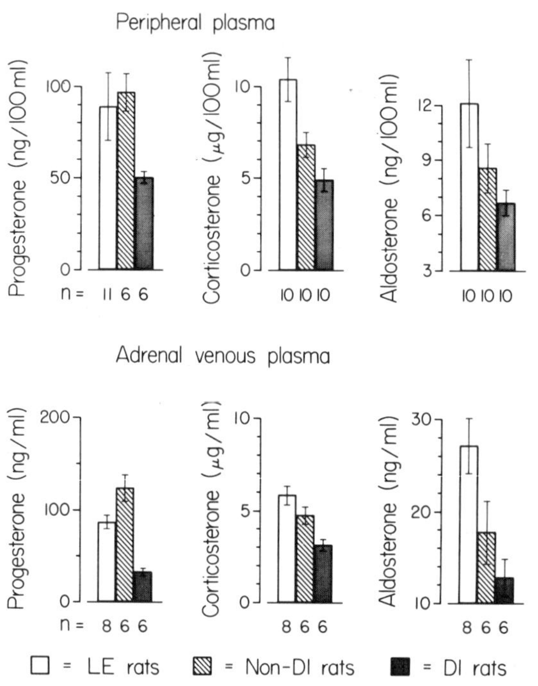

FIGURE 2. Progesterone, corticosterone, and aldosterone concentrations in peripheral (upper panel) and adrenal venous (lower panel) plasma of Brattleboro DI and HZ (i.e., non-DI), and LE rats. Each bar represents the mean ± SE.

STATISTICAL COMPARISONS (p)

	Progesterone	Corticosterone	Aldosterone
Peripheral Plasma			
LE vs. HZ	NS	<0.05	NS
LE vs. DI	NS	<0.01	<0.05
HZ vs. DI	<0.01	NS	NS
Adrenal Venous Plasma			
LE vs. HZ	<0.05	NS	NS
LE vs. DI	<0.01	<0.001	<0.01
HZ vs. DI	<0.01	<0.02	NS

TABLE 2

PERIPHERAL PLASMA CONCENTRATIONS, PLASMA CLEARANCE, AND PRODUCTION RATES OF CORTICOSTERONE AND ALDOSTERONE IN DI AND HZ BRATTLEBORO AND LE RATS

		Corticosterone			Aldosterone		
		Plasma Concentration (μg/100 ml)	Plasma Clearance (ml/100 g BW/min)	Production Rate (μg/100 g BW/min)	Plasma Concentration (ng/100 ml)	Plasma Clearance (ml/100 g (ml/100 g	Production Rate (ng/100 g BW/min)
	LE	12.6 ± 1.8	3.8 ± 0.7	0.46 ± 0.09	16.0 ± 1.5	2.7 ± 0.1	0.44 ± 0.05
	HZ	15.6 ± 1.9	3.9 ± 0.3	0.59 ± 0.05	12.0 ± 2.5	3.7 ± 0.2	0.37 ± 0.05
	DI	4.3 ± 0.7	2.3 ± 0.2	0.10 ± 0.02	10.0 ± 2.0	4.6 ± 0.5	0.40 ± 0.07
Statistical	LE vs. HZ	NS	NS	NS	NS	<0.001	NS
comparisons	LE vs. DI	<0.001	NS	<0.001	<0.05	<0.01	NS
(p)	HZ vs. DI	<0.001	<0.01	<0.001	NS	NS	NS

Values are means ± SE, $N = 8$ in each group.

The differential development of the zona glomerulosa in the DI rat is of interest in view of its relative independence of pituitary influences.[22] The high plasma renin activity reported in DI animals may be responsible for this enhanced zona glomerulosa growth and maintenance.[23] In keeping with this morphological development, the estimated rate of aldosterone production in these animals is within the normal range and not depressed, as previously suggested by data on peripheral hormone concentrations.[2] This also concurs with the similar zona glomerulosa *in vitro* aldosterone biosynthetic capacities in DI and HZ rats.[9] The lower plasma aldosterone concentration of the DI rat seems to result from an elevated rate of plasma hormone clearance, perhaps from increased renal[24] or hepatic extraction or from greater tissue 'usage.' Thus it would appear that although general adrenal steroidogenesis is depressed in the vasopressin-deficient rat, aldosterone secretion is maintained within the normal range. The altered adrenal steroidogenesis in the absence of vasopressin, may be understood in terms of the disturbed hypothalamic-pituitary-adrenal axis although the cause of associated alterations in steroid clearances remains obscure.

Acknowledgments

We are indebted to Dr. M. R. A. Morgan for the measurement of progesterone. We thank Professor V. H. T. James for the aldosterone antiserum.

References

1. McCann, S. M., J. Antunes-Rodrigues, R. Naller & H. Valtin. 1966. Pituitary-adrenal function in the absence of vasopressin. Endocrinology **79:** 1058–1064.
2. Möhring, B., J. Möhring, G. Dauda & D. Haack. 1974. Potassium deficiency in rats with hereditary diabetes insipidus. Am. J. Physiol. **227:** 916–920.
3. Buckingham, J. C. & J. H. Leach. 1980. Hypothalamo-pituitary-adrenocortical function in rats with inherited diabetes insipidus. J. Physiol. (Lond.) **305:** 395–404.
4. Arimura, A., T. Saito, C. Y. Bowers & A. V. Schally. 1967. Pituitary-adrenal activation in rats with hereditary hypothalamic diabetes insipidus. Acta Endocr. (Copnh.) **54:** 155–165.
5. Wiley, M. K., A. F. Pearlmutter & R. E. Miller. 1974. Decreased adrenal sensitivity to ACTH in the vasopressin-deficient (Brattleboro) rat. Neuroendocrinology **14:** 257–270.
6. Yates, F. E., S. M. Russell, M. F. Dallman, G. A. Hedge, S. M. McCann & A. P. S. Dhariwal. 1971. Potentiation by vasopressin of corticotrophin release induced by corticotrophin-releasing factor. Endocrinology **88:** 3–15.
7. Krieger, D. T., A. Liotta & M. J. Brownstein. 1977. CRF distribution in normal and Brattleboro rat brain and effect of deafferentation, hypophysectomy and steroid treatment in normal animals. Endocrinology **100:** 227–237.
8. Gillies, G. & P. J. Lowry. 1980. Corticotrophin releasing activity in extracts of the stalk median eminence of Brattleboro rats. J. Endocrinol. **84:** 65–73.
9. Kenyon, C. J., G. Hargreaves & I. W. Henderson. 1978. Adrenocortical function in rats with inherited hypothalamic diabetes insipidus. J. Steroid Biochem. **9:** 345–348.

10. MÖHRING, J., B. MÖHRING, A. SCHOMIG, H. SCHOMIG-BREKNER & D. HAACK. 1974. Acute effects of vasopressin on potassium and water balance in rats with diabetes insipidus. Am. J. Physiol. **227:** 921–926.
11. MÖHRING, J., G. KOHRS, B. MÖHRING, M. PETRI, E. HOMSY & D. HAACK. 1978. Effects of prolonged vasopressin treatment in Brattleboro rats with diabetes insipidus. Am. J. Physiol. **234:** F106–F111.
12. GUTMAN, Y. & F. BENZAKEIN. 1974. Antidiuretic hormone and renin in rats with diabetes insipidus. Eur. J. Pharmacol. **28:** 114–118.
13. BALMENT, R. J., I. W. HENDERSON & J. A. OLIVER. 1975. The effects of vasopressin on pituitary oxytocin content and plasma renin activity in rats with hypothalamic diabetes insipidus (Brattleboro strain). Gen. Comp. Endocrinol. **26:** 468–477.
14. MOSES, A. H. & M. MILLER. 1970. Accumulation and release of pituitary vasopressin in rats heterozygous for hypothalamic diabetes insipidus. Endocrinology **86:** 34–41.
15. TAIT, J. F., B. LITTLE, S. A. S. TAIT & C. FLOOD. 1962. The metabolic clearance rate of aldosterone in pregnant and non-pregnant subjects estimated by both single injection and constant infusion methods. J. Clin. Invest. **41:** 2093–3000.
16. KIME, D. E. 1977. Measurement of 1α-hydroxycorticosterone and other corticosteroids in elasmobranch plasma by radioimmunoassay. Gen. Comp. Endocrinol. **33:** 344–351.
17. MORGAN, M. R. A., P. G. WHITTAKER, B. P. FULLER & P. D. G. DEAN. 1980. A radioimmunoassay for Equilin in post-menopausal plasma: plasma levels of Equilin determined after oral administration of conjugated equine oestrogens (Premarin). J. Steroid Biochem. **13:** 551–555.
18. ARIMURA, A., S. SAWANO, T. W. REDDING & A. V. SCHALLY. 1968. Studies on the retarded growth in rats with hypothalamic diabetes insipidus. Neuroendocrinology **3:** 187–192.
19. VALTIN, H. 1967. Hereditary hypothalamic diabetes insipidus in rats (Brattleboro strain). Am. J. Med. **42:** 814–827.
20. SAPIRSTEIN, L. A. & H. GOLDMAN. 1959. Adrenal blood flow in the albino rat. Am. J. Physiol. **196:** 159–162.
21. PORTER, J. C. & M. S. KLAIBER. 1965. Corticosterone secretion in rats as a function of ACTH input and adrenal blood flow. Am. J. Physiol. **209:** 811–814.
22. GIROUD, C. J. P., J. STACHENKO & E. H. VENNING. 1956. Secretion of aldosterone by the zona glomerulosa of rat adrenal glands incubated *in vitro*. Proc. Soc. Expt. Biol. Med. **92:** 154–158.
23. REBUFFAT, P., A. S. BELLONI, G. MAZZOCCHI, P. VASSANELLI & G. G. NUSSDORFER. 1979. A stereological study of the trophic effects of the renin-angiotensin system on the rat adrenal zona glomerulosa. J. Anat. **129:** 561–570.
24. KINTER, L. B., W. FLAMENBAUM, J. C. MELBY & R. BEEUWKES. 1980. Renin and aldosterone in conscious diabetes insipidus rats: a paradox? Proc. Intl. Union Physiol. Sci., XIV. 28th Congress, Budapest, 1980. Abstract 2026.

DISCUSSION OF THE PAPER

H. W. SOKOL (*Dartmouth Medical School, Hanover, N.H.*): Our results are really rather different. We didn't see an increase in zona glomerulosa size, rather a slight decrease. We also found a significant reduction in the combined width of the zona fasciculata and zona reticularis. I am puzzled that you found a marked decrease in ACTH and corticosterone production at a time when the cortex was increased.

C. M. MILNE: This is a problem, I agree, and the zona fasciculata and reticularis might be expected to decrease. However, it should be remembered that adrenal secretory capacities in response to trophic stimuli may be tempered by dietary intake of Na and K. Some of the apparent differences between the adrenocortical profiles in our two laboratories may simply reflect differences in the compositions of our animal diets.

SOKOL: Your finding that aldosterone production in normal and DI rats was not very different is consistent with the morphology of the zona glomerulosa in my studies.

MILNE: For the DI rat there is an enhanced rate of aldosterone production relative to the secretion of corticosterone, which is the major steroid product in the rat. Such preferential stimulation of aldosterone production would be consistent with our observed expansion of the zona glomerulosa in the DI animals.

L. B. KINTER (*Smith Kline and French Laboratories, Philadelphia, Pa.*): Obviously I am very pleased to see these results. Dr. Rosen and I at the Beth Israel Hospital in Boston have just completed measurements in DI and Long-Evans rats that confirm your observations. We see up to a 50% increase in zona glomerulosa thickness. We were very careful to keep the plane of sectioning constant in all adrenals; several of us (without knowing the identity of the rats) made measurements that were averaged. I would offer an explanation for low plasma aldosterone levels accompanying high clearance rates and an increase in the zona glomerulis. All of us are measuring plasma steroid levels during the day, acutely in unconscious or unanesthetized animals. It may be at night, when all of these animals are active, that those levels are very much higher.

MILNE: We have examined circadian variations in plasma steroid levels in Long-Evans rats, and peaks in aldosterone and corticosterone concentration occur just prior to the dark period, that is about 6 p.m. All the experiments we have described were carried out between 10 a.m. and 4 p.m.

W. BAILEY (*Rockefeller University, New York, N.Y.*): Do you consider the possibility that the reduced secretion of adrenal corticosteroids could be a result of the hypokalemia?

MILNE: Indeed our DI rats are hypokalemic and this may contribute to the altered adrenal steroid dynamics in these animals.

SOKOL: In fact, zona glomerulosa cells in particular have been shown to secrete less aldosterone in a low potassium environment.

OXYTOCIN RELEASE AND RENAL ACTIONS IN NORMAL AND BRATTLEBORO RATS *

R. J. Balment

Department of Zoology
University of Manchester
Manchester M13 9PL, England

M. J. Brimble

Department of Pharmacy
Brighton Polytechnic
Brighton BN2 4GJ, England

M. L. Forsling

Department of Physiology
Middlesex Hospital Medical School
London W1P 6DB, England

Introduction

The presence of oxytocin-secreting neurons in the male[1] has increased speculation that oxytocin might have functions outside its established role in female reproductive activity. Circumstantial evidence in favor of an osmoregulatory role is provided by the observation that dehydration results in progressive depletion of pituitary oxytocin content, in parallel with the depletion of pituitary vasopressin.[2] Plasma oxytocin concentrations are also elevated in dehydrated animals.[3] We have also shown in the male rat that the progressive increase in plasma osmolality, following salt loading, is positively correlated with the coincident rise in plasma oxytocin concentration.[4] Thus, it would appear that an elevation of plasma osmolality might be an effective stimulus for oxytocin release. This hypothesis is supported by the observation that in the lactating female rat, in which oxytocin neurons may be identified,[5] elevation of plasma osmolality by salt loading results in activation of oxytocin neurons and elevation of plasma oxytocin levels.[6]

There have been numerous reports that oxytocin is capable of influencing renal function.[7] Oxytocin when given in sufficient quantities has been consistently found to increase renal Na and Cl excretion, though this may depend upon the presence of circulating vasopressin.[8] Most investigators have also reported that the hormone causes an increase in urine production,[9-11] though an antidiuretic action has also been observed in the normal rat[12, 13] and in the vasopressin-deficient Brattleboro rat.[12] Whilst it is not clear what plasma oxytocin levels were achieved in these studies, as oxytocin is rapidly removed from the circulation,[14] the amounts of oxytocin employed were relatively large compared with an effective antidiuretic dose of vasopressin.

In the present study pituitary content, osmotic release, and renal actions of oxytocin are compared in Long-Evans (LE) and vasopressin-deficient Brattle-

* Supported by the Science Research Council (Grant GR/A/624/5).

0077-8923/82/0394-0241 $1.75/0

boro (DI) rats. The renal actions of oxytocin are also examined in LE rats in which endogenous vasopressin secretion has been suppressed by either ethanol anesthesia or hypophysectomy.

METHODS

Animals

Adult male LE (350–500 g) and Brattleboro DI (250–350 g) were bred in the Department of Zoology, University of Manchester. Animals were maintained on a 12 h light/12 h dark regime and allowed free access to food (Labsure PMD diet) and water.

Measurement of Pituitary Oxytocic Activity

Pituitaries were removed from five LE and eight DI rats, and from seven DI rats which had been injected s.c. daily for seven days with 1.0 U vasopressin (Pitressin tannate in oil; Parke Davis & Co.) to restore normal water turnover.[15] Glands were acetone-dried and extracted into 5 ml of 0.05 M acetic acid and 0.5% chlorobutanol. At least five separate four-point bioassays were carried out on each pituitary extract against the Third International Standard for Oxytocin, using the isolated rat uterus[16] preparation from stilbestrol-treated rats.[17]

Estimation of Plasma Oxytocin Concentration

Trunk blood was collected from decapitated rats into cooled heparinized containers within 30 sec of stunning the animals. An aliquot was reserved for plasma osmolality determination and the remaining separated plasma was stored at —20° C until radioimmunoassayed for oxytocin as described by Chard and Forsling.[18] Blood was collected in this manner from eight untreated LE and seven DI rats, and from six DI rats, which were allowed to mildly dehydrate under Inactin anesthesia to raise their plasma osmolality. Values obtained by radioimmunoassay have been previously shown to correlate well with bioassay estimates.[19]

Renal Actions of Oxytocin

The renal effects of i.v. administered oxytocin were examined in groups of five to seven DI and LE rats. Animals were anesthetized with i.p. injection of Inactin, 0.11 g/kg (5-ethyl-5-(1′-methyl-propyl)-2-thiobarbiturate; Byk Gulden) and a tracheostomy performed. The right jugular vein was cannulated (Portex PP50) for hypotonic saline infusion, the left carotid artery cannulated (Portex PP50) for monitoring blood pressure (Elcomatic EM 750 pressure transducer) and urinary bladder cannulated (Portex PP90) for urine collection. Animals were maintained on a continuous i.v. infusion of 0.077 M NaCl at 150 μl/min (Harvard Model 935 infusion pump) and, after an initial 3.5 h equilibration period, 5-min urine collections were taken into preweighed vials to determine flow rate. Urinary Na concentrations were determined by flame photometry (Corning-EEL 430) and C1 by conductivity meter (Corning-EEL 920).

After four control urine collections (20 min) the infusate was changed for 20 min to one of identical ionic composition but containing synthetic oxytocin (Sigma Grade III); hormone-free infusate was returned for a final 30 min. Oxytocin was added to the infusate at different concentrations to allow delivery of hormone at 0.05, 0.15, or 1.5 mU/min. The plasma oxytocin concentrations induced by these rates of hormone infusion were determined in a parallel series of animals ($N = 6$ in each group). Blood samples were taken via the venous cannula at the end of the 20 min period of oxytocin administration, after discarding the initial 0.2 ml of collected blood. This procedure obviated any detectable (by radioimmunoassay) contamination of the plasma sample by the oxytocin infusion.

Further studies employed the highest rate of hormone administration (1.5 mU/min) only but followed the same basic protocol. In groups of five LE and six DI rats [^{14}C]inulin (Amersham International Ltd.) was added to the infusate (0.3 μCi/ml) throughout the experiment, including the initial 3.5 h equilibration period. Plasma ^{14}C counts had stabilized prior to the start of urine collections. Mean plasma ^{14}C counts were obtained from arterial blood samples (300 μl) withdrawn before the initial control period and at the end of the experiment. 50-μl aliquots of each urine collection were counted for ^{14}C to allow calculation of 5-min renal clearances.

The renal actions of oxytocin were also examined in LE rats in which endogenous vasopressin secretion was suppressed by cither ethanol anesthesia or acute hypophysectomy. A group of five LE rats were anesthetized with 12% ethanol in deionized water (60 ml/kg), given by stomach tube in two equal doses 20 min apart. This was supplemented as necessary during surgery by i.p. Brietal, 12 mg (Methohexitone sodium, Eli Lilly & Co.), and for the remainder of the experiment by addition of 3% ethanol to the infusate. A group of eight LE rats were hypophysectomized, by the parapharyngeal approach, immediately prior to the standard cannulation and infusion procedures. Completeness of pituitary removal was checked visually at post mortem. In both groups the renal response to a 20 min period of oxytocin administration (1.5 mU/min) was assessed as before.

Statistics

Values are presented as means ± SEM. In the renal studies the mean excretory rate of the last two control collections was used as a baseline for comparison with subsequent collections. The significance of any deviations from this baseline was assessed using the paired *t*-test. All other comparisons were by the unpaired *t*-test.

RESULTS

Pituitary Oxytocic Activity

The pituitary oxytocin content of Brattleboro DI rats (FIGURE 1) was very much lower than that in LE rats ($p < 0.001$). Vasopressin replacement

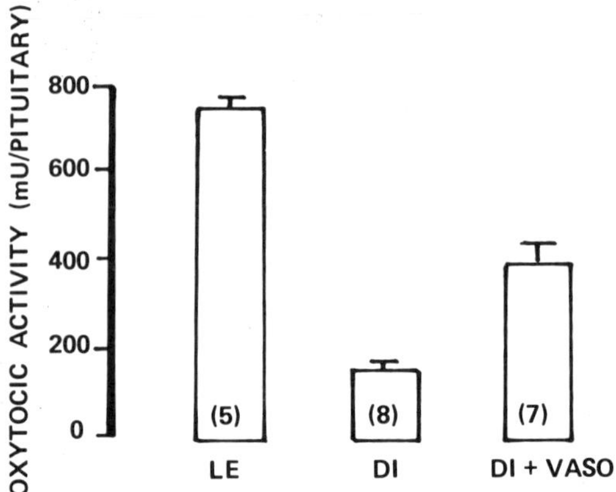

FIGURE 1. Pituitary oxytocin content in LE and DI rats, and in DI rats given 1.0 U vasopressin for seven days to correct their water turnover. Values are means, vertical bars indicate SEM and *N* values are shown in parenthesis.

in the DI rat, sufficient to restore normal water turnover,[15] significantly increased $p < 0.001$) the pituitary store of oxytocic activity but not to the level seen in LE animals.

Plasma Oxytocin Concentration

In association with the reduced pituitary oxytocin store, plasma oxytocin concentration was higher in DI than in LE rats. While oxytocin was not detectable (<2 μU/ml) in the plasma of untreated LE rats (plasma osmolality 299 ± 2 mOsmole/kg), untreated DI rats (plasma osmolality 303 ± 2 mOsmole/kg) showed plasma levels of 2.2 ± 0.5 μU/ml. In DI rats which were allowed to become mildly dehydrated under Inactin anesthesia (plasma osmolality 316 ± 4 mOsmole/kg) plasma oxytocin concentration rose to 8 ± 4 μU/ml. FIGURE 2 presents the mean plasma oxytocin concentrations and coincident plasma osmolalities of these two groups of DI rats. The values from neither of these groups differed significantly from the superimposed regression line, which describes the previously established relationship between plasma oxytocin concentration and plasma osmolality obtained in LE rats.[4]

Renal Actions of Oxytocin

By the end of the 3.5 h equilibration period the rates of renal excretion of fluid (150 ± 12 μl/min), Na (10.5 ± 2.6 μmole/min), and Cl (11.5 ± 1.5 μmole/min) were not significantly different in LE rats from the rates of infusion. Oxytocin administration at 0.05, 0.15, and 1.5 mU/min for 20 min produced plasma hormone concentrations of 5 ± 1, 16 ± 2, and 179 ± 29 μU/ml, respectively. While changes in renal excretion were equivocal with the lowest dose, significant increases in the rate of urine flow and Na and Cl excretion were induced with the higher doses (FIGURE 3). The transient increases in Na and Cl excretion reached a peak during the period of hormone administration, but the rise in urine flow produced by the highest dose was delayed to reach a maximum 5–10 min after oxytocin infusion had ceased (see FIGURE 4). These oxytocin-induced increases in renal water and electrolyte excretion are clearly related to the dose of oxytocin employed and the resultant plasma hormone concentration.

In DI rats at the end of the 3.5 h equilibration period urine flow (200 ± 15 μl/min) was considerably higher than the rate of infusion, while Na (8.5 ± 1.5 μmole/min) and Cl (9 ± 1.6 μmole/min) excretory rates were lower than infusion levels (11.6 μmole/min). In contrast to LE rats, oxytocin administration at 0.15 mU/min failed to induce any significant change in renal excretion in DI rats, despite producing a plasma hormone level of 28 ± 3 μU/ml, which was significantly higher ($p < 0.01$) than in LE rats. Only oxytocin administration at the highest rate (1.5 mU/min) produced significant changes in renal excretion in the DI rat. By the end of the period of oxytocin infusion Na and Cl excretion (μmole/min) had increased by 4.9 ± 0.97 ($p < 0.01$) and 5.6 ± 0.95 ($p < 0.01$), respectively. However, contrary to the diuresis seen in the LE rats urine production fell 44 ± 12.7 μl/min ($p < 0.02$) in DI rats during the period of hormone infusion (FIGURE 5). Oxytocin administration at 1.5 mU/min produced similar plasma hormone levels in LE and DI (148 ± 24 μU/ml) rats.

Mean arterial blood pressure (mm Hg) in anesthetized LE (120 ± 7, $N = 7$) and DI (128 ± 5, $N = 6$) rats was unaffected by oxytocin administra-

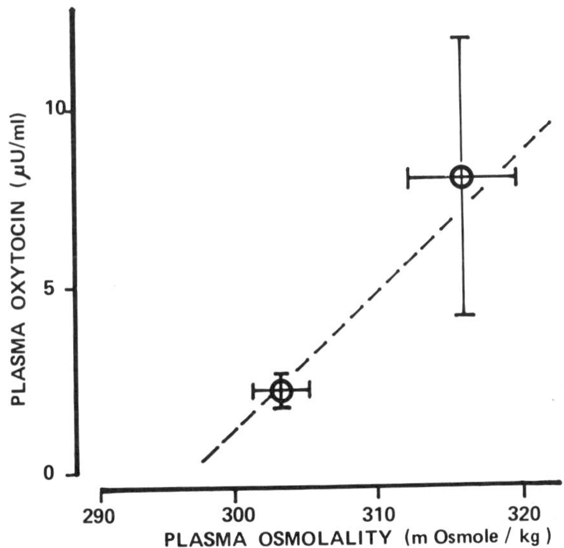

FIGURE 2. Comparison of plasma oxytocin concentration with plasma osmolality for groups of seven untreated and six mildly dehydrated DI rats. The regression line (dotted) describes the relationship between plasma oxytocin concentration and plasma osmolality previously established in LE rats (4) and is defined by the equation: Plasma oxytocin (μU/ml) = 0.37 × [Plasma osmolality (mOsmole/kg) − 297]. Plotted values are means ± SEM.

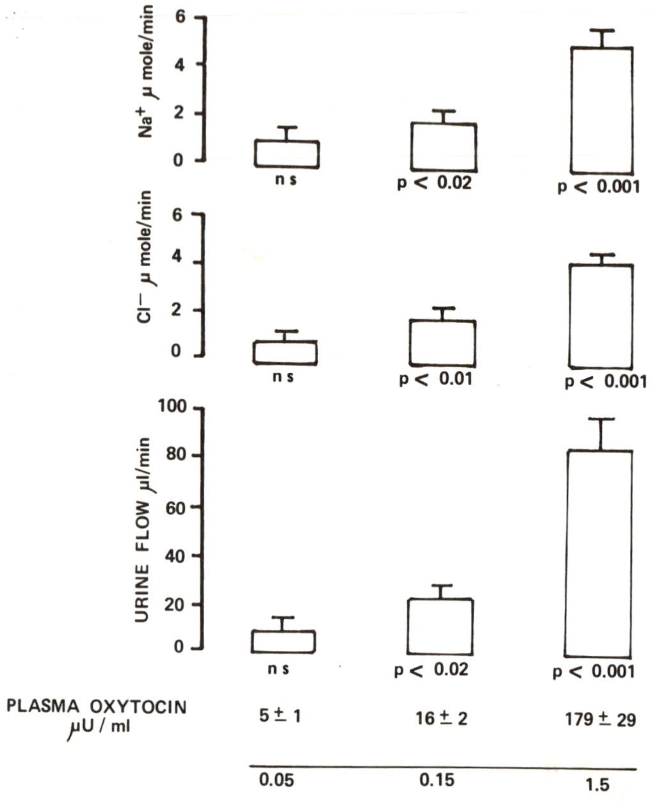

FIGURE 3. Plasma oxytocin concentrations and increases in renal excretion induced by oxytocin administration at 0.05, 0.15, and 1.5 mU/min in anesthetized, hypotonic saline infused LE rats. The increases in Na and Cl excretion and urine production are compared with initial control values by the paired *t*-test. Plasma oxytocin concentrations were obtained in a series of rats run parallel to the renal study groups. Values are means ± SEM from six or seven animals.

tion at any of the rates employed. Oxytocin administration at 1.5 mU/min also failed to produce any significant change in the rate of glomerular filtration in LE (FIGURE 4) or DI (FIGURE 5) rats. Rather, the diuresis associated with oxytocin administration in LE rats was coincident with a fall in the urine/plasma inulin ratio ($p < 0.05$), indicating a reduction in renal tubular fluid reabsorption. The contrasting antidiuretic response to oxytocin in DI rats was coincident with a significant increase in urine/plasma inulin ratio ($p < 0.05$), indicative of an increased rate of renal tubular fluid reabsorption.

The higher glomerular filtration rate of LE rats reflects the greater mean body weight of these animals; when related to body weight filtration rates (ml/100 g/min) were similar in LE (0.82 ± 0.07) and DI (0.79 ± 0.1) animals. However, the urine/plasma inulin ratio was significantly lower in DI (11 ± 1.8) compared with LE (31 ± 4.6) rats ($p < 0.001$), reflecting the lower fractional water reabsorption and consequent faster urine flow in these animals.

Suppression of endogenous vasopressin secretion in LE rats, by water loading and ethanol anesthesia or by hypophysectomy (FIGURE 6), was associated with an antidiuretic response to administered oxytocin similar to that previously seen in DI rats (FIGURE 5). Urine production fell by 79 ± 13 $\mu l/min$ ($p < 0.01$) in the ethanol anesthetized group and by 77 ± 13 $\mu l/min$ ($p < 0.001$) in the hypophysectomized group.

DISCUSSION

Vasopressin deficiency in the Brattleboro DI rat was associated with a depletion of pituitary oxytocin content.[20] In normal rats elevated plasma

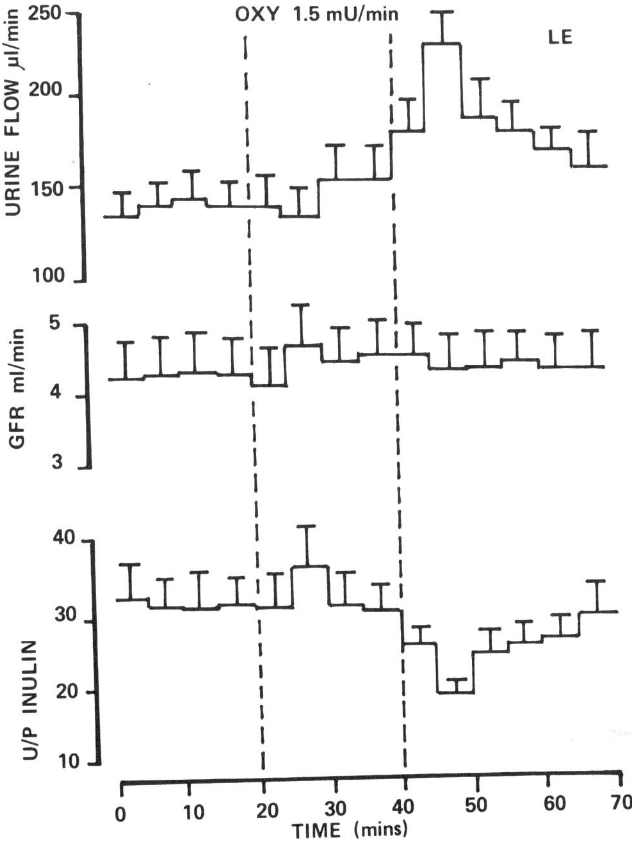

FIGURE 4. Effect of oxytocin administration (1.5 mU/min) on urine flow, glomerular filtration rate (GFR), and urine/plasma (U/P) inulin ratio in anesthetized, hypotonic saline infused LE rats ($N = 5$). Values are means, vertical bars represent SEM.

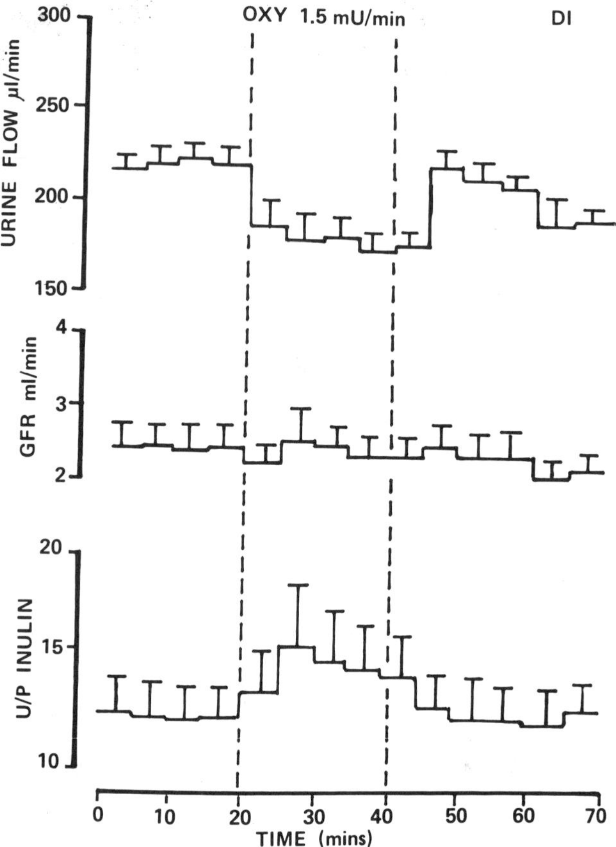

FIGURE 5. Effect of oxytocin administration (1.5 mU/min) on urine flow, glomerular filtration rate (GFR), and urine/plasma (U/P) inulin ratio in anesthetized, hypotonic saline infused DI rats ($N=6$). Values are means, vertical bars represent SEM.

osmolality is known to activate oxytocin-releasing neurons,[21] raise plasma oxytocin concentration,[4, 6] and, if prolonged, may lead to depletion of pituitary oxytocin stores.[2] The raised plasma osmolality of DI rats is associated with hypertrophy of the hypothalamo-neurohypophyseal system,[22] increased firing rate of neurohypophyseal neurons,[23] and elevated plasma oxytocin levels.[3] The correction of water turnover in DI rats by vasopressin administration led to a restoration of pituitary oxytocin store toward normal levels.[24]

In the LE rat the concentration of oxytocin in plasma is positively correlated with plasma osmolality.[4] The regression line describing this relationship is shown in FIGURE 2. Accordingly oxytocin was not detectable in the plasma of normal LE rats but was evident in the plasma of untreated DI rats which have a somewhat higher plasma osmolality. Plasma hormone concentrations in untreated and mildly dehydrated DI rats were not significantly different from the levels expected on the basis of this relationship between plasma oxytocin concentration and plasma osmolality. Thus, it would appear that the reduced pituitary oxytocin store and raised plasma hormone concentration, in DI animals, reflect enhanced hormone release in response to chronic dehydration rather than a defect in oxytocin synthesis or release.

The saliuretic and diuretic effects of oxytocin in LE rats were similar to those previously reported,[7] the size of the response being clearly dependent on the plasma oxytocin concentration. Significant renal responses were achieved when plasma hormone concentration was raised to 16 μU/ml, a level that can be induced in the conscious animal by increasing plasma osmolality to 340 mOsmole/kg.[4] The vasopressin-deficient DI rats appeared to be less sensitive than normal LE rats to administered oxytocin, despite their reported greater sensitivity to vasopressin.[25] Oxytocin administration at 0.15 mU/min, though inducing plasma hormone levels of 28 μU/ml, failed to produce significant changes in renal excretion. Experiments in the neurohypophysectomized dog[8] suggest that circulating vasopressin may be required for the full expression of the renal actions of oxytocin. Indeed, we have been able to confirm the presence

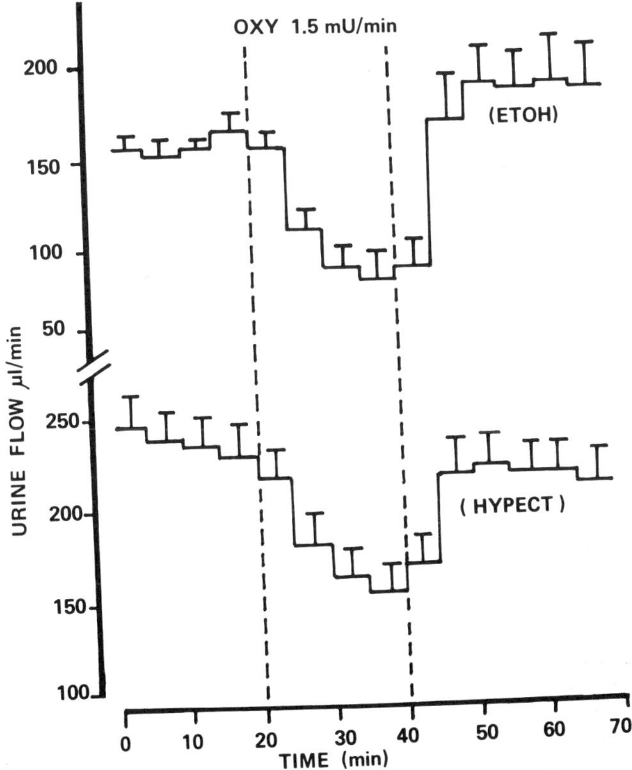

FIGURE 6. Effect of oxytocin administration (1.5 mU/min) on urine flow in five ethanol-anesthetized (ETOH) and eight hypophysectomized (HYPECT) LE rats receiving hypotonic saline infusion at 150 μl/min. Values are means, vertical bars represent SEM.

of vasopressin (unpublished observations) in the blood of the LE (1.0 ± 0.25 μU/ml, $N = 6$) but not the DI rat preparations as used in the renal studies. Although vasopressin may enhance the renal response to oxytocin, it is clearly not essential, as oxytocin administration at the higher rate of 1.5 mU/min produced a natriuresis and chloruresis in the DI rat similar to those seen in LE animals.

Despite evoking similar saliuretic responses in both groups, oxytocin administration at 1.5 mU/min produced an antidiuresis in DI rats in contrast to the diuresis seen in LE animals.[12] These differing effects on urine production were related to hormone-induced changes in tubular fluid reabsorption rather than altered rates of glomerular filtration. Ethanol anesthesia and acute hypophysectomy, both recognized procedures for suppression of endogenous vasopressin secretion, were associated with an antidiuretic response to vasopressin. Chan,[26] on the evidence of molecular modification studies, proposed that the effects of oxytocin on urine flow were largely due to oxytocin acting as a competitive antagonist at the renal vasopressin receptor. However, the antidiuretic action of oxytocin, seen when vasopressin is congenitally absent or its release is suppressed, suggests that oxytocin does possess some intrinsic antidiuretic activity, though it is much less potent than vasopressin. When vasopressin is present in the circulation, the diuretic action of oxytocin may be due to the displacement of the more potently antidiuretic vasopressin from the renal receptor allowing urine flow to increase. The results of the present study are thus consistent with oxytocin acting as a partial agonist at the renal vasopressin receptor.

In conclusion it would appear that increased plasma osmolality is an effective stimulus for oxytocin release, which may be responsible for the depletion of pituitary oxytocin and the high plasma oxytocin titers in DI rats. Plasma oxytocin concentrations, similar to those induced by raising plasma osmolality to the levels produced by severe dehydration,[2] are effective in increasing renal Na and Cl excretion and urine flow. The level of vasopressin in the circulation may determine the pattern of oxytocin-induced changes in urine flow and may enhance the saliuretic response to oxytocin. While the present study provides no evidence that oxytocin plays any major role in the day-to-day regulation of salt or water balance in the water-replete animal, it could be of physiological significance under conditions of water deprivation or salt loading.

References

1. Swaab, D. F., F. Nijveldt & C. W. Pool. 1975. Distribution of oxytocin and vasopressin in the rat supraoptic and paraventricular nucleus. J. Endocr. **67:** 461–462.
2. Jones, C. W. & B. T. Pickering. 1969. Comparison of the effects of water deprivation and sodium imbibition on the hormone content of the neurohypophysis of the rat. J. Physiol. (Lond.) **203:** 449–458.
3. Dogterom, J., Tj. B. van Wimersma Greidanus & D. F. Swaab. 1977. Evidence for the release of vasopressin and oxytocin into cerebrospinal fluid: Measurements in plasma and CSF of intact and hypophysectomised rats. Neuroendocrinology **24:** 108–118.
4. Balment, R. J., M. J. Brimble & M. L. Forsling. 1980. Release of oxytocin induced by salt loading and its influence on renal excretion in the male rat. J. Physiol. (Lond.) **308:** 439–449.

5. Poulain, D. A., J. B. Wakerley & R. E. J. Dyball. 1976. Electrophysiological differentiation of oxytocin- and vasopressin-secreting neurones. Proc. Roy. Soc. Lond. Ser. B **196:** 367–384.
6. Brimble, M. J., R. E. J. Dyball & M. L. Forsling. 1978. Oxytocin release following osmotic activation of oxytocin neurones in the paraventricular and supraoptic nuclei. J. Physiol. (Lond.) **278:** 69–78.
7. Peters, G. & R. Roche-Ramel. 1970. Renal effects of posterior pituitary peptides and their derivatives. *In* Pharmacology of the Endocrine System and Related Drugs: the Neurohypophysis. H. Heller & B. T. Pickering, Eds.: 229–278. Pergamon Press. New York, N.Y.
8. Brooks, F. P. & M. Pickford. 1958. The effect of posterior pituitary hormones on the excretion of electrolytes in dogs. J. Physiol. (Lond.) **142:** 468–493.
9. Jacobson, H. N. & R. H. Kellogg. 1956. Isotonic NaCl diuresis in rats. Antidiuresis and chloruresis produced by posterior pituitary extracts. Am. J. Physiol. **184:** 376–389.
10. Rosas, R., L. Barnafi, T. Pereda & H. Croxatto. 1962. Effect of oxytocin structural changes on rat renal excretion of Na, K and water. Am. J. Physiol. **202:** 901–904.
11. Lees, P. & M. F. Lockett. 1964. The influence of hypophysectomy and of adrenalectomy on the urinary changes induced by oxytocin in rats. J. Physiol. (Lond.) **171:** 403–410.
12. Sawyer, W. H. & H. Valtin. 1967. Antidiuretic responses of rats with hereditary hypothalamic diabetes insipidus to vasopressin, oxytocin and nicotine. Endocrinology **80:** 207–210.
13. Chan, W. Y. 1965. Effects of neurohypophysial hormones and their deamino analogues on renal excretion of Na^+, K^+ and water in rats. Endocrinology **77:** 1097–1104.
14. Fabian, M., M. L. Forsling, J. J. Jones & J. Lee. 1969. The release, clearance and plasma protein binding of oxytocin in the anaesthetized rat. J. Endocr. **43:** 175–189.
15. Balment, R. J., I. W. Henderson & J. A. Oliver. 1975. The effects of vasopressin on pituitary oxytocin content and plasma renin activity in rats with hypothalamic diabetes insipidus (Brattleboro strain). J. Gen. Comp. Endocrinol. **26:** 468–477.
16. Holton, P. 1948. A modification of the method of Dale and Laidlaw for standardization of posterior pituitary extracts. Brit. J. Pharmacol. **3:** 328–334.
17. Follett, B. K. & P. J. Bentley. 1963. The bioassay of oxytocin: increased sensitivity of the rat uterus in response to serial injections of stilboestrol. J. Endocr. **29:** 277–282.
18. Chard, T. & M. L. Forsling. 1976. Bioassay and radioimmunoassay of oxytocin and vasopressin. *In* Hormones in Human Plasma. H. N. Antoniades, Ed.: 488–516. Harvard Press. Cambridge, Mass.
19. Chard, T., M. L. Forsling, M. A. R. James, M. J. Kitau & J. Landon. 1970. The development of a radioimmunoassay for oxytocin. Sensitivity of the assay in aqueous buffer solution, specificity, and dissociation of immunological and biological activity. J. Endocr. **46:** 533–542.
20. Valtin, H., W. H. Sawyer & H. W. Sokol. 1965. Neurohypophysial principles in rats homozygous and heterozygous for hypothalamic diabetes insipidus (Brattleboro strain). Endocrinology **77:** 701–706.
21. Brimble, M. J. & R. E. J. Dyball. 1977. Characterization of the responses of oxytocin- and vasopressin-secreting neurones in the supraoptic nucleus to osmotic stimulation. J. Physiol. (Lond.) **271:** 253–271.
22. Orkand, P. M. & S. Palay. 1967. Effect of treatment with exogenous vasopressin on the structural alterations in the hypothalamo-neurohypophysial system of rats with hereditary diabetes insipidus. Anat. Rec. **157:** 295.

23. DYBALL, R. E. J. 1974. Single unit activity in the hypothalamo-neurohypophysial system of Brattleboro rats. J. Endocr. **60:** 135–143.
24. SUNDE, D. A. & H. W. SOKOL. 1975. Quantification of rat neurophysins by polyacrylamide gel electrophoresis (PAGE): Application to the rat with hereditary hypothalamic diabetes insipidus. Ann. N.Y. Acad. Sci. **248:** 345–361.
25. VIERLING, A. F., J. B. LITTLE & E. P. RADFORD, JR. 1967. Antidiuretic hormone bioassay in rats with hereditary hypothalamic diabetes insipidus (Brattleboro strain). Endocrinology **80:** 211–214.
26. CHAN, W. Y. 1976. An investigation of the natriuretic, antidiuretic and and oxytocic actions of neurohypophysial hormones and related peptides: delineation of separate mechanisms of action and assessment of molecular requirements. J. Pharmac. Exp. Ther. **196:** 746–757.

DISCUSSION OF THE PAPER

T. BENNETT (*University of Nottingham Medical School, Nottingham, England*): In the experiments you did on the Long-Evans rat with oxytocin, what effect did it have on free-water clearance? You showed a diuresis, but I wasn't clear whether that was a mobilization of water or solutes.

R. J. BALMENT: The diuresis in the Long-Evans rat reached a peak after the period of hormone infusion, whereas the saluretic response was almost over by this time. There was an increase in free water clearance coincident with a fall in urine osmolality during this diuretic period.

M. D. LINDHEIMER (*University of Chicago, Chicago, Ill.*): Have you tried any other usual stimuli of vasopressin secretion to see what it does to oxytocin in your rats—like nicotine or hypotension?

BALMENT: This, as well as volume depletion, obviously should be done. We are presently looking at oxytocin release and renal actions during milk ejection.

B. R. EDWARDS (*Dartmouth Medical School, Hanover, N.H.*): Perhaps I should comment since I have a poster that deals with very similar experiments. Our overall results are very similar. At the highest dose that we used, which was very similar to your highest dose, we get almost identical plasma concentrations, and an antidiuresis just as you do. Our lower dose is equivalent to your intermediate dose and again we get almost identical plasma concentrations. However, whereas you found essentially no effect in the DI rat, we found no effect on urine flow, but there was a slight increase in urine osmolality, brought about by an increase in sodium excretion. But that, I think is the only difference.

BALMENT: We observed a slight increase in salt excretion in the DI rats, but this was not statistically significant, contrasting with the pronounced saluresis produced by this level of oxytocin in Long-Evans rats.

EDWARDS: Overall, the results of both our studies would suggest that, in the physiological range, oxytocin has little renal effect. For example, our own results indicate that oxytocin probably has little role to play in the increased concentrating ability that is seen in the dehydrated DI rat.

D. M. GASH (*University of Rochester, Rochester, N.Y.*): We have also looked at infusions of oxytocin, but in a somewhat different way. Osmopumps were implanted subcutaneously. At every concentration, prolonged infusion

(2–7 days) produced an antidiuretic effect in conscious animals housed in metabolism cages. Although oxytocin can act on the kidney, even at the highest doses, urine osmolalities were not greater than 1,000 mOsmole/kg.

BALMENT: That would be in accord with one possible explanation for our results, that is that oxytocin is a very weak antidiuretic agent, and may only be a partial agonist at the renal vasopressin receptor.

G. VALIQUETTE (*Columbia University, New York, N.Y.*): As far as the electrolytes are concerned, oxytocin seems to act as a pure agonist. Would you care to comment on that?

BALMENT: The saluretic effects of oxytocin and its effects on water balance seem to be totally divorced. This has been very clearly shown by molecular modification studies, particularly by Chan and du Vigneaud. They have shown that modification of the oxytocin molecule can result in the retention of its saluretic properties in the absence of effects on water excretion.

EFFECT OF POTASSIUM ON PLASMA RENIN CONCENTRATION IN THE PRESENCE AND ABSENCE OF ADH (BRATTLEBORO RAT MODEL)

Emma Fernández-Repollet, Susan Opava-Stitzer, and Manuel Martínez-Maldonado

Department of Physiology
University of Puerto Rico School of Medicine
San Juan, Puerto Rico

Rats with hereditary hypothalamic diabetes insipidus (so-called DI rats) have elevated plasma renin levels.[1, 2] Although the mechanism responsible for this condition has not been elucidated, it seems reasonable to postulate that the absence of ADH [3] and/or the hypokalemia [4] previously reported in these rats might contribute to the elevation of plasma renin concentration (PRC). Evidence in favor of this hypothesis emerges from studies in which both ADH [5] and potassium [6] have been shown to inhibit renin release. In an attempt to examine the relative roles of ADH and potassium in the regulation of renin secretion, PRC was measured in DI rats maintained on a potassium-free, normal potassium, or high potassium diet in the presence and absence of ADH treatment.

Male and female DI rats were used in all experiments. Body weights ranged from 250–350 g. As depicted in TABLE 1, balance studies and PRC determinations were performed during a control period in four groups of untreated DI rats fed with a normal diet. Twenty-four–hour urine samples were collected for the measurement of urine volume, osmolality, and sodium and potassium concentrations. Plasma renin concentration was measured by radioimmunoassay of angiotensin I (AI, New England Nuclear). Following the control period, subgroups of six rats were placed for three weeks on the respective dietary regimen and/or ADH treatment; balance studies and PRC measurements were performed as previously described.

Our data indicate that in the absence of ADH treatment PRC of DI rats maintained on a K^+-free diet (97 ± 11 ng AI/ml/h) was not significantly different from that of DI rats maintained on a normal diet (98 ± 20 ng AI/ml/h). As illustrated in FIGURE 1, ADH treatment significantly diminished PRC of DI rats maintained on a low or a normal potassium intake. Additionally, no significant difference between PRC of ADH-treated rats maintained on a normal or a low K^+ diet (48 ± 6 vs. 71 ± 13 ng AI/ml/h) was detected. FIGURE 2 depicts the effect of a high potassium diet on PRC of DI rats in the absence of ADH treatment. It is evident from the results that a high K^+ intake significantly reduced PRC of DI rats (74 ± 7 vs. 44 ± 6 ng AI/ml/h, $p < 0.01$). This value of PRC was also significantly lower than that observed in control DI rats maintained concurrently on a normal potassium intake (109 ± 15, $p < 0.01$). As shown in FIGURE 3, PRC was found to be significantly lower in rats treated with both ADH and a high K^+ diet (34 ± 3 ng AI/ml/h) than in rats treated with ADH alone (53 ± 5 ng AI/ml/h, $p < 0.01$), suggesting an additive effect of potassium and ADH on PRC. As summarized in TABLE 2,

0077–8923/82/0394–0254 $1.75/0

TABLE 1

EXPERIMENTAL PROTOCOL

		Control (Two weeks) Balance Study Renin	Control (Three weeks)	Experimental (Two weeks) Balance Study Renin
Group 1	6	N	N	N
	6	N	High K^+	High K^+
Group 2	6	N	OIL+N	OIL+N
	6	N	ADH+N	ADH+N
	6	N	ADH+High K^+	ADH+High K^+
Group 3	6	N *	N *	N *
	6	N *	K^+-Free	K^+-Free
Group 4	6	N *	ADH+N *	ADH+N *
	6	N *	ADH+K^+-Free	ADH+K^+-Free

N = Purina laboratory chow (K^+: 0.92%); High K^+ = Purina laboratory chow + 15 g KCl/100 g of food; N * = K^+-free diet + 1.2 g KCl/100 g of food; and K^+-Free = K^+-free diet (General Biochemicals).

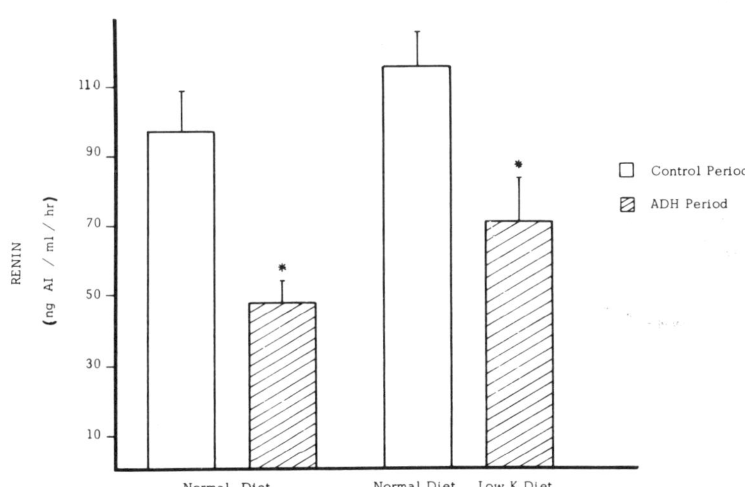

FIGURE 1. Effect of ADH and a low potassium diet on plasma renin concentration in DI rats. Bars to the left represent values in a control group maintained on a normal diet throughout the experiment. Bars to the right are values in the experimental group before and after a low K^+ diet. ADH was administered to all rats in the experimental period. * Significantly different from values in the control period ($p < 0.01$).

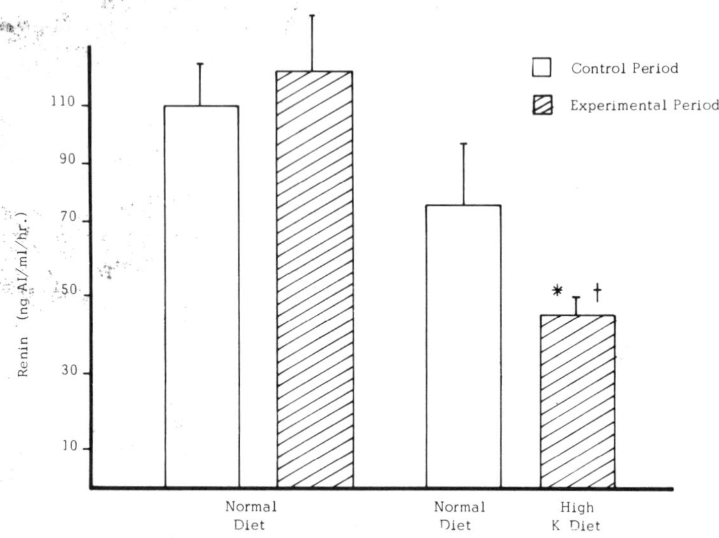

FIGURE 2. Effect of a high potassium diet on plasma renin concentration in DI rats. Format is the same as in FIGURE 1. * Significantly different from values during the control period in the same rats or † from the control group ($p < 0.01$).

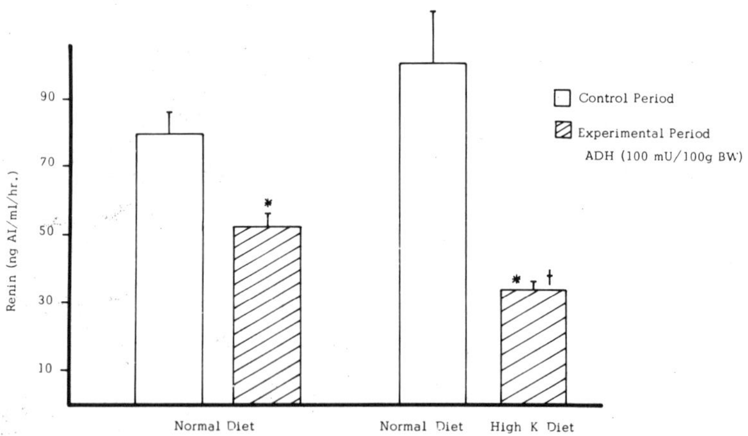

FIGURE 3. Effects of ADH and a high potassium diet on plasma renin concentration in DI rats. For explanation see legend to FIGURE 1. * Significantly different from values obtained during a control period in the same rats or † from the control group ($p < 0.01$).

ADH treatment always reduced PRC in DI rats regardless of the potassium intake.

The present study demonstrates that a K^+-free diet did not increase PRC in DI rats; however, PRC of ADH-treated rats on a K^+-free diet was higher than that of ADH-treated rats on a normal diet. Although the difference was not significant, such a tendency suggests that, in the presence of ADH, potassium deficiency might stimulate PRC in DI rats as in normal rats. In contrast, PRC in DI rats was significantly reduced on a high K^+ intake. This finding is consistent with the inhibitory effect of potassium on PRC observed in other experimental animals.[7, 8] Since PRC of ADH-treated rats on a high potassium diet was significantly lower than that of ADH-treated rats on a normal diet, one might speculate that the effects of these two inhibitors on renin release are additive and probably have different mechanisms of action. As in other studies [1, 2] ADH administration significantly decreased PRC of DI rats on a normal diet.

Although the present study has not dealt with the mechanism by which ADH or potassium inhibits renin release, several actions might be postulated, e.g. a direct action of either ADH or potassium on the granular cells or the renal baroreceptor, an indirect action of potassium mediated through changes

TABLE 2

PLASMA RENIN CONCENTRATION IN ADH-TREATED DI RATS ON VARYING POTASSIUM INTAKES

	Renin (ng AI/ml/h)		
	K^+-free	Normal K^+	High K^+
DI	97±11	74±16	44±5
DI + ADH	71±13 *	53±5 *	34±3 *

Values presented as mean ± SEM; * $p < 0.01$.

in sodium delivery to the macula densa, or indirect effects of ADH as a result of water or sodium retention.

Finally, the fact that ADH treatment consistently diminished PRC in DI rats regardless of the potassium intake strongly suggests that the absence of ADH either directly or indirectly is the principal factor involved in the elevated PRC characteristic of DI rats.

REFERENCES

1. GROSS, F., G. DAUDA, S. KAZDA, J. KYNČL, J. MÖHRING & H. ORTH. 1970. Increased fluid turnover and the activity of the renin-angiotensin system under various experimental conditions. Circ. Res. **30–31** (Suppl. II): 173–181.
2. GUTMAN, Y. & F. BENZAKEIN. 1974. Antidiuretic hormone and renin in rats with diabetes insipidus. Eur. J. Pharmacol. **28:** 114–118.
3. VALTIN, H. & H. A. SCHROEDER. 1964. Familial hypothalamic diabetes insipidus in rats (Brattleboro strain). Am. J. Physiol. **206:** 425–430.
4. MÖHRING, B., J. MÖHRING, G. DAUDA & D. HAACK. 1974. Potassium deficiency in rats with hereditary diabetes insipidus. Am. J. Physiol. **297:** 916–920.

5. Oparil, S. & E. Harber. 1974. The renin-angiotension system. N. Engl. J. Med. **291:** 389–401.
6. Davis, J. O. & R. H. Freeman. 1976. Mechanisms regulating renin release. Physiol. Rev. **56**(1): 1–56.
7. Sealey, J. E., I. Clark, M. B. Bull & J. H. Laragh. 1970. Potassium balance and the control of renin secretion. J. Clin. Invest. **49:** 2119–2127.
8. Flamenbaum, W., J. G. Kleinman, J. S. McNeil, R. J. Hamburger & T. A. Kotchen. 1975. Effect of KCl infusion on renin secretion and aldosterone excretion in dogs. Am. J. Physiol. **229:** 370–375.

THE ROLE OF RENIN IN THE PARADOXICAL ANTIPOLYURIC EFFECTS OF CHLOROTHIAZIDE IN THE HOMOZYGOUS BRATTLEBORO RAT

C. A. de Groot

University of Leiden
2300 RA Leiden
the Netherlands

Chlorothiazide (C) added to the food (16 g/kg food) induces a sustained reduction of about 50% in the polyuria of mature female Brattleboro (DI) rats and increases plasma renin activity (PRA) by about 250%.[1] In order to evaluate the role of PRA in C antidiuresis, two sets of experiments were carried out: (a) with the angiotensin antagonist saralasin (S) and (b) with the β-adrenergic antagonist propranolol (P), which interferes with renin release.[2] Female rats (b.w. about 200 g) were given food ($\pm$C) and water ad libitum for at least five days; diurnal water, food (and C) intake, and urine production were measured at 10 a.m. each day.

S is ineffective orally and rapidly broken down after injection; its effect was extended to 8 h by injecting it twice s.c., at 9 a.m. and at 1 p.m. in a 12% solution of gelatin in a dose of 10 mg/kg b.w. In this way S was given to rats pretreated for six days $\pm$C and adapted to a reversed day-night scheme, i.e. during their active period. On this same day they were killed at 5 p.m. by decapitation. In the other experimental series P was added to the food (1.8 g/kg) $\pm$C during five days, without day-night reversal. They were decapitated at 10 a.m. on day 6. At decapitation, carotid blood was collected in heparinized tubes. The plasma was used to measure PRA by radioimmunoassay for angiotensin I (New England PRA kit). In other plasma samples Na and K concentrations were determined (Flapho flame photometer) as well as osmolality (freezing-point determination, Knauer apparatus), urea (Merck kit) and creatinine (BDH kit) concentrations. Similar measurements were made in the urine collected during the experimental period (8 h in S experiments and 24 h in P experiments). The standard ground laboratory food of Hope Farms, which contains about 0.4% Na and 0.8% K was used. C was kindly provided by Merck, Sharp & Dohme, propranolol by ICI, and saralasin acetate by Dr. K. Ellis of Norwich Pharmacal Company. "Free" water was calculated as

$$\text{Urine Volume} - \frac{\text{Urine Volume} \times \text{Urine Osmolality}}{\text{Plasma Osmolality}}.$$

Statistical evaluation was done by analysis of variance followed by Duncan's new multiple comparison test.[3]

Results

Food intake was not affected by any treatment (8 ± 1 g/animal/8 h; 16 ± 1 g/24 h). As is shown in Figure 1, C halves DI polydipsia and polyuria,

0077–8923/82/0394–0259 $1.75/0 © 1982, NYAS

even in the short period of 8 h. S and P are each also effective in this respect, but not to the same extent; P increases the action of C. S and P diminish "free" water excretion, but C reverses it.

The data of TABLES 1 and 2 show that C does not affect urinary Na^+ levels significantly, but reduces the total Na^+ loss, whereas K^+ concentration is increased; urine osmolality is nearly doubled. S also reduces Na^+ loss, but it differs from C in that it reduces total K^+ excretion and does not affect urine osmolality. The combined treatment leads to increased osmolality and urea excretion, but a reduction of the increase in urinary K^+ concentration caused

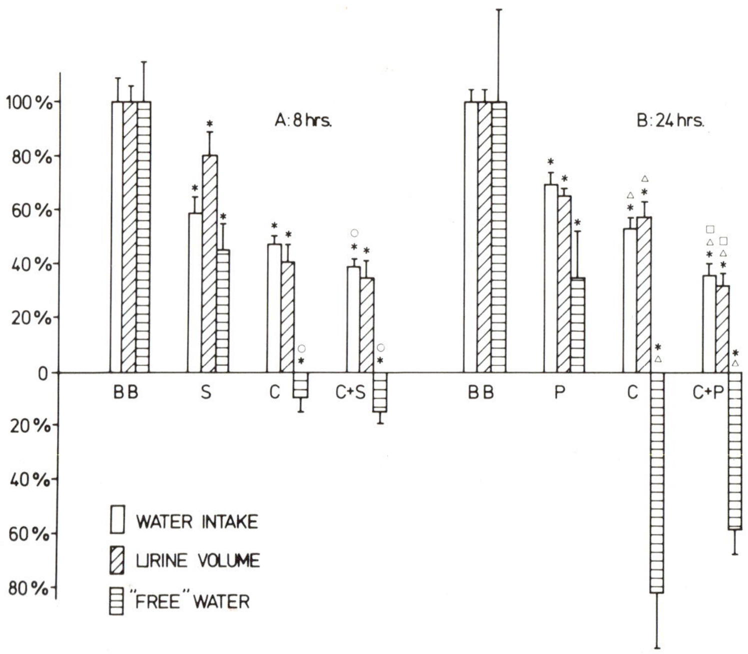

FIGURE 1. A. 8 h values (N = 11) and B. 24 h values (N = 9) of water intake, urinary volume, and calculated "free" water as compared to untreated Brattleboro (BB) homozygotes. S: homozygotes injected with 20 mg/kg b.w. saralasin acetate; and C: homozygotes treated for at least five days with chlorothiazide 16 g/kg food. P: homozygotes with propranolol, 1.8 g/kg food during the same time. C + S and C + P: combined treatments. ★ is significant as compared to group BB, ○ to group S, □ to group C and △ to group P.

by C. On the whole, P imitates S, except for the effects on K^+ excretion. In FIGURE 2 PRA data are presented. C markedly increases PRA; S also induces an increase, but the combined treatment does not significantly alter the effect obtained by C alone. Here, P does not mimic S results: PRA is not increased by P, and P added to C *decreases* the high PRA level significantly. FIGURE 2 gives creatinine clearances. C as well as S lower this clearance, while P does not do so and even slightly diminishes, in the combined treatment, the significant reduction obtained by C.

TABLE 1

SARALASIN EXPERIMENTS *

	H	HS	C	CS
Na^+ (mEq/L)	10.8±1.4	12.5±3.8	12.8±1.3	12.6±3.5
K^+ (mEq/L)	19.1±1.7	18.4±1.7	47.4±6.1 [hs]	32.7±3.3 [hcs]
Total Urinary				
Na ((mmol)	0.55±0.09	0.29±0.09 [h]	0.29±0.03 [h]	0.17±0.02 [hs]
K (mmol)	0.84±0.07	0.55±0.05 [h]	0.92±0.10	0.59±0.05 [hc]
Urine Osmolality (mOsm/L)	158±12	198±32	276±17 [h]	329±47 [hs]
Urine Urea (conc. g/L)	5.8±0.5	8.1±1.3	9.3±1.0	12.3±1.9 [hs]

* Eight-hour values of levels of Na^+ and K^+ (mEq/L); mmol Na and K in the urine, urine osmolality, and urea concentration.

Note: H = Brattleboro homozygotes; HS = Brattleboro homozygotes given 10 mg/kg b.w. S acetate at 9 p.m. and again at 1 p.m. on day of experiment; C = Brattleboro homozygotes pretreated for six days with C in food (16 g/kg); and CS = C rats given S. $N = 11$ per group. Entries represent mean ± SEM. [h] is significant as compared to group H; [s] to group HS; and [c] to group C.

TABLE 2

PROPRANOLOL EXPERIMENTS *

	H	HP	C	CP
Na^+ (mEq/L)	15.4±1.3	11.2±1.9	17.3±3.2	22.0±3.7
K^+ (mEq/L)	16.8±0.9	18.9±2.7	39.2±2.7 [hp]	43.3±2.5 [hp]
Total Urinary				
Na (mmol)	2.32±0.21	1.11±0.22 [h]	1.31±0.35 [h]	1.09±0.26 [h]
K (mmol)	2.30±0.14	1.83±0.17	2.73±0.32 [p]	2.19±0.24
Urine Osmolality (mOsm/L	198±19	230±14	364±30 [hp]	379±16 [hp]
Urine Urea (conc. g/L)	5.5±0.5	6.8±0.2	8.1±1.0	13.2±1.7 [hpc]

* 24-hour values for Na^+ and K^+ (mEq/L); mmol Na and K excreted in the urine; and urine osmolality and urea concentration.

Note: H = Brattleboro homozygotes; HP = Brattleboro homozygotes given P, 1.8 g/kg food during five days; C = Brattleboro homozygotes treated for five days with C in food (16 g/kg); and CP = Brattleboro homozygotes given a combined treatment of C and P. $N = 9$ per group. Entries represent mean ± SEM; [h] is significant ($p < 0.05$) as compared to group H; [p] to HP rats; and [c] to C rats.

Discussion

In these experiments we set out to evaluate the role of renin in the antidiuresis caused by high doses of C, which, as in previous work,[1] halves the polyuria of hereditary diabetes insipidus and increases PRA to about 250%. The new data indicate the C antidiuresis does not evolve from renin-dependent processes, since both the angiotensin antagonist S, as well as the β-adrenolytic P (which should interfere with renin-dependent processes), do not diminish C antidiuresis. They even enhance the effects of C to some degree (Figure 1). S is indeed an effective antagonist of angiotensin, since K^+ output is significantly reduced after it is increased by C (Table 1). Also, P is effective in decreasing elevated PRA (Figure 2) and, accordingly, the effect of P on K^+ elimination is marginal (Table 2). Each antidiuretic drug regimen *reduces* total Na^+ output in DI homozygotes (Tables 1 and 2); apparently this reduction is independent of PRA, which is high after C or S and nearly normal after P. This reduction during stabilized antidiuresis is a confirmation of earlier C data.[1] High PRA (Figure 2) is correlated with a decreased creatinine clearance (Figure 3), i.e. reduced glomerular filtration rate, and might be caused by decreased perfusion pressure in the kidney.[4]

Thirst

C antidiuresis was originally discovered in rats with diabetes insipidus induced by hypothalamic lesioning,[5] where it is far more effective than in

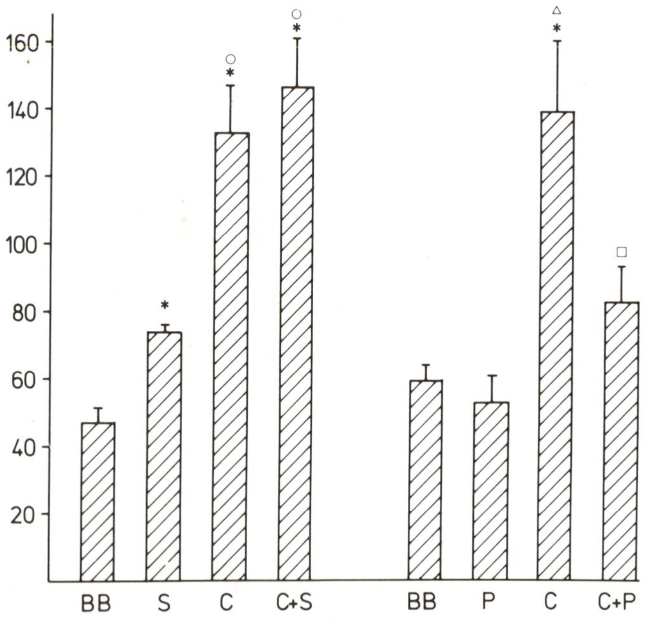

Figure 2. PRA values (ng Angiotensin I/ml plasma/h incubation) of the rats of Figure 1. Same symbols are used.

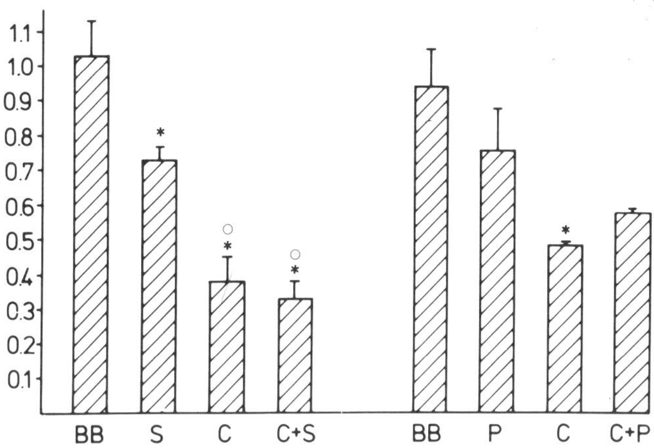

FIGURE 3. Creatinine clearance, ml/min of the rats of FIGURE 1. Same symbols are used.

DI rats. The difference might be explained by the role of vasopressin in the regulation of thirst. Angiotensin stimulates thirst [6] as well as vasopressin release in normal rats; [7] vasopressin can suppress thirst.[8] Hence, lack of vasopressin could prolong thirst in DI rats, but not in lesioned rats, where vasopressin is still present; and hence, C is more effective. In DI rats, disruption of the renin-angiotensin system should enhance effectiveness, as the stimulus to drink is interfered with. This is indeed the case: P as well as S enhance the anti-polyuric effects of C (FIGURE 1). Moreover, they even reduce water intake in DI rats not treated with C. Captopril, which blocks the formation of angiotensin II, does the same.[9] In what way C produces antidiuresis in DI rats is not yet clear; it is not correlated with increased Na^+ elimination, nor with high PRA. On the contrary, it exists in spite of the increased drinking behavior that high PRA should induce in DI rats. Perhaps a vascular effect is involved in C antidiuresis.

REFERENCES

1. DE GROOT, C. A. & A. M. I. TIJSSEN. 1979. The effects of long-term oral treatment with chlorothiazide or furosemide on hereditary diabetes insipidus in rats. Arch. Int. Pharmacodyn. Thér. **237:** 75–87.
2. FORMAN, B. H. & P. J. MULROW. 1974. Suppression of plasma renin activity by propranolol: correlation with plasma propranolol level. Proc. Soc. Exp. Biol. Med. **146:** 530–533.
3. STEEL, R. G. D. & J. H. TORRIE, Eds. 1960. Principles and Procedures of Statistics. McGraw-Hill. New York, N.Y.
4. HOFBAUER, K. G., H. ZSCHIEDRICH & F. GROSS. 1975. Regulation of renin release and intrarenal formation of angiotensin. Clin. Exp. Pharmacol. Phys. **3:** 73–93.
5. CRAWFORD, J. D. & G. C. KENNEDY. 1959. Chlorothiazide in diabetes insipidus. Nature **183:** 89–90.

6. Epstein, A. N., T. Fitzsimons & B. J. Rolls. 1970. Drinking induced by injection of angiotensin into the brain of the rat. J. Physiol. **210:** 457–474.
7. Bonjour, J. P. & R. L. Malvin. 1970. Stimulation of ADH release by the renin-angiotensin system. Am. J. Physiol. **218:** 1555–1559.
8. Rolls, B. 1971. The effect of intravenous infusion of antidiuretic hormone on water intake in the rat. J. Physiol. **219:** 333–340.
9. Henderson, I. W., A. McKeever & C. J. Kenyon. 1979. Captopril depresses drinking and aldosterone in rats lacking vasopressin. Nature **281:** 569–570.

ACTH AND INDOMETHACIN-INDUCED ANTIDIURESIS IN LOW SALT BRATTLEBORO STRAIN RATS: INDUCTION AND ESCAPE

S. Khosla,* L. B. Kinter, R. Zusman, and R. Beeuwkes III

Department of Physiology
Harvard Medical School
and
Cardiac Unit
Massachusetts General Hospital
Boston, Massachusetts

Introduction

Previous studies have shown that ACTH and indomethacin administration are associated with antidiuresis in rats and humans maintained on chronic low salt diets.[1, 2] Such diets could lead to volume depletion and elevation of vasopressin. Thus, the observed antidiuresis could be due to direct antidiuretic effects of the administered agents or could be secondary to potentiation of the action of endogenous vasopressin. *In vitro* studies in the toad urinary bladder have shown that inhibitors of prostaglandin biosynthesis, such as indomethacin, do enhance vasopressin-stimulated water permeability.[3] In addition, adrenal steroids have been shown to increase vasopressin-stimulated water flow in the toad urinary bladder.[4] On the other hand, both prostaglandin synthesis inhibitors and adrenal steroids are known to have significant effects on GFR and tubular function and could produce an antidiuresis independent of endogenous vasopressin.[5, 6] Finally, previous studies have demonstrated that the renal response to ACTH and indomethacin depends on the salt state of the animal.[1, 5] The purpose of this investigation was thus to separate direct antidiuretic effects of these agents from effects due to potentiation of endogenous vasopressin and, in addition, to examine the response as a function of chronic adaptation to low salt states in rats. For this purpose, we used heterozygous (HZ) and homozygous diabetes insipidus (DI) Brattleboro strain rats.[7]

Methods

Studies were performed on 36 HZ rats weighing 200–350 grams and 22 DI Brattleboro strain rats weighing 150–300 grams purchased from Blue Spruce Farms (Altamont, N.Y.). The rats were divided into eight groups (Table 1) and were maintained on a sodium-deficient diet (Sodium-Deficient Diet Rat Modified, ICN Nutritional Biochemicals, Cleveland, Ohio) and distilled water *ad libitum* for either two weeks (Groups 1–4) or more than five weeks (Groups 5–8). All animals were housed individually in metabolism cages and maintained in a 12-hour light/dark photoperiod in an isolated animal room

* Address all correspondence to S. Khosla, 402 Rindge Ave. #12A, Cambridge, Mass. 02140.

0077-8923/82/0394-0265 $1.75/0

TABLE 1

Group	Type	N	Diet	Regimen
1	HZ	12		ACTH
2	HZ	11	2 weeks	Indomethacin
3	DI	6	low salt	ACTH
4	DI	7		Indomethacin
5	HZ	12		ACTH
6	HZ	6	5 weeks	Indomethacin
7	DI	4	low salt	ACTH
8	DI	5		Indomethacin

with controlled temperature and near constant humidity. As outlined in TABLE 1, at the end of the appropriate time period, Groups 1, 3, 5, and 7 were administered ACTH (ACTHAR-Gel, Armour Biochemicals, Scottsdale, Ariz.) 8 units/day subcutaneously for two successive days. Twenty-four–hour urines were collected during three days prior to and three days after first drug administration. Similarly, Groups 2, 4, 6, and 8 were administered indomethacin (Merck, Sharp and Dohme, West Point, Pa.) dissolved in drinking water at a concentration giving an estimated daily intake of 3 mg/d·kg.[8] Twenty-four–hour urines were collected for three days prior to and during the three days of drug administration. Urine osmolalities (U_{osm}) were determined by freezing-point depression (Osmette Precision Systems Inc., Sudbury, Mass).

Statistical Analysis

The control value for each of the study groups was the mean U_{osm} during the three days prior to drug administration. Similarly, the experimental value was the mean value during the three days after first drug administration. Experimental values were compared to control values by Student's *t*-test for paired observations.[9]

RESULTS

Two Weeks Low Salt

FIGURE 1 summarizes the observations after two weeks on a sodium-deficient diet. ACTH increased U_{osm} in the HZ group from 779 ± 87 mOsm/kg·H_2O to 1104 ± 116 mosm/kg·H_2O ($p < .001$) (FIGURE 1A). In contrast, ACTH had no significant effect on U_{osm} in the DI group (123 ± 8 mOsm/kg·H_2O to 128 ± 5 mOsm/kg·H_2O (NS)) (FIGURE 1C). Indomethacin, however, increased U_{osm} in the HZ group from 927 ± 128 mOsm/kg·H_2O to 1166 ± 158 mOsm/kg·H_2O ($p < .005$) (FIGURE 1B) as well as in the DI group from 153 ± 14 mOsm/kg·H_2O to 182 ± 17 mOsm/kg·H_2O ($p < .02$) (FIGURE 1D). Body weight was maintained in all groups during drug treatment, suggesting that the observed changes in U_{osm} were not due to dehydration resulting from morbidity or avoidance of the indomethacin-containing drinking solution.

Five Weeks Low Salt

After five weeks on the sodium-deficient diet, the HZ rats failed to increase U_{osm} in response to either ACTH (801 ± 53 mOsm/kg·H_2O to 885 ± 53 mOsm/kg·H_2O (NS) or indomethacin (962 ± 105 mOsm/kg·H_2O to 825 ± 81 mOsm/kg·H_2O, $p < 0.01$) (FIGURES 2A and B). In fact, the fall in U_{osm} with indomethacin was significant. In contrast, the DI rats continued to have an antidiuretic response to indomethacin, increasing U_{osm} from 141 ± 12 mOsm/kg·H_2O to 168 ± 12 mOsm/kg·H_2O ($p < .005$) (FIGURE 2D).

DISCUSSION

Brattleboro strain rats provide a convenient model with which to separate direct mechanisms of drug-induced antidiuresis from indirect mechanisms involving potentiation of endogenous vasopressin. In the present experiments, after two weeks of low salt conditioning, ACTH increased U_{osm} in the HZ rats but not in the vasopressin-deficient DI group. This finding indicates that vasopressin is required for the ACTH effect. This is reasonable since ACTH administration would be expected to elevate endogenous adrenal steroid levels and a synergism between vasopressin and adrenal steroids, resulting in enhanced vasopressin-stimulated water flow, has long been recognized both *in vivo* [10, 11]

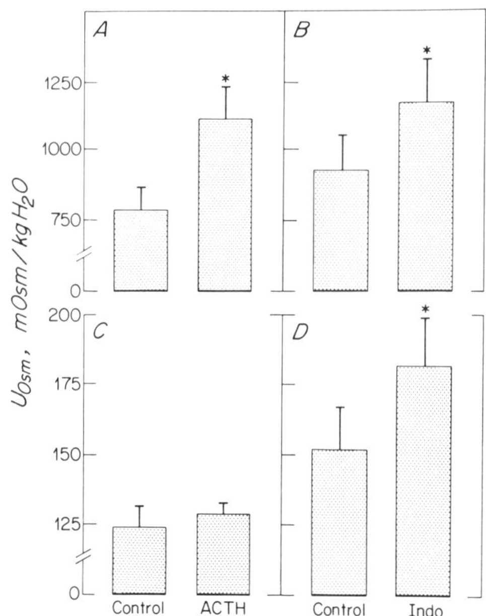

FIGURE 1. Effect of ACTH and indomethacin on U_{osm} (mOsm/kg·H_2O) after two weeks of sodium depletion. A: HZ given ACTH (Group 1); B: HZ given indomethacin (Group 2): C: DI given ACTH (Group 3); D: DI given indomethacin (Group 4). Asterisk (*) indicates statistical significance ($p < .05$) compared to the control period.

and *in vitro*.[4] It is notable, however, that HZ rats maintained on the low salt regimen for more than five weeks failed to increase U_{osm} in response to the ACTH challenge. Such escape has to our knowledge not been previously reported, perhaps because few investigators maintain their rats on low salt for more than two weeks. The mechanism of this phenomenon is unknown at the present time.

Indomethacin increased U_{osm} in both the HZ and DI groups maintained on low salt for two weeks. In the DI rat this effect must be independent of endogenous vasopressin and is presumably due to another mechanism. It is possible that reductions of inner medullary blood flow[12] and/or a change in

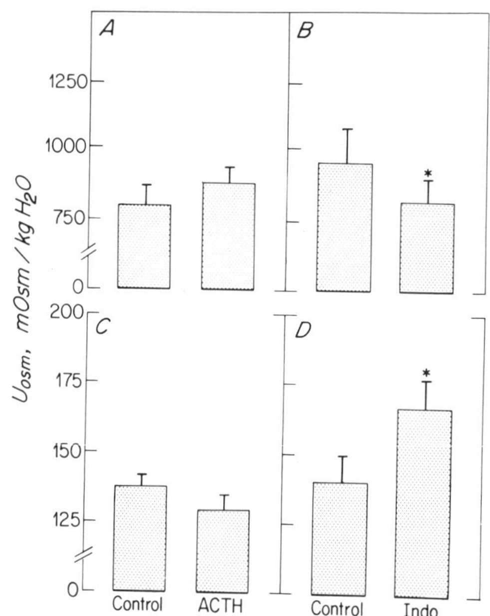

FIGURE 2. Effect of ACTH and indometharin on U_{osm} (mOsm/Kg·H_2O) after 5 weeks of sodium depletion. A: HZ given ACTH (Group 5); B: HZ given indomethacin (Group 6); C: DI given ACTH (Group 7) and D: DI given indomethacin (Group 8). Asterisk indicates statistical significance ($p < .05$) compared to the control period.

intramedullary tonicity[13] caused by indomethacin may be responsible for the increase in U_{osm} observed in the DI rat. This vasopressin independence in the DI rats does not, however, exclude the possibility that indomethacin raised U_{osm} in the HZ group at least in part by potentiation of the effects of endogenous vasopressin.[3, 14]

Again, after more than five weeks on the low salt regimen, the HZ rats failed to increase U_{osm} in response to indomethacin. Strikingly, however, the DI rats continued to demonstrate an antidiuresis in response to an indomethacin challenge. These results suggest that vasopressin either directly or indirectly promoted escape from indomethacin-induced antidiuresis in HZ rats. Such a phenomenon has also not to our knowledge been previously reported and the mechanism is as yet unknown.

In summary, both ACTH and indomethacin induce antidiuresis when administered to conscious rats maintained on a sodium-deficient diet for at least two weeks. The present studies show that the ACTH-induced antidiuresis requires the presence of endogenous vasopressin whereas the indomethacin-induced antidiuresis apparently does not. Finally, endogenous vasopressin may play a role in the chronic adaptation to imposed salt states in rats.

References

1. Khosla, S., L. B. Kinter, R. M. Zusman & R. Beeuwkes. 1980. Effect of adrenocorticotropic hormone and prostaglandins on renal water reabsorption. Clin. Res. **28:** 452A (Abstract).
2. Zusman, R. M., J. M. Vinci, R. E. Bowden, D. Horwitz & H. R. Derser. 1979. Effect of indomethacin and adrenocorticotrophic hormone on renal function in man: An experimental model of inappropriate anditidiuresis. Kidney Int. **15:** 62–70.
3. Zusman, R. M., H. R. Keiser & J. S. Handler. 1977. Vasopressin-stimulated prostaglandin E biosynthesis in the toad urinary bladder: effect on water flow. J. Clin. Invest. **60:** 1339–1347.
4. Zusman, R. M., H. R. Keiser & J. S. Handler. 1978. Effect of adrenal steroids on vasopressin-stimulated PGE synthesis and water flow. Am. J. Physiol. **234**(6): F532–F540.
5. Schor, N., I. Ichikawa & B. M. Brenner. 1980. Glomerular adaptations to chronic dietary salt restriction on excess. Am. J. Physiol. **238**(7): F428–F436.
6. Green, H. H., A. R. Harrington & H. Valtin. 1970. On the role of antidiuretic hormone in the inhibition of acute water diuresis in adrenal insufficiency and the effects of gluco- and mineralocorticoids in reversing the inhibition. J. Clin. Invest. **49:** 1724–1736.
7. Valtin, H. 1967. Hereditary hypothalamic diabetes insipidus in rats (Brattleboro Strain). A useful experimental model. Am. J. Med. **42:** 814–827.
8. Kinter, L. B., M. J. Dunn, T. R. Beck, R. Beeuwkes III & A. Hassid. 1982. The interactions of prostaglandins and vasopressin in the kidney. Ann. N.Y. Acad. Sci. (This volume.)
9. Snedecor, G. W. & W. G. Cochran. 1967. The comparison of two samples. *In* Statistical Methods. 91–119. Iowa State University Press. Ames, Iowa.
10. Crabbe, J. 1962. The role of aldosterone in the renal concentration mechanism in man. Clin. Sci. **23:** 9–46.
11. Goonaratna, C. F. W. & O. M. Wrong. 1975. Assessment of urine-concentrating ability in man: effect of fluorocortisone and urea in enhancing response to vasopressin. Clin. Sci. Molec. Med. **48:** 269–278.
12. Solez, K., J. A. Fox, M. Miller & R. H. Heptinstall. 1974. Effects of indomethacin on inner medullary plasma flow. Prostaglandins **7:** 91–98.
13. Ganguli, M., L. Tobian, S. Azar & M. O'Donnell. 1977. Evidence that prostaglandin synthesis inhibitors increase the concentration of sodium and chloride in rat renal medulla. Circ. Res. **40** (Suppl. 1): 135–139.
14. Berl, T., W. Czaczkes & C. Kleeman. 1975. Effect of prostaglandin (PG) synthesis inhibition on the renal action of vasopressin (ADH). Studies in man and the rat. Am. Soc. Nephrology **8:** 72 (Abstract).

BASIC ALTERATIONS IN SERUM LEVELS OF SEVERAL CHEMICAL SUBSTANCES IN BRATTLEBORO RATS

Helen M. Murphy and Cyrilla H. Wideman

Department of Psychology
John Carroll University
Cleveland, Ohio 44118

INTRODUCTION

Studies have been conducted on blood parameters in homozygous and heterozygous male and female Brattleboro rats. Valtin and Schroeder [1] reported that serum Na concentrations are significantly higher in homozygous male and female Brattleboro (DI) rats than in heterozygous Brattleboro (HZ) animals. There were no significant differences in the hematocrit between the two groups. Barnett *et al.*[2] reported that there were no significant changes with age or between genotypes when the hematocrit and hemoglobin concentrations were measured in DI and HZ rats. Both DI and HZ male rats had significantly higher hematocrits and hemoglobin concentrations than their female counterparts. Using plasma measurements, the following observations were reported: (1) glucose levels were significantly lower in DI males and females than in HZ animals; (2) lipids were significantly lower in DI males and females than in HZ females; (3) protein was significantly lower in DI animals than in HZ controls and protein was significantly lower in DI males than in DI females; and (4) no significant differences in Na and K were found between DI and HZ animals. In 1976, Balment *et al.*[3] reported that DI females and HZ females had lower concentrations of plasma Na than their male counterpart. DI females also had a higher concentration of K than the other groups. By indirect measurements, Galton *et al.*[4] determined that thyroid function was normal in both DI and HZ male Brattleboro rats.

The present study was conducted in an attempt to further examine blood constituents in homozygous Brattleboro rats. Previously measured parameters such as Na, K, Ca, Cl, glucose, total protein, hematocrit, and hemoglobin were reexamined employing different techniques than those previously utilized.[1-4] In addition, several other blood parameters were examined. Three variables were of particular concern in the study: (1) Long-Evans hooded rats (LE) may serve as better controls than heterozygous animals of the Brattleboro strain, (2) the sex of the animals may influence the level of blood constituents, and (3) the age of the animals may be critical when assessing blood parameters. These variables were examined in two experiments.

EXPERIMENT 1

Materials and Methods

Twenty-four female DI rats served as the experimental group and eight female LE rats served as the control group. At 35 weeks of age the animals

0077-8923/82/0394-0270 $1.75/0 © 1982, NYAS

were sacrificed by decapitation and the following tests were run on the serum: Na, K, Ca, Cl, P, glucose, blood urea nitrogen, creatinine, total protein, albumin, globulin, alkaline phosphatase, serum glutamic-oxaloacetic transaminase (SGOT), serum glutamic-pyruvic transaminase (SGPT), lactate dehydrogenase (LDH), cholesterol, triglycerides, Fe, bilirubin, and uric acid. These substances were analyzed in the Technicon SMAC system.[5] Serum thyroid stimulating hormone (TSH), total thyroxin (T_4), and prolactin were analyzed by radioimmunoassay. Rat TSH and prolactin standards were provided by Dr. Parlow through National Institute of Arthritis, Metabolism, and Digestive Diseases.

Results

The results of Experiment 1 are given in TABLE 1. The DI female rats had significantly lower levels of Ca ($p < .05$), triglycerides ($p < .001$), albumin ($p < .01$), and total protein ($p < .01$). The same animals had significantly higher levels of Na ($p < .001$), Cl ($p < .001$), alkaline phosphatase ($p < .05$), and Fe ($p < .05$). There were no significant differences between experimental and control animals in levels of K, P, glucose, blood urea nitrogen, creatinine, globulin, SGOT, SGPT, LDH, cholesterol, bilirubin, uric acid, TSH, T_4, and prolactin.

EXPERIMENT 2

Materials and Methods

This experiment utilized 43 animals. There were 10 DI males (DI-M), 10 DI females (DI-F), 11 LE males (LE-M), and 12 LE females (LE-F). All animals were sacrificed at 8 weeks of age. All of the tests performed in Experiment 1 were repeated except for TSH, T_4, and prolactin. In addition, tests were conducted on the hematocrit, hemoglobin, and red and white blood cells.

Results

The results of Experiment 2 are given in TABLE 2. Genotypic and sex differences were found in a number of the parameters. The following comparisons were made: DI-M vs. DI-F, DI-M vs. LE-M, DI-F vs. LE-F, and LE-M vs. LE-F. Significant differences were noted in the following parameters. DI-F rats were higher than LE-F rats in Na ($p < .001$). DI-M were lower than LE-M animals in K ($p < .05$). The Ca level of the DI-F group was lower than the LE-F group. Cl levels were different in the DI-M vs. LE-M ($p < .001$) and DI-F vs. LE-F ($p < .001$) groups. In both instances the Brattleboro rats were higher than the controls. With P, the DI-M vs. LE-M ($p < .001$) and LE-M vs. LE-F ($p < .05$) groups showed differences. The DI-M animals were lower than the LE-M animals and the LE-F rats were also lower than the LE-M rats. Glucose was lower in DI-M than in LE-M animals ($p < .001$). Creatinine levels were higher in DI-F rats as compared to LE-F rats ($p < .01$). There were differences in total protein between DI-M vs. DI-F ($p < .001$) and

DI-F vs. LE-F animals ($p < .001$). In both instances the DI-F group was higher. The same picture emerged with serum albumin. DI-F rats had higher levels of this constituent than DI-M rats ($p < .001$) and LE-F rats ($p < .01$). DI-M animals had lower levels of globulin than DI-F animals ($p < .05$). With alkaline phosphatase, DI-F rats were lower than LE-F rats ($p < .05$). The opposite was true with LDH. The DI-F group was higher than the LE-F group ($p < .01$). Cholesterol levels were higher in DI-F animals when compared with LE-F animals ($p < .05$). In this constituent the LE-F group was higher than the LE-M group ($p < .01$). The DI-F group was lower than the LE-F group in triglycerides ($p < .01$) and the LE-M group was also lower than the

TABLE 1

SERUM LEVELS OF CHEMICAL SUBSTANCES IN HOMOZYGOUS BRATTLEBORO AND LONG-EVANS FEMALE RATS AT 35 WEEKS OF AGE (MEANS ± SEM)

Blood Constituent	Brattleboro Rats (DI)	Long-Evans Rats (LE)
Na (mEq/L)	147.83±0.72	140.38±0.53
K (mEq/L)	7.03±0.15	7.41±0.11
Ca (mg/dl)	10.50±0.08	10.90±0.17
Cl (mEq/L)	108.50±0.64	102.50±0.73
P (mg/dl)	5.84±0.18	5.65±0.13
Glucose (mg/dl)	133.00±2.21	140.75±3.45
Blood urea nitrogen (mg/dl)	20.46±0.72	20.50±0.86
Creatinine (mg/dl)	0.60±0.03	0.58±0.07
Total Protein (g/dl)	6.44±0.08	6.98±0.21
Albumin (g/dl)	3.71±0.06	4.09±0.14
Globulin (g/dl)	2.73±0.05	2.89±0.11
Alkaline phosphatase (I.U./L)	344.29±23.96	229.25±26.34
SGOT (I.U./L)	236.53±13.39	197.00±15.74
SGPT (I.U./L)	100.67±8.56	74.88±3.85
LDH (mμ/ml)	940.48±55.65	886.14±109.39
Cholesterol (mg/dl)	68.42±2.06	71.00±3.26
Triglycerides (mg/dl)	84.25±5.77	276.00±46.04
Fe (mg/dl)	369.00±17.94	285.75±15.28
Bilirubin (mg/dl)	0.13±0.01	1.00±0.01
Uric acid (mg %)	1.12±0.06	0.96±0.02
TSH (ng/ml)	313.06±28.86	326.13±40.06
T_4 (μg/dl)	3.05±0.20	2.60±0.23
Prolactin (ng/ml)	21.41±3.62	15.34±4.72

LE-F group ($p < .01$). DI-M animals had lower levels of Fe than DI-F animals ($p < .05$). With both bilirubin and uric acid, DI-M animals had lower levels of these constituents than DI-F animals ($p < .05$). Differences in hemoglobin were found between DI-M vs. Le-M ($p < .001$) and DI-F vs. LE-F rats ($p < .01$). The Brattleboro animals were higher in both instances. The same results were noted with the hematocrit. The DI-M animals were higher than the LE-M animals ($p < .001$) and DI-F rats were higher than LE-F rats ($p < .001$). Red blood cell counts were also higher in Brattleboro animals (DI-M vs. LE-M, $p < .001$ and DI-F vs. LE-F, $p < .001$). DI-M animals had fewer white blood cells than LE-M animals ($p < .001$). No differences were found among the groups in levels of blood urea nitrogen, SGOT, and SGPT.

TABLE 2

BLOOD PARAMETERS IN HOMOZYGOUS BRATTLEBORO AND LONG-EVANS MALE AND FEMALE RATS AT 8 WEEKS OF AGE (MEANS ± SEM)

Blood Constituent	Brattleboro Male (DI–M)	Brattleboro Female (DI–F)	Long-Evans Male (LE–M)	Long-Evans Female (LE–F)
Na (mEq/L)	153.60 ± 4.24	152.00 ± 1.52	147.00 ± 3.32	142.20 ± 1.43
K (mEq/L)	7.81 ± 0.24	8.02 ± 0.45	8.43 ± 0.17	8.37 ± 0.28
Ca (mg/dl)	11.10 ± 0.22	10.65 ± 0.20	11.47 ± 0.11	11.33 ± 0.12
Cl (mEq/L)	113.89 ± 1.79	114.40 ± 1.89	104.78 ± 1.73	103.08 ± 0.83
P (mg/dl)	8.63 ± 0.29	9.14 ± 0.35	10.65 ± 0.25	9.78 ± 0.30
Glucose (mg/dl)	160.50 ± 2.49	160.56 ± 4.06	179.00 ± 4.20	168.60 ± 3.18
Blood urea nitrogen (mg/dl)	21.40 ± 1.65	19.20 ± 1.19	19.36 ± 0.71	20.58 ± 0.88
Creatinine (mg/dl)	0.50 ± 0.09	0.60 ± 0.09	0.42 ± 0.02	0.37 ± 0.02
Total Protein (g/dl)	5.38 ± 0.07	6.10 ± 0.13	5.49 ± 0.12	5.52 ± 0.10
Albumin (g/dl)	3.25 ± 0.06	3.76 ± 0.11	3.19 ± 0.07	3.30 ± 0.06
Globulin (g/dl)	2.13 ± 0.07	2.34 ± 0.07	2.30 ± 0.06	2.22 ± 0.05
Alkaline phosphatase (I.U./L)	571.40 ± 45.33	519.90 ± 33.67	596.45 ± 44.30	609.00 ± 23.19
SGOT (I.U./L)	296.78 ± 29.93	288.10 ± 34.67	234.09 ± 21.68	224.42 ± 18.71
SGPT (I.U./L)	78.00 ± 6.77	74.00 ± 9.53	69.82 ± 5.67	59.73 ± 3.83
LDH (mμ/ml)	1090.90 ± 113.00	1163.40 ± 126.67	854.73 ± 76.90	759.75 ± 50.30
Cholesterol (mg/dl)	81.00 ± 4.37	91.70 ± 3.70	71.64 ± 2.59	82.92 ± 2.44
Triglycerides (mg/dl)	178.10 ± 16.83	131.89 ± 18.30	135.64 ± 13.83	199.83 ± 15.48
Fe (μg/dl)	302.20 ± 14.70	358.20 ± 24.43	355.18 ± 21.27	381.08 ± 17.38
Bilirubin (mg/dl)	0.08 ± 0.02	0.17 ± 0.04	1.00 ± 0.03	1.00 ± 0.18
Uric acid (mg %)	1.05 ± 0.05	1.70 ± 0.27	1.00 ± 0.01	1.00 ± 0.01
Hematocrit (%)	38.49 ± 0.83	39.12 ± 0.87	34.12 ± 0.39	34.52 ± 0.67
Hemoglobin (gm %)	14.43 ± 0.31	14.60 ± 0.33	12.99 ± 0.12	13.13 ± 0.28
Red blood cells (m/mm^3)	6.51 ± 0.15	6.71 ± 0.15	5.73 ± 0.07	5.83 ± 0.13
White blood cells (m/mm^3)	6170.00 ± 358.33	6580.00 ± 856.00	8300.00 ± 364.00	7425.00 ± 416.86

Discussion and Conclusion

Interesting observations can be made when comparing the data from Experiment 1 on older female rats (35 wks.) and Experiment 2 on younger female rats (8 wks.). Older DI-F animals can be contrasted with younger DI-F animals and older LE-F rats with younger LE-F rats. For the DI rats, the younger animals had higher levels of Na ($p < .01$), K ($p < .01$), Cl ($p < .001$), P ($p < .001$), glucose ($p < .001$), alkaline phosphatase ($p < .001$), cholesterol ($p < .001$), triglycerides ($p < .01$), and uric acid ($p < .01$). The younger DI-F animals had lower levels of total protein ($p < .05$) and globulin ($p < .001$). There were no significant differences in any of the other parameters. For the LE rats, the younger animals had significantly higher levels of K ($p < .05$), P ($p < .001$), glucose ($p < .001$), alkaline phosphatase ($p < .001$), cholesterol ($p < .01$), and Fe ($p < .001$). The younger LE-F animals had lower levels of creatinine ($p < .001$), total protein ($p < .001$), albumin ($p < .001$), globulin ($p < .001$), and SGPT ($p < .05$). There were no significant differences in any of the other parameters.

Previous studies on blood parameters in Brattleboro rats compared the DI and HZ animals of the strain.[1–4] The LE strain, from which the Brattleboro rats mutated[1] was not examined. The present study compared DI rats and LE rats because these are the commonly used groups in studies on memory and other psychological and physiological phenomena. The use of LE animals as controls may explain why our data are not in agreement with Valtin and Schroeder[1] with respect to hematocrit and with Barnett *et al.*[2] with respect to hematocrit and hemoglobin concentrations. Age differences may explain discrepancies between the present study and that of Barnett *et al.* with respect to total protein. The rats in the present study (Experiment 2) were 8 wks. old, whereas those utilized by Barnett *et al.* were 37 to 138 wks. of age. The same can be said for the results obtained on levels of Na, Ca, and K in the Balment *et al.* study.[3] Their subjects were 18 to 35 wks. of age. The present study confirms, by direct measurement, the results of Galton *et al.*[4] that thyroid function is normal in Brattleboro rats.

It is suggested that in experiments utilizing Brattleboro rats a control group of LE rats be employed. It is further suggested that the sex and age of the animals be noted since several of the blood constituents presented in this study show marked variations when correlated with these two parameters.

References

1. Valtin, H. & H. A. Schroeder. 1964. Familial hypothalamic diabetes insipidus in rats (Brattleboro strain). Am. J. Physiol. **206:** 425–430.
2. Barnett, J. L., P. Cheeseman, J. Cheeseman, J. M. Douglas & J. G. Phillips. 1974. Changes in organ weights and blood parameters in ageing Brattleboro rats with hereditary diabetes insipidus. Age & Ageing **3:** 229–239.
3. Balment, R. J., I. C. Jones, I. W. Henderson & J. A. Oliver. 1976. Effects of adrenalectomy and hypophysectomy on water and electrolyte metabolism in male and female rats with inherited hypothalamic diabetes insipidus (Brattleboro strain). J. Endocr. **71:** 193–217.
4. Galton, V. A., H. Valtin & D. G. Johnson. 1966. Thyroid function in the absence of vasopressin. Endocrinology **78**: 1224–1229.
5. The Technicon SMAC System is the automated system utilized by Consolidated Biomedical Laboratories, Columbus, Ohio.

THE RENIN-ANGIOTENSIN SYSTEM IN THE ABSENCE OF ANTIDIURETIC HORMONE

J. Ann Oliver, Ian W. Henderson,* Annette McKeever, and Richard J. Balment †

Department of Zoology
University of Sheffield
Sheffield S10 2TN, England

and

† *Department of Zoology*
University of Manchester
Manchester M13 9PL England

The renin-angiotensin system and antidiuretic hormone are key factors in the maintenance of mammalian water balance. Although exogenous vasopressin suppresses plasma renin activity and angiotensin can stimulate vasopressin release, a simple negative feedback between these renal and neurohypophyseal hormones is unlikely.[1, 2] This is largely because these secretions act upon internal body fluid distribution, upon vascular smooth muscle, upon the adrenal cortex (in particular aldosterone), and upon sodium homeostasis.

Comparison of Brattleboro rats with (DI) and without (non-DI) hypothalamic diabetes insipidus, revealed that DI animals had much more active renin-angiotensin systems.[3-5] Such data are compatible with vasopressin inhibition of renin release. Moreover, blockade of angiotensin I-converting enzyme reduces vasopressin excretion in the spontaneously hypertensive rat.[6] The relationship becomes more complex, for earlier studies[4, 5] suggested that males displayed a greater activation of the renin-angiotensin system than female DI Brattleboro rats. Exogenous vasopressin administered daily for one week to reduce water turnover of DI rats by about 60% provoked a fall in PRA of male rats, but actually increased that of females.[5] This sexual difference in the PRA response to exogenous vasopressin also occurs acutely in anesthetized DI rats (FIGURE 1), and although more marked in DI rats, is not unique to them.[7]

The underlying mechanisms responsible for these sex differences in renin-vasopressin interactions are unclear but certain data are pertinent: in male DI rats castration or treatment with cyproterone acetate blocks the vasopressin-induced reduction in PRA; ovariectomy of female DI rats does not qualitatively change the vasopressin-induced increase in PRA; hypophysectomized or adrenalectomized male and female DI rats give reductions in PRA after vasopressin. Patently, many factors could account for these significant qualitative differences between male and female rats with respect to the renin-angiotensin system, among them the known sex differences in angiotensinogen metabolism,[8] androgen- and/or estrogen-dependent alterations in vascular reactivity,[9, 10] and gonadal influences upon neurohypophyseal function.[11]

Inhibition of the hyperactive renin-angiotensin system in DI rats produces obvious, if not as yet clearly defined, changes in the syndrome.[12, 13] Thus

* Correspondence to: Dr. Ian W. Henderson, Department of Zoology, University of Sheffield, Sheffield S10 2TN, England.

0077-8923/82/0394-0275 $1.75/0 © 1982, NYAS

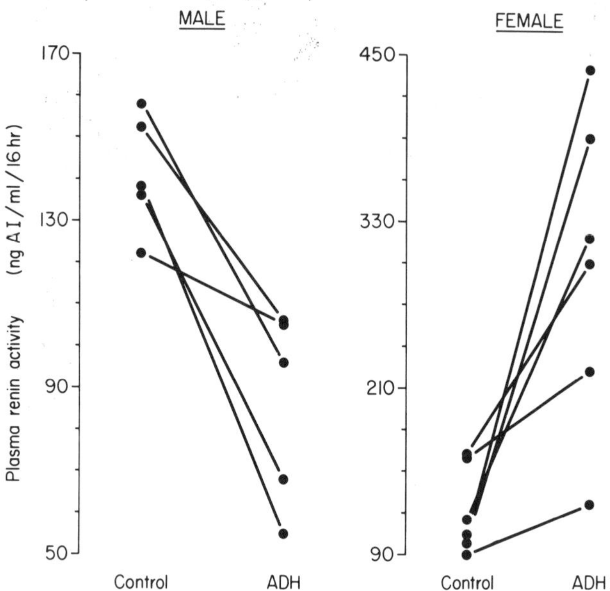

FIGURE 1. Plasma renin activities of male and female DI Brattleboro rats before and after intravenous injection of antidiuretic hormone (ADH). (From Henderson *et al.*[7])

administration of Captopril (SQ 14225), an inhibitor of angiotensin I-converting enzyme reduces the intake and urinary output of water; water balance itself is unaltered and it is not clear whether the primary effect is on renal or central nervous mechanisms. Natriuretic and ontikaliuretic actions of Captopril also occur alongside reduced plasma aldosterone and unchanged corticosterone concentrations[14] (TABLE 1). Structural changes in the adrenal cortex include a narrowing of the zona glomerulosa.

TABLE 1

EFFECTS OF ORAL SQ 14225 ON PLASMA ELECTROLYTE AND CORTICOSTEROID CONCENTRATIONS AND RENIN ACTIVITY IN DI MALE BRATTLEBORO RATS

	Control Rats	Rats Drinking SQ 14225 For 6 Days
Aldosterone (ng %)	8.4 ± 2.0(5)	2.8 ± 0.4(6)†
Corticosterone (μg %)	15.9 ± 2.8(5)	12.4 ± 2.2(9)
Potassium (mmol/L)	3.82 ± 0.15(10)	4.24 ± 0.21(9)*
Sodium (mmol/L)	133.8 ± 3.2(5)	141.6 ± 4.7(9)
Osmolarity (mosmol/L)	313.8 ± 6(5)	307.8 ± 5.4(9)
Plasma renin activity (ng equivalents angiotensin I/ml/16 h)	127.3 ± 7.3(5)	429.3 ± 110.5(5)

Means ± SE of numbers of rats given parentheses, * and †, $p < 0.1$ and < 0.02, respectively.[12]

It is concluded that certain facets of the syndrome of diabetes insipidus of the Brattleboro rat reflect the actions of an unchecked hyperactive renin-angiotensin system; the primary and secondary consequences upon central and peripheral nervous, renal, and general endocrine systems await clarification.

References

1. Davis, J. O. & R. H. Freeman. 1976. Mechanisms regulating renin release. Physiol. Rev. **56:** 1–56.
2. Keeton, T. K. & W. B. Campbell. 1980. The pharmacologic alteration of renin release. Pharmacol. Rev. **32:** 81–227.
3. Gutman, Y. & F. Benzakein. 1971. Effect of an increase and a lack of antidiuretic hormone on plasma renin activity in the rat. Life Sci. **10:** 1081–1085.
4. Gutman, Y. & F. Benzakein. 1974. Antidiuretic hormone and renin in rats with diabetes insipidus. Eur. J. Pharmacol. **28:** 114–118.
5. Balment, R. J., I. W. Henderson & J. A. Oliver. 1975. The effects of vasopressin on pituitary oxytocin content and plasma renin activity in rats with hypothalamic diabetes insipidus (Brattleboro strain). Gen. Comp. Endocrinol. **26:** 468–477.
6. Crofton, J. T., L. Share & Z. P. Horovitz. 1979. The effect of SQ14,225 on systolic blood pressure and urinary excretion of vasopressin in the developing spontaneously hypertensive rat. Hypertension **1:** 462–467.
7. Henderson, I. W., R. J. Balment & J. A. Oliver. 1978. Vasopressin effects on plasma renin activity in male and female rats. Clin. Sci. Mol. Med. **55:** 301–307.
8. Hasegawa, H., A. Nasjletti, K. Rice & G. M. C. Masson. 1973. Role of pituitary and adrenals in the regulation of plasma angiotensinogen. Am. J. Physiol. **225:** 1–6.
9. Baker, P. J., E. R. Ramey & P. W. Ramwell. 1978. Androgen-mediated sex differences of cardiovascular responses in rats. Am. J. Physiol. **235:** H242–H246.
10. Hoeg, J. M., L. R. Willis & M. H. Weinberger. 1977. Estrogen attenuation of development of hypertension in spontaneously hypertensive rats. Am. J. Physiol. **233:** H369–H373.
11. Skowsky, W. R., L. Swan & P. Smith. 1979. Effects of sex steroid hormones on arginine vasopressin in intact and castrated male and female rats. Endocrinology **104:** 105–108.
12. Henderson, I. W., A. McKeever & C. J. Kenyon. 1979. Captopril (SQ-14225) depresses drinking and aldosterone in rats lacking vasopressin. Nature **281:** 569–570.
13. Stamoutsos, B. A., R. G. Carpenter & S. P. Grossman. 1981. Role of angiotensin-II in the polydipsia of diabetes insipidus in the Brattleboro rat. Physiol. Behav. **26:** 691–693.
14. McKeever, A., C. J. Kenyon, J. A. Oliver & I. W. Henderson. 1980. Effects of angiotensin I converting enzyme inhibitor SQ14225 on water and electrolyte balance and adrenocortical function in rats with inherited diabetes insipidus (Brattleboro strain). J. Endocrinol. **85:** 16P–17P.

EFFECT OF LITHIUM AND ANTIDIURETIC HORMONE ON PLASMA RENIN CONCENTRATION IN DIABETES INSIPIDUS RATS (BRATTLEBORO RAT MODEL)

S. Opava-Stitzer

Department of Physiology
University of Puerto Rico School of Medicine
San Juan, Puerto Rico 00936

Antidiuretic hormone (ADH) is known to inhibit renin secretion in many species,[1-6] but the mechanism of this inhibition and its importance in the control of renin secretion are unknown. Measurements of plasma renin concentration (PRC) and plasma renin activity (PRA) in the diabetes insipidus (DI) rat suggest that physiological levels of ADH may exert a tonic inhibitory effect on renin release. PRC and PRA are increased in the DI rat[7-9] and reduced following treatment with ADH.[9-11] Nevertheless it is unclear to what extent these alterations in renin are due to the presence or absence of ADH *per se*. In the absence of ADH, plasma volume,[12] blood pressure,[7] and glomerular filtration rate[13] are reduced, all of which might stimulate renin secretion. Conversely, ADH might reduce plasma renin levels in the DI rat indirectly, secondary to its effects on water balance, as well as by a direct effect on renin secretion. To differentiate between direct and indirect effects of ADH on renin secretion, plasma renin concentration was measured in DI rats, before and during ADH administration, with or without LiCl treatment. Using different doses of ADH, combined with LiCl treatment, it was possible to raise the titer of antidiuretic hormone in the plasma of the DI rat while blocking to a varying extent the action of ADH on tubular water reabsorption.[14] The effect of ADH on PRC of LiCl-treated and control (NaCl-treated) DI rats was compared.

All rats were maintained in individual metabolism cages for the entire study. As shown in TABLE 1, the protocol consisted of pretreatment with 0.1 ml/100 g BW/day of isotonic saline ($N = 37$) or LiCl (3.0 mEq/kg BW/day) ($N = 38$), i.p., for 11 days, followed by two weeks of balance study and bleeding for determination of PRC. Forty-three rats (half LiCl-, half NaCl-treated) were then injected with 10 mU ($N = 12$), 25 mU ($N = 11$), or 200 mU ($N = 20$) of ADH (Parke-Davis, Pitressin tannate in oil) per 100 g BW per day, or an equal volume of sesame oil ($N = 22$), s.c., for one week, followed by two weeks of balance study and bleeding. Once begun, LiCl, NaCl, ADH, and oil injections were continued throughout the study. Ten rats received neither ADH nor oil but only LiCl ($N = 5$) or NaCl ($N = 5$) injections. Blood samples were obtained by tailcutting in the conscious rat. PRC was determined by radioimmunoassay (New England Nuclear)[15] of the amount of angiotensin I generated in the presence of excess substrate.[16] Balance studies consisted of daily measurement of body weight, water intake, urine volume, osmolality, and sodium and potassium concentrations. The effects of an intermediate dose of ADH (20 mU/100 g BW/day) on hematocrit, plasma osmolality, plasma protein, and sodium and potassium concentrations were also determined in 10 LiCl-treated and 10 NaCl-treated DI rats. Sodium and potassium concentra-

0077-8923/82/0394-0278 $1.75/0

TABLE 1

PROTOCOL FOR THE STUDY OF THE EFFECTS OF ADH TREATMENT ON PLASMA RENIN CONCENTRATION IN LiCl-TREATED AND CONTROL (NaCl-TREATED) DI RATS

	Time (days) 1–11	12–18	19–25	26–32	33–39	40–44
Procedures	1. NaCl or LiCl injections	1. NaCl or LiCl injections 2. 2-Day balance study (Days 13–15) 3. Renin (Day 16)	1. NaCl or LiCl injections 2. 2-Day balance study (Days 20–22) 3. Renin (Day 23)	1. NaCl or LiCl injections 2. ADH or oil injections	1. NaCl or LiCl injections 2. ADH or oil injections 3. 2-Day balance study (Days 34–36) 4. Renin (Day 37)	1. NaCl or LiCl injections 2. ADH or oil injections 3. 2-Day balance study (Days 41–43) 4. Renin (Day 44)

tions in urine and plasma were determined by flame photometry; osmolality by freezing-point depression.

The effect of ADH on urine osmolality was greater during the second week of study and only these results have been included in FIGURES 1 and 2. As shown in FIGURE 1, ADH had widely differing effects on urine osmolality of LiCl- and NaCl-treated rats. At doses of 10 and 25 mU/100 g BW/day, LiCl-treated rats continued to excrete hypotonic urine while the urine of NaCl-treated rats was hypertonic. At the highest dose of ADH both groups were able to concentrate their urine, although to differing degrees. Despite these disparate effects on water excretion, ADH affected PRC similarly in both groups. As shown in FIGURE 2, at any dose of ADH, there was no difference in the PRC of LiCl- and NaCl-treated DI rats. These data also suggest that a dose-response relationship may have existed between ADH and renin secretion with maximal effects obtained at 25 mU/100 g BW/day. There was no difference in the effect of this dose and the 200 mU dose on PRC. Although the 10 mU dose of ADH

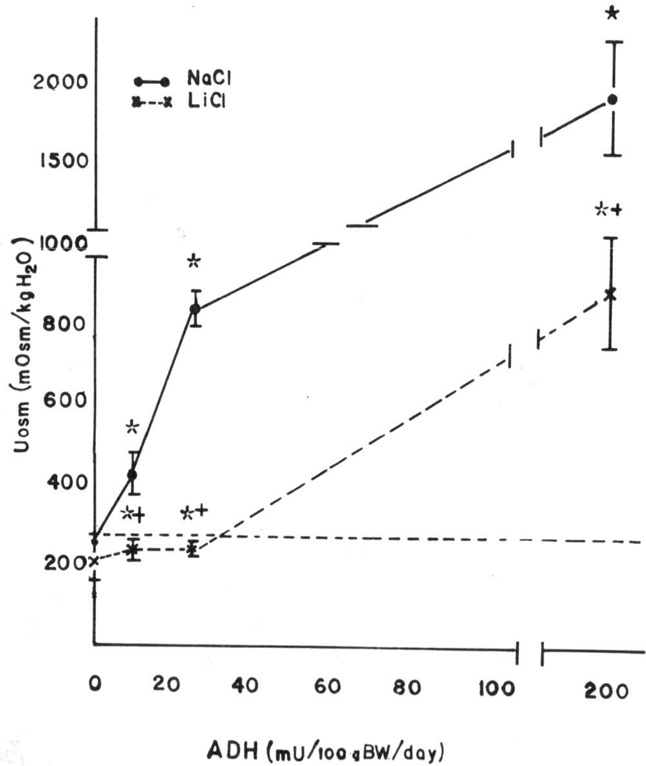

FIGURE 1. Effect of different doses of ADH or oil (ADH = 0) on urine osmolality (U_{Osm}) of LiCl-treated and NaCl-treated DI rats. * Significantly different from values prior to ADH or oil treatment, $p < 0.05$. + Significantly different from NaCl-treated rats, $p < 0.05$. Values are means ± SEM.

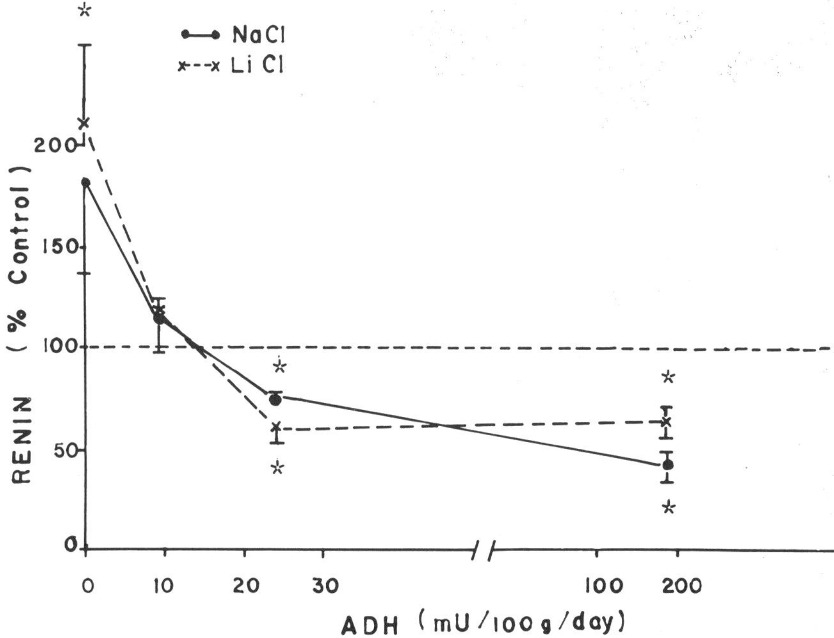

FIGURE 2. Effect of different doses of ADH or oil (ADH = 0) on plasma renin concentration of LiCl-treated and NaCl-treated DI rats. * Significantly different from values prior to ADH or oil treatment, $p < 0.05$. Values are means ± SEM.

did not alter PRC, oil injections alone (ADH = 0, FIGURE 2) increased PRC, as did multiple bleedings of LiCl- and NaCl-treated rats with neither ADH nor oil injections superimposed. These results suggest that the lowest dose of ADH may have prevented such a rise in PRC.

As shown in FIGURE 3, when all data were pooled, there was no significant correlation between plasma renin concentration and urine osmolality. Moreover, there was no difference in PRC when the urine was hypertonic (78 ± 4 ng AI/ml/h, $N = 65$) or hypotonic (82 ± 5, $N = 84$).

As shown in TABLE 2, ADH treatment significantly affected only plasma osmolality, which was decreased by ADH in control (NaCl) rats, but increased in LiCl-treated rats, probably not as a result of, but despite ADH treatment.

These data indicate that the inhibition of renin secretion in the DI rat during ADH treatment is unrelated to effects of ADH on urine concentration. Thus systemic effects on renin release, secondary to changes in water balance, seem an unlikely explanation for the observed inhibition of renin. Rather a direct action of ADH on the renin-producing cell or baroreceptor is suggested.

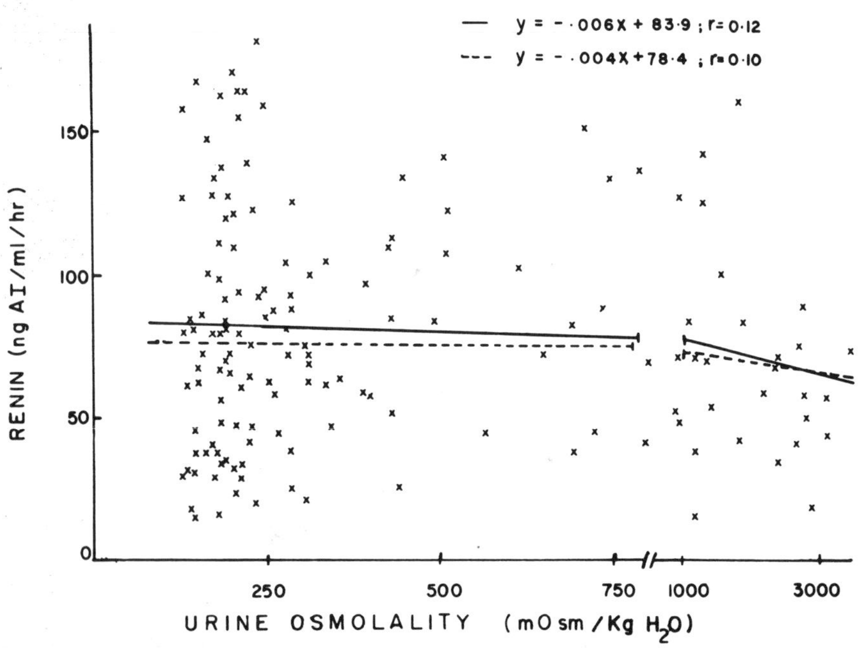

FIGURE 3. Linear regression analysis of the relationship between plasma renin concentration and urine osmolality in LiCl-treated and NaCl-treated DI rats. The solid line is the regression line for values during ADH treatment only; the dotted line is the regression line for all values.

TABLE 2

EFFECTS OF ADH TREATMENT ON HEMATOCRIT (HCT), PLASMA OSMOLALITY (P_{Osm}), AND PLASMA PROTEIN, SODIUM (P_{Na}) AND POTASSIUM (P_K) CONCENTRATION IN LiCl- AND NaCl-TREATED DI RATS

		HCT (%)	Protein (g %)	P_{Osm} (mOsm/L)	P_{Na} (mM)	P_K (mM)
LiCl	Control	39 ± 0.6	7.5 ± 0.17	294 ± 4	142 ± 0.7	3.6 ± 0.09
	ADH	40 ± 0.7	7.4 ± 0.09	305 ± 3 †	143 ± 0.8	3.6 ± 0.11
NaCl	Control	39 ± 0.9	7.7 ± 0.12	313 ± 2	142 ± 1.3	3.4 ± 0.11
	ADH	41 ± 0.7	7.8 ± 0.17	308 ± 1 *	142 ± 1.4	3.1 ± 0.12

Values are means ± SEM.
* Significantly different from control, $p < 0.05$; † $p < 0.001$.

References

1. Buñag, D. D., I. H. Page & J. W. McCubbin. 1967. Inhibition of renin release by vasopressin and angiotensin. Cardiovasc. Res. **1:** 67–73.
2. Vander, A. J. 1968. Inhibition of renin release in the dog by vasopressin and vasotocin. Circ. Res. **23:** 605–609.
3. Smith, M. J., A. W. Cowley, Jr., A. C. Guyton & R. D. Manning, Jr. 1979. Acute and chronic effects of vasopressin on blood pressure, electrolytes, and fluid volumes. Am. J. Physiol. **237:** F232–F240.
4. Vandongen, R. 1975. Inhibition of renin secretion in the isolated rat kidney by antidiuretic hormone. Clin. Sci. Mol. Med. **49:** 73–76.
5. Konrads, A., G. Hofbauer, U. Werner & F. Gross. 1978. Effects of vasopressin and its deamino-D-arginine analog on renin release in the isolated perfused rat kidney. Pflügers Arch. **377:** 81–85.
6. Hesse, B. & I. Nielsen. 1977. Suppression of plasma renin activity by intravenous infusion of antidiuretic hormone in man. Clin. Sci. Mol. Med. **52:** 357–360.
7. Gross, F., G. Dauda, S. Kazda, J. Kynčl, J. Möhring & H. Orth. 1972. Increased fluid turnover and the activity of the renin-angiotensin system under various experimental conditions. Circ. Res. **30,31** (Suppl. II): 173–181.
8. Gutman, Y. & F. Benzakein. 1971. Effect of an increase and lack of antidiuretic hormone on plasma renin activity in the rat. Life Sci. **10:** 1081–1085.
9. Balment, R. J., I. W. Henderson & J. A. Oliver. 1975. The effects of vasopressin on pituitary oxytocin content and plasma renin activity in rats with hypothalamic diabetes insipidus (Brattleboro strain). Gen. Comp. Endocrinol. **26:** 468–477.
10. Gutman, Y. & F. Benzakein. 1974. Antidiuretic hormone and renin in rats with diabetes insipidus. Eur. J. Pharmacol. **28:** 114–118.
11. Möhring, J., G. Kohrs, B. Möhring, M. Petri, E. Homsy & D. Haack. 1978. Effects of prolonged vasopressin treatment in Brattleboro rats with diabetes insipidus. Am. J. Physiol. **234:** F106–F111.
12. Möhring, J., B. Möhring, A. Schömig-Brekner & D. Haack. 1974. Acute effects of vasopressin on potassium and water balance in rats with diabetes insipidus. Am. J. Physiol. **227:** 921–926.
13. Gellai, M. & H. Valtin. 1979. Chronic vascular constrictions and measurements of renal function in conscious rats. Kidney Int. **15:** 419–426.
14. Myers, J. B., T. O. Morgan, S. L. Carney & C. Ray. 1980. Effects of lithium on the kidney. Kidney Int. **18:** 601–608.
15. Haber, E., T. Koerner, L. B. Page, B. Kliman & A. Purnode. 1969. Application of a radioimmunoassay for angiotensin I to the physiologic measurements of plasma renin activity in normal human subjects. J. Clin. Endo. Metab. **29:** 1349–1355.
16. Churchill, P. C., M. C. Churchill & F. McDonald. 1973. Renin release in anesthetized rats. Kidney Int. **4:** 273–279.

EFFECTS OF ANTI-OXYTOCIN SERUM IN BRATTLEBORO RATS

I. C. A. F. Robinson, R. G. Clark, K. M. Fairhall, P. M. Jones,* and the late J. A. Parsons

Laboratory for Endocrine Physiology and Pharmacology
The National Institute for Medical Research
London, NW7 1AA, England

Introduction

Although oxytocin (OT) possesses both natriuretic [1] and antidiuretic activities [2] the physiological relevance of these activities is obscure. Homozygous Brattleboro rats present an interesting model in which to study this since they lack vasopressin and secrete oxytocin at a high rate. In this study we have investigated the effects of oxytocin on salt and water excretion in conscious Brattleboro rats before and after passive immunization with an antiserum to OT.

Materials and Methods

Homozygous Brattleboro rats (200–220 g) were equipped with indwelling jugular venous cannulae under halothane anesthesia, and housed singly in metabolic cages with food and distilled water *ad lib.* The venous cannulae were connected via swivel joints to infusion lines through which 0.9% NaCl was continuously delivered at 0.3 ml/h. After several days, timed urine samples were collected, the volumes noted, Na^+ and K^+ concentrations determined by flame photometry, and the oxytocin-like immunoreactivity ('OT') estimated by radioimmunoassay.[3] For passive immunization, each rat received 0.2 ml i.v. of a high titer (3×10^5) anti-oxytocin serum (anti-OT), described in detail elsewhere.[3, 4] Oxytocin (Syntocinon, Sandoz Ltd.) was diluted in saline just before use.

Results

In one experiment involving 16 rats, groups of four animals received OT infusions at 3, 1, 0.3, or 0 pmol/min for 5 h. Figure 1 illustrates the effects of OT infusion at the highest rate. Urine flow falls, accompanied by rises in both Na^+ and K^+ output. The control group also showed a fall in urine volume at this time of day although Na^+ and K^+ outputs remained steady. Large amounts of OT-like material appeared in the urine during and after the infusion. Table 1 shows the values obtained in four collection periods during and immediately after infusion, expressed as a percentage of the outputs in the two pre-infusion periods. Table 2 shows a similar comparison of the last three periods post-infusion as a percentage of the pre-infusion values. Note that the

* P. M. Jones is a Medical Research Council Scholar.

0077–8923/82/0394–0285 $1.75/0

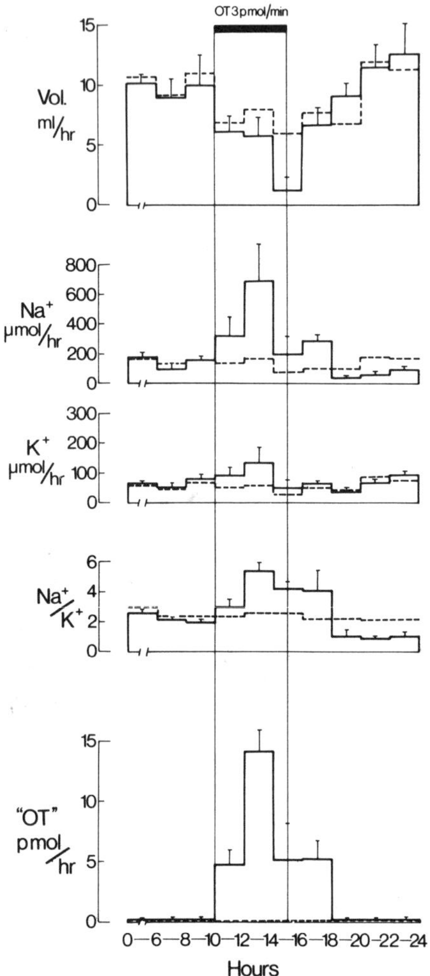

FIGURE 1. Oxytocin infusion in Brattleboro rats. The animals were infused with 0.9% NaCl with (——) or without (– – –) the addition of oxytocin (OT) to the infusate for a five-hour period. Urinary outputs of volume (Vol) sodium (Na^+), potassium (K^+), and oxytocin-like immunoreactivity ('OT') were measured on 2-h collections of urine. Results shown are Mean ± SEM ($N = 4$).

TABLE 1

INFUSION OUTPUTS AS A PERCENTAGE OF PRE-INFUSION OUTPUTS

	Volume	Na^+	K^+	Na^+/K^+	'OT'
Control *					
A	58±15	242±60 †	124±35	201±10 §	6510±980 §
B	46±7 †	78±7	67±7	120±13	654±253
C	63±18	97±17	90±24	117±12	348±76 †
D	70±6	86±18	79±13	106±9	77±6
Immunized *					
A	44±5	113±34	72±19	153±10 §	2800±815 †
B	83±14	132±45	141±51	99±10 †	478±130 †
C	95±22	105±29	126±36	84±9	235±33 †
D	86±18	69±21	107±22	64±10	135±23

* OT infusion rates: A=3, B=1, C=0.3, and D=0 pmol/min. Immunized animals received 0.2 ml antioxytocin serum prior to OT infusion (see text). Differences, group D vs. A, B or C; † $p < 0.05$, § $p < 0.001$, ± = SEM, N = 4 in all cases.

Na^+/K^+ ratio rose during infusion but fell significantly below control values post-infusion. Two days later the rats were passively immunized with anti-OT and the OT infusions repeated as before. Oxytocin again produced a fall in urine flow and a rise in Na^+ output (FIGURE 2) but these effects were smaller, and both delayed and prolonged when compared to the results obtained before the animals were immunized (FIGURE 1). This is clearly seen in TABLE 2 where the Na^+/K^+ ratio again rose during OT infusion at the higher rates, but remained at or above control levels for several hours post-infusion. Immunization also delayed and prolonged the excretion of 'OT' (FIGURE 3).

In a second experiment, six rats received a 3-h infusion of OT at 3 pmol/min on consecutive days, before and after passive immunization. OT infusion

TABLE 2

POST-INFUSION OUTPUTS AS A PERCENTAGE OF PRE-INFUSION OUTPUTS

	Volume	Na^+	K^+	Na^+/K^+	'OT'
Control *					
A	119±5	44±9 †	102±10	43±7 †	180±30
B	108±10	39±12 †	94±20	38±8 †	99±22
C	127±14	103±26	145±31	67±7	127±20
D	98±11	107±17	122±19	89±12	98±13
Immunized *					
A	50±6 †	71±18	84±15	88±23	3220±795 ‡
B	65±11	94±19	100±17	93±10 †	1025±390
C	119±26	116±38	164±43	72±16	194±94
D	111±23	106±35	160±36	63±6	123±23

* OT infusion rates: A=3, B=1, C=0.3, and D=0 pmol/min. Immunized animals received 0.2 ml antioxytocin serum prior to OT infusion (see text). Differences, group D vs. A, B or C; † $p < 0.05$, ‡ $p < 0.01$, § $p < 0.001$, ± = SEM, N = 4 in all cases.

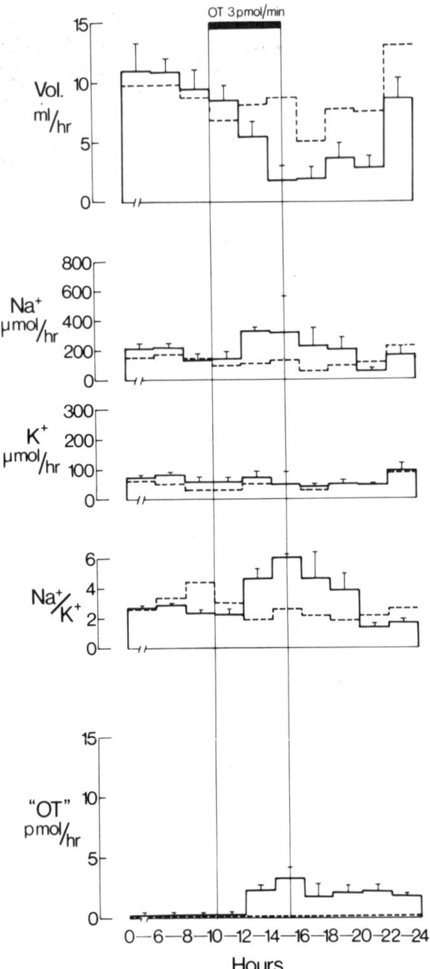

FIGURE 2. Oxytocin infusion in Brattleboro rats passively immunized with anti-OT. Experimental conditions as in FIGURE 1 except that both control group (– – – –) and OT-infused group (——) were passively immunized with 0.2 ml anti-OT at 6 a.m. Urine samples were collected at 2 h intervals. Results shown are Mean ± SEM ($N = 4$).

doubled Na^+ and K^+ output and increased 'OT' excretion, all of which returned to control levels 4 h post-infusion. After immunization, urinary 'OT' excretion was reduced and no increase occurred during OT infusion. However urinary Na^+ output rose as before and remained high 4 h post-infusion.

In a further experiment, 10 rats were passively immunized without receiving continuous saline infusions. Anti-OT had no effect on daily urine volume or Na^+ output, even though these animals had high titers of anti-OT ($>10^3$) more than two days after immunization.

DISCUSSION

It has long been known that water deprivation or salt-loading depletes pituitary oxytocin [5] and that elevations in plasma osmolality increase OT secretion in normal rats.[6,7] Since OT is both natriuretic and antidiuretic when given acutely to Brattleboro rats [2,7] we were interested to study the effectiveness of longer-term infusions of this peptide and to passively immunize these rats against OT whilst monitoring salt and water output.

The most striking effect of OT in our study was an increase in Na^+ output at a time when urine flow fell, the Na^+/K^+ ratio increasing in a dose-dependent manner. After the infused oxytocin had been excreted, sodium was retained, the Na^+/K^+ ratio falling below control values. Passive immunization reduced but did not abolish these effects of infused OT. The clearance of OT from plasma was delayed and prolonged, judging by the appearance of OT-like material in the urine over a much longer period. It is likely that significant amounts of free oxytocin would be present in the plasma of immunized animals for a period much longer than that of the infusion.[4] Although passive immunization reduced the excretion of endogenous OT-like material, it had no effect on daily urine volume or Na^+ output.

We therefore conclude that endogenous OT does not play a major role in regulating salt and water excretion in Brattleboro rats, but that pharmacological doses of OT produce their known natriuretic effect.

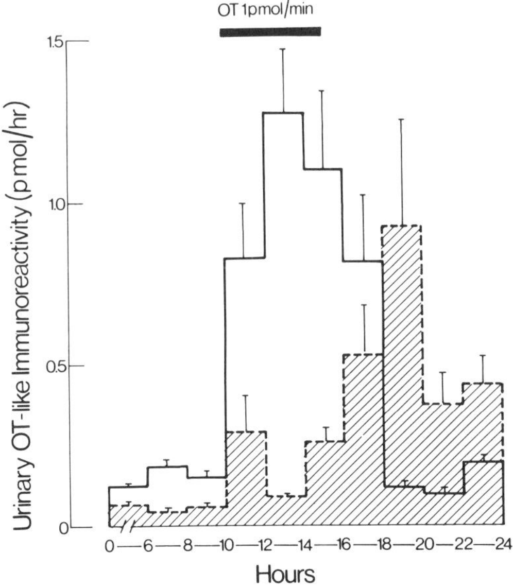

FIGURE 3. Urinary excretion of oxytocin-like immunoreactivity. Oxytocin was infused for 5 h (solid bar) and the urinary excretion of OT-like immunoreactivity estimated by RIA (open columns). Two days later, the rats were passively immunized with anti-OT, and the OT infusion repeated (hatched columns). Values shown are Mean ± SEM ($N = 4$).

Summary

Brattleboro rats were housed singly in metabolic cages and continuously infused with saline. Timed urine collections were analyzed for Na^+ and K^+; oxytocin (OT)-like material was estimated by RIA. OT infusions reduced urine flow and increased Na^+ output. Urinary Na^+/K^+ ratios rose during OT infusion, falling below control values after infusion. Passive immunization with an OT antiserum reduced but did not prevent these renal effects of OT. In a second experiment, immunization had no effect on daily urine volume or Na^+ output, even though high titers were achieved and excretion of endogenous OT-like material was reduced. We conclude that endogenous OT has little effect on Na^+ excretion in these rats, although pharmacological doses of OT produced their known natriuretic effect.

References

1. Chan, W. Y. 1965. Effects of neurohypophysial hormones and their deamino analogues on renal excretion of Na^+, K^+ and water in rats. Endocrinology **77:** 1097–1104.
2. Sawyer, W. H. & H. Valtin. 1967. Antidiuretic responses of rats with hereditary hypothalamic diabetes insipidus to vasopressin, oxytocin and nicotine. Endocrinology **80:** 207–210.
3. Robinson, I. C. A. F. 1980. The development and evaluation of a sensitive and specific radioimmunoassay for oxytocin in unextracted plasma. J. Immunoassay **1:** 323–347.
4. Robinson, I. C. A. F. & J. A. Parsons. 1981. Suckling in the guinea pig: the effects of passive immunization with an antiserum to oxytocin. J. Endocr. **90:** (In press.)
5. Jones, C. W. & B. T. Pickering. 1969. Comparison of the effects of water deprivation and sodium chloride imbibition on the hormone content of the neurohypophysis of the rat. J. Physiol. **203:** 449–458.
6. Dogterom, J., T. J. B. Van Wimersma Greidanus & D. F. Swaab. 1977. Evidence for the release of vasopressin and oxytocin into cerebrospinal fluid: measurements in plasma and CSF of intact and hypophysectomized rats. Neuroendocrinology **24:** 108–118.
7. Balment, R. J., M. J. Brimble & M. L. Forsling. 1980. Release of oxytocin induced by salt loading and its influence on renal excretion in the male rat. J. Physiol. **308:** 439–449.

MORPHOLOGICAL CORRELATES OF RENIN AND ALDOSTERONE SECRETION IN THE BRATTLEBORO RAT *

Hilda Weyl Sokol and Jan Möhring

Department of Physiology
Dartmouth Medical School
Hanover, New Hampshire 03755

There is a paradoxical relationship between angiotensin and aldosterone in the homozygous Brattleboro rat, which lacks vasopressin (VP) and therefore has diabetes insipidus. Compared to normal animals, higher levels of plasma renin and angiotensin II are accompanied by lower concentrations of aldosterone in the blood in the DI rat.[1] This finding prompted a comparative morphological study (by light and electron microscopy) of the structures responsible for the secretion of renin and aldosterone in normal Long-Evans (LE), heterozygous (HZ), and homozygous (DI) Brattleboro rats.

The renal juxtaglomerular apparatus (JGA) and the adrenal cortex were examined in the three different genotypes. Some DI rats were injected with vasopressin, which corrects fluid turnover as well as the aberrant circulating levels of angiotensin II and aldosterone.[2] To assess the responsiveness of the renin-angiotensin-aldosterone system to a severe stimulus in both the presence and absence of vasopressin, both HZ and DI Brattleboro rats were placed on a salt-deficient diet.

The results of these studies permit the broad generalization that the granularity of the JGA and the width of the zona glomerulosa reflect the plasma levels of renin-angiotensin II and aldosterone, respectively.

Materials and Methods

Animals

Long-Evans and Brattleboro rats were bred and maintained in the Pharmacological Institute of the University of Heidelberg, Germany. Male rats approximately three months old were used. There were six groups of eight rats each: (1) normal Long-Evans rats, (2) heterozygous Brattleboro rats, (3) DI rats (four of which received daily subcutaneous injections of 0.1 ml peanut oil for two weeks), (4) DI rats treated with vasopressin (100 mU VP tannate in oil, Parke-Davis) daily for two weeks, (5 & 6) DI and HZ rats on sodium-deficient diets for 30 days.

Preparation of Tissues: Kidney Cortex and Adrenal Glands

After the rats were anesthetized with Inactin (100 mg/kg) the abdominal aorta was cannulated for whole body perfusion. Blood was first washed out

* These studies were performed while H. W. S. held a Senior Fogarty International Fellowship (N.I.H.), 1975–76, under the auspices of Dr. Wilhelm Kriz, Director of the Anatomical Institute, University of Heidelberg.

0077-8923/82/0394-0291 $1.75/0 © 1982, NYAS

with a modified Ringer's solution to which heparin had been added, followed by perfusion with 300–400 ml 3.0% glutaraldehyde in 0.1 M cacodylate buffer. After kidneys and adrenal glands were removed, small (ca. 2.5 mm^3) pieces of the superficial cortex of the kidney (where most JGA are located) were fixed overnight in the glutaraldehyde buffer mixture. Subsequently the small blocks of tissue were rinsed in 0.1 M phosphate buffer, followed by further fixation with osmium tetroxide, dehydration in alcohol and propylene oxide, and final embedding in Epon. Semi-thin (0.5–1.0 μm) sections stained with toluidine blue were examined under a light microscope to localize areas containing juxtaglomerular apparatuses. Favorable areas were subsequently sectioned with an MT-2B Sorval ultramicrotome, stained with uranyl acetate and lead citrate, and examined under a Philips 301 microscope. Upon removal, the adrenal glands were fixed in formalin for several days, dehydrated in alcohol, embedded in paraffin, sectioned at 6 μm, and stained with hematoxylin and eosin for light microscopy.

Analysis of Tissues

Adrenal cortex. Sections through the middle of the adrenal gland were photographed under low power (16×). The width of the entire cortex and the width of the zona glomerulosa, separately, were measured in both glands of 30 rats on photographic enlargements (×80).

Superficial cortex of kidney. Stained semi-thin sections (2.5 mm^2) were examined under the light microscope. At a magnification of 63×, every JGA in a section was photographed and the number of *granulated* cells appearing in each JGA was counted on ×300 photographic enlargements. The total JGA granularity in each group of rats was assessed by multiplying the average number of granulated cells per JGA by the average number of JGAs per 2.5 mm^2 of cortex.

Ultrathin sections were viewed under the electron microscope and every JGA that was visible (i.e., not masked by the grid) was photographed. These electron micrographs were not subjected to morphometric analysis, since an earlier study [3] revealed no significant morphological alterations (e.g., granularity, and the volumes of the Golgi apparatus and rough endoplasmic reticulum) in the JGA of DI rats in comparison to LE rats.

Although 48 animals were used in this study, a good perfusion was not achieved in all. In addition, hydronephrosis occurred in 20% of the rats (excluding those on the sodium-deficient diet) appearing in the non-DI as frequently as in the DI rats. This reduced the number of animals from which adequate tissue could be examined.

Results

JGA

Examination of juxtaglomerular apparatuses in the superficial cortex (albeit only small fragments) under the light microscope, in contrast to the electron microscope, affords the viewer an overall impression of the degree of granularity. While the number of granulated cells discernible in each JGA is not significantly different in the three genotypes, there are more JGAs per unit of tissue in the homozygous Brattleboro rat (TABLE 1). DI rats given VP replacement therapy for two weeks exhibited a slight reduction in total granularity (i.e., number of cells per JGA multiplied by the number of JGAs per 2.5 mm^2 field of cortex.) On the other hand, a pronounced increase not only in number of recognizable JGAs, but in the number of cells per JGA, were seen in rats placed on a sodium-deficient diet.

At the ultrastructural level, the putative renin-containing granules appeared to be more abundant, and the rough endoplasmic reticulum and Golgi apparatus somewhat better developed, in JGA cells of DI rats than in non-DI rats (FIGURE 1 a, b; FIGURE 2 a, b, c). Although a more precise morphometric analysis by Peter and Möhring [3] did not show significant differences in these cellular organelles, these investigators did remark on the trend toward enlargement of these structures in the DI rat. However, hypertrophy and hyperplasia of these elements was striking in the sodium-deficient HZ and DI rats (FIGURE 1 c, d; FIGURE 2 d) where large JGA cells were packed with dense granules of varying sizes apparently due to the coalescence of granules. In addition, the cisternae of the rough endoplasmic reticulum were conspicuously dilated in rats on a salt-free diet, a condition seldom seen in normal rats (FIGURE 2 b, d).

TABLE 1

GRANULARITY OF THE JUXTAGLOMERULAR APPARATUS (JGA) IN NORMAL AND BRATTLEBORO RATS

Rat Type	Number of Granular Cells per JGA	Total Granularity *
Normal LE (8)†	2.41±0.73 ‡	9.9
Heterozygote (4)	2.62±0.80	7.8
Homozygote, DI (6)	2.78±0.49	15.9
DI+Vasopressin (5)§	2.45±0.34	13.0
DI, Na^+ deficient (17)¶	3.65±0.84 **	21.3

* Number of granular cells per JGA multiplied by the number of JGAs seen in a 2.5-mm^2 section of superficial cortex.

† (n) number of sections analyzed.

‡ Mean±1 SEM.

§ VP tannate in oil (100 mU/day) for 10 days.

¶ Rats on a sodium-free diet ad libitum for 30 days.

** $p < 0.03$ compared to untreated DI.

Zona Glomerulosa

The width of the aldosterone-secreting zone of the adrenal cortex is somewhat narrower in the DI rat compared to LE and HZ rats (TABLE 2). In contrast to a slightly reduced zona glomerulosa, the inner glucocorticosteroid-producing zonae fasciculata and reticularis are markedly smaller in DI rats than in non-DI rats. Thus, the zona glomerulosa of the homozygous Brattleboro rat actually occupies a larger proportion of the adrenal cortex than it does in normal Long-Evans and heterozygous Brattleboro rats. Vasopressin replace-

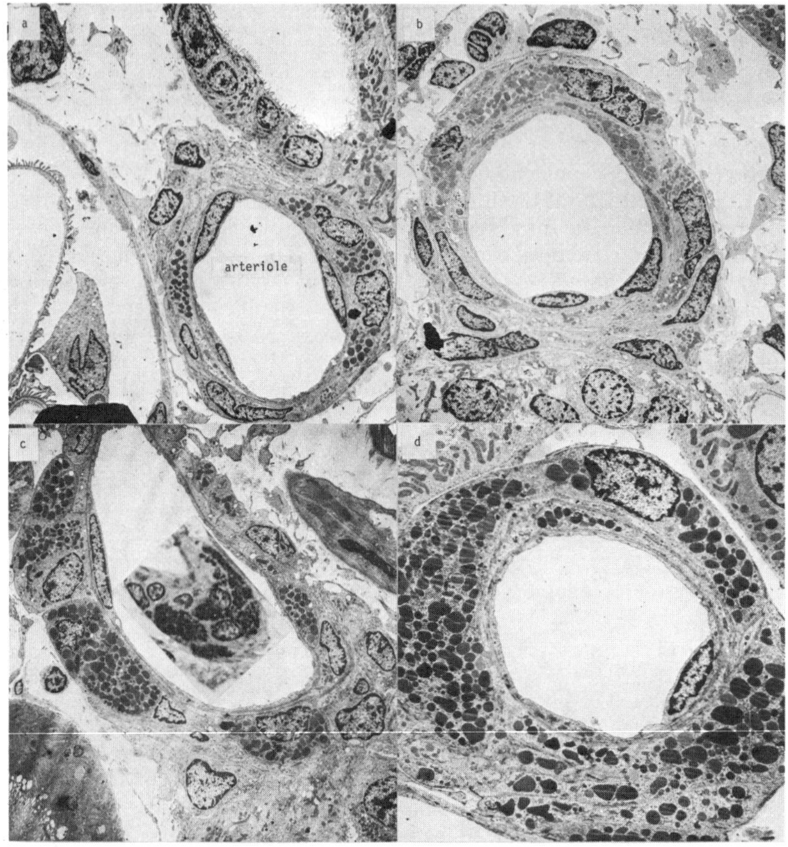

FIGURE 1. The juxtaglomerular apparatus (JGA) in the Brattleboro rat: (a) heterozygote (HZ); (b) homozygote (DI); (c, d) DI on a sodium-deficient diet. The granular epithelioid JGA cells in the media of the arteriole appear to be more numerous and granulated in the untreated DI (b) than in the HZ (a) rat, although there is great variation among JGAs even in the same tissue section. Hypertrophy and hyperplasia of the JGA cells and an increase in size and number of their granules is evident in the salt-deficient state (c, d). Magnifications: a, b, c × 1330; d × 2375; inset in (c) × 144.

ment therapy restored the zona glomerulosa to normal dimensions and caused an even greater enlargement of the zonae fasciculata-reticularis in the DI rat, resulting in a zona glomerulosa: adrenal cortex ratio approaching the normal condition (TABLE 2). This suggests that VP may be enhancing ACTH secretion since the fascicular and reticular zones are more dependent upon ACTH than the zona glomerulosa. However, aldosterone-producing cells of the zona glomerulosa are more responsive to low sodium than other cortical cells. This is seen in the marked hypertrophy of this zone with a compensatory decrease in the glucocorticoid-secreting zones in animals on a sodium-deficient diet (TABLE 2).

DISCUSSION

In general, there is a correspondence between the levels of circulating renin, angiotensin, and aldosterone and the morphology of the tissues responsible for their synthesis, although the structural differences between DI and non-DI rats is slight. The trend is nevertheless consistent with the concept that VP may act directly on JGA cells to inhibit the synthesis and release of renin.[4] Renin concentration in the kidney has been reported to be significantly different between LE and DI rats,[3] which is reflected in the increased total granularity (TABLE 1) seen in the DI rats. This may mean that in the absence of VP dormant or potential JGA epithelioid cells are recruited from smooth muscle cells surrounding the afferent glomerular arteriole.[5-7] That renin-producing

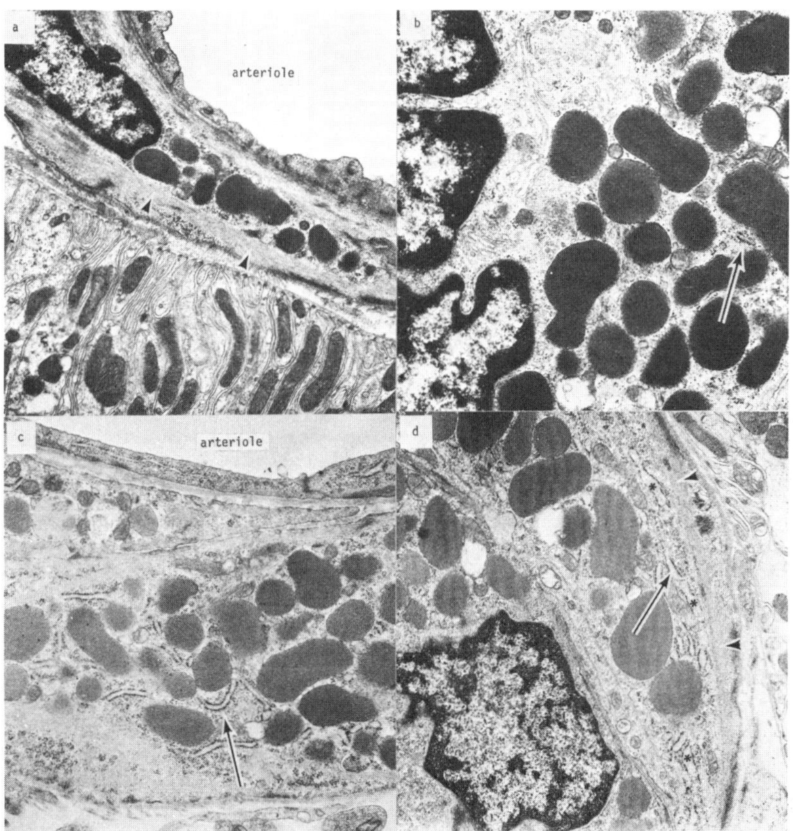

FIGURE 2. Granular epithelioid JGA cells in nomal LE (a, b) and DI (c, d) rats. Myofilaments (arrowheads) are readily seen in both LE (a) and DI (d) rats. Rough endoplasmic reticulum (arrows) is well developed in all cells but most pronounced in the sodium-deficient DI rat (d) which displays very dilated cisternae (*). Magnifications: a, d × 10,450; b, c × 13,300.

cells are derived from smooth muscle cells is suggested by the presence of myofilaments (leiomyofilaments, myofibrils) in granulated JGA cells (FIGURE 2 a, d). Well-defined rough endoplasmic reticulum, the site of protein synthesis, is also more readily seen in DI than in normal rat JGA cells (FIGURE 2 b, c).

It has been suggested that the lower plasma aldosterone concentration in the DI rat is a function of its higher urinary excretion rate in these animals.[8, 9] This may even be accompanied by an increase in zona glomerulosa width, suggesting an increase in the secretion of aldosterone.[9] The opposite finding in this study of a slightly reduced zona glomerulosa, which was restored to normal size in DI rats injected with vasopressin (TABLE 2), suggests that *reduced* synthesis of aldosterone may contribute to the lower plasma levels. Earlier studies showed a significantly smaller adrenal gland in DI rats even when expressed relative to body weight.[1] Usually, elevated plasma angiotensin II would result in increased aldosterone secretion. Hyperaldosteronism in a DI rat whose serum osmolality is already elevated [1] could exacerbate the problem. Instead hypokalemia, an abnormality also present in the DI rat,[1] may actually be advantageous since zona glomerulosa cells secrete less aldosterone in a low potassium environment.[10] One might also expect elevated blood pressure to accompany the higher angiotensin II levels in the blood. Since DI rats are not hypertensive [11] it is tempting to speculate that the absence of vasopressin may be a contributing factor. Recent studies [12] suggest that VP may play a role as a physiological pressor agent even at "low" concentrations.

TABLE 2

ADRENAL CORTEX AND ZONA GLOMERULOSA WIDTHS IN NORMAL AND BRATTLEBORO RATS

Rat Type	Number of Rats	Body Weight (g)	Adrenal Cortex (width μm)	Zona Glomerulosa (zG) (width μm)	ZG: Adrenal Cortex (%)
Normal LE	3 ‡	281	825.0±13.8 *	61.3±3.5 †	7.5
Heterozygote (HZ)	2	316	766.6±24.8	57.9±3.5	7.6
Homozygote (DI)	5	239	650.8±16.5	54.4±1.1	8.6
DI+vasopressin	7	258	746.6±22.5 *	60.0±1.5 *	8.2
DI, Na⁺ deficient	8	174	615.4±17.5	136.1±3.0	22.0
HZ, Na⁺ deficient	5	268	759.8±24.9	148.1±5.8	19.4

* $p < 0.001$ compared to untreated DI.
† $p < 0.02$ compared to untreated DI.
‡ 6 measurements per gland.

The renin-angiotensin-aldosterone complex in the homozygous Brattleboro rat represents a viable adaptation to the lack of vasopressin with altered homeostatic mechanisms.

REFERENCES

1. MÖHRING, B., J. MÖHRING, G. DAUDA & D. HAACK. 1974. Potassium deficiency in rats with hereditary diabetes insipidus. Am. J. Physiol. **227:** 916–920.
2. MÖHRING, J., G. KOHRS, B. MÖHRING, M. PETRI, E. HOMSY & D. HAACK. 1978. Effects of prolonged vasopressin treatment in Brattleboro rats with diabetes insipidus. Am. J. Physiol. **234** (Renal Fluid Electrolyte Physiol. 3): F106–F111.
3. PETER, S. & J. MÖHRING. 1978. The juxtaglomerular apparatus of rats with hereditary hypothalamic diabetes insipidus. Cell Tiss. Res. **188:** 335–339.
4. KEETON, T. K. & W. B. CAMPBELL. 1980. The pharmacologic alteration of renin release. Pharmacological Rev. **32:** 81–227.
5. LATTA, H. & A. MAUNSBACH. 1962. The juxtaglomerular apparatus as studied electron microscopically. J. Ultrastruct. Res. **6:** 547–561.
6. PETER, S., J. LAZAR, F. GROSS & W. G. FORSSMANN. 1974. Studies on the juxtaglomerular apparatus. II. Quantitative morphology after adrenalectomy. Cell Tiss. Res. **151:** 457–469.
7. PETER, S., J. LAZAR, F. GROSS & W. G. FORSSMANN. 1974. Studies on the juxtaglomerular apparatus. III. Quantitative morphology after treatment with deoxycorticosterone (DOC). Cell Tiss. Res. **151:** 471–480.
8. KINTER, L. B. 1978. Homeostatic mechanisms in the diabetes insipidus rat. Ph.D. Thesis, Harvard University, Cambridge, Mass. 323 pp.
9. MILNE, C. M., R. J. BALMENT, I. W. HENDERSON, W. MOSLEY & I. C. JONES. 1982. Adrenocortical function in the Brattleboro rat. Ann. N.Y. Acad. Sci. (This volume.)
10. FREDLUND, P., S. SALTMAN, T. KONDO, J. DOUGLAS & K. J. CATT. 1977. Aldosterone production by isolated glomerulosa cells: Modulation of sensitivity to angiotensin II and ACTH by extracellular potassium concentration. Endocrinology **100:** 481–486.
11. GELLAI, M. & H. VALTIN. 1982. The Brattleboro rat and the trained, conscious, chronically catheterized rat. *In* Research Animals and Concepts of Applicability to Clinical Medicine. K. Gartner & H. Stolte, Eds. Karger. Basel. (In press.)
12. MÖHRING, J., J. KINTZ, J. SCHOUN, R. ARBOGAST, J. F. LIARD, J. P. MONTANI, J. A. MACIEL, K. GLÄNZER & R. DÜSING. 1981. The antidiuretic hormone and arterial hypertension: Recent observations in rats. Adv. Nephrol. **10:** 75–87.

In Memoriam

Jan Möhring died on December 26, 1979 at the age of thirty-nine. With his untimely departure we lost a true Renaissance man. He was much more than an able and productive scientist with over 40 publications, in less than a decade, in the fields of vasopressin and hypertension. He was a philosopher, an excellent pianist, a linguist, a mountain climber (in the figurative sense as well), and a compassionate human being.

The Brattleboro rat was the catalyst for our friendship. We first met in Washington, D.C. in 1972 at the IV International Congress of Endocrinology. A trip to Hanover immediately after these meetings made Möhring a "household word" among the renal physiologists and neuroendocrinologists in the Physiology Department of Dartmouth Medical School. Jan also became a personal friend

of my family. Through his efforts, I (plus family) spent a sabbatical year (1975–1976) at the University of Heidelberg in the Department of Anatomy under the direction of Dr. Wilhelm Kriz. Möhring, at that time a member of the Department of Pharmacology, and I collaborated on the work reported here.

Jan Möhring leaves behind a legacy of excellent work and many good (and some sad) memories for those who knew him. We mourn the early death of both a friend and scientist and dream of what might have been. . . .

H. W. S.

ROLE OF VASOPRESSIN IN THE CONTROL OF SYSTEMIC HEMODYNAMICS—LESSONS LEARNED FROM THE BRATTLEBORO RAT

Gary Aisenbrey and Tomas Berl

Department of Medicine
University of Colorado Health Sciences Center
Denver, Colorado 80262

In 1895, Oliver and Schafer were the first to note that the infusion of whole pituitary extract was associated with a marked increase in blood pressure.[1] Three years later, Howell localized this effect to the posterior lobe of the pituitary[2] but it was not until 1928 that Kamm *et al.*[3] determined that the posterior pituitary substance responsible for the rise in blood pressure is distinct from the one responsible for uterine contractions. Despite the longstanding recognition that the hormone can exert changes in blood pressure, pulse rate, and/or cardiac output not only in experimental animals but also in man,[4-6] most of the focus on this hormone has been directed at its role in the control of renal water excretion. The last few years, however, have witnessed renewed interest in the elucidation of a role for vasopressin in the control of systemic hemodynamics. In this review we will discuss the current knowledge of the systemic vascular effects of this hormone in both normal and pathologic states.

DOES VASOPRESSIN PLAY A ROLE IN THE CONTROL OF SYSTEMIC HEMODYNAMICS IN BASAL CONDITIONS?

In 1937, Page and Sweet observed that animals that were otherwise normal had a decrease in blood pressure following hypophysectomy,[7] and soon thereafter Sattler *et al.* noted that section of the median eminence in the dog was followed by a period of hypotension.[8] While these early observations were by no means conclusive they appeared to support a role for antidiuretic hormone (ADH) in blood pressure regulation. Since then, however, a number of arguments have been put forth in support of the contention that vasopressin is not an important component in the control of blood pressure. Thus, concentrations of circulating vasopressin required to produce a significant pressor response are rarely reached under physiologic conditions and are, in fact, several fold higher than those needed to achieve maximal hydroosmotic effect.[9] Further evidence against a role for vasopressin is provided by studies with sustained infusion of physiologic levels of the hormone in the dog.[10] In these experiments, mean arterial pressure was unaffected acutely but rose significantly to a peak on day 9. This increase, however, appeared to correlate with increased plasma volume. In fact, as the infusion was continued, compensatory adjustments of volume were noted as were simultaneous declines in arterial pressure. The authors concluded that ADH produces a transient hypervolemic hypertension. These data are also supported by the observation that patients with the syndrome of inappropriate ADH secretion as well as the infusion of exogenous vasopressin does not increase blood pressure in man.[11]

0077–8923/82/0394–0299 $1.75/0

The interpretation of these experiments in which blood pressure remains unaltered as supporting the view that vasopressin plays no role in the control of systemic hemodynamics must be made with caution, because reflex mechanisms could counteract an effect of the hormone on cardiac output and/or peripheral resistance and maintain systemic pressure unchanged. In this regard the observation of Wagner and Braunwald in patients with autonomic insufficiency is of great interest.[12] These authors found that the infusion of very small doses of vasopressin, which possibly approach the physiologic range, resulted in a pressor response and significant hemodynamic changes in these patients. The loss of the normal vasoregulatory reflexes that normally act to counter the vasoconstrictor effects of vasopressin therefore made its effect on pressure evident. This interpretation was corroborated by experiments performed by Cowley *et al.* in 1974.[13] The hemodynamic consequences of physiologic levels of intravenous vasopressin (less than 1 mU/kg/min) were studied in conscious intact dogs and dogs with sinoaortic baroreceptor denervation. An average increase in arterial pressure of 5 mm Hg was noted in intact dogs while a 33 mm Hg increase was seen in dogs with baroreceptor denervation as the threshold sensitivity was increased eleven-fold. These investigations were complemented by the recent studies of Montani *et al.* who examined cardiovascular events associated with one-hour vasopressin infusions (40–5,000 fmol/kg/min) in conscious dogs.[14] These doses were chosen to produce plasma concentrations that would be expected in physiologic situations. They noted that in dogs with intact baroreceptor reflexes, increases of 2–20 fmol/ml in plasma ADH concentration did not affect mean arterial pressure, although cardiac output fell and total peripheral resistance was increased. After carotid sinus baroreceptor denervation, however, ADH infusion rates as low as 40 fmol/kg/min were associated with increases in mean arterial pressure even though cardiac output was not affected until higher infusion rates were used. These observations strongly suggest that vasopressin may have significant hemodynamic effects but that these effects are not ordinarily associated with blood pressure changes in animals with an intact baroreceptor reflex mechanism. It could be proposed therefore that earlier studies arguing against a pressor role of ADH in blood pressure regulation may indeed have been obscured by reflex action of the sympathetic nervous system and the functioning baroreceptors.

Does Vasopressin Play A Role in the Pathogenesis of Hypertension?

Studies performed both *in vivo* and *in vitro* have documented that vasopressin is a potent vasoconstrictor. Thus, the infusion of the hormone in conscious animals causes a greater increment in total systemic iliac and mesenteric resistance for any increase in blood pressure than either a catecholamine agonist or angiotensin II.[15] Likewise, on a molar basis 8-arginine vasopressin is a more potent vasoconstrictor of rat blood vessels than either norepinephrine or angiotensin II.[16] Despite this potency, the role of vasopressin in the pathogenesis of hypertensive states has remained controversial.

A possible role for vasopressin in hypertensive states was first suggested by Page and Sweet in 1937 when dogs with experimental hypertension of the Goldblatt type decreased their blood pressure[17] following hypophysectomy. However, in these studies an effect due to loss of the anterior lobe could not be excluded. Subsequently Sattler and Ingram, however, reported that section of the supraopticohypophyseal tract in the median eminence in dogs with Goldblatt hypertension was associated with significant decreases in blood pressure in five out of eight dogs studied.[8] They were not able to unequivocally implicate ADH, however, because postsurgical injection of Pitressin in oil did not reverse the section-induced fall in blood pressure. Suggestive evidence was also obtained from human studies by Keedy[13] who reported that division of the hypophyseal stalk in hypertensive patients lowered blood pressure in most patients.

Although these early studies appeared to incriminate ADH as a potential vascular hormone that could be important in the pathogenesis of hypertension, much of the research effort to further the field remained neglected for a number of years. The late Dr. Jan Möhring was probably most responsible for rejuvenating an interest in the vascular effects of ADH. In a study published in 1978, he examined the role of vasopressin in rats with two-kidney Goldblatt hypertension.[19] Animals with both benign, and even more so with malignant, hypertension had elevated vasopressin levels. In approximately 50 per cent of the latter rats, blood pressure decreased transiently toward normal after administration of an arginine vasopressin (AVP) antiserum. On the other hand, studies by Pullan *et al.*, performed in the one-kidney Goldblatt model, disputed the role of ADH as pathogenetic in the development of the hypertension.[20] These authors noted that mean arterial pressure rose by nearly 30 mm Hg at a time when plasma AVP had approximately doubled. Since vasopressin infusions in another group of animals increased plasma AVP by nearly five times with only a 9 mm Hg change in blood pressure, it seemed unlikely that vasopressin played a significant role. In addition, it has been shown that rats with congenital diabetes insipidus develop two-kidney Goldblatt hypertension despite the absence of circulating ADH.[21] It appears therefore that, as relates to the Goldblatt model of hypertension, the results obtained with the AVP antiserum are at variance with other experimental data. In assessing this controversy it may be appropriate to quote Möhring himself who stated "the results obtained so far should be interpreted with caution until they are controlled and confirmed by comparable experiments using specific AVP antagonists or purified AVP antibody, neither of which is available to date."[19] Since that time, analogues of AVP with specific antivasopressor activity have become available through the elegant research efforts of Manning and Sawyer.[22,23] Recently, Rabito *et al.* employed such a pressor antagonist in rats with two-kidney one-clip hypertension.[24] These authors failed to see any change in systemic pressure and therefore could not support an AVP-independent mechanism in the pathogenesis of this form of hypotension.

The AVP antiserum was also employed by Möhring *et al.* to evaluate the role of ADH in the pathogenesis of DOC hypertension in the rat.[25] In rats that developed the malignant phase of the disease, the AVP antiserum transiently lowered blood pressure to normal or subnormal levels. These studies appeared to support earlier findings in which a markedly attenuated increase in blood pressure was seen in congenital[26] or surgically induced diabetes insipidus rats[27] treated with DOC. Saito *et al.*, however, have recently published studies in which DOC-treated, unilaterally nephrectomized diabetes insipidus rats became hypertensive when dDAVP, which is devoid of pressor activity, was included in the saline drinking water.[28] These studies suggest an important role for volume expansion as the mechanism whereby AVP contributes to the hypertension in this model. Studies employing vasopressin inhibitors in rats with DOC-saline hypertension have unfortunately yielded conflicting results. Thus, while Crofton *et al.*[26] found that the antagonists lowered systemic pressure by at least 14 mm Hg, Rabito *et al.*[24] found no such effect. The cause for this discrepancy remains unresolved.

The role of vasopressin has also been studied in spontaneously hypertensive rats. Möhring reported that plasma AVP concentrations are elevated in these animals.[29] AVP antiserum normalized blood pressure in these animals and thus it appeared attractive to postulate that AVP was involved in the hypertension in these animals. Crofton *et al.* carried these studies further and demonstrated that the urinary excretion of AVP in spontaneously hypertensive rats was twice that of controls.[30] However, when attempts were made to associate urinary AVP excretion with the development of the hypertension, the correlation was poor. Furthermore, Matsaguchi *et al.* observed that despite increased levels of vasopressin, the AVP pressor antagonist failed to significantly lower blood pressure in either salt-sensitive or salt-resistant genetically hypertensive rats.[31] Another disorder in which hypertension is commonly seen and in which the role of vasopressin has been examined is in acute renal failure. Hofbauer *et al.* induced renal failure in rats by intramuscular injection of glycerol and demonstrated that intravenous vasopressin antiserum lowered blood pressure.[32] Hence, the forty-fold increase of plasma AVP noted in these animals was considered to have exerted significant systemic vasoconstrictor activity. These observations will have to be confirmed with the more specific AVP pressor antagonists. Also, the contribution of the hormone to the hypertension in chronic renal failure, a setting in which its levels are somewhat elevated, also remains to be elucidated.

As a result of the evidence that vasopressin has an important role in some forms of animal hypertension, the question of whether circulating AVP can cause or contribute to hypertension in humans has also been entertained. Cowley *et al.* have noted elevated levels of the hormone in hypertensive subjects, particularly when the diastolic pressure was above 95 mm Hg.[33] Padfield *et al.* evaluated plasma concentrations of AVP in forty patients with benign essential hypertension and in twelve patients with malignant phase hypertension.[34] Although the values tended to be higher in the malignant phase group, when five normal volunteers were infused with exogenous AVP to produce concentrations five times higher than the highest recorded in the malignant phase group, no change in blood pressure was seen. Padfield also examined the relationship of plasma vasopressin to blood pressure in patients with malignant hypertension and in patients with the syndrome of inappropriate ADH secretion.[11] In the malignant hypertensive group AVP was found to be elevated, but the average 13 pg/ml concentration did not correlate with arterial blood pressure. Furthermore, as in his previous study, when ADH was infused into normal volunteers to levels beyond those seen in malignant hypertensives, no change in mean arterial pressure was noted. In addition, no correlation was noted with either systolic or diastolic pressure in the patients with syndrome of inappropriate ADH in whom plasma AVP levels averaged 39 pg/ml. These results argued against a significant vasoconstrictor role of AVP in these diseases, but definitive evidence could only be accomplished by injecting an AVP pressor antagonist, as compensatory mechanisms in normals but not hypertensives may have altered the response to increased vasopressin levels and/or infusion. Until such studies are performed it will be difficult to entirely exclude AVP as a pressor hormone in human hypertension.

In assessing the contribution of vasopressin to elevations in blood pressure an awareness of the possible interaction of the hormone with other vasoconstrictors or alterations in sodium balance need to be considered. Thus, what in absolute terms could be normal or even subpressor levels of AVP could have significant vascular effects. The interaction between vasopressin and catecholamines is of particular interest. It has been suggested for a long time that catecholamines sensitize the vascular system to a hormone liberated from the pituitary.[35] In 1965, Bartelstone demonstrated that subpressor doses of vasopressin enhance the pressor responses to injected or endogenously liberated catecholamines.[36] When added to an *in vitro* bath containing aortic strip preparations, vasopressin also potentiated both the isometric and isotonic responses of the muscle. *In vivo* evidence for such an interaction was noted in the studies of Cowley *et al.* in decapitated dogs.[13] When such dogs received norepinephrine the vascular sensitivity to exogenous AVP was increased by a factor of 8,000. It is also possible that the greatly enhanced response to vasopressin observed in the spontaneously hypertensive rat[29] could involve a catecholamine interaction since these have been reported to be increased in this strain of rats.[37] The experiments of Johnson *et al.*[38] suggest that a similar interaction may exist between vasopressin and angiotensin II. In addition to its interaction with vasoconstrictors, the vascular response to vasopressin may be altered by changes in total body, as well as serum, sodium concentration. *In vitro* studies have suggested that neurohypophyseal peptide-induced contractions in arterial smooth may be independent of external sodium concentrations.[17] However, studies in which serum sodium concentration is increased with hypertonic saline suggest that the release of vasopressin that presumably accompanies this maneuver is an important mediator of the increase in systemic pressure.[39] Since the AVP pressor antagonist decreased pressure more in hypertonic sodium chloride treated rats than an infusion of hypertonic mannitol, which is known to have an equally vasopressin-stimulating effect, it is possible to postulate that the increase in serum sodium concentration was a determinant of the pressor response to the hormone.[40] Finally, the role of increased extracellular fluid volume on vasopressin responsiveness needs to be considered. The experiments of Manning *et al.* involving chronic infusions of subpressor doses of vasopressin suggested that these infusions can have hypertensive effects, but this is mediated primarily by its fluid retaining properties, despite the attendant hyponatremia.[41] These studies argued against beliefs that salt per se, independent of the volume effects, is important as a hypertensive agent. It is also possible that the increased responsiveness to vasopressin in DOC-salt hypertension[42] may be related to volume expansion although a role for interaction with catecholamines cannot be ruled out in this model.[43] In summary, the effects of both changes in serum sodium and total body sodium on the action of vasopressin will require further investigation.

Does Vasopressin Play a Role in the Maintenance of Systemic Hemodynamics in Volume Depletion?

Volume depletion and circulatory stress states characteristically bring forth a number of reflex neurohumoral responses designed to maintain the integrity of the circulation. In view of its vasoconstrictive properties, it is attractive to postulate that vasopressin could also contribute to such a homeostatic response. Evidence that this may indeed be the case has steadily accrued over the last decade. In 1973 Szczepanska-Sadowska[44] demonstrated that the infusion of vasopressin to levels observed in non-hypotensive hemorrhage exerts significant hemodynamic effects in conscious dogs as peripheral resistance increased and cardiac output decreased. Subsequently, Khokar and Slater[45] showed that even mild hydropenia (24-hour fluid deprivation) increases the rate of urinary vasopressin excretion in man. Further evidence for a role for vasopressin in hypotensive hemorrhage was provided by Laycock *et al.*[46] These investigators demonstrated that blood loss of two percent body weight in Brattleboro rats was associated with a significantly lower blood pressure than an identical blood loss in Long-Evans rats with an intact source of AVP. It was concluded from these studies that vasopressin plays a significant role in maintaining arterial blood pressure during hemorrhage. An assessment of the quantitative importance of AVP in the regulation of arterial pressure during hypotensive hemorrhage in the dogs was performed by Cowley *et al.*[47] Studies performed in dogs with complete interruption of the efferent sympathetic nervous system and bilateral nephrectomy (to eliminate compensatory responses of the renin-angiotensin system) demonstrated that a decrease in arterial pressure to 50 mm Hg was accompanied by a five-fold increase in plasma AVP concentration and a 71% compensatory increase in arterial pressure. This compensation was abolished by the administration of an inhibitor of the pressor action of AVP. In addition, dogs that were also decapitated and lost AVP secretory ability could muster only a 10% compensatory response. Recently specific pressor antagonists have been employed to establish the role of vasopressin in the maintenance of systemic hemodynamics in the water-deprived rat. In experiments performed in our laboratory,[48] when the AVP antagonist was given to rats after 24 hours of fluid deprivation (at which time their mean plasma AVP levels averaged 21.6 pg/ml) a significant fall in blood pressure occurred. The effect on blood pressure occurred primarily due to a fall in peripheral vascular resistance. Specificity of the antagonist was demonstrated by its blood pressure lowering effect in Brattleboro rats receiving exogenous AVP and its failure to alter pressor responses to exogenous angiotensin II or norepinephrine. Similar results were obtained by Andrews and Brenner[49] who employed a different antagonist. They also showed that the AVP antagonist-induced fall in blood pressure during fluid deprivation was transient and that saralasin, an angiotensin II antagonist, not only prevented the return of blood pressure to normal but also increased the hypotensive response seen following injection of the AVP antagonist. These important studies provide strong evidence that AVP is important not only as ADH, but also in circulatory homeostasis during fluid deprivation in the rat. In addition, when the pressor effects of AVP are blocked in fluid-deprived animals, angiotensin II may assume a significant role in blood pressure control.

In conclusion, much interest in the role of AVP as a vascular hormone has evolved over the years stemming from the early work of Oliver and Howell. It has been difficult, until recently, to positively ascribe a significant vascular role to AVP. This has, in part, been due to interference of compensatory baroreflex phenomenon, interaction or compensatory changes of other endogenous vasoactive hormones, and the lack of availability of a specific AVP pressor antagonist. Many of these initial problems have been overcome through intense collaborative research efforts in the fields of physiology, biochemistry, and pharmacology. The interest in AVP as a pressor hormone has not only been significant with respect to better understanding the role of AVP in normal and abnormal pathophysiologic states, but also has opened new doors in the understanding of other components involved in blood pressure regulation. Finally, AVP clearly may possess significant vascular activity in plasma concentrations that were previously thought high enough only for antidiuretic activity. A comprehensive understanding of its pressor role will undoubtedly require extensive research efforts in the years to come but we now appear to be embarking on some of the most provocative aspects of its role in blood pressure control.

References

1. Oliver, G. & E. A. Schaefer. 1895. On the physiological action of extracts of pituitary body and certain other glandular organs. J. Physiol. (Lond) **18:** 277–279.
2. Howell, W. H. 1898. The physiological effects of extracts of the hypophysis cerebri and infundibular body. J. Exp. Med. **3:** 245–258.
3. Kamm, O., T. B. Aldrich, F. W. Corote, L. W. Rowe & E. P. Bugbee. 1928. Active principles of the posterior lobe of the pituitary gland. I. The demonstration of the presence of two active principles. II. The separation of the two principles and their concentration in the form of potent solid preparations. J. Am. Chem. Soc. **50:** 573–580.
4. Grollman, A. & E. M. K. Geilrug. 1932. The cardiovascular and metabolic reactions of man to the intramuscular injection of posterior pituitary liquid (Pituitrin), Pitressin, and Pitocin. J. Pharmacol. Exp. Ther. **46:** 447–456.
5. Graybiel, A. & R. E. Glendy. 1941. Circulatory effects following the intravenous administration of pitressin in normal persons and in patients with hypertension and angina pectoris. Am. Heart J. **2:** 481–489.
6. Cushing, H. 1930. Neurohypophysial mechanisms from clinical standpoint. (Lister Memorial Lectures.) Lancet **2:** 219–224.
7. Page, I. H. & J. E. Sweet. 1937. The effect of hypophysectomy on arterial blood pressure of dogs with experimental hypertension. Am. J. Physiol. **120:** 238–247.
8. Sattler, D. G. & W. R. Ingram. 1941. Experimental hypertension and the neurohypophysis. Endocrinol. **29:** 952–957.
9. Sawyer, W. H. 1961. Neurohypophyseal hormones. Pharmacol. Rev. **13:** 225–277.
10. Smith, W. M., A. W. Cowley, A. C. Guyton & R. D. Manning. 1979. Acute and chronic effects of vasopressin on blood pressure, electrolytes, and fluid volumes. Am. J. Physiol. **237:** F232–F238.
11. Padfield, P. L., J. J. Brown, A. F. Lever, J. J. Morton & J. I. S. Robertson. 1981. Blood pressure in acute and chronic vasopressin excess. N. Engl. J. Med. **304:** 1067–1071.
12. Wagner, H. N. & E. Braunwald. 1956. The pressor effect of the antidiuretic principle of the posterior pituitary in orthostatic hypotension. J. Clin. Invest. **35:** 1412–1417.
13. Cowley, A. W., E. Monos & A. C. Guyton. 1974. Interaction of vasopressin and the baroreceptor reflex system in the regulation of arterial blood pressure in the dog. Circ. Res. **34:** 505–511.
14. Montani, M. P., J. F. Liard, J. Schoun & J. Möhring. 1980. Hemodynamic effect of exogenous and endogenous vasopressin at low plasma concentrations in conscious dogs. Circ. Res. **47:** 346–352.
15. Heyndrickx, G. R., D. H. Boettcher & S. F. Vatner. 1976. Effects of angiotensin, vasopressin, and methoxamine on cardiac function and blood flow distribution in conscious dogs. Am. J. Physiol. **231:** 1579–1586.
16. Page, I. H. 1938. The effect of bilateral adrenalectomy on arterial blood pressure of dogs with experimental hypertension. Am. J. Physiol. **122:** 352–359.

17. ALTURA, B. M. & B. T. ALTURA. 1977. Vascular smooth muscle and neurohypophyseal hormones. Fed. Proc. **36:** 1853–1860.
18. KEEDY, C. 1957. Effect of division of the hypophyseal stalk in the hypertensive patient. Am. Surg. **23:** 439–444.
19. MÖHRING, J., B. MÖHRING, M. PETRI & D. HAACK. 1978. Plasma vasopressin concentrations and effects of vasopressin antiserum on blood pressure in rats with malignant two-kidney Goldblatt hypertension. Circ. Res. **42:** 17–22.
20. PULLAN, P. T., C. I. JOHNSTON, W. P. ANDERSON & P. I. KORNER. 1978. The role of vasopressin in blood pressure control and in experimental hypertension. Clin. Sci. Mol. Med. **55:** 251s–254s.
21. JOHNSTON, C. I., P. PULLAN, R. WOODS, D. CUSLEY & K. SOWARDS. 1981. Role of vasopressin in experimental renal hypertension in the rat. Hypertension. (In press.)
22. MANNING, M., J. LOWBRIDGE, C. T. STIER, J. HALDAR & W. H. SAWYER. 1977. [1-deaminopenicillamine, 4-valine]-8-D-arginine vasopressin, a highly potent inhibitor of the vasopressor response to arginine vasopressin. J. Med. Chem. **20:** 1228–1231.
23. LOWBRIDGE, J., M. MANNING, J. HALDAR & W. H. SAWYER. 1978. [1-(β-mercapto-β, β-cyclopentamethylenepropionic acid), 4-valine, 8-D-arginine] vasopressin, a potent and selective inhibitor of the vasopressor response to arginine vasopressin. J. Med. Chem. **21:** 313–317.
24. RABITO, S. F., O. A. CARRETERO & A. G. SCICLI. 1981. Evidence against a role of vasopressin in the maintenance of high blood pressure in mineralocorticoid and renovascular hypertension. Hypertension **3:** 34–38.
25. MÖHRING, J., B. MÖHRING, M. PETRI & D. HAACK. 1977. Vasopressor role of ADH in the pathogenesis of malignant DOC hypertension. Am. J. Physiol. **232:** F260–F264.
26. CROFTON, J. T., L. SHARE, R. E. SHADE, W. J. LEE KWON, M. MANNING & W. H. SAWYER. 1979. The importance of vasopressin in the development and maintenance of DOC-salt hypertension in the rat. Hypertension **1:** 31–37.
27. FREIDMAN, S. M., C. L. FREIDMAN & M. NAKASHIMA. 1960. Accelerated appearance of DOCA hypertension in rats treated with pitressin. Endocrinology **67:** 752–759.
28. SAITO, Y., Y. YAJIMA & T. WATANABE. 1980. Involvement of AVP in the development and maintenance of hypertension in rats. *In* Antidiuretic Hormone. S. Yoshida, L. Share & K. Yagi, Eds.: 215–225. University Park Press. Baltimore, Md.
29. MÖHRING, J., J. KINTZ & J. SCHOUN. 1978. Role of vasopressin in blood pressure control of spontaneously hypertensive rats. Clin. Sci. Mol. Med. **55:** 247s–250s.
30. CROFTON, J. T., L. SHARE, R. E. SHADE, C. ALLEN & D. TARNOWSKI. 1978. Vasopressin in the rat with spontaneous hypertension. Am. J. Physiol. **235:** H361–H366.
31. MATSUOUCHI, H., P. SCHMID, D. VANORDEN & A. L. MARK. 1981. Does vasopressin contribute to salt-induced hypertension in the Dahl strain? Hypertension **3:** 174–181.
32. HOFBAUER, K. G., A. KONRADS, K. BAVEREISS, B. MÖHRING, J. MÖHRING & F. GROSS. 1977. Vasopressin and renin in glycerol-induced acute renal failure in the rat. Circ. Res. **40:** 424–428.
33. COWLEY, A. W., W. C. CUSHMAN, E. W. QUILLEN, M. M. SKELTON & H. G. LANGFORD. 1981. Vasopressin elevation in essential hypertension and increased responsiveness to sodium intake. Hypertension **3:** I93–I100.
34. PADFIELD, P. L., A. F. LEVER, J. J. BROWN, J. J. MORTON & J. I. S. ROBERTSON. 1976. Changes of vasopressin in hypertension: cause or effect? Lancet **1:** 1255–1257.
35. CHENOWETH, M. B., G. L. ELLMAN, R. C. REYNOLDS & P. J. SHEA. 1958. A

secondary response to pressor stimuli caused by sensitization to an endogenous pituitary hormone. Circ. Res. **6:** 334–342.
36. Bartelstone, H. J. & P. A. Nasmyth. 1965. Vasopressin potentiation of catecholamine actions in dog, rat, cat, and rat aortic strip. Am. J. Physiol. **208:** 754–762.
37. Grobecker, H., M. F. Roizen, V. Weise, J. M. Scavedra & I. J. Kopin. 1975. Sympathoadrenal medullary activity in young, spontaneously hypertensive rats. Nature **258:** 267–269.
38. Johnson, A. J., S. Ichikawa, K. Kutz, W. Fowler & C. Payne. 1981. Pressor response to vasopressin in rabbits within a 3-day renal artery stenosis. Am. J. Physiol. **240:** H862–H867.
39. Hatzinikolaou, P., H. Gavras, H. R. Brunner & I. Gavras. 1981. Role of vasopressin, catecholamines, and plasma volume in hypertonic saline-induced hypertension. Am. J. Physiol. **240:** H827–H831.
40. Hatzinikolaou, P., H. Gavras, H. R. Brunner & I. Gavras. 1980. Sodium-induced elevation of blood pressure in the anephric state. Science **209:** 935–936.
41. Manning, R. D., A. C. Guyton, T. G. Coleman & R. E. McCaa. 1979. Hypertension in dogs during antidiuretic hormone and hypotonic saline infusion. Am. J. Physiol. **236:** H314–H319.
42. Crofton, J. T., L. Share, B. C. Wang & R. E. Shade. 1980. Pressor responsiveness to vasopressin in the rat with DOC-salt hypertension. Hypertension **2:** 424–431.
43. Reid, J. L., J. A. Zivin & I. J. Kopin. 1975. Central and peripheral adrenergic mechanisms in the development of deoxycorticosterone-saline hypertension in rats. Circ. Res. **37:** 569–576.
44. Szezepanska-Sadowska, E. 1973. Hemodynamic effects of a moderate increase of the plasma vasopressin level in conscious dogs. Pflugers Arch. **338:** 313–322.
45. Khokhar, A. M. & J. D. H. Slater. 1976. Increased renal excretion of arginine vasopressin during mild hydropenia in young men with mild essential benign hypertension. Clin. Sci. Mol. Med. **51:** 691s–694s.
46. Laycock, J. F., W. Penn, D. G. Shirley & S. J. Walter. 1979. The role of vasopressin in blood pressure regulation immediately following acute haemorrhage in the rat. J. Physiol. **296:** 267–275.
47. Cowley, A. W., S. J. Switzer & M. M. Guinn. 1980. Evidence and quantification of the vasopressin arterial pressure control system in the dog. Circ. Res. **46:** 58–65.
48. Aisenbrey, G. A., W. A. Handelman, P. E. Arnold, M. Manning & R. W. Schrier. 1981. Vascular effects of arginine vasopressin during fluid deprivation in the rat. J. Clin. Invest. **67:** 961–968.
49. Andrews, C. E. & B. M. Brenner. 1981. Relative contributions of arginine vasopressin and angiotensin II to maintenance of systemic arterial pressure in the anesthetized water-deprived rat. Circ. Res. **48:** 254–258.

Discussion of the Paper

M. Miller (*Veterans Administration Medical Center, Syracuse, N.Y.*): In the last data, you have presented only one point in time demonstrating the lowering of the blood pressure following administration of pressor antagonists. Could you comment as to how long that fall can be maintained?

T. BERL: That is a very transient phenomenon, similar to that achieved by Möhring with AVP antiserum. Andrews and Brenner have carried the observation further and have given an angiotensin II blocker. The hypotensive response was not only more profound but also maintained for a much longer period of time. It appears that the return in pressure is partially due to the renin-angiotensin system and perhaps other compensatory mechanisms.

C. I. JOHNSTON (*Prince Henry's Hospital, Melbourne, Australia*): There is one interaction that you alluded to that occurs in the normal state and that is the relationship to the baroreceptors. Vasopressin when infused causes a profound bradycardia and in normal animals this probably prevents them from becoming hypertensive. I am interested to know why it doesn't seem to happen when one is volume depleted. If an animal is volume depleted vasopressin does not cause a bradycardia, which may explain why there is no fall in blood pressure.

BERL: I suspect that in the volume-depleted state catecholamine levels are high and that this and other compensatory mechanisms try to maintain normal cardiac output after the blood pressure falls. These mechanisms may obscure the direct effect of vasopressin to decrease heart rate and cardiac output. The stroke volume is also down but the lowering of cardiac output is primarily due to decrease in heart rate.

JOHNSTON: The other problem is that the sum of the vasoconstrictor effects of vasopressin and angiotensin seem to be additive, so that if you remove one, the other compensates for it. Remove the angiotensin system and vasopressin will compensate for it as in hemorrhage, for example.

BERL: Yes, that's the one setting in which it has been shown.

DEVELOPMENT OF DOCA-SALT HYPERTENSION IN THE BRATTLEBORO RAT

Toshikazu Saito and Yukiko Yajima

Division of Endocrinology and Metabolism
Jichi Medical School
Tochigi, 320–04, Japan

and

Tokyo Metropolitan Institute of Medical Science
Tokyo 113, Japan

INTRODUCTION

The importance of vasopressin in the maintenance of experimental hypertension in rats has been suggested by the fact that when vasopressin antiserum is intravenously injected into rats with DOC-salt hypertension endogenous vasopressin is inactivated and the blood pressure is thereby lowered.[1] In their report, Crofton *et al.*[2] came to a similar conclusion based on the results of their experiment that indicated that when an analogue of vasopressin, which specifically hinders the action of vasopressin on the blood vessels, is injected into rats having DOC-salt hypertension, lowering of the blood pressure is observed. These findings suggest that vasoconstrictive action of vasopressin plays an important role in the maintenance of DOC-salt hypertension. They also reported that in rats with hereditary diabetes insipidus even when loading of DOC and salt is performed after uninephrectomy the onset of hypertension is not observed. This seems to suggest that vasopressin plays an important role not only in the maintenance but also in the onset of DOC-salt hypertension in rats.

Regarding the hypertension induced by the stenosis added to the unilateral renal artery, Möhring *et al.*[3] reported that vasopressin is important for the maintenance of high blood pressure at the malignant phase. Pullan *et al.*,[4] on the other hand, reported that, in renovascular hypertension, vasopressin may play some role in regulating the cardiovascular system, but that its vasoconstrictive action does not perform a significant function. In the studies of rats with spontaneous hypertension, Möhring *et al.*[5] reported that vasoconstrictive action of vasopressin plays an important role in the maintenance of this hypertension. Crofton *et al.*,[6] on the other hand, confirmed an increase of the secretion of vasopressin in rats with this hypertension, but the levels of vasopressin in plasma seem to be too low to cause hypertension.

TABLE 1 shows the effect of intravenous infusion of 0.4 ml of highly specific vasopressin antiserum on arterial blood pressure in hypertensive rats.[7] The most striking fall in blood pressure was observed in the rats with DOCA-salt hypertension in response to inactivation of endogenous vasopressin after infusion of vasopressin antiserum. No significant effect on the blood pressure was observed in the hypertensive rats with unilateral renal artery stenosis.

In order to further clarify the roles of vasopressin in the development or maintenance of experimental hypertension in rats, the present studies were

0077–8923/82/0394–0309 $1.75/0 © 1982, NYAS

TABLE 1

EFFECT OF INFUSION OF VASOPRESSIN ANTISERUM ON BLOOD PRESSURE IN RATS

		Blood Pressure (mm Hg, Mean ± SE)			
Hypertension	*N*	Before	After Normal Rabbit Serum	After Antiserum	p
DOCA-Salt	11	176 ± 5	184 ± 5	159 ± 12	<.05
Renal Artery Stenosis	6	189 ± 9	185 ± 9	185 ± 11	NS
SHR	9	184 ± 6	185 ± 6	171 ± 9	<.05
SHR-SP	11	195 ± 5	196 ± 6	183 ± 9	<.05
Normal	7	116 ± 2	113 ± 2	117 ± 3	NS

SHR: Spontaneously hypertensive rats. SHR-SP: Spontaneously hypertensive rats, stroke prone substratin.

performed in the rats with hereditary hypothalamic diabetes insipidus, Brattleboro homozygotes.

MATERIALS AND METHODS

The Brattleboro homozygotes and the Long-Evans rats 4 to 6 months old were uninephrectomized on the left side. Food containing 5.5% sodium chloride was given to the rats and subcutaneous injection of DOCA were carried out at the dose of 20 mg/kg BW weekly. The rats with diabetes insipidus were divided into five groups, each consisting of five to six rats. Either 0, 0.1, 0.4, or 1.6 μg/100 g BW/day of synthetic arginine vasopressin (Sigma, Grade VI) was subcutaneously and continuously infused for 10 weeks using an osmotic micropump (Alza) implanted subcutaneously. After this, administration of vasopressin was terminated, while loading of DOCA and salt was continued, and observations were made under this condition for one more week. Daily urine output, urinary osmolality and urinary excretion of electrolytes were also measured weekly for each rat. In another group of the Brattleboro homozygotes, 0.6 ng/100 g BW/day of desmopressin was infused subcutaneously as done in the groups infused with arginine vasopressin. In five groups of untreated rats with diabetes insipidus, continuous and subcutaneous infusions of vasopressin or desmopressin as were employed for the DOCA-salt loaded rats were undertaken and daily urine output and urinary osmolality were measured.

In the experiments of pressor response to vasopressin the Brattleboro homozygotes and the Long-Evans rats with or without DOCA-salt hypertension were anesthetized with α-chloralose (0.06 g/kg BW) and urethane (0.6 g/kg BW). One mg/kg BW of dibenzyline was further given and vasopressin was infused intravenously. The blood pressure was continuously recorded via a catheter inserted into the abdominal aorta via a femoral artery. The DOCA-salt hypertension was induced in the Brattleboro homozygotes by giving 1.0% saline containing 1 μg/ml of desmopressin as drinking water and injection of 5 μg/kg BW of DOCA every week.[7]

In the experiment of salt loading without DOCA administration the Brattleboro homozygotes, the Brattleboro heterozygotes, and the Long-Evans rats were given the usual food containing 0.3% sodium chloride until week 14 after uninephrectomy on the left side. From this time on the rats were given food containing 5.5% salt as "high salt diet" for an additional nine weeks. Blood pressure was determined weekly during the whole experimental period.

Statistical analysis of the data was performed by the paired or unpaired Student's *t*-tests.

Results and Discussion

Table 2 shows the urinary output and its osmolality of the untreated rats having diabetes insipidus, after continuous infusion of vasopressin through an osmotic micropump implanted subcutaneously, and of Long-Evans rats. The osmolality of the daily urine of rats into which 0.1 μg/100 g BW/day of vasopressin was infused was significantly higher than that of rats infused with control solution ($p < .02$). The osmolaiity of the urine of rats infused with 1.6 μg/100 g BW/day was not significantly different from those of Long-Evans rats.

Figure 1 shows the blood pressure of rats with diabetes insipidus during the 10-week period during which either the control fluid or vasopressin solution at three different concentrations was continuously infused, as well as the blood pressure of the Long-Evans rats during the same period. In the rats into which the vehicle was infused, loading of DOCA and salt did not result in elevation of the blood pressure. In the Long-Evans rats, on the other hand, rapid elevation of the blood pressure was observed from the commencement of the loading of DOCA and salt at week 1, and at week 3 the blood pressure had reached 148 ± 6 mm Hg, which was significantly higher than that prior to salt loading. At week 10, a maximum blood pressure of 182 ± 4 mm Hg had been obtained in them. In the rats having diabetes insipidus into which three different concentration levels of vasopressin solution were infused, sustained elevation of the blood pressure was noted from week 3 on. By week 7, the average blood pressure in each group had risen to beyond 150 mm Hg, which was significantly higher than that of the rats into which control fluid had been injected. However, there was no significant difference in the blood pressures among three groups of rats with diabetes insipidus that were given three different concentration levels of vasopressin solution; and their blood pressures were lower than those of Long-Evans rats. From week 8 on, the elevations of the blood pressure depended on the dose of vasopressin infused. In week 8, the blood pressures of the rats that were infused with 1.6 μg/100 g BW/day of vasopressin were significantly higher than those of rats that were infused with 0.1 μg/100 g BW/day of vasopressin. During the period from week 8 to week 10 the blood pressures of the rats that were given 1.6 μg/100 g BW/day of vasopressin were not significantly different from those of Long-Evans rats, whereas those of rats that were given 0.1 μg/100 g BW/day of vasopressin were significantly lower than those of Long-Evans rats ($p < .02$). No significant difference was found in the urinary excretion of sodium and potassium among the groups at week 8 except that potassium excretion in the rats infused with control solution was significantly lower than that of the Long-Evans rats. Urinary output and urinary osmolality decreased and increased, respectively, depending upon the dose of vasopressin infused (Table 3). One week after vasopressin infusions were terminated at week 10 under the treatment of DOCA and salt loading, the blood pressure fell significantly in rats into which 0.4 or 1.6 μg/100 g BW/day of

TABLE 2

URINARY VOLUME AND OSMOLALITY IN RATS WITH DIABETES INSIPIDUS SUBCUTANEOUSLY INFUSED WITH SYNTHETIC ARGININE VASOPRESSIN FOR 24 HOURS

Rate of Vasopressin Infusion (μg/100 g BW/day)	Urinary Volume * (ml/100 g BW/day)	Urinary Osmolality * (mOsm/L)
0	104.0 ± 6.5	99.8 ± 8.4
0.1	51.3 ± 11.0	193.5 ± 66.5
0.4	7.9 ± 1.6	1437.7 ± 179.0
1.6	6.3 ± 0.4	1878.8 ± 111.9
Long-Evans	7.2 ± 1.2	1737.0 ± 225.1

* Mean ± SE, $N = 6$.

vasopressin was infused (FIGURE 1). The infusion of 0.6 ng/100 g BW/day of desmopressin into the untreated Brattleboro homozygotes resulted in the elevation of their urinary osmolality to 201 ± 50 mOsm/L, which is not significantly different from the urinary osmolality of the Brattleboro homozygotes infused with 0.1 μg/100 g BW/day of vasopressin, 194 ± 67 mOsm/L. The courses of blood pressures of two groups of the Brattleboro homozygotes receiving the replacement of arginine vasopressin or desmopressin, whose antidiuretic activities were identical, are shown in FIGURE 2. The blood pressure of the rats

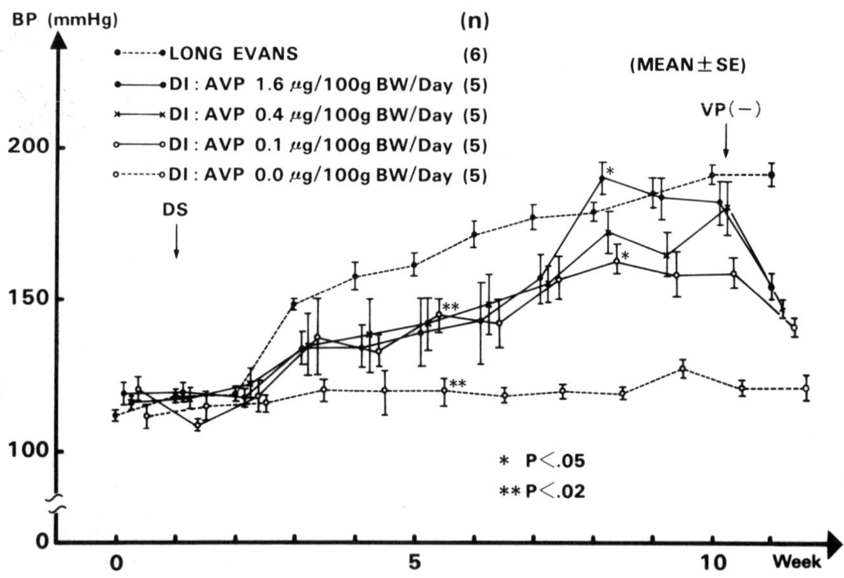

FIGURE 1. Course of blood pressure in rats treated with deoxycorticosterone acetate (DOCA) and sodium chloride (salt). "DI" and "AVP" represent the Brattleboro homozygotes and synthetic arginine vasopressin, respectively. The rats were uninephrectomized on the left side. After one week, subcutaneous infusion of vasopressin was started one week before measuring the initial level of blood pressure at week 0. Administration of DOCA and salt were commenced as shown by "DS". "VP (−)" indicates the point of cessation of vasopressin infusion.

TABLE 3

BLOOD PRESSURE, URINE VOLUME, URINARY OSMOLALITY, AND EXCRETION OF Na AND K IN DOCA-SALT TREATED RATS AT WEEK 8

Rate of Vasopressin Infusion (μg/100 g BW/day)	(n)	BP * (mm Hg)	U_V * (ml/100 g BW)	U_{Osm} * (mOsm/L)	U_{Na} * (mEq/100 g BW/day)	U_K * (mEq/100 g BW/day)
0	(5)	119.6 ± 2.0	104.0 ± 11.2	124.7 ± 20.4	2.89 ± 0.99	1.98 ± 1.17
0.1	(5)	163.8 ± 5.3	78.4 ± 12.4	198.6 ± 38.1	3.18 ± 1.05	0.58 ± 0.22
0.4	(5)	172.0 ± 6.9	24.6 ± 4.6	578.1 ± 107.8	3.41 ± 0.95	0.85 ± 0.22
1.6	(5)	190.5 ± 5.0	19.0 ± 4.2	663.3 ± 196.4	2.78 ± 0.86	0.91 ± 0.28
Long-Evans	(8)	185.4 ± 5.5	28.6 ± 1.7	640.6 ± 67.3	4.66 ± 0.55	1.19 ± 0.42

* Mean ± SE.

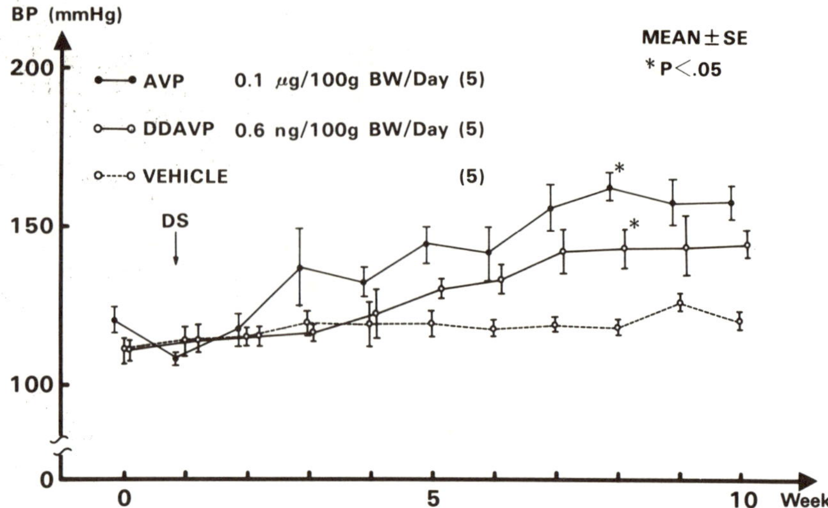

FIGURE 2. Course of blood pressure in rats treated with DOCA and salt under the replacement with 0.1 μg/100 g BW/day of arginine vasopressin or 0.6 ng/100 g BW/day of desmopressin (DDAVP.)

infused with desmopressin was lower than the rats infused with arginine vasopressin during the whole period.

The finding that the loading of DOCA and salt after removal of one kidney in rats having diabetes insipidus does not result in a significant elevation of blood pressure indicated that vasopressin is an essential factor causing the onset of this type of hypertension. This seems to support the reports made by Share's group [2] as well as by the present authors.[8] Wisgerhof *et al.*[9] reported that hypertension was observed after loading of DOCA and salt when Pitressin, which is a mixture of lysine vasopressin and arginine vasopressin, was administered as a supplement to rats with diabetes insipidus. The present authors [8] succeeded in creating the onset of DOCA-salt hypertension by maintaining the antidiuresis through supplementary oral administration of desmopressin to rats having diabetes insipidus. The present experiments established the induction of DOCA-salt hypertension in rats with diabetes insipidus by supplementing arginine vasopressin, which is the endogenous antidiuretic hormone of rats.

It is interesting that the amount of vasopressin necessary to cause the onset of DOCA-salt hypertension in rats having diabetes insipidus is 0.1 μg/100 g BW/day and that, when this amount of vasopressin is infused into untreated rats having diabetes insipidus, the osmolality of the urine was only 17% of that of Long-Evans rats. Plasma levels of vasopressin produced at this rate of continuous subcutaneous infusion would be much lower than the physiological level in the Long-Evans rats. Thus it is indicated that a plasma concentration of vasopressin below physiological level would be one of the factors required for development of the DOCA-salt hypertension in rats. Moreover the fact that the infusion of desmopressin, which has no vascular effect, resulted in less of an increase in the blood pressure, compared with the infusion of arginine vasopressin with a comparable antidiuretic activity, seems to indicate the involvement of vasopressor activity as well as antidiuretic activity of vasopressin in the development of DOCA-salt hypertension in rats. The fall in blood pressure after withdrawal of vasopressin supplement at week 10 confirms the concept that vasopressin is one of the factors which maintain the blood pressure in this hypertension.

It has been generally believed that vasopressin must be present at a high concentration of pharmacological level in order for it to manifest vasoconstrictive action. Recent reports indicate, however, that vascular sensitivity to vasopressin seems to be enhanced by baroreceptor denervation[10, 11] or presence of DOCA-salt hypertension.[12]

The pressor response to vasopressin in the present study was significantly enhanced in the Brattleboro homozygotes as compared with the Long-Evans rats and in the hypertensive rats as compared with the normotensive group of its control (FIGURE 3). This finding seems to suggest that the endogenous vasopressin may modify the vascular function at its physiological concentration.

In the experiments of salt loading without DOCA administration, the blood pressure in the three groups of rats remained unchanged during the period of administration of normal food (FIGURE 4). There was no significant difference among the blood pressures of the three groups of rats except at week 9. From week 14 when the high salt diet was started, the blood pressures in the Long-Evans rats and in the Brattleboro heterozygotes increased continuously while the blood pressure in the Brattleboro homozygotes still remained unchanged. At week 23, the blood pressure in the Long-Evans rats is significantly higher than that in the Brattleboro heterozygotes, which was still higher than the level of blood pressure in the Brattleboro homozygotes. As clearly shown here, the

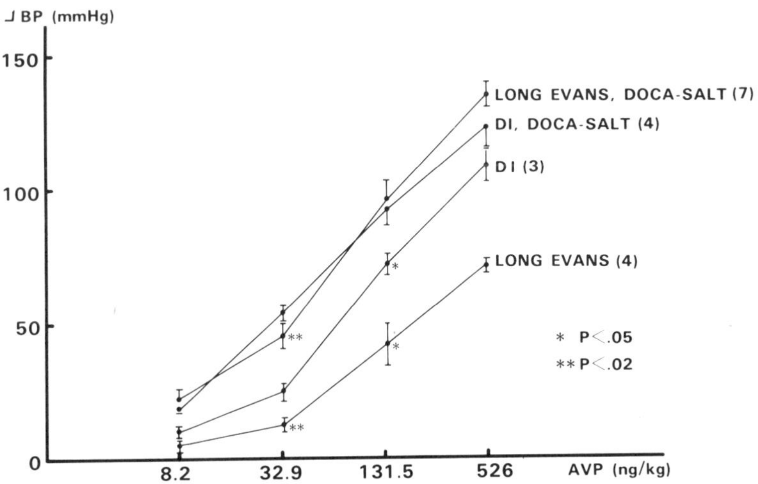

FIGURE 3. Increase in blood pressure in rats in response to infusion of vasopressin. "DI" and "DOCA-SALT" represent the Brattleboro homozygotes and rats with DOCA-salt hypertension, respectively. After anesthesia with α-chloralose and urethane and treatment with dibenzyline, four doses of synthetic arginine vasopressin were infused in 5 min. Numbers in parentheses show number of animals.

load of salt without administration of DOCA also induces hypertension in uninephrectomized rats only in the presence of endogenous secretion of vasopressin.

In conclusion, it may be said that the presence of endogenous vasopressin is necessary for the onset and maintenance of salt-dependent hypertension in rats through its antidiuretic and vascular action. The degree of elevation in blood pressure in DOCA-salt hypertension depends on the concentration of vasopressin in plasma below the physiological level.

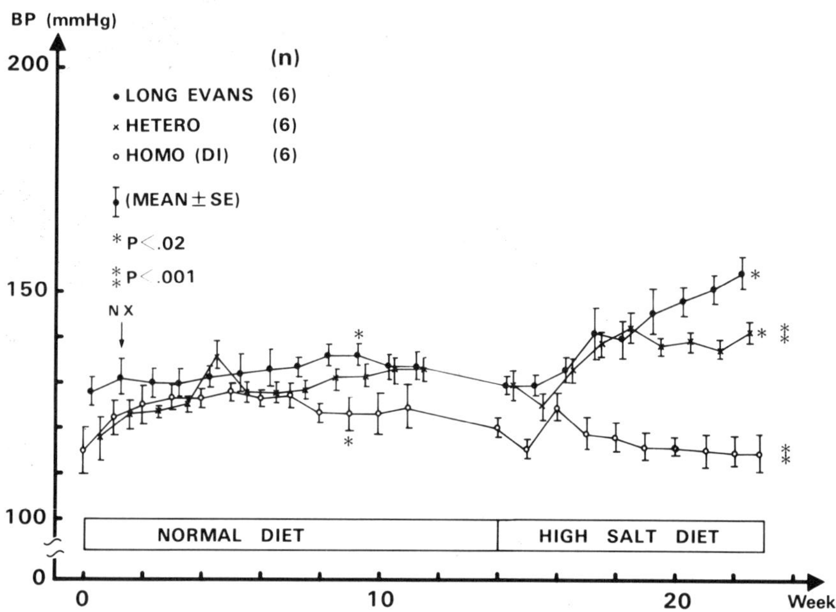

FIGURE 4. Course of blood pressure in uninephrectomized rats. "Hetero" and "Homo" represent the Brattleboro heterozygotes and the Brattleboro homozygotes, respectively. The left kidney was removed one week after the start of the experiment as shown by "NX".

REFERENCES

1. MÖHRING, J., B. MÖHRING, M. PETRI & D. HAACK. 1977. Vasopressor role of ADH in the pathogenesis of malignant DOC hypertension. Am. J. Physiol. **232:** F260–269.
2. CROFTON, J. T., L. SHARE, R. E. SHADE, W. J. LEE-KWON, M. MANNING & W. H. SAWYER. 1979. The importance of vasopressin in the development and maintenance of DOC-salt hypertension in the rats. Hypertension **1:** 31–38.
3. MÖHRING, J., B. MÖHRING, M. PETRI & D. HAACK. 1978. Plasma vasopressin concentrations and effects of vasopressin antiserum on blood pressure in rats with malignant two-kidney Goldblatt hypertension. Circ. Res. **42:** 17–22.
4. PULLAN, P. T., C. T. JOHNSTON, W. P. ANDERSON & P. I. KORNER. 1980. Plasma vasopressin in blood pressure homeostasis and in experimental renal hypertension. Am. J. Physiol. **239:** H81–H87.

5. MÖHRING, J., J. KINTZ & J. SCHOUN. 1978. Role of vasopressin in blood pressure control of spontaneously hypertensive rats. Clin. Sci. Molecular Med. **55:** 247–250.
6. CROFTON, J. T., L. SHARE, R. E. SHADE, C. ALLEN & D. TARNOWSKI. 1978. Vasopressin in the rat with spontaneous hypertension. Am. J. Physiol. **235:** H361–336.
7. SAITO, T., Y. YAJIMA, K. SATO & T. WATANABE. 1979. Role of vasopressin in the pathogenesis of experimental hypertension in rats. Japanese Heart J. **20** (Suppl. 1): 322–324.
8. SAITO, T., Y. YAJIMA & T. WATANABE. 1980. Involvement of AVP in the development and maintenance of hypertension in rats. *In* Antidiuretic Hormone. S. Yoshida, L. Share & K. Yagi Eds.: 215–225. University Park Press. Baltimore, Md.
9. WISGERHOF, M. V., T. E. NORTHRUP, D. M. HEUBELEIN, R. D. BROWN & T. P. DOUSA. 1977. Role of antidiuretic hormone (ADH) in the elevation of blood pressure (B.P.) caused by deoxycorticosterone (DOC) and NaCl. Fed. Proc. **36:** 491.
10. COWLEY, A. W. JR., E. MONOS & A. C. GUYTON. 1974. Interaction of vasopressin and the baroreceptor reflex system in the regulation of arterial blood pressure in the dog. Circ. Res. **34:** 505–514.
11. MONTANI, J. P., J. F. LIARD, J. SCHOUN & J. MÖHRING. 1980. Hemodynamic effects of exogenous and endogenous vasopressin at low plasma concentration in conscious dogs. Circ. Res. **47:** 346–355.
12. CROFTON, J. T., L. SHARE, B. C. WANG & R. E. SHADE. 1980. Pressor responsiveness to vasopressin in the rat with DOC-salt hypertension. Hypertension **2:** 424–431.

DISCUSSION OF THE PAPER

T. BERL (*University of Colorado Medical Center, Denver, Col.*): I understand that when you give dDAVP you don't attain the same increase in blood pressure as when you give AVP. Is that correct?

SAITO: Yes.

BERL: Would you like to summarize your data and tell us what you think about the relative contribution of the vasoconstrictor effect of AVP, the antidiuretic effect of AVP, and perhaps AVP-independent mechanisms to the hypertension in these animals?

SAITO: According to my infusion experiments, approximately 80% of the vasopressin effect in causing the high blood pressure is due to its antidiuretic effect, while less than 20% seems to be dependent on the vascular effect. Infusion of antiserum led to a 15% decrease in blood pressure. This seems to be due to the vascular effect of AVP, because no antidiuresis was seen within 5–10 minutes after the infusion. This is compatible with results of experiments in DI rats chronically infused with AVP or dDAVP.

BERL: So you think an inhibitor of the vascular *and* antidiuretic effects would completely normalize blood pressure in the DOCA-salt hypertensive rat?

SAITO: If an inhibitor is administered to block the initial effect of vasopressin completely, it probably would be very difficult to produce DOCA-salt hypertension in rats. Once hypertension is developed, however, I am not sure blood pressure would be normalized.

Berl: To summarize, in the generation process AVP may be very important, but once the hypertension is established, there may be other factors involved.

K. Berecek (*University of Alabama, Birmingham, Ala.*): What I do not understand about the volume dependency of DOCA hypertension is why we find a positive sodium balance and a positive water balance within the first three days of DOCA administration (Silastic implant containing 100 mg DOCA/kg) after which the animals undergo DOCA escape. Water turnover at that stage is very great, and the animals are neither in positive sodium balance nor positive water balance. Five days after DOCA the animals are again in water balance; however another four or more days elapse before arterial pressure begins to rise. If the antidiuretic effect of vasopressin produces some sort of volume dependency, I have not found it to be linked with the rise in arterial pressure. Could you comment whether there might be an action common to both dDAVP and AVP other than a direct pressor or antidiuretic effect? Perhaps both molecules modify the central nervous system in some way, or have an indirect action in the periphery.

Saito: We cannot completely exclude the possibility that these two peptides may affect the central nervous system. However, the antidiuretic effect does not increase the plasma volume significantly. I measured plasma volume by the Evans blue method and with radioiodinated albumin in rats with and without treatment with dDAVP, and detected no difference in plasma volume. The water content of the vascular system may increase the resistance of arterioles and thus affect volume regulation.

B. M. Altura (*Downstate Medical Center, SUNY, New York, N.Y.*): We have some direct microcirculatory evidence that dDAVP can actually sensitize vascular receptors in arterioles to other hormones that are normally constrictor substances. Perhaps this is the missing link to this whole story.

VASOPRESSIN AND DEOXYCORTICOSTERONE HYPERTENSION IN BRATTLEBORO RATS *

K. H. Berecek † and M. J. Brody ‡

Cardiovascular Center
Department of Pharmacology
University of Iowa
Iowa City, Iowa 52242

Increasing evidence suggests that vasopressin (VP) may play a role in the development of DOCA/salt hypertension. Friedman[1] found that surgical ablation of the median eminence producing diabetes insipidus prevented the development of DOCA/salt hypertension while administration of large doses of VP hastened its onset. Subsequently, elevations in plasma and urinary VP levels have been reported in DOCA/salt-treated rats in both early and chronic stages of hypertension.[2,3] Moreover, administration of VP antagonists or antisera produced an acute reduction in arterial pressure in DOCA/salt hypertensive rats.[2-4] Of great importance has been the finding that Brattleboro rats homozygous for diabetes insipidus (DI rats), which totally lack VP, fail to develop hypertension when treated with DOCA/salt.[3-5] These rats, however, did become hypertensive when Pitressin® (arginine + lysine VP)[5,6] or deamino-8-D-arginine VP (DDAVP), a synthetic analogue that has an antidiuretic effect but no pressor action was replaced.[4]

The mechanisms by which VP might participate in DOCA/salt hypertension are not understood nor is it certain whether VP is necessary for the maintenance of hypertension. Earlier experiments on the DOCA/salt hypertensive rat suggested that changes in vascular reactivity, per se, may participate in the development of hypertension.[7] We, therefore, studied DI rats to determine (a) whether VP replacement could restore the hypertensive properties of DOCA/salt in these rats, (b) whether alterations in vascular responsiveness in these rats would parallel changes in arterial pressure, and (c) whether the maintenance of DOCA/salt hypertension was dependent upon a VP-mediated mechanism.

Materials and Methods

Studies were carried out on 12-week old male DI and Long-Evans (LE) rats (Blue Spruce Farms). DI rats at this age weighed 305 ± 10 g whereas LE rats weighed 396 ± 14 g. Rats were housed individually in metabolic cages and fed a standard laboratory chow (Purina, 100 μEq NaCl/g). We examined changes in arterial pressure and renal vascular responsiveness to vasoconstrictor

* Supported in part by U.S. Public Health Service grants HLB–14388 and HLB–07121 and a gift from the Searle Family Trust.

‡ Address all correspondence to: Michael J. Brody, Department of Pharmacology, Bowen Science Building, University of Iowa, Iowa City, IA 52242.

† Present address: Cardiovascular Research and Training Center, University of Alabama in Birmingham, 1012 Zeigler Building, Birmingham, AL 35294.

0077–8923/82/0394–0319 $1.75/0 © 1982, NYAS

agents in (1) DI rats lacking VP, (2) DI rats replaced with VP, and (3) normal LE rats after unilateral nephrectomy and treatment with DOCA (100 mg/kg s.c.) and 0.3% NaCl. Two additional control groups were included: untreated DI rats and DI rats treated with VP alone. DI rats replaced with VP were given the pituitary extract vasopressin tannate (Pitressin® tannate, Parke-Davis) at a daily dose of 100 mU/100 g rat, s.c. VP replacement in DI rats was begun two weeks prior to DOCA treatment.

Daily fluid intake was monitored. Systolic blood pressure was determined twice weekly using an indirect tail cuff method in conscious restrained rats pre-warmed at 37° C for 5–10 min (automated cuff inflator pulse detector, IITC, Inc.).

Perfusion of the kidney was performed 6–8 weeks after DOCA. In an additional study kidneys from DOCA-treated DI rats and DI rats replaced with VP were perfused 3–4 days post-DOCA. We also studied the effect of removal of VP on arterial pressure in separate groups of DOCA-treated DI rats replaced with VP. Vasopressin treatment was stopped at five and nine weeks post-DOCA. Arterial pressure in these rats was monitored for an additional five weeks following removal of VP.

Renal perfusion experiments were carried out under sodium pentobarbital anesthesia (Nembutal®, Abbott labs, 50 mg/kg i.p.) in isolated kidneys perfused at constant flow with a modified Krebs-Henseleit solution containing Ficoll (Pharmacia AB 35 g/L).[7] Perfusion pressure was measured from the side arm of the perfusion cannula and perfusate flow from a renal venous catheter. Vasoconstrictors employed for the reactivity studies were norepinephrine (NE, Levophed® bitartrate, Breon Lab., Inc.) and vasopressin (VP, Pitressin®, Parke-Davis). Perfusate flow was approximately 5.5 ml/g/min. At this flow rate the renal vascular bed was maximally dilated as bolus injections of papaverine HCl (5–10 mg) produced no further fall in perfusion pressure. Cumulative dose-response curves were obtained to NE and then VP with a period of 25 min between drugs. The drugs were injected into the system in bolus amounts from subthreshold to maximum doses.

All values presented in the text and figures are means ± standard error (SE). One-way analysis of variance was used to evaluate weekly measurements of arterial pressure, daily measurement of fluid intake, and hemodynamic characteristics of the rats at perfusion. Vascular sensitivity was characterized by the ED_{50}, which was determined by probit analysis of the dose-response curves.[8] Differences in sensitivity were determined among the various groups through use of one-way analysis of variance.

In all cases of multiple comparisons when a significant ($p < .05$) F-ratio was obtained the Newman Keuls test was used to determine which of the comparisons were significantly different.

Results

Arterial Pressure

LE rats treated with DOCA/salt showed an increase in systolic blood pressure that was significant at two weeks and averaged 180 mm Hg at four weeks (Table 1). DI rats lacking VP showed no rise in arterial pressure after DOCA, however DI rats replaced with VP developed hypertension after DOCA,

TABLE 1

SYSTOLIC ARTERIAL PRESSURE *

Rat Groups	Control	Weeks 1	2	3	4	5	6	7	8
DI control untreated ($N = 7$)		115 ± 4	119 ± 5	123 ± 5	126 ± 4	126 ± 4	123 ± 7	124 ± 7	118 ± 6
DI + VP ($N = 7$)		123 ± 5	119 ± 6	117 ± 3	126 ± 4	125 ± 7	124 ± 7	123 ± 7	120 ± 6
		Weeks After DOCA or Silastic 1	2	3	4	5	6	7	8
DI + DOCA/salt ($N = 7$)	122 ± 5	124 ± 4	118 ± 2	124 ± 6	129 ± 5	128 ± 5	119 ± 5	126 ± 4	123 ± 5
DI + DOCA/salt + VP ($N = 7$)	120 ± 2	133 ± 2	150 ± 2 †‡	167 ± 4 †‡	179 ± 4 †‡	187 ± 2 †‡	192 ± 6 †‡	206 ± 7 †‡	196 ± 11 †‡
LE control ($N = 6$)	122 ± 1	120 ± 3	122 ± 4	125 ± 5	126 ± 6	122 ± 5	124 ± 5		
LE + DOCA/salt ($N = 6$)	123 ± 4	132 ± 5	154 ± 2 ‡§	177 ± 5 ‡§	190 ± 5 ‡§	192 ± 3 ‡§	201 ± 5 ‡§	196 ± 6 ‡§	196 ± 4 ‡§

Responses are mean ± the standard error.
* Systolic arterial pressure measured in conscious, restrained rats with an indirect tail cuff method (mm Hg).
† $p < .01$ significantly different from all other DI groups.
‡ $p < .01$ significantly different from control measurements.
§ $p < .01$ significantly different from LE control group.

the rise in arterial pressure following a course and obtaining a magnitude similar to that obtained in LE rats (TABLE 1). Note that VP treatment alone in DI rats for 6–8 weeks produced no effect on arterial pressure.

Effectiveness of VP replacement was evaluated by monitoring daily fluid intake in DI rats. DI rats treated with DOCA/salt had an average daily fluid intake of 0.3% NaCl of approximately 350 ml/rat prior to VP replacement. Within 24 hours after the initial injection of VP, intake fell to an average of 90 ml/rat which was significantly less than that seen in DI rats treated with DOCA/salt alone and was comparable to the intake of LE rats. The decrease in fluid intake in the DI rats was maintained as long as the VP replacement continued.

Renal Vascular Resistance

Renal vascular beds of DI rats treated with VP alone and DI rats treated with DOCA/salt alone showed a resistance at maximal vasodilation that was comparable to that of untreated DI rats (FIGURE 1). In contrast renal vascular beds of DI rats replaced with VP who developed hypertension in response to DOCA/salt showed significantly greater ($p < .05$) renal vascular resistance than the other DI groups (FIGURE 1). Kidneys from LE rats treated with DOCA/salt that became hypertensive also showed increased renal vascular resistance when compared to LE control rats (FIGURE 1).

Vascular Sensitivity

At eight weeks post-DOCA renal vascular beds from hypertensive LE rats showed an increase in sensitivity to NE as indicated by a significant decrease in ED_{50} (TABLE 2B). The average ED_{50} of the NE response curves in hyper-

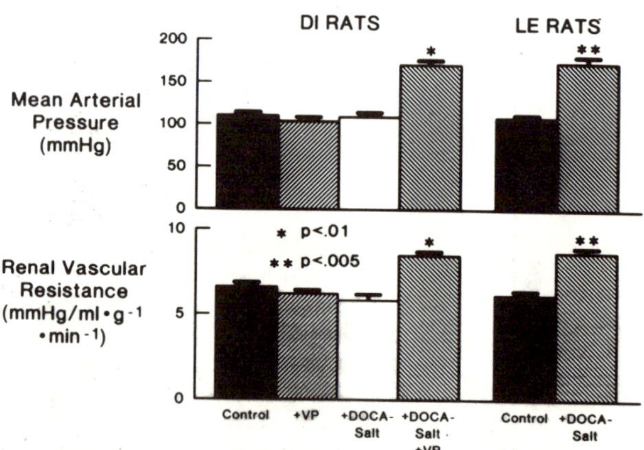

FIGURE 1. Mean arterial pressure (direct measurement) and renal vascular resistance (at maximal vasodilation with papaverine (HC1) in anesthetized rats at the time of perfusion. Asterisks indicate significant differences in arterial pressure and renal vascular resistance determined by one-way analysis of variance among the group responses and the Newman-Keul's test. Bars represent mean ± SE.

TABLE 2

RENAL VASCULAR SENSITIVITY AS DETERMINED BY ED_{50}*

	Norepinephrine (10^{-8} g)	Vasopressin (10^{-9} g)
A. Chronic DI Rats		
DI + DOCA/salt + VP ($N=7$)	1.65 ± 0.25 †	0.75 ± 0.05 †
DI + DOCA/salt ($N=7$)	46.0 ± 13.6 †	11.0 ± 2.1 †
DI + VP ($N=7$)	10.1 ± 1.98	2.82 ± 0.41
DI Control ($N=7$)	14.05 ± 5.8	3.01 ± 0.18
B. Chronic LE Rats		
LE + DOCA/salt ($N=6$)	2.27 ± 0.24 ‡	1.97 ± 0.56 ‡
LE Control ($N=8$)	11.2 ± 0.20	19.0 ± 7.6
C. DI Rats—3 Days Post-DOCA		
DI + DOCA/salt + VP ($N=6$)	1.61 ± 0.2 §	1.48 ± 0.17 §
DI + DOCA/salt ($N=6$)	25.4 ± .3 §	12.5 ± 2.7 §
DI Control ($N=6$)	16.2 ± .4	3.6 ± 1.1

Responses are expressed as mean ± standard error (SE). The values are × 10^{-8} g for norepinephrine and × 10^{-9} g for vasopressin.

* The dose at the half-maximal response is achieved; determined by probit analysis (see text).

Significance of differences determined by one-way analysis of variance and the Newman-Keuls test.

† Different from all other chronic DI groups, $p < .01$.

‡ Different from chronic LE control, $p < .01$.

§ Different from all other DI groups, $p < .05$.

tensive LE rats was five-fold less than that of LE control rats. DI rats replaced with VP who developed hypertension with DOCA also showed enhanced sensitivity to NE (TABLE 2A). The ED_{50} of hypertensive DI rats was 8.3-fold less than untreated DI control rats. Vasopressin replacement alone was not responsible for the enhanced sensitivity to NE as the ED_{50} of this group was similar to that of untreated DI control rats. DI rats treated with DOCA/salt alone that did not show a rise in AP also did not develop enhanced vascular sensitivity. Rather, these rats showed a decreased sensitivity as the ED_{50} was 3.3-fold greater than control and 28-fold greater than hypertensive DI rats.

The pattern of changes in vascular sensitivity in hypertensive LE and DI rats to VP was similar to that seen with NE. Hypertensive LE rats had an ED_{50} that was 9.5-fold less than LE control rats (TABLE 2B). Hypertensive DI rats had an ED_{50} that was four-fold less than untreated DI control rats. Again, DI rats treated with DOCA/salt alone that failed to develop hypertension also failed to develop increased vascular sensitivity. The ED_{50} of DI rats treated with DOCA/salt alone in response to VP was 3.5-fold greater than control and 14.7-fold greater than hypertensive rats. In addition to a decrease in ED_{50} in the hypertensive rats, dose-response curves to NE and VP were also characterized by parallel leftward shifts, decreased thresholds, and increased maximum responses.

Changes in vascular sensitivity were also assessed in DI rats and DI rats replaced with VP three days after the start of DOCA/salt treatment, prior to a rise in arterial pressure or change in vascular resistance. At three days post-DOCA the ED_{50} of dose-response curves to NE and VP in DI rats replaced with VP was significantly less than that of untreated DI control rats, whereas the ED_{50} of DI rats treated with DOCA/salt alone was greater.

Maintenance of Hypertension

In separate groups of DI rats treated with DOCA/salt + VP we stopped VP treatment at five weeks post-DOCA (systolic blood pressure = 185 mm Hg, $N = 6$) and at nine weeks post-DOCA (systolic blood pressure = 198 mm Hg, $N = 6$) and followed the rats for an additional five weeks. At both times removal of VP produced a significant ($p < .05$) decline in arterial pressure by five weeks after stopping VP treatment, but rats remained hypertensive (FIGURE 2). In DI rats treated for five weeks with DOCA/salt + VP blood pressure decreased to 165 mm Hg while in DI rats treated for nine weeks blood pressure fell to 169 mm Hg.

DISCUSSION

Our current studies suggest that VP participates in the development and maintenance of DOCA/salt hypertension although the maintenance was not solely VP-dependent. In confirmation of the findings of Crofton *et al.*[3] and Saito *et al.*,[4] we have found that DI rats totally devoid of VP do not become hypertensive with DOCA/salt. We also confirm the preliminary report of Wisgerhof *et al.*[6] who found that VP (Pitressin®) replacement restored the hypertensive properties of DOCA/salt. Moreover, restoration by VP of hypertension in DI rats was associated with an induction of increased renal vascular sensitivity to NE and VP and that this induction occurred rapidly, preceding the rise in arterial pressure. The pattern of changes in arterial pressure and vascular sensitivity found in DI rats treated with VP and DOCA was similar to that seen in DOCA-treated LE rats with normal hypophyseal function.

A potential role for VP in the pathogenesis of DOCA/salt hypertension was suggested by Möhring *et al.*,[2] who demonstrated that plasma concentrations of VP were increased three- to ten-fold in chronic DOCA-treated rats. Furthermore, Crofton *et al.* identified increased plasma concentration[9] and urinary excretion of VP[3] in early stages of development of DOCA-hypertension. Möhring *et al.* have also shown that acute administration of a specific VP antiserum resulted in a transient large reduction in mean arterial pressure in chronic DOCA hypertensive rats. Saito *et al.*[4] also showed a blood pressure reduction in DOCA hypertensive rats after administration of VP antiserum, which preceded the appearance of diuresis, suggesting involvement of a vasoconstrictor action of VP in the maintenance of hypertension. Use of VP antiserum or VP antagonists to determine its role in the maintenance of hypertension has not produced consistent results. Unlike the substantial drop in arterial pressure to near normal levels that occurred in the Möhring study using VP antiserum, Saito *et al.* found a significant but much smaller (14%) decrease in arterial pressure. We found that when removing VP from DI rats there was an 11% decrease in AP in rats treated for five weeks with DOCA/salt + VP and a 15% decrease in AP in rats treated for nine weeks, results similar to those of Saito *et al.*[4] Crofton *et al.*[3] used two antagonists of the pressor effect of VP, [1-deamino-pencillamine, 4-valine, 8-D-arginine]VP (dPVDAVP) and [1-(β mercapto-β,β-cyclopentamethylene-proprionic acid), 4-valine, 8-D arginine] VP (cyclo dVDAVP). They found that both antagonists produced a substantial reduction in arterial pressure of chronic DOCA hypertensive LE rats. In contrast, Rabito *et al.*[10] found that these same antagonists did not induce significant changes in mean arterial pressure in DOCA-hypertensive Sprague-Dawley rats. Whether these differences are due to differences in the compounds, the methods used to employ the compounds, or the strains of rats in which they are used remains to be determined. In our study administration of VP in an amount sufficient to produce normal levels of plasma AVP, water intake, hematocrit, serum Na^+, serum K^+, plasma renin activity, and plasma aldosterone in DI rats[11] restored the ability of these rats to develop hypertension with DOCA/salt. After establishment of hypertension, removal of VP led to only a 10–15% decrease in arterial pressure suggesting that in these rats hypertension was not maintained solely through a VP pressor mechanism.

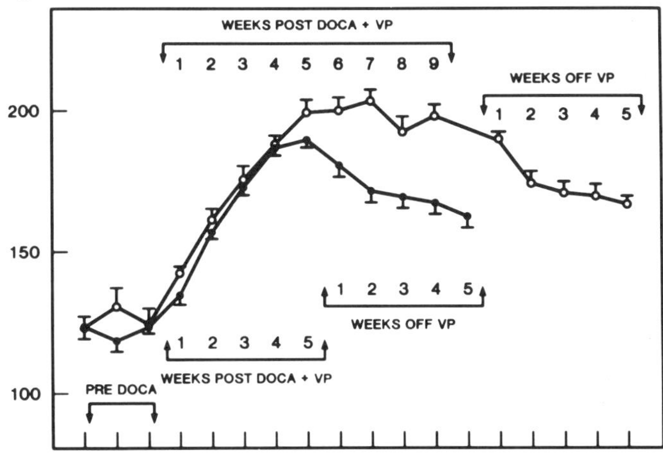

FIGURE 2. Effect of removal of vasopressin on systolic arterial pressure (determined by an indirect tail cuff method in conscious restrained rats) in DI rats treated with DOCA/salt and vasopressin. Vasopressin was removed after five weeks of treatment (●) and after nine weeks of treatment (○). Values represent mean ± SE. Statistics are indicated in the text.

Few studies have addressed the question of the mechanism of action of VP in DOCA/salt hypertension. Saito *et al.*[4] were able to induce DOCA/salt hypertension in DI rats by replacing them with deamino-8-D-arginine VP (DDAVP), which has an antidiuretic effect but no vasoconstrictor effect. This finding suggests that either both the vasoconstrictor and antidiuretic actions of endogenous VP may be involved in DOCA/salt hypertension in rats or there is an indirect action common to both arginine VP and DDAVP.

Plasma concentrations of VP in DOCA/salt hypertensive rats are elevated but to a level below that sufficient to induce a rise in arterial pressure in the normal rat.[12, 13] *In vitro* studies in normal rats indicate that normal plasma concentrations of VP are capable of constricting resistance vessels.[14] However, plasma concentrations of VP have to be increased to supraphysiological levels before a pressor response can be detected. These findings have led to the hypothesis that an increase in sensitivity to the vasoconstrictor effects of VP must occur before VP at physiological concentrations can exert a pressor effect.[15] A marked increase in pressor sensitivity to VP has been demonstrated in baroreceptor-denervated dogs,[16, 17] in decapitated, spinal-anesthetized dogs,[16] and in humans with primary autonomic insufficiency.[18] In these preparations there was a pressor response to physiological concentrations of VP. These data suggest that under normal circumstances the direct pressor effect of VP is buffered by the baroreflex or other central reflexes or inhibited by the sympathetic nervous system. However, in conditions such as hypertension where the function of these reflexes may be altered, physiological concentrations of VP might begin to become pressor.

Recent studies of Montani *et al.*[17] suggest an explanation for potentiation of the pressor response to VP after baroreceptor denervation. These workers found that infusion of physiological concentrations of VP in dogs produced a peripherally mediated increase in total peripheral resistance and a centrally mediated decrease in cardiac output resulting in an unchanged arterial pressure. Removal of baroreceptor afferents suppressed the central cardiac inhibition of VP, leaving the peripheral resistance effect. Hence, the response to VP was not found to be an increase in arterial pressure. There have been few studies on the pressor sensitivity to VP in DOCA hypertensive rats. An increase in systemic pressor sensitivity to VP was observed in chronic DOCA/salt hypertensive rats, but this appeared late in the course of the hypertension and was not a consistent finding in all rats.[9]

Vasopressin may participate in the development of DOCA/salt hypertension by indirectly modulating vascular smooth muscle responses to other vasoconstrictors. VP enhances blood pressure responses[19] and potentiates catecholamine-induced contractions in the microvasculature[14] by unknown mechanisms. Potentiation of the catecholamine response appears not to be due to a VP-induced alteration in the function of the adrenergic nerve terminal.[20] Our studies suggest that the presence of VP in DI rats does modulate the function of the vascular smooth muscle cell by inducing an increase in vascular sensitivity and that the ability of VP to induce such a change precedes the rise in arterial pressure. Moreover, there appears to be a unique interaction between VP and DOCA since vascular reactivity was either unchanged or reduced in DI rats treated with DOCA or VP alone.

From our findings it is evident that VP plays a primary role in DOCA hypertension and that its mechanism of action may, in part, involve an indirect, peripheral effect of inducing an increase in vascular smooth muscle sensitivity to endogenous vasoconstrictor agents.

ACKNOWLEDGMENTS

We gratefully thank Dr. Melvyn Gluckman, Director, Department of Pharmacology, Warner-Lambert Pharm., Research Division, Ann Arbor, Mich. for generously supplying Pitressin® tannate for these studies.

REFERENCES

1. FRIEDMAN, S. M., C. L. FRIEDMAN & M. NAKASHIMA. 1960. Accelerated appearance of DOCA hypertension in rats treated with pitressin. Endocrin. **67:** 752.
2. MÖHRING, J., B. MÖHRING, M. PETRI & D. HAACK. 1977. Vasopressor role of ADH in the pathogenesis of malignant DOC hypertension. Am. J. Physiol. **232:** F260–F269.
3. CROFTON, J. T., L. SHARE, R. E. SHADE, W. J. LEE-KWON, M. MANNING & W. H. SAWYER. 1979. The importance of vasopressin in the development and maintenance of DOC-salt hypertension in the rat. Hypertension **1:** 31–38.
4. SAITO, T., Y. YAJIMA & T. WATANABE. 1980. Involvement of AVP in the development and maintenance of hypertension in rats. *In* Antidiuretic Hormone. S. Yoshida, L. Share & K. Yagi, Eds.: 215–225. University Park Press. Baltimore, Maryland.
5. BERECEK, K. H., R. D. MURRAY, F. GROSS & M. J. BRODY. 1980. Vasopressin (VP) and vascular reactivity in the development of deoxycorticosterone hypertension in rats with hereditary diabetes insipidus. Circulation **62:** III–221.
6. WISGERHOF, M. V., T. E. NORTHRUP, D. M. HEUBELEIN, R. D. BROWN & T. P. DOUSA. 1977. Role of antidiuretic hormone (ADH) in the elevation of blood pressure (B.P.) caused by deoxycorticosterone (DOC) and NaCl. Fed. Proc. **36:** 491.
7. BERECEK, K. H., R. D. MURRAY & F. GROSS. 1980. Significance of sodium, sympathetic innervation and central adrenergic structures on renal vascular responsiveness in DOCA-treated rats. Circ. Res. **47:** 675–683.
8. GOLDSTEIN, A. 1975. Biostatistics: An Introductory Text. Macmillan Company. New York, N.Y.
9. CROFTON, J. T., L. SHARE, B. C. WANG & R. E. SHADE. 1980. Pressor responsiveness to vasopressin in the rat with DOC-salt hypertension. Hypertension **2:** 424–431.
10. RABITO, S. F., O. A. CARRETERO & A. G. SCICLI. 1981. Evidence against a role of vasopressin in the maintenance of high blood pressure in mineralocorticoid and renovascular hypertension. Hypertension **3:** 34–38.
11. MÖHRING, J., G. KOHRS, B. MÖHRING, M. PETRI, E. HOMSY & D. HAACK. 1978. Effects of prolonged vasopressin treatment in Brattleboro rats with diabetes insipidus. Am. J. Physiol. **234:** F106–F111.
12. HAACK, D. & J. MÖHRING. 1978. Vasopressin mediated blood pressure response to intraventricular injection of angiotensin II in the rat. Pflugers Arch. **373:** 1670.
13. MÖHRING, B. & J. MÖHRING. 1975. Plasma ADH in normal Long-Evans rats and in Long-Evans rats heterozygous and homozygous for hypothalamic diabetes insipidus. Life Sci. **17:** 1307.
14. ALTURA, B. M. & B. T. ALTURA. 1977. Vascular smooth muscle and neurohypophyseal hormones. Fed. Proc. **36:** 1853–1860.
15. MÖHRING, J. 1978. Neurohypophyseal vasopressor principle: vasopressor hormone as well as antidiuretic hormone? Klinwochenschr. **56** (Suppl I): 71–79.
16. COWLEY, A. W., E. MONOS & A. C. GUYTON. 1974. Interaction of vaso-

pressin and the baroreceptor reflex system in the regulation of arterial pressure in the dog. Circ. Res. **34:** 505–514.

17. Montani, J.-P., J.-F. Liard, J. Schoun & J. Möhring. 1980. Hemodynamic effects of exogenous and endogenous vasopressin at low plasma concentrations in conscious dogs. Circ. Res. **47:** 346.
18. Möhring, J., K. Glanzer, J. A. Macill, Jr., R. Dusing, H. J. Kramer, R. Arbogast & J. Koch-Weser. 1980. Greatly enhanced pressor response to antidiuretic hormone in patients with impaired cardiovascular reflexes due to idiopathic orthostatic hypotension. J. Cardiovasc. Pharmacol. **2:** 367.
19. Bartelstone, J. H. & P. A. Nasmyth. 1965. Vasopressin potentiation of catecholamine actions in dog, rat, cat and rat aortic strip. Am. J. Physiol. **208:** 754–762.
20. Starke, K., U. Werner, R. Hellerforth & H. J. Schumann. 1970. Influence of peptides on the output of noradrenaline from isolated rabbit hearts. Eur. J. Pharmacol. **9:** 136.

Discussion of the Paper

H. Valtin (*Dartmouth Medical School, Hanover, N.H.*): I would like to invite a comment from Dr. Sawyer. Repeatedly throughout the meeting, we have heard about the use of dDAVP as a strictly antidiuretic agent, with the assumption that it has no vascular effect whatsoever. How valid and how absolute is this assumption?

W. H. Sawyer (*Columbia University, New York, N.Y.*): How absolute is absolute? dDAVP does have inherent vascular activity. It is a vasopressor agent, there is no doubt about it; but, it is very, very weak compared to AVP.

Valtin: But then it is not absolute, and that's my point. It follows that we must be careful to observe the dose of dDAVP when we invoke the assumption. I do not mean this point to be a criticism of any of the papers that we have heard; I raise it as a general consideration for these proceedings.

S. Opava-Stitzer (*University of Puerto Rico, San Juan, P.R.*): I think that is an extremely important point. In the mesenteric artery preparation in the rat, it has been shown that lithium blocks the potentiating effect of vasopressin on the response to other vasoconstrictor substances, but does not block the direct pressor effect. The fact that this potentiating action of AVP is blocked by lithium, as is the antidiuretic effect, suggests that dDAVP may also act in this way despite a very weak direct pressor effect.

Berecek: Another possible mechanism of action of vasopressin in hypertension is that it may interfere with the baroreceptor reflex. There is evidence that suggests that vasopressin may exert an inhibitory effect on the baroreflex (Ciriello, J. & F. Calaresu. *Am. J. Physiol.* **239:**R137, 1980.) We have done some baroreceptor sensitivity work and have found that DI rats show much greater heart-rate changes to a given rise in arterial pressure. This suggests that a lack of vasopressin increases the sensitivity of the baroreceptor reflex. Vasopressin in DOCA/salt hypertension may depress the baroreceptor reflex and allow a greater rise in arterial pressure to occur.

M. Gellai (*Dartmouth Medical School, Hanover, N.H.*): How old are your rats when you start the treatment?

BERECEK: The first group of rats consisted of four-month-old males. The second group was two months of age.

I. W. HENDERSON (*University of Sheffield, Sheffield, England*): Is there any difference in the time of escape from DOCA between DI and normal rats?

BERECEK: Our preliminary studies don't show any difference, but we have to do far more studies in terms of their metabolism and body fluid volumes. Some additional studies in four-month-old rats suggest that it is occurring at the same time, within the first couple of days.

T. SAITO (*Jichi Medical School, Tochigi, Japan*): There seem to be some differences in our results concerning the effect of withdrawal of vasopressin in the hypertensive rats. Did you continue the administration of DOCA and salt when vasopressin was withdrawn?

BERECEK: Yes.

SAITO: In my experiment dDAVP, DOCA, and salt were all withdrawn, resulting in a decrease in hypertension in the DI rats.

BERECEK: I am surprised that you find such a dramatic decrease in arterial pressure since you removed DOCA and salt after 10 weeks of treatment. The severity of hypertension in your group of rats, however, was not as great as in our group. I would anticipate that blood pressures of 200 mm Hg for a duration of five weeks or so, would produce structural changes in the vascular beds and the heart. Thus, in spite of DOCA and salt withdrawal, irreversible, or only slowly reversible, structural changes in the vasculature may sustain the hypertension.

AGE-DEPENDENT SALT HYPERTENSION IN BRATTLEBORO RATS: A HEMODYNAMIC ANALYSIS

J. Zicha, J. Křeček, and H. Dlouhá

Institute of Physiology
Czechoslovak Academy of Sciences
Prague 4, CS 142 20, Czechoslovakia

A mild to moderate elevation of systolic blood pressure was described in uninephrectomized Brattleboro rats homozygous (DI) for diabetes insipidus that drank 0.6% NaCl solution.[6, 11] The blood pressure response to uninephrectomy plus 0.6% saline intake was augmented in rats in which this hypertensive stimulus was applied before sexual maturation.[6] A high salt intake from prepuberty was essential for the hypertensive action of renal-mass reduction. Uninephrectomy of adult DI rats elicited hypertension only in those rats that drank saline from prepuberty.[7] A similar dependence of the blood pressure response on the age at which consumption of 8% NaCl diet started and the importance of high salt intake in prepuberty for induction of late salt hypertension were already reported in salt hypertensive rats of Dahl S (susceptible) substrain.[5]

In the present study, an attempt was made to evaluate the hemodynamics of conscious hypertensive DI Brattleboro rats. A biphasic pattern of systolic blood pressure development was demonstrated in uninephrectomized DI females drinking saline from youth,[7] while this pattern was blunted in heterozygous rats (FIGURE 1). With respect to this blood pressure (BP) development, four age periods (10–12, 13–15, 17–18, and 20–31 weeks of age) were selected in which the hemodynamics of unoperated and uninephrectomized DI rats both drinking saline from the fourth week of age were studied. In order to study the role of age at which rats began to drink saline, the hemodynamics of uninephrectomized and unoperated DI rats both drinking saline from adulthood (12th week of age) were compared with both rats drinking saline from youth and unoperated, water-drinking DI rats at two age periods (13–15 and 20–31 weeks).

MATERIAL AND METHODS

102 DI female Brattleboro rats were weaned at the age of 30 days and they were fed a diet containing 1% NaCl. Twenty-three unoperated rats (group NACL30) and 27 rats uninephrectomized at the age of 25 days (group NACL30UNX25) started to drink 0.6% saline instead of water from the age of 30 days. At the age of 80 days 0.6% saline was administered to additional 14 unoperated rats (group NACL80) and to 15 rats uninephrectomized at the age of 80 days (group NACL80UNX80). Twenty-three unoperated water drinking DI females were used as controls.

Systolic blood pressure (SBP), mean arterial pressure (MAP), diastolic blood pressure (DBP), pulse pressure (PP), cardiac output (CO), total peripheral resistance (TPR), heart rate (HR), stroke volume (SV), and central blood volume (CBV) were studied in conscious unrestrained animals as described in

0077–8923/82/0394–0330 $1.75/0 © 1982, NYAS

detail elsewhere.[19] Arterial rigidity was estimated from pulse pressure-to-stroke volume ratio (PP/SV).[1] Volume-dependent parameters were also expressed per unit of body surface in order to minimize the age-dependent changes of cardiac output and systemic resistance.

Differences between means of groups were evaluated either by Student's t test or by Walsh t test, which was employed for comparison of groups with different variance ($p < .05$). Linear regression analysis was used to find the relationship between the age of animals and individual hemodynamic parameters as well as between MAP and CO or TPR, respectively.

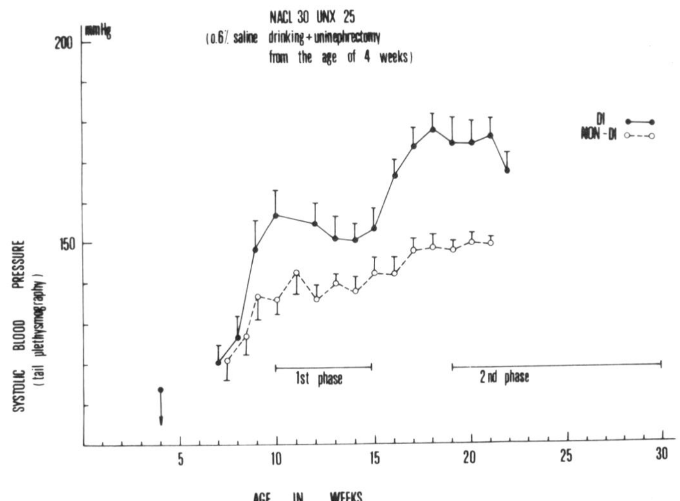

FIGURE 1. The development of systolic blood pressure in uninephrectomized homozygous (DI) and heterozygous (NON-DI) Brattleboro female rats drinking 0.6% saline from the 4th week of age (unpublished indirect blood pressure measurements). Vertical bars indicate SEM.

RESULTS

The Development of Blood Pressure, Cardiac Output, and Total Peripheral Resistance in Conscious NACL30 and NACL30UNX25 Rats

Separate age subgroups of NACL30 rats did not differ in blood pressure (SBP, MAP, DBP). The cardiac output was high in rats aged 10–12 and especially 13–15 weeks in comparison with 20 to 30-week-old NACL30 rats in which total peripheral resistance was elevated (FIGURES 2 and 3).

On the other hand, two blood pressure peaks were demonstrated in NACL30UNX25 rats. In two age subgroups (13 to 15- and 20 to 30-week-old rats) blood pressure (SBP, MAP, and DBP) of NACL30UNX25 rats was higher than values observed in corresponding age subgroups of NACL30 rats. Blood pressure (SBP, MAP, and DBP) of 13 to 15-week-old rats also exceeded values of NACL30UNX25 rats aged 10–12 and 17–18 weeks. In rats aged

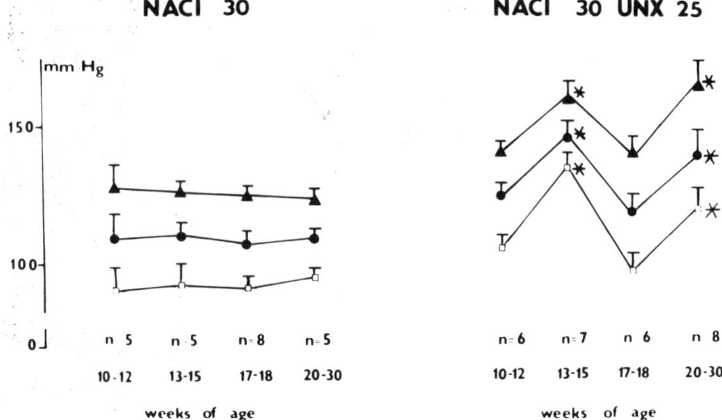

FIGURE 2. Blood pressure of rats aged 10–12, 13–15, 17–18, and 20–30 weeks. Systolic BP (triangles), mean arterial pressure (dots), and diastolic BP (open squares). Asterisks indicate the differences ($p < .05$) between the corresponding age subgroups of unoperated (NACL30) and uninephrectomized (NACL30UNX25) rats both drinking 0.6% saline from the age of 4 weeks.

20–30 weeks only systolic BP was elevated significantly. At the age of 13–15 weeks cardiac output of NACL30UNX25 rats was lower and total peripheral resistance was higher than values of NACL30 rats. In contrast to this, both CO and TPR of NACL30UNX25 rats aged 20–30 weeks were insignificantly elevated as compared to NACL30 rats of this age (FIGURES 2 and 3).

A significant negative correlation ($r = -.491$, $p < .001$, 93 determinations)

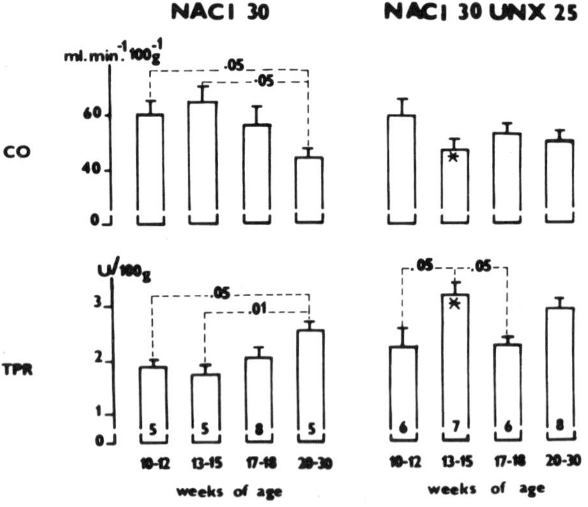

FIGURE 3. Cardiac output (CO in ml·min^{-1}·100 g^{-1}) and total peripheral resistance (TPR in mm Hg·ml^{-1}·min·100 g). Groups and asterisks as in FIGURE 2.

between cardiac output (expressed per unit of body surface) and the age of animals was observed in NACL30 rats aged 10–30 weeks. This was due to the age-dependent decrease of both heart rate ($r = -.346$, $p < .001$) and stroke volume ($r = -.277$, $p < .01$). Systemic resistance rose with age ($r = .577$, $p < .001$). Neither MAP nor SBP and DBP correlated with age but pulse pressure decreased with age of NACL30 rats ($r = -.347$, $p < .001$). Contrary to NACL30 rats, systolic BP and pulse pressure of NACL30UNX25 rats correlated positively with the age of rats ($r = .324$ and $r = .415$ respectively, $p < .001$, 106 determinations). Neither MAP and DBP nor CO and TPR correlated with the age of NACL30UNX25 rats.

The Influence of Saline Consumption from the Age of 4 or 12 Weeks on the Hemodynamics of Unoperated and Uninephrectomized DI Rats

Rats Aged 13–16 Weeks

Systolic blood pressure and mean arterial pressure of NACL30 rats were significantly lower in comparison with rats drinking water (TABLES 1 and 2). Cardiac output and stroke volume of NACL30 rats were elevated, while their systemic resistance was depressed. Short-term saline drinking did not influence BP of NACL80 rats except for pulse pressure increase. While cardiac output did not differ from water-drinking rats, total peripheral resistance of NACL80 rats was significantly lower. Thus SBP and TPR were more influenced in NACL30 than in NACL80 rats. In both NACL30 and NACL80 rats central blood volume tended to be increased but the difference between water and saline drinking rats was significant for NACL80 rats only.

At this age period blood pressure (SPB, MAP, and DBP) of NACL30UNX25 rats differed significantly from both NACL30 and water drinking rats. Cardiac output and stroke volume of hypertensive NACL30UNX25 rats were equal to values of water drinking rats and they were lower than in NACL30 rats. Systemic resistance of NACL30UNX25 rats was highly elevated in comparison with NACL 30 rats but its 18% increase above values of water drinking rats was not significant. CBV and PP/SV of NACL30UNX25 rats did not differ from water drinking or NACL30 rats.

On the other hand, no hemodynamic differences between NACL80 and NACL80UNX80 rats were observed in this age period. Increased central blood volume (per unit of body surface) was the only difference between NACL80UNX80 and water drinking rats. Consequently blood pressure (SBP, MAP, and DBP) as well as systemic resistance were significantly lower in NACL80UNX80 than in NACL30UNX25 rats, while the opposite was true for central blood volume and PP/SV ratio (TABLES 1 and 2).

Rats Aged 20–31 Weeks

At this age period the difference in the duration of high salt intake between rats drinking saline from youth or adulthood was diminished (TABLES 3 and 4). Blood pressure (SBP, MAP and DBP) of both NACL30 and NACL80 rats was significantly lower in comparison with water drinking rats but their pulse pres-

TABLE 1

THE HEMODYNAMICS OF CONSCIOUS DI FEMALE RATS AGED 13–16 WEEKS: WATER DRINKING RATS (H_2O), RATS DRINKING 0.6% SALINE FROM THE 4TH WEEK OF AGE (NACL30), OR FROM THE 12TH WEEK OF AGE (NACL80)

Group	H_2O	NACL30	NACL80
Number of rats	9	5	7
Age of animals (weeks)	13.2±0.15	13.6±0.24	14.4±0.20**, †
High salt intake (weeks)	0	9.6±0.24	2.4±0.20††
Body weight (g)	158.7±3.58	143.6±8.05	163.0±3.87†
Body surface (cm^2)	292.9±4.43	273.8±10.17	298.3±4.73†
Hematocrit (%)	45.6±0.81	43.8±1.36	43.6±0.71
SBP (mm Hg)	138.0±3.07	125.6±4.13*	139.3±2.54†
MAP (mm Hg)	124.5±3.30	111.2±4.94*	122.1±3.45
DBP (mm Hg)	109.5±4.09	92.3±7.75	103.2±3.61
PP (mm Hg)	28.4±3.15	33.3±5.61	36.6±1.57*
CO ($ml \cdot min^{-1} \cdot 100\ g^{-1}$)	47.9±3.62	64.7±5.66*	56.1±2.07
($ml \cdot min^{-1} \cdot 100\ cm^{-2}$)	25.8±1.83	33.8±2.91*	30.6±1.19
TPR ($mm\ Hg \cdot ml^{-1} \cdot min \cdot 100\ g$)	2.75±0.233	1.77±0.155*	2.20±0.109*, †
($mm\ Hg \cdot ml^{-1} \cdot min \cdot 100\ cm^2$)	5.07±0.406	3.38±0.281*	4.04±0.197*
HR (beats/min)	439±16.8	427±9.0	440±7.2
SV ($\mu l/100\ g$)	111±11.0	153±15.0*	128±5.1
($\mu l/100\ cm^2$)	60±5.8	80±7.9	70±3.1
CBV (ml/100 g)	1.38±0.128	1.71±0.177	1.76±0.102*
($ml/100\ cm^2$)	0.75±0.070	0.89±0.094	0.96±0.058*
PP/SV ($mm\ Hg \cdot \mu l^{-1} \cdot 100\ g$)	261±26.2	218±35.6	288±12.6
($mm\ Hg \cdot \mu l^{-1} \cdot 100\ cm^2$)	483±48.4	422±73.6	527±24.8

Data are given as means ± SEM.
*, ** Significantly different ($p < .05$, $p < .01$) from H_2O rats.
†, †† Significantly different ($p < .05$, $p < .01$) from NACL30 rats.

sure was not influenced. However, BP was lowered more in NACL30 than in NACL80 rats. In NACL80 rats stroke volume was higher while heart rate and PP/SV ratio were lower than values of water drinking rats. PP/SV ratio of NACL80 rats was also lower as compared to NACL30 rats (TABLES 3 and 4).

Although even in this late age period NACL30UNX25 rats differed from NACL30 rats by elevated blood pressure (SBP, MAP, and DBP) (FIGURE 2), only systolic BP of NACL30UNX25 rats significantly exceeded values of water drinking rats. Neither cardiac output nor total peripheral resistance of NACL30UNX25 rats were significantly elevated in comparison with NACL30 or water drinking rats of this age. Central blood volume, pulse pressure, and PP/SV ratio of NACL30UNX25 rats were increased as compared to both water drinking and NACL30 rats as well as to NACL30UNX25 rats aged 13–16 weeks (TABLE 2).

Blood pressure of NACL80UNX80 rats was also elevated in comparison with water drinking rats (SBP) and NACL80 rats (SBP and MAP). Similarly as in NACL80 rats heart rate of NACL80UNX80 rats was lower and stroke volume was higher than in water drinking rats. Central blood volume and pulse pressure but not PP/SV ratio of NACL80UNX80 rats were increased in comparison with both NACL80 and water drinking rats (TABLES 3 and 4).

Due to the low number and great scatter of NACL80UNX80 rats of this age it was difficult to evaluate the role of age at which the hypertensive stimulus was applied. Increased arterial rigidity (PP/SV) of NACL30UNX25 rats was the principal hemodynamic peculiarity of rats subjected to the hypertensive regimen from youth (TABLES 3 and 4).

The Relationship of CO and of TPR to MAP

The relations of cardiac output and of total peripheral resistance to mean arterial pressure were analyzed in two age subgroups (10–15 and 17–30 weeks of age) of water drinking, NACL30, and NACL30UNX25 rats (TABLE 5). Mean values for each rat ($\overline{MAP}$, $\overline{CO}$, $\overline{TPR}$), individual determinations (iMAP, iCO, iTPR) as well as individual determinations expressed in percentages of mean values (%MAP, %CO, %TPR) were used for linear regression analysis.

At the age of 10–15 weeks there was a negative correlation of $\overline{CO}$ and $\overline{MAP}$ and a positive correlation of $\overline{TPR}$ and $\overline{MAP}$ in water drinking rats. There was

TABLE 2

THE HEMODYNAMICS OF CONSCIOUS DI FEMALE RATS AGED 13–16 WEEKS: RATS UNINEPHRECTOMIZED AND DRINKING 0.6% SALINE FROM THE 4TH WEEK OF AGE (NACL30UNX25) OR FROM THE 12TH WEEK OF AGE (NACL80UNX80)

Group	NACL30UNX25	NACL80UNX80
Number of animals	7	10
Age of animals (weeks)	14.6±0.30**, †	15.5±0.17**, ‡, §§
High salt intake (weeks)	10.6±0.30†	3.5±0.17††, §§
Body weight (g)	149.7±3.69	173.9±5.27*, ‡
Body surface (cm²)	281.8±4.68	311.3±6.36*, ‡
Hematocrit (%)	45.8±0.58	41.6±1.05**, ‡‡
SBP (mm Hg)	160.5±5.84**, ††	140.7±3.69‡‡
MAP (mm Hg)	147.2±5.42**, ††	124.0±2.79‡‡
DBP (mm Hg)	135.2±4.90**, ††	106.1±2.27‡‡
PP (mm Hg)	25.3±2.18	34.5±2.72‡
CO ($ml \cdot min^{-1} \cdot 100\ g^{-1}$)	47.0±2.90†	52.5±2.66
($ml \cdot min^{-1} \cdot 100\ cm^2$)	25.0±1.60†	29.3±1.54
TPR ($mm\ Hg \cdot ml^{-1} \cdot min \cdot 100\ g$)	3.20±0.217††	2.42±0.105‡‡
($mm\ Hg \cdot ml^{-1} \cdot min \cdot 100\ cm^2$)	6.04±0.410††	4.34±0.188‡‡
HR (beats/min)	418±21.2	421±13.2
SV (μl/100 g)	114±7.7†	126±8.3
($\mu l/100\ cm^2$)	60±4.2†	70±5.0
CBV (ml/100 g)	1.50±0.057	1.65±0.045‡
($ml/100\ cm^2$)	0.79±0.028	0.92±0.031*, ‡
PP/SV ($mm\ Hg \cdot \mu l^{-1} \cdot 100\ g$)	223±15.5	277±17.1‡
($mm\ Hg \cdot \mu l^{-1} \cdot 100\ cm^2$)	421±32.8	497±30.6

*, ** Significantly different ($p < .05$, $p < .01$) from H_2O rats (TABLE 1).
†, †† Significantly different ($p < .05$, $p < .01$) from NACL30 rats (TABLE 1).
‡, ‡‡ Significantly different ($p < .05$, $p < .01$) from NACL30UNX25 rats.
§, §§ Significantly different ($p < .05$, $p < .01$) from NACL80 rats (TABLE 1).

TABLE 3

THE HEMODYNAMICS OF CONSCIOUS DI FEMALE RATS AGED 20–31 WEEKS: WATER DRINKING RATS (H_2O), RATS DRINKING 0.6% SALINE FROM THE 4TH WEEK OF AGE (NACL30) OR FROM THE 12TH WEEK OF AGE (NACL80)

Group	H_2O	NACL30	NACL80
Number of rats	14	5	7
Age of animals (weeks)	23.1±0.87	25.6±1.69	30.6±0.20**, †
High salt intake (weeks)	0	21.6±1.69	18.6±0.20
Body weight (g)	181.8±4.80	169.0±2.53*	164.3±7.19*
Body surface (cm^2)	320.6±5.56	305.6±3.06*	299.6±8.63*
Hematocrit (%)	47.0±0.37	44.5±1.07*	48.1±0.20*, †, ^{W}
SBP (mm Hg)	138.1±1.91	124.0±1.68**	131.1±1.44*, ††, ^{W}
MAP (mm Hg)	124.5±2.08	109.8±2.31**	116.4±2.14*
DBP (mm Hg)	110.2±2.18	94.8±2.74**	101.4±1.50*, †
PP (mm Hg)	28.0±0.99	29.1±1.78	29.7±2.25
CO ($ml \cdot min^{-1} \cdot 100\ g^{-1}$)	45.7±2.52	44.0±3.06^{w}	51.0±3.08
($ml \cdot min^{-1} \cdot 100\ cm^2$)	25.8±1.37	24.3±1.66^{w}	28.0±2.02
TPR ($mm\ Hg \cdot ml^{-1} \cdot min \cdot 100\ g$)	2.91±0.241	2.55±0.143^{W}	2.34±0.150
($mm\ Hg \cdot ml^{-1} \cdot min \cdot 100\ cm^2$)	5.13±0.411	4.61±0.253^{w}	4.30±0.314
HR (beats/min)	409±10.3	403±28.0	328±22.7**, ^{W}
SV ($\mu l/100\ g$)	112±5.9	111±6.1^{w}	164±20.1*
($\mu l/100\ cm^2$)	63±3.2	61±3.4	91±12.0*
CBV (ml/100 g)	1.46±0.062	1.37±0.100	1.54±0.093
($ml/100\ cm^2$)	0.82±0.031	0.76±0.057	0.84±0.056
PP/SV ($mm\ Hg \cdot \mu l^{-1} \cdot 100\ g$)	258±12.7	270±26.4	191±16.7**, †
($mm\ Hg \cdot \mu l^{-1} \cdot 100\ cm^2$)	456±21.9	489±47.3	350±33.8*, †

*, ** Significantly different ($p < .05$, $p < .01$) from H_2O rats.
†, †† Significantly different ($p < .05$, $p < .01$) from NACL 30 rats.
^{w}, ^{W} Significantly different ($p < .05$, $p < .01$) from rats aged 13–16 weeks (TABLE 1).

no correlation between %CO and %MAP, but %TPR correlated positively with %MAP. Similarly in NACL30UNX25 rats there was a positive correlation of $\overline{TPR}$ and $\overline{MAP}$ and of %TPR and %MAP. Cardiac output tended to correlate negatively with mean arterial pressure at this age ($\overline{CO}$ and $\overline{MAP}$: $p = .072$, iCO and iMAP: $p < .01$). Hence, mean arterial pressure of individual NACL30UNX25 rats was set to different levels by the changes of systemic resistance, while cardiac output was depressed proportionally to BP elevation. Nevertheless, a positive correlation of %CO and %MAP indicated that at each level of mean arterial pressure it could be elevated by the increase of cardiac output. On the other hand, NACL30 rats differed from both preceding groups since a positive correlation between total peripheral resistance and mean arterial pressure was absent in both analyzed age periods. However, their cardiac output correlated positively with mean arterial pressure. This was true only for iCO and iMAP in rats aged 10–15 weeks, while at the age of 17–30 weeks this also applied for $\overline{CO}$ and $\overline{MAP}$ and for %CO and %MAP. In NACL30UNX25 rats aged 17–30 weeks, there was also a positive correlation of $\overline{TPR}$ and $\overline{MAP}$ and of %TPR and %MAP as in preceding age periods but no correlation between cardiac output and mean arterial pressure was observed.

Discussion

Direct blood pressure measurements disclosed that elevated systolic blood pressure of uninephrectomized salt hypertensive DI Brattleboro rats was accompanied by an elevation of both mean arterial pressure and diastolic blood pressure. Hence, salt hypertension could be induced in rats lacking vasopressin.[6, 7, 11] Nevertheless, the high salt intake in prepuberty was demonstrated to be of critical importance for eliciting salt hypertension in DI rats. The failure to induce DOC-salt hypertension in uninephrectomized DI male rats[4] could be explained by the fact that adult animals had been used. The data presented do not support the hypothesis on the key role of vasopressin in the pathogenesis of salt hypertension.[4, 15] Recent experiments with selective antagonists of vasopressor effects of vasopressin also speak against this hypothesis, at least in salt-hypertensive Dahl S rats,[14] in adrenal-regeneration hypertension and in malignant DOC-salt hypertension.[16]

The present hemodynamic study indicated the existence of two phases in the

Table 4

The Hemodynamics of Conscious DI Female Rats Aged 20–31 Weeks: Rats Uninephrectomized and Drinking 0.6% Saline from the 4th Week of Age (NACL30UNX25) or from the 12th Week of Age (NACL80UNX80)

Group	NACL30UNX25	NACL80UNX80
Number of rats	8	5
Age of animals (weeks)	22.6±0.42	28.2±0.20*, ‡‡, §§
High salt intake (weeks)	18.6±0.42	16.2±0.20‡‡, §§
Body weight (g)	171.8±7.11	177.6±8.41
Body surface (cm^2)	308.6±8.62	315.6±10.04
Hematocrit (%)	44.7±1.46	42.6±2.15
SBP (mm Hg)	165.4±8.88*, ††	154.0±14.00*, §
MAP (mm Hg)	139.8±8.78†	135.0±10.79§
DBP (mm Hg)	119.1±9.19†	114.3±8.34
PP (mm Hg)	46.3±3.72**, ††, W	39.6±9.10*, §
CO ($ml \cdot min^{-1} \cdot 100\ g^{-1}$)	50.0±3.45	55.6±6.14
($ml \cdot min^{-1} \cdot 100\ cm^2$)	27.9±2.22	31.3±3.44
TPR ($mm\ Hg \cdot ml^{-1} \cdot min \cdot 100\ g$)	2.93±0.310	2.58±0.328
($mm\ Hg \cdot ml^{-1} \cdot min \cdot 100\ cm^2$)	5.31±0.600	4.59±0.586
HR (beats/min)	410±21.5	339±17.1**, ‡, W
SV (μl/100 g)	128±13.9	167±22.6*
($\mu l/100\ cm^2$)	72±8.5	94±12.2*
CBV (ml/100 g)	1.85±0.093**, ††, W	2.02±0.243*
($ml/100\ cm^2$)	1.02±0.041**, ††, W	1.13±0.128*, §, w
PP/SV ($mm\ Hg \cdot \mu l^{-1} \cdot 100\ g$)	386±33.7**, †, W	254±55.5‡
($mm\ Hg \cdot \mu l^{-1} \cdot 100\ cm^2$)	700±67.3**, †, W	448±95.0‡

*, ** Significantly different ($p < .05$, $p < .01$) from H_2O rats (Table 3).
†, †† Significantly different ($p < .05$, $p < .01$) from NACL30 rats (Table 3).
‡, ‡‡ Significantly different ($p < .05$, $p < .01$) from NACL30UNX25 rats.
§, §§ Significantly different ($p < .05$, $p < .01$) from NACL80 rats (Table 3).
w, W Significantly different ($p < .05$, $p < .01$) from rats aged 13–16 weeks (Table 2).

development of salt hypertension in uninephrectomized DI rats. In the early phase (13–15 weeks of age) cardiac output of hypertensive NACL30UNX25 animals was low, while their systemic resistance was elevated. The role of elevated systemic resistance, however, was not so clear-cut in the late phase of hypertension (20–30 weeks of age). When NACL30UNX25 rats were compared with NACL30 rats, a 27% elevation of mean arterial pressure in hypertensive rats was accompanied by a 15% elevation of both cardiac output and total peripheral resistance. In this late phase only systolic blood pressure of

TABLE 5

THE RELATIONSHIP OF THE CARDIAC OUTPUT (CO) AND OF THE TOTAL PERIPHERAL RESISTANCE (TPR) TO THE MEAN ARTERIAL PRESSURE (MAP) IN CONSCIOUS DI FEMALE RATS DRINKING EITHER WATER (H_2O) OR 0.6% SALINE FROM THE 4TH WEEK OF AGE (NACL30) OR UNINEPHRECTOMIZED AND DRINKING 0.6% SALINE FROM THE SAME AGE (NACL30UNX25): LINEAR REGRESSION ANALYSIS

Group	H_2O	NACL30	NACL30UNX25
	Rats Aged 10–15 Weeks		
Number of rats	19	10	13
$\overline{CO}$ and $\overline{MAP}$	−.675^{S}	.458**	−.515†
$\overline{TPR}$ and $\overline{MAP}$	.814^{S}	.255	.788^{S}
Number of determinations	88	45	48
iCO and iMAP	−.540^{S}	.388^{S}, **	−.430^{S}, ††
iTPR and iMAP	.742^{S}	.240**	.736^{S}, ††
%CO and %MAP	.060	.239	.510^{S}, **
%TPR and %MAP	.419^{S}	.228	.327^{s}
	Rats Aged 17–30 Weeks		
Number of rats	14	13	14
$\overline{CO}$ and $\overline{MAP}$	−.046	.722^{s}, *	.017†
$\overline{TPR}$ and $\overline{MAP}$	.452	−.382*	.664^{s}, ††
Number of determinations	67	48	58
iCO and iMAP	−.019^{W}	.645^{S}, **	.033††, ^{w}
iTPR and iMAP	.397$^{S, W}$	−.270**, ^{w}	.654^{S}, *, ††
%CO and %MAP	.252^{s}	.353^{s}	.210
%TPR and %MAP	.065^{w}	.059	.306^{s}

Data are given as correlation coefficients *r*.

r coefficient significantly different ($p < .05$, $p < .01$): ^{s}, ^{S}—from zero; *, **—from H_2O rats; †, ††—from NACL30 rats; ^{w}, ^{W}—from rats aged 10–15 weeks.

$\overline{MAP}$, $\overline{CO}$, $\overline{TPR}$—average values of MAP, CO and TPR for each rat.

iMAP, iCO, iTPR—individual determinations of MAP, CO and TPR.

%MAP, %CO, %TRP—iMAP, iCO and iTPR in percentages of $\overline{MAP}$, $\overline{CO}$, and $\overline{TPR}$.

hypertensive rats was significantly elevated in comparison with water drinking rats. The increase of pulse pressure as well as of estimated arterial rigidity indicated the importance of changes of arterial vasculature for systolic blood pressure elevation in the late phase of hypertension. The central blood volume of NACL30UNX25 rats was also increased only in the late phase. This increase could not be explained either by an increase of cardiac output or by blood volume expansion (early phase: 7.1 ± 0.13 ml/100 g, $N = 9$; late phase:

7.1 ± 0.11, $N = 18$; unpublished data) and therefore it had to be considered as the redistribution of circulating blood into the cardiopulmonary vascular bed in the late development of salt hypertension.

The dominant role of increased systemic resistance in the early phase of hypertension and the contribution of other factors in the late phase were already evident in experiments in which neonatally sympathectomized (guanethidine-treated) DI rats were uninephrectomized and drank 0.6% saline from youth. The early increase of systolic blood pressure was completely abolished, while the late hypertension occurred. The late elevation of systolic blood pressure in sympathectomized DI NACL30UNX25 rats was accompanied by a further increase of blood volume.[13, 18]

A biphasic pattern of systolic blood pressure increase was also observed in Dahl S rats fed a high salt diet from the sixth week of age, i.e. from prepuberty.[12] If the high salt diet was supplemented with thiazide, only the early increase of systolic blood pressure was observed and no hypertension occurred after the age of 16 weeks.[12] Both systolic and mean arterial pressures were elevated in conscious 14-week-old Dahl S rats fed a high salt diet from the eighth week of age.[14] Normal cardiac output but increased systemic resistance were observed in anesthetized Dahl S rats in which the elevation of mean arterial pressure was induced by seven-day feeding of a high salt diet in prepuberty.[10] Neurogenic vasoconstriction accounted for 50% of increased vascular resistance observed in 14-week-old Dahl S rats consuming a high salt diet from the age of 8–10 weeks.[17] Neonatal guanethidine sympathectomy abolished the early rise of systolic blood pressure in conscious Dahl S rats fed a high salt diet from the 4th week of age. However, these animals were only followed until the 10th week of age.[9]

Unfortunately, no data on the hemodynamics of the late phase were reported in rats in which salt-dependent hypertension was induced in youth. Nevertheless, a decrease of functional distensibility of the aorta was observed in 20-week-old uninephrectomized Wistar male rats that were DOC-salt treated from the age of 4 weeks.[2] Low aortic distensibility was ascribed to both the increase of relative wall thickness and altered aortic wall composition. This was in agreement with our observation of elevated arterial rigidity (estimated *in vivo* as PP/SV) in hypertensive NACL30UNX25 rats aged 20–30 weeks. Hence, in the late phase of salt-dependent hypertension, abnormalities of the arterial system[2] must be considered besides body fluid changes indicated by the effects of thiazides on salt hypertensive Dahl S rats.[12]

The most interesting hemodynamic finding was the decrease of systemic resistance in unoperated, saline drinking DI female rats aged 13–15 weeks. This decrease was more pronounced in rats drinking saline from youth in which it was accompanied by a significant increase of cardiac output. A similar decrease of systemic resistance was previously described in female rats of salt-resistant Dahl substrain after a short-term feeding a high salt diet in prepuberty.[10] These changes were even more prominent in normotensive Wistar-Kyoto females rats exposed to long-term drinking of 1% saline in adulthood.[3] Even if the elevation of cardiac output and the decrease of systemic resistance in NACL30 DI rats disappeared during later development, the dependence of mean arterial pressure on cardiac output was even strengthened.

The high mortality of unoperated DI rats drinking 0.6% saline from youth was reported earlier[20] while no death occurred during 14 weeks of the experiment in rats that started to consume saline in adulthood. The question remains open whether the high mortality of NACL30 rats was related to peculiarities of their circulation (low resistance, high flow, relative independence of mean arterial pressure on systemic resistance changes) or to their sodium metabolism (hypernatremia).[8] The former possibility was supported by the low mortality of NACL30UNX25 DI rats [20] in which the mechanisms regulating mean arterial pressure were more similar to water drinking DI rats than to NACL30 DI rats. Minor circulatory changes induced by saline consumption in adult rats of the NACL80 group did not influence their good survival on a high salt regimen.[13, 20]

Summary

The hemodynamic effects of 0.6% saline, consumed either from youth (4th week of age) or from adulthood (12th week of age), were studied in unanesthetized, unoperated, and uninephrectomized homozygous female Brattleboro rats. Long-term saline drinking induced a general decrease of blood pressure in unoperated rats which was more pronounced in rats drinking it from youth. The relation of low systemic resistance and high cardiac output (observed at the age of 10–15 weeks) to the high mortality of these rats was discussed.

Two phases were recognized in the development of salt hypertension in uninephrectomized rats drinking saline from youth. The increased systemic resistance played a major role during the early phase (13–15 weeks), while changes of body fluids as well as altered arterial compliance contributed to the elevation of systolic blood pressure in the late phase of salt hypertension (20–30 weeks of age).

In uninephrectomized rats drinking saline from adulthood, the late blood pressure response was only slightly attenuated in comparison with uninephrectomized rats drinking saline from youth. The absence of increased arterial rigidity in the former group was the only major hemodynamic difference between these two groups of uninephrectomized rats aged 20–30 weeks.

References

1. Alicandri, C. L., R. Fariello, E. Agabiti-Rosei, G. Romanelli & G. Muiesan. 1980. Influence of the sympathetic nervous system on aortic compliance. Clin. Sci. **59** (Suppl. 6): 279–282.
2. Berry, L. C. & S. E. Greenwald. 1976. Effects of hypertension on the static mechanical properties and chemical composition of the rat aorta. Cardiovasc. Res. **10:** 437–451.
3. Chrysant, S. G., G. M. Walsh, D. C. Kem & E. D. Frohlich. 1979. Hemodynamic and metabolic evidence of salt sensitivity in spontaneously hypertensive rats. Kidney Int. **15:** 33–37.
4. Crofton, J. T., L. Share, R. E. Shade, W. J. Lee-Kwon, M. Manning & W. H. Sawyer. 1979. The importance of vasopressin in the development and maintenance of DOC-salt hypertension in the rat. Hypertension **1:** 31–38.
5. Dahl, L. K., K. D. Knudsen, M. A. Heine & G. J. Leitl. 1968. Effects of chronic excess salt ingestion. Modification of experimental hypertension in the rat by variations in the diet. Circulation Res. **22:** 11–18.
6. Dlouhá, H., J. Křeček & J. Zicha. 1977. Hypertension in rats with hereditary diabetes insipidus. The role of age. Pflügers Arch. **369:** 177–182.

7. Dlouhá, H., J. Křeček & J. Zicha. 1979. Effect of age on hypertensive stimuli and the development of hypertension in Brattleboro rats. Clin. Sci. **57:** 273–275.
8. Dlouhá, H., J. Křeček & J. Zicha. 1980. Sodium metabolism in rats with hereditary defect of vasopressin synthesis. *In* Hormonal Regulation of Sodium Excretion. B. Lichardus, R. W. Schrier & J. Ponec, Eds.: 129–134. Elsevier/North Holland Biomedical Press. Amsterdam.
9. Friedman, R., L. M. Tassinari, M. Heine & J. Iwai. 1979. Differential development of salt-induced and renal hypertension in Dahl hypertension-sensitive rats after neonatal sympathectomy. Clin. Exp. Hypertension **1:** 779–799.
10. Ganguli, M., L. Tobian & J. Iwai. 1979. Cardiac output and peripheral resistance in strains of rats sensitive and resistant to NaCl hypertension. Hypertension **1:** 3–7.
11. Hall, C. E., S. Ayachi & O. Hall. 1973. Spontaneous hypertension in rats with hereditary hypothalamic diabetes insipidus (Brattleboro strain). Texas Rep. Biol. Med. **31:** 471–487.
12. Iwai, J., E. V. Ohanian & L. K. Dahl. 1977. Influence of thiazide on salt hypertension. Circulation Res. **40** (Suppl. 1): 131–134.
13. Křeček, J., S. Doležel, H. Dlouhá & J. Zicha. 1979. Blood pressure increase in Brattleboro rats with diabetes insipidus—a protective mechanism against age-dependent toxic effects of salt? Physiol. Bohemoslov. **28:** 449.
14. Matsuguchi, H., P. G. Schmid, D. Van Orden & L. Mark. 1981. Does vasopressin contribute to salt-induced hypertension in the Dahl strain? Hypertension **3:** 174–181.
15. Möhring, J., B. Möhring, M. Petri & D. Haack. 1977. Vasopressor role of ADH in the pathogenesis of malignant DOC hypertension. Am. J. Physiol. **232:** F260–F269.
16. Rabito, S. F., O. A. Carretero & A. G. Scicli. 1981. Evidence against a role of vasopressin in the maintenance of high blood pressure in mineralocorticoid and renovascular hypertension. Hypertension **3:** 34–38.
17. Takeshita, A. & A. L. Mark. 1978. Neurogenic contribution to hindquarters vasoconstriction during high salt intake in Dahl strain of genetically hypertensive rat. Circulation Res. **43** (Suppl. 1): 86–91.
18. Zicha, J., J. Křeček & H. Dlouhá. 1981. The hemodynamic analysis of two phases in the development of salt hypertension in homozygous (DI) Brattleboro rats. Physiol. Bohemoslov. **30:** 189.
19. Zicha, J., P. Karen, V. Krpata, H. Dlouhá & J. Křeček. 1982. Hemodynamics of conscious Brattleboro rats. Ann. N.Y. Acad. Sci. (This volume.)
20. Dlouhá, H., J. Křeček & J. Zicha. 1981. Hypertension in Brattleboro rats and the injurious influence of salt in youth. Physiol. Bohemoslov. **30:** 531–538.

Discussion of the Paper

T. Bennett (*University of Nottingham Medical School, Nottingham, England*): Would you care to comment about the effectiveness of the technique you use for sympathectomy? I ask because in our hands it doesn't work very well at all.

Zicha: We have verified the status of the peripheral sympathetic nervous system in surviving rats (aged 20–30 weeks) by means of the histochemical fluorescent technique (Falck, *Acta Physiol. Scand* Suppl. **197:**1, 1962). No signs of sympathetic regeneration were observed. But it is well known (Johnson *et al. Eur. J. Pharmacol.* **37:**45, 1976) that early guanethidine treatment does not eliminate sympathetic activity in either central nervous system or adrenal medulla.

I. L. SCHWARTZ (*Mt. Sinai School of Medicine, New York, N.Y.*): If Dr. Lewis Dahl were here today, he would be very pleased with your paper because he had the idea a good many years ago, that we are producing hypertensives in our population by putting salt in baby foods.

ZICHA: I think that our data on salt hypertension in Brattleboro rats do confirm the earlier observation of Dr. Lewis K. Dahl (Dahl *et al. Circ. Res.* **22:** 11, 1968). They also observed that a relatively short period of high salt intake in youth was of critical importance for the eliciting of severe hypertension in adult animals.

THE ROLE OF VASOPRESSIN IN THE CONTROL OF RENAL HEMODYNAMICS. THE BRATTLEBORO RAT AS AN EXPERIMENTAL MODEL *

Miklos Gellai

Department of Physiology
Dartmouth Medical School
Hanover, New Hampshire 03755

In 1976, reviewing information available at that time, K. Aukland stated: "There is no convincing evidence that physiological variation in ADH secretion causes marked changes in total or regional RBF." [1] The data that have been gathered since 1976 concerning the effect of vasopressin on *total* renal blood flow and glomerular filtration (GFR) do not justify changing that conclusion.

Lately, most research in this field has dealt with the possible influence of vasopressin on intrarenal distribution of blood flow and nephron filtration rate. However, in spite of the availability of a wide variety of techniques designed to measure blood flow in different regions of the kidney, none of them have been sufficiently validated to date. As a consequence, data resulting from their use should be interpreted with some caution.[2]

In contrast, data on the glomerular effect of vasopressin appear more convincing. Cytochemical methods were used by Sato *et al.*[3] to show the presence of adenylate cyclase localized on the basement membrane and the mesangial matrix of the glomeruli. The vasopressin sensitivity of the adenylate cyclase was demonstrated in isolated glomeruli from rabbits,[4] but not in those from rats.[5] Ausiello *et al.* presented evidence that rat glomerular mesangial cells, grown in culture, contract when exposed to vasopressin.[6] Consistent with these findings is the report by Ichikawa and Brenner that infusion of subpressor doses of vasopressin effected a decrease in the ultrafiltration coefficient (K_f).[7] It has yet to be demonstrated whether or not such reduction in K_f plays a role in the control of GFR. Ichikawa and Schor have proposed that the decrease in K_f is offset by an increase in transcapillary hydraulic pressure difference as a result of vasopressin-induced antidiuresis; the net effect may be no change in GFR.[8]

THE USE OF ANIMALS WITH DIABETES INSIPIDUS

Most experiments designed to study the effect of vasopressin on renal hemodynamics have been carried out in animals or persons in whom a water diuresis was induced. However, excessive water loading may lead to a large increase in extracellular fluid volume, and in turn to reflex changes in renal hemodynamics. Furthermore, it is usually not safe to assume that plasma vasopressin concentration is reduced to zero during water diuresis. For these reasons, increasing numbers of investigators are selecting animals with diabetes insipidus as the experimental model.

It was in such a model—dogs with diabetes insipidus produced by hypothalamic-hypophyseal tract section—that the first convincing evidence regarding

* Supported principally by U.S. Public Health Service Research Grant AM-08469.

0077–8923/82/0394–0343 $1.75/0 © 1982, NYAS

the rôle of vasopressin in the intrarenal distribution of blood flow was presented.[9] Using the ^{85}Kr washout technique, Fisher *et al.* demonstrated that, in the absence of vasopressin, outer cortical blood flow was greater, whereas juxtamedullary cortical flow and outer medullary flow were lower than normal. Additional data obtained by the retrograde venous injection of silicon rubber—resulting in filling of venous vasculature and glomeruli in the DI dogs, but not in normal dogs—led these authors to conclude that vasopressin exerts an important control on postglomerular vascular resistance. Recently, Akatsuka *et al.*[10] and Johnson *et al.*[11] have confirmed these effects of vasopressin on the distribution of renal cortical blood flow in dogs in water diuresis.

That the Brattleboro homozygous (DI) rat may serve as a useful model for studying the effect of vasopressin on glomerular dynamics was demonstrated by Davis and Schnermann in 1971.[12] Using the Hanssen ferrocyanide method, as modified by de Rouffignac,[13] to measure regional differences in single nephron filtration rate, these workers showed that infusion of vasopressin resulted in a significant and selective increase of juxtamedullary nephron filtration rate. Using the same technique, Bankir and associates [14] demonstrated that the well known heterogeneity of single nephron filtration rate is absent in the DI rat, but could be restored by treatment with vasopressin [pitressin tannate or (desamino)D-arginine vasopressin (dDAVP)]. This restoration was the sole result of a selective increase in juxtamedullary nephron filtration rate. However, whether or not this event was due to the vascular or antidiuretic action of vasopressin cannot be determined from these experiments. As was pointed out by Davis and Schnermann,[12] the rise in urine osmolality during vasopressin treatment could lead to an increase in the viscosity of blood in the vasa recta and that this in turn, may be followed by increased postglomerular resistance and effective filtration pressure. That the heterogeneity of single nephron filtration is a consequence of the ability to concentrate urine, and does not depend on a direct vasoconstrictor action of vasopressin was demonstrated recently by data from experiments performed on nephrogenic DI mice, with normal or high plasma levels of vasopressin.[15]

Complicating Factors in the Study of Vasopressin and Renal Hemodynamics

The studies presented above exemplify the difficulties of interpreting experimental data that deal with the role of hormones (including vasopressin) in the control of renal hemodynamics. Some of the most nagging problems are discussed next.

(1) It is difficult, if not impossible, to dissect the direct tubular influence of vasopressin in causing antidiuresis from its possible (and simultaneous) hemodynamic effect.

(2) There is a complex interaction between vasopressin and many other hormones, especially those of renal origin: renin-angiotensin, prostaglandins, kallikrein-kinin; [16–18] and these interactions are often further complicated by the intervention of the autonomic and adenylate cyclase/cyclic AMP systems.[8]

(3) Anesthesia and acute surgical stress have been shown to severely affect glomerular filtration and renal blood flow, as well as the nervous and hormonal systems.[19, 20] Of necessity, most experiments are performed under

anesthesia. Caution should be applied when interpreting such experiments as they may not be representative of truly physiological conditions.

(4) Uncontrolled changes in the state of hydration of the experimental subjects may directly alter renal hemodynamics. Dehydration of the already volume-contracted DI rat results in decreased glomerular filtration.[21] On the other hand, infusion of vasopressin for a prolonged period, when accompanied by volume expansion, may lead to the syndrome of inappropriate ADH secretion.[22]

The Conscious, Trained, Chronically Catheterized DI Rat

The development by us of this new experimental approach[23] greatly enhanced our ability to study changes in renal functions during acute or chronic replacement with vasopressin. This model allows us to collect blood and urine samples, infuse drugs, measure blood pressure and heart rate, and regulate renal perfusion pressure in trained, conscious rats for weeks or months following surgical preparation. The advantage of having the same animal serve as its own control during a prolonged experimental protocol (38 days) is illustrated in Figure 1.

In the past few years, we have systematically examined the effects of oxytocin and arginine vasopressin on GFR and effective renal plasma flow (ERPF) in conscious DI rats. Detailed results from these experiments have been presented elsewhere.[24, 25] Briefly, infusion of physiological doses of oxytocin for 60 minutes resulted in an approximately 25% increase in GFR but no change in ERPF. Acute infusion of three different doses of vasopressin (2.5, 25, and 100 pg/min, yielding P_{AVP} of < 1.25, 2.3, and 8.0 pg/ml) had no effect on these variables; chronic (10 days) replacement of vasopressin on the other hand resulted in marked increases in GFR and ERPF. Additional observations during the chronic treatment included an increase in body weight, decreased hematocrit, increase in plasma potassium concentration, and decrease in plasma sodium concentration. An important difference between the acute and chronic experiments—and one that might serve to explain the contrasting findings—was in the control of fluid intake. Whereas in the acute experiments care was taken to keep the body weight of the rat constant, no such attempt was made during the chronic treatment when the rats were allowed to eat and drink ad libitum. Other investigators have shown that the DI rat differs from its parent strain, the Long-Evans normal rat, in many ways. For example, high renin-angiotensin levels,[26] decreased renal synthesis of prostaglandins,[27] decreased plasma corticosterone,[28] hypokalemia, and lower plasma volume[26] have all been observed in the DI rat; all of these conditions were corrected by prolonged vasopressin treatment.

These findings illustrate the difficulty of trying to identify the possible mechanism(s) by which vasopressin influences glomerular filtration and renal blood flow. It is possible that the increase in GFR and ERPF during chronic vasopressin replacement are secondary to changes in fluid or hormonal homeostasis brought on by the primary action of vasopressin, namely, of changing the water permeability of the distal tubules and collecting ducts. In order to better understand the physiological processes resulting from vasopressin replacement in the DI rat, experiments should be performed to assess the effect of increased fluid volume, increased plasma potassium concentration, decreased plasma sodium concentration, reducing renin-angiotensin levels, increasing prostaglandin synthesis, etc. on renal hemodynamics, all in the absence of vasopressin.

SUMMARY

It has been demonstrated through the use of new techniques that the action of vasopressin on the kidneys is not limited to changing the water permeability of distal tubules and collecting ducts. However, it has yet to be established whether these additional actions, such as lowering K_f (possibly by contracting

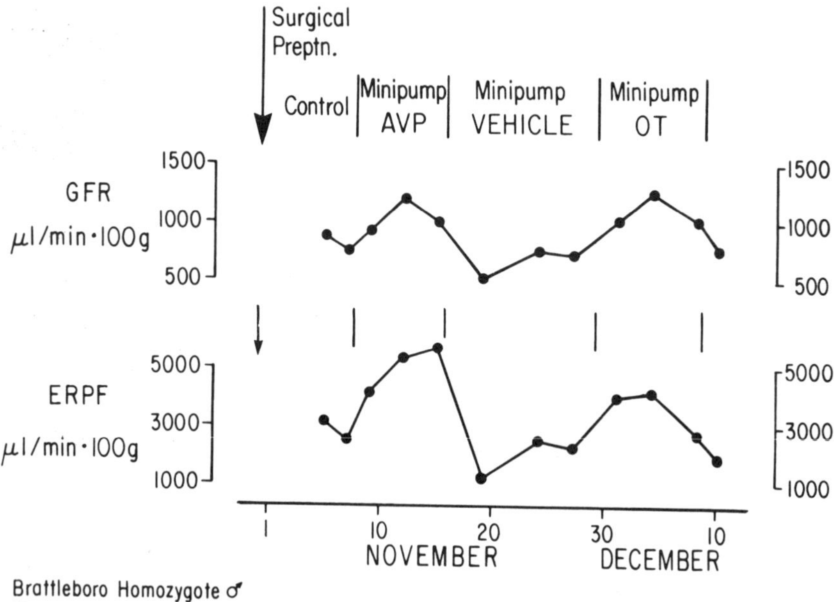

FIGURE 1. Changes in GFR and effective renal plasma flow (ERPF) in a conscious, trained DI rat, consecutively: during a four-day control period, vasopressin (AVP) treatment (for 10 days), recovery (12 days), oxytocin (OT) treatment (for 10 days), and recovery (two days). Subcutaneously implanted osmotic minipumps (Alzet) were used to deliver the hormones.

mesangial cells), or increasing postglomerular vascular resistance, are important factors in the control of GFR and renal blood flow.

The use of animals with diabetes insipidus, particularly the Brattleboro homozygous (DI) rat, may help to circumvent a number of methodological problems and provide a useful model for assessing the role of vasopressin in the control of renal hemodynamics. Although that role may be exerted through a direct effect on the vascular tone, it may be an indirect effect in which the antidiuretic action of vasopressin alters fluid balance and elicits secondary changes in other vasoactive hormones. The complexity of this latter possibility suggests that other methodological problems (in the measurement and/or control of the related variables) may complicate the final resolution of this issue for some time to come.

References

1. Aukland, K. 1976. Renal blood flow. Int. Rev. Physiol. **11:**23–79.
2. Aukland, K. 1980. Methods for measuring renal blood flow: total flow and regional distribution. Ann. Rev. Physiol. **42:**543–555.
3. Sato, T., R. Garcia-Bunuel & D. Brendes. 1974. Ultrastructural cytochemical localization of adenylate cyclase in the rat nephron. Lab. Invest. **30:** 222–229.
4. Imbert, M., D. Charbardés & F. Morel. 1974. Hormone-sensitive adenylate cyclase in isolated rabbit glomeruli. Mol. Cell. Endocrinol. **1:**295–304.
5. Sraer, J., R. Ardaillou, N. Loseau & J. D. Sraer. 1974. Evidence for parathyroid hormone sensitive adenylate cyclase in rat glomeruli. Mol. Cell. Endocrinol. **1:**285–294.
6. Ausiello, D. A., J. I. Kreisberg, C. Roy & M. J. Karnovsky. 1980. Contraction of cultured rat glomerular mesangial cells after stimulation with angiotensin II and arginine vasopressin. J. Clin. Invest **65:**754–760.
7. Ichikawa, I. & B. M. Brenner. 1977. Evidence for glomerular actions of ADH and dibutyryl cyclic AMP in the rat. Am. J. Physiol. **233:**F102–F117.
8. Ichikawa, I. & N. Schor. 1981. Humoral regulation of glomerular filtration rate: in vivo evidence. Proc. 8th Int. Congr. of Nephrol. Athens. 109–116.
9. Fisher, R. D., J.-P. Grünfeld & A. C. Barger. 1970. Intrarenal distribution of blood flow in diabetes insipidus: role of ADH. Am. J. Physiol. **219:**1348–1358.
10. Akatsuka, N., W. H. Moran, M. L. Morgan & M. F. Wilson. 1977. Effects of steady state plasma vasopressin levels on the distribution of intrarenal blood flow on electrolyte excretion. J. Physiol. (London) **266:**567–586.
11. Johnson, M. D., C. S. Park & R. L. Malvin. 1977. Antidiuretic hormone and the distribution of renal cortical blood flow. Am. J. Physiol. **232:**F111–F116.
12. Davis, J. M. & J. Schnermann. 1971. The effect of antidiuretic hormone on the distribution of nephron filtration rates in rats with hereditary diabetes insipidus. Pflügers Arch. **330:**323–334.
13. deRouffignac, C., S. Deiss & J. P. Bonvalet. 1970. Determination du taux individuel de filtration glomérulaire des nephrons accessibles et inaccessibles à la microponction. Pflügers Arch. **315:**272–290.
14. Trinh-Trang-Tan, M. M., M. Diaz, J.-P. Grünfeld & L. Bankir. 1981. ADH-dependent nephron heterogeneity in rats with hereditary hypothalamic diabetes insipidus. Am. J. Physiol. **240:**F372–F380.
15. Trinh-Trang-Tan, M. M., H. W. Sokol, L. Bankir & H. Valtin. 1982. Homozygous Brattleboro rats lack the normal nephron heterogeneity as a consequence of their urine concentrating defect. Ann. N.Y. Acad. Sci. (This volume.)
16. Dunn, M. J. & V. L. Hood. 1977. Prostaglandins and the kidney. Am. J. Physiol. **233:**F169–F184.
17. Levinsky, N. G. 1979. The renal kallikrein-kinin system. Circ. Res. **44:**441–451.
18. Levens, N. R., M. J. Peach & R. M. Carey. 1981. Role of the intrarenal renal-angiotensin system in the control of renal function. Circ. Res. **48:**157–167.
19. Gellai, M. & H. Valtin. 1981. Autoregulation of Glomerular Filtration and Renal Blood Flow in Conscious Rats. *In* Advances of Physiological Sciences. L. Takács, Ed. **11:**217–221. Akadéimai Kiadó. Budapest.

20. PEFFINGER, W. A., K. TANAKA, K. KEETON, W. B. CAMPBELL & S. N. BROOKS. 1975. Renin release, an artifact of anesthesia and its implications in rats. Proc. Soc. Exp. Biol. Med. **148:**625–630.

21. GELLAI, M., B. R. EDWARDS & H. VALTIN. 1979. Urinary concentrating ability during dehydration in the absence of vasopressin. Am. J. Physiol. **237:** F100–F104.

22. SMITH, M. J., JR., A. W. COWLEY, JR., A. C. GUYTON & R. D. MANNING, JR. 1979. Acute and chronic effects of vasopressin on blood pressure, electrolytes, and fluid volumes. Am. J. Physiol. **237:**F232–F240.

23. GELLAI, M. & H. VALTIN. 1979. Chronic vascular constrictions and measurements of renal function in conscious rats. Kidney Int. **15:**419–426.

24. GELLAI, M., F. T. LAROCHELLE, JR. & H. VALTIN. 1980. Effect of oxytocin on renal function in conscious Brattleboro homozygous rats. A model for studying renal regulation by hormones. *In* Hormonal Regulation of Sodium Excretion. B. Lichardus, R. W. Schrier & J. Ponec, Eds.: 121–128. Elsevier/ North Holland. Amsterdam.

25. LAROCHELLE, F. T., M. GELLAI, J.-C. HWANG & H. VALTIN. 1982. Effect of acute versus chronic treatment with vasopressin on renal function in conscious Brattleboro homozygotes. Ann. N.Y. Acad. Sci. (This volume.)

26. MÖHRING, J., G. KOHRS, B. MÖHRING, M. PETRI, E. HOMSY & D. HAACK. 1978. Effects of prolonged vasopressin treatment in Bratttleboro rats with diabetes insipidus. Am. J. Physiol. **234:**F106–F111.

27. WALKER, L. A., A. R. WHORTON, M. SMIGEL, R. FRANCE & J. FROLICH. 1978. Antidiuretic hormone increases renal prostaglandin synthesis in vivo. Am. J. Physiol. **235:**F180–F185.

28. MCCANN, S. M., J. ANTUNES-RODRIGUES, R. NALLAR & H. VALTIN. 1966. Pituitary-adrenal function in the absence of vasopressin. Endocrinol. **79:** 1058–1064.

DISCUSSION OF THE PAPER

T. BENNETT (*University of Nottingham Medical School, Nottingham, England*): You said that chronic administration of vasopressin to DI rats lowers blood pressure. Is that what you meant?

GELLAI: Yes. We cannot make a conclusive statement at this point because we have only studied three animals thus far. The blood pressure in those three animals in the control state was around 110 mm Hg, and during treatment decreased to around 98–100 mm Hg. This moderate decrease probably was caused by other factors, secondary to expanded volume.

I. W. HENDERSON (*University of Sheffield, Sheffield, England*): In the changes induced by vasopressin in superficial versus juxtamedullary single nephron filtration rates, where do you view its site of action? Is it an effect on the ultrafiltration coefficient, is it on the glomerular capillary, or is it on the afferent and/or efferent glomerular arteriole?

GELLAI: In the absence of reliable experimental evidence, one can only speculate regarding the site of action. Perhaps Dr. Bankir could comment on the subject.

L. BANKIR (*Hôpital Necker, Paris, France*): We only have hypotheses as to how this effect is triggered—no experimental data to confirm it. We suspect, as proposed 10 years ago by Davis and Schnermann, that it might be an increased resistance in the postglomerular vessels of the deep nephrons, i.e., the vasa recta, induced by the increase in osmolality in the papilla. Another hypothesis could be that after VP administration the feedback mechanism would be activated to a greater extent in deep nephrons than in superficial ones due to the effect of AVP on the thick ascending limbs, which are shorter in deep nephrons than in superficial nephrons. There might be other mechanisms, e.g. prostaglandins might be involved, but no one has been able to test this yet.

T. BERL (*University of Colorado Medical Center, Denver, Col.*): Is the plasma flow changed during the chronic infusion of vasopressin in these DI rats?

GELLAI: Yes, there is a marked increase as measured by PAH clearance.

BERL: The rats have a renal vein catheter. Are you doing extractions too?

GELLAI: Not in these rats. With the present state of the art, renal venous catheters can be successfully maintained patent only about 10 days; thus, we have not implanted them for these long-term studies. We are working on solving this problem.

BERL: Well, it's an unbelievable technical feat anyway.

IMPORTANCE OF AVP FOR BLOOD PRESSURE CONTROL DURING DEVELOPMENT: A STUDY IN THE BRATTLEBORO RAT

Anita Aperia, Peter Herin, Ann-Christine Eklöv, and Vivianne Johnsson

Department of Pediatrics
Karolinska Institute
St. Göran's Children's Hospital
Stockholm, Sweden

Introduction

The doses needed for obtaining a blood pressure effect with arginine vasopressin (AVP) are generally reported to be 10^3–10^4 times higher than the doses needed for an antidiuretic effect with AVP. The physiological importance of AVP in the control of blood pressure may therefore be questioned. Studies on animals subjected to moderate or large bleedings, however, imply that AVP is of physiological importance for blood pressure control during blood volume contraction.[1] Bleeding is one of the most potent stimuli for the endogenous release of AVP.[2–4] Bleeding results in an immediate large increase in plasma AVP concentration in the fetus [5, 6] although, at this stage of development, the antidiuretic effect of AVP is relatively unimportant.[7] The highest plasma levels of AVP occur in the perinatal period,[8–10] when they also show a correlation with the arterial blood pressure.[8]

In the present study we have re-examined the question of whether AVP exerts a physiological effect on blood pressure control during blood volume contraction, and whether this effect may be more pronounced during infancy. In order to demonstrate the physiological vasoactive effect of AVP, most of the studies were performed in rats of the Brattleboro strain, i.e., so-called diabetes insipidus (DI) rats, which lack the capacity to produce AVP.[11] For reference purposes, we also studied Sprague-Dawley (SD) rats, which had an intact hypothalamic pituitary system. We compared the effect of bleeding 0.5 ml/100 g on mean arterial blood pressure (MAP) in young and adult rats, and also determined the effect of continuous infusion of AVP during control conditions and during hemorrhagic hypotension. Since the results suggested that the pressor effect of AVP depends on the blood pressure level before AVP administration, we have examined more closely the relationship between the blood pressure effect of AVP and the initial MAP in experiments where bolus doses of AVP were administered.

Material and Methods

The experiments were performed on rats of the Brattleboro strain aged 20–22 days and 50–80 days and in Sprague-Dawley rats aged 20–22 days and 50–80 days. The animals were anesthetized with Inactin (Promonta) (8 mg/100 g bw in infant and adult rats).

* Supported by grants from the Swedish Medical Research Council (No. 3644).

0077–8923/82/0394–0350 $1.75/0 © 1982, NYAS

An endotracheal tube was inserted and a stream of 95% oxygen was passed in front of the endotracheal tube during the entire experiment. This resulted in a stable P_{O_2} around 12 kPa (1 kPa ≈ 7.52 mm Hg) and a stable pH and P_{CO_2}.[12] One carotid artery and one jugular vein were catheterized with polyethylene tubing of appropriate size for the vessels.

During the experiments the animals were given a continuous intravenous infusion of normal saline at a rate of 4 ml/100 g bw/h in DI rats and 1 ml/100 g bw/h in SD rats. With these infusion rates, the hematocrits were 30 ± 4.5 in infant DI rats, 36 ± 2.6 in adult DI rats, 32 ± 4.7 in infant SD rats, and 41 ± 1.1 in adult SD rats. Values are mean ± 1 S.D. We have previously reported that hematocrit in rats increases from infancy to adulthood.[13]

The animals were bled 0.5–1.0 ml/100 g from the carotid artery catheter for a period of 1.5–2 min. MAP was recorded with a Statham transducer connected to a Grass recorder. Towards the end of the experiments the hematocrit was determined by centrifuging microcaps at 5,000 rpm for 5 min.

Bleeding

Following 10 min of stable MAP recordings the animals were bled 0.5–1.0 ml/100 g for approximately 90 sec. The MAP was recorded for 5–10 min after the bleeding was terminated. The animals were then retransfused. Bleeding was usually repeated 10 min later.

AVP Administration

During most of the experiments the rats were given intravenous arginine vasopressin (Ferring) in varying doses. AVP was dissolved in 0.9% NaCl and 0.1% BSA (1 μU AVP = 2.5 pg AVP). In some of the experiments, AVP, 200–10,000 pg/100 g/min, was injected continuously. In bleeding experiments, the infusion was started 10 min before bleeding. In other experiments, AVP, 20–2,000 pg/100 g, was given as a bolus dose. The injection was given either during control conditions or 2 min after the bleeding was terminated.

Angiotensin

In some of the adult animals, angiotensin I in arterial blood was determined with a standard radioimmunoassay (NEN). The blood samples were taken during control conditions and 5 min after bleeding 1 ml/100 g was terminated.

Student's *t*-test and paired *t*-test have been used for statistical analysis. The values are given as mean ± 1 S.D.; p values less than 0.05 are considered significant.

Results

During control conditions, the MAP does not differ significantly in infant DI and SD rats (Table 1). As reported previously in studies on SD rats, the MAP increases significantly from the age of 20 days to adulthood.[14] The increase is more pronounced in DI rats than in SD rats, and the adult DI rats have a significantly higher MAP than the adult SD rats.

MAP During Bleeding

The effect of a moderate bleeding, e.g. 0.5 ml/100 g, on the MAP is shown in TABLE 2. For comparison between the groups, the MAP following bleeding has been recorded in percent of the MAP immediately before bleeding was started. In all groups bleeding resulted in a fall in the MAP, but this fall was significantly greater in infant DI rats than in adult DI rats, infant SD rats and adult SD rats. Between the latter three groups there was no significant difference in the MAP change at any time after the bleeding. When the adult DI rats were bled 1.0 ml/100 g there was a fall in the MAP similar to that recorded in infant DI rats bled 0.5 ml/100 g.

TABLE 1

MAP IN INFANT AND ADULT DI AND SD RATS DURING CONTROL (NON-BLED) CONDITIONS

	MAP (mm Hg)
Infant DI N = 34	106 ± 19*
Adult DI N = 34	164 ± 20†
Infant SD N = 10	104 ± 12*
Adult SD N = 13	141 ± 12

Values are mean ± 1 S.D.
* Significantly different from adult DI rats.
† Significantly different from adult SD rats.

MAP Following AVP Administration

Continuous infusion of AVP, 200 pg/100 g/min, significantly blunted the hemorrhagic hypotension in infant DI rats bled 0.5 ml/100 g (FIGURE 1) and in adult DI rats bled 1.0 ml/100 g (FIGURE 2). AVP, 10,000 pg/100 g/min, completely abolished the fall in the MAP during bleeding in infant DI rats and almost abolished it in adult DI rats.

The effect in infant DI rats of a bolus dose of AVP, 20 pg/100 g, given 2 min after bleeding was terminated, is shown in FIGURE 3. There was an immediate significant increase in the MAP. When a bolus dose of AVP, 200 pg/100 g, was given the MAP was restored to control values.†

† The control rats in FIGURE 3 (upper panel) were not identical with those in FIGURE 1.

TABLE 2

CHANGES IN MAP DURING THE FIRST TEN MINUTES AFTER BLEEDING

Experiment	MAP (% of Initial Value)				
	15 sec after bleeding	1 min after bleeding	2 min after bleeding	5 min after bleeding	10 min after bleeding
Infant DI, bled 0.5 ml/100 g N = 11	57.8 ± 2.3*	58.4 ± 3.5*	57.5 ± 4.2*	61.5 ± 5.9*	91.2 ± 3.7
Adult DI, bled 0.5 ml/100 g N = 9	81.9 ± 4.7	85.4 ± 4.9	88.5 ± 3.1	91 ± 5.5	97 ± 2.4
Infant SD, bled 0.5 ml/100 g N = 10	72.6 ± 5.4	82.2 ± 4.7	86.9 ± 5.2	96.2 ± 1.3	99.4 ± 0.2
Adult SD, bled 0.5 ml/100 g N = 5	81.3 ± 5.8	87.3 ± 4.7	92.8 ± 4.6	98 ± 3.5	100 ± 0
Adult DI, bled 1.0 ml/100 g N = 12	54.3 ± 4.73†	61.9 ± 4.43†	67.3 ± 4.16†	77.2 ± 3.75†	89.4 ± 2.41
Adult SD, bled 1.0 ml/100 g N = 9	52.9 ± 1.55‡	53.3 ± 6.20‡	54.3 ± 6.68‡	66.3 ± 6.66‡	76.7 ± 6.03

Values are mean ± 1 S.D.

* Significantly different from adult DI, infant SD, and adult SD rats bled 0.5 ml/100 g at corresponding times after bleeding.

† Significantly different from adult DI rats bled 0.5 ml/100 g but not significantly different from adult SD rats bled 1.0 ml/100 g at corresponding times after bleeding.

‡ Significantly different from adult SD rats bled 0.5 ml/100 g at corresponding times after bleeding.

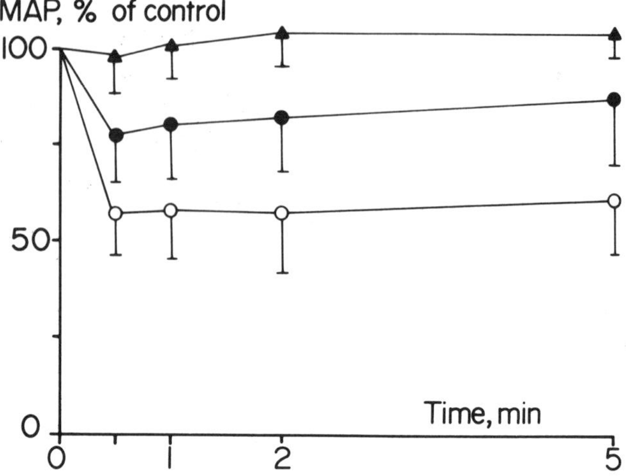

FIGURE 1. MAP in infant DI rats bled 0.5 ml/100 g during control conditions (open circles), during continuous infusion of AVP, 200 pg/100 g/min (closed circles), and 10,000 pg/100 g/min (closed triangles).

During control, i.e. non-bleeding conditions, continuous infusion of AVP, 200 pg/100 g/min, had no significant effect on the MAP in infant rats. Continuous infusion of AVP, 10,000 pg/100 g/min, increased the MAP 5–20% in infant DI rats. In adult and non-bled DI rats, the MAP response to continuous infusion of AVP ranged from −30 to +15%. Negative changes were generally noted in the rats with the most pronounced hypertension.

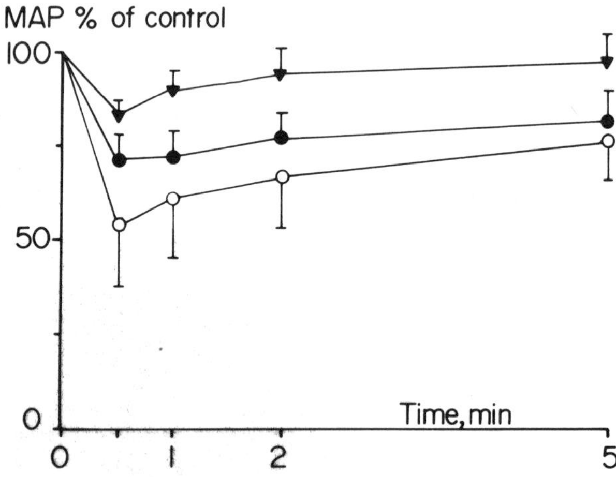

FIGURE 2. MAP in adult DI rats bled 1 ml/100 g during control conditions (open circles) and during the continuous infusion of AVP, 200 pg/100 g/min (closed circles) and 10,000 pg/100 g/min (closed triangles).

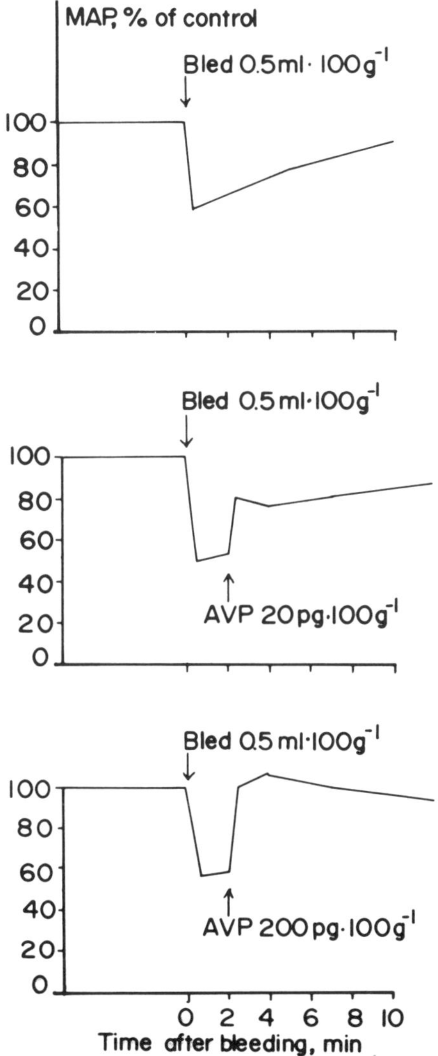

FIGURE 3. MAP in eleven infant DI rats bled 0.5 ml/100 g (upper panel). In seven studies (middle panel) a bolus dose of 20 pg/100 g AVP was given 2 min after the bleeding was terminated and in six studies (lower panel) a bolus dose of AVP 200 pg/ 100 g was given 2 min after the bleeding was terminated.

Relationship Between Pressure Effect of AVP and Initial MAP

Since the effect of AVP on the MAP appeared to depend on the initial MAP, this relationship was more closely examined by giving bolus doses of AVP to infant and adult DI rats during non-bleeding conditions, as well as following different degrees of bleeding in order to get a wide range of initial MAP (FIGURES 4 and 5). The effect of AVP on the MAP ranged from +55 to −28%. In the hypertensive adult DI rats, AVP generally produced a fall in blood pressure. Different doses of AVP were tested to establish whether the maximal pressure effect of AVP was recorded. A highly significant corre-

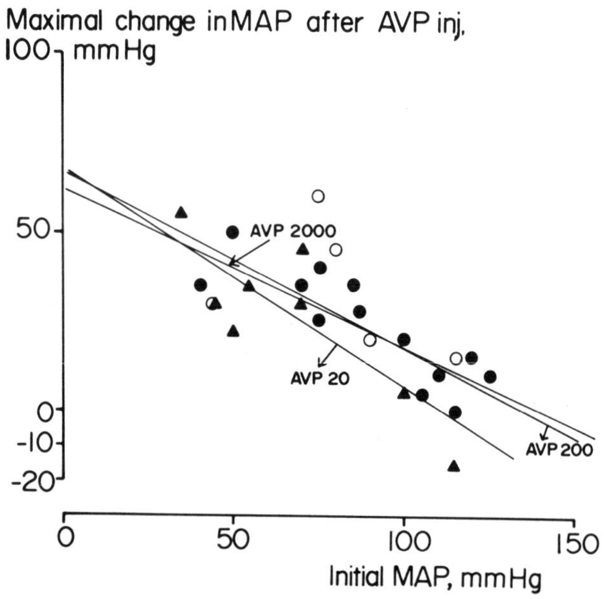

FIGURE 4. Relationship in infant DI rats between initial MAP and change in MAP following a bolus dose of AVP, 20 pg/100 g (closed triangles), 200 pg/100 g (closed circles), and 2000 pg/100 g (open circles).

lation between the initial MAP and the pressure effect of AVP was found in both infant and adult DI rats. The *r*-values were −0.90 for 20 pg AVP/100 g, −0.92 for 200 pg AVP/100 g, and −0.93 for 2,000 pg AVP/100 g. There was no significant difference between the correlation coefficients obtained for doses of 20, 200, and 2,000 pg/100 g, respectively. In contrast, the effect of AVP on the MAP in adult SD rats was unpredictable (FIGURE 6). In the majority of bled rats, AVP had little or no effect on MAP. The correlation between the initial MAP and the change in the MAP following 200 pg/100 g was 0.002.

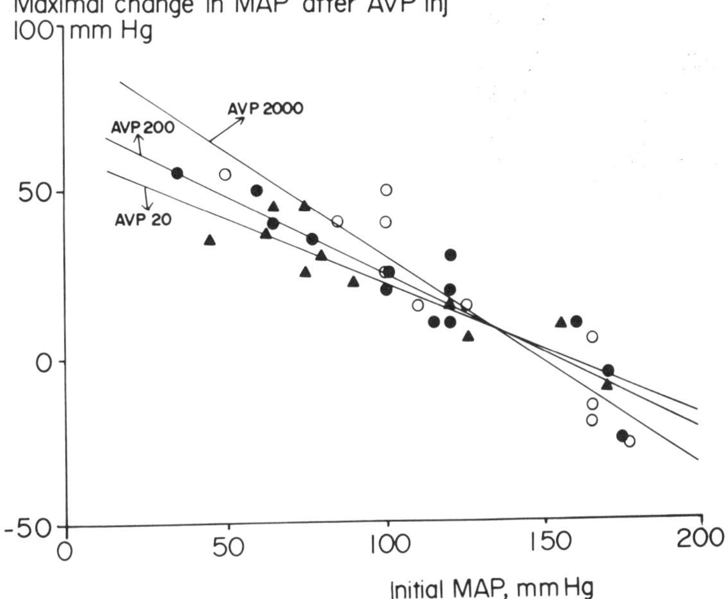

FIGURE 5. Relationship in adult DI rats between initial MAP and change in MAP following a bolus dose of AVP, 20 pg/100 g (closed triangles), 200 pg/100 g (closed circles), and 2,000 pg/100 g (open circles).

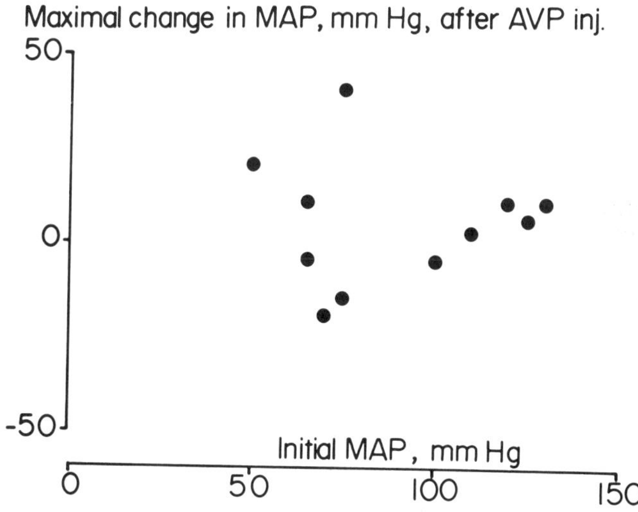

FIGURE 6. Relationship in adult SD rats between MAP and change in MAP following a bolus dose of AVP, 200 pg/100 g (closed circles).

Angiotensin Levels

The angiotensin levels in the adult DI and SD rats were determined in blood taken under control conditions and after bleeding, 1 ml/100 g (TABLE 3). There was no difference in angiotensin levels between DI and SD rats either during control conditions or following bleeding. In both DI and SD rats, bleeding significantly increased the angiotensin values.

DISCUSSION

Two findings in this study provide further evidence that AVP is of physiological importance in the control of blood pressure and that this effect is most apparent during blood volume contraction.

(1) In infant normotensive DI rats, the pressure fall following bleeding was more pronounced than in SD rats of corresponding age. The difference in pressure response can most likely be attributed to the absence or the presence

TABLE 3

ANGIOTENSIN I IN ADULT DI AND SD RATS DURING CONTROL CONDITIONS AND 5 MIN FOLLOWING BLEEDING (1 ml/100 g)

	Angiotensin I (ng/ml/h)	
	Control	Bled
Adult DI N = 4	10.4 ± 4.7	45.3 ± 16.2*
Adult SD N = 4	13.4 ± 6.0	42.0 ± 10.9†

Volues are mean ± 1 S.D.

* Significantly different from control DI rats, but not significantly different from bled SD rats.

† Significantly different from control SD rats.

of AVP. AVP is so far the only pressor factor that has been found lacking in DI rats of the Brattleboro strain.[11]

(2) Low doses of AVP, 200 pg/100 g/min, significantly increased the MAP during hemorrhagic hypotension in both infant and adult DI rats. This dose is ten times more than the threshold dose needed for the antidiuretic effect of AVP.[15] On the other hand, it has been reported that in unanesthetized dogs continuous infusion of 200 pg/100 g AVP results in a serum level of AVP around 30 pg/ml, and that serum AVP levels increases from 6.6 ± 2.3 to 48 ± 21 pg/ml [16] following moderate hemorrhage (20% of blood volume).

The adult DI rats were more resistant to hemorrhagic hypotension than the infant DI rats. Yet the sensitivity to AVP was the same in adult and infant DI rats when hypotension was well established. This implies that adult DI rats have developed a compensatory mechanism for maintaining blood pressure during mild blood volume contraction. This may reflect an increase in the circulating angiotensin levels, since moderate hemorrhage is followed by immediate increases not only in plasma AVP levels but also in plasma angiotensin levels.[2] This possibility, however, could be excluded. Angiotensin levels were the same in DI and SD rats both during control conditions and following bleeding.

The most significant new finding in this study was that the effect of AVP on the MAP in DI rats depended more on the initial MAP than on the AVP dose. At high MAP, AVP even resulted in a MAP fall. This implies that AVP has a dual effect on blood pressure control. It has already been suggested by studies on baroreceptor-denervated dogs that AVP can act both peripherally by causing vasoconstriction and centrally by enhancing the baroreceptor stimulus to the depressor vasomotor center.[17] The existence of peripheral vascular AVP receptors is well documented and the hormone receptor interaction has also been shown to result in smooth muscle contraction.[18] In baroreceptor-denervated dogs, the MAP rises and the vasopressor effect of AVP is greatly enhanced.[17,19,20] The results of the present study are in accord with the hypothesis that AVP has a predominantly peripheral vasoconstrictive action at low MAP and potentiates the baroreceptor stimulus to the depressor vasomotor center with increasing blood pressure. If the MAP is sufficiently increased, the central vasodilatation effect outweighs the peripheral vasoconstrictive effect of AVP, and this could explain why we generally observed a blood pressure fall when AVP was administered to hypertensive adult DI rats.

The hypertension in adult DI rats may well be due to resetting of the baroreceptor stimulus to the vasomotor center; but we did not examine this possibility. The finding that adult DI rats develop hypertension with advancing age has been documented previously.[21,22] Hall *et al.*[22] have shown that the hypertension in DI rats is probably attributable to lack of AVP. The greater tolerance to bleeding that we found in the adult than in the infant DI rats may also be related to hypertension. Laycock *et al.*[23] previously studied the effect of bleeding in adult *normotensive* DI rats and found that the fall in blood pressure following hemorrhage, 0.5–2 ml/100 g was significantly greater than the fall in blood pressure in Long-Evans rats having intact AVP production.

An almost maximal pressure effect of AVP appeared to be achieved by a bolus dose of 20 pg/100 g. This is a remarkably low dose; and it is possible that the lack of effect of AVP on the MAP in the SD rats with hemorrhagic hypotension was due to the fact that endogenously produced AVP already exerted a maximal vasoconstrictive action.

In summary, this study shows that one of the physiological effects of AVP is to balance the arterial blood pressure around a level that may be optimal for the organism by increasing the MAP during blood volume contraction and decreasing the MAP during hypertension. The importance of AVP for the prevention of hypotension during blood volume contraction seems greatest during infancy. A lack of AVP may reset other mechanisms responsible for blood volume control and thereby cause hypertension in adult life.

References

1. Szczepanski-Sadowska, E. 1973. Hemodynamic effects of a moderate increase of the plasma vasopressin level in conscious dogs. Pflügers Arch. **338:**313–322.
2. Claybaugh, J. R. & L. Share. 1973. Vasopressin, renin and cardiovascular response to continuous slow hemorrhage. Am. J. Physiol. **234**(3):519–523.
3. Robertson, G. L. 1977. The regulation of vasopressin function in health and disease. Rec. Prog. Horm. Res. **33:**333–385.

4. WEITZMAN, R. E., A. REVICZKY, T. H. ODDIE & D. A. FISCHER. 1980. Effect of osmolality on arginine vasopressin and renin release after hemorrhage. Am. J. Physiol. **238**(1):E62–68.
5. DRUMMOND, M. H., A. M. RUDOLPH, L. C. KEIL, P. D. GLÜCKMAN, A. A. MACDONALD & M. A. HEYMANN. 1980. Arginine vasopressin and prolactin after hemorrhage in fetal lamb. Am. J. Physiol. **238**(3):E214–219.
6. ROBILLARD, J. E., R. E. WEITZMAN, D. A. FISHER & F. G. SMITH, JR. 1979. The dynamics of vasopressin release and blood volume regulation during fetal hemorrhage in the lamb fetus. Pediat. Res. **13:**606–610.
7. SCHLONDORFF, D., H. WEBER, W. TRIZNA & L. G. FINE. 1978. Vasopressin responsiveness of renal adenylate cyclase in newborn rats and rabbits. Am. J. Physiol. **234**(1):F16–21.
8. POHJAVUORI, M. & F. FYHRQUIST. 1980. Hemodynamic significance of vasopressin in the newborn infant. J. Pediat. **97:**462–465.
9. POLIN, R. A., M. K. HUSAIN, L. S. JAMES & A. G. FRANTZ. 1977. High vasopressin concentrations in human umbilical cord blood—Lack of correlation with stress. J. Perinat. Med. **5:**114–119.
10. REES, L., M. L. FORSLING & C. G. D. BROOK. 1980. Vasopressin concentrations in the neonatal period. Clin. Endocrin. **12:**357–362.
11. VALTIN, H. 1976. Animal model of human disease. Hereditary hypothalamic diabetes insipidus in the Brattleboro strain of rat. Am. J. Pathol. **83:**633–636.
12. ELINDER, G. & A. APERIA. 1981. Renal oxygen consumption and sodium reabsorption during isotonic volume expansion in the developing rat. Pediat. Res. (In press.)
13. APERIA, A. & G. ELINDER. 1981. Distal tubular Na reabsorption in the developing rat kidney. Am. J. Physiol. **240:**F487–491.
14. APERIA, A. & P. HERIN. 1975. Development of glomerular perfusion rate and nephron filtration rate in rats 17–60 days old. Am. J. Physiol. **228:**1319–1325.
15. HARMANCI, M. C., P. STERN, W. A. KACHADORIAN, H. VALTIN & V. A. DISCALA. 1980. Vasopressin and collecting duct intramembranous particle clusters: a dose-response relationship. Am. J. Physiol. **239:**F561–564.
16. PULLAN, P. T., C. I. JOHNSTON, W. P. ANDERSSON & P. I. KORNER. 1980. Plasma vasopressin in blood pressure homeostasis and in experimental renal hypertension. Am. J. Physiol. **239:**H81–87.
17. MONTANI, J.-P., J.-F. LIARD, J. SCHOUN & J. MÖHRING. 1980. Hemodynamic effects of exogenous and endogenous vasopressin at low plasma concentrations in conscious dogs. Circ. Res. **47:**346–355.
18. AUSIELLO, D. A., J. I. KREISBERG, C. ROY & M. J. KARNOVSKY. 1980. Contraction of cultured rat glomerular cells of apparent mesangial origin after stimulation with angiotensin II and arginine vasopressin. J. Clin. Invest. **65:** 754–760.
19. COWLEY, JR., A. W., E. MONOS & A. C. GUYTON. 1974. Interaction of vasopressin and the baroreceptor reflex system in the regulation of arterial blood pressure in the dog. Circ. Res. **34:**505–514.
20. COWLEY, JR., A. W., S. J. SWITZER & M. M. GUINN. 1980. Evidence and quantification of the vasopressin arterial pressure control system in the dog. Circ. Res. **46:**59–67.
21. DLOUHÁ, H. J., KŘEČEK & J. ZICHA. 1977. Hypertension in rats with hereditary diabetes insipidus. The role of age. Pflüfiers Arch. **369:**177–182.
22. HALL, C. E., S. AYACHI & O. HALL. 1973. Spontaneous hypertension in rats with hereditary hypothalamic diabetes insipidus (Brattleboro Strain). Texas Rep. Biol. Med. **31:**471–487.
23. LAYCOCK, J. F., W. PENN, D. G. SHIRLEY & S. J. WALTER. 1979. The role of vasopressin in blood pressure regulation immediately following acute haemorrhage in the rat. J. Physiol. **296:**267–275.

Discussion of the Paper

B. M. Altura (*Downstate Medical Center, SUNY, New York, N.Y.*): I would just like to make a couple of comments for the people in the audience who may not be aware of certain things about vasopressin in the vascular system. Vasopressin at 10^{-6} to 10^{-7} nM will preferentially constrict venules in many vascular beds, long before the threshold of the arterioles is obtained. Therefore, we must think in terms of capacitance actions of vasopressin, i.e. what they may do to hemodynamics. It is not necessary to have a blood pressure increase to observe constrictor effects of vasopressin.

The second point is that in your studies you have utilized animals that are 20 days old. In piglets aged 1–12 days, vasopressin in low concentration will contract the vasculature in the renal bed until about 5 or 6 days after birth. I think this is very important, since the vascular receptors have not really been developed by that time. This is very early—earlier than your 20-day study. It is important to keep this in mind when discussing this area.

Herin: I think from a renal physiological point of view the 20-day old rat corresponds fairly well to the 35–36th gestational week in man.

T. Bennett (*University of Nottingham Medical School, Nottingham, England*): We have recently found that if you hemorrhage DI rats (and others have found this also) they show a poor recovery of blood pressure following the hemorrhage, compared to Long-Evans rats. If you pretreat them with naloxone, recovery is very much improved. Do you have any evidence that would support that in your infant rats?

The other point is that I believe Sjostrand showed quite some time ago that Sprague-Dawley rats respond quite differently to hemorrhage from various other strains of rats. I believe you used Sprague-Dawley rats as your controls.

Herin: I can agree that Sprague-Dawleys are not the best control for these studies, and I think we will have to start using Long-Evans rats. Concerning the naloxone and recovery from hemorrhage, I have no information.

T. Berl (*University of Colorado Medical Center, Denver, Col.*): Do you have any ideas as to how the naloxone may work?

Bennett: These are very preliminary experiments. Intuitively one is imagining that the absence of vasopressin in the homozygote DI rat unveils the depressor effect of endogenous opiates. But the action of naloxone may be much less specific than that.

C. I. Johnston (*Prince Henry's Hospital, Melbourne, Australia*): I wonder if you could tell us how specific the vasoconstrictor response was for vasopressin? Did you give angiotensin II and see if the blood pressure could be brought back to normal as well as change the position of the baroreceptor curves?

Herin: We didn't give angiotensin, although we measured it in control and bleeding conditions. The values were the same in DI and normal rats, which surprised us somewhat. There was a significant increase, of course, during hemorrhage, but the levels of angiotensin in both types of rats rose similarly.

J. Haldar (*Columbia University, New York, N.Y.*): In regard to naloxone, did you do this only in DI rats, or in normal rats also?

BENNETT: We have done it in Long-Evans rats and DI rats.

HALDAR: Naloxone also helps in Long-Evans rats, doesn't it?

BENNETT: Not in our hands. The Long-Evans rat doesn't, in fact, show the impairment of blood pressure recovery.

HALDAR: But I think Holaday has reported that in spinal and hemorrhagic shock, pretreatment with naloxone, intravenous or intraventricular, reverses the cardiovascular changes due to shock.

BENNETT: That's in Sprague-Dawley, not Long-Evans rats.

THE CONTROL OF HEART RATE IN RATS WITH HEREDITARY HYPOTHALAMIC DIABETES INSIPIDUS (BRATTLEBORO STRAIN)*

Sheila M. Gardiner and T. Bennett

Department of Physiology and Pharmacology
University of Nottingham Medical School
Queen's Medical Centre
Nottingham NG7 2UH, England

Introduction

There is increasing evidence to show that vasopressin impinges on cardiovascular control mechanisms in several ways other than through its effects on the renal handling of water and urea.

For instance, the quantity of vasopressin released during hemorrhage is sufficient to elicit a pressor response in baroreceptor-denervated dogs [1,2] and even sub-pressor levels can induce vasoconstriction in certain vascular beds.[3–5] Furthermore, the pressor action of vasopressin is important in maintaining arterial blood pressure in states of water deprivation [6] and in certain forms of hypertension.[7,8]

Brattleboro rats have a congenital inability to synthesize vasopressin and are therefore an interesting model in which to study the involvement of vasopressin in cardiovascular homeostasis. Arterial blood pressure is within the normal range in Brattleboro rats but they are reported to have a resting tachycardia.[9,10] Since, in the absence of vasopressin, there is a profound diuresis, Hall *et al.*[9] suggested that the tachycardia was a reflex response to a chronic state of volume depletion. Other workers [11] observed increased activity of dopamine-β-hydroxylase in the serum of Brattleboro rats which was corrected by administration of vasopressin; this they took as evidence for sympathetic hyperactivity due to volume depletion. If the resting tachycardia was due to baroreflex responses to volume depletion, one would expect a concurrent reduction in vagal tone. We have therefore studied the relative importance of sympathetic and parasympathetic activity in the control of heart rate in Brattleboro rats compared to rats of the parent strain (Long-Evans).

Vasopressin can influence heart rate by mechanisms independent of changes in fluid balance. Pullan *et al.*[12] and Montani *et al.*[3] found that the infusion of vasopressin into dogs caused a reduction in heart rate, even when the concentration of the peptide was too low to affect arterial blood pressure. Montani and his co-workers [3] suggested that vasopressin may interact with the barorecepor feedback system to give "greater inhibition of sympathetic tone and/or greater stimulation of vagal tone for a given pressure at the receptor level." It is possible, therefore, that the tachycardia in Brattleboro rats is related directly to a lack of vasopressin. In the present work we have assessed the baroreceptor reflex control of heart rate in Brattleboro rats and Long-Evans

* Supported by the British Heart Foundation and the Medical Research Council.

0077-8923/82/0394-0363 $1.75/0 © 1982, NYAS

rats and have studied the effects of acute administration of vasopressin on blood pressure, heart rate, and baroreceptor reflex sensitivity.

Methods

Male homozygous Brattleboro rats (body weight 260–330 g) and age-matched, male Long-Evans rats (body weight 320–380 g) were used in this study.

Rats were anesthetized with sodium methohexitone (Brietal, Lilly; 60 mg/kg intraperitoneally). Catheters, filled with weakly heparinized saline (12.5 U/ml), were implanted in the right jugular vein for drug administrations and in the abdominal aorta via the caudal artery for arterial blood pressure recording (for full details of the design of catheter and recording systems see Reference 13). The catheters were fed subcutaneously through a hollow needle, exteriorized at the back of the neck, and led through a fine spring (attached to a harness worn by the rat) and finally out of the cage. Systolic and diastolic blood pressures were recorded on a U.V. recorder (SE, 3006) using a Bell and Howell pressure transducer (Type 4-442) and an E.M.M.A. (Electromedical Multi-channel amplifier) system. Heart rate was also recorded on the U.V. recorder by way of a heart-rate meter (Devices) triggered by the arterial pulse. The animals were allowed to recover from the anesthetic for at least 5 h before any measurements were made and were given free access to water throughout the day. Drugs were dissolved in 0.9% saline, and administered in a volume of 0.1 ml flushed in with a further 0.1 ml saline.

Measurements were made for 30 min in the resting state and the mean levels recorded during the last 15 min were taken as the baseline values.

Experiment 1. The Effects of Autonomic Blockade on Blood Pressure and Heart Rate in Brattleboro Rats (N = 5) and Long-Evans Rats (N = 5)

Intravenous injections of propranolol (1 mg/kg) and atropine (1 mg/kg) were given at 20 min intervals; measurements were made when the responses had stabilized.

Experiment 2. The Effects of Vasopressin on Blood Pressure and Heart Rate in Brattleboro Rats and in Long-Evans Rats

Blood pressures and heart rates were measured before and during a slow infusion of vasopressin (10 mU/ml; 1.5 ml/h for 40 min) in Long-Evans rats (N = 6) and Brattleboro rats (N = 10).

Experiment 3. The Influence of Autonomic Blockade on the Heart Rate Response to Vasopressin in Brattleboro Rats

In this experiment, two catheters were implanted in the jugular vein. Three groups of Battleboro rats (N = 5 in each) were used. *Group 1* was given an intravenous infusion of vasopressin (10 mU/ml; 1.5 ml/h) for 40 min, followed by bolus injections of propranolol (1 mg/kg), and then atropine (1 mg/kg) at 15 min intervals while the vasopressin infusion was continued.

Group II received propanolol (1 mg/kg) 15 min before the start of the vasopressin infusion, and atropine (1 mg/kg) 40 min later. *Group III* were given atropine (1 mg/kg) first, followed by vasopressin infusion, and then propranolol (1 mg/kg) 40 min later.

Experiment 4. Baroreflex Control of Heart Rate in Brattleboro Rats and Long-Evans Rats

Baroreceptor reflex sensitivity was assessed by relating systolic blood pressure to the pulse interval of the succeeding beat[14] during an increase in pressure induced by infusion of methoxamine (0.4 mg/ml; 0.2 ml/min for 15 sec) or during a decrease in pressure induced by infusion of glyceryl trinitrate (0.8 mg/ml; 0.2 ml/min). Different groups of rats were used for the pressor (12 Brattleboro rats; 6 Long-Evans rats) and depressor (5 Brattleboro rats; 5 Long-Evans rats) tests.

Experiment 5. The Effects of Vasopressin on Baroreceptor Reflex Sensitivity in Brattleboro Rats

Two catheters were implanted in the jugular vein, one for the administration of a vasoactive agent for the baroreflex test, and one for the simultaneous infusion of vasopressin. Baroreflex sensitivity was assessed as in Experiment 2 in the control state. Then a slow infusion of vasopressin was begun (10 mU/ml; 1.5 ml/h), blood pressures and heart rates were recorded continuously, and 40 min later the baroreflex test was repeated. As before, different groups of rats were used for the pressor ($N = 5$) and depressor ($N = 5$) tests. (Note: the same Brattleboro rats were used in Experiments 2 and 5)

Drugs

The following drugs were used: vasopressin (Pitressin; Parke, Davis and Company); glyceryl trinitrate (Macarthays); methoxamine hydrochloride (Vasoxyl; Burroughs, Wellcome and Company); propranolol hydrochloride (Inderal; Imperial Chemical Industries Ltd.); and atropine sulfate (Sigma).

Statistics

Values are expressed as the mean ± 1 SE; N is the number of animals. Differences were tested for statistical significance using Students' paired or unpaired *t*-test.

Results

Experiment 1. The Effects of Autonomic Blockade on Blood Pressure and Heart Rate in Brattleboro Rats and Long-Evans Rats

In the resting state, there were no significant differences between the arterial blood pressures of Brattleboro rats ($151 \pm 1.8/96 \pm 4.5$ mm Hg; systolic/diastolic) and Long-Evans rats ($154 \pm 2/104 \pm 4$ mm Hg) but the heart rate of the Brattleboro rats (393 ± 7 beats/min) was significantly higher ($p < 0.001$) than that of the Long-Evans rats (337 ± 15 beats/min; Figure 1).

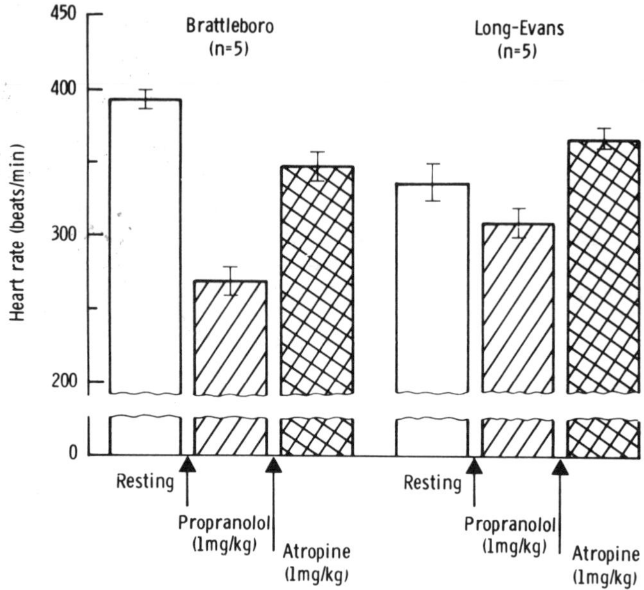

FIGURE 1. The effect of autonomic blockade on heart rate (mean ± SEM) in Brattleboro rats and Long-Evans rats. Resting heart rate was significantly ($p < 0.001$) higher in the Brattleboro rats than in the Long-Evans rats. Propranolol caused a greater fall in heart rate in the Brattleboro rats than in the Long-Evans rats whereas atropine caused a similar tachycardia in both groups.

Propranolol caused a reduction in heart rate in both groups but the change was much greater in the Brattleboro rats (—128 ± 12 beats/min) than in the Long-Evans rats (—27 ± 5 beats/min; FIGURE 1). Subsequent administration of atropine increased the heart rate to a similar extent in both groups (Brattleboro rats, +76 ± 6 beats/min; Long-Evans rats, +58 ± 6 beats/min, FIGURE 1). Thus in the presence of propranolol and atropine, the heart rate of the Brattleboro rats (346 ± 11 beats/min) was less than the resting value whereas that of the Long-Evans rats (368 ± 7 beats/min) was higher (FIGURE 1).

Neither propranolol nor atropine significantly affected the arterial blood pressures of the Brattleboro rats or the Long-Evans rats.

Experiment 2. The Effects of Vasopressin on Blood Pressure and Heart Rate in Brattleboro Rats and Long-Evans Rats

Forty minutes after the start of a vasopressin infusion in Long-Evans rats, systolic blood pressure had increased by 8 ± 0.8 mm Hg and heart rate had fallen by 23 ± 3 beats/min. In contrast, vasopressin infusion in Brattleboro rats increased systolic blood pressure by 18 ± 8 mm Hg and lowered heart rate by 87 ± 13 beats/min; the onset of the change in heart rate in the Brattleboro rats preceded the onset of any change in blood pressure (FIGURE 2).

Experiment 3. The Influence of Autonomic Blockade on the Heart Rate Response to Vasopressin in Brattleboro Rats

Group I (FIGURE 3). Vasopressin infusion caused a large fall in heart rate (—81 ± 9 beats/min) with a relatively small increase in systolic blood pressure (+20 ± 8 mm Hg; cf. Exp. 2). Subsequent administration of propranolol caused only a small further reduction in heart rate (—15 ± 5 beats/min), markedly less than in Experiment 1 when propranolol was given alone. Finally, atropine increased the heart rate by 64 ± 5 beats/min, a change similar to that observed in Experiment 1 in the absence of vasopressin.

Group II (FIGURE 4). Propranolol given before any other drug lowered the heart rate by 74 ± 8 beats/min (cf. Experiment 1). Infusion of vasopressin in the presence of propranolol increased systolic blood pressure by 13 ± 7 mm Hg but caused only a small further reduction in heart rate (—5 ± 2 beats/min.) As before, the heart rate response to atropine in the presence of propranolol and vasopressin (+57 ± 8 beats/min) was similar to that observed in Experiment 1 in the absence of vasopressin.

Group III (FIGURE 5). Intravenous administration of atropine prior to any other drug increased the heart rate by 75 ± 6 beats/min. During the vasopressin infusion, systolic blood pressure increased from 144 ± 5 mm Hg

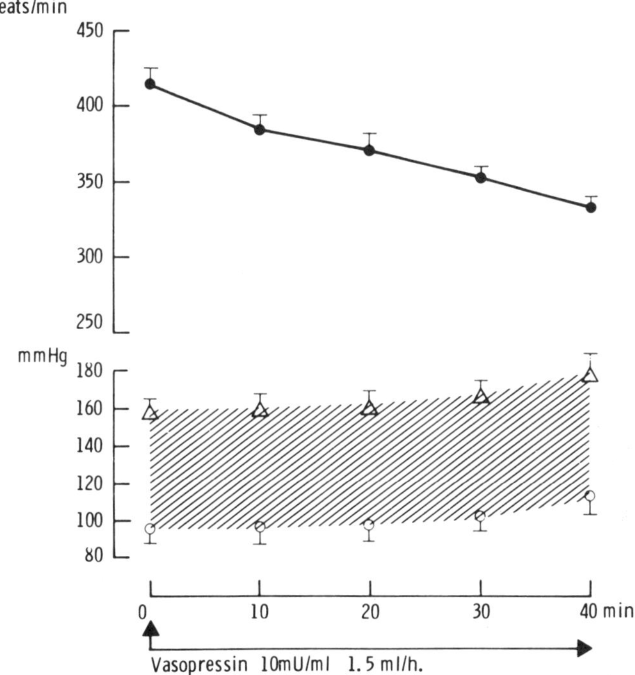

FIGURE 2. The effects of vasopressin infusion on systolic (△) and diastolic (○) blood pressures and heart rate (●) in Brattleboro rats. Values represent the mean ± 1 SEM. Note, the onset of the bradycardia occurred at the start of the vasopressin infusion, prior to any change in arterial blood pressure.

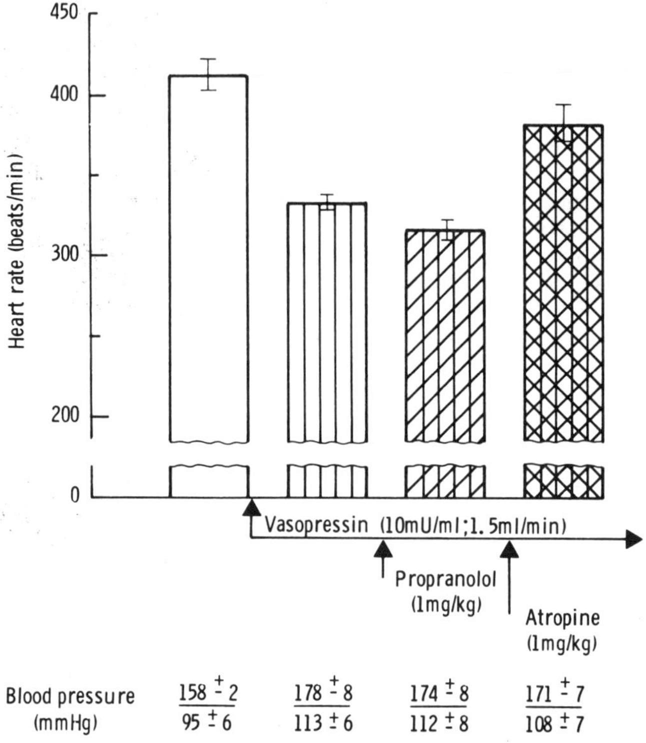

FIGURE 3. Blood pressure and heart rate responses to vasopressin infusion followed by autonomic blockade in Brattleboro rats (mean ± SEM, $N = 5$). Vasopressin infusion caused a large fall in heart rate with a relatively small increase in blood pressure. Subsequent administration of propranolol caused only a small further reduction in heart rate (markedly less than in the absence of vasopressin). The heart rate response to atropine was unaffected by vasopressin.

to 154 ± 6 mm Hg but heart rate fell by 101 ± 20 beats/min. Finally propranolol, in the presence of atropine and vasopressin, caused only a small further reduction in heart rate (—16 ± 8 beats/min), which was not significant.

Experiment 4. Baroreflex Control of Heart Rate in Brattleboro Rats and Long-Evans Rats

In the text, the slope of the line relating systolic blood pressure to the pulse interval of the succeeding beat during drug-induced changes in blood pressure is termed the baroreflex sensitivity.

The baroreflex sensitivity obtained from the pulse-interval response to an increase in pressure induced by methoxamine was significantly ($0.05 > p > 0.02$) less in the Brattleboro rats (1.02 ± 0.09 msec/mm Hg) than in the Long-Evans rats (1.41 ± 0.15 msec/mm Hg).

There was no significant difference between the response to a reduction in blood pressure induced by glyceryl trinitrate in Brattleboro rats and Long-Evans rats. (Brattleboro rats = 0.98 ± 0.13 msec/mm Hg; Long-Evans rats = 1.18 ± 0.09 msec/mm Hg).

Experiment 5. The Effects of Vasopressin on Baroreceptor Reflex Sensitivity in Brattleboro Rats

Vasopressin infusion did not alter the pulse interval response to an increase in blood pressure induced by methoxamine in Brattleboro rats (before vasopressin = 1.10 ± 0.16 msec/mm Hg; during vasopression infusion = 1.04 ± 0.10 msec/mm Hg) but enhanced the pulse interval response to a depressor stimulus (before vasopressin =0.96 ± 0.14 msec/mm Hg; during vasopressin infusion = 1.44 ± 0.16 msec/mm Hg; $0.05 > p > 0.02$).

Discussion

The present results have confirmed that Brattleboro rats have a normal arterial blood pressure but a resting tachycardia. In the presence of propranolol

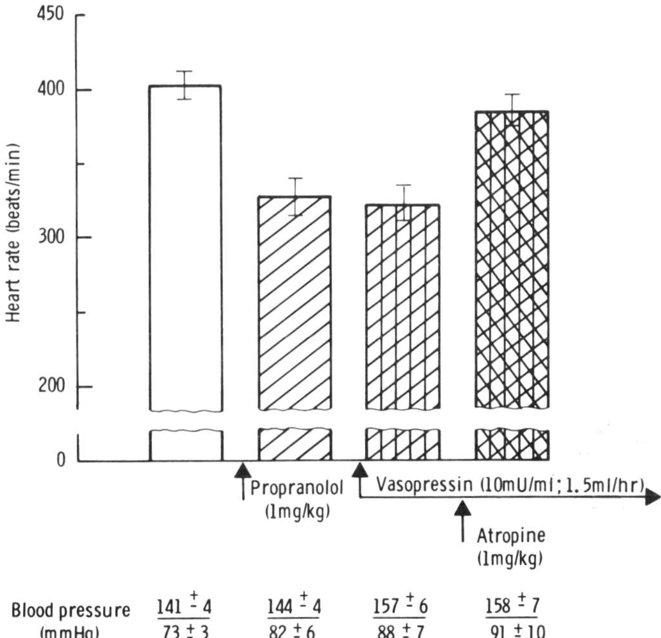

FIGURE 4. Blood pressure and heart rate responses to vasopressin after prior administration of propranolol to Brattleboro rats (mean ± SEM, $N = 5$). Propranolol caused a pronounced bradycardia. Subsequent administration of vasopressin caused no further change in heart rate although systolic blood pressure increased by 13 mm Hg. The presence of vasopressin did not affect the heart rate response to atropine.

and atropine, the 'intrinsic' heart rate of the Brattleboro rats is lower than that of the Long-Evans rats; thus the tachycardia does not appear to be due to a fundamental change in the myocardium.

Hall *et al.*[9] suggested that the high heart rate in the Brattleboro rats was a reflex response to volume depletion but, on the basis of the present work, this seems unlikely for the following reasons.

Firstly, we have shown that sympathetic hyperactivity is largely responsible for the tachycardia, with no evidence of a concurrent reduction in vagal tone; the latter would be expected to be part of the reflex response to volume depletion.

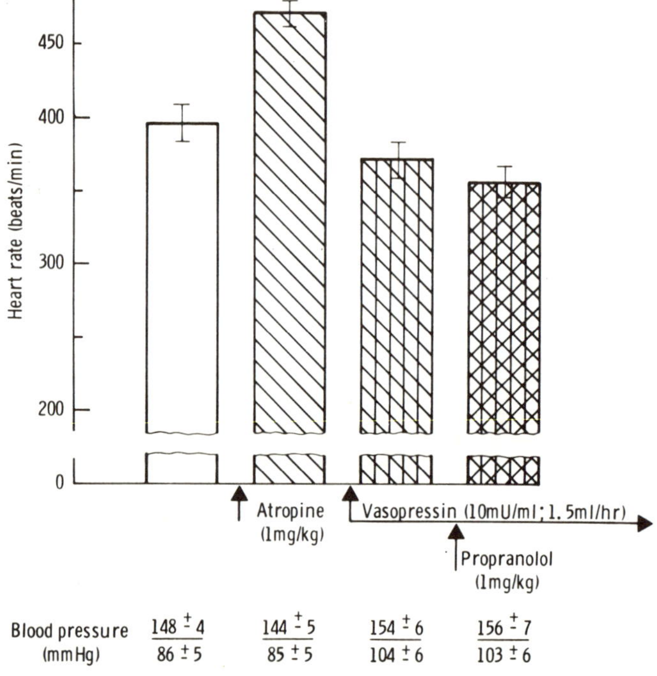

FIGURE 5. Blood pressure and heart rate responses to vasopressin after prior administration of atropine to Brattleboro rats (mean ± SEM; $N=5$). Atropine caused an increase in heart rate with no change in arterial blood pressure. The bradycardia in response to vasopressin in the presence of atropine was similar to that observed in the control state. Vasopressin markedly reduced the heart rate response to propranolol.

Secondly, if the cardiac sympathetic hyperactivity was an attempt to sustain arterial blood pressure then β-adrenoceptor antagonism should be accompanied by hypotension. But in the present work, propranolol caused a profound bradycardia in Brattleboro rats with no significant change in arterial blood pressure. It could be argued that in the presence of propranolol, arterial blood pressure was sustained by a reflex increase in peripheral vascular resistance in response to a fall in cardiac output.[15] If this was the case, in the presence of propranolol, an α-adrenoceptor antagonist ought to cause a marked lowering of blood pressure. But we have shown (unpublished data) that in the presence of propranolol and atropine, α-adrenoceptor antagonism with phenoxybenzamine causes a similar lowering of blood pressure in Long-Evans rats (—24 mm Hg) and Brattleboro rats (—20 mm Hg).

Thirdly, the present results demonstrate that the tachycardia can be reversed by the acute administration of vasopressin. This effect occurred within 40 minutes of the onset of treatment and therefore is unlikely to have been due to volume expansion as a result of an acute antidiuresis.

The bradycardia that occurred in response to vasopressin in the Long-Evans rats can largely be explained by a reflex response to the increase in pressure. Thus, the slope of the line joining the baseline values of systolic blood pressure and pulse interval to those at the peak of the pressure change induced by vasopressin is 1.63 msec/mm Hg; when methoxamine is the pressor agent, the corresponding value is 1.41 msec/mm Hg. But, in the Brattleboro rats, although vasopressin caused a greater increase in arterial blood pressure, the reflex bradycardia was disproportionately large. The pulse interval response to the increase in pressure induced by vasopressin was 1.90 msec/mm Hg but when methoxamine was used the corresponding value was only 0.96 msec/mm Hg.

Other workers have shown that acute administration of vasopressin causes a bradycardia that cannot be accounted for by any change in blood pressure.[3, 12, 16, 17] Pullan *et al.*[12] demonstrated that the bradycardia persisted after total autonomic blockade and concluded that it was due to a direct negative chronotropic action of the peptide. Varma *et al.*[17] also concluded that, in the absence of any change in blood pressure, part of the bradycardia with vasopressin was a direct myocardial depressant effect, but suggested, in addition, that vasopressin lowered heart rate by directly stimulating the cardioinhibitory neurons in the region of the vagal nuclei in the medulla. However Yourmans *et al.*[18] demonstrated that vasopressin still lowered heart rate in animals in which vagal cardioinhibitory responses had been completely abolished. They suggested that vasopressin may exert "a blocking action somewhere in the cardioaccelerator pathway within the nervous system or at the level of the cardioaccelerator center".[18] In the present study, propranolol abolished the heart rate response to vasopressin whereas atropine was without effect. Thus it is unlikely that the bradycardia was due to direct myocardial depression or that stimulation of the vagal cardioinhibitory center was involved. The results we obtained are most readily accounted for by vasopressin having an inhibitory action on sympathetic activity. Since vasopressin enhances peripheral sympathetic effects[19] we suggest that the present findings are best explained by vasopressin acting centrally to inhibit sympathetic tone.

The observation that vasopressin had much less marked effects on Long-Evans rats is in line with the lower levels of cardiac sympathetic tone in these animals (see Experiment 1). Our findings are consistent with those of Bohus[20] who demonstrated a suppressant effect of vasopressin on brain stem centers controlling autonomic function, and with reports of altered catecholamine concentrations and turnover in medullary and hypothalamic regions involved in cardiovascular control in Brattleboro rats.[21]

Altered catecholamine metabolism in the nucleus tractus solitarius[21] may also account for the abnormalities in the baroreflex control of heart rate in Brattleboro rats, which we have presently reported. However, two findings that are difficult to reconcile are that vasopressin, which in the control state appears to be inhibiting sympathetic drive, enhanced the baroreflex response to a decrease in pressure, and did not influence the baroreflex response to a pressor stimulus.

References

1. Rocha e Silva, M. Jr. & M. Rosenberg. 1969. The release of vasopressin in response to haemorrhage and its role in the mechanism of blood pressure regulation. J. Physiol. (London) **202:**535–557.
2. Szczepanski-Sadowska, E. 1973. Hemodynamic effects of a moderate increase of the plasma vasopressin level in conscious dogs. Pflügers Arch. **338:**313–322.
3. Montani, J-P., J-F. Liard, J. Schoun & J. Möhring. 1980. Hemodynamic effects of exogenous and endogenous vasopressin at low plasma concentrations in conscious dogs. Circ. Res. **47:**346–355.
4. Schmid, P. G., F. M. Abboud, M. G. Wendling, E. S. Ramberg, A. L. Mark, D. D. Heistad & J. W. Eckstein. 1974. Regional vascular effects of vasopressin: plasma levels and circulatory responses. Am. J. Physiol. **227:**998–1004.
5. Hoffman, W. E. 1980. Regional vascular effects of antiduiretic hormone in normal and sympathetic blocked rats. Endocrinology **107:**334–341.
6. Andrews, C. E. Jr. & B. M. Brenner. 1981. Relative contributions of arginine vasopressin and angiotensin II to maintenance of systemic arterial pressure in the anesthetized water-deprived rat. Circ. Res. **48:**254–258.
7. Möhring, J., B. Möhring, M. Petri & D. Haack. 1977. Vasopressor role of ADH in the pathogenesis of malignant DOC hypertension. Am. J. Physiol. **232:** F260–F269.
8. Möhring, J., B. Möhring, M. Petri & D. Haack. 1978. Plasma vasopressin concentrations and effects of vasopressin antiserum on blood pressure in rats with malignant two-kidney Goldblatt hypertension. Circ. Res. **42:** 17–22.
9. Hall, C. E., S. Ayachi & O. Hall. 1973. Spontaneous hypertension in rats with hereditary hypothalamic diabetes insipidus (Brattleboro strain). Tex. Rep. Biol. Med. **31:** 471–487.
10. Laycock, J. F., W. Penn, D. G. Shirley & S. J. Walter. 1979. The role of vasopressin in blood pressure regulation immediately following acute haemorrhage in the rat. J. Physiol. (London) **296:** 267–275.
11. Wooten, G., T. Hanson & F. Lamprecht. 1975. Elevated serum dopamine-β-hydroxylase activity in rats with inherited diabetes insipidus. J. Neural Transm. **36:** 107–112.
12. Pullan, P. T., C. I. Johnston, W. P. Anderson & P. I. Korner. 1980. Plasma vasopressin in blood pressure homeostasis and in experimental renal hypertension. Am. J. Physiol. **239:** H81–H87.
13. Gardiner, S. M., T. Bennett & P. A. Kemp. 1980. Systemic arterial hypertension in rats exposed to short-term isolation; intra-arterial systolic and diastolic blood pressure and baroreflex sensitivity. Med. Biol. **58:** 232–239.
14. Smyth, H. S., P. Sleight & G. W. Pickering. 1969. Reflex regulation of arterial pressure during sleep in man: A quantitative method of assessing baroreflex sensitivity. Circ. Res. **24:** 109–121.
15. Struyker-Boudier, H. A. J., J. F. Smits & H. Van Essen. 1979. The role of the baroreceptor reflex in the cardiovascular effects of propranolol in the conscious spontaneously hypertensive rat. Clin. Sci. **56:** 163–167.
16. Nakano, J. 1974. Cardiovascular responses to neurohypophysial hormones. Handb. Physiol. Sect. 7, Vol. IV, Part 1: 395–442.
17. Varma, S., B. P. Jaju & K. P. Bhargava. 1969. Mechanism of vasopressin-induced bradycardia in dogs. Circ. Res. **24:** 787–792.

18. Yourmans, W. B., H. V. Good & A. F. Hewitt. 1952. Inhibitory effect of vasopressin on cardio-accelerator mechanism after sino-aortic denervation. Am. J. Physiol. **168:** 182–188.
19. Bartelstone, H. & P. A. Nasmyth. 1965. Vasopressin potentiation of catecholamine actions in dog, rat, cat and rat aortic strips. Am. J. Physiol. **208:** 754–762.
20. Bohus, B. 1974. The influence of pituitary peptides on brain centres controlling autonomic responses. Prog. Brain Res. **41:** 175–183.
21. Versteeg, D. H. G., M. Tanaka & E. R. De Kloet. 1978. Catecholamine concentration and turnover in discrete regions of the brain of the homozygous Brattleboro rat deficient in vasopressin. Endocrinology **103:** 1654–1661.

Discussion of the Paper

B. M. Altura (*Downstate Medical Center, SUNY, New York, N.Y.*): Have you tried to do these experiments on isolated perfused hearts? I don't think you have eliminated the strong possibility that there is a direct effect on the myocardium due to the inherent genetic changes in the Brattleboro rat. Other beta adrenergic antagonists should also be tried.

Gardiner: We have not looked at that.

M. Gellai (*Dartmouth Medical School, Hanover, N.H.*): Did you try lower concentrations of vasopressin? Do you have any idea of what plasma concentration you were achieving?

Gardiner: No, that is the obvious experiment to do next. We chose the infusion rate and concentration of vasopressin from the work of Laycock, who measured vasopressin levels in Long-Evans rats, to produce an equivalent plasma concentration. We haven't actually measured the plasma concentrations, but we have since shown (in agreement with work presented by others this morning) that the pressor sensitivity to vasopressin is increased in Brattleboro rats compared to Long-Evans rats; we obviously need to try a lower concentration.

K. Berecek (*University of Alabama, Birmingham, Ala.*): I don't really know how to explain the differences in our studies. One major difference is that your work has been done on conscious animals while my experiments were done in the anesthetized animal. We measured arterial pressure for prolonged time intervals using an indirect method and spent three weeks conditioning the rats to the restraining devices. We noted the DI rats showed an increased heart rate and arterial pressure, compared to the Long-Evans rats, during the conditioning period. However, once the DI rats were conditioned to the restraining devices, their heart rates were 25–50 beats per minute less than Long-Evans rats. Did you do any preconditioning of your Long-Evans or DI rats prior to putting them in their cages to run the experiment?

Gardiner: Not in these experiments, but we have done experiments with Brattleboro rats using tail-cuff measurements. We also condition those animals before measurements are made, but we still find a tachycardia compared to the Long-Evans rats.

T. Berl (*University of Colorado Medical Center, Denver, Col.*): Has anybody measured catecholamines in DI rats? Is there any known effect of AVP on actual myocardial contractility?

GARDINER: I don't think plasma catecholamines have been measured, but central catecholamine half-life and turnover is altered. Serum dopamine beta hydroxylase activity has been shown to be increased in Brattleboro rats. Vasopressin does affect heart rate directly by depressing myocardial contractility, I believe. However, I don't think that accounts for our results since propanolol can block the effect, although I accept your point.

M. JOHNSON (*West Virginia University Medical Center, Morgantown, W. Va.*): I have measured, in a preliminary experiment, renal catecholamine content in DI rats and weight-matched normal rats. The DI rat has a renal catecholamine content that is almost double that of the normal rat. We intend to pursue that and we also plan to measure plasma catecholamines in the future.

H. W. SOKOL (*Dartmouth Medical School, Hanover, N.H.*): Very preliminary data for a Ph.D. thesis in our department show elevated plasma catecholamine levels in the DI rat, but this must be repeated.

BERL: So everything is decreased in these rats except perhaps catecholamines.

MICROCIRCULATORY AND VASCULAR SMOOTH MUSCLE BEHAVIOR IN THE BRATTLEBORO RAT: RELATIONSHIP TO RETICULOENDOTHELIAL SYSTEM FUNCTION AND RESISTANCE TO SHOCK AND TRAUMA

Burton M. Altura

Department of Physiology
State University of New York
Downstate Medical Center
Brooklyn, New York 11203

INTRODUCTION

Several lines of investigation have revealed that hemorrhage, fluid loss, and trauma are strong stimuli for the release of vasopressin from the pituitary gland.[1–5] This naturally occurring pituitary hormone is a potent constrictor of vascular smooth muscle cells,[6–8] including arterioles and precapillary sphincters in the microcirculation of the normal animal.[6–13] A little over 15 years ago, it was clearly demonstrated, independently, by two different groups that synthetic analogues and homologues of vasopressin could be successfully employed as therapeutic regimens to promote survival in experimental forms of circulatory shock and trauma.[14, 15] During the past 15 years, this concept has been verified and extended to other synthetic analogues of vasopressin, including studies in man.[16–27] It is not known, however, whether the endogenous release of vasopressin in circulatory shock is necessary for survival nor is it known whether the absence of released and circulating vasopressin is necessary to maintain vascular tone, reactivity, and homeostasis. With these points in mind, together with the background data on vasopressin analogues, microcirculation, and shock (discussed below), we thought the Brattleboro rat might represent an ideal animal to examine some of these questions.[28] The results presented herein indicate that release of endogenous vasopressin, particularly in low-flow states, may be critical in maintenance of circulatory homeostasis.

INFLUENCE OF VASOPRESSIN ANALOGUES ON MICROCIRCULATION AND SURVIVAL AFTER CIRCULATORY SHOCK AND TRAUMA

Among other aspects of therapy for the late (agonal) stages of circulatory shock, the use of vasopressor drugs has been unfavorable. This experience is based on a large body of data and must be considered a 'working' assumption. The key factor in the indictment of vasopressors derives from indirect and clinical evidence that responses of these pressor agents are accompanied by excessive vasoconstriction, sufficiently occlusive to invalidate their pressor effects on local tissue blood flow. As a result, the vasopressors now in use—i.e., catecholamines, sympathomimetics, and angiotensin—abet rather than relieve tissue ischemia.[16, 18, 20, 23, 26, 29, 30] Since the principle of vaso-excitor-pressor therapy late in shock is logical, it must be that the vasopressor drugs on which most of our experience is based are inappropriate.

Pharmacologic manipulation of organ systems at the microcirculatory level

0077–8923/82/0394–0375 $1.75/0

is by no means a novel or impractical concept. However, in thinking seriously about using vasopressors late in shock, our approach can be considered somewhat at variance with the current antipressor environment. When the phase of the low-flow state syndrome is characterized by a decompensating peripheral vasculature—that is, marked hypotension and pooling and sequestration of blood in the microscopic capacitance vessels—we seek vasoexcitor-pressor agents that exert selective microvascular constrictor effects, primarily on the microscopic muscular venules. Fundamental to this rational hemodynamic approach is the fact that evaluation and choice of vasoactive agents should be based not only on their effects on the cardiovascular components proximal to the terminal vascular bed but also on their actions on the various types of microvessels that regulate capillary blood flow and distribution. Using such a direct, *in vivo* microcirculatory approach with a closed-circuit television microscope recording system (which can resolve 0.02 μm and record up to 30 measurements of lumen size and wall thickness per minute), we have been examining, designing, and testing (with some success) synthetic analogues of the neurohypophyseal peptides vasopressin and oxytocin.[6–13, 15–18, 20, 22–27, 31–44]

Until recently, vasopressin was given little serious thought either as a useful drug or as an endogenous humoral regulator of the peripheral circulation in shock because of its supposed coronary constrictor action.[45] However, certain new evidence has questioned the latter assumption. (1) Studies on a variety of isolated bovine and canine coronary arteries (both large and small diameter vessels) have revealed that both arginine-vasopressin and lysine-vasopressin either are extremely weak vasoconstrictors (eliciting no more than 10% of a maximal contractile response) or (surprisingly) relax, in a dose-dependent manner, coronary arterial tone and tension.[34, 46] (2) At least two different synthetic analogues of vasopressin may have significant antiarrhythmic properties in man.[16, 21, 47] One of these two analogues, namely [2-phenylalanine, 8-lysine]-vasopressin (i.e., PLV-2), has been reported by our group to be beneficial in the treatment of several forms of experimental and clinical circulatory shock syndromes in several species, including humans.[15, 16, 18, 20, 22–27, 32, 33, 39, 43, 44] It is thought that the selective, nonocclusive venular constrictor action of some of these analogues, together with their ability to maintain normotensive arterial blood pressure levels and restore vascular reactivity to normal in the microvasculature, are primarily responsible for their beneficial effects (e.g., FIGURE 1, TABLE 1). Vasopressin analogues that occlude either arterioles, venules, or precapillary sphincters and metarterioles are detrimental when used as therapy in experimental shock (e.g., TABLE 1).[22, 26, 44] In view of such findings, (1) we have been screening several old, as well as new, analogues of vasopressin to determine whether structural changes in the vasopressin molecule will produce analogues that exert more highly selective microvascular constrictor actions. We are seeking, especially, molecules with submaximal arteriolar and venular constrictor actions. (2) One must entertain the strong possibility that endogenously released vasopressin molecules may be critical in maintenance of circulatory homeostasis.

ROLE OF RETICULOENDOTHELIAL SYSTEM IN HOST DEFENSE AND THE EFFECTS OF VASOPRESSIN MOLECULES ON PHAGOCYTIC FUNCTION IN SHOCK AND TRAUMA

During the past 15 years, evidence has been gathered that suggests that tolerance to various types of circulatory shock, trauma, and surgery is associated with the functional capacity of the phagocytic cells of the reticuloendothelial system.[26, 27, 44, 48–52] This body of data, both in animals and humans, suggests that reticuloendothelial system (RES) phagocytic indices (i.e., *K* values, or half-times of clearance) may provide diagnostic and prognostic measurements of shock syndromes and their response to therapy. Using a variety of test colloids for RES phagocytic clearance measurements (including colloidal carbon, lipid

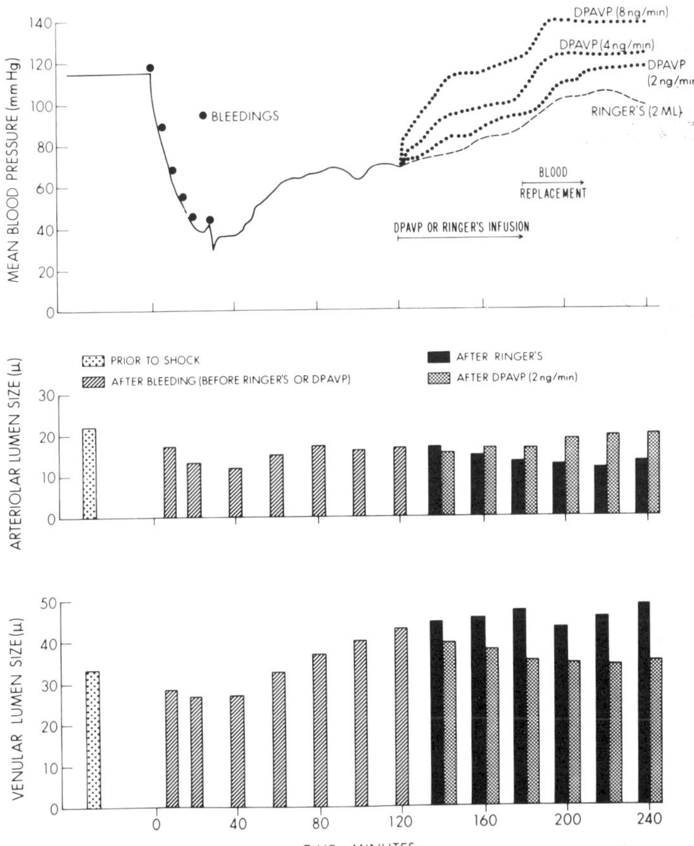

FIGURE 1. Composite of mean blood pressure, precapillary arteriolar responses, and venular responses in 42 rats subjected to hemorrhagic shock before and after intravenous Ringer's solution and DPAVP (i.e., 1-deamino-[2-phenylalanine, 8-arginine]-vasopressin. Twenty-six rats were used for the data on arteriolar and venular diameters. The SEM's for the arteriolar and venular measurements ranged from 0.8 to 2.8 μm. The SEM's for the arterial blood pressure points ranged from 4.2 to 9.4 mm Hg in the 42 rats that were examined.

radiolabeled emulsions, and denatured radiolabeled human serum albumin), and irrespective of the form of shock or its etiology, it has been demonstrated that the magnitude of the RES phagocytic index parallels the degree of shock or trauma; that is, the greater the degree of injury or shock the *lower the K value.* Animals that eventually die from circulatory shock exhibit a depressed RES up until death.[26, 27, 44, 48, 51, 52] In marked contrast to what happens in nonsurvivors, animals and humans that survive these lethal shock syndromes show, with time, progressively improved phagocytic indices. Furthermore, usually within 24–96 hours, such subjects exhibit K values that are increased by more than 100% over normal unshocked subjects.[26, 27, 44, 48–56] These hyperfunctional RE systems usually return spontaneously to normal within 96–144 hours. K values thus appear to reflect accurately the degree of tissue injury and indicate whether a mammal will survive the shock syndrome or succumb.

Substances (e.g., certain colloids, chemicals, hormones, and drugs) and microorganisms that depress the phagocytic powers of RE cells increase mortality, while materials that stimulate RES phagocytic activity are, in most instances, associated with increased tolerance to many forms of circulatory shock, trauma, and systemic stress.[26, 27, 44, 48–54, 57–66]

TABLE 1

RELATIVE HORMONE-RECEPTOR AFFINITIES ON RAT MICROVESSELS, INFLUENCE OF NEUROHYPOPHYSEAL PEPTIDE ANALOGS ON RE FUNCTION, AND SURVIVAL OF RATS SUBJECTED TO HEMORRHAGE

Peptide (—vasopressin)	Preferential Microvessel Affinity	Influence on Survival After Shock	RES Function (K values)
[Orn^8]-	Arteriole	—*	decreased
[Phe^2, Orn^8]-	Venule	+++	increased
[Phe^2, Lys^8]-	Venule	++	increased
[Phe^2, $Ileu^3$, Orn^8]-	Precapillary Sph.	——	decreased
1-Deamino-[Arg^8]-	Arteriole	—	decreased
1-Deamino-[Phe^2, Arg^8]-	Venule	+++	increased
[$Ileu^3$, Arg^8]-	Arteriole	——	decreased
[$Ileu^3$, Orn^8]-	Metarteriole	——	decreased
[$Ileu^3$, Leu^8]-	Metarteriole-Precap. Sph.	——	decreased
[N^{ϵ}-For-Lys^8]-	Arteriole	—	decreased

* Symbols: — =no effect on survival; —— =exacerbation of mortality; ++ =40–50% increase in survival; +++ =70–100% increase in survival.

Since RE cells are constituents of the vascular system, they can be affected, primarily, by only two agencies—microcirculatory vasomotor behavior or vasoactive substances in the blood or lymph. Elsewhere, I have reviewed some direct microcirculatory evidence to indicate that RE cells probably participate in regulation of blood flow not only in the liver, spleen, and lymph nodes but also in other tissues. Other more recent studies also point to an interaction between microvascular integrity and RE cell function.[44] To illustrate, we have shown that antibiotics that induce RE cell depression (viz., tetracyclines, polymixin B, cephalothin, kanamycin)[41, 51, 66, 67] produce some constriction of splanchnic microvessels and an increased sensitivity to constrictor catecholamines. Treatment of animals with doses of estrogen or glucocorticoids, which results in enhanced survival and RE cell phagocytosis after shock and trauma,[61, 64] prevents the tissue ischemia noted early in low-flow states, thereby maintaining a *vis a tergo* and venular tone;[64, 68] The latter could, in large measure, be responsible for both enhanced RE cell phagocytosis and survival. Vasoactive drugs that improve survival after different forms of circulatory shock and trauma not only favorably alter microcirculatory dynamics (i.e., decrease reactivity to endogenous and exogenous constrictors, increase capillary inflow and outflow, and restore vasomotion and microvascular tone) but also result in stimulation of RE cell phagocytosis, restoration of arterial blood pressure, and hemodilution (as opposed to hemoconcentration).[26, 27, 44, 48, 51, 61, 64, 68–70] As could be predicted from such data, therapeutically effective doses of vasopressin analogues with selective microvascular actions (see above), concomitantly produce stimulation of RES phagocytic function within three hours after intravenous infusion to severely (irreversibly) shocked animals (e.g., TABLE 1).[24, 26, 27, 44, 48, 51, 71–74] In fact, infusion of very small, therapeutically effective doses (i.e., 4 ng/min) of some of these vasopressin molecules (e.g., DPAVP) increases carbon clearance rates by almost 400% over untreated shocked controls.[24] Infusion of equi-pressor doses of catecholamines or angiotensin not only fails to enhance survival after shock and trauma but results in a further depression in RES phagocytic indices.[48, 65, 71–74] Collectively, such data suggest that (1) RES phagocytic function parallels the course of the shock syndrome and its response to therapy; (2) it is possible to treat so-called refractory shock and trauma with vasopressin analogues which possess special microvascular properties; and (3) release of endogenous vasopressin and/or neurohypophyseal peptides may be extremely important in regulating the microvasculature, circulatory homeostasis, and RES function.

BRATTLEBORO RATS ARE SENSITIVE TO BLOOD AND FLUID LOSS: IMPORTANCE OF ENDOGENOUS VASOPRESSIN AND RES

In view of the importance of RE cells in host defense, and the question as to whether a lack of circulating vasopressin in shock alters RES function, experiments were performed, in our laboratory, using Brattleboro, Long-Evans, and Wistar rats.[28] Using mild (25–30% mortality in controls) hemorrhagic and bowel ischemic forms of circulatory shock, we found that male homozygous Brattleboro rats are extremely sensitive to loss of blood and/or fluid. The quantitative survival data clearly showed that these rats, which lack vasopressin in their pituitary glands, are at least seven times more susceptible to blood and fluid loss than are normal Long-Evans or Wistar control rats of the same age, weight, and sex.[28] The mild forms of shock used in our experiments lowered arterial blood pressure only 20–30 mm Hg 60 minutes post-shock in the control (Long-Evans and Wistars) rats, but resulted in marked hypotension in the Brattleboro strain (TABLE 2). In addition, the results in TABLE 2 indicate that the hematocrit values 60 minutes post-shock in both the Wistar and Long-Evans rats exhibit hemodilution, while those of the Brattleboro animals exhibit hemoconcentration. Although RES phagocytic function prior to induction of hemorrhage or bowel ischemia is not different in the three different strains of rats, virtually no particulate matter is cleared by the RES in Brattleboro rats three hours post hemorrhage or bowel ischemia (TABLE 2). The 30–50% depression of RES phagocytic function observed in Wistar and Long-Evans control rats post-shock (TABLE 2) is in concert with previous findings.[24, 44, 48–55] These experiments thus support the idea that release of endogenous vasopressin from the posterior pituitary gland into the circulation during shock syndromes is probably an intrinsic, homeostatic regulator of vascular tone. In addition,

TABLE 2

MEAN ARTERIAL BLOOD PRESSURES, ARTERIAL HEMATOCRITS, AND RETICULOENDOTHELIAL SYSTEM PHAGOCYTIC FUNCTION IN WISTAR, LONG-EVANS, AND BRATTLEBORO RATS AFTER MILD HEMORRHAGE AND BOWEL ISCHEMIA *

	Hemorrhage			Bowel Ischemia		
Rat Strain	Mean Blood Pressure (mm Hg ± SEM)	Hct (% ± SEM)	*K* Value (mean ± SEM)	Mean Blood Pressure (mm Hg ± SEM)	Hct (% ± SEM)	*K* Value (mean ± SEM)
Wistar						
Control	122 ± 6	42.2 ± 2.2	0.043 ± 0.005	124 ± 6	41.8 ± 1.6	0.046 ± 0.004
Post-shock	95 ± 4†	36.0 ± 1.6†	0.025 ± 0.004†	92 ± 5†	37.3 ± 1.4†	0.033 ± 0.005†
Long-Evans						
Control	120 ± 7	43.0 ± 1.8	0.048 ± 0.006	118 ± 5	42.2 ± 1.4	0.045 ± 0.003
Post-shock	97 ± 4†	35.5 ± 1.9†	0.027 ± 0.007†	90 ± 6†	36.4 ± 1.8†	0.031 ± 0.005†
Brattleboro						
Control	128 ± 6	44.2 ± 2.5	0.045 ± 0.006	126 ± 7	43.5 ± 1.5	0.051 ± 0.005
Post-shock	30 ± 6‡	52.5 ± 2.2‡	0.002 ± 0.002‡	32 ± 8‡	50.8 ± 2.1‡	0.001 ± 0.001‡

* $N = 8$–12 different animals for each group. All postshock values recorded here are at 60 min post-hemorrhage or post-release of the superior mesenteric arterial occlusion.

† Significantly different from paired controls ($p < 0.01$).

‡ Significantly different from all other values ($p < 0.001$).

our findings suggest that RES function in shock syndromes may also be due to an intrinsic regulation by vasopressin. The hematocrit patterns noted here in normal animals, containing vasopressin in their posterior pituitary glands, seem to closely parallel what would be expected in mild forms of shock (i.e., a hemodilution) characteristic of vascular compensation. However, Brattleboro rats, lacking endogenous vasopressin, exhibit progressive hemoconcentration, which is characteristic of circulatory decompensation. An inability of an animal's peripheral vasculature to undergo compensatory vasoconstriction after bleeding and/or fluid loss (e.g., in the case of an animal lacking endogenous vasopressin) would, at least in theory, result in marked hypotension, progressive hemoconcentration, and marked RES phagocytic depression, all of which take place in Brattleboro rats subjected to only mild episodes of hemorrhage and intestinal ischemia.

These findings in Brattleboro rats, when viewed in light of the therapeutic value of vasopressin molecules (*vide supra*) in circulatory shock, could be indicative of the possibility that the microvasculature and vascular smooth muscle cells may be altered in the diabetes insipidus state. It must also be recalled that, normally, many types of arterial and arteriolar smooth muscle cells are normally more sensitive to the constrictor actions of endogenous circulating vasopressin than they are to other vasoactive peptides, amines, or prostanoids.[8–12, 38, 42] If the latter were altered in peripheral blood vessels of animals exhibiting diabetic insipidus, it could aid in explaining why Brattleboro rats are sensitive to blood and fluid loss and are unable to clear particulate matter from their RE systems when faced with only minimal loss of blood and/or fluid.

Responsiveness of Arteriolar Smooth Muscle Cells of Brattleboro Rats to Neurohypophyseal Peptides and Catecholamines

An examination of mesenteric terminal arterioles of Wistar, Long-Evans, and Brattleboro rats, using our television-microscope recording system (*vide supra*), revealed several interesting findings. As expected,[75] arterioles from female rats were more sensitive (based on threshold and EC_{50} concentrations) to the contractile effects of vasopressin, oxytocin, norepinephrine, and epinephrine than were comparable vessels in male animals; this was noted in all strains of rats (Table 3–5). This sex difference is, however, more pronounced in the control animals than in the Brattleboro rat. Possibly more importantly, arterioles of the Brattleboro rat are much more sensitive (one to five orders of magnitude depending upon agonist and gender; see Tables 3–5) to the constrictor actions of neurohypophyseal peptides and catecholamines than are either of the two control strains. The latter is suggestive of the possibility that the receptors for these agonists demonstrate a greater affinity on arteriolar smooth muscle of the Brattleboro animal. It should also be pointed out that the *intrinsic activity* (i.e., in this case the maximal contractile response) of the arteriolar smooth muscle cells of the Brattleboro strain for these agonists is greater than for comparable smooth muscle cells in either the Wistar or Long-Evans strains.

Enhanced receptor affinity and intrinsic activity of arterioles for circulating oxytocin and catecholamines in the diabetes insipidus state could play important

TABLE 3

SENSITIVITY AND CONTRACTILITY OF INTACT MESENTERIC ARTERIOLES FROM DIFFERENT RAT STRAINS TO LYSINE-VASOPRESSIN

Sex, Strain	Threshold Concentration (M)	EC_{50} (M)	% Maximal Response
Male			
Wistar	$3.7 \pm 0.4 \times 10^{-10}$	$2.2 \pm 0.3 \times 10^{-8}$	83.0 ± 3.5
Long-Evans	$2.3 \pm 0.3 \times 10^{-9}$ *	$3.6 \pm 0.5 \times 10^{-8}$ *	82.5 ± 4.2
Brattleboro	$1.6 \pm 0.2 \times 10^{-14}$ †	$1.2 \pm 0.2 \times 10^{-10}$ †	100 ± 0 †
Female			
Wistar	$1.9 \pm 0.3 \times 10^{-11}$ ‡	$3.5 \pm 0.4 \times 10^{-9}$ ‡	100 ± 0
Long-Evans	$3.7 \pm 0.6 \times 10^{-11}$ ‡	$1.6 \pm 0.2 \times 10^{-8}$ ‡	100 ± 0
Brattleboro	$3.7 \pm 0.4 \times 10^{-13}$ ‡	$8.8 \pm 0.8 \times 10^{-10}$ ‡	100 ± 0

$N = 6$–14 each.
* Significantly different from Wistar ($p < 0.01$).
† Significantly different from Wistar and Long-Evans ($p < 0.01$).
‡ Significantly different from all other values ($p < 0.01$).

roles in stabilizing hemodynamics and blood pressure, and, thus, may explain why animals (and possibly humans) do not ordinarily lose homeostatic control under normal circumstances; in other terms the microvasculature appears to undergo a compensatory adaptation in diabetes insipidus. The latter could also explain why animals (and humans) in such a diseased state are extremely sensitive to blood and/or fluid loss. The presence of an enhanced reactivity of arterioles and metarterioles in diabetes insipidus could result in intense tissue

TABLE 4

SENSITIVITY AND CONTRACTILITY OF INTACT MESENTERIC ARTERIOLES FROM DIFFERENT RAT STRAINS TO OXYTOCIN

Sex, Strain	Threshold Concentration (M)	EC_{50} (M)	% Maximal Response
Male			
Wistar	$9.6 \pm 0.8 \times 10^{-6}$	$6.8 \pm 0.5 \times 10^{-5}$	60.0 ± 2.0
Long-Evans	$4.8 \pm 0.5 \times 10^{-5}$ *	$2.1 \pm 0.3 \times 10^{-4}$ *	52.2 ± 2.5
Brattleboro	$1.9 \pm 0.3 \times 10^{-7}$ †	$9.3 \pm 1.0 \times 10^{-6}$ †	100 ± 0 †
Female			
Wistar	$9.8 \pm 0.9 \times 10^{-8}$ ‡	$3.0 \pm 0.3 \times 10^{-5}$ ‡	100 ± 0
Long-Evans	$9.8 \pm 1.0 \times 10^{-7}$ §	$4.7 \pm 0.5 \times 10^{-5}$ §	100 ± 0
Brattleboro	$7.8 \pm 0.8 \times 10^{-8}$ ‡	$2.4 \pm 0.3 \times 10^{-5}$ ‡	100 ± 0

$N = 6$–16 each.
* Significantly different from Wistar ($p < 0.01$).
† Significantly different from Wistar and Long-Evans ($p < 0.01$).
‡ Significantly different from paired values for male ($p < 0.01$).
§ Significantly different from female Wistar and Brattleboro, and paired male ($p < 0.01$).

ischemia, hypoxia, and anoxia upon only slight loss of fluid and/or blood. The end results of the latter would be increased susceptibility of subjects with diabetes insipidus to hemorrhage and/or fluid loss, loss of blood pressure control, and failure of the RES to clear particulate matter—all of which we have indeed noted in Brattleboro rats.[28]

But are the vascular smooth muscle cells really more sensitive to constrictor agonists as our microcirculatory data suggest or is it the presence (or absence) of certain endogenous circulating substances, metabolites, etc. possibly brought about by the diabetic insipidus state which accounts for the apparent changes in hormone receptor affinities and intrinsic activities? In order to test such an hypothesis, we proceeded to examine excised blood vessels obtained from adult male and female rats (of the same age) from all three strains.

TABLE 5

SENSITIVITY AND CONTRACTILITY OF INTACT MESENTERIC ARTERIOLES FROM DIFFERENT RAT STRAINS TO NOREPINEPHRINE

Sex, Strain	Threshold Concentration (M)	EC_{50} (M)	% Maximal Response
Male			
Wistar	$6.5 \pm 0.7 \times 10^{-8}$	$8.2 \pm 0.7 \times 10^{-7}$	80.0 ± 3.6
Long-Evans	$7.4 \pm 0.8 \times 10^{-8}$	$1.3 \pm 0.2 \times 10^{-6}$	82.0 ± 2.5
Brattleboro	$2.8 \pm 0.3 \times 10^{-9}$ *	$8.2 \pm 0.8 \times 10^{-8}$ *	100 ± 0 *
Female			
Wistar	$2.7 \pm 0.3 \times 10^{-8}$ †	$5.3 \pm 0.5 \times 10^{-7}$ †	95.0 ± 1.5
Long-Evans	$4.1 \pm 0.4 \times 10^{-8}$ †	$7.0 \pm 0.8 \times 10^{-7}$ †	93.0 ± 2.0
Brattleboro	$2.6 \pm 0.3 \times 10^{-9}$ *	$6.5 \pm 0.6 \times 10^{-8}$ *	100 ± 0

$N = 6$–10 each.
* Significantly different from Wistar and Long-Evans ($p < 0.01$).
† Significantly different from paired male values ($p < 0.03$).

REACTIVITY OF AORTIC SMOOTH MUSCLE EXCISED FROM WISTAR, LONG-EVANS, AND BRATTLEBORO RATS TO PEPTIDES, AMINES, AND POTASSIUM

Using isolated aortic strips, and isometric recording techniques similar to those published previously,[6–8, 34, 38] it can clearly be seen that in all cases, irrespective of gender, both vasopressin and oxytocin elicit greater contractile responses on aortas obtained from Brattleboro rats than on comparable tissues obtained from control Long-Evans rats (FIGURES 2 and 3); these differences are significant at the 1% level. It is also interesting to note that the aortas of Brattleboro animals, irrespective of gender, are more sensitive to the contractile actions of vasopressin than are comparable tissues of either Wistar or Long-Evans rats. As in the case of the intact arterioles (TABLES 3 and 4), aortic smooth muscle obtained from female animals exhibit, quantitatively, greater contractile responses to both vasopressin and oxytocin (FIGURES 2 and 3), thus supporting the notion[75] that the activity of neurohypophyseal peptides on vascular smooth muscle is modulated by sex hormones.

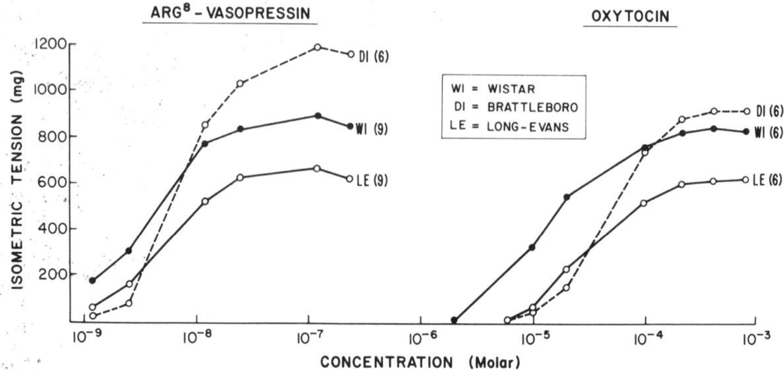

FIGURE 2. Comparative contractile effects of vasopressin and oxytocin on isolated aortas of female Wistar (WI), Long-Evans (LE), and Brattleboro rats (DI). The numbers in parentheses indicate the number of different animals examined.

Collectively, these data and those on the intact arterioles seemed to indicate that some (or all) arterial and arteriolar smooth muscle cells in Brattleboro animals might exhibit a nonspecific supersensitivity to vasoactive substances that are capable of evoking contraction. If this is true, one might expect to see potentiation of potassium chloride-induced contractions as well. The data shown in FIGURE 4 do, indeed, seem to support this contention.

Recently, Haack and Möhring [76] demonstrated that intraventricular injection of angiotensin II into Brattleboro rats homozygous for hereditary hypothalamic diabetes insipidus, where plasma arginine-vasopressin was undetectable, resulted in a smaller (50–80%) rise in arterial blood pressure when compared to what is seen in Long-Evans rats. They suggested that the response is less because there is no arginine-vasopressin in the posterior pituitary gland of the

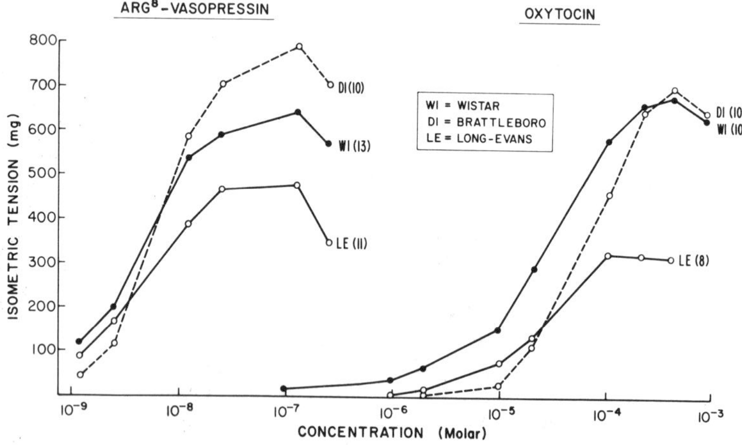

FIGURE 3. Comparative contractile effects of vasopressin and oxytocin on isolated aortas of male Wistar (WI), Long-Evans (LE), and Brattleboro rats (DI).

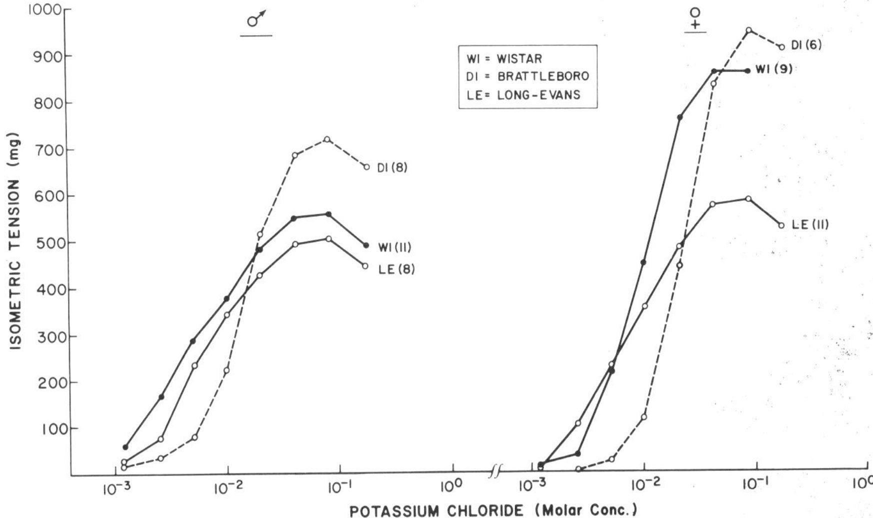

FIGURE 4. Comparative contractile effects of potassium chloride on isolated aorta of male (*left panel*) and female (*right panel*) Wistar (WI), Long-Evans (LE), and Brattleboro rats (DI). Note that maximum contractile responses generated on DI aortas are much greater than those attained on aortas excised from Long-Evans rats, irrespective of gender.

Brattleboro rat, and it is the vasopressin, per se, which is needed for most of the angiotensin II-induced blood pressure response. Another possibility must, however, be entertained, *viz.*, that certain arterial and arteriolar smooth muscle cells of the Brattleboro rat may be hyposensitive to angiotensin II. The data shown in FIGURE 5, on isolated aortic smooth muscle, do, somewhat surprisingly, demonstrate that certain vascular muscle in the Brattleboro rat may, indeed, be hyposensitive to angiotensin II stimulation; whether a similar phenomenon is seen on intact peripheral arterioles remains to be investigated. In any event, our

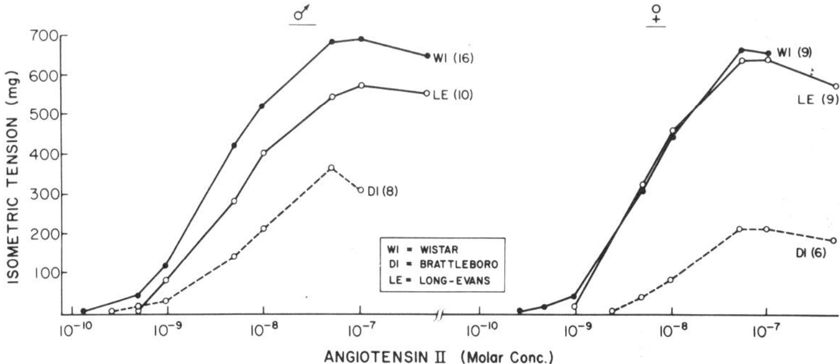

FIGURE 5. Isolated aortic smooth muscle from male (*left panel*) and female (*right panel*) Brattleboro rats is much less reactive to angiotensin II amide than aortic tissue of other rat strains, male or female.

findings are suggestive of the possibility that some arterial and arteriolar blood vessels in animals with hereditary diabetes insipidus may be supersensitive to certain endogenous vasoactive substances, including the neurohypophyseal peptides and catecholamines, whereas these same vessels may exhibit a hyposensitivity to certain endogenous peptide hormones like angiotensin II.

In view of these data, it must be entertained that the physiological and/or pharmacological behavior of peripheral vascular smooth muscle cells in mammals with diabetes insipidus may be altered; such vascular alterations may help to explain why (1) on the one hand, blood pressure is stabilized in mammals exhibiting diabetes insipidus; and (2) on the other hand, such diseased animals may be very susceptible to systemic stress such as blood and/or fluid loss.

Conclusions

The data reviewed herein demonstrate that release of endogenous vasopressin, particularly in low-flow states, may be critical in maintenance of circulatory homeostasis and reticuloendothelial system function. Synthetic analogues of vasopressin can be designed that have selective microcirculatory actions and that can be highly effective as therapeutic agents in the treatment of both experimental and clinical circulatory shock states. The beneficial effects of vasopressin molecules appear to reside in their ability to selectively restore venular tone and arteriolar vasomotor tone, and to stimulate the phagocytic elements of the reticuloendothelial system in states of circulatory shock and trauma.

Experiments with intact arterioles (18–25 μm i.d.) and isolated aortas of Brattleboro rats indicate that the smooth muscle cells, per se, of peripheral blood vessels in diabetes insipidus appear to have undergone some adaptation due to the diseased state. Such vascular smooth muscle cells are supersensitive (e.g., increased receptor affinity and intrinsic activity) not only to vasopressin and oxytocin, but to catecholamines and potassium ions as well. Peripheral vessels from female animals exhibit more pronounced supersensitivity than do blood vessels of male animals. Although micro- and macrovascular smooth muscle cells from Brattleboro rats may be more sensitive to most endogenous, circulating vasoactive substances than are comparable cells from Wistar and Long-Evans strains, they are hyposensitive to the potent pressor peptide, angiotensin II. These marked alterations in responsiveness of peripheral blood vessels may aid in explaining why mammals with hereditary diabetes insipidus are able to regulate and maintain blood pressure and organ blood flows. In addition, it is possible that these bizarre alterations in micro- and macrovessel reactivity could explain why rats exhibiting hereditary homozygous diabetes insipidus are extremely sensitive to blood and fluid loss, and are not able to clear particulate matter from their RE systems. Future work in this area should determine why the absence of circulating vasopressin produces direct and marked contractility alterations of vascular smooth muscle cells to a variety of vasoactive peptides and amines.

References

1. Ginsburg, M. & L. M. Brown. 1956. Effect of anaesthetics and hemorrhage on the release of neurohypophysial antidiuretic hormone. Brit. J. Pharmacol. **14:** 327–333.

2. BARATZ, R. A. & R. C. INGRAHAM. 1960. Renal haemodynamics and anti-diuretic hormone release associated with volume regulation. Am. J. Physiol. **198:** 565–570.
3. BELESIN, D., G. W. BISSETT, J. HALDAR & R. L. POLAK. 1967. The release of vasopressin without oxytocin in response to haemorrhage. Proc. Roy. Soc. B **166:** 443–458.
4. FORSLING, M. L. 1977. Anti-diuretic Hormone. Eden Press. Montreal, Canada.
5. ROCHA E SILVA, JR., M., W. CELSO DE LIMA & E. M. CASTRO DE SOUZA. 1978. Vasopressin secretion in response to haemorrhage. Mathematical modelling of the factors involved. Pflüg. Arch. **376:** 185–190.
6. ALTURA, B. M. 1970. Significance of amino acid residues in vasopressin on contraction in vascular muscle. Am. J. Physiol. **219:** 222–229.
7. ALTURA, B. M. 1972. Significance and interaction of the amino acid residues in positions 1,2,3 and 8 of vasopressins on contractile activity in vascular smooth muscle. *In* Chemistry and Biology of Peptides. J. Meienhofer, Ed.: 441–447. Ann Arbor Science Publ., Inc. Ann Arbor, Mich.
8. ALTURA, B. M. & B. T. ALTURA. 1977. Vascular smooth muscle and neuro-hypophyseal hormones. Fed. Proc. **36:** 1853–1860.
9. ALTURA, B. M. 1967. Evaluation of neurohumoral substances in local regulation of blood flow. Am. J. Physiol. **212:** 1447–1454.
10. ALTURA, B. M. 1971. Chemical and humoral regulation of blood flow through the precapillary sphincter. Microvasc. Res. **3:** 361–384.
11. ALTURA, B. M. 1973. Selective microvascular constrictor actions of some neurohypophyseal peptides. Eur. J. Pharmacol. **24:** 49–60.
12. ALTURA, B. M. 1973. Significance of amino acid residues in position 8 of vasopressin on contraction of rat blood vessels. Proc. Soc. Exp. Biol. Med. **142:** 1104–1110.
13. ALTURA, B. M. 1974. Neurohypophyseal hormones and analogues: Rat pressor potency versus contractile activity on rat arterioles and arteries. Proc. Soc. Exp. Biol. Med. **146:** 1054–1060.
14. CORT, J. M., J. HAMMER, M. ULRYCH, Q. PISA, T. DOUSA & J. RUDINGER. 1964. Synthetic extended-chain analogues of vasopressin and oxytocin in treatment of experimental hemorrhagic shock. Lancet **2:** 840–841.
15. ALTURA, B. M., R. HSU, V. D. B. MAZZIA & S. G. HERSHEY. 1965. Influence of vasopressors on survival after traumatic, intestinal ischemia and endotoxin shock in rats. Proc. Soc. Exp. Biol. Med. **119:** 389–393.
16. HERSHEY, S. G. & B. M. ALTURA. 1966. Behandlung des Schocks durch Beeinflussung der peripheren Zirkulation mit Vasoaktiven Wirkstoffen: eine mikrozirkulatorische Basis fur die Therapie. Schweiz. Med. Wschr. **96:** 1467–1471, 1516–1522 (2 parts).
17. ALTURA, B. M. & S. G. HERSHEY. 1967. Pharmacology of neurohypophyseal hormones and their synthetic analogues in the terminal vascular bed: Structure-activity relationships. Angiology **18:** 428–439.
18. ALTURA, B. M. & S. G. HERSHEY. 1968. Structure-activity relationships of neurohypophyseal polypeptides in the microcirculation: A molecular basis for shock therapy. *In* Intermedes Proceeding 1968: Combined Injuries and Shock. B. Schildt & L. Thoren, Eds.: 247–259. Amqvist and Wiksell. Stockholm.
19. SCHLAG, G. 1969. Klinische Untersuchung mit einen vasoaktiven Polypeptid POR-8 (Ornithin^8-Vasopressin). Arzneim-Forschung **19:** 1521–1524.
20. ALTURA, B. M., S. G. HERSHEY & B. T. ALTURA. 1970. Microcirculatory actions of vasoactive polypeptides and their use in the treatment of experimental shock. Adv. Exp. Med. Biol. **8:** 239–247.
21. NIELSEN, O. V. & N. VALENTIN. 1970. Ornithine-8-vasopressin, a new vasoconstrictor used for haemostasis during operation for genital prolapse. Acta Obstet. Gynecol. Scand. **49:** 45–48.

22. ALTURA, B. M. & S. G. HERSHEY. 1972. A structure-activity basis for vasotropic peptide therapy in shock. Adv. Exp. Med. Biol. **21:** 399–408.
23. HERSHEY, S. G. & B. M. ALTURA. 1973. Vasopressors and low-flow states. *In* Pharmacology of Adjuvant Drugs. H. L. Zauder, Ed.: 31–76. F. A. Davis Co. Philadelphia, Pa.
24. ALTURA, B. M. 1976. DPAVP: A vasopressin analog with selective microvascular and RES actions for the treatment of circulatory shock in rats. Eur. J. Pharmacol. **37:** 155–168.
25. ALTURA, B. M. 1976. Microcirculatory approach to the treatment of circulatory shock with a new analog of vasopressin, [2-phenylalanine, 8-ornithine]-vasopressin. J. Pharmacol. Exp. Ther. **197:** 187–196.
26. ALTURA, B. M. 1980. Reticuloendothelial and neuro-endocrine stimulation in shock therapy. Adv. Shock Res. **3:** 3–25.
27. ALTURA, B. M. 1980. Recent progress in patho-physiology of shock: Reticuloendothelial and neuro-endocrine stimulation. J. Clin. Anesth. **4:** 745–758.
28. ALTURA, B. M. 1980. Evidence that endogenous vasopressin plays a protective role in circulatory shock. Role for reticuloendothelial system using Brattleboro rats. Experientia **36:** 1080–1081.
29. THOMPSON, W. L. 1979. Drug therapy of shock. *In* Recent Advances in Clinical Pharmacology. P. Turner & D. G. Shand, Eds. Churchill-Livingstone. New York, N.Y.
30. GUYTON, A. C. 1981. Medical Physiology. W. B. Saunders. Philadelphia, Pa.
31. ALTURA, B. M., S. G. HERSHEY & B. W. ZWEIFACH. 1965. Effects of a synthetic analogue of vasopressin on vascular smooth muscle. Proc. Soc. Exp. Biol. Med. **119:** 258–261.
32. HERSHEY, S. G., V. D. B. MAZZIA, B. M. ALTURA & L. GYURE. 1965. Effects of vasopressors on the microcirculation and on survival in hemorrhagic shock. Anesthesiol. **26:** 179–189.
33. ALTURA, B. M., S. G. HERSHEY & V. D. B. MAZZIA. 1966. Microcirculatory approach to vasopressor therapy in intestinal ischemic (SMA) shock. Am. J. Surg. **111:** 186–192.
34. ALTURA, B. M. 1966. Differential actions of polypeptides and other drugs on coronary inflow vessels. Am. Heart J. **72:** 709–711.
35. ALTURA, B. M. 1971. Pharmacology of neurohypophyseal hormones and analogs on isolated vascular smooth muscle and in the terminal vascular bed. *In* Physiology and Pharmacology of Vascular Neuroeffector Systems. J. A. Bevan, R. F. Furchgott, R. A. Maxwell & A. P. Somlyo, Eds.: 274–290. Karger. Basel, Switzerland.
36. ALTURA, B. M. 1972. Structure-activity relationships of neurohypophyseal peptides on different types of isolated mammalian blood vessels. Adv. Exp. Med. Biol. **21:** 187–196.
37. ALTURA, B. M. 1972. Can metarteriolar vessels occlude their lumens in response to vasoactive substances? Proc. Soc. Exp. Biol. Med. **140:** 1270–1274.
38. ALTURA, B. M. 1975. Dose-response relationships for arginine vasopressin and synthetic analogs on three types of rat blood vessels: Possible evidence for regional differences in vasopressin receptor sites within a mammal. J. Pharmacol. Exp. Ther. **193:** 413–423.
39. ALTURA, B. M. 1976. Microcirculatory approach to the treatment of shock using a new analogue of vasopressin. *In* Microcirculation Vol. 2. J. Grayson & W. Zingg, Eds.: 193–194. Plenum. New York, N.Y.
40. ALTURA, B. M. 1978. Pharmacology of venular smooth muscle: New insights. Microvasc. Res. **16:** 91–117.
41. LASSOFF, S. & B. M. ALTURA. 1980. Do pial terminal arterioles respond to local perivascular application of the neurohypophyseal peptide hormones, vasopressin and oxytocin? Brain Res. **196:** 266–269.
42. ALTURA, B. M. 1981. Pharmacology of the microcirculation. *In* Microcircula-

tion. E. M. Effros, H. Schmid-Schonbein & J. Ditzel, Eds.: 51–105. Academic Press. New York, N.Y.

43. ALTURA, B. M. 1981. Pharmacology of venules: Some current concepts and clinical potentials. J. Cardiovasc. Pharmacol. **3:** 1413–1428.
44. ALTURA, B. M. 1981. Relationship of reticuloendothelial cell function to microcirculatory blood flow and low-flow states. *In* Pathophysiology of the Reticuloendothelial System. B. M. Altura & T. M. Saba, Eds.: 159–223. Raven Press. New York, N.Y.
45. SAAMELI, K. 1968. The circulatory actions of the neurohypophyseal hormones. Handb. Exp. Pharmacol. **23:** 748–801.
46. TURLAPATY, P. D. M. V. & B. M. ALTURA. 1982. Effects of neurohypophyseal peptide hormones on isolated coronary arteries: role of magnesium ions. Experientia. (In press.)
47. KATZ, R. L. 1965. Epinephrine and PLV-2: Cardiac rhythm and local vasoconstrictor effects. Anesthesiol. **26:** 619–623.
48. ALTURA, B. M. & S. G. HERSHEY. 1973. Reticuloendothelial function in experimental injury and tolerance to shock. Adv. Exp. Med. Biol. **33:** 545–569.
49. SABA, T. M. 1975. Reticuloendothelial systemic host defense after surgery and traumatic shock. Circulatory Shock **2:** 91–108.
50. SCHILDT, B. 1976. The present view of RES and shock. Adv. Exp. Med. Biol. **73A:** 375–387.
51. ALTURA, B. M. 1980. Reticuloendothelial cells and host defense. Adv. Microcirculation **9:** 252–294.
52. ALTURA, B. M. & T. M. SABA. 1981. Pathophysiology of the Reticuloendothelial System. Raven Press. New York, N.Y.
53. ALTURA, B. M. & S. G. HERSHEY. 1968. RES phagocytic function in trauma and adaptation to experimental shock. Am. J. Physiol. **215:** 1414–1419.
54. ALTURA, B. M. & S. G. HERSHEY. 1971. Acute intestinal ischemia and reticuloendothelial system function. J. Reticuloendothelial Soc. **10:** 361–371.
55. ALTURA, B. M. & S. G. HERSHEY. 1972. Sequential changes in reticuloendothelial system function after hemorrhage. Proc. Soc. Exp. Biol. Med. **139:** 935–939.
56. HERSHEY, S. G. & B. M. ALTURA. 1969. Function of the reticuloendothelial system in experimental shock and combined injury. Anesthesiol. **30:** 138–144.
57. ALTURA, B. M., S. G. HERSHEY & C. HYMAN. 1965. Influence of choline on the reticuloendothelial system and on survival after experimental shock. J. Reticuloendothelial Soc. **3:** 57–64.
58. HERSHEY, S. G. & B. M. ALTURA. 1966. Effect of pretreatment with aggregate albumin on reticuloendothelial system activity and survival after experimental shock. Proc. Soc. Exp. Biol. Med. **122:** 1195–1199.
59. ALTURA, B. M. & S. G. HERSHEY. 1970. Influence of glyceryl trioleate on the RES and survival after experimental shock. J. Pharmacol. Exp. Ther. **175:** 555–564.
60. ALTURA, B. M. 1974. Hemorrhagic shock and reticuloendothelial system phagocytic function in pathogen-free animals. Circulatory Shock **1:** 295–300.
61. ALTURA, B. M. 1975. Glucocorticoid-induced protection in circulatory shock: Role of reticuloendothelial system function. Proc. Soc. Exp. Biol. Med. **150:** 202–206.
62. REICHARD, S. M. 1967. Factors that diminish radiation lethality. Radiology **89:** 501–508.
63. REICHARD, S. M. 1972. RES stimulation and transfer of protection against shock. J. Reticuloendothelial Soc. **12:** 604–617.
64. ALTURA, B. M. 1976. Sex and estrogens in protection against circulatory stress reactions. Am. J. Physiol. **231:** 842–847.
65. HERSHEY, S. G. 1980. The reticuloendothelial system: Relationship to shock

and host defense. *In* Microcirculation. G. Kaley & B. M. Altura, Eds.: **3:** 69–105. University Park Press. Baltimore, Md.
66. ALTURA, B. M. & A. GEBREWOLD. 1980. Prophylactic administration of antibiotics compromises reticuloendothelial system function and exacerbates shock mortality in rats. Brit. J. Pharmacol. **68:** 19–21.
67. ALTURA, B. M., S. G. HERSHEY, M. ALI & C. THAW. 1966. Influence of tetracycline on phagocytosis, infection, and resistance to experimental shock: Relationship to microcirculation. J. Reticuloendothelial Soc. **3:** 447–457.
68. ALTURA, B. M. & B. T. ALTURA. 1974. Peripheral vascular actions of glucocorticoids and their relationship to protection in circulatory shock. J. Pharmacol. Exp. Ther. **190:** 300–315.
69. ALTURA, B. M. & S. HALEVY. 1978. Circulatory shock, histamine and antihistamines: Therapeutic aspects. Handb. Exp. Pharmacol. **18:** 575–602.
70. ALTURA, B. M. 1979. Reticuloendothelial system (RES) phagocytic depression in shock is ameliorated by H_1-receptor antihistamines. Eur. J. Pharmacol. **59:** 165–167.
71. ALTURA, B. M. & S. G. HERSHEY. 1967. Use of reticuloendothelial phagocytic function as an index in shock therapy. Bull. N.Y. Acad Med. **43:** 259–266.
72. ALTURA, B. M. & S. G. HERSHEY. 1968. Influence of vasopressor drugs on reticuloendothelial phagocytic function in experimental shock. *In* Intermedes Proceedings 1968: Combined Injuries and Shock. B. Schildt & L. Thoren, Eds.: 185–193. Almqvist and Wiksell. Stockholm.
73. HERSHEY, S. G. & B. M. ALTURA. 1969. The effects of vasoactive drugs on reticuloendothelial function in experimental shock and combined injury. Anesthesiol. **30:** 144–149.
74. ALTURA, B. M. 1980. Reticuloendothelial function and humoral factors in shock and trauma. *In* 11th Symposium on Shock and Reanimation. Tokyo, Japan. T. Oyama, Ed. **11:** 73–91.
75. ALTURA, B. M. 1975. Sex and estrogens and responsiveness of terminal arterioles to neurohypophyseal hormones and catecholamines. J. Pharmacol. Exp. Ther. **193:** 402–412.
76. HAACK, D. & J. MÖHRING. 1978. Vasopressin-mediated blood pressure response to intraventricular injection of angiotensin II in the rat. Pflügers Arch. **373:** 167–173.

DISCUSSION OF THE PAPER

K. BERECEK (*University of Alabama, Birmingham, Ala.*): Have you done any of your experiments in DI rats that have been chronically replaced with vasopressin to see if any of these changes were reversed?

ALTURA: We have done those studies and can reverse many of the abnormalities in the rat. (1) We have given pitressin tannate oil at various doses for 3–7 days and have been able to restore almost total reticuloendothelial phagocytic function. (2) We have used vasopressin analogues and have achieved even more striking improvement. In terms of microvasculature we have not looked at other vessels in other beds. I would like to examine renal vessels in the rat. We have looked at canine vessels but, of course, we don't have a canine model with good diabetes insipidus to compare them with. I think that some of the critical beds, i.e. the renal and coronary beds, have to be looked at certainly in view of the cardiac effects of vasopressin that we heard about this morning.

L. B. KINTER (*Smith Kline and French Laboratories, Philadelphia, Pa.*): Brattleboro rats are known to have elevated plasma renin activity. In our laboratory we have shown not only a higher baseline plasma renin activity in conscious DI rats, but also a larger response of plasma renin levels to stimuli, including dehydration and hemorrhage, than in conscious heterozygous control rats. I wonder whether some of the effects of blood loss you described in DI rats might be due to their high levels of plasma angiotensin II. I realize that you have shown a striking decreased sensitivity to angiotensin II in the DI rat; however, this may be a compensatory mechanism protecting the rat from high levels of angiotensin II.

ALTURA: I think what you are saying has great validity. We must seriously contend with what you are raising; but I have no answers, I have no data. Those are tricky experiments. We have thought of using something like captopril and doing some studies along those lines.

M. GELLAI (*Dartmouth Medical School, Hanover, N.H.*): On your isolated strip studies, did you compare passive length-tension relationships of vessels from DI rats, with those from the normal rats? How did you decide what resting tension to apply?

ALTURA: That is a very good question, and I think a fair one as well. All of our studies are always done under what we consider optimal rest length condition, since we are looking at isometric contractions. But, what does that mean when we are dealing with an animal that for some reason has different arteriolar wall thicknesses, as we have shown in these DI rats. How does one know we are really dealing with the same kind of set point? We use a variety of tensions and a comparison of the data shows very similar results at the different tensions.

J. HALDAR (*Columbia University, New York, N.Y.*). Do you always see oxytocin as a kind of vasoconstrictor or in some conditions does it act as a vasodilator?

ALTURA: For years my views on what oxytocin does and doesn't do in terms of the vasculature have been somewhat controversial. Eighty-five percent of all the studies with oxytocin, up until this time, have utilized a preparation that contains a preservative (usually chlorobutanol). In most cases the amount of chlorobutanol in such preparations is so overwhelmingly potent that it produces a vasodilation. Removal of chlorobutanol and careful use of preservative-free oxytocin (its bioactivity can be checked with uterotonic assay) reveals that in most vascular beds, oxytocin is a constrictor. However, there are vascular beds where oxytocin may indeed be a vasodilator; in fact, we believe, that endothelial cells may be very important in some of the oxytocin relaxation responses. In general, however, pure oxytocin is a constrictor in many vascular beds.

RELATIONSHIP BETWEEN VASOPRESSIN AND THE ANTEROVENTRAL THIRD VENTRICLE REGION IN DEOXYCORTICOSTERONE/SALT HYPERTENSION (BRATTLEBORO RAT MODEL) *

K. H. Berecek,† K. W. Barron, R. L. Webb, and M. J. Brody ‡

Cardiovascular Center and
Department of Pharmacology
University of Iowa
Iowa City, Iowa 52242

DOCA/salt hypertension does not develop in rats with hereditary deficiency in vasopressin (VP, DI rats) [1, 2] nor does it develop in normal rats after lesion of the anteroventral region of the third ventricle (AV3V).[3] The purpose of the present studies was to determine whether alterations in VP mechanisms mediated through the AV3V region were involved in the protective effect of the lesion. We examined, therefore, DOCA/salt hypertension in AV3V-lesioned rats and in AV3V-lesioned rats given vasopressin. Sham-lesioned rats treated with DOCA/salt served as controls. We also conducted these experiments in VP-deficient Brattleboro rats (DI rats) given DOCA/salt and VP.

MATERIALS AND METHODS

Lesion of the AV3V region was made in male Sprague-Dawley rats (180–200 g) by passing an anodal current of 2 mA for 15–20 sec through a stereotaxically placed lesioning electrode (24 gauge nichrome wire, insulated except at the tips). The coordinates for placement of the electrode were on the midline, 0.3 mm posterior from bregma and 7.5 mm below the dura. The same procedure was used for male DI rats (150–200 g); however, these rats were given VP replacement beginning two weeks prior to the lesion and continuing for the duration of the experiment. Deoxycorticosterone acetate (Sigma, 100 mg/kg) was given to unilaterally nephrectomized rats in a single silastic implant four weeks after AV3V or sham lesions. All rats received 0.9% NaCl as drinking water. All rats given VP received daily subcutaneous injections of VP-tannate (Pitressin® Parke-Davis, 100 mU/100 g rat). In AV3V-lesioned Sprague-Dawley rats VP treatment began on the day of DOCA implantation. Systolic arterial pressure was monitored one to two times per week in conscious restrained rats using an indirect tail-cuff method. At the termination of the study, samples of rats from each group were rapidly decapitated and trunk blood was collected for measurement of plasma arginine vasopressin.[4]

* Supported in part by U.S. Public Health Service grants HLB-14388 and HLB-07121, and a gift from the Searle Family Trust.

† Present address: Cardiovascular Research and Training Center, University of Alabama in Birmingham, 1012 Zeigler Building, Birmingham, AL 35294.

‡ Address all correspondence to: Michael J. Brody, Department of Pharmacology, Bowen Science Building, University of Iowa, Iowa City, IA 52242.

0077–8923/82/0394–0392 $1.75/0 © 1982, NYAS

RESULTS

Sham-lesioned Sprague-Dawley rats showed a significant rise in arterial pressure two weeks after DOCA/salt and averaged 195 mm Hg at five weeks (TABLE 1). AV3V-lesioned rats showed no rise in arterial pressure after DOCA/salt; however, when given VP their arterial pressure increased and averaged 155 mm Hg at five weeks (TABLE 1). Although not shown here, a rise in arterial pressure also occurred in AV3V-lesioned rats given DOCA/salt when VP treatment was begun five weeks post-DOCA. Plasma arginine vasopressin levels were elevated (threefold) in chronic DOCA/salt hypertensive rats (FIGURE 1). AV3V-lesioned rats treated with DOCA also showed a rise in plasma VP (1.5-fold) but the rise was not as great as that seen in sham-lesioned rats. AV3V-lesioned rats given VP showed a plasma VP level that was similar to that seen in sham-lesioned DOCA-treated rats.

AV3V lesion in DI rats was found to completely prevent the development of DOCA hypertension induced by vasopressin (TABLE 2). Five weeks post-DOCA, intact DI rats replaced with VP showed a rise in arterial pressure that averaged 195 mm Hg. In contrast, the arterial pressure of lesioned DI rats treated with DOCA + VP was 120 mm Hg and was not significantly different from the pressure prior to DOCA treatment.

DISCUSSION

The AV3V region is a critical receptor site for central actions of angiotensin II and hyperosmotic NaCl.[3] The pressor action of centrally administered AII and hypertonic NaCl, mediated via the AV3V region, depends in part upon elaboration of vasopressin.[3] There is also anatomical and functional evidence that this region exerts direct control over release of VP.[3] Importantly, lesion of this area in rats completely prevents the development of DOCA/salt hypertension.[3] Based on these findings we decided to examine the relationship between the AV3V region and VP in DOCA/salt hypertension. Since DOCA hypertension is associated with elevated blood levels of VP[5] we hypothesized that the protective effect of the AV3V lesion might be due to failure of lesioned rats to increase VP levels in response to DOCA/salt. Similar to previous studies[3] we found that AV3V lesion in normal rats prevented the development of DOCA/salt hypertension. AV3V-lesioned rats given VP did show an increase in arterial pressure after DOCA but the increase was attenuated. Although the amount of VP given to AV3V-lesioned rats was sufficient to raise the plasma level to an amount comparable to that seen in the sham-lesioned rats, VP treatment alone did not fully restore DOCA/salt hypertension in the lesioned rats. These data suggested that the protective effect of the AV3V lesion in DOCA/salt hypertension was not simply due to prevention of increased VP secretion.

The protective effect may also involve a disruption of pathways required for a central mechanism of action of VP. This hypothesis is supported by our findings in DI rats that after AV3V lesion VP could no longer induce DOCA/salt hypertension. Furthermore, the vasoconstrictor effects of electrical stimulation of the AV3V region are attenuated in DI rats.[6]

In conclusion, our data suggest that the ability of the AV3V lesion to prevent DOCA hypertension may be due to (1) an alteration in the release of VP as well as (2) interference with pathways required for the mechanism of a potential central action of vasopressin.

TABLE 1

THE EFFECT OF AV3V LESION AND VP TREATMENT ON SYSTOLIC ARTERIAL PRESSURE * IN DOCA/SALT-TREATED SPRAGUE-DAWLEY RATS

	Systolic Arterial Pressure (mm Hg)					
		Weeks Post-DOCA				
	Pre-DOCA †	1	2	3	4	5
Sham-lesion rats (N=10)	118 ± 2	124 ± 2	147 ± 3 §	177 ± 4 ‡	187 ± 4 ‡	197 ± 7 ‡
AV3V-lesion rats (N=9)	121 ± 4	124 ± 3	133 ± 4	134 ± 3	120 ± 6	119 ± 5
AV3V-lesion rats + VP (N=10)	118 ± 3	135 ± 3	144 ± 2 §	156 ± 6 ‡	155 ± 3 ‡	161 ± 2 ‡

* Arterial pressure measured in conscious, restrained rats with an indirect tail cuff.
† Average of measurements taken on three separate days.
‡ Significantly different from other groups, $p < .05$.
§ Significantly different from AV3V-lesioned rats, $p < .05$.
Differences determined by one-way analysis of variance and the Newman-Keuls ranking test.

TABLE 2

THE EFFECT OF AV3V LESION ON VP-REPLACED DI RATS ON SYSTOLIC ARTERIAL PRESSURE *
AFTER DOCA/SALT TREATMENT

	Systolic Arterial Pressure (mm Hg)					
		Weeks Post-DOCA				
	Pre-DOCA	1	2	3	4	5
Intact DI rats + VP (N=14)	123 ± 2	138 ± 3 †	159 ± 2 †	174 ± 2 †	187 ± 2 ‡	194 ± 2 †
AV3V-lesion DI rats + VP (N=10)	122 ± 3	120 ± 3	124 ± 3	115 ± 4	121 ± 2	123 ± 5

* Systolic arterial pressure measured in conscious, restrained rats with an indirect tail cuff method.

† Significantly different from AV3V-lesioned DI rats $p < .01$ determined by one-way analysis of variance and the Newman-Keuls ranking test.

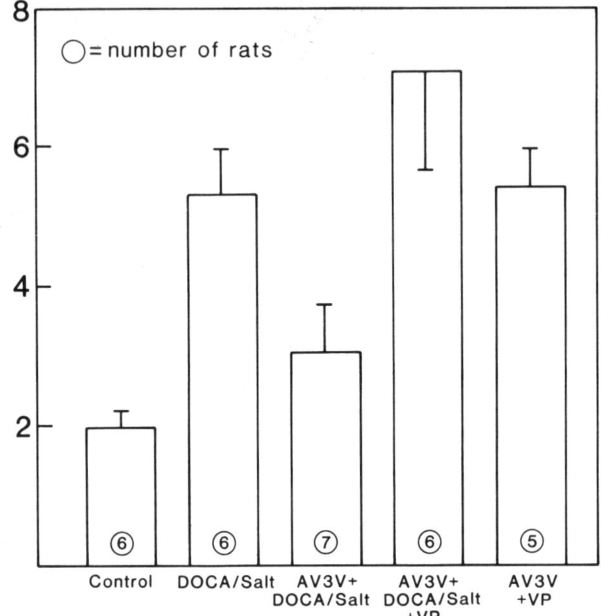

FIGURE 1. Effects of DOCA/salt treatment, DOCA/salt + AV3V lesion, DOCA/salt + AV3V lesion + VP, and AV3V lesion + VP on plasma arginine-vasopressin levels.

ACKNOWLEDGMENTS

We gratefully thank Dr. Melvyn Gluckman, Director, Department of Pharmacology (Warner-Lambert, Ann Arbor, Mich.) for generously supplying Pitressin® tannate for these studies, and to Dr. Dianna Van Orden and Kevin Taylor for measurement of plasma vasopressin.

REFERENCES

1. CROFTON, J. T., L. SHARE, R. E. SHADE, W. J. LEE-KWAN, M. MANNING & W. H. SAWYER. 1979. The importance of vasopressin in the development and maintenance of DOC-salt hypertension in the rat. Hypertension **1:** 31–38.
2. SAITO, T., Y. YAJIMA & T. WATANABE. 1980. Involvement of AVP in the development and maintenance of hypertension in rats. *In* Antidiuretic Hormone. S. Yoshida, L. Share & K. Yagi, Eds. : 215–225. University Park Press. Baltimore, Md.
3. BRODY, M. J. & A. K. JOHNSON. 1980. Role of the anteroventral third ventricle region in fluid and electrolyte balance, arterial pressure regulation and hypertension. Front. Neuroendocrinol. **6:** 249–292.

4. MATSUGUCHI, H., P. G. SCHMID, D. VAN ORDEN & A. L. MARK. 1981. Does vasopressin contribute to hypertension in the Dahl strain? Hypertension **3:** 174–181.
5. MÖHRING, J., B. MÖHRING, M. PETRI & D. HAACK. 1977. Vasopressor role of ADH in the pathogenesis of malignant DOC hypertension. Am. J. Physiol. **232:** F260–F269.
6. BERECEK, K. H., R. L. WEBB, K. W. BARRON & M. J. BRODY. 1982. Vasopressin projections and central control of cardiovascular function. (This volume.)

RENAL HYPERTENSION IN THE BRATTLEBORO DIABETES INSIPIDUS RAT *

C. I. Johnston, J. M. Abrahams, and R. L. Woods

Monash University Department of Medicine
Prince Henry's Hospital
Melbourne, Victoria
Australia

INTRODUCTION

Arginine vasopressin (AVP), the antidiuretic hormone, is also a potent vasoconstrictor and raises the blood pressure in both animals and humans. This vasopressor response has in the past been regarded solely as a pharmacological action. More recently however, evidence has been accumulated that vasopressin may also play an important role as a vasoconstrictor in response to dehydration and hemorrhage, and in the pathogenesis of hypertension.[1, 2] Plasma vasopressin levels have been reported to be increased in experimental DOCA-salt hypertension,[3] in the spontaneously hypertensive rat,[4] in experimental renal hypertension[5] and in human hypertension.[6] Furthermore, Möhring[1] has shown that administration of a specific vasopressin antibody transiently lowers the blood pressure in a variety of experimental hypertensive states. Crofton *et al.*[7] have demonstrated that specific vasopressin vascular receptor antagonists will also lower the blood pressure in hypertension. However a recent report[8] failed to confirm the hypotensive action of vasopressin vascular receptor antagonists in DOCA-salt hypertension, in adrenal regeneration hypertension, or in experimental renal hypertension in rats.

The Brattleboro diabetes insipidus (DI) rat, genetically deficient in vasopressin, provides an ideal model for elucidating the role of vasopressin in the pathogenesis of hypertension. Crofton *et al.*[7] have reported that the Brattleboro (DI) rat does not develop hypertension when given DOCA and salt to drink. However this experimental model of hypertension is a very volume-dependent form of hypertension and thus the failure of the DI rat to develop this form of hypertension may be due to the lack of the antidiuretic properties of vasopressin rather than the vasoconstrictor properties of the hormone.

We have therefore attempted to produce experimental renal hypertension in the Brattleboro rat, using both the two-kidney, one-clip (2K-1C) Goldblatt hypertensive model and one-kidney, one-clip (1K-1C) Goldblatt hypertensive model. The importance of the antidiuretic action of vasopressin in these models of experimental renal hypertension was assessed by administering 1-desamino-8-D-arginine vasopressin (DDAVP), the antidiuretic but nonpressor synthetic analogue of AVP, to the animals.

* Supported by a Grant in Aid by the National Health and Medical Research Council of Australia. R. L. Woods is an N.H. and M.R.C. Biomedical Scholar.

0077–8923/82/0394–0398 $1.75/0

Renal Hypertension in the Brattleboro Diabetes Insipidus Rat

Long-Evans rats and homozygous Brattleboro DI rats (110–140 g) were made hypertensive by placing a 0.2 mm silver clip around the left renal artery and either leaving the right kidney untouched (2K-1C) or removing the right kidney (1K-1C). Systolic blood pressure by tail plethysmography and body weight were measured every 5 days for 35 days. At day 35 the rats were decapitated and the first 2.5 ml of blood collected into 2,3-dimercapto-1-propanol (BAL)-EDTA for plasma renin activity and plasma angiotensin II estimations while the remainder was collected for the measurements of hematocrit, plasma sodium, potassium, vasopressin, and DDAVP. Hearts were removed and ventricular weights expressed as a percentage of body weight.

Both 2K-1C Long-Evans and 2K-1C Brattleboro rats developed hypertension at the same rate and to the same degree (Figure 1A). Also, these two groups of rats had similar degrees of cardiac hypertrophy. Brattleboro rats with 1K-1C Goldblatt renal hypertension developed progressive increases in the blood pressure at the same rate as 1K-1C Long-Evans rats. However the absolute level of blood pressure achieved between 30–35 days was significantly less. The degree of cardiac hypertrophy in the hypertensive 1K-1C Brattleboro rats was also less than in the 1K-1C Long-Evans hypertensive rats (Figure 1B).

The plasma osmolality and sodium were also significantly higher in the Brattleboro rats than the Long-Evans rats, as was the hematocrit.

Effect of DDAVP on One-Kidney, One-Clip (1K-1C) Goldblatt Renal Hypertensive Rats

To determine whether in the 1K-1C Brattleboro rats the failure to reach the same hypertensive level of blood pressure as the 1K-1C Long-Evans rats was due to the lack of the vasoconstrictor properties or to the antidiuretic action of AVP, rats were administered DDAVP ("Minirin") in their drinking water. Rats were placed in individual metabolic cages and DDAVP was given in a 2 μg/ml concentration between days 18–35 following renal artery constriction.[9] The intake of DDAVP averaged 27.8 $\pm$ 2.8 μg/100 g/day and the plasma levels reached 403 $\pm$ 36 pg/ml.

DDAVP had little effect on the water intake, urine volume, body weight, hematocrit, or blood pressure of the hypertensive Long-Evans rats. DDAVP reduced fluid intake and urine flow in the Brattleboro rats to levels not significantly different from those of the control Long-Evans rats.[10] Also, body weight increased and hematocrit fell in the 1K-1C Brattleboro rats treated with DDAVP. Systolic blood pressure rose significantly by the second day of treatment and was maintained for seven days (Figure 2) at levels similar to those in hypertensive 1K-1C Long-Evans rats. Cardiac hypertrophy also increased in the DDAVP-treated Brattleboro rats.

These results clearly show that correcting the water balance and hence volume in these Brattleboro rats, using a vasopressin analogue without pressor properties, elevated the blood pressure to the same hypertensive levels as hypertensive Long-Evans rats. This suggests that the antidiuretic properties of AVP may be important in maintaining blood pressure in volume-dependent forms of hypertension.

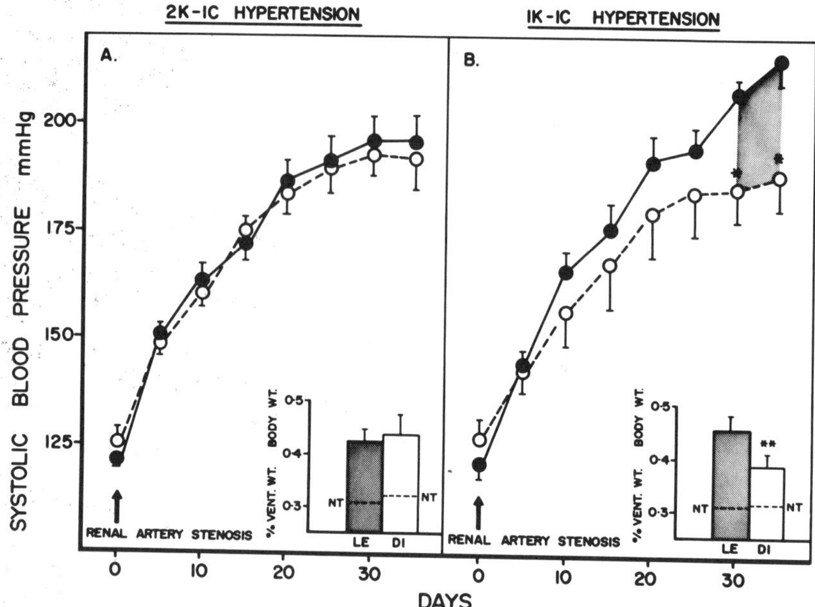

FIGURE 1. A. Systolic blood pressure in two-kidney, one-clip (2K–1C) Long-Evans (●), $N = 20$, and homozygous Brattleboro (○), $N = 31$, rats following unilateral renal artery stenosis. B. Systolic blood pressure in one-kidney, one-clip (1K–1C) Long-Evans (●), $N = 23$, and homozygous Brattleboro (○), $N = 16$, rats following left renal artery stenosis with right kidney excision. Shading and * $p < 0.01$ Brattleboro blood pressure < Long-Evans blood pressure.

Inserts show cardiac hypertrophy expressed as % ventricular weight to body weight. ** $p < 0.005$.

Renin-Angiotensin System in Hypertensive Brattleboro DI Rats

There is evidence in dehydration [11] and hemorrhage [12] to suggest that vasopressin and the renin-angiotensin system act as interrelated vasoconstrictor systems. It has been suggested that DI rats may maintain their blood pressure and become hypertensive by compensation through excessive activity of the renin-angiotensin system. We decided to investigate this postulate.

In a further series of studies plasma renin activity (PRA) and plasma angiotensin II (PAII) were measured at day 35 following renal artery stenosis in 2K-1C Goldblatt and 1K-1C Goldblatt hypertensive Long-Evans and Brattleboro rats. 2K-1C Goldblatt hypertensive Long-Evans rats showed significant increases in PRA and PAII compared to normotensive Long-Evans controls. PRA and PAII were elevated to comparable levels in 2K-1C hypertensive Brattleboro rats (FIGURE 3). Neither 1K-1C Long-Evans hypertensive rats nor 1K-1C Brattleboro hypertensive rats had significant changes in the PRA or PAII compared to normotensive controls.

These results do not support the contention that in the vasopressin-deficient animal, hypertension is maintained by excessive increases in the renin-angiotensin system.

Discussion and Conclusions

Brattleboro rats, deficient in vasopressin, were able to become hypertensive after renal artery stenosis. The blood pressure in the 2K-1C Goldblatt renal hypertensive model in the acute stages is dependent on the renin-angiotensin system and it is not surprising therefore that the Brattleboro rat develops hypertension at the same rate and to the same degree as the Long-Evans rats. The 1K-1C Goldblatt renal hypertensive model is thought to be a more volume-dependent form of hypertension. The findings in the 1K-1C Goldblatt renal hypertensive Brattleboro rats are in keeping with this contention. Brattleboro rats are relatively chronically volume-depleted, as shown by a higher hematocrit and plasma sodium concentration. These rats developed hypertension in response to renal artery stenosis and contralateral nephrectomy, but the absolute level of blood pressure reached and the degree of cardiac hypertrophy were lower than in the hypertensive 1K-1C Long-Evans rats. The lower blood pressure in the 1K-1C Brattleboro rats may reflect their volume depletion. Their blood pressure levels and cardiac hypertrophy could however be made to reach hypertensive levels comparable to Long-Evans rats by correcting their water balance. This was achieved with administration of the synthetic antidiuretic vasopressin analogue, DDAVP, which has little vasoconstrictor activity. These results suggest that vasopressin may play an important role via volume homeostasis in this form of hypertension.

Studies in the hypertensive Brattleboro rat do not support a role for vasopressin as a direct vasoconstrictor in the pathogenesis of hypertension but do provide evidence that its antidiuretic properties may contribute via volume homeostasis to the absolute level of blood pressure reached in hypertension.

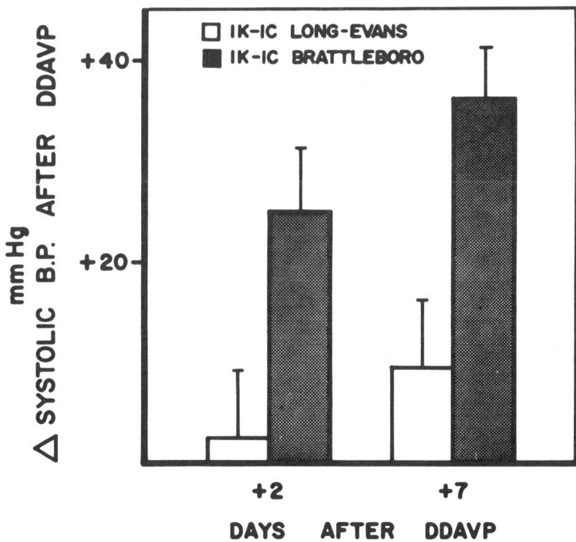

FIGURE 2. Changes in systolic blood pressure at days 2 and 7 after administration of DDAVP in the drinking water (2 μg/ml) in 1K–1C hypertensive Long-Evans $N = 6$, and in 1K–1C hypertensive Brattleboro $N = 11$, rats.

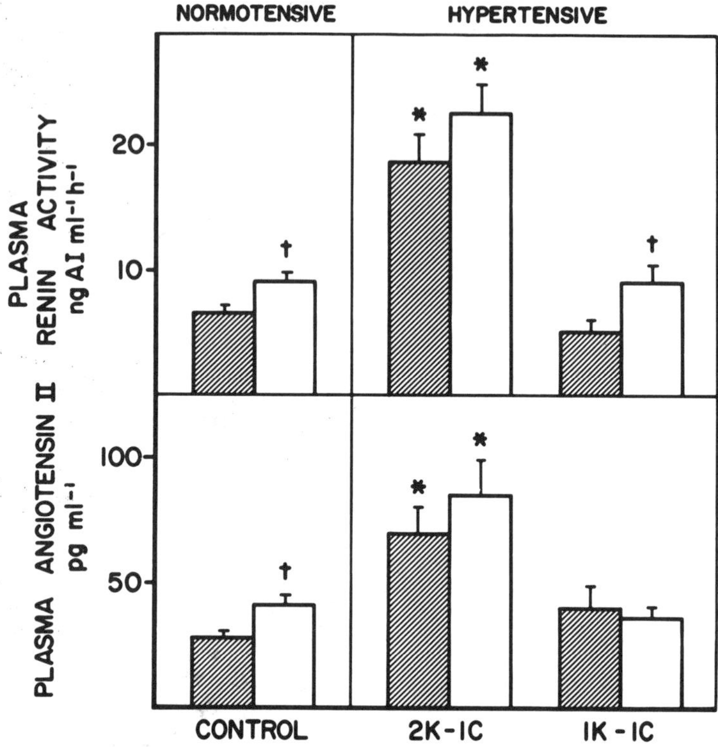

FIGURE 3. *Top Panel:* Plasma renin activity (PRA) in normotensive, $N = 19$, hypertensive 2K–1C, $N = 15$, and hypertensive 1K–1C, $N = 9$, Long-Evans rats (■) and in normotensive, $N = 18$, hypertensive 2K–1C, $N = 9$, and hypertensive 1K–1C, $N = 8$, Brattleboro rats (□). *Lower Panel:* Plasma angiotensin II (PAII) levels in rats as in top panel. * $p < 0.001$ Hypertensive rats compared to normotensive controls. † $p < 0.05$ Brattleboro rats compared to Long-Evans rats.

REFERENCES

1. MÖHRING, J., R. ARBOGAST, R. DUSING, K. GLANZER, J. KINTZ, J-F. LIARD, J. A. MACIEL, JR., J. P. MONTANI & J. SCHOUN. 1980. Vasopressor role of vasopressin in hypertension. Brain and Pituitary Peptides, Ferring Symposium, pp. 157–167. Munich 1979. Karger. Basel.
2. JOHNSTON, C. I., M. NEWMAN & R. L. WOODS. 1981. Role of vasopressin in cardiovascular homeostasis and hypertension. Clin. Sci. **61:** 129s–139s.
3. MÖHRING, J., B. MÖHRING, M. PETRI & D. HAACK. 1977. Vasopressor role of ADH in the pathogenesis of malignant DOC hypertension. Am. J. Physiol. **232:** F260–F269.
4. CROFTON, J. T., L. SHARE, R. E. SHADE, C. ALLEN & D. TARNOWSKI. 1978. Vasopressin in the rat with spontaneous hypertension. Am. J. Physiol. **135:** H361–H365.
5. JOHNSTON, C. I., P. T. PULLAN & N. M. A. WALTER. 1978. Vasoactive pep-

tides in experimental renal hypertension. Klin. Wochenschr. **56** (suppl. I): 81–85.
6. PADFIELD, P. L., J. J. BROWN, A. F. LEVER, J. J. MORTON & J. I. S. ROBERTSON. 1976. Changes of vasopressin in hypertension: cause or effect? Lancet **1:** 1255–1257.
7. CROFTON, J. T., L. SHARE, R. E. SHADE, W. J. LEE-KWON, M. MANNING & W. H. SAWYER. 1979. The importance of vasopressin in the development and maintenance of DOC-salt hypertension in the rat. Hypertension **1**(1): 31–38.
8. RABITO, S. F., O. A. CARRETERO & A. G. SCICLI. 1981. Evidence against a role of vasopressin in the maintenance of high blood pressure in mineralocorticoid and renovascular hypertension. Hypertension **3:** 34–38.
9. WOODS, R. L. & C. I. JOHNSTON. 1981. Importance of antidiuretic properties of vasopressin in experimental renal hypertension. Clin. Exp. Pharm. Physiol. (In press.)
10. WOODS, R. L. & C. I. JOHNSTON. 1982. The role of vasopressin in hypertension: studies using the Brattleboro diabetes insipidus rat. Am. J. Physiol. (In press.)
11. ANDREWS, C. E., JR. & B. M. BRENNER. 1981. Relative contributions of arginine vasopressin and angiotensin II to maintenance of systemic arterial pressure in the anaesthetised water-deprived rat. Circ. Res. **48:** 254–258.
12. COWLEY, A. W. JR., S. J. SWITZER & M. M. GUINN. 1980. Evidence and quantification of the vasopressin arterial pressure control system in the dog. Circ. Res. **46:** 58–66.

ALCOHOL AND BLOOD PRESSURE REGULATION AFTER HEMORRHAGE (BRATTLEBORO RAT MODEL)

S. J. Walter and J. F. Laycock

Department of Physiology
Charing Cross Hospital Medical School
London W6 8RF England

Vasopressin and Blood Pressure Regulation

Although high concentrations of vasopressin can cause an increase in blood pressure, it has been generally accepted in the past that physiological levels of vasopressin do not play any significant role in blood pressure regulation. However, some recent experiments[1,2] suggest that the importance of the hemodynamic effects of vasopressin may have been underestimated. Brattleboro rats, which lack the ability to synthesize vasopressin and consequently suffer from diabetes insipidus (DI) provide an ideal animal model for research in this field, and, in experiments that have already been reported in detail,[3] we have attempted to provide direct evidence for the involvement of vasopressin in blood pressure regulation. The effect of a 2% hemorrhage was studied in three groups of animals: Long-Evans (LE) rats, Brattleboro (DI) rats, and Brattleboro rats infused with vasopressin in a manner calculated to induce plasma antidiuretic activities, both prior to and in the period immediately following a 2% hemorrhage, similar to those measured in Long-Evans rats. The results of these experiments are shown in Figure 1. Mean values for the mean arterial blood pressure during the control (pre-hemorrhage) period and *immediately* after the 2% hemorrhage do not differ significantly in the three groups of animals. However, both 5 and 10 minutes after the hemorrhage, the blood pressure observed in DI rats is significantly lower than that of LE rats. In contrast, in vasopressin-infused DI rats the blood pressure observed after a 2% hemorrhage does not differ significantly from the equivalent pressure in LE rats. These results strongly suggest that vasopressin released following an acute hemorrhage in normal anesthetized rats influences the rate of recovery of arterial blood pressure.

Alcohol and Blood Pressure Regulation

It has long been known that alcohol has a diuretic action and it is now generally accepted that this diuresis results from an inhibitory influence of alcohol on the release of vasopressin from the neurohypophysis. We have attempted to determine whether administration of alcohol reduces the rate of recovery of arterial blood pressure following a hemorrhage in anesthetized rats, and whether inhibition of vasopressin is implicated in any alcohol-induced impairment of blood pressure regulation.

Brattleboro and Long-Evans rats were anesthetized with Inactin and after a control period of at least 10 minutes, during which the arterial blood pressure

0077–8923/82/0394–0404 $1.75/0 © 1982, NYAS

remained stable, infused intravenously over a 10-minute period with either 1 ml of 75 mmol/L NaCl (saline) or 1 ml of a 50:50 mixture of ethanol and 150 mmol/L NaCl (alcohol). After a further 10 minutes had elapsed, each animal was bled via a catheterized femoral artery into pots containing trace amounts of heparin. The amount of blood removed was equivalent to 2% of the body weight, the process being completed in approximately two minutes. The experiment was terminated either 5, 10, or 20 minutes later and a final blood sample obtained. Arterial blood pressure was monitored throughout the

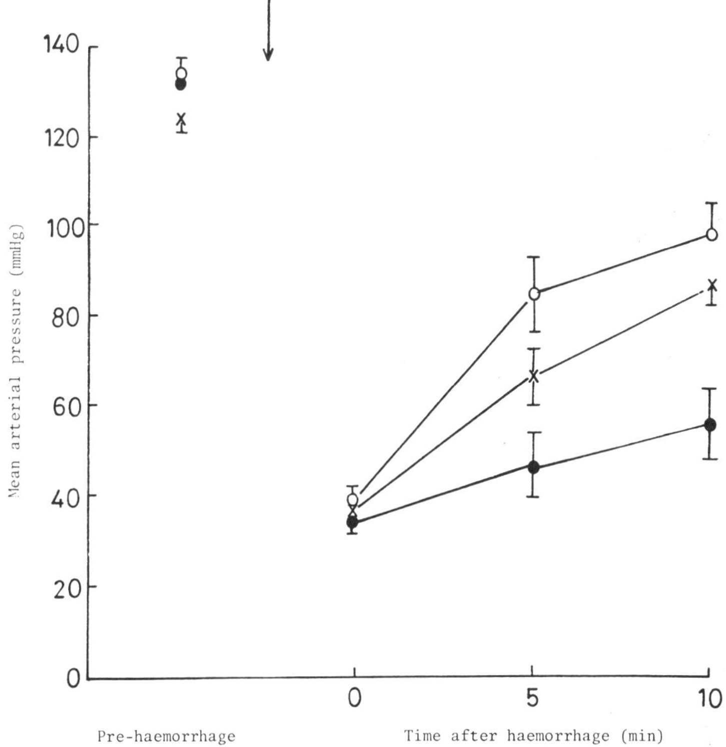

FIGURE 1. Mean arterial blood pressures (± standard errors of the means (SE)) in Long-Evans rats [×], Brattleboro rats [●], and Brattleboro rats given vasopressin replacement [○], measured before and after (0, 5, and 10 min) a single 2% hemorrhage which is represented by the vertical arrow.[3]

experiment. Antidiuretic activity was measured on plasma taken from the first 1.2 ml of the final blood sample from each LE rat, using a bioassay technique.[4] The results shown in FIGURE 2 confirm our previous findings. Thus the initial measurements of mean arterial blood pressure in the two strains of rats do not differ significantly, whereas the rate of recovery after hemorrhage is significantly lower in Brattleboro rats than in the parent Long-Evans strain ($p < 0.001$). This difference between the two strains of rats is associated with a difference in the response of the heart to the acute hemorrhage. The mean heart rate in

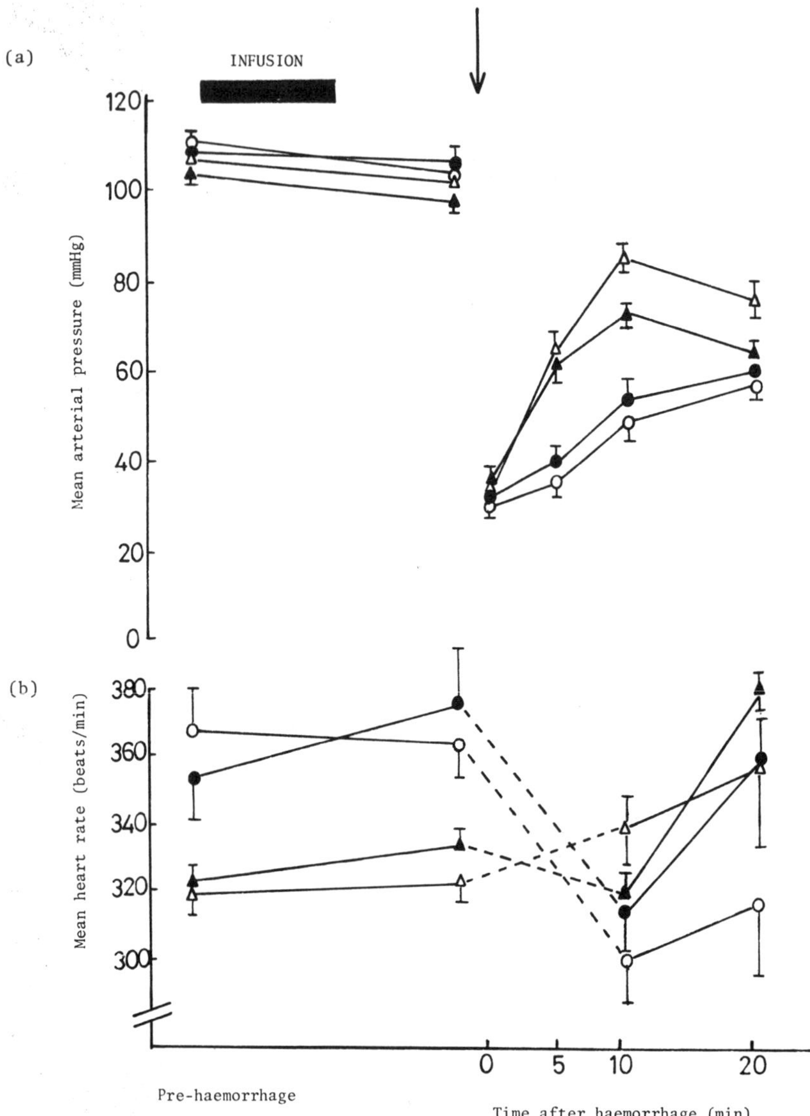

FIGURE 2. (a) Mean arterial blood pressure (±SE) and (b) mean heart rate (±SE) in saline-infused LE rats [△], alcohol-infused LE rats [▲], saline-infused DI rats [○] and alcohol-infused DI rats [●] measured before and after the infusion (shown by the horizontal bar) of either 1 ml of 75 mmol/L NaCl or 1 ml of a 50:50 mixture of ethanol and 150 mmol/L NaCl, and after a single 2% hemorrhage (represented by the vertical arrow).

saline-infused DI rats measured before hemorrhage is significantly greater ($p < 0.005$) than that of LE rats. This may indicate a reduced vascular smooth muscle tone in the rats lacking vasopressin. Furthermore, whereas in Brattleboro rats there is a marked fall in heart rate in response to the hemorrhage, Long-Evans rats show a slight but not significant rise (FIGURE 2). This difference between the two groups of rats is highly significant ($p < 0.001$).

Both the plasma alcohol concentration (LE rats 52.7 ± 1.8 mmol/L; DI rats 52.6 ± 2.4 mmol/L) and the consequent increase in plasma osmolality (LE rats 52.5 mosmol/kg H_2O; DI rats 53.0 mosmol/kg H_2O) were similar in Long-Evans and Brattleboro rats infused with 0.5 ml of ethanol. This concentration of alcohol brought about an immediate and significant rise in the heart rate in both strains of animals ($p < 0.05$ in LE rats, $p < 0.005$ in DI rats). There is little immediate effect on arterial blood pressure as neither the slight fall before the hemorrhage nor the profound fall brought about by the hemorrhage itself are significantly influenced by the infusion of alcohol. However, in LE rats it is clear that a high plasma alcohol concentration leads to a reduction in the rate of recovery of arterial pressure. The rise in pressure measured in the first 10 minutes following the hemorrhage is significantly greater ($p < 0.005$) in saline-infused LE rats than in LE rats infused with alcohol prior to the hemorrhage. Although there is no equivalent difference between the two groups of LE rats in the change in pressure between 10 and 20 minutes after the hemorrhage, the final (20 min) pressure remains significantly lower ($p < 0.05$) in the alcohol-infused rats.

TABLE 1

PLASMA ANTIDIURETIC ACTIVITIES

Time after hemorrhage (min)	5	10	20
Saline-infused Long-Evans rats			
Number of rats	4	7	5
Antidiuretic activity (μU/ml)	53: <7.5–147	56: 0–114	<65: <8–182
Alcohol-infused Long-Evans rats			
Number of rats	5	10	3
Antidiuretic activity (μU/ml)	91: <20–296	78: <22–224	<50: <50–78

Plasma antidiuretic activities (medians and ranges) measured 5, 10, and 20 minutes after a single 2% hemorrhage in saline-infused and alcohol-infused (see text) Long-Evans rats. Some samples contained less antidiuretic activity than could be detected with the bioassay preparation, and are therefore expressed as "less than" (<) the assay sensitivity. There were no significant differences between the medians (using the non-parametric median test) for the two groups of LE rats at either 5, 10, or 20 min post-hemorrhage.

In marked contrast, the rate of recovery of arterial blood pressure following an acute 2% hemorrhage is almost identical in alcohol-infused and saline-infused Brattleboro rats. This difference between the two strains of rats suggests that an inhibition of vasopressin release may be an important element in the reduced rate of recovery of arterial blood pressure in alcohol-infused Long-Evans rats. However, results obtained from the bioassay of antidiuretic activity in blood samples from LE rats (TABLE 1) do not confirm this hypothesis. We were unable to demonstrate any significant difference between the antidiuretic activity in the blood taken from alcohol-infused and saline-infused animals. These results are of interest in that they provide direct evidence in support of the suggestion [5] that hemorrhage provides a stimulus for vasopressin release sufficiently profound to override any influence resulting from a relatively high plasma alcohol concentration.

In conclusion, we have shown that in Long-Evans, but not Brattleboro rats, a high plasma alcohol concentration is associated with a reduction in the rate of recovery of arterial blood pressure following an acute hemorrhage. However, preliminary bioassay results suggest that an inhibition of vasopressin release may not provide the mechanism whereby alcohol exerts its effect.

References

1. Rocha e Silva, M. Jr. & M. Rosenberg. 1969. The release of vasopressin in response to haemorrhage and its role in the mechanism of blood pressure regulation. J. Physiol. **202:** 535–557.
2. Cowley, A. W. Jr., E. Monos & A. C. Guyton. 1974. Interaction of vasopressin and the baroreceptor reflex system in the regulation of arterial blood pressure in the dog. Circ. Res. **34:** 505–514.
3. Laycock, J. F., W. Penn, D. G. Shirley & S. J. Walter. 1979. The role of vasopressin in blood pressure regulation immediately following acute haemorrhage in the rat. J. Physiol. **296:** 267–275.
4. Penn, W. 1966. An assessment of the methods for the concentration of vasopressin from the blood and urine. Part 1. J. Sci. Tech. **12:** 151–169.
5. Tata, P. S. & R. Byzalkov. 1966. Vasopressin studies in the rat. III. Inability of ethanol anaesthesia to prevent ADH secretion due to pain and haemorrhage. Pflügers Arch. **290:** 294–297.

HEMODYNAMICS OF CONSCIOUS BRATTLEBORO RATS

J. Zicha, P. Karen, V. Krpata, H. Dlouhá, and J. Křeček

Institute of Physiology
Czechoslovak Academy of Sciences
Prague 4, CS 142 20 Czechoslovakia

Vasopressin exerts distinct cardiovascular effects in many species including the rat.[8] However, its role in blood pressure regulation is still not clear.[6] Brattleboro rats homozygous for diabetes insipidus (DI) completely lack vasopressin,[9] while 50% of normal vasopressin synthesis was demonstrated in their heterozygous (HZ) littermates.[7] Adult female DI and HZ rats were compared in order to ascertain whether the inborn absence of vasopressin could influence their hemodynamics.

Material and Methods

Systolic, mean arterial and diastolic blood pressures, pulse pressure, cardiac output, stroke volume, heart rate, total peripheral resistance, and central blood volume were studied in conscious, unrestrained Brattleboro rats (23 HZ and 33 DI) aged 10–38 weeks. Rats (weaned at the age of 30 days) were given tap water and consumed a diet containing 1% NaCl. Rats with a daily water intake greater than 50% of body weight were considered to be DI.

The tip of a heparinized PE 50 catheter was inserted through the carotid artery into the aortic arch and a PE 10 catheter was placed through jugular vein into the right atrium under pentobarbital anesthesia (40 mg/kg) 24–48 hours before the experiment. Implanted catheters were exteriorized behind the rat's ears. After a one-hour adaptation of animals to the measuring conditions, 4–6 determinations of cardiac output, arterial blood pressure, and heart rate were made at 5-min intervals.

The cardiac output was measured by dye dilution using a single-channel microcuvette densitometer.[1] The response of its photocell was linear up to 15 mg of indocyanine green per liter. A single dose (9 μl) of 0.25% indocyanine green solution (Cardiogreen, HWD) was injected into the right atrium. Aortic blood samples for the determination of dye concentration were withdrawn at a rate of 1.1 ml/min (less than 2% of cardiac output). The withdrawn blood (0.6 ml) was reinfused after the measurement. The low perfusion rate of the cuvette (18 μl/sec), as related to its internal volume (25 μl), caused substantial alteration of the measured changes of dye concentration. The time course of the indicator concentration in the aortic blood entering the cuvette ($C_1(t)$) was registered as the time course of optical density changes that were proportional to the mean indicator concentration in the cuvette ($C_2(t)$). The dependence of these two functions was given by the differential equation $dC_2(t)/dt = a[C_1(t) - C_2(t)]$, where a was a constant given by the properties of the mixing system. By rearranging this equation as $C_1(t) = C_2(t) + [dC_2(t)/dt]/a$, it was possible to reconstruct the time course of aortic blood

0077-8923/82/0394-0409 $1.75/0

dye concentration. Constant a (0.80/sec) was estimated from the time course of the optical density changes caused by the entry of the pulsating blood front to the saline-filled cuvette. The numerical estimate of the first derivative of $C_2(t)$ was calculated from the points of the recorded dilution curve $dC_2(t)/dt = [C_{2,\,i+1} - C_{2,\,i-1}]/[t_{i+1} - t_{i-1}]$. The recorded dilution curves were digitized and the original time course of aortic blood dye concentration was reconstructed (FIGURE 1) using a Hewlett Packard 21 calculator with digitizer. We used a new method of correcting the indicator recirculation.[4] A third-degree polynomial was fitted in all points of the S-shaped semilog transformed downslope of the dilution curve by a least-square procedure. The exponential decay constant was evaluated from the slope of the tangent at the inflection point of the fitted polynomial. The cardiac output (CO in ml/min) was

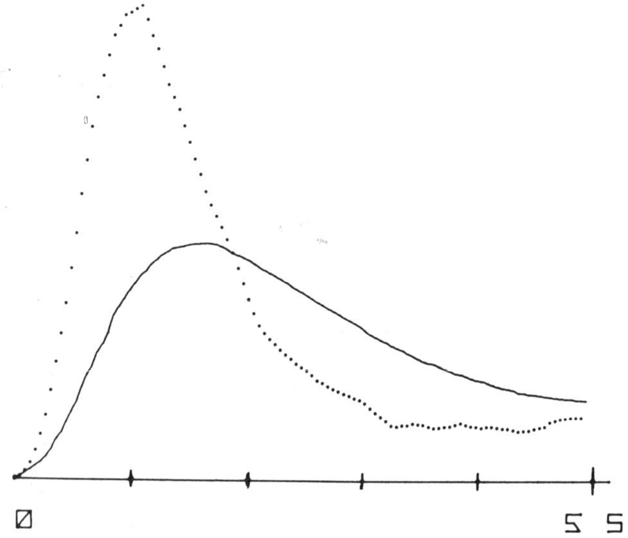

FIGURE 1. A comparison of the recorded dye dilution curve (solid line) with a reconstruction of the time course of dye concentration in the rat aortic blood (dotted line). Abscissa: time (sec).

calculated as $CO = K \cdot D \cdot V \cdot c/S$, where K was a constant correcting the plasma calibration value for hematocrit of rat blood, D was a plasma calibration value, i.e. mm of recording pen deflection (from plasma baseline) caused by measuring the solution containing one dose of injected indicator dissolved in a known volume (V in ml) of human plasma, c was the recorder chart speed (mm/min), S was the area (mm^2) under the corrected original dilution curve. The central blood volume was calculated from each dilution curve as cardiac output multiplied by mean transit time, i.e. the time that had elapsed between indicator injection and the moment at which half the area under the corrected curve was reached. The blood pressure was recorded on dual-channel HP 321 amplifier with MP-15 transducers. The use of the same arterial catheter artificially lowered the recorded pressure during blood withdrawal. Because the systemic pressure of anesthetized rats (recorded from the femoral artery) was not influenced by a pump action, cardiac output was related to blood pressure and heart rate recorded during 30 sec prior to cardiac output measurement.

Student's t test was used to evaluate the group differences.

RESULTS

The greater body weight and body surface and the higher values of hematocrit were observed in HZ rats as compared to DI rats (TABLE 1). There were no differences in the mean arterial pressure and diastolic blood pressure, but systolic blood pressure and pulse pressure of DI rats exceeded those of HZ rats.

TABLE 1

THE HEMODYNAMICS OF CONSCIOUS HOMOZYGOUS (DI) AND HETEROZYGOUS (HZ) BRATTLEBORO FEMALE RATS

Genotype	HZ	DI	
Number of animals	23	33	
Age (weeks)	21.5±2.37	16.8±1.04	NS
Body weight (g)	202.3±6.33	165.3±3.75	p < .001
Body surface * (cm^2)	343.8±7.16	300.7±4.54	p < .001
Hematocrit (%)	47.3±0.32	46.2±0.36	p < .05
Systolic blood pressure (mm Hg)	132.8±2.39	138.7±1.64	p < .05
Mean arterial pressure (mm Hg)	121.8±2.18	124.5±1.52	NS
Diastolic blood pressure (mm Hg)	108.3±2.10	110.3±1.66	NS
Pulse pressure (mm Hg)	24.6±1.11	28.4±1.19	p < .05
Cardiac output			
($ml \cdot min^{-1} \cdot 100\ g^{-1}$)	38.1±1.22	48.4±1.73	p < .001
($ml \cdot min^{-1} \cdot 100\ cm^{-2}$)	22.2±0.69	26.4±0.89	p < .001
Total peripheral resistance			
($mm\ Hg \cdot ml^{-1} \cdot min \cdot 100\ g$)	3.30±0.141	2.73±0.135	p < .001
($mm\ Hg \cdot ml^{-1} \cdot min \cdot 100\ cm^2$)	5.62±0.224	4.97±0.233	NS
Heart rate (beats/min)	398±8.3	415±7.8	NS
Stroke volume (μl/100 g)	97±3.9	118±4.5	p < .002
(μl/100 cm^2)	57±2.2	64±2.4	p < .05
Central blood volume (ml/100 g)	0.96±0.033	1.42±0.048	p < .001
(ml/100 cm^2)	0.57±0.021	0.78±0.025	p < .001

* Body surface was calculated from body weight (g) as $10 \cdot \text{body weight}^{0.667}$.
NS—non-significant ($p \geq .05$).

A significantly greater cardiac output of DI rats (due to greater stroke volume) was accompanied by lower total peripheral resistance. Only in DI rats cardiac output correlated significantly with mean arterial pressure ($r = -.440$, $p < .02$, $N = 33$). The central blood volume of DI rats was about 35–40% greater than that of HZ rats.

DISCUSSION

The observation of slightly higher systolic blood pressure in DI rats (as compared to HZ females) confirmed previously reported data.[10] These findings do not support the idea that vasopressin plays an important role in the setting of blood pressure level. However, its absence in DI rats was accompanied by pronounced hemodynamic abnormalities (low systemic resistance, high stroke volume and cardiac output, enlarged central blood volume). The hemodynamic parameters of HZ rats were closely analogous to the data reported in conscious albino rats.[3, 5] The central blood volume of HZ rats was 16% of their total blood volume, while it represented 24% in DI rats (calculated from the average values of blood volume—HZ: 5.97 ± .15 ml/100 g b.w., $N = 16$; DI: 6.03 ± .09 ml/100 g b.w.; unpublished observations). The hemodynamic effects of vasopressin[8] would explain the major part of the observed hemodynamic differences between DI and HZ rats. It has been reported that the administration of vasopressin decreased the cardiac output in conscious rats.[2] Further experiments are therefore necessary to elucidate the question of whether the observed hemodynamic peculiarities of DI rats are caused merely by the absence of vasopressin or whether they are the consequence of the development of rat hemodynamics in the absence of this hormone.

Summary

The hemodynamics of conscious adult homozygous (vasopressin deficient) and heterozygous Brattleboro female rats were studied. The systolic blood pressure and pulse pressure of rats homozygous for diabetes insipidus were greater in comparison with their heterozygous littermates, but there were no differences in their mean arterial or diastolic blood pressures. The stroke volume and cardiac output of homozygotes were greater and the total peripheral resistance was lower in comparison with heterozygotes. Enlarged central blood volume indicated the redistribution of blood towards the cardiopulmonary circulation in homozygous rats.

References

1. Albrecht, I., M. Hällback, S. Julius, Y. Lundgren, L. Stage, L. Weiss & B. Folkow. 1975. Arterial pressure, cardiac output and systemic resistance before and after pithing in normotensive and spontaneously hypertensive rats. Acta Physiol. Scand. **94:** 378–385.
2. Goldman, H. 1966. Vasopressin modulation of the distribution of blood flow in the unanaesthetized rat. Neuroendocrinology **1:** 23–30.
3. Goldman, H. & L. A. Sapirstein. 1973. Brain blood flow in the conscious and anaesthetized rat. Am. J. Physiol. **224:** 122–128.
4. Karen, P. & J. Zicha. 1981. Computations of cardiac output determined by single injection indicator dilution method: polynomial regression method to correct for recirculation of indicator. Physiol. Bohemoslov. **30:** 172–173.
5. Lundin, S., B. Folkow, P. Friberg & B. Rippe. 1980. Central blood volume in spontaneously hypertensive and Wistar-Kyoto normotensive rats. Clin. Sci. **59** (Suppl. 6): 389–391.
6. Montani, J. P., J. F. Liard, J. Schoun & J. Möhring. 1980. Hemodynamic effects of exogenous and endogenous vasopressin at low plasma concentrations in conscious dogs. Circ. Res. **47:** 346–355.
7. Moses, A. M. & M. Miller. 1970. Accumulation and release of pituitary vasopressin in rats heterozygous for hypothalamic diabetes insipidus. Endocrinology **86:** 34–41.

8. NAKANO, J. 1974. Cardiovascular response to neurohypophysial hormones. *In* The Pituitary Gland and Its Neuroendocrine Control. E. Knobil & W. H. Sawyer, Eds. **4:** 395–442. Handbook of Physiology. American Physiological Society. Bethesda, Md.
9. VALTIN, H. 1967. Hereditary hypothalamic diabetes insipidus in rats (Brattleboro strain). Am. J. Med. **42:** 814–827.
10. ZEBIDI, A., B. MARTIN, D. GRAFMEYER, M. VINCENT & C. GHARIB. 1976. Pression artérielle et système renine-angiotensine-aldostérone chez le Rat à diabète insipide héréditaire (Brattleboro). J. Physiol. Paris. **72:** 59A.

WATER BALANCE IN THE BRATTLEBORO RAT: SINGLE OR MULTIPLE DEFECTS? *

Brian R. Edwards

Department of Physiology
Dartmouth Medical School
Hanover, New Hampshire 03755

The very existence of this symposium speaks to the growing areas of physiology, pathophysiology, and related disciplines in which the Brattleboro rat has become a useful tool. Nevertheless, for most people in the field, I think it safe to say that the term "Brattleboro rat" is first associated with the concept of disorders of water balance. During the first decade following the report of their discovery,[1] most of the work performed on Brattleboro rats centered on the elucidation of the mechanisms underlying their characteristic polyuria and polydipsia, as well as on the genetics of their disorder. During the second decade, and with the advent of several successful breeding colonies throughout the world, many different aspects of the biology of the Brattleboro rat have commanded increasing attention. Nevertheless, interest in the problems of water balance of the Brattleboro rat has remained strong. As a model of hereditary hypothalamic diabetes insipidus, the Brattleboro homozygote is a most useful tool in the study of the many facets of water balance, ranging from neuroendocrinology to the renal countercurrent system.

On the one hand, the wealth of published information concerning water balance in the Brattleboro rat (hereinafter referred to as the DI rat) affords ample material for any review; on the other hand, limitations of time and space prevent justice being done to all aspects of (and all workers in) this broad field. Consequently, I shall confine my comments to a few restricted areas. The theme that I shall pursue is the following: Is the disorder of the urine-concentrating mechanism related solely to the absence of vasopressin-induced changes in water permeability of the collecting ducts, or might other defects contribute? If there are other defects, are they discrete, or are they simply secondary to the absence of vasopressin?

An indication that the concentrating defect might entail more than the absence of vasopressin-induced osmotic equilibration across the collecting ducts arose from the studies of Harrington and Valtin [2] (FIGURE 1). They administered daily subcutaneous doses of one unit of pitressin tannate in oil to DI rats for up to eight weeks, and showed that urine osmolality continued to rise throughout this period. Concentrating ability equivalent to untreated normal rats was attained only after four to five weeks; it took the full eight weeks for urine osmolality of treated DI animals to achieve values similar to those of vasopressin-treated normal rats. Osmotic equilibration across the collecting duct of the DI rats was observed on the third day of treatment, and the gradual rise in urine osmolality was therefore attributed to a progressive increase in the osmolality of the papillary interstitium. Reduction of water content and increase of urea content were primarily responsible for the restoration of papillary

* Supported by Research Grant AM 26553 from the National Institutes of Health.

0077-8923/82/0394-0414 $1.75/0 © 1982, NYAS

osmolality to normal. However, there was no explanation offered for the time delay involved. Does the slow time course for correction reflect gradual changes in other variables that impinge upon the concentrating mechanism? In addressing this question, I shall focus on three principal areas: differences in renal morphology of the DI rat that pertain to the concentrating mechanism; the effect of vasopressin on the countercurrent multiplier; and, finally, differences in the renal prostaglandin system of the DI rat.

Renal Morphology

Tisher and coworkers have examined the influence of vasopressin on the ultrastructure of both medullary [3] and cortical [4] segments of the collecting ducts of

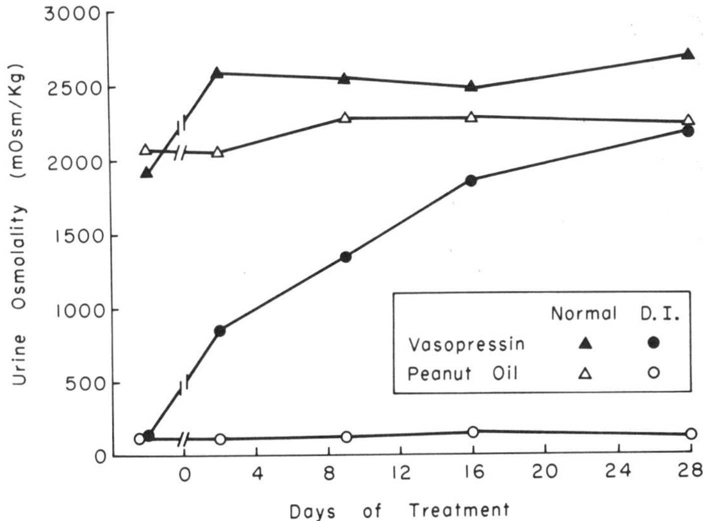

Figure 1. Urine osmolality in normal Long-Evans rats and in DI rats given either 1 U pitressin tannate in oil s.c. daily, or peanut oil as a control. (Reproduced with permission from Harrington, A. R. & H. Valtin. 1965. *Proc. Soc. Exp. Biol. Med.* **118:** 448–450.)

DI rats. In the presence of a favorable osmotic gradient, these authors noted that vasopressin led to cell swelling and dilatation of the intercellular spaces throughout the collecting duct and in the late distal tubule. In other studies on the permeability of the tight junctions to lanthanum, Tisher and Yarger [5] showed that these junctions resisted lanthanum penetration in the cortical and outer medullary segments of the collecting duct, whereas the inner medullary and papillary segments were freely permeable, thus suggesting the presence of a paracellular pathway for solute and water movement. However, these characteristics were independent of the presence or absence of vasopressin.

It should be noted that in all the foregoing studies, little mention was made of any morphological characteristics of the DI rat that were uniquely different from other strains. In contrast, Sun and coworkers [6] were able to distinguish several unique features in the DI rat. For the papillary collecting duct, these included large intracellular vacuoles and dilated intercellular spaces with less interdigitation between cells. Mitochondria showed focal swelling and degeneration. The connective tissue of the papilla contained less mucopolysaccharide than normal rats. This last finding has also been confirmed in a recent article by McAuliffe.[7] Not only was the volume of papillary interstitium in DI rats less than in normal animals, but the intensity of staining of glycosaminoglycans was greatly reduced (FIGURE 2). The interstitial cells lacked the characteristic branch-like structure as well as the typical arrangement of being perpendicular to the tubular elements. In general, the appearance of the intracellular organelles of the interstitial cells indicated that these cells were relatively "inactive" as compared to interstitial cells of normal rats. McAuliffe postulated that the deficiency in interstitial glycosaminoglycans reflected the altered ultrastructure of the interstitial cells. Prolonged treatment of DI rats with dDAVP essentially restored the normal appearance (McAuliffe, personal communication).

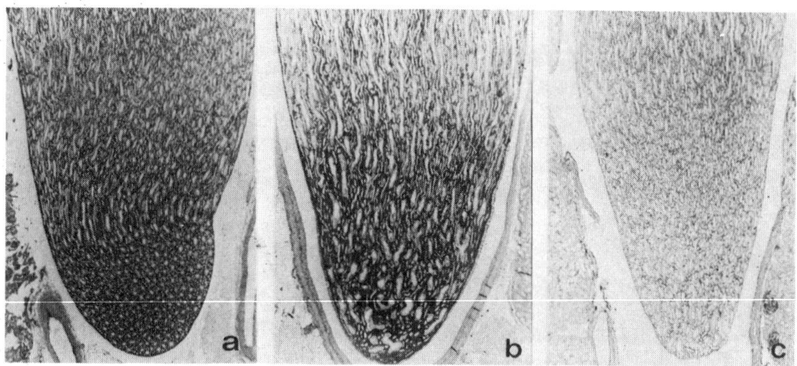

FIGURE 2. Renal papillae from (A) normal Long-Evans rat, (B) Brattleboro heterozygote, and (C) Brattleboro homozygote. Glycosaminoglycans are darkly stained with colloidal iron. Note almost complete absence of glycosaminoglycans in homozygote. Distribution in heterozygote is similar to that in normal rat. Colloidal iron-PAS; 21×. (Reproduced with permission from McAuliffe.[7])

As reported elsewhere in this symposium, Ivanova and colleagues[8] have also found decreased levels of glycosaminoglycans in the renal papillary interstitium of DI rats. While treatment with dDAVP resulted in partial restoration of glycosaminoglycans, full restoration, particularly within the interstitium, required treatment with hydrocortisone. It should be noted that in normal rats the opposite condition holds; that is, water loading is associated with high levels of interstitial mucopolysaccharides and vasopressin causes their polymerization and subsequent disappearance.[9] Other differences also exist. For example, in this symposium, Sundelin and Bohman reported that the medullary interstitial cells of DI rats do not show atrophic changes. Moreover, these cells disappeared under the influence of prolonged vasopressin treatment.[10] These observations appear to contradict those of McAuliffe.[7]

The role of the interstitial cells and of the glycosaminoglycans have still to be resolved. It has been postulated that the latter influence the degree of water permeability of the collecting ducts,[11] or that they serve to stabilize the corticopapillary solute gradient.[7] The former may be one site of the synthesis of prostaglandins which, in turn, may modulate the renal concentrating mechanism. It is clear, therefore, that the ultimate resolution of the properties and functions of the renal interstitial tissue may be of great importance to our understanding of water balance in the DI rat.

Recently, another interesting anatomical difference between DI and normal rats has been reported—a difference that may be important to the operation of the concentrating mechanism. This concerns the concept of nephron heterogeneity. Trinh-Trang-Tan and coworkers [12] have observed that for proximal tubular length and for glomerular volume, the customary heterogeneity between superficial and juxtamedullary nephrons is absent (or greatly reduced) in the DI rat (TABLE 1). This observation was also extended to the ratio of superficial/juxtamedullary (S/JM) single nephron glomerular filtration rate (SNGFR). For all three variables the S/JM ratio was close to unity, compared to values of 0.50 to 0.73 in control animals. The higher ratios in DI rats were due exclusively to differences in the juxtamedullary nephrons. Daily treatment of DI rats with vasopressin or dDAVP for eight weeks (beginning at age two

TABLE 1

DIMINISHED NEPHRON HETEROGENEITY IN THE DI RAT *

	Superficial/Juxtamedullary ratio of		
	Glomerular Volume	Proximal Tubular Length	SNGFR
Heterozygotes (Control)	0.50	0.73	0.71
DI rats	0.77	0.90	1.04
DI rats + dDAVP for 8 weeks	0.59	0.78	0.78

* Data from Trinh-Trang-Tan *et al.*[12]

weeks) restored the nephron heterogeneity toward normal (TABLE 1). The authors emphasized the positive correlation that appears to exist between concentrating ability and the degree of nephron heterogeneity across a broad range of species, the inference being that the lack of nephron heterogeneity in the DI rat may contribute to the concentrating defect of this strain. These authors have reported in this symposium [13] an extension of their studies to the nephrogenic DI mouse. The concentrating defect in this species coexists with normal or elevated circulating levels of endogenous vasopressin, yet evidence of a similar deficiency of nephron heterogeneity was found. Therefore, the lack of nephron heterogeneity in DI rats might follow from some aspect of the high water turnover in these animals, rather than from the deficiency of vasopressin *per se.*

LOOP OF HENLE AS A TARGET FOR VASOPRESSIN

Dousa and coworkers reported that the basal level and the vasopressin-stimulated activity of adenylate cyclase in medullary homogenates from DI rats were significantly lower than in control animals.[14] Pretreatment of DI rats with one unit of vasopressin daily for 30 days increased the vasopressin-stimulated activity of adenylate cyclase toward normal. Similar results have been reported by others.[15]

With the advent of more sophisticated techniques it has now become possible to measure the activities of the enzymes associated with the cyclic AMP system within individual segments of the nephron. It is now known, for example, that in the rat about 80% of the adenylate cyclase response to vasopressin of renal medullary homogenates is accounted for by action at the level of the thick ascending limb.[16] When applied to the DI rat, these techniques have revealed that the adenylate cyclase response to vasopressin was the same as in normal rats for the medullary collecting duct, but was reduced by about 50% below normal in the medullary thick ascending limb [17, 18] (FIGURE 3). These

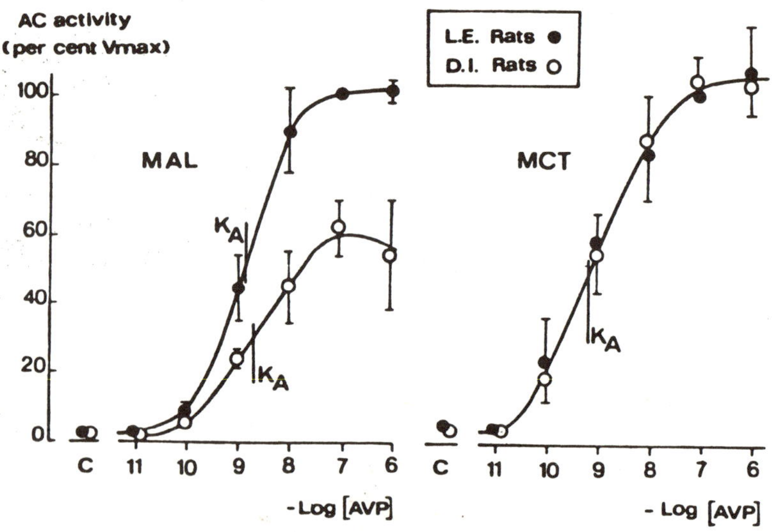

FIGURE 3. Vasopressin-dependent adenylate cyclase activity (as % of value obtained with 10^{-7} M vasopressin) in medullary collecting ducts (MCT) and medullary thick ascending limbs (MAL) of DI and Long-Evans (LE) rats. K_A is dose for half-maximal activation. (Reproduced with permission from Morel *et al.*[18])

results suggest that part of the concentrating defect of the DI rat—and, more specifically, the time required to correct this defect with vasopressin replacement therapy—may have to do with an impaired function of the ascending limb. On the other hand, it should be pointed out that the threshold of response for the ascending limb is higher than for the collecting duct, suggesting that the *in vivo* concentration of vasopressin needed to elicit any biological response could also be greater for the ascending limb.[16]

What is the "biological response" induced in the thick ascending limb? It is tempting to suggest an increase in NaCl reabsorption and hence a strengthening of the countercurrent multiplier system. No direct evidence exists to support this notion for the DI rat. Elsewhere in this symposium, Kriz and Bankir have reported vasopressin-induced morphological changes in the thick ascending limb of DI rats that are at least consistent with some functional change.[19] Using the isolated perfused tubule technique, evidence for vasopressin-stimulated chloride transport in the thick ascending limb is strongest for the mouse,[20, 21] but more equivocal for the rat and rabbit.[21] The evidence for vasopressin increasing the NaCl content of the renal medulla is conflicting (see Reference 16 for review). However, since active chloride transport in the thick ascending limbs is the step that provides the driving force for passive reabsorption of urea from inner medullary collecting ducts—and hence for the medullary accumulation of urea—it is conceivable that vasopressin primarily influences the urea component of the corticopapillary interstitial osmotic gradient. This would be consistent with the findings of Harrington and Valtin that it is primarily the medullary content of urea that gradually rises with prolonged vasopressin treatment.[2]

It is interesting that Dousa's group has reported a qualitatively similar picture for the nephrogenic DI mouse, which has high circulating levels of vasopressin, as described above for the DI rat. That is, comparing nephrogenic DI mice with control mice, vasopressin-stimulated adenylate cyclase activity was only slightly lower in the medullary collecting duct, but was markedly reduced in the medullary thick ascending limb.[22] Moreover, this group also showed that, in the rat, cyclic AMP levels increase in the inner medullary collecting duct in response to increasing osmolality of the incubation medium.[23] Whether this also holds for the thick ascending limb is not yet known. From these results it appears that whatever physiological function may ultimately be ascribed for vasopressin to serve in the thick ascending limb *in vivo*, the defect associated with the lack of vasopressin in the DI rat may, once again, be secondary to the increased water turnover and the reduced tonicity of the medullary interstitium.

Renal Prostaglandins

There is increasing evidence that vasopressin and prostaglandins (PG)—particularly of the E series—are linked in a negative feedback relationship. In this scheme, vasopressin stimulates the production of PGs in the renal medulla, and these compounds then reduce vasopressin's action to increase water permeability of the collecting duct.[24, 25] Consistent with this arrangement is the finding that DI rats have low levels of urinary excretion (and hence, of renomedullary synthesis) of PGE_2 and $PGF_{2\alpha}$.[26, 27] Administration of vasopressin or of dDAVP increases PG excretion in a dose-related fashion.[26–29] However, this situation for the DI rat appears to conflict with reports in water-loaded dogs and humans where increased PG excretion is associated with increased urine flow.[25] Whether this is a species difference, or whether the difference reflects the acute versus the chronic state of polyuria, is not known. In favor of the latter possibility is a recent report [30] on patients with central diabetes insipidus in whom urinary PG excretion was suppressed, but could be markedly increased with dDAVP. Acute water-loading of normal volunteers slightly enhanced PG excretion.

Thus, hypothalamic diabetes insipidus appears to be associated with reduced medullary synthesis of PGs. If this characteristic is to influence water balance in the DI rat, we must look for areas of action of PGs other than that associated with vasopressin-mediated changes in water permeability. Some possibilities are shown in Figure 4. The possible actions of PGs on water and sodium balance have attracted a great deal of interest in recent years and the interested reader is referred to more up-to-date and complete reviews.[24, 25, 29, 31] Although disagreement persists, most investigators have found that PGE_2 and $PGF_{2\alpha}$ (the principal medullary PGs) inhibit NaCl transport in the medullary thick ascending limb.[24, 29] In addition, PGs are believed to enhance medullary blood flow and thereby contribute to increased wash-out of the corticopapillary interstitial gradient.[24, 29] In the DI rat, therefore, the decreased level of PG synthesis should serve to enhance urinary concentrating ability (albeit mildly) through the opposite chain of events. Consistent with this notion are observations in the DI rat that inhibitors of cyclooxygenase activity—which reduce medullary PG synthesis still further—enhance papillary solute concentration [32] and urinary osmolality.[29, 32, 33] However, the increase was slight and urine remained hypotonic to plasma.

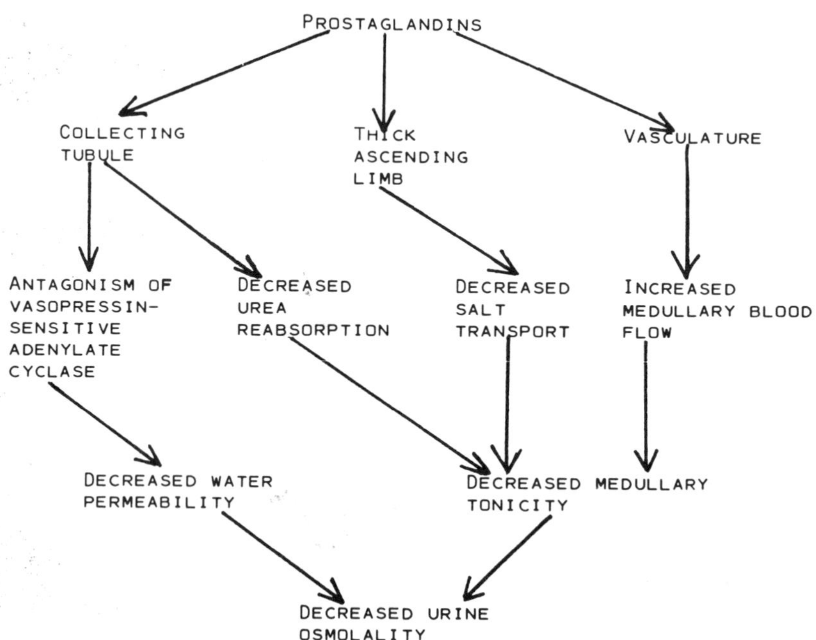

FIGURE 4. Interrelationships of renal prostaglandins and the urinary concentrating mechanism. The left-hand limb is absent in the DI rat. (Reproduced with permission from Kinter *et al.*[29])

With respect to the disorder of water balance in the DI rat, while the reduced level of medullary PG synthesis seems well established (and may well follow from the absence of vasopressin), this condition cannot strictly be construed as a defect since, if anything, it works in favor of improved concentrating efficiency. However, returning to Harrington and Valtin's description of the protracted recovery of full concentrating ability in vasopressin-treated DI rats,[2] this may in part be explained by the simultaneous stimulation of renal medullary PG synthesis which serves to modulate the corrective action of vasopressin.

Summary

The total restoration of urinary concentrating ability of the DI rat given daily injections of vasopressin takes several weeks, although complete osmotic equilibrium across the collecting duct is manifest within hours. This suggests that there may be other deficiencies of the renal concentrating mechanism that, if corrected by vasopressin treatment, are corrected more slowly. I have focussed on just three possibilities. First, the morphology of the medullary interstitium is different from normal rats. Perhaps associated with this finding are alterations in the levels of medullary glycosaminoglycans which may have a role to play in water balance. Functional and morphological changes in the juxtamedullary nephrons are also evident. Second, the possibility exists that the countercurrent multiplier of the DI rat operates less efficiently than in the normal animal. Finally, reduced synthesis of PGs in the renal medulla of DI rats may also influence the concentrating mechanism, although in a favorable direction. While most (if not all) of these differences are secondary to the lack of vasopressin, in some instances it appears that it is the high water turnover (possibly the altered chemical composition of the medullary interstitium) that is the primary culprit.

While the DI rat remains an excellent model for the study of water balance and the action of vasopressin, the presence of multiple defects within the system should be borne in mind. This is particularly true when comparing data obtained following acute treatment with vasopressin versus that following chronic treatment.[34]

Acknowledgment

Dr. Lewis B. Kinter provided many helpful suggestions during the preparation of this manuscript.

References

1. Valtin, H., H. A. Schroeder, K. Benirschke & H. W. Sokol. 1962. Familial hypothalamic diabetes insipidus in rats. Nature **196:** 1109–1110.
2. Harrington, A. R. & H. Valtin. 1968. Impaired urinary concentration after vasopressin and its gradual correction in hypothalamic diabetes insipidus. J. Clin. Invest. **47:** 502–510.
3. Tisher, C. C., R. E. Bulger & H. Valtin. 1971. Morphology of renal medulla in water diuresis and vasopressin-induced antidiuresis. Am. J. Physiol. **220:** 87–94.
4. Woodhall, P. B. & C. C. Tisher. 1973. Response of the distal tubule and cortical collecting duct to vasopressin in the rat. J. Clin. Invest. **53:** 3095–3108.
5. Tisher, C. C. & W. E. Yarger. 1975. Lanthanum permeability of tight junctions along the collecting duct of the rat. Kidney Int. **7:** 35–43.
6. Sun, C. N., H. J. White & E. J. Towbin. 1972. Histochemistry and electron microscopy of the renal papilla in a genetic strain of rats with diabetes insipidus. Nephron **9:** 308–317.
7. McAuliffe, W. G. 1980. Histochemistry and ultrastructure of the interstitium of the renal papilla in rats with hereditary diabetes insipidus (Brattleboro strain). Am. J. Anat. **157:** 17–26.

8. Ivanova, L. N., T. E. Goryunova, L. F. Nikiforovskaya & N. I. Tishchenko. 1982. Hyaluronate hydrolase activity and glycosaminoglycans in the Brattleboro rat kidney. Ann. N.Y. Acad. Sci. (This volume.)
9. Ivanova, L. N. & V. V. Vinogradov. 1963. Histochemical features of mucopolysaccharides in interstitial tissue of renal medulla. Fed. Proc. **22:** T931–T934.
10. Sundelin, B. & S.-O. Bohman. 1982. Interstitial cells in the renal medulla of Brattleboro rats—effects of long term vasopressin treatment. Ann. N.Y. Acad. Sci. (This volume.)
11. Ginetzinsky, A. G. 1958. Role of hyaluronidase in the reabsorption of water in renal tubules: the mechanism of action of the antidiuretic hormone. Nature **182:** 1218–1219.
12. Trinh-Trang-Tan, M.-M., M. Diaz, J.-P. Grünfeld & L. Bankir. 1981. ADH-dependent nephron heterogeneity in rats with hereditary hypothalamic diabetes insipidus. Am. J. Physiol. **240** (Renal Fluid Electrolyte Physiol. 9): F372–F380.
13. Trinh-Trang-Tan, M.-M., H. W. Sokol, L. Bankir & H. Valtin. 1982. Homozygous Brattleboro rats lack normal nephron heterogeneity as a consequence of their urine concentrating defect. Ann. N.Y. Acad. Sci. (This volume.)
14. Dousa, T. P., Y. F. S. Hui & L. D. Barnes. 1975. Renal medullary adenylate cyclase in rats with hypothalamic diabetes insipidus. Endocrinology **97:** 802–807.
15. Rajerison, R. M., D. Butlen & S. Jard. 1977. Effects of in vivo treatment with vasopressin and analogues on renal adenylate cyclase responsiveness to vasopressin stimulation in vitro. Endocrinology **101:** 1–12.
16. Imbert-Teboul, M., D. Chabardès, M. Montégut, A. Clique & F. Morel. 1978. Vasopressin-dependent adenylate cyclase activities in the rat kidney medulla: Evidence for two separate sites of action. Endocrinology **102:** 1254–1261.
17. Imbert-Teboul, M., D. Chabardès, M. Montégut, A. Clique & F. Morel. 1978. Impaired response to vasopressin of adenylate cyclase of the thick ascending limb of Henle's loop in Brattleboro rats with diabetes insipidus. Renal Physiol. **1:** 3–10.
18. Morel, F., M. Imbert-Teboul & D. Chabardès. 1980. Cyclic nucleotides and tubule function. Adv. Cyclic Nucleotide Res. **12:** 301–313.
19. Kriz, W. & L. Bankir. 1982. ADH-induced changes in the epithelium of the thick ascending limb in Brattleboro rats with hereditary hypothalamic diabetes insipidus. Ann. N.Y. Acad. Sci. (This volume.)
20. Hall, D. A. & D. M. Varney. 1980. Effect of vasopressin on electrical potential difference and chloride transport in mouse medullary thick ascending limb of Henle's loop. J. Clin. Invest. **66:** 792–802.
21. Sasaki, S. & M. Imai. 1980. Effects of vasopressin on water and NaCl transport across the in vitro perfused medullary thick ascending limb of Henle's loop of mouse, rat and rabbit kidneys. Pflügers Arch. **283:** 215–221.
22. Jackson, B. A., R. M. Edwards, H. Valtin & T. P. Dousa. 1980. Cellular action of vasopressin in medullary tubules of mice with hereditary nephrogenic diabetes insipidus. J. Clin. Invest. **66:** 110–122.
23. Edwards, R. M., B. A. Jackson & T. P. Dousa. 1981. ADH-sensitive cAMP system in papillary collecting duct: Effect of osmolality and PGE_2. Am. J. Physiol. **240** (Renal Fluid Electrolyte Physiol. 9): F311–F318.
24. Stokes, J. B. 1981. Integrated actions of renal medullary prostaglandins in the control of water excretion. Am. J. Physiol. **240** (Renal Fluid Electrolyte Physiol. 9): F471–F480.
25. Handler, J. S. 1981. Vasopressin-prostaglandin interactions in the regulation of epithelial cell permeability to water. Kidney Int. **19:** 831–8438.

26. WALKER, L. A., A. R. WHORTON, M. SMIGEL, R. FRANCE & J. C. FRÖLICH. 1978. Antidiuretic hormone increases renal prostaglandin synthesis in vivo. Am. J. Physiol. **235** (Renal Fluid Electrolyte Physiol. 4): F180–F185.
27. DUNN, M. J., H. P. GREELY, H. VALTIN, L. B. KINTER & R. BEEUWKES III. 1978. Renal excretion of prostaglandins E_2 and $F_{2\alpha}$ in diabetes insipidus rats. Am. J. Physiol. **235** (Endocrinol. Metab. Gastrointest. Physiol. 4): E624–E627.
28. WALKER, L. A. & J. C. FRÖLICH. 1981. Dose-dependent stimulation of renal prostaglandin synthesis by deamino-8-D-arginine vasopressin in rats with hereditary diabetes insipidus. J. Pharmacol. Exp. Ther. **217:** 87–91.
29. KINTER, L. B., M. J. DUNN, T. R. BECK, R. BEEUWKES III & A. HASSID. 1981. The interactions of prostaglandins and vasopressin in the kidney. Ann. N.Y. Acad. Sci. **372:** 163–179.
30. DÜSING, R., R. HERRMANN, K. GLÄNZER, H. VETTER, A. OVERLACK & H. J. KRAMER. 1981. Renal prostaglandins and water balance: Studies in normal volunteer subjects and in patients with central diabetes insipidus. Clin. Sci. **61:** 61–67.
31. FRÖLICH, J. C., A. S. NIES & R. W. SCHRIER, Eds. 1981. Prostaglandins and the kidney. Kidney Int. **19:** 755–880.
32. STOFF, J. S., R. M. ROSA, P. SILVA & F. H. EPSTEIN. 1981. Indomethacin impairs water diuresis in the DI rat: Role of prostaglandins independent of ADH. Am. J. Physiol. **241** (Renal Fluid Electrolyte Physiol. 10): F231–F237.
33. PONEC, J. & B. LICHARDUS. 1980. Decreased free water excretion after indomethacin in the absence of antidiuretic hormone in saline loaded hypophysectomized Wistar and hydropaenic Brattleboro rats. Endokrinologie **75:** 67–76.
34. KINTER, L. B. 1982. Water balance in the Brattleboro rat: Considerations for hormone replacement therapy. Ann. N.Y. Acad. Sci. (This volume.)

DISCUSSION OF THE PAPER

T. BERL (*University of Colorado Medical Center, Denver, Col.*): How long does it take for the biochemical defect to reverse?

EDWARDS: So far as I am aware, that question has not been specifically addressed. Dousa's group has shown that vasopressin treatment for 30 days will restore vasopressin-stimulated adenylate cyclase activity in the whole medulla (Dousa, T. P., Y. F. S. Hui & L. D. Barnes. *Endocrinology* **97:**802–807, 1975). However, they did not investigate the minimum duration or dose of vasopressin required to reverse the defect. Perhaps Dr. Bankir has more information on this.

L. BANKIR (*Hôpital Necker, Paris, France*): I don't know of any systematic determination of the time needed to reverse this defect. Jard and coworkers (Rajerison, Butlen & Jard, *Endocrinology* **101:**1–12, 1977) showed that the AVP-sensitive adenylate cyclase activity in medullary membrane fractions of DI rats was 30% lower than in control rats. On the other hand, chronic treatment with dDAVP for 9 days in normal rats increased by 30% the maximal response to vasopressin in medullary membrane preparations. Chronic water diuresis induced by drinking 5% glucose in normal rats induced a progressive decrease in the maximal response to AVP, apparent after one week, maximum (20–25%) after two weeks. Thus, AVP administration for one to two weeks is able to modify vasopressin-sensitive adenylate cyclase activity in normal rats.

ADH-INDUCED CHANGES IN THE EPITHELIUM OF THE THICK ASCENDING LIMB IN BRATTLEBORO RATS WITH HEREDITARY HYPOTHALAMIC DIABETES INSIPIDUS *

Wilhelm Kriz and Lise Bankir

Anatomisches Institut der Universität Heidelberg
6900 Heidelberg, Federal Republic of Germany

and

Institut National de la Sauté et de la Recherche Médicale
Unité 90
Hôpital Necker
75730 Paris, Cedex 15, France

Introduction

In addition to its well known effect on water permeability in the renal collecting duct, ADH has recently been shown to stimulate adenylate cyclase activity in thick ascending limb.[1] The functional relevance of this effect appears to be a stimulation of chloride transport.[2,3] In homozygous Brattleboro rats, which are unable to synthesize ADH,[4] an impairment of the ADH-induced adenylate cyclase stimulation in the thick ascending limb has been reported.[5]

Since it has been demonstrated that chronic hormonal treatment modifies the morphology of the target cells in a transporting epithelium,[6] we studied the structural organization of the thick ascending limb (TAL) in Brattleboro diabetes insipidus rats (DI) and in Brattleboro-DI rats after a prolonged treatment with ADH (TDI). Additional investigations were done in Brattleboro heterozygotes (HZ), which have a subnormal ADH production,[4] and in Long-Evans rats (LE) from which the Brattleboro strain had been derived. The results of these studies unequivocally demonstrate that ADH stimulates the growth of the medullary thick ascending limb: the diameter of the tubule as well as epithelial thickness increased far beyond the values of all three control groups.

Materials and Methods

Two experiments were done. In experiment (A), DI rats were chronically treated with antidiuretic hormone and compared to untreated DI littermates. Experiment (B) was designed to compare DI rats with HZ and LE controls. HZ and DI rats were bred at Necker Hospital; LE rats were bought from Lessieux (France). They were fed a rat pellet diet and drank water *ad libitum*. Sexes and body weights are given in Tables 1 and 2.

* Supported by "Deutsche Forschungsgemeinschaft," Sonderforschungsbereich 90, Heidelberg and by Delegation Générale á la Recherche Scientifique et Technique, ACC BRD 80 E 0809, Paris.

0077–8923/82/0394–0424 $1.75/0 © 1982, NYAS

Experiment (A)

Nine rats with diabetes insipidus from one litter (both parents DI) were used in this experiment. Five of these rats received, from 17 days of age, daily injections of dDAVP in oil while the four other rats received oil alone. Details of the treatment are reported in Trinh-Trang-Tan *et al.*[7] The vasopressin analogue dDAVP (de-amino-D-arginine-vasopressin) has a potent antidiuretic effect and a longer biological half-life than AVP and has no pressor effect. Rats were weaned at four weeks of age without discontinuing the treatment. Doses were increased with age from 500 ng to 2 μg per day. Rats were placed in metabolic cages for 24 hours once every two weeks and urinary output and urine osmolality (U_{Osm}) were measured. After 6.5 weeks of treatment kidneys were fixed by *in vivo* perfusion, according to the procedure described below.

Experiment (B)

Four rats with diabetes insipidus and four heterozygotes were used. These rats were not littermates but were of the same age (11 weeks old). Three control Long-Evans rats were also studied. Kidneys were fixed *in situ* according to the procedure described below.

Fixation and Processing of the Kidneys

Food but not water was withheld for the 18 hours preceding the experiment. Rats were anesthetized with Inactin (10 mg/100 g BW i.p.). In experiment (B) a sample of urine was taken from the bladder with a small syringe for determination of urine osmolality. Kidneys were perfusion-fixed *in situ* according to the previously described procedure.[8] Briefly, the abdominal aorta was clamped below the renal artery, and a cannula was inserted into the lower segment and connected to the perfusion apparatus. When perfusion was started, the clamp was removed and the inferior vena cava was opened. Perfusion was started with a washing solution for 20 sec followed by a 3-min perfusion of the fixative at a pressure of roughly 150 mm Hg. Both solutions were warmed at 37° C.

The washing solution consisted of an oxygenated Ringer solution added with 0.33 g/L $CaCl_2 + 2\ H_2O$, 5 g/L procain-HCl, 25 g/L PVP (polyvinylpyrrolidone) (MW $\sim$ 40,000) and 5,000 I.U./L Heparin; osmolality 350 mOsm/L; pH 7.3.

The fixative solution contained 3% glutaraldehyde in 0.1 M cacodylate buffer, supplemented with 0.66 g/L $CaCl_2 + 2\ H_2O$, 0.05 g/L picric acid, and 25 g/L PVP. The carrier osmolality was 200 mOsmol/L H_2O, the total osmolality 540 mOsmol/L H_2O.

After perfusion, the kidneys were removed, decapsulated, weighed, and cut into slices, which stayed for at least 12 h in the same fixative solution as used for the perfusion. After washing in 0.1 M cacodylate buffer (osmolality increased with sucrose to 350 mOsmol) representative pieces of cortex, outer and inner stripe of the outer medulla, and inner medulla of the left kidney were processed for electron microscopy following standard procedures.[8] Semithin transverse sections (1 μm thick) were cut with an ultramicrotome and stained with Methylene blue and Azur II.

Measurements, Calculations, and Statistics

From each kidney, three inner stripe sections from three blocks taken from three different levels of the inner stripe (upper, middle, and lower level) were photographed with a Zeiss photomicroscope (100×). Thick ascending limb measurements were made on pictures enlarged 8.1 times.

Seven points were drawn at random on a sheet of transparent paper. This sheet was superimposed to each picture and the seven thick ascending limb closest to the points were measured. Other ascending limbs were ignored. This resulted in 21 TAL measurements per rat. The diameter of each tubule was measured. When the tubule did not appear quite circular, the smallest diameter was considered. The height of the epithelium was measured in two diametrically opposite points and averaged. Measured lengths were converted into microns according to the final magnification (810×). Epithelium volume per mm length of tubule was calculated as the epithelium cross-sectional area expressed in μm^2 multiplied by 1 mm, assuming tubules were cylindrical. Values given for each rat are the means of the 21 measured tubules.

Results obtained for the different groups of rats were compared by the Student's *t*-test.

Results

Body weights and urine osmolalities are given in Tables 1 and 2. Kidney weights, only measured in the second experiment, are expressed as a function of body weight.

In experiment (A), the dDAVP treatment resulted in a very good correction of the diabetes insipidus, treated rats having a urine osmolality 8.3 times higher than their untreated littermates (Table 1) and similar to the value observed in normal Long-Evans rats (Table 2).

No urine could be obtained from one of the four heterozygotes studied and two other had very low U_{Osm}. This has sometimes been observed in our laboratory on short time urine samples, but, in general, the mean urine osmolality of HZ rats on 24-hour collection in metabolic cages is above 1,200 mOsmol/kg H_2O.

In the TAL measurements (Tables 1 and 2), a striking difference was observed in both diameter and epithelium height between DI rats treated with dDAVP and control DI rats as can be seen on Figures 1 and 2. This resulted in a more than two-fold increase in the volume of the epithelium per unit length of the thick ascending limb.

A significant difference in epithelium height and tubular diameter was also observed between DI and LE rats in experiment (B). The TAL epithelium was slightly thicker in HZ than in DI rats but the difference did not reach significant values.

We do not know if the thicker epithelium of DI rats in experiment (B) compared to that of DI rats in experiment (A) is simply due to differences in age and body weight between the two groups. It is important to know however that tubular diameter and epithelial height increased as a consequence of the chronic ADH treatment far beyond the values observed in DI rats, heterozygous rats, and normal Long-Evans rats.

The values concern only the part of the thick ascending limb lying in the

TABLE 1

EXPERIMENT A

			Body Weight (g)	Urine Osmolality * (mOsmol/kg H_2O)	Tubule Diameter (μm)	Epithelium Height (μm)	Epithelium Volume per mm Tubular Length ($\mu m^3 \times 10^{-3}$)
DI	1	♂	143	185	30.83	4.65	384
	2	♂	143	180	27.20	4.94	346
	3	♀	108	335	25.84	4.65	309
	4	♀	102	170	25.77	5.64	357
Mean			124	223	27.41	4.97	349
SEM			11	48	1.19	0.23	16
Treated DI	1	♂	164	2090	38.15	7.81	740
	2	♂	160	1465	35.19	8.23	697
	3	♀	139	1930	37.28	9.96	854
	4	♀	119	1350	36.16	9.75	805
	5	♀	134	2445	36.33	9.22	777
Mean			143	1856	36.62	8.99	774
SEM			10	255	0.51	0.42	27
t-test			NS	$p < 0.001$	$p < 0.001$	$p < 0.001$	$p < 0.001$

* 24-h Urine collection in metabolic cages.

TABLE 2

EXPERIMENT B

			Body Weight (g)	Kidney Weight (mg/100 g BW)	Urine Osmolality * (mOsmol/kg H_2O)	Tubule Diameter (μm)	Epithelium Height (μm)	Epithelium Volume per mm Tubular Length $\mu m^3 \times 10^{-3}$
DI	1	♂	190	1078	175	26.81	6.73	424
	2	♂	255	964	362	27.75	6.30	422
	3	♀	155	1240 †	240	26.56	6.24	398
	4	♀	145	1088	230	27.16	6.49	422
Mean			186	1092	252	27.07	6.44	416
SEM			25	56	40	0.26	0.11	6
HZ	1	♂	215	924	—	29.05	8.47	550
	2	♂	245	952	1220	25.36	6.50	386
	3	♀	150	1054	275	25.79	7.76	441
	4	♀	175	908	480	25.47	7.11	412
Mean			196	959		26.42	7.46	447
SEM			21	33		0.88	0.42	36
LE	1	♂	205	1168	1990	30.46	7.70	550
	2	♂	200	1200	1310	28.79	7.87	518
	3	♂	195	1186	2110	28.65	8.11	523
Mean			200	1185	1803	29.30	7.89	530
SEM			2	9	247	0.58	0.12	10
t-test DI vs. HZ				NS		NS	NS	NS
t-test HZ vs. LE				$p < 0.001$		NS	NS	NS
t-test DI vs. LE				NS	$p < 0.001$	$p < 0.025$	$p < 0.001$	$p < 0.001$

* Last urine formed before anesthesia.

† Right kidney had hydronephrosis; left kidney was slightly hypertrophied.

inner stripe, i.e. the beginning portion of this tubular segment. The outer stripe and cortical portion of the thick ascending limb were not measured since a simple inspection by eye did not show any difference between the different experimental groups. Even within the inner stripe the measurements were made

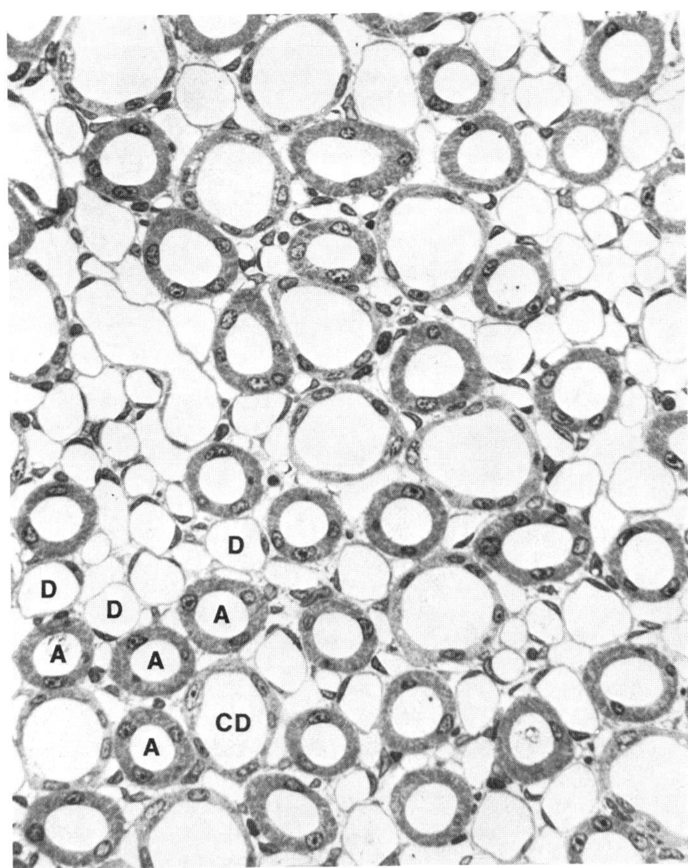

FIGURE 1. DI rat (Experiment A). Semithin (1 μm) cross section through the lower level of the inner stripe. Descending thin limbs (D), thick ascending limbs (A), and collecting ducts (CD) are found. Compare with FIGURE 2: both figures are equally enlarged. ~ 370×.

in sections taken from three different levels and averaged. This was done because it is known that the thick ascending limb reduces in diameter and epithelial thickness when approaching to the cortex.[8] This reduction was obvious in all experimental groups (not shown).

Discussion

Homozygous Brattleboro rats have diabetes insipidus of hypothalamic origin. They cannot synthesize antidiuretic hormone. When given exogenous ADH they are able to increase significantly their urine osmolality, but Harrington and Valtin [9] showed that the full restoration of a normal concentrating ability was

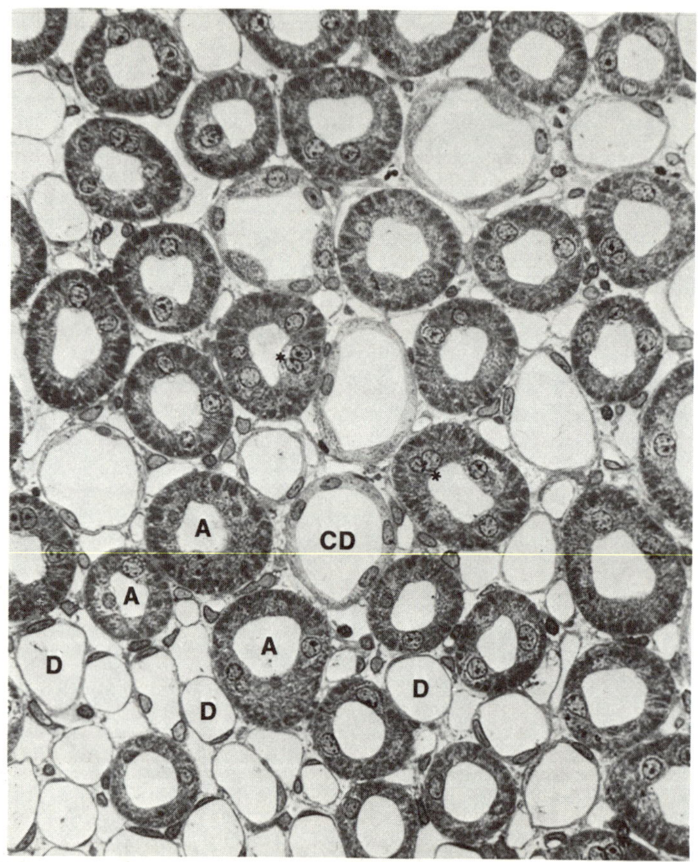

FIGURE 2. DI rat after ADH treatment (Experiment A). Semithin cross section through the lower level of the inner stripe. In comparison with an untreated DI rat (FIGURE 1) the thick ascending limbs are strikingly increased in diameter and cellular thickness. Many binuclear cells are found in the thick ascending limb epithelium.* Symbols as in FIGURE 1. ~ 370×.

obtained only after several weeks of treatment. This observation suggests that some long-term–induced process must develop in the kidney to make possible the achievement of the highest urine osmolalities.

The effect of ADH on stimulation of adenylate cyclase activity in plasma membrane fractions of the medulla of DI rats have been studied by two different groups.[10, 11] They both showed that ADH induced a lower stimulation of adenylate cyclase in DI rat medulla membrane fractions than in normal rat. A normal activity could be restored after several days of vasopressin administration.[10] Subsequently, Imbert *et al.*[1] showed that, in addition to its well known effect on collecting duct, ADH also acts on the thick ascending limb of Henle's loop especially in the rat and the mouse. In response to ADH, the adenylate cyclase activity of the cells in the thick ascending limb is increased several fold.[1] The same authors found that the response to ADH in homozygous DI rats was impaired in TAL but not in collecting ducts.[5] Thus the decreased response to ADH observed in medullary membrane preparations with DI rats was likely attributable to TAL. Imbert *et al.*[5] found no difference in diameter of the medullary thick ascending limb between DI and LE rats. We found a slight difference (8%) in tubular diameter, but a clear difference in epithelial thickness (18%). The most surprising difference, however, occurred in tubular diameter and epithelial thickness between the chronically ADH-treated DI rats and all other groups. Comparing treated to untreated DI rats (Experiment A) the tubular diameter increased by 34% and the epithelial height by 81%.

The growth effect of ADH on the thick ascending limb is comparable to the action of desoxycorticosterone acetate (DOCA) on the cortical collecting duct. Adaptation of rats to a chronic stimulation with DOCA enhances Na and K transport in the cortical collecting duct due to an increase in basolateral cell membrane area of principal cells.[6] In a recent study it was shown that a high concentration of endogenous aldosterone provokes a similar increase in cell membrane area in the principal cells of the cortical collecting duct.[12] Thus, a hormone, DOCA or aldosterone, that stimulates a specific ion transport at a specific tubular site, elicits a growth effect on the transporting membrane. ADH is known to acutely stimulate Cl transport in the thick ascending limb[2, 3] and chronic treatment provokes a growth effect on the epithelium of the TAL. Even if the survey pictures (FIGURES 1 and 2) only demonstrate the overall thickening of the epithelium, we know from a short inspection in the electron microscope that the transporting basolateral membranes together with their associated mitochondria have increased proportionally. Judging from the structural finding the transport capacity of the thick ascending limb should be drastically increased.

There are additional similarities of our results to those obtained by other chronic stimulations to other nephron segments. After adaptation to low Na and high K intake the epithelium of the connecting tubule increases markedly, most pronounced in the very beginning and approaching control values near the end of this tubular segment.[12] The thick ascending limb has, a priori, such a configuration with the thickest epithelium at its beginning and a thin epithelium at its end in the cortex.[8] ADH treatment enhances the difference by a further increase of the beginning portions without any obvious alterations at the terminal portions. Thus, it is the inner stripe portion of the thick ascending limb that is primarily affected and, within the inner stripe portion, the beginning parts (i.e. those parts lying towards the inner zone) that develop the thickest epithelium. An explanation for this axial differences along the thick ascending limb would be a gradual decline in ADH sensitivity along the segment. That the cortical parts are less sensitive to ADH with respect to the response of the cAMP-system than the medullary parts has been shown in several species.[13]

From a teleological point of view it appears reasonable that the transport capacity is increased most at the site where it is most effective. With respect to the urine concentrating ability the very beginning portions of the thick ascending limb must be regarded as the most effective ones. One is tempted to speculate that the observed increase in transport capacity of the thick ascending limb might represent an adaptation of general importance that would allow animals to adapt to changes in water intake.

Acknowledgments

The authors are grateful to Ms. B. Koura, Ms. I. Hartmann, and Ms. M. Douté for technical assistance, to Ms. I. Ertel for photographic help, and to Ms. M. Goos for typing the manuscript.

References

1. Imbert-Teboul, M., D. Chabardes, M. Montegut, A. Clique & F. Morel. 1978. Vasopressin-dependent adenylate cyclase activities in the rat kidney medulla: evidence for two separate sites of action. Endocrinology **102:** 1254–1261.
2. Hall, D. A. & D. M. Varney. 1980. Effect of vasopressin on electrical potential difference and chloride transport in mouse medullary thick ascending limb of Henle's loop. J. Clin. Invest. **66:** 792–802.
3. Sasaki, S. & M. Imai. 1980. Effects of vasopressin on water and NaCl transport across the in vitro perfused medullary thick ascending limb of Henle's loop of mouse, rat and rabbit kidneys. Pflügers Arch. **383:** 215–221.
4. Valtin, H. & H. A. Schroeder. 1964. Familial hypothalamic diabetes insipidus in rats (Brattleboro strain). Am. J. Physiol. **206:** 425–430.
5. Imbert-Teboul, M., D. Chabardes, M. Montegut, A. Clique & F. Morel. 1978. Impaired response to vasopressin of adenylate cyclase of the thick ascending limb of Henle's loop in Brattleboro rats with diabetes insipidus. Renal Physiol. **1:** 3–10.
6. Wade, J. B., R. G. O'Neil, J. L. Pryor & E. L. Boulpaep. 1979. Modulation of cell membrane area in renal collecting tubules by corticosteriod hormones. J. Cell Biol. **81:** 439–445.
7. Trinh-Trang-Tan, M. M., M. Diaz, J. P. Grünfeld & L. Bankir. 1981. ADH-dependent nephron heterogeneity in rats with hereditary diabetes insipidus. Am. J. Physiol. **240:** F372–F380.
8. Kaissling, B. & W. Kriz. 1979. Structural analysis of the rabbit kidney. Adv. Anat. Embryol. Cell Biol. **56:** 1–123. Springer Verlag. Heidelberg.
9. Harrington, A. R. & H. Valtin. 1968. Impaired urinary concentrations after vasopressin and its gradual corrections in hypothalamic diabetes insipidus. J. Clin. Invest. **47:** 502–510.
10. Dousa, T. P., Y. F. S. Hui & L. D. Barnes. 1975. Renal medullary adenylate cyclase in rats with hypothalamic diabetes insipidus. Endocrinology **97:** 802–807.
11. Rajerison, R. M., D. Butlen & S. Jard. 1977. Effects of in vivo treatment with vasopressin and analogues on renal adenylate cyclase responsiveness to vasopressin stimulation in vitro. Endocrinology **101:** 1–12.
12. Kaissling, B. & M. Le Hir. 1982. Analysis of distal tubular segments in the rabbit kidney after adaptation to altered Na- and K-intake. Cell Tissue Res. (In press.)
13. Morel, F. 1981. Sites of hormone action in the mammalian nephron. Am. J. Physiol. **240:** F159–F165.

Discussion of the Paper

S.-O. Bohman (*Karolinska Institute, Huddinge, Sweden*): I want to congratulate you on this very important paper and I trust that you will now measure the actual surface area per cell or surface area per unit length of tubule. As you know, it has been shown in parts of the distal nephron that adaptive changes may occur in these membrane surface areas, for example, in response to potassium depletion and mineralocorticoid treatment.

I. W. Henderson (*University of Sheffield, Sheffield, England*): Some of the changes that you have noted are reminiscent of those seen in normal postnatal development of the mammalian kidney. Thus, there is an initial failure to concentrate urine in the neonate and there then ensues a renal maturation towards the preparation of urine hyperosmotic to plasma. Are there parallel morphological bases when your observations are compared with these immediate postnatal changes?

Bankir: I don't know if it has been studied. In the rat the kidney is not mature at birth, that is, the full complement of nephrons has not been achieved and the medulla is not well developed. Dr. Edwards and coworkers of Dr. Valtin's group have shown that during kidney maturation, the ability to concentrate urine in young rats appears and develops in parallel with the progression of the loops of Henle in the medulla. When the loops of superficial nephrons enter the medulla, the concentrating ability appears and develops. Maybe Dr. Edwards could explain this better than I can.

B. R. Edwards (*Dartmouth Medical School, Hanover, N.H.*): We found that the superficial loops enter the outer medulla at about 10 days of age and achieve maximal penetration of this zone by approximately 16 days. We also found that while papillary and urinary osmolalities increased throughout the first 3–4 postnatal weeks, the rate of increase of these variables accelerated dramatically after the superficial loops had fully penetrated the outer medulla. However, we did not perform any histological studies that would answer Dr. Hendersen's question. Perhaps Dr. Kriz would comment as I know that several years ago he and Dr. Ricciarelli analyzed the histological changes in the loops of Henle of developing rats.

W. Kriz (*University of Heidelberg, Heidelberg F.R.G.*): The renal medulla is not fully developed 3 weeks after birth at which time we started AVP treatment. We administered AVP during that period of time when tubules are normally growing; specific stimulation of the thick ascending limb occurred, since this tubular segment developed far beyond control values.

I. L. Schwartz (*Mt. Sinai School of Medicine, New York, N.Y.*): Since these morphological changes go far beyond the normal, is there a reversion back to normal with continued treatment?

Bankir: We have not studied this, but it might be interesting to look at.

P. Stern (*Dartmouth Medical School, Hanover, N.H.*): Do you know if there is any correlation between the anatomic changes you have described and changes in sodium content or concentration, particularly in the outer medulla, in the DI rat? That is, do they occur in parallel?

Bankir: I have no data of my own on this, but it is known that there are changes in the composition of the medulla when you begin to treat DI rats, so it should be compared with other studies where these parameters have been measured.

J. ZICHA: I would like to mention a study by Deveny (*Acta Morph. Acad. Sci. Hung.* **12**:365–366, 1964). In these experiments Wistar rats received 1.6 IU of pitressin tannate for 4 days either alone or in combination with 5 ml of saline given intraperitoneally six times per day. In the outer medulla of the former group (VP alone) a mild increase (4×) in the number of mitotic cells was observed, while in the latter group (VP + saline) mitoses were increased 50 times. No increase in mitoses was observed in cortical tubules. Sixty-eight percent of the mitoses in the outer medulla were in collecting ducts, 21% in thick ascending limbs and 9% in the thin descending limbs. Binucleated cells were also described. This study shows: (1) that the growth-promoting effects of vasopressin can occur in normal rats in the presence of endogenous vasopressin; and (2) that it does not seem to be proportional to the sensitivity of the ADH-dependent adenylate cyclase in the various segments.

BANKIR: Thank you for this interesting information. I have not read this paper but, from what you say, the growth-promoting effect in this case seems to be due to the high amount of saline rather than to the vasopressin. Actually, normal rats receiving 6 × 5 ml saline i.p. per day, probably undergo a marked saline diuresis in spite of the vasopressin. The high number of mitoses in the collecting ducts could be due to the increased urine flow rate. Nevertheless, it would be interesting to investigate this problem further.

INTERSTITIAL CELLS IN THE RENAL MEDULLA OF BRATTLEBORO RATS: EFFECTS OF LONG TERM VASOPRESSIN TREATMENT *

Birgitta Sundelin and Sven-Olof Bohman †

Karolinska Institutet
Department of Pathology
Huddinge Hospital
S-141 86 Huddinge, Sweden

Introduction

During recent years evidence has accumulated that the lipid-laden interstitial cells type 1, of the renal medulla have endocrine functions and are of importance in the regulation of blood pressure (for review, see Reference 1). The chemical nature of the postulated hormone, as well as the mechanisms that control the activity of the interstitial cells, are not yet defined. Previous ultrastructural studies have shown that the amount of lipid stored in the interstitial cells is greatly influenced by the state of diuresis [2] and recent biochemical studies have indicated that the cells are sensitive to vasopressin.[3,4] Consistent with this, McAuliffe [5] recently reported that the interstitial cells showed ultrastructural signs of reduced activity and atrophy in homozygous Brattleboro rats and that interstitial cells type 2, a lymphocyte-like cell [6] was present in abnormal numbers and locations in the medulla. Our study was undertaken to confirm these observations and to study the possible effects of vasopressin treatment on interstitial cell ultrastructure. In order to make our observations as objective as possible, we used a quantitative, stereological method.

Materials and Methods

Sixteen male homozygous Brattleboro (DI) rats (TNO, Holland) and four normal untreated Long-Evans hooded (LE) rats, all weighing 150–200 g, were studied. The animals were fed standard Rostock diet and tap water ad libitum. They were intermittently kept in metabolic cages to check urine volume and osmolality. The rats were divided into groups and treated with vasopressin as shown in Table 1.

After the period of treatment the animals were anesthetized with sodium pentobarbital and the kidneys were perfusion-fixed according to methods previously described.[6] The rinsing solution contained 0.2% procaine hydrochloride and 2% dextran T40 in 0.08 M sodium cacodylate buffer (pH 7.3) and the fixation solution consisted of 3% glutaraldehyde and 2% dextran T40 in a 0.08 M sodium cacodylate buffer. To adjust the osmolality of these solutions

* Supported by the Swedish Medical Research Council Grant No. B82–12X–05937–02.

† Correspondence to Dr. S.-O. Bohman, Karolinska Institutet, Department of Pathology F42, Huddinge Hospital, S–141 86 Huddinge, Sweden.

0077–8923/82/0394–0435 $1.75/0 © 1982, NYAS

TABLE 1

EXPERIMENTAL GROUPS OF DIABETES INSIPIDUS (DI) AND LONG-EVANS HOODED (LE) RATS

Group	Rat	*n*	Treatment	Period of Treatment	Final Urine Osmolality (mean, mOsm/kg water)
1	DI	4	none		193
2	DI	3	daily injections of peanut oil s.c.	19 days	174
3	DI	3	1.0 U/day vasopressin tannate in oil s.c.*	2 days	1239
4	DI	3	"	19 days	2274
5	DI	3	1.0 U/day synthetic AVP, continous infusion s.c.†	19 days	1944
6	LE	4	none		1800

* Pitressin Tannate in oil (Parke, Davis & Co., USA).

† Synthetic arginine vasopressin (Ferring AB Sweden) in 0.9% sodium chloride was given in subcutaneously implanted capsules (Alzet osmotic minipumps, Alza Corp., USA).

to the osmotic conditions in the renal papilla of each of the experimental groups, 0–4.7% sodium chloride was added to both the rinsing and fixative solutions.

A 0.5-mm thick tissue slice was then excised exactly half-way between the papilla tip and the border between the inner and outer zones of the medulla. One part of the tissue slice was cut out separately and embedded flat for light microscopy. The rest of the tissue slice was chopped into small pieces and embedded with random orientation for electron microscopic stereology. The tissue was post-fixed in osmium tetroxide, stained en bloc with uranyl acetate and embedded in EM-bed-812 (E.M.S., Fort Washington, Pa., USA).

Sections from three blocks from each animal were then cut for electron microscopy, taken up on 200-mesh coordinate grids (TEBRA), and stained with uranyl acetate and lead citrate. Four micrographs were taken from each section at random orientation using the grid coordinates. The electron-optical magnification was 3,300× and the negatives were enlarged 3.5×.

The following tissue parameters were determined by point-counting according to standard stereological procedures[7] and using a double-lattice square test system ($d = 5$ and 50 mm, respectively): (1) Volume fraction or volume density (V_v) of interstitial cells type 1 in whole tissue; (2) V_v of interstitial cells type 2 in whole tissue; (3) V_v of mitochondria in interstitial cells type 1; (4) V_v of lipid droplets in interstitial cells type 1; (5) Surface-to-volume ratio or surface density (S_v) of plasma membrane of interstitial cells type 1.

Results

Interstitial Cells in Untreated Homozygous Brattleboro Rats

The light- and electron microscopic appearance of interstitial cells in Long-Evans hooded (LE) rats (Figure 1) was essentially similar to that described in Wistar and Sprague Dawley rats.[8] Cell shapes varied considerably, some cells had a rounded shape while most of them had irregular shapes with long cytoplasmic processes making contact with the outer aspect of thin limbs and capillaries. Cytoplasmic lipid inclusions or lipid droplets were numerous and were found in all parts of the cytoplasm. Rough surfaced ER was prominent and mitochondria occurred in moderate numbers. In accordance with previous studies[2] the interstitial cells type 1 made up about 10% of the whole tissue volume (Table 2).

At the level of the light microscope the interstitial cells type 1 in untreated Brattleboro (DI) rats appeared very similar to those in LE rats (Figure 2a, b). Most interstitial cells had irregular, often stellate shapes with long cytoplasmic processes. Lipid droplets were numerous. In contrast to the situation in LE rats and other normal rat strains,[6] interstitial cells type 2 were sometimes abundant in the mid-level of the papilla in DI rats. They were sometimes interposed between the epithelial cells of collecting ducts. However, their number varied very much between animals and therefore the mean volume density (V_v) of type 2 cells was not increased by more than 50% (Table 2).

The electron microscope investigation largely confirmed the impression obtained in the light microscope. Subjectively, we did not detect any differences in interstitial cell ultrastructure between LE and DI rats (Figure 4). The

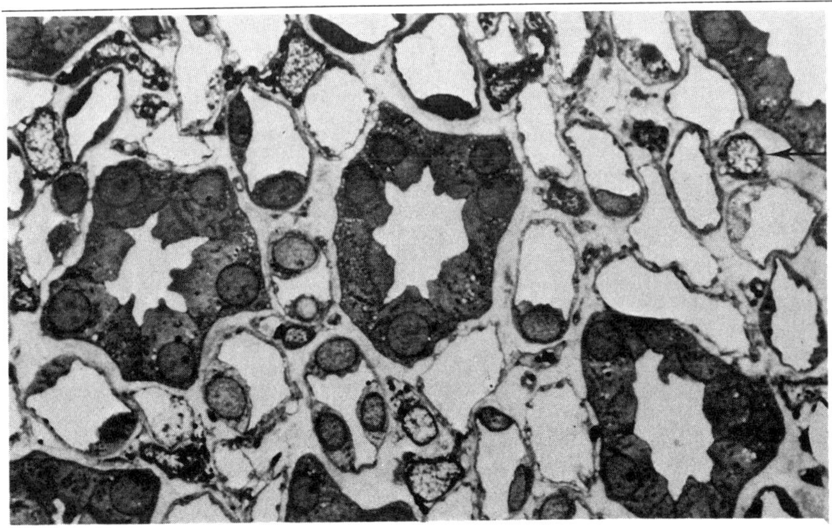

Figure 1. Light micrograph from renal medulla of Long-Evans rat. Interstitial cells are abundant and contain numerous lipid droplets. Some interstitial cells are rounded (arrow) while most of them have irregular shapes. 330×.

stereological data (TABLE 2) showed only small differences between the two groups with respect to the mean value of volume density and surface-to-volume ratio of type 1 cells. The mean volume density of mitochondria was somewhat lower and that of lipid droplets somewhat higher in the DI rats. Standard deviations of the latter figures were high, however.

Effect of Vasopressin Treatment

After two days (55 h) of vasopressin treatment in DI rats, light microscopic investigation showed a reduced number of interstitial cells in the renal medulla as compared to untreated DI rats. The volume fraction of interstitial cells type 1 (TABLE 3) was also considerably reduced but no other structural changes were apparent from the stereological data. Type 2 interstitial cells were not detected in DI rats after two days of vasopressin treatment (TABLE 3).

TABLE 2

STEREOLOGICAL PARAMETERS OF INTERSTITIAL CELLS IN RENAL PAPILLA OF BRATTLEBORO (DI) RATS AND LONG-EVANS HOODED (LE) RATS *

	Groups †	
	6 LE Rats	1 DI Rats
V_v of interstitial cells type 1 in whole tissue (%)	9.3 ± 2.7	11.6 ± 0.9
S_v of cell membrane of interstitial cells type 1 ($\mu m^2/\mu m^3$)	1.2 ± 0.1	1.0 ± 0.2
V_v of lipid droplets in interstitial cells type 1 (%)	1.7 ± 0.5	2.9 ± 1.3
V_v of mitochondria in interstitial cells type 1 (%)	2.7 ± 1.1	2.0 ± 0.1
V_v of interstitial cells type 2 in whole tissue (%)	0.2 ± 0.4	0.3 ± 0.6

* All figures indicate mean ± standard deviation. V_v = volume density, volume fraction; S_v = surface density, surface-to-volume ratio.

† See TABLE 1.

The reduction in the number of interstitial cells was dramatic in animals treated with vasopressin for 19 days. In the light microscope, large areas of the medulla were seen to be essentially devoid of interstitial cells, leaving a wide, rather homogeneous interstitium of moderate density (FIGURE 3a, b). Occasional capillaries showed a marginal rim of platelets but no occluding thromboses were seen. The stereological analysis showed a decrease in the volume fraction of interstitial cells to about 6% of the control value (TABLE 3). Electron microscopic examination showed an interstitium of approximately normal width containing numerous strands of basement membrane-like material and a rather irregularly distributed fine flocculent material. In addition, cellular debris and fragments probably derived from disintegrated interstitial cells were

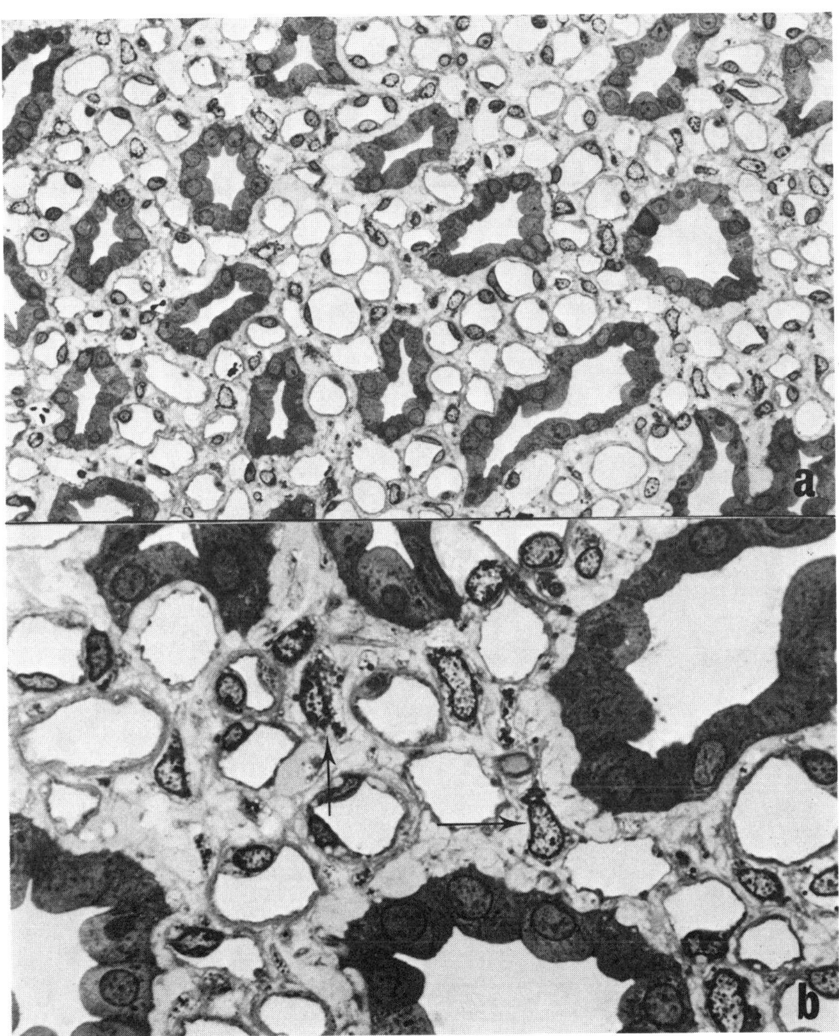

FIGURE 2. (a) Light micrograph from renal medulla of untreated homozygous Brattleboro rat. Note the abundance of interstitial cells. 330×. (b) Higher magnification showing the irregularly shaped interstitial cells (arrows) which contain numerous lipid droplets. 830×.

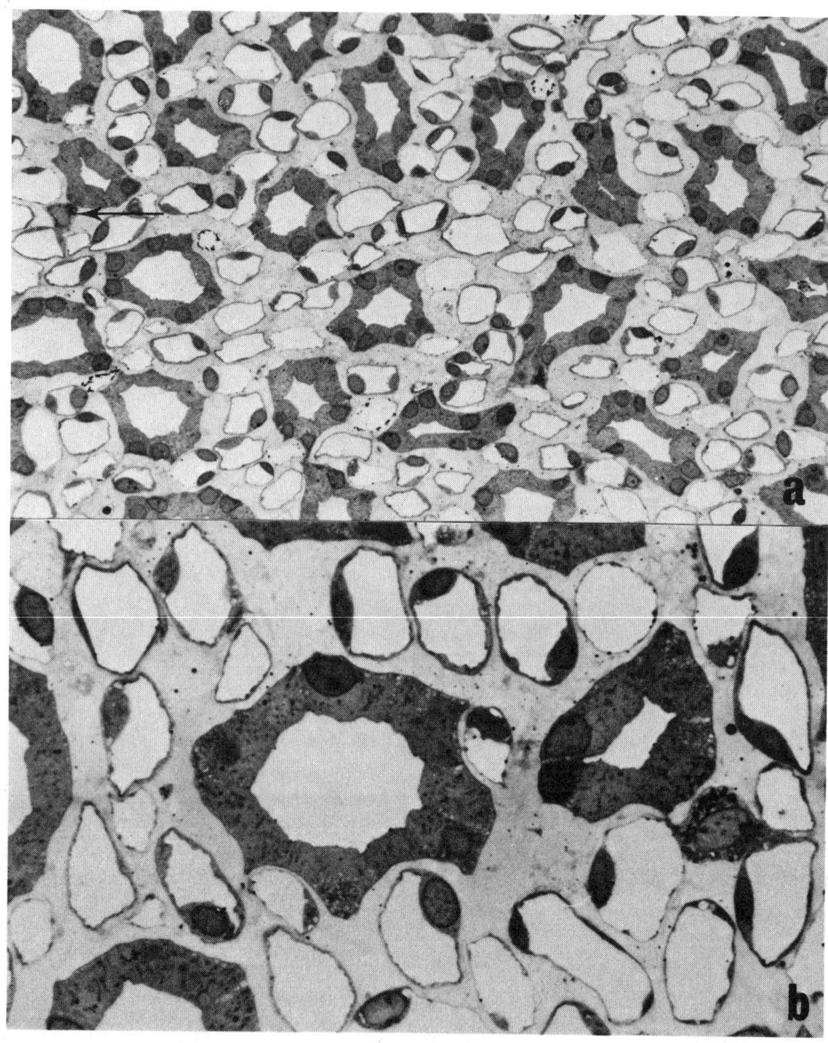

FIGURE 3. (a) Light micrograph from renal medulla of Brattleboro rat, treated with 1.0 U vasopressin for 19 days. The wide interstitial space is practically empty. Very few interstitial cells are seen (arrow). Compare FIGURE 2A 330×. (b) Higher magnification showing the empty interstitium containing only one interstitial cell. Other structures appear well preserved. Compare FIGURE 2b 830×.

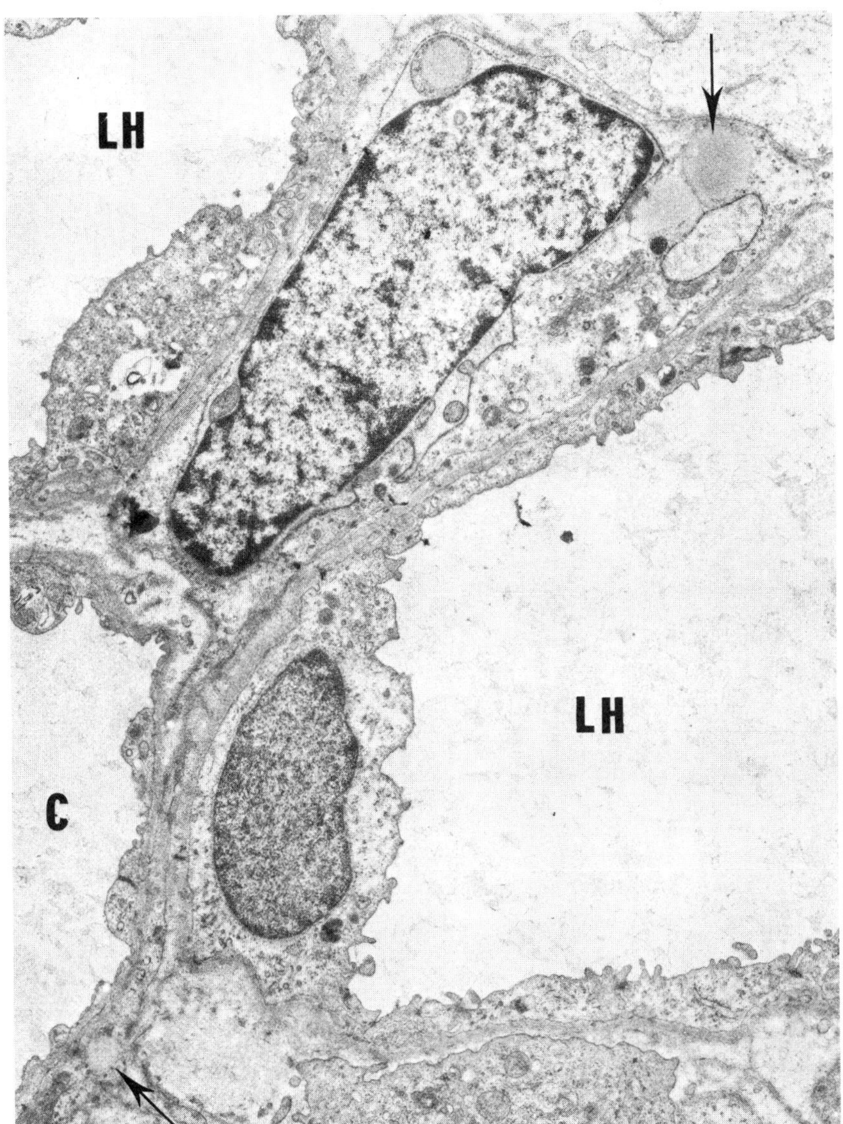

FIGURE 4. Electron micrograph showing an interstitial cell type 1 in the renal medulla of control Brattleboro rat. The cell has an irregular shape and is in close contact with surrounding thin limbs (LH) and capillaries (C). Several lipid droplets are seen, also in thin cytoplasmic projections (arrows). 8,600×.

often seen (FIGURE 5). Some capillaries showed areas of endothelial damage with platelet aggregation along the denuded basal lamina (FIGURE 6). Collecting ducts and thin limbs appeared generally well preserved.

When the vasopressin treatment was given in the form of continuous subcutaneous infusion from an implanted capsule (osmotic minipump) the effect on the interstitial cells was similar to that described above but less pronounced (TABLE 3). These animals also did not reach the same high urine osmolalities as the animals given vasopressin tannate in oil for the same period of time (TABLE 1).

TABLE 3

CHANGES IN INTERSTITIAL CELLS IN RENAL PAPILLA OF CONTROL AND VASOPRESSIN-TREATED BRATTLEBORO RATS *

	Groups †			
	2 Control, peanut oil 19 days	3 Pitressin tannate in oil 1.0 U/day, 2 days	4 Pitressin tannate in oil 1.0 U/day, 19 days	5 AVP, continuous infusion 1.0 U/day, 19 days
V_v of interstitial cells type 1 in whole tissue (%)	9.3 ± 4.9	5.0 ± 1.1	0.6 ± 1.0	5.6 ± 1.1
S_v of cell membrane of interstitial cells type 1 ($\mu m^2/\mu m^3$)	1.6 ± 0.3	1.7 ± 0.4	0.8 ± 1.1	2.0 ± 0.2
V_v of lipid droplets in interstitial cells type 1 (%)	2.0 ± 0.4	1.0 ± 1.0	2.4	1.9 ± 1.5
V_v of mitochondria in interstitial cells type 1 (%)	1.6 ± 0.4	1.8 ± 0.7	2.8	2.3 ± 0.4
V_v of interstitial cells type 2 in whole tissue	0.6 ± 0.2	0.0	0.0	0.0

* All figures indicate mean ± standard deviation.
† See TABLE 1.

DISCUSSION

Our light- and electron microscopic investigation of the interstitial cells in the renal medulla of homozygous Brattleboro rats failed to confirm the previous report by McAuliffe,[5] who found an almost total absence of cytoplasmic processes as well as a pronounced reduction of the number of mitochondria, lipid droplets, and other cell organelles in interstitial cells of DI rats. These degenerative or atrophic changes were interpreted to be a result of the lack of vasopressin stimulation of the interstitial cells.

In addition to a subjective, mainly qualitative investigation we made a quantitative analysis of the volume fraction of interstitial cells as well as their lipid droplets and mitochondria. In addition, we determined the surface-to-volume ratio of interstitial cells as a parameter sensitive to changes in cell

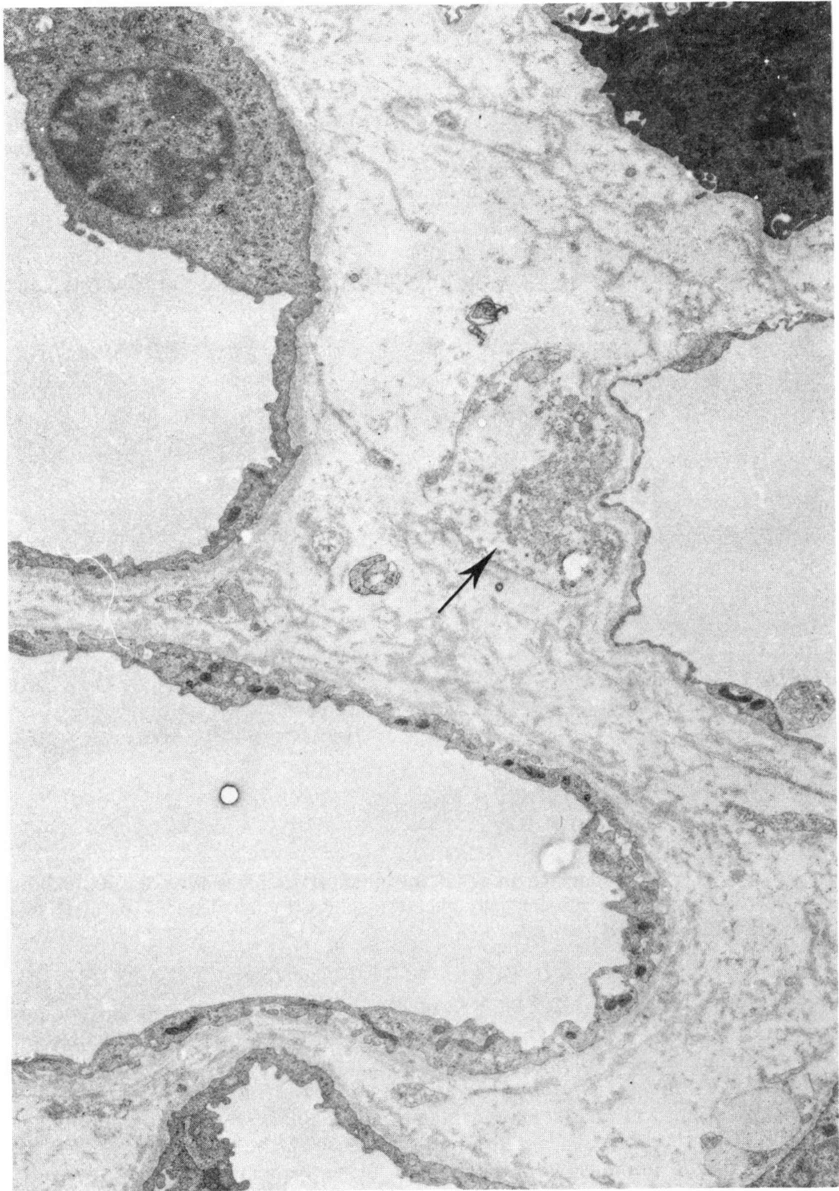

FIGURE 5. Electron micrograph from renal medulla of Brattleboro rat treated with vasopressin for 19 days. The interstitium contains only strands of basement membrane-like material and cellular debris (arrow). 8,100×.

shape, e.g. the loss of long cytoplasmic processes. The differences found between LE and DI rats were not impressive and although our material is too limited to exclude minor differences, we find a clear discrepancy between our results and those of McAuliffe.[5] In our opinion, the evidence for an inactivity of interstitial cells in DI rats is weak. The reason for the above discrepancy between our results is not clear to us at the moment.

In addition to the reported changes in interstitial cells of DI rats,[5] a reduced ability of the medullary interstitial substance to stain with Alcian blue, colloidal iron, and other mucopolysaccharide stains[5, 10] has been found. We have not yet studied this alteration in our laboratory.

The vasopressin treatment in some of our experimental groups was with a high (supramaximal) dose and for a long period of time (19 days). Previously, acute effects of vasopressin on the medullary ultrastructure in DI rats have

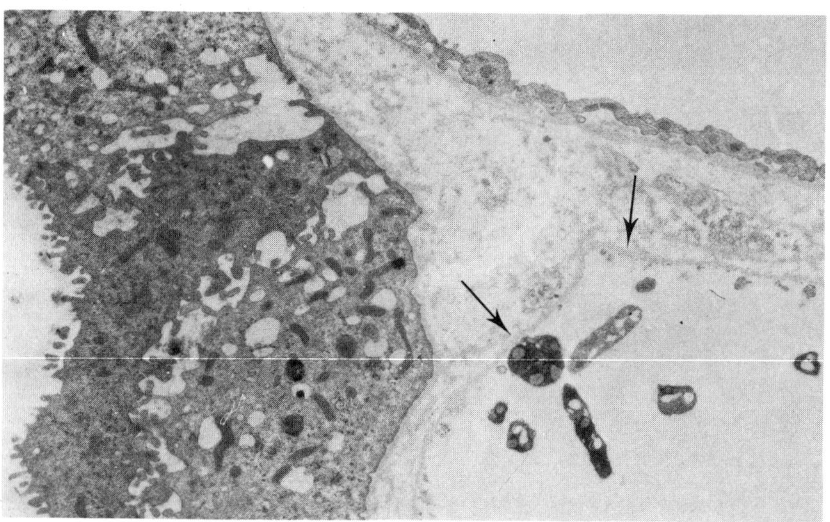

FIGURE 6. Capillary damage in renal medulla of vasopressin-treated Brattleboro rat. Endothelial layer is missing and platelets adhere to the denuded basal lamina (arrows). 8,100×.

been studied by Tisher *et al.*[9] They found no obvious differences in morphology between LE and DI rats. One to two hours after vasopressin injection the interstitial space appeared widened but no change in interstitial cells were reported. The effects of prolonged vasopressin treatment was studied by Sun[10] but no mention was made about interstitial cells. In our hands, prolonged vasopressin treatment caused a dramatic decrease in the number of interstitial cells and in the animals with the highest urine osmolality they were nearly absent. The presence of some cell debris in the interstitial space suggests that the interstitial cells had undergone necrosis although whole necrotic cells were rarely encountered. This cell damage appeared rather selective for the interstitial cells. However, in the animals with the most pronounced changes some endothelial damage was also seen.

At present, the mechanism behind the observed changes of the interstitial cells is not known. Although the changes may be due to a direct effect of the hormone on the interstitial cells, it cannot be excluded that they are secondary to the increase in papillary interstitial solute concentration that occurs gradually during the first two weeks of vasopressin treatment.[12]

Interestingly, Shimamura and Trojanowski,[13] who studied the effect of repeated deprivation of drinking water on the morphology of the renal medulla of normal rats, found a loss of interstitial cells after 5 to 16 weeks of treatment. It is not known whether this was an effect of high vasopressin stimulation or other, local changes in the medulla. However, interstitial cells are considered to be rather resistant to a hypertonic milieu.[14,15] In a preliminary study on Wistar rats, we could not see any obvious changes of interstitial cells after 19 days of vasopressin treatment (1.0 U/day).

While rare in the inner part of the papilla of normal rats, interstitial cells type 2[6] were very abundant in some DI rats. This was normalized already after two days of vasopressin treatment when type 2 cells were virtually absent. Interstitial cells of type 2 are morphologically similar to lymphocytic or monocytic cells[6,16] and may be found in close apposition to the cell membrane of type 1 cells.[6] However, since the function of type 2 cells in the renal medulla and their exact relation to type 1 cells is unknown, the interpretation of the present findings with respect to type 2 cells is difficult.

Although pronounced effects of vasopressin treatment on interstitial cells have been demonstrated, this study does not clarify the exact relationship between the antidiuretic hormone and the interstitial cells type 1. The hypothesis that the interstitial cells have vasopressin receptors remains an attractive one but further studies are necessary to substantiate it. Maybe this could give a further clue to the understanding of the role of the kidney, sodium, and volume in hypertension.

Acknowledgments

We are indebted to Miss Erika Eliasson and Mr. Ulf Nilsson for competent technical assistance. Synthetic arginine vasopressin was a generous gift from Ferring AB, Malmö, Sweden.

References

1. Mandal, A. K. & S.-O. Bohman, Eds. 1980. The Renal Papilla and Hypertension. Plenum Publ. Corp. New York, N.Y.
2. Bohman, S.-O. & P. K. A. Jensen. 1976. Morphometric studies on the lipid droplets of the interstitial cells of the renal medulla in different states of diuresis. J. Ultrastruct. Res. **55:** 182–192.
3. Zusman, R. M. & H. R. Keiser. 1977. Prostaglandin biosynthesis by rabbit renomedullary interstitial cells in tissue culture: Stimulation by angiotensin II, bradykinin and arginine vasopressin. J. Clin. Invest. **60:** 215–223.
4. Beck, T. R., A. Hassid & M. J. Dunn. 1980. The effect of arginine vasopressin and its analogs on the synthesis of prostaglandin E_2 by rat renal medullary interstitial cells in culture. J. Pharmacol. Exp. Ther. **215:** 15–19.
5. McAuliffe, W. G. 1980. Histochemistry and ultrastructure of the interstitium of the renal papilla in rats with hereditary diabetes insipidus (Brattleboro strain). Am. J. Anat. **157:** 17–26.

6. Bohman, S.-O. 1974. The ultrastructure of the rat renal medulla as observed after improved fixation methods. J. Ultrastruct. Res. **47:** 329–360.
7. Weibel, E. R. 1970. Stereological techniques for electron microscopic morphometry. *In* Principles and Techniques of Electron Microscopy. M. A. Hayat, Ed. Vol. **3:** 237–296. Van Nostrand Reinhold Co. New York, N.Y.
8. Bohman, S.-O. 1980. The ultrastructure of the renal medulla and the interstitial cells. *In* The Renal Papilla and Hypertension. A. K. Mandal & S.-O. Bohman, Eds.: 7–33. Plenum Publ. Corp. New York, N.Y.
9. Tisher, C. C., R. E. Bulger & H. Valtin. 1971. Morphology of renal medulla in water diuresis and vasopressin-induced antidiuresis. Am. J. Physiol. **220:** 87–94.
10. Sun, C. N., H. J. White & E. J. Towbin. 1972. Histochemistry and electron microscopy of the renal papilla in a genetic strain of rats with diabetes insipidus. Nephron **9:** 308–317.
11. Sun, C. N. 1980. Effect of antidiuretic hormone to the connective tissues of rat renal papilla. Exp. Path. **18:** 469–473.
12. Valtin, H. 1967. Hereditary hypothalamic diabetes insipidus in rats (Brattleboro Strain). Am. J. Medicine **42:** 814–827.
13. Shimamura, T. & S. Trojanowski. 1976. Effects of repeated deprivation of drinking water on the structure of renal medulla of rats. Am. J. Pathol. **84:** 87–92.
14. Kuroda, M., H. Ueno, S. Sakato, N. Funaki & R. Takeda. 1979. A unique affinity and adaptation of renomedullary interstitial cells for hypertonic medium. Prostaglandins **18:** 209–220.
15. Bulger, R. E., R. Beeuwkes III & A. J. Saubermann. 1981. Application of scanning electron microscopy to X-ray analysis of frozen-hydrated sections. III. Elemental content of cells in the rat renal papillary tip. J. Cell Biol. **88:** 274–280.
16. Bulger, R. E. & R. B. Nagle. 1973. Ultrastructure of the interstitium in the rabbit kidney. Am. J. Anat. **136:** 183–204.

Discussion of the Paper

W. Kriz (*University of Heidelberg, Heidelberg, F.R.G.*): You presented very striking findings. Do you have an explanation for the difference between your and Dr. McAuliffe's results? How stable is the shape of these cells? Might these cells be very mobile in shape; furthermore, might the osmolality of the fixative alter the shape of the cells?

Bohman: The question about fixation artifacts is a relevant one, especially in a study of the renal papilla, which is notoriously difficult to preserve in an optimal way. But I cannot imagine, and I say this from some experience in this area, that these differences are due to fixation effects, either in Dr. McAuliffe's work or in mine. Hearing the discussion yesterday about the differences that may occur in different Brattleboro rat colonies around the world, I am starting to get worried that one of us has obtained rats that differ from normal rats, not only with respect to the lack of vasopressin but also with respect to the structure of the renal medulla, although this seems somewhat remote. Our rats were obtained from the TNO in Holland and Dr. McAuliffe, I think, got his rats from two different breeders in the United States.

A. M. MOSES (*State University Hospital, Syracuse, N.Y.*): Another possibility is that hypokalemia is a major stimulus to prostaglandin production in these cells. After the discussions yesterday about potassium balance in rats and the response to AVP treatment, etc., there may be indirect effects on morphology as well as on prostaglandin synthesis related to potassium levels.

BOHMAN: Yes, that is possible. We would expect different potassium balances in untreated and treated rats.

S. OPAVA-STITZER (*University of Puerto Rico, San Juan, P.R.*): Dr. Michael Dunn has studied the effects of hypokalemia on prostaglandin synthesis in normal rats; a degree of hypokalemia very similar to what is found in Brattleboro rats (when they are hypokalemic) had no effect on prostaglandin synthesis.

WATER BALANCE IN THE BRATTLEBORO RAT: CONSIDERATIONS FOR HORMONE REPLACEMENT THERAPY *

Lewis B. Kinter

Department of Pharmacology
Smith Kline and French Laboratories
Philadelphia, Pennsylvania 19101

Introduction

Brattleboro strain rats homozygous for diabetes insipidus (DI rats) lack detectable arginine vasopressin (AVP, antidiuretic hormone) and drink and excrete enormous quantities of water, as compared to vasopressin-replete Brattleboro heterozygotes (HZ) and Long-Evans (LE) strain controls (Figure 1). Discovered in 1961, DI rats have become an increasingly popular experimental model of vasopressin insufficiency.[1] This chapter will summarize the mechanisms that permit net water balance to be achieved in both the presence and absence of antidiuretic hormone (ADH), and describe the "water state" of the DI rat. Antidiuretic hormone replacement in DI rats will be discussed briefly in terms of integration of exogenously administered hormone with endogenous feedback-controlled mechanisms.

Water Balance in the DI Rat

Water balance describes the condition in which the rates of total water intake and total water loss are equal. As a result, total body water (TBW) is maintained at a constant, although undefined, level. (For a detailed statement of water balance see Appendix.) Water balance is achieved by the interaction of feedback-controlled mechanisms regulating water intake and water loss (Figure 2). Water intake is regulated by thirst and environmental water availability. Water losses include both a large insensible component and renal water losses regulated by antidiuretic hormone. Feedback control of TBW, and hence water balance, is based upon two primary receptor systems, one based on circulating blood volume, the other based on plasma osmolality.[2, 3] Since both thirst and ADH release result from activation of either of these systems, a single "ADH-thirst drive" has been found sufficient to model the control of TBW.[4] Thus a pure volume decrease such as hemorrhage causes both a reduction in the rate of renal water loss and stimulation of thirst while a pure osmolar increase (ingestion of salt) leads to both thirst and renal water conservation. In normal animals, ADH-thirst drive results in water intake and ADH release proportional to the slopes of their respective response curves (Figure 2). These slopes depend upon species, environmental constraints, and physiologic

* Supported in part by the National Institutes of Health Grant AM 18294 and by a gift from R.J. Reynolds Industries. L.B.K. is the recipient of the National Institutes of Health fellowship award AM05938.

0077–8923/82/0394–0448 $.1.75/0

state. The ratio of these slopes determines the relative magnitudes of the thirst and ADH responses to a unit of total drive. Thus, in a desert environment, or during experimental water deprivation, little or no water intake can occur despite a high level of total drive and the slope of the water intake response approaches zero. Animals maintained in such conditions must maintain their TBW primarily through water conservation. In certain desert mice conservation is so efficient that daily fluid losses are substantially replaced with water derived from metabolism alone.[5] In the DI rat, on the other hand, ADH is absent and the slope of the ADH response is zero (FIGURE 3). Renal water conservation is reduced and copious quantities of dilute urine are formed

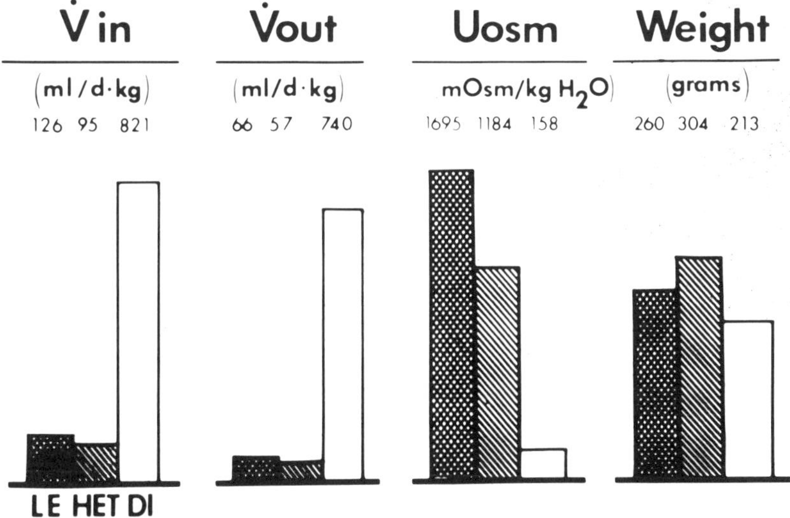

FIGURE 1. 24 h water intake ($\dot{V}_{in}$), urine output ($\dot{V}_{out}$), and urine osmolality (U_{osm}) in Brattleboro strain rats homozygous (DI, di/di) and heterozygous (HET, +/di) for diabetes insipidus, and Long-Evans rats (LE, +/+).

continuously. In DI rats TBW must be maintained entirely through a high rate of water intake driven by the thirst response to total drive (FIGURE 3).

ADH-THIRST DRIVE IN DI RATS

High rates of water intake in DI rats could result from either greater levels of total drive or a greater response (e.g., steeper slope of the water intake response curve) to the same level of drive. Morphological, biochemical, and physiological evidence from studies of the DI rat posterior pituitary-hypothalamic axis strongly supports the former. Hypertrophy of neurosecretory neurons, parikarya, nuclei, and nucleoli in supraoptic and paraventricular nuclei; increased protein synthetic activity; and increased electrical activity within the neurohypophyseal tracts of DI rats are consistent with greater stimulation and thus greater total drive.[6-10]

Greater total drive could result from greater input from either osmolality- or volume-dependent components, or both (FIGURE 3). DI plasma osmolality, sodium, and chloride levels have been measured by many investigators and are consistently reported elevated (~5%) above those of antidiuretic controls (TABLE 1). These levels, if measured in ADH-replete animals would indicate a significant elevation of osmolality-dependent drive. Direct evidence for elevation of the volume component of total drive in DI rats has been elusive. Valtin

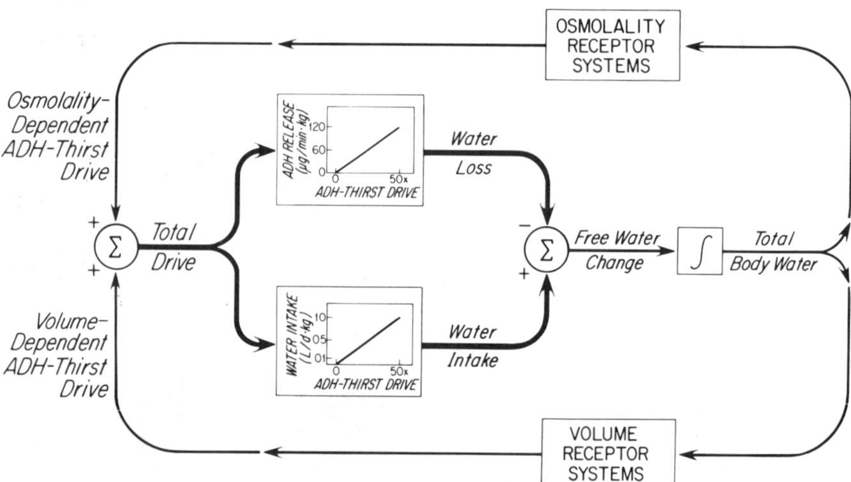

FIGURE 2. Simplified diagram showing feedback control of total body water through volume and osmoreceptor systems. Drives resulting from high plasma osmolality and decreased effective circulating volume and summed to create a total ADH-thirst drive. The rate of ADH release depends upon the slope of the ADH response curve. Increased ADH reduces the rate of water loss. If water is freely available, the rate of water intake depends on the slope of the thirst response curve. The net rate of change in body water is given by the difference between rates of intake and water loss. Integration of this rate with respect to time yields the net change resulting from the compensatory mechanisms. The constant of integration (not shown) defines the initial volume state.

et al. first proposed that elevation of the concentrations of the plasma solutes in DI rats indicated mild dehydration and plasma volume constriction.[11] This concept is supported by reports of increased body weight, expansion of plasma volume, and dilution of hematocrit associated with acute ADH replacement in DI rats.[12, 13] In addition, during chronic ADH replacement the growth curve for DI rats was displaced, first upwards (~5%) with initiation of therapy, then downwards (~5%) when therapy was withdrawn (FIGURE 4). Such displacement is consistent with expansion of TBW with ADH replacement and thus contraction of TBW with ADH withdrawal. Paradoxically, the only reported measurements of body fluid volumes in Brattleboro rats suggest that

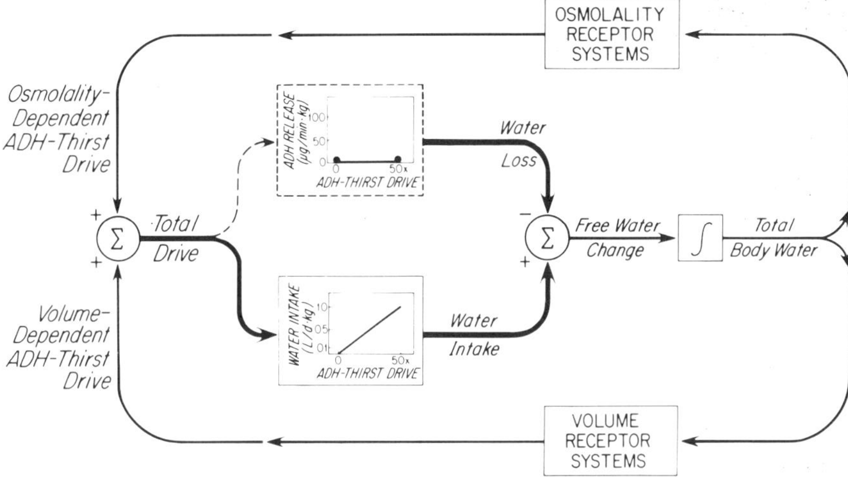

FIGURE 3. Simplified diagram showing feedback control of total body water in the vasopressin-deficient DI rat. As shown in FIGURE 2, total ADH-thirst drive results from input from both volume and osmoreceptor mechanisms. In hypothalamic diabetes insipidus the ADH response to total drive is absent (broken lines) and a large obligatory water loss results. The thirst response, however, is intact and water intake is increased in response to greater levels of drive to balance water losses.

TABLE 1

PLASMA SOLUTE PROFILES IN DI RATS

Antidiuretic Control *	DI	(p)	Δ%	Reference
Plasma Osmolality (mOsm/kg H_2O)				
302	323	(<.01)	+7.0%	12
308	321	(<.01)	+4.2%	13
307	320	(<.001)	+4.2%	1
303	320	(<.01)	+5.6%	23
304	321	(<.01)	+5.6%	24
Plasma Sodium (mEq/L)				
145	152	(<.05)	+4.8	1
140	147	(<.01)	+4.7	12
141	144	(<.01)	+2.1	13
140	146	(<.01)	+4.3	22
141	147	(<.01)	+4.1	23
137	146	(<.01)	+6.6	24
Plasma Chloride (mEq/L)				
103	108	(<.001)	+4.9	24

* Brattleboro heterozygotes or Long-Evans rats.

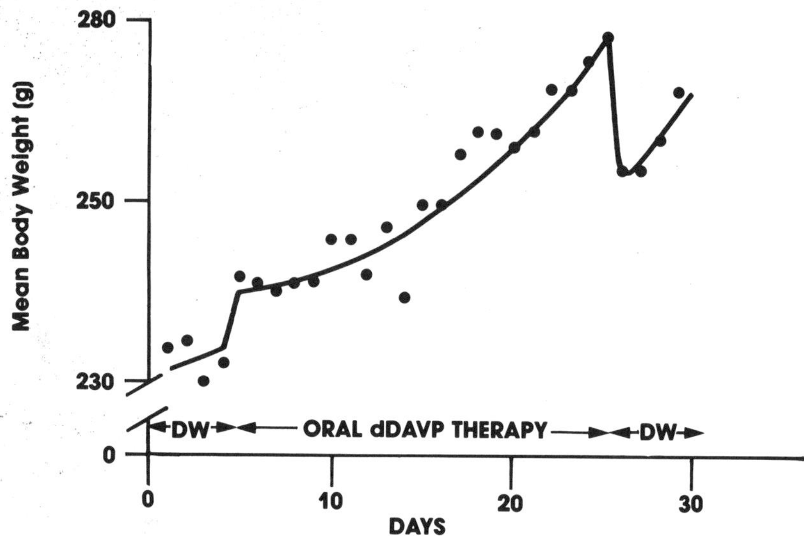

FIGURE 4. Mean body weight in DI rats before, during, and after treatment with dDAVP. Six male DI rats drank distilled water for four days then increasing concentration of dDAVP (first 5, then 10, 100, 1000, and 2000 μg/L for at least four days at each concentration). A final four day posttreatment control terminated the experiment. Daily U_{osm} and intake and excretion of water for this experiment are presented in FIGURE 9. The data suggest an upward displacement of the growth curve during dDAVP therapy. The changes in body weight associated with initiation and termination of drug treatment were statistically significant ($p < .01$).

extracellular fluid volume (ml/100 g bw) is slightly greater in DI rats than in heterozygotes.[14] However, DI rats tend to be much leaner than heterozygotes (personal observation) and thus comparison of fluid volumes expressed as a function of total body weight may be misleading.

Recently much attention has been focused on hormonal factors that might modulate total drive. Growing evidence suggests that peripheral levels of angiotensin II (AII) may stimulate both thirst and ADH release. Elevated plasma renin activity and AII levels have been consistently reported in DI rats.[15-19] Reduction of plasma AII levels with SQ14225 is reported to lower water intake in DI rats,[20] although others have failed to show this effect.[21] Thus it may be appropriate to include a secondary endocrine component to total drive (FIGURE 3, not shown) in the DI rat.

Conclusion

DI rats lack ADH, drink and excrete large quantities of water daily, and achieve water balance solely by means of the thirst response to elevated levels of total drive. The physiological significance of elevated drive in DI rats extends beyond its role in maintaining water balance. The factors contributing to that drive almost certainly influence other regulatory mechanisms (e.g., sympathetic nervous system activity) and thus constitute separate considerations when evaluating physiological and pharmacological responses in DI rats.

Antidiuretic Replacement Therapy

Replacement therapy with ADH causes increased renal water reabsorption and results in expansion of plasma volume, dilution of plasma osmolality, reduction of total drive, and finally, reduction of water intake. It is significant that systemic ADH levels apparently do not modulate thirst directly, but only indirectly via drive.[4] Since TBW depends upon the appropriate ratio of ADH and thirst responses to total drive (Figure 2), appropriate ADH replacement must insure not only adequate plasma ADH levels, but also rapid and adequate adjustments in water intake. The remainder of this paper will discuss briefly three modes of ADH replacement with respect to this principle.

Bolus Injection

The most widely used mode of antidiuretic therapy is the daily subdermal injection of semipurified pituitary extracts, synthetic vasopressins, or vasopressin analogues, in either aqueous solution or oil suspension (Table 2, Figure 5). Valtin and colleagues first reported use of this therapy in DI rats and showed that while polyurea and polydipsia were decreased, stable levels were only

Table 2

Summary of Antidiuretic Hormone Replacement in DI Rats

Drug	Dose	Reference
Daily Subdermal Bolus Injection		
Pitressin	500 mU/rat	1
"	500–2,500 mU/rat	25
"	1,000 mU/rat	18
"	40–400 mU/100 g	15, 17
"	50–500 mU/rat	13
"	250–500 mU/rat	26
"	1,000 mU/rat	27
"	1,000 mU/rat	28
AVP	100 mU/rat	23
Continuous Infusion		
Pitressin	330 μU/min	29
AVP	80 ng/h	30
LVP	20 U/kg·week	31
dDAVP	2 ng/h	30, 24
dDAVP	5–300 ng/kg·day	32
Gastric Gavage		
dDAVP	100–400 ng/rat	33
Oral Self-Administration		
dDAVP	5–2,000 μg/L	24, 34–37

achieved after several weeks of daily injections at a dose of 0.5–1.0 U/rat.[1] This observation, confirmed by others[24, 26] (FIGURE 6), has promoted much speculation concerning possible additional renal defects in DI rats.[25, 37] With the advent of radioimmunoassay techniques for AVP it was shown that daily injection of as little as 100 mU/rat resulted in physiological plasma AVP levels, while injection of 500 mU/rat resulted in plasma levels higher than those observed in severe hydropenia.[23] In addition, only five days of treatment were required to achieve a stable antidiuretic state at the 100 mU/rat dose.[23] Thus, while the possibility of additional defects cannot be excluded, it seems likely that the impaired urine concentrating ability observed at doses greater than 100 mU/rat is primarily due to an inappropriate ratio of hormone delivery and

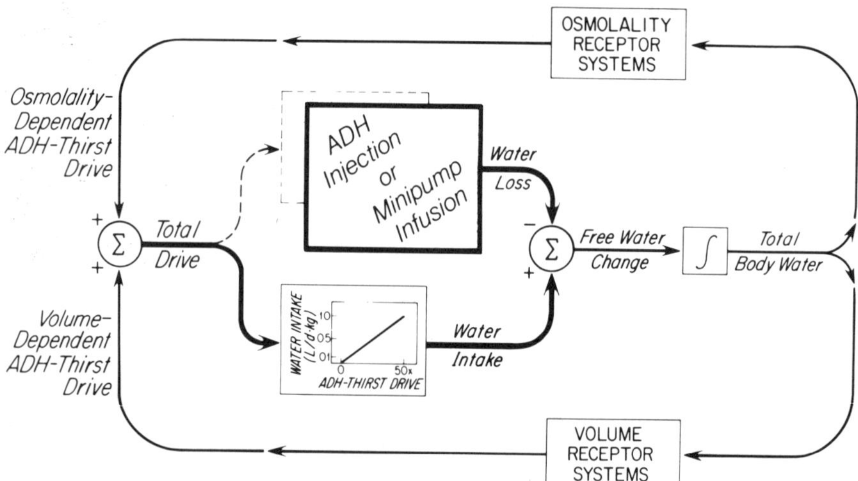

FIGURE 5. Simplified diagram showing feedback control of total body water in DI rats given ADH by bolus injection or minipump infusion. Exogenous hormone is imposed on the endogenous feedback control mechanism. Renal water reabsorption is increased, plasma osmolality is diluted and plasma volume expanded. Total drive is reduced. Water intake is reduced in response to total drive. Significant points include: (1) ADH delivery by bolus injection or by minipump infusion is not feedback controlled and (2) rising titers of ADH apparently do not inhibit thirst directly.

water intake. The secretion of inappropriate amounts of endogenous AVP is a recognized clinical syndrome (Syndrome of Inappropriate ADH Secretion, SIADH) and is also characterized by impaired urine concentrating ability.[38]

Chronic Infusion

The development of miniature implantable infusion pumps (Alzet, Alza Corp., Palo Alto, Calif.) has provided a relatively simple means for maintaining a constant infusion of ADH in experimental animals.[39] These pumps can be implanted subdermally or intraperitoneally and can deliver ADH to be absorbed

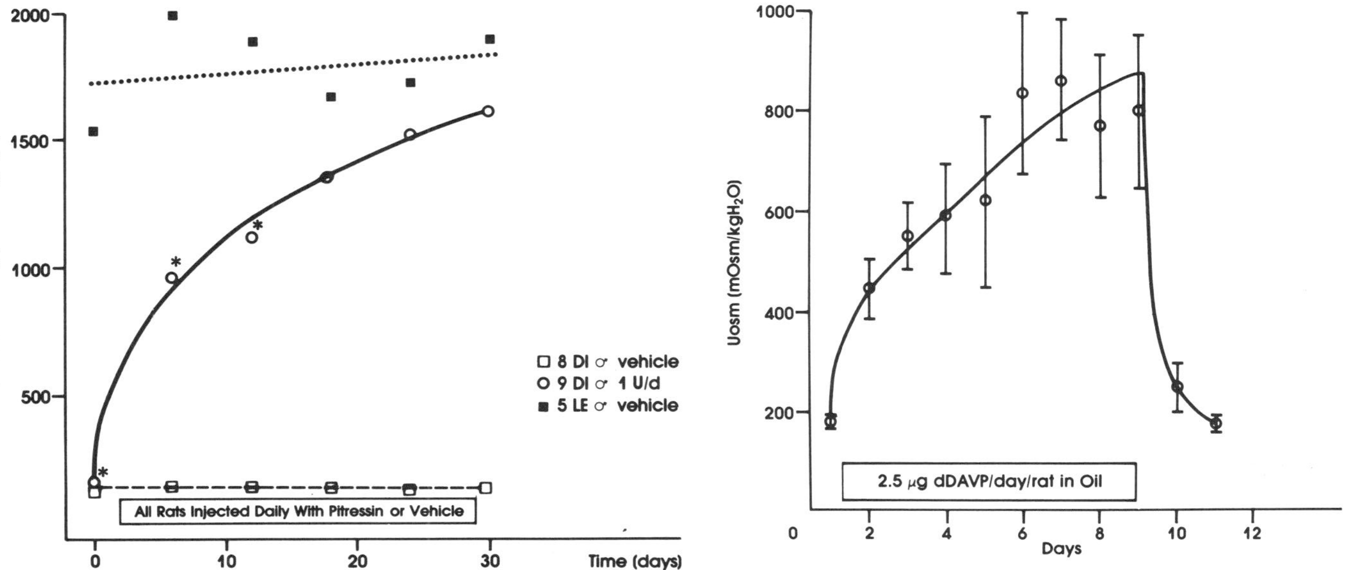

FIGURE 6. Antidiuretic responses of DI rats given Pitressin (1 U/day in oil, $N = 9$, (left) or dDAVP (2.5 μg/day in oil, $N = 6$) (right). More than two weeks of daily Pitressin injections were required before mean U_{osm} in DI rats was not significantly different from those of vehicle injected LE rats (*, $p < .01$). After eight days of treatment with a large dose of dDAVP mean U_{osm} in six additional DI rats was not above 1,000 mOsm, similar to the effect seen with Pitressin. It is likely that the impaired urine concentrating ability in DI rats titrated with large doses of Pitressin or dDAVP is due to the presence of supraphysiological amounts of antidiuretic activity. Very large amounts of exogenous antidiuretic activity may produce a Syndrome of Inappropriate ADH Secretion-like condition.

locally or, with the addition of a catheter, can deliver ADH intravenously for up to two weeks. However, like bolus injection, ADH infusion by minipump is independent of total drive (FIGURE 5). Surprisingly, experience with minipumps in DI rats is limited (TABLE 2). Extravascular delivery of between 20 and 100 ng/h AVP appears to produce reasonable levels of antidiuresis. Plasma concentrations associated with these levels of infusion have not been reported. Achievement of similar levels of antidiuresis are reported with the infusion of the vasopressin agonist desamino-8-D-arginine vasopressin (dDAVP) at only 2 to 10 ng/h. In contrast to previous bolus injection studies, stable levels of antidiuresis are reported achieved well within the seven-day infusion period, although that stability appears again to be somewhat dose-related [32] (FIGURE 7). Results with minipump infusion replacement in DI rats support the concept that the impaired urine concentrating ability previously observed with supramaximal ADH doses is related to delivery of inappropriate amounts of hormone. More experience is needed with this unique delivery system especially as regards chronic intravenous infusions.

Oral Self-dosing Therapy

The discovery that dDAVP was orally active [34] led to the development of an entirely new approach to antidiuretic therapy in DI rats, in which hormone delivery was coupled via thirst to total drive (FIGURE 8). Addition of dDAVP

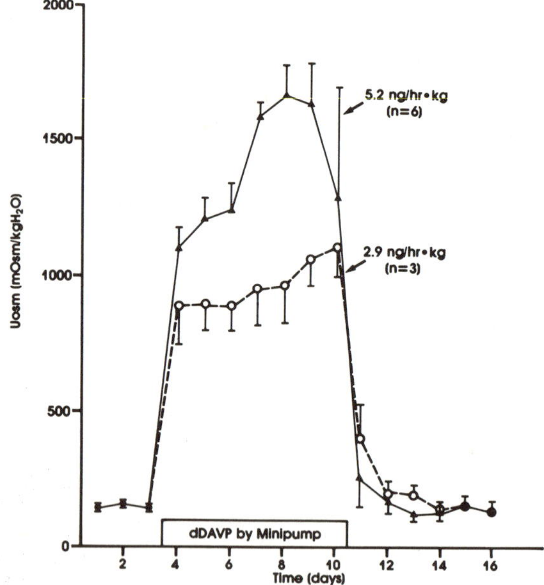

FIGURE 7. Effect of chronic infusion of dDAVP on U_{osm} in DI rats. Two groups of DI rats were given minipumps (Alzet #1701, Alza Corp.) implanted i.p. and loaded to infuse 2.9 or 5.2 ng dDAVP/h·kg respectively for seven days. In contrast to bolus injection studies (FIGURE 6), stable high levels of U_{osm} were achieved within the seven-day pump infusion period.

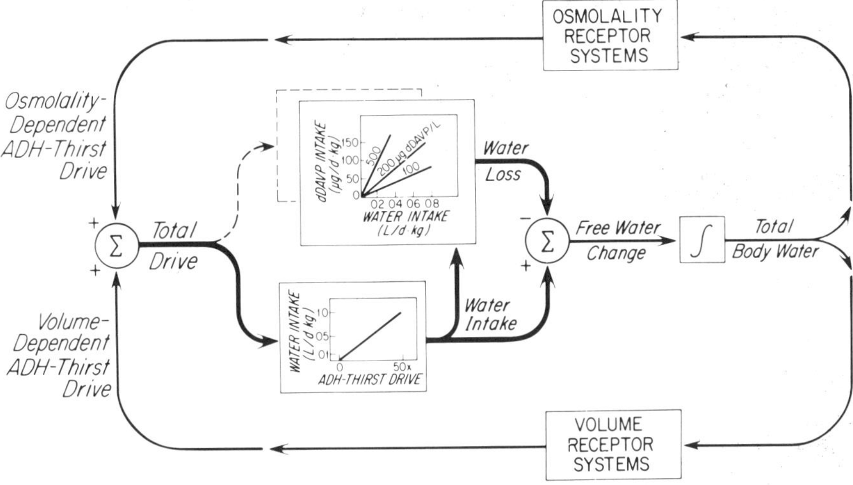

FIGURE 8. Simplified diagram showing feedback control of total body water in hypothalamic diabetes insipidus as achieved by addition of dDAVP to drinking water. As shown in FIGURE 2, total ADH-thirst drive results from the operation of volume and osmoreceptor mechanisms. In hypothalamic diabetes insipidus however, the ADH response is absent (dotted lines), while the thirst response is maintained. The presence of an orally active antidiuretic agent (dDAVP) in the water ingested results in drive-proportional hormone administration. Increasing the concentration of agent in the water increases the slope of the new response curve, and changes the proportion between water-intake and antidiuretic responses to a given level of total ADH-thirst drive.

(5–2,000 μg/L) to the drinking water results in rapid achievement (within 24 hours) of dose-dependent antidiuretic states (FIGURE 9) in spite of the large doses ingested. dDAVP intake (μg/d·kg) increases as a nearly linear function of the concentration of dDAVP in the drinking water (FIGURE 10), in spite of progressive reductions in water intake (FIGURE 9). Plasma dDAVP levels achieved are not available, but levels of dDAVP in urine suggest that only a small fraction of the oral dose is actually absorbed in an active form.[35] The ratio of dDAVP intake to water intake is determined by the concentration of dDAVP in the drinking water and is conveniently controlled by the investigator (FIGURE 8). Each ratio is characterized by a unique level of U_{osm}. The addition of an orally active ADH analogue to the drinking water of DI rats sets up a unique situation in which both water and hormone are self-administered in an appropriate ratio in response to total drive. Clearly the usefulness of such oral therapy is limited to chronic studies in conscious animals. It is important to emphasize that drive-dependent dDAVP intake in the DI rat is in series with drive-dependent water intake (FIGURE 8) whereas drive-dependent ADH release in the normal rat is in parallel with water intake (FIGURE 2). The consequence of series delivery of water and dDAVP is that as drinking behavior becomes more periodic, greater hourly fluctuations in urine concentrating ability will occur. This is in contrast to parallel delivery where reduced water intake is compensated for by greater endogenous hormone release (FIGURE 2).

Conclusion

Antidiuretic hormone replacement in DI rats must rapidly produce an appropriate ratio of hormone delivery and water intake in order to maintain net water balance. Low-dose, continuous ADH infusion (via minipumps) and drive-dependent dDAVP self-administration appear to meet these criteria and permit the rapid achievement of stable antidiuretic states. Prolonged unstable antidiuretic responses in DI rats appear to be primarily associated with administration of inappropriately large doses of hormone. Reduced renal concentrating ability associated with such administration is remarkably similar to that associated with inappropriate secretion of endogenous ADH and is thus probably not the result of other proposed defects in the DI rat kidney.

Acknowledgments

I wish to thank Mr. Dan Silkiss (European Chemical Co., New York) and Dr. Jan Mulder (Ferring AB, Sweden) for their generous contributions of dDAVP, Ms. Greer Geiger, Ms. Janet Shahood, and Mr. David Shier for their expert technical assistance, and Dr. Reinier Beeuwkes for his valuable suggestions concerning the manuscript.

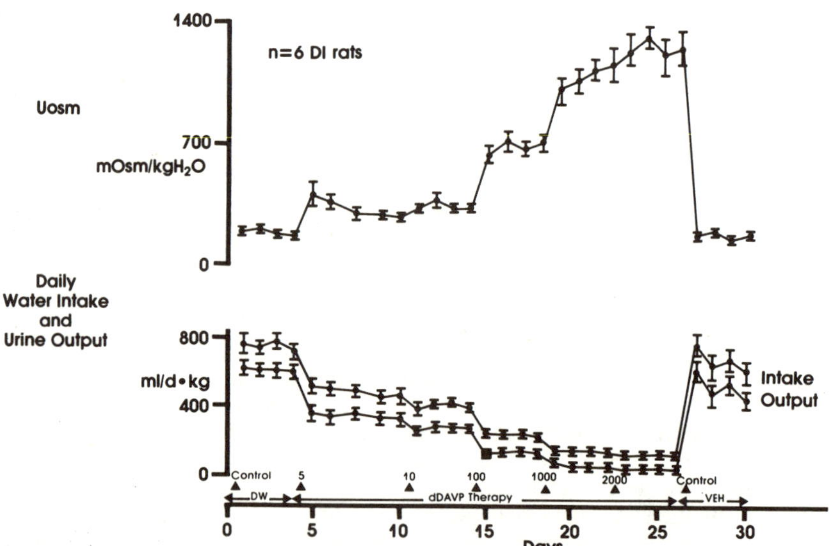

FIGURE 9. Effect of drinking increasing concentrations of dDAVP on U_{osm}, water intake, and urine output in DI rats. Six DI rats were first given distilled water (control, four days) then increasing concentrations of dDAVP (5–2,000 μg/L) at intervals of at least four days, followed an additional four days of vehicle (VEH). Progressive and dose-dependent changes in U_{osm}, water intake, and urine output were observed. Mean body weight increased during the treatment period (FIGURE 4) suggesting maintenance of positive metabolic balance and net expansion of plasma volume.

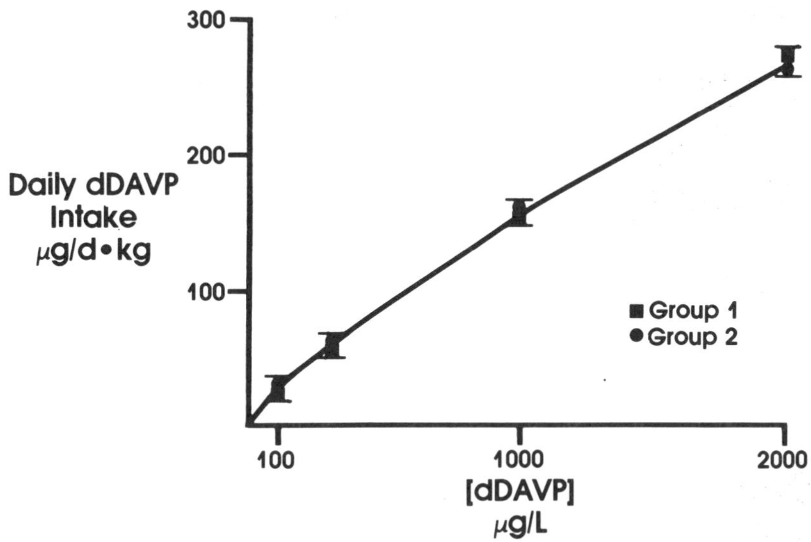

FIGURE 10. Relationship between daily intake (μg/d·kg) and concentration (μg/L) of dDAVP ingested when administered orally in the drinking water. Two groups of DI rats drank increasing concentrations of dDAVP for at least three days at each concentration and the daily dDAVP dose calculated from the water intake. Despite a progressive fall in water intake associated with ingestion of increasing concentrations of dDAVP (FIGURE 9), dDAVP dose increased as a nearly linear function of concentration administered in the drinking water.

REFERENCES

1. VALTIN, H. & H. A. SCHROEDER. 1964. Familial hypothalamic diabetes insipidus in rats (Brattleboro strain) Am. J. Physiol. **206:** 425–430.
2. GAUER, O. H. & J. P. HENRY. 1976. Neurohormonal control of plasma volume. *In* Cardiovascular Physiology II. A. C. Guyton & A. W. Cowley, Jr., Eds.: 145–190. University Park Press. Baltimore, Md.
3. PETERS, G., J. T. FITZSIMONS & L. PETERS-HAEFEL. 1975. Control Mechanisms of Drinking. Springer-Verlag. New York.
4. COWLEY, A. W. 1975. Role of thirst and vasopressin in control of body fluid osmolality and volume. *In* Circulatory Physiology II: Dynamics and Control of the Body Fluids. A. C. Guyton, A. E. Taylor & H. J. Granger, Eds. Ch. 9. Saunders Co. Philadelphia, Pa.
5. MACMILLEN, R. E. & A. K. LEE. 1967. Australian desert mice: Independence of exogenous water. Science **158:** 383–385.
6. SOKOL, H. W. & H. VALTIN. 1965. Morphology of the neurosecretory system in rats homozygous and heterozygous for hypothalamic diabetes insipidus (Brattleboro strain). Endocrinology **77:** 692–700.
7. SCOTT, D. E. 1968. Fine structural features of the neural lobe of the hypophysis of the rat with homozygous diabetes insipidus (Brattleboro strain). Neuroendocrinology **3:** 156–163.
8. KALIMO, H. & U. K. RINNE. 1972. Ultrastructural studies on the hypothalamic neurosecretory neurons of the rat. II. The hypothalamic-neurohypophyseal

system in rats with hereditary diabetes insipidus. Z. Zellforsch. Mikrosk. Anat. **134:** 205–217.

9. Granger, F. & J. C. Sloper. 1974. Correlation between microtubular number and transport activity of hypothalamo-neurohypophyseal secretory neurons. Cell Tissue Res. **153:** 101–108.
10. Dyball, R. E. 1974. Single unit activity in the hypothalamo-neurohypophyseal system of Brattleboro rats. J. Endocrinol. **60:** 135–144.
11. Valtin, H. & A. R. Harrington. 1968. Accumulation of medullary urea during prolonged treatment of hypothalamic diabetes insipidus with vasopressin (ADH). *In* Urea and the Kidney. B. Schmidt-Nielsen & D. W. S. Kerr, Eds.: 430–442. Excerpta Medica Foundation. Amsterdam, the Netherlands.
12. Möhring, B., J. Möhring, G. Dauda & D. Haack. 1974. Potassium deficiency in rats with hereditary diabetes insipidus. Am. J. Physiol. **227**(4): 916–920.
13. Möhring, J., B. Möhring, A. Schomig, H. Schomig-Brekner & D. Haack. 1974. Acute effects of vasopressin on potassium and water balance in rats with diabetes insipidus. Am. J. Physiol. **227**(4): 921–926.
14. Friedman, S. M. & C. L. Friedman. 1965. Salt and water distribution in hereditary and in induced hypothalamic diabetis insipidus in the rat. Can. J. Physiol. Pharmacol. **43:** 699–705.
15. Gutman, Y. & F. Benzakein. 1971. Effect of an increase and a lack of antidiuretic hormone on plasma renin activity in the rat. Life Sci. **10**(1): 1081–1085.
16. Kyncl, J., L. Miksche, M. Khayyal, J. Möhring & F. Gross. 1971. Renin-angiotensin system in rats with hereditary hypothalamic diabetes insipidus. Physiologist **14:** 177A.
17. Gutman, Y. & F. Benzakein. 1974. Antidiuretic hormone and renin in rats with diabetes insipidus. Eur. J. Pharmacol. **28:** 114–118.
18. Henderson, I. W., R. J. Balment & J. A. Oliver. 1978. Vasopressin effects on plasma renin activity in male and female rats. Clin. Sci. Mol. Med. **55:** 301–307.
19. Kinter, L. B., D. N. Shier, A. C. Barger, W. Flamenbaum & R. Beeuwkes III. 1979. Comparison of plasma renin activity (PRA) in conscious and anesthetized Brattleboro strain rats. Fed. Proc. **38:** 892A.
20. Henderson, I. W., A. McKeever & C. J. Kenyon. 1979. Captopril (SQ-14225) depresses drinking and aldosterone in rats lacking vasopressin. Nature **281:** 569–570.
21. Mann, J. F. E., W. Rascher, A. Schömig & R. Dietz. 1978. Inhibition of the renin-angiotensin-system in Brattleboro rats with hereditary hypothalamic diabetes insipidus. Klin. Wschr. 56 Suppl. **1:** 67–70.
22. Gross, F., G. Dauda, S. Kazda, J. Kyncl, J. Möhring & H. Orth. 1972. Increased fluid turnover and the activity of the renin-angiotensin system under various experimental conditions. Circ. Res. (Supp. II) **30** & **31:** II–173–II–181.
23. Möhring, J., G. Kohrs, B. Möhring, M. Petri, E. Homsy & D. Haack. 1978. Effects of prolonged vasopressin treatment in Brattleboro rats with diabetes insipidus. Am. J. Physiol. **234**(2): F106–F111.
24. Kinter, L. B. 1978. Homeostatic Mechanisms in the Diabetes Insipidus Rat. Unpublished doctoral thesis. Harvard University. Cambridge, Mass.
25. Harrington, A. V. & H. Valtin. 1968. Impaired urinary concentration after vasopressin and its gradual correction in hypothalamic diabetes insipidus. J. Clin. Invest **47:** 502–510.
26. Sokol, H. W. & J. Sise. 1973. The effect of exogenous vasopressin and growth hormone on growth of rats with hereditary hypothalamic diabetes insipidus. Growth **37:** 127–142.
27. Lee, J. & P. G. Williams. 1972. The effect of vasopressin (Pitressin) ad-

ministration and dehydration on the concentration of solutes in renal fluids of rats with and without hereditary hypothalamic diabetes insipidus. J. Physiol. **220:** 729–743.
28. Laycock, J. F. & P. G. Williams. 1973. The effect of vasopressin (Pitressin) administration on sodium, potassium, and urea excretion in rats with and without diabetes insipidus (DI), with a note on the excretion of vasopressin in the DI rat. J. Endocrinol. **56:** 111–120.
29. Licht, A., J. D. Silver, G. S. Levey, D. C. Lehotay & J. J. Bourgoignie. 1978. The effect of continuous low dose administration of vasopressin (VP) to rats with and without diabetes insipidus. Kidney Int. **12:** 565A.
30. Fujimoto, S. & G. A. Hedge. 1981. Pituitary-thyroid function in the Brattleboro rat with hereditary diabetes insipidus. Fed. Proc. **40**(3): 457.
31. Fisher, M. E. & Kupchella, C. E. 1981. Antidiuretic hormone induced kidney glycosaminoglycans in the Brattleboro rat. Fed. Proc. **40**(3): 554.
32. Walker, L. A. & J. C. Frölich. 1981. Dose-dependent stimulation of renal prostaglandin synthesis by deamino-8-D-arginine vasopressin in rats with hereditary diabetes insipidus. J. Pharmacol. Exp. Ther. **217:** 87–91.
33. Vavra, I., A. Machoua & I. Krejci. 1974. Antidiuretic action of 1-desamino-[8-D-arginine]-vasopressin in unanesthetized rats. J. Pharmacol Exp. Ther. **188:** 241–247.
34. Kinter, L. B., G. L. Geiger & R. Beeuwkes III. 1977. Oral antidiuretic therapy in diabetes insipidus rats. Kidney Int. **12:** 563A.
35. Kinter, L. B., M. J. Dunn, T. R. Beck, R. Beeuwkes III & A. Hassid. 1981. The interactions of prostaglandins and vasopressin in the kidney. Ann. N.Y. Acad. Sci. **372:** 163–179.
36. Dunn, M. J., H. P. Greely, H. Valtin, L. B. Kinter & R. Beeuwkes III. 1978. Renal excretion of prostaglandins E_2 and $F_{2\alpha}$ in diabetes insipidus rats. Am. J. Physiol. **235:** E624–E627.
37. Imbert-Teboul, M., D. Chabardes, M. Montegut, A. Clique & F. Morel. 1978. Vasopressin-dependent adenylate cyclase activities in the rat kidney medulla: evidence for two separate sites of action. Endocrinology **102:** 1254–1261.
38. Fanestil, D. D. 1977. Hyposmolar syndromes. *In* Disturbances in Body Fluid Osmolality. T. E. Andreoli, J. J. Grantham & F. C. Rector, Jr., Eds.: 267–284. American Physiological Society. Bethesda, Md.
39. Capozza, R. 1978. An implantable, self-powered pump for drug delivery use in laboratory animals. *In* Polymeric Delivery Systems. Midland Macromolecular Monographs. R. J. Kostelnick, Ed. **5:** 261–267. Gordon and Breach. New York, N.Y.

Appendix

Water Balance

Water balance is described by the mass action statement

$$\frac{dH_2O_{in}}{dt} - \frac{dH_2O_{out}}{dt} = O \tag{1}$$

where dH_2O_{in}/dt and dH_2O_{out}/dt are the respective rates of total water intake and output with respect to time.

In growing animals (with or without ADH) total body water (TBW) increases with increasing body mass. Hence, the mass action statement is modified to include the TBW term:

$$\frac{dH_2O_{in}}{dt} - \frac{dH_2O_{out}}{dt} = \frac{dTBW}{dt} \tag{2}$$

Integration of **2** with respect to time yields the net change in TBW

$$\int_0^t \frac{dTBW}{dt} dt = \Delta \text{ TBW} \tag{3}$$

and

$$\Delta \text{ TBW} + TBW_0 = TBW' \tag{4}$$

where TBW_0 is the level total body water at $t = 0$ and TBW′ is the new absolute level of TBW.

Discussion of the Paper

L. Bankir (*Hôpital Necker, Paris, France*): It could appear that oral dDAVP therapy is the best one for AVP replacement in Brattleboro DI rats. However, I think it is not and would like to present some strong reservations concerning this route of administration.

My most important reservation is that, by this route, the antidiuretic hormone is given exactly at the wrong time. The concomitant intake of water and antidiuretic hormone is absolutely anti-physiologic. In normal animals, water intake is most likely followed by a decrease in endogenous AVP level. Simultaneous water intake and increased AVP (dDAVP) level will result in periods of relative volume expansion, the duration would depend on the amount of dDAVP ingested. We studied DI rats receiving various concentrations of dDAVP in drinking water. Urine collections were made every four hours for several consecutive days. Urine volume and osmolality showed large variations over the 24 hours. The pattern of these variations differed according to the dDAVP concentration in the drinking water and were all different from the circadian pattern observed in normal rats. Alzet osmotic minipumps provide constant AVP delivery, which is also not similar to spontaneous AVP release in normal rats, but it is certainly less antiphysiologic than the oral therapy.

Other reservations are: (1) the high cost of the treatment, due to the large amounts of dDAVP needed; (2) only total intake can be determined, not the effective dose. According to the observed plasma levels or to the resulting antidiuretic effect, less than 1/10,000th of the dDAVP is actually absorbed. As small peptides are broken down in the stomach, it might well be that the effective dDAVP is absorbed in the early digestive tract (mouth and esophagus). In human therapy (nasal spray) dDAVP is absorbed by the nasal mucosa; (3) only dDAVP can be administered by this route. Comparison with AVP or other AVP analogues, not produced in such large scale, is impossible.

The only real advantage of the oral therapy appears to be its ease of administration.

* * *

General problems encountered with various methods of vasopressin replacement therapy were also discussed:

T. Saito (*Jichi Medical School, Tochigi, Japan*): We have recently observed that acute replacement with Pitressin causes a larger increase in urinary excretion of sodium in comparison to replacement with synthetic vasopressin in man. Did you compare the effect of synthetic vasopressin to that of Pitressin? Pitressin may contain some impurities.

F. T. LaRochelle (*Northrop Services Inc., Houston, Tex.*): Vasopressin clearly can have natriuretic activity at high enough doses. In experiments using minipumps and lower doses of synthetic vasopressin, we don't see a natriuresis. Dr. Valtin may have injected pitressin tannate and have shown a natriuresis in earlier studies.

H. Valtin (*Dartmouth Medical School, Hanover, N.H.*): I have done it, and I have not been impressed by a natriuresis.

C. I. Johnston (*Monash University, Melbourne, Austr.*): May I just ask whether the minipump infusions were given intravenously or subcutaneously?

LaRochelle: Subcutaneously. And I think that is where part of the problem is. You get adhesions around the spot where the pump is implanted.

Johnston: With several other peptides, it has recently been shown that actually one needs to give them intravenously rather than subcutaneously. With subcutaneous administration there is a variable degree of breakdown and this may account for the low and variable plasma levels.

L. B. Kinter (*Smith Kline and French Laboratories, Philadelphia, Pa.*): An alternative route for chronic minipump infusions is to place them into the peritoneal cavity via small incisions. We find that this works very well (for AVP and related peptides) and we have never observed any morbidity, nor do we see any adhesions.

Johnston: We have also done that. The rates needed to attain the same plasma levels of angiotensin, vasopressin, or bradykinin are about 10 times higher if you administer the substance subcutaneously or via the peritoneal cavity than if you give it intravenously. So it is suggested that quite a lot of the peptides are destroyed before they actually get into the circulation.

EFFECTIVE ANTAGONISTS OF THE ANTIDIURETIC ACTION OF VASOPRESSIN IN RATS *

W. H. Sawyer and M. Manning

Department of Pharmacology
College of Physicians and Surgeons
Columbia University
New York, New York 10032

and

Department of Biochemistry
Medical College of Ohio
Toledo, Ohio 43699

Specific antagonists can be potent pharmacological tools. Effective antagonists of the antidiuretic response to vasopressin could be used to produce diabetes insipidus without surgery. This could provide animal models without nonspecific hypothalamic injury, avoiding unrelated genetic defects that may exist in Brattleboro rats and also permit pharmacological induction of diabetes insipidus in species other than rats. Antagonists could also be clinically useful for treating the syndrome of inappropriate ADH release and, possibly, in other pathological situations involving excessive fluid retention. Thus the design of specific antagonists acting at the level of the renal antidiuretic receptors has long been a goal of chemists and pharmacologists studying neurohypophyseal peptides. We will attempt to summarize here the limited progress that we have made toward this goal and to describe some of the many problems that remain to be solved.

Mammalian vasopressor receptors are of at least two types, clearly defined by differing specificities for agonistic analogues of vasopressin. Receptors on vascular smooth muscle and liver cells have been classified as "V_1" receptors.[1] Their activation does not involve stimulation of adenylate cyclase. Renal tubular "V_2" receptors manifest their effects by stimulating adenylate cyclase. Vascular receptors are rather easily blocked *in vivo* while antidiuretic receptors have proved to be much more of a problem.[2]

Many effective antagonists of vascular responses to vasopressin have been synthesized in recent years. These have been widely exploited to characterize receptors and to probe the contribution of vasopressin to cardiovascular regulation under a variety of circumstances.[3-6] We hope that similarly useful antagonists of the antidiuretic response will soon become available.

The Design of Antagonistic Peptides

Antagonists of vasopressor responses were first developed by du Vigneaud's group. These resulted from three types of structural change involving the first

* Unpublished work by the authors cited in this paper has been largely supported by Research Grants AM 01940, HL 12738, and GM 25280 from the National Institutes of Health.

0077-8923/82/0394-0464 $1.75/0

two amino acids in vasopressin or oxytocin. These were: (1) alkylation of the hydroxyl group of the 2-tyrosine,[7] (2) acetylation of the free amino group on the 1-hemicystine,[8] and (3) dimethyl substitutions on the β-carbon of the 1-hemicystine, to make 1-penicillamine or 1-deaminopenicillamine analogues.[9] None of these changes, alone or in combination, however, produced effective antagonists of the antidiuretic response.

Our original attempts to design antagonists of the antidiuretic response were based on the naive assumption that potent and specific agonists could be appropriately modified to make potent and specific antagonists. This arose from the belief that binding and activating functions of vasopressin reside in distinct structural features, and that some of the specificity of receptors involves the recognition of specific binding elements on the peptide. If this belief was sound it should be possible to preserve specific binding while eliminating activation, to preserve affinity while removing or blocking those structures responsible for intrinsic activity or "efficacy," thus producing a classical competitive antagonist.

Although this reasoning was overly simplistic, analogues designed from this approach led us to a number of good antagonists,[2] including the most potent known antagonists acting at vascular, myometrial, and mammary receptors, and, more recently, the first truly effective antagonists of the antidiuretic response.[2, 10]

Antagonists of the Antidiuretic Response

We chose to explore the possibility that one of four most potent and specific antidiuretic agonists, 1-deamino-[4-valine,8-D-arginine]-vasopressin[11] or "dVDAVP" could be modified into a specific antagonist of the antidiuretic response. Initially this led to a series of useful antagonists of the vasopressor response[2] but not of the antidiuretic response (Table 1). One of this series of analogues, [1-(β-mercapto-β,β-cyclopentamethylenepropionic acid), 4-valine, 8-D-arginine]-vasopressin, or "d$(CH_2)_5$VDAVP" provided an important lead. This analogue acts as a competitive antagonist of the activation of adenylate cyclase in rat renal medullary membrane preparations *in vitro*.[12] In itself it causes scarcely detectable activation of cyclase in these preparations. But d$(CH_2)_5$VDAVP acts as a weak partial antidiuretic agonist without evident antagonistic activity *in vivo*.[13] This may reflect the presence of an excess of vasopressin-sensitive adenylate cyclase in rat renal tubular cells. Activation of a small fraction of the available pool may suffice to cause antidiuresis.[12] Considering these findings a knowledgeable colleague told us quite flatly that it would be impossible to design an effective antagonist of the antidiuretic response *in vivo*.

Undaunted by this sage advice we forged ahead. From earlier experience with antagonists of the oxytocic response[14] we realized that combining changes in the first and second amino acids of neurohypophyseal peptides often resulted in additive or more than additive enhancement of antagonistic potency (Table 1). Meanwhile, Larsson *et al.*[15] reported that 2-*O*-ethyltyrosine analogues of deamino-lysine-vasopressin were partial antagonists of the antidiuretic response, although antagonism was inconsistent and not dose-related. These analogues themselves were antidiuretic agonists at higher doses. We thus proceeded to try 2-*O*-alkyltyrosine substitutions in d$(CH_2)_5$VDAVP. The resulting analogues were promising.[10] Although they still caused an initial antidiuresis when in-

TABLE 1

ADDITIVE EFFECTS OF SUBSTITUTIONS IN THE 1- AND 2-POSITIONS OF dVDAVP ON VASOPRESSOR ANTAGONISM AND ANTIDIURETIC ACTIVITY

Analogues *	Anti-vasopressor Effective Doses † (nmoles/kg)	Anti-diuretic Activity (U/mg)	References
dVDAVP	7.2	1230	20
dPVDAVP	1.0	123	20
$d(CH_2)_5$VDAVP	1.5	0.1	13
dTyr(Me)VDAVP	8.4	2000	unpublished data
dPTyr(Me)VDAVP	1.0	3.2	unpublished data
$d(CH_2)_5$Tyr(Me)VDAVP	0.3	inhibits	16

† The "effective dose" is defined as the i.v. dose of antagonist that reduces the response to 2x units of agonist to equal the response to 1x units in the absence of antagonist.

* Structures: dVDAVP = 1-deamino[4-valine, 8-D-arginine]-vasopressin. dPVDAVP = [1-deaminopenicillamine, 4-valine, 8-D-arginine]-vasopressin. $d(CH_2)_5$VDAVP = [1-(β-mercapto-β,β-cyclopentamethylenepropionic acid), 4-valine, 8-D-arginine]-vasopressin. dTyr(Me)VDAVP = 1-deamino[2-*O*-methyltyrosine, 4-valine, 8-D-arginine]-vasopressin. dPTyr(Me)VDAVP = [1-deaminopenicillamine, 2-*O*-methyltyrosine, 4-valine, 8-D-arginine]-vasopressin. $d(CH_2)_5$Tyr(Me)VDAVP = [1-(β-mercapto-β,β-cyclopentamethylenepropionic acid), 2-*O*-methyltyrosine, 4-valine, 8-D-arginine]-vasopressin.

jected i.v. into rats this was followed by a period lasting for an hour or more in which antidiuretic responses to ADH were significantly inhibited (FIGURE 1). When injected i.p. in conscious rats these analogues caused a marked diuresis and a fall in urinary osmolality (FIGURE 2). This was not secondary to polydipsia, since it occurred when water was withheld. This seems to be, in essence, instant pharmacologically induced diabetes insipidus.

Since our initial report of four effective antagonists of the antidiuretic response [10] we have synthesized about thirty additional analogues that are also effective antagonists, and many other related analogues, that are not. We shall spare you the details here and list only a few of these in TABLE 2. Although following the trail from a specific and potent agonist, dVDAVP, via $d(CH_2)_5$VDAVP, an *in vitro* antagonist, to its 2-*O*-alkyltyrosine analogues that are *in vivo* antagonists, we had been lucky in that we had carried along at least two molecular features that appear essential to antagonism. These are the cyclopentamethylene substitution on the β-carbon of the 1-hemicystine and the valine in position 4. Other substitutions in these positions eliminate antagonism of the antidiuretic response (unpublished data). The presence of an 8-D-arginine was important in making the parent analogue, dVDAVP, a specific antidiuretic agonist [11] but the presence of this residue was unnecessary, actually deleterious, in antidiuretic antagonists.[10] 8-L-Arginine analogues were more effective (TABLE 2). A surprising finding was that D- rather than L-(*O*-alkyl) tyrosines were more effective in position 2 (TABLE 2). Although 2-L-(*O*-ethyl) tyrosine analogues appeared more effective than their *O*-methyl, *O*-isopropyl, *O*-propyl, or *O*-butyl counterparts,[16] differences seemed to disappear when the same alkyl substituents were introduced into D-tyrosine in the 2-position. In fact, unalkylated 2-D-tyrosine itself [17] was consistent with substantial antagonistic activity (TABLE 2). Thus we have been able to discard some of the apparently unneeded substitutions that we had accumulated on the circuitous trail toward more active antagonists. The last analogue shown in TABLE 2 is actually the simplest and the most easily synthesized peptide of this series. It will be interesting to see if further exploration will reveal other parts of these molecules that are expendable without significant loss of antagonistic activity.

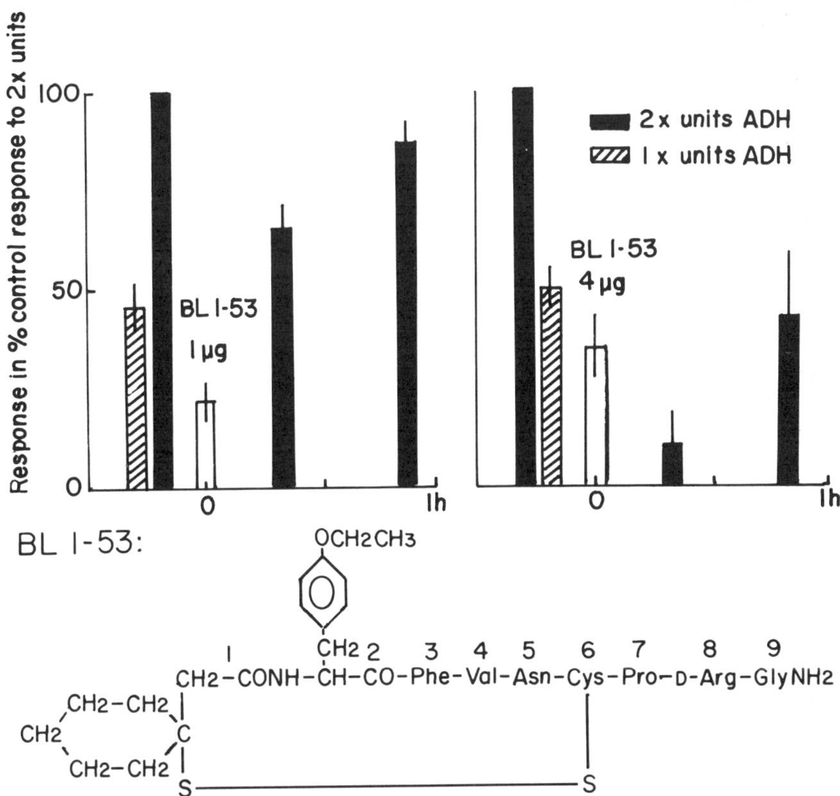

[1-(β-mercapto-β,β-cyclopentamethylenepropionic acid),2-*O*-ethyltyrosine, 4-valine,8-D-arginine]-vasopressin or "d(CH2)5Tyr(Et)VDAVP"

FIGURE 1. Responses to an analogue, BL I–53, that has both agonistic and antagonistic activities on the rat intravenous antidiuretic assay. Each rat received a high (2x) and a low (1x) dose of the USP Posterior Pituitary Reference Standard, usually 30 and 15 μU. The antagonist was then injected and it caused a brief partial antidiuretic response, indicated by the open bars. Rats were again injected with the high (2x) dose of standard at 20 and 50 min after the antagonist and the responses are indicated by the solid bars. Four female rats, weighing 210 to 240 g were used. They were water-loaded and anesthetized with ethanol. Two rats received the 1 μg dose of BL I–53 first and two, the 4 μg dose. Using responses at 20 min after injection of BL I–53 the effective dose was estimated to be 6.5 μg/kg or 5.7 ± 0.5 nmoles/kg.[10] The structure of BL I–53 is also shown.

Potency, Specificity, and Possible Side-Effects

Although an antagonist such as $d(CH_2)_5$D-Tyr(Et)VAVP has an "effective dose" of about 1 μg/kg when given i.v. (Table 2) it would be a great advantage to find even more potent antagonists. Doses several times the "effective" dose

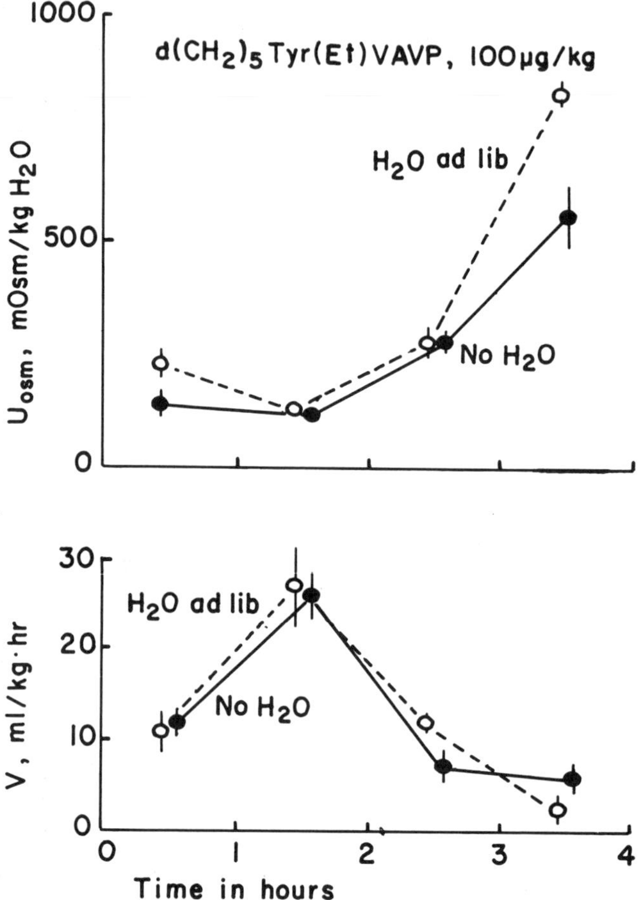

Figure 2. Responses by four conscious rats to the i.p. injection of $d(CH_2)_5$Tyr(Et) VAVP at time 0, when water bottles were removed from the cages (closed circles). Urine osmolalities of these rats injected with the solvent only averaged 1111 ± 215 mOsm/kg H_2O and urine flow rates, 1.8 ± 0.8 ml/kg per h. Responses by a different group of rats to the same dose of the antagonist when water remained available are shown for comparison (open circles).

are needed to block the action of endogenous ADH and these antagonists are available only in small quantities. Thus more potent antagonists would permit experiments in larger animals and chronic experiments that are not now practical.

TABLE 2

SOME ANTAGONISTS OF ANTIDIURETIC, VASOPRESSOR, AND OXYTOCIC RESPONSES TO NEUROHYPOPHYSEAL PEPTIDES

Analogues *	Effective Doses † (nmole/kg) *in vivo*			References
	Anti-Antidiuretic	Anti-Vasopressor	Anti-Oxytocic	
$d(CH_2)_5$Tyr(Et)VDAVP	5.7 ± 0.5	0.34 ± 0.04	—	10, 16
$d(CH_2)_5$Tyr(Et)VAVP	1.9 ± 0.2	0.49 ± 0.11	13 ± 2	10, 16, unpublished data
$d(CH_2)_5$Tyr(Et)AVP	agonist	0.23 ± 0.02	—	unpublished data
$d(CH_2)_5$VAVP	agonist	0.76 ± 0.10	26 ± 7	unpublished data
$d(CH_2)_5$D-Tyr(Et)VAVP	1.1 ± 0.2	0.45 ± 0.11	1.6 ± 0.4	17, unpublished data
$d(CH_2)_5$D-TyrVAVP	2.2 ± 0.2	0.29 ± 0.09	6.9 ± 1.3	17, unpublished data

† See TABLE 1 for definition of "effective dose." For these analogues 1 nmole equals about 1.1 μg.

* Structures: $d(CH_2)_5$Tyr(Et)VDAVP = [1-(β-mercapto-β,β-cyclopentamethylenepropionic acid), 2-*O*-ethyltyrosine, 4-valine, 8-D-arginine]-vasopressin. $d(CH_2)_5$Tyr(Et)VAVP = [1-(β-mercapto-β,β-cyclopentamethylenepropionic acid), 2-*O*-ethyltyrosine, 4-valine]-arginine-vasopressin. $d(CH_2)_5$Tyr(Et)AVP = [1-(β-mercapto-β,β-cyclopentamethylenepropionic acid), 2-*O*-ethyltyrosine]-arginine-vasopressin. $d(CH_2)_5$VAVP = [1-(β-mercapto-β,β-cyclopentamethylenepropionic acid), 4-valine]-arginine-vasopressin. $d(CH_2)_5$D-Tyr(Et)VAVP = [1-(βmercapto-β,β-cyclopentamethylenepropionic acid), 2-D-(*O*-ethyl)tyrosine, 4-valine]-arginine-vasopressin. $d(CH_2)_5$D-TyrVAVP = [1-(β-mercapto-β,β-cyclopentamethylenepropionic acid), 2-D-tyrosine, 4-valine]-arginine-vasopressin.

In theory one might expect that potency could be enhanced by eliminating residual agonistic activity. All the peptides in TABLE 1 cause a transient antidiuresis when injected i.v. Preliminary experiments indicate that it is possible to design effective antagonists devoid of detectable agonistic activity, although at a slight loss of antagonistic potency. We hope that we will be able to develop further analogues that are both potent and pure antagonists.

The antagonists shown in TABLE 2 are far from specific for antidiuretic receptors. They also block vasopressor and oxytocic responses. More specific antagonists would make more incisive pharmacological probes and hold greater therapeutic promise.

The design of more specific antagonists would be impossible if the several types of tissue receptors have similar binding characteristics and if antagonism results simply from the deletion or modification of active elements that endow peptides with specific agonistic properties. Studies on the binding of vasopressin analogues to renal antidiuretic receptors do not appear to support this pessimistic outlook, since antidiuretic activities *in vivo* appear correlated with the affinities of agonists for these receptors *in vitro*.[12] We have recently found several prototypic analogues that show substantially greater potency as antagonists of the antidiuretic response than of vasopressor or oxytocic responses. Thus the design of even more specific antagonists of the antidiuretic response appears to be a reasonable possibility.

More potent and specific antagonists would not only provide sharper experimental tools but would certainly seem desirable when one considers possible clinical applications. Both enhanced potency and specificity should reduce the incidence and severity of undesired side-effects.

We can speculate on the properties of presently known antagonists of the antidiuretic response that could be troublesome in clinical situations. These concern blockade of receptors other than those mediating antidiuresis. It is unlikely that antagonism of vasopressor responses would have noticeable consequences except, perhaps, under extraordinary circumstances, such as hypovolemic shock.[4] Blockade of myometrial and mammary gland receptors, however, could be a serious problem in women near term and during lactation. A more generally disturbing type of side-effect could derive from interference of antagonists with the still poorly defined behavioral actions of vasopressin and oxytocin within the central nervous system. Koob *et al.*[18] recently reported that [1-deaminopenicillamine, 2-*O*-methyltyrosine]-arginine-vasopressin (dPTyr(Me) AVP), an analogue that antagonizes vasopressor responses but remains a full antidiuretic agonist, can hasten the extinction of conditioned avoidance behavior in rats. It can also prevent the delay of extinction by vasopressin injected s.c. or into the cerebral ventricles. The behavioral effects of the antagonist occur after s.c. injection indicating that the peptide gains access to sites within the brain at which vasopressin affects behavior. These observations raise the gloomy prospect that an antagonist administered to treat hyponatremia might also have deleterious effects on short-term memory.

A more frightening possibility arises from the observations of Pederson and Prange [19] suggesting that oxytocin is important in promoting maternal behavior in rats. The possible ramifications of antagonizing this central action of oxytocin in human beings are certainly unpleasant to contemplate.

Conclusions

Although synthetic analogues of the neurohypophyseal hormones that can antagonize some of their pharmacological actions have been known for over twenty years it is only within the past two years that we have demonstrated that it is possible to design analogues of vasopressin that act as effective antagonists of its antidiuretic action *in vivo*. This represents a promising start but it is important that we keep trying to improve on both the potency and specificity of antagonists to provide more useful pharmacological probes and to avoid potentially deleterious side-effects that would limit their potential therapeutic value.

References

1. Michell, R. H., C. J. Kirk & M. M. Billah. 1979. Hormonal stimulation of phosphatidylinositol breakdown, with particular reference to the hepatic effects of vasopressin. Biochem. Soc. Trans. **7:** 861–865.
2. Sawyer, W. H., Z. Grzonka & M. Manning. 1981. Neurohypophysial peptides: Design of tissue-specific agonists and antagonists. Mol. Cell. Endocrin. **22:** 117–134.
3. Crofton, J. T., L. Share, W. J. Lee-Kwon, M. Manning & W. H. Sawyer. 1979. The importance of vasopressin in the development and maintenance of DOC-salt hypertension in the rat. Hypertension **1:** 31–38.
4. Cowley, A. W., Jr., S. J. Switzer & M. M. Guinn. 1980. Evidence and quantification of the vasopressin arterial pressure control system in the dog. Circ. Res. **46:** 58–67.
5. Andrews, C. E., Jr. & B. M. Brenner. 1981. Relative contributions of arginine vasopressin and angiotensin II to maintenance of systemic arterial pressure in the anesthetized water-deprived rat. Circ. Res. **48:** 254–258.
6. Aisenbrey, G. A., W. A. Handelman, P. Arnold, M. Manning & R. W. Schrier. 1981. Vascular effects of arginine vasopressin during fluid deprivation in the rat. J. Clin. Invest. **67:** 961–968.
7. Law, H. D. & V. du Vigneaud. 1960. Synthesis of 2-*p*-methoxyphenylalanine oxytocin (*O*-methyl-oxytocin) and some observations on its pharmacological behavior. J. Am. Chem. Soc. **82:** 4579–4581.
8. Cash, W. D. & B. L. Smith. 1963. Synthesis and biological properties of 1-acetyl-8-lysine-vasopressin. J. Biol. Chem. **238:** 994–997.
9. Schulz, H. & V. du Vigneaud. 1966. Synthesis of 1-L-penicillamine-oxytocin, 1-D-penicillamine-oxytocin, and 1-deaminopenicillamine-oxytocin, potent inhibitors of the oxytocic response to oxytocin. J. Med. Chem. **9:** 647–650.
10. Sawyer, W. H., P. K. T. Pang, J. Seto, M. McEnroe, B. Lammek & M. Manning. 1981. Vasopressin analogs that antagonize antidiuretic responses by rats to the antidiuretic hormone. Science **212:** 49–51.
11. Sawyer, W. H., M. Acosta, L. Balaspiri, J. Judd & M. Manning. 1974. Structural changes in the arginine vasopressin molecule that enhance antidiuretic activity and specificity. Endocrinology **94:** 1106–1115.
12. Butlen, D., G. Guillon, R. M. Rajerison, S. Jard, W. H. Sawyer & M. Manning. 1978. Structural requirements for activation of vasopressin-sensitive adenylate cyclase and hormone binding: Effects of highly potent antidiuretic analogues and competitive inhibitors. Mol. Pharmacol. **14:** 1006–1017.
13. Lowbridge, J., M. Manning, J. Haldar & W. H. Sawyer. 1979. [1-(β-mercapto-β,β-cyclopentamethylenepropionic acid), 4-valine, 8-D-arginine]-vaso-

pressin, a potent and selective inhibitor of the vasopressor response to arginine vasopressin. J. Med. Chem. **21:** 313–315.

14. Sawyer, W. H., J. Haldar, D. Gazis, J. Seto, K. Bankowski, J. Lowbridge, A. Turan & M. Manning. 1980. The design of effective *in vivo* antagonists of rat uterus and milk-ejection responses to oxytocin. Endocrinology **106:** 81–91.
15. Larsson, L.-E., G. Lindeberg, P. Melin & V. Pliška. 1978. Synthesis of O-alkylated lysine-vasopressins, inhibitors of the antidiuretic response to lysine-vasopressin. J. Med. Chem. **21:** 352–356.
16. Manning, M., B. Lammek, A. Kolodziejczyk, J. Seto & W. H. Sawyer. 1981. Synthetic antagonists of *in vivo* antidiuretic and vasopressor responses to arginine vasopressin. J. Med. Chem. (In press.)
17. Manning, M., A. Olma, W. A. Klis, A. Kolodziejczyk, J. Seto & W. H. Sawyer. 1981. Design of more potent antagonists of the antidiuretic action of arginine vasopressin. Proc. 7th Am. Peptide Symp. (In press.)
18. Koob, G. F., M. Le Moal, O. Gaffori, M. Manning, W. H. Sawyer, J. Rivier & F. E. Bloom. 1981. Arginine vasopressin and a vasopressin antagonist peptide: Opposite effects on extinction of active avoidance in rats. Regulatory Peptides **2:** 153–163.
19. Pedersen, C. A. & A. J. Prange, Jr. 1979. Induction of maternal behavior in virgin rats after intracerebroventricular administration of oxytocin. Proc. Natl. Acad. Sci. USA **76:** 6661–6665.
20. Manning, M., J. Lowbridge, C. T. Stier, Jr., J. Haldar & W. H. Sawyer. 1977. [1-Deaminopenicillamine, 4-valine, 8-D-arginine]-vasopressin, a highly potent inhibitor of the vasopressor response to arginine vasopressin. J. Med. Chem. **20:** 1228–1230.

REPLACEMENT THERAPY WITH ARGININE VASOPRESSIN IN HOMOZYGOUS BRATTLEBORO RATS *

Savio W. T. Cheng, W. G. North, and M. Gellai

Department of Physiology
Dartmouth Medical School
Hanover, New Hampshire 03755

INTRODUCTION

Since their discovery nearly twenty years ago, Brattleboro homozygotes with hereditary hypothalamic diabetes insipidus (DI) have become a useful animal model for studying the physiological role of arginine vasopressin, the naturally occurring antidiuretic hormone in most mammalian species. In most of the early studies, vasopressin was replaced in DI rats by daily subcutaneous injection of 100 mU to 1,000 mU of vasopressin tannate (Pitressin Tannate, Parke, Davis & Co) in oil.[1-3] However, there are problems in interpreting data obtained using this experimental procedure.

First, impure vasopressin was used in these studies. According to the manufacturer, Pitressin Tannate is supplied as an extract from bovine and porcine pituitaries and contains 50% arginine vasopressin (AVP), the native antidiuretic hormone in rats and 50% lysine vasopressin (LVP), which has an antidiuretic activity 60% of that of AVP.[4] Accurate measurement of plasma AVP concentration by radioimmunoassay in Pitressin Tannate-treated DI rats is hampered by the presence of LVP because the cross-reactivity between AVP and LVP is generally high.[5]

Secondly, there is good reason to believe that the subcutaneous injection of Pitressin Tannate failed to provide a steady rate of replacement. Möhring *et al.* have shown that the release of Pitressin Tannate from subcutaneous sites into the circulation decreases with time.[6]

Recent pharmacological and technical advances may help to circumvent these problems. Synthetic AVP, the naturally occurring antidiuretic hormone in rats, is now available and should be a more effective reagent than the impure Pitressin Tannate, making interpretation easier. In addition, according to the manufacturer (Alza, Palo Alto, Calif.), osmotic minipumps can provide a means of steady release of AVP up to two weeks. Steady rate of release minimizes the variations of plasma AVP level, which may be essential in studies of the effects of vasopressin on renal electrolytes and water excretion as well as other balance studies.

The present study was designed to determine the optimal AVP replacement dosage in DI rats to modulate levels of antidiuresis. Some of the physiological parameters in DI rats after chronic AVP treatment were also assessed, such as plasma AVP concentration, osmolality, electrolyte concentrations, and urinary excretion of AVP.

* Supported in part by Research Grants AM-08469, CA-19613 from the National Institutes of Health; and by The Ryan Foundation.

0077-8923/82/0394-0473 $1.75/0

Methods and Materials

Male DI rats (Blue Spruce Farms Inc., Altamont, N.Y.) weighing 180–320 g were placed singly in metabolic cages for an adjustment period of 2–7 days. The animals were housed in a room with constant temperature (76° F) and 12-hour intervals of light and darkness. Standard laboratory diet and tap water were provided ad libitum throughout the experiment.

After the animals were acclimated to the metabolic cages, osmotic minipumps (Alzet 2002) containing different concentrations of synthetic AVP (Pierce Chemical Co., Rockford, Ill.) were implanted subcutaneously or intraperitoneally under light ether anesthesia. The delivery rates of AVP were calculated to range from 2 to 3,000 ng per day. Synthetic AVP was dissolved in 0.05 M acetic acid and diluted to the desired volume with bacteriostatic saline. Minipumps were weighed before and after AVP was loaded to ensure proper filling. Care was taken not to introduce any air bubbles into the minipumps. They were soaked overnight in bacteriostatic saline at room temperature before implantation on the following day. This was recommended by the manufacturer as a means of expediting the onset of the delivery of AVP.

Twenty-four-hour urine collections were made daily before and after AVP treatment and were collected under oil to prevent evaporation. Urine output was expressed in ml per 100 g body weight per 24 hours. Body weight was also measured daily. Seven to 10 days after minipump implantation, an indwelling catheter (medical grade Tygon® tubing S-54-HL, Norton Co., Akron, Ohio) was placed into the femoral or carotid artery of each rat using ketamine hydrochloride (60 mg/kg body weight) and pentobarbital sodium (21 mg/kg body weight) anesthesia. Two days later, approximately 1 ml of blood (heparinized) was taken from the conscious rat in restraining cage for plasma AVP, osmolality, and electrolyte measurements. Plasma and urinary AVP concentrations were measured by radioimmunoassay; osmolalities by vapor pressure osmometer (Wescor Inc., Logan, Utah); electrolytes by flame photometer (Instrumentation Lab. Inc., Lexington, Mass.); and plasma protein by refractometry.

A total of 45 male DI rats, 7 in a pilot study and 38 in four experimental series, were used. Only male rats were used in this study to avoid the effects of the estrus cycle and because female rats are claimed to be less sensitive to AVP.[7]

Radioimmunoassay (RIA) for AVP

AVP was first extracted from plasma using columns of ODS-silica[8] (C18 Sep-Paks, Waters Assoc., Inc., Milford, Mass.) and assayed by previously described methods.[5] The same lot of synthetic AVP used in the minipumps was used as standard in the RIA. The sensitivity of the assay was to 0.2 pg AVP.

Results and Discussion

All rats showed an increase in urine osmolality on the first day following minipump implantation. The fact that this increase was observed even in the control group (AVP minipump release rate = 2 ng/day) suggests that it was partially due to post-surgical stress. On the second day urine osmolality dropped slightly and from then on remained relatively stable, except in the group that received supramaximal dose of AVP (release rate = 3000 ng/day). Their urine concentrating ability continued to increase for six to seven days and became stabilized thereafter (FIGURE 1).

An AVP dose-dependent antidiuretic response was generated in this study (FIGURE 2a). The threshold AVP release rate for initiation of antidiuresis was between 20 to 50 ng per day, with plasma AVP concentrations of 3.6 ± 1.0 to 6.8 ± 1.8 pg/ml, respectively (FIGURE 3a). Measured plasma AVP concentration at maximal antidiuresis was 26.7 ± 4.6 pg/ml. Others have shown that plasma AVP concentration in Long-Evans normal rats to be in the range of 2–4 pg/ml (LaRochelle, F. T. & B. R. Edwards, unpublished data). Only mild antidiuresis (U_{osm} = 500 mOsm/kg H_2O) was achieved in the present study with the same level of plasma AVP (FIGURE 3b). It is not understood at this time why relatively high levels of plasma AVP were needed to attain a certain

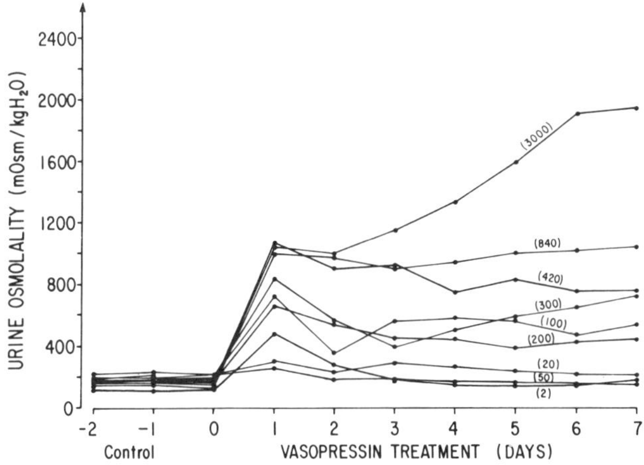

FIGURE 1. Urine osmolality before and after AVP administration in nine groups of DI rats. Numbers in parentheses indicate the assumed AVP minipump release rates (ng/day). (Values are averages from all animals in each group.)

level of antidiuresis. One possible explanation is that inbreeding has given us a subpopulation that is more severe in diabetes insipidus or less sensitive to AVP.

With vasopressin treatment, urine output of all rats decreased except those in the control group (FIGURE 2c). Water was retained as indicated by the percentage increase in body weight (TABLE 1). Plasma osmolality and sodium and protein concentrations were lowered because of an increase in plasma volume as reflected by the gain in body weight (FIGURE 2b and TABLE 1). Möhring and coworkers have shown that plasma volume of Pitressin Tannate-treated DI rats increased by about 15% in 21 hours.[9] The hematocrit values were less revealing: they are not significantly different before and after AVP treatment (TABLE 1). Contrary to previous work by Möhring,[10] hypokalemia was not observed in the present study. Fernandez-Repollet and collaborators have suggested that mild dehydration may be responsible for hypokalemia in DI rats through changes in renal tissue potassium concentration and conse-

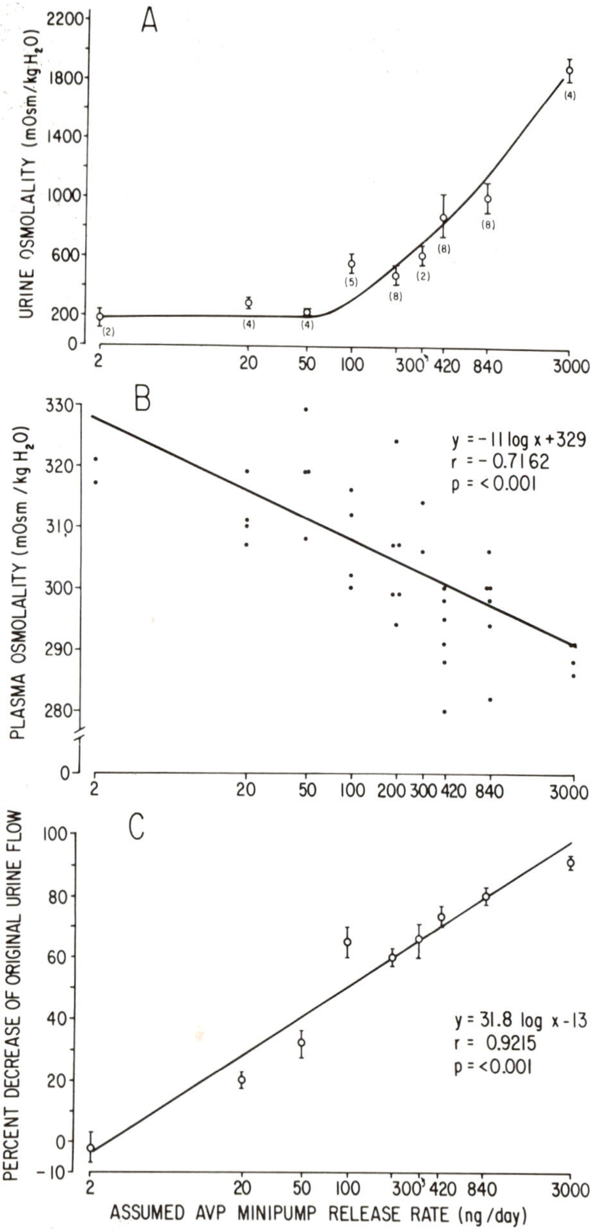

FIGURE 2. (A) Dose-dependent antidiuretic response plotted on a semi-log scale. Values are mean ± SE, calculated by averaging the urine osmolalities from day 1 of vasopressin treatment to the day when the animals were catheterized. Figures in parentheses indicate number of animals used in each group. (B) Relationship between plasma osmolality and the assumed AVP minipump release rate of individual animals plotted on a semi-log scale. (C) Dose-dependent percentage decrease of original urine output plotted on a semi-log scale.

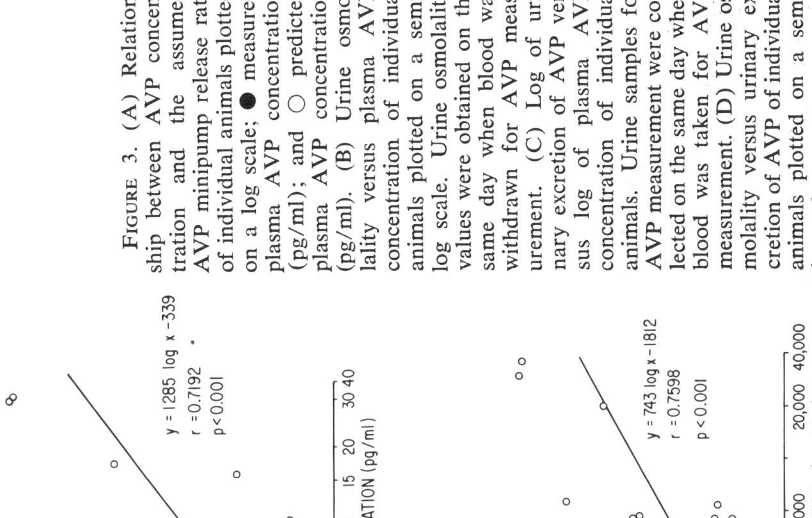

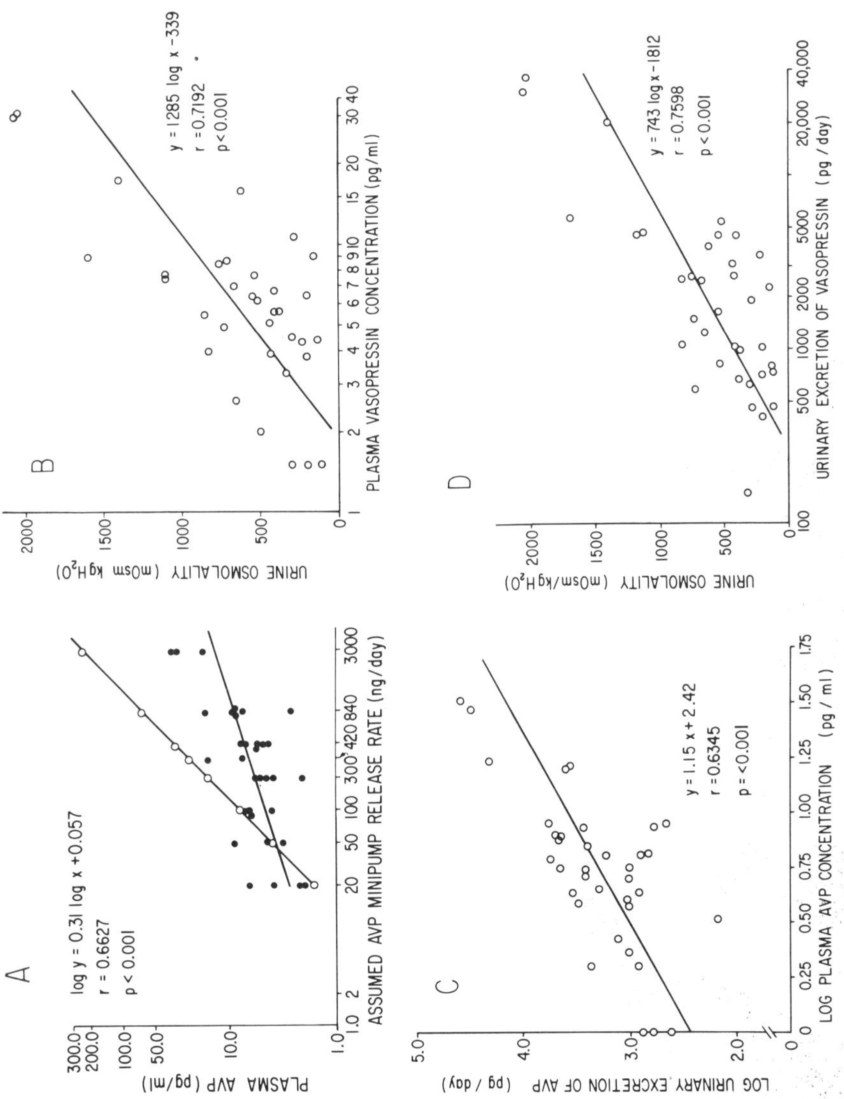

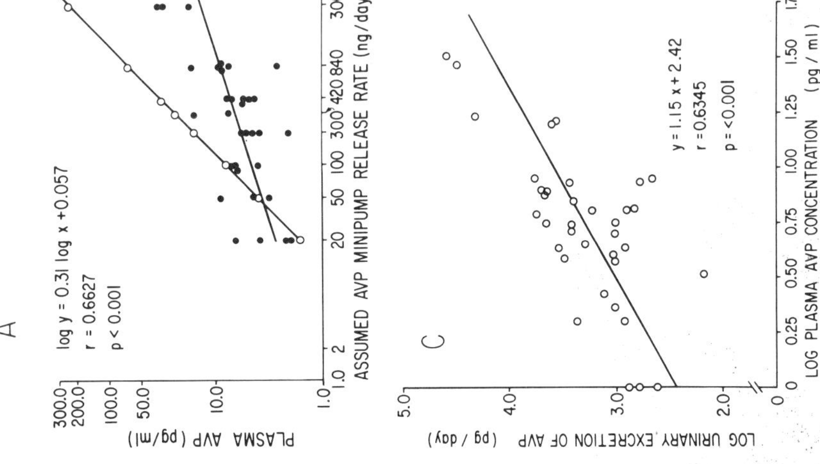

FIGURE 3. (A) Relationship between AVP concentration and the assumed AVP minipump release rate of individual animals plotted on a log scale; ● measured plasma AVP concentration (pg/ml); and ○ predicted plasma AVP concentration (pg/ml). (B) Urine osmolality versus plasma AVP concentration of individual animals plotted on a semi-log scale. Urine osmolality values were obtained on the same day when blood was withdrawn for AVP measurement. (C) Log of urinary excretion of AVP versus log of plasma AVP concentration of individual animals. Urine samples for AVP measurement were collected on the same day when blood was taken for AVP measurement. (D) Urine osmolality versus urinary excretion of AVP of individual animals plotted on a semi-log scale.

TABLE 1

EFFECT OF VASOPRESSIN TREATMENT ON THE INDICES OF FLUID BALANCE

	Assumed AVP Minipump Release Rate (ng/day)								
	2	20	50	100	200	300	420	840	3000
Plasma Na^+ (mEq/L)	149.9±0.6 (2)*	160.9±2.8 (4)	149.5±1.1 (4)	150.9±3.2 (4)	147.0±1.9 (6)	155.8±2.8 (2)	145.8±1.4 (6)	147.4±1.5 (6)	140±0.2 (4)
Plasma K^+ (mEq/L)	4.8±0.1 (2)	5.2±0.5 (4)	4.2±0.1 (4)	4.8±0.2 (4)	4.6±0.1 (6)	5.3±0.2 (2)	4.4±0.1 (6)	5.3±0.8 (6)	3.8±0.1 (4)
HCT %	43±1.0 (2)	45.3±1.5 (3)	41.8±2.0 (4)	42.9±1.2 (4)	41.3±1.1 (6)	42.3±2.0 (2)	42.8±0.8 (5)	41.3±1.0 (6)	42.8±0.4 (3)
Plasma protein (g/100 ml)		7.6±0.8 (2)	6.5±0.2 (4)	6.6±0.1 (2)	6.4±0.2 (4)	6.4 (1)	6.3±0.2 (4)	6.2±0.2 (4)	6.8±0.2 (4)
% Increase body weight	0.25±2.8 (2)	5.2±2.4 (4)	0.9±1.3 (4)	10.5±2.9 (5)	6.8±2.9 (9)	5.4±0.7 (2)	7.1±2.4 (8)	5.7±1.0 (8)	4.3±0.3 (4)

* Figure in parentheses indicates numbers of animals in each group.

quently in urinary potassium excretion. Correction of dehydration during prolonged periods in metabolic cages may correct the potassium deficiency spontaneously.[11]

Urinary excretion of AVP was calculated by multiplying urinary AVP concentration and urine output. FIGURE 3c showed that the amount of immunologic intact AVP excreted in urine correlates positively with the level of AVP in the plasma. When urine osmolality was plotted against urinary excretion of AVP (FIGURE 3d), the relationship was similar to that of urinary osmolality versus plasma AVP concentration (FIGURE 3b). This indicates that urinary excretion of AVP can serve as an important piece of information if plasma AVP values are unavailable.

The actual release rates of AVP were calculated using the scheme provided by Valtin and coworkers,[12] assuming a half-life of two minutes and a volume of distribution of 10% of body weight.[13] The fraction of intact AVP cleared by the kidneys (calculated by dividing the urinary excretion of AVP by the actual release rate of AVP) at various ranges of plasma AVP concentrations (pg/ml) of 2–4, 4–6, 6–8, 8–16, and 16–33 were 4.1 ± 1.1, 3.5 ± 0.7, 3.4 ± 0.8, 2 ± 0.7, and 8.8 ± 0.5 percent, respectively.

Plasma AVP concentrations measured by radioimmunoassay were less than those predicted from the release rate of AVP by the minipump (FIGURE 3a). This difference may be explained by (1) the route of release of AVP. The predicted plasma AVP concentration was calculated by assuming that AVP was released directly into the blood stream, whereas the actual subcutaneous release may have led to partial degradation of AVP along its path to the circulation. (2) The release rate of the minipump was assumed from information provided by the manufacturer and was not verified by us.

It is anticipated that future studies can be improved if AVP is delivered intravenously by attaching a small catheter to the pump as well as determining the actual release rate.

There were no signs of past bleeding, inflammation, or scarring of tissue when minipumps were removed at the end of the experiments. The osmotic minipumps, continuously delivering solutions for an extended period without the need of frequent animal handling provide a convenient tool for prolonged vasopressin replacement in DI rats.

SUMMARY

We studied the replacement therapy of different doses of AVP in DI rats. It is surprising that relatively high levels of plasma AVP were needed to achieve significant antidiuretic effects. Measured plasma AVP concentrations are less than those predicted from the release rates of AVP by the minipumps. This difference may be due to the subcutaneous mode of release of AVP. The study also provides evidence that urinary excretion of AVP is a good indicator of the plasma AVP level.

REFERENCES

1. MCCANN, S. M., J. ANTUNES-RODRIGUES, R. NALLAR & H. VALTIN. 1966. Pituitary adrenal function in the absence of vasopressin. Endocrinology **79:** 1058–1064.
2. HARRINGTON, A. R. & H. VALTIN. 1965. Vasopressin effect on urinary concentration in rats with hereditary hypothalamic diabetes insipidus (Brattleboro strain). Proc. Soc. Exp. Biol. Med. **118:** 448–450.
3. LEE, J. & P. G. WILLIAMS. 1972. The effect of vasopressin (Pitressin) administration and dehydration on the concentration of solutes in renal fluids of rats with and without hereditary hypothalamic diabetes insipidus. J. Physiol. **220:** 729–743.
4. WALTER, R., J. RUDINGER & I. L. SCHWARTZ. 1967. Chemistry and structure-activity relations of the antidiuretic hormones. Am. J. Med. **42:** 653–677.
5. NORTH, W. G., F. T. LAROCHELLE, JR., J. HALDAR, W. H. SAWYER & H. VALTIN. 1978. Characterization of an antiserum used in a radioimmunoassay for arginine-vasopressin: Implications for reference standards. Endocrinology **103:** 1976–1984.
6. MÖHRING, J., G. KOHRS, B. MÖHRING, M. PETRI, E. HOMSY & D. HAACK. 1978. Effects of prolonged vasopressin treatment in Brattleboro rats with diabetes insipidus. Am. J. Physiol. **234**(2): F106–F111.
7. BALMENT, R. J., I. W. HENDERSON & J. A. OLIVER. 1975. The effects of vasopressin on pituitary oxytocin content and plasma renin activity in rats with hypothalamic diabetes insipidus (Brattleboro strain). Gen. Comp. Endocrinology **26:** 468–477.
8. LAROCHELLE, JR., F. T., W. G. NORTH & P. STERN. 1980. A new extraction of arginine vasopressin from blood: The use of Octadecasilyl-Silica. Pflügers Arch. **387:** 79–81.
9. MÖHRING, J., B. MÖHRING, A. SCHÖMIG, H. SCHÖMIG-BREKNER & D. HAACK. 1974. Acute effects of vasopressin on potassium and water balance in rats with diabetes insipidus. Am. J. Physiol. **227**(4): 921–926.
10. MÖHRING, B., J. MÖHRING, G. DAUDA & D. HAACK. 1974. Potassium deficiency in rats with hereditary diabetes insipidus. Am. J. Physiol. **227:** 916–920.
11. FERNANDEZ-REPOLLET, E., M. MARTINEZ-MALDONADO & S. OPAVA-STITZER. 1980. Role of water balance in the enhanced potassium excretion and hypokalaemia of rats with diabetes insipidus. J. Physiol. **305:** 97–108.
12. VALTIN, H., J. STEWART & H. W. SOKOL. 1974. Genetic control of the production of posterior pituitary principles. Handb. Physiol. Sect. 7, Vol. IV, Part 1: 131–171.
13. LAUSON, H. D. 1974. Metabolism of the neurohypophysial hormones. Handb. Physiol. Sect. 7, Vol. IV, Part 1: 328–393.

OSMOREGULATION IN THE PREGNANT BRATTLEBORO RAT *

Jacques A. Dürr, Barbara A. Stamoutsos, William M. Barron, and Marshall D. Lindheimer †

Departments of Obstetrics and Gynecology and Medicine
The Pritzker School of Medicine
The University of Chicago
Chicago, Illinois 60637

INTRODUCTION

Plasma osmolality (P_{osm}) and the osmotic threshold for vasopressin (AVP) secretion decrease approximately 10 mOsm/kg during gestation in Sprague-Dawley rats.[1] To test the hypothesis that this new level of body tonicity can only be achieved and maintained if rats decrease their osmotic threshold for thirst as well, osmoregulation was studied in Brattleboro rats, since homozygous animals in this strain produce no AVP. Results demonstrate a significant decrement in P_{osm} during gestation in homozygous rats, suggesting that the osmotic threshold for thirst is decreased during pregnancy.

METHODS

Studies were performed on gravid homozygous and heterozygous Brattleboro rats, gravid Long-Evans animals, and their respective age-matched virgin controls. Except where noted otherwise, Brattleboro rats were bred by us between their 8–11th week of life by exposure to a homozygous male for three successive days. Long-Evans animals were bred by the supplier (Blue Spruce Farms, Altamont, N.Y.) during their 10th week of life.

Evaluation of Basal Conditions

Rats were studied 20 days after mating. The animals were weighed the evening prior to the experiment and then left undisturbed until they were sacrificed rapidly by decapitation the next morning. In addition at least six gravid and six virgin animals were placed in separate metabolic cages (Nalgene®, Rochester, N.Y.) and their urine was collected under mineral oil during the 24 hours prior to sacrifice.

Metabolic Balance Studies

Homozygous and heterozygous Brattleboro animals were placed in individual metabolic cages and once acclimated to their environment, food intake and

* Supported by generous grants from the National Institutes of Health (HD 5572) and the Mothers' Aid Research Fund of Lying-in Hospital.

† Send all correspondence to: Marshall D. Lindheimer, Department of Medicine and Obstetrics and Gynecology, 950 E. 59th Street (Box 83), Chicago, Illinois 60637.

0077–8923/82/0394–0481 $1.75/0 © 1982, NYAS

water balance were measured for five days. Studies were then interrupted for one week during which time the animal was mated with a homozygous male and reacclimated to the cage. Balance measurements were resumed for the next 19 days, the study ending one to two days prior to delivery. One month later the protocol was repeated after the same females were mated with heterozygous breeders.

Experiments Designed to Alter P_{osm} and U_{osm}

Homozygous animals were studied shortly after midpregnancy. After control measurements each animal received Pitressin in oil (s.c.) twice daily for six days at two dose schedules (40 mU and 80 mU BID), the last injections given two to three days before labor (in animals who were not sacrificed). Simultaneously studied virgin animals served as controls. In other experiments water was withheld for periods ranging from 24–48 h, timed to end between the 18–20th gestational day in heterozygote rats and the morning of day 20 in Long-Evans animals.

The methodology used has been previously detailed.[1] An AVP antiserum provided by Dr. F. H. Katz (Denver, Colo.) was used in the initial studies and our own antiserum #10169 in subsequent experiments. These antisera can usually detect 0.1 pg of AVP per assay tube, and always 0.2 pg/assay tube. Further information on these highly specific assays are described elsewhere.[1-3] Urinary prostaglandin (PGE_2) was assayed by a modification of Dray's method,[4] using an antiserum obtained from the Pasteur Institute (Paris, France).

RESULTS

Basal values in 18–20-day gravid animals and their respective age-matched controls are summarized in TABLES 1 and 2. P_{osm} was lower near term in the gravidas of each group ($p < 0.001$) and in each instance the difference could be accounted for primarily by decrements in P_{Na} and its attendant anions ($p < 0.001$). Comparisons between heterozygote and Long-Evans animals revealed several differences between the strains (TABLE 2). Basal P_{osm}, P_{Na}, and U_{osm} were greater in virgin heterozygote Brattleboro rats compared to nonpregnant Long-Evans animals ($p < 0.005$) and P_{osm} was higher in gravid heterozygotes compared to gravid Long-Evans ($p < 0.05$). Of note was the continued inability of gravid homozygous animals to concentrate their urines, while U_{osm} remained at or above 950 mOsm/kg in other rats. Also while water ingestion and urine volumes were increased in all three pregnant groups, these increments were massive in the pregnant homozygous rats, fluid intake and urinary output increasing $204 \pm$ SD 36 and 189 ± 22 ml/day, respectively (TABLE 1).

The results of metabolic balance studies before and during two successive gestations in six homozygous rats are depicted in FIGURE 1. Water intake and urine volumes increased during gestation, the greatest increments occurring during the final week of pregnancy. U_{osm} also increased slightly but significantly ($p < 0.025$) early in pregnancy returning to preconception values during the final week of gestation. However even at the highest values obtained, the urine remained hypotonic to plasma (FIGURE 1). There were small differences in the patterns recorded during homozygous-homozygous mating compared to homozygous-heterozygous mating, due possibly to the fact that the latter breeding took place when the rats were one month older. Of importance, however, is that U_{osm} were similarly dilute during the final week of both studies, the period when one would expect that any contribution of fetal AVP to maternal concentrating ability should become manifest. As expected, food intake and solute excretion increased during gestation. Water-to-food ratio (ml/g), 13.0 ± 0.8 during control measurements, averaged 13.6 ± 1 during early and midpregnancy and increased slightly averaging 15.6 ± 1 ($p < 0.01$) during the final five days of both study periods. Differences between fluid intake and urinary output also increased during this period suggesting that the small increment in the water-to-food ratio might be due to fluid retained in the mother and products of conceptus, or increases in insensible loss.

The results from balance studies performed on four gravid heterozygous rats were more variable (FIGURE 2). In most instances increments in fluid intake and urine volumes were small compared to that in homozygous animals and U_{osm} were often 950 mOsm/kg or greater. However, some dams periodically manifested increments in fluid ingestion and urine volume resembling those noted in homozygous animals (FIGURE 3) bringing to mind a syndrome in humans called transient diabetes insipidus of pregnancy.[5] Also, and in contrast to the pattern described in the homozygous strain, U_{osm} did not increase during early or midgestation; values either decreased slightly or remained similar to basal measurements (FIGURE 2).

TABLE 1

BASAL VALUES OF PLASMA OSMOLALITY AND WATER BALANCE IN NEAR TERM HOMOZYGOUS BRATTLEBORO GRAVID AND VIRGIN RATS

	Virgin	Pregnant	p
Plasma Osm (mOsm/kg)	310 ± 6.0 * (13)†	292 ± 3.5 (11)	<0.001
Hct (%)	40.3 ± 2.2 (13)	33 ± 1.8 (11)	<0.001
Plasma Urea N (mg/100 ml)	21.9 ± 5.2 (6)	19.5 ± 5.6 (6)	NS
Plasma Na (mEq/L)	146 ± 1.5 (6)	139 ± 1.1 (6)	<0.001
Water Intake (ml/24 h)	169 ± 23 (6)	375 ± 49 (6)	<0.001
Urine Volume (ml/24 h)	138 ± 18 (6)	336 ± 46 (6)	<0.001
Urinary Osm (mOsm/kg)	145 ± 45 (6)	135 ± 41 (6)	NS

* ± SD.
† *N*.

Pitressin (80 mU daily) administered for 6 days near term to gravid homozygous rats had no effect on P_{osm} (292 ± 3 mOsm/kg in treated vs. 292 ± 4 mOsm/kg in controls), while treatment of virgin animals resulted in a small but significant reduction (304 ± 3 mOsm/kg in treated vs. 310 ± 6 mOsm/kg in controls, $p < 0.05$). Unfortunately only one of the animals housed in metabolic cages conceived, but its U_{osm} had increased from 166 mOsm/kg before to 863 mOsm/kg with treatment, and its P_{osm} was 289 mOsm/kg at sacrifice. (U_{osm} in six virgins similarly studied increased from 152 ± 20 mOsm/kg to 876 ± 213 mOsm/kg). In parallel experiments urinary PGE_2 was measured before and on the sixth day of Pitressin treatment. Despite a greater urinary volume, PGE_2 was similar in gravid (18 ± 9 ng/24 h) and virgin (15 ± 8 ng/24 h) rats, and the increment in excretion evoked by Pitressin was similar: urinary PGE_2 rising to 48 ± 25 and 126 ± 33 ng/24 h

TABLE 2

COMPARISON OF OSMOREGULATION IN PREGNANCY BETWEEN HETEROZYGOUS BRATTLEBORO AND LONG-EVANS RATS

	Heterozygote			Long-Evans		
	Virgin	p	Pregnant	Virgin	p	Pregnant
Plasma Osm (mOsm/kg)	300 ± 3(12)	++	288 ± 5(11)	292 ± 2(13)	++	284 ± 3(13)
Hct (%)	44 ± 2(12)	++	38 ± 2(11)	42 ± 2(13)	++	35 ± 2(13)
Plasma Urea N (mg %)	18 ± 2(12)	NS	19 ± 4(11)	19 ± 2(13)	++	17 ± 3(13)
Plasma Na (mEq/L)	141 ± 1(12)	++	135 ± 2(11)	138 ± 1(13)	++	133 ± 2(13)
Plasma AVP (pg/ml)	3.8 ± 2.8(12)	NS	4.1 ± 3.0(11)	2.2 ± 1.5(13)	NS	2.1 ± 1.4(13)
Water Intake (ml/24 h)	39 ± 5(8)	**	54 ± 14(16)	32 ± 5(17)	++	50 ± 7(18)
Urine Volume (ml/24 h)	23 ± 4(8)	*	31 ± 11(16)	17 ± 4(17)	++	26 ± 8(18)
Urine Osm (mOsm/kg)	1234 ± 221(8)	NS	1168 ± 318(16)	1721 ± 337(17)	NS	1455 ± 556(18)
Maximal U_{osm} (mOsm/kg) (48 h dehydration)	3446 ± 538(15)	NS	3465 ± 628(13)	4188 ± 356(6)	NS	3960 ± 814(6)

( ) = *N;* ± = 1 SD; significance; p .05 = *, .01 = **, .005 = +, .001 = ++.

when gravid rats were treated with 80 mU and 160 mU daily, respectively, while comparable values in controls were 49 ± 32 and 135 ± 36 ng/24 h at each dose schedule.

Results of dehydration protocols in heterozygote and Long-Evans animals

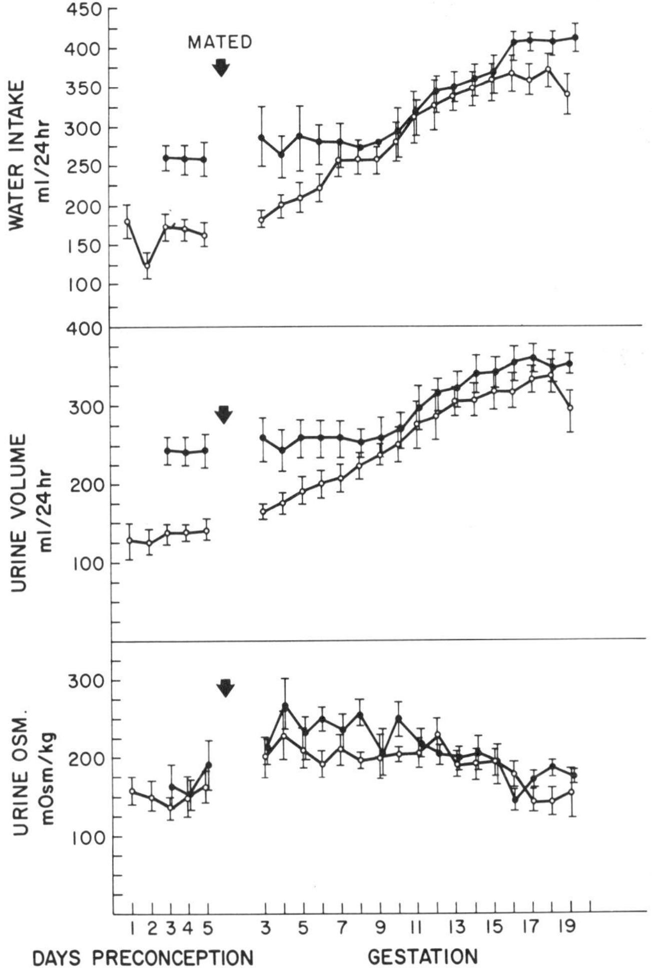

FIGURE 1. Water balance studies during two successive gestations in six homozygous Brattleboro animals, first after mating with homozygous males (○), and again following exposure to heterozygous breeders (●). Gestational days were derived by assuming delivery to be 21 days after conception (see text for details). I = SEM.

are summarized in TABLE 2 and FIGURE 4. The virgin Long-Evans animals' maximal U_{osm} were slightly higher than all others ($p < 0.01$) but of importance is that gravid and control rats of each strain were able to concentrate their urines to 3,000 mOsm/kg or greater. Water restriction resulted in gradual

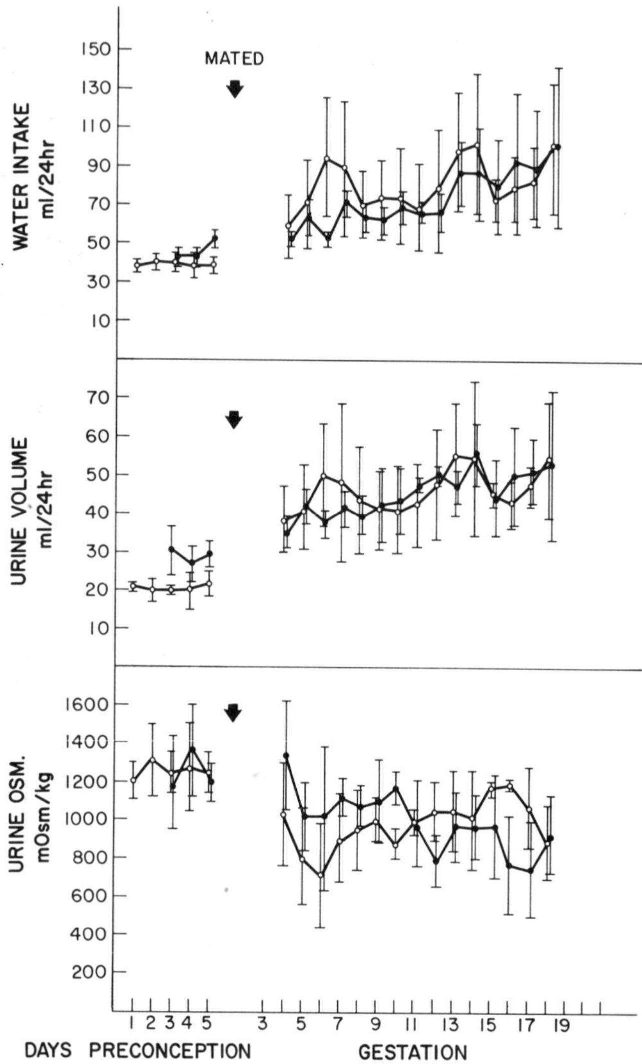

FIGURE 2. Water balance studies in heterozygous Brattleboro rats during two successive pregnancies: homozygous breeders; (○) heterozygous breeders; (●). I = SEM.

increments in both body tonicity and AVP levels, and the correlation of P_{AVP}* vs. P_{osm} was significant (range of *r*'s 0.66–0.92) for each group, however,

* However, even here as in the balance studies, the data for heterozygotes could be quite variable. For instance in one experiment the mean P_{AVP} at 48 h increased to 28±4 and 26±1 pg/ml in gravid and virgin heterozygotes, respectively, while in the study depicted in FIGURE 4 (where all four groups of animals were studied simultaneously) the highest value in virgin heterozygotes dehydrated for 48 h was only 15 pg/ml.

the slopes of the regression lines in the heterozygotes were less than those of the Long-Evans. The regression lines also defined the apparent osmotic thresholds for AVP secretion which were 282 and 281 mOsm/kg in gravid heterozygote and Long-Evans rats compared to 292 and 290 mOsm/kg in virgin heterozygote and Long-Evans rats, respectively.

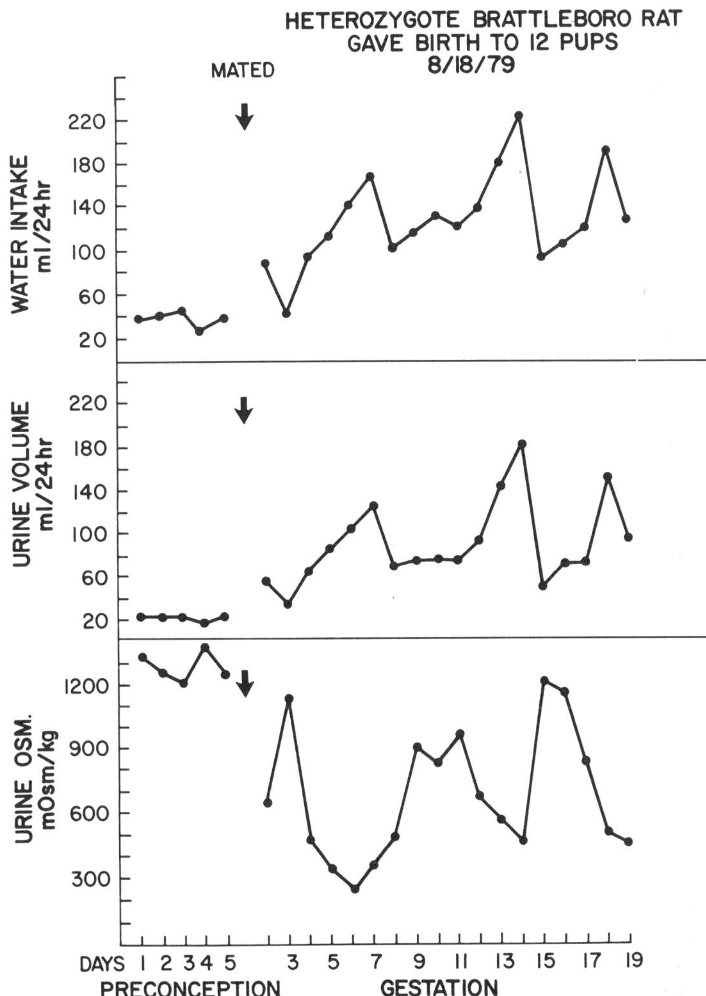

FIGURE 3. Water balance study during pregnancy in a single heterozygous Brattleboro rat, who periodically manifested increments in fluid ingestion, urine volume, and decrements in U_{osm} to values resembling those noted during gestation in homozygous animals. These increases could not have been artifactual as the metabolic cages are so designed that the water bottle spigot is outside the cage and spillage is caught and recorded in an underlying container.

Comment

The observations described above represent our preliminary results from studies still in progress when this symposium was convened. Data demonstrate that P_{osm} decrease significantly during gestation in both homozygous and heterozygous Brattleboro rats and the parent Long-Evans strain. The decrease in body tonicity in the pregnant Brattleboro rats is all the more remarkable since these animals have no circulating AVP, observations supporting our hypothesis that the threshold for water ingestion is also decreased during pregnancy.

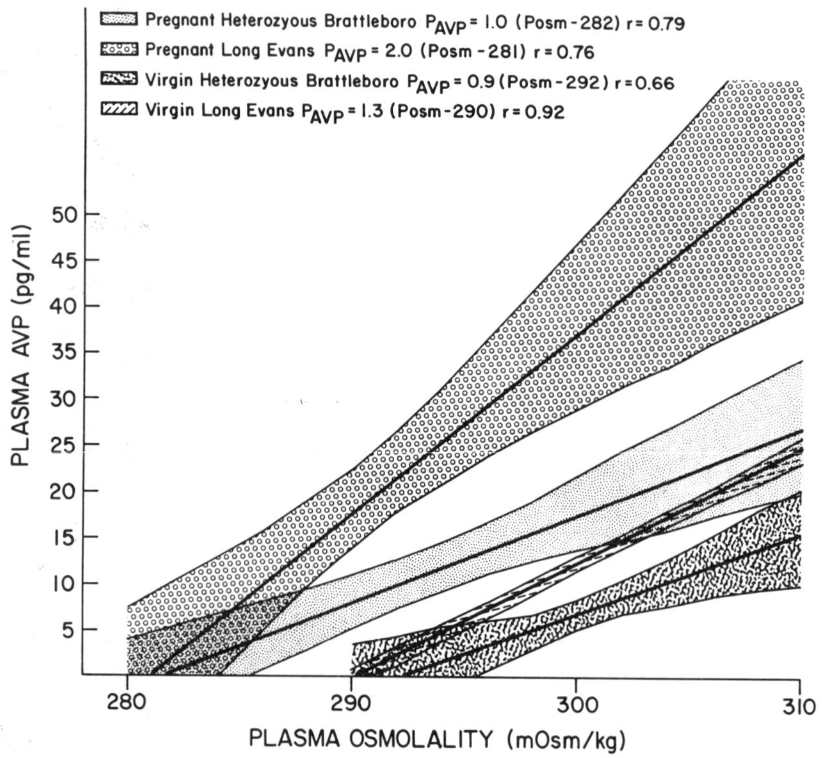

FIGURE 4. Regression lines with 95% confidence bands P_{AVP} vs. P_{osm} during dehydration experiment #2, in which pregnant and virgin heterozygous Brattleboro and Long-Evans rats were studied simultaneously.

The homozygous animals demonstrated a small increase in U_{osm} noted during the first week following conception and present throughout the first half of pregnancy. Since the highest values achieved, (252 ± SE 47 and 226 ± 43 mOsm/kg during the first and second matings) were still considerably below the P_{osm} normally present in nonpregnant homozygous rats, the increments in U_{osm} alone cannot be responsible for the decrement in body tonicity that occurs in pregnancy. One must therefore assume that in these gravid rats thirst continued to be stimulated to a marked degree despite a considerable decrease in P_{osm}.

The reason why U_{osm} increases early in gestation is not apparent from our data. U_{osm} may rise in the absence of AVP when solute excretion increases, but U_{osm} was similarly dilute in the near-term gravid compared to virgin animals at a time when solute excretion in the pregnant rats was twice that of the controls (TABLE 1). Other nonosmotic factors, such as alterations in renal hemodynamics, the influence of several hormones such as oxytocin, prolactin, and vasotocin need to be evaluated.

In addition to findings in the homozygous animals, our results also suggest a resetting of the osmotic threshold for AVP secretion in gravid heterozygote Brattleboro and Long-Evans rats. This is because despite their lower P_{osm}, these animals concentrated their urines and had basal P_{AVP} levels similar to those of the nonpregnant controls. Also, P_{AVP} increased immediately when increases in P_{osm} were evoked by fluid restriction (FIGURE 4). One should be aware, however, that water restriction in rats not only increases P_{osm} but also decreases intravascular volume [1] and the latter change too may stimulate AVP secretion. Thus we are currently increasing P_{osm} without concomitant volume depletion by injecting hypertonic saline i.p. into heterozygote Brattleboro animals and are again observing immediate increments in P_{AVP} (unpublished results).

Reasons why the osmotic thresholds for thirst and AVP secretion decrease in pregnant rats are not apparent from our data, and are discussed in more detail elsewhere.[1] We have been unable to demonstrate roles for estrogen, progesterone, or angiotensin II [6] and are currently investigating if prolactin or endorphins affect P_{osm} in the rat.

It was actually the marked polydypsia, polyuria, and decrements in U_{osm}, which occurred periodically during pregnancy in certain heterozygote dams, that led to our studies comparing gestation in the heterozygote Brattleboro and Long-Evans strains. We postulated that in pregnancy there might be increased metabolism or resistance to AVP, leading to secretory 'exhaustion' of antidiuretic hormone in the gravid Brattleboro animals. Both pregnant and virgin heterozygote rats increased their P_{AVP} and concentrated their urine remarkably well during 48 h of total water restriction (TABLE 2) but as seen in FIGURE 4 the increment in P_{AVP} as P_{osm} rises is less in both gravid and virgin heterozygotes than in Long-Evans animals. Obviously these results do not completely resolve the questions posed above concerning the existence of AVP resistance or exhaustion, and studies currently in progress in our laboratory comparing the effects of Pitressin on U_{osm} in gravid and virgin homozygous animals, and the influence of i.p. hypertonic saline on P_{AVP} in pregnant and control heterozygote rats may clarify these issues further.

In another publication [1] we suggested that parallel changes in the osmotic thresholds for thirst and AVP secretion must occur if the pregnant animal is to achieve and maintain its new steady-state osmolality within a narrow range. For example P_{osm} should rise even though AVP is secreted at a lower body tonicity if the rat isn't stimulated to drink, and considerable polydipsia is required to maintain any P_{osm} in the absence of vasopressin. The data presented here demonstrating an enormous water turnover in the gravid homozygous compared to the minimal increments in drinking and urinary output in the gravid heterozygous and Long-Evans strains, which maintain a high U_{osm} when P_{osm} is low, support these predictions.

In conclusion (1) data demonstrating marked decrements in P_{osm} in gravid homozygous Brattleboro rats suggest that the osmotic threshold for thirst decreases during pregnancy. (2) Vasopressin produced in fetuses does not enhance the ability of the homozygous mothers to concentrate their urines during gestation. (3) P_{osm} also falls during pregnancy in heterozygous Brattleboro and Long-Evans rats and their osmotic thresholds for AVP secretion decreases as well.

Acknowledgments

We thank Ms. Catherine Regovic for her excellent secretarial assistance. We are grateful to Dr. David W. Feigal for his statistical advice.

References

1. Dürr, J. A., B. Stamoutsos & M. D. Lindheimer. 1981. Osmoregulation during pregnancy in the rat: Evidence for resetting of the threshold for vasopressin secretion during gestation. J. Clin. Invest. **68:** 337–346.
2. Katz, F. H., J. A. Smith, J. P. Lock & D. E. Loeffel. 1979. Plasma vasopressin variation and renin activity in normal active humans. Hormone Res. **10:** 289–302.
3. Rascher, W., R. E. Lang, Th. Unger, D. Ganten & F. Gross. Vasopressin in the brain of spontaneously hypertensive rats. (Submitted for publication.)
4. Dray, F., B. Charbonnel & J. Maclouf. 1975. Radioimmunoassay of prostaglandins $F_{2\alpha}$, E_1 and E_2 in human plasma. Eur. J. Clin. Invest. **5:** 311–318.
5. Millar, D. G., E. G. Robertson & R. Hall. 1970. Transient diabetes insipidus in pregnancy: case report. *In* Reproductive Endocrinology. W. J. Irvine, Ed.: 152–156. Livingstone Ltd. Edinburgh, Scotland.
6. Dürr, J. A., B. A. Stamoutsos & M. D. Lindheimer. 1981. Plasma osmolality (P_{osm}) in pregnant rats in the absence of vasopressin (AVP) and during angiotensin blockade. Kidney Int. **19:** 238 (Abst.).

BRATTLEBORO HOMOZYGOTES CAN CONCENTRATE THEIR URINE DURING DEHYDRATION WITHOUT A CHANGE IN GFR *

Brian R. Edwards

Department of Physiology
Dartmouth Medical School
Hanover, New Hampshire 03755

INTRODUCTION

The observation that Brattleboro homozygotes (DI rats) produce increasingly concentrated urine when deprived of drinking water is an old one.[1] The mechanism of this response has not been fully elucidated, however. With our development of techniques that allow clearance experiments to be performed in unanesthetized, trained rats,[2] we were able to measure changes in glomerular filtration rate (GFR) and concentrating ability in fully conscious DI rats that were deprived of drinking water for up to 24 h.[3] These experiments clearly showed a progressive decline in GFR as urine became concentrated during dehydration in the absence of vasopressin.

As a possible explanation for this finding we turned to the mechanism that was first postulated by Berliner and Davidson.[4] In this scheme, the fall in GFR, by reducing delivery of solutes and water to the loops of Henle, results in decreased generation (and ultimate excretion) of free water. This effect, together with a reduced intratubular flow rate, would allow increased fractional reabsorption from a smaller load of water in the collecting ducts, sufficient to ultimately render the urine hyperosmotic to plasma.

In our previous studies,[3] the first measurements of GFR and concentrating ability were made after 3 h of dehydration had elapsed. However, additional preliminary observations had suggested that during the early hours of dehydration, urine osmolality begins to rise before any change in GFR is detectable. The present experiments were designed to more completely investigate this possibility. For, if true, the Berliner-Davidson mechanism could not be invoked—at least for the initial stages of dehydration. We also sought to expand upon our previous studies by including measurements of effective renal plasma flow (ERPF), and of the renal handling of urea and sodium during dehydration. A decrease in the fractional excretion of both substances would be predicted.

METHODS

Experiments were performed on nine DI rats of either sex weighing between 178 and 332 g. By methods previously described,[2, 3] catheters were placed in a jugular vein (for infusion of solutions), a carotid artery (for blood sampling and pressure measurement), and the urinary bladder. Following surgery the rats were trained to the restraining cage, and experiments performed from 4 to 24 days later.

* Supported by Research Grant AM 26553 from the National Institutes of Health.

0077–8923/82/0394–0491 $1.75/0

For the experiment, the rat was weighed, placed in the restraining cage, and an i.v. infusion of 10% Inutest (for GFR) and 1% PAH (for ERPF) in Ringer's solution started at 8 μl/min·100 g BW. Urine flow was immediately measured and an i.v. infusion of 2.5% dextrose in ¼-strength Ringer's was begun at the same rate. The infusion rate was adjusted in order to balance urine flow at approximately 10-min intervals throughout the 1-h equilibration. Thereafter, three 10-min control urine collections with midpoint blood samples were obtained following which the hypotonic Ringer's infusion was discontinued. This marked the beginning of the phase of dehydration. Then, consecutively, three 10-min, three 20-min, and three 30-min urine samples (with midpoint blood samples) were collected.

The sources of our analytical methods have been published previously,[3, 5] with the exception of urea.[6] In the following figures and table, the experimental values are compared by paired *t*-test with the mean of the three control values.

Results and Discussion

Table 1 summarizes some of the more important data that do not appear in graphic form. Three hours of dehydration elicited an hypertonic volume contraction as judged from the weight loss (6.6 ± 0.5 (SE) %) and the rise in plasma osmolality (from 321 ± 2 to 332 ± 3 mOsm/kg H_2O, p = 0.0009). Mean systemic arterial pressure was well maintained, and even increased slightly during the first 50 min. Heart rate decreased progressively.

As shown in Figure 1, urine flow, which began decreasing almost immediately after the start of the dehydration, fell rapidly during the first hour but more slowly thereafter. A significant increase in urine osmolality was evident by 20 min, and hypertonicity was reached by 60 min. For the latter half of the dehydration, urine osmolality remained within the range 384 to 412 mOsm/kg H_2O. This finding is similar to our previous report where urine osmolality increased from 155 to 376 mOsm/kg H_2O after 3 h of dehydration.[3]

The large decrease in urine flow resulted principally from the reduction of free-water clearance, although the clearance of total solute also decreased significantly during the final 2 h (Figure 2). Over the 3 h, the fall in osmolar excretion (from 15.9 ± 1.8 to 8.6 ± 0.4 μOsm/min·100 g, p = 0.007) could be accounted for, almost entirely, by decreases in the excretion of urea (4.8 ± 0.4 to 3.0 ± 0.2 μM/min·100 g, p = 0.002), sodium (3.1 ± 0.9 to 1.4 ± 0.2 μEq/min·100 g, p = 0.127), potassium (1.6 ± 0.2 to 1.2 ± 0.1 μEq/min·100 g, p = 0.018), plus accompanying anions. Note that the fall in sodium excretion was not significant, largely because of considerable inter-animal variation in the control values. The significant increase in fractional reabsorption of urea is presumably a result of increased fractional reabsorption of water in the collecting ducts. Dehydration of DI rats has been shown to increase the urea content of the papillary interstitium.[1, 7]

The surprising finding in this study was the relative stability of GFR throughout the period of dehydration (Figure 3). Apart from a transient reduction in the first 10 min, GFR did not fall significantly, but in fact tended to rise toward the end of the 3 h. This is in marked contrast to our previous findings where the formation of hypertonic urine after 3 h of dehydration was associated with an 18% reduction in GFR.[3] No compelling arguments can be marshalled to reconcile this difference, although a couple of observations are

TABLE 1

SUMMARY OF EFFECTS OF THREE HOURS OF DEHYDRATION IN DI RATS

	Control	Minutes of Dehydration								
		0–10	10–20	20–30	30–50	50–70	70–90	90–120	120–150	150–180
Plasma Osmolality ($mOsm/kg\ H_2O$)	321.6 ±2.3	324.9 * ±2.3	328.0 * ±2.6	329.5 * ±2.6	331.0 * ±2.8	329.3 * ±2.9	328.7 * ±2.2	330.4 * ±2.9	333.6 * ±2.9	332.4 * ±3.1
Mean Arterial Pressure (mm Hg)	114 ±2	116 * ±2	118 * ±2	118 * ±2	118 * ±2	117 ±3	116 ±3	116 ±3	115 ±3	116 ±3
Heart Rate (beats/min)	434 ±14	420 * ±11	411 * ±10	401 * ±11	392 * ±13	388 * ±13	385 * ±15	385 * ±14	398 * ±18	374 * ±20
Urine Osmolality ($mOsm/kg\ H_2O$)	130.6 ±6.9	137.1 ±6.6	168.1 * ±7.7	199.8 * ±7.0	262.3 * ±11.1	353.4 * ±9.1	404.4 * ±30.6	412.3 * ±38.8	348.3 * ±26.3	394.2 * ±31.2
GFR ($\mu l/min \cdot 100$ g BW)	827 ±36	765 * ±35	819 ±33	801 ±31	765 ±45	780 ±38	825 ±60	868 ±68	964 * ±37	861 ±50
C_{PAH} (ERPF) ($\mu l/min \cdot 100$ g BW)	2947 ±216	2546 * ±195	2824 ±236	2700 ±228	2730 ±305	2843 ±248	3084 ±311	3196 ±273	3288 * ±176	3574 * ±264
GFR/ERPF	0.29 ±0.02	0.31 * ±0.03	0.31 ±0.03	0.31 ±0.03	0.30 ±0.03	0.28 ±0.02	0.28 ±0.02	0.28 ±0.01	0.30 ±0.02	0.25 ±0.02
$\dot{V}$/GFR (%)	14.8 ±1.4	14.6 ±1.5	11.8 * ±1.3	9.9 * ±1.0	7.2 * ±0.9	4.4 * ±0.5	3.2 * ±0.3	2.8 * ±0.3	2.6 * ±0.3	2.7 * ±0.2
C_{Osm}/GFR (%)	6.0 ±0.6	6.0 ±0.5	5.9 ±0.5	5.9 ±0.5	5.5 ±0.6	4.6 * ±0.5	3.8 * ±0.3	3.3 * ±0.3	2.8 * ±0.2	3.1 * ±0.2
C_{H_2O}/GFR (%)	8.8 ±0.9	8.6 ±1.0	6.0 * ±0.9	4.0 * ±0.6	1.6 * ±0.4	−0.3 * ±0.1	−0.6 * ±0.2	−0.5 * ±0.2	−0.3 * ±0.1	−0.4 * ±0.2
C_{Na}/GFR (%)	2.47 ±0.65	2.67 ±0.52	2.81 ±0.43	3.15 ±0.44	2.92 ±0.43	2.26 ±0.37	1.66 ±0.22	1.26 ±0.18	1.02 ±0.16	1.06 ±0.13
C_K/GFR (%)	54.0 ±6.9	57.3 ±6.5	56.2 ±6.2	52.6 ±5.8	49.1 ±5.6	44.4 * ±6.0	37.9 * ±5.1	34.5 * ±5.5	31.5 * ±4.3	36.3 * ±4.0
C_{urea}/GFR (%)	74.4 ±1.8	78.1 ±2.7	73.5 ±2.4	72.9 ±1.9	72.1 ±2.3	67.7 * ±2.5	64.8 * ±2.0	62.6 * ±3.0	57.7 * ±3.5	64.0 * ±2.0

Mean ± SE. * Significantly different from Control at $p < 0.05$ or better.

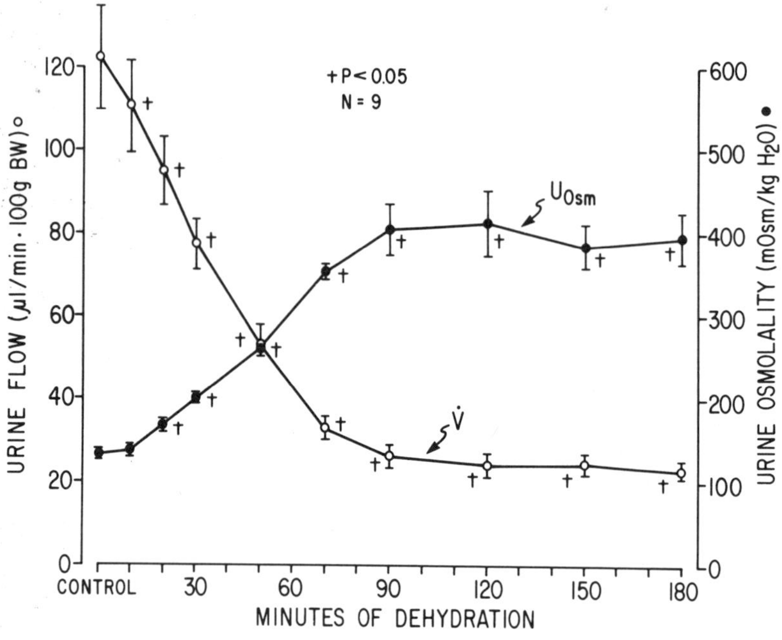

FIGURE 1. Urine osmolality and urine flow before and during 3 h of dehydration in DI rats. Mean ± SE are shown.

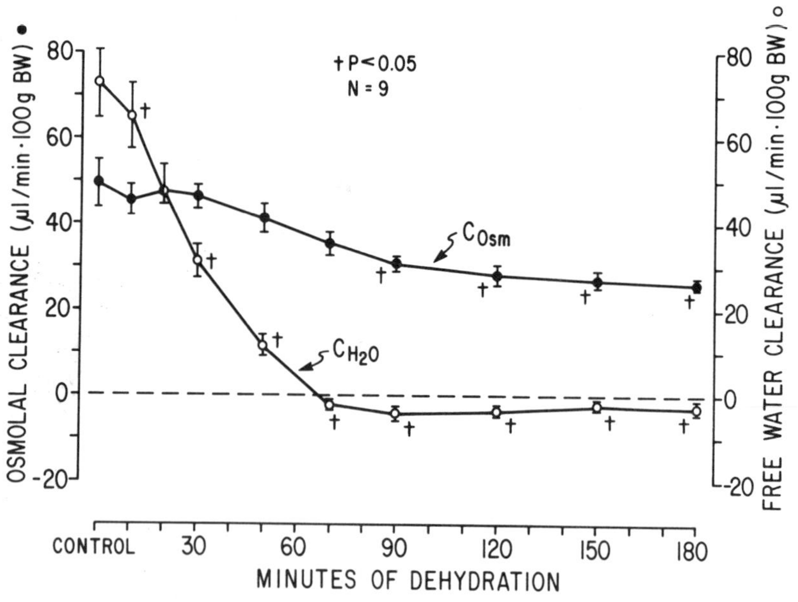

FIGURE 2. Osmolal clearance and free water clearance before and during 3 h of dehydration in DI rats. Mean ± SE are shown.

worthy of note. First, in our earlier study,[3] the DI rats were provided from our own colony, whereas a commercial supplier (Blue Spruce Farms, Inc., Altamont, N.Y.) was used for the present study. Conceivably, prolonged inbreeding might result in subtle differences among animals from different colonies that go largely unrecognized. Certainly the baseline values for urine flow in our purchased animals are 40% higher than those we were accustomed to finding in our own stock.[3] Secondly, in the present study, each rat remained in the restraining cage for the duration of the experiment. This contrasts with our earlier experiments where the rat was removed from the restraining cage following the control period, and then replaced in the cage some two hours later for further clearance determinations.[3] It is possible that this additional handling stress was sufficient to influence renal function.

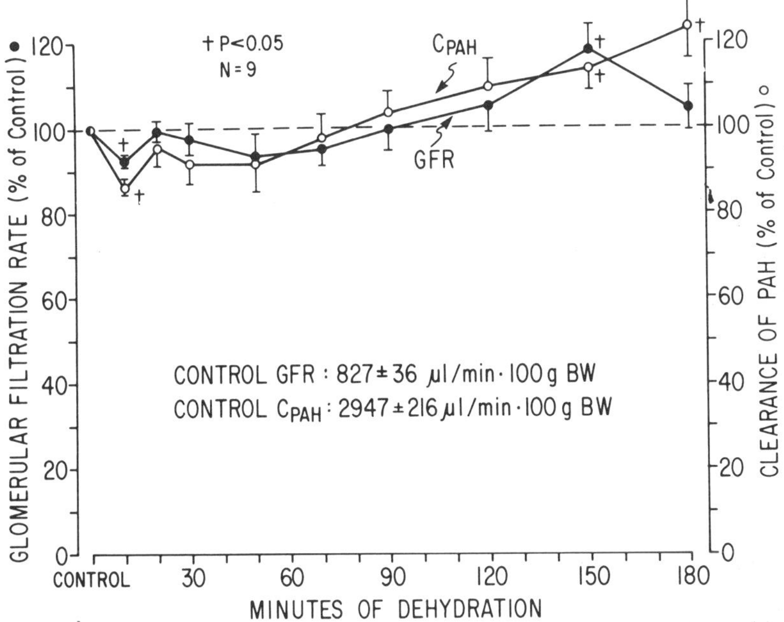

FIGURE 3. GFR and clearance of PAH (ERPF) as a % of control values during 3 h of dehydration in DI rats. Mean ± SE are shown.

ERPF followed an almost identical pattern to that of GFR (FIGURE 3)—that is, a transient decrease at 10 min, no significant change for most of the study, but a tendency to rise during the final hour. Given this parallelism, GFR/ERPF (or filtration fraction) was not significantly altered during dehydration. A rise in filtration fraction, by increasing fractional reabsorption in the proximal tubules, would be another means of reducing delivery of solutes and water to the loops, thereby initiating the Berliner-Davidson effect. However, this possibility is excluded by the present data, just as is a primary decrease in GFR. Our present findings are more in line with our previous studies in normally hydrated DI rats in which aortic constriction induced a similar rise in urine osmolality while GFR remained unchanged.[5] We were unable to provide a conclusive explanation for these results.

For the present study, we are forced to the conclusion that DI rats can concentrate their urine during dehydration (at least the early hours of dehydration), in the absence of vasopressin, without a decrease in GFR. The mechanism of this response cannot be determined from the information currently available, but the Berliner-Davidson mechanism appears not to be a viable candidate—at least for the initial stages of dehydration. During more prolonged dehydration, when GFR clearly falls,[3] the Berliner-Davidson mechanism is likely to play a major role.

Acknowledgment

The expert assistance of Ms. Betsy Craig is gratefully acknowledged.

References

1. Valtin, H. 1966. Sequestration of urea and nonurea solutes in renal tissues of rats with hereditary hypothalamic diabetes insipidus: effect of vasopressin and dehydration on the countercurrent mechanism. J. Clin. Invest. **45:** 337–345.
2. Gellai, M. & H. Valtin. 1979. Chronic vascular constrictions and measurements of renal function in conscious rats. Kidney Int. **15:** 419–426.
3. Gellai, M., B. R. Edwards & H. Valtin. 1979. Urinary concentrating ability during dehydration in the absence of vasopressin. Am. J. Physiol. **237:** F100–F104.
4. Berliner, R. W. & D. G. Davidson. 1957. Production of hypertonic urine in the absence of pituitary antidiuretic hormone. J. Clin. Invest. **36:** 1416–1427.
5. Edwards, B. R., M. Gellai & H. Valtin. 1980. Concentration of urine in the absence of ADH with minimal or no decrease in GFR. Am. J. Physiol. **239:** F84–F91.
6. Beale, R. N. & D. Croft. 1961. A sensitive method for the colorimetric determination of urea. J. Clin. Pathol. **14:** 418–424.
7. Williamson, G. L. & B. R. Edwards. 1982. Changes in the corticopapillary osmotic gradient during prolonged dehydration of the Brattleboro rat. Ann. N.Y. Acad. Sci. (This volume.)

CONCENTRATION OF URINE BY DEHYDRATED BRATTLEBORO HOMOZYGOTES: IS THERE A ROLE FOR OXYTOCIN? *

Brian R. Edwards, Frederick T. LaRochelle, Jr.,† and Miklos Gellai

Department of Physiology
Dartmouth Medical School
Hanover, New Hampshire 03755

†*Life Sciences Laboratory*
Northrop Services Inc.
Houston, Texas 77034

Introduction

Experiments in rats have shown that during prolonged water deprivation, pituitary stores of oxytocin become depleted [1] and plasma levels of this hormone increase.[2] Although Brattleboro homozygotes (DI rats) cannot synthesize vasopressin, they are capable of producing and releasing oxytocin. We have previously shown that DI rats are capable of producing hypertonic urine when deprived of drinking water,[3] and we therefore wondered whether oxytocin might have a role to play in this response.

Previous reports on the effects of oxytocin on the kidney are difficult to resolve. Both diuretic and antidiuretic responses have been noted, and similarly conflicting data abound for other renal functions.[4] Beyond species differences, the explanation for these conflicting reports may relate to the widely differing doses of oxytocin that have been used. In many instances, these doses have been far beyond the physiological range.

Therefore, in the present study, we first sought to determine the circulating levels of endogenous oxytocin before and after dehydration of DI rats, and then, based upon these results, to investigate some of the renal effects of appropriate dosages of exogenously administered synthetic oxytocin.

Methods

Metabolic Cage Study

Nineteen DI rats (163–367 g) and 20 Long-Evans (LE) rats (172–239 g) of either sex were acclimated to individual metabolic cages. On the day of experiment, 10 DI and 10 LE rats were deprived of drinking water for 24 h. The remaining (control) animals continued to have free access to food and water. The rats were weighed at the beginning and at the end of the 24-h period. Urine was collected under oil for the final 4 h (i.e., from 20 to 24 h) and its volume and osmolality recorded. At 24 h the rats were quickly decapi-

* Supported in part by Research Grants AM 08469, AM 26553, and HD 12123 from the National Institutes of Health.

0077-8923/82/0394-0497 $1.75/0

tated and trunk blood collected in chilled, heparinized containers. The plasma oxytocin levels were measured using an extraction procedure and radioimmunoassay (RIA) techniques that have been previously reported.[5] The antiserum had a cross-reactivity with either arginine vasopressin or arginine vasotocin of less than 1:10,000. The sensitivity of the RIA was 1.25 pg/ml of plasma when 0.5 ml plasma was extracted.

Infusion of Synthetic Oxytocin

Experiments were performed on 16 conscious DI rats (200–240 g) of either sex. By methods previously described,[3] catheters were placed in the superior vena cava (for infusion of solutions), the descending aorta (for blood sampling and pressure measurement), and the urinary bladder. Following surgery the rats were trained to the restraining cage, and experiments run 4 to 7 days later. During each experiment, ¼-strength Ringer's solution was infused at a rate that was continuously adjusted in order to balance the prevailing urine flow. By this means the weight of the animal was held constant for the duration of the experiment. Following a 45-min equilibration period, three 10-min control urine collections (bracketed by arterial blood samples) were obtained; then, synthetic oxytocin (Vega Biochemicals, Tucson, Ariz.) was infused continuously for 1 h at one of two doses: 250 or 2,500 pg/min·100 g body weight (BW). Five minutes after starting this infusion, six additional 10-min urine collections (with appropriate blood samples) were obtained. At the end of each experiment, an additional blood sample was drawn for subsequent determination of plasma oxytocin by RIA. Urine and plasma samples were analyzed for osmolality and for sodium and potassium concentrations.

Results and Discussion

The results of the metabolic cage study are summarized in Figure 1 and Table 1. In the LE rats, the increase in urine osmolality in response to 24 h of dehydration was not accompanied by any significant change in plasma oxytocin concentration. The latter finding contrasts with that of Dogterom *et al.*,[2] who reported a continuous rise of plasma oxytocin levels of Wistar rats subjected to four days of water deprivation. We are not aware of any analogous studies in LE rats. However, it has been reported that, in this strain, plasma oxytocin levels increase following intraperitoneal injection of hypertonic saline, although there was considerable variability in this response.[6] Plasma osmolality was not measured in the present study but, given the weight loss (Table 1), an increase would be predicted. We are unable to explain the apparent conflict between our results and the previous findings[2] except to postulate that, in our study, the degree of volume contraction—and hence the rise in plasma osmolality—was less than that required to provide a positive stimulus for neurohypophyseal release of oxytocin.[6]

Comparing the normally hydrated animals, the basal level of circulating oxytocin in DI rats was higher than in normal rats. This conforms with previous findings,[2,6] although the absolute levels reported by Dogterom *et al.*[2] for DI rats were exceptionally high (50 pg/ml). As previously reported,[3] the DI rats concentrated their urine to hypertonic levels following 24 h of dehydration. This was accompanied by a significant rise in plasma oxytocin levels (Figure 1). Moreover, there was a clear separation of the ranges of values encountered in the control and the dehydrated DI rats (Table 1).

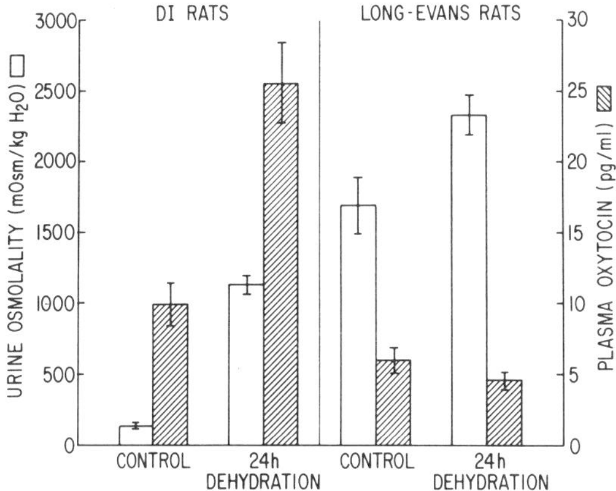

FIGURE 1. Urine osmolality and plasma oxytocin concentration before and after 24 h dehydration of DI and Long-Evans rats. Means ± SE are shown.

The second series of experiments was therefore designed to test the possibility that the rise in plasma oxytocin concentration induced by dehydration of DI rats might contribute to the observed increase in concentrating ability. The lower dose of oxytocin (250 pg/min · 100 g BW) was calculated to achieve a plasma concentration similar to that observed following 24 h of dehydration. The reasonableness of our calculations was confirmed by the fact that the average oxytocin level measured at the end of the infusion in eight experiments was 46.3 ± 4.7 pg/ml; this value falls within the range observed after dehydration (TABLE 1). However, when this dose of oxytocin was infused into normally hydrated DI rats, there was no significant effect on urine flow, and the rise in urine osmolality was not impressive (FIGURE 2A). There was a clear natriuresis, however, and the rise in sodium excretion (plus accompanying anion) appeared to almost completely account for the increased urine osmolality (FIGURE 3A). There was no change in potassium excretion.

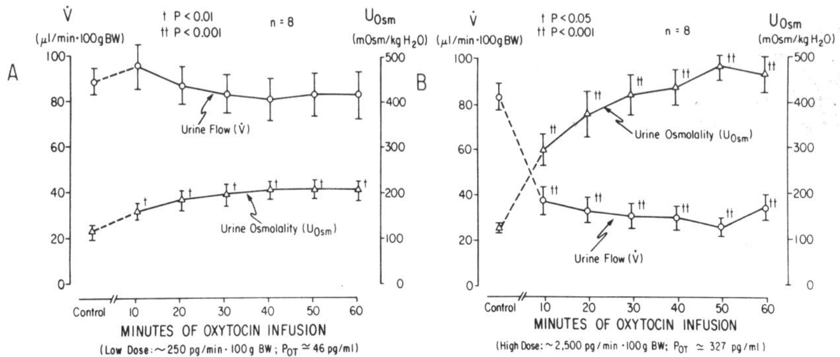

FIGURE 2. Urine flow and urine osmolality in normally hydrated DI rats before (control) and during infusion of synthetic oxytocin at (A) 250 pg/min · 100 g BW or (B) 2,500 pg/min · 100 g BW. Plasma oxytocin concentrations (P_{OT}) were determined by RIA. p-values refer to comparisons with control. Means ± SE are shown.

TABLE 1

SUMMARY OF METABOLIC CAGE STUDY

	DI Rats		Long-Evans Rats	
	Controls	Dehydrated (24 h)	Controls	Dehydrated (24 h)
Number of rats	9	10	10	10
0 to 24 h:				
Water intake (ml/24 h · 100 g BW)	91.2 ± 8.1	0 †	14.5 ± 1.1	0 †
Change in body weight (%)	+ 2.3 ± 1.1	− 15.0 ± 0.6 †	− 0.7 ± 0.7	− 9.7 ± 0.3 †
20 to 24 h:				
Urine flow (ml/4 h · 100 g BW)	9.0 ± 1.5	1.0 ± 0.1 †	1.1 ± 0.2	0.2 ± 0.03 ‡
Urine osmolality (mOsm/kg H_2O)	141.8 ± 17.1	1,129.5 ± 58.5 †	1,683.2 ± 199.7	2,322.0 ± 139.6 §
24 h:				
Plasma oxytocin (pg/ml)	9.9 ± 1.5	25.5 ± 2.8 †	6.0 ± 0.9	4.6 ± 0.6
{range}	{4.3–18.8}	{19.3–47.0}	{3.5–10.6}	{2.3–7.8}

Mean ± S.E. † $p < 0.001$; ‡ $p < 0.005$; and § $p < 0.02$ compared with corresponding controls.

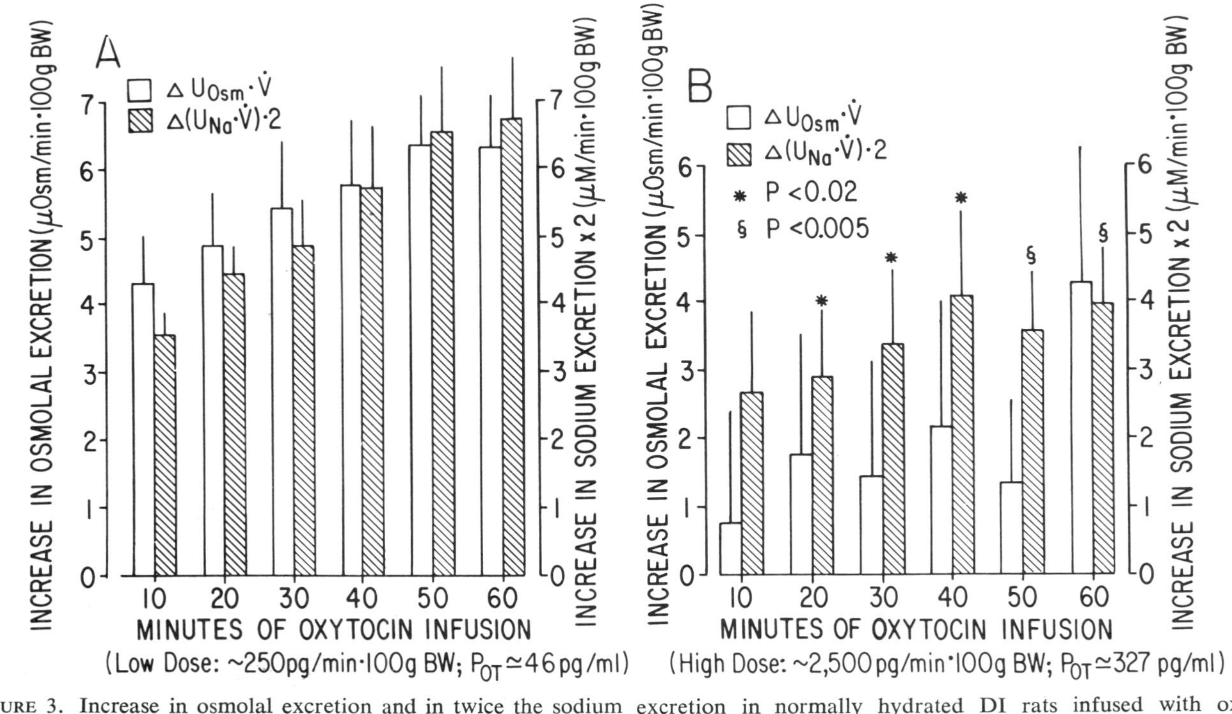

FIGURE 3. Increase in osmolal excretion and in twice the sodium excretion in normally hydrated DI rats infused with oxytocin. Means ± SE are shown. (A) Dose of oxytocin = 250 pg/min · 100 g BW. All changes significant at $p < 0.001$. Control $U_{Osm} \cdot \dot{V} = 10.2 \pm 1.1$ μOsm/min · 100 g BW; Control $(U_{Na} \cdot \dot{V}) \cdot 2 = 2.7 \pm 0.8$ μM/min · 100 g BW. (B) Dose of oxytocin = 2,500 pg/min · 100 g BW. Significant changes are indicated. Control $U_{Osm} \cdot \dot{V} = 10.4 \pm 0.9$ μOsm/min · 100 g BW; Control $(U_{Na} \cdot \dot{V}) \cdot 2 = 2.7 \pm 0.7$ μM/min · 100 g BW.

Despite the fact that oxytocin, at a dose that mimicked post-dehydration levels of this hormone, did not exert a major effect on concentrating ability, the DI rat *is* capable of responding to higher doses of oxytocin. Thus, in eight additional rats, the infusion of a ten-fold higher dose resulted in a clear antidiuretic response (FIGURE 2B). The plasma concentration achieved in these experiments, 327 ± 29 pg/ml, is close to (or even above) the upper limit of the physiological range—perhaps resembling levels seen during parturition, coitus, and lactation.[7] A somewhat smaller natriuresis was observed, with no significant change in overall solute excretion (FIGURE 3B), or potassium excretion. The rise in urine osmolality to hypertonic levels therefore reflects increased osmotic equilibration across the collecting ducts, presumably as a result of an increase in water permeability mediated via oxytocin.

Balment and coworkers have infused similar doses of oxytocin into anesthetized DI rats and achieved comparable plasma levels.[6] They found that while the lower dose was without any effect, the higher dose increased the excretion of sodium and chloride, reduced urine flow, and increased urine osmolality (but barely to hypertonicity). It is likely that the anesthetized, acutely instrumented DI rat is a less satisfactory model to demonstrate small changes in concentrating ability than is the conscious, trained animal.

In summary, plasma oxytocin levels increase in the DI rat following prolonged dehydration. However, when such elevated levels are reproduced in the normally hydrated DI rat, there ensues only a slight rise in urine osmolality that appears to result from the natriuretic properties of the hormone. Although this finding suggests that oxytocin may have little or no role to play in the increased concentrating ability of dehydrated DI rats, the possibility cannot be excluded that the same level of oxytocin has a greater influence on the renal concentrating mechanism in the volume-contracted state than in the euvolemic condition.

References

1. JONES, C. W. & B. T. PICKERING. 1969. Comparison of the effects of water deprivation and sodium chloride imbibition on the hormone content of the neurohypophysis of the rat. J. Physiol. **203:** 449–458.
2. DOGTEROM, J., T. B. VAN WIMERSMA GREIDANUS & D. F. SWAAB. 1977. Evidence for the release of vasopressin and oxytocin into cerebrospinal fluid: measurements in plasma and CSF of intact and hypophysectomized rats. Neuroendocrinology **24:** 108–118.
3. GELLAI, M., B. R. EDWARDS & H. VALTIN. 1979. Urinary concentrating ability during dehydration in the absence of vasopressin. Am. J. Physiol. **237:** F100–F104.
4. THORN, N. A. 1968. The influence of the neurohypophysial hormones and similar polypeptides on the kidneys. Handb. Exp. Pharmacol. **23:** 372–442.
5. GELLAI, M., F. T. LAROCHELLE, JR. & H. VALTIN. 1980. Effect of oxytocin on renal function in conscious Brattleboro homozygous rats. A model for studying renal regulation by hormones. *In* Hormonal Regulation of Sodium Excretion. B. Lichardus, R. W. Schrier & J. Ponec, Eds.: 121–128. Elsevier/North-Holland Biomedical Press. Amsterdam.
6. BALMENT, R. J., M. J. BRIMBLE & M. L. FORSLING. 1980. Release of oxytocin induced by salt loading and its influence on renal excretion in the male rat. J. Physiol. **308:** 439–449.
7. TINDAL, J. S. 1974. Stimuli that cause the release of oxytocin. Handb. Physiol. Sec. 7, Vol. IV, Part 1: 257–267.

HYALURONATE HYDROLASE ACTIVITY AND GLYCOSAMINOGLYCANS IN THE BRATTLEBORO RAT KIDNEY

L. N. Ivanova, T. E. Goryunova, L. F. Nikiforovskaya, and N. I. Tishchenko

Institute of Cytology and Genetics
USSR Academy of Sciences Siberian Department
630090 Novosibirsk 90, USSR

More than twenty years ago it was found that the renal medullary interstitium of the mammals contained abundant amounts of glycosaminoglycans (GAG) and that the urine of hydropenic animals had a high hyaluronidase activity. Histochemical and physiological observations led to the hypothesis that passage of water out of the renal collecting duct following antidiuretic hormone (ADH) might be facilitated by the release of hyaluronidase with subsequent depolymerization of GAG in the interstitium.[1] But the literature contains contradictory reports as to the influence of ADH on the GAG of the renal medullary interstitium,[3] and up to now the effect of ADH on the interstitial structures, which, as supposed, plays an important role in the urinary concentration, remains disputable. The homozygous Brattleboro rats with diabetes insipidus seemed to offer unique advantages for solving this problem. Recently, it has been found that the renal papilla interstitium of homozygous rats contains greatly reduced amounts of GAG when compared with normal or heterozygous ones.[3] We have obtained the same data and undertook a research to investigate the renal enzyme system of hyaluronate hydrolases and GAG in detail.

Experiments were carried out on the homozygous (DI) and heterozygous (HZ) Brattleboro rats bred from the stock supplied by the Institute of Physiology of the Czechoslovak Academy of Sciences. The total hyaluronate hydrolases (HH) activity of the renal papilla tissue was determined according to Bollet *et al.*[4] To investigate the hyaluronidase (Hy) activity, the exohydrolases' effect on the substrate was blocked by mucic acid.[5] For the evaluation of β-glucuronidase and *N*-acetylhexosaminidase activities we used as substrates PNP-β-D-glucuronide, PNP-2-acetamido-2-deoxy-β-D-glucopyranoside, respectively (Koch-Light, England). Urine and plasma osmolality was determined by freezing-point osmometry. For the histochemical study, the kidneys were fixed in the 10% solution of neutral formalin. The slices were stained by several methods for identification of GAG.[6] To avoid some possible faults when staining, the slices of control and experimental rats were processed simultaneously.

The study of HH activity in the papilla tissue showed that the basal level of the total HH activity and the Hy of two exohydrolases were significantly less in DI rats than in HZ ones (TABLE 1); but the activity was nearly equal to that of Wistar rats.[7] The treatment of the rats with ADH (Pituitrin P, 500 μU/100 g b.wt., subcutaneously twice in 30 min) increased the urinary osmolality as well as the HH activity. Although the level of the enzyme activity in the DI rat kidney remained lower than in HZ rat, the percentage of

0077-8923/82/0394-0503 $1.75/0 © 1982, NYAS

TABLE 1

EFFECT OF PITUITRIN P ON THE URINE OSMOLALITY AND GLYCANO-HYDROLASE ACTIVITY IN THE RENAL PAPILLA OF HOMOZYGOUS (DI) AND HETEROZYGOUS (HZ) BRATTLEBORO RATS

	Experimental Conditions	Urine Osmolality (mOsm/L)	Total Activity of Glycano-hydrolases (μg *N*-acetylglycosamine/mg of protein)	Hyaluronidase Activity (μg *N*-acetylglycosamine)	Glucuronidase Activity (μM *N*-nitrophenol/mg of protein)	*N*-acetylglycosaminooxidase (μM *N*-nitrophenol per mg of protein)
DI	Control	188 ± 21	65.1 ± 2.4 *** (3)	16.4 ± 0.1 *** (3)	16.0 ± 1.1 (3)	43.6 ± 3.7 *** (4)
	Pituitrin P	562 ± 32	80.8 ± 2.3 **‡ (3)	21.3 ± 2.5 **‡ (3)	19.6 ± 0.3 † (3)	55.2 ± 4.1 *† (4)
HZ	Control	657 ± 27	87.4 ± 2.2 (6)	23.1 ± 1.3 (6)	17.8 ± 2.6 (3)	64.8 ± 1.6 (5)
	Pituitrin P	1540 ± 45	101.2 ± 3.1 † (7)	30.4 ± 1.8 ‡ (7)	20.8 ± 1.0 † (3)	74.0 ± 4.8 † (5)

The numbers in parentheses indicate the number of experiments in which the materials on three animals are summed.
Significance of differences in comparison with control in each group of animals: † $p < 0.05$; ‡ $p < 0.01$.
Significance of enzyme activity differences between homozygotes and heterozygotes: * $p < 0.02$; ** $p < 0.01$; *** $p < 0.001$.

its rise was approximately the same. Close correlation between the urine osmolarity and the renal tissue enzyme activity was observed ($r = 0.95$, $p < 0.01$, for the total HH activity, $r = 0.99$, $p < 0.01$, for the Hy, $r = 0.85$, $p < 0.05$, for the β-glucuronidase, and $r = 0.92$, $p < 0.05$ for *N*-acetylhexosaminidase).

The GAG content in the DI rat papilla interstitium differed significantly from that of HZ rat. While the interstitium of HZ rat renal papilla and the apical coat of collecting tubule cells were intensively stained, those of DI rats were characterized by a striking reduction of staining (FIGURE 1a, 2a).

The reduction of GAG staining in the DI rats papilla might be the result of a decrease in synthesis because of ADH deficiency. However, the treatment of DI rats with DDAVP (Adiuretin-SD, Spofa, ČSSR, 20 μU/100 g b.wt., i.p. during six days) didn't eliminate the GAG defect, although it caused the urine osmolarity to increase from 185 ± 23 to 1594 ± 38 mOsm/L. At this highest dose of DDAVP, only a slight increase of staining was observed in the interstitium of the tip and the middle zone of the papilla. The granules' appearance in the collecting duct epithelial cells and the intensification of cell-coat staining was seen as well. Nevertheless, the histochemistry was not entirely restored to that observed in the HZ rats (FIGURE 1b).

One can suppose that DDAVP stimulated the renal prostaglandin system and the latter, in its turn, influenced the GAG synthesis.[8-10] Simultaneously, DDAVP affected the HH activity and GAG hydrolysis prevented GAG from increasing in the interstitium.

Glucocorticoids have been recently found to cause the renal HH activity decrease.[11] This observation prompted us to investigate the influence of corticosteroids on the renal GAG in the DI rats. The treatment of DI rats with hydrocortisone (Hydrocortisone, Gedeon Richter, Hungary, 0.25 mg/100 g b.wt. during six days) did not increase the urine osmolarity (U_{osm} was 282 ± 26 mOsm/L by the end of the experimental period). At the same time a moderate increase of GAG in the renal papilla interstitium, in the collecting duct epithelial cells, and in their apical coat was found (FIGURE 1c).

The influence of hydrocortisone has been still more clearly revealed in another set of experiments. A water balance close to that of DI rats could be reached in HZ rats by giving them a 4% sucrose solution instead of drinking water (food was removed). The rats drank the volume of the solution equal to 80–90% of their body weight daily. In six days the cortico-papillar gradient of sodium and urea decreased considerably, and the urine osmolarity dropped to 56 ± 15 mOsm/L.[12] It appeared that the HZ rats' papilla GAG were stained rather weakly as well as those of DI rats (FIGURE 2b). The daily injection of hydrocortisone (0.25 mg/100 g b.wt.) accompanied by sugar solution drinking did not affect either the cortico-papillar gradient of sodium and urea or the urine osmolality. However, it almost completly preserved GAG staining in the papilla interstitium (FIGURE 2c).

The results suggest to us that the display of HH activity and the GAG staining in DI rats are due to some complex alterations of hormonal balance. Lower HH activity in the DI rats' papilla as compared to that of HZ seems to be closely connected with the ADH deficiency. Meanwhile, enzyme activity is not suppressed completely, the system's ability to be stimulated by exogenous ADH being preserved. The GAG staining decrease might be explained not only by the decrease in biosynthesis of these compounds but also their hydrolysis due to HH, their activity remaining comparatively high and the hydrolysis

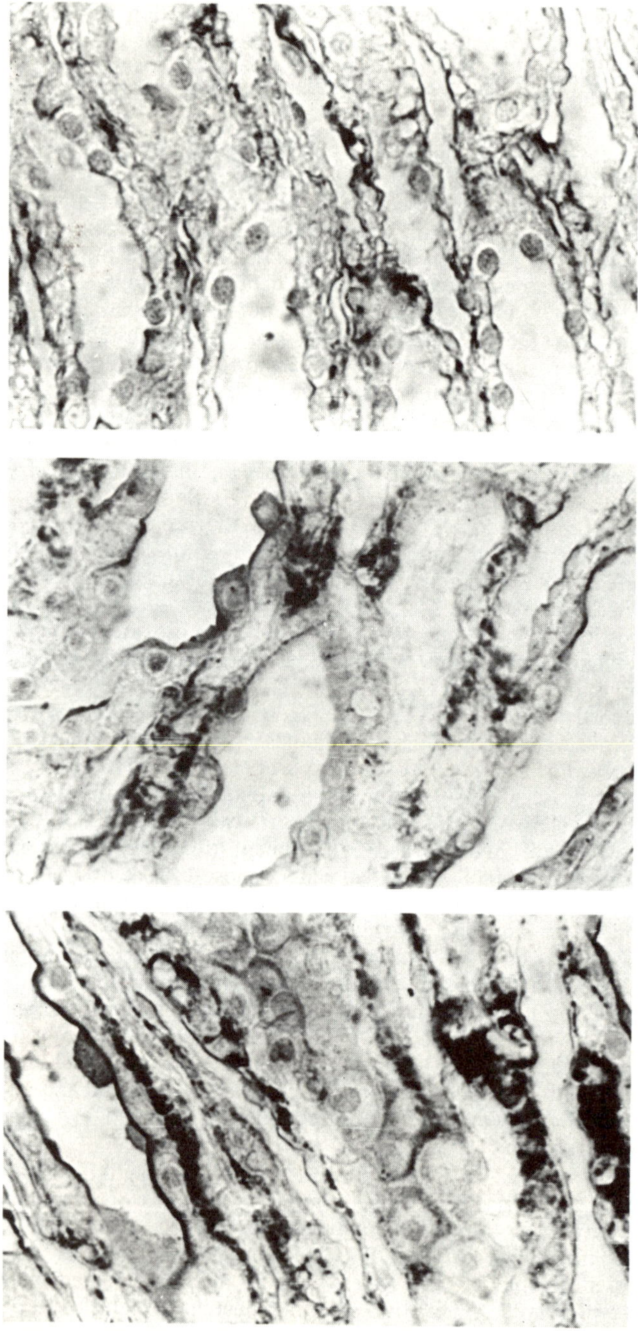

FIGURE 1. Tip of the renal papilla of homozygous Brattleboro rats: (top) control, (middle) DDAVP treatment, and (bottom) hydrocortisone treatment. GAG stained with colloidal iron: $\times$ 175.

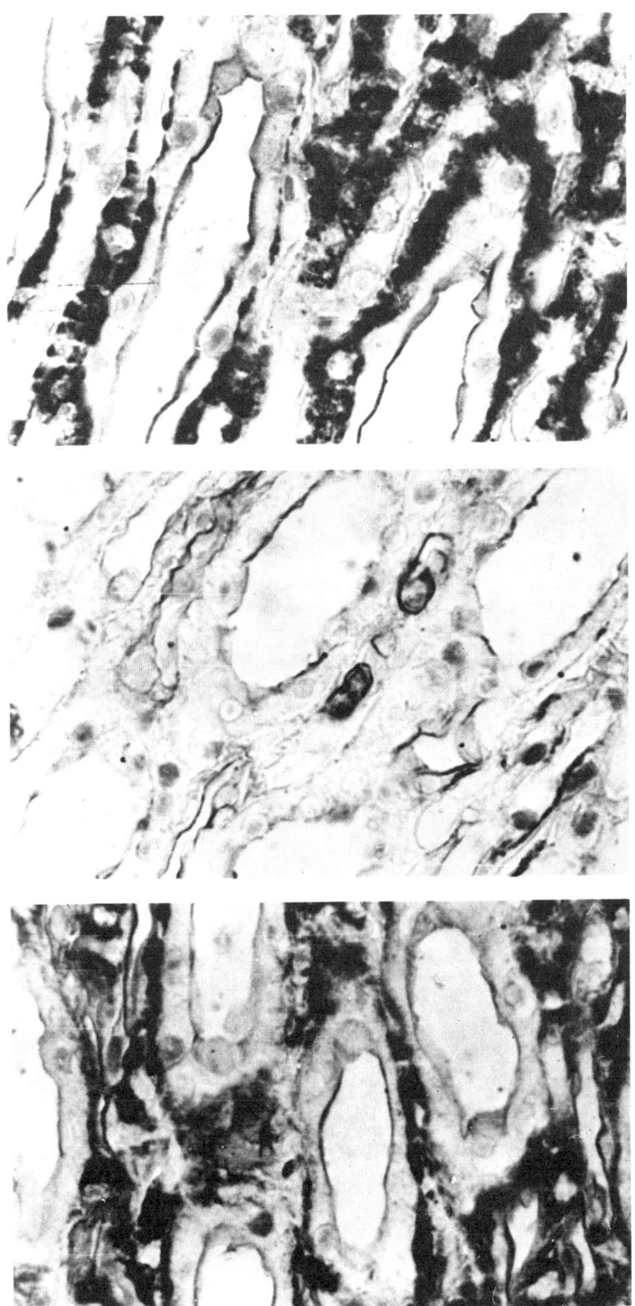

FIGURE 2. Tip of the renal papilla of heterozygous Brattleboro rats: (top) control, (middle) drinking of the 4% sucrose solution during 6 days, (bottom) drinking of the 4% sucrose solution and treatment with hydrocortisone during 6 days. GAG stained with colloidal iron: ×175.

products being washed out. The GAG synthesis change in the DI rats may also be the result of potassium balance alteration as is the case with induced hypokalemia.[13] The changes in the HH and GAG in the DI rats' papilla may have several possible explanations, but regardless of the interpretation these data demonstrate the significant role of HH and GAG in the ADH effect.

References

1. Ginetzinsky, A. G. 1958. Role of hyaluronidase in the reabsorption of water in renal tubules. Nature **182:** 1218–1219.
2. Dicker, S. E. 1970. *In* Mechanisms of Urine Concentration and Dilution in Mammals. Chap. 7. Monographs of Physiol. Soc. Edward Arnold, Ltd. London.
3. McAuliffe, W. G. 1980. Histochemistry and ultrastructure of the interstitium of the renal papilla in rats with hereditary diabetes insipidus (Brattleboro strain). Am. J. Anat. **157:** 17–26.
4. Bollet, A. J., W. M. Bonner, Jr. & J. L. Nance. 1963. The presence of hyaluronidase in various mammalian tissues. J. Biol. Chem. **238:** 3522–3527.
5. Vikha, I. V., M. N. Privalenko & A. E. Khormen. 1973. The determination of serum hyaluronidase, β-glucuronidase and N-acetyl-D-glucosaminidase. Vopr. Med. Chim. **19:** 90–96.
6. Pearse, A. G. E. 1960. Histochemistry—theoretical and applied. J. & A. Churchill Ltd. London.
7. Ivanova, L. N. & T. E. Goryunova. 1981. Mechanism of the renal hyaluronate-hydrolase activation in response to ADH. Adv. Physiol. Sci. **11:** 587–591. Kidney and Body Fluids. L. Takács, Ed. Akademiai Kiado, Budapest.
8. Dunn, M., L. B. Kinter, R. Beeuwkes III, D. Shier, H. P. Greeley & H. Valtin. 1980. Interaction of vasopressin and renal prostaglandins in the homozygous diabetes insipidus rat. *In* Advances in Prostaglandin and Thromboxane Research. B. Samuelsson, P. W. Ramwell & R. Paoletti, Eds. **7:** 1009–1015. Raven Press. New York, N.Y.
9. Walker, L. A., A. R. Whorton, M. Smigel, R. France & J. C. Frölich. 1978. Antidiuretic hormone increases renal prostaglandin synthesis *in vivo*. Am. J. Physiol. **235:** F180–F185.
10. Schönhöfer, P. S., H. D. Peters, A. Wasmus, B. A. Pesker, R. von Figura & I. Klappstein. 1978. Prostaglandins, cyclic nucleotides and glycosaminoglycan biosynthesis in cultured fibroblasts. Pol. J. Pharmacol. Pharm. **30:** 183–193.
11. Chocha, A. 1980. Wplyw octanu kortyzonu na zawartošć kransynch mukopolisacharydow (glikozoaminoglikuronianow) w nerkach krolika. III Zmiany w Zakresie aktunownošci beta-N-acetylo-D-glukozaminidazy. Endocrinol. pol. **31:** 267–271.
12. Melidi, N. N. & V. A. Lavrinenko. 1981. Peculiarities of volume- and ionoregulation in different strains of rats. (In press.)
13. Zaks, M. G., T. V. Krestinskaya & M. M. Sokolova. 1964. Antidiuretic hormone effect in hypokalemic rats. Sechenov physiol. J. USSR **50:** 1490–1495.

SPONTANEOUS FORMATION OF BLADDER STONES IN DIABETES INSIPIDUS RATS (BRATTLEBORO STRAIN) *

Lewis B. Kinter,† Jeanne McDonald,‡ Andrea Harris,† Reinier Beeuwkes III,† and Rubin Gittes‡

† *Department of Physiology*
Harvard Medical School
Boston, Massachusetts 02115

‡ *Department of Urology*
Brigham and Women's Hospital
Boston, Massachusetts 02115

Introduction

Elevation of urine flow is generally thought to protect against urolithiasis, whereas oliguria is often associated with stone formation. However, we have detected spontaneously occurring struvite ($MgNH_4PO_4 \cdot 6H_2O$) and apatite ($(Ca_{10}(PO_4 \cdot CO_3OH_6)(OH)_2)$) bladder stones in Brattleboro strain rats homozygous for diabetes insipidus (DI rats) in which urine excretion exceeds 200 ml/day (70% of body weight) and total urinary solute levels are less than 150 mOsm.[1] In investigating this apparent paradox we sought to determine: first, whether DI rat urine was saturated *in vivo* with respect to struvite and apatite, and second, what effect chronic urinary bladder infection with a urea-splitting organism (*Proteus mirabilis*) would have on *in vivo* stone formation in diabetes insipidus.

Methods

Male DI rats were selected at random from the same Brattleboro colony (Harvard Medical School, Boston, Mass.) in which spontaneous stone formation was first observed. Age-matched control antidiuretic Long-Evans strain rats were purchased from a local breeder (Charles River, Medford, Mass.). Animals were housed communally (up to four rats/cage) on soft bedding, and fed tap water and the same diet (Charles River #RMH2000) previously associated with stone formation. Three groups of animals were studied.

Group I (*Colony Control*). Spontaneous stone formation was monitored in 13 DI and 25 LE controls.

Group II (*Nidus Control*). *In vivo* saturation with respect to stone-forming solutes in diuretic and antidiuretic rat urine was determined by the method first described by Vermeulen *et al.*[2] Tared sterile zinc disks (weighing ~100

* Supported in part by National Institutes of Health Grant AM 18249, Grants from the Brigham Surgical Group, Inc. and the Thornberg Medical Foundation, and by a gift from R. J. Reynolds Industries. L.B.K. was the recipient of National Institutes of Health Fellowship Award AM 05938.

0077-8923/82/0394-0509 $1.75/0 © 1982, NYAS

mg) were implanted during ether anesthesia into the urinary bladders of 12 DI and 9 LE rats. These disks provided nucleation sites for crystal growth.

Group III (*Nidus + Proteus*). Urease-containing *Proteus mirabilis* has been associated with urinary-tract stone formation in experimental animals and man.[3] After disk implantation the urinary bladders of an additional 11 DI and 5 LE rats were injected with a live *Proteus* culture ($\sim 2.2 \times 10^6$ cells/bladder). Care was taken to prevent peritoneal *Proteus* infection.

After six months all bladder contents was removed, dried, weighed, and analyzed for crystallographic composition (Urolithiasis Laboratory, Baylor College of Medicine, Division of Urology, Houston, Texas). Empty urinary bladders were blotted dry and weighed. Statistically significant differences in bladder weights and rates of stone formation were determined by application of the Student's t test for unpaired observations.

Results

Implanted zinc disks were always associated with the formation of stone in LE rats but not in DI rats (Table 1, $p < 0.001$). In fact, in 5 of 12 DI rats the zinc disks were partially dissolved. All stones were mixtures of struvite and apatite, similar to the composition of stones formed spontaneously in DI rats. No stone formation was observed in DI and LE control groups, nor have any additional spontaneously formed stones been observed in the Brattleboro colony.

Proteus infection (+disk) was also associated with the formation of large amounts of stone in LE rats. However in contrast to colony-control and nidus-control groups *Proteus*-infected DI rats formed copious quantities of stone. All calculi contained mixtures of struvite and apatite. Spot checks of free-catch urine samples during the experiment showed consistent *Proteus* levels in infected animals, and undetectable *Proteus* levels in nidus and colony controls.

Urinary bladders in colony-control DI rats were significantly heavier than those of LE controls ($p < .001$, Table 2). Urinary bladders constituted 0.095% of body weight in the DI group and 0.032% in the LE group $P < .001$).

Discussion

Urolithiasis is one of the oldest diseases in recorded medicine.[4] Hippocrates (460–370 B.C.) described both renal and bladder calculi and was well aware of an association between stone formation and putrid urine.[5] He recommended against surgical removal of these stones (in apparent recognition of the high mortality of these early operations), favoring treatment with diuretics and increased fluid intake.[4, 5] Hippocrates recognized that elevation of fluid intake resulted in increased fluid output that he predicted would "wash away" the stones.[4, 5] This treatment, still practiced effectively today,[6] is in agreement with the present results. DI rat urine is undersaturated with respect to struvite and apatite. Little or no stone formation occurs in such urine, even when a foreign nidus is provided. Preliminary results suggest that sterile struvite stones slowly dissolve when implanted in the bladders of DI rats. On the other hand concentrated antidiuretic LE rat urine is saturated with respect to struvite and apatite as these solutes readily precipitate *in vivo* when a foreign nidus is provided to initiate crystal growth.

TABLE 1

BLADDER STONE FORMATION IN DI AND LE RATS

	Weight of Stone Formed (mg/6 months)		
	Nidus Control	Infected	p
DI (diuretic)	1.8 ± 1.7 (12)	949 ± 377 (11)	$p < .01$
LE (antidiuretic)	267 ± 84 (9)	575 ± 202 (5)	$p < .03$
	$p < .001$	ns	

Values are means (± SE) for groups containing (*N*) rats.

Struvite precipitation is commonly associated with urinary tract infection in humans and most experimental animals with the exception of the rat.[2, 3, 6] The present studies support previous findings of struvite stone formation in antidiuretic rats in the absence of frank infection. While a possible low level infection can not be excluded high urine concentration (1500–2000 mOsm) and relatively alkaline pH (>7.0, personal observation) are probably the most significant factors resulting in struvite precipitation in the LE nidus-control group.[6–8]

Chronic infection of bladder urine with *Proteus* markedly alters struvite and apatite solubility even in dilute urine. Large amounts of stone were formed in both infected DI and LE groups. Bacterial urease activity has been specifically linked to the formation of such "infection stones" in rats and man.[3] The splitting of urea increases both urinary pH and NH_4^+ levels, conditions that markedly reduce struvite and apatite solubility.[3, 7, 8] Preliminary studies suggest that *Proteus* infection alone (in the absence of a foreign nidus) can result in stone formation in DI rats (personal observation). We suspect a similar infection may have been responsible for spontaneous stone formation first observed in our colony.

Enlarged bladders in DI rats may also potentiate bacterial urolithiasis. *In vivo* these bladders may contain up to 10 ml of urine and may prolapse during voiding, resulting in incomplete emptying, and thus providing a continuous

TABLE 2

WEIGHTS OF URINARY BLADDERS IN DI AND LE RATS

	N	Wet Weight (mg)	% Weight (%)
DI (diuretic)	13	216 ± 13 *	0.095 ± 0.005 *
LE (antidiuretic)	25	111 ± 7	0.037 ± 0.003

Values are means (± SE) for freshly excised empty urinary bladders blotted dry prior to weighing (wet weight) and expressed as a percentage of body weight ((mg bladder ÷ mg body) × 100). Statistically significant differences between diuretic and antidiuretic groups are *, $p < .001$.

environment for both bacterial and stone growth. The etiology of bladder hypertrophy in DI rats is unknown. It probably results from attempts to maintain periodic voiding behavior in spite of the diabetes insipidus.

In conclusion the present studies illustrate the insidious stone-forming potential of urinary tract *Proteus* infection even in diabetes insipidus. In addition a new role for the DI rat is indicated as a useful model in experimental studies of urolithiasis.

Acknowledgments

We wish to thank Dr. D. P. Griffith (Urolithiasis Laboratory, Division of Urology, Baylor College of Medicine) for the crystallographic analyses reported in this paper. We also thank Mr. David Shier and Mr. Steven Kerr for their expert technical assistance, and Ms. Ada Luz for her assistance in the preparation of the manuscript.

References

1. Valtin, H. & H. A. Schroeder. 1964. Familial hypothalamic diabetes insipidus in rats. (Brattleboro Strain) Am. J. Physiol. **206:** 425–430.
2. Vermeulen, C. W., W. J. Grove, R. Goetz, H. D. Ragins & N. O. Correll. 1950. Experimental urolithiasas. I. Development of calculi upon foreign bodies surgically introduced into bladders of rats. J. Urol. **64:** 541–548.
3. Griffith, D. P. 1978. Struvite stones. Kidney Int. **13:** 372–382.
4. Butt, A. J. 1956. Historical survey of etiologic factors in renal lithiasis. *In* Etiologic Factors in Renal Lithiasis. A. J. Butt, Ed.: 3–47. Charles C. Thomas. Springfield, Ill.
5. Adams, F. 1929. The Genuine Works of Hippocrates. William Wood and Co., New York, N.Y.
6. Grove, W. J., C. W. Vermeulen, R. Goetz & H. D. Ragins. 1950. Experimental urolithiasis II. The influence of urine volume upon calculi experimentally produced upon foreign bodies. J. Urol. **64:** 549–554.
7. Elliot, J. S., R. F. Sharp & L. Lewis. 1959. The solubility of struvite in urine. J. Urol. **81**(3): 366–368.
8. Elliot, J. S., H. E. Todd & L. Lewis. 1961. Some aspects of calcium phosphate solubility. J. Urol **85**(4): 428–431.

USE OF THE BRATTLEBORO RAT IN STUDIES OF THE MECHANISM BY WHICH CHLORPROPAMIDE ENHANCES THE ACTION OF VASOPRESSIN *

Arnold M. Moses,† Rosemary Fenner, Edward T. Schroeder, and Richard Coulson

Veterans Administration Medical Center
and *Departments of Medicine and Pharmacology*
Upstate Medical Center
State University of New York
Syracuse, New York 13210

The antidiuretic action of arginine vasopressin (AVP) is enhanced by chlorpropamide (CP) treatment of the toad bladder,[1] rat,[2] and man.[3]

Using the homozygous Brattleboro rat, we have demonstrated that the injection of CP increases the antidiuretic responses to injected dDAVP[4] (TABLE 1). Since Brattleboro rats are devoid of AVP, the augmentation phenomena observed must be due to the enhanced activity of dDAVP and not to additive effects of endogenous AVP and injected dDAVP.[5]

Since the antidiuretic response to vasopressin is mediated by cyclic AMP, we determined the concentration of cyclic AMP in extracts of kidneys taken from Brattleboro rats and perfused *in vitro* for one hour with 10^{-8} M dDAVP. dDAVP significantly increased tissue content of cyclic AMP, and this was further increased by pretreatment with CP (TABLE 1).[4] Since tissue cyclic AMP content is increased by adenylate cyclase activity and reduced by phosphodiesterase, studies were then conducted on the effect of CP treatment on the activities of these enzymes. Injection of the rats with CP resulted in significantly greater adenylate cyclase activation by dDAVP (FIGURE 1). In contrast, phosphodiesterase activities of both supernatant and particulate fractions of renal homogenates were unaltered by CP.[4] The combination of increased dDAVP-stimulated adenylate cyclase activity in CP-treated rats with the lack of effect of CP on phosphodiesterase apparently accounts for the observations we made on the renal concentration of cyclic AMP. Since there is a vasopressin-sensitive adenylate cyclase in both the mammalian collecting tubule and in the thick ascending limb of the loop of Henle,[5] the effect of CP on enhancing vasopressin-stimulated adenylate cyclase could occur at either or both loci. CP treatment augmented adenylate cyclase activity of submaximal and maximal dDAVP concentrations, although its augmentation of the antidiuretic action of dDAVP occurred only at submaximal concentrations.[4] This probably reflects a rate-limiting step beyond the generation of cyclic AMP. CP or its metabolites may act directly on the kidney or indirectly by stimulating endogenous vasopressin release.[3] Since chronically elevated levels of vasopressin have been shown to augment the response of medullary adenylate cyclase to vasopressin

* Supported by Veterans Administration research funds and U.S. Public Health Service Grant 2RO1–AM–14401.

† Reprint requests should be addressed to: Dr. Arnold M. Moses, State University Hospital, 750 E. Adams St., Syracuse, N.Y. 13210.

0077–8923/82/0394–0513 $1.75/0 © 1982, NYAS

TABLE 1

EFFECT OF CHLORPROPAMIDE TREATMENT

	No Rx	CP	dDAVP	CP + dDAVP
U_{osm} (mOsm/kg)	180 ± 18	203 ± 14	326 ± 43 *	450 ± 31 ¶
$UPGE_2$ (ng/mg creatinine)	8.04 ± 2.55	6.67 ± 1.02	20.15 ± 3.90 *	39.63 ± 6.24 ‡
Renal cyclic AMP (nmol/g dry wt)	5.57 ± 0.68	5.25 ± 0.60	9.16 ± 0.16 †	14.0 ± 1.8 §

Effect of chlorpropamide treatment on six-hour urine osmolality and prostaglandin excretion following 1 ng dDAVP injection into homozygous Brattleboro rats. Renal cyclic AMP was determined after an hour of *in vitro* perfusion of kidneys with 10^{-8} M dDAVP. Chlorpropamide dosage was 20 mg/100 g/day for seven days. Chlorpropamide alone had no significant effects. dDAVP alone increased all three measurements, while dDAVP injected into rats treated with chlorpropamide resulted in higher values than dDAVP alone. Some of these data are from Moses & Coulson.[4]

* $p < .01$ vs. no treatment, † $p < .001$ vs. no treatment, ‡ $p < .05$ vs. dDAVP, § $p < .025$ vs. dDAVP, and ¶ $p < .02$ vs. dDAVP.

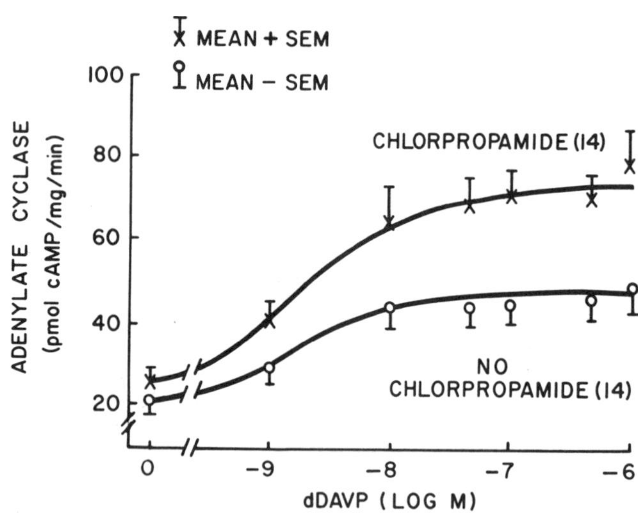

FIGURE 1. Dose response of adenylate cyclase activity to dDAVP stimulation in renal medullary cell membrane preparations obtained from homozygous Brattleboro rats, either untreated or injected witnh chlorpropamide, 20 mg/100 g/day for seven days. The chlorpropamide effect was statistically significant. (From Moses & Coulson.[4] With permission from *Endocrinology*.)

added *in vitro,*[6] this could have been a factor in our results if we had not used Brattleboro rats homozygous for diabetes insipidus.

In subsequent experiments, also using the Brattleboro rat, the *in vitro* addition of the post-receptor stimuli GTP, Gpp(NH)p, and F^- increased adenylate cyclase activity of renal medullary membranes, but these stimulated activities were not augmented by the injection of CP (FIGURE 2).

Vasopressin stimulates PGE synthesis in the toad bladder and the prostaglandins in turn inhibit the water permeability response to vasopressin.[7] In the same preparation, the *in vitro* addition of CP inhibited vasopressin-stimulated prostaglandin synthesis.[7] This was proposed as a mechanism for the CP-induced enhancement of water flow in response to vasopressin. Diabetes insipidus rats

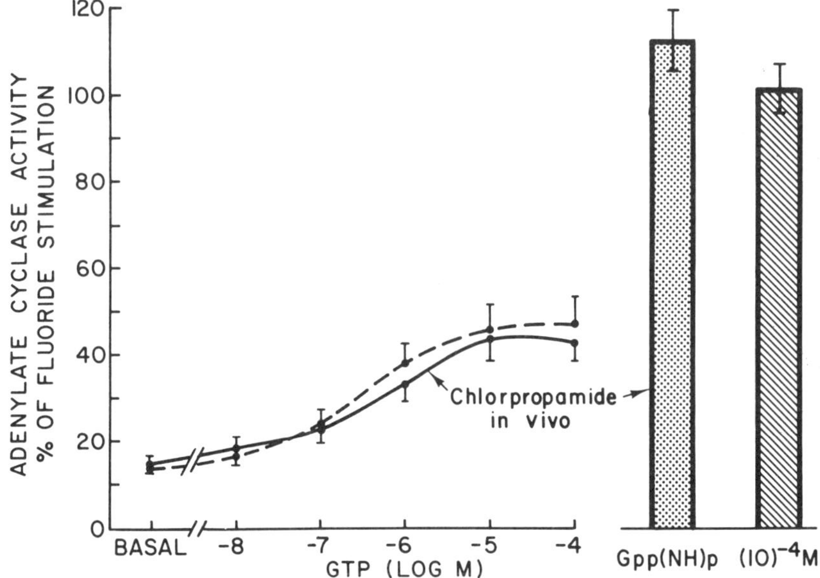

FIGURE 2. Lack of effect of injected chlorpropamide on renal medullary adenylate cyclase activity stimulated by addition of GTP or Gpp(NH)p. The F^- stimulated adenylate cyclase activities were 184 ± 12 in chlorpropamide-treated rats, and 162 ± 9 in controls. ATP concentration was 1 mM.

have a decreased urinary prostaglandin excretion that increases after the injection of Pitressin or dDAVP.[8] However, Dunn *et al.* were unable to show that treatment with CP altered the PGE_2 response to dDAVP, although the antidiuretic response was increased.[9] In these experiments both the CP and the dDAVP were administered in the drinking water. We again utilized the Brattleboro rat, injecting either CP, dDAVP, or control substances and confirmed previous observations that dDAVP increases urinary PGE_2 excretion (TABLE 1). However, in contrast to the observations of Dunn *et al.*, we found that injection of diabetes insipidus rats with CP significantly augmented the PGE_2 response to dDAVP (TABLE 1).

We believe, therefore, that CP enhances the renal responses to dDAVP in terms of the cyclic AMP and prostaglandin systems, while not altering responses to post-receptor stimuli of adenylate cyclase. Since the biochemical pathways for the cyclic AMP and PGE_2 systems are different, our observations are consistent with the thesis that CP acts at the receptor level to augment the action of vasopressin. Studies using radioactive vasopressin to determine the effect of the drug on hormone binding have not yet been conducted, but Olefsky and Reaven have shown that insulin binding to mononuclear cells of diabetic patients is increased when the patients are treated with CP.[10]

The increased PGE_2 synthesis by dDAVP serves a negative feedback function by modulating the action of vasopressin in the renal tubule.[9, 11] In our experiments the CP-treated rats developed a higher urine osmolality after dDAVP when treated with CP. This occurred with a greater activation of both the cyclic AMP and prostaglandin systems indicating that functionally the activation of the former predominated over the latter.

In conclusion, the injection of CP into Brattleboro rats with diabetes insipidus enhances the responses to dDAVP in terms of antidiuresis, activation of renal medullary adenylate cyclase activity, content of cyclic AMP, and PGE_2 excretion. In contrast, the injection of CP does not alter either GTP, Gpp(NH)p, or F^- stimulated adenylate cyclase activity of these renal membranes. CP, therefore, enhances the renal responses to dDAVP in terms of the cyclic AMP and prostaglandin systems, while not altering responses to post-receptor stimuli. These observations support the concept that CP (chlorpropamide) acts at the receptor level to augment the renal responses to vasopressin.

References

1. Mendoza, S. A. & C. F. Brown. 1974. Effect of chlorpropamide on osmotic water flow across toad bladder and the response to vasopressin, theophylline and cyclic AMP. J. Clin. Endocrinol. Metab. **38:** 883–889.
2. Miller, M. & A. M. Moses. 1970. Potentiation of vasopressin action by chlorpropamide in vivo. Endocrinology **86:** 1024–1027.
3. Moses, A. M., P. Numann & M. Miller. 1973. Mechanism of chlorpropamide induced antidiuresis in man: Evidence for release of ADH and enhancement of peripheral action. Metabolism **22:** 59–66.
4. Moses, A. M. & R. Coulson. 1980. Augmentation by chlorpropamide of 1-deamino-8-D-arginine vasopressin-induced antidiuresis and stimulation of renal medullary adenylate cyclase and accumulation of adenosine 3′,5′-monophosphate. Endocrinology **106:** 967–972.
5. Jard, S., D. Butlen, R. Rajerison & C. Roy. 1977. The vasopressin-sensitive adenylate cyclase from the mammalian kidney. *In* Neurohypophysis. A. M. Moses & L. Share, Eds.: 211–219. Karger. Basel.
6. Rajerison, R. M., D. Butlen & S. Jard. 1977. Effects of in vivo treatment with vasopressin and analogues on renal adenylate cyclase responsiveness to vasopressin stimulation in vitro. Endocrinology **101:** 1–12.
7. Zusman, R. M., H. R. Keiser & J. S. Handler. 1977. Inhibition of vasopressin-stimulated prostaglandin E biosynthesis by chlorpropamide in the toad urinary bladder. Mechanism of enhancement of vasopressin-stimulated water flow. J. Clin. Invest. **60:** 1348–1353.
8. Walker, L. A., A. R. Whorton, M. Smigel, R. France & J. C. Frölich. 1978. Antidiuretic hormone increases renal prostaglandin synthesis in vivo. Am. J. Physiol. **235:** F180–F185.

9. DUNN, M. J., L. B. KINTER, R. BEEUWKES, D. SHIER, H. P. GREELEY & H. VALTIN. 1980. Interaction of vasopressin and renal prostaglandins in the homozygous diabetes insipidus rat. *In* Advances in Prostaglandin and Thromboxane Research. B. Samuelsson, P. W. Ramwell & R. Paoletti, Eds.: 1009–1015. Raven Press. New York, N.Y.
10. OLEFSKY, J. M. & G. M. REAVEN. 1976. Effect of sulfonylurea therapy on insulin binding to mononuclear leukocytes of diabetic patients. Am. J. Med. **60:** 89–95.
11. EDWARDS, R. M., B. A. JACKSON & T. P. DOUSA. 1981. ADH-sensitive cAMP system in papillary collecting duct: effect of osmolality and PGE_2. Am. J. Physiol. **240:** F311–F318.

VASOPRESSIN AND INTRAMEMBRANOUS PARTICLE CLUSTERS IN COLLECTING DUCT CELLS OF BRATTLEBORO AND LONG-EVANS RATS *

Paul Stern, Mehmet C. Harmanci, and Brian R. Edwards

Departments of Maternal and Child Health and Physiology
Dartmouth Medical School
Hanover, New Hampshire 03755

and

Renal Service
Public Health Service Hospital
Staten Island, New York 10304

Introduction and Methods

Intramembranous particle clusters (IPC), revealed by freeze-fracture electron microscopy, are induced by vasopressin (VP) in the luminal membrane of collecting duct cells (CDLM) of the rat.[1] This is the membrane where permeability to water increases in response to the hormone.[2] We studied the relationship between VP, IPC, and urine osmolality (U_{Osm}) in Brattleboro homozygous (DI) and Long-Evans (LE) rats. In DI rats, the effect of infusion of VP, and of changes in water balance in the absence of VP, were examined. In LE rats, maneuvers to either suppress or release endogenous VP were employed.

Three experimental protocols were carried out. The first one, which we have described in detail previously,[3] involved infusion of a range of doses of VP (Arginine Vasopressin, Pierce Chemical Co., Rockford, Ill.) into a jugular vein of DI rats anesthetized with Inactin (Henley and Co., New York, N.Y.). The doses of VP ranged from subminimal to supramaximal, as assessed by U_{Osm}. From each rat, a kidney was removed before starting the infusion of VP, and the contralateral kidney was removed during infusion, once U_{Osm} had reached an apparent plateau. In other rats, the first kidney was removed 20 min after starting infusion of VP, and the contralateral kidney was removed after stopping VP, when U_{Osm} had returned to pre-infusion values.

In the second protocol, both kidneys were removed one hour after an oral water load equivalent to 3% of body weight had been given to unanesthetized DI or LE rats. In the third protocol, DI and LE rats were kept in metabolic cages without food or water for 24 hours before removal of the kidneys. Rats in the second and third protocols were induced to void, and then killed by a sharp blow to the head immediately prior to nephrectomy.

Urine osmolalities were measured by vapor pressure osmometry (Wescor Osmometer, model 5100B, Wescor, Inc., Logan, Utah). The freeze-fracture procedure has been described in detail before.[1, 3] In brief, a kidney removed after a given control or experimental period was immediately sliced to expose

* Supported in part by Grants HD–12123, AM–26553, AM–18710, and AM–08469 from the National Institutes of Health.

0077–8923/82/0394–0518 $1.75/0

the papilla and fixed in buffered glutaraldehyde. A small portion of the papilla was frozen and fractured in a Balzers freeze-etch unit (BAF 301, Balzers Corp., Nashua, N.H.). The exposed fracture surfaces were coated with platinum at an angle (shadowing), followed by perpendicular coating with carbon. Electron micrographs of the resulting replicas were obtained. Replicas from 20 collecting duct cells of each kidney were examined and the results averaged; the number of IPC was counted, and the area of membrane occupied by the IPC was measured. The data reported are from examination of the E-face replicas (clusters induced by VP in CDLM occupy both complementary fracture faces: exoplasmic or E-face, and protoplasmic or P-face).

The electron micrographs of the replicas were examined without knowledge of the particular experimental maneuver involved.

Results and Discussion

The morphology of the IPC can be seen in Figure 1b. The IPC seen in LE rats had a similar conformation. We have previously reported the results obtained with infusion of VP into DI rats.[3] In brief, a dose-related increase in frequency of IPC (number of IPC/100 μm^2 CDLM) was observed that paralleled the increase in U_{Osm}. A plateau in frequency of, and area occupied by, IPC was reached at VP doses calculated to resemble rates of pituitary release of VP in dehydrated normal rats (~36 pg/min·100 g body weight).[4] The relationship between the frequency of IPC and area of CDLM occupied by them can be seen in Figure 2. The maximum area of CDLM occupied by all the clusters was between 6 and 7%. While there was a direct relationship between frequency and total area occupied, the mean size of individual clusters did not vary with dose of VP. This is consistent with what has been observed in amphibian urinary bladder, where IPC, although structurally different from those in rat CDLM,[1] appear in luminal membranes of granular cells upon stimulation by VP.[5] In the toad bladder there is evidence that the IPC are "pre-formed." They seem to be inserted in the luminal bilayer by fusion of cytoplasmic membrane structures—that contain the pre-formed IPC—with the luminal membrane after stimulation by VP.[6]

The change in frequency of IPC in the DI rat infused with VP not only paralleled the change in U_{Osm}, but did so following a similar time course, as can be seen in Figure 3. Whether this implies that the VP-induced change in water permeability of collecting duct cells is dependent on the IPC response produced by the hormone cannot be established by these experiments. In toad bladder, however, the IPC in granular cells have been specifically linked to the change in water permeability produced by VP in that epithelium.[7]

The results of the water loading and dehydration experiments can be seen in Figure 4. After 24 hours of dehydration, LE rats increased U_{Osm} (to 2419 ± 105 mosmol/kg H_2O, $N = 4$) and IPC frequency (to 115 ± 12). In comparison, the maximum IPC frequency observed in DI rats infused with VP was 111. LE rats given an oral water load equivalent to 3% of body weight had a U_{Osm} of 218 ± 30 ($N = 4$) after one hour, and IPC frequency of 4.6 ± 0.6. In DI rats, U_{Osm} increased to 993 ± 149 mosmol/kg H_2O ($N = 4$) after 24 hours of dehydration. DI rats had a weight loss of $17.8 \pm 0.6\%$, significantly more than the weight loss in LE rats after 24 hours of dehydration ($10.0 \pm 1.2\%$). IPC frequency in the dehydrated DI rats was 5.9 ± 0.7, a

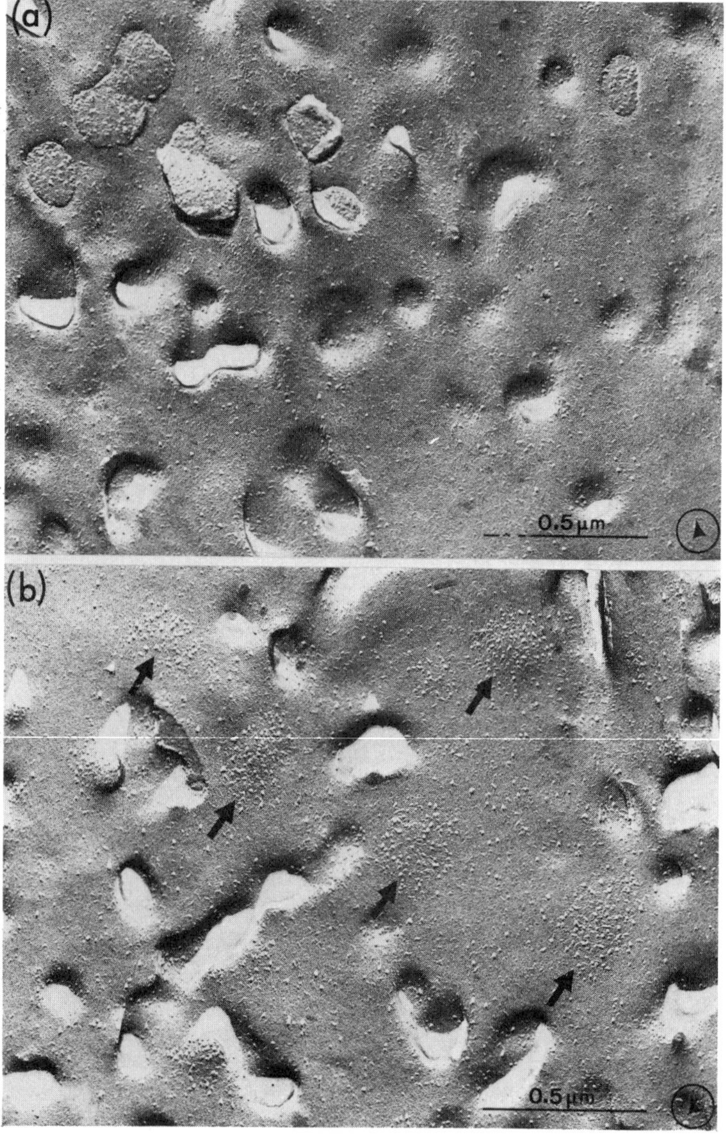

FIGURE 1. (a) E-face replica of luminal membrane of a collecting duct cell (CDLM) of a kidney removed from an untreated Brattleboro homozygous (DI) rat. No intramembranous particle clusters (IPC) are present. (b) E-face replica of CDLM from DI rat treated with vasopressin. Numerous IPC can be seen, indicated by arrows. Arrowheads point in the direction of shadowing.

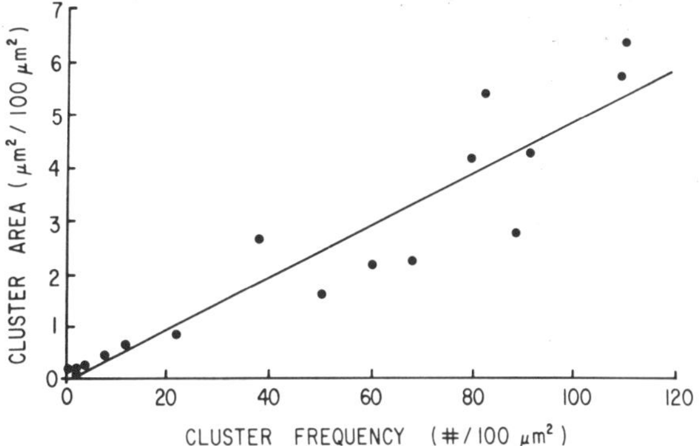

FIGURE 2. Relationship between the frequency of IPC observed in CDLM of kidneys removed while the DI rats received different infusion rates of VP, and the total area of CDLM occupied by the clusters ($y = 0.127 + 0.049x$; $r = 0.94$; $p < 0.001$; $N = 17$).

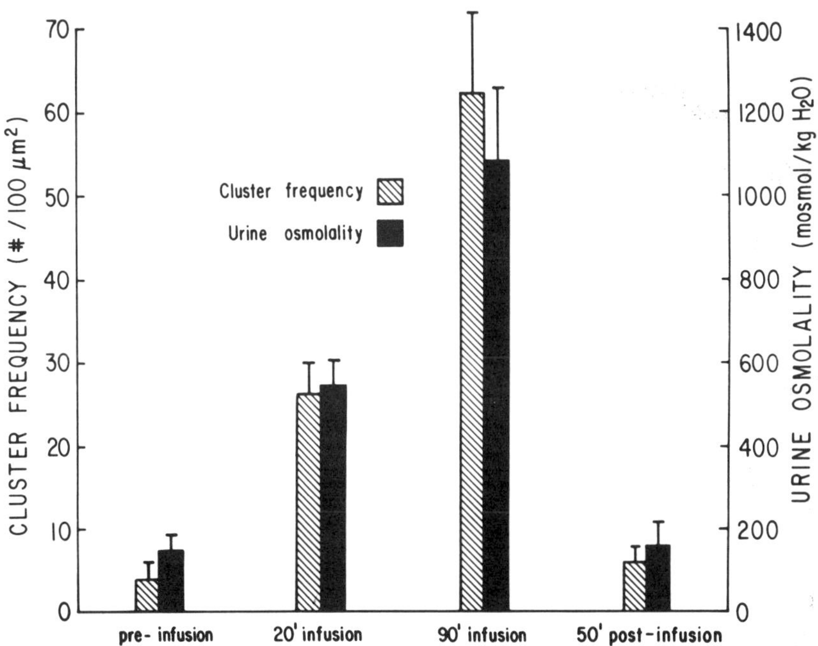

FIGURE 3. Cluster frequency and urine osmolality obtained just prior to removal of the kidney, observed in DI rats before starting an infusion of VP (column 1), after 20 min and 90 min of an infusion of 9 pg of VP/min·100 g body weight (columns 2 and 3, respectively), and 50 min after stopping VP infusion (column 4). Columns 1 and 3 correspond to one set of rats ($N = 3$), and columns 2 and 4 to another ($N = 5$).

value not different from DI rats given a water load (2.9 ± 1, $N = 4$) or water ad libitum (3 ± 1, $N = 4$).

In conclusion, endogenous VP release (dehydrated LE rats) elicited the same morphological response in CDLM as did infusion of VP in DI rats. There is a close association between the morphological change (IPC frequency) and the functional change (U_{Osm}) induced by VP in the LE rat in different conditions of water balance, and in the DI rat given different doses of VP. The DI rat can concentrate urine during dehydration[8] by a mechanism not involving VP or the IPC associated with this hormone.

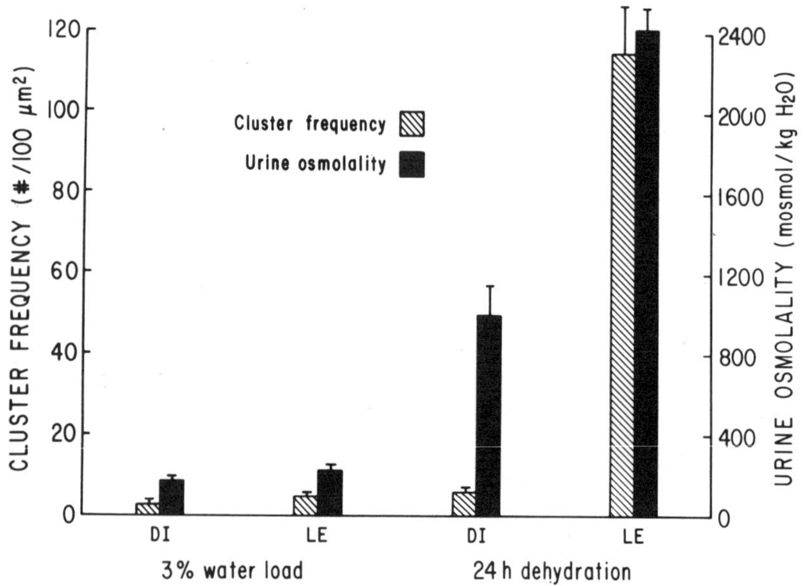

FIGURE 4. Cluster frequency and urine osmolality in kidneys of Brattleboro homozygous rats (DI) or Long-Evans rats (LE) given a water load equivalent to 3% body weight or dehydrated for 24 hours.

ACKNOWLEDGMENT

We thank Ethel B. Garrity for secretarial assistance.

REFERENCES

1. HARMANCI, M. C., W. A. KACHADORIAN, H. VALTIN & V. A. DISCALA. 1978. Antidiuretic hormone-induced intramembranous alterations in mammalian collecting ducts. Am. J. Physiol. **235:** F440–F443.
2. GRANTHAM, J. J. 1974. Action of antidiuretic hormone in the mammalian kidney. *In* Kidney and Urinary Tract Physiology. I. K. Thurau, Ed. Vol. **6:** 247–272. University Park Press. Baltimore, Md.

3. HARMANCI, M. C., P. STERN, W. A. KACHADORIAN, H. VALTIN & V. A. DISCALA. 1980. Vasopressin and collecting duct imtramembranous particle clusters: a dose-response relationship. Am. J. Physiol. **239:** F560–F564.
4. VALTIN, H., J. STEWART & H. W. SOKOL. 1974. Genetic control of the production of posterior pituitary principles. Handb. Physiol. Sec. 7, Vol. IV, Part 1: 131–171.
5. KACHADORIAN, W. A., J. B. WADE & V. A. DISCALA. 1975. Vasopressin-induced structural change in toad bladder luminal membrane. Science **190:** 67–69.
6. MULLER, J., W. A. KACHADORIAN & V. A. DISCALA. 1980. Evidence that ADH-stimulated intramembrane particle aggregates are transferred from cytoplasmic to luminal membranes in toad bladder epithelial cells. J. Cell Biol. **85:** 83–95.
7. KACHADORIAN, W. A., S. D. LEVINE, J. B. WADE, V. A. DISCALA & R. M. HAYS. 1977. Relationship of aggregated intramembranous particles to water permeability in vasopressin-treated toad urinary bladder. J. Clin. Invest. **59:** 576–581.
8. GELLAI, M., B. R. EDWARDS & H. VALTIN. 1979. Urinary concentrating ability during dehydration in the absence of vasopressin. Am. J. Physiol. **237:** F100–F104.

HOMOZYGOUS BRATTLEBORO RATS LACK NORMAL NEPHRON HETEROGENEITY AS A CONSEQUENCE OF THEIR URINE CONCENTRATING DEFECT *

Marie-Marcelle Trinh-Trang-Tan,† Hilda Weyl Sokol,‡
Lise Bankir,† and Heinz Valtin ‡

† *Institut National de la Santé et de la Recherche Médicale U90*
Hôpital Necker F 75730
Paris Cedex 15 France

‡ *Department of Physiology*
Dartmouth Medical School
Hanover, New Hampshire 03755

INTRODUCTION

Numerous studies have shown that in most mammalian species all nephrons are not identical. Several anatomical and functional differences exist between superficial and deep (or juxtamedullary) nephrons. In particular, deep nephrons have a larger glomerulus, a longer proximal tubule, and a higher glomerular filtration rate than do superficial nephrons.[1, 2] Moreover, a few studies have shown that these differences are particularly pronounced in desert-adapted rodents, which exhibit a high urine-concentrating ability.[3, 4]

We have recently studied nephron heterogeneity in Brattleboro rats.[5] Homozygous rats exhibit hereditary diabetes insipidus (DI) due to their inability to synthesize antidiuretic hormone (ADH), while heterozygous (HZ) rats have a nearly normal concentrating ability. DI rats were found to have considerably lower nephron heterogeneity than HZ rats. The synthetic analogue of ADH, dDAVP, given daily for 6–8 weeks, restored normal urine osmolality and induced nearly normal nephron heterogeneity in DI rats (see Reference 5 and FIGURE 1). It cannot be determined from these observations whether the absence of nephron heterogeneity is due to the lack of ADH per se or to its consequences, namely, the defective urine-concentrating ability.

Mice with nephrogenic DI are a good model to study this question. Despite normal or high vasopressin plasma levels, these mice are unable to concentrate urine because of their renal unresponsiveness to the hormone.[6–8] We, therefore, studied the degree of heterogeneity between superficial and juxtamedullary nephrons in this strain of mice. If vasopressin is responsible for the nephron heterogeneity found in normal mammals, this heterogeneity should be present in DI mice. If, on the other hand, this heterogeneity depends on the high osmotic pressure occurring in the papilla during antidiuresis, DI mice should lack nephron heterogeneity, as do Brattleboro DI rats.

* This work was supported in part by The Institut National de la Santé et de la Recherche Médicale (A.T.P. No. 79110); The Délégation Générale à la Recherche Scientifique et Technique (A.C. No. 77.7.1395); The Hitchcock Foundation; U.S. Public Health Service Research Grant AM–08469; and the Philippe Foundation.

0077–8923/82/0394–0524 $1.75/0 © 1982, NYAS

MATERIAL AND METHODS

Animals were adult mice of both sexes, 2–3 months old. The study included six DI +/+ severe and six DI +/+ non-severe, which served as controls.[7]

Experimental procedure was that described by Hanssen, using sodium ferrocyanide as a glomerular filtration rate marker.[9] After urine collection, mice were anesthetized with pentobarbital and a 50 μl bolus of 15% ^{14}C-labeled sodium ferrocyanide (25 μCi) was injected via the jugular vein (New England Nuclear). Fifteen seconds after this injection, the mice were sacrificed by

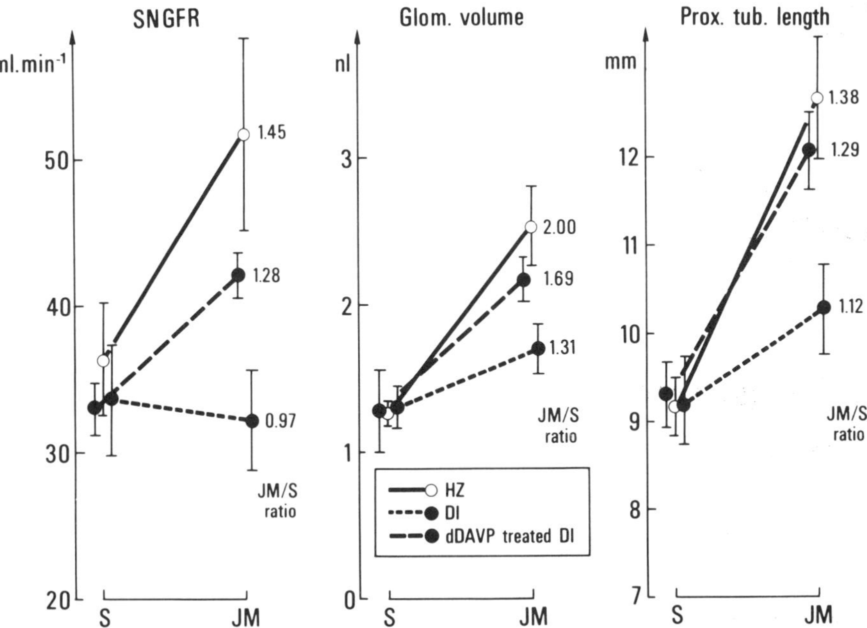

FIGURE 1. Mean single nephron glomerular filtration rate (SNGFR), glomerular volume, and proximal tubular length in HZ (heavy line, $N = 4$), DI (dotted line, $N = 6$) and chronically dDAVP-treated DI Brattleboro rats (broken line, $N = 6$). Vertical bars represent 1 SE. JM/S ratios are indicated on the right side of each graph. (Modified from Reference 5.)

cutting the body between the heart and the renal vessels, and the kidneys were quickly removed and frozen in a mixture of dry ice and acetone. Precipitation of ferrocyanide into Prussian Blue was performed overnight at −20° C in 15% ferric chloride in absolute ethanol. Kidneys were then macerated in 20% hydrochloric acid for 150 min at 37° C. They were stored until microdissection in 0.2% ferric chloride in 1% acetic acid. Proximal tubules and the attached glomeruli were microdissected from 10 superficial and 10 juxtamedullary nephrons in each mouse. They were spread on plastic strips and drawn with a known magnification. Proximal tubular length (×107) and glomerular diameter (×430) were measured on the drawings. Assuming spherical glomeruli, glomerular volumes were calculated from glomerular diameters. The radio-

TABLE 1

PROXIMAL TUBULAR LENGTH, GLOMERULAR VOLUME, AND GLOMERULAR FILTRATION RATE IN SUPERFICIAL (S) AND JUXTAMEDULLARY (JM) NEPHRONS OF DI +/+ NON-SEVERE AND DI +/+ SEVERE MICE

	Proximal Tubule Length mm			Glomerular Volume nl			Glomerular Filtration Rate ^{14}C Counts		
Mice	S	JM	JM/S	S	JM	JM/S	S	JM	JM/S
				DI +/+ Non-Severe					
1	6.21	9.46	1.52	0.52	0.67	1.29	2806	3429	1.22
2	5.50	8.92	1.62	0.45	0.78	1.73	2340	3178	1.36
3	6.87	10.32	1.50	0.71	1.04	1.46	2468	3748	1.52
4	8.90	10.90	1.22	1.15	1.53	1.33	2740	3280	1.20
5	5.45	8.54	1.57	0.63	1.00	1.58	2580	3395	1.32
6	6.45	9.26	1.44	0.34	0.71	2.09	2884	4564	1.58
Means	6.56	9.57	1.48	0.63	0.95	1.58	2636	3599	1.37
± SE	0.52	0.36	0.07	0.12	0.13	0.12	86	208	0.06
				DI +/+ Severe					
1	6.96	8.25	1.19	0.41	0.54	1.36	3357	3404	1.01
2	6.68	8.74	1.31	0.45	0.65	1.46	3065	2945	0.96
3	6.74	8.56	1.27	0.64	1.03	1.68	3060	4509	1.47
4	5.01	7.25	1.45	0.57	0.97	1.71	2655	3065	1.15
5	6.53	7.71	1.18	0.92	1.36	1.47	2476	2236	0.90
6	6.42	7.75	1.21	0.56	0.66	1.18	2616	2700	1.03
Means	6.39	8.04 *	1.27 *	0.59	0.87	1.46	2872	3143	1.09 *
± SE	0.29	0.23	0.04	0.08	0.13	0.08	138	316	0.08

* Indicates significant difference between DI +/+ severe and DI +/+ non-severe (Student's *t* test $p < 0.02$).

Each value is the mean of 10 microdissected nephrons.

activity of each tubule was counted for 50 min in a Packard Tri-Carb liquid scintillation counter. Urine osmolalities were measured with a vapor pressure Wescor osmometer (model 5100B).

RESULTS

For DI +/+ non-severe and DI +/+ severe, respectively, mean body weights were 27.3 ± 1.4 (SE) g and 23 ± 1.6 g, and mean urine osmolalities were 1,769 ± 267 mOsm/kg H_2O and 296 ± 63 mOsm/kg H_2O.

The lengths of proximal tubules and the glomerular volumes of the superficial and juxtamedullary microdissected nephrons are given in TABLE 1. The amounts of radioactivity found in these nephrons are shown in the same table. They represent a relative measurement of the glomerular filtration rate. In control mice (DI +/+ non-severe), marked differences were observed between superficial and juxtamedullary nephrons for both anatomical and functional parameters. JM/S ratios, considered as an index of nephron heterogeneity, were 1.48 ± .07, 1.58 ± .14, and 1.37 ± .06 for proximal tubular length, glomerular volume, and glomerular filtration rate, respectively. In contrast, in mice with severe DI, the difference between S and JM nephrons was less marked. The JM/S ratios were significantly lower than in control mice for the length of the proximal tubule and for the glomerular filtration rate. The JM/S ratio for the volume of the glomeruli was also lower, although the difference between the two groups of mice did not reach significance (TABLE 1).

DISCUSSION

Our previous study on DI Brattleboro rats showed that lack of ADH and/or the diuretic status is accompanied by absence of both anatomical and functional nephron heterogeneity. This was seen to be mainly due to the small size and low filtration of juxtamedullary nephrons (FIGURE 1). The purpose of the present study was to assess the role of antidiuretic hormone in the causation of nephron heterogeneity.

Like DI Brattleboro rats, DI mice have a very high flow rate of hypo-osmotic urine, but unlike these rats, they do have endogenous ADH. The results we obtained in these mice show that, like DI Brattleboro rats, they lack glomerular filtration rate heterogeneity and have reduced anatomical heterogeneity. As in the rats, the proximal tubular length of the juxtamedullary nephrons in these mice is significantly reduced (TABLE 1). FIGURE 2 summarizes results obtained in both rats and mice. In all DI animals, whether

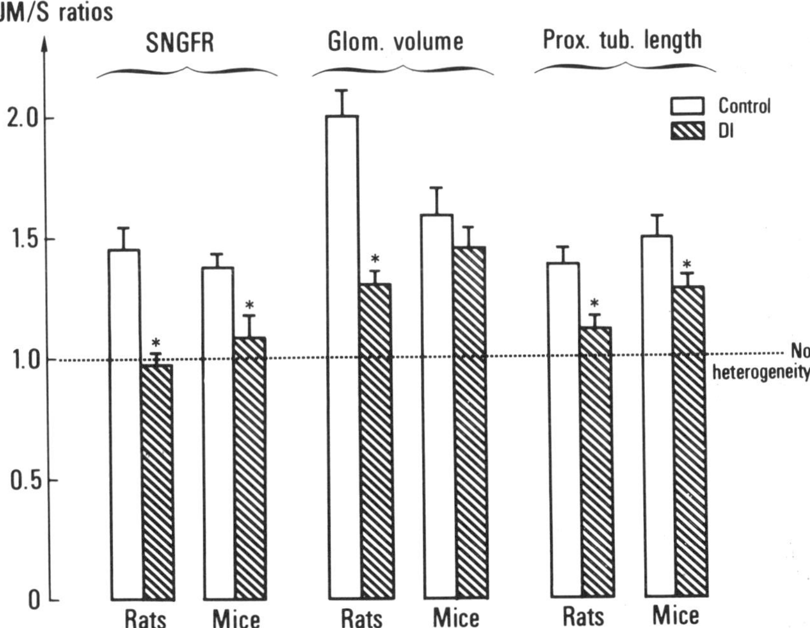

FIGURE 2. Mean JM/S ratios for SNGFR, glomerular volume, and proximal tubular length in control rats or mice (open bar) and in DI rats or mice (striped bar). Asterisks indicate a significant difference between DI animals versus control animals (Student's *t* test $p < 0.05$ or less).

with ADH (mice) or without ADH (rats), JM/S ratios are clearly lower than in control animals. It can be concluded that ADH does not act directly on juxtamedullary nephrons to induce nephron heterogeneity. Rather, the high osmotic pressure that occurs in the inner medulla of animals with normal urine concentrating ability may determine nephron heterogeneity in mammals, as suggested by Davis and Schnermann.[10]

References

1. Lameire, N. H., M. D. Lifschitz & J. H. Stein. 1977. Heterogeneity of nephron function. Ann. Rev. Physiol. **39:** 159–184.
2. Valtin, H. 1977. Structural and functional heterogeneity of mammalian nephrons. Am. J. Physiol. **233:** F491–F501.
3. Munkacsi, I. & M. Palkovits. 1965. Volumetric analysis of glomerular size in kidneys of mammals living in desert, semidesert or water rich environment in the Sudan. Circ. Res. **17:** 303–311.
4. Altschuler, E. M., R. B. Nagle, E. J. Braun, S. L. Lindstedt & P. H. Krutzsch. 1979. Morphological study of the desert Heteromyid kidney with emphasis on the Genus Perognathus. Anat. Rec. **194:** 461–463.
5. Trinh-Trang-Tan, M. M., J. P. Grünfeld, M. Diaz & L. Bankir. 1981. ADH-dependent nephron heterogeneity in rats with hereditary hypothalamic diabetes insipidus. Am. J. Physiol. **240:** F372–F380.
6. Naik, D. V. & H. W. Sokol. 1970. The hypothalamo-hypophysial neurosecretory system in mice mith vasopressin-resistant urinary concentrating defects. Gen. Comp. Endocrinol. **15:** 59–69.
7. Kettyle, W. M. & H. Valtin. 1972. Chemical and dimensional characterization of the renal countercurrent system in mice. 1972. Kidney Int. **1:** 135–144.
8. Sokol, H. W. 1979. Diabetes Insipidus. *In* Spontaneous Animal Models of Human Disease. E. J. Andrews, B. C. Ward & N. H. Altman, Eds. Vol. 2: 293–295. Academic Press. New York, N.Y.
9. Hanssen, O. D. 1961. The relationship between glomerular filtration and length of the proximal convoluted tubule in mice. Acta Pathol. Microbiol. Scand. **53:** 265–279.
10. Davis, J. M. & J. Schnermann. 1971. The effect of antidiuretic hormone on the distribution of nephron filtration rates in rats with hereditary diabetes insipidus. Pflügers Arch. **330:** 323–334.

CHANGES IN THE CORTICOPAPILLARY OSMOTIC GRADIENT DURING PROLONGED DEHYDRATION OF THE BRATTLEBORO RAT *

Gail L. Williamson and Brian R. Edwards †

Department of Physiology
Dartmouth Medical School
Hanover, New Hampshire 03755

INTRODUCTION

Because of the absence of vasopressin, there normally exists a considerable osmotic disequilibrium between the urine and the papillary interstitium of the Brattleboro homozygote (DI rat).[1, 2] Nevertheless, when these animals are deprived of drinking water, they demonstrate substantial increases in urine osmolality.[1, 2] Gellai and coworkers[2] found that glomerular filtration rate (GFR) of DI rats decreased progressively during prolonged dehydration, and they suggested the so-called Berliner-Davidson hypothesis[3] as a possible explanation for the concomitant rise in urine osmolality.

Whatever the step initiating the increase in concentrating ability of DI rats during dehydration, ultimately the increase in urine osmolality must follow from a relative increase in reabsorption of water from the collecting ducts. In the absence of vasopressin, there are limited possibilities to account for this increased reabsorption. These may include, first, an augmented driving force for water reabsorption as a result of an increase in the magnitude of the corticopapillary interstitial osmotic gradient. Second, there may be increased osmotic equilibration across the collecting ducts as a result of prolonged contact time, without any alteration in water permeability. The major objective of the present study was to determine the relative importance of these two mechanisms during different stages of dehydration of the DI rat.

Gellai and coworkers found that while urine osmolality rose progressively throughout 24 hours of dehydration, papillary osmolality increased only during the first nine hours.[2] The nature of the solutes in the medullary interstitium of DI rats has only been examined after 12 hours of dehydration;[1] at this time the rise in medullary osmolality was principally the result of increased concentration and content of urea.

Thus, in the present study, changes in the entire corticopapillary osmotic gradient were quantified after 12, 24, and 48 hours of dehydration, and the relative contributions of urea and non-urea solute were assessed. Also, the degree of osmotic equilibration between collecting duct fluid and papillary interstitium was determined at each time point by calculation of the urine/papillary tip osmolality ratio. Gellai and colleagues reported a substantial increase in osmotic equilibration by the 24th hour of dehydration and suggested this resulted from decreased flow rate of fluid in the collecting ducts.[2] This

* Supported by Research Grant AM 26553 and by National Research Service Award AM 07301 from the National Institutes of Health.

† Correspondence to B. R. Edwards.

posed the following question: could dehydration for 48 hours, when presumably GFR and collecting duct fluid flow rate are even lower than at 24 hours, result in complete osmotic equilibration, even in the absence of vasopressin?

Methods

Experiments were performed on 23 DI rats of both sexes that were between 12 and 16 weeks of age (163–359 g). Animals were placed in metabolism cages (with free access to food) and were deprived of drinking water for 0, 12, 24, or 48 hours. After each period of dehydration, urine samples were obtained and the rats then sacrificed by cervical dislocation. The renal hila were clamped with hemostats and both kidneys were excised immediately. A column of tissue was cut from the core of each kidney so that pieces of cortex, outer medulla, inner medulla, papillary base, and papillary tip could be sliced easily and rapidly. The tissue slices were ultracentrifuged under water-equilibrated paraffin oil at 15,000 *g* and $-1°$ C for 30 minutes. This procedure yielded approximately 0.5–2 μl of fluid that had exuded from the tissue. Osmolality of this fluid was determined with a freezing-point depression nanoliter osmometer (Clifton Technical Physics, New York, N.Y.) and urea concentrations of this fluid and of urine were determined by a colorimetric assay.[4] A vapor pressure osmometer (Wescor, Logan, Ut.) was used to measure urine osmolality.

Tissue water contents were determined on similar slices obtained from kidneys of a separate, but identically treated, series of 12 rats. The slices were quickly weighed in pre-tared aluminum foil envelopes, dried to constant weight, and the percent water content calculated. As judged from the percent body weight loss and from the increase in urine osmolality, both series of rats reacted to identical periods of dehydration in a similar fashion. Therefore, the values of percent water content were applied to the results from the first series of rats in order to estimate the tissue content of the different solutes.

Results and Discussion

The entire corticopapillary interstitial osmotic gradient increased dramatically during the first 12 hours of dehydration (Figure 1A). In agreement with Valtin,[1] we found that this was largely attributable to a rise in the concentration and content of urea (Figures 1B and 2B), since the percent of total tissue solute contributed by urea increased significantly in each tissue region during this time period (Figure 3). The accumulation of urea was greatest in the inner medulla and papilla. Presumably, as a consequence of decreased GFR and intratubular flow rate, fractional reabsorption of water in earlier segments of the nephron increases, raising the concentration of urea in collecting duct fluid, and thus enhancing the diffusion of urea into the medullary interstitium. Indeed, Edwards found that fractional reabsorption of urea was elevated in DI rats dehydrated for just three hours.[5]

The concentration and content of non-urea solute in inner medulla and papilla rose substantially during the first 12 hours of dehydration (Figures 1C and 2C). Although differing from the findings of Valtin,[1] these data are consistent with the passive model of countercurrent multiplication in the inner medulla.[6] Thus, following an increase in interstitial urea concentration in the

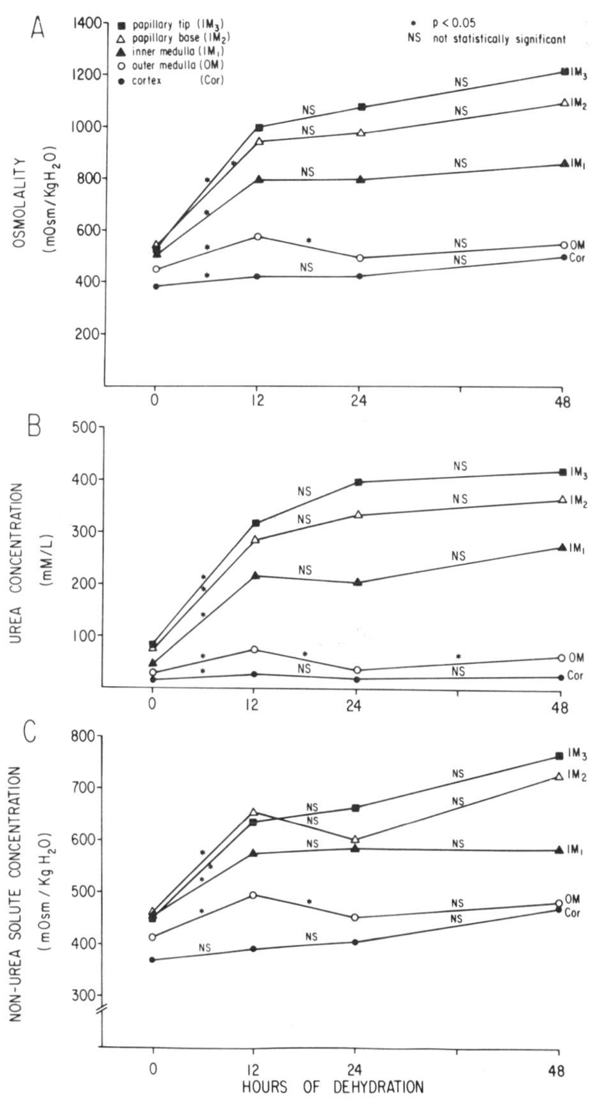

FIGURE 1. (A) Tissue osmolality, (B) tissue urea concentration, and (C) tissue non-urea solute concentration in DI rats dehydrated for 0, 12, 24, and 48 hours.

inner medulla, there should be greater osmotic extraction of water from the descending limbs of the loops of Henle of deep nephrons, concentrating the NaCl within the loops and thereby increasing the driving force for passive reabsorption of NaCl from the thin ascending limbs into the inner medullary interstitium. In addition, there may be a decrease in medullary blood flow during dehydration that would enhance the concentration of both urea and non-urea solute.

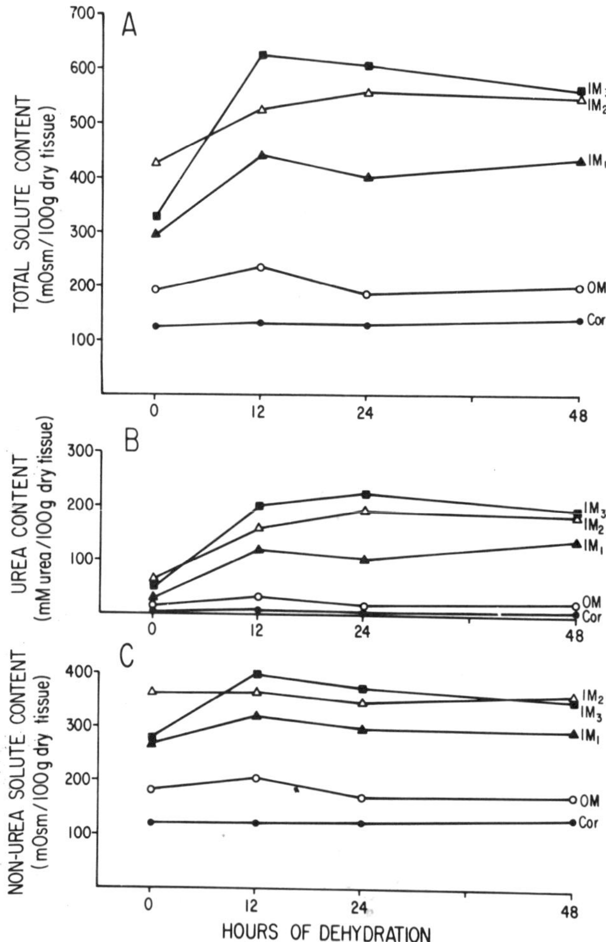

FIGURE 2. (A) Tissue solute content, (B) tissue urea content, and (C) tissue non-urea solute content in DI rats dehydrated for 0, 12, 24, and 48 hours.

Levinsky and coworkers reported that, in the presence of vasopressin, large decrements in GFR compromised the development of the corticopapillary gradient, presumably because of decreased delivery of NaCl to the loops of Henle.[7] This might explain our finding that the concentrations of total solute, urea, and non-urea solute failed to rise significantly between the 12th and 24th

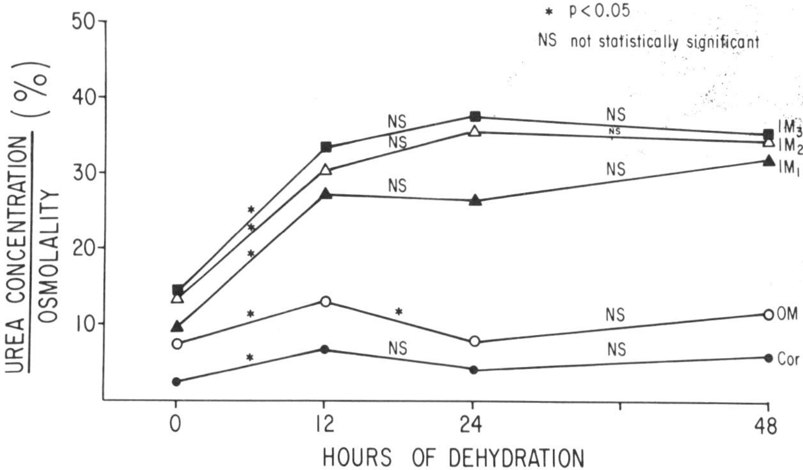

FIGURE 3. Change in the tissue urea concentration/osmolality ratio with increasing length of dehydration.

and between the 24th and 48th hours of dehydration (FIGURES 1A, B and C). GFR is known to fall markedly between the 12th and 24th hours of dehydration of DI rats,[2] and thus might counterbalance the forces favoring further enhancement of the corticopapillary gradient.

The degree of osmotic equilibration of collecting duct fluid with papillary interstitium increased dramatically after 12, 24, and 48 hours of dehydration (FIGURE 4B). Presumably the enhanced equilibration resulted from decreased flow rate in the collecting ducts, which would increase the contact time of

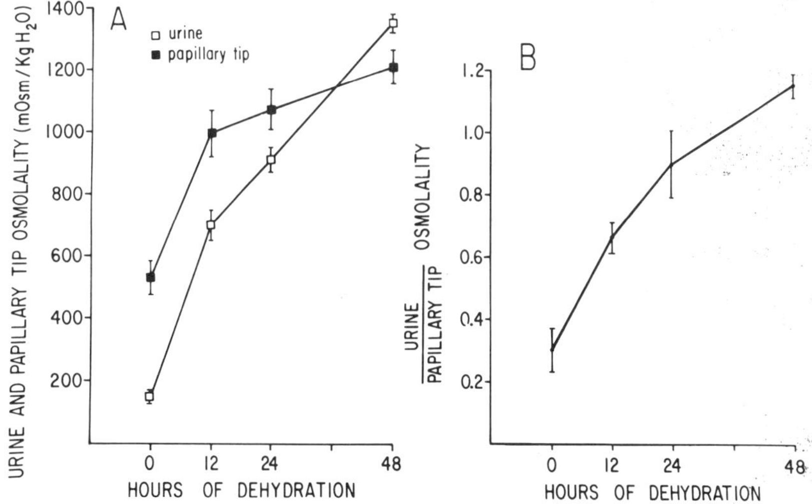

FIGURE 4. (A) Papillary tip and urinary osmolalities, and (B) the urine/papillary tip osmolality ratio in DI rats dehydrated for 0, 12, 24, and 48 hours.

tubular fluid with surrounding interstitium. Indeed, by the 48th hour of dehydration, when urine osmolality averaged 1,348 mOsm/kg H_2O, osmotic equilibration between collecting duct fluid and papillary interstitium was essentially complete.

In conclusion, it appears that the increase in urine osmolality during the first 12 hours of dehydration was due to both an increase in the corticopapillary interstitial osmotic gradient and to increased osmotic equilibration between collecting duct fluid and papillary interstitium, while the increase in urine osmolality from the 12th to 48th hour of dehydration was due primarily to increased osmotic equilibration, probably as a result of decreased tubular flow rate.

Acknowledgment

The expert assistance of Ms. Betsy Craig is gratefully acknowledged.

References

1. Valtin, H. 1966. Sequestration of urea and nonurea solutes in renal tissues of rats with hereditary hypothalamic diabetes insipidus: effect of vasopressin and dehydration on the countercurrent mechanism. J. Clin. Invest. **45:** 337–345.
2. Gellai, M., B. R. Edwards & H. Valtin. 1979. Urinary concentrating ability during dehydration in the absence of vasopressin. Am. J. Physiol. **237:** F100–F104.
3. Berliner, R. W. & D. G. Davidson. 1957. Production of hypertonic urine in the absence of pituitary antidiuretic hormone. J. Clin. Invest. **36:** 1416–1427.
4. Beale, R. N. & D. Croft. 1961. A sensitive method for the colorimetric determination of urea. J. Clin. Pathol. **14:** 418–424.
5. Edwards, B. R. 1982. Brattleboro homozygotes can concentrate their urine during dehydration without a change in GFR. Ann. N.Y. Acad. Sci. (This volume.)
6. Jamison, R. L. & C. R. Robertson. 1979. Recent formulations of the urinary concentrating mechanism: a status report. Kidney Int. **16:** 537–545.
7. Levinsky, N. G., D. G. Davidson & R. W. Berliner. 1959. Effects of reduced glomerular filtration rate on urine concentration in the presence of antidiuretic hormone. J. Clin. Invest. **38:** 730–740.

THE HORMONAL STATUS OF THE BRATTLEBORO RAT *

Hilda Weyl Sokol

Department of Physiology
Dartmouth Medical School
Hanover, New Hampshire 03755

Earl A. Zimmerman

Department of Neurology
College of Physicians and Surgeons
Columbia University
New York, New York 10032

Although the primary defect in the homozygous (DI) Brattleboro rat is the inability to synthesize vasopressin, other endocrine abnormalities are manifested in this animal as well. It is believed that such differences from the normal condition are probably secondary effects attributable to the lack of vasopressin and/or the state of diabetes insipidus itself. Replacement therapy with exogenous vasopressin corrects the high fluid turnover in these rats as well as some, but apparently not all, of the accompanying abnormalities. The purpose of this paper is to review what is currently known or surmised about the hormonal status of this valuable experimental model. Unless its endocrine and other physiological deviations are known and taken into account, the experimental results obtained with this animal could be misleading.

NEUROHYPOPHYSEAL PRINCIPLES

The first endocrinopathy to be described in the Brattleboro rat was its deficiency in the neurohypophyseal hormones. Not only was vasopressin absent in the neural lobes of the pituitary gland of Brattleboro homozygotes but storage levels of oxytocin were markedly less than in normal rats.[1, 2] Results from subsequent experiments involving dehydration or chronic vasopressin replacement, which corrected fluid turnover and plasma osmolality in DI rats,[3-8] confirmed that vasopressin (VP) and its associated neurophysin (VP-NP) were not present; oxytocin (OT) and its associated neurophysin (OT-NP) were shown to be produced and released in the untreated DI rat in amounts greater than normal.[9] A partial vasopressin deficiency was reported in heterozygous (HZ) Brattleboro rats in their neural lobe stores, circulating plasma levels, and urinary excretion rates.[1, 2, 8, 10-13] The hormones and their respective neurophysins show a consistent 1:1 molar relationship in the pituitary glands in all three genotypes,[5, 7] i.e., in normal Long-Evans (LE), HZ, and DI rats (TABLE 1). The corresponding molar ratios of these neurohypophyseal principles in the plasma remain to be determined, although the hormone levels are now known (TABLE 1).

* Supported by U.S. Public Health Service Research Grants AM–08469, AM–22032, AM–20337, and HL–24105.

0077–8923/82/0394–0535 $1.75/0

All studies thus far support the concept that homozygous Brattleboro rats are unable to synthesize vasopressin and vasopressin neurophysin and that heterozygotes are deficient in the production of these substances.[14–17] It is surprising that vasopressin neurons in HZ rats do not seem to be able to compensate for their gene deficiency by enhanced synthesis, since supraoptic neurons do exhibit nucleolar hypertrophy.[1] During dehydration, HZ rats release less VP and VP-NP than LE controls, and subsequent repletion of neural lobe stores also progresses more slowly in HZ rats than in normal rats.[5, 8, 10, 11] On the other hand, oxytocin synthesis and release appear to be normal in the Brattleboro heterozygote.[5]

TABLE 1

NEUROHYPOPHYSEAL PRINCIPLES IN THE PITUITARY GLAND AND PLASMA OF NORMAL AND BRATTLEBORO RATS

		Normal	Brattleboro Rats	
		Long-Evans (LE)	Heterozygous (HZ)	Homozygous (DI)
Pituitary Gland (pMoles/neural lobe)*				
	VP	808 ± 74	437 ± 31	undetectable
	VP-NP	828 ± 53	394 ± 36	undetectable
	OT	739 ± 69	807 ± 26	347 ± 48
	OT-NP	736 ± 70	674 ± 75	312 ± 0
Plasma (pg/ml)				
	VP	3.0 ± 0.3 5pm †	2.0 ± 0.3 5pm	
		4.4 ± 0.3 10am †	3.2 ± 0.5 10am	undetectable
		2.6 ± 0.6 ‡	2.0 ± 0.2	<1
	OT	6.0 ± 0.9 §		9.9 ± 1.5

* Valtin *et al.*[7]
† Möhring & Möhring.[11]
‡ Krieger *et al.*[13]
§ Edwards *et al.*[9]

The enhanced synthesis and release of oxytocin in the DI rat may be responsible for some of the difficulty DI rats appear to experience at the time of parturition (as discussed in the first session of this symposium). It is possible that the elevated (2× normal) circulating oxytocin levels make the uterine musculature less sensitive to oxytocin.[18] Such uterine insensitivity may be related to the reduced number of oxytocin receptors in the DI rat.[19] Lessened uterine responsiveness coupled with lower oxytocin reserves in the neural lobe could be responsible for inadequate contractions during parturition. This in turn could lead to higher incidences of stillbirths, runts, or neonates that die soon after birth.

Adenohypophyseal Hormones

Abnormalities, other than diabetes insipidus, have been observed in the homozygous Brattleboro rat which strongly implicate altered adenohypophyseal function (Table 2). Growth retardation, subnormal responses to certain stresses, and reduced fertility were some of the aberrations recognized soon after the discovery of this mutant rat. Surprisingly little work has been continued in these areas since the early initial studies were carried out. In brief, it has been reported that DI rats have essentially normal thyroid activity;[20, 21] have somewhat reduced adrenal activity;[22-24] have a pronounced deficit in body weight, length, and pituitary GH content;[2, 21, 25] and can be bred more successfully if heterozygous females are mated with DI males, although matings between two DI rats do produce viable litters.[26]

Growth Hormone (GH)

Despite growth retardation, early studies showed that homozygous Brattleboro rats had normal plasma GH levels as measured by radioimmunoassay (unpublished observations of J. B. Martin, S. Z. Burday, D. S. Schlach, and S. Reichlin). These findings have been corroborated by more recent investigations.[27] However, estimations of GH concentration in the anterior pituitary by bioassay revealed considerably less GH stores in the DI rat than normal[21] and more recently we have observed a similar trend by radioimmunoassay[27] (Table 2). While Arimura *et al.*[21] found differences in pituitary GH content between normal and DI rats, this could not be ascribed to a deficiency in hypothalamic GH releasing factor activity. Long-term injections of growth hormone, but not vasopressin, enhanced body length in DI rats which indicated that their target tissues were sensitive to GH.[25] In addition, these studies showed that although the correction of the diabetes insipidus syndrome with exogenous VP did not improve skeletal growth, it did result in increased body weight.[25] The possibility that all the immunoreactive circulating GH in DI rats may not be biologically active could perhaps account for their subnormal growth. It is also possible that there is an impairment of the more subtle regulation of the episodic secretory pattern of GH release shown to exist in normal rats.[28] Perhaps VP plays such a modulatory role within the central nervous system (CNS).

Thyroid Stimulating Hormone (TSH)

Although TSH content of the pituitary gland has been reported to be normal,[21] plasma levels appear to be significantly elevated in both heterozygous and homozygous Brattleboro rats[29, 30] (Table 2). The paper that follows reports that the thyroid gland in DI rats appears to be less sensitive to TSH resulting in significantly lower circulating thyroid hormones.[29]

Adrenocorticotropic Hormone (ACTH)

Much conflicting evidence exists concerning the status of the pituitary-adrenal axis in the Brattleboro rat. Although pituitary ACTH concentration

TABLE 2

PITUITARY AND PLASMA CONCENTRATIONS OF ANTERIOR PITUITARY HORMONES IN NORMAL LONG-EVANS (LE) AND HOMOZYGOUS (DI) BRATTLEBORO RATS

Hormone	Anterior Pituitary ($\mu g/mg$)		Plasma (ng/ml)		Reference
	LE	DI	LE	DI	
GH	84.7 * (63.7–112.5)	35.4 (21.2–59.5)			21
	33.5 ± 5.9	20.3 ± 6.1	61.4 ± 24.0	78.3 ± 46.9	27
PRL	1.15 ± 0.20 *	0.64 ± 0.04	15.7 ± 2.3	18.7 ± 3.2	37
		1.68 ± 0.5	19.5	18.2	38
	0.412 ± 0.118	0.472 ± 0.231	18.2 ± 3.7	19.5 ± 5.3	27
			95 ± 11 *	54 ± 10	30
ACTH	20.4 mU (2.9–30.2)	19.5 mU (14.0–26.9)			22
	33 mU *	23 mU	0.56 mU/100 ml *	0.16 mU/100 ml	48
TSH			305 ± 20 *	598 ± 100	29
LH	25.8 ± 3.5	24.7 ± 2.8	62.2 ± 7.4	51.9 ± 21.1	39
FSH	46.4 ± 6.0 *	62.2 ± 7.4	450 ± 36 *	794 ± 110	39

* $p < 0.05$ Normal vs. DI.

appears to be normal,[22, 24, 31, 32] a recent study found the storage and release of ACTH in the DI rat to be deficient.[48] Significantly lower circulating plasma ACTH has also been reported in Brattleboro rats,[48] which could provide a partial answer to the etiology of corticosteroid deficiency exhibited by the DI rat. Both normal[24, 32] and lower[12, 33, 34, 48] basal levels of corticosterone have been reported in DI rats. Similarly the response to stress has been found to be normal by some[24] and deficient by others.[22, 23, 32, 35] The adrenal gland in the DI rat has been found to be smaller than normal, whether expressed in absolute weight or in relation to total body weight by most investigators.[22, 32, 33, 48] The inner glucocorticosteroid producing zones (Z. fasciculata and Z. reticularis) of the adrenal cortex have been found to be relatively less developed in the DI rat, a deficiency which was corrected by prolonged treatment with arginine vasopressin.[36] This is consistent with the finding that isolated zonae fasciculata/reticularis cells from DI rats are less responsive to ACTH than those in non-DI rats.[35] Thus, the reduced response to stress in the DI rat could partially reside in its decreased adrenal sensitivity to ACTH.

Gonadotropins (Prolactin, PRL; Follicle Stimulating Hormone, FSH; Luteinizing Hormone, LH)

Little information is available to date on prolactin,[37, 38] FSH, and LH[39] in the Brattleboro rat. Fortunately some of the gaps will be filled in by recent research to be presented at this session. One might assume that profound differences are not likely to exist since Brattleboro rats are capable of reproducing themselves. The precisely timed and titered secretions of gonadotropins and sex hormones that are required for successful ovulation, fertilization, implantation, gestation, and parturition probably precludes serious deviations from the normal condition. However, since Brattleboro rats have thus far proved themselves to be amazingly adaptive (by establishing a new homeostasis in many systems) we must approach their reproductive function in an unbiased manner, being prepared to find the unexpected. Preliminary data from several laboratories on the prolactin status of DI rats are not in complete agreement, since both normal[37, 38] and reduced[30] levels of circulating PRL have been reported. The apparent conflict may reside in the much higher concentrations of PRL found in *both* normal and DI rats in the study[30] reporting a difference between the two types of rat. The relatively lower PRL levels in the DI rat compared to the normal in that study may reflect a deficiency in the acute release of PRL rather than a difference in basal plasma levels.

Improvements in radioimmunoassay and the availability of purified rat pituitary hormones, coupled with a renewed interest in the total endocrine status of the Brattleboro rat, mean that more definitive results are on the horizon. This is already in evidence by the contributions to this symposium. It is likely, although certainly not proved, that all the endocrine abnormalities manifest in the DI rat are secondary to the lack of vasopressin. Deficiencies remaining even after normalization of fluid turnover by exogenous vasopressin do not necessarily mean, however, that these defects are not dependent upon VP. The distribution of vasopressinergic (and oxytocinergic) pathways from the paraventricular nuclei[40] of the hypothalamus, and possibly from the supraoptic nuclei as well,[41] to many other parts of the CNS (i.e. not exclusively terminating in the neural lobe of the pituitary gland) strongly suggests that VP may act directly at nerve endings.[42–44] In addition, the important vasopressin pathway from the paraventricular nucleus to the hypophyseal portal capillary system in the median eminence insures the delivery of VP directly to the anterior pituitary at concentrations up to a thousand times higher than in the peripheral circulation.[45, 50]

Hypothalamic Releasing Factors

Corticotropin Releasing Factor—CRF

(Or, "Is vasopressin CRF?) If vasopressin is *the* corticotropin releasing factor, just as it is *the* antidiuretic hormone, then the Brattleboro rat lacking vasopressin should exhibit profound deficits in ACTH and corticosteroid functions—which it does not. That is not to say, however, that VP is not *a* CRF—which it is. It is possible, as suggested in numerous recent studies excellently reviewed by Gillies and Lowry,[46] that VP acting in synergy with one or more other hypothalamic factors constitutes an essential element in that elusive "entity," CRF, which stimulates the secretion of ACTH. The homozygous Brattleboro rat is an excellent model in which to pursue the question: "Does VP account for all the activity ascribed to the yet unidentified CRF?" Since the answer to that question is "No," the questions that remain are "What other substances qualify as CRF?" and "What is their relation to vasopressin?"

Most investigators agree that hypothalamic extracts from Brattleboro homozygotes contain CRF activity and that some of the physical-chemical properties (e.g. stability during storage) are different from normal. The current controversy centers around how much CRF activity resides in the DI rat hypothalamus, different researchers reporting normal,[47] reduced,[13, 48] or minimal activity.[31]

Gillies and Lowry[31] who found markedly reduced CRF activity in the homozygous Brattleboro rat have been able to restore DI hypothalamic CRF potency to normal by the addition of an amount of VP equivalent to that found in normal rat hypothalamic extracts. Chromatographic evidence from this laboratory strengthens the concept of a critical CRF role for VP. When normal hypothalamic extracts are run over a BioGel P_2 column, three peaks with CRF activity are clearly differentiated, the major peak co-eluting with immunoreactive VP. Although each peak had some ACTH-releasing capabilities (using a dispersed cell perfusion system), the sum of the individual responses fell far short of the CRF potency of crude extracts; however, when the three peaks were recombined full CRF activity was restored. In contrast, chromatography of DI hypothalamic extracts yielded two minor peaks whose minimal CRF potency was enhanced and brought up to normal levels by the addition of synthetic AVP (Figure 1). Thus, these workers conceptualize a CRF that is multifactorial, its several components acting synergistically. Such multiplicity of substances could account for the elusiveness of this hypothalamic releasing factor, which was the first one hypothesized to exist almost 30 years ago.†

Results from a variety of early *in vivo* studies on the brain-pituitary-adrenal axis formed the foundation for the more recent *in vitro* studies on CRF and

† Two weeks after this symposium the purification and synthesis of a hypothalamic peptide containing 41 amino acids with CRF activity was reported in *Science*.[68]

VP. Hedge *et al.*[49] suggested that VP may act as a CRF-releasing factor. Yates *et al.*[23] created a schema in which VP, in addition to stimulating the release of CRF in the hypothalamus, could function directly as a CRF, and could also potentiate the action of CRF at the level of the anterior pituitary gland. The detection of high plasma AVP levels in the hypophyseal portal blood[45, 50] and the anatomical evidence for increased immunoreactive VP in the outer zone of the median eminence after adrenalectomy[51, 52] gave strong support to the concept of VP involvement in the normal regulation of adrenal activity. Thus, it is not at all surprising that an altered brain-pituitary-adrenal axis exists in the DI rat, the details of which still need to be investigated.

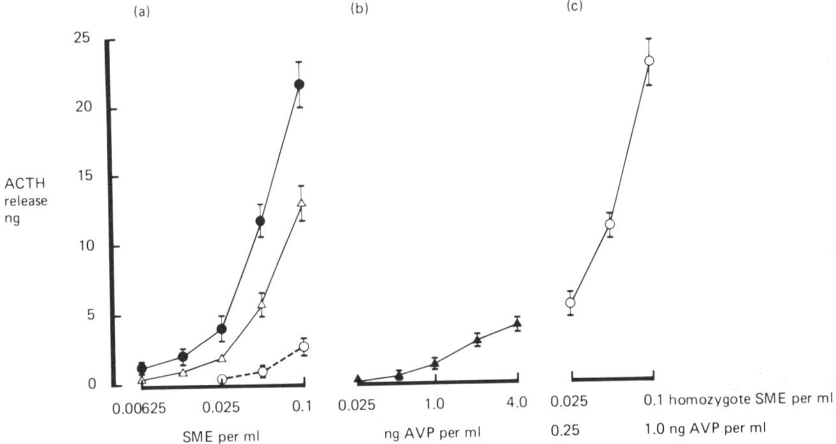

FIGURE 1. Comparison of CRF bioactivity (A) of stalk median eminence (SME) from normal (Wistar or Long Evans) rats (●—●), heterozygous Brattleboro rats (△—△), and homozygous Brattleboro rats (○- -○); (B) synthetic AVP and (C) homozygous Brattleboro SME with AVP added in concentrations equivalent to those found in normal SME. Identical results are obtained when SME is tested on perfused anterior pituitary cell columns prepared from normal or Brattleboro rats. (Reproduced with permission from Gillies and Lowry.[39])

Other Releasing Factors

Little is known about the other hypophysiotropic regulating factors in the homozygous Brattleboro rat. Information available usually constitutes only a small part of a study whose primary focus was in another area. Thus, it has been reported that the DI rat does not seem to be deficient in growth hormone-releasing factor (GH-RF),[21] nor was impairment in luteinizing hormone-releasing hormone (LH-RH or LRH) activity observed in DI stalk-median eminence extracts.[31]

RENIN-ANGIOTENSIN-ALDOSTERONE

The contributions of the late Jan Möhring[11, 12, 33, 53] have provided valuable data on the relationship between vasopressin and the renin-angiotensin-aldosterone (R-A-A) complex in the Brattleboro rat (TABLE 3). A more recent comprehensive study of body fluid homeostatic mechanisms in the DI rat by Kinter[54] has elucidated some of the paradoxical relationships in its R-A-A system. Elevated plasma renin activity and angiotensin II (AII) occur simultaneously with lowered plasma aldosterone in the untreated homozygous Brattleboro rat, aberrations rectified by prolonged vasopressin treatment.[12, 33, 54, 55]

The data consolidated in TABLE 3 were obtained from measurements made in a variety of laboratories.[33, 34, 54, 55] Although the absolute values may differ, there is consensus in the relative differences between normal (or heterozygous) and DI rats. In general, DI rats have at least two times the plasma renin activity of normal rats, although much lower absolute amounts are obtained when blood is collected from unanesthetized rats bearing chronic catheters for the withdrawal of blood.[54] Their plasma AII levels reflect plasma renin activity,

TABLE 3

LEVELS OF CIRCULATING RENIN, ANGIOTENSIN II, AND ADRENAL CORTICOSTEROIDS IN NORMAL LONG-EVANS (LE), AND HOMOZYGOUS (DI) BRATTLEBORO RATS

Substance (concentration)	LE Rats	DI Rats	Reference †
Plasma Renin Activity			
(ng AI/ml·h)	25.5 ± 9.1	92.5 ± 14.2	1(a)
	5.2 ± 0.6 *	10.1 ± 1.8	1(b)
(ng AI/ml·16 h)	99.1 ± 11.9 *	231.9 ± 43.0	2
Angiotensin II			
(pg/ml)	85 ± 5	267 ± 44	3(a)
Aldosterone	8.0 ± 2.4	3.6 ± 0.9	3(b)
(ng/100 ml)	12.1 ± 2.4	6.7 ± 0.8	4
	10.3 ± 1.2	6.9 ± 1.6	6
Corticosterone	10.1 ± 1.4	4.9 ± 1.2	3(b)
(μg/100 ml)	10.4 ± 1.2	4.9 ± 0.6	4
	15 ± 6	8 ± 6	5

* Heterozygous Brattleboro rats.

†1. Kinter[54] (a) anesthesia (b) no anesthesia. 2. Balment *et al.*[55] 3. Möhring *et al.*[33] (a) anesthesia (b) no anesthesia. 4. Milne *et al.*[34] 5. Buckingham & Leach.[48] 6. Opava-Stitzer *et al.*[67]

while the concentrations of the corticosteroids, both corticosterone and aldosterone, were only half of the normal condition. The enhanced renin secretion that occurs in the DI rat may be ascribed directly to the lack of vasopressin that normally inhibits renin release from the juxtaglomerular cells in the kidney.[56–58]

The unusual inverse relationship between plasma AII and aldosterone is probably related to the heightened urinary excretion of aldosterone in the DI rat.[54] In this symposium Milne and coworkers[34] corroborate Kinter's findings of an elevated aldosterone clearance rate which they, too, suggest may account for the lower plasma levels of aldosterone in the DI rat (TABLE 3). These investigators also report that the zona glomerulosa is hypertrophied in the DI rat in comparison with the normal LE rat, a finding not consistent with the data Sokol and Möhring[36] presented at these meetings showing a slight reduction in the width of the aldosterone-producing zone. In this study[36] the hypokalemic state of the DI rat was invoked as a possible explanation for the lower circulating aldosterone despite elevated AII levels.[58] Zona glomerulosa cells are known to be less sensitive to stimulation in a medium low in potassium,[59] a teleologically sound adaptation.

It is difficult at this time to account for some of these divergent results. Perhaps, as discussed in an earlier session, the possibility that some genetic drift has occurred in the different colonies from which we obtain our experimental animals cannot be ruled out. Sex differences must also be taken into account.[55]

Angiotensin in the CNS

The question of the existence of a renin-angiotensin (R-A) system, which generates AII within the brain, remains controversial.[60, 61] Although there is good evidence for most of the components of the R-A system in the CNS, including AII receptors, identification and quantitation of endogenous AII have been problematic.[61] It may be that brain AII remains elusive because of its rapid destruction during extraction. That all the R-A components could reside in a single nerve cell is suggested by the discovery of all the essential elements of this system in a homogeneous neuroblastoma-glioma cell line[62] and in juxtaglomerular cells of the kidney.[63]

Angiotensin II-like immunoreactivity has been detected in supraoptic and paraventricular neurons in the hypothalamus of the normal rat by immunocytochemistry.[64, 65] AII-like material was localized in vasopressin neurons and was also found to increase in median eminence terminals after adrenalectomy[65] reminiscent of vasopressin accumulation under similar conditions.[51, 52] Despite the elevated concentrations of circulating AII in peripheral blood of DI rats,[33] immunocytochemical probes reveal only a rare supraoptic or paraventricular perikaryon with AII-like immunoreactivity; nor does correction of the diabetes insipidus with exogenous vasopressin result in an enhancement of AII-like material in the hypothalamus.[66] Although the significance of these observations is not clear, they at least suggest that AII-like material visualized in the brain in these studies does not come from the peripheral blood. Its presence in normal vasopressin-producing cells combined with its rarity in magnocellular neurons in the homozygous Brattleboro rat may mean that VP and AII-like material (1) share a common precursor molecule, (2) are unrelated molecules but share some common antigenic determinants, or that (3) AII synthesis in the brain is dependent upon the presence of vasopressin in the cell. Internalization of extracellular AII must also be considered.

Thus, interactions between vasopressin and the renin-angiotensin system are probably not limited to the periphery but may play a vital role in the integration of neural activity within the CNS. The significance of the numerous central hypothalamic projections of vasopressin (and oxytocin) fibers to cortex, brainstem, and spinal cord[42–44] is just beginning to be understood. The Brattleboro rat, lacking functional vasopressin projections, will continue to help us find some of the answers.

Conclusion

It is obvious that the endocrine status of homozygous and heterozygous Brattleboro rats is largely unknown. Not only do we have little information on the circulating and tissue levels of most of the anterior pituitary hormones, but few of their target organ hormones have been explored, except perhaps for the corticosteroids. Gaps in our knowledge include the sex hormones, parathyroid hormone, insulin, glucagon, the catecholamines, and the whole gamut of peptides (including hypothalamic releasing factors) found in both the gastrointestinal tract and the central nervous system. Much interesting work and probably unexpected results lie ahead!

Table 4

Summary: Hormone Levels Reported in the DI Rat Relative to the Amounts Found in the LE Rat

Hormone (or associated substance)	Plasma	Tissue
Vasopressin	not detectable	
Vasopressin neurophysin	not detectable	
Oxytocin	↑	↓
Oxytocin neurophysin	↑	↓
Growth Hormone (GH)	=	↓
Prolactin (PRL)	=	↓
Thyrotropin (TSH)	↑	=
Luteinizing Hormone (LH)	=	=
Follicle Stimulating Hormone (FSH)	↑	↑
Corticotropin (ACTH)	↓	=,↓
ACTH-releasing factor (CRF)		=,↓
GH-releasing factor (GHRF)		=
LH-releasing hormone (LRH)		=
Renin	↑	↑
Angiotensin II	↑	
Aldosterone	↓	
Corticosterone	↓,=	
Thyroid hormones: T_3, T_4	↓	

References

1. Sokol, H. W. & H. Valtin. 1965. Morphology of the neurosecretory system in rats homozygous and heterozygous for hypothalamic diabetes insipidus (Brattleboro strain). Endocrinology **77:** 692–700.
2. Valtin, H., W. H. Sawyer & H. W. Sokol. 1965. Neurohypophysial principles in rats homozygous and heterozygous for hypothalamic diabetes insipidus (Brattleboro strain). Endocrinology **77:** 701–706.
3. Sokol, H. W. & H. Valtin. 1967. Evidence for the synthesis of oxytocin and vasopressin in separate neurons. Nature **214:** 314–316.
4. Valtin, H. 1967. Hereditary hypothalamic diabetes insipidus in rats (Brattleboro strain). A useful experimental model. Am. J. Med. **42:** 814–827.

5. SUNDE, D. A. & H. W. SOKOL. 1975. Quantification of rat neurophysins by polyacrylamide gel electrophoresis (PAGE): Application to the rat with hereditary hypothalamic diabetes insipidus. Ann. N.Y. Acad. Sci. **248:** 345–364.
6. VALTIN, H., H. W. SOKOL & D. SUNDE. 1975. Genetic approaches to the study of the regulation and actions of vasopressin. Rec. Prog. Horm. Res. **31:** 447–486.
7. VALTIN, H., W. G. NORTH, F. T. LAROCHELLE, JR., H. W. SOKOL & J. F. MORRIS. 1978. Biochemical and anatomical aspects of ADH production. Proc. VIIth Int. Congr. Nephrology. Montreal, June 18–23, 1978. S. Karger. Basel, pp. 313–320.
8. MILLER, M. & A. M. MOSES. 1971. Radioimmunoassay of urinary antidiuretic hormone with application to study of the Brattleboro rat. Endocrinology **88:** 1389–1396.
9. EDWARDS, B. R., F. T. LAROCHELLE, JR. & M. GELLAI. 1982. Concentration of urine by dehydrated Brattleboro homozygotes: Is there a role for oxytocin? Ann. N.Y. Acad. Sci. (This volume.)
10. MOSES, A. M. & M. MILLER. 1970. Accumulation and release of pituitary vasopressin in rats heterozygous for hypothalamic diabetes insipidus. Endocrinology **86:** 34–41.
11. MÖHRING, B. & J. MÖHRING. 1975. Plasma ADH in normal Long-Evans rats and in Long-Evans rats heterozygous and homozygous for hypothalamic diabetes insipidus. Life Sci. **17:** 1307–1314.
12. MÖHRING, J., G. KOHRS, B. MÖHRING, M. PETRI, E. HOMSY & D. HAACK. 1978. Effects of prolonged vasopressin treatment in Brattleboro rats with diabetes insipidus. Am. J. Physiol. **234:** F106–F111.
13. KRIEGER, D. T., A. LIOTTA & M. J. BROWNSTEIN. 1977. Corticotropin releasing factor distribution in normal and Brattleboro rat brain and effect of deafferentation, hypophysectomy and steroid treatment in normal animals. Endocrinology **100:** 227–237.
14. BROWNSTEIN, M. J. & H. GAINER. 1977. Neurophysin biosynthesis in normal rats and in rats with hereditary diabetes insipidus. Proc. Natl. Acad. Sci. USA **74:** 4046–4049.
15. BROWNSTEIN, M. J., J. T. RUSSELL & H. GAINER. 1980. Synthesis, transport and release of posterior pituitary hormones. Science **207:** 475–478.
16. RUSSEL, J. T., M. J. BROWNSTEIN & H. GAINER. 1980. Biosynthesis of vasopressin, oxytocin and neurophysins: Isolation and characterization of two common precursors (propressophysin and prooxyphysin). Endocrinology **107:** 1880–1891.
17. SOKOL, H. W., E. A. ZIMMERMAN, W. H. SAWYER & A. G. ROBINSON. 1976. The hypothalamic-neurohypophysial system of the rat: Localization and quantitation of neurophysin by light microscopic immunocytochemistry in normal rats and in Brattleboro rats deficient in vasopressin. Endocrinology **98:** 1176–1188.
18. HALDAR, J., L. KUPFER & H. W. SOKOL. 1982. Decreased sensitivity to oxytocin of uteri from homozygous diabetes insipidus rats. Ann. N.Y. Acad. Sci. (This volume.)
19. GEONZON, R. M., H. J. GOREN, M. D. HOLLENBERG, K. LEDERIS & D. MORGAN. 1980. Oxytocin binding and responsiveness in the uterus of the Brattleboro rat. J. Physiol. **303:** 52P.
20. GALTON, V. A., H. VALTIN & D. G. JOHNSON. 1966. Thyroid function in the absence of vasopressin. Endocrinology **78:** 1224–1229.
21. ARIMURA, A., S. SAWANO, T. W. REDDING & A. V. SCHALLY. 1968. Studies on retarded growth of rats with hereditary hypothalamic diabetes insipidus. Neuroendocrinology **3:** 187–192.
22. MCCANN, S. M., J. ANTUNES-RODRIGUES, R. NALLAR & H. VALTIN. 1966.

Pituitary-adrenal function in the absence of vasopressin. Endocrinology **79:** 1058–1064.

23. YATES, F. E., S. M. RUSSELL, M. F. DALLMAN, G. A. HEDGE, S. M. MCCANN & A. P. S. DHARIWAL. 1971. Potentiation by vasopressin of corticotropin-releasing factor. Endocrinology **88:** 3–15.
24. ARIMURA, A., T. SAITO, C. Y. BOWERS & A. V. SCHALLY, 1967. Pituitary-adrenal activation in rats with hereditary hypothalamic diabetes insipidus. Acta Endocrinol. (Copenhagen) **54:** 155–165.
25. SOKOL, H. W. & J. SISE. 1973. The effect of exogenous vasopressin and growth hormone on the growth of rats with hereditary hypothalamic diabetes insipidus. Growth **37:** 127–142.
26. SAUL, G. B. II, E. B. GARRITY, K. BENIRSCHKE & H. VALTIN. 1968. Inherited hypothalamic diabetes insipidus in the Brattleboro strain of rats. J. Hered. **59:** 113–117.
27. ADLER, R. A. & H. W. SOKOL. 1982. Prolactin and growth hormone in the unstressed homozygous Brattleboro rat. Ann. N.Y. Acad. Sci. (This volume.)
28. MARTIN, J. B., P. BRAZEAU, G. S. TANNENBAUM, J. O. WILLOUGHBY, J. EPELBAUM, L. C. TERRY & D. DURAND. 1978. Neuroendocrine organization of growth hormone regulation. *In* The Hypothalamus. S. Reichlin, R. J. Baldessarini & J. B. Martin, Eds.: 329–357. Raven Press. New York, N.Y.
29. FUJIMOTO, S. & G. A. HEDGE. 1982. Abnormalities in thyrotropin and thyroxine secretion in Brattleboro rats. Ann. N.Y. Acad. Sci. (This volume.)
30. FUJIMOTO, S. & G. A. HEDGE. 1982. Alterations in pituitary function in rats with hypothalamic diabetes insipidus (Brattleboro strain). Ann. N.Y. Acad. Sci. (This volume.)
31. GILLIES, F. & P. J. LOWRY. 1980. Corticotrophin-releasing activity in extracts of the stalk median eminence of Brattleboro rats. J. Endocrinol. **84:** 65–73.
32. WILEY, M. K., A. F. PEARLMUTTER & R. E. MILLER. 1974. Decreased adrenal sensitivity to ACTH in the vasopressin-deficient rat (Brattleboro). Neuroendocrinology **14:** 257–270.
33. MÖHRING, B., J. MÖHRING, G. DAUDA & D. HAACK. 1974. Potassium deficiency in rats with hereditary diabetes insipidus. Am. J. Physiol. **227:** 916–920.
34. MILNE, C. M., R. J. BALMENT & I. W. HENDERSON. 1982. Adrenal steroidogenesis in vasopressin deficiency. Ann. N.Y. Acad. Sci. (This volume.)
35. KENYON, C. J., G. HARGREAVES & I. W. HENDERSON. 1978. Adrenocortical function in rats with inherited hypothalamic diabetes insipidus (Brattleboro strain). J. Steroid Biochem. **9:** 345–348.
36. SOKOL, H. W. & J. MÖHRING. 1982. Morphological correlates of renin and aldosterone secretion in the homozygous Brattleboro rat. Ann. N.Y. Acad. Sci. (This volume.)
37. COCULESCU, M., M. OPRESCU & V. DIMITRIU. 1976. Prolactin determination in hypothalamic diabetes insipidus. VIth Int. Congr. Endocrinology. Hamburg, Fed. Rep. Germany, July 18–24, 1976. Abstract No. 22, p. 9.
38. ADLER, R. A., S. DOLPHIN, M. SZEFLER & H. W. SOKOL. 1979. The effects of elevated circulating prolactin in rats with hereditary hypothalamic diabetes insipidus. (Brattleboro strain). Endocrinology **105:** 1001–1006.
39. OPRESCU, M., L. SIMIONESCU & M. COCULESCU. 1982. Gonadotropins in rats with hereditary diabetes insipidus. Ann. N.Y. Acad. Sci. (This volume.)
40. SWANSON, L. W. & P. E. SAWCHENKO. 1980. Paraventricular nucleus: A site of integration of neuroendocrine and autonomic mechanisms. Neuroendocrinology **31:** 410–417.
41. BUIJS, R. M. 1980. Vasopressin and oxytocin innervation of the rat brain. A light- and electronmicroscopical study. Ph.D. Thesis. University of Amsterdam, Amsterdam, The Netherlands.

42. BUIJS, R. M. 1978. Intra- and extrahypothalamic vasopressin and oxytocin pathways in the rat. Cell Tiss. Res. **192:** 423–435.
43. NILAVER, G., E. A. ZIMMERMAN, J. WILKINS, J. MICHAELS, D. HOFFMAN & A.-J. SILVERMAN. 1980. Magnocellular hypothalamic projections to lower brainstem and spinal cord of the rat. Neuroendocrinology **30:** 150–158.
44. ZIMMERMAN, E. A. 1981. The organization of oxytocin and vasopressin pathways. *In* Neurosecretion and Brain Peptides, J. B. Martin, S. Reichlin & K. L. Bick, Eds.: 63–75. Raven Press. New York, N.Y.
45. RECHT, L. D., D. HOFFMAN, J. HALDAR, A.-J. SILVERMAN & E. A. ZIMMERMAN. 1981. Vasopressin concentrations in hypophysial portal plasma: Insignificant reduction following removal of the posterior pituitary gland. Neuroendocrinology **33:** 88–90.
46. GILLIES, G. & P. J. LOWRY. 1982. Corticotropin releasing hormone and its vasopressin component. *In* Frontiers in Neuroendocrinology. W. F. Ganong & L. Martini, Eds. Vol. 7. Raven Press. New York, N.Y. (In press.)
47. PEARLMUTTER, A. F., L. DOKAS, A. KONG, R. MILLER & M. SAFFRAN. 1980. Is corticotropin releasing factor modulated vasopressin? Nature **283:** 697–698.
48. BUCKINGHAM, J. C. & J. H. LEACH. 1980. Hypothalamo-pituitary-adrenocortical function in rats with inherited diabetes insipidus. J. Physiol. **305:** 397–404.
49. HEDGE, G. A., M. B. YATES, R. MARCUS & F. E. YATES. 1966. Site of action of vasopressin in causing corticotrophin release. Endocrinology **79:** 328–340.
50. ZIMMERMAN, E. A., P. W. CARMEL, M. K. HUSAIN, M. FERIN, M. TANNENBAUM, A. G. FRANTZ & A. G. ROBINSON. 1973. Vasopressin and neurophysins: High concentrations in monkey hypophysial portal blood. Science **182:** 925–927.
51. STILLMAN, M. A., L. D. RECHT, S. L. ROSARIO, S. M. SEIF, A. G. ROBINSON & E. A. ZIMMERMAN. 1977. The effects of adrenalectomy and glucocorticoid replacement on vasopressin and vasopressin neurophysin in the zona externa of the median eminence of the rat. Endocrinology **101:** 42–49.
52. BURLET, A., M. CHATEAU & P. CZERNICHOW. 1979. Infundibular localization of vasopressin, oxytocin and neurophysins in the rat; its relationship with corticotrope function. Brain Res. **168:** 275–286.
53. PETER, S. & J. MÖHRING. 1978. The juxtaglomerular apparatus of rats with hereditary hypothalamic diabetes insipidus. Cell Tiss. Res. **188:** 335–339.
54. KINTER, L. B. 1978. Homeostatic mechanisms in the diabetes insipidus rat. Ph.D. Thesis. Harvard University. Cambridge, Mass.
55. BALMENT, R. J., I. W. HENDERSON & J. A. OLIVER. 1975. The effects of vasopressin on pituitary oxytocin content and plasma renin activity in rats with hypothalamic diabetes insipidus (Brattleboro strain). Gen. Comp. Endocrinol. **26:** 468–477.
56. GUTMAN, Y. & R. BENZAKEIN. 1974. Antidiuretic hormone and renin in rats with diabetes insipidus. Eur. J. Pharmacol. **28:** 114–118.
57. GANONG, W. F., J. SHINSAKO, I. A. REID, L. C. KEIL, D. L. HOFFMAN & E. A. ZIMMERMAN. 1982. Role of vasopressin in the renin and ACTH responses to intraventricular angiotensin II. Ann. N.Y. Acad. Sci. (This volume.)
58. KOTCHEN, T. A. & G. P. GUTHRIE, JR. 1980. Renin-angiotensin-aldosterone and hypertension. Endocrine Reviews **1:** 78–99.
59. FREDLUND, P., S. SALTMAN, T. KONDO, J. DOUGLAS & K. J. CATT. 1977. Aldosterone production by isolated glomerulosa cells: Modulation of sensitivity to angiotensin II and ACTH by extracellular potassium concentration. Endocrinology **100:** 481–486.
60. GANTEN, D., G. SPECK, P. SCHELLING & T. H. UNGER. 1981. The brain renin-angiotensin system. *In* Neurosecretion and Brain Peptides. J. B. Martin, S. Reichlin & K. L. Bick, Eds.: 359–372. Raven Press. New York, N.Y.

61. RAMSAY, D. J. 1982. Effects of circulating angiotensin II on the brain. *In* Frontiers in Neuroendocrinology. W. F. Ganong & L. Martini, Eds. Vol. 7. Raven Press. New York, N.Y. (In press.)
62. FISHMAN, M. C., E. A. ZIMMERMAN & E. E. SLATER. 1981. Renin and angiotensin. The complete system within the neuroblastoma X glioma cell. Science **214:** 921–923.
63. CELIO, M. R. & T. INAGAMI. 1981. Angiotensin II-immunoreactivity coexists with renin in the juxtaglomerular cells of the kidney. 32nd Annu. Meeting of the Histochemical Society. Abstract No. 62.
64. PHILLIPS, M. I., J. WEYHENMEYER, D. FELIX, D. GANTEN & W. E. HOFFMAN. 1979. Evidence for an endogenous brain renin-angiotensin system. Fed. Proc. **38:** 2260–2266.
65. KILCOYNE, M. M., D. L. HOFFMAN & E. A. ZIMMERMAN. 1980. Immunocytochemical localization of angiotensin II and vasopressin in rat hypothalamus: Evidence for production in the same neuron. Clin. Sci. **59:** 57S–60S.
66. HOFFMAN, D. L., L. KRUPP, D. SCHRAG, G. VALIQUETTE, G. NILAVER, M. M. KILCOYNE & E. A. ZIMMERMAN. 1982. Angiotensin immunoreactivity in vasopressin cells in rat hypothalamus and its relative deficiency in homozygous Brattleboro rats. Ann. N.Y. Acad. Sci. (This volume.)
67. OPAVA-STITZER, S., E. FERNANDEZ-REPOLLET, C. RODRIGUEZ-SARGENT, J. L. CANGIANO & M. MARTINEZ-MALDONADO. 1982. Exaggerated natriuretic response of diabetes insipidus (DI) rats to extracellular volume expansion. Ann. N.Y. Acad. Sci. (This volume.)
68. VALE, W., J. SPIESS, C. RIVIER & J. RIVIER. 1981. Characterization of a 41-residue ovine hypothalamic peptide that stimulates secretion of corticotropin and β-endorphin. Science **213:** 1394–1397.

DISCUSSION OF THE PAPER

J. F. MORRIS (*Oxford University, Oxford, England*): It struck me that, from basic cell biology, a peptide that attaches to a receptor on the surface of a cell can get to the Golgi apparatus and might therefore be packaged. So that, if magnocellular neurons have enkephalin and angiotensin II, which are active on the surface of these cells, it would be possible for these peptides to reach the Golgi in small amounts and perhaps be co-packaged with the neurosecretory material. Thus, the cells might not make these peptides but they could still pick up a small amount from afferent synapses. That would provide an explanation for some of the data that we have discussed earlier in this conference.

ABNORMALITIES IN THYROTROPIN AND THYROXINE SECRETION IN BRATTLEBORO RATS

Seigo Fujimoto and George A. Hedge

Department of Physiology
West Virginia University Medical Center
Morgantown, West Virginia 26506

Although there are numerous reports of direct stimulatory effects of vasopressin (VP) on thyroid secretion [1-4] and on pituitary secretion of thyrotropin (TSH),[5, 6] we are aware of only one paper in which the pituitary-thyroid axis has been studied in the Brattleboro rat, which is devoid of endogenous VP. Using several contemporary indices of thyroid function, such as urinary excretion of radioactive iodide, basal metabolic rate (BMR), and protein-bound iodide (PBI), Galton *et al.*[7] did indeed find some abnormalities in the thyroid axis of the homozygous (DI) rat. However, these abnormalities were reversed when the excess urine flow was normalized by the administration of exogenous VP. Because of the definitive nature of these results, there has been rather little interest in further examination of other aspects of structure and function of the pituitary-thyroid axis in the Brattleboro rat.

The present work was initiated by our suspicion in another study that VP might be affecting the anterior pituitary sensitivity to TRH, a notion that was not testable at the time of the earlier study mentioned above.[7] We then recognized the value of the Brattleboro rat in such an inquiry, and after initiating this specific investigation, we were quickly drawn into a more extensive description of the function (and some of the structure) of this neuroendocrine axis in these rats with diabetes insipidus. Although our study has some conceptual similarities to the earlier work of Galton *et al.*,[7] we have had the very great advantage of being able to benefit from some technological advances of the past 15 years, such as the availability of: (1) radioimmunoassays (RIA) for all of the hormones of interest, (2) synthetic TRH to test pituitary sensitivity, and (3) implantable minipumps to allow continuous physiological administration of VP preparations.

MATERIALS AND METHODS

Normal Long-Evans rats and rats of the Brattleboro homozygous (DI) and heterozygous (HZ) strain (180–200 g at the beginning of the study) were used. Urine volume and osmolality were measured for each animal during 24 h collection periods.

For RIA of plasma TSH and thyroxine (T_4), blood was obtained by cardiac puncture under light ether anesthesia, before and 10 min after the i.v. injection of thyrotropin releasing hormone (TRH) at a mid-range dose of 250 ng/100 g BW. Similar samples were obtained before and 4 h after the i.v. injection of bovine TSH using a midrange dose of 100 mU/100 g BW. Plasma (100 μl for TSH and 25 μl for T_4) was separated and stored frozen until assayed as described previously.[8, 9]

0077-8923/82/0394-0549 $1.75/0

The disappearance rates of plasma TSH and T_4 were determined in rats bearing atrial catheters (Silastic®) implanted under sodium pentobarbital anesthesia (4 mg/100 g BW, i.p.). This preparation allowed the administration of labeled hormones and collection of blood samples in unstressed, unanesthetized rats, although an occasional sample was collected by cardiac puncture under light ether anesthesia. Experiments were performed 1–4 days after catheterization. Blood was collected at 5, 10, 20, 30, 45, 60, and 90 min after [^{125}I]TSH injection (10^6 cpm/0.25 ml/rat; rat TSH was iodinated using the chloramine T method) and 6, 12, 20, 28, and 36 h after [^{125}I]T_4 injection (5.5 μCi/0.25 ml/rat). Plasma (50 μl) was separated and stored frozen until immunoprecipitation of all samples.

For the determination of TSH disappearance rate, plasma was added to 450 μl of 2% normal rabbit serum in 0.01 M phosphate buffered saline (NRS-PBS), and then 100 μl of 0.1 M EDTA in PBS and 200 μl of first antibody (NIH kit, diluted 1:1,000 with 2% NRS-PBS) were added to each tube, then gently mixed. After 24 h incubation at room temperature, sheep anti-rabbit gamma globulin (40 μl) was added and mixed and then incubated at 4° C for another 24 h. The pellet precipitated by centrifugation for 20 min at $575 \times g$ was counted in a gamma counter (Beckman 9000). The "blank tube" was treated in the same way with the exception that the first antibody was not added. The results for each animal were plotted as ln cpm of immunoprecipitated plasma radioactivity as a function of time. Two exponentials were found in the disappearance curve, so a single meaningful half-life ($t_{1/2}$) could not be calculated. The last three data points were used to obtain the rate constant for the slower component, which was then subtracted from the total disappearance curve to obtain the faster one. The initial volume distribution (V_d^I) and metabolic clearance rate (MCR) were calculated from the dose administered and the slopes and intercepts of these two exponential terms.[10, 11]

The T_4 disappearance rate was determined as described in our previous study[8] except for slight modification in which 0.1% bovine gamma globulin in barbital buffer and 30% polyethylene glycol (to separate bound from free) were used. Standard procedures were used to calculate $t_{1/2}$, V_d, and MCR.

Drugs

[^{125}I]T_4 (100 mCi/mg) and carrier free ^{125}I for iodination of rat TSH were purchased from Industrial Nuclear Co. and Amersham Corp., respectively. Synthetic TRH, L-T_4, bovine TSH and gamma globulin were obtained from Sigma Chemical Corp. (St. Louis, Mo.). TSH dissolved in phosphate buffer as a stock solution was diluted with saline and the other drugs were dissolved in 1–2% bovine serum albumin (BSA) for i.v. injection. Desamino-D-arginine vasopressin (DDAVP, Ferring Pharmaceuticals, Inc.) diluted with saline was infused at a rate of 48 ng/24 μl/day for seven days from osmotic minipumps (Alzet, Alza Corp.) implanted s.c. in DI and HZ (but not the control rat). Synthetic AVP purchased in a solution (Sigma Chemical Corp.) was diluted with saline and infused for seven days in the same manner as DDAVP. Infusion rates were 1.92 and 0.96 μg/24 μl/day in DI and HZ, respectively.

Schedules for VP replacement were as follows: After blood samples were obtained for basal and TRH-induced plasma TSH levels (pre-DDAVP control period), osmotic minipumps loaded with DDAVP were implanted s.c. Seven days later, the TSH response to TRH was determined (post-DDAVP period). The basal and TRH-induced TSH levels were again determined 26 days after termination of the DDAVP infusion (pre-AVP control period), and then osmotic minipumps loaded with AVP were implanted s.c. Seven days later, the TSH response to TRH was determined once again (post-AVP period).

The significance of a difference between two means was ascertained by the paired or unpaired Student's *t* test, as appropriate.

Thyroid glands for histological study were rapidly extirpated, fixed in Bouin's solution, sliced in 9 μm sections, and stained with hematoxylin and eosin.

Results

The effects of continuous infusion of DDAVP or AVP on urine volume and osmolality were determined in DI and it was found that both volume and osmolality were returned completely to the control levels (Figure 1, panels A and B) after seven days of DDAVP infusion (48 ng/day, s.c.). Polyuria and hypo-osmolality were again evident 26 days after termination of the DDAVP infusion. Both of these variables were only partially (but significantly) restored by the seven-day infusion of AVP (1.92 μg/day, s.c.). The time course of the normalization of urine volume and osmolality by DDAVP is shown in Figure 2.

Since the abnormalities of the pituitary-thyroid axis in HZ were similar to those in DI (described below), we also determined the effects of VP on urine production in HZ (Figure 1). The volume and osmolality of urine produced by HZ were quite normal before the initiation of the infusion of DDAVP. In response to this infusion, urine volume decreased and osmolality increased compared to the levels seen in control rats. Twenty-six days after the termination of the DDAVP infusion, control data were again collected (Figure 1, panel C). At this time, urine volume was slightly greater, and osmolality slightly less, in HZ than in control rats, although neither was radically different from the initial control values in HZ (i.e., before DDAVP). Such differences have been reported by others,[13] but this is the only occasion on which our HZ rats have exhibited significant differences in this regard; we cannot rule out the possibility that this is a result of the rats having previously been treated with DDAVP for seven days. In any case, both urine volume and osmolality were completely normal relative to control rats during a seven-day infusion of AVP (0.96 μg/day).

We then compared plasma levels of TSH and T_4 among the control rat, DI and HZ without VP treatment. As shown in Figure 3 (Panel A), basal plasma TSH levels in DI and HZ were similar to each other and were significantly higher than the control level. The TSH increments induced by TRH (250 ng/100 g BW, i.v.) in DI and HZ were also greater than the control increment. On the other hand, the plasma T_4 level in DI was significantly lower than that in the control rat, while the level in HZ was not different from the control level.

We also determined the effects of VP on the plasma levels of these hormones in the Brattleboro rats. In DI, neither DDAVP nor AVP had any effect on the levels of T_4 or TSH (basal or stimulated). In HZ, however, although the plasma T_4 level remained unchanged during the s.c. infusion of either DDAVP or AVP, the DDAVP did reduce the basal level of TSH, and both DDAVP and AVP restored the TSH responsiveness to TRH to the control level.

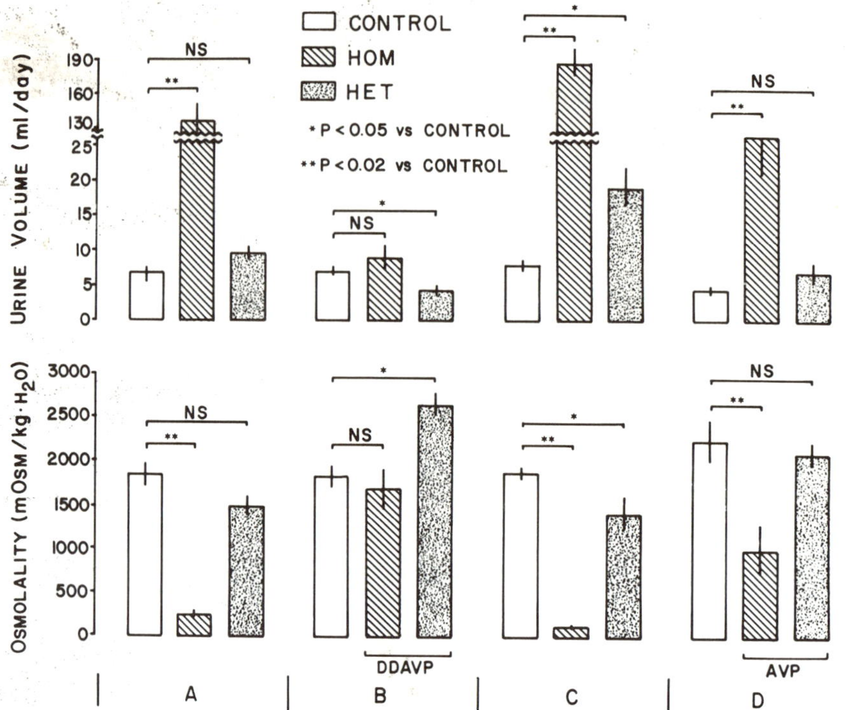

FIGURE 1. Urine volume and osmolality in control, homozygous (HOM), and heterozygous (HET) rats. Control data are presented in Panel A. Data in Panel B were obtained after seven days of DDAVP infusion. Control data in Panel C were collected 26 days after stopping DDAVP, and Panel D data were obtained seven days after AVP infusion. Results are presented as means ± SEM in 4–9 animals. The data from the DDAVP portion of this study (FIGURES 1 and 3) have been redrawn from our previous paper.[12]

In this study, we have used plasma hormone concentrations as an index of the rate of hormone secretion. However, since the former is also a function of hormone clearance rate, we compared the disappearance rates of radioactive T_4 and TSH in control, DI, and HZ. The results for T_4 are presented in FIGURE 4; the MCR's calculated from these data ranged from 1.5 to 1.6 ml/h/100 g BW and there were no significant differences among the three values. There were also no differences in the $t_{½}$ or the V_d values among the three groups. Multiexponential analysis of the TSH disappearance curves (see MATERIALS AND METHODS) revealed that the V_d^I and MCR for TSH are also completely normal in both DI and HZ.

In comparing the thyroid response to TSH among the three groups of rats, it was found that the plasma T_4 level increased significantly in response to bovine TSH (100 mU/100 g BW, i.v.) in the control rat, but not in either of the types of Brattleboro rats (FIGURE 5).

Since the above findings suggest that the thyroid responsiveness to circulating TSH is suppressed in DI and HZ, the thyroid glands from all three groups were subjected to histological examination (FIGURE 6). Control tissues were characterized by cuboidal follicular cells enclosing moderate amounts of colloid. In contrast, the cells in both DI and HZ were columnar and domed in association with a "scalloped" appearance of the colloid borders, which is typical of a stimulated thyroid gland.

DISCUSSION

Since Valtin *et al.*[14] reported the discovery of a strain of rats with familial hypothalamic diabetes insipidus (DI) in 1962, which was later named the Brattleboro strain,[15] many aspects of the physiology and pathology of these rats have been examined. Lacking the ability to synthesize and secrete VP, these rats provide an excellent model for the investigation of the physiological roles for endogenous VP. Shortly after the discovery of this strain, Galton *et al.*[7] studied several characteristics of thyroid hormone metabolism and thyroid function in the DI rat of this strain. These authors reported that the urinary excretion of ^{131}I following i.v. injections of [^{131}I]T_4 or Na^{131}I was greater in the DI rat than in the control rat. However, the greater excretion was due to excess urine production rather than to enhanced deiodination. Using the indices available at that time, there was no evidence of any abnormalities in thyroid function itself. Since that time, the thyroid axis of the Brattleboro rats has been virtually ignored. Thus, in the present studies we have used techniques developed since the studies of Galton *et al.* to re-examine the pituitary-thyroid axis in DI, HZ, and Long-Evans control rats, and to determine whether abnormalities that might exist can be reversed by physiological treatment with exogenous VP.

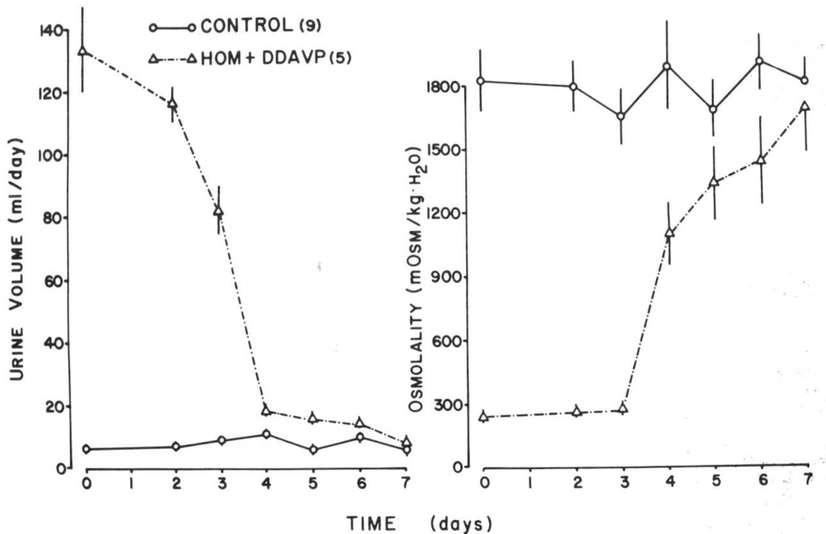

FIGURE 2. Time-dependent normalization by DDAVP of urine volume (left) and osmolality (right) in HOM. The DDAVP infusion was started at time zero in HOM, but the Long-Evans control rats were not treated with DDAVP. Results are expressed as means ± SEM. Numbers in parentheses indicate number of animals.

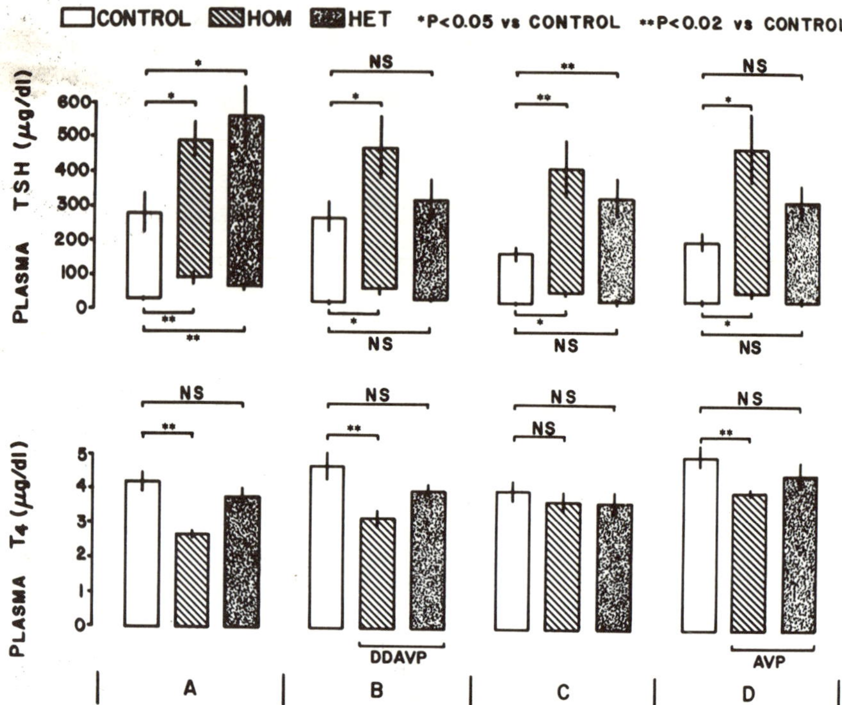

FIGURE 3. Plasma TSH (basal and TRH-induced) and T_4 concentrations in the same rats whose water balance data are presented in FIGURE 1. In top panels, the bottom and top ends of each column indicate basal and TRH-induced TSH levels, respectively.

Although implantable osmotic minipumps are being used increasingly in a variety of endocrine investigations, they are not yet widely used for the replacement of vasopressin preparations in Brattleboro rats. We feel that the use of such pumps is particularly advantageous in a study such as the present one since VP has a relatively short biological half life. Thus, we have accomplished a relatively long-term, steady-state return to normal urine production and osmolality, rather than the transients associated with pulsatile injections of large doses. We chose to test DDAVP in addition to the rats' natural hormone (AVP) because of the greater specificity of the DDAVP for antidiuretic rather than pressor activity. From our results in both DI and HZ it is clear that the antidiuretic effect of our DDAVP treatment is greater than that of our AVP treatment. The former treatment normalized water balance in DI and overcompensated in HZ, whereas the latter brought HZ to normal, but was only partially effective in DI. Since normalization of water balance was in no case able to normalize endocrine function, we conclude that the disruptions in the pituitary-thyroid axis in Brattleboro rats can not be attributed to excess water turnover. The independence of these endocrine effects from the state of water balance is analogous to the recent findings of Groblewski *et al.*[16] that VP infusion with osmotic minipumps (1 IU or 2.72 μg/day for seven days) reversed the polydipsia but did not affect certain behavioral abnormalities in DI.

The fact that there were no differences in disappearance rates of either plasma TSH or plasma T_4 between the control and the Brattleboro rats, makes us confident that the hormone concentrations that we have measured in plasma reflect rates of hormone secretion, not metabolism. With this information in hand, we suggest that the most parsimonious explanation for the increased plasma TSH levels in Brattleboro rats (at least in DI) is that there is a reduction in negative feedback inhibition of TSH secretion resulting from lowered levels of plasma thyroid hormones. Although not presented here, we have also found that plasma T_3 concentrations in both DI and HZ are lower than the control level.[12] The low plasma concentrations of thyroid hormones in the presence of high TSH levels in these rats suggest that the thyroid can not respond adequately to TSH. This contention was confirmed by our finding that the thyroids of both DI and HZ exhibit reduced responses to a challenge of exogenous TSH. Finding no previous histological study of the thyroid glands of Brattleboro rats, we proceeded to examine the tissues from our rats and found further evidence that the glands were obviously in a stimulated state, but were apparently unable to respond by releasing hormones.

Although the interpretation of our results from DI is rather straightforward, the results from HZ are a bit more complicated. The HZ rat has a partial VP deficiency, and at least in our hands, has urine production and osmolality values

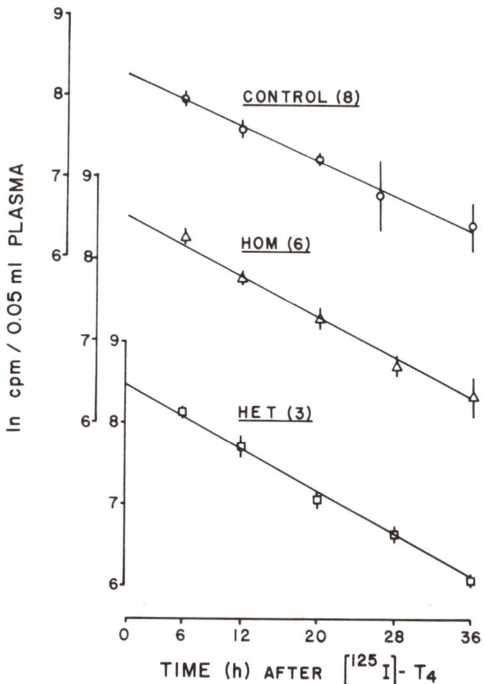

FIGURE 4. Disappearance of radioactive T_4 from plasma after bolus injection at time zero. Numbers of animals used are indicated in parentheses.

that do not differ from the Long-Evans control. In spite of this, the TSH levels in HZ are abnormally high to the same extent as observed in DI. The T_4 levels in HZ are not different from control, yet their elevated TSH levels are indicative of a hypothyroid state as was seen in DI. Certainly other explanations are possible, but it seems most likely that this situation is due to the fact that T_3 levels are reduced in HZ.[12] Thus, it appears that there are multiple abnormalities in the pituitary-thyroid axis of Brattleboro rats, some of which may be present only in DI or in HZ. However, it is equally apparent from our thyroid stimulations and histological findings that the abnormality within the thyroid gland is common to both DI and HZ.

In summary, our re-examination of the pituitary-thyroid axis in Brattleboro rats has revealed physiological and histological evidence of thyroid gland abnormalities. Although the results in HZ remain somewhat enigmatic, the resultant decrease in thyroid hormone concentrations in DI causes a compensatory increase in TSH secretion, and the restoration of water balance by continuous treatment with either AVP or DDAVP fails to reverse these endocrine abnormalities.

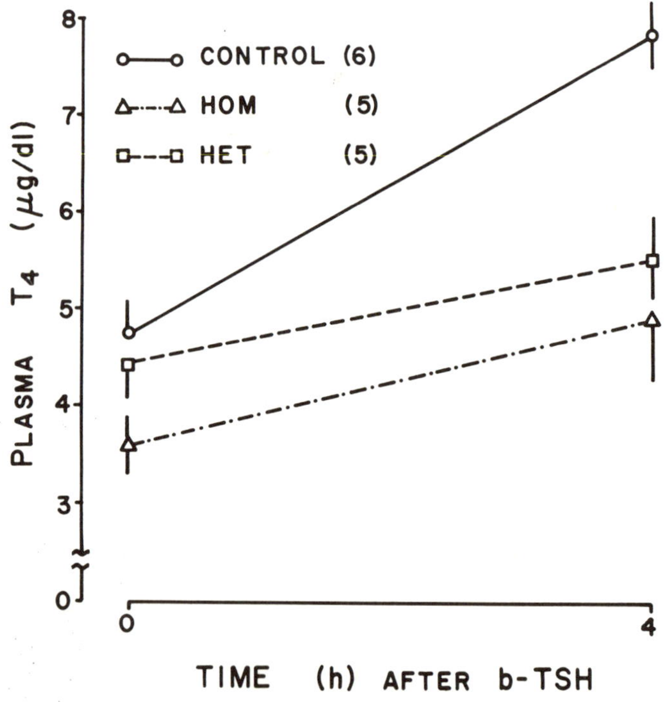

FIGURE 5. Plasma T_4 concentration before and 4 h after i.v. injection of bovine TSH (100 mU/100 g BW). One percent bovine serum albumin in saline as vehicle did not significantly change plasma T_4 levels in any of the three groups.

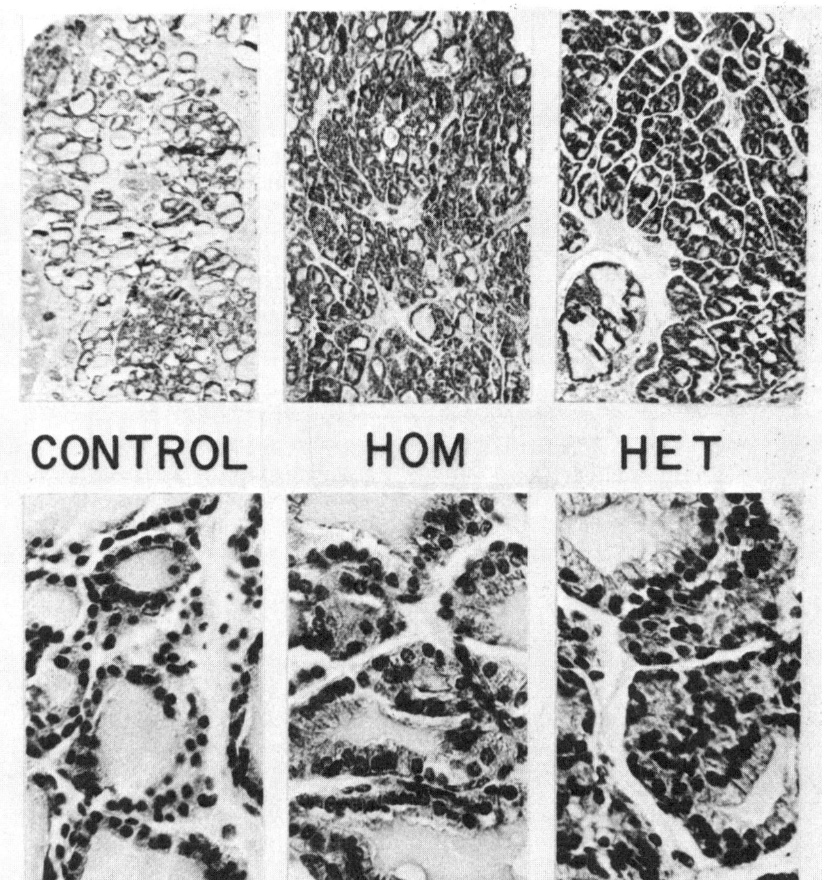

FIGURE 6. Low power (top, ×34) and higher power (bottom, ×340) photomicrographs of thyroid tissue from untreated control, HOM, and HET rats.

ACKNOWLEDGMENTS

We appreciate the very valuable consultations of Dr. Lewis B. Kinter. Smith Kline & French Laboratories, Philadelphia, in choosing vasopressin replacement rates, and of Dr. David E. Hinton, West Virginia University Medical Center, in interpreting the histological results. The experimental assistance of Marilyn Howton and Avonell Painter is also gratefully acknowledged.

REFERENCES

1. OGAWA, E., K. ARAI & K. SHIBATA. 1956. Studies on the thyroid uptake of ^{131}I. I. Effect of pituitary hormones. Endocrinol. Japon. **3:** 211–216.
2. PETERSON, R. E., R. F. CECH & R. W. JOHNSON. 1960. An effect of vasopressin on thyroid function in man. J. Lab. Clin. Med. **56:** 934–935.

3. LIPSCOMB, H., D. HATHAWAY & D. GARD. 1961. Direct effect of vasopressin on thyroid. Clin. Res. **9:** 29.
4. GARCIA, J., G. W. HARRIS & W. J. SCHINDLER. 1964. Vasopressin and thyroid function in the rabbit. J. Physiol. **170:** 487–515.
5. LABELLA, F. S. 1964. Release of thyrotrophin *in vivo* and *in vitro* by synthetic neurohypophysial hormones. Can. J. Physiol. Pharmacol. **42:** 75–83.
6. KRASS, M. E., F. S. LABELLA & S. R. VIVIAN. 1968. Thyrotropin release *in vitro:* The role of metabolism in the secretory response to vasopressin, oxytocin and epinephrine. Endocrinology **82:** 1183–1189.
7. GALTON, V. A., H. VALTIN & D. G. JOHNSON. 1966. Thyroid function in the absence of vasopressin. Endocrinology **78:** 1224–1239.
8. THOMPSON, M. E., G. P. ORCZYK & G. A. HEDGE. 1977. *In vivo* inhibition of thyroid secretion by indomethacin. Endocrinology **100:** 1060–1067.
9. CHOPRA, I. J. 1972. A radioimmunoassay for measurement of thyroxine in unextracted serum. J. Clin. Endocrinol. Metab. **34:** 938–947.
10. KANEKO, M., T. HIROSHIGE, J. SHINSAKO & M. F. DALLMAN. 1980. Diurnal changes in amplification of hormone rhythms in the adrenocortical system. Am. J. Physiol. **239:** R309–R316.
11. JACKSON, I. V., P. D. PAPAPETROU & S. REICHLIN. 1979. Metabolic clearance of thyrotropin-releasing hormone in the rat in hypothyroid and hyperthyroid states: Comparison with serum degradation *in vitro.* Endocrinology **104:** 1292–1298.
12. FUJIMOTO, S. & G. A. HEDGE. 1982. Altered pituitary-thyroid function in the Brattleboro rat with diabetes insipidus. Endocrinology. (In press.)
13. VALTIN, H., W. H. SAWYER & H. W. SOKOL. 1965. Neurohypophysial principles in rats homozygous and heterozygous for hypothalamic diabetes insipidus (Brattleboro strain). Endocrinology **77:** 701–706.
14. VALTIN, H., H. A. SCHROEDER, K. BENIRSCHKE & H. W. SOKOL. 1962. Familial hypothalamic diabetes insipidus in rats. Nature **196:** 1109–1110.
15. VALTIN, H. & H. A. SCHROEDER. 1964. Familial hypothalamic diabetes insipidus in rats (Brattleboro strain). Am. J. Physiol. **206:** 425–430.
16. GROBLEWSKI, T. H., A. A. NUNEZ & R. M. GOLD. 1981. Circadian rhythms in vasopressin deficient rats. Brain Res. Bull. **6:** 125–130.

DISCUSSION OF THE PAPER

K. STITZER (*University of Puerto Rico, San Juan, P.R.*): In your slide presentation you said that you have an increased thyroidal response to TSH but you have a decrease in the secretion of T_3 and T_4. That means the thyroidal response is decreased hormonal output.

HEDGE: It is the magnitude of the thyroid's responsiveness to stimulation by TSH that we feel is reduced in the Brattleboro rat. This notion was suggested by the occurrence of low T_3 and T_4 levels in the presence of elevated TSH, and it was confirmed by demonstrating that the T_3 and T_4 responses to an exogenous TSH challenge were lower in Brattleboro rats than in controls. The nature of the thyroidal defect is still unknown, but there is a whole series of intrathyroidal events that could be involved. So far, we only know (from our histological findings) that the very first step, the recognition of the presence of elevated TSH, is not likely to be compromised. Thus, it appears that the Brattleboro thyroid is constantly exposed to high levels of TSH and it is "aware" of this stimulation, but it cannot secrete the normal amount of thyroid hormone in response to it.

M. MILLER (*Veterans Administration Medical Center, Syracuse, N.Y.*): The set of findings that you present are quite typical of what a clinician would recognize in a patient with primary hypothyroidism. I would like to know if these animals express, clinically if you will, any manifestation of hypothyroidism. This is really quite important because, as you are well aware, a hypothyroid human or animal may have significant alterations in water handling ability. When studying water metabolism in these animals, one then has to keep in mind what potential contribution defective thyroid function might be making to the abnormal water regulation so as to separate this effect from alterations that are due directly to disturbances of vasopressin itself.

HEDGE: I completely agree with your assessment that the picture here is virtually identical to that of primary hypothyroidism. However, we have not yet examined any of the common clinical indices of hypothyroidism other than those from which the present data have been derived.

FETAL AND POSTNATAL MATURATION OF CORTICOTROPE FUNCTION IN THE VASOPRESSIN-DEFICIENT RAT (BRATTLEBORO STRAIN): A RADIOIMMUNOLOGICAL, IMMUNOCYTOCHEMICAL, AND MORPHOMETRIC STUDY *

P. Tankosic,† A. Burlet,† S. Jegou,§ M. Chateau,‡ H. Vaudry,§ C. Burlet,† and M. Boulangé ‡

† Laboratoire d'Histologie A
Faculté de Médecine
B.P. 184 54500 Vandoeuvre-lès-Nancy

‡ Laboratoire de Physiologie B
Faculté de Médecine
B.P. 184 54500 Vandoeuvre-lès-Nancy

§ Laboratoire d'Endocrinologie
Faculté des Sciences de Rouen
76130 Mont-Saint-Aignan
France

INTRODUCTION

The antidiuretic hormone, vasopressin (VP) is not exclusively located at the cerebral level in the hypothalamo-neurohypophyseal tract. VP has been found in the fibers of the external layer of the median eminence in contact with the portal vessels [1–4] and its presence has also been suggested in the anterior lobe of the hypophysis (AL) in man,[5] monkey,[6] and rat.[7] The biochemical and biological methods used by Chateau *et al.*[7] had shown that chronic and experimental stimulations of ACTH secretion overload the VP infundibular endings and increase the VP adenohypophyseal content. Dexamethasone injected into adrenalectomized rats prevents VP accumulation in both median eminence and AL.[4] These data are in agreement with many reports on participation of VP in the corticotrope function, but its exact role is still under discussion. Nevertheless, VP has been distinguished from hypothalamic corticotrophin-releasing factor (CRF).[8–13] Suggestions have been made that VP might potentiate the response of the adenohypophysis to hypothalamic CRF.[14]

Studies on the Brattleboro rat, genetically devoid of VP and its carrier protein,[15, 16] should give more information on the physiological role of VP. Data on the corticotrope function and its regulation in the Brattleboro rats have demonstrated some anomalies characterized by a weakened sensitivity to ACTH,[17, 18] a diminished corticosteroidogenic response to stress,[14, 17, 19, 20] and lower,[13] equal [17, 19–21] or higher [18] ACTH concentration at AL level. The present study was carried out to re-examine some aspects of the regulation of the corticotrope function in the adult Brattleboro rat and see if the perturbations observed exist in the developing rat. A role for VP in the maturation of the ACTH cells is discussed.

* Supported by the Délégation Générale à la Recherche Scientifique et Technique (Grant No. 80.7.0352.).

0077–8923/82/0394–0560 $1.75/0 © 1982, NYAS

Materials and Methods

Animals

Experiments were performed on animals bred in our animal facilities. Rats of the Long-Evans hooded strain (LE) were purchased from R. Janvier (Le Genest, France). A Brattleboro rat colony was established in the laboratory from male adult DI rats, which were kindly supplied by Dr. Rossbach (Switzerland). Our usual breeding program consisted in crossing DI males first with LE female rats and then with HZ females. At six and eight weeks of age, rats were classified on the basis of daily urine output and urinary osmolality. Young rats, too small for measurements in metabolism cages, were classified according to the presence, at the pituitary level, of VP detected by a radioimmunoassay (RIA) or immunocytochemistry. For young rats, the date of birth was considered as "Day 1." DI fetuses were obtained by crossing DI males with DI females. Animals were caged overnight from 6.00 p.m. to 9.00 a.m. If spermatozoa were found in the vaginal smears the next morning, this day was noted as "Day 1" of gestation and females were then put into individual cages.

Nychthemeral Rhythm

Nychthemeral rhythm of plasma ACTH was studied on adult DI males and females synchronized by a 14 Light (8.00)–10 Dark (6.00) photoperiod in a temperature-controlled room ($24 \pm 1°$ C).

Adrenalectomy

A bilateral adrenalectomy was performed on adult DI females by dorsal surgical approach and under ether anesthesia. Animals received a 0.9% NaCl solution as drinking water. They were killed 14 days after surgery. Intact rats were sham-operated and drank tap water.

Management with 1-Desamino-8-D-Arginine Vasopressin (DDAVP)

Two experimental groups, intact and adrenalectomized DI females received intramuscular injections of DDAVP (2×20 ng) twice a day.

RIA Procedures

74 LE, 98 HZ, and 98 DI infant rats were killed by decapitation at 1, 4, 8, 10, 12, 15, 18, 21, and 30 days of age, and adult rats at two months. The peptides were extracted from hypophyseal tissues according to the technique of Amatruda *et al.*[22] AL and neurohypophysis (for VP determination) were removed and immersed in ice-cold acetone for 24 hours. The tissues were vacuum-dried. After weighing on a Cahn-gram electro-balance, the tissue samples were homogenized in 0.04 M acetic acid solution. After centrifugation, the supernatant was frozen to $-30°$ C.

The ACTH content in adult rats was determined using a standard RIA kit (C.E.A., Gif-sur-Yvette, France) allowing a detection of 10 pg ACTH/ml. In young rats, the ACTH content was determined using an antibody raised against the 1–24 fraction of the ACTH molecule. For the RIA, an aliquot of 100 μl was incubated with 50 μl of ^{125}I-labeled ACTH (6,000 cpm) and 300 μl of the antibody at a final dilution of 1/400,000. A standard curve was set up with unlabeled human ACTH, dilutions ranging from 2 to 625 pg/tube. The sensitivity of the antibody was 4 pg/tube. All the assays were carried out in duplicate and at two different dilutions of the sample. The presence of VP in neurohypophyseal extracts was sought using the previously described anti-AVP serum.[23]

Immunocytochemical Procedures

Pregnant rats were anesthetized and fetuses of 15–21 days of gestation (41 LE, 39 DI) were removed by caesarean section. 21 LE, 25 HZ, and 42 DI postnatal rats were killed by decapitation at 1, 4, 8, 12, 15, 21, and 30 days of age. Fetus heads and dissected brains plus hypophysis were fixed with Gerard fixative (80% Bouin, 10% neutral formaldehyde, 10% saturated $HgCl_2$), dehydrated, and embedded in paraffin. Sagittal sections, 8 μm thick, were performed on fetuses. ACTH cells were stained using an anti- $^{1\text{-}24}$ACTH serum (72 hours at 4° C) at 1/400 dilution and a sheep anti-rabbit γ-globulin conjugated with fluorescein isothiocyanate (Institut Pasteur, Paris) for 45 min. Sections were counterstained with Evans Blue (0.01%) and mounted in buffered glycerin. Postnatal maturation of corticotrope function was carried out on serial sagittal sections, 5 μm thick, using the same antibody and an immunoenzymatic technique using peroxidase-anti-peroxidase complex (PAP), according to the method of Sternberger *et al.*[24] The reaction product was visualized with 3,3′-diamino-benzidine tetrahydrochloride in Tris-buffered saline (pH 7.6), the filtered solution was used with 0.01% hydrogen peroxide. The same method and specific anti-AVP serum previously described[4] were used for VP determination.

Specificity of the Anti $^{1\text{-}24}$ACTH Sera

The specificity of the antisera was sought using several natural or synthetic polypeptides related to ACTH. For the RIA, the cross-reactivities with the specific antiserum were calculated from the ratios of the inhibition curves on a weight basis. The specificity of the immunocytochemical reaction was tested as usual: the specific antiserum was replaced by a phosphate-buffered saline, a normal pre-immune serum or the specific serum previously adsorbed (1 hour at 37° C) with graded concentrations of ACTH-related polypeptides. All the data were reported in TABLE 1.

Morphometric and Stereologic Analysis

A morphometric study was performed on the serial 5 μm thick sections using an electronic image analyzer Quantimet 720 (QTM) (Cambridge Instrument, Ltd). The method is based on the principle of Delesse[25] who proved that the volume density (V_V) of a component can be estimated on random sections by measuring the relative areas of their profiles, also called the areal density (A_A) of the profiles on section. This method depends on the localization and distribution of ACTH cells in the tissue, i.e. anisotropy. The tissue components must be randomized, by an adequate method of sampling, so that the samples studied are representative of the whole organ. We have used a systematic sampling, measuring several sections spaced by a constant thickness. Two parameters were measured on the QTM at a constant magnification: the area of the AL section (a_T) and the area of the ACTH cells (a_O). This analytic step was done by sorting out the image points on the basis of gray level values. The discriminator selected a range of gray values within which all picture points belonging to ACTH-stained cells should fall. The measure was taken into account when the original image (FIGURE 1a) was superimposed with the electronic image (FIGURE 1b) displayed on the screen of the QTM. The gray level threshold was also selected so that a cell with a lesser ACTH content was detected in the same way as a heavily loaded cell. The ratio of the sum of cellular areas ($\Sigma\, a_O = A_O$) to the sum of AL areas ($\Sigma\, a_T = A_T$) was evidently the areal density of ACTH cells, A_A. According to the principle of Delesse, if the sample was representative of the cell population, we evidently had $V_V = A_A$.

TABLE 1
SPECIFICITY OF THE ANTI$^{1-24}$ACTH SERA

Peptides	RIA Cross-Reactivities	Inhibition of the Immunostainings
$^{1-24}$ACTH (synacthen)	1	0.5 μg/ml
Porcine ACTH	3.7×10^{-1}	–
Rat ACTH	1.3×10^{-1}	–
$^{1-10}$ACTH	10^{-6}	> 100 μg/ml
$^{1-16}$ACTH amide	2×10^{-4}	75 μg/ml
$^{4-10}$ACTH	10^{-6}	> 100 μg/ml
$^{11-19}$ACTH amide	3.8×10^{-1}	> 100 μg/ml
$^{11-24}$ACTH	6×10^{-3}	50 μg/ml
$^{17-39}$ACTH	10^{-6}	100 μg/ml
α-MSH	10^{-5}	100 μg/ml
Human β MSH	10^{-6}	–
Bovine β MSH	10^{-5}	–

For the RIA, the cross-reactivities of peptides were calculated from the ratios of the inhibition curves on a weight basis.

For the immunocytochemical procedure, specific antiserum (dilution: 1/400) was absorbed with synthetic or natural peptides for 1 hour at 37° C.

Total AL volume was obtained in multiplying the sum of AL areas (A_T) by the distance between two equidistant serial sections. Finally, the product between V_V and V_T gave the volume of the ACTH cells. It could be considered as a reflection of the number of corticotrope cells.

Statistics

Data presented as means are accompanied by their standard errors. Statistical evaluation of differences in mean values were calculated using Student's *t* test.

RESULTS

Regulation of Corticotrope Function in the Adult Brattleboro Rat Nychthemeral Rhythm (TABLE 2)

TABLE 2 shows the nychthemeral variations of plasma ACTH in male and female DI rats. The amplitude of the fluctuations was remarkably high since

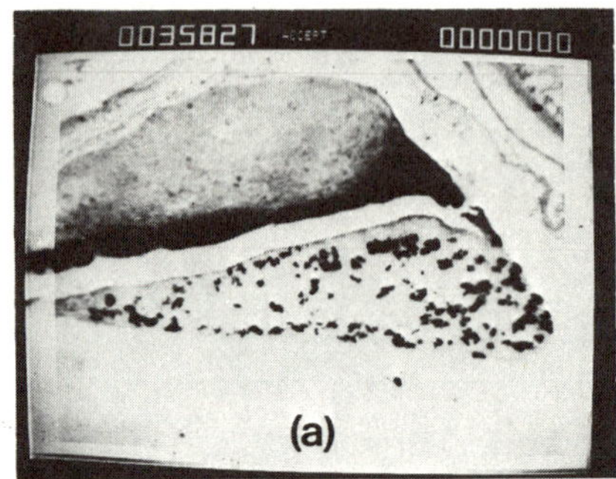

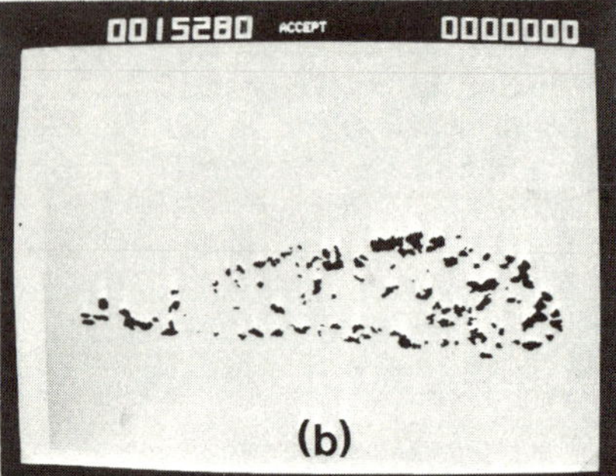

FIGURE 1. Morphometric study of a sagittal section of the pituitary treated with an anti $^{1-24}$ACTH serum. Real (a) and electronic (b) images are visualized on the screen of a QTM 720 image analyzer.

the values, in female DI, rose from a minimum of 54 to a maximum of 664 pg ACTH/ml, this latter value being observed during the dark period (02.00)

Effects of Adrenalectomy (TABLE 3)

Adrenalectomy did not induce any compensatory hypertrophy of the AL structures. We observed a significant increase in the AL ACTH content (×1.6) and especially in the plasma ACTH concentration (×17).

Management with DDAVP (TABLE 3)

In the intact DI rat, the twice daily injections of DDAVP induced a weak increase in the plasma ACTH and a similar increase in the pituitary ACTH content when compared to adrenalectomized rats. DDAVP treatment did not significantly modify the effects induced by adrenalectomy.

TABLE 2

CHANGES IN THE PLASMA ACTH CONTENT OF THE BRATTLEBORO RAT DURING THE NYCHTHEMERE (14L-10D)

Time of Measurements		Plasma ACTH (pg/ml)	
		DI ♀	DI ♂
Light period	10 h	385 ± 99 (7)	191 ± 93 (4)
	14 h	54 ± 3 (6)	56 ± 6 (6)
	18 h	90 ± 13 (5)	357 ± 117 (4)
Dark period	22 h	272 ± 98 (8)	453 ± 36 (6)
	2 h	664 ± 116 (6)	–
	6 h	288 ± 78 (7)	224 ± 76 (6)

Values are means ± SEM; numbers between parentheses indicate the number of rats.

TABLE 3

EFFECT OF ADRENALECTOMY ON THE CORTICOTROPE FUNCTION OF THE BRATTLEBORO RAT-MANAGEMENT WITH DDAVP

	Experimental Groups of DI			
	Intact	DDAVP	Adrenalectomy	Adrenalectomy + DDAVP
Pituitary (mg)	0.992 ± 0.180 (5)	1.101 ± 0.114 (7)	1.018 ± 0.078 (13)	0.863 ± 0.269 (7)
Pituitary ACTH (ng/AL)	457 ± 29 (5)	705 ± 106 * (7)	730 ± 113 † (13)	647 ± 170 * (7)
Plasma ACTH (pg/ml)	54 ± 3 (5)	183 ± 34 ‡ (7)	926 ± 86 ‡ (13)	797 ± 76 ‡ (7)

Intact rats were sham-operated and fed a normal diet.
Adrenalectomized rats were given a 0.9% NaCl solution as drinking water. They were killed 14 days after surgery.
Treated rats received twice a day intramuscular injections of 1-desamino-8-D-arginine vasopressin (2 × 20 ng).
Values are means ± SEM; numbers between brackets indicate the number of rats.
* Compared to intact rats $p < 0.02$, † Compared to intact rats $p < 0.01$, and ‡ Compared to intact rats $p < 0.001$.

Maturation of Corticotrope Function

Ontogenesis of ACTH Cells (FIGURE 2)

The first ACTH cells appeared in the LE rat on the morning of the 16th day of gestation (A), at the basal part of the AL. A few immunoreactive cells were always found in the LE at this period. Eight out of the nine DI rats studied were devoid of any ACTH-containing cell (G). They were observed in the afternoon of the 16th day (H), four to five hours later than in the LE rats. Thereafter, they increased in number in the two rat strains and were scattered throughout the whole AL. However, cell proliferation, and especially fluores-

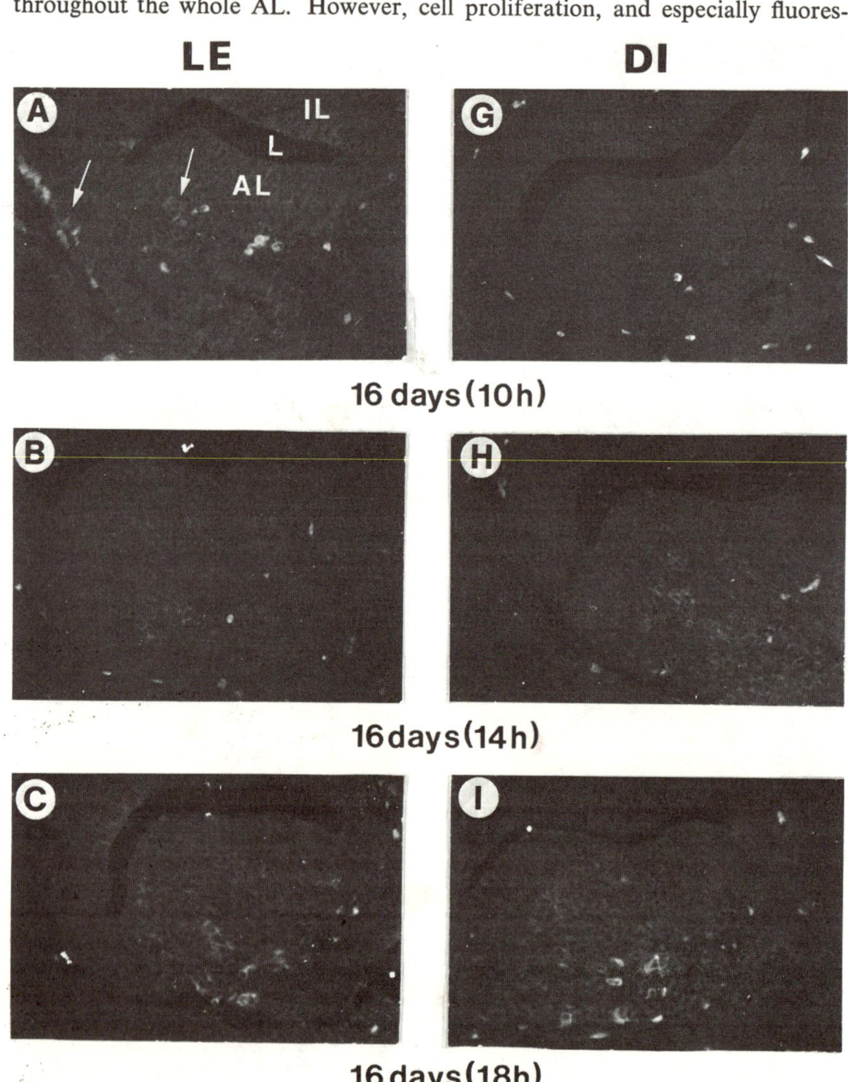

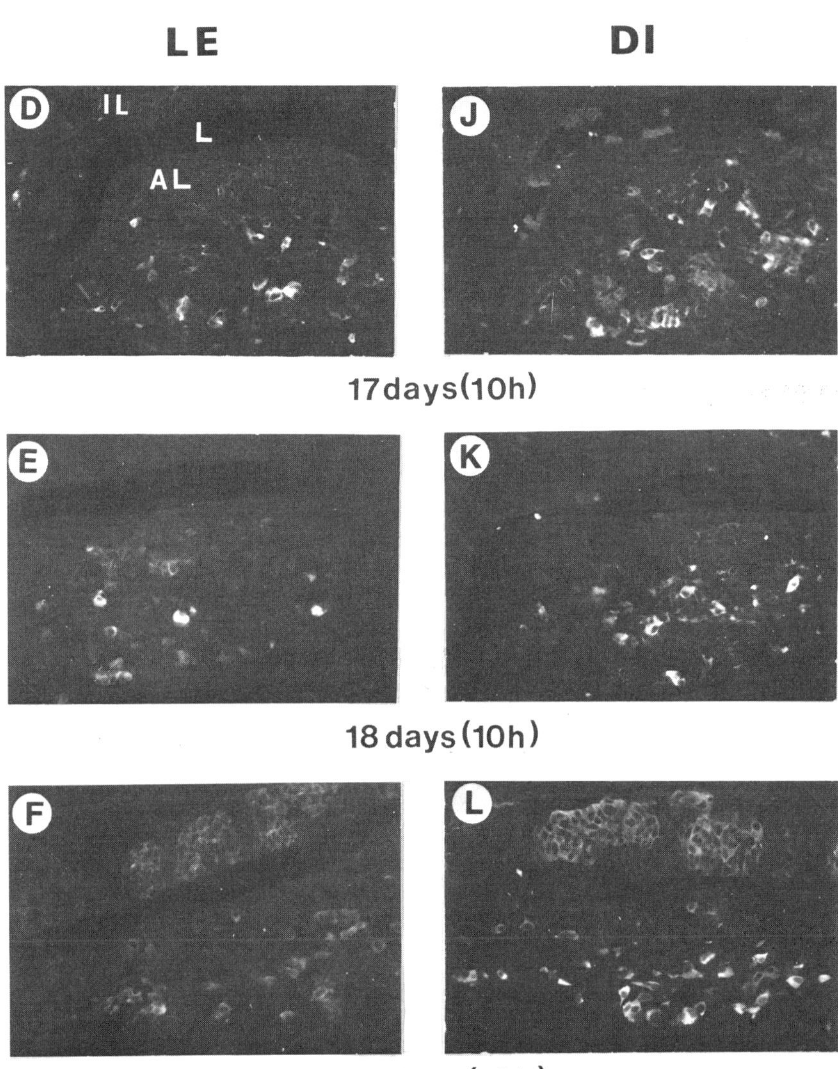

FIGURE 2. Ontogenesis of the corticotroph cells, stained by an anti $^{1-24}$ACTH serum, in the hypophysis of fetal LE and DI Brattleboro rats. ACTH cells appeared in the LE rats on the morning of the 16th day of gestation (A) (→) and were absent in the DI rats (G). The immunoreactive content is less in the LE than in the DI rats until the 18th day of gestation. AL: anterior lobe; IL: intermediate lobe; L: residual lumen. (150×).

cence intensity, seemed more significant in the DI rats, this fact being observable as early as the evening of day 16 (I), but also on day 17, and to a lesser extent on the 18th day of gestation. During the subsequent steps of development, such a disparity tended to diminish and the hormonal load seemed similar in LE and DI rats.

Postnatal Maturation of Corticotrope Function (FIGURES 3–5)

The RIA ACTH content (FIGURE 3a) was weak at birth (0.1 ng ACTH/AL). We noted a low increase until the middle of the second week of postnatal life, followed by a flat period, and then an accelerated hormonal synthesis at weaning. Between the 1st and the 10th day of life, the ACTH content was identical in LE, HZ, and DI rats (10th day: 41 to 49 ng ACTH/AL). In 21 day-old rats, the ACTH content was six-fold in LE, five-fold in HZ, and three-fold in DI rats. Therefore, a significant decrease of about 30% was observed in DI rats. The rate of increase for ACTH rapidly accelerated after 3 weeks of age: LE 995 ± 68 ng ACTH/AL; HZ 872 ± 46; and DI 618 ± 67 ng ACTH/AL. If data were expressed in ng ACTH/mg AL (FIGURE 3b), the observed differences were reduced (15% compared with LE rats) because of the AL hypotrophy in the DI.

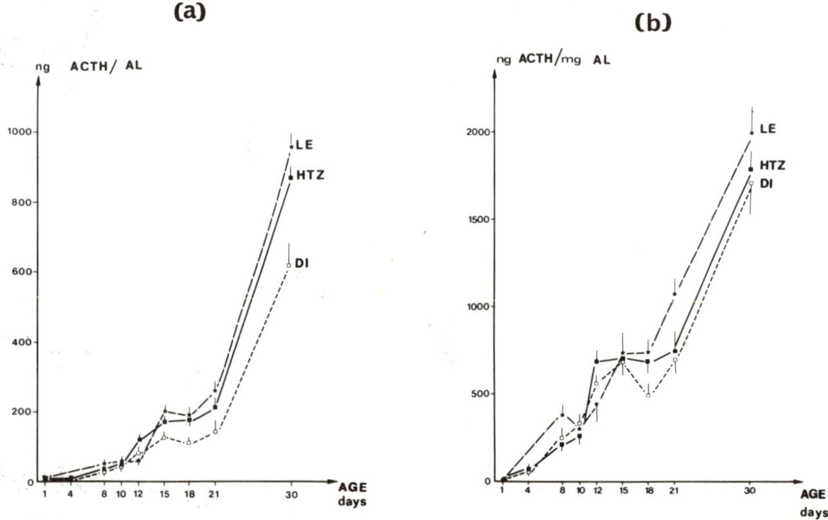

FIGURE 3. Postnatal evolution of the ACTH content (a) and concentration (b) in the LE, HZ, and DI Brattleboro rats. Vertical bars indicate standard errors of the means.

The morphometric data showed a heterogeneous distribution of ACTH cells (FIGURE 4), which were more concentrated in the lateral regions of AL. A good estimation for V_V and V_T was obtained by measuring one section every 200 μm. The data showed a high value for V_V at birth (FIGURE 5a): LE 21.7 ± 1.5%; HZ 16.1 ± 2.6%; and DI 15.3 ± 0.9%. Thereafter V_V decreased and reached a constant value during the second week of age. Significant differences were observed from the end of the first week between LE and HZ rats, but as soon as birth between LE and DI rats. No significant differences were noted between HZ and DI rats. The postnatal evolution of the ACTH cellular volume was similar between LE and HZ rats during the whole period of study. Significant differences in the corticotrope volume existed between LE and DI rats. Between the 1st and the 21st day, values rose from 26.8 ± 2.5 to

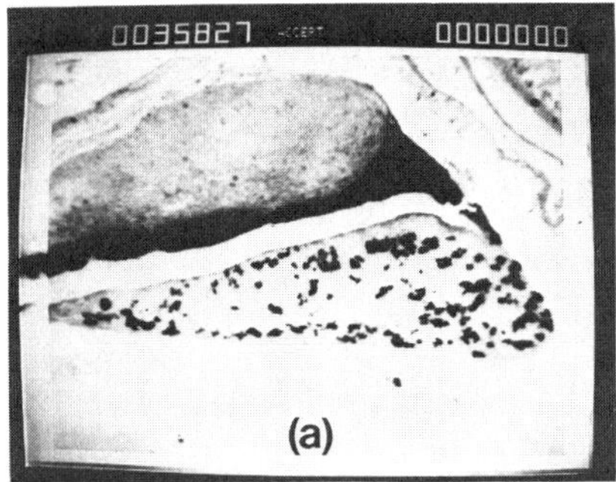

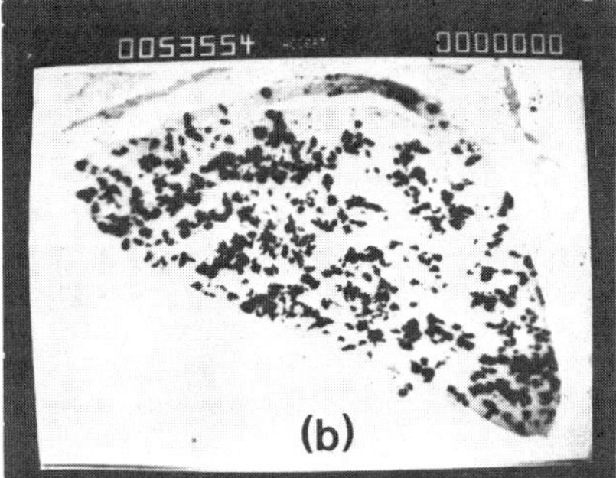

FIGURE 4. Sagittal sections of the pituitary treated with an anti [1-24]ACTH serum. ACTH cells were less concentrated in the medial (a) than in the lateral region.

$121.4 \pm 15.4 \times 10^{-3}$ mm^3 in LE; 20.2 ± 3.4 to $105.1 \pm 9.1 \times 10^{-3}$ mm^3 in HZ, and 17.6 ± 2.1 to $52.1 \pm 4.2 \times 10^{-3}$ mm^3 in DI rats. The differences were reduced after 3 weeks of age.

DISCUSSION

These experiments show that, in the DI rat, the nychthemeral rhythm exists with a chronology similar to that reported by several investigators in other rat strains.[26-29] However, the amplitude of fluctuations seems to be greater in the DI Brattleboro strain.

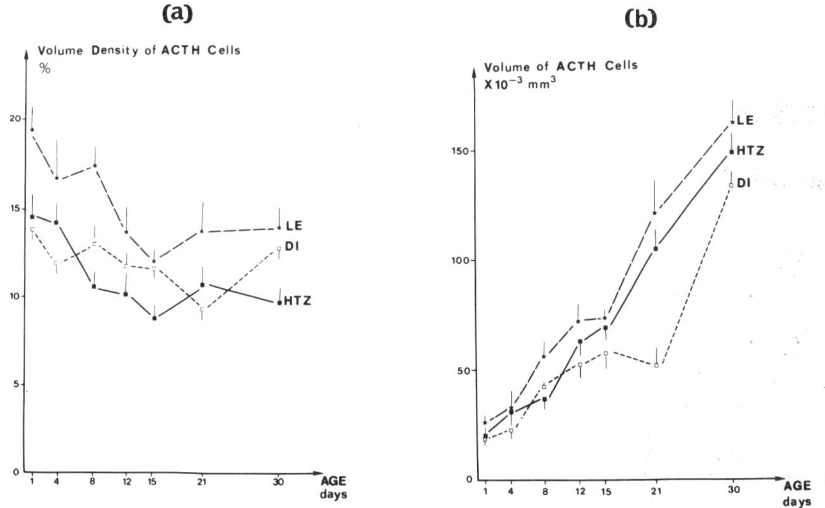

FIGURE 5. Postnatal evolution of the volume density and total volume of ACTH cells in the LE, HZ, and DI Brattleboro rats. Vertical bars indicate standard errors of the means.

Adrenalectomy induces an increase in the pituitary ACTH content and in the plasma ACTH concentration, similar to those demonstrated in the DI [21] or in the Wistar [13, 21] rat. In normal rats, without VP deficiency, adrenalectomy induces a parallel increase in the adenohypophyseal VP concentration [7] and a specific overloading in the VP infundibular fibers.[21] In the DI rat, DDAVP treatment, which restores a quite normal renal physiology (decreased urine flow, increased urinary osmolarity), does not modify the effects of adrenalectomy. In addition to an apparent nychthemeral release of ACTH, the DI rat presents a regulated corticotrope function even in the absence of VP.

The most marked anomalies are observed during the early stages of development. The first ACTH cells appear in the LE rat on the morning of the 16th day of gestation. These data are in agreement with the observations of Watanabe and Daikoku [30] in the Sprague-Dawley rat. Other investigators [31, 32] have found a later appearance of the ACTH cells in the Wistar rat (day 17 in our chronology). The reason for this discrepancy remains unknown. Thereafter, immunoreactive cells multiply but the fluorescence intensity, which reflects the hormonal load, is scarce from the 16th to the 18th day of gestation. This fact maybe reflects an ACTH release rather than a slower hormonal synthesis, since Dupouy [33] observed at this stage a decrease in the pituitary ACTH and a parallel increase in the plasma corticosterone.

The DI rat shows a delay in the appearance of the first ACTH cell and an overloading of the hormonal content between the 16th and the 18th day of gestation. In Dupouy's hypothesis, this fact could be interpreted as a slower or a delayed ACTH release. The DI rat presents a similar postnatal evolution of the corticotrope function when compared to Wistar rat,[34] the curves showing a plateau during the same period. However, the DI rat is characterized by a lower ACTH content than in the LE or HZ rats. In terms of concentration, the differences are reduced but they still exist. These observations tend to demonstrate that the decreased ACTH content is not only induced by the AL hypotrophy. The diminished ACTH levels which are found in the adult DI rat [13, 18] occur in the earlier developmental stages. The morphometric data on the corticotrope volume suggest that the decreased pituitary ACTH may be due to a lesser number of ACTH cells rather than a decreased hormonal synthesis.

It is now obvious that VP plays a role in ACTH release. VP found in the infundibular endings and in the AL of hypophysis is related to corticotrope function.[4, 7] VP has a corticotropin-releasing function when it is being released in the primary portal circulation.[19] The corticotropin-releasing activity of VP has been demonstrated *in vivo* [35] and *in vitro* [36] and its physiological action may be a potentiation of the CRF. Nevertheless, the lack of VP in the DI rat does not dramatically affect the mechanism of corticotrope function—although diminished, it remains regulated. If the decreased ACTH release in the fetus and the decreased pituitary ACTH content can be explained by a failure of VP to potentiate the CRF action, we believe that the delayed appearance of ACTH cells in the fetal DI rat and the reduced cell concentration and number in the postnatal DI rat seem to be due to a slower or a delayed cellular development. These data may suggest a role of VP in the maturation of the corticotrope cell. Has VP a direct action on the corticotrope cell, on eventual membrane receptors,[37] or is it mediated by growth hormone, since the DI rat shows a deficit in this hormone? [38] Some further investigations are necessary to clear up this action.

ACKNOWLEDGMENTS

We gratefully acknowledge the technical assistance of Mrs. E. Mary, B. Fernette, and E. Angel, and the secretarial work of Miss J. Bara.

REFERENCES

1. BURLET, A., J. MARCHETTI & J. DUHEILLE. 1974. Immunohistochemistry of vasopressin: study of the hypothalamo-neurohypophyseal system of normal, dehydrated and hypophysectomized rats. *In* Neurosecretion, the Final Endocrine Pathway. F. Knowles & L. Vollrath, Eds.: 24. Springer Verlag. Berlin.
2. BURLET, A., M. CHATEAU & J. MARCHETTI. 1976. Contribution of immunoenzymatic techniques to the study of diencephalic localization of vasopressin. *In* Immunoenzymatic Techniques. G. Feldman, P. Druet, J. Bignon & S. Avrameas, Eds.: 333. Elsevier/North-Holland. Amsterdam.
3. STILLMAN, M. A., D. L. RECHT, S. L. ROSARIO, S. M. SEIF, A. G. ROBINSON & E. A. ZIMMERMAN. 1977. The effects of adrenalectomy and glucocorticoid replacement on vasopressin-neurophysin in the zona externa of the median eminence of the rat. Endocrinology **101:** 42–49.
4. BURLET, A., M. CHATEAU & P. CZERNICHOW. 1979. Infundibular localization of vasopressin, oxytocin and neurophysins in the rat: its relationship with the corticotrope function. Brain Res. **168:** 275–286.
5. BARNAFI, L. & H. CROXATTO. 1965. Vasopressor and oxytocic activities in human hypothalamus, posterior and anterior lobes in the pituitary glands. Acta Endocr. **48:** 177–185.
6. ZIMMERMAN, E. A., P. W. CARMEL, M. K. HUSAIN, M. FERIN, M. TANNENBAUM, A. G. FRANZ & A. G. ROBINSON. 1973. Vasopressin and neurophysin: high concentrations in monkey hypophyseal portal blood. Science **182:** 925–927.
7. CHATEAU, M., J. MARCHETTI, A. BURLET & M. BOULANGE. 1979. Evidence of vasopressin in adenohypophysis: research into its role in corticotrope activity. Neuroendocrinology **28:** 25–35.
8. SCHALLY, A. V. & C. Y. BOWERS. 1964. Corticotrophin releasing factor and other hypothalamic peptides. Metabl. Clin. Exptl. **13:** 1190–1205.
9. RAMIREZ, V. D. & S. M. MCCANN. 1964. Thioglycollate stable luteinizing hormone and corticotrophin releasing factors. Am. J. Physiol. **207:** 441–445.
10. CHAN, L. T., S. M. SCHAAL & M. SAFFRAN. 1969. Properties of the corticotrophin releasing factor of the rat median eminence. Endocrinology **85:** 644–651.
11. MULDER, A. H., J. J. GEUZE & D. DE WIED. 1970. Studies on the subcellular localization of corticotrophin releasing factor (CRF) and vasopressin in the median eminence of the rat. Endocrinology **87:** 61–79.
12. PORTANOVA, R. & G. SAYERS. 1973. Isolated pituitary cells: CRF like activity of neurohypophysial and related polypeptides. Proc. Soc. Exp. Biol. Med. **143:** 661–666.
13. BUCKINGHAM, J. C. & J. H. LEACH. 1980. Hypothalamo-pituitary-adrenocortical function in rats with inherited diabetes insipidus. J. Physiol. (London) **305:** 397–404.
14. YATES, F. E., S. M. RUSSELL, M. L. DALLMAN, G. A. HEDGE, S. M. MCCANN & A. P. S. DHARIWAL. 1971. Potentiation by vasopressin of corticotrophin release induced by CRF. Endocrinology **88:** 3–15.
15. VALTIN, H., H. A. SCHROEDER, K. BENIRSCHKE & H. W. SOKOL. 1962. Familial hypothalamic diabetes insipidus in rats. Nature **196:** 1109–1110.
16. VALTIN, H. 1967. Hereditary hypothalamic diabetes insipidus in rats (Brattleboro strain). A useful experimental model. Am. J. Med. **42:** 814–827.

17. WILEY, C., F. PEARLMUTTER & R. E. MILLER. 1974. Decreased adrenal sensitivity to ACTH in the vasopressin deficient (Brattleboro) rat. Neuroendocrinology **14:** 257–270.
18. KRIEGER, D. T. & A. LIOTTA. 1977. Pituitary ACTH responsiveness in the vasopressin deficient rat. Life Sci. **20:** 327–377.
19. MCCANN, S. M., J. A. RODRIGUEZ, R. NALLAR & H. VALTIN. 1966. Pituitary adrenal function in the absence of vasopressin. Endocrinology **19:** 1058–1064.
20. ARIMURA, A., T. SAITO, C. Y. BOWERS & A. SCHALLY. 1967. Pituitary adrenal activation in rats with hereditary diabetes insipidus. Acta Endocrinol. **54:** 155–165.
21. SCHLEIFFER, R., C. MIALHE, B. BRIAUD; B. LUTZBUCHER & B. KOCH. 1979. Effects of adrenalectomy and hypercorticism on the ACTH content of the anterior and the posterior pituitary in rats with inherited diabetes insipidus (Brattleboro strain). Hormone & Metab. Res. **11**(2): 130–135.
22. AMATRUDA, T. T., P. J. MULROW, J. C. GALLAGHER & W. H. SAWYER. 1963. Carcinoma of the lung with inappropriate antidiuresis. N. Eng. J. Med. **269:** 544–549.
23. TANKOSIC, P., M. CHATEAU & A. BURLET. 1980. Evolution postnatale du contenu en hormones neurohypophysaires chez le rat Brattleboro. C. R. Soc. Biol. (Nancy) **174:** 21–27.
24. STERNBERGER, L. A. & P. H. HARDY. 1970. The unlabeled antibody enzyme method of immunochemistry. Preparation and properties of soluble antigen-antibody complex (Horseradish peroxydase-antihorseradish peroxydase) and its use in identification of Spirochetes. J. Histochem. Cytochem. **18:** 315–333.
25. DELESSE, M. A. 1847. Procédé mécanique pour déterminer la composition des roches. C. R. Acd. Sci. (Paris) **25:** 544–545.
26. REES, L. A., D. M. COOK, J. W. KENDALL, C. F. ALLEN, R. M. KRAMER, J. C. RATCLIFFE & R. A. KNIGHT. 1971. A radioimmunoassay for rat plasma ACTH. Endocrinology **89:** 254–261.
27. USATEGUI, R., C. OLIVER, H. VAUDRY, G. LOMBARDI, I. ROZENBERG & A. M. MOURRE. 1976. Immunoreactive α-MSH and ACTH in rat plasma and pituitary. Endocrinology **98:** 189–196.
28. CHIAPPA, S. A. & G. FINK. 1977. Hypothalamic luteinizing hormone releasing factor and corticotrophin releasing activity in relation to pituitary and plasma hormone levels in male and female rats. J. Endocrinol. **72:** 195–210.
29. SZAFARCZYK, A., M. HERY, E. LAPLANTE, G. IXART, I. ASSENMACHER & C. KORDON. 1980. Temporal relationships between the circadian rhythmicity in plasma levels of pituitary hormones and in hypothalamic concentrations of releasing factors. Neuroendocr. **30:** 369–376.
30. WATANABE, Y. G. & S. DAIKOKU. 1979. An immunohistochemical study on the cytogenesis of adenohypophysial cells in fetal rats. Dev. Biol. **68:** 557–567.
31. SETALO, G. & P. K. NAKANE. 1972. Studies on the functional differentiation of cells in fetal anterior pituitary glands of rats with peroxidase labelled antibody method. Anat. Rec. **172:** 403–404.
32. CHATELAIN, A., J. P. DUPOUY & M. P. DUBOIS. 1979. Ontogenesis of cells producing polypeptide hormones (ACTH, MSH, LPH, GH, Prolactin) in the fetal hypophysis of the rat: influence of the hypothalamus. Cell Tissue Res. **196:** 409–427.
33. DUPOUY, J. P. 1976. Evolution de la teneur de l'hypophyse foetale en ACTH. Etude chez le rat en fin de gestation. C. R. Acad. Sci. (Paris) **282:** 211–214.
34. CHIAPPA, S. A. & G. FINK. 1977. Releasing factor and hormonal changes in the hypothalamic-pituitary-gonadotrophin and adrenocorticotrophin systems before and after birth and puberty in male, female and androgenized female rats. J. Endocrinol. **72:** 211–224.
35. GONZALES-LUQUE, A., M. L'AGE, P. S. DHARIWAL & F. E. YATES. 1970.

Stimulation of corticotrophin release by corticotrophin-releasing factor (CFR) or by vasopressin following intrapituitary infusions in unanesthetized dogs: inhibition of the responses by dexamethasone. Endocrinology **86:** 1134–1142.
36. HEDGE, G. A. & P. G. SMELIK. 1969. The action of dexamethasone and vasopressin on hypothalamic CRF-production and release. Neuroendocrinology **4:** 242–253.
37. LUTZ, B., B. KOCH & C. MIAHLE. 1974. Présence et mode d'action de l'hormone antidiurétique au niveau de l'antéhypothyse du Rat. C. R. Acad. Sci. (Paris) **279:** 1903–1906.
38. VALTIN, H., H. W. SOKOL & D. SUNDE. 1975. Genetic approaches to the study of the regulation and actions of vasopressin. Rec. Prog. Horm. Res. **35:** 447–486.

DISCUSSION OF THE PAPER

T. BENNETT (*University of Nottingham Medical School, Nottingham, England*): In the developmental study, how much was the difference between Long-Evans and DI rats a function of your sampling time? Did you look at any points between 10 AM and 2 PM?

TANKOSIC: We began our study from the 15th day of gestation. We were surprised to note a lack of ACTH-containing cells on the 16th day, and then greater amounts of immunoreactive ACTH on the 17th day in the DI rat than in the Long-Evans rat. These data seemed to show some differences in the ontogenesis of the ACTH cells between Long-Evans and DI rats. To get more precise data on changes occurring during development we obtained samples every 4 hours. Further reduction of sampling time would probably not improve the data because a limiting factor exists, namely the exact time of fertilization. Jost *et al.* have assumed that it occurs around 01:00, but it is quite possible that a variation exists.

R. G. ALLEN (*Veterans Administration Hospital, Portland, Ore.*): The Long-Evans rat may be completely different than the Sprague-Dawley rat, in which we find a couple of hundred nanograms of ACTH and β-endorphin on the 20th day of fetal development. Unless I misinterpreted your slide on neonatal development, the content of ACTH in the anterior lobe was around zero, is that right?

TANKOSIC: At birth, we observe a drop in the pituitary ACTH content (1.1 ng). However, we have no data on the radioimmunoassayable ACTH content in the fetuses of Long-Evans and DI rats at time of birth. In the Wistar rat, Chiappa and Fink (1977) have found 2–3 ng ACTH in the anterior lobe during the last days of gestation. Although a drop in the pituitary ACTH content is possible at the time of parturition, I think that the differences between your results and ours may be due to differences in our antisera, in addition to possible differences between the two rat strains. Some variations in the molecular forms of ACTH may also be taken into account, since it has been shown that a "big" form of ACTH, with high radioimmunological activity, prevails during the fetal period.

MOLECULAR FORMS AND ANTERIOR PITUITARY CONTENT OF ACTH, β-LIPOTROPIN, AND β-ENDORPHIN IN DIABETES INSIPIDUS di/di (BRATTLEBORO) RATS *

Richard G. Allen,†‡ John C. Crabbe,‡ Julianne Stack,† and N. Donna Gaudette ‡

† *Department of Medicine*
Division of Endocrinology and Metabolism
Oregon Health Sciences University
Portland, Oregon 97201

‡ *Veterans Administration Medical Center*
Portland, Oregon 97201

Introduction

The biosynthesis and processing of corticotropin (ACTH), β-lipotropin (β-LPH), and β-endorphin are well characterized. The amino acid sequences of these and several other pituitary hormones are contained in a single common precursor molecule that has been tentatively named pro-opiomelanocortin (POMC). In the Sprague-Dawley rat, differential processing of POMC by anterior lobe or intermediate-posterior lobe cells gives rise to different hormone end products. Anterior lobe processing produces relatively large amounts of 1–39 ACTH and β-LPH, while the intermediate lobe completes the processing of the precursor molecule to 1–39 ACTH derivatives, α-melanocyte-stimulating hormone (α-N-acetyl (1–13) ACTH-NH_2) and CLIP (corticotropin-like intermediate lobe peptide, which is 18–39 ACTH), and the β-LPH derivatives, β-endorphin (61–91 β-LPH) and γ-LPH.[1]

Vasopressin has long been implicated as a potential corticotropin-releasing factor (CRF), but it has just as clearly been shown that there are other factors besides vasopressin that have CRF activity.[2] Since di/di (diabetes insipidus Brattleboro rats) rats have been shown to lack vasopressin, we thought it would be of interest to ascertain whether the defect alters anterior pituitary content of ACTH, β-LPH, or β-endorphin, as well as to investigate the distribution of molecular forms of β-endorphin in di/di rats. We found that in Long-Evans normals (+/+), heterozygotes (di/+), or di/di genotypes the anterior lobe content of ACTH, β-LPH, and β-endorphin was not different, nor was the distribution of molecular forms of β-endorphin. We also performed the same studies on anterior lobe tissue obtained from Roman High Avoidance (RHA) rats that have had the vasopressin defect backcrossed from di/di (Brattleboro) rats.[3] Again, the +/+, di/+, and di/di RHA anterior lobes did not differ in their content of ACTH, β-LPH, and β-endorphin, or the distribution of molecular forms of β-endorphin.

* Supported by the Veterans Administration and National Institute for Drug Abuse grant DA 02799.

0077–8923/82/0394–0574 $1.75/0 © 1982, NYAS

Materials and Methods

Tissue Extraction and Preparation

Long-Evans and RHA male rats of the genotype +/+ (normal); di/+ heterozygote); and di/di (diabetes insipidus), as well as Sprague-Dawley males and females of the same age (6–7 wk.) were killed. Their pituitaries were quickly removed and the anterior lobes were separated from the neurointermediate lobes. The tissue was extracted in 5 N acetic acid containing 1 mM phenylmethylsulfanylfluoride (PMSF) and iodoacetic acid (IAA) to inhibit proteolysis. The samples were centrifuged, and supernatants lyophilized and treated as previously described.[4]

Radioimmunoassays and Antisera

Radioimmunoassays (RIAs) were performed as described by Allen *et al.*[5] Two different RIAs were used to measure ACTH. One antiserum is specific for the 11–24 region of 1–39 ACTH, the other is our "C"-terminal, 25–39 antiserum. The β-endorphin antiserum recognizes the "N"-terminal of β-endorphin (61–91 β-LPH), and reacts with β-LPH and β-endorphin on an equimolar basis. All antisera were raised in rabbits by injecting carbodiimide-coupled β-endorphin or 1–39 ACTH to bovine serum albumin. All antisera recognize the higher molecular weight forms of ACTH and β-endorphin.

Sodium Dodecyl Sulfate/Polyacrylamide Gel Electrophoresis; SDS/PAGE

SDS/PAGE (12% polyacrylamide; Biorad) was performed as previously described.[6] Dansylated yeast alcohol dehydrogenase (YADH; MW 37,000), and myoglobin (Mb; MW 16,000) were included in all gel runs as internal molecular weight markers.

Results

Other workers have shown that the ACTH content in anterior lobes of di/di rats is less than the amount of ACTH found in anterior lobes of +/+ (normal Long-Evans rats).[2] We found that there was no significant difference in the ACTH, β-LPH, and β-endorphin content in the anterior lobes of the +/+, di/+, and di/di genotypes (Figure 1A). Nor was there a significant difference in the total content of these hormones compared to the amounts found in male Sprague-Dawley anterior lobes.

As a further test we obtained RHA rats; normals (+/+), heterozygotes (di/+), and diabetes insipidus (di/di) from Dr. Carl Hansen (National Institutes of Health). Figure 1B shows that the same result was obtained as that described above. There was no significant difference in anterior lobe content of ACTH, β-LPH, or β-endorphin in the three genotypes studied. In this particular experiment Sprague-Dawley females were included because we are very familiar with their anterior lobe content of ACTH, β-LPH, and β-endorphin. All of the Brattleboro and RHA rats were males in this study. It seems likely that

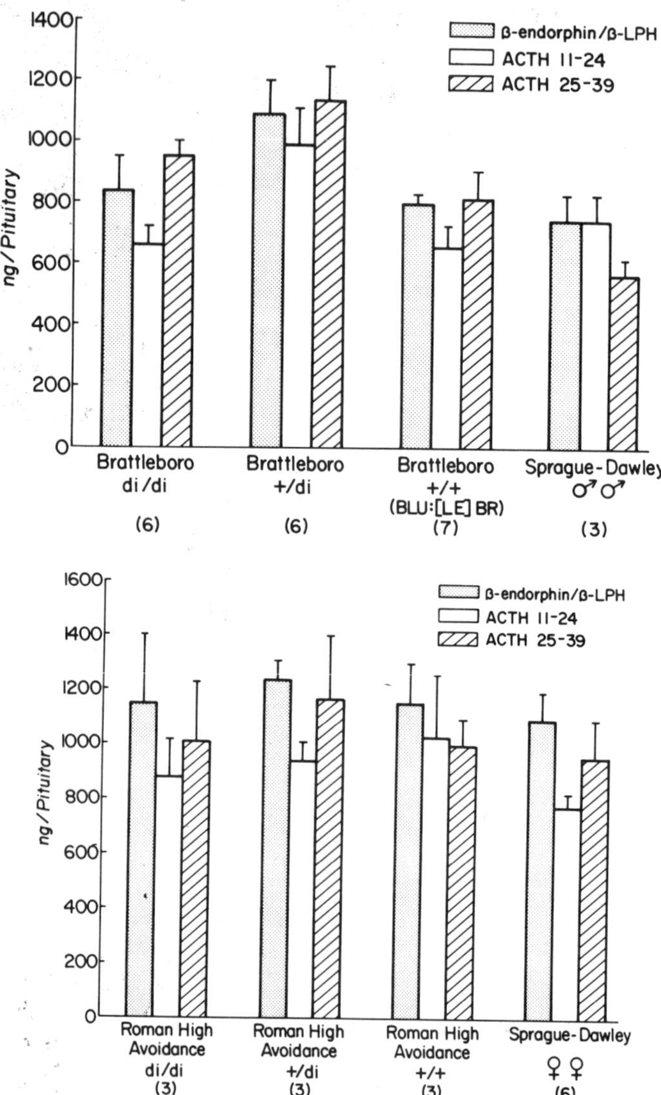

FIGURE 1. The amounts of ACTH, β-LPH, and β-endorphin found in anterior pituitary lobes of (top) +/+, di/+, di/di, Brattleboro and Sprague-Dawley male rats and (bottom) +/+, di/+, di/di RHA males and Sprague-Dawley female anterior lobes. Data are shown as RIA ACTH, β-LPH, and β-endorphin in ng/anterior pituitary ± SEM. The specificity of each antiserum employed is noted in each panel.

the 10–20% difference seen in hormone content of Brattleboro (+/+, di/+, di/di), RHA (+/+, di/+, di/di) when compared with Sprague-Dawley males (+/+) is within the error expected in this type of study, and is not really statistically different.

Since the immunodeterminants of ACTH and β-endorphin are contained in several molecular forms, total RIA does not give a complete picture of how these determinants are distributed among the forms of each hormone. In order to assess the distribution of forms of each hormone we fractionated representative samples of anterior pituitary extracts on 12.5% polyacrylamide gels (in the presence of SDS) of each Brattleboro genotype examined for total RIA. The gels were sliced, eluted, and assayed for ACTH, β-LPH, and β-endorphin (FIGURE 2). An examination of these gel profiles reveals that all these genotypes contain very similar amounts of POMC (ca. 35K). There does,

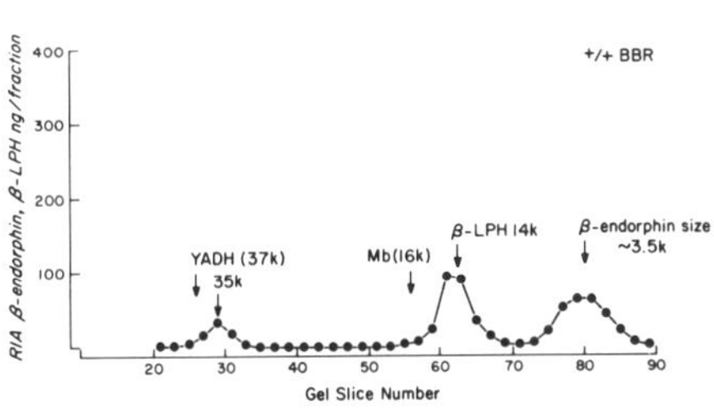

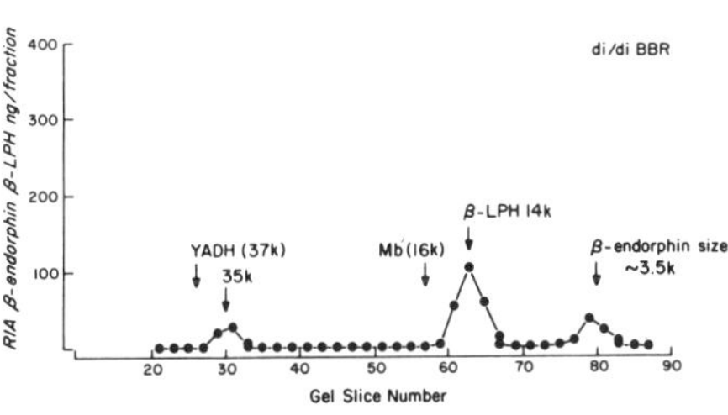

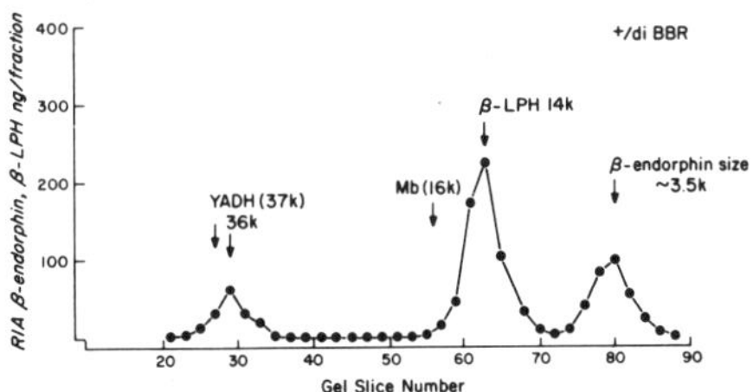

FIGURE 2. The distribution of forms of β-endorphin in anterior pituitary tissue of Brattleboro rats. Anterior pituitaries were removed and homogenized in 30% acetic acid containing 1 mM phenylmethyl sulfonylfluoride and iodoacetic acid to inhibit proteolysis. The homogenates were freeze-thawed and diluted to 5% acetic acid. The extracts were then lyophilized, resuspended in gel sample buffer, and fractionated on 12% Biophore (Richmond, Calif.) gels as previously described.[5] The gels were sliced, eluted, and the eluates assayed for β-LPH and β-endorphin immunoreactivity as previously described.[6] The β-endorphin antiserum recognizes β-endorphin and β-LPH on an equimolar basis, and also recognizes the high molecular weight forms of β-endorphin. Dansylated yeast alcohol dehydrogenase (YADH) and myoglobin (Mb) were included in each gel run as internal molecular weight standards. 3,500 β-endorphin, and 13,000 β-LPH shown here co-migrate with pure standards. β-LPH (sheep) was a generous gift of M. Chretien. β-endorphin (synthetic; human) was purchased from Peninsula Labs, Inc. The type of radioimmune assay and source of tissue are noted in each panel. The recovery of each hormone (β-LPH, β-endorphin), is always 90–105% in this gel system. The total RIA applied to the gels is always equal to the sums of each form recovered on the gels. (top) +/+, normal Long-Evans. (middle) di/di, diabetes insipidus Brattleboro rat. (bottom) di/+, heterozygous Brattleboro rat. All animals obtained from Blue Spruce Farms.

however, appear to be a difference in the amount of β-LPH found in di/+ compared to +/+ and di/di, as well as differences in the amounts of β-endorphin found in all three genotypes. The reader should keep in mind that the profiles shown here are static pictures of dynamic processes (i.e. material is always moving from the precursor pool to smaller forms) and that if one averages several gel profiles, the apparent differences become small or non-existent. Very similar results were obtained when we examined the distribution of forms of ACTH in anterior lobes of +/+, di/+, and di/di Brattleboro rats (data not shown), as well as RHA rats of the same three genotypes. All four major size classes of ACTH (35,000, 20,000, 13,000, and 4,500 MW) were seen in the gel profiles, and all were present in appropriate amounts.

Discussion

It appears that lack of vasopressin does not cause a decrease or increase in anterior lobe content of ACTH, β-LPH, or β-endorphin. There is one possible problem with this interpretation however, and that is the fact that researchers in this area still are in a quandary about whether the di/di rat really lacks the vasopressin gene.[7] It is quite possible that the di/di rat has very small amounts of vasopressin, undetectable by RIA, enough to maintain its contribution to pituitary function, but not enough to function at the kidney. Until this question is answered, all studies involving the Brattleboro rat or other strains of rats lacking a functional vasopressin will be difficult to interpret.

References

1. Eipper, B. A. & R. E. Mains. 1980. Structure and biosynthesis of adrenocorticotropin/endorphin and related peptides. Endocrine Reviews **1:** 1–31.
2. Buckingham, J., & J. H. Leach. 1979. Vasopressin and hypothalamo-pituitary-adrenocorticotrophic activity. Physiological Society. July 1979. p. 87. (b) This volume.
3. The Roman High Avoidance rats were characterized and kindly provided by Dr. Carl Hansen; National Institutes of Health.
4. Hinman, M. B. & E. Herbert. 1980. Processing of the precursor to ACTH and β-lipotropin in monolayer cultures of mouse anterior pituitary. Biochemistry **19:** 5395–5402.
5. Allen, R. G., E. Herbert, M. Hinman, H. Shikuya & C. B. Pert. 1978. Coordinate control of ACTH, β-lipotropin and β-endorphin release in mouse pituitary cell cultures. Proc. Natl. Acad. Sci. USA **75:** 4972–4976.
6. Allen, R. G., E. Orwoll, J. W. Kendall, E. Herbert & H. Paxton. 1980. The distribution of ACTH and β-endorphin in normal, tumorous, and autopsy pituitary tissue: Virtual absence of 13K ACTH. J. Clin. Endocrinol. Metab. **51:** 376–380.
7. This volume.

DISCUSSION OF THE PAPER

P. J. LOWRY (*St. Bartholomew's Hospital, London*): There are some experiments with purified human pro-γ-MSH (Al Dujaili *et al., Nature* **291:**156, 1981). We are using two perfused adrenal cell systems in parallel, one of which is perfused with pro-γ-MSH. With increasing doses of ACTH, increasing amounts of corticosterone are generated in both systems; but, in the MSH-containing system there is marked potentiation of steroidogenesis, so that it's not quite true that the 16K fragment does not have any true activity.

ALLEN: It has also been reported that the 16K fragment potentiated aldosterone and also some glucocorticoid secretion in the adrenals. What you are saying here is that the 16K fragment is a glucocorticoid stimulator at the level of the adrenal gland.

LOWRY: When we start adding γ-MSH to our system there is no increase in steroidogenesis at all until ACTH is put in. This is the native polypeptide, which is inactive by itself. I think that Pedersen and Brownie generally increased activity with the trypsinized 16K fragment. We use the whole peptide.

J. R. HODGES (*Royal Free Hospital Medical School, London*): Dr. Hedge referred to the remarkable advantages of radioimmunoassay methods and Dr. Sokol to discrepancies in the literature. I believe that many of these discrepancies, particularly in the ACTH field, are due to the indiscriminate use of radioimmunoassay methods. Many new techniques for the biological standardization of hormones have been developed recently. They are very sensitive, precise, and specific and, whenever possible, radioimmunoassay methods should be validated by comparison with them.

ALLEN: Since a hormone may have several molecular forms which have different biological activities, I think the state of the art demands not only bioassays and radioimmunoassays, but also requires fractionation chemistry.

K. PASS (*New York State Department of Health, Albany, N.Y.*): With respect to your comment of no target organ for β-endorphin, Russell recently presented evidence to indicate that renal ornithine decarboxylase (ODC) is stimulated following injection of β-endorphin. The response was not greatly different from that which we have seen after injection of arginine vasopressin.

POTENTIATION OF HYPOTHALAMIC CORTICOTROPIN RELEASING ACTIVITY BY VASOPRESSIN: STUDIES IN THE BRATTLEBORO RAT *

Julia C. Buckingham

Academic Department of Pharmacology
Royal Free Hospital School of Medicine
London NW3 2QG England

It is firmly established that the secretion of corticotropin (ACTH) is controlled by a corticotropin releasing factor (CRF) produced by neurons in the hypothalamus and conveyed to the adenohypophysis by the hypothalamo-hypophyseal portal vessels. However, the chemical nature of the hypothalamic hormone is still not known. Over twenty-five years ago it was shown that rats with lesions in the median eminence develop diabetes insipidus together with an inability to secrete ACTH in response to stressful stimuli.[1] Vasopressin was effective in evoking ACTH release in such animals[2] and it was suggested that the octapeptide may be the physiological corticotropin releasing factor. The findings of others do not support this hypothesis. For example, de Wied *et al.*[3] showed a great deal of corticotropin releasing activity but little pressor activity in hypothalamic extracts prepared from hypophysectomized rats while McDonald *et al.*[4] found a complete lack of correlation between the release of these two hormones in response to nicotine, water loading and water deprivation in man. Furthermore, according to Saffran & Saffran[5] the doses of vasopressin used to induce ACTH secretion in hypothalamic-lesioned rats were several thousand times greater than the dose needed to inhibit diuresis maximally. The ability of vasopressin to stimulate pituitary-adrenocorticotropic activity has been studied extensively in several laboratories using a variety of CRF assay systems.[6-9] There can be no doubt that vasopressin stimulates ACTH secretion but only in high, non-physiological concentrations. Furthermore, in all the systems employed, the relationship between the doses of vasopressin and their biological responses differed markedly from those of hypothalamic extracts, indicating that the corticotropin releasing activity in the hypothalamus is not identical with vasopressin. Nevertheless, a substantial amount of evidence suggests that vasopressin may be involved in the sequence of events which leads to ACTH secretion.[10, 11] Attempts to investigate this aspect of hypothalamo-pituitary-adrenocortical (HPA) physiology in rats congenitally lacking vasopressin (Brattleboro strain) have yielded conflicting data[10, 12] probably because only indirect or insensitive methods were employed to assess pituitary-adrenocorticotropic function. Recently, sensitive and precise methods have been developed for the determination of ACTH[13] and CRF.[8] These techniques have been employed to re-examine the functional activity of the HPA system in rats with inherited diabetes insipidus in an attempt to clarify the physiological role of vasopressin in the control of corticotropin secretion.

Male homozygous and heterozygous Brattleboro rats and normal (Long-Evans) controls weighing between 150–200 g were used. The animals were

* Supported by the Medical Research Council.

0077–8923/82/0394–0580 $1.75/0

housed two per cage for at least one week before each experiment in a room in which the temperature was maintained at 22° C. They were handled regularly[14] and food and water were available *ad libitum*. Rats were killed by decapitation. Blood was collected from the trunks into chilled plastic tubes and centrifuged immediately. The plasma was separated and stored at —70° C for ACTH and corticosterone determination. Anterior pituitary glands and hypothalami were also removed from the decapitated rats and homogenized in 0.1 M HCl. The ACTH and CRF contents respectively of the homogenates were determined. Corticosterone was assayed fluorimetrically[15] and corticotropin by cytochemical bioassay. CRF activity was determined using a sensitive and precise ($\lambda = 0.046 \pm 0.002$ for 10 dose-response lines) bioassay method that depends upon the ability of segments of anterior pituitary tissue to secrete ACTH *in vitro*.[8] The capacities of adenohypophyseal segments and hypothalami removed from Brattleboro and Long-Evans rats to secrete ACTH and CRF respectively in response to tropic stimuli were also studied using the methods of Buckingham & Hodges.[8, 10]

The concentrations of corticosterone in the plasma and ACTH in the plasma and adenohypophysis were all significantly ($p < 0.01$) reduced in rats congenitally lacking vasopressin (homozygotes) compared with normal controls (TABLE 1) as also was the ability of segments of their anterior pituitary tissue

TABLE 1

	Plasma Corticosterone	Plasma ACTH	Pituitary ACTH	Hypothalamic CRF
Homozygotes	7.4 ± 2.6	0.16 ± 0.03	23.3 ± 1.6	422 ± 36
Heterozygotes	10.0 ± 2.5	0.26 ± 0.07	29.8 ± 2.1	615 ± 43
Controls	19.4 ± 5.3	0.55 ± 0.12	33.3 ± 2.6	773 ± 45

Concentrations of corticosterone (μg/100 ml) and ACTH (mU/100 ml) in the plasma, ACTH (mU/mg) in the adenohypophysis and CRF (expressed as ACTH (μU/ml/mg pituitary tissue)) in the hypothalamus in homozygous and heterozygous Brattleboro rats and in normal (Long-Evans) controls. Each value is the mean of 10 determinations and is shown with its standard error.

to secrete ACTH in response to tropic stimuli. Hypothalamic extracts prepared from homozygous Brattleboro rats stimulated pituitary ACTH production but their activity was significantly ($p < 0.001$) less than those from controls (TABLE 1). Moreover, the capacity of hypothalami removed from the homozygotes to secrete CRF *in vitro* in response to acetylcholine or 5-hydroxytryptamine was also impaired.

These results show that the functional activity of the hypothalamo-pituitary-adrenocortical system is impaired in rats with inherited diabetes insipidus. The marked reduction in pituitary-adrenocorticotropic activity probably results from lack of tropic stimuli from the hypothalamus. Certainly the CRF content of the hypothalamus is reduced as also is the capacity of the organ to secrete the releasing hormone in response to tropic stimuli. The data suggest that vasopressin plays an important role in the control of ACTH secretion but they do not explain its site or mode of action. The finding that hypothalami from rats congenitally lacking vasopressin are capable of secreting considerable amounts of CRF *in vitro* does not support the recent proposal[17] that vasopressin is identical with corticotropin releasing factor and that it requires a synergistic factor to express its full biological activity. However, the results do not preclude the possibilities that vasopressin acts either by stimulating the release of CRF [18] or by potentiating the action of CRF at the pituitary level.[11] Of these, the latter seems more feasible as vasopressin does not affect the secretion of CRF by isolated hypothalami *in vitro*.[19]

The possibility that vasopressin acts synergistically with CRF has been investigated by studying the effects of arginine vasopressin and specific arginine vasopressin antiserum on the corticotropin releasing activity of hypothalamic extracts prepared from homozygous Brattleboro rats and control (Long-Evans) animals. Batches of hypothalamic extract were prepared from Brattleboro and control rats as described by Buckingham & Hodges.[8] The CRF activity of pure extracts and those treated with either arginine vasopressin or arginine vasopressin antiserum were determined.

Hypothalamic extracts prepared from both groups of rats caused dose-related increases in corticotropin production by pituitary segments *in vitro* but both the activity and the slope of the dose-response line of the Brattleboro extract were significantly ($p < 0.001$) less than that of the control. Addition of arginine vasopressin to the control extract, in a concentration (16.4 μU/ml) approximately 1,000 times less than that required to stimulate ACTH production directly by pituitary segments *in vitro,* did not affect its corticotropin releasing activity (FIGURE 1). On the other hand, in the same concentration it rendered the dose-response line of the Brattleboro extract identical with that of the control (FIGURE 1). Treatment with arginine vasopressin antiserum did not affect the corticotropin releasing activity of the Brattleboro extract. However, it not only diminished markedly the corticotropin releasing activity of the control extract but also made the slope of its dose-response line parallel with that of the Brattleboro extract. These results are shown in FIGURE 2.

The results suggest that while arginine vasopressin is not the corticotropin releasing factor, it is involved in the sequence of events within the pituitary gland that leads to ACTH secretion. It seems unlikely that it acts by directly stimulating ACTH release as the concentration of the octapeptide in hypothalamic extracts from control animals is not sufficient to evoke directly ACTH secretion. However it may act synergistically with CRF and potentiate its action at the pituitary level. It has been shown before that relatively high concentrations of vasopressin [11, 21, 22] and vasopressin-related peptides (Buckingham & van Wimersma Greidanus, unpublished observations) potentiate the activity of CRF both *in vivo* and *in vitro.* The present data demonstrate clearly that minute, physiological concentrations of the hormone fully restore the biological activity of hypothalamic extracts prepared from Brattleboro rats. The hypothesis that vasopressin is necessary for the full expression of CRF activity is further supported by the finding that treatment of hypothalamic extracts from control rats with arginine vasopressin antiserum makes their corticotropin releasing activity identical with that of Brattleboro rats. These data are not in accord with those of Gillies & Lowry [17] who found that hypothalamic extracts from control rats treated with vasopressin antiserum were completely devoid of corticotropin releasing activity and who, on the basis of which, suggested that vasopressin is the major component of CRF. The reason for this discrepancy is not clear. It may be associated with differences in the specificity of the antiserum employed. The antibody used in the present work appears to be specific for arginine vasopressin. Certainly it abolishes the

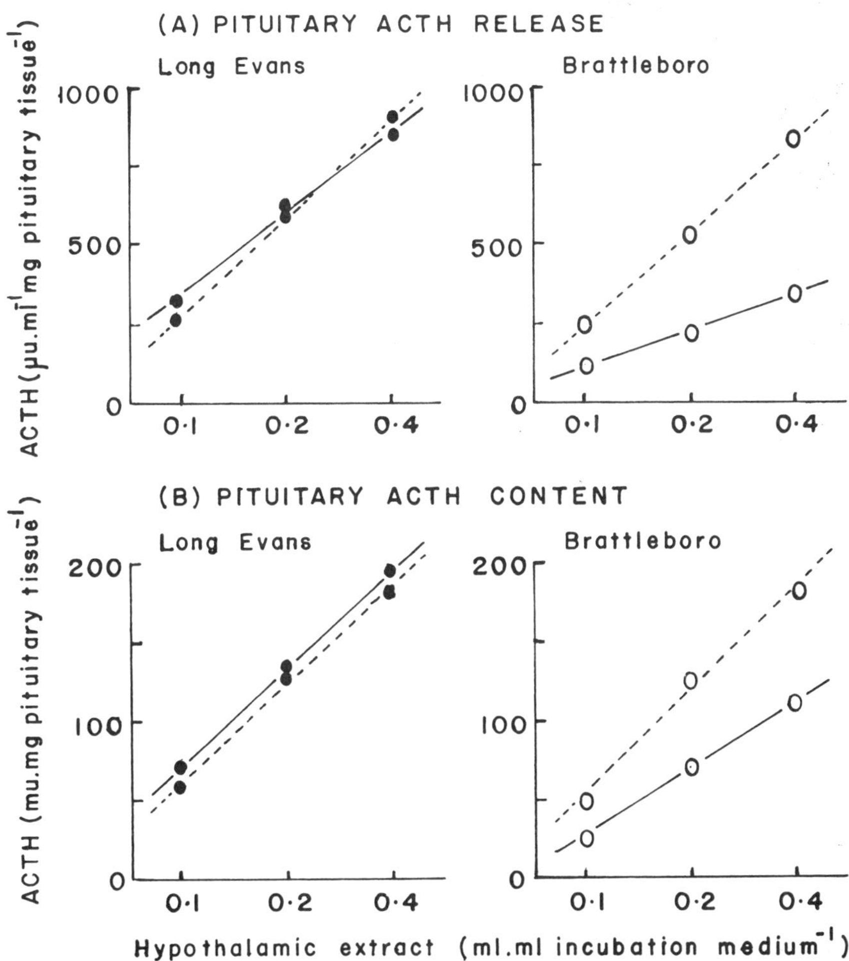

FIGURE 1. Influence of arginine vasopressin on the corticotropin releasing activity of hypothalamic extracts prepared from homozygous Brattleboro rats (○) and from Long-Evans controls (●). ———, hypothalamic extracts; – – –, hypothalamic extracts with arginine vasopressin (16.4 μU/ml). 1.0 ml hypothalamic extract contains the equivalent of one hypothalamus. Each point is the mean of five determinations. For clarity, standard errors are not shown but in every case they were within ± 10% of the mean. (From Buckingham.[20] Reprinted with permission from the *Journal of Physiology.*)

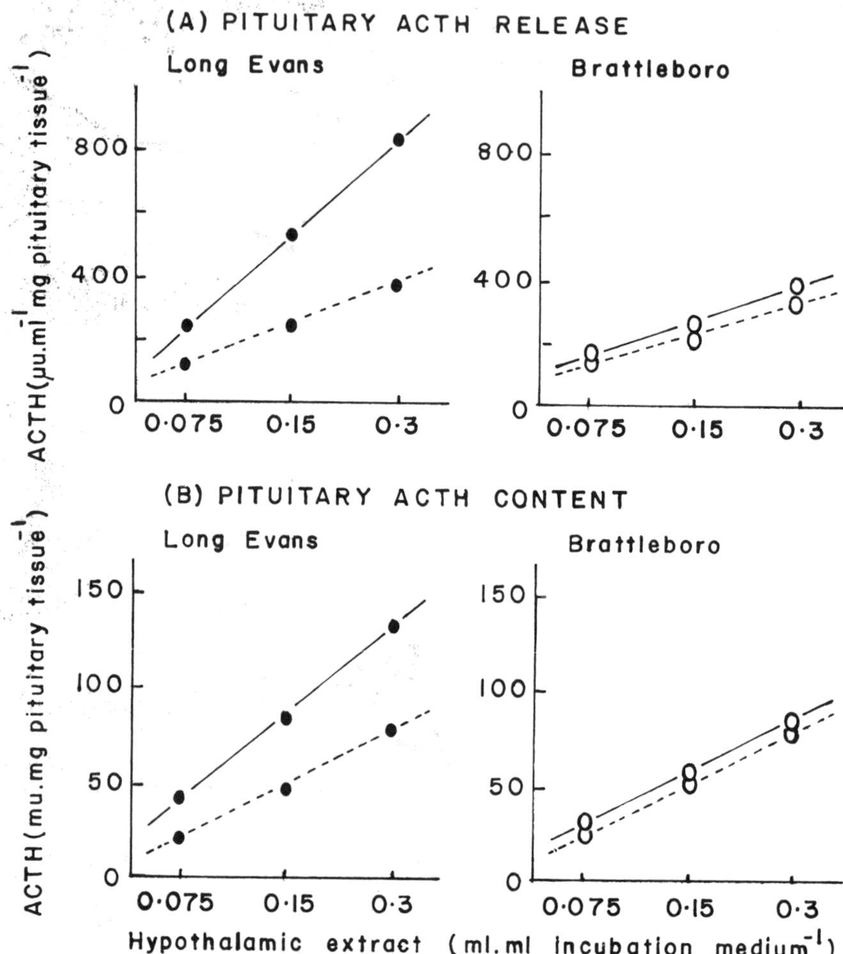

FIGURE 2. Influence of arginine vasopressin antiserum on the corticotropin releasing activity of hypothalamic extracts prepared from homozygous Brattleboro rats (○) and from Long-Evans controls (●). ———, hypothalamic extracts alone; – – –, hypothalamic extracts with arginine vasopressin antiserum. 1.0 ml hypothalamic extract contains the equivalent of one hypothalamus. Each point is the mean of five determinations. For clarity, standard errors are not shown but in every case they were within ± 10% of the mean. (From Buckingham.[20] Reprinted with permission from the *Journal of Physiology*.)

biological activity of arginine vasopressin but does not influence that of arginine vasotocin, a peptide of which the structure and biological activity closely resemble that of vasopressin. The discrepancies may also be due to differences in the "assay" systems employed to assess CRF activity. Since no satisfactory standard preparation of CRF is available it is essential that any new bioassay for the hypothalamic hormone should be validated, as far as is possible, by comparing the activity of the same samples with other older, well-established systems. Comparative studies have shown a good correlation between the results obtained using pituitary segments *in vitro* (the method used in this study) and those achieved with *in vivo* systems using rats in which the endogenous release of CRF was prevented either surgically by the placement of lesions in the hypothalamus [23] or pharmacologically by treatment with pentobarbitone and chlorpromazine [24] or pentobarbitone and morphine (Buckingham, unpublished observations). No such comparative studies appear to have been done for the isolated cell assay used by Gillies & Lowry [17] and, in this context, it is significant that Jones and his colleagues (personal communication) recently found marked discrepancies between their assay results obtained with isolated pituitary cells and those achieved with either hypothalamic-lesioned rats or pituitary segments *in vitro*.

Although the results described do not throw any further light on the chemical nature of the corticotropin releasing factor they add further support to the growing mass of evidence that vasopressin is not the CRF but is involved in the sequence of events leading to ACTH secretion. They suggest that vasopressin acts synergistically with CRF and is essential for the full expression of pituitary-adrenocorticotropic activity.

Acknowledgments

I am grateful to Professor J. R. Hodges for his valuable advice, to the World Health Organization for IIIrd corticotropin I.W.S. and for arginine vasopressin, to Organon Laboratories Ltd. for corticosterone, to Dr. L. Kiel for arginine vasopressin antiserum and to Mrs. J. Dobson for excellent technical assistance.

References

1. McCann, S. M. & J. R. Brobeck. 1954. Evidence for the role of the supraoptico-hypophysial system in the regulation of adrenocorticotrophin secretion. Proc. Soc. Exp. Biol. Med. **87:** 318–324.
2. McCann, S. M. & A. Fruit. 1957. Effect of vasopressin on release of adrenocorticotrophin in rats with hypothalamic lesions. Proc. Soc. Exp. Biol. Med. **96:** 566.
3. de Wied, D., P. G. Smelik, J. Moll & P. R. Bouman. 1964. On the mechanisms of ACTH release. *In* Major Problems in Neuroendocrinology. E. Bajusz & G. Jasmin, Eds.: 156–176. Williams & Wilkins. Baltimore, Md.
4. McDonald, R. K., H. N. Wagner & V. K. Weise. 1957. Relationships between exogenous antidiuretic hormone activity and ACTH release in man. Proc. Soc. Exp. Biol. Med. **96:** 566–567.
5. Saffran, M. & J. Saffran. 1959. Adenohypophysis and the adrenal cortex. Ann. Rev. Physiol. **21:** 403–444.

6. DE WIED, D. 1961. An assay of corticotrophin-releasing principles in hypothalamic lesioned rats. Acta Endocr. (Copenhagen) **37:** 288–297.
7. PEARLMUTTER, A. F., E. RAPINO & M. SAFFRAN. 1974. A semi-automated *in vitro* assay for CRF: activities of peptides related to oxytocin and vasopressin. Neuroendocrinology **15:** 106–119.
8. BUCKINGHAM, J. C. & J. R. HODGES. 1977. The use of corticotrophin production by adenohypophysial tissue *in vitro* for the detection and estimation of potential corticotrophin releasing factors. J. Endocr. **72:** 187–193.
9. GILLIES, G., T. B. VAN WIMERSMA GREIDANUS & P. J. LOWRY. 1978. Characterisation of rat stalk median eminence vasopressin and its involvement in adrenocorticotropin release. Endocrinology **103:** 528–534.
10. MCCANN, S. M., J. ANTUNES-RODRIGUES, R. NALLAR & H. VALTIN. 1966. Pituitary-adrenal function in the absence of vasopressin. Endocrinology **79:** 1058–1064.
11. YATES, F. E., S. M. RUSSELL, M. F. DALLMAN, G. A. HEDGE, S. M. MCCANN & A. P. S. DHARIWAL. 1971. Potentiation by vasopressin of corticotrophin release induced by corticotrophin-releasing factor. Endocrinology **88:** 3–15.
12. ARIMURA, A., T. SAITO, C. Y. BOWERS & A. V. SCHALLY. 1967. Pituitary adrenal activation in rats with hereditary hypothalamic diabetes insipidus. Acta Endocr. (Copenhagen). **54:** 155–165.
13. ALAGHBAND-ZADEH, J., J. R. DALY, L. BITENSKY & J. CHAYEN. 1974. The cytochemical section assay for corticotrophin. Clin. Endocr. **3:** 319–327.
14. HODGES, J. R. & S. MITCHLEY. 1970. The effect of 'training' on the release of corticotrophin in response to minor stressful procedures in the rat. J. Endocr. **47:** 253–254.
15. ZENKER, N. & D. E. BERNSTEIN. 1958. The estimation of small amounts of corticosterone in rat plasma. J. Biol. Chem. **231:** 695–701.
16. BUCKINGHAM, J. C. & J. R. HODGES. 1977. Production of corticotrophin releasing hormone by the isolated hypothalamus of the rat. J. Physiol. **272:** 469–479.
17. GILLIES, G. & P. J. LOWRY. 1979. Corticotrophin releasing factor may be modulated vasopressin. Nature (London) **278:** 463–464.
18. HEDGE, G. A., M. B. YATES, R. MARCUS & F. E. YATES. 1966. Site of action of vasopressin in causing corticotrophin release. Endocrinology **79:** 328–340.
19. BURDEN, J., D. J. HARRISON, E. W. HILLHOUSE, M. R. IRONMONGER & M. T. JONES. 1975. Effects of chlorpromazine, pentobarbitone, vasopressin, angiotensin II, bradykinin and corticotrophin on the secretion of corticotrophin releasing hormone from the hypothalamus *in vitro*. J. Endocr. **67:** 45P.
20. BUCKINGHAM, J. C. 1981. The influence of vasopressin on hypothalamic corticotrophin releasing activity in rats with inherited diabetes insipidus. J. Physiol. **312:** 9–16.
21. BUCKINGHAM, J. C. & J. H. LEACH. 1980. Hypothalamo-pituitary-adrenocortical function in rats with inherited diabetes insipidus. J. Physiol. **350:** 397–404.
22. GILLIES, G. & P. J. LOWRY. 1980. Corticotrophin releasing activity in extracts of the stalk median eminence of Brattleboro rats. J. Endocr. **84:** 65–73.
23. JONES, M. T., E. W. HILLHOUSE & J. BURDEN. 1976. Secretion of corticotrophin-releasing hormone *in vitro*. *In* Frontiers in Neuroendocrinology. L. Martini & W. F. Ganong, Eds. **4:** 195–226. Raven Press. New York, N.Y.
24. BUCKINGHAM, J. C. & TJ. B. VAN WIMERSMA GREIDANUS. 1977. Neurohypophysial hormones and their analogues on corticotrophin release. J. Endocr. **72:** 9P.

(DISCUSSION OF THIS PAPER BEGINS ON PAGE 602.)

CORTICOTROPHIN RELEASING FACTOR ACTIVITY IN THE BRATTLEBORO RAT MEDIAN EMINENCE *

A. Frances Pearlmutter, Linda A. Dokas,
Bonnie Loeser, and Murray Saffran

Department of Biochemistry
Medical College of Ohio
Toledo, Ohio 43699

Introduction

There have been conflicting claims in the literature about the role of vasopressin as a releasing factor for ACTH.[1] The most potent corticotrophin releasing factor (CRF) among peptide hormones of *known* structure is arginine vasopressin.[2] However, the amount of vasopressin required to stimulate the release of ACTH from rat anterior pituitary tissue is about 2,500-fold more than is found in a rat median eminence.[3] To probe further the relationship between CRF and vasopressin, we used the Brattleboro rat as a source of CRF. By comparing the characteristics of CRF from normal and Brattleboro rats in which vasopressin and its corresponding neurophysin are absent, we hoped to delineate more exactly the role of vasopressin as a CRF and to purify partially a vasopressin-free CRF from Brattleboro rats.

In the posterior pituitary, the neurohypophyseal hormones, oxytocin and vasopressin, are stored in granules. Within the granules, the hormones are associated in a noncovalent complex with rat neurophysin I (for vasopressin) and rat neurophysin II (for oxytocin) *in vivo*. Either neurophysin can bind oxytocin or vasopressin with nearly identical affinity *in vitro*.[4, 5] Neurophysin affinity chromatography has been used to purify oxytocin and/or vasopressin.[6] Because of a possible similarity in structure between AVP and CRF, we attempted to use a neurophysin affinity column to purify CRF from the Brattleboro rat. This CRF would, presumably, be free of contamination by vasopressin.

We had shown previously that the CRF activity in a median eminence extract of normal rats is due to at least two factors of large and small molecular weight, which must be combined to obtain full activity.[7] As an additional comparison between normal and Brattleboro CRF, we tested Brattleboro rat median eminence extract to determine if the same two factors were necessary for ACTH release.

Materials and Methods

CRF activity was determined with a semi-automated *in vitro* bioassay.[2] In this bioassay, rat pituitary halves are incubated with the presumed CRF material. The ACTH released from the pituitary halves is measured using

* Supported by National Institutes of Health grants AM18383, T32–AM7316, and BRS 5–SO1–RR05700.

0077–8923/82/0394–0587 $1.75/0

adrenal quarters in a continuous superfusion system that gives a minute-to-minute reading of corticosterone production. For each CRF determination, a control is used to account for basal ACTH secretion by the pituitary as well as for any endogenous ACTH activity in the sample. The same set of pituitaries and the same adrenal tissue are used for each sample and its control. This bioassay has a linear dose-response relationship between 0.04 and 2.0 median eminence equivalents.

Rat median eminence fragments were collected in 0.1 N HCl (25 μl per fragment) and stored in the freezer prior to processing. A crude extract was prepared by crushing the tissue with a glass rod and then removing the particulate material by centrifugation at 1,500 rpm. After rinsing the precipitate, the final volume was made up to 25 μl per fragment.

Brattleboro rats were obtained from Blue Spruce Farms (Altamont, N.Y.) and some hypothalamic fragments from Brattleboro rats were collected and sent to us by Dr. Hilda Sokol (Dartmouth Medical School). Normal Sprague-Dawley rats were bred at Medical College of Ohio (Toledo, Ohio) and were donated by Dr. Melvyn Soloff and Mats Fernstrom. All animals were kept on a 12 hour light/dark cycle with water and food freely available.

The neurophysin affinity column was made following the procedure of Robinson and Walker.[6] Bovine neurophysins I and II were purified from freeze-dried posterior pituitaries (Pel-Freez) as previously described.[8] Prior to its use, the neurophysin affinity column was tested for binding affinity using [^{3}H]oxytocin (~30 Ci/mM) as a tracer with 100 μg of nonradioactive hormone. The oxytocin bound to the column at pH 5.8 could be eluted completely from the affinity column with 0.1 M formic acid.

Results

Median eminence (ME) extracts from normal Long-Evans, normal Sprague-Dawley and homozygous (di/di) Brattleboro rats were prepared in 0.1 N HCl as described above. The CRF activity of all three groups was tested in the bioassay. A pool of ME extract prepared in the laboratory from normal Sprague-Dawley rats was used as a reference standard. Stalk ME extract of Brattleboro rats contains 98% of the CRF-like activity of normal Sprague-Dawley rats[1] and 96% of that of normal Long-Evans rats (Figure 1).

The usefulness of neurophysin affinity chromatography as a tool to purify CRF activity was tested first with median eminence extract from normal (Sprague-Dawley) rats. The ME extract was applied to the column in ammonium acetate buffer at pH 5.8, conditions which are optimal for the binding of neurophysin to the neurohypophyseal hormones. After the unbound material was collected, the bound material was eluted from the column with 0.1 M formic acid. An aliquot of the ME extract was treated in an identical fashion, but without affinity chromatography (unfractionated ME). Virtually all of the CRF activity in the normal ME extract bound to the neurophysin affinity column (Figure 1). When Brattleboro ME extract was run over the neurophysin affinity column, with an experimental protocol identical to that for normal ME, 65% of the CRF activity did *not* bind to the column; the material that bound to the affinity column contained only 8% of the total activity (Figure 1).

By Sephadex chromatography, normal rat hypothalamic CRF can be separated into two components that must be combined to produce full CRF activity.[7] To determine whether Brattleboro ME extract contains the same components required for full CRF activity in normal rat ME extract, we chromatographed Brattleboro ME extract on Sephadex G-25 (2.5/30) and eluted with 0.2 M acetic acid. Unlike normal rat ME extract, the chromatographed Brattleboro ME extract had virtually all of the CRF activity in the larger molecular weight fraction and no potentiation of CRF activity was observed when the larger and smaller molecular weight fractions were combined (FIGURE 1).

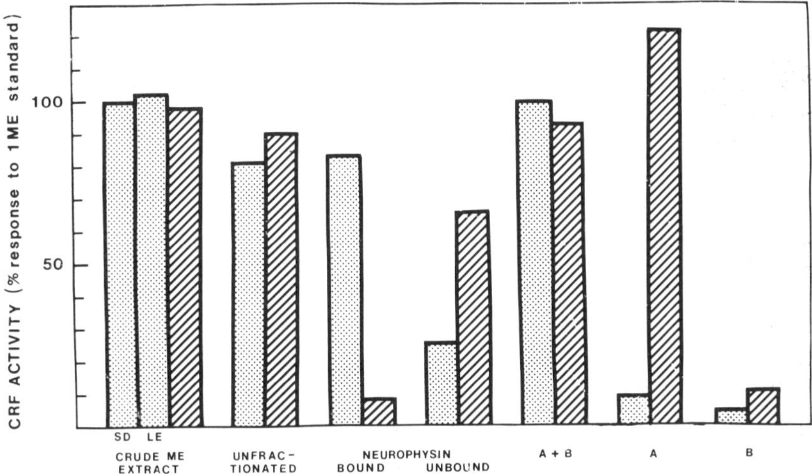

FIGURE 1. CRF activity of median eminence extracts from normal (dotted bars) and Brattleboro rats (striped bars). Brattleboro ME extract has been compared with Sprague-Dawley (SD) and Long-Evans (LE) rats. The ME extracts have been subjected to neurophysin affinity chromatography and both bound and unbound fractions tested. The controls for these experiments are the unfractionated samples. Fractions from the Sephadex G–25 chromatography have been divided into a large molecular weight pool, A, and a small molecular weight pool, B. A and B is a recombination of these pools.

DISCUSSION

A comparison of the properties of CRF activity in ME extract of Brattleboro rats with that of normal rats shows that the amount of CRF-like activity contained in Brattleboro rat ME extract is identical to that found in normal rats. Unlike normal rat ME extract, only one factor contains all of the CRF-like activity in Brattleboro ME extract. In addition, the factor which is responsible for CRF activity in Brattleboro ME extract, unlike normal rat CRF, does not bind to a neurophysin affinity column. These observations led us to conclude that the substance responsible for CRF activity in the Brattleboro ME extract is different from that in normal rats. One possible explanation for this difference is that CRF may be processed from a larger precursor; and, in the Brattleboro rat, there may be some defect in this processing mechanism, which results in an altered form of CRF. This suggestion is in agreement with the results of other workers who have found differences in neurophysin precursor processing in Brattleboro rats compared with normal rats.[9]

References

1. Pearlmutter, A. F., L. Dokas, A. Kong, R. Miller & M. Saffran. 1980. Is corticotropin releasing factor modulated vasopressin? Nature **283:** 697–698.
2. Pearlmutter, A. F., E. Rapino & M. Saffran. 1974. A semi-automated *in vitro* assay for CRF activities of peptides related to oxytocin and vasopressin. Neuroendocrinology **15:** 106–119.
3. Pearlmutter, A. F., L. A. Dokas, B. K. Loeser, A. Kong, M. Saffran & W. Simmons. 1981. Properties of CRF from normal and Brattleboro rat median eminence. Neuroendocrinology **34:** 99–103.
4. Breslow, E. 1979. Chemistry and biology of the neurophysins. Ann. Rev. Biochem. **48:** 251–274.
5. Soloff, M. & A. F. Pearlmutter. 1979. Biochemical actions of neurohypophyseal hormones and neurophysin. *In* Biochemical Actions of Hormones. G. Litwack, Ed. **6:** 265–333. Academic Press. New York, N.Y.
6. Robinson, A. F. & P. Walter. 1979. Extraction of small amounts of oxytocin from biological fluids by means of agarose-bound neurophysin. J. Endocrinol. **80:** 191–202.
7. Pearlmutter, A. F., E. Rapino & M. Saffran. 1975. The ACTH-releasing hormone of the hypothalamus requires a co-factor. Endocrinology **97:** 1336–1339.
8. Pearlmutter, A. F. & C. McMains. 1977. Interaction of bovine neurophysin with oxytocin and vasopressin measured by temperature-jump relaxation. Biochemistry **16:** 628–633.
9. Russell, J. T., M. J. Brownstein & H. Gainer. 1980. Biosynthesis of vasopressin, oxytocin, and neurophysins: isolation and characterization of two common precursors (propressophysin and prooxyphysin). Endocrinology **107:** 1180–1891.

(Discussion of This Paper Begins on Page 602.)

CENTRAL CONTROL OF ACTH SECRETION IN DIABETES INSIPIDUS BRATTLEBORO RATS *

Alex J. Baertschi and Jean-Louis Bény

Department of Animal Biology
University of Geneva
1211 Geneva 4
Switzerland

The vasopressin-deficient Brattleboro rat[1] has been a popular subject for the study of corticotrophin releasing factor. Already before the discovery that hypothalamic extracts evoke the release of ACTH from pituitary tissue *in vitro*,[2, 3] McCann and colleagues[4-6] showed that electrical lesions of the supra-optico-hypophysial tract inhibited the ACTH release to stressful stimuli, thus implicating vasopressinergic neural pathways in the control of the adrenal cortex. When the Brattleboro rat became available, McCann, Antunes-Rodrigues, Nellar, and Valtin undertook what appears to be the first study concerning adrenocortical control in these animals.[7] They showed that the corticosterone secretory response to ether or mild restraint was significantly impaired in Brattleboro diabetes insipidus rats (DI) compared to controls. However, the content of the medial basal hypothalamus in corticotrophin releasing factor (CRF), as measured by the adrenal ascorbic acid depletion in response to the i.v. injection of extracts into median eminence lesioned rats, was not significantly different for DI compared to control rats. These findings helped spark a controversy, which has carried on for more than a decade,[8, 9] concerning the physiological role of vasopressin as a CRF. In the present study we raise the questions if oxytocin may also be a physiological CRF, where in the hypothalamus other CRFs are synthesized, and if intermediate lobe ACTH can participate in the overall adrenocortical control.

Release of CRF from the Neural Lobe *In Vivo* and *In Vitro*

With a few exceptions,[10-12] most previous studies on CRF have been performed using *extracts* from either the neural lobe or median eminence. We asked the question if *released* CRF possesses characteristics similar to those of extracted CRF. Since large amounts of CRF have been found in the neural lobe,[13] we measured the effect of electrical stimulation of the neural lobe in normal and DI rats on plasma ACTH *in vivo*,[14] and on medium CRF activity *in vitro*.[15]

In vivo experiments were carried out on urethane-anesthetized lactating rats by placing a bipolar stimulating electrode into the neural lobe. The hypothalamo-hypophyseal (HHT) tract was lesioned in order to prevent antidromic action potentials from reaching the hypothalamus (Figure 1A). Blood samples of 2 ml, collected at a rate of 1.5 ml/min from a carotid artery and replaced

* Supported in part by the Swiss National Science Foundation (Grant 3.581.79) and Sandoz-Stiftung.

0077–8923/82/0394–0591 $1.75/0 © 1982, NYAS

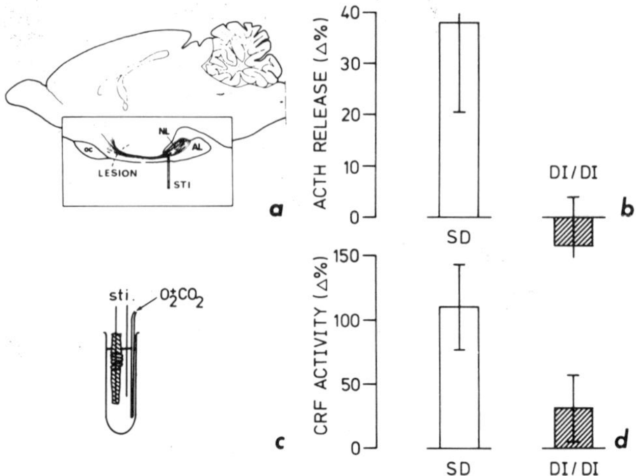

FIGURE 1. Electrical stimulation of rat neural lobes *in vivo* and *in vitro.* a and b: *In vivo* experiment. Neural lobes (NL) were electrically stimulated with a bipolar electrode (STI) in animals where the hypothalamo-neurohypophyseal tract was transsected (LESION). Columns represent mean ± SEM of changes in ACTH plasma concentrations over basal ACTH concentrations (SD: Sprague-Dawley, $N = 4$; DI/DI: Diabetes insipidus, $N = 7$) during the three minutes following the neural lobe stimulation. Increase for SD rats is significant at 0.025 level. c and d: *In vitro* experiments. Ten posterior pituitaries were stacked on a platinum needle and placed in 1 ml medium gassed with 95% O_2 and 5% CO_2. Electrical stimulation was applied between the needle that carried the posterior pituitaries and another platinum needle in the medium. Duplicate samples of 100 μl were tested for CRF activity in batches of one ml of anterior pituitary cell suspension. CRF activity (mean ± SEM, $N = 5$ experiments) is expressed as percentage increase of ACTH secretion over basal ACTH release to KRBGA. ACTH content of posterior pituitary medium was measured separately and subtracted to yield the net CRF effect. Increase for SD significant at 0.025 level. When CRF activity was expressed as percentage increase of ACTH secretion over basal ACTH secretion to 100 μl of medium from non-stimulated posterior pituitaries, then increase for DI was significant at 0.05 level.

continuously by donor blood, were immediately centrifuged at 4° C and the plasma stored at −20° C pending radioimmunoassay of ACTH.[16] Following a 3-min stimulation of the neural lobe with biphasic constant current pulses at 30 Hz (5 sec on and off), there was a 40% rise in plasma ACTH levels of Sprague-Dawley (SD) rats (FIGURE 1B). However, no such increase of plasma ACTH was observed in DI rats. These results suggested that vasopressin had mediated the increased ACTH secretion to neural lobe stimulation in SD animals.

The *in vitro* experiments were carried out by stacking 10 posterior pituitaries on a platinum needle (FIGURE 1C) and incubating them in 1 ml of a Krebs-Ringer bicarbonate solution containing 2 mg glucose and 1 mg beef serum albumin (KRBGA). The medium was collected every 20 minutes and replaced. The CRF activity and ACTH content of the medium was determined by the Sayers dispersed pituitary and adrenocortical cell assay.[15, 17] During the third

20 min incubation period, the neural lobes were stimulated electrically for 15 min (5 sec on and off) at 30 Hz with biphasic constant current pulses. For SD posterior pituitaries, there was a large increase in CRF release (white column, FIGURE 1D). The amount of vasopressin which was released simultaneously into the medium was estimated to be about 7 mU/lobe (based on bioassayable [18] oxytocin content and a ratio of vasopressin/oxytocin release of one). The same amount of exogenous vasopressin had CRF activity (FIGURE 2) not significantly different from that shown by the white column of FIGURE 1D. The experiment, carried out with DI posterior pituitaries, showed that only very little CRF was released during neural lobe stimulation. The amount of oxytocin released simultaneously was 3.2 mU/lobe and could entirely explain the small CRF effect (FIGURE 2).

The results of FIGURE 1 show that vasopressin in SD and oxytocin in DI can account for all the CRF activity *released* from the posterior pituitary. Therefore, it was of interest to measure with the same bioassay the amount of CRF activity in median eminence and neural lobe *extracts* that could be accounted for by AVP and oxytocin.

CRF EXTRACTS FROM MEDIAN EMINENCE AND NEURAL LOBE

The median eminences and neural lobes were homogenized in 50 μl of 0.1 N HCl and frozen. The intermediate lobe of the posterior pituitary was aspirated beforehand so that its ACTH content could not interfere with the assays. Before the CRF assay, the homogenates were neutralized with 5 μl of 1 N NaOH, supplemented with 445 μl of KRBGA and submitted to centrifugation. The CRF content of 0.2 median eminences (0.2 ME) was at least two times lower in DI than in SD rats ($p < 0.05$) (FIGURE 3, top panel). In another series of DI neural lobe extracts (FIGURE 4), the median eminence CRF content was even lower, that is three times less than the amounts shown

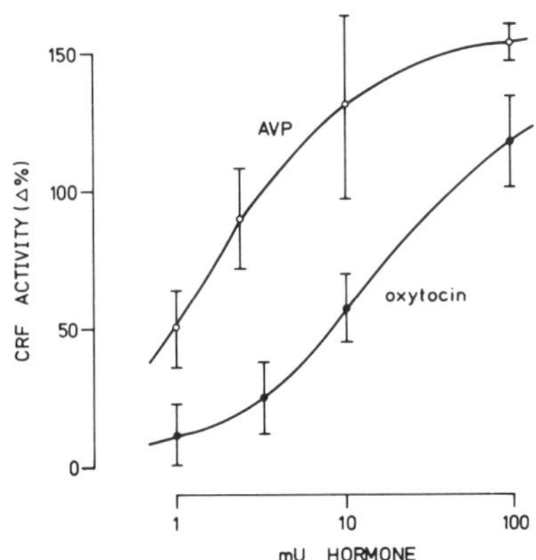

FIGURE 2. CRF activity of vasopressin (AVP) and oxytocin. CRF activity is expressed as percent increase of ACTH secretion over basal ACTH release to KRBGA. Hormone doses are in mU added to 1 ml of anterior pituitary cell suspensions ($N = 3$ to 7 experiments).

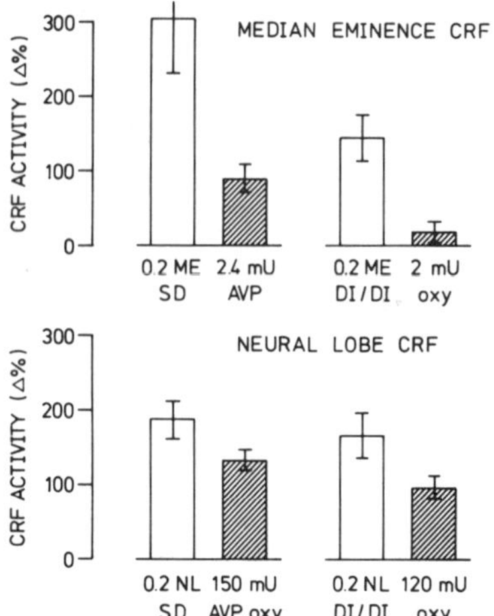

FIGURE 3. CRF activity for median eminence and neural lobe extracts compared to CRF activity of AVP and oxytocin contained in extracts. CRF activity (mean $\pm$ SEM, $N = 8$ experiments for all extracts, $N = 4$ experiments for AVP and oxytocin) was measured and expressed as in FIGURE 1d. ACTH content of extracts was measured separately and subtracted to yield net CRF effect.

for SD neural lobes in FIGURE 3. Median eminences contain about 5 to 12 mU AVP in SD rats[19, 20] and at most 10 mU oxytocin in DI animals (Bény and Baertschi, unpublished). The CRF activity of vasopressin and oxytocin contained in 0.2 ME was, however, at least three times less than that of 0.2 ME for both SD and DI rats (hatched columns, top panel, FIGURE 3).

The CRF activity of 0.2 neural lobe extracts (0.2 NL) of DI rats was not significantly different from that of SD animals (white columns in bottom panel, FIGURE 3). Neural lobes contain 750 mU AVP and 750 mU oxytocin in SD rats,[21] and at the most 600 mU in DI animals.[1] The CRF activity of 150 mU AVP plus 150 mU oxytocin accounts for about 70% of CRF content in 0.2 NL of SD, and that of 120 mU oxytocin for about 60% of CRF content in 0.2 NL of DI rats.

Thus a CRF distinct from either AVP or oxytocin accounts for 70 to 90% of the CRF content of SD and DI median eminences, and for 30 to 40% of CRF content of SD and DI neural lobes. These results differ from those shown in FIGURE 1, where no CRF distinct from AVP or oxytocin could be detected in released CRF. We have recently shown that AVP accounts for all CRF released from the SD median eminence *in vitro,* but that CRFs distinct from AVP account for about 30% of total released CRF when ascorbic acid is added to the medium.[22] Ascorbic acid either enhances the release of CRF distinct from AVP or protects CRF from degradation[23] in the medium.

ORIGIN AND NATURE OF MEDIAN EMINENCE CRF IN DI RATS

In view of the large proportion of CRF, distinct from vasopressin and oxytocin, that we detected in DI median eminence extracts (FIGURE 3), we investigated the origin of the CRF axons.

Previous experiments have suggested that the paraventricular nucleus (PV) may be involved in adrenocortical control (see discussion). In collaboration with G. Makara, we [24] lesioned the PV region of male DI rats with a specially designed rotating knife. Histological examination of 10 μm thick paraffin sections revealed that at least 90% of the PV was destroyed. The lesion also extended to other hypothalamic areas, such as half of the paraventricular nucleus and small parts of the ventromedial, dorsomedial and anterior hypothalamic nuclei, but the basal hypothalamus remained intact (insert of FIGURE 4). In control animals, the knife was also lowered into the hypothalamus, but was not rotated. Histology revealed minimal damage, usually a track left by the shaft of the knife. Four days later, the animals were stressed with ether for 1 min and the first blood sample of 1.5 ml was taken from a jugular vein. The animals were decapitated 30–40 min later, 1.5 ml of trunk blood was collected for the second blood sample, and the median eminence (ME) dissected, homogenized, stored at —20° C and assayed not more than three days later for CRF activity as shown previously. Plasma corticosterone of the first (C_1) and second (C_2) blood sample was determined by fluorimetry and the corticosterone response to ether was quantified by $(C_2 - C_1)/\frac{1}{2}(C_2 + C_1)$.

In sham-lesioned animals (closed circles, FIGURE 4), the ME CRF activity

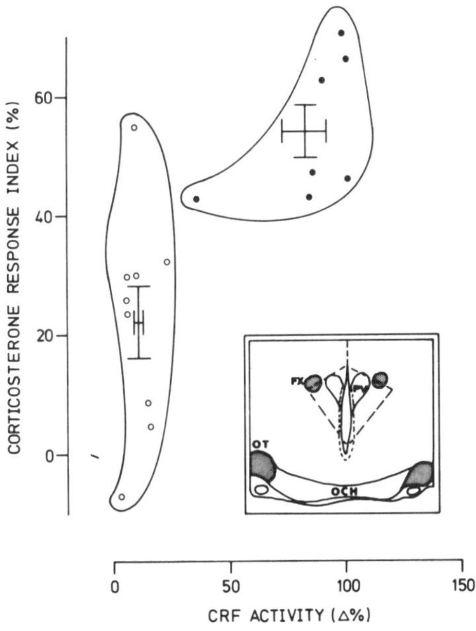

FIGURE 4. Effect of paraventricular nucleus lesion on median eminence CRF and corticosterone response to ether in DI rats. CRF activity of 0.2 ME was measured as in FIGURE 1d. Individual data points are from seven sham-lesioned (closed circles) and nine PV-lesioned animals (open circles). Brackets indicate mean ± SEM. Insert shows frontal section through the hypothalamus (PV: paraventricular nucleus; Fx: fornix; OT optic tract; OCH: optic chiasm). Heavy stippled line indicates extent of histologically checked lesion. Thin stippled line around third ventricle marks position of paraventricular nucleus.

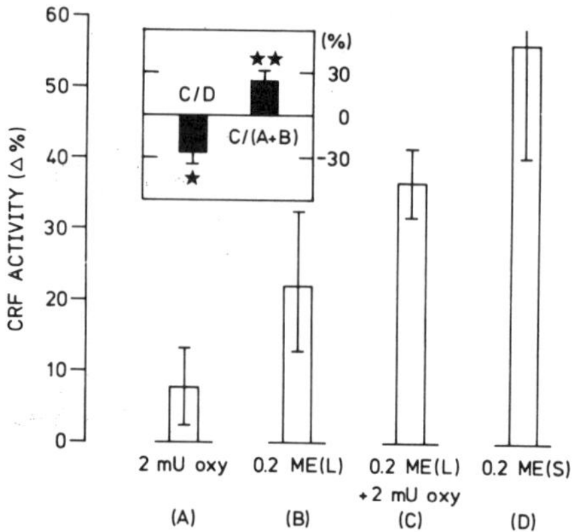

FIGURE 5. Potentiation by oxytocin of median eminence CRF from PV-lesioned [ME(L)] DI rats. CRF activity (mean ± SEM, $N = 4$ to 5 experiments) was measured as in FIGURE 1d. Insert shows percentage changes from 100% of the mean ± SEM of individual ratios (* $p < 0.05$). ME(S) = sham lesioned.

was only slightly lower than shown in FIGURE 3, and these animals responded with a large increase in corticosterone secretion to ether stress. In PV-lesioned animals (open circles, FIGURE 4), the ME CRF was about 90% lower, and in another series (FIGURE 5) 61% lower than in the sham-operated group. The corticosterone response to ether stress was much smaller though still significant, as shown by the individual data points each representing one animal. One exception can be seen where a low CRF content is not accompanied by a reduced corticosterone response.

The results clearly show that a CRF is transported through or synthesized in the PV region. As Gillies and Lowry suggested that CRF may be modulated vasopressin [23] and that AVP can potentiate the DI CRF activity [25] of the median eminence, we wondered if DI CRF was modulated oxytocin. Thus the loss of CRF activity in PV-lesioned DI could have been due to a loss of oxytocin, or of CRF distinct from oxytocin and originating in PV.

To explore these possibilities, DI rats were subjected to a PV lesion as described previously, and the ME was dissected four days later, homogenized and stored at —20° C pending CRF assay. A part of the ME of lesioned rats (0.2 ME(L)) was assayed for CRF content either alone or together with 2 mU oxytocin. ME of PV-lesioned rats had 61% less CRF activity than ME of sham-lesioned rats (FIGURE 5). Oxytocin (2 mU) by itself had hardly any CRF effect, but increased the CRF activity of ME from lesioned rats. Because of the large variability of the data, these changes did not achieve a high statistical significance. Therefore, the data were expressed by the mean ratio of the CRF effect of 2 mU oxytocin plus 0.2 ME(L) (=C) divided by the sum of the individual CRF effects of oxytocin (=A) and 0.2ME(L) (=B). C was significantly ($p < 0.025$) larger than A plus B, demonstrating potentiation of the CRF activity of oxytocin by 0.2 ME(L). Furthermore, 2 mU oxytocin alone were not sufficient to raise the CRF activity of 0.2 ME(L) to the level found in sham-lesioned rats ($p < 0.05$, C vs. D), suggesting that a CRF distinct from oxytocin must also originate from the PV region.

Release of Bioactive ACTH from the Posterior Lobe of SD and DI Rats

The previous results demonstrate that the DI rat may partially compensate for its AVP deficiency by releasing oxytocin. Another compensation mechanism may operate through the intermediate lobe. A previous study[15] had shown that the release of bioactive ACTH was under a neural inhibitory control, since electrical stimulation of the posterior lobe *in vitro* inhibited ACTH release (Figure 6, left panel) in SD rats. The same experiment, repeated in DI animals (Figure 6, right panel), shows: (1) initial ACTH release is almost twice that found in SD rats ($p < 0.05$), and (2) the release of ACTH is not inhibited by the electrical stimulation of the DI posterior lobe (hatched columns, Figure 6).

Discussion

Our study clearly indicates that CRF axons originate or pass through the PV region. Previous histochemical studies on normal rats showed that vasopressinergic pathways, originating in the paraventricular nucleus and projecting to the ME[26] are involved in adrenocortical control.[27, 28] In normal animals, the lesion of the paraventricular nuclei is followed by a decrease of median eminence CRF content,[29] electrical stimulation of the PV region elicits the release of ACTH *in vivo*,[30] and PV induces a ten-fold increase of ACTH release from anterior pituitary in co-culture.[31] In DI, significant amounts of CRF were detected in extracts of microdissected PV.[32]

These findings are in apparent contradiction with classic concepts that CRF is synthesized in the basal hypothalamus.[33] The latter studies were based on experiments involving hypothalamic deafferentation. However, recent evidence suggests that a complete hypothalamic island is difficult to perform, since usually a small fiber bundle that carries CRF axons is left intact in the antero-

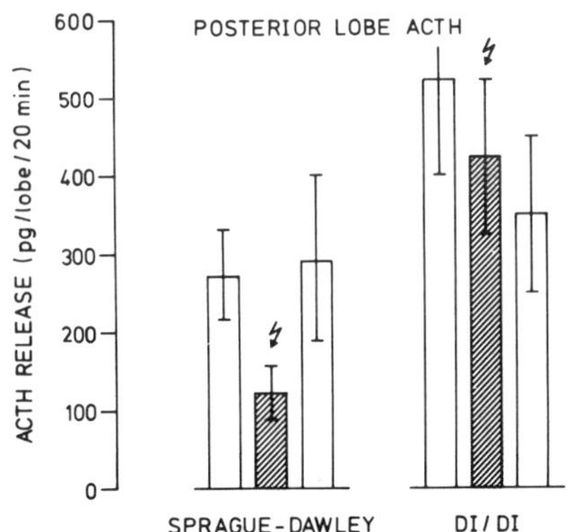

Figure 6. Release of bioactive ACTH from posterior pituitaries *in vitro*. ACTH content of medium (mean ± SEM, $N = 5$ to 8 experiments) was measured as in Figure 1d by the output of corticosterone from adrenocortical cells. Same protocol as in Figure 1c and d. Electrical stimulation was applied during incubation indicated by hatched column. Open columns = pre- and poststimulated states.

lateral basal hypothalamus.[29, 34] This may also explain why ME CRF remained unchanged following 'complete' hypothalamic deafferentation.[32]

The results presented in FIGURES 1 to 5 may shed some light on the nature and origin of DI CRF. In TABLE 1, we propose several hypotheses. Factors controlling ACTH release may include oxytocin, a potentiating factor for oxytocin that originates in the PV region (PV-PF), another potentiating factor for oxytocin that originates outside the PV region (X-PF), a CRF distinct from oxytocin that originates in the PV region (PV-CRF), and a CRF distinct from oxytocin that originates outside the PV-region (Y-CRF). Although the possibility of another type of potentiating factor as well as of an inhibiting factor cannot be excluded,[35] we have not considered these aspects. The existence of potentiating factors for oxytocin can be anticipated from the finding that vasopressin potentiates the CRF activity of DI median eminence extracts[35] and of a high molecular weight fraction of median eminence extracts from normal[23] and DI[25] rats. Out of 19 possibilities listed in TABLE 1, only three are compatible with all the results of this study (Hypothesis 17, 18, 19). Thus,

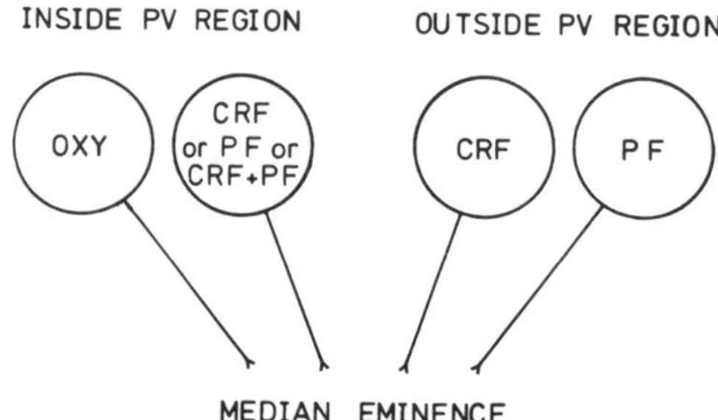

FIGURE 7. Control of ACTH release from anterior pituitary in DI rats. Model is based on data presented in FIGURES 1 to 5 and on TABLE 1. PF: potentiating factors for oxytocin; CRF: CRF distinct from oxytocin and not PF. PF is defined as having no CRF activity by itself, but as being capable of magnifying the CRF activity of oxytocin. Model remains consistent if PF and CRF are one and the same molecule.

oxytocin appears to play an important role, not so much because of its inherent ACTH releasing potency (FIGURE 2), but because its action can be potentiated by a factor outside the PV region (X-PF) (FIGURE 5). Since a significant CRF activity remains in median eminence extracts of PV-lesioned DI rats (FIGURES 4 and 5), another CRF distinct from oxytocin must be synthesized outside the PV region (Y-CRF). The addition of oxytocin to ME extracts of lesioned DI rats does not bring about the full CRF effect encountered in ME extracts of sham-lesioned DI rats (FIGURE 5). Thus, another factor must originate in the PV region, either a potentiating factor (PV-PF), or a CRF distinct from oxytocin (PV-CRF), or both. These concepts have been summarized in FIGURE 7. The quantitative aspects of this scheme remain to be elucidated, though the relative importance of each factor can be roughly estimated (in decreasing order): oxytocin plus X-PF, PV-PF or PV-CRF, Y-CRF.

The role of the DI rat intermediate lobe in adrenocortical control has rarely been investigated. Our results show that DI intermediate lobes secrete about

TABLE 1

COMPATIBILITY OF RESULTS WITH VARIOUS HYPOTHESES CONCERNING THE NATURE AND ORIGIN OF CRF IN DI RATS

	Factors controlling ACTH					Compatible †	
Hypothesis	Oxytocin	PV-PF	X-PF	PV-CRF	Y-CRF	Yes	No
1	0	0	0	0	1	–	1,2,3,4,5
2	0	0	0	1	0	–	1,2,3,4,5
3	0	0	0	1	1	4	1,2,3,5
4	1	0	0	0	0	1,2	3,4,5
5 to 11	1	1 or 0 *	1 or 0 *	1 or 0 *	0	1,2,3	4,5
12	1	0	0	0	1	1,2,3	4,5
13 to 15	1	1 or 0 *	0	1 or 0 *	1	1,2,3,4	5
16	1	0	1	0	1	1,2,3,4	5
17 to 19	1	1 or 0 *	1	1 or 0 *	1	1,2,3,4,5	–

* At least one alternative in a row must be different from 0, 1: factor is necessary, and 0: factor is not necessary.
† Numbers refer to FIGURES in text.

two times more ACTH than those of SD, and that this secretion cannot be inhibited, as in SD rats, by electrical stimulation of the posterior pituitary. Other investigators found that adrenalectomy in DI, but not in normal rats, is followed by a decrease of ACTH content in the intermediate lobe.[36] These and our results would suggest that DI rats may partially compensate for their vasopressin deficiency by secreting increased amounts of intermediate lobe ACTH, and by releasing oxytocin as a substitute for vasopressin to control the release of ACTH from the anterior pituitary.

Acknowledgments

We are indebted to Ms. M. Friedli and A. Aebi for technical assistance, and to Ms. A. Dupré for drawings. We thank Dr. R. Mathison for helpful discussion.

References

1. Valtin, H., W. H. Sawyer & H. W. Sokol. 1965. Neurohypophysial principles in rats homozygous and heterozygous for hypothalamic diabetes insipidus (Brattleboro strain). Endocrinology **77:** 701–706.
2. Saffran, M. & A. V. Schally. 1955. The release of corticotrophin by anterior pituitary tissue *in vitro.* Can. J. Biochem. Physiol. **33:** 408–415.
3. Guillemin, R. & B. Rosenberg. 1955. Humoral hypothalamic control of anterior pituitary: study with combined tissue cultures. Endocrinology **57:** 599–607.
4. McCann, S. M. 1953. Effect of hypothalamic lesions on adrenal cortical response to stress in the rat. Am. J. Physiol. **175:** 13–20.
5. McCann, S. M. & J. R. Brobeck. 1954. Evidence for the role of the supraopticohypophysial system in the regulation of adrenocorticotrophin secretion. Proc. Soc. Exp. Biol. Med. **87:** 318–324.
6. McCann, S. M. & K. L. Snydor. 1954. Blood and pituitary adrenocorticotrophin in adrenalectomized rats with hypothalamic lesions. Proc. Soc. Exp. Biol. Med. **87:** 369–373.
7. McCann, S. M., J. Antunes-Rodrigues, R. Nallar & H. Valtin. 1966. Pituitary-adrenal function in the absence of vasopressin. Endocrinology **79:** 1058–1064.
8. Pearlmutter, A. F., L. Dokas, A. Kong, R. Miller & M. Saffran. 1980. Is corticotrophin releasing factor modulated vasopressin? Nature **283:** 697–698.
9. Gillies, G. E. & P. J. Lowry. 1980. Reply. Nature **283:** 698.
10. Bradbury, M. W. M., J. L. Burden, E. W. Hillhouse & M. T. Jones. 1974. Stimulation electrically and by acetylcholine of the rat hypothalamus *in vitro.* J. Physiol. (Lond.) **239:** 269–283.
11. Jones, M. T., E. W. Hillhouse & J. L. Burden. 1976. Effect of various putative neurotransmitters on the secretion of corticotrophin-releasing hormone from the rat hypothalamus in vitro. Endocrinology **69:** 1–10.
12. Fehm, H. L., K. M. Voigt, R. I. Lang & E. F. Pfeiffer. 1980. Effects of neurotransmitters on the release of corticotrophin releasing hormone (CRH) by rat hypothalamic tissue in vitro. Exp. Brain Res. **39:** 229–234.
13. Yasuda, N. & M. A. Greer. 1976. Distribution of corticotrophin releasing factor(s) activity in neural and extraneural tissues of the rat. Endocrinology **99:** 944–948.
14. Baertschi, A. J., P. Vallet, J. B. Baumann & J. Girard. 1980. Neural lobe

of pituitary modulates corticotrophin release in the rat. Endocrinology **106:** 878–882.
15. BÉNY, J.-L. & A. J. BAERTSCHI. 1980. Release of corticotrophin and corticotrophin releasing factors from rat posterior pituitary in vitro. Neuroendocrinology **30:** 108–112.
16. GIRARD, J., J. B. BAUMANN, U. BÜHLER, K. ZUPPINGER, H. G. HAAS, J. J. STAUB & H. I. WYSS. 1978. Cyproteroneacetate and ACTH adrenal function. J. Clin. Endocr. Metab. **47:** 581–586.
17. SAYERS, G. 1977. Bioassay of ACTH using isolated adrenal cortex cells. Ann. N.Y. Acad. Sci. **297:** 220–241.
18. BISSET, G. W., B. J. CLARK, J. HALDAR, M. C. HARRIS, G. P. LEWIS & M. ROCHA E SILVA, JR. 1967. The assay of milk-ejecting activity in the lactating rats. Br. J. Pharmac. Chemother **31:** 537–549.
19. GILLIES, G., T. B. VAN WIMERSMA GREIDANUS & P. J. LOWRY. 1978. Characterization of rat stalk median eminence vasopressin and its involvement in adrenocorticotropin release. Endocrinology **103:** 528–534.
20. BENY, J. L. & A. J. BAERTSCHI. 1981. Release of vasopressin from the microdissected median eminence of the rat. Brain Res. **206:** 469–473.
21. YOUNG, T. K. & M. B. VAN DYKE. 1968. Repletion of vasopressin and oxytocin in the posterior lobe of the pituitary gland of the rat. J. Endocr. **40:** 337–342.
22. BÉNY, J. L. & A. J. BAERTSCHI. 1981. Corticotrophin releasing factors (CRF) secreted by the rat median eminence *in vitro* in the presence or absence of ascorbic acid: quantitative role of vasopressin and catecholamines. Endocrinology **109:** 813–817.
23. GILLIES, G. & P. J. LOWRY. 1979. Corticotrophin releasing factor may be modulated vasopressin. Nature **278:** 463–464.
24. BAERTSCHI, A. J., J.-L. BÉNY & G. B. MAKARA. 1980. Corticotrophin releasing factor of diabetes insipidus rats: effects of paraventricular nuclei lesion. Adv. Physiol. Sci. **14:** 57–61.
25. GILLIES, G. & P. J. LOWRY. 1980. Corticotrophin releasing activity in extracts of the stalk median eminence of Brattleboro rats. J. Endocr. **84:** 65–73.
26. VANDESANDE, F., K. DIERICKX & J. DE MEY. 1977. The origin of the vasopressinergic and oxytocinergic fibres of the external region of the median eminence of the rat hypophysis. Cell Tiss. Res. **180:** 443–452.
27. ZIMMERMAN, E. A., M. A. STILLMAN, L. D. RECHT, J. L. ANTUNES & P. W. CARMEL. 1977. Vasopressin and corticotrophin releasing factor: an axonal pathway to portal capillaries in the zona externa of the median eminence containing vasopressin and its interaction with adrenal corticoids. Ann. N.Y. Acad. Sci. **297:** 405–417.
28. BURLET, A., M. CHATEAU & P. CZERNICHOW. 1979. Infundibular localization of vasopressin, oxytocin and neurophysins in the rat; its relationships with corticotrope function. Brain Res. **168:** 275–286.
29. MAKARA, G. B., E. STARK, M. KARTESZI, M. PALKOVITS & G. RAPPAY. 1981. The effect of paraventricular lesion on stimulated ACTH release and CRF in stalk/median eminence of the rat. Am. J. Physiol. **240:** E441–E446.
30. DORNHORST, A., D. E. CARLSON, S. M. SEIF, A. G. ROBINSON, E. A. ZIMMERMAN & D. S. GANN. 1981. Control of release of adrenocorticotrophin and vasopressin by the supraoptic and paraventricular nuclei. Endocrinology
31. BAERTSCHI, A. J., J.-L. BENY, M. FRIEDLI & B. GÄHWILER. 1981. Vasopressin and ACTH in long-term tissue cultures of hypothalamic paraventricular nuclei and adenohypophysis. J. Physiol. (Lond.).
32. KRIEGER, D. T., A. LIOTTA & M. J. BROWNSTEIN. 1977. Corticotrophin releasing factor distribution in normal and Brattleboro rat brain, and effect of

deafferentation, hypophysectomy and steroid treatment in normal animals. Endocrinology **100:** 227–237.
33. Szentagothai, J., B. Flerko, B. Mess & B. Halasz. 1972. Hypothalamic Control of the Anterior Pituitary. 3rd edit. Akadémiai Kiado. Budapest.
34. Palkovits, M., G. B. Makara & F. Stark. 1976. Hypothalamic region and pathways responsible for adrenocortical response to surgical stress in rats. Neuroendocrinology **21:** 280–288.
35. Buckingham, J. C. 1981. The influence of vasopressin on hypothalamic corticotrophin releasing activity in rats with inherited diabetes insipidus. J. Physiol. **312:** 9–16.
36. Schleiffer, R., C. Miahle, B. Briaud, B. Lutz-Bucher & Koch. 1979. Effects of adrenalectomy and hypercorticism on the ACTH content of the anterior and the posterior pituitary in rats with inherited diabetes insipidus (Brattleboro strain). Horm. Metab. Res. **11:** 130–135.

Discussion of the Three Preceding Papers

R. G. Allen (*Veterans Administration Hospital, Portland, Ore.*): Everyone talks about hypothalamic extract and yet no one clearly specifies how it was purified, what was thrown away, what was kept, etc. Since Vale is soon going to reveal to the world the structure of CRF, (*Science* **213:**1394, 1981), it would be nice to know what *is* hypothalamic extract. I have prepared some myself and I am sure mine differs from all three of yours with regard to CRF activity.

A. F. Pearlmutter (*Medical College of Ohio, Toledo, Ohio*): Ours is a crude median eminence extract. We expose the ventral surface of the brain, and with curved eyelet scissors press down on the mammillary body and cut out a fragment, which contains the pituitary stalk, median eminence, and a small amount of ventral hypothalamus (about 3 mg) and place that in 25 μl of 0.1 N HCl per fragment. Sometimes we collect 3 or 4 fragments at a time, put them in the freezer and continue to add to that on subsequent days. The Brattleboro and Long-Evans material was made up in an identical fashion.

Allen: Don't you desalt it or do anything else to it? Is it just a crude prep?

Pearlmutter: When we use this extract in our bioassay, we add 25 μl of extract to 1.0 ml of Krebs-Ringer solution. Thus, whatever salts and other things that may be in there probably would not play a large role. In addition, we run a parallel control to correct for any effects that might be due to the extract acting on the adrenal.

J. C. Buckingham (*Royal Free Hospital Medical School, London, England*): Our preparation procedure is very similar to Dr. Pearlmutter's except that we use the whole hypothalamus and not just the median eminence. Our extraction method is identical. However, we never freeze our extracts but dilute and assay them within one hour.

A. J. Baertschi (*University of Geneva, Geneva, Switzerland*): Our extraction procedure is similar to those in the two previous papers, however, the region we take now is a 100-μg piece of the microdissected median eminence (*Brain Res.* **206:**469, 1981). The first thing we do is take out the whole medial basal hypothalamus, which weighs about 2 mg, then with a scalpel dissect out just the median eminence region which only weighs about 100 μg. From that piece the median eminence extracts are prepared as mentioned by others.

J. R. Hodges (*Royal Free Hospital Medical School, London, England*): I should like to ask Dr. Pearlmutter why she thought that her results were very different from those of Dr. Buckingham for they seem to me to be remarkably similar. There are differences in your methods for the quantification of corticotrophin releasing activity but, of course, neither of you is using a valid bioassay since there is no CRF standard and your results are expressed in terms of the ACTH released from pituitary tissue. It is, therefore, impossible to make quantitative comparisons and this is well shown by the differences between the dose-response lines of Dr. Buckingham's Brattleboro rats and her Long-Evans controls.

Pearlmutter: We do not have dose-response curves of Brattleboro median eminence extracts because we did not have enough material to do this. Secondly, the response to one median eminence is fairly high on the dose-response curve that we obtained previously with normal rats. So that at least until that point, the dose-response curves would be identical. However, you are correct. If Dr. Buckingham's dose-response curve were to go higher, perhaps then we would no longer see a dissociation between the normal and the Brattleboro CRF content.

P. J. Lowry (*St. Bartholomews Hospital, London, England*): I think the big differences in all these assays is the fact that freshly excised anterior pituitaries actually contain endogenous vasopressin. In fact, it takes several hours for the endogenous vasopressin to disappear from pituitary halves cultured *in vitro*. So that when one uses pituitary halves, one has to take into account the actual endogenous vasopressin that is there. I think this may explain, first of all, why Dr. Pearlmutter's experiments are different from Dr. Buckingham's. Five hours is apparently the time it takes for all the vasopressin to disappear from the anterior pituitary *in vitro*.

Secondly, in our hands CRF is quite unstable. When one starts fractionating it becomes unstable; that is why we first started using ascorbic acid because to add an antioxidant seemed a sensible thing to do. In addition, it appears that vasopressin actually stabilizes the CRF complex, indicating that there was probably some association. If there is an association of vasopressin with the CRF complex, that would perhaps explain Dr. Pearlmutter's experiment: when the crude median eminence extract is passed through a neurophysin column, the association of vasopressin to the CRF complex would then associate the complex with the neurophysin on the resin.

Thirdly, there are problems with preparations of oxytocin and vasopressin. The only good vasopressin preparation that Dr. Glenda Gillies and I have is one that is about 12 years old, which we got from Ferring. This vasopressin was synthesized classically and not by the modern solid-phase automated machines. We got very confused at one stage because the Ferring material was running out, and we started switching over to the more modern synthetic preparations. We did not, however, get the consistent results that we previously observed. We, therefore, purified some vasopressin from posterior pituitaries and our old results returned. Quite recently we checked several preparations by HPLC (ultrasphere propyl-CN). Our purified material, which we have been using for many years, gives quite a nice, single peak with slight contamination. The synthetic vasopressin, however, gave multiple peaks, at least four of equal abundance. The same thing happened with purified oxytocin, and purified lysine vasopressin. Some of these could be partial agonists or antagonists. We have to reveal more about our peptides, e.g. where we get them, who made them, how they are purified, etc., because these major contaminants could have marked effects in any bioassay system.

ALLEN: The last slide I showed was a profile of human 31β-endorphin.

LOWRY: Oh yes, β-endorphin is great, enkephalin is great and any straight chain peptide is alright. I think the problem has something to do with the disulfide bridges in synthetic peptides.

PEARLMUTTER: I just want to respond to the first thing that you said about the vasopressin being present in the anterior pituitary tissue that we use in our CRF bioassay. Were that the case, we never would have seen the decrease in CRF activity with peak A alone, the larger molecular weight peak found in our Sephadex chromatography of median eminence extract. By your reasoning we always would have had the vasopressin, which you think is the small molecular weight material. Thus, peak A should have always shown full activity and we would have never been able to see this striking potentiation when the larger and the smaller molecular weight materials from the Sephadex column were combined. Also, we have done some bioassays with Brattleboro pituitaries and they look fairly normal. Because of financial limitations we were only able to do one set of experiments. But that would be the way to test your hypothesis.

LOWRY: I think Dr. Buckingham has done some work with Brattleboro rats.

BUCKINGHAM: We thought the Brattleboro material was very much *less* sensitive to hypothalamic extracts than normal tissue, but perhaps very surprisingly, they are equally sensitive to lysine vasopressin or arginine vasopressin.

LOWRY: Using isolated cells, there is no difference in the response of Brattleboro pituitary cells versus normal to stalk-median eminence or AVP. Once again the sequestered vasopressin makes the difference in pituitary halves in culture.

B. T. PICKERING (*University of Bristol, Bristol, England*): Following on from what you have been saying about possible association with vasopressin, has any one ever looked at the median eminence extracts from homozygous Brattleboro rats that have been restored to normal by vasopressin?

BUCKINGHAM: I have not done this yet but I intend to when we breed enough Brattleboro rats.

PICKERING: Would you suspect, Dr. Lowry, that such animals might then have more normal median eminences in terms of their CRF activity?

LOWRY: If you add the appropriate amount of vasopressin you can restore it to normal.

PICKERING: I am not talking about doing the thing *in vitro*. I am wondering whether an animal that has been living in an environment of vasopressin, and that perhaps can take up vasopressin, would give rise to a median eminence extract that would behave more like that from a normal animal.

LOWRY: What you are now invoking is neural peptide uptake into the median eminence.

PICKERING: I am not invoking anything, I am just asking if the experiment has been done.

LOWRY: No we haven't done that.

H. W. SOKOL (*Dartmouth Medical School, Hanover, N.H.*): I think Gillies and Lowry have shown that the amount of CRF in the hypothalamus, i.e. in the two smaller peaks, is really almost quantitatively identical in the DI and normal rat. Is that correct?

LOWRY: Yes.

SOKOL: One might increase CRF by treating rats with vasopressin. Is that what you are suggesting Dr. Pickering?

PICKERING: Not increase the CRF *per se:* one's talking in terms of biological activity rather than in terms of chemical entities. What I am wondering is whether the biological activity of the median eminence of a vasopressin-treated Brattleboro rat might have the same pharmacological properties as the median eminence of a normal rat; in other words, have the same sort of dose-response curve.

LOWRY: I think also that there is some problem with vasopressin in the CRF complex. I really don't understand the way the whole complex is stabilized. But I think there must be some physicochemical association.

P. STERN (*Dartmouth Medical School, Hanover, N.H.*): I think there is evidence that peripheral infusion of vasopressin would not necessarily restore median eminence vasopressin.

HODGES: I am delighted, of course, that the last three speakers have all been using bioassay techniques, but none of them are really suitable bioassays or very valuable because of the lack of the standard. I think that one of the main reasons for the differences between findings by Dr. Lowry's and Dr. Buckingham's groups is the fact that one group uses pituitary segments, the other group uses isolated pituitary cells for their assays of CRF. It seems to me that in the preparation of isolated pituitary cells there is a change in the receptors: receptors to vasopressin seem to be upgraded and receptors to CRF downgraded.

LOWRY: I think one has to take into account many different aspects depending upon whether one prepares cells *or* tissue in culture. One could criticize one method against the other indefinitely, but I think everyone wants to know what CRF is. I think that researchers should use as many tools that they have at their disposal.

HODGES: The point I am trying to make is that we will only get meaningful results when we use parallel quantitative assays. I think we all need to go to a course in pharmacology.

G. A. HEDGE (*West Virginia University Medical Center, Morgantown, W. Va.*): In response to the comments by Dr. Hodges, I would like to mention some of our experience with a dispersed pituitary cell preparation very similar to Dr. Lowry's except we have been concentrating on TSH and prolactin secretion. We have encountered several situations in which we have had to determine whether or not these enzymatically dispersed cells are likely to be giving us the same answers as we would get with fragments of pituitaries. Somewhat to our surprise, the only case in which we have found what we would call an abnormality of the dispersed cell regards the negative feedback effects of thyroid hormones on TSH secretion. In contrast to pituitary fragments, the cells do not decrease their TSH secretion upon exposure to thyroid hormones. We were surprised by this because we would have expected the enzyme to disturb surface receptors for peptide hormones more readily than the intracellular mechanisms involved in thyroid hormone action. In every other way the cells function quite normally, and we find them to be of particular value in the study of secretory dynamics using a continuous superfusion system.

A. G. ROBINSON (*University of Pittsburgh, Pittsburgh, Pa.*): I would like to make a comment about physiology. Most of this discussion is about *in vitro* work and biochemistry. I would like to tell you of some studies Dr. Donald Gann has done in collaboration with Earl Zimmerman and me. Dr. Gann has stimulated areas of the cat anterior hypothalamus and measured coincident release of ACTH and the vasopressin. When we stimulated in the area of the paraventricular nucleus (PVN) there was a good correlation between release of vasopressin in the peripheral blood and release of ACTH. Since then, in work which we presented at The Endocrine Society, we have tried to block the release of ACTH in response to PVN stimulation by prior administration of antisera to vasopressin. When we preadminister vasopressin antisera we block the release of ACTH in response to the stimulation of the PVN, but if we stimulate other areas in the hypothalamus, there is no change in ACTH release in response to those stimulations. The last point, in terms of physiology, we can't block with antiserum to oxytocin.

ALLEN: This is directed to all four speakers. We have worked with AT20 tumor cells which are grown in monolayer culture, are stable, and have been alive for 15 years. I would like to know why these cells do not respond to vasopressin under most conditions.

LOWRY: The fact that you need to keep on stimulating the pituitary gland with LHRH to keep its receptors, and that down-regulation of receptors is observed when the gland experiences long-term absence of a stimulating factor, could explain your observation. Also, they are tumor cells and maybe that mutation is linked to the vasopressin stimulation of ACTH release.

ALLEN: They apparently still have a CRF receptor.

LOWRY: Other things that you may have to take into account are the biological effects of not only the whole 16K, or even the γ-MSH peptide, but some of the peptides from the N-terminal part of pro-opiocortin.

J. J. DREIFUSS (*University of Geneva School of Medicine, Geneva, Switzerland*): My question is addressed to Dr. Hedge. You showed that the TSH from pituitaries was elevated in di/di rats, and that this was not corrected by dDAVP in doses where dDAVP corrected the renal defect. Is this because the receptors somewhere higher up are different from the renal receptor, or what is the explanation?

HEDGE: We have concluded from our findings that the pituitary-thyroid abnormality is not a direct function of disturbed water balance or *circulating* vasopressin levels. This leaves open the interesting possibility of a local effect of vasopressin somewhere other than the thyroid. We have recently started some experiments in which we are administering vasopressin into the hypothalamus to test for a local interaction that one would not detect by peripheral replacement of vasopressin.

SOKOL: The difficulty with that explanation is that one might expect TRH secretion to be interfered with, but since you did show elevated TSH release in the DI rat it doesn't appear that the hypothalamic-pituitary axis is at fault.

HEDGE: You are correct in that our data indicate that the thyroid status in these rats is not totally determined by the activity of the hypothalamic-pituitary axis. In fact, we are thinking along other lines in the experiments involving vasopressin replacement into the region of the brain from which endogenous vasopressin is derived. We are intrigued by the possibility that the thyroid nerves might be involved here since there is good evidence that the adrenergic innervation of the thyroid can to some extent "tune" the sensitivity of the thyroid to circulating TSH. Thus, what we are trying to get to is the possibility of a relationship between the vasopressin and the nervous system that affects the thyroid gland via some mechanism other than the TRH-TSH pathway.

PROLACTIN AND GROWTH HORMONE IN THE UNSTRESSED BRATTLEBORO RAT *

Robert A. Adler and Hilda Weyl Sokol

Departments of Medicine and Physiology
Dartmouth Medical School
Hanover, New Hampshire 03755

INTRODUCTION AND METHODS

Because Brattleboro (DI) rats are smaller than aged-matched Long-Evans (LE) rats, we investigated anterior pituitary function in unstressed, unanesthetized rats fitted with indwelling intra-aortic catheters. Male DI and weight-matched LE rats received intra-arterial injections of thyrotropin-releasing hormone (TRH) (400 ng/100 g BW), insulin (1 unit/100 g BW), or the alpha-adrenergic agent, clonidine (0.1 mg/100 g BW), at approximately weekly intervals. Blood samples for prolactin (PRL), growth hormone (GH), and glucose (in the insulin test) were obtained through the catheter just before and at specific intervals after the injection. Serum separated from whole blood was frozen for later radioimmunoassay using reagents kindly provided by the National Institute of Allergy, Metabolism, and Digestive Diseases (National Institute of Arthritis, Diabetes, and Digestive and Kidney Diseases) Hormone Distribution Program.

At sacrifice, anterior pituitaries were removed, and halves were frozen for later radioimmunoassay of GH and PRL concentration or preserved in Bouin's solution for immunocytochemical analysis. The frozen halves were thawed, weighed, and homogenized in phosphosaline buffer just before assay. Pituitary glands from female homozygous (DI), heterozygous (HZ), and Long-Evans (LE) rats as well as male HZ rats were prepared at the same time as the preserved hemi-pituitaries from this study. Transverse 6 μm sections of Bouin's-fixed, paraffin-imbedded glands were incubated with the same antisera to rPRL and rGH as used in the radioimmunoassay of these hormones. Subsequent application of peroxidase-antiperoxidase and enzyme-substrate[1-3] resulted in the deposition of brown reaction products at specific antigenic sites.

RESULTS

Serum Levels of PRL and GH

Basal and stimulated PRL secretion. Basal serum PRL (average of three weekly determinations/rat) was 19.5 ± 5.3 ng/ml (SEM) in six male DI rats compared to 18.2 ± 3.7 ng/ml in six LE male rats (p = NS). There was a small response to TRH but it was similar in the two groups. Insulin-induced hypoglycemia was followed by a decrease in PRL to less than half of the baseline level in both groups of rats. The ratio of the nadir of PRL to baseline was

* Supported by National Institutes of Health Grants AM 08469 and AM 22032.

0077-8923/82/0394-0607 $1.75/0 © 1982, NYAS

0.42 in DI rats and 0.50 in LE rats (p = NS). Clonidine increased PRL in both DI and LE rats to about twice baseline. There was no difference between the two groups.

Basal GH secretion. There was considerable day-to-day change in basal GH in these unstressed male rats. Serum GH levels in DI and LE rats were generally in the range of 10 to 25 ng/ml. However, we also observed very high GH levels reflecting the pulsatile nature of the secretion of this hormone. For example, DI rat #75–1 had the following serum GH concentrations under the same conditions and at the same time of day at three weekly blood drawings: 59.7, 12.2, and 158.5 ng/ml. Corresponding PRL levels were 8.2, 9.0, and 10.6 ng/ml. Similarly LE rat #75–5 had basal serum GH levels of 207.8, 210.9, and 44.1 ng/ml with corresponding serum PRL levels of 9.4, 12.7, and 10.8 ng/ml. The large variations in serum GH make comparisons difficult, but the average basal GH of six DI rats was 78.3 ± 46.9 ng/ml compared to 61.4 ± 24.0 ng/ml in six LE rats (p = NS).

GH response to TRH. There was no evidence that serum GH was increased by the administration of TRH. If the basal GH levels were high that particular morning, post-TRH samples would decrease. If the basal levels were in the 10 to 20 ng/ml range, the post-TRH levels varied in no discernible pattern in either the DI or LE group.

GH response to insulin-induced hypoglycemia. The intra-arterial injection of 1 unit/100 g BW of regular insulin produced a brisk hypoglycemic response. The serum glucose was lower in the DI rats at all points, but only the 30-minute sample was statistically significant (54.3 ± 1.4 vs. 67.3 ± 4.5 mg/dl, $p < 0.05$). The mean nadir of serum glucose was 42.2 ± 2.5 mg/dl in the DI group and 51.3 ± 4.9 mg/dl in the LE group (p = 0.1). In five of six DI rats the serum GH was higher at 60 minutes after insulin than at baseline. The one animal not demonstrating this phenomenon had a very high basal GH (422.4 ng/ml). In the LE group, four of six rats had a higher serum GH 60 minutes after insulin, but this was usually preceded by serum GH below baseline levels at 30 and/or 45 minutes. This pattern was not observed in any of the DI rats. The higher GH level in DI rats at 45 minutes after the injection of insulin just barely misses statistical significance when compared to LE rats (p = 0.07).

GH response to clonidine. There was a clear increase in serum GH after clonidine administration to three DI and three LE rats. However, the ratio of peak GH to basal GH ranged from 1.85 to 23.38. This variability of response was seen in both groups and no qualitative or quantitative differences were observed.

Anterior Pituitary Content of PRL and GH

Pituitary prolactin concentrations. PRL concentrations in hemi-pituitaries were similar in four male DI and four male LE rats (0.472 ± 0.231 μg/mg wet weight vs. 0.412 ± 0.118 μg/mg wet weight).

Lactotrope morphology. A comparison between male and female pituitaries (FIGURE 1) revealed a striking sex difference in the number and morphology of PRL-secreting cells. In general, lactotropes are polygonal or angular and coarsely granulated, with processes extending between other cells. The lactotropes from male pituitaries were not only distributed more sparsely throughout the anterior pituitary, but the cells were considerably less granulated. A com-

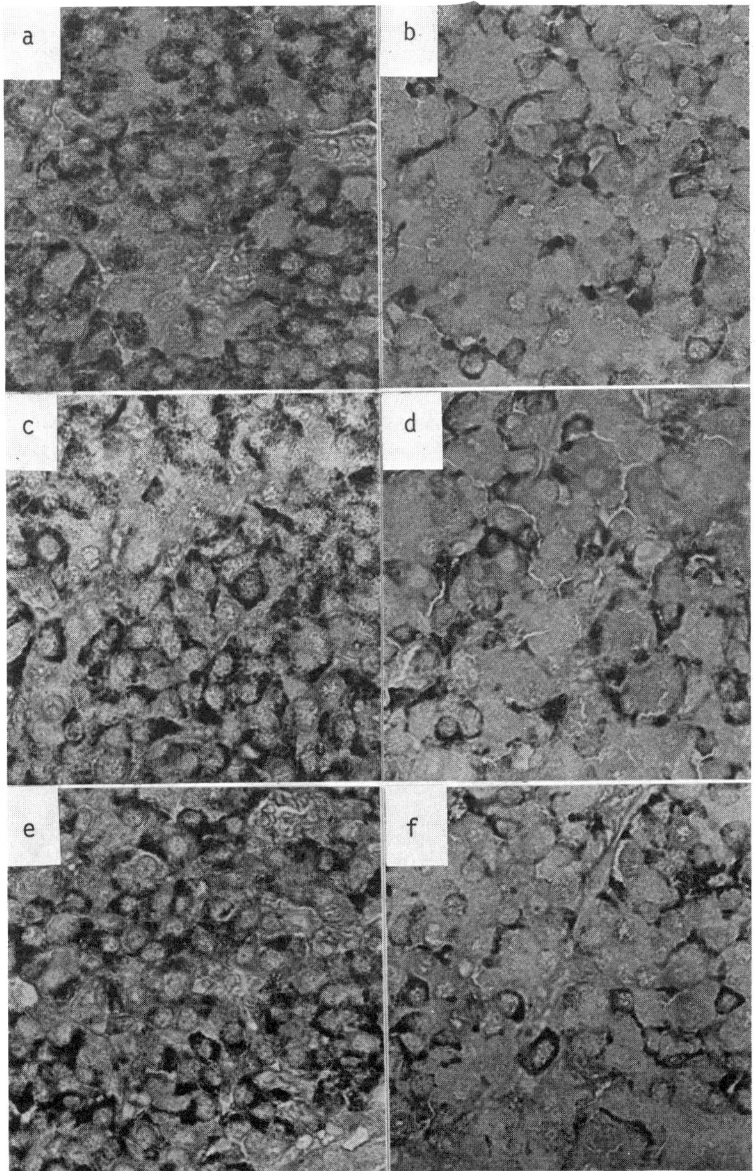

FIGURE 1. Lactotropes (visualized using an immunoenzyme technique with anti-rPRL) in normal LE (a,b), heterozygous (c,d), and homozygous (e,f) Brattleboro rats of both sexes (females, left; males, right). PRL cells in males are sparsely distributed in the anterior pituitary and are noticeably less granulated than in females. PRL cells in normal and heterozygous rats were larger and more conspicuously granulated than in DI rats, a difference most easily seen in females. Magnification 482×.

parison of the pituitary glands in the three genotypes revealed that PRL cells in LE and HZ rats were larger and more conspicuously granulated than in the DI pituitary, the difference being most striking in females.

Pituitary growth hormone concentration. The concentration of GH in four male hemi-pituitary glands from DI rats was 20.3 ± 6.1 μg/mg wet weight as contrasted with 33.5 ± 5.9 μg/mg wet weight in four male LE half-glands. Considerable variation within groups precluded statistical significance.

Somatotrope morphology. Somatotropes, the most numerous of anterior pituitary cells, are widely distributed throughout the lobe and are easily distinguishable from lactotropes. The ovoid somatotrope contains granules of fine texture. Interestingly, the GH-secreting cells of male rats are much larger and less densely stained, suggesting enhanced synthesis. A comparison of GH cells in the three rat genotypes revealed somewhat smaller cells in the DI rats, especially noticeable in the LE versus DI males. Somatotropes from DI female rats had less dense cytoplasm than LE or HZ females (FIGURE 2). The sex differences in somatotropes are consistent with higher GH concentrations in anterior pituitary glands (40 vs. 17 μg/mg wet weight) in males compared to females by analytical polyacrylamide gel electrophoresis (Sokol, unpublished). The difference in pituitary GH concentration between LE and DI males, as measured by radioimmunoassay in this study and by bioassay,[4] is consistent with the cytological differences observed in the GH cells.

DISCUSSION

Our data suggest that neither PRL nor GH secretion in DI rats is reduced when compared to that of weight-matched LE control rats. PRL has been claimed to have antidiuretic properties in mammals by some investigators,[5,6] but the weight of evidence is against such a role.[2,7] If PRL were antidiuretic, serum PRL levels might be expected to be higher in rats lacking the potent antidiuretic substance, arginine vasopressin. However, Brattleboro rats do not have elevations of basal serum PRL levels, and PRL concentrations and lactotrope morphology of the male DI and LE pituitaries are similar. The PRL response of male DI rats to insulin, TRH, and clonidine is appropriate and comparable to the response in LE males. Since male and female rats may have different PRL responses to a variety of stimuli such as TRH[8] and clonidine,[9] studies comparing female DI and LE rats will be necessary to fully characterize dynamic PRL synthesis and secretion in the DI rat.

Growth is retarded in Brattleboro rats, and large doses of bovine GH treatment significantly enhance growth in the DI rat.[10] However, the efficacy of injected heterologous GH does not necessarily prove that altered endogenous GH secretion contributes to the growth retardation in these rats. Unfortunately, the wide range of serum GH in normal rats,[11] makes it difficult to assess basal GH secretion. Episodic GH release, stress, and anesthesia can account for wide variations in serum GH concentrations.[12] There are also considerable sex differences in GH secretion. For example, in rats androgens increase and estrogens decrease serum GH.[13] We attempted to avoid some of these problems by using rats fitted with indwelling intra-aortic catheters allowing blood withdrawal without acute stress or anesthesia. Furthermore, we infused TRH, insulin, and clonidine to stimulate or inhibit hormone release and compared the responses in DI and LE controls. No essential differences in response

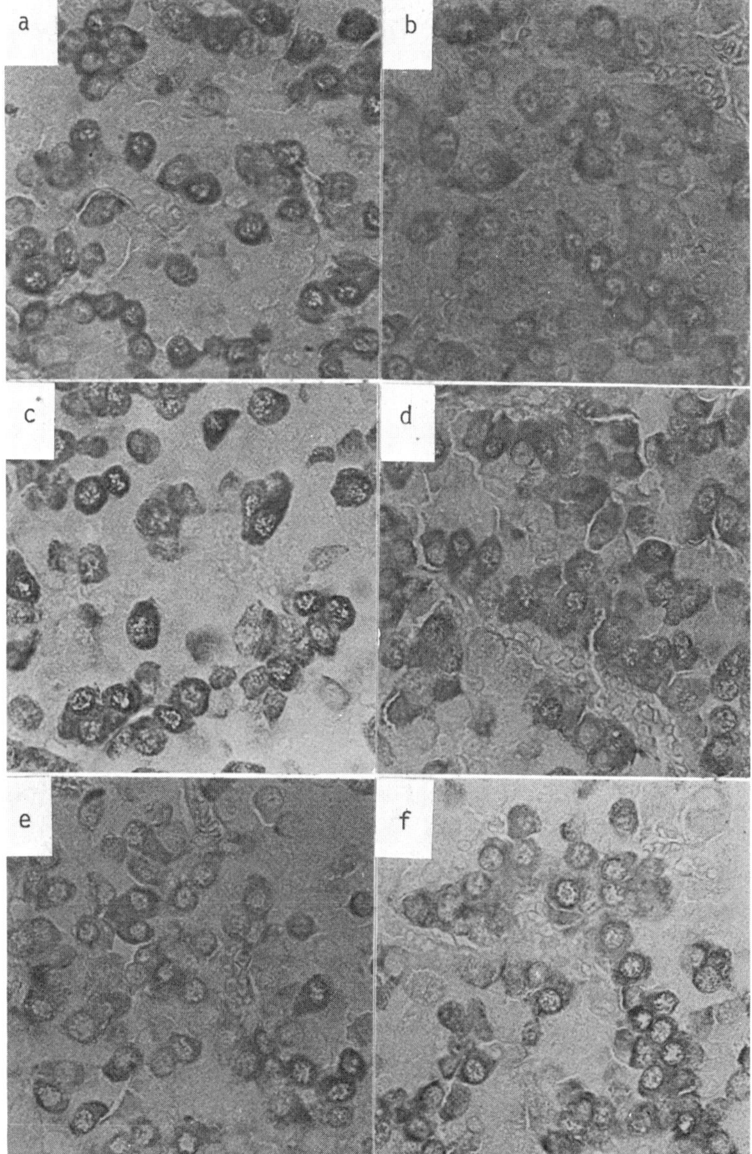

FIGURE 2. Somatotropes (visualized using an immunoenzyme technique with anti-rGH in normal LE (a,b), heterozygous (c,d), and homozygous (e,f) Brattleboro rats of both sexes (females, left; males, right). GH cells in males are much larger and less densely stained than in females. GH cells in male DI rats are smaller than in normal or heterozygous controls; in the female DI rat they are slightly smaller and less immunoreactive than in non-DI females. Magnification 482×.

between normal and DI rats were observed. If anything, post-insulin plasma GH was slightly elevated in the DI rats despite a normal PRL response to insulin. It is also possible that the drop in serum glucose to slightly lower levels in the DI rats compared to LE rats is a manifestation of inadequate catecholamine or corticosterone response to hypoglycemia. At any rate, there is no evidence for inadequate GH release in the DI rat.

Striking differences between males and females can be seen in the morphology of their lactotropes and somatotropes. Prolactin cells are more numerous and better granulated in females while growth hormone cells show signs of greater activity in males. Thus, it will be important to repeat the studies using female rats. Differences in these hormone-producing cells are also discernible between normal and DI rats, although not as striking as the sexual dichotomy. The lower GH content, as measured by RIA, of anterior pituitary glands in DI rats is reflected in the cytology of somatotropes.

Although there is a correspondence between pituitary hormone content and its cellular morphology, the plasma levels of GH do not reflect these differences. Therefore, it must be emphasized that we measured only *immunoreactive* GH and PRL. The GH of the DI rat could include a molecular form having immunoreactivity but not biological activity. Further studies are planned to determine the molecular size of pituitary and serum GH in DI rats. Although both GH and PRL may stimulate the production of somatomedin-C,[14] GH is a much more potent stimulus and probably exerts its growth-promoting effects through somatomedin. It will therefore be necessary to measure somatomedin-C in the DI rat. Although we can conclude that secreted immunoreactive GH and PRL are not decreased in DI rats, the mystery of their retarded growth remains unsolved.

References

1. Petrali, J. P., D. M. Hinton, G. C. Moriarty & L. A. Sternberger. 1974. The unlabeled antibody enzyme method of immunocytochemistry. Quantitative comparison of sensitivities with and without peroxidase-antiperoxidase complex. J. Histochem. Cytochem. **22:** 782–801.
2. Adler, R. A., S. Dolphin, M. Szefler & H. W. Sokol. 1979. The effects of elevated circulating prolactin in rats with hereditary hypothalamic diabetes insipidus (Brattleboro strain). Endocrinology **105:** 1001–1006.
3. Adler, R. A., S. J. Brown & H. W. Sokol. 1980. Characteristics of prolactin secretion from anterior pituitary implants. Prog. Reproductive Biol. **6:** 24–30.
4. Arimura, A., S. Sawano, T. W. Redding & A. V. Schally. 1969. Studies on retarded growth of rats with hereditary diabetes insipidus. Neuroendocrinology **3:** 187–192.
5. Horrobin, D. F., I. J. Lloyd, A. Lipton, P. G. Burstyn, N. Durkin & K. L. Muiruri. 1971. Actions of prolactin on human renal function. Lancet **2:** 352–354.
6. Miller, M. & M. Van Gemert. 1976. Antidiuretic action of prolactin in the rat with diabetes insipidus. Horm. Res. **7:** 319–332.
7. Carey, R. M., A. J. Johanson & S. M. Seif. 1977. The effects of ovine prolactin on water and electrolyte excretion in man are attributable to vasopressin contamination. J. Clin. Endocrinol. Metab. **44:** 850–858.
8. Piercy, M. & S. H. Shin. 1980. Comparative studies of prolactin secretion in estradiol-primed and normal male rats induced by ether stress, pimozide and TRH. Neuroendocrinology **31:** 270–275.

9. Stevens, R. W. & D. M. Lawson. 1977. The influence of estrogen on plasma prolactin levels induced by thyrotrophin releasing hormone (TRH), clonidine and serotonin in ovariectomized rats. Life Sci. **20:** 261–266.
10. Sokol, H. W. & J. Sise. 1973. The effect of exogenous vasopressin and growth homone on the growth of rats with hereditary hypothalamic diabetes insipidus. Growth **37:** 127–142.
11. Schalch, D. S. & S. Reichlin. 1966. Plasma growth hormone concentration in the rat determined by radioimmunoassay: Influence of sex, pregnancy, lactation, anesthesia, hypophysectomy and extrasellar pituitary transplants. Endocrinology **79:** 275–280.
12. Martin, J. B., G. Tannenbaum, J. O. Willoughby, L. P. Renaud & P. Brazeau. 1975. Functions of the central nervous system in regulation of pituitary GH secretion. *In* Hypothalamic Hormones. M. Motta, P. G. Crosignani & L. Martini, Eds.: 217–235. Academic Press. New York, N.Y.
13. Birge, C. A., G. T. Peake, I. K. Mariz & W. H. Daughaday. 1967. Radioimmunoassayable growth hormone in the rat pituitary gland: Effects of age, sex and hormonal state. Endocrinology **81:** 195–204.
14. Holder, A. T. & M. Wallis. 1977. Actions of growth hormone, prolactin and thyroxine on serum somatomedin-like activity and growth in hypopituitary dwarf mice. J. Endocrinol. **74:** 223–229.

ALTERATIONS IN PITUITARY FUNCTION IN RATS WITH HYPOTHALAMIC DIABETES INSIPIDUS (BRATTLEBORO STRAIN) *

Seigo Fujimoto and George A. Hedge

Department of Physiology
West Virginia University Medical Center
Morgantown, West Virginia 26506

Many studies have focused on the possible role(s) of vasopressin (VP) in regulating the secretion of anterior pituitary hormones. VP-secreting neurons have been discovered in hypothalamic areas known to be involved in regulating adenohypophyseal secretion. Also, this peptide and neurophysin are present in the external layer of the median eminence, the portal blood, and the anterior pituitary. However, the physiological significance of VP in regulating anterior pituitary secretion has not as yet been established.

VP is thought not to be the physiological corticotropin-releasing hormone, but it has corticotropin (ACTH)-releasing and ACTH-like activities.[1] There is also evidence that, at least under certain circumstances, VP is a stimulant for the secretion of growth hormone,[2] prolactin (PRL),[3] thyrotropin (TSH),[4] and luteinizing hormone.[5]

The Brattleboro rat seems to be a good model for clarifying the role of VP on pituitary secretion. Some investigators have reported that plasma concentrations of corticosterone (used as an index of ACTH secretion) are lower in the Brattleboro rat than in the control rat,[6] but others have not observed this.[7] On the other hand, there is agreement that the elevation of plasma corticosterone (B) concentration during certain types of stress is less in the Brattleboro rat than in the control rat.[7, 8] Thus, there has been considerable attention devoted to the pituitary-adrenal axis in these rats, but by comparison, there is still relatively little known about other anterior pituitary hormones in the Brattleboro rat. Therefore, we have assessed TSH, PRL, and B levels in homozygous (HOM) and heterozygous (HET) Brattleboro rats before and during infusion of VP.

Methods

All of the methods and materials involved in the *in vivo* experiments are as described in our previous paper in this volume.[9] Plasma PRL was assayed by RIA, and B was measured by a competitive protein binding assay.

The *in vitro* pituitary response to TRH was compared among the control, HOM, and HET, using the following two preparations: (1) Anterior pituitary cells from three animals were enzymatically dispersed and superfused as described in detail previously.[10] Superfusate was collected every 5 min, and TRH was superfused for 10 min. (2) Each hemisected anterior pituitary was cut into nine fragments, pooled, and preincubated in 2 ml of Krebs-Ringer bicarbonate

* Supported by the National Institutes of Health Grant AM 21348.

0077-8923/82/0394-0614 $1.75/0 © 1982, NYAS

buffer containing 0.5% bovine serum albumin for 30 min at 37° C. The fragments were then incubated in the buffer for 30 min (control period) and then in the buffer containing TRH at a final concentration of 10^{-6} M for another 30 min. The fragments from contralateral hemipituitaries were used as a "vehicle-treated group" and treated in the same manner except that the TRH stimuli were omitted.

Statistical comparisons were made using the paired or unpaired Student's *t* test as appropriate.

RESULTS

As described in our previous paper,[9] basal levels of plasma TSH in HOM and HET were higher than the control level and the pituitary response to TRH (250 ng/100 g BW, i.v.) was greater in HOM and HET than the control rat. The present experiments confirm these findings (FIGURE 1), and they also provide information on PRL and ACTH secretion in these same animals. It was found that plasma PRL levels in HOM and HET were similar to each other and significantly lower than that in the control rat. In contrast, no differences were found in B levels between the control and the Brattleboro rats.

To determine if the abnormal TSH and PRL levels could be corrected by the normalization of water balance, DDAVP was infused s.c. at a rate of 48 ng/day for seven days, a treatment that restored normal water balance in HOM.[9] In HOM, basal plasma TSH levels and TSH responses to TRH were not significantly changed, but plasma PRL levels were restored to the control level by this treatment (FIGURE 1).

On the other hand, the same dose of DDAVP, which was antidiuretic and increased urine osmolality above the normal range in HET,[9] restored basal TSH and PRL levels, as well as TRH-induced TSH increments, to the control levels.

Since the exaggerated TSH response to TRH observed in HOM and HET could be caused by the low plasma concentration of thyroid hormone,[9] we studied the TRH responsiveness of isolated pituitary tissue. Dispersed pituitary cells from HOM spontaneously secreted more TSH than did the control cells (FIGURE 2). When TRH was superfused, TSH secretion was increased in the three groups, but the variation in baseline in this preparation prevented quantification of the amounts of TSH secreted in response to TRH. Consequently, we next determined the pituitary responsiveness to TRH in a static incubation experiment. The basal level of TSH in the medium in HOM was again higher than those of control, but the magnitude of the TSH response to TRH was similar in all groups (FIGURE 3).

DISCUSSION

The present experiments confirm our previous findings that basal and TRH-induced plasma TSH levels were abnormally high in HOM and HET.[9] Although we suggested that the increase in the TSH secretion might be secondary to the simultaneous reduction in plasma thyroid hormones,[9] these *in vivo* experiments could not rule out the possibility of an inherent difference in pituitary sensitivity in the Brattleboro rats. Thus, we have used two types of tissue preparations in an attempt to determine whether the pituitary response to TRH in the absence of thyroid hormones is altered in the Brattleboro rat. In both cases, we found that the spontaneous secretion of TSH in HOM was higher than in the control rat. Although the physiological significance of this phenomenon is not apparent, it is presumably related to the high level of plasma TSH in HOM. However, the present evidence that there is no difference among the three groups regarding the TSH response to TRH *in vitro*, suggests that in the Brattleboro rats the increased *in vivo* response to TRH was due to the endocrine environment of the pituitary rather than to an inherent sensitivity difference.

The restoration of water balance by DDAVP[9] was not followed by normalization of the basal or TRH-induced TSH levels in HOM. Nevertheless, DDAVP restored these variables to the control levels in HET, while it reduced

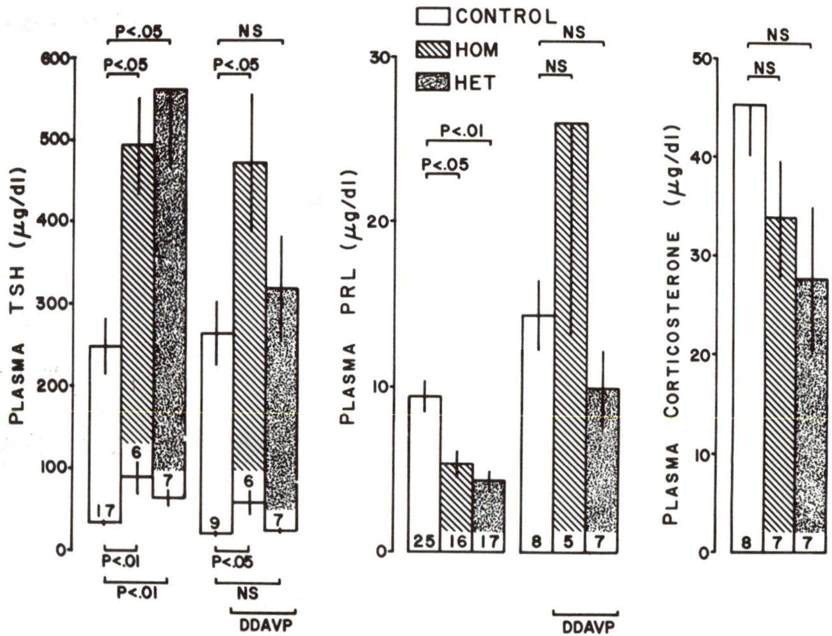

FIGURE 1. Basal and TRH-induced plasma TSH, and basal plasma PRL and B levels before and after seven-day infusion of DDAVP. In left panel, bottom and top lines of each bar indicate basal and TRH-induced TSH levels, respectively. Vertical lines indicate SE of means. Numbers within bars represent numbers of animals.

urine production and increased osmolality. These results indicated clearly that these abnormalities in HOM were not simply due to alterations in water metabolism.

Since TRH can stimulate PRL as well as TSH secretion, and since there is a similarity between the functional receptors involved,[3] we also measured the plasma PRL levels in the Brattleboro rats. As expected, PRL levels in the control rat were higher than commonly observed in unstressed rats because our blood samples were obtained by cardiac puncture under ether anesthesia. However, these stress-induced PRL levels in HOM and HET were significantly lower than in controls. The experiment with DDAVP suggested that abnormal water

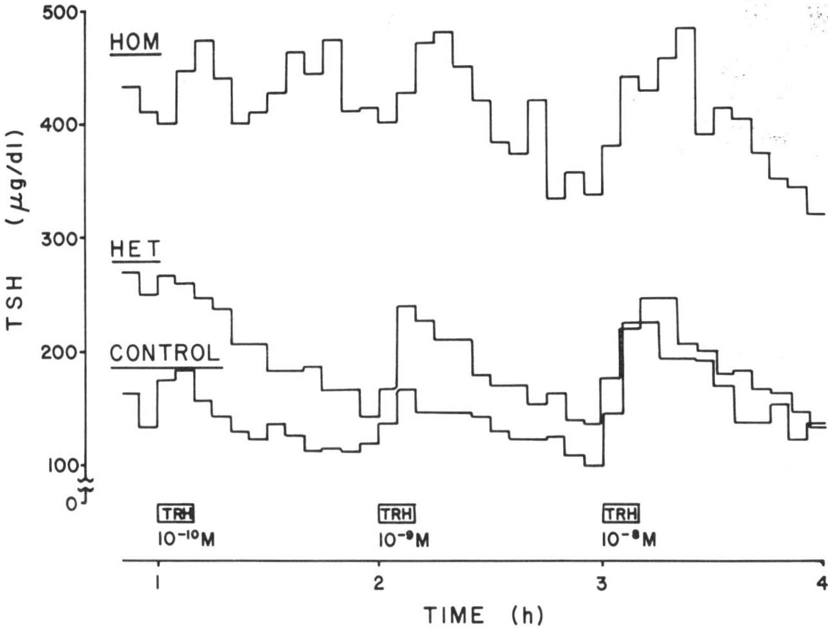

FIGURE 2. TSH secretion from superfused dispersed pituitary cells. Concentrations of TRH in the buffer are indicated below the horizontal bars.

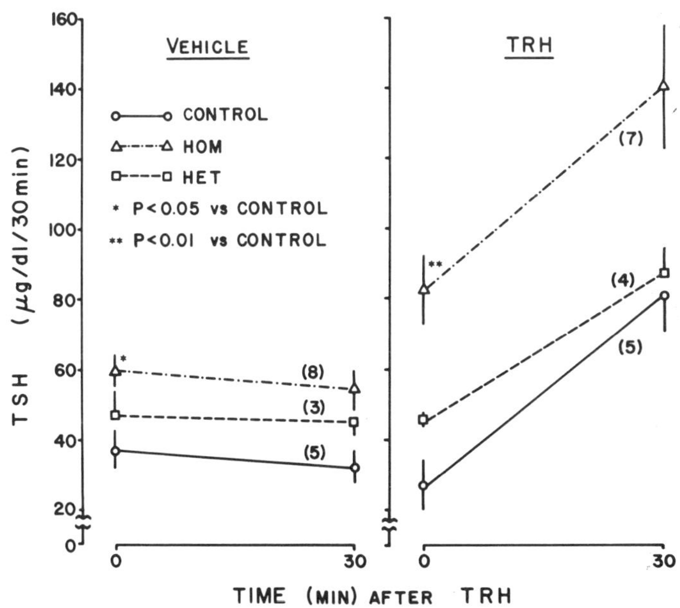

FIGURE 3. Spontaneous and TRH-induced secretion of TSH from pituitary fragments.

balance or plasma VP concentrations might be responsible for the diminished stress response in plasma PRL level in HOM.

In summary, we have found that plasma TSH levels are elevated and those of PRL are depressed in Brattleboro rats. These rats also have an increased pituitary response to TRH, which appears to be due to reduced negative feedback by thyroid hormones. The abnormalities in plasma PRL (but not TSH) levels can be reversed by replacement of vasopressin and restoration of water balance.

Acknowledgment

We acknowledge the consultation of Dr. L. Kinter.

References

1. Edwards, C. R. W. 1977. Vasopressin. *In* Clinical Neuroendocrinology. L. Martini & G. M. Besser, Eds.: 527–567. Academic Press. New York, N.Y.
2. Meyer, V. & E. Knobil. 1966. Stimulation of growth hormone secretion by vasopressin in the rhesus monkey. Endocrinology **79:** 1016–1018.
3. Rivier, C. & W. Vale. 1974. *In vivo* stimulation of prolactin secretion in the rat by thyrotropin releasing factor, related peptides and hypothalamic extracts. Endocrinology **95:** 978–983.
4. Krass, M. E., F. S. LaBella & S. R. Vivian. 1968. Thyrotropin release *in vitro:* The role of metabolism in the secretory response to vasopressin, oxytocin and epinephrine. Endocrinology **82:** 1183–1189.
5. Endroczi, E. & J. Hilliard. 1965. Luteinizing hormone releasing activity in different parts of rabbit and dog brain. Endocrinology **77:** 667–673.
6. Möhring, J., G. Kohrs, B. Möhring, M. Petri, E. Homsy & D. Haack. 1978. Effects of prolonged vasopressin treatment in Brattleboro rats with diabetes insipidus. Am. J. Physiol. **234:** F106–F111.
7. Yates, F. E., S. M. Russel, M. F. Dallman, G. A. Hedge, S. M. McCann & A. P. S. Dhariwal. 1971. Potentiation by vasopressin of corticotropin release induced by corticotropin-releasing factor. Endocrinology **88:** 3–15.
8. McCann, S. M., J. Antunes-Rodrigues, R. Nallar & H. Valtin. 1966. Pituitary-adrenal function in the absence of vasopressin. Endocrinology **79:** 1058–1064.
9. Fujimoto S. & G. A. Hedge. 1982. Abnormalities in thyrotropin (TSH) and thyroxine (T_4) secretion in Brattleboro rats. (This volume.)
10. J. M. Connors, K. C. Wright, A. M. Judd, C. M. Liu & G. A. Hedge. 1981. Dynamics and regulation of TSH secretion by superfused anterior pituitary cells. Horm. Res. **14:** 1–17.

ROLE OF VASOPRESSIN IN THE RENIN AND ACTH RESPONSES TO INTRAVENTRICULAR ANGIOTENSIN II *

William F. Ganong, Jeannette Shinsako, and Ian A. Reid

Department of Physiology
University of California
San Francisco, California 94143

Lanny C. Keil

NASA Ames Research Center
Moffett Field, California 94035

Donald L. Hoffman and Earl A. Zimmerman

Department of Neurology
College of Physicians & Surgeons
Columbia University
New York, New York 10032

INTRODUCTION

When administered into the cerebral ventricles, angiotensin II has at least six endocrine or endocrine-related effects: it increases the secretion of ACTH and vasopressin, it decreases renin secretion, and it increases blood pressure, water intake, and salt intake.[1] The increase in blood pressure is due at least in part to increased sympathetic discharge.[2] Since increased sympathetic discharge increases renin secretion by way of circulating catecholamines and increased renal nerve activity,[2] one would expect intraventricular angiotensin II to increase rather than decrease renin secretion. However, the increase in plasma vasopressin could overcome the stimulatory effect, since vasopressin inhibits renin secretion.[3] This possibility can be investigated, and the more general problem of the role of vasopressin secretion in the day-to-day regulation of renin secretion can be explored by comparing the effects of intraventricular angiotensin II in homozygous Brattleboro rats to the effects in suitable controls.

Comparison of the effects of intraventricular angiotensin II in normal and Brattleboro rats also provides information about the role of vasopressin in the ACTH response to the angiotensin II. Vasopressin has corticotropin-releasing hormone activity,[4] and is released into the portal hypophyseal vessels.[5] It is not the only corticotropin-releasing hormone, but it appears to be one of the corticotropin-releasing hormones in the hypothalamus, and may potentiate the action of one or more other corticotropin-releasing hormones.[4] If the increase in ACTH secretion produced by intraventricular angiotensin II is due to increased secretion of vasopressin, it should be absent in homozygous Brattleboro rats deficient in vasopressin.

Consequently, we injected angiotensin II intraventricularly in homozygous

* Supported by U.S. Public Health Service Grants AM06704, HL24105, and the National Aeronautics and Space Administration.

0077-8923/82/0394-0619 $1.75/0

Brattleboro rats and normal Long-Evans controls, and measured plasma vasopressin, renin activity, ACTH, and 11-oxycorticoids. An abstract reporting preliminary results of the experiments has been published.[6]

Methods

Adult homozygous Brattleboro rats and normal Long-Evans rats (Blue Spruce Farms, Altamont, N.Y.) weighing 150–250 grams were used. Two experiments were carried out, one employing only male animals and the second employing male and female animals. Cannulas made of 22 gauge stainless steel tubing were implanted in a lateral ventricle under ether anesthesia, and held in place with dental cement. Ten days later, a 28 gauge needle was inserted through the guide cannula without anesthesia, and 50 ng of angiotensin II in 10 microliters of saline, or saline alone was injected into the ventricle. Rats were sacrificed by decapitation 15 minutes later. Trunk blood was collected in 0.3 M ethylenediaminetetraacetic acid (1 vol/10 vol blood) and the plasma separated and frozen. Plasma vasopressin was subsequently measured, using a radioimmunoassay.[7] With this assay, values of 0.3 pg/ml or less are not significantly different from zero. Plasma renin activity was determined by radioimmunoassay of generated angiotensin I.[8] Plasma ACTH was measured by a radioimmunoassay described elsewhere.[9] In the first experiment, 11-oxycorticoids were measured by a protein binding method,[10] and in the second experiment, they were measured by radioimmunoassay employing an antibody that measured corticosterone (Dallman *et al.*, unpublished observations).

Statistical analysis of the data was carried out using Student's *t* test.

Results

The effects of intraventricular angiotensin II and saline on plasma vasopressin, renin activity, ACTH, and 11-oxycorticoids in normal and Brattleboro rats are summarized in Tables 1 and 2. In the first experiment (Table 1), intraventricular angiotensin II caused a statistically significant increase in plasma vasopressin in normal rats. In association with this increase, there was a statistically significant decrease in plasma renin activity. In the Brattleboro rats, plasma vasopressin values were at or near the limit of sensitivity of the method. Plasma renin activity in the Brattleboro rats receiving intraventricular saline was more than twice as great as it was in the saline-injected controls, and it was not reduced by angiotensin II. Results in the second experiment (Table 2) were generally comparable, although the plasma vasopressin and renin activity were somewhat lower in the control rats.

Plasma ACTH and 11-oxycorticoids were elevated in the normal rats injected with saline in both experiments, in all probability because of the stress of the injection. Plasma 11-oxycorticoids measured by protein binding were higher in experiment 1 than 11-oxycorticoids measured by the more specific radioimmunoassay in experiment 2. Plasma ACTH was markedly elevated in the normal rats injected with angiotensin II. Corticoids also increased, but to a lesser degree. The stress of injecting saline probably produced near maximum corticoid output, and there was little opportunity for further increase. When maximum corticoid output is reached, further increases in plasma ACTH are

TABLE 1

EFFECTS OF INTRAVENTRICULAR ANGIOTENSIN II IN HOMOZYGOUS BRATTLEBORO RATS AND NORMAL LONG-EVANS CONTROLS. EXPERIMENT 1

	Normal Rats		Brattleboro Rats	
Plasma Concentration	Saline	Angiotensin II	Saline	Angiotensin II
Vasopressin (pg/ml)	1.1 ± 0.3	4.6 ± 1.2 *	0.7 ± 0.2	0.2 ± 0.1 §
Renin activity (ng AI/ml/3 h)	7.0 ± 0.8	4.3 ± 0.4 †	15.8 ± 4.6 §	17.6 ± 7.3
ACTH (pg/ml)	173 ± 36	372 ± 29 *	199 ± 86	194 ± 21 §
11-oxycorticoids (μg/dl)	41.1 ± 7.2	63.4 ± 5.6 ‡	39.6 ± 4.1	50.2 ± 1.1 * ¶

* $p < 0.01$ vs. saline.
† $p < 0.05$ vs. saline.
‡ $p = 0.06$ vs. saline.
§ $p < 0.01$ vs. comparable value in normal rats.
¶ $p < 0.05$ vs. comparable value in normal rats.

not mirrored by comparable increases in plasma corticoids.[11] In the Brattleboro rats treated with angiotensin II, corticoids were also higher than they were in the Brattleboro rats treated with saline. In the first experiment, there was one very high plasma ACTH value in a Brattleboro rat injected with saline. Consequently, the mean was as high as in the angiotensin II-injected rats, and the standard error was large. Without this value, the mean was 113.8 ± 17.0 pg/ml. In the Brattleboro rats in experiment 2, plasma ACTH was significantly greater after angiotensin II than it was after saline. In both experiments, ACTH and corticoid values were lower in the angiotensin II-treated Brattleboro rats than in the angiotensin II-treated normal rats.

TABLE 2

EFFECTS OF INTRAVENTRICULAR ANGIOTENSIN II IN HOMOZYGOUS BRATTLEBORO RATS AND NORMAL LONG-EVANS CONTROLS. EXPERIMENT 2

	Normal Rats		Brattleboro Rats	
Plasma Concentration	Saline	Angiotensin II	Saline	Angiotensin II
Vasopressin (pg/ml)	0.7 ± 0.3	1.3 ± 0.5	0.4 ± 0.2	0.1 ± 0.1 §
Renin activity (ng/AI/ml/3 h)	4.9 ± 1.2	2.4 ± 0.4	9.9 ± 1.3 ‡	10.3 ± 2.7 ‡
ACTH (pg/ml)	101 ± 16	279 ± 43 *	104 ± 19	180 ± 23 † ¶
11-oxycorticoids (μg/dl)	31.9 ± 6.7	41.5 ± 4.4	24.6 ± 3.3	25.3 ± 2.5 ‡

* $p < 0.01$ vs. saline.
† $p < 0.05$ vs. saline.
‡ $p < 0.01$ vs. comparable value in normal rats.
§ $p < 0.05$ vs. comparable value in normal rats.
¶ $p = 0.06$ vs. comparable value in normal rats.

Discussion

The failure of angiotensin II to decrease plasma renin activity in Brattleboro rats supports the hypothesis that the decrease in normal rats is due to the increase in plasma vasopressin. Similar data have been obtained in dogs. In this species, the decrease in plasma renin activity produced by intraventricular angiotensin II is also associated with an increase in plasma vasopressin. When circulating vasopressin is reduced to very low, constant levels by surgical removal of the pituitary, intraventricular angiotensin II no longer lowers plasma renin activity.[12]

Although plasma renin activity failed to fall in our angiotensin II-treated Brattleboro rats, it also failed to show the rise that would be predicted if vasopressin were the only factor inhibiting the stimulatory effect of increased sympathetic discharge. The failure to rise could have been due to the direct inhibitory effect of an increase in blood pressure on renin secretion.[13] Blood pressure was not measured in our experiments, but in hypophysectomized dogs, the pressor effect of intraventricular angiotensin II was still present and plasma renin activity did not rise significantly.[12] However, only a slight blood pressure response was reported to very large doses of intraventricular angiotensin II in Brattleboro rats.[14]

The increased plasma renin activity in Brattleboro rats given intraventricular saline is consistent with other reports of increased plasma renin activity at rest in untreated Brattleboro rats.[15, 16] This increase could be due to chronic hypovolemia; plasma osmolality and total protein are elevated in Brattleboro rats, although hematocrit is normal.[17] The increase could also be due to the lack of the direct inhibitory effect of vasopressin on renin secretion. One piece of evidence supporting this conclusion is the fact that relatively small doses of vasopressin produce prompt decreases in plasma renin activity in Brattleboro rats [15] before there is time for expansion of extracellular fluid volume. However, plasma vasopressin was not measured in those studies, and it is important to demonstrate that plasma renin activity is reduced by normal plasma levels of vasopressin.

The fact that the ACTH and corticoid responses to intraventricular angiotensin II were reduced in Brattleboro rats suggests that the responses are partly due to vasopressin in normal animals. However, there was a statistically significant elevation in plasma ACTH in the angiotensin II-treated Brattleboro rats in experiment 2, and plasma corticoids rose significantly in experiment 1. Consequently, angiotensin II appears to also produce an increase in ACTH secretion that is independent of vasopressin and mediated by some other mechanism. Others have reported that pituitary-adrenal responses to other stresses are diminished but still present in Brattleboro rats.[4] There is also some evidence that angiotensin II has corticotropin-releasing activity.[18, 19] Whether the ACTH-stimulating activity of angiotensin II is direct, indirect, or both remains a question for future research.

Summary

In conclusion, the present data support the hypothesis that the decrease in plasma renin activity produced by intraventricular injection of angiotensin II in normal animals is due to increased plasma vasopressin. The high levels of renin in Brattleboro rats suggest but do not prove that there is tonic inhibition of renin secretion by vasopressin in normal animals. The data also demonstrate that the increase in plasma ACTH produced by intraventricular angiotensin II is due in part to increased release of vasopressin, but is also due to some other mechanism.

REFERENCES

1. REID, I. A. 1980. Interactions between the renin-angiotensin system and the brain. Adv. Exp. Med. Biol. **130:** 257–291.
2. GANONG, W. F., C. D. RUDOLPH & H. ZIMMERMAN. 1979. Neuroendocrine components in the regulation of blood pressure and renin secretion. Hypertension **1:** 207–218.
3. GANONG, W. F. & C. BARBIERI. 1982. Neuroendocrine components in the regulation of renin secretion. Front. Neuroendocrinol. Vol. 7. (In press.)
4. GILLIES, G. & P. J. LOWRY. 1982. Corticotropin releasing hormone and its vasopressin component. Front. Neuroendocrinol. **7:** 45–75.
5. ZIMMERMAN, E. A., P. W. CARMEL, M. K. HUSAIN, M. FERIN, M. TANNENBAUM, A. G. FRANTZ & A. G. ROBINSON. 1973. Vasopressin and neurophysin: high concentration in monkey hypophyseal portal blood. Science **182:** 925–927.
6. GANONG, W. F., E. A. ZIMMERMAN, D. L. HOFFMAN, L. C. KEIL, J. SINSAKO & I. A. REID. 1981. The role of vasopressin (AVP) in the renin and ACTH responses to intraventricular angiotensin II (AII). Endocrinology **108:** 272A (abstract).
7. KEIL, L. C. & W. B. SEVERS. 1977. Reduction in plasma vasopressin levels of dehydrated rats following acute stress. Endocrinology **100:** 30–38.
8. STOCKIGT, J. R., R. D. COLLINS & E. G. BIGLIERI. 1971. Determination of plasma renin concentration by angiotensin I radioimmunoassay. Circ. Res. **28–29** (Suppl. II) I184–I189.
9. DALLMAN, M. F., D. DEMANICOR & J. SHINSAKO. 1974. Diminishing corticotrope capacity to release ACTH during sustained stimulation. The 24 hours after bilateral adrenalectomy in the rat. Endocrinology **95:** 65–73.
10. MURPHY, B. E. P. 1967. Some studies on the protein binding of steroids and their application to the routine micro and ultramicro measurement of various steroids in body fluids by competitive protein binding radioimmunoassay. J. Clin. Endocrinol. **27:** 973–990.
11. GANONG, W. F. 1963. The central nervous system and the synthesis and release of ACTH. *In* Advances in Neuroendocrinology. A. V. Nalbandov, Ed.: 92–149. University of Illinois Press. Urbana, Illinois.
12. MALAYAN, S. A., L. C. KEIL, D. J. RAMSAY & I. A. REID. 1979. Mechanism of suppression of plasma renin activity by centrally administered angiotensin II. Endocrinology **104:** 672–675.
13. DAVIS, J. O. & I. FREEMAN. 1976. Mechanisms regulating renin release. Physiological Rev. **56:** 1–56.
14. HUTCHINSON, J. E., P. SCHELLING, J. MÖHRING & D. GANTEN. 1976. Pressor action of centrally perfused angiotensin II in rats with hereditary diabetes insipidus. Endocrinology **99:** 819–823.
15. GUTMAN, Y. & F. BENZAKEIN. 1974. Antidiuretic hormone and renin in rats with diabetes insipidus. Eur. J. Pharmacol. **28:** 114–118.
16. ZERBE, R. L., G. FEUERSTEIN, D. K. MEYER & I. J. KOPIN. 1981. Cardiovascular, catecholamine and renin responses to hemorrhage in Brattleboro rats. Neurosci. Abstr. (In press.)

17. EDWARD, B. R., M. GELLAI & H. VALTIN. 1980. Concentration of urine in the absence of ADH with minimal or no decrease in GFR. Am. J. Physiol. **239:** F84–F91.
18. MARAN, J. W. & F. E. YATES. 1977. Cortisol secretion during intrapituitary infusion of angiotensin II in conscious dogs. Am. J. Physiol. **232:** E273–E285.
19. SOBEL, D. O., A. H. VAGNUCCI & E. EVANS. 1981. Angiotensin mediated secretion of ACTH from adenohypophyseal cells in monolayer culture. Endocrinology **108:** 291A (abstract).

OXYTOCIN ACTION IN ISOLATED ADIPOCYTES FROM BRATTLEBORO RATS *

H. Joseph Goren,† Khawar Hanif, Morley D. Hollenberg, and Karl Lederis

† Department of Medical Biochemistry
Department of Pharmacology and Therapeutics
University of Calgary
Calgary, Alberta
Canada T2N 4N1

Circulating levels of oxytocin in the Brattleboro rat (homozygous diabetes insipidus strain, HoDI) are believed to be above normal.[1,2] With many hormones, an excess leads to a decrease in tissue receptor number and/or response.[3,4]

A significantly lower number of receptors (one-seventh of normal) in the uterus of the HoDI rats together with unchanged uterine response to oxytocin has recently been reported.[5] Glucose oxidation, which is stimulated by oxytocin in isolated fat cells of normal rats, does not occur in the adipocytes of HoDI rats, although these cells have unaltered oxytocin binding activity.[5] Thus, epididymal adipocytes from HoDI rats are defective in coupling of oxytocin-receptor occupancy to the initiation of a response. This report describes some investigations into the source of the defect.

Materials and Methods

Porcine zinc insulin (exp. 491–4) and synthetic oxytocin were gifts from Dr. J. Clement (Connaught Laboratories, Toronto, Canada) and Dr. G. Moore (University of Calgary), respectively. Other chemicals used were bovine serum albumin (fraction V, Armour Pharmaceutical, lot T 13904); collagenase (Worthington, CLS 48M201); Pitressin Tannate (Parke Davis); and [U-^{14}C] glucose and [1-^{14}C]glucose (New England Nuclear). All laboratory reagents were of highest quality commercially available.

Preparation of isolated adipocytes, their number in suspension, and glucose oxidation have been described previously.[5] Glucose transport activity as measured by the oxidation of [1-^{14}C]glucose in the presence of phenazine methosulfate has also been described.[6,7] Measurement of lipogenesis followed the procedure of Gliemann,[8] and measurements of inhibition of epinephrine-stimulated lipolysis were done using the method of Taylor *et al.*[9] with glycerol being measured enzymatically.[10]

* Supported by the Medical Research Council of Canada (MA 7271 (H. J. G.)), DG183 (M. D. H.), and MT3911 (K. L.)) and the Canadian Diabetes Association (H. J. G.). K. L. is a Career Investigator of the Canadian Medical Research Council.

0077–8923/82/0394–0625 $1.75/0 © 1982, NYAS

Results and Discussion

In isolated adipocytes from Long-Evans rats, the parent strain to HoDI rats, oxytocin stimulated glucose oxidation and triglyceride synthesis (Figures 1A, B). Fat cells from HoDI rats, however, did not respond to 100 nM oxytocin, a dose sufficient to give maximal glucose oxidation and lipogenesis (K. Hanif, H. J. Goren, K. Lederis, and M. D. Hollenberg, unpublished observation), but did respond to insulin (Figures 1A, B). These results suggest that biochemical processes that are stimulated by insulin and result in conversion of glucose to CO_2 and fatty acids are operative in HoDI rats but that these processes are not activated despite oxytocin receptor occupation.

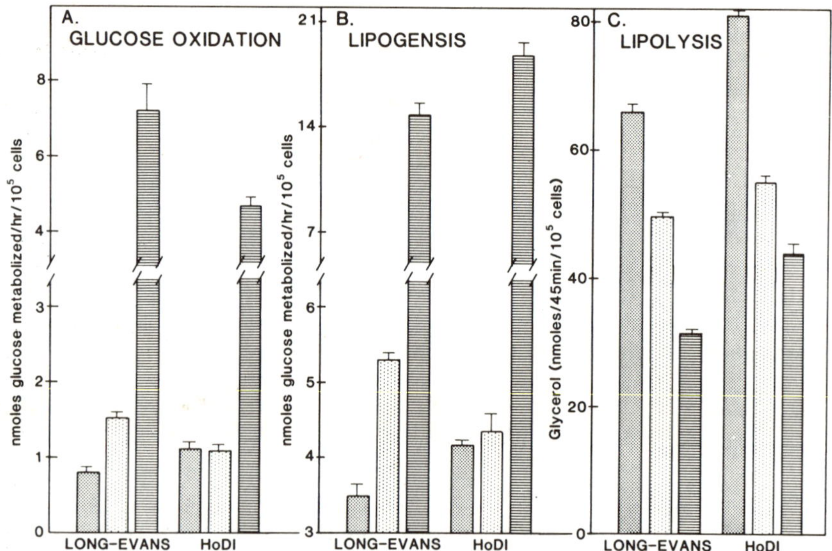

Figure 1. Glucose oxidation, lipogenesis, and lipolysis in isolated adipocytes of Long-Evans and of homozygous diabetes insipidus (Brattleboro strain) rats. Adipocytes were isolated from Long-Evans or HoDI rats and were used in a glucose oxidation (A), lipogenesis (B), or lipolysis (C) assay. Buffers contained no peptide hormone (▩), or 100 nM oxytocin (▢), or 1.7 nM insulin (▤). During lipolysis, 0.3 μM norepinephrine was present in the incubation buffers. Results are the mean ± SEM ($N = 4$).

In contrast, oxytocin inhibited the epinephrine-stimulated lipolysis in adipocytes from normal and HoDI rats (Figure 1C), suggesting that oxytocin binding to its receptor in adipocytes of HoDI rats elicits the elaboration of a second messenger. Assuming the oxytocin-receptor coupling results in a single second messenger, then this messenger is recognized by some biochemical processes (e.g., inhibition of catecholamine-stimulated lipolysis) but not by others (e.g., glucose metabolism to CO_2 or to triglycerides).

Calcium has been suggested to have a role in oxytocin action in adipocytes.[11] Table 1 illustrates that lowering or raising calcium concentrations in Krebs-

TABLE 1

EFFECT OF CALCIUM CONCENTRATION ON GLUCOSE OXIDATION IN ISOLATED ADIPOCYTES OF HoDI BRATTLEBORO RATS *

Calcium (mM)	Glucose Oxidized (nmoles/h/10^5 cells) Basal	Oxytocin	Insulin
0.63	1.06 ± .01	1.12 ± .04	5.08 ± .16
1.25	1.50 ± .03	1.35 ± .07	5.25 ± .19
2.50	1.25 ± .06	1.39 ± .15	5.19 ± .04

* Isolated adipocytes were incubated with [U-^{14}C]glucose in Krebs-Ringer bicarbonate buffer containing 0.63, 1.25, or 2.50 mM $CaCl_2$ in the absence (basal) or presence of oxytocin (100 nM) or insulin (1.7 nM). Results are the means ± SEM. ($N = 4$). There is no statistical difference (Student *t*-test, $p < 0.1$) between the basal and the oxytocin-stimulated rate at any calcium concentration.

Ringer bicarbonate buffer was not sufficient to change the inability of oxytocin to stimulate glucose oxidation.

Urinary flow and the osmolarity of urine in HoDI rats are improved upon vasopressin injection.[12] When these animals were injected with one unit of arginine vasopressin for 28 days, the responsiveness of adipocytes to oxytocin did not improve (TABLE 2). Thus, the absence of vasopressin and presumably excess oxytocin is not the cause of lack of oxytocin-receptor–response coupling in HoDI rat adipocytes.

The rate-limiting step in glucose metabolism is glucose transport. In the presence of insulin, the phosphorylation of glucose becomes the rate-limiting step.[13] Insulin, therefore, stimulates glucose transport. FIGURE 2 illustrates that oxytocin did not stimulate [1-^{14}C]glucose oxidation in the presence of phenazine methosulfate, in either Long-Evans or HoDI Brattleboro rats. Since these measurements reflect glucose transport,[6] oxytocin does not stimulate this biochemical event in this strain of rats.

In conclusion, our findings suggest that in HoDI rats insulin and oxytocin, by binding to their own receptors: (1) elicit second-messengers; (2) these

TABLE 2

GLUCOSE OXIDATION IN ISOLATED ADIPOCYTES OF VASOPRESSIN-TREATED HoDI BRATTLEBORO RATS *

Treatment	Glucose Oxidized (nmoles/h/10^5 cells) Basal	Oxytocin	Insulin
1. Peanut Oil	1.16 ± .08	1.17 ± .03	4.62 ± .13
2. Arginine Vasopressin	1.07 ± .09	1.30 ± .13	5.46 ± .86

* Three HoDI rats were injected intramuscularly daily for 28 days with 1 unit Pitressin tannate (Arginine vasopressin) in peanut oil. Control HoDI rats were injected with peanut oil for 28 days. The rats were sacrificed, adipocytes were isolated and used in a glucose oxidation assay; basal—no additive, oxytocin—30 nM, and insulin—1.7 nM. Results are the means ± SEM. ($N = 4$). There is no statistical difference (student *t*-test, $p < 0.1$) between the basal and oxytocin-stimulated rates.

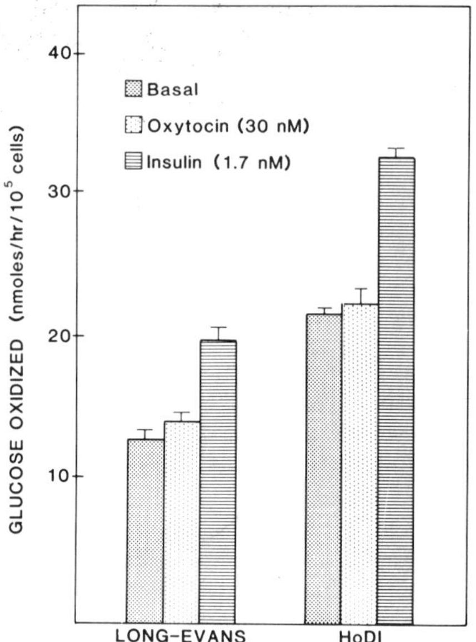

FIGURE 2. Glucose transport in isolated adipocytes of Long-Evans and of homozygous diabetes insipidus (Brattleboro strain) rats. Adipocytes were isolated from Long-Evans or HoDI rats. Glucose transport was measured indirectly as [1-^{14}C] glucose oxidized in the presence of 20 μM phenazine methosulfate where the incubation buffer contained no peptide hormone (basal, ▩), 30 nM oxytocin (☐), or 1.7 nM insulin (▤). Results are the mean ± SEM ($N = 4$).

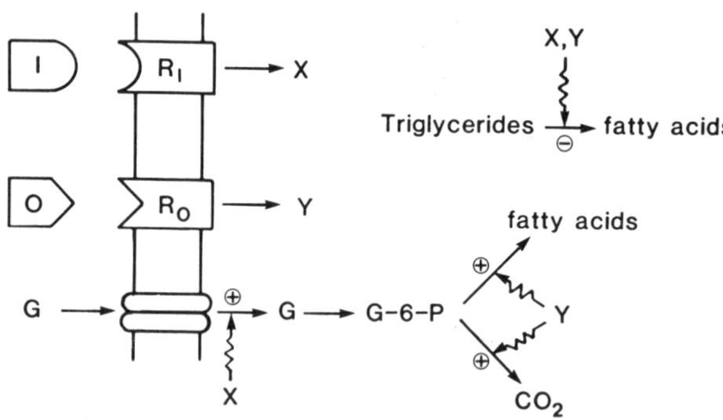

FIGURE 3. Proposed mechanism of oxytocin and insulin action in intracellular adipocyte metabolism. Insulin (I) binds to its receptor, R_I, resulting in the formation of second messenger, X. Oxytocin (O) binds to its receptor, R_o, yielding its second messenger, Y. Both X and Y inhibit catecholamine stimulated lipolysis. X stimulates glucose (G) transport, whereas Y stimulates conversion of glucose-6-phosphate (G-6-P) to fatty acids and CO_2.

second-messengers are probably not the same for the two hormones; (3) the two hormones affect glucose metabolism but at different biochemical points; (4) the second-messenger to oxytocin-receptor occupancy is probably mediated in HoDI rat adipocytes but it is not recognized by some of the cell's enzymes; and (5) the excess of oxytocin or absence of vasopressin is not a cause of oxytocin resistance. Some of these conclusions are summarized in FIGURE 3.

REFERENCES

1. VALTIN, H., W. H. SAWYER & H. W. SOKOL. 1965. Neurohypophysial principles in rats homozygous and heterozygous for hypothalamic diabetes insipidus (Brattleboro Strain). Endocrinol. **77:** 701–706.
2. DOGTEROM, J., C. M. F. VON RHEENEN-VERBERG, TJ. B. VAN WIMERSMA GREIDANUS & D. F. SWAAB. 1977. Vasopressin and oxytocin in cerebrospinal fluid of rats. J. Endocrinol. **72:** 74–75.
3. KAHN, C. R. 1976. Membrane receptors for hormones and neurotransmitters. J. Cell Biol. **70:** 261–286.
4. CATT, K. J., J. P. HARWOOD, G. A. AQUILERA & M. L. DUFAU. 1979. Hormonal regulation of peptide receptors and target cell responses. Nature **280:** 109–116.
5. GOREN, H. J., R. M. GEONZON, M. D. HOLLENBERG, K. LEDERIS & D. O. MORGAN. 1980. Oxytocin action: Lack of correlation between receptor number and tissue responsiveness. J. Supramol. Struct. **14:** 129–138.
6. HALPERIN, M. L., M. L. MAK & W. M. TAYLOR. 1978. Control of glucose transport in adipose tissue of the rat: Role of insulin, ATP, and intracellular metabolites. Can. J. Biochem. **56:** 708–712.
7. GOREN, H. J. & S. H. ROTH. 1981. Glucose oxidation, glucose transport and insulin binding in isolated rat epididymal adipocytes in the presence of anesthetics. Life Sci. **28:** 2151–2159.
8. GLIEMANN, J., S. GAMMELTOFT & J. VINTEN. 1975. Time course of insulin-receptor binding and insulin-induced lipogenesis in isolated rat fat cells. J. Biol. Chem. **250:** 3368–3374.
9. TAYLOR, W. M., M. D'COSTA, A. ANGEL & M. L. HALPERIN. 1977. Insulin-like effects of fluoroacetate on lipolysis and lipogenesis in adipose tissue. Can. J. Biochem. **55:** 982–987.
10. DAVIDSON, M. B. & R. KARJALA. 1970. Simplified method for the determination of plasma glycerol. J. Lipid Res. **11:** 609–612.
11. BONNE, D., O. BELHADJ & P. COHEN. 1977. Modulation by calcium of the insulin action and the insulin-like effect of oxytocin on isolated rat lipocytes. Eur. J. Biochem. **75:** 101–105.
12. HARRINGTON, A. R. & H. VALTIN. 1965. Vasopressin effect on urinary concentration in rats with hereditary hypothalamic diabetes insipidus (Brattleboro strain). Proc. Soc. Exp. Biol. Med. **118:** 448–450.
13. CROFFORD, O. B. & A. E. RENOLD. 1965. Glucose uptake by incubated rat epididymal adipose tissue. Rate-limiting steps and site of insulin action. J. Biol. Chem. **240:** 14–21.

DEMONSTRATION OF CORTICOTROPIN-RELEASING FACTOR (CRF) ACTIVITY IN THE NEUROHYPOPHYSIS OF BRATTLEBORO RATS

M. Kárteszi, G. B. Makara, F. A. László, G. Rappay, and E. Stark

Institute of Experimental Medicine
Hungarian Academy of Sciences
H-1450 Budapest POB 67

Department of Endocrinology
School of Medicine
Szeged, Hungary

Brattleboro rats with hereditary diabetes insipidus (DI rats), which lack the capability of synthesizing vasopressin, provide a good model to study the relative importance of vasopressin versus other CRFs or potentiating factors. There is no doubt that DI rats are able to respond to various types of stress with an increase in ACTH release,[1] however, the hypothalamo-pituitary-adrenocortical axis of these animals is evidently compromised to some extent.[12,16]

Since the findings on ACTH-stimulating action of vasopressin are controversial,[16,17] it seemed important to study the CRF activity of brain regions in DI rats. Using an *in vitro* CRF bioassay,[10] both stalk-median eminence and neurohypophyseal extracts from homozygous Brattleboro rats were shown to contain the same amount of CRF-like activity as those of the controls.[11,12] Here we present data on the CRF-like activity in an *in vivo* test of neurohypophyseal extracts from heterozygous and homozygous Brattleboro rats.

Materials and Methods

Male rats of CFY origin or heterozygous and homozygous Brattleboro rats were housed at 24 ± 1° C and 55–75% humidity with fluorescent lights on between 06:00 and 18:00. The water consumption of the DI rats was studied, and those rats that drank more than 200 ml or 50–100 ml/day were considered as homozygous or heterozygous, respectively.

Preparation of Brain Extracts

The animals were killed by decapitation and the posterior lobe of the pituitary was dissected. The intermediate lobe tissue was removed by vacuum aspiration, and the neural lobe was homogenized in 10 μl 0.1 N HCl. For reference, cerebral cortical extract was prepared the same way. Before the intrapituitary injection of the test materials, the extracts were neutralized by an appropriate amount of 1 M NaOH.

0077–8923/82/0394–0630 $1.75/0 © 1982, NYAS

In Vivo CRF Bioassay

For testing, CFY rats were pretreated with dexamethasone-morphine-pentobarbitone according to the protocol of Dunn and Critchlow.[4] A volume of 0.5 μl of test materials was injected as a bolus into the anterior pituitary via a stereotaxically guided glass capillary[8] just after taking a small amount (300–400 μl) of venous blood for determination of the initial level of corticosterone. 20 min after completing the intrapituitary injection, about 0.5 μl of Pontamine sky blue (0.1%) was infused through the glass microcapillary in order to mark the site of the injection, and then the animals were killed by decapitation. Trunk blood was collected and plasma corticosterone was measured by radioimmunoassay[7] using our own antiserum. The increment in corticosterone secretion following the injection of brain extracts was considered an index of CRF activity of the test materials.

Results and Discussion

It was found that intrapituitary injection of cortical cerebral extracts from CFY rats failed to stimulate corticosterone secretion in pharmacologically pretreated rats (Figure 1). In contrast, neurohypophyseal extracts from both heterozygous and homozygous Brattleboro rats caused a significant increase in corticosterone secretion. There was no quantitative difference in ACTH-releasing potency between the two Brattleboro extracts.

These findings are in a good agreement with our previous results obtained with an *in vitro* CRF bioassay, suggesting that the stalk-median eminence and the neurohypophysis of DI rats contain as much CRF-like activity as similar samples from control animals.[11] On the basis of these results, one can conclude that the CRF activity of rat brain extracts is mainly distinct from vasopressin. This conclusion is also supported by our findings that even an excess amount of exogeneous vasopressin failed to potentiate the ACTH-releasing effect of brain extracts from homozygous Brattleboro rats.[12]

On the other hand, our results are at variance with measurements of CRF activity in MBH fragments from Brattleboro rats using either isolated pituitary cells[5, 13] or incubated pituitary segments[3] as the test system. Other studies also indicated that vasopressin present in the hypothalamic extracts accounted for 21%,[15] 41%,[14] or 80%[6] of overall CRF activity. The explanation for these discrepancies is not clear. However, it seems reasonable that the differences in sensitivity of the CRF bioassays toward vasopressin may be responsible —at least in part—for the great variation of the results from different laboratories.

Although it is possible that vasopressin may be involved in the acute response to stress, it is unlikely that vasopressin would be the physiological CRF.[18] It is important to note that in the experiment of Hedge *et al.*[8] intrapituitary injection of vasopressin was relatively ineffective in releasing ACTH, while systemic administration produced a good release of ACTH. The results of this elegant study emphasize that the mechanism of action of substances with ACTH-releasing activity ought to be investigated with meticulous care.

In a very recent experiment[2] oxytocin was shown to be secreted *in vitro* as a major corticotropin-releasing factor from the posterior pituitary of DI rats. However, it is unlikely that the present results of the CRF test using the *in vivo* pituitary microinjection could be due to the oxytocin content of the neural lobe, since this test system is known to be relatively insensitive to both vasopressin and oxytocin.[9] Nevertheless, the possibility that oxytocin may be an effective modulator of CRF release cannot be excluded.

ACKNOWLEDGMENTS

The authors are grateful to Ms. V. Garamvölgyi and Mrs. I. Schay for their excellent technical assistance, and to the Hormone Laboratory of the Institute of Experimental Medicine (headed by Dr. Zs. Acs) for the hormone measurements.

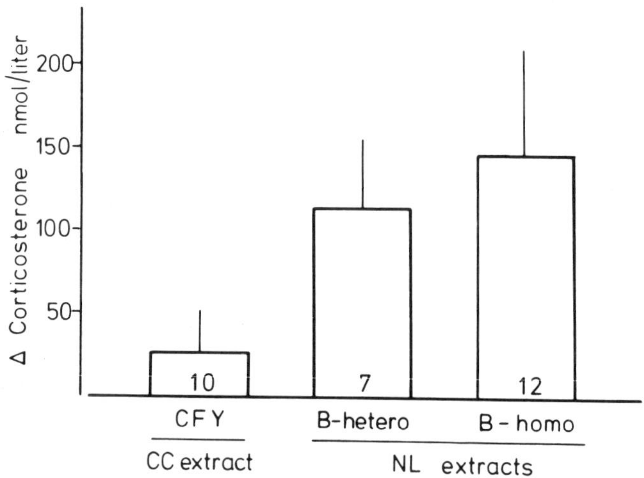

FIGURE 1. Effect of intrapituitary injection of cortical cerebral extract (CC) and neurohypophyseal extracts (NL) from heterozygous and homozygous Brattleboro rats (B hetero and B homo, respectively) on corticosterone secretion in pharmacologically pretreated normal rats. The columns represent mean ± SE increment in corticosterone secretion induced by the test materials. The number of the animals with a successful injection in the anterior lobe of the pituitary is indicated in the columns.

REFERENCES

1. ARIMURA, A., T. SAITO, C. BOWERS & A. V. SCHALLY. 1967. Pituitary-adrenal activation in rats with hereditary hypothalamic diabetes insipidus. Acta Endocrinologica **54:** 155–165.
2. BÉNY, J.-L. & A. J. BAERTSCHI. 1980. Oxytocin: Major corticotropin-releasing factor secreted from diabetes insipidus rat posterior pituitary *in vitro.* Neuroendocrinology **31:** 261–264.
3. BUCKINGHAM, J. C. & J. H. LEACH. 1979. Corticotrophin secretion in the Brattleboro rat (Abstract). J. Endocrinol. **81:** 126P.
4. DUNN, J. & V. CRITCHLOW. 1973. Electrically stimulated ACTH release in pharmacologically blocked rats. Endocrinology **93:** 835–842.

5. GILLIES, G. & P. J. LOWRY. 1978. Perfused rat isolated anterior pituitary cell column as bioassay for factor(s) controlling release of adrenocorticotropin: validation of a technique. Endocrinology **103:** 521–537.
6. GILLIES, G. & P. J. LOWRY. 1980. Corticotrophin-releasing activity in extracts of the stalk-median eminence of Brattleboro rats. J. Endocrinol. **84:** 65–73.
7. GOMEZ-SANCHEZ, C., B. A. MURRY, D. C. KEM & N. M. KAPLAN. 1975. A direct radioimmunoassay of corticosterone in rat serum. Endocrinology **96:** 796–798.
8. HEDGE, G. A., M. B. YATES, R. MARCUS & F. E. YATES. 1966. Site of action of vasopressin in causing corticotropin release. Endocrinology **79:** 328–340.
9. HIROSHIGE, T., H. KUNITA, C. OGURA & S. ITOH. 1968. Effect on ACTH release of intrapituitary injection of posterior pituitary hormones and several amines in the hypothalamus. Jap. J. Physiol. **18:** 609–619.
10. KÁRTESZI, M., E. STARK, G. B. MAKARA, I. FAZEKAS & GY. RAPPAY. 1978. Corticoliberin (CRF) activity of the rat neurohypophysis. Endocrinol. Exptl. **12:** 204–208.
11. KÁRTESZI, M., E. STARK, F. A. LÁSZLÓ, G. B. MAKARA & GY. RAPPAY. 1980. Possible role of neurohypophyseal corticotropin releasing factor (NH-CRF) in the regulation of the pituitary-adrenal axis. Adv. Physiol. Sci. **14:** 63–68.
12. KÁRTESZI, M., E. STARK, GY. RAPPAY, F. A. LÁSZLÓ & G. B. MAKARA. 1981. Corticotrophin releasing activity of the neurohypophysis is distinct from vasopressin: *in vivo* and *in vitro* studies in normal and Brattleboro rats. Am. J. Physiol. **240:** 689–693.
13. KRIEGER, D. T., A. LIOTTA & M. BROWNSTEIN. 1977. Corticotropin-releasing factor distribution in normal and Brattleboro rat brain, and effect of deafferentation, hypophysectomy and steroid treatment in normal animals. Endocrinology **100:** 227–237.
14. LUTZ-BUCHER, B., B. KOCH, C. MIALHE & B. BRIAUD. 1980. Involvement of vasopressin in corticotropin-releasing effect of hypothalamic median eminence extract. Neuroendocrinology **30:** 178–182.
15. MCCANN, S. M., J. ANTUNES-RODRIQUES, R. NALLAR & H. VALTIN. 1966. Pituitary-adrenal function in the absence of vasopressin. Endocrinology **79:** 1058–1064.
16. MCCANN, S. M. 1980. Control of anterior pituitary hormone release by brain peptides. Neuroendocrinology **31:** 355–363.
17. SAFFRAN, M. & A. V. SCHALLY. 1977. The status of the corticotropin releasing factor (CRF). Neuroendocrinology **24:** 259–375.
18. VALE, W., J. SPIESS, C. RIVIER & J. RIVIER. 1981. Characterization of a 41-residue ovine hypothalamic peptide that stimulates secretion of corticotropin and β-endorphin. Science **213:** 1394–1397.

INFLUENCE OF POSTERIOR PITUITARY HORMONES ON THE PITUITARY-ADRENOCORTICAL RESPONSE TO NEUROGENIC STRESS IN THE BRATTLEBORO RAT *

B. Lutz-Bucher and B. Koch

Laboratoire de Physiologie
rue René Descartes
67084 Strasbourg Cedex France

Reports that in the homozygous diabetes insipidus (DI) or Brattleboro rat, adrenocortical response to various stressors may be impaired,[1-4] suggested that vasopressin (VP) may play a role in activation of the hypothalamo-hypophyseal axis. However, although it is well established that VP may exhibit CRF activity (see review[5]), the mechanism of action of the neuropeptide with regard to control of ACTH secretion is still unclear.

The aim of the present study was to investigate the possible influence of VP, as well as the parent hormone oxytocin (OT), on the ACTH response of DI rats to a neurogenic stimulus. Moreover, the interference of naturally occurring high levels of VP, such as those reported in plasma of spontaneously hypertensive rats (SHR[6]), was also examined. Finally, in view of recent findings pointing to OT as being a major CRF,[7] we also focused attention on OT concentrations in DI rats pretreated with VP.

We used female DI and SHR rats, as well as Long-Evans and Wistar Kyoto controls. The corticotropic response to sound stress (frequency of 600 cycles/sec for 10 min) of these animals was assessed by measuring plasma ACTH (using a bioassay) and corticosterone in both plasma and adrenal glands. In some experiments, animals were injected (s.c.) with lysine-VP or OT (a generous gift from Sandoz, Basel) at a dose of 10 IU (b.i.d.) for three weeks. By measuring water consumption, we found this dose of VP necessary to restore water intake to normal levels. OT and VP was assayed by RIAs developed in our laboratory (Guerné, unpublished[8]). These assays were highly specific for each neuropeptide. Statistical significance of results was evaluated using a nonparametric test.

Results displayed in FIGURE 1 clearly show that the corticotropic response to a neurogenic stimulus, as measured in terms of plasma ACTH and corticosterone levels, was reduced in DI rats compared with Long-Evans controls. Chronic treatment of DI rats with not only VP, but interestingly with OT as well, was found to significantly enhance corticosterone response to stress (FIGURE 2). However, this did not seem to be due to an altered responsiveness of the brain-pituitary axis, as could be expected to result from chronic treatment with these neuropeptides (i.e. CRF agents). Control rats, indeed, when treated under identical conditions failed to show a potentiated corticosterone response to the stimulus (TABLE 1).

Since VP deficiency appeared to be associated with a reduced capacity of the pituitary-adrenocortical axis to respond to stress, we thought, on the con-

* Supported by the Centre de la Recherche Scientifique (LA 309) and the Institut National de la Santé et de la Recherche Medicale (CRL 781.227.4).

0077–8923/82/0394–0634 $1.75/0 © 1982, NYAS

trary, to test the influence of endogenous, high circulating levels of VP. SHR rats were therefore exposed to sound stress, but no significant differences in increments of adrenal corticosterone concentrations were observed, as compared to normotensive controls: steroid contents increased from 2.6 ± 1.4 to 35.0 ± 2.4 and from 2.5 ± 0.5 to 30.7 ± 1.4 $\mu g/g$, respectively ($N = 6$).

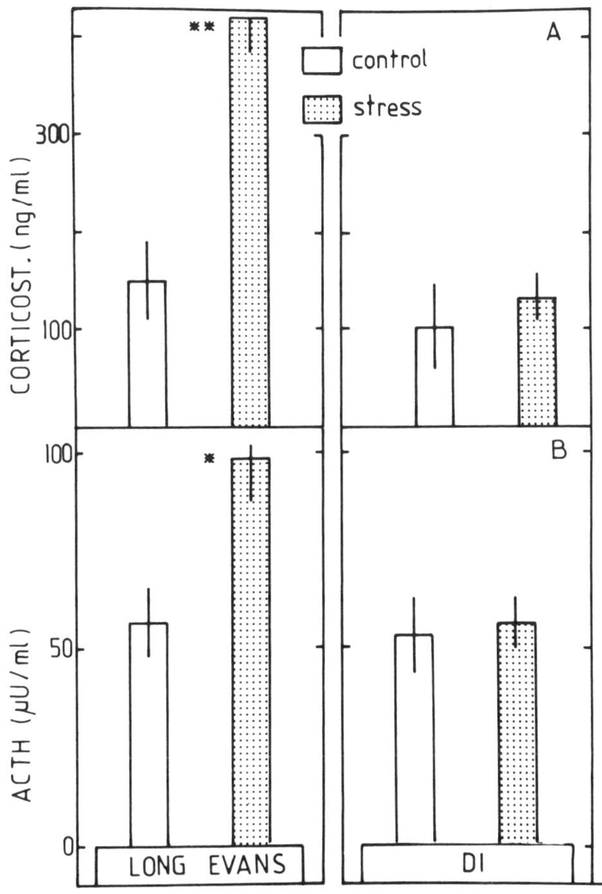

FIGURE 1. Effect of sound stress (10 min) on plasma corticosterone (A) and ACTH levels (B) in female DI rats and Long-Evans controls. Data are means ± SEM of six animals. ACTH is given in terms of i.v. IU of the 3rd International Working Standard. * $p < 0.05$; ** $p < 0.01$, as compared to nonstress conditions.

Observations regarding the effect of VP and OT treatments, as well as stress, on OT concentrations in pituitary neuro-intermediate lobe and plasma are seen in TABLE 2. Our data confirm that VP caused an increment in pituitary OT content,[9] and show, in addition, that OT failed to exert such an effect. Also, the fact that sound triggered a significant increase of plasma OT levels in both untreated and VP-treated DI rats (that is in stress-nonresponsive and responsive

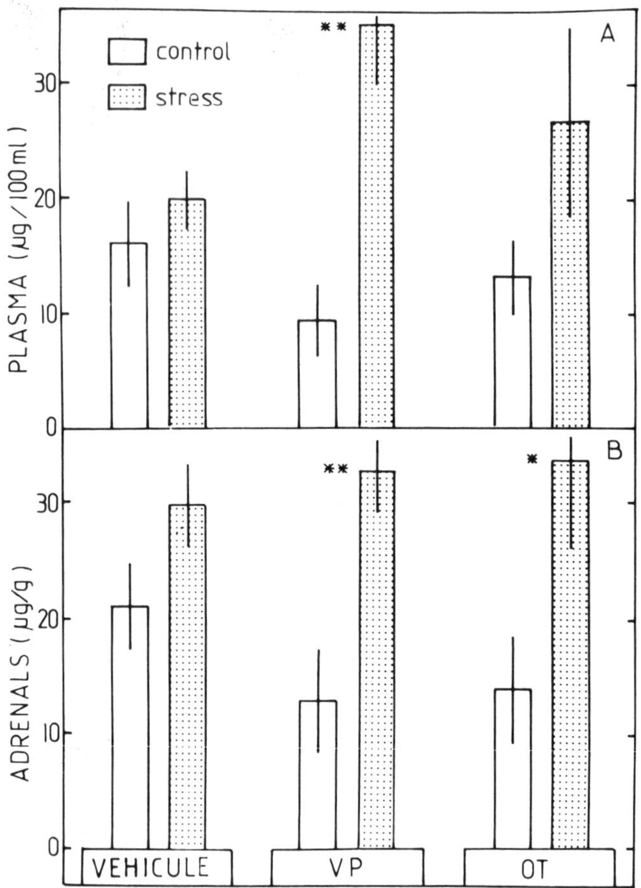

FIGURE 2. Effect of sound stress on plasma (A) and adrenal gland (B) concentrations of corticosterone in DI rats treated with vehicle, VP, or OT. Results are means ± SEM of 12 animals. * $p < 0.02$; ** $p < 0.01$, as compared to controls killed under basal conditions.

TABLE 1

EFFECT OF PRETREATMENT OF LONG-EVANS RATS WITH VP, OT, OR VEHICLE ON CORTICOTROPIC RESPONSE TO SOUND STRESS *

Treatment	Stress	Corticosterone/Adrenals
vehicle	—	1.9 ± 0.3
vehicle	+	36.0 ± 3.9
VP	+	41.7 ± 6.0
OT	+	35.3 ± 7.6

* Values are means ± SEM from six animals and are given as μg/g adrenal gland. The sound stimulus was applied for up to 10 min.

animals), does not seem, at first sight, to provide evidence for OT being a CRF.

These data not only support the view of De Wied[10] that VP may play a role in the corticotropic response to neurogenic stimuli, but interestingly, also suggest that OT may be active as well. Furthermore, we present evidence that VP only needs to be present in a critical amount, as abnormal high levels of the peptide in plasma of hypertensive rats did not further enhance the responsiveness of the pituitary-adrenal axis to stress.

The stimulatory effect of posterior pituitary hormones on the corticotropic response to neurogenic stress may be explained in several ways. First, one may argue that the effect may be indirect and mediated via factors controlling the hydromineral metabolism.[11, 12] Second, it may be suggested that the neuropeptides are acting on the CNS at sites related to the control of synthesis/release process of CRF. With regard to this, it must be noted that VP is thought to alter catecholamine neurotransmission,[13] which is partly implicated in response to neurogenic stimuli.[10]

TABLE 2

EFFECT OF NEUROPEPTIDES ADMINISTRATION AND STRESS ON PITUITARY NEUROINTERMEDIATE LOBE AND PLASMA OT CONCENTRATIONS OF FEMALE DI RATS *

Treatment	Stress	Pituitary OT	Plasma OT
vehicle	—	622 ± 72	5.9 ± 0.4
	+	265 ± 39 †	8.0 ± 0.6 *
VP	—	1620 ± 150	5.6 ± 0.7
	+	1474 ± 237	9.6 ± 2.1 *
OT	—	624 ± 86	22.6 ± 5.0
	+	620 ± 61	31.5 ± 5.1

* Data are given as μg/gland and pg/ml plasma and are means ± SEM from 8–10 animals. * $p < 0.02$; † $p < 0.01$.

ACKNOWLEDGMENTS

We thank Drs. H. Weidmann and H. Friedli (Sandoz, Basel) for the generous gift of lysine-vasopressin and oxytocin. We are also indebted to Mrs. L. Fraquet, F. Herzog, and A. Hoeft for skillful technical assistance.

REFERENCES

1. MCCANN, S. M., J. ANTUNES-RODRIGUES, R. NALLAR & H. VALTIN. 1966. Pituitary-adrenal function in the absence of vasopressin. Endocrinology **79:** 1058–1064.
2. ARIMURA, A., T. SAITO, C. Y. BOWERS & A. V. SCHALLY. 1967. Pituitary-adrenal activation in rats with hereditary hypothalamic diabetes insipidus. Acta Endocrin. **54:** 155–165.
3. YATES, F. E., S. M. RUSSEL, M. F. DALLMAN, G. A. HEDGES, S. M. MCCANN & A. P. S. DHARIWAL. 1971. Potentiation by vasopressin of corticotropin releasing factor. Endocrinology **88:** 3–15.

4. Wiley, M. K., A. F. Pearlmutter & R. E. Miller. 1974. Decreased adrenal sensitivity to ACTH in the vasopressin-deficient rat. Neuroendocrinology **14:** 257–270.
5. Lutz-Bucher, B., B. Briaud, B. Koch & G. Schmitt. 1980. Spécificité de l'action de la vasopressine au niveau de l'antéhypophyse du rat. J. Physiol. (Paris) **76:** 243–247.
6. Möhring, J., B. Möhring, M. Petri & D. Haack. 1978. Plasma vasopressin and effects of vasopressin antiserum on blood pressure in rats with malignant two-kidney Goldblatt hypertension. Circulation Res. **42:** 17–22.
7. Bény, J. L. & A. J. Baertschi. 1980. Release of corticotropin and corticotropin-releasing factors from rat posterior pituitary in vitro. Neuroendocrinology **31:** 261–264.
8. Lutz-Bucher, B., B. Koch & C. Mialhe. 1977. Comparative in vitro studies on corticotropin releasing activity of vasopressin and hypothalamic extract. Neuroendocrinology **23** 181–192.
9. Balment, R. J., I. W. Henderson & J. A. Oliver. 1975. The effects of vasopressin on pituitary oxytocin content and plasma renin activity in rats with hypothalamic diabetes insipidus. Gen. Comp. Endocrinol. **26:** 468–477.
10. DeWied, D. 1960. The significance of the posterior pituitary in the release of corticotropin following noxious stimuli. First Int. Congr. Endocrinol. Copenhagen, p. 119.
11. Valtin, H., W. H. Sawyer & H. W. Sokol. 1965. Neurohypophysial principles in rats homozygous and heterozygous for hypothalamic diabetes insipidus (Brattleboro strain). Endocrinology **77:** 701–706.
12. Jones, C. W. & B. T. Pickering. 1969. Comparison on the effects of water deprivation and sodium chloride inhibition on the hormone content of the neurohypophysis of the rat. J. Physiol. (London) **203:** 449–458.
13. Versteeg, D. H. G., M. Tanaka & E. R. de Kloet. 1978. Catecholamine concentrations and turnover in discrete regions of the brain of the homozygous Brattleboro rat deficient in vasopressin. Endocrinology **103:** 1654–1661.

GONADOTROPINS IN RATS WITH HEREDITARY DIABETES INSIPIDUS (BRATTLEBORO STRAIN)

M. Oprescu, L. Simionescu, and M. Coculescu

Institute of Endocrinology
Department of Endocrinology
Faculty of Medicine
Bucharest, Romania

In the Brattleboro strain of rats, the genetic defect in vasopressin synthesis [1] seems to be accompanied by a defect in the synthesis of other structurally related nonapeptides, such as the pineal arginine vasotocin [2, 3] which has antigonadotropic properties in animals.[4] It was therefore tempting to investigate gonadotropin secretion in rats with hereditary hypothalamic diabetes insipidus (DI rats).

Materials and Methods

A Brattleboro rat colony with hereditary hypothalamic diabetes insipidus (the Brattleboro strain) was established in the Institute of Endocrinology of the Academy of Medical Sciences (Bucharest) in 1976 from a breeding nucleus supplied by Dr. Jan Möhring (Department of Pharmacology, Heidelberg). The usual breeding program was to cross homozygous males with heterozygous females. A Long-Evans colony of normal rats was established simultaneously from the same source.

Three groups of animals, each composed of 20 male rats aged between 12 and 16 weeks and weighing 180–270 g were used in our study.

The rats were sacrificed by decapitation last spring during the interval 8:00 a.m.–10:00 a.m. Serum samples were frozen and stored at —20° C until time for assay of gonadotropins. The pituitary glands were removed immediately after decapitation, weighed and homogenized separately in 2.0 ml of 0.05 M phosphate buffer with 0.1% bovine serum albumin and centrifuged for 40 minutes (4,000 r.p.m.) at 4° C. The supernatants were stored at —20° C until time for assay. After thawing, the pituitary extract was diluted 1:1,000 in 0.04 M phosphate buffer and 0.5% bovine serum albumin, pH 7.4.

The Assay Method

The double-antibody radioimmunoassay (RIA) technique was applied, using the reagents kindly supplied by the Rat Pituitary Hormone Distribution Program (National Institute of Arthritis, Metabolism, and Digestive Diseases, Bethesda, Md.), and the second antibody was prepared in our laboratory by immunizing pigs with rabbit or monkey γ-globulin, as described elsewhere.[5]

Assay of rat LH was made using for radioiodination a NIAMDD-rat-LH-I_3 solution with a biological activity of approximately $1.0 \times$ NIH-LH-S_1 (OAAD assay), a FSH contamination less than $0.01 \times$ NIH-FSH-S_1 (HCG-augmenta-

0077–8923/82/0394–0639 $1.75/0 © 1982, NYAS

tion assay), and TSH contamination less than 0.03 USP units/mg (McKenzie assay). The NIAMDD-rabbit anti-rat LH serum-1 and the reference preparation NIAMDD-rat-LH-RP-1 with a biological activity of 0.03 × NIH-LH-S_1 (OAAD assay), 0.54 × NIH-FSH-S_1 (HCG-augmentation assay), as well as 0.22 USP bovine TSH Units/mg (McKenzie assay) were also used.

The FSH-RIA technique was applied using the following NIAMDD reagents: for radioiodination, rat-FSH-I_3 with a biological activity of approximately 100 × NIH-FSH-S_1 (HCG-augmentation assay) and LH contamination less than 0.002 × NIH-LH-S_1 (OAAD assay). The anti-rat FSH serum-6 was used at an initial dilution of 1:2,500. The reference preparation also supplied by NIAMDD, was rat-FSH-RP-1 with a biological activity of 2.1 × NIH-FSH-S_1 (HCG augmentation assay), LH contamination of 0.02 × NIH-LH-S_1 (OAAD assay), and TSH contamination of 0.3 USP U/mg (McKenzie assay).

Only the intra-assay data are included in our study. The RIA for the serum samples and pituitary extracts were performed separately. The sensitivity of the method was 3.8 ng/ml for LH and 39.0 ng/ml for FSH. For statistical significance the Student's *t* test was used.

TABLE 1

THE CHARACTERISTICS OF THE RIA SYSTEMS

Parameter	Hormone	
	LH	FSH
Ab * dilution	1:30,000	1:2,500
Percent B_o †	55.99 ± 1.12 ¶	30.05 ± 2.15 ¶
% CV ‡	2.00	7.15
Sensitivity ng/ml	3.8	39.0
Slope	0.99	0.71
r §	0.99	0.87

* Antibody; † tracer bound at the zero point of the reference curve; ‡ coefficient of variability; § coefficient of the linear correlation between the experimental and the calculated reference curve; and ¶ mean ± standard deviation.

Incubation volume: 500 μl (200 μl buffer-BSA, 100 μl standard or sample, 100 μl Ab, 100 μl ^{125}I-labeled hormone. Incubation time: 24 h second Ab). Incubation at room temperature.

RESULTS

The characteristics of the RIA system for LH and FSH are presented in TABLE 1.

As shown in TABLE 2, there were no significant differences between the serum and pituitary LH concentrations in the three groups of male rats, i.e., homozygous and heterozygous rats of the Brattleboro strain with hereditary diabetes insipidus and normal Long-Evans strain rats.

Serum FSH levels were significantly lower in normal Long-Evans male rats as compared either to the homozygous or the heterozygous Brattleboro group

TABLE 2

THE SERUM AND PITUITARY LEVELS OF RAT-LH AND RAT-FSH IN THE BRATTLEBORO MALE RATS BY COMPARISON TO THE LONG-EVANS MALE GROUP

Rat Strain Group	Number of Animals	Serum Level (ng/ml) $\overline{X} \pm SE$	p	Pituitary Level No.	μg/gland $\overline{X} \pm SE$	p	No.	μg/mg $\overline{X} \pm ESM$	p
Rat LH									
Homozygous Brattleboro rats	20	51.58 ± 21.05		20	101.37 ± 9.37		20	24.67 ± 2.81	
Heterozygous Brattleboro rats	20	36.83 ± 5.48		20	88.97 ± 7.68		20	25.66 ± 12.08	
Normal Long-Evans rats	20	62.16 ± 7.35	p * = NS	19	103.43 ± 13.53	p * = NS	19	25.84 ± 3.46	p * = NS
Rat FSH									
Homozygous Brattleboro rats	20	794.4 ± 109.6		20	256.91 ± 24.25		20	62.22 ± 7.4	
Heterozygous Brattleboro rats	19	617.9 ± 53.65		20	215.29 ± 23.9		20	64.09 ± 12.2	
Normal Long-Evans rats	20	449.8 ± 35.95	p * < 0.01	19	184.6 ± 19.9	p * < 0.05	19	46.42 ± 6.01	p * < 0.05

p * was calculated separately between each group of Brattleboro rats and the group of normal Long-Evans rats. NS = nonsignificant.

($p < 0.01$). The pituitary concentrations of FSH were also lower in normal male rats (the Long-Evans strain) than in the homozygous or heterozygous rats with hereditary diabetes insipidus ($p < 0.05$).

Discussion

Rats with hereditary diabetes insipidus were studied by many investigators as a quasi-experimental model for the effect of lack of vasopressin synthesis on animal physiology. There have been reports of a significantly diminished stress-induced response [6, 7] of homozygous Brattleboro rats, due to a specific defect in corticotrophin-releasing activity at hypothalamic level, linked to vasopressin deficiency,[8] and a smaller amount of arginine-vasotocin, the closest structurally related nonapeptide in the pineal gland of the DI rats [2] or, even to absence of pineal vasotocin [3] in DI rats in contrast to normal rats.

However, some major peculiarities of the DI rat biology, namely, growth retardation [9] and some difficulty in reproduction could not be explained as a direct or indirect effect of vasopressin.

Thus, it is well known that retarded growth, observed in DI rats, is due to a diminished pituitary growth hormone level (GH content) in spite of normal growth-hormone–releasing activity in the hypothalamus of DI rats.[9] Chronic treatment of DI rats with vasopressin also failed to restore growth retardation.[10] Both serum and pituitary LH level have the same concentration in the DI rats as in the control rats (Table 2). FSH concentration was higher in DI rats than in normal rats, but insignificantly different between the homozygous rats with manifest diabetes insipidus and the heterozygous rats with partial deficit of vasopressin (Table 2).

The dissociated variation of LH and FSH in the pituitary gland and serum of DI rats is accompanied by normal bio- and immunoactivities of the hypothalamic LH-releasing hormone.[8] Therefore, the influence of the decreased amount of pineal arginine-vasotocin, known to act as an antigonadotropic factor at the hypothalamic level, in determining the gonadotropin level in Brattleboro rats is improbable. A diminished content of pituitary prolactin was also reported for DI rats [11] but the serum concentrations of prolactin were comparable to those in the control animals.[11, 12] It is tempting to consider the dissociated variation of LH and FSH as an independent genetic peculiarity of the DI rats.

References

1. Valtin, H., W. H. Sawyer & H. Sokol. 1965. Neurohypophysial principles in rats homozygous and heterozygous for hypothalamic diabetes insipidus (Brattleboro strain). Endocrinology **77:** 701–706.
2. Coculescu, M., V. Matulevicius, R. Goldstein & S. Pavel. 1978. Presence and synthesis of vasotocin in the pineal gland of Brattleboro rats. J. Endocr. (London) **77:** 145–146.
3. Rosenbloom, A. A. & D. A. Fisher. 1975. Radioimmunoassayable AVT and AVP in adult mammalian brain tissue: comparison of normal and Brattleboro rats. Neuroendocrinology **17:** 354–361.
4. Pavel, S. Arginine vasotocin as a pineal hormone. 1978. J. Neurol. Transmiss. (Suppl. 13): 135–156.

5. SIMIONESCU, L. 1979. Metode de analiză in vitro folosind compuşi marcaţi. *In* Compuşi Marcaţi şi Radiofarmaceutici cu Aplicaţii în Medicina Nucleară. A. T. Balaban, L. Gălăţeanu, G. Georgescu & L. Simionescu, Eds. Academiei Republicii Socialiste România. Bucureşti, Romania.
6. MCCANN, S. M., J. ANTUNES-RODRIGUES, R. MALLAR & H. VALTIN. 1966. Pituitary adrenal function in the absence of vasopressin. Endocrinology **79:** 1058–1064.
7. WILEY, M. K., A. F. PEARLMUTTER & R. E. MILLER. 1974. Decreased adrenal sensitivity to ACTH in the VP-deficient (Brattleboro) rat. Neuroendocrinology **14:** 257–270.
8. GILLIES, G. & P. J. LOWRY. 1980. Corticotrophin releasing activity in extracts of the stalk median eminence of Brattleboro rats. J. Endocr. (London) **84:** 65–73.
9. ARIMURA, A., S. SAWANO, T. W. REDDING & A. V. SCHALLY. 1968. Studies on retarded growth of rats with hereditary hypothalamic diabetes insipidus. Neuroendocrinology **3:** 187–192.
10. REICHLIN, S. 1960. Growth and the hypothalamus. Endocrinology **66:** 760–773.
11. COCULESCU, M., M. OPRESCU & V. DIMITRIU. 1976. Prolactin determination in hypothalamic diabetes insipidus. Abstract volume. V Int. Congr. Endocrinol. Hamburg. Abstract 22.
12. ADLER, R. A., S. DOLPHIN, M. SZEFLER & H. W. SOKOL. 1979. The effects of elevated circulating prolactin in rats with hereditary hypothalamic diabetes insipidus (Brattleboro strain). Endocrinology **105:** 1001–1006.

ROLE OF POLYAMINES IN THE REDUCED GROWTH OF BRATTLEBORO RATS

K. A. Pass and J. J. Postulka

New York State Department of Health
Division of Laboratories and Research
Birth Defects Institute
Albany, New York 12201

Introduction

Rats homozygous for the recessive allele for hypothalamic diabetes insipidus (DI) grow more slowly than their heterozygous (HZ) siblings.[1] Daily injections of vasopressin from weaning[2] or birth[3] were ineffective in reversing this growth retardation, although urine osmolalities were restored to normal. Analysis of pituitary growth hormone (GH) content revealed a much lower concentration in DI than HZ littermates.[4] Treatment of DI rats with GH resulted in gains in body weight and length comparable to control HZ animals, while having no effect on water turnover,[2] thus suggesting the stunted growth in DI rats was the result of reduced GH synthesis and/or release. However, it is not yet clear if there is normal utilization of GH by target tissues of the DI rat, since large supraphysiologic doses of GH were used in those experiments.

The polyamines are a group of aliphatic nonprotein bases present as normal cellular components of almost all living species. Polyamine levels are elevated dramatically in rapidly growing tissues of all types and appear to be correlated with changes in nucleic acid metabolism.[5] A key enzyme in the synthesis of the polyamines, ornithine decarboxylase (ODC, EC 4.1.1.17), is increased in numerous tissues by various growth stimuli, such as hormones or drugs.[5] In that regard, removal of a portion of the liver is a strong stimulus for GH release, leading to regrowth of the organ and increased polyamine synthesis.[6] The actions of GH in stimulating ODC have also been described in the adrenals and kidney.[7, 8] ODC activity has been reported to peak in the chick embryo during the period of limb bud formation and to decline thereafter.[5] Thus there is abundant evidence that the polyamines play a role in growth and development.

The present studies were undertaken to investigate the activity of ODC in Brattleboro rats. The answers to several questions were sought: (1) Are there any differences in basal ODC activity between DI and HZ rats? (2) Are the effects of GH on ODC similar in these models? (3) Does renal ODC in DI and HZ rats respond differently to hormonal stimulation?

Materials and Methods

Rats used in these experiments were from a colony maintained in our laboratory (from breeding stock obtained from Blue Spruce Farms, Altamont, New York). Animals were used at 21 to 26 weeks of age. Animal quarters were controlled for temperature (22° C) and light (14 h on) and except as noted in the dehydration studies, were provided with food and water ad libitum. All

0077–8923/82/0394–0644 $1.75/0 © 1982, NYAS

experiments were completed before noon to avoid any diurnal variation in ODC activity.[9]

The assay for ODC was performed as originally described by Janne and Williams-Ashman[10] and later modified.[11] Briefly, *in vivo* ODC activity was estimated from the *in vitro* release of $^{14}CO_2$ from ^{14}C-ornithine. Results are expressed as pmole $^{14}CO_2$ per hour per mg tissue. DL-(1-^{14}C) ornithine monohydrochloride was obtained from New England Nuclear, deamino-8-D-arginine vasopressin (dDAVP) from Ferring, and other reagents from Sigma Chemical Company (St. Louis, Mo.). Urine osmolality was measured by freezing-point depression on a Model 2007 Osmette (Precision Systems).

The data was analyzed by Student's *t* test or two-way ANOVA. Results were considered significant at $p < 0.05$.

Results and Discussion

There is evidence that pituitary content of GH is reduced in DI rats[4] and that GH treatment restores toward normal the retarded growth of DI rats.[2] Therefore, it has been concluded that GH deficiency is responsible for the reduced growth of DI rats.[2] However, the question of proper utilization of GH by the tissues of DI rats remains unanswered, since large amounts of GH were used in these treatments. Since polyamine metabolism and growth are inextricably linked,[5] and GH administration stimulates the activity of ODC, a key enzyme in polyamine biosynthesis, we examined the activity of ODC in Brattleboro rats in an attempt to answer this question. As shown in Table 1, basal levels of ODC were no different in DI and HZ rats in any of the tissues examined. This suggests that under tonic conditions the apparent lack of GH in the DI rats has no effect on polyamine biosynthesis in those organs examined. However, it should be noted that a reduced pituitary content of GH[4] is not necessarily indicative of a reduced serum concentration of that hormone and indeed, basal serum levels of GH may be the same.

Partial hepatectomy results in a profound increase (up to 10-fold) in hepatic ODC within 4 h of the procedure and a further increase thereafter.[12] Hypophysectomy abolishes the ODC response in regenerating rat liver, while treatment with GH restores the increase in ODC.[12] Therefore, this procedure serves as an ideal index of cellular response to GH in actively growing tissues. As shown in Figure 1, the response of DI and HZ rats to partial hepatectomy was identical. There was no difference either in absolute response or the percent change from controls. Although it is difficult to compare reported values for enzyme activity, it is interesting to note that the ODC response to partial hepatectomy in Brattleboro rats (both DI and HZ) was below that reported in other rat strains, suggesting a possible defect in the utilization of GH (or other hormones) by the tissues of these animals.

We have previously reported changes in renal ODC of mice following administration of vasopressin (ADH).[11] In that study, we showed that acute injection of ADH produced an increase in renal ODC, whereas chronic treatment with ADH for three days caused a fall in enzyme activity. This latter result mimicked the effects of dehydration, which also caused a decline in renal ODC.[11] Thus, it was of particular interest to examine renal ODC in an animal devoid of ADH. As shown in Table 1, basal levels of renal ODC were not different in HZ and DI, thus suggesting that, although the kidney is a primary

TABLE 1

BASAL ODC ACTIVITY IN BRATTLEBORO RATS

	HZ	DI	p Value
Heart	1.67 ± 0.22	1.20 ± 0.13	NS
Adrenal	1.40 ± 0.39	0.69 ± 0.12	NS
Kidney	6.21 ± 0.60	5.70 ± 0.56	NS
Liver	4.95 ± 0.85	4.45 ± 0.42	NS

Values (pmoles·h^{-1}·mg $tissue^{-1}$) are mean ± SEM for eight rats.

target organ for ADH and has one of the highest basal levels of ODC activity of any organ, ADH is not responsible for tonic control of ODC in this organ. Treatment of Brattleboro rats with ADH produced increases in renal ODC of comparable magnitude in both DI and HZ (TABLE 2). Injection of dDAVP likewise produced similar responses in both DI and HZ, but of a much smaller magnitude.

Dehydration of Brattleboro rats for 24 h led to comparable declines in renal ODC in DI and HZ (33% vs. 30%), while the DI rats exhibited a much larger (60%) loss in body weight (FIGURE 2). After 48 h dehydration, the DI rats had lost almost 25% of their body weight, compared with 14% for HZ rats; yet renal ODC activity had declined similar amounts in both. Measurements of urine osmolality during dehydration (data not shown) indicated that the DI rat is capable of concentrating urine, apparently by mechanisms other

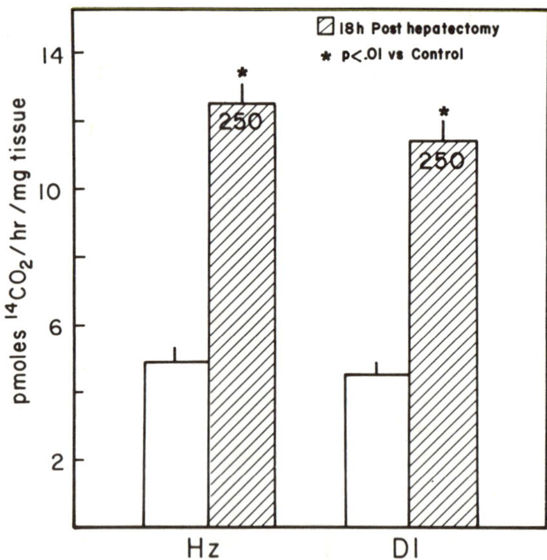

FIGURE 1. Effects of partial hepatectomy on liver ODC activity in Brattleboro rats. Values are mean ± SEM for nine rats. Values in striped bar represent percent increase from control.

TABLE 2

EFFECTS OF VASOPRESSIN ON RENAL ODC IN BRATTLEBORO RATS

	HZ	DI	p Value
Saline	7.43 ± 0.46	8.31 ± 0.58	NS
AVP *	22.72 ± 2.73	21.01 ± 2.76	NS
DDAVP †	11.0 ± 0.95	12.37 ± 1.33	NS

* Arginine vasopressin (3 U/100 g).
† DDAVP (0.32 μg/rat).

than those initiated by ADH,[13] and that the fall in ODC activity had no effect on this capacity. In an animal presumably never exposed to ADH, it is surprising that receptor mechanisms and intracellular sequelae (e.g., ODC activation) are still in place and functioning. More importantly, however, these results suggest that renal ODC may be involved in functions other than protein synthesis and water metabolism as we[14] and others[15] have suggested.

Orchidectomy of adult rats leads to a decrease in kidney weight within one week,[16] whereas treatment of castrated rats with testosterone brings about an augmentation in kidney weight.[17] Along with the increase in weight, testosterone administration also produces dramatic increases in renal ODC activity while having no effect on renal concentrating mechanisms.[18] Using testosterone as a probe, we examined its effect on renal ODC in Brattleboro rats. In HZ rats receiving testosterone, renal ODC was increased over that of oil-injected controls (FIGURE 3). Testosterone also caused an increase in renal ODC of DI rats, but

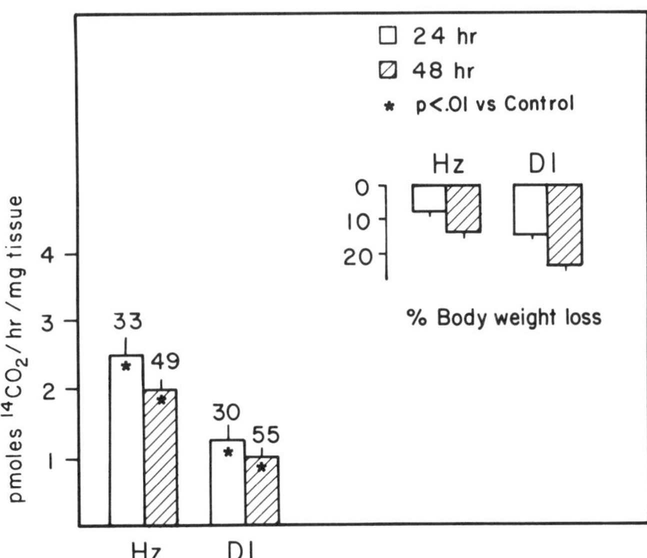

FIGURE 2. Effects of dehydration on body weight and renal ODC in Brattleboro rats. Values are mean ± SEM for 7–10 rats. Values above bars represent percent decrease from control.

the magnitude of this response was significantly lower than in HZ rats. Adrenal response to testosterone also differed significantly in the two groups (FIGURE 3), with a much greater increase in the HZ animals. Since testosterone has both androgenic and anabolic properties, it is not clear which facet of its action is represented here. However, it is interesting to note that ADH is a membrane-active hormone relying on cAMP mechanisms for its antidiuretic effect,[17] whereas testosterone exerts its effect primarily on the nucleus.[19] Thus, the mechanisms by which these two hormones increase renal ODC may be quite different, as has been demonstrated in rat liver.[20] If the primary effect of testosterone on the kidney is anabolic, and it may well be,[21] then the reduced response of the DI rats may be indicative of a general tissue characteristic of these animals and could account in part for their reduced growth rate.

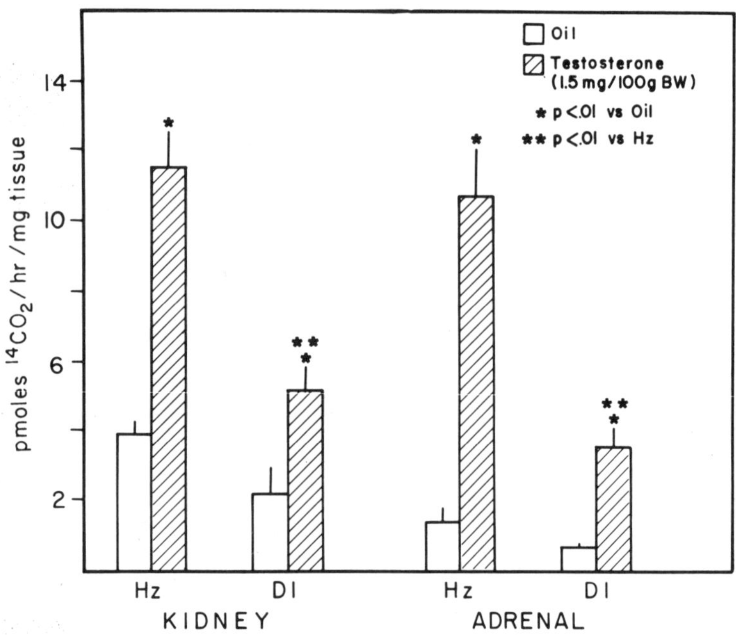

FIGURE 3. Effects of testosterone injection on ODC activity in the kidney and adrenal of Brattleboro rats. Values are mean ± SEM for eight rats.

CONCLUSION

As indicated from measurements of basal levels of ODC in several organs and enzyme activity following perturbations such as dehydration and partial hepatectomy, there are no significant differences in ODC activity between DI and HZ rats. Thus, the hypothesis that reduced polyamine activity could account for the retarded growth of DI rats would seem disproved. However, there were significant differences in renal enzyme activity in the two models when testosterone was used as a stimulus for enzyme activity. Thus, the possibility exists that testosterone and perhaps other steroids may be involved in growth retardation of the DI animals. We are currently pursuing this avenue of investigation.

References

1. Valtin, H., W. H. Sawyer & H. W. Sokol. 1965. Neurohypophyseal principles in rats homozygous for hypothalamic diabetes insipidus. Endocrinology **77:** 701–706.
2. Sokol, H. W. & J. Sise. 1973. The effect of exogenous-vasopressin and growth hormone on the growth of rats with hereditary hypothalamic diabetes insipidus. Growth **37:** 127–142.
3. Wright, W. A. & C. L. Kutscher. 1977. Vasopressin administration in the first month of life: effects on growth and water metabolism in hypothalamic diabetes insipidus rats. Pharmac. Biochem. Behav. **6** (5): 505–509.
4. Arimura, A., S. Sawano, T. W. Redding & A. V. Schally. 1968. Studies on retarded growth of rats with hereditary hypothalamic diabetes insipidus. Neuroendocrinology **3:** 187–192.
5. Russell, D. H. & B. M. G. Durie. 1978. Polyamines as biochemical markers of normal and malignant growth. *In* Progress In Cancer Research and Therapy. **8:** 1–45. Raven Press. New York, N.Y.
6. Higgins, H. M. & R. M. Anderson. 1931. Experimental pathology of the liver. Arch. Path. **12:** 186–202.
7. Brandt, J. T., K. A. Pierce & N. Fausto. 1972. Ornithine decarboxylase activity and polyamine synthesis during kidney hypertrophy. Biochim. Biophys. Acta **279:** 184–193.
8. Levine, J. H., W. E. Nicholson & D. N. Orth. 1972. Evidence for an adrenal growth factor other than ACTH. Clin. Res. **20:** 431.
9. Nicholson, W. E., J. H. Levine & D. N. Orth. 1976. Hormonal regulation of renal ornithine decarboxylase activity in the rat. Endocrinology **98:** 123–128.
10. Janne, J. & H. G. Williams-Ashman. 1971. On the purification of L-ornithine decarboxylase from rat prostate and effects of thiol compounds on the enzyme. J. Biol. Chem. **246:** 1725–1732.
11. Pass, K. A., H. L. Vallet, J. E. Bintz & J. J. Postulka. 1980. Effect of dehydration and arginine vasopressin on renal ornithine decarboxylase activity in mice. Life Sci. **26:** 1913–1918.
12. Russell, D. H. & S. H. Snyder. 1969. Amine synthesis in regenerating rat liver: Effect of hypophysectomy and growth hormone on ornithine decarboxylase. Endocrinology **84:** 223–228.
13. Gellai, M., B. R. Edwards & H. Valtin. 1979. Urinary concentrating ability during dehydration in the absence of vasopressin. Am. J. Physiol. **237**(2): F100–F104.
14. Pass, K. A., J. E. Bintz, J. J. Postulka & H. L. Vallet. 1981. Effect of hormonal status on renal ornithine decarboxylase activity. Proc. Soc. Exp. Biol. Med. **167:** 270–274.
15. Russell, D. H., C. V. Byns & Manem. 1976. Proposed model of major sequential biochemical events of a trophic response. Life Sci. **19:** 1297–1306.
16. Basinger, G. T. & R. F. Gittes. 1974. Effects of testosterone propionate on compensatory renal hypertrophy. Endocrinology **94:** 599–601.
17. Rajerison, R. M., D. Butler & S. Jard. 1977. Effect of *in vivo* treatment with vasopressin and analogues on renal adenylate cyclase responsiveness to vasopressin stimulation *in vitro*. Endocrinology **101:** 1–12.
18. Pass, K. A., J. E. Bintz & J. J. Postulka. 1981. Effects of testosterone on renal ornithine decarboxylase and kidney function. Enzyme (In press.)
19. O'Malley, B. W. & L. Birnbaumer, Eds.. 1978. Receptors and Hormone Action. Vol. 2 Academic Press. New York, N.Y.
20. Combest, W. L. & D. H. Russell. 1980. Separate mechanisms of induction for ornithine decarboxylase by triiodothyroxine and aminophylline. Life Sci. **27:** 651–658.
21. Nowinski, W. W. & R. J. Gross. 1969. Compensatory renal hypertrophy. Academic Press. New York, N.Y.

THE HYPOTHALMIC MEDIAN EMINENCE OF THE HOMOZYGOUS BRATTLEBORO RAT IS DEFICIENT IN CALCITONIN-SPECIFIC BINDING SITES *

Mark van Houten,† Arturo J. Rizzo, David Goltzman, and Barry I. Posner

Department of Medicine
McGill University
Montreal, Quebec
Canada H2A 2B2

Calcitonin, a 32 amino-acid polypeptide hormone secreted by the thyroid, has been shown to interact directly with the central nervous system.[1, 2] Recently, we have localized specific binding sites for blood-borne calcitonin to the circumventricular organs of the rat brain.[3] These highly specialized regions of the central nervous system may serve as the brain's main receptive system in mediating the direct action of blood-borne peptide hormones on brain visceral function.[4] However, it is unclear as to how and what brain functions are influenced specifically by circulating calcitonin.

Classical studies have implicated calcitonin in the regulation of renal function and electrolyte balance.[5] Thus, we have considered the possibility that calcitonin action in the central nervous system may serve to mobilize brain hydromineral regulatory functions to complement this action in kidney. To examine this hypothesis we sought to quantify the calcitonin-binding capacity of the circumventricular organs of the Brattleboro rat brain, because if calcitonin action is related to hydromineral regulation, profound disturbance of this regulation may disrupt calcitonin–brain binding interactions.

Methods and Materials

Iodination

Synthetic salmon calcitonin (sCt) (Sandoz Co. Ltd., Basel, Switzerland) was moniodinated by chloramine T to a specific activity of 328–742 μCi/μg, as described.[1]

Specimen

Age-matched adult Long-Evans rats, homozygous and heterozygous Brattleboro rats (Blue Spruce Farms, Altamont, N.Y.) were maintained *ad libitum*

* Supported by grants from the Canadian Diabetic Association, the Juvenile Diabetes Foundation, the Medical Research Council of Canada (MT–4182 and MA–5775), and the U.S. Public Health Service (1 RO1 AM 19573).

† Address correspondence to: Mark van Houten, McGill University, Polypeptide Hormone Laboratory, Strathcona Anatomy Bldg., rm 2–11, 3640 University Street, Montreal, Quebec H2A 2B2.

0077–8923/82/0394–0650 $1.75/0 © 1982, NYAS

for 10 days, three rats per metabolic cage, before commencement of *in vivo* and *in vitro* binding studies and plasma calcitonin determinations. For 19 consecutive days three homozygous rats received daily injections of vasopressin (Pitressin tannate; Parke, Davis and Co., Brockville, Ontario; 120 mU/100 g/day) in peanut oil (1:1) and three received vehicle only. During this time daily records of water consumption, urinary output, and urine osmolality were collected on each group, including the Long-Evans controls, until the day of sacrifice.

In Vivo *Binding Studies*

Rats were anesthetized and given an intracardiac injection of [^{125}I]sCt (153×10^6 cpm) alone or with 100 μg of unlabeled sCt, all by methods published previously.[6] Five minutes after injection rats were perfused sequentially with lactated Ringer's solution, followed by Bouin's fixative.

Light Microscope Radioautography

Perfusion-fixed blocks of median eminence were prepared for light microscope radioautography, as described.[6] Five-micron thick sections were exposed for four to seven days before development. The intensity of radioautographic reactions over the external zone of the median eminence of the Brattleboro rats and Long-Evans controls were compared quantitatively by methods published.[6]

In Vitro *Binding Studies*

Competitive binding studies were performed on membranes extracted from hypothalami of normal Long-Evans and homozygous Brattleboro rats by *in vitro* methods previously reported.[1]

Determination of Plasma Calcitonin Concentration

Peripheral blood from six Long-Evans controls and six homozygous Brattleboro rats was obtained by cardiac puncture. Serum was tested at multiple dilutions in the radioimmunoassay, as previously described.[7]

Results

The histological distribution of calcitonin-binding sites in the median eminence and other circumventricular organs of the Long-Evans rat brain resembled that previously observed in the Sprague-Dawley rat brain.[3] Calcitonin-binding sites were concentrated in the neuropil of the external zone in the vicinity of capillaries of the portal plexus. Two major binding-rich regions were observed in the external zone: they occurred bilaterally in the centrolateral aspect of the zona externa, and were separated by a binding-poor region in the center of the median eminence. These binding sites were calcitonin-specific, because the binding of [^{125}I]sCt was blocked by unlabeled sCt in proportion to concentration, and to a lesser extent by the structurally homologous human calcitonin, but not by structurally dissimilar polypeptides.[3]

Quantitative radioautographic analysis showed that the capacity of the external zone of the homozygous Brattleboro rat median eminence to specifically bind blood-borne calcitonin was diminished within a range of 27–42%, as compared to the median eminence of the Long-Evans controls (FIGURE 1). No significant reduction in median eminence calcitonin-binding capacity was observed in heterozygous Brattleboro rats.[3]

Assessment of calcitonin binding to hypothalamic membranes *in vitro* in normal Long-Evans and homozygous Brattleboro rats confirmed the binding deficit observed *in vivo* (FIGURE 2). Scatchard analysis of the homozygous hypothalamus suggested the absence of at least one-fifth of the specific binding sites with normal affinity.

Daily vasopressin replacement treatment of the homozygous Brattleboro rat normalized water consumption and urine volumes, and greatly improved urine concentration, as expected.[8] The capacity of the homozygous median eminence to bind blood-borne calcitonin, however, was not normalized by vasopressin treatment (FIGURE 1), nor was any obvious improvement observed.

Plasma levels of calcitonin in homozygous Brattleboro rats (0.53 ± 0.25, mean ± SD ng-equivalents/ml) were not statistically different from values in the Long-Evans rats.

DISCUSSION

The inability of vasopressin treatment to correct the calcitonin-binding deficit of the homozygous median eminence and the absence of elevated plasma

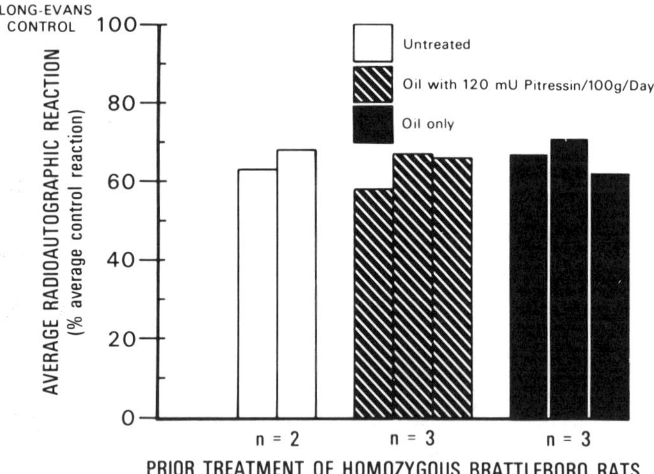

FIGURE 1. Average calcitonin-binding capacity of the median eminence of homozygous Brattleboro rats receiving vasopressin, vehicle, or no treatment prior to binding study is expressed as a percent of the average binding capacity in Long-Evans controls.

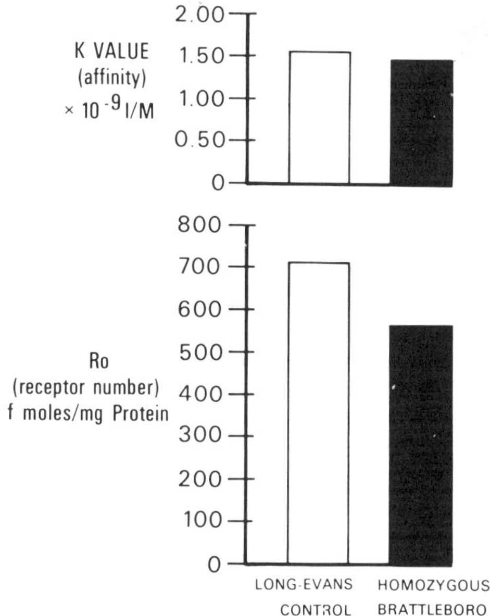

FIGURE 2. The affinity constants and number of calcitonin receptors in hypothalamic homogenates from homozygous Brattleboro rats and Long-Evans controls, as reported by van Houten *et al.*[3]

calcitonin levels in these rats suggest that this receptor deficiency may represent a genetic lesion, which could relate directly to their complex endocrinopathy. Indeed, a calcitonin receptor deficiency may be a second disturbance in vasopressinergic neurons that project to the median eminence.

Interestingly, the distribution of calcitonin binding sites in median eminence corresponds closely to the distribution of median eminence somatostatin.[9] Calcitonin has been shown to influence growth hormone [10] and TSH [11] secretion *in vivo,* but we have found no evidence for specific binding sites for blood-[^{125}I]sCt in adenohypophysis (unpublished observations). Thus, it may well be that calcitonin influences GH and TSH secretion indirectly via a direct action on the median eminence, perhaps involving somatostatin. Furthermore, a disruption in median eminence calcitonin-binding capacity may relate by this action to the marked growth retardation of the homozygous Brattleboro rat, which, like calcitonin-specific binding capacity, is not corrected by chronic vasopressin therapy.[12]

REFERENCES

1. RIZZO, A. J. & D. GOLTZMAN. 1981. Calcitonin receptors in the central nervous system of the rat. Endocrinology **108:** 1672–1677.
2. FREED, W. J., M. J. PERLOW & R. J. WYATT. 1979. Calcitonin: inhibitory effects on eating in rats. Science **206:** 850–852.
3. VAN HOUTEN, M., D. GOLTZMAN & B. I. POSNER. 1980. Calcitonin receptors

in brain: Localization in normals and deficiency in hereditary diabetes insipidus. Clinical Res. **28:** 676A.

4. VAN HOUTEN, M. & B. I. POSNER. 1981. Cellular basis of direct insulin action in the central nervous system. Diabetologia **20** (Suppl.): 255–267.
5. ALDRED, J. P., R. R. KLESZYNSKI & J. W. BASTIAN. 1970. Effects of acute administration of porcine and salmon calcitonin on urine electrolyte excretion in rats. Proc. Soc. Exp. Biol. Med. **134:** 1175–1180.
6. VAN HOUTEN, M., B. I. POSNER, B. M. KOPRIWA & J. R. BRAWER. 1979. Insulin binding sites in rat brain: *in vivo* localization to the circumventricular organs by quantitative radioautography. Endocrinology **105:** 666–672.
7. GOLTZMAN, D. & A. S. TISCHLER. 1978. Characterization of the immunochemical forms of calcitonin released by a medullary thyroid carcinoma in tissue culture. J. Clin. Invest. **61:** 502–510.
8. HARRINGTON, A. R. & H. VALTIN. 1968. Impaired urinary concentration after vasopressin and its gradual correction in hypothalabic diabetes insipidus. J. Clin. Invest. **47:** 502–510.
9. BAKER, B. L. & Y. Y. YU. 1976. Distribution of growth hormone release-inhibiting hormone (somatostatin) in the rat brain as observed by immunohistochemistry. Anat. Rec. **186:** 343–356.
10. CANTALAMESSA, L., A. CATANIA, E. RESCHINI & M. PERACCHI. 1978. Inhibitory effects of calcitonin on growth hormone and insulin secretion in man. Metabolism **27:** 987–992.
11. LEICHT, E., G. BIRO & K. F. WENIGES. 1974. Inhibition of releasing-hormone-induced secretion of TSH and LH by calcitonin. Horm. Metab. Res. **6:** 410–414.
12. VALTIN, H., H. W. SOKOL & D. SUNDE. 1975. Genetic approaches to the study of the regulation and actions of vasopressin. Rec. Prog. Horm. Res. **31:** 447–486.

DISTURBED BEHAVIOR AND MEMORY OF THE BRATTLEBORO RAT

Tj. B. van Wimersma Greidanus

Rudolf Magnus Institute for Pharmacology
Medical Faculty
University of Utrecht
3521 GD Utrecht
the Netherlands

The neurohypophyseal hormones vasopressin and oxytocin have been shown to play an important role in brain processes related to memory function. Vasopressin induces a long-term inhibition of extinction of a conditioned avoidance response (CAR) in rats, which extends far beyond the actual presence of the injected peptide in the organism. This suggests that vasopressin triggers a long-term effect on the maintenance of a learned response, probably by facilitating storage and subsequent retrieval of information in the central nervous system by enhancement of consolidation processes.[1-4] Vasopressin not only affects maintenance of conditioned avoidance behavior, it also affects acquisition of this behavior. This latter effect however, may depend on the level at which the animals tend to respond to relevant stimuli.[5]

The long-term preservation of an acquired response following vasopressin administration was not only found using active avoidance behavior as paradigm, but also in a passive avoidance situation.[6-9] In addition it was found that the effects of vasopressin on avoidance of shock are similar to those on avoidance of attack.[10]

Further evidence for the effects of vasopressin on memory processes was obtained when it appeared that vasopressin antagonizes amnesia induced by puromycin, by pentylenetetrazol, by CO_2, or by electroconvulsive shock.[9, 11-13] In fact it appeared that vasopressin is able to prevent as well as to reverse amnesia. These results suggest that vasopressin, in addition to affecting memory consolidation,[3] also promotes its retrieval.[13, 14]

In addition several data indicate that vasopressin can affect not only avoidance behavior, but performance of positively rewarded tasks as well, such as sexually motivated approach behavior[15] and appetitive discrimination tasks.[16] Besides, vasopressin and/or its congeners facilitate the development of resistance to the analgesic action of morphine,[17] which may be regarded as a form of learning or memory.[18, 19]

That vasopressin may be involved physiologically in memory processes was indicated when picogram amounts were found to be behaviorally active after intracerebroventricular (icv) administration.[20]

In addition, it appeared that oxytocin exerts an effect on behavior that is generally opposite to the effect induced by vasopressin. Upon icv administration oxytocin impairs passive avoidance behavior and facilitates extinction of an active avoidance response.[21, 22] It has been suggested that oxytocin may be a naturally occurring amnesic peptide. Generally it is assumed that vasopressin and oxytocin modulate memory function in an opposite manner by a direct action on the brain.[19]

0077-8923/82/0394-0655 $1.75/0 © 1982, NYAS

Further evidence for a physiological role of vasopressin and oxytocin in brain processes related to memory function was obtained from experiments with rats with a reduced amount of bio-available hormone in the brain, due to central administration of specific antisera to vasopressin or oxytocin.[21, 23–27] From the disturbed behavior of rats treated with anti-vasopressin serum it was suggested that centrally available vasopressin is of importance for memory consolidation as well as for retrieval processes. The results of the experiments in which anti-vasopressin serum is applied also reinforce the notion that learning can take place in the temporary absence of vasopressin, but that the maintenance of an acquired behavioral response is disturbed.[19, 26] Since neutralization of centrally available oxytocin by antiserum to oxytocin generally results in effects opposite to those obtained after treatment with anti-vasopressin serum it has been suggested that brain function needed for optimal adaptation to environmental stimuli depends on a subtle equilibrium between the bio-available amounts of these hormones in the brain. Disturbances in this brain endocrine homeostatic mechanism may result in alterations in brain functions which may be reflected in behavioral changes. In fact these disturbances may interfere with the flexibility or adaptive capacity of the organism to cope with environmental changes when it is confronted with situations previously experienced.

Further evidence for a physiological role of vasopressin in brain processes related to memory function was obtained from experiments with Brattleboro rats. However, interpretation of the peculiarities in the behavioral performance of homozygous diabetes insipidus rats of the Brattleboro strain as a primary consequence of their vasopressin deficiency is probably too simple. The chronic vasopressin deficiency results in a variety of physiological disturbances that may affect brain function and behavior. Nevertheless the homozygous diabetes insipidus (HO-DI) rat of the Brattleboro strain is a useful model to study the effects of neurohypophyseal functioning on brain and behavior, if one is aware of the multiple effects of chronic vasopressin deficiency.

In addition to the virtual absence of vasopressin in blood, pituitary, and hypothalamus of HO-DI rats, the oxytocin levels in blood and CSF are higher in HO-DI rats than in heterozygous animals,[28, 29] whereas the oxytocin levels in pituitaries of HO-DI rats were lower (approximately 50%) than in heterozygous or normal animals.[30, 31] HO-DI rats have serum levels of potassium, which are lower than normal[32] and high levels of renin and angiotensin II.[30, 33, 34] HO-DI rats also show a diminished response of the pituitary-adrenocortical system to stress.[35, 36]

Memory function of HO-DI rats is impaired in one-trial passive avoidance behavior, when retention is tested 24 h or more after training. Vasopressin, administered immediately after the single acquisition trial restores the disturbed behavior.[37] This favors the idea that memory rather than learning is disturbed in the absence of vasopressin. Indeed, full retention of passive avoidance behavior is obtained in HO-DI rats when retention is tested shortly after the learning (acquisition) trial. Thus, the main disturbance is in the maintenance of an acquired behavioral response and not in the acquisition itself. This is also illustrated by the observation that HO-DI rats are able to acquire a conditioned avoidance response in multiple trial paradigms, albeit the rate of acquisition is slightly slower than in heterozygous rats.[38, 39]

The fact that open field activity and responsiveness to electric foot shock (EFS) are approximately the same in HO-DI and heterozygous rats[36, 38] excludes the possibility that differences in exploratory or motor activity or in sensitivity to a painful stimulus underlie the differences in avoidance behavior between HO-DI and heterozygous animals.

In fact even the maintenance of acquired behavior, although severely disturbed, has not disappeared completely in the absence of vasopressin, since HO-DI rats display a certain level of passive avoidance response even at 24 h after acquisition, although the retention of the passive avoidance response is extremely weak.[37]

This lack of an absolute retention deficit in passive avoidance behavior is more clearly illustrated by the experiments of Bailey and Weiss.[40] The results of their experiments show that passive avoidance behavior of HO-DI rats is poorer than that of heterozygous animals, but that the absence of vasopressin is not associated with a total impairment of passive avoidance. They point to factors other than the total deficiency of vasopressin that might explain the behavioral characteristics of HO-DI rats, such as the chronic elevation of plasma oxytocin levels[28, 29] and the alterations in adrenocortical function,[35, 41] in growth and metabolism[42, 43] and in electrolyte balance,[44] etc.

In fact HO-DI rats display a pituitary-adrenal response during passive avoidance performance which shows a marked relationship with the avoidance behavior. This suggests that the impairment of the pituitary-adrenal response in HO-DI rats depends primarily on a disturbance in the psychological response to stress.[38]

Nevertheless, the observation that a single injection of vasopressin completely normalizes the disturbed behavior of HO-DI rats as observed by De Wied *et al.*[37] does not support the idea that secondary long-term consequences of the chronic vasopressin deficiency underlie the "memory loss" of HO-DI rats.

Data obtained by Bohus *et al.*[38] in multiple learning trial paradigms also show that HO-DI rats are able to acquire fear-motivated responses. Acquisition of a pole jump avoidance response (one-way active conditioning) is approximately similar in HO-DI and heterozygous rats. HO-DI rats display avoidance behavior in a shuttle box as well, although their rate of acquisition in this two-way active avoidance behavior is slightly slower than that of heterozygous rats. However, again the maintenance of the acquired response seems disturbed in the absence of vasopressin, witness the facilitated extinction of the CAR in HO-DI rats.

In contrast, Celestian *et al.*[45] reported an unimpaired maintenance of a multiple trial CAR in HO-DI rats, although differences in behavioral performance were observed between HO-DI rats and heterozygous animals. Heterozygous rats showed a greater increase in avoidance acquisition during most stages of learning in contrast to the HO-DI rats, which did not achieve the high levels reached by the heterozygous ones. However, HO-DI rats that did reach the learning criterion of conditioned avoidance responding retained more conditioned avoidance responses, reflected in a retarded rate of extinction of the response.[45]

Miller *et al.*[39] also confirmed Bohus' data in that acquisition of an avoidance response in a multiple-trial paradigm is inferior in HO-DI rats. They also report that HO-DI animals make fewer responses in extinction than heterozygous rats, but they suggest a greater resistance to extinction in HO-DI rats than in heterozygous animals, if rates of extinction are estimated from the slopes of the extinction curves. If they take into account the terminal performance in acquisition the apparent differences in extinction of HO-DI and heterozygous rats vanish.[39] However, one has to take into account that in the experiments of

Bohus *et al.*[38] the behavioral procedure is slightly different from the ones of Celestian *et al.*[45] and of Miller *et al.*[39] One of these differences is the duration of presentation and the intensity of the unconditioned stimulus (US) of electric footshock (EFS) applied during acquisition of the CAR. In Bohus' experiments EFS is applied until the rat makes a response, whereas Celestian terminates the US of shock after 10 sec in case the animals are not responding to the conditioning stimulus. Furthermore extinction of the CAR is studied by Celestian in a group of animals selected by reaching the acquisition criterion, whereas in Bohus' experiments all animals were included for extinction.

Recently it has been shown[46] that HO-DI rats adapted more slowly than normal animals in a T maze and that they made significantly fewer correct responses in a visual discrimination task than normal rats. Apparently HO-DI rats were learning the visual discrimination, but the acquisition rate was inferior to that of normal rats. A similar inferiority of HO-DI rats was observed when the animals were trained for an olfactory discrimination task. Concerning performance of a contingently reinforced task of delayed alteration in a T maze, HO-DI rats and normal animals displayed a more or less similar gradual improvement in performance, at least when short-intertrial intervals were applied.[46] At long intertrial intervals HO-DI rats might be slightly impaired in T maze alternation.[46] However, according to these authors HO-DI rats, although in some aspects deficient in behavior, show better retention of the punishment effect in the approach-avoidance conflict test than normal animals. In fact, their data support the notion that HO-DI rats differ in their learning ability from normal rats, but they also suggest that not all aspects of learning are equally affected. According to Brito *et al.* HO-DI rats may have defects in reference-memory mechanisms and adaptability (temperment-related tasks) rather than in working-memory processes.[46, 52]

Since development of tolerance might be regarded as a form of learning or memory[18] and similar mechanisms as in learning and memory might be involved in development of tolerance, it is worthwhile to mention that HO-DI rats show a delayed development of tolerance to the antinociceptive action of morphine.[47] However, the fact that the rate of development of tolerance was slower in HO-DI rats than in heterozygous animals does not necessarily mean that the endpoint of tolerance development is different in these animals. These data correlate well with reduced tolerance to the hypothermic effect of ethanol in HO-DI rats as observed by Pittman *et al.*[53]

Although no concordance exists between the observations and interpretations of the data from experiments with Brattleboro rats performed by the various investigators, it generally may be concluded that learning itself, in terms of acquisition of a response, is under certain circumstances disturbed in the absence of vasopressin as in HO-DI rats. Memory function in terms of the maintenance of an acquired response is also deficient under some conditions, although not completely absent. That this deficiency in memory is not absolute may also be concluded from the observation that $ACTH_{4-10}$, a neuropeptide that enhances processes related to arousal, attention, motivation, or retrieval is able to improve passive avoidance behavior in HO-DI rats.[48] If in HO-DI rats information storage would have been absolutely absent, $ACTH_{4-10}$ would never be able to improve passive avoidance behavior by increased arousal or retrieval. Interestingly, $ACTH_{4-10}$ is able to normalize disturbed passive avoidance behavior in rats with impaired retrieval, but not with impaired information storage (consolidation), due to treatment with antiserum to vasopressin.[49] However, the effects of $ACTH_{4-10}$ in HO-DI rats are not completely understood and reports in the literature are not in agreement.[36, 48, 50] It has been reported[36, 50] that $ACTH_{4-10}$ may impair passive avoidance behavior of HO-DI rats. This effect has been interpreted[26] as a consequence of a different arousal state of HO-DI rats in combination with the U-shape dose-response curve of ACTH.[51] Nevertheless, the data obtained with the neuropeptide $ACTH_{4-10}$ generally illustrate that HO-DI rats may be deficient in memory function, but that this deficiency is not absolute and that at least some of the behavioral differences between HO-DI rats and heterozygous or normal animals may be of secondary nature rather than being primary consequences of the chronic absence of vasopressin.

That the absence of vasopressin, as in HO-DI rats, does not result in an absolute memory deficiency may also be explained by the fact that neuropeptides such as ACTH, MSH, and the endorphins are present in the brain of HO-DI rats. These peptides also affect active and passive avoidance behavior and may enhance processes related to memory. Therefore it seems likely that the presence of the mentioned pro-opiomelanocortin fragments and of related peptides in the brain of HO-DI rats may have consequences for consolidation and retrieval processes in these animals and enable them to display a certain degree of memory.

The precise mechanism of action by which vasopressin normally affects brain functions related to memory is still unknown and remains to be elucidated. Nevertheless it may be concluded that the HO-DI Brattleboro rat has a disturbed behavior due to the absence of vasopressin, depending on the behavioral paradigm used.

References

1. De Wied, D. 1971. Long term effect of vasopressin on the maintenance of a conditioned avoidance response in rats. Nature **232:** 58–60.
2. De Wied, D., B. Bohus & Tj. B. van Wimersma Greidanus. 1974. The hypothalamo-neurohypophyseal system and the preservation of conditioned avoidance behavior in rats. Prog. Brain Res. **41:** 417–428.
3. De Wied, D., Tj. B. van Wimersma Greidanus, B. Bohus, I. Urban & W. H. Gispen. 1976. Vasopressin and memory consolidation. Prog. Brain Res. **45:** 181–194.
4. van Wimersma Greidanus, Tj. B., B. Bohus & D. De Wied. 1973. Effects of peptide hormones on behaviour. *In* International Congress Series No. **273:** 197–201. Excerpta Medica. Amsterdam.
5. King, A. R. & D. De Wied. 1974. Localized behavioral effects of vasopressin on maintenance of an active avoidance response in rats. J. Comp. Physiol. Psychol. **86:** 1008–1018.
6. Ader, R. & D. De Wied. 1972. Effects of lysine vasopressin on passive avoidance learning. Psychon. Sci. **29:** 46–48.
7. Wang, S. S. 1972. Synthesis of desglycinamide lysine vasopressin and its behavioral activity in rats. Biochem. Biophys. Res. Commun. **48:** 1511–1515.
8. Lissák, K. & B. Bohus. 1972. Pituitary hormones and avoidance behavior of the rat. Int. J. Psychobiol. **2:** 103–115.
9. Bookin, H. B. & W. D. Pfeifer. 1977. Effect of lysine vasopressin on pentylenetetrazol-induced retrograde amnesia in rats. Pharmacol. Biochem. Behav. **7:** 51–54.
10. Leshner, A. I. & K. E. Roche. 1977. Comparison of the effects of ACTH and lysine vasopressin on avoidance of attack in mice. Physiol. Behav. **18:** 879–883.

11. LANDE, S., J. B. FLEXNER & B. FLEXNER. 1972. Effect of corticotropin and desglycinamide-lysine vasopressin on suppression of memory by puromycin. Proc. Natl. Acad. Sci. USA **69:** 558–560.
12. PFEIFER, W. D. & H. B. BOOKIN. 1978. Vasopressin antagonizes retrograde amnesia in rats following electroconvulsive shock. Pharmacol. Biochem. Behav. **9:** 261–263.
13. RIGTER, H., H. VAN RIEZEN & D. DE WIED. 1974. The effects of ACTH- and vasopressin-analogues on CO_2-induced retrograde amnesia in rats. Physiol. Behav. **13:** 381–388.
14. RIGTER, H. & H. VAN RIEZEN. 1978. Hormones and memory. *In* Psychopharmacology: A Generation of Progress. M. A. Lipton, A. DiMascio & K. F. Killam, Eds.: 677–689. Raven Press. New York, N.Y.
15. BOHUS, B. 1977. Effect of desglycinamide-lysine vasopressin (DG-LVP) on sexually motivated T-maze behavior of the male rat. Horm. Behav. **8:** 52–61.
16. HOSTETTER, G., S. L. JUBB & G. P. KOZLOWSKI. 1977. Vasopressin affects the behavior of rats in a positively-rewarded discrimination task. Life Sci. **21:** 1323–1328.
17. KRIVOY, W. A., E. ZIMMERMAN & S. LANDE. 1974. Facilitation of development of resistance to morphine analgesia by desglycinamide[9]-lysine-vasopressin. Proc. Natl. Acad. Sci. USA **71:** 1852–1856.
18. COHEN, M., A. S. KEATS, W. A. KRIVOY & G. UNGAR. 1965. Effect of actinomycin D on morphine tolerance. Proc. Soc. Exp. Biol. Med. **119:** 381–384.
19. VAN WIMERSMA GREIDANUS, TJ. B. & D. H. G. VERSTEEG. 1981. Neurohypophyseal hormones: their role in endocrine function and behavioral homeostasis. *In* Behavioral Neuroendocrinology. Ch. B. Nemeroff & J. Dunn, Eds. Spectrum Publ. Jamaica, N.Y. (In press.)
20. DE WIED, D. 1976. Behavioral effects of intraventricularly administered vasopressin and vasopressin fragments. Life Sci. **19:** 685–690.
21. BOHUS, B., I. URBAN, TJ. B. VAN WIMERSMA GREIDANUS & D. DE WIED. 1978. Opposite effects of oxytocin and vasopressin on avoidance behaviour and hippocampal theta rhythm in the rat. Neuropharmacology **17:** 239–247.
22. BOHUS, B., G. L. KOVÁCS & D. DE WIED. 1978. Oxytocin, vasopressin and memory: opposite effects on consolidation and retrieval processes. Brain Res. **157:** 414–417.
23. VAN WIMERSMA GREIDANUS, TJ. B. J. DOGTEROM & D. DE WEID. 1975. Intraventricular administration of anti-vasopressin serum inhibits memory consolidation in rats. Life Sci. **16:** 637–644.
24. VAN WIMERSMA GREIDANUS, TJ. B., B. BOHUS & D. DE WIED. 1975. The role of vasopressin in memory processes. Prog. Brain Res. **42:** 135–141.
25. VAN WIMERSMA GREIDANUS, TJ. B., B. BOHUS & D. DE WIED. 1981. Vasopressin and oxytocin in learning and memory. *In* Endogenous Peptides and Learning and Memory Processes. J. L. Martinez, R. A. Jensen, R. B. Messing, H. Rigter & J. L. McGaugh, Eds. Academic Press. San Diego, Calif. (In press.)
26. VAN WIMERSMA GREIDANUS, TJ. B. & D. DE WIED. 1976. Modulation of passive avoidance behavior of rats by intracerebroventricular administration of antivasopressin in serum. Behav. Biol. **18:** 325–333.
27. LESHNER, A. I., R. HOFSTEIN, D. SAMUEL & TJ. B. VAN WIMERSMA GREIDANUS. 1978. Intraventricular injection of antivasopressin serum blocks learned helplessness in rats. Pharmacol. Biochem. Behav. **9:** 889–892.
28. VALTIN, H., W. H. SAWYER & H. W. SOKOL. 1965. Neurohypophyseal principles in rats homozygous and heterozygous for hypothalamic diabetes insipidus (Brattleboro strain). Endocrinology **77:** 701–706.
29. DOGTEROM, J., TJ. B. VAN WIMERSMA GREIDANUS & D. F. SWAAB. 1977. Evidence for the release of vasopressin and oxytocin into cerebrospinal fluid:

Measurements in plasma and CSF of intact and hypophysectomized rats. Neuroendocrinology **24:** 108–118.

30. Balment, R. J., I. W. Henderson & J. A. Oliver. 1975. The effects of vasopressin on pituitary oxytocin content and plasma renin activity in rats with hypothalamic diabetes insipidus (Brattleboro strain). Gen. Comp. Endocrinol. **26:** 468–477.
31. Van Wimersma Greidanus, Tj. B. & D. de Wied. 1977. The physiology of neurohypophyseal system and its relation to memory processes. *In* Biochemical Correlates of Brain Structure and Function. A. N. Davison, Ed.: 215–248. Academic Press. London.
32. Möhring, J., G. Dauda, D. Haack, E. Homsy, G. Kohrs & B. Möhring. 1972. Increased potassium intake and kaliopenic nephropathy in rats with genetic diabetes insipidus. Life Sci. **11:** 679–683.
33. Möhring, J., G. Kohrs, B. Möhring, M. Petri, E. Homsy & D. Haack. 1978. Effects of prolonged vasopressin treatment in Brattleboro rats with diabetes insipidus. Am. J. Physiol. **234:** F106–F111.
34. Hoffman, W. E., U. Ganten, P. Schelling, M. I. Phillips, P. G. Schmid & D. Ganten. 1978. The renin and isorenin-angiotensin system in rats with hereditary hypothalamic diabetes insipidus. Neuropharmacology **17:** 919–923.
35. Wiley, M. K., A. F. Pearlmutter & R. E. Miller. 1974. Decreased adrenal sensitivity to ACTH in the vasopressin-deficient (Brattleboro rat). Neuroendocrinology **14:** 280–288.
36. Bailey, W. H. & J. M. Weiss. 1981. Avoidance conditioning and endocrine function in Brattleboro rats. *In* Endogenous Peptides and Learning and Memory Processes. J. L. Martinez Jr., R. A. Jensen, R. B. Messing, H. Rigter & J. L. McGaugh, Eds. Academic Press. San Diego, Calif. (In press.)
37. De Wied, D., B. Bohus & Tj. B. Van Wimersma Greidanus. 1975. Memory deficit in rats with hereditary diabetes insipidus. Brain Res. **85:** 152–156.
38. Bohus, B., Tj. B. Van Wimersma Greidanus & D. De Wied. 1975. Behavioral and endocrine responses of rats with hereditary hypothalamic diabetes insipidus (Brattleboro strain). Physiol. Behav. **14:** 609–615.
39. Miller, M., E. G. Barranda, M. C. Dean & F. R. Brush. 1976. Does the rat with hereditary hypothalamic diabetes insipidus have impaired avoidance learning and/or performance? Pharmacol. Biochem. Behav. **5:** 35–40.
40. Bailey, W. H. & J. M. Weiss. 1979. Evaluation of a "memory deficit" in vasopressin-deficient rats. Brain Res. **162:** 174–178.
41. Vinson, G. P., C. Goddard & B. J. Whitehouse. 1977. Steroid profiles formed by adrenocortical tissues from rats with hereditary diabetes insipidus (Brattleboro strain) and from normal female Wistar rats under different conditions of stimulation. J. Endocrinol. **75:** 31P–32P.
42. Dlouhá, H., J. Krecek & J. Zicha. 1977. Growth and urine osmolarity in young Brattleboro rats. J. Endocrinol. **75:** 329–330.
43. Sokol, H. W. & J. Sise. 1973. The effect of exogenous vasopressin and growth hormone on the growth of rats with hereditary hypothalamic diabetes insipidus. Growth **37:** 127–142.
44. Laycock, J. F. 1977. The Brattleboro rat with hereditary hypothalamic diabetes insipidus. Gen. Pharmacol. **8:** 297–302.
45. Celestian, J. F., R. J. Carey & M. Miller. 1975. Unimpaired maintenance of a conditioned avoidance response in the rat with diabetes insipidus. Physiol. Behav. **15:** 707–711.
46. Brito, G. N., G. J. Thomas, S. I. Gingold & D. M. Gash. 1981. Behavioral characteristics of vasopressin-deficient rats (Brattleboro strain). Brain Res. Bull. **6:** 71–75.
47. De Wied, D. & W. H. Gispen. 1976. Impaired development of tolerance to morphine analgesia in rats with hereditary diabetes insipidus. Psychopharmacologie **46:** 27–29.

48. Bohus, B. 1979. Inappropriate synthesis and release of vasopressin in rats: behavioral consequences and effects of neuropeptides. Neurosci. Lett. (suppl.) **3:** 329.
49. Van Wimersma Greidanus, Tj. B. 1979. Neuropeptides and avoidance behavior; with special reference to the effects of vasopressin, ACTH and MSH on memory processes. *In* Central Nervous System Effects of Hypothalamic Hormones and Other Peptides. R. Collu, A. Barbeau, J.-R. Ducharme & J.-G. Rochefort.: 177–187. Raven Press. New York, N.Y.
50. Bailey, W. H. & J. M. Weiss. 1978. Effect of $ACTH_{4-10}$ on passive avoidance of rats lacking vasopressin (Brattleboro strain). Horm. Behav. **10:** 22–29.
51. Gold, P. E. & R. Van Buskirk. 1976. Effects of posttrial hormone injections on memory processes. Horm. Behav. **7:** 509–517.
52. Brito, G. N. O., G. J. Thomas, D. M. Gash & J. H. Kitchen. 1982. The behavior of Brattleboro rats. Ann. N.Y. Acad. Sci. (This volume.)
53. Pittman, Q. J., J. Rogers & F. E. Bloom. 1982. Deficits in tolerance to ethanol in Brattleboro rats. Ann. N.Y. Acad. Sci. (This volume.)

(Discussion of This Paper Begins on Page 681.)

ETHANOL PREFERENCE IN HOMOZYGOUS DIABETES INSIPIDUS (BRATTLEBORO) RATS: EFFECT OF VASOPRESSIN FRAGMENTS *

Henk Rigter † and John Crabbe ‡§

† *CNS Pharmacology Department*
Scientific Development Group
Organon International BV
Oss, the Netherlands

‡ *VA Medical Center*
Departments of Medical Psychology and Pharmacology
Oregon Health Sciences University
Portland, Oregon 97201

INTRODUCTION

Recently, peptides related to Lys-vasopressin (LVP) have been reported to affect the acceptance of alcohol solutions in normal and hypophysectomized rats.[1] Rats were forced to accept as their sole drinking fluid concentrations of alcohol in water. Concentrations were raised every two days if the animal accepted the previous concentration, or lowered if the animal refused to drink. A final acceptance concentration (FAC) was determined for each animal during 24 days of testing. All animals were then tested for their choice between water and their individual FAC of alcohol for an additional 24 days. Three doses of des-Gly-[Lys]-vasopressin (DGLVP) were given every two days.

Normal rats given DGLVP attained twice the FAC levels that control rats reached during the training period and consequently showed higher daily g/kg alcohol intake. DGLVP treatment did not affect hypophysectomized animals; they initially accepted alcohol, but within six days began to reject all but the lowest alcohol concentrations. During the free-choice testing period, alcohol intake of all groups declined but relative group positions remained similar. This experiment suggested that DGLVP treatment might have increased the rate of acquisition of alcohol drinking and, additionally, might have delayed the rate of extinction during the choice conditions.[1]

Using intact rats and a very similar design, this group subsequently studied the effects of daily s.c. injections of DGLVP, LVP, and Pro-Leu-Gly-NH_2 (PLG), the tripeptide side chain of oxytocin, on alcohol acceptance. Both DGLVP and PLG, but not LVP, enhanced ethanol acceptance when given during acquisition. In a separate study, DGLVP failed to maintain ethanol intake when given after FAC had been reached. In neither experiment were there striking effects after the animals were shifted to a free choice condition.[2]

Thus, vasopressin fragments with reduced endocrine activity [3, 4] appear to

* J. Crabbe is supported by the Veterans Administration and by National Institute of Drug Abuse Grant DA 02799.

§ Send all correspondence to J. Crabbe, Research Service (151), VA Medical Center, Portland, Oregon 97201

0077–8923/82/0394–0663 $1.75/0 © 1982, NYAS

enhance ethanol drinking under some circumstances. We sought to relate these effects to the role that endogenous vasopressin might play in the control of ethanol drinking. We chose to study rats of the Brattleboro strain homozygous for a recessive gene for diabetes insipidus. These rats are totally lacking in vasopressin.[5, 6] We report that diabetes insipidus rats showed altered ethanol drinking and that treatment with vasopressin analogues returned ethanol drinking to control levels.

METHODS

Male Brattleboro rats were purchased from TNO-Zeist, the Netherlands. The rats were 3–4 months old at the time of the experiments. Animals were homozygous for the diabetes insipidus trait (di/di), heterozygous at this locus (di/+), or homozygous-normal (+/+).

During the experiments, rats were individually housed and were provided with two drinking bottles, one containing an ethanol solution and the other tap water. Food was available *ad libitum.* Each day at noon the rats and the bottles were weighed to the nearest gram and the position of the bottles was changed. To correct for spillage of fluid during the weighing procedure, fluid loss from tap water or ethanol solution containing bottles from empty cages was averaged and subtracted from the experimental data.

In one testing paradigm, rats were offered a choice between a particular concentration of ethanol and tap water for several days. In another paradigm, rats were offered a choice between 2.2% ethanol and water for days 1 and 2, 4.6% ethanol versus water for days 3 and 4, and 10% ethanol and water for the remaining test days. The ethanol preference ratio (PR) was defined as:

$$\frac{\text{intake of ethanol solution in ml}}{\text{intake of (ethanol solution + tap water) in ml}} \times 100.$$

The following measures were calculated: (a) the average daily total fluid intake (ml); (b) the average daily ethanol intake (g); and (c) the average daily ethanol preference ratio.

In some experiments, rats were anesthetized with tribromoethanol and an Alzet® osmotic minipump (model 1701; volume = 170 μl) was implanted s.c. in the nape of the neck. The pumps were filled with a sterilized 0.9% saline solution containing LVP or des-Gly-[Arg]-vasopressin acetate (DGAVP). Peptides were synthesized by Drs. H. M. Greven and Jan van Nispen (Organon, Oss). According to factory specifications, osmotic minipumps are operational for seven days with a release rate of 1 μl/h. However, in the present studies the minipumps apparently released peptides for approximately 8–9 days. Estimated rates of release were based on this time period. Control groups were implanted with Silastic® tubing of the same size as osmotic minipumps, or minipumps containing saline. At the end of the experimental period, the rats were again anesthetized and the minipumps were removed. All implants were free of scar tissue.

Data were subjected to statistical analyses by means of Kruskal-Wallis one-way analyses of variance, followed, where appropriate, by two-tailed randomization tests for two independent samples.[7]

RESULTS AND DISCUSSION

We first tested the development of ethanol preference in the three genotypes. Five to seven rats of each genotype were offered a choice between 2.2%, 4.6%, or 10.0% ethanol and tap water for seven days. Results are given in TABLE 1. Ethanol preference scores varied considerably among groups ($H = 44.9$, $df = 8$, $p < 0.001$). The di/di rats had lower preference scores than either heterozygotes or normals, especially under the 4.6% and 10% conditions (TABLE 1). Ethanol preference was affected by ethanol concentration (all $p < 0.01$) and appeared to be inversely related to concentration within each genotype.

Total fluid intake varied significantly among groups ($H = 47.4$, $df = 8$, $p < 0.001$). Rats with diabetes insipidus displayed excessive fluid intake under

TABLE 1

THE EFFECT OF GENOTYPE ON TOTAL FLUID INTAKE, ETHANOL INTAKE, AND ETHANOL PREFERENCE

Ethanol Concentration (%)	Genotype	Average Daily Fluid Intake (ml)	Median Daily Ethanol Preference (%)	Average Daily Ethanol Intake (g)
	di/di	166.8 ± 8.3	68.2	1.9 ± 0.2
2.2	di/+	44.6 ± 4.7 ‡	81.8 *	0.6 ± 0.1 †
	+/+	38.6 ± 2.3 ‡	77.6	0.5 ± 0.1 †
	di/di	158.8 ± 7.4	12.2	0.7 ± 0.1
4.6	di/+	39.1 ± 0.4 ‡	38.1 ‡	0.7 ± 0.1
	+/+	37.2 ± 1.0 ‡	28.5 †	0.6 ± 0.2
	di/di	173.2 ± 7.2	3.6	0.6 ± 0.1
10.0	di/+	39.3 ± 1.7 †§	18.9 †	0.7 ± 0.1
	+/+	32.9 ± 0.5 †	19.9 †	0.6 ± 0.1

Mean ± SEM, except for median preference scores.

*, †, ‡: significant differences with respect to corresponding di/di group; $p < 0.05$, $p < 0.01$, $p < 0.001$, respectively, by two-tailed randomization test.

§ $p < 0.05$ versus +/+ group.

all conditions. Heterozygotes ingested slightly more fluid than normal animals but only the di/+ 10% group differed significantly from the corresponding +/+ group. Ethanol concentration affected total fluid intake in +/+ rats ($H = 7.7$, $df = 2$, $p < 0.05$) but did not influence intake in di/di or the di/+ rats. A more detailed analysis revealed that +/+ animals drank less under the 10% condition than under the 2.2% or 4.6% conditions (TABLE 1).

Groups similarly differed in average daily amount of ethanol ingested ($H = 19.3$, $df = 8$, $p < 0.02$). Although they had reduced relative ethanol preference, the excessive drinking of diabetes insipidus rats apparently led to their drinking absolute amounts of ethanol that were either higher (2.2% condition) or similar to those ingested by the di/+ or +/+ rats (TABLE 1).

Ethanol concentration did not influence the absolute amount of ethanol ingested by di/+ or +/+ animals.

In contrast, the absolute ethanol intake of di/di rats varied inversely with ethanol concentration ($H = 12.8$, df $= 2$, $p < 0.01$). The di/di rats ingested markedly more ethanol under the 2.2% condition than under the 4.6% or 10% conditions (TABLE 1). Thus, the hypothesis that ethanol preference is controlled by endogenous vasopressin was supported by the lowered preference of vasopressin-deficient di/di rats. The tendency for di/di rats to drink more alcohol on an absolute basis (although only at the lowest concentration) was surprising. We next sought to replace vasopressin in di/di rats to see if this treatment would result in a pattern of development of preference for ethanol that resembled normal rats.

Homozygotes and heterozygotes were implanted with minipumps containing 0.83 ng/μl LVP. One milligram of peptide was equivalent to 288 IU. Estimated release rate was 65 ng/h. Three hours after surgery, the animals were individually housed and fluid intake was measured for nine days. The rats were offered a choice between tap water and ethanol solution. During days 1 and 2 the ethanol concentration was 2.2%, during days 3 and 4, 4.6% and during days 5 through 9, 10%.

Median preference scores for the first six days are shown in FIGURE 1. Results for the di/di control and di/+ groups were similar to those reported above. Most importantly, di/di controls had significantly lower preference scores at all concentrations than di/+ rats. For both groups, preference declined as concentration increased (FIGURE 1). Treatment of di/di rats with LVP increased preference for ethanol at all concentrations. LVP-treated di/di rats did not differ from di/+ rats with respect to ethanol preference; both groups of rats had significantly higher preference for any of the ethanol concentrations than di/di control animals.

LVP treatment also blocked the increase in absolute ethanol consumption by di/di rats during days 1 and 2 (i.e., at the 2.2% concentration). Untreated diabetes insipidus rats consumed significantly more ethanol in absolute terms than LVP-treated di/di or di/+ rats, who did not differ from each other.

Thus far, we concluded that rats with inherited vasopressin deficiency showed low preference for ethanol. An unexplained result was the increased absolute intake of ethanol by diabetes insipidus rats at the lowest concentration studied. Ethanol drinking (preference and absolute intake) similar to normal rats was seen during chronic treatment with LVP. Since ethanol drinking disturbances in di/di rats and their reversal by LVP treatment paralleled the polydipsia in these animals, we next sought to study the effect of a vasopressin fragment with reduced antidiuretic activity.

Groups of five di/di rats received a minipump with des-Gly-[Arg]-vasopressin acetate (DGAVP) in concentrations of 0.1 μg/μl and 1 μg/μl in sterilized saline or vehicle. Two groups of 10 di/+ rats were implanted with a minipump containing saline or 1 μg/μl DGAVP. Three hours after surgery the rats were individually housed and fluid intake was recorded for 11 days. The solutions offered were ethanol (2.2% vol/vol) and tap water. Assuming that the minipumps were operational for about 8–9 days, the estimated release rates of DGAVP were 78 and 780 ng/h for the low (0.1 μg/μl) and the high (1 μg/μl) concentration of the peptide, respectively.

In the doses used, DGAVP exerted antidiuretic activity. Overall fluid intake of the di/di rats treated with the 1 μg/μl concentration of the peptide was comparable to the intake of di/+ placebo animals. The 0.1 μg/μl concentration of DGAVP was slightly less effective in reducing fluid intake. DGAVP also decreased drinking levels in di/+ rats. Ethanol preference data are given in FIGURE 2. As expected, di/di placebo rats had lower ethanol preference scores than rats from the corresponding di/+ group. DGAVP increased ethanol preference in di/di animals. From the beginning of the study, di/di rats treated with the 0.1 μg/μl concentration of the peptide had significantly higher preference scores than di/di control rats.

Preference scores tended to decline somewhat after the minipumps had ceased to release DGAVP (FIGURE 2). The median preference score for the

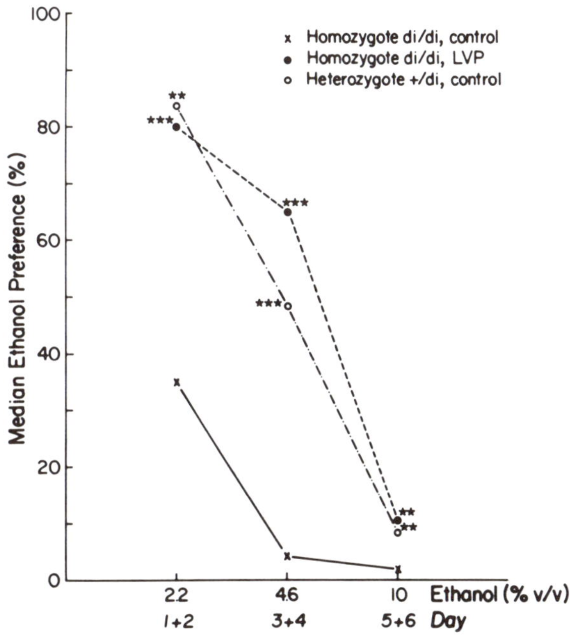

FIGURE 1. Lowered preference for ethanol in di/di rats and its reversal with LVP. Median preference ratios for ethanol are shown for successive two-day test periods. Asterisks indicate significance by two-tailed randomization test with respect to di/di control group (**, $p < 0.01$; ***, $p < 0.001$). The di/+ and di/di-LVP groups did not differ from each other.

0.1 μg/μl DGAVP di/di group for days 9–11 was 72% versus 29.5% for the di/di placebo group; this difference was no longer significant. The 1 μg/μl concentration of the peptide also enhanced ethanol preference in di/di rats, although this increase tended to be delayed in comparison with the effect produced by the low dose of the peptide. However, due to the limited number of animals and to the high variability within these groups, di/di control and 1 μg/μl DGAVP groups differed significantly only during days 6–9. The increased ethanol preference of the di/di group receiving the 1 μg/μl concentration of DGAVP markedly decreased on cessation of peptide release (FIGURE 2). In contrast to its effect in di/di rats, DGAVP tended to reduce ethanol preference in di/+ animals. This tendency was significant during days 1–5 (p's < 0.05) but not thereafter (p's > 0.10; FIGURE 2), and the di/+ groups did not differ in overall preference over the eight-day course of minipump effectiveness.

In the previous experiments it was found that di/di rats consumed a larger volume of a 2.2% ethanol solution, and therefore higher absolute amounts of ethanol than di/+ rats. This finding was confirmed in the present study. DGAVP decreased the absolute ethanol intake of di/di rats in a dose-dependent manner. Homozygous (di/di) rats treated with the 1 μg/μl concentration of

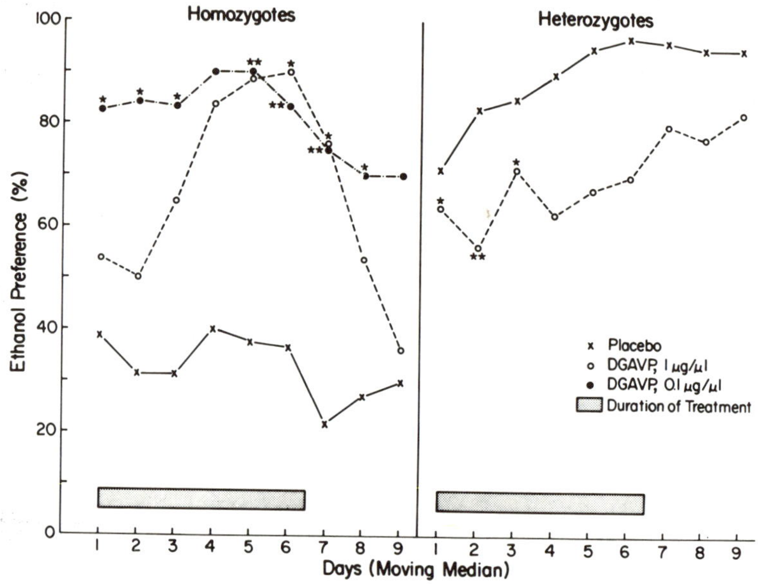

FIGURE 2. Enhancement of ethanol preference drinking in di/di, but not di/+, rats by DGAVP. Median ethanol preference for di/di (left panel) and di/+ (right panel) rats offered a choice between 2.2% ethanol and tap water for eleven days. Medians were averaged over three-day blocks. Normal duration of peptide release is shown by crosshatched bars. The di/di given 0.1 μg/μl DGAVP differed significantly from its placebo group during days 1–5 ($p < 0.05$), 5–9 ($p < 0.01$), and 8–10 ($p < 0.05$). The difference at time points 4 and 9 were not significant. The di/di group given 1 μg/μl differed from its placebo group only during days 6–9 ($p < 0.05$). The di/+ group given DGAVP differed ($p < 0.05$) from its placebo group only during days 1–5. All differences were tested with two-tailed randomization tests.

DGAVP did not differ from di/+ control rats with respect to the amount of ethanol ingested. Since others had reported that DGLVP had no effect on ethanol acceptance in the rat after acceptance had been established, we next tested DGAVP in animals that had already developed stable ethanol preference. Homozygotes and heterozygotes were given a choice between tap water and 2.2% ethanol (vol/vol). After an 11-day period during which ethanol preference scores stabilized, the animals were subjected to surgery and implanted with a minipump containing either sterilized saline or 1 μg/μl DGAVP in sterilized saline. After implantation of the minipumps, the rats were tested for ethanol preference for another 11 days.

DGAVP in a concentration of 1 $\mu g/\mu l$ normalized overall fluid intake of di/di rats: i.e., intake levels of the peptide-treated di/di and the placebo-treated di/+ groups were similar. Again, DGAVP also reduced overall fluid intake in the di/+ animals. DGAVP did not affect ethanol preference in either di/di or di/+ rats: ethanol preference scores remained low in di/di animals and high in the di/+ animals. As seen previously, however, the peptide reduced the absolute amounts of ethanol ingested. Even though DGAVP enhanced the development of ethanol preference in di/di rats in the previous experiment, it failed to affect an established preference in this study.

Finally, we wondered whether the reduced ethanol preference and increased absolute intake of ethanol in di/di rats could be idiosyncratic to the Brattleboro strain we studied. We obtained Roman High Avoidance (RHA) rats of similar age from Dr. Carl Hansen of the National Institutes of Health. Through a regimen of repeated backcrossing, he had introduced the *di* gene into this rat strain, which had been selectively bred for superior avoidance learning. We tested RHA/N-di/di and RHA/N-+/+ rats (homozygous diabetes insipidus and normal, respectively) for preference for 2.2% ethanol, 4.6% ethanol, and 10% ethanol during days 1–2, 3–4, and 5–9, respectively. Half the animals in each genotype were implanted with minipumps containing DGAVP (1 $\mu g/\mu l$) and half with minipumps containing saline. Results were virtually identical to those reported for Brattleboro rats. RHA/N-di/di rats had lower preference and higher absolute ethanol intake (at 2.2%) than normal RHA rats. DGAVP increased preference and reduced ethanol g/day intake. We conclude that the effects of the *di* gene on preference and the effects of replacement therapy with DGAVP are to at least some degree independent of the genetic background on which the gene is placed.

Others have found that ethanol acceptance was enhanced in normal rats by treatment with DGLVP.[3] However, acceptance was not modified by daily injections of LVP.[4] Further, the effectiveness of peptide treatment was limited to the period during development of ethanol acceptance.[2] We found a similar result with DGAVP in our studies of ethanol preference. Mucha and Kalant[2] suggested several possible reasons why peptide treatment was only effective during the period of initial exposure to ethanol in their study, including habituation to the aversive effects of alcohol, learning processes during the development of drinking, and positive reinforcement. These factors could plausibly have been altered in the studies we report by treatment with peptides related to vasopressin. All of these factors related to learning, motivation, and memory are known to be affected by vasopressin and related peptides.[5] However, neither their studies nor ours offer clear indications of which, if any, of these factors may have been affected.

The increased absolute ethanol intake by di/di rats at the 2.2% concentration of ethanol is an intriguing finding. Inability to taste such a low ethanol concentration may be rejected as a possible explanation because the lowered preference of di/di rats argues for their ability to discriminate between the two offered fluids. It is possible that di/di rats will accept ethanol in absolute quantities up to a certain threshold value. At this level, taste or the intoxicating effects could signal the animals to stop drinking. If very low concentrations of ethanol provide insufficient cue to di/di rats, or if they fail to learn of ethanol's aversive properties quickly enough to avoid drinking ethanol during the initial presentation of the drug, higher absolute intake could result. Such an hypothesis remains to be rigorously tested.

Our results with vasopressin-deficient di/di rats would seem to indicate a role for endogenous vasopressin in the control of intake of and preference for ethanol. This interpretation is ventured with caution, however. As many papers at this conference have amply documented, the endocrine status of Brattleboro rats is considerably more complicated than simply the absence of vasopressin in di/di rats. Brattleboro di/di rats also lack arginine vasotocin [8] and store less oxytocin in posterior pituitary than di/+ and +/+ rats.[9] Homozygous diabetes insipidus rats also have elevated plasma levels of oxytocin.[10]

That exogenously administered vasopressin-related peptides reversed the impaired ethanol drinking of di/di rats seems strong support for a role for endogenous vasopressin. However, we were unable to increase preference at doses of DGAVP that did not also reduce total fluid consumption to normal levels, presumably through an antidiuretic action. Restoration of normal water balance in di/di rats with exogenous vasopressin also normalizes the oxytocin content of the neurohypophysis.[11] Thus, we cannot ascribe the enhancing effects of LVP or DGAVP on preference drinking to an effect of these agents within the central nervous system. Indeed, the failure of others [1] to affect impaired ethanol acceptance drinking in hypophysectomized rats with DGAVP also cannot be used to support a specific role for vasopressin. The possibility remains that both the impairments and reversals of impairments by vasopressin treatment so far reported are due to realignments of the complex neuroendocrine balance that have yet to be identified. On the other hand, in the studies we report here, when preference was already established, peptide replacement effective in restoring water balance was not able to enhance ethanol preference. This suggests that further work may allow us to dissociate vasopressin effects on fluid balance from those on ethanol drinking.

References

1. Finkelberg, F., H. Kalant & A. Le Blanc. 1978. Effect of vasopressin-like peptides on consumption of ethanol by the rat. Pharmac. Biochem. Behav. **9:** 453–458.
2. Mucha, R. & H. Kalant. 1979. Effects of desglycinamide (9)-Lysine (8)-vasopressin and Prolyl-Leucyl-Glycinamide on oral ethanol intake in the rat. Pharmac. Biochem. Behav. **10:** 229–234.
3. De Wied, D., H. Greven, S. Lande & A. Witter. 1972. Dissociation of the behavioural and endocrine effects of lysine vasopressin by tryptic digestion. Brit. J. Pharmacol. **45:** 118–122.
4. Walter, R., P. Hoffman, J. Flexner & L. Flexner. 1975. Neurohypophyseal hormones, analogs and fragments: their effect on puromycin-induced amnesia. Proc. Natl. Acad. Sci. USA **72:** 4180–4184.
5. Rigter, H. & J. Crabbe. 1979. Modulation of memory by pituitary hormones and related peptides. Vitamins & Hormones **37:** 153–241.
6. Valtin, H., H. Schroeder, K. Bernischke & H. Sokol. 1962. Familial hypothalamic diabetes insipidus in rats. Nature **196:** 1109–1110.
7. Siegel, S. 1956. Nonparametric Statistics. McGraw-Hill Book Co., Inc. New York, N.Y.
8. Rosenbloom, A. & D. Fisher. 1975. Radioimmunoassayable AVT and AVP in adult mammalian brain tissue: comparison of normal and Brattleboro rats. Neuroendocrinol. **17:** 354–361.

9. VALTIN, H., W. H. SAWYER & H. W. SOKOL. 1965. Neurohypophysial principles in rats homozygous and heterozygous for hypothalamic diabetes insipidus (Brattleboro strain). Endocrinol. **77:** 701–706.
10. DOGTEROM, J., T. J. VAN WIMERSMA GREIDANUS & D. SWAAB. 1977. Evidence for the release of vasopressin and oxytocin into cerebrospinal fluid: measurements in plasma and CSF of intact and hypophysectomized rats. Neuroendocrinol. **24:** 108–118.
11. VALTIN, H., H. W. SOKOL & D. SUNDE. 1975. Genetic approaches to the study of the regulation and actions of vasopressin. Rec. Prog. Horm. Res. **31:** 447–479.

(DISCUSSION OF THIS PAPER BEGINS ON PAGE 681.)

BEHAVIORAL MODIFICATION IN BRATTLEBORO RATS DUE TO VASOPRESSIN ADMINISTRATION AND NEURAL TRANSPLANTATION *

Don M. Gash,† Philip H. Warren, Leslie B. Dick, John R. Sladek, Jr., and James R. Ison

Department of Anatomy
Center for Brain Research
Department of Psychology
University of Rochester
Rochester, New York 14642

INTRODUCTION

Three distinct anatomical pathways are known by which vasopressin may directly influence the central nervous system (CNS). Blood-borne vasopressin circulates through the choroid plexuses, the brain capillaries and larger blood vessels. Cerebrospinal fluid, circulating through and around the brain and spinal cord, contains moderate levels of vasopressin.[1] Axons containing vasopressin project from magnocellular neurons in the hypothalamus to many extra hypothalamic sites, including the central amygdala, nucleus of the solitary tract, dorsal motor nucleus of the vagus, and spinal cord.[2] Buijs and Swaab have shown,[3] that at least for the amygdala, these axons terminate in peptide-containing synapses, which do not differ in their ultrastructure from the classical neurotransmitter-containing synapses.

Vasopressin, due to its wide distribution throughout the CNS, is in a position to influence many neural activities. Studies over the past decade have implicated vasopressin involvement in functions ranging from the regulation of brain fluid and electrolyte homeostasis[4] to facilitation of memory and learning.[5, 6] However, by no means has a complete listing of the role(s) of vasopressin in CNS function been formulated. Comparison of behavior between normal animals and rats homozygous for diabetes insipidus (DI or Brattleboro rats) affords an excellent opportunity to dissect out those behaviors affected by vasopressin. Our research group has conducted an extensive series of tests to determine behavioral deficits in Brattleboro rats. Our studies on learning and memory are reported elsewhere[7] (see also Brito, Thomas, Gash, and Kitchen, this volume). This present paper discusses modification of drinking behavior and hyperresponsiveness by vasopressin in homozygous Brattleboro rats.

DRINKING BEHAVIOR

Drinking is a complex behavior that requires an extensive CNS involvement and interaction with the oral-pharyngeal region and kidneys.[8, 9] The major

* Supported by National Institutes of Health Grants NS 15901 and NS 17543, and a Biomedical Research Support Grant from the University of Rochester School of Medicine and Dentistry.

† Send correspondence to: Don M. Gash, Department of Anatomy, Box 603, University of Rochester, Rochester, New York 14642.

0077–8923/82/0394–0672 $1.75/ © 1982, NYAS

stimulus for drinking seems to be cellular dehydration, which stimulates osmotically sensitive neurons in the preoptic area and hypothalamus. The paraventricular preoptic-hypothalamic region has been shown by the experiments of Johnson and Buggy [10] to be very important in drinking behavior. Stereotaxic lesions in this region produce adipsia without affecting other behavior. Lesioned animals show an increase in plasma osmolality to about 440 mOsm on the third day after surgery but fail to reduce their urine volume or substantially increase urine osmolality. Other studies have shown (for a review, see Reference 8) that water passing through the oral-pharyngeal region stimulates neurons in the preoptic and lateral hypothalamus, and the responsiveness of these neurons seems to depend on the extent of water deprivation. These studies indicate that the preoptic area and hypothalamus play a critical role in the initiation and maintenance of drinking. One response to water deprivation is the release of vasopressin by the posterior pituitary into the systemic circulation. Vasopressin then acts on the distal convoluted tubules of the kidney to allow free water reabsorption into the blood and thus concentration of the urine.[11]

We have examined the ability of chronically infused vasopressin to alleviate the symptoms of diabetes insipidus in homozygous Brattleboro rats. Mini-osmopumps (Alzet 2001, Alza, Palo Alto, Calif.) containing either arginine vasopressin (Bachem, Torrance, Calif.) or physiological saline were subcutaneously implanted above the scapula of young adult male rats of the Brattleboro strain (Blue Spruce Farms, Altamont, N.Y.). In the first study, test animals received titers of 0.001 μg/μl ($N = 6$), 0.01 μg/μl ($N = 6$), 0.1 μg/μl ($N = 10$), or 1.0 μg/μl ($N = 7$) of vasopressin dissolved in saline and continuously delivered at the rate of 1 μl/h. Control animals received physiological saline. Water consumption and urine osmolality were monitored for the three days prior to and again on the third day after pump implantation. The results were statistically evaluated by an analysis of variance. Vasopressin caused a dose-dependent decrease in water consumption (FIGURE 1A) and increase in urine osmolality. A maximum response to vasopressin was seen with the infusion of 0.1 μg vasopressin/h. Increasing the titer of infused vasopressin tenfold did not result in a significant additional effect on water consumption or urine osmolality.

It was noted that the water intake was only reduced to .15 ml/g body weight and urine osmolality increased to 1450 mOsm/l in the groups in which vasopressin was most effective, compared to values of .11–.17 ml/g body weight and 2130–2159 mOsm/l in normal adult rats. Because even the highest titers of vasopressin did not bring the animals up to normal levels of urine osmolality within the three-day time period, a longer term study was conducted. Seven male Brattleboro rats received constant infusions of 1 μg/μl vasopressin at the rate of 1 μl/h for 10 days (FIGURE 2). While water consumption fell to within normal limits by the second day after implantation, it took six days for urine osmolality to approach normal levels. Urine osmolality of the treated DI rats did not quite reach the levels seen in our Long-Evans rats, used as controls, for normal urine osmolality levels during the 10-day period of infusion. In an early study on the DI rat, Harrington and Valtin [12] found that it took up to 28 days for vasopressin tannate (a vasopressin analogue with a long half-life in the blood) injected subcutaneously at the rate of 1 unit/day to bring urine osmolality up to normal levels. However, using a similar protocol, Balment, Henderson, and Oliver [13] found that daily subcutaneous injections of vasopressin tannate could increase urine osmolality to normal levels within four to five days.

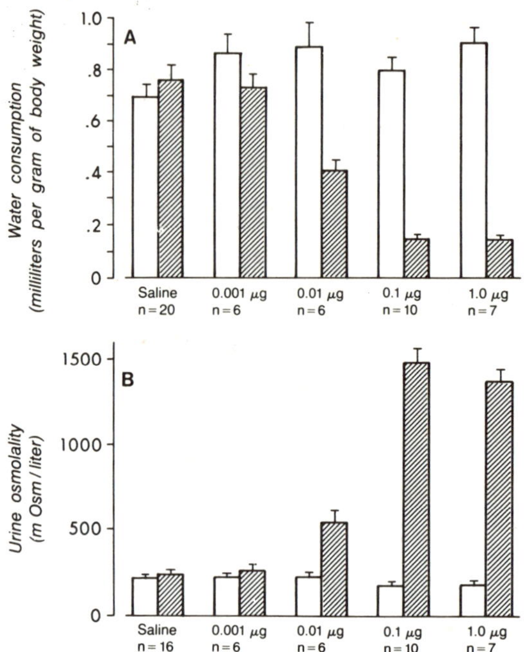

FIGURE 1. (A) The dose-dependent increase in water consumption due to the chronic subcutaneous administration of vasopressin is demonstrated here. Comparing water consumption in the three-day period prior to pump implants (clear bars) to consumption on the third day of infusion (dark bars) shows that titers of 0.01 μg/μl, 0.1 μg/μl, and 1.0 μg/μl of vasopressin significantly reduced consumption ($p \leq 0.001$). The standard error of the mean is shown for each group. (B) Changes in urine osmolality were inversely correlated with the decreases in water consumption ($r = -.848$). Average urine osmolality for the three-day period prior to pump implantation (clear bars) compared to osmolality on the third day of infusion (dark bars) shows that titers of 0.01 μg/μl, 0.1 μg/μl, and 1.0 μg/μl vasopressin led to significantly ($p \leq 0.001$) more concentrated urine. The standard error of the mean is shown for each group.

The data reported in the latter paper are qualitatively and quantitatively similar to our results from chronic vasopressin infusion. Collectively these studies indicate that the kidney, after a long term absence of vasopressin, cannot immediately restore normal water balance upon stimulation by even high titers of exogenous vasopressin; the process requires several days to weeks.

DRINKING BEHAVIOR FOLLOWING NEURAL GRAFTS

Our research group has examined[14-16] the antidiuretic influences of transplanted CNS tissue upon young adult DI rats. When the third ventricle is used as a site for transplantation, grafts of anterior hypothalami from 17- and 19-day post-coitus normal rat fetuses significantly ameliorate the symptoms of diabetes insipidus (FIGURE 3). In approximately 20% of these graft recipients sustained decreases in water consumption and increases in urine osmolality up to 1,000 mOsm/kg were observed. In contrast, the surgical procedures and transplantation of neural tissue from the fetal occipital cortex exerted only transient effects on drinking and urine concentration.

Magnocellular vasopressin neurons have been immunocytochemically identified within the transplants of the fetal anterior hypothalamic tissue, with their axonal processes coursing to blood vessels within the graft and the host median eminence (14, 15 and Sladek, Schöler, Notter and Gash, this volume). Vasopressin, at about 3% of the levels seen in normal animals,[16] has been measured by radioimmunoassay of Palkovits-type punches[17] taken from transplants.

The site of transplantation appears to be important. When the anterior hypothalami from 19- and 21-day normal fetuses were transplanted into the lateral ventricle of young adult male rats ($N = 12$), no marked change in host urine osmolality or water consumption was observed (FIGURE 4). A slight decrease in water intake in the 48 hours following grafting was attributed to the effects of surgery. Transplants taken for histological and immunocytochemical evaluation 20 days after implantation were found associated with the ventricular wall and choroid plexus of the lateral ventricle. The grafts were well vascularized. Most neural perikarya were of the parvocellular size range, that is 12–14 μm in diameter. Magnocellular neurons were rare. Only a few neurophysin-containing neurons and fibers were seen and the immunocytochemically stained axonal processes remained within the boundaries of the graft. These data suggest that features/factors found within the median eminence region of the hypothalamus are important for the functional development of vasopressin neurons. It may be that the presence of numerous fenestrated

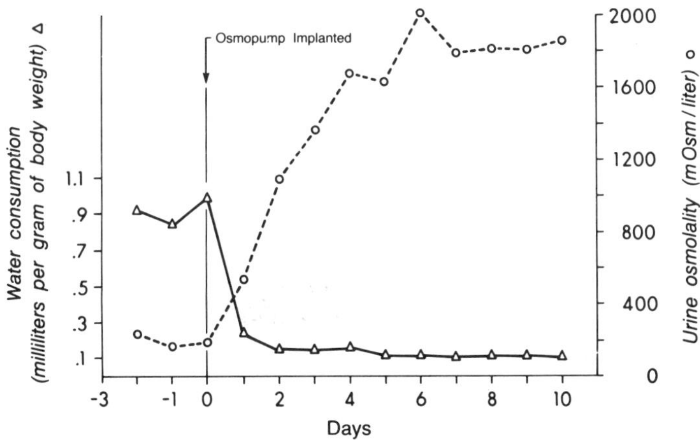

FIGURE 2. Changes in water consumption and urine osmolality are shown for seven Brattleboro rats receiving chronic infusions of 1 μg/μl vasopressin. Intragroup differences after osmopump implantation were minimal, with the standard error of the mean ≤ 0.02 ml/g body weight for water consumption and ≤ 82 mOsm/kg for urine osmolality for each day studied. Ranges for water consumption and urine osmolality in age-matched young male Long-Evans rats in our colony are, respectively, 0.11 to 0.17 ml H_2O/g body weight and 2130 to 2559 mOsm/kg.

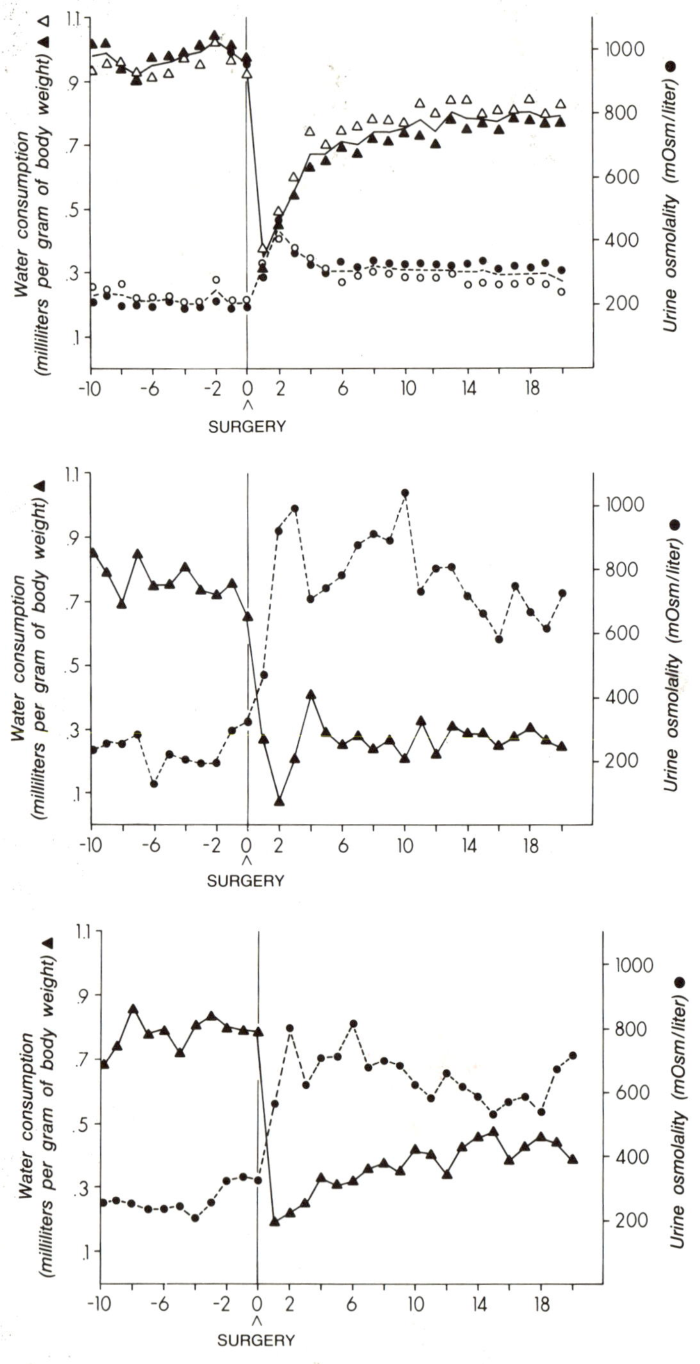
Water consumption (milliliters per gram of body weight) ▲ △
Urine osmolality (mOsm/liter) ●
1.1
.9
.7
.5
.3
.1
1000
800
600
400
200
-10
-6
-2
0
2
6
10
14
18
SURGERY
Water consumption (milliliters per gram of body weight) ▲
Urine osmolality (mOsm/liter) ●
SURGERY
Water consumption (milliliters per gram of body weight) ▲
Urine osmolality (mOsm/liter) ●
SURGERY
DAYS

capillaries of the hypothalamo-neurohypophyseal portal system in the median eminence provides an optimal environment for the establishment of appropriate neural-hemal contacts by the grafted neurons. Other factors may be involved and it is clear that this area of transplantation neurobiology requires further study.

The antidiuresis observed following transplantation into the third ventricle along with the immunocytochemical and radioimmunoassay identification of vasopressin within the graft is consistent with the interpretation that the transplanted neurons are releasing biologically active vasopressin into the host's circulation and affecting kidney function. This then indicates that the transplants are capable of classical neuroendocrine activity. Direct interactions between the grafts and the host CNS may also influence drinking. Transplants in the third ventricle are juxtaposed with the paraventricular hypothalamic preoptic region which is intimately involved in drinking behavior and thirst. The boundary between grafted tissue and the host brain is often hard to distinguish.[14, 16] Axons have been identified coursing through the graft-host interface: vasopressin-containing fibers from the transplant and catecholamine-containing fibers of host origin. Vasopressin neurons in the transplant are contacted by catecholamine varicosities.[16] Thus an anatomical basis for host CNS control of the transplant exists. Similarly, drinking and regulation of the water balance may be affected by neural input from the graft with the establishment of neural connections between the graft and host being extremely important in development of an appropriate transplant function.

Reflex Behavior: The Startle Response

While drinking is the most obvious behavior affected in DI rats, there have been many reports that animals without vasopressin show other altered be-

FIGURE 3. Effects of surgery and transplantation in the control and experimental rats. (A) The average daily consumption of water by the sham-operated rats (△, $N = 11$) and the control tissue-implanted rats (▲, $N = 18$) is indicated by the solid line. The average osmolality of the urine is plotted on the same graph. Since we did not measure osmolality for all the controls, the sample size is smaller; $N = 18$ for the sham-operated rats (○) and $N = 10$ for the rats with occipital cortex implants (●). The standard error for each daily average was ± 0.04 ml for water consumption and ± 43 milliosmoles for urine osmolality. In some of the experimental animals [(B) a rat that received vasopressin neurons from a 17-day-old fetus and (C) a rat that received such tissue from a 19-day-old fetus], there were sustained decreases in water consumption and sustained increases in urine osmolality. Changes of these magnitudes were never observed in the control animals. (From Gash *et al.*[16] With permission from *Science*. Copyright 1980 by the American Association for the Advancement of Science.)

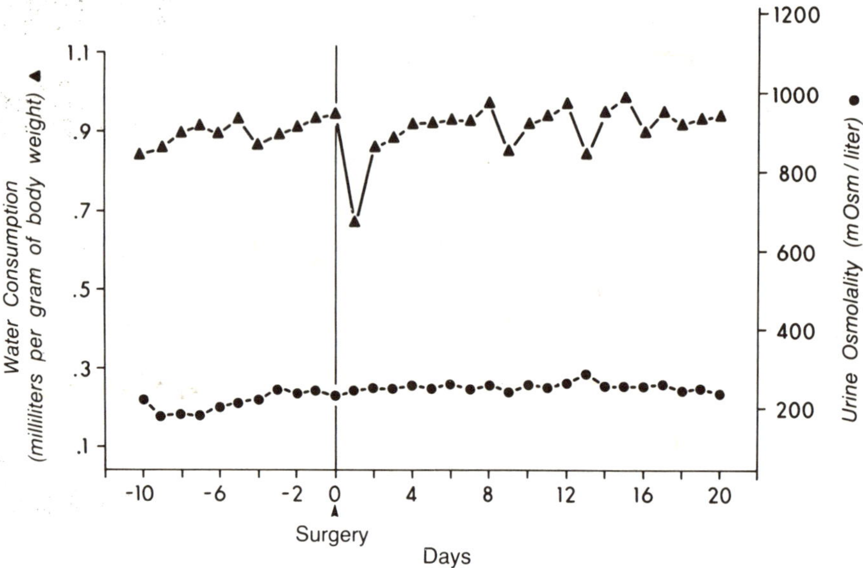

FIGURE 4. Changes in water consumption and urine osmolality are shown for twelve adult male Brattleboro rats that received transplants of fetal hypothalamic tissue into the lateral ventricle ($N = 12$). Urine osmolality appeared unaffected by the surgery; water consumption data showed a small post-surgical effect but returned to pre-surgical levels within three days of tissue implantation.

havioral patterns.[5-7] Since most of these studies have employed stressful stimuli such as electrical shock, an issue to be addressed is the responsiveness of DI rats to noxious stimuli as compared to the responsiveness of other animals. We have addressed this issue by using a sensitive quantitative test, based on the startle response,[18-20] to measure responsiveness to tail shock in DI, heterozygous for diabetes insipidus (HZ), and Long-Evans rats (LE).

In these tests, animals are placed in a restraining cage on a semi-rigid shelf attached to a Statham accelerometer. Cutaneous electrodes are attached to the tail with tail resistance uniformly reduced to 20K. The amplitude of the startle jump reflex elicited by an electrical shock is recorded, measured as amplified and rectified accelerometer output integrated over a 100 msec time period following the shock with the reading displayed on a digital voltmeter. Data are evaluated by analysis of variance. In the first study,[21] 11 DI and 10 HZ age-matched young male rats were presented with shock intensities ranging from 200 μA to 1600 μA in 200 μA increments, these intensities randomly intermixed with two trials at each intensity. The DI animals showed a significantly ($p \leq 0.01$) higher startle amplitude to the middle four shock-intensities of 600, 800, 1000, and 1200 μA. These results indicated that animals lacking vasopressin are hyperresponsive to electrocutaneous stimulation.

To determine if exogenous vasopressin could reverse the hyperresponsiveness, the following experiment was conducted using young male DI ($N = 12$) and LE ($N = 12$) rats. In the first test, half of each of the DI and LE rats received s.c. injections of 1 μg vasopressin (Bachem, Torrance, Calif.); the other half received injections of the vehicle, 0.2 ml physiological saline. All animals were tested 1 h later for their startle response to 10 repetitions of 800 μA shock. One week later, the animals were retested under reverse conditions so that animals which had received vasopressin in the first trial were given saline for the second test and saline-treated animals given vasopressin. In contrast to the powerful effects of vasopressin treatment in water consumption and urine osmolality, this experiment failed to find any effect of the injection in startle behavior in DI rats. The saline-treated DI rat was again hyperresponsive compared to the normal LE, and vasopressin did not alter its behavior.

These results indicate that Brattleboro male rats are hyperresponsive to electric shock in comparison to HZ and LE animals. Our findings suggest caution in evaluating those behavioral tests which are designed to measure memory consolidation and retrieval in these animals (such as conditioned active avoidance and conditioned passive avoidance) but which employ noxious stimulation to motivate or reinforce animal behavior. The results may be a reflection of some performance variable other than memory, such as increased responsiveness to the stimuli involved in the experiment.

Conclusions

The mutant DI rat exhibits a number of behavioral differences from HZ and LE rats. The present paper discusses two of these behavioral variances,

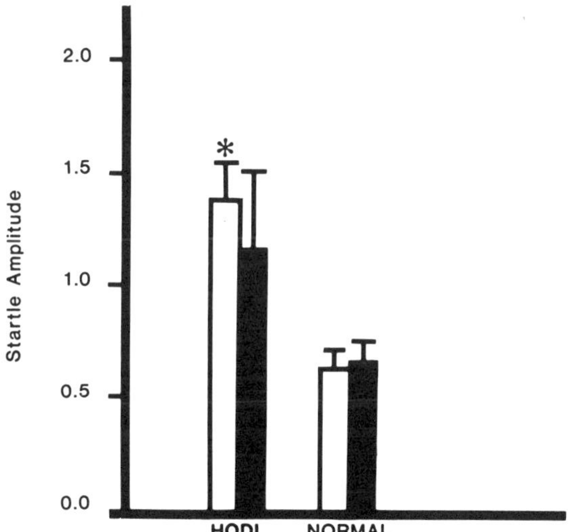

FIGURE 5. The startle response to electrical shock after acute vasopressin administration. The amplitude of the startle response for DI and normal Long-Evans rats is shown one hour following either saline (open bars) or 1 μg vasopressin (black bars) administration. The startle amplitude in saline-treated DI rats (*) was significantly different ($p \leq 0.005$) from saline and vasopressin-treated normal animals. The standard error of the mean is shown for each group.

drinking and startle responsiveness to electrical shock. Exogenous vasopressin injections normalize drinking behavior, but, at least in the initial studies, does not repair the exaggerated startle response. Additionally, a new surgical technique for investigating the role of vasopressin in behavior is described. Vasopressin neurons can be grafted into areas of the brain believed to be involved in vasopressin-modified behavior. The neural connections of the transplant with the host CNS may be determined by immunocytochemistry and fluorescence. Our previous studies have shown that vasopressin neurons, grafted into the third ventricle of adult DI rats ameliorate the symptoms of diabetes insipidus. The influence of transplants on other behaviors remains to be investigated.

Acknowledgments

We thank Susan Connor, Arnold Campo, and Robert Coopersmith for technical assistance, and Louise Behrens for typing the manuscript. We also thank Dr. Garth Thomas for his helpful insights into Brattleboro rat behavior.

References

1. Luerssen, T. G. & G. L. Robertson. 1980. Cerebrospinal fluid vasopressin and vasotocin in health and disease. *In* Neurobiology of Cerebrospinal Fluid. J. H. Wood, Ed. **1:** 613–623. Plenum Publishing Co. New York, N.Y.
2. Sofroniew, M. V., A. Weindl, U. Schrell & R. Wetzstein. 1981. Immunohistochemistry of vasopressin, oxytocin, and neurophysin in the hypothalamus and extrahypothalamic regions of the human and primate brain. Acta Histochemica Suppl. **24:** 79–95.
3. Buijs, R. M. & D. F. Swaab. 1979. Immuno-electron microscopic demonstration of vasopressin and oxytocin synapses in the limbic system of the rat. Cell Tissue Res. **204:** 355–365.
4. Raichle, M. E. 1981. Hypothesis: a central neuroendocrine system regulates brain ion homeostasis and volume. *In* Neurosecretion and Brain Peptides. J. B. Martin, S. Reichlin & K. L. Bick, Eds.: 329–336. Raven Press. New York, N.Y.
5. de Wied, D., B. Bohus & T. J. B. van Wimersma Greidanus. 1975. Memory deficit in rats with hereditary diabetes insipidus. Brain Res. **85:** 152–156.
6. Rigter, H. & J. C. Crabbe. 1980. Modulation of memory by pituitary hormones and related peptides. Vitamins & Hormones **37:** 153–241.
7. Brito, G. N. O., G. J. Thomas, S. I. Gingold & D. M. Gash. 1981. Behavioral characteristics of vasopressin-deficient rats (Brattleboro strain). Brain Res. Bull. **6:** 71–75.
8. Rolls, B. J., R. J. Wood & E. T. Rolls. 1980. Thirst: the initiation, maintenance and termination of drinking. Prog. Psychobiol. Physiol. Psych. **9:** 263–321.
9. Ramsay, D. J. & W. F. Ganong. 1980. CNS regulation of salt and water intake. *In* Neuroendocrinology. D. T. Kreiger & J. C. Hughes, Eds.: 123–129. Sinauer Associates. Sunderland, Mass.
10. Johnson, A. K. & J. Buggy. 1978. Periventricular preoptic-hypothalamus is vital for thirst and normal water economy. Am. J. Physiol. **234:** R122–R129.
11. Sawyer, W. H. 1974. The mammalian antidiuretic response. Handb. Physiology Sec. 7, Vol. IV, Part 1: 443–468.
12. Harrington, A. R. & H. Valtin. 1965. Vasopressin effect on urinary con-

centration in rats with hereditary hypothalamic diabetes insipidus (Brattleboro strain) Proc. Soc. Exp. Biol. Med. **118:** 448–450.

13. Balment, R. J., I. W. Henderson & J. A. Oliver. 1975. The effects of vasopressin on pituitary oxytocin content and plasma renin activity in rats with hypothalamic diabetes insipidus (Brattleboro strain). Gen. Comp. Endocrinology **26:** 468–477.
14. Gash, D. M. & J. R. Sladek, Jr. 1980. Vasopressin neurons grafted into Brattleboro rats. Viability and activity. Peptides **1:** 11–14.
15. Gash, D. M., J. R. Sladek, Jr. & C. D. Sladek. 1980. Functional development of grafted vasopressin neurons. Science **210:** 1367–1369.
16. Gash, D. M., C. D. Sladek & J. R. Sladek, Jr. 1980. A model system for analyzing functional development of transplanted peptidergic neurons. Peptides **1** (Suppl. 1): 125–134.
17. Palkovits, M. 1973. Isolated removal of hypothalamic and other brain nuclei of the rat. Brain Res. **59:** 449–450.
18. Davis, M. 1980. Neurochemical modulation of sensory-motor reactivity: acoustic and tactile startle reflexes. Neurosci. Biobehav. Rev. **4:** 241–263. 1980.
19. Hoffman, H. S. & J. R. Ison. 1980. Reflex modification in the domain of startle: I. Some empirical findings and their implications for how the nervous system processes sensory input. Psychol. Rev. **87:** 175–189.
20. Warren, P. H. & D. M. Gash. 1981. Vasopressin rats are hyperresponsive. (Submitted for publication.)
21. Gash, D. M. & P. H. Warren. 1981. Brattleboro rats, which congenitally lack vasopressin, are hyperresponsive to electrical shock and acoustical stimulation. Soc. Neurosci. Abstr. **7:** 30.

Discussion of the Three Preceding Papers

L. B. Kinter (*Smith Kline and French Laboratories, Philadelphia, Pa.*): I am curious about your suggestion that the application of pituitary hormones in the third ventricle might perhaps reduce thirst. Is that the impression you were trying to create?

Gash: What I was referring to is that lesions in the paraventricular preoptic and hypothalamic area have an effect on drinking. We were putting tissue in there that could have an effect upon drinking just by being there and perhaps interacting with this area of the hypothalamus, which is very strongly involved with drinking behavior. It is not the application of vasopressin as such, but the fact that neural tissue was there and could be interacting with the area that regulates drinking.

Kinter: That qualifies my concern. It has been suggested in the past that perhaps these rats overdrink, and if somehow water availability were curtailed one could get them to come into balance at a lower level of water intake. I tried to do that and failed, and I want to hear if anybody else has ever demonstrated that these animals could come into balance with reduced fluid turnover.

Gash: Dr. Valtin and others have reported that DI rats can concentrate their urine, but that they also lose tremendous amounts of weight, when deprived of water. We don't find significant weight loss in rats with transplants and reduced water consumption.

KINTER: Can you implant these cells, perhaps in the peritoneal cavity, where they would be closer to the peripheral circulation?

GASH: Yes, we are now placing tissue under the kidney capsule. We really don't have enough data to report on these studies yet.

H. W. SOKOL (*Dartmouth Medical School, Hanover, N.H.*): I was actually very intrigued with the differential functional response you observed depending upon whether the implants were placed in the lateral ventricle or in the third ventricle. It seemed to indicate to me that the implantation site determined the function of that implant. When placed in the third ventricle the cells are at least a little closer to their normal location. Taking into consideration the neural synapses (as demonstrated earlier by J. R. Sladek) occurring between transplant and host tissue, it is perhaps possible that normal stimuli that induce the release of AVP may be able to stimulate these transplanted neurons.

GASH: We are not the only group that is looking at the lateral ventricle. At the time we were beginning to do transplants into the third ventricle Dr. Perlow and Dr. Wyatt at the National Institute of Mental Health were also implanting hypothalamic tissue into Brattleboro rats. They chose the lateral ventricle as their site and found no antidiuretic effects in any of the 36 animals bearing grafts. We were fortunate in choosing the third ventricle as a site for transplantation in our initial experiments. Fenestrated capillaries of the hypothalamic-neurohypophyseal portal system are located there, so that in many ways it's the ideal site for the correct neuro-hemal contacts to be established. This may be why our transplants are functional.

R. BODNAR (*Queens College, SUNY, New York, N.Y.*): The hyperalgesic effects we have shown (in the poster session) in the DI rat seem to be consistent with the increase in the amplitude of the startle jump reflex that you get following electric shock. However, you said that AVP has no effect on the amplitude of the shock response. If AVP or dDAVP is given repeatedly to a DI rat so as to reduce water turnover, we found the reinstatement of the normal pain response on flinch jump tests. What was your dose regimen of AVP?

GASH: We have tested chronic infusions of vasopressin (0.1 μg/h and 1.0 μg/h) released by Alzet osmopumps implanted subcutaneously into the Brattleboro for at least a 3–5 day period. This certainly brings the animal back to approximately normal water balance. The 0.1 μg dose showed a trend toward reducing the hyperresponsiveness; the 1.0 μg dose actually potentiated the response so that the animals were even more hyperresponsive. We are now trying to interpret that data.

M. MILLER (*Veterans Administration Medical Center, Syracuse, N.Y.*): I would like to preface my comment with the statement that the animals used in our studies have all been produced by mating DI males with heterozygous females. This may perhaps account for some of the differences in my results compared to those reported by Dr. Greidanus. FIGURE 1 shows data from a passive avoidance model using three different foot shock intensities, 0.5–1.5 mAmp, in which homozygous DI rats are compared to normal Long-Evans rats. The latency responses do indeed vary with the intensity of the foot shock; however, there is no difference at any given shock intensity between the latency response of normal animals and that of animals with vasopressin deficiency.

FIGURE 2 shows data from a study based on an active avoidance situation, again comparing normal Long-Evans rats from our colony with rats with diabetes insipidus. What I wish to point out here is a performance that once again, if anything, is better in the vasopressin-deficient rats than in the normal rats. The basis for the difference in the observations that we have made in contrast to those from Dr. Greidanus and his colleagues is difficult to resolve except that there may well be significant differences in the DI rats that have been bred in separate colonies over the years. This may be a consequence of variations in breeding techniques and perhaps in some other factors that are less identifiable. In any case, I think this raises some question as to the physiologic role vasopressin may play in a variety of behavioral phenomena.

W. BAILEY (*Rockefeller University, New York, N.Y.*): One of the things that I would like to point out is the need for the distinction in the behavioral studies between primary and secondary effects of the deficiency of vasopressin. By primary, I mean those effects that might be attributed to a direct effect of vasopressin at some hypothetical CNS site. By secondary factors, I mean those

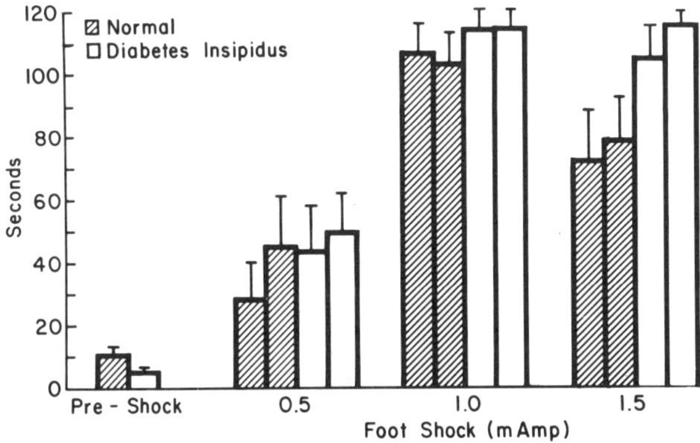

FIGURE 1. Passive avoidance performance following prior exposure to increasing intensity of foot shock. The latency interval (seconds) for movement of the rats from an illuminated chamber to a dark chamber is represented by the height of the vertical bars. Each pair of bars at each shock level indicates the values for the first and second retention tests. There is progressive lengthening of the latency interval with increase in shock intensity and the interval for DI rats is significantly greater than that for normal rats on the second retention test ($p < 0.05$, two-way ANOVA).

indirect and often compensatory changes that are secondary to the chronic loss of vasopressin. Among the factors that might be expected to affect the behavior of the animals in avoidance tests are those that we have reviewed at this conference, like the synthesis and release of oxytocin, alterations in adrenal steroid synthesis, alterations in DI rat metabolism, electrolyte balance, increased sympathetic activity, and the development of potassium-dependent renal pathology. Some of the differences between the deWied, Bohus, and van Wimersma Greidanus groups and our own with regard to passive avoidance might be that we use older animals in which you would expect the development of kaliopenic nephropathy in the kidney to be greater than in younger rats. Another possibility, although it appears not to be supported by some of the data presented at this meeting, is that DI rats are affected by chronic undernutrition in the perinatal period and by the long term effects of mild dehydration.

Besides these points, I think that if we look at the behavioral performance of DI rats, we really have to admit that it is very difficult to validate the concept of vasopressin involvement in memory processes with the paradigms that have been used by all investigators so far. Realistically, I think we are talking about performance and thinking about memory rather than talking about memory directly. A final possibility that I would like to present to clarify our understanding of the behavior of Brattleboro rats relates to concepts like fear and arousal level. For instance, you would expect a relatively powerful effect of differences in fear and arousal on paradigms like passive avoidance, but a really insignificant effect on more strenuous paradigms like active avoidance. Similar to data just shown by Dr. Greidanus we found no difference between het-

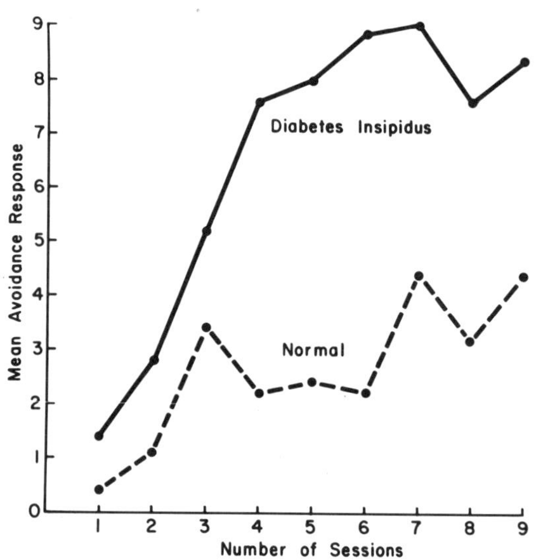

FIGURE 2. Active avoidance response in a two-way shuttle box system. The avoidance performance of the DI group is superior to that of the normal group ($p < 0.05$, two-way ANOVA).

erozygous and diabetes insipidus rats during the acquisition of active avoidance. We also considered that there may be differences in the behavioral patterns of these animals that may affect their performance. For instance, the animals' performance on the open field might be interesting to look at. In such tests there are no differences between the heterozygous and the diabetes insipidus rats, but their activity level in terms of numbers of crossings during a three-minute test is significantly lower than that of normal Long-Evans rats. If the differences between Long-Evans rats and Brattleboro rats in these tests are generalizable to other behavioral experiments, you might expect to see this. Brattleboro rats, both HZ and DI, treated with saline vehicle show greater passive avoidance performance, interpretable perhaps by some people in terms of greater memory retention, than vehicle-injected Long-Evans rats with normal vasopressin capabilities. We have also looked at the effect of ACTH 4–10 on the passive avoidance of these animals and have found that the effect of ACTH depends on the initial arousal state of the animals. Among those animals who initially show very good performance, latencies are decreased by the administration of ACTH 4–10, whereas the latencies of animals that usually demonstrate poor passive avoidance are increased. Under some circumstances, I think it's possible to demonstrate that vasopressin-deficient rats exhibit better passive avoidance than normal rats. Data also support this hypothesis. The passive avoidance data of vehicle-injected HZ and DI rats together with that of animals who were chronically maintained on 0.5 U of pitressin tannate for 10 days before testing were compared. Animals given oil vehicle showed passive avoidance latencies greater than those given vasopressin replacement! In fact, the administration of vasopressin to DI rats almost totally abolished their passive avoidance performance in these tests. Granted that pitressin tannate is a quite impure preparation, nevertheless, it is still intriguing that chronic replacement with a vasopressin mixture like this did in fact abolish passive avoidance, which is exactly the opposite of what we would have expected from reports of acute therapy with vasopressin.

J. J. DREIFUSS (*University of Geneva Medical School, Geneva, Switzerland*): May I just say, Dr. Bailey raised some very important questions as to how far the passive avoidance behavior paradigm was really capable of testing memory, and I think I would like to have Dr. Greidanus continue along that line.

T. B. VAN WIMERSMA GREIDANUS (*Rudolf Magnus Institute for Pharmacology, Utrecht, the Netherlands*): With respect to the use of the word memory, in our studies on the effects on memory, we are affecting processes that are *related* to memory; that states it a little more carefully. I think that so many processes are involved in memory, such as storage, retrieval, alertness, concentration, or vigilance, or what have you, that I agree that you cannot simplify the matter by nominating one of these processes as "memory." Nevertheless, we think that vasopressin is involved in storage processes, as well as in retrieval processes, and therefore, affects memory.

The passive avoidance situation is a suitable paradigm to study effects on storage and/or retrieval. That there are differences in the findings obtained with Brattleboro rats in the United States and in the Netherlands by our "Utrecht group" is clear. However, our "Utrecht group" is not the only group that found differences between homozygous DI rats and controls. Marked differences in passive avoidance behavior between homozygous DI rats and controls have also been found by Rigter and Crabbe from the Organon Company in Oss (the Netherlands), similar to our results. That kind of compensation for the vasopressin deficiency may have taken place in the brain of the DI rats, and whether or not that compensation is different in the colonies bred in the United States and Europe, we really don't know. In this respect it is important to consider the presence of ACTH in the brain, as well as all the other pro-opiomelanocortin fragments (such as β-endorphin, α-endorphin, and γ-endorphin) that affect active as well as passive avoidance behavior. To avoid possible differences in compensatory mechanisms, investigators may have to use animals that are obtained from the same colony in order to solve the problem of discrepancies in our experimental results. I would be very well willing to collaborate in such studies.

G. THOMAS (*University of Rochester, Rochester, N.Y.*): It is quite clear, as Dr. Greidanus has indicated, that behavioral results are quite variable from lab to lab. I want to make a comment because I think the phenomenon might give behavior a bad name. The problem is one that has also occurred in other areas of study of the Brattleboro rat. It stems, in my opinion, from the fact that the various colonies in different labs started from limited matings of rats that came from the original Brattleboro strain. People refer to this as "genetic drift" or, more likely, it would also be what evolutionists called the "founder effect," i.e., there could be varying behavioral (and other) results from rats in these different colonies because of the happenstance concentration of other genes in these rats—in addition to the *di* gene(s). For instance, we intended to use the Utrecht idea of a simple memory test that we could use as an assay (in this case to study the effect of the transplant) but we didn't find an effect so we used a battery of behavioral tests. We used DI rats that Dr. Gash was maintaining as a colony, and we got results similar to those that Dr. Greidanus presented. We repeated the study with more tests and other rats and didn't replicate our previous findings, particularly with regard to olfactory and visual discrimination. Now I am testing a third group (obtained from Blue Spruce Farms) and they are also a little different. I am testing Blue Spruce Brattleboro rats (both HZ and DI) on the runway, and they are even more timid than our earlier groups. This timid disposition in animals from Blue Spruce Farms and Dr. Gash's colony shows up on these behavioral tests and may affect all kinds of performance and could result in misleading conclusions about vasopressin-dependent cognitive capabilities.

S. GARDNER (*University of Nottingham Medical School, Nottingham, England*): May I ask you a very general question, and that is whether between the groups, the housing conditions and handling procedures between the Long-Evans rats and the Brattleboro rats are exactly the same, since for instance individual housing can affect behavior. In our experience, our Brattleboro rats are handled more than our Long-Evans rats simply because of cleaning procedures.

VAN WIMERSMA GREIDANUS: I do not think there are many differences due to that. We generally house the DI rats in similar cages as the controls. However, I think how the animals are handled throughout the whole behavioral procedure is relevant. When you are using the passive avoidance paradigm, for instance, the innate response of preference for a dark compartment is used. If in the adaptation trial there is an animal that does not show this innate response to prefer darkness, it is not used in such experiments. This all accounts for differences, I think; housing may be just one of the factors, and handling and the way the behavioral procedure is performed are other important variables.

J. C. CRABBE (*Veterans Medical Center, Portland, Ore.*): I would like to suggest a way out of this, as far as the behavioral heterogeneity that's being reported in DI rats is concerned. The solution to the problem of identifying specific behavioral effects of the *di* gene is going to come from the several different genotypes that Dr. Hanson is breeding for us all. RHA-DI rats are in fact timid and jumpy, but RHA rats are timid and jumpy before you insert the *di* gene. Osborne and Mendel rats, on which backgrounds the *di* gene is also being inserted, are not particularly jumpy animals. We are going to have a choice of four or five very different genetic backgrounds on which to examine any proposed behavioral deficit due to the *di* gene, and I think that that is where the answer will come from. Examining such a collection of genotypes, all of which are available with or without the *di* gene, will allow each of us to isolate the gene effects from effects due to inbreeding and handling that are characteristic of our own colony.

K. P. CONRAD (*Dartmouth Medical School, Hanover, N.H.*): This is a general question. In your experiments, how do you know if it is the direct effects of AVP-deficiency on the brain that alters the performance and memory of Brattleboro rats? It seems to me that the indirect effects of AVP-deficiency, namely frequent urination and drinking, in themselves, might alter performance and memory. If sorting out the direct (brain) and indirect (water turnover) effects of AVP-deficiency on performance and memory is a difficulty, would not the nephrogenic DI mouse which, unlike the DI rat, has ample AVP, but like the DI rat experiences high water turnover, be a good model to study in comparison with the Brattleboro rat?

VAN WIMERSMA GREIDANUS: When we measure vasopressin in relation to behavior, we generally do not see a relationship between behavioral performance and vasopressin levels in the peripheral blood, but see such a relationship only in blood taken from the eye-plexus. We think that one is only dealing with the vasopressin that is present in the brain. For instance, if an animal is water-deprived for 24 hours, it will have high vasopressin levels in the blood; but, that does not necessarily mean that the vasopressin levels in the brain have been altered. We have not seen any effect of water deprivation on the maintenance of learned behavior. I think this is possibly due to the fact that there is no change in centrally available vasopressin, as we have demonstrated by collecting cerebrospinal fluid from water-deprived rats and measuring vasopressin by radioimmunoassay.

R. ELFONT (*Dartmouth Medical School, Hanover, N.H.*): Dr. Crabbe, could you tell us about the ethanol preference of normal dehydrated rats as compared to the diabetes insipidus rats?

CRABBE: These rats were not dehydrated. As far as I know a moderate degree of dehydration doesn't markedly influence ethanol preference.

G. BRITO (*University of Rochester, Rochester, N.Y.*): It seems to me that we should question the usefulness of the Brattleboro rat as a model to study the behavioral functions of vasopressin because lack of AVP is only one aspect of the biochemical pathology of the DI rat. Intracerebral administration of VP antiserum in normal rats may provide more definitive answers in regard to the behavioral functions of vasopressin.

VAN WIMERSMA GREIDANUS: In fact, with all the discrepancies between the behavior of the different Brattleboro rats in mind, the method of introducing specific antiserum to vasopressin into the lateral ventricle of the brain, or locally in certain brain sites that we think are the sites of behavioral action of vasopressin, has two advantages: (1) one is dealing with a normal animal in which vasopressin can be specifically neutralized; (2) one can manipulate vasopressin levels in the brain *temporarily* avoiding problems due to chronic vasopressin deficiency. Nevertheless, I think the Brattleboro rat is a very interesting animal, but for the purpose of finding out the physiological role of vasopressin in behavior, the use of antiserum may be an adequate approach.

SOKOL: There are obviously many useful approaches to the study of the functions of vasopressin, use of the DI rat being one of them. It has been proposed in this symposium that conflicting data obtained from studies in the DI rat may be attributable to genetic variation between colonies. Lest we leave this meeting with the impression that this proposal is fact, let me remind you of the data presented here by Drs. Crabbe and Allen on the Roman High Avoidance rats that carry the *di* gene. Their results suggest that the genetic background on which the *di* gene is placed is not that critical since the Brattleboro and the RHA rats responded similarly with regard to alcohol preference and characteristics of the ACTH family of hormones.

G. VALIQUETTE (*Columbia University, New York, N.Y.*): Just a very brief comment to come back to the effect of thirst on behavior. Quite a few patients with diabetes insipidus will tell you that they never feel really comfortable in a normal environment until they have located a bathroom and a source of fluid. This may be the same for the DI rat, so that the test chamber or environment may be more anxiety provoking for the DI than the normal rat.

VAN WIMERSMA GREIDANUS: Maybe we didn't solve all the problems about the behavior and the behavioral performance of the Brattleboro rat. This might be a reason to encourage collaborations between people who are studying behavior in these rats.

ELECTROPHYSIOLOGY OF VASOPRESSIN IN NORMAL RATS AND IN RATS OF THE BRATTLEBORO STRAIN *

J. J. Dreifuss,† M. Mühlethaler, and B. H. Gähwiler

Department of Physiology
University of Geneva Medical School
1211 Geneva 4, Switzerland
and
Preclinical Research, Sandoz Ltd.
Basel, Switzerland

Brattleboro rats, with their absolute deficiency in vasopressin biosynthesis,[1-4] provide a unique opportunity to assess the consequences of permanent vasopressin deprivation on brain development,[5] on the expression of various behaviors,[6] as well as on neural mechanisms at the level of single neurons.

Surprisingly few investigators have used Brattleboro rats to assess the consequences of the complete absence of vasopressin on the functioning of single nerve cells. The few studies that have appeared in print showed that the mechanisms of propagation and of patterning of action potentials in hypothalamic magnocellular endocrine neurons, i.e. in the cells primarily affected in the diabetes insipidus mutant, do not differ to any great extent from those of normal rats.[7, 8]

The idea that vasopressin might be a neurotransmitter or a neuromodulator in the mammalian brain has been strengthened by the discovery of vasopressinergic axon endings in various brain locations outside of the hypothalamus.[9-11] Except for studies performed on invertebrate neurons,[12, 13] the electrophysiological evidence favoring a direct action of vasopressin on single nerve cells was deduced, until recently, from the recording of neural firing during intracerebroventricular or iontophoretic applications of the peptide.

The development of suitable *in vitro* preparations of brain tissue has permitted the study of the effects of vasopressin and related peptides on single neurons in the hypothalamus and hippocampus. The latter is a region of the central nervous system known to receive a vasopressinergic innervation. Vasopressin, at low concentrations, markedly increased the rate of firing of hippocampal pyramidal cells. In the final part of this article, we report that hippocampal neurons in Brattleboro rats possess the same sensitivity to exogenous vasopressin as those of normal rats.

Hypothalamo-Neurohypophyseal Neurons in Normal Rats and in Brattleboro Rats

The axons that terminate in the neural lobe of the pituitary gland originate from the cell body of hypothalamic magnocellular neurons and are deemed to

* Supported in part by grants from the Swiss National Science Foundation.

† Send correspondence to: Prof. J. J. Dreifuss, Département de Physiologie, Centre Médical Universitaire, 1211 Genève 4.

0077-8923/82/0394-0689 $1.75/0 © 1982, NYAS

be either oxytocinergic or vasopressinergic.[4, 14] Electrophysiological criteria can be used to identify hypothalamo-neurohypophyseal neurons in the living animal.[15] The principles of this identification procedure are briefly outlined below (FIGURE 1).

By applying an electrical stimulus to the neurohypophysis or to the pituitary stalk, an action potential can be generated that travels up the axon to its parent cell body located in the hypothalamus. These antidromically conducted spikes can be recorded by an electrode located in the hypothalamus. The detection of an action potential at a *fixed latency* after the stimulus constitutes the first indication that the activity of a neuron projecting to the neural lobe is being monitored. A second criterion is that antidromically conducted spikes should be elicited in response to *high frequency* stimuli in a one-to-one manner. Third, if an antidromically conducted spike is evoked shortly after the generation of a spontaneous action potential at the cell body, the two spikes *collide* somewhere along the hypothalamo-neurohypophyseal tract and cancel each other; therefore, the antidromically conducted spike never reaches the cell body where the recording electrode is located.

Three groups of investigators have applied these criteria to identify magnocellular endocrine neurons and to study their firing characteristics in rats with hereditary hypothalamic diabetes insipidus.

Conduction Velocity

From a knowledge of (a) the approximate distance between stimulation and recording electrodes and (b) the latency for antidromic invasion, the conduction velocity of the hypothalamo-neurohypophyseal axons can be estimated. The mean latency for antidromic invasion of supraoptic neurons was much shorter in homozygous Brattleboro rats than in Long-Evans rats of similar weight,[8, 16] indicating a faster conduction velocity in the diabetes insipidus rats (FIGURE 2). This increase is probably explained by the fact that cell body and axon of hypothalamo-neurohypophyseal neurons are hypertrophic in Brattleboro rats.[17] It is well established that conduction velocity depends on axon diameter.

Rate of Spontaneous Firing

Supraoptic and paraventricular neurons fire spontaneously at a higher mean rate in Brattleboro rats than in Long-Evans rats.[8, 16] This difference probably reflects the sustained increase in plasma osmolality in the diabetes insipidus rats, which is known to accelerate the firing of hypothalamo-neurohypophyseal neurons in normal rats.[18-20]

Interaction Between Magnocellular Neurons

In his intracellular study of hypothalamic magnocellular neurons in the goldfish, Kandel[21] observed that antidromic volleys applied to the posterior lobe of the pituitary gland elicited a hyperpolarizing postsynaptic potential with a latency of onset only slightly longer than that of the antidromically conducted spike and with characteristics typical of a chemically mediated event.

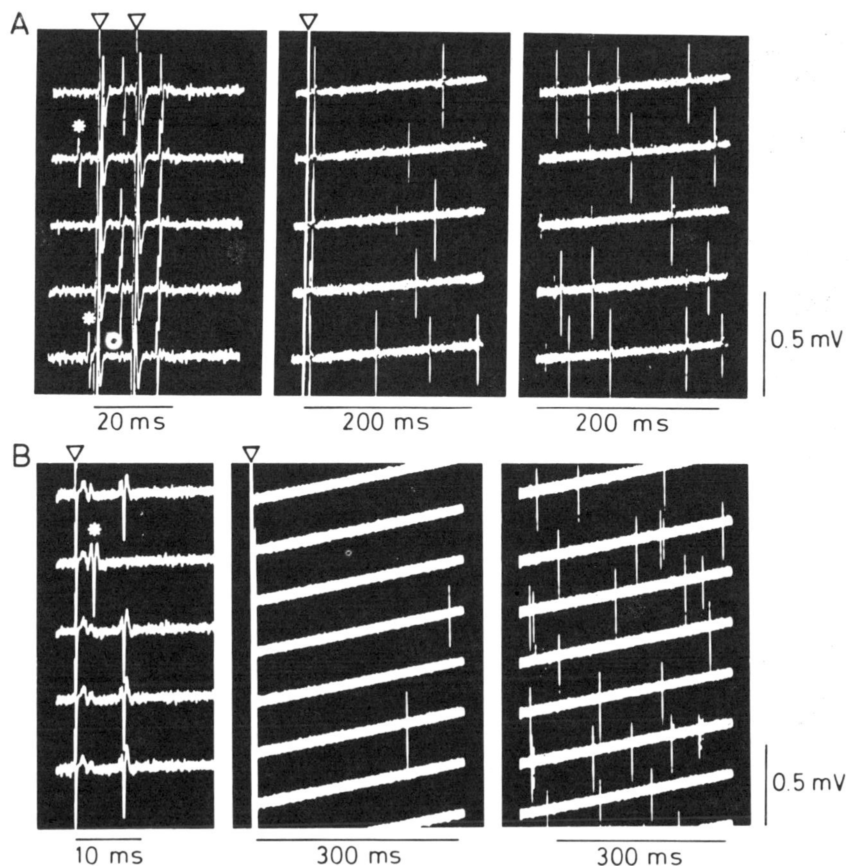

FIGURE 1. Effect of electrical stimulation of the pituitary stalk in Brattleboro rats on two supraoptic neurons. Left-hand panels illustrate criteria used to show that the cells projected to the pituitary. Antidromically conducted spikes occurred at fixed latency (7.5 msec for both cells) following application of stimuli (at ▽), except when spontaneously occurring action potentials (*) had cancelled the antidromically conducted spike In A, each of two stimuli applied at an interval of 11 msec was followed by an antidromic action potential. Middle panels show that, in addition to, and following the antidromically conducted spike, stimulation reduced the firing probability for some 100 (in A) to 250 msec (in B). This is best seen by comparing the records photographed in the absence of electrical stimulation (right-hand panels) with those obtained during pituitary stalk stimulation (middle panels).

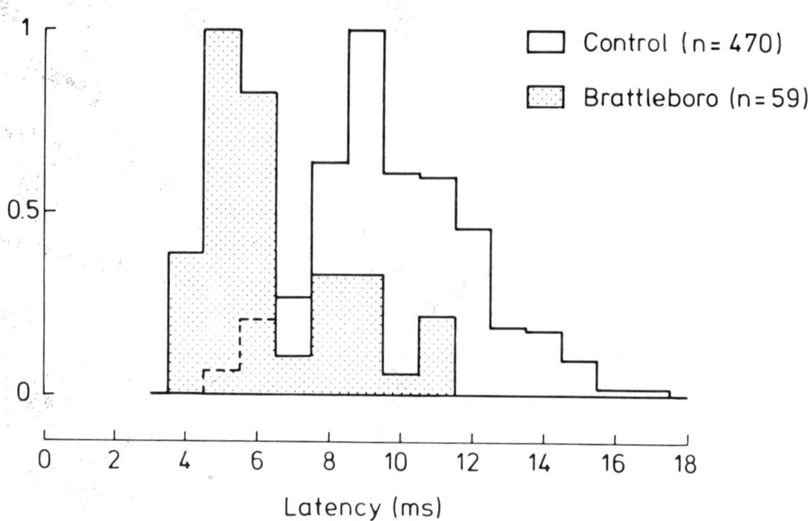

FIGURE 2. Latency for antidromic invasion of neurons in the rat supraoptic nucleus in response to electrical stimulation of the pituitary stalk. Distribution of latencies is shown for 59 neurons from homozygous Brattleboro rats and for 470 neurons from control rats (Sprague-Dawley and Long-Evans). The modal value for both distributions has been given the arbitrary value of one on the ordinate. For normal rats, the average conduction velocity of hypothalamo-neurohypophyseal axons was approximately 0.5 m/sec. It was about twice as fast in the Brattleboro rats.

He suggested that this inhibitory response was probably carried by a system of recurrent axon collaterals and argued that this finding indicated either that neurohypophyseal hormones could serve as transmitter substances at central synapses or that the axon of neuroendocrine cells was capable of releasing a transmitter substance in addition to its neurohormone.

A pause in firing following 'antidromic' stimulation has also been observed in hypothalamo-neurohypophyseal neurons in rat and cat brains and has been generally interpreted to be mediated by a recurrent inhibitory pathway.[15, 22]

In cats, a correlation had been found between those supraoptic neurons exhibiting this 'recurrent inhibition' and those whose rate of firing was depressed by iontophoretically applied vasopressin. Nicoll and Barker[23] have therefore proposed that vasopressin might serve as neurotransmitter along the presumptive recurrent inhibitory pathway.

This hypothesis was contradicted by the observation that 'recurrent inhibition' existed in homozygous Brattleboro rats and had a similar time course as that seen in non-diabetic rats.[7, 8] FIGURE 1 shows a reduction in firing probability following antidromic invasion of supraoptic magnocellular neurons in the Brattleboro rat. A pause in firing in response to 'antidromic' stimulation was recently described to occur both in vasopressinergic and in oxytocinergic neurons.[22] Since in the previous studies, no distinction was made between these two neuronal populations, it could still be envisioned that vasopressin and/or oxytocin can act as neurotransmitters mediating recurrent inhibition in normal rats. This function could be mediated by oxytocin in Brattleboro rats.

Firing Patterns: Phasic Cells

Electrophysiological studies have indicated that a proportion of hypothalamo-neurohypophyseal neurons discharges in an intermittent pattern that is characterized by bursts of activity interrupted by periods devoid of action potentials.[24] Initially noticed in the rat hypothalamus, this pattern of firing has also been observed in hypothalamo-neurohypophyseal neurons of other mammalian species. It is now widely accepted that, at least in the rat, this pattern of activity is characteristic of vasopressinergic neurons.[24, 25] Oxytocinergic neurons, i.e. neurons involved in reflex milk-ejection, do not discharge phasically.[25, 26] Although Brattleboro rats are incapable of vasopressin biosynthesis, phasically firing neurons are readily observed in this strain (FIGURE 3). This observation indicates that the impairment of peptide biosynthesis in Brattleboro rats does not markedly alter the electrical properties of the affected neurons. Actually, a somewhat higher proportion (49%) of supraoptic neurons fire phasically in this strain than in Long-Evans rats (33%), which is not unexpected when considering the osmotic stimulus to which Brattleboro rats are steadily exposed.[16]

Leng[27] has recently described a 'constant-collision stimulation technique' in which each recorded action potential from a supraoptic neuron was quickly

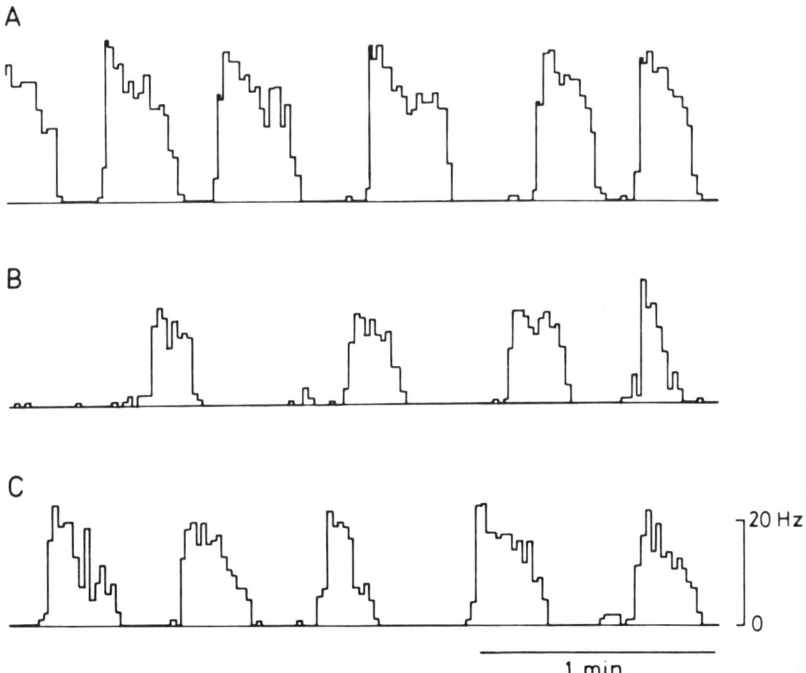

FIGURE 3. Ratemeter records to show the pattern of firing of 'phasic' supraoptic neurons in Brattleboro rats. (This pattern is similar to that observed in normal rats). Latency for antidromic invasion was 5 msec for cell A. Cells B and C were from another Brattleboro rat and had antidromic latencies of 7.0 and 5.5 msec, respectively.

followed by an electrical stimulus applied to the hypothalamo-neurohypophyseal tract. In normal rats, this stimulation pattern consistently shortened the burst duration of phasically firing cells, without changing the rate of firing during the bursts. In Brattleboro rats, however, this stimulation paradigm was found to be rather ineffective in shaping the pattern of firing of phasic neurons.[16] It was conjectured that this relative inefficiency reflected the lack of vasopressin in this strain. According to Leng and Wiersma,[16] vasopressin or a related substance might therefore, in normal rats, act as a neurotransmitter or a neuromodulator in the area of the hypothalamic supraoptic nucleus.

Effects of Posterior Pituitary Hormones on Neuronal Firing in the Hypothalamus

Evidence is indeed available to suggest that vasopressin and oxytocin can alter the rate of firing of hypothalamic neurons. To assess the effects of neurohypophyseal hormones, the blood-brain barrier must be bypassed. Toward this end, vasopressin and oxytocin can be either injected intraventricularly or applied locally.

Intracerebroventricular Injection

An early report stated that the global neural activity of the rabbit hypothalamus was affected by intraventricular vasopressin and oxytocin.[28] Following the injection of nanogram quantities of vasopressin, Bhargava and coworkers[29] noted a dose-related decrease in the level of antidiuretic hormone in venous blood of dogs. No electrophysiological recordings were performed in this study. Pharmacological evidence suggested that this vasopressin-induced reduction in vasopressin secretion was indirect and probably mediated via catecholamine receptors located within the central nervous system.

Excitatory effects of intraventricularly injected oxytocin on neuronal firing were reported by Freund-Mercier & Richard.[30] In lightly anesthetized lactating rats, these investigators recorded the activity of magnocellular endocrine neurons located in the hypothalamic paraventricular nucleus. When less than 1 ng oxytocin was injected, the frequency of the characteristic bursts of activity seen at milk ejection increased in presumptive oxytocinergic neurons. This facilitation was not seen following the injection of similar amounts of vasopressin.

Iontophoretic Application

Responses with a faster time course and with a more localizable site of action have been obtained following the local ejection of vasopressin or oxytocin by iontophoresis from fine glass pipettes filled with concentrated solutions of these peptides. Controlled ejection of small quantities of these peptides was attempted by applying positive current to the pipettes. No tests were performed, however, to check how much vasopressin or oxytocin was actually released. Only approximately half of the neurons tested responded to vasopressin and to oxytocin (Table 1). Reactive hypothalamo-neurohypophyseal neurons located in the supraoptic nucleus were said to be predominantly inhibited by vaso-

TABLE 1

EFFECTS OF IONTOPHORETICALLY APPLIED VASOPRESSIN AND OXYTOCIN

Area	Species	Vasopressin			Oxytocin			Reference
		+	~	−	+	~	−	
Supraoptic nucleus	rat	7	22	29		n.d.		23
Paraventricular nucleus	rabbit	0	16	0	19	1	0	31
Paraventricular nucleus	rat	0	11	2	13	4	0	31
Cerebral cortex	rat	44	35	6		n.d.		23
Caudal medulla	rat		n.d.		5	15	46	39
Locus coeruleus	rat	27	4	0		n.d.		37

Explanation of symbols: +, increase in firing rate; ~, no effect on firing rate; —, decrease in firing rate; and n.d., not determined.

pressin,[23] whereas another group reported strong excitations induced by oxytocin in a majority of hypothalamic paraventricular neurons.[31]

Bath Application

The development of *in vitro* preparations of brain tissue that can be perfused with solutions containing known concentrations of peptide, has permitted more quantitative studies of vasopressin action. During the past few years, we have cultured for periods of several weeks small explants taken from the area of the supraoptic nucleus from young rats. In these cultures, intracellular recordings can be obtained from large nerve cells that are visible under phase contrast microscopy.[32, 33] In preliminary studies, we found vasopressin added to the perfusion medium at micromolar concentrations to accelerate the firing of approximately half of some thirty cells tested. The remainder were not affected, inhibitory effects were never observed. FIGURE 4 shows recordings from a cultured supraoptic neuron stimulated by vasopressin in a dose-dependent manner at concentrations in excess of 10^{-8} M.

EFFECTS OF POSTERIOR PITUITARY HORMONES ON HIPPOCAMPAL NEURONS IN NORMAL RATS AND IN BRATTLEBORO RATS

Traditionally, the hypothalamo-neurohypophyseal system was regarded as forming a system of 'final common' neurons whose function is to release vasopressin and oxytocin into blood vessels of the neural lobe of the pituitary gland.[34] Numerous new data indicate that these peptides have a wider distribution and additional, non-endocrine effects. Vasopressin, oxytocin, and their associated neurophysin have been shown by immunoassay and immunocytochemistry to be present in different brain regions outside of the hypothalamus.[35] Axon terminals, containing 'vesicles' that reacted with vasopressin antiserum, have been detected by immunoelectron microscopy in several locations within the brain.[36] Oxytocin and vasopressin, applied by iontophoresis, affected neuronal firing in extrahypothalamic brain areas (TABLE 1). Thus, vasopressin increased the rate of the firing of neurons located in the region of locus coeruleus [37] a nucleus known to receive a vasopressinergic innervation.[5, 38] In the caudal brain stem, a region rich in oxytocinergic axon terminals, neuronal firing was predominantly depressed following the iontophoretic application of oxytocin.[39]

In addition, much evidence has accumulated to suggest an involvement of vasopressin (and oxytocin) in various behaviors. In particular, a role of vasopressin in memory in rat, mice, and man has been considered.[40-44] Although it remains somewhat controversial, the association of vasopressin and memory consolidation is supported by a growing number of experimental findings. For example, the injection of vasopressin into cerebrospinal fluid [45, 46] or into discrete brain regions in rats,[47] has been shown to facilitate the retention of learned behavior. This central effect of vasopressin may even be of physiological significance since the injection of either vasopressin antiserum,[48] or vasopressin antagonist [49] inhibited the retention of learned behavior in rats. Oxytocin, on the other hand, induced maternal behavior in virgin female rats.[50]

Since the hippocampus is believed by many investigators to be involved in memory consolidation, it appeared worthwhile to assess whether vasopressin might affect the firing of neurons located in this part of the brain. The ventral hippocampus is known to receive a vasopressinergic innervation[10] and there are some indications in the literature suggesting that vasopressin might affect the bioelectric activity of hippocampal neurons.[46, 51–53]

To test whether such effects may be explained by a local action, we prepared

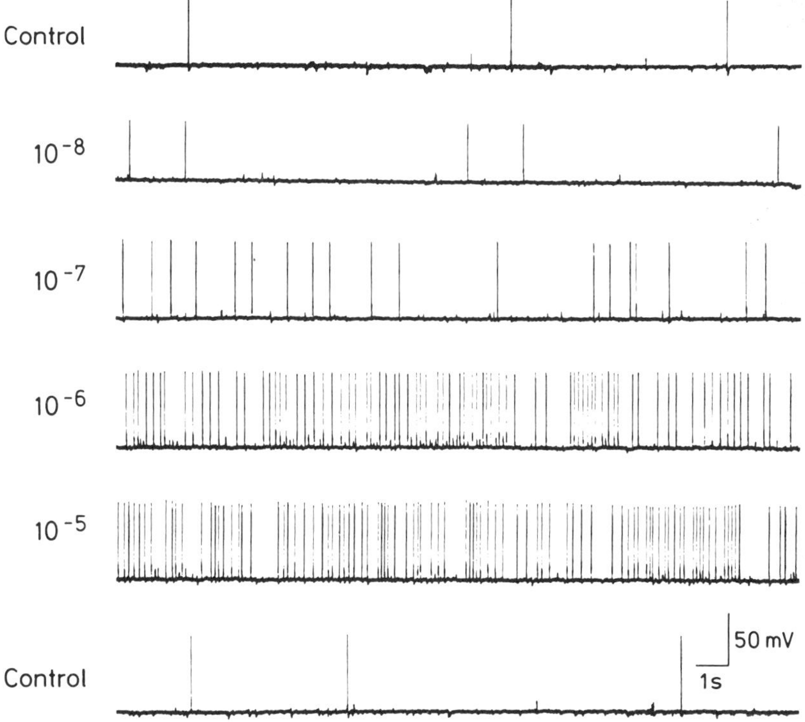

FIGURE 4. Intracellular records from a neuron in a cultured explant of the area of the hypothalamic supraoptic nucleus prepared from a 7-day-old rat. Approximately 20 sec of recording are shown in each tracing, obtained in the absence (control) or in the presence of lysine-vasopressin (LVP), applied to the perfusion medium at the molar concentrations indicated. The peptide caused a dose-dependent increase in rate of firing from 0.3 Hz in control solution to 0.6 (10^{-7} M), 7.0 (10^{-6} M), and 7.5 Hz (10^{-5} M). Between each LVP application, the rate of firing returned to the control level, but only the final control has been illustrated. Oxytocin had no detectable effect even when applied at higher concentration. Spike amplitude, 75 mV.

transverse slices of rat hippocampus and recorded extracellularly the effects of vasopressin and oxytocin on single cell activity.[54]

Vasopressin consistently induced a sustained and fully reversible increase in firing of neurons located in the CA1 area. The threshold concentration of vasopressin, which accelerated cell firing was 10^{-8} M, maximal stimulation of firing being observed at $\sim 10^{-6}$ M. The stimulatory effect of vasopressin was fully antagonized by a synthetic vasopressin analogue known to antagonize the pressor effects of vasopressin.[55, 56] Oxytocin also excited hippocampal cells. It is not yet clear whether separate binding sites exist for vasopressin and oxytocin or to what extent these two neurohypophyseal peptides interact with a single 'receptor'.

Memory is partially impaired in Brattleboro rats, a deficit that can be alleviated by the injection of vasopressin.[6] Therefore, Brattleboro rats might possess in their CNS the same binding sites for neurohypophyseal peptides as normal, nondiabetic rats. To test this conjecture we compared the effects of exogenous vasopressin on hippocampal slices obtained from Long-Evans rats or from homozygous Brattleboro rats. In Brattleboro rats, 10^{-7} M vasopressin accelerated neuronal firing and 10^{-6} M did so more powerfully (FIGURE 5). Dose-response relations varied to an appreciable extent among cells from Brattleboro rats and among cells from Long-Evans rats, and did not differ significantly between the two strains (FIGURE 6). These observations show that hippocampal neurons of Brattleboro rats retain a normal sensitivity to exogenous vasopressin, although these rats are incapable of producing endogenous vasopressin. As in normal rats, the stimulatory effect of vasopressin was reversibly abolished by a synthetic vasopressin antagonist.

Studies are in progress to characterize the sites in the hippocampus with which vasopressin reacts in Brattleboro rats and to assess whether these sites also interact with oxytocin.

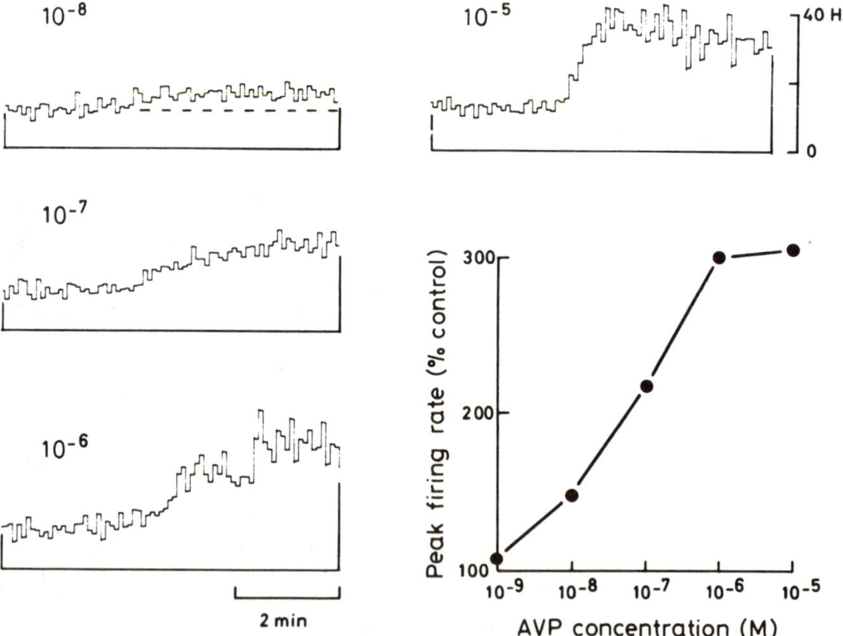

FIGURE 5. Effect of synthetic arginine-vasopressin (AVP) on the firing rate of a neuron in a transverse hippocampal slice cut from the brain of a homozygous Brattleboro rat. Ratemeter records illustrate the effects of AVP applied to the perfusion medium at the molar concentrations indicated. Lower right graph shows the dose-response curve obtained.

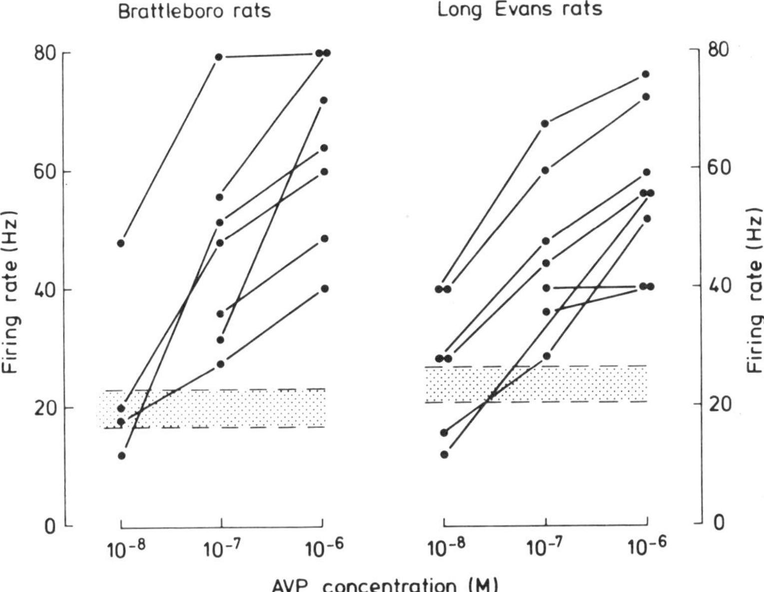

FIGURE 6. Dose-response of arginine-vasopressin (AVP) effects as obtained in hippocampal slices from homozygous Brattleboro rats ($N = 7$) and from Long-Evans rats ($N = 8$). Dotted areas shown mean ($\pm$ SEM) firing rate in the absence of AVP. Note that in both strains vasopressin had a clear stimulatory effect at concentrations of 10^{-7} M or greater.

Summary and Conclusions

Electrophysiological investigations performed on hypothalamo-neurohypophyseal neurons revealed no major, unexpected alterations in firing characteristics in Brattleboro rats, as compared to normal rats. This suggests that the biosynthetic defect of these cells in Brattleboro rats is not mirrored by any marked abnormality of their excitable membrane.

Iontophoretic applications of oxytocin and vasopressin in normal rats yielded ambiguous results, perhaps owing to uncertainties about peptide delivery from the pipettes used. In contrast, clear-cut quantitative results can be obtained with the use of *in vitro* preparations. Vasopressin, applied to the bathing solution in minute concentration ($<10^{-6}$ M) consistently excited hippocampal, as well as hypothalamic neurons. This effect could be blocked by an analogue known to antagonize the pressor action of vasopressin.

Vasopressin was found to be as effective in stimulating hippocampal pyramidal cells in Brattleboro rats as in normal rats. In the former strain, it is conceivable either that vasopressin interacts with "oxytocin receptors," or that the hippocampus retains its normal set of binding sites, including the putative "vasopressin receptors."

Acknowledgments

The Brattleboro rats were kindly supplied by the Institut für Tierzucht, Füllinsdorf, Switzerland and the vasopressin antagonists by Dr. M. Manning, Toledo, Ohio.

REFERENCES

1. VALTIN, H., W. H. SAWYER & H. W. SOKOL. 1965. Neurohypophysial principles in rats homozygous and heterozygous for hypothalamic diabetes insipidus (Brattleboro strain). Endocrinology **77:** 701–706.
2. MÖHRING, B. & J. MÖHRING. 1975. Plasma ADH in normal Long-Evans rats and in Long-Evans rats heterozygous and homozygous for hypothalamic diabetes insipidus. Life Sci. **17:** 1307–1314.
3. BROWNSTEIN, M. J. & H. GAINER. 1977. Neurophysin biosynthesis in normal rats and in rats with hereditary diabetes insipidus. Proc. Natl. Acad. Sci. USA **74:** 4046–4049.
4. RHODES, C. H., J. I. MORELL & D. W. PFAFF. 1981. Immunohistochemical analysis of magnocellular elements in rat hypothalamus: distribution and numbers of cells containing neurophysin, oxytocin and vasopressin. J. Comp. Neurol. **198:** 45–64.
5. BOER, G. J., R. M. BUIJS, D. F. SWAAB & G. J. DE VRIES. 1980. Vasopressin and the developing rat brain. Peptides **1** (Suppl. 1): 203–209.
6. VAN WIMERSMA GREIDANUS, TJ. B. 1982. Disturbed behavior and memory of the Brattleboro rat. Ann. N.Y. Acad. Sci. This volume.
7. DREIFUSS, J. J., J. J. NORDMANN & J. D. VINCENT. 1974. Recurrent inhibition of supraoptic neurosecretory cells in homozygous Brattleboro rats. J. Physiol. (London) **237:** 25–27P.
8. DYBALL, R. E. J. 1974. Single unit activity in the hypothalamo-neurohypophysial system of Brattleboro rats. J. Endocr. **60:** 135–143.
9. BUIJS, R. M., D. SWAAB, J. DOGTEROM & F. VAN LEEUWEN. 1978. Intra- and extrahypothalamic vasopressin and oxytocin pathways in the rat. Cell Tissue Res. **186:** 423–433.
10. BUIJS, R. M. 1978. Intra- and extrahypothalamic vasopressin and oxytocin pathways in the rat. Pathways to the limbic system, medulla oblongata and spinal cord. Cell Tissue Res. **192:** 423–435.
11. SOFRONIEW, M. V. 1980. Projections from vasopressin, oxytocin, and neurophysin neurones to neural targets in the rat and human. J. Histochem. Cytochem. **28:** 475–478.
12. BARKER, J. L., M. IFSHIN & H. GAINER. 1975. Studies on bursting pacemaker potential activity in molluscan neurons. III. Effects of hormones. Brain Res. **84:** 501–513.
13. BARKER, J. L. & T. C. SMITH. 1976. Peptide regulation of neuronal membrane properties. Brain Res. **103:** 167–170.
14. VANDESANDE, F. & K. DIERICKX. 1975. Identification of the vasopressin-producing and of the oxytocin-producing neurons in the hypothalamic magnocellular system of the rat. Cell Tissue Res. **164:** 153–162.
15. DREIFUSS, J. J. & J. S. KELLY. 1972. Recurrent inhibition of antidromically identified rat supraoptic neurones. J. Physiol. (London) **220:** 87–103.
16. LENG, G. & J. WIERMSA. 1981. Effects of neural stalk stimulation on phasic discharge of supraoptic neurones in Brattleboro rats devoid of vasopressin. J. Endocr. **90:** 211–220.
17. SOKOL, H. W. & H. VALTIN. 1965. Morphology of the neurosecretory system in rats homozygous and heterozygous for hypothalamic diabetes insipidus (Brattleboro strain). Endocrinology **77:** 692–700.
18. ARNAULD, E., J. D. VINCENT & J. J. DREIFUSS. 1974. Firing patterns of hypothalamic supraoptic neurones during water deprivation in monkeys. Science **185:** 535–537.
19. BRIMBLE, M. J. & R. E. J. DYBALL. 1977. Characterization of the responses of oxytocin- and vasopressin-secreting neurones in the supraoptic nucleus to osmotic stimulation. J. Physiol. (London) **271:** 253–271.

20. WAKERLEY, J. B., D. A. POULAIN & D. BROWN. 1978. Comparison of firing patterns in oxytocin- and vasopressin-releasing neurones during progressive dehydration. Brain Res. **148:** 425–440.
21. KANDEL, E. 1964. Electrical properties of hypothalamic neuroendocrine cells. J. Gen. Physiol. **47:** 691–717.
22. POULAIN, D. A., F. ELLENDORF & J. D. VINCENT. 1980. Septal connections with identified oxytocin and vasopressin neurones in the supraoptic nucleus of the rat. An electrophysiological investigation. Neuroscience **5:** 379–387.
23. NICOLL, R. A. & J. L. BARKER. 1971. The pharmacology of recurrent inhibition in the supraoptic neurosecretory system. Brain Res. **35:** 501–511.
24. DREIFUSS, J. J., M. C. HARRIS & E. TRIBOLLET. 1976. Excitation of phasically firing hypothalamic supraoptic neurones by carotid occlusion in rats. J. Physiol. (London) **257:** 337–354.
25. DREIFUSS, J. J., E. TRIBOLLET & M. MÜHLETHALER. 1981. Temporal patterns of neural activity and their relation to the secretion of posterior pituitary hormones. Biol. Reprod. **24:** 51–72.
26. LINCOLN, D. W. & J. B. WAKERLEY. 1974. Electrophysiological evidence for the activation of supraoptic neurones during the release of oxytocin. J. Physiol. (London) **242:** 533–554.
27. LENG, G. 1981. The effects of neural stalk stimulation upon firing patterns in rat supraoptic neurones. Exp. Brain Res. **41:** 135–145.
28. SCHULZ, H., H. UNGER, H. SCHWARZBERG, G. POMMRICH & R. STOLZE. 1971. Neuronenaktivität hypothalamischer Kerngebiete von Kaninchen nach intraventrikulärer Applikation von Vasopressin und Oxytocin. Experientia **27:** 1482–1483.
29. BHARGAVA, K. P., U. K. KULSHRESTHA & Y. P. SRIVASTAVA. 1977. Central mechanisms of vasopressin-induced changes in antidiuretic hormone release. Brit. J. Pharmacol. **60:** 77–81.
30. FREUND-MERCIER, M. J. & PH. RICHARD. 1981. Excitatory effects of intraventicular injections of oxytocin on the milk-ejection reflex in the rat. Neurosci. Lett. **23:** 193–198.
31. MOSS, R. L., R. E. J. DYBALL & B. A. CROSS. 1972. Excitation of antidromically identified neurosecretory cells of the paraventricular nucleus by oxytocin applied iontophoretically. Exp. Neurol. **34:** 95–102.
32. GÄHWILER, B. H. & J. J. DREIFUSS. 1979. Phasically-firing neurones in long-term cultures of the rat hypothalamic supraoptic area: pacemaker and follower cells. Brain Res. **177:** 95–103.
33. GÄHWILER, B. H. & J. J. DREIFUSS. 1980. Transition from random to phasic firing induced in neurones cultured from the hypothalamic supraoptic area. Brain Res. **193:** 415–423.
34. KNOWLES, F. & L. VOLLRATH, Editors. 1974. Neurosection—The Final Neuroendocrine Pathway. Springer Verlag. Berlin.
35. BROWNSTEIN, M. J. 1980. Peptidergic pathways in the central nervous system. Proc. R. Soc. (London), Ser. B **210:** 79–90.
36. BUIJS, R. M. & D. F. SWAAB. 1979. Immuno-electron microscopical demonstration of vasopressin and oxytocin synapses in the limbic system of the rat. Cell Tissue Res. **204:** 353–365.
37. OLPE, R. R. & V. BALTZER. 1981. Vasopressin activates noradrenergic neurones in the rat locus coeruleus: a microiontophoretic investigation. Eur. J. Pharmacol. **77:** 377–378.
38. ROSSOR, M. N., L. L. IVERSEN, J. HAWTHORN, V. T. Y. ANG & J. S. JENKINS. 1981. Extrahypothalamic vasopressin in human brain. Brain Res. **214:** 349–355.
39. MORRIS, R., T. E. SALT, M. V. SOFRONIEW & R. G. HILL. 1980. Actions of microiontophoretically applied oxytocin, and immunohistochemical localiza-

tion of oxytocin, vasopressin and neurophysin in the rat caudal medulla. Neurosci. Lett. **18:** 163–168.

40. DE WIED, D. 1971. Long term effect of vasopressin on the maintenance of a conditioned avoidance response in rats. Nature (London) **232:** 58–60.
41. DE WIED, D. 1980. Behavioral actions of neurohypophysial peptides. Proc. R. Soc. (London) Ser. B **210:** 183–195.
42. WALTER, R., P. L. HOFFMAN, J. B. FLEXNER & L. B. FLEXNER. 1975. Neurohypophysial hormones, analogs and fragments: their effects on puromycin-induced amnesia. Proc. Natl. Acad. Sci. USA **72:** 4180–4184.
43. ROCHE, K. E. & A. J. LESHNER. 1979. ACTH and vasopressin treatments immediately after a defeat increase future submissiveness in male mice. Science **204:** 1343–1344.
44. WEINGARTNER, H., P. W. GOLD, J. C. BALLENGER, S. A. SMALLBERG, R. SUMMERS, D. R. RUBINOW, R. M. POST & F. K. GOODWIN. 1981. Effects of vasopressin on human memory functions. Science **211:** 601–603.
45. DE WIED, D. 1976. Behavioral effects of intraventricularly administered vasopressin and vasopressin fragments. Life Sci. **19:** 685–690.
46. BOHUS, B., I. URBAN, TJ. B. VAN WIMERSMA GREIDANUS & D. DE WIED. 1978. Opposite effects of oxytocin and vasopressin on avoidance behavior and hippocampal theta rhythm in the rat. Neuropharmacology **17:** 239–247.
47. KOVACS, C. L., B. BOHUS, D. H. G. VERSTEEG, E. R. DE KLOET & D. DE WIED. 1979. Effect of oxytocin and vasopressin on memory consolidation: sites of action and catecholaminergic correlates after local microinjection into limbic midbrain structures. Brain Res. **175:** 303–314.
48. VAN WIMERSMA GREIDANUS, TJ. B., J. DOGTEROM & D. DE WIED. 1975. Intraventricular administration of antivasopressin serum inhibits memory consolidation in rats. Life Sci. **16:** 637–644.
49. KOOB, C. F., M. LE MOAL, O. GAFFORI, M. MANNING, W. H. SAWYER, J. RIVIER & F. E. BLOOM. 1981. Arginine vasopressin and a vasopressin antagonist peptide: opposite effects on extinction of active avoidance in rats. Regul. Peptides **2:** 153–163.
50. PEDERSEN, C. A. & A. J. PRANGLE. 1979. Induction of maternal behavior in virgin rats after intracerebroventricular administration of oxytocin. Proc. Natl. Acad. Sci. USA **76:** 6661–6665.
51. HUSTON, J. P. & L. JACOBARTH. 1977. Evidence for selective susceptibility of hippocampus to spreading depression induced by vasopressin. Neurosci. Lett. **33:** 69–72.
52. URBAN, I. & D. DE WIED. 1978. Neuropeptides: effects on paradoxical sleep and theta rhythm in rats. Pharmacol. Biochem. Behav. **8:** 51–59.
53. ABOOD, L. G., R. KNAPP, T. MITCHELL, H. BOOTH & L. SCHWAB. 1980. Chemical requirements of vasopressin for barrel rotation convulsions and reversal by oxytocin. J. Neurosci. Res. **5:** 191–199.
54. MÜHLETHALER, M., J. J. DREIFUSS & B. H. GÄHWILER. 1981. Vasopressin causes excitation of hippocampal neurones. Nature (London). (In press.)
55. KRUSZYNCKI, M., B. LAMMEK, M. MANNING, J. SETO, J. HALDAR & W. H. SAWYER. 1980. [1-(β-Mercapto-β, β-cyclopentamethylenepropionic acid), 2-(O-methyl) tyrosine] arginine-vasopressin and [1-(β-Mercapto-β, β-cyclopentamethylene-propionic acid)] arginine-vasopressin, two highly potent antagonists of the vasopressor response to arginine-vasopressin. J. Med. Chem. **23:** 364–368.
56. SAWYER, W. H., Z. GRZONKA & M. MANNING. 1981. Neurohypophysial peptides. Design of tissue specific agonists and antagonists. Molec. Cell. Endocr. **22:** 117–134.

(DISCUSSION OF THIS PAPER BEGINS ON PAGE 728.)

THE REGIONAL IMPAIRMENT OF BRAIN DEVELOPMENT IN THE BRATTLEBORO DIABETES INSIPIDUS RAT; SOME VASOPRESSIN SUPPLEMENTATION STUDIES

G. J. Boer, H. B. M. Uylings, A. J. Patel,* K. Boer, and R. Kragten

Netherlands Institute for Brain Research
IJdijk 28, 1095 KJ Amsterdam
the Netherlands

* *MRC Developmental Neurobiology Unit*
33 John's Mews
London WC1N 2NS, England

Introduction

In the neonate diabetes insipidus (DI) Brattleboro rat, body growth and the development of the brain are impaired.[1, 2] In the homozygous (HOM) form of this genetically recessive, vasopressin-deficient mutancy,[3] a lower brain weight persists throughout life, the severity of which depends upon the breeding scheme for raising the HOM-DI versus the heterozygous (HET) DI controls not deficient for vasopressin. Earlier reports did not regard the missing vasopressin as the causal factor in the disturbed body growth because of the finding that in early pedigrees of the Brattleboro strain the growth retardations of HOM-DI animals appeared more severe than in later ones.[4] Additionally, vasopressin treatments neither improved early postnatal body growth[5] nor restored the weight deficits in more adult stages.[6] Development of the brain however, was never investigated in these studies.

Vasopressin is not only known as a neurohormone with an antidiuretic action, but also acts as a centrally active neuropeptide. In the adult it influences such processes as memory consolidation,[7] monoamine metabolism,[8] and body temperature regulation.[9] Vasopressin seems to act centrally as a neurotransmitter rather than as a hormone. From the sites of synthesis in the cell bodies of the supraoptic, paraventricular, and suprachiasmatic nuclei of the hypothalamus, an extensive network of vasopressinergic nerve fibers emerges, which extends to the olfactory bulb as well as to the spinal cord,[10–13] synaptically innervating extrahypothalamic brain areas.[14] The (a) total absence of this exohypothalamic vasopressin system in the HOM-DI Brattleboro, and the (b) early prenatal development of this system in normal rats,[15–17] taken together with the (c) effects upon brain development of other neurotransmitter and endocrine systems,[18–21] led us to re-examine a possible role of vasopressin in growth, especially in that of the brain.[22]

The present state of biochemical knowledge of the disturbances of regional brain growth was achieved under strictly controlled conditions of Brattleboro breeding. The results suggest that the way in which mutation affects brain growth is different from other known circumstances of disturbed brain development. The most dramatically affected area appears to be the cerebellum, which

0077–8923/82/0394–0703 $1.75/0 © 1982, NYAS

is why this part of the bain has been the primary subject of further investigations in vasopressin supplementation studies.

Design of Breeding

HOM- and HET-DI (control) Brattleboro rats are raised under standard conditions in genetically homogeneous litters by mating HOM-DI females with HOM-DI and HOM-N (normal) males respectively. The Brattleboro parents were obtained from the Dutch Central Institute for Breeding of Laboratory Animals (TNO-CPB, Zeist, the Netherlands). The day of birth is called postnatal day 0. Litters were nursed at constant litter size, usually seven or eight pups. Care was taken to keep the litters dry on ample bedding, which was daily changed sawdust. Water nipples were made accessible for the pups from the beginning of cage exploratory behavior. The prenatal as well as the postnatal environmental conditions for both experimental and control animals were essentially the same in this breeding scheme and intra-litter competition between HET and HOM animals was excluded as well, both prerequisites for a developmental study.[2, 23]

Whole litters were sacrificed during the postnatal brain development period and the genetics of the animals were always checked afterwards either by the radioimmunological detection of the presence or absence of vasopressin in the isolated pituitary[24] or, for animals older than 30 days, by the measurements of diuresis in metabolism cages.[25]

Regional Impairments of Brain Growth

Total brain weight as measured between postnatal day 8 and 30, i.e., including the second peak of neurogenesis in the brain, was 1–9% less in the HOM-DI animals as compared to HET-DI and became about 5% less in adulthood for both sexes.[2] In order to determine the extent to which the brain was regionally affected, selected brain areas were dissected out and weighed.[26] Differences during development could only be demonstrated in the HOM-DI cerebral cortex, cerebellum, and medulla oblongata, whereas the weight of five other areas—olfactory bulb, colliculi, hippocampus, hypothalamus, and a residue (thalamus and basal ganglia mainly)—appeared unaffected (cf. for males Figure 1). However, the severity of the deficit appeared to be quite different for the three areas, as were regional differences apparent in the adult (Table 1, first two lines). For the cerebral cortex, there was only a slight retardation in the early postnatal period, which, at 180 days of age, reached HET-DI levels. A similar observation was made for the more affected medulla, which indicates that significant catch-up of growth exists in the HOM-DI brain after day 30, as is also seen for total body weight.[22] However the lower weight of the cerebellum persists into adulthood, making this area the most severely and consistently affected area in the HOM-DI Brattleboro brain. Tissue parameters such as dry weight and protein and DNA content showed similar impairments in the cerebellum (Figure 2), while average cell density (weight or protein/DNA) at the end of the brain growth spurt did not appear to differ from control.[2] From yet unpublished morphological studies it appears that the cerebellar lobes are differentially affected during the postnatal period and that no uniform growth retardation exists in the different layers.[27]

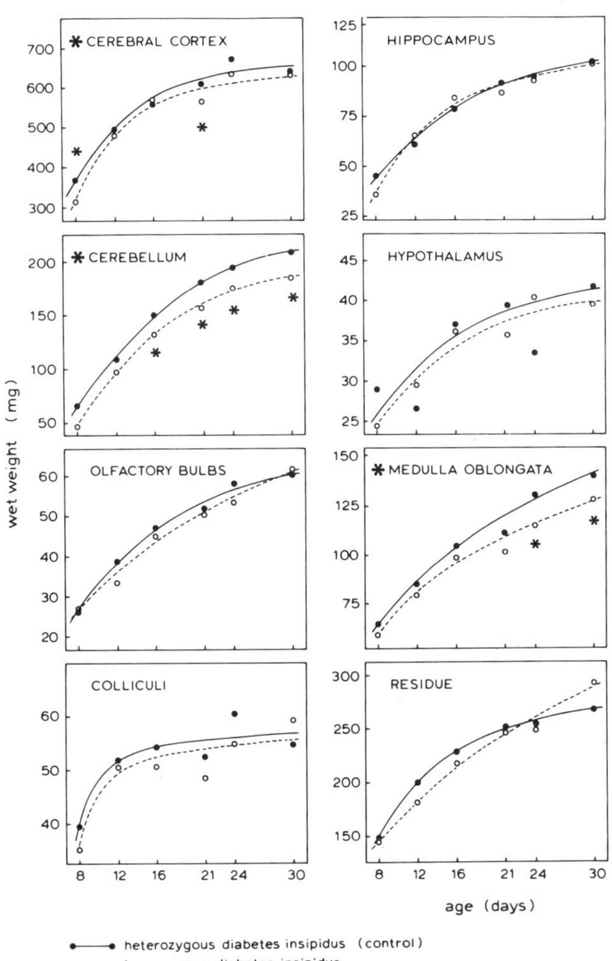

FIGURE 1. Regional postnatal growth of the brain of HET-(control) and HOM-DI Brattleboro male rats. Litters, being either completely homozygous or heterozygous by genotype and brought to 7–8 pups from day 3 onwards, were raised and nursed by a HOM-DI mother. Data are obtained from three litters per age in both groups. For the first month significant differences ($p \leq 0.05$) between the two genotype groups are only present for cerebral cortex, cerebellum, and medulla oblongata (analysis of variance, large asterisks). For these areas statistically significant differences were additionally tested for the various age groups using the Student *t*-test (small asterisks). (Data from Boer *et al.*[2])

TABLE 1

REGIONAL WEIGHT CHANGES IN THE BRAIN OF HOM-DI ADULT BRATTLEBORO RATS UPON ONE MONTH OF PERIPHERAL (A) AND/OR CENTRAL (B) VASOPRESSIN ADMINISTRATION

	Number of Animals	Total Brain (g)	Cerebral Cortex (g)	Cerebellum (mg)	Olfactory Bulbs (mg)	Hippocampus (mg)	Hypo-thalamus (mg)	Colliculi (mg)	Medulla Oblongata (mg)	Residue (mg)	
Males											
HET-DI	9	2.04 ± 0.03 †	0.76 ± 0.02	277 ± 6 †	63 ± 5 †	118 ± 7	43 ± 1	77 ± 6	218 ± 10	362 ± 9	
HOM-DI	10	1.94 ± 0.03	0.74 ± 0.02	259 ± 4	77 ± 5	124 ± 4	40 ± 2	68 ± 3	220 ± 9	341 ± 6	A
HOM-DI O	5	1.92 ± 0.04	0.70 ± 0.02	252 ± 10	76 ± 5	128 ± 8	43 ± 2	65 ± 4	206 ± 14	354 ± 13	A
HOM-DI P	6	1.92 ± 0.03	0.73 ± 0.03	256 ± 9	73 ± 4	131 ± 6	42 ± 2	68 ± 4	211 ± 12	394 ± 53	A
HOM-DI P/S	4	1.96 ± 0.04	0.72 ± 0.04	236 ± 12	66 ± 6	143 ± 8 *†	46 ± 4	58 ± 5	173 ± 15 †	361 ± 19	B
HOM-DI P/VP	7	1.94 ± 0.04	0.70 ± 0.03	251 ± 8	61 ± 3 †‡	125 ± 5	51 ± 5 †	61 ± 2	190 ± 13	376 ± 26	B
HOM-DI O/VP	5	2.09 ± 0.05 *†‡	0.78 ± 0.05	251 ± 8	63 ± 5	143 ± 2 *†	53 ± 4 †	63 ± 3	196 ± 5	406 ± 25 †	B
Females											
HET-DI	10	1.90 ± 0.02 †	0.71 ± 0.02	261 ± 3 †	59 ± 3 †	117 ± 4	40 ± 2	75 ± 6	217 ± 6	330 ± 14	
HOM-DI	7	1.78 ± 0.02	0.70 ± 0.02	245 ± 5	80 ± 4	119 ± 4	41 ± 2	68 ± 3	200 ± 12	340 ± 7	A
HOM-DI O	7	1.76 ± 0.02	0.67 ± 0.01	230 ± 9	65 ± 4 *†	120 ± 4	42 ± 3	62 ± 2	179 ± 10	328 ± 5	A
HOM-DI P	7	1.74 ± 0.02	0.66 ± 0.02	228 ± 7	65 ± 2 *†	117 ± 5	40 ± 2	60 ± 3	185 ± 8	328 ± 11	A
HOM-DI P/S	8	2.03 ± 0.05 †‡	0.84 ± 0.03 †	248 ± 4	51 ± 2 †‡	135 ± 7	43 ± 4	72 ± 5	203 ± 6	349 ± 18	B
HOM-DI P/VP	2	1.91	0.72	240	54	125	55	62	173	358	B
HOM-DI O/VP	7	1.92 ± 0.06 †‡	0.76 ± 0.02 †	234 ± 4	47 ± 4 †‡	121 ± 10	45 ± 4	68 ± 6	197 ± 10	335 ± 14	B

HET-DI rats came out of three litters. The HOM-DI animals were from two series of seven- to eight-pup litters (A and B) and were randomly distributed over the experimental groups. Daily vasopressin administration starting at day 150 of life and continuing for one month consisted of injections of Pitressin tannate (P; 5 U/ml) or arachis oil (O) given subcutaneously (0.1 ml/100 g body weight) and/or 50 ng vasopressin (VP; Sigma grade VIII) in 2 μl saline (S) via a polyethylene canula implanted in the right lateral ventricle according to De Wied.[44] In series B, 40% of the animals did not survive the treatment and occasionally barrel rotation was observed upon the intracerebroventricular injection.[43]

Tissue preparation has been described previously[2] and data given as mean ± SEM. Differences were regarded as statistically significant using Student t-test at the level of ≤ 0.05 and were compared within a series (*) as well as with the untreated HOM-DI group (†). Additionally those HOM-DI data that approach the HET-DI level by the vasopressin treatment are indicated (‡).

Differences in tissue water were absent during the postnatal period of cerebellar growth.[2] However, a 0.7% higher water content was found in the adult cerebellum of HOM-DI rats, which was also present in the cerebral cortex (TABLE 2, first two lines) and colliculi, but not in any other part of the brain.[2]

Normally the size of the rat olfactory bulb does not increase after day 30 of age,[28] which was confirmed for the HET-DI controls. However, in HOM-DI animals the bulbs continued to develop well beyond this period, eventually reaching weights above those of controls (cf. FIGURE 1 and TABLE 1), and with a concomitant increased cell content (TABLE 3).

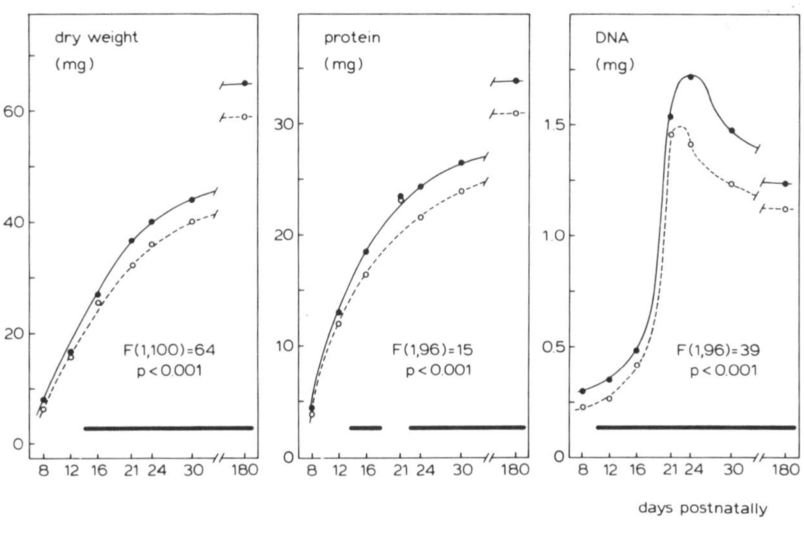

FIGURE 2. Tissue parameters of the developing cerebellum of HET-(control) and HOM-DI male Brattleboro rats. Data are of cerebelli from the animals of FIGURE 1. The difference of the growth curves of the genotypes are all statistically significant differences. Horizontal black bars indicate significance using the Student *t*-test for the various ages. (Adapted from Boer *et al.*[2])

RATE OF CELL ACQUISITION IN THE BRAIN

Using Long-Evans instead of HOM-N Brattleboro males for siring the HET-DI control litters, and with HOM-DI animals obtained from two different sources (Dept. of Anatomy, University of Bristol and Charing Cross Hospital, London, respectively), bred separately in London, the rate of [^{3}H]thymidine incorporation into the DNA of forebrain, cerebellum, and olfactory bulbs was investigated to elucidate the apparent disturbances in cell acquisition of the mutant.[29] To eliminate the influences of changes in the availability of labeled precursor, the results are expressed in terms of the [^{3}H]DNA formed per brain part corrected for the concentration of radioactivity of the acid-soluble fraction freed of tritiated water, i.e. the radioactivity in thymidine and its phosphate derivates (FIGURE 3).

A small but consistent decrease in cell acquisition rates, measured in terms of amount of DNA per brain part, was observed in all the three brain regions of HOM-DI rats, but, in agreement with previous studies described above (cf. FIGURES 1 and 2), the differences between HOM-DI and HET-DI rats were significant only for the cerebellum (FIGURE 3). A similar small reduction in [^{3}H]thymidine incorporation into DNA was also observed throughout the brain; however, in contrast to total cell number results, the differences were significant only for forebrain and olfactory bulbs. As expected, the reduction in DNA

TABLE 2

BRAIN WATER CONTENT OF THE HOM-DI BRATTLEBORO ADULT RATS UPON PERIPHERAL (A) AND/OR CENTRAL (B) VASOPRESSIN ADMINISTRATION

	One day Urine Production (ml/100 g b.w.)	Tissue Water (%)		
		Cerebral Cortex	Cerebellum	
Males				
HET-DI	5.5 ± 0.6 †	76.9 ± 0.1 †	76.7 ± 0.1 †	
HOM-DI	53 ± 5	77.6 ± 0.04	77.3 ± 0.2	A
HOM-DI O	45 ± 3	77.4 ± 0.2	77.2 ± 0.1	A
HOM-DI P	3.1 ± 0.5 *†‡	77.5 ± 0.2	77.2 ± 0.2	A
HOM-DI P/S	7.4 ± 2.7 *†‡	77.7 ± 0.4	76.9 ± 0.3	B
HOM-DI P/VP	5.9 ± 0.4 *†‡	77.8 ± 0.2	77.1 ± 0.1	B
HOM-DI O/VP	51 ± 5	77.7 ± 0.4	77.0 ± 0.2	B
Females				
HET-DI	5.8 ± 0.6 †	76.9 ± 0.1 †	76.4 ± 0.1 †	
HOM-DI	91 ± 6	77.5 ± 0.1	77.2 ± 0.2	A
HOM-DI O	78 ± 6	77.5 ± 0.1	77.3 ± 0.1	A
HOM-DI P	19 ± 5 *†‡	77.7 ± 0.1	77.3 ± 0.1	A
HOM-DI P/S	11 ± 3 *†‡	78.6 ± 0.3 †	77.1 ± 0.1	B
HOM-DI P/VP	8.5	78.0	77.1	B
HOM-DI O/VP	83 ± 8	78.1 ± 0.2 †	77.0 ± 0.2	B

The groups of animals are the same as described in TABLE 1 (see also for abbreviations and statistics).

Daily urine production data are the averages ± SEM of values obtained one week after the onset of daily injections (day 157 of age) and before sacrifice (day 180). Tissue water percentage was calculated after wet and dry weight measurements.[2]

labeling in HOM-DI rats preceded the decrease in brain cell number. In the cerebellum, in contrast to the forebrain and olfactory bulbs, the rate of acquisition of cells (DNA deposition) for the same mitotic activity ([^{3}H]DNA labeling) was very large. Therefore, it is possible that while the consistent decrease in DNA labeling was associated with a significant reduction in cerebellar cell number, in forebrain and olfactory bulbs the significant decrease in [^{3}H]DNA formation was accompanied by a reduction in the amount of DNA that did not reach statistical significance. However, possible differences in thymidine metab-

TABLE 3

DNA CONTENT OF CEREBRAL CORTEX, CEREBELLUM, AND OLFACTORY BULBS OF HOM-DI MALE BRATTLEBORO RATS AFTR VASOPRESSIN THERAPY

	Cerebral Cortex	Cerebellum	Olfactory Bulb	
HET-DI	1.09 ± 0.02	1.24 ± 0.06	0.23 ± 0.02 †	
HOM-DI	1.10 ± 0.04	1.13 ± 0.07	0.32 ± 0.01	A
HOM-DI O	1.01 ± 0.02	1.06 ± 0.06	0.32 ± 0.02	A
HOM-DI P	1.02 ± 0.03	0.95 ± 0.09	0.30 ± 0.02	A
HOM-DI P/S	1.05 ± 0.18	0.91 ± 0.11	0.27 ± 0.02 †‡	B
HOM-DI P/VP	0.91 ± 0.09	1.10 ± 0.08	0.21 ± 0.01 *†‡	B
HOM-DI O/VP	1.14 ± 0.21	1.14 ± 0.09	0.27 ± 0.02 †‡	B

The group of animals are the same as those described in TABLE 1 (see also for abbreviations and statistics).

DNA was measured in the frozen-dried tissues using the ethidium bromide method [45] and given in μg ± SEM.

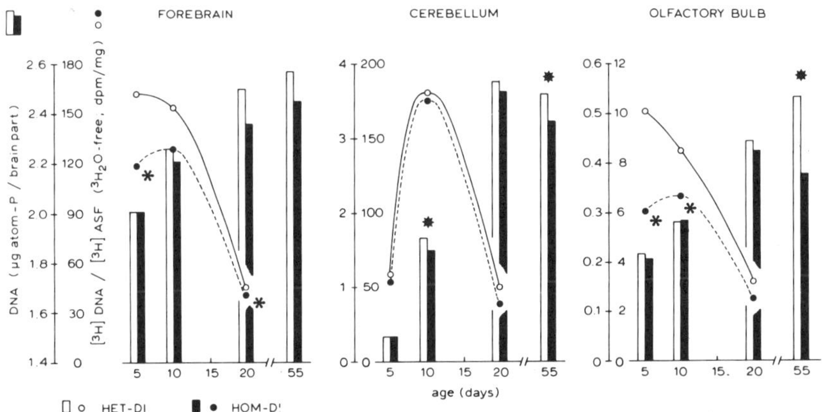

FIGURE 3. Cell acquisition in brain of HET-(control) and HOM-DI Brattleboro rats. Homogeneously genotyped litters were raised by HOM-DI females after mating with HOM-DI or Long-Evans males. The respective HOM- and HET-DI litters were given subcutaneous [^{3}H]thymidine injections on the day of investigation (one litter per age) and killed 30 min later in order to determine its incorporation into the DNA. No distinction of sex was made. [^{3}H]DNA data are the mean and are given as corrected values by expressing them in terms of concentration of labeled thymidine plus its phosphate derivates, i.e. the acid-soluble fraction freed of [^{3}H]water. Total DNA content was also determined (vertical bars). Statistically significant differences for the various ages using Student *t*-test ($p \leq 0.05$) are given for both parameters by asterisks and stars, respectively. (Adapted from Boer & Patel.[29])

olism and in postnatal cell death between HET-DI and HOM-DI may also have contributed to these apparent discrepancies.

Another puzzling observation was a 30% reduction in the amount of DNA in the olfactory bulbs of 55-day old HOM-DI rats (FIGURE 3). This result is exactly opposite to that in the earlier series reported in TABLE 3. Possible differences in the Brattleboro breeding stocks used in the two experiments as well as the use of Long-Evans instead of HOM-N Brattleboro males to sire HET-DI pups in the present experiments may have contributed to the discrepancy. However, the influences of the environment and breeding conditions cannot be excluded (see below). Apart from these discrepancies, the present results do at least reveal disturbances in both [^{3}H]thymidine incorporation into DNA and in total cell number in the brain of the mutant rat.

Etiology of the Impairment

A number of environmental, nutritional, hormonal, and neurotransmitter factors are known to influence the development of the brain, causing persistent deficits. Although the present breeding scheme used to raise control and experimental Brattleboro groups should largely exclude the first two factors, changes in the endogenous factors, whether related to the vasopressin deficiency or not, may be responsible for the HET-/HOM-DI differences. So far, however, the mutation in the Brattleboro strain appears to have its own particular etiology among the brain growth disturbances caused by several known factors that have been investigated biochemically. Taking together (a) the absence of a marked effect on neocortical growth, (b) the rather specific and persistent defect in the cerebellum, (c) the unusual alterations for the olfactory bulb weights, (d) the unaffected average cell size (weight or protein/DNA) in the adult cerebellum, and (e) the relatively less reduction in the postnatal [^{3}H]thymidine incorporation into DNA in the cerebellum than in forebrain and olfactory bulbs, along with the observation of (f) an earlier opening of the eyes in HOM-DI Brattleboro rats,[15] one is led to the conclusion that neither malnutrition, thyroid hormone imbalance, nor increased corticosteroid levels can account for the present results.[22] Elsewhere it has been argued that growth hormone deficiency is also not likely to be involved.[2, 22]

Although other factors are not definitely excluded, the particular expression of the Brattleboro mutation upon brain development is assumed to depend directly or indirectly upon the absence of vasopressin in HOM-DI animals. This is supported by the observations that vasopressin is a mitogenic agent *in vitro* on mouse fibroblasts,[30] on rat chondrocytes,[31] and on rat thymic lymphocytes;[32] and *in vivo* on the short-term post-hemorrhage growth response in red bone marrow.[33] However, it was hoped that results would come from vasopressin supplementation experiments.

Vasopressin Supplementation

Vasopressin supplementation in HOM-DI Brattleboro rats has been reported not to restore the impaired body growth. However the therapies were based upon peripheral administrations of either Pitressin tannate, a mixture of the tannates of vasopressin and ^{8}lysine-vasopressin,[6] or ^{8}lysine-vasopressin alone.[5]

Since only a small percentage of vasopressin given peripherally reaches the brain,[34] we have also administered the actual missing (^{8}arginine-)vasopressin intracerebrally, both in adulthood and in the early postnatal period.

Adulthood

Two series of age-matched 150-day-old HOM-DI animals of both sexes handled completely similarly [2] received a one-month vasopressin therapy, either peripherally by subcutaneous injections (series A), or centrally using a canula in the right lateral ventricle, or a combination of both routes (series B) (see legends TABLE 1).

Although the *peripheral* Pitressin tannate injections largely normalized diuresis (TABLE 2), there were no significant effects, either for male and female mutants, on total and regional brain weights (A, TABLE 1) or on brain tissue water (A, TABLE 2). Moreover, the DNA content of the areas known to be affected did not change in the male HOM-DI adult (A, TABLE 3). The only change observed was a lower olfactory bulb weight in the females given arachis oil as well as to Pitressin tannate treatment, and this cannot therefore be regarded as an effect of the injected vasopressin.

Males that received *intracerebroventricularly* 50 ng of vasopressin daily (O/VP group of series B) had significantly greater total brain weights, approaching the levels of the HET-DI animals (TABLE 1). The underlying regional weight changes, however, cannot be interpreted as a HOM-DI brain assuming HET-DI characteristics, since the cerebellar weight remained unaffected, whereas hippocampal, hypothalamic, and 'residual' weights are in fact enlarged as compared to the untreated HOM-DI males of series A (TABLE 1). Only the olfactory bulbs attained the HET-DI size, but within series B this effect could not significantly be related to the intraventricularly administered vasopressin, whereas generally the figures for the bulb weights already appeared to be at the HET-DI level (TABLE 1). However, with Brattleboro males given vasopressin centrally as well as peripherally, when no change was seen for total brain weight, on the basis of DNA content, a significant change diminishing the HOM/HET difference was apparent for the olfactory bulbs (P/VP group, TABLE 3). Moreover, this group has a significantly lower hippocampal weight as compared to the others in the series (TABLE 1).

The results of central application of vasopressin in females appeared less differentiated. No increase in total brain weight as well as in the isolated areas was found (TABLE 1). However, for the series B females, brain weights all appeared to be indistinguishable from the HET-DI reference value. The even more enlarged water content of (at least) the cerebral cortex, which was not seen for the males in this series B (TABLE 2), is consistent with this finding. It might have been caused (female-specifically) by the repeated intraventricularly injections. As in males, the olfactory bulb in the series was smaller as compared to the untreated HOM-DI group of series A (TABLE 1).

In view of these results, the present findings appear to be inconsistent with the idea that vasopressin might be able to restore the brain deficits of the Brattleboro mutant in the adult. The finding that olfactory bulb weights in series B were not different from the HET-DI figure, as seen in series A (FIGURE 1), may point to the sensitivity of the olfactory bulb to environmental changes, since the two series were injected at different periods of the year.

Neonatal Period

Vasopressin treatment has recently been performed postnatally using two sets of three eight-pup litters.[35] The pups were taken from six or seven HOM-DI nests at day 5 of age in order to obtain a small body weight range together with a 50/50 composition for both sexes. Each litter, nursed by a HOM-DI mother, was randomly divided into three groups from which one received control treatment.

Pitressin tannate given subcutaneously in a daily dosage of 0.25 U from day 5 until day 10 and thereafter 0.5 U until day 28 of life, decreased brain weight as measured on day 32, but did not significantly affect cerebellar weight and DNA content (FIGURE 4). Vasopressin suspended in arachis oil and given as similar dosages failed to have any apparent effect on growth when compared to control. However a lower cerebellar DNA content was found as compared to the Pitressin tannate group.

Centrally applied vasopressin was given daily under light ether anesthesia by direct injection alternately into the right and left lateral ventricle of the brain. The site of the injection was drawn on the head-skin using experimentally obtained coordinates and fixed depths depending on the age of the pups (FIGURE 5). One or 100 ng of vasopressin in 5 μl saline was administered daily by a perpendicular injection through the skull from day 5 until 28. All the pups withstood the short daily stress very well. In general, however, these litters grew somewhat less than those used for the peripheral treatment study. This may be partly due to the injection procedure, or determined by the body weight differences at the onset of treatment (10.7 ± 0.1 g versus 11.0 ± 0.1 g for the 24 selected pups of both series). From FIGURE 4 it will be clear that neither dose of the missing vasopressin, centrally applied in the neonatal period, did not influence the selected brain parameters.

DISCUSSION AND CONCLUSION

The reduced size of the cerebellum throughout life is the most characteristic and obvious defect of the HOM-DI Brattleboro rats. It can be expected to have its consequences for these animals in behavioral tasks in addition to those on the vasopressin-related memory processes.[36]

Assuming that the absence of vasopressin in the mutant is involved in the stunted growth, it may well be that vasopressin is acting directly on the brain itself. Three possible routes may be mentioned in this respect. Firstly, the blood circulation for the neurohypophyseal-released vasopressin. The hypothalamo-neurohypophyseal system is already present prenatally at day 17,[16, 37] and becomes functional shortly after birth,[38] but probably even prenatally.[39] Moreover the blood-brain barrier is not well developed perinatally, and for rat cerebellum even not completely established before the end of the third week of postnatal life.[40] The relatively frequent occurrence of vasopressin nerve endings along the third ventricle in fetal and early postnatal stages of the rat [15, 16, 22] might point to the cerebrospinal fluid as a second route of transport at the time of the period of neurogenesis in the brain. Finally, there exist vasopressinergic neurites, many of which reach extrahypothalamic brain areas long before neurogenesis virtually stops [15–17] and these may provide a direct route of transfer.

So far, substitution of vasopressin in early neonatal life or in the adult given either peripherally or centrally did not restore the developmental disturbances of the HOM-DI brain. The negative finding in adults seem to exclude

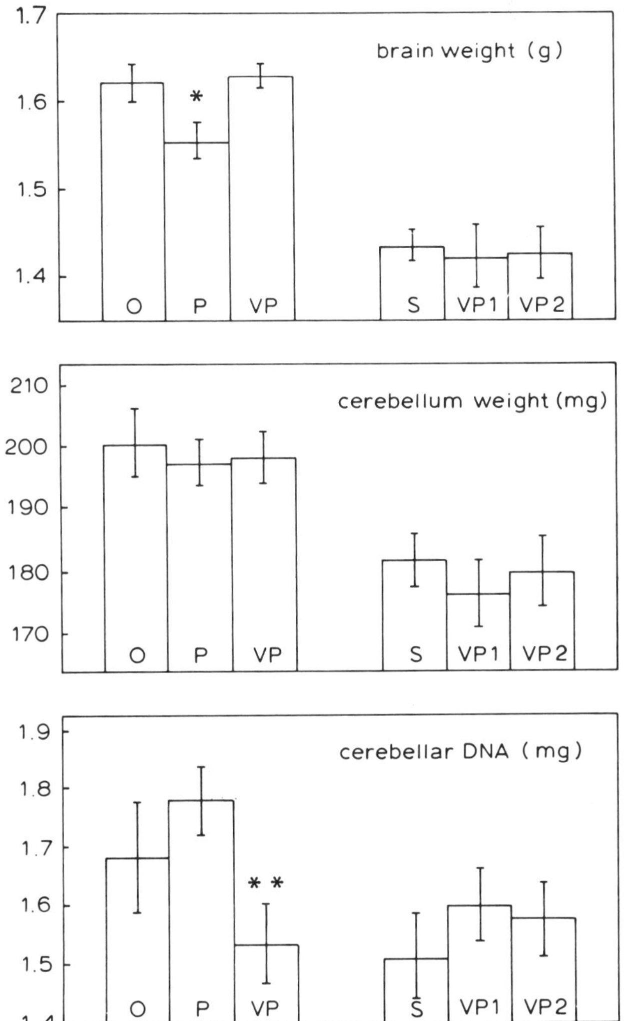

FIGURE 4. Brain and cerebellum weight of young HOM-DI Brattleboro rats after vasopressin therapy during the first month of life. In the first series (left bars), the pups of three eight-pup litters of small weight range received either a daily subcutaneous injection of arachis oil as control (O), Pitressin tannate (P), or vasopressin suspended in oil (VP). Dosage from day 5 to 10 was 0.25 U/0.25 ml, and thereafter until day 28 of age 0.5 U/0.5 ml. In a similar second series (right bars), the young received a daily 5 μl vasopressin injection throughout the same period directly into the lateral ventricle either containing 1 (VP 1) or 100 ng (VP 2) in saline (S). Vasopressin was obtained from Sigma (grade VIII). In both series pups were sacrificed at the age of 32 days and data are given as mean ± SEM for the eight pups per experimental group. (Sex of pups not determined.) A single asterisk denotes statistically significant difference from control, whereas two asterisks indicate significance with other experimental group. (Student *t*).

a possible relationship between vasopressin and glial proliferation, as has been suggested previously.[1] In spite of the absence of an effect on brain growth postnatally, permanent changes after vasopressin treatment have been described and were observed in the present experiments. Prenatal as well as postnatal treatment with vasopressin or its analogues results in a long-lasting increase in body water turnover in later life.[5, 35, 41, 42] For HOM-DI Brattleboro rats no indication was found for the induction of antibodies against vasopressin as an explanation for this phenomenon and response seems to be due to a central effect and not determined on the level of the kidney.[42]

On account of the hitherto negative findings we now propose that the possible influence of vasopressin on brain growth requires continuous instead of discontinuous (central) treatment or earlier replacement of the missing vasopressin. The doses injected into the ventricular system were the highest possible. Adult rats given more than 50 ng of vasopressin all at once frequently showed barrel rotation behavior,[43] while in the young the 100 ng dose sometimes caused head tremor. Continuous application might avoid these toxic phenomena. On the other hand the stunted growth may already be determined prenatally. HOM-DI brain weight measured the day after birth, shows a 10% deficit as compared to HET-DI controls both for males and females (TABLE 4). Immunocytochemically identifiable vasopressin cells first appear by day 16 of prenatal age and signs of exohypothalamic pathways reaching their final target brain areas were seen from fetal day 17 onwards.[15, 16] An effect of vasopressin on the formation of Purkinje cells may for instance have its repercussion in postnatal cerebellar development. In order to test this hypothesis it will be necessary to apply vasopressin during the intrauterine period, a difficult task!

SUMMARY

The mutation of the HOM-DI Brattleboro rat appears to result in an impairment of brain growth. These deficits cannot be accounted for by any other

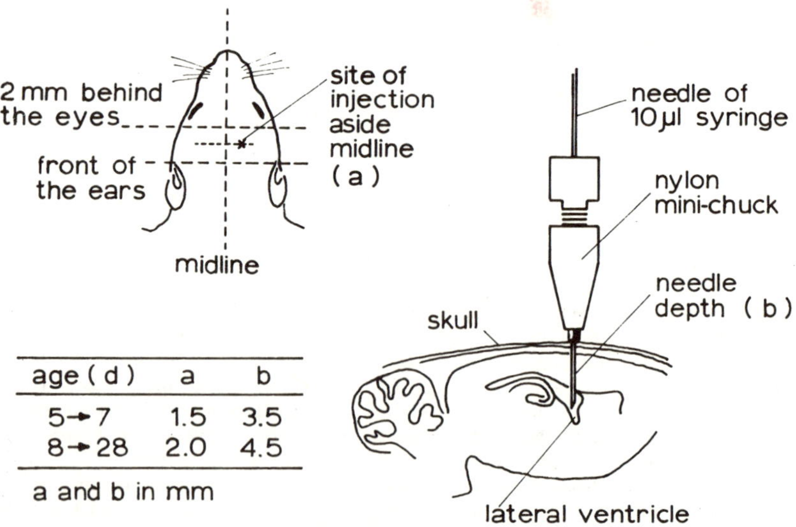

FIGURE 5. Direct intracerebroventricular injection in neonatal rats.

TABLE 4

DAY 1 BRAIN WEIGHT OF HET-(CONTROL) AND HOM-DI BRATTLEBORO RATS

		N	Body Weight (g)	Brain Weight (mg)
HET-DI	♂	21	7.13 ± 0.11	296 ± 4
	♀	24	6.99 ± 0.12	285 ± 3
HOM-DI	♂	27	6.50 ± 0.10 *	266 ± 3 *
	♀	27	6.19 ± 0.10 *	260 ± 3 *

Data are the mean ± SEM of pups derived from five litters in each group. Only litters having at birth between 8 and 11 young were sacrificed; average litter size was 9.0 in both control and experimental group. Asterisk denotes statistically significant difference at the level of $p < 0.05$ (Student *t*-test) as compared to HET-DI.

known circumstance of disturbed brain development. The cerebellum in particular appears to be persistently lower in weight and cell content.

The absence of vasopressin in these animals is assumed to be involved directly or indirectly in the stunted brain growth. However, daily injections of vasopressin peripherally and centrally, both in the adult as well as between day 5 and day 28 of postnatal life, did not reserve the abnormal brain development in homozygote DI Brattleboro rats. It is possible that the vasopressin may exert its effect on brain development already before birth.

ACKNOWLEDGMENTS

The authors wish to thank Prof. Dr. D. F. Swaab and Dr. R. Balàzs for their advice and encouragement, Dr. R. Baker and Dr. B. W. L. Brooksbank for revising the English style and Mrs. A. van der Velde and I. Kemna for their typing contributions. This work is supported, in part, by a grant from the Wellcome Trust, London, England.

REFERENCES

1. BOER, G. J., H. B. M. UYLINGS, C. M. F. VAN RHEENEN-VERBERG & B. FISSER. 1978. Postnatal brain development in rats with hereditary diabetes insipidus (Brattleboro Strain). *In* Hormones and Brain Development. G. Dörner & M. Kawakami, Eds.: 253–258. Elsevier. Amsterdam.
2. BOER, G. J., C. M. F. VAN RHEENEN-VERBERG & H. B. M. UYLINGS. 1982. Impaired brain development of the diabetes insipidus Brattleboro rat. Dev. Brain Res. (In press.)
3. VALTIN, H. & H. A. SCHROEDER. 1964. Familial hypothalamic diabetes insipidus in rats (Brattleboro strain). Am. J. Physiol. **206:** 425–430.
4. SAUL, G. B., E. B. GARRITY, K. BENIRSCHKE & H. VALTIN. 1968. Inherited hypothalamic diabetes insipidus in the Brattleboro strain of rats. J. Hered. **59:** 113–117.
5. WRIGHT, W. A. & C. L. KUTSCHER. 1977. Vasopressin administration in the first month of life: effects on growth and water metabolism in hypothalamic diabetes insipidus rats. Pharm. Biochem. Behav. **6:** 505–509.
6. SOKOL, H. W. & J. SISE. 1973. The effect of exogenous vasopressin and growth hormone on the growth of rats with hereditary hypothalamic diabetes insipidus. Growth **37:** 127–142.

7. DE WIED, D., B. BOHUS & TJ. B. VAN WIMERSMA GREIDANUS. 1974. The hypothalamoneurohypophyseal system and the preservation of conditioned avoidance behavior in rats. Progr. Brain Res. **41:** 417–428.
8. KOVÁCS, G. L., B. BOHUS & D. H. G. VERSTEEG. 1980. The interaction of posterior pituitary neuropeptides with monoaminergic neurotransmission: significance in learning and memory processes. Progr. Brain Res. **53:** 123–140.
9. COOPER, K. E., N. W. KASTING, K. LEDERIS & W. L. VEALE. Evidence supporting a role for endogenous vasopressin in natural suppression of fever in the sheep. J. Physiol. **295:** 33–45.
10. WEINDL, A., M. V. SOFRONIEW & I. SCHINKO. Psychotrope Wirkungen hypothalamischer Hormone: Immunohistochemische Identifikation extrahypophysärer Verbindungen neuroendokriner Neurone. Arzeneimittelforsch. **26:** 1191–1194.
11. BROWNFIELD, M. S. & G. P. KOZLOWSKI. 1977. The hypothalamo-choroidal tract. I. Immunohistochemical demonstration of neurophysin pathways to telencephalic plexuses and cerebrospinal fluid. Cell Tiss. Res. **178:** 111–127.
12. BUIJS, R. M. 1978. Intra- and extrahypothalamic vasopressin and oxytocin pathways in the rat. Pathways to the limbic system, medulla oblongata and spinal cord. Cell Tiss. Res. **192:** 423–435.
13. STERBA, G., W. NAUMANN & G. HOHEISEL. 1980. Exohypothalamic axons of the classic neurosecretory systems and their synapses. Prog. Brain Res. **53:** 141–158.
14. BUIJS, R. M. & D. F. SWAAB. 1979. Immunoelectronmicroscopical demonstration of vasopressin and oxytocin synapses in the rat limbic system. Cell Tiss. Res. **204:** 355–365.
15. BOER, G. J., R. M. BUIJS, D. F. SWAAB & G. J. DE VRIES. 1980. Vasopressin and the developing rat brain. Peptides **1**(Suppl. 1): 203–209.
16. BUIJS, R. M., D. N. VELIS & D. F. SWAAB. 1980. Ontogeny of vasopressin and oxytocin in the fetal rat: early vasopressinergic innervation of the fetal brain. Peptides **1:** 315–324.
17. DE VRIES, G. J., R. M. BUIJS & D. F. SWAAB. 1981. Ontogeny of the vasopressinergic neurons of the nucleus suprachiasmaticus and their exohypothalamic projections in the rat brain. Presence of a sex-linked difference in the lateral septum. Brain Res. **218:** 67–78.
18. BALÁZS, R. 1974. Hormonal influence on brain development. *In* Biochemistry and Mental Disorders. L. L. Iversen & S. P. R. Rose, Eds.: 39–57. Biochemical Society. London.
19. SWAAB, D. F. 1980. Neuropeptides and brain development—a working hypothesis. *In* Multidisciplinary Approach to Brain Development. C. Di Benedetta, R. Balázs, G. Gombos & P. Porcellati, Eds.: 181–196. Elsevier. Amsterdam.
20. PATEL, A. J. & P. D. LEWIS. 1981. Effects on cell proliferation of pharmacological agents acting on the central nervous system. *In* Mechanisms of Actions of Neurotoxic Substances. A. Vernadakis & K. N. Prasad, Eds.: 181–218. Raven Press. New York, N.Y.
21. PATEL, A. J., O. BAROCHOVSKY & P. D. LEWIS. 1981. Psychotropic drugs and brain development: effects on cell replication *in vivo* and *in vitro*. Neuropharmacology. (In press.)
22. BOER, G. J., D. F. SWAAB, H. B. M. UYLINGS, K. BOER, R. M. BUIJS & D. N. VELIS. 1980. Neuropeptides in rat brain development. Prog. Brain Res. **53:** 207–227.
23. BOER, G. J., K. BOER & D. F. SWAAB. 1982. On the reproductive and developmental differences within the Brattleboro strain. Ann. N.Y. Acad. Sci. (This volume.)
24. DOGTEROM, J., TJ. B. VAN WIMERSMA GREIDANUS & D. F. SWAAB. 1977. Evidence for the release of vasopressin and oxytocin into cerebrospinal fluid: measurements in plasma and CSF of intact and hypophysectomized rats. Neuroendocrinology **24:** 108–118.

25. SWAAB, D. F., G. J. BOER & J. W. L. NOLTEN. 1973. The hypothalamo-neurohypophysial system (HNS) of the Brattleboro rat. Acta Endocrin. Suppl.**177:** 41–53.
26. PATEL, A. J., M. DEL VECCHIO & D. J. ATKINSON. 1978. Effect of undernutrition on the regional development of transmitter enzymes: glutamate decarboxylase and choline acetyltransferase. Devel. Neurosci. **1:** 41–53.
27. UYLINGS, H. B. M., W. VAN NORDE & G. J. BOER. Unpublished results.
28. ROSELLI, L. & J. ALTMAN. 1979. The postnatal development of the main olfactory bulb of the rat. J. Develop. Physiol. **1:** 295–313.
29. BOER, G. J. & A. J. PATEL. Disorders of cell acquisition in the postnatal developing brain of Brattleboro rats deficient for vasopressin. (In preparation.)
30. ROZENGURT, E., A. LEGG & P. PETTICAN. 1979. Vasopressin stimulation of mouse 3T3 cell growth. Proc. Natl. Acad. Sci. USA **76:** 1284–1287.
31. MILLER, R. P., F. HUSAIN, M. SVENSSON & S. LOHIN. 1977. Enhancement of (^{3}H-methyl)thymidine incorporation and replication of rat chondrocytes grown in tissue culture by plasma, tissue extracts and vasopressin. Endocrinology **100:** 1365–1375.
32. WHITFIELD, J. F., A. D. PERRIS & T. YOUDALE. 1969. The calcium mediated promotion of mitotic activity in rat thymocyte populations by growth hormone, neurohormones, parathyroid hormone and prolactin. J. Cell Physiol. **73:** 203–212.
33. HUNT, N. H., A. D. PERRIS & P. A. SANDFORD. 1977. Role of vasopressin in the mitotic response of rat bone marrow cells to haemorrhage. J. Endocrin. **72:** 5–16.
34. WILLUMSEN, N. B. S. & P. BIE. 1969. Tissue to plasma ratios of radioactivity in the rat hypothalamo-hypophyseal system after intravenous injection of ^{3}H-Lysine^{8}-vasopressin and ^{3}H-mannitol. Acta Endocrin. **60:** 389–400.
35. KRAGTEN, R. & G. J. BOER. Unpublished results.
36. BRITO, G. N. O., G. J. THOMAS, S. I. GINGOLD & D. M. GASH. 1981. Behavioral characteristics of vasopressin-deficient rats (Brattleboro strain). Brain Res. Bull. **6:** 71–75.
37. WOLF, G. & G. STERBA. 1978. Development of the hypothalamo-neurohypophysial system in rats. *In* Hormones and Brain Development. G. Dörner & M. Kawakami, Eds.: 217–222. Elsevier. Amsterdam.
38. SINDING, C., A. G. ROBINSON & S. M. SEIF. 1980. Levels of neurohypophyseal peptides in the rat during the first month of life. II. Response to physiological stimuli. Endocrinology **107:** 755–760.
39. GALABOV, P. & T. H. SCHIEBLER. 1978. The ultrastructure of the developing neural lobe. Cell Tiss. Res. **189:** 313–329.
40. JOHANSON, C. E. 1980. Permeability and vascularity of the developing brain: cerebellum vs cerebral cortex. Brain Res. **190:** 3–16.
41. LICHARDUS, B., J. PONEC & A. BRLICOVÁ. 1978. The application of dDAVP in pregnancy interferes with ontogenesis of osmoregulation in rats. *In* Hormones and Brain Development. G. Dörner & M. Kawakami, Eds.: 247–252. Elsevier Amsterdam.
42. BOER, G. J. & D. F. SWAAB. Unpublished results.
43. KRUSE, H., TJ. B. VAN WIMERSMA GREIDANUS & D. DE WIED. 1977. Barrel rotation induced by vasopressin and related peptides in rats. Pharmac. Biochem. Behav. **7:** 311–313.
44. DE WIED, D. 1976. Behavioral effects of intraventricularly administered vasopressin and vasopressin fragments. Life Sci. **19:** 685–690.
45. BOER, G. J. 1975. A simplified microassay of DNA and RNA using ethidium bromide. Anal. Bioch. **65:** 225–231.

(DISCUSSION OF THIS PAPER BEGINS ON PAGE 728.)

AN ALTERED NORADRENERGIC INNERVATION OF THE BRATTLEBORO RAT SUPRAOPTIC NUCLEUS *

Jan Schöler † and John R. Sladek, Jr.

Department of Anatomy
University of Rochester School of Medicine
Rochester, New York 14642

INTRODUCTION

The magnocellular neurosecretory peptides oxytocin (OX) and vasopressin (VP) are synthesized in and transported by neurons of the paraventricular (PVN) and supraoptic (SON) nuclei of the hypothalamus.[1] Classically these substances have been known to play an important role in the onset of parturition and milk ejection and to help regulate the water and electrolyte balance within intra- and extra-cellular fluid compartments. More recently however, it has become apparent that these peptides also function to integrate autonomic and endocrine mechanisms by means of the apparent reciprocal connection between hypothalamus and brain stem.[2, 3] Nilaver *et al.*[3] demonstrated that although the caudal projection from the paraventricular nucleus is largely oxytocinergic, a VP projection innervates the nucleus tractus solitarius (NTS) and the dorsal motor nucleus of the vagus. Ward and Gann[4] demonstrated these two regions of the medulla to be involved in the reflexive release of ACTH in response to hemodynamic stimuli arising from cardiac receptors. Although the relation between VP and corticotropin-releasing factor (CRF) is still not fully resolved,[5] the presence of VP fibers in the region of NTS and the dorsal motor nucleus of the vagus raised the possibility of VP modulating ACTH secretion by influencing afferent input to the brain stem[3] in addition to its apparent involvement of ACTH secretion at the pituitary level.[6]

The SON and PVN receive a dense input from brain stem noradrenergic neurons.[7–9] The densest accumulation of fluorescent varicosities within the SON is localized to its ventral regions, where the majority of neurons are vasopressinergic.[10] A recent developmental study has demonstrated the onset of peptide staining within the SON and PVN to precede the presence of fluorescent varicosities in the same regions.[11] Based on this finding, it was postulated that the neurosecretory products might act as neurotrophic agents providing a target for ingrowing, noradrenergic axons during the establishment of their functional interaction. To further test this hypothesis, the Brattleboro rat, genetically devoid of a specific target neuron peptide, was employed to determine if the absence of VP is correlated with an altered noradrenergic innervation. Our initial studies have demonstrated an altered noradrenergic input into the SON of the Brattleboro rat,[12] which raised the issue of whether the apparently decreased innervation was the result of an increased turnover of transmitter

† Present address: Department of Anatomy, LSU School of Medicine, New Orleans, Louisiana.

* This work was supported by Public Health Service grants NS 15816, AG 00847 and National Institute of Mental Health training grant MH 14577.

0077–8923/82/0394–0718 $1.75/0 © 1982, NYAS

within the terminals or whether the terminals were no longer present. The present study describes the altered noradrenergic input to the SON as well as preliminary data on the blocking of monoamine oxidase activity within noradrenergic terminals of the SON.

Materials and Methods

For the morphological study of the noradrenergic input into the SON, six adult, male homozygous Brattleboro rats and five normal, Long-Evans rats of the same age were processed for formaldehyde-induced fluorescence of catecholamines.[13] All the blocks processed in this manner were serially sectioned at 6 μm and every tenth section was mounted and stained with Luxol-fast blue and Cresyl violet for anatomical orientation. Sections were chosen for fluorescence microscopy throughout rostral, middle, and caudal levels of the SON; adjacent sections were mounted on gelatin-coated slides and stained immunocytochemically for rat neurophysin using the PAP procedure.[14] Photomontages were assembled of the SON at each level. All neurons of the SON were traced onto acetate overlays marking (1) those that stained positively for neurophysin and (2) in the Brattleboro rat, those that stained negatively for NP (i.e., the majority of which are the VP-deficient neurons). The acetate overlay was then transferred to the photomontage of the adjacent section processed for fluorescence microscopy at which time the varicosities abutting both NP-positive and NP-negative perikarya were marked. Utilizing this approach, the pattern of varicosities within the SON of normal and Brattleboro rats was assessed quantitatively. A total of 1,415 neurons was analyzed; the number of varicosities in apposition to each neuron was counted. To determine if a statistically significant difference existed between the Brattleboro and control rats, one-way analysis of variance was performed on the means of the number of varicosities in apposition to perikarya per animal.

To allow more precise identification of individual fluorescent varicosities, the second phase of the study employed an aqueous aldehyde method (FAGLU) for fluorescent histochemical localization of catecholamines (CA).[15] To increase catecholamine fluorescence, two homozygous Brattleboro rats and two normal controls were injected with pargyline (100 mg/kg). Four hours later the treated rats and an equal number of saline-injected controls were perfused with 4% paraformaldehyde, 0.5% gluteraldehyde in 0.1 M phosphate buffer pH 7.0. After overnight fixation, the brains were processed through increasing concentrations of a polyethelene glycol-fixative mixture and finally, embedded in 100% polyethylene glycol (molecular weight 1,000).[16] Sections were cut at two microns on a rotary microtome at five different levels through the SON and photographed. Again, photomontages were constructed and the number of varicosities in apposition to neuronal cell bodies as well as the overall fluorescence within the SON was judged independently by four observers.

Results

Immunocytochemical staining for NP in the Brattleboro rat identifies those cells that contain oxytocin and leaves the vasopressin-deficient cells unstained. In examining the noradrenergic input into the nucleus, it is important to recall the pattern of neurophysin staining within the nucleus since each population of cells occupies a specific portion of the nucleus (FIGURE 1).

The SON of control animals contained a dense input of catecholamine fibers. The distribution was densest in the ventral regions of the nucleus and it was progressively less dense more dorsally (FIGURE 2a). This pattern of noradrenergic input clearly favored the VP neurons in the normal animal as reported for other rat strains.[10, 17]

The catecholamine fluorescence in the SON of the Brattleboro rat, however, appeared different. The ventral regions of the nucleus which contained the

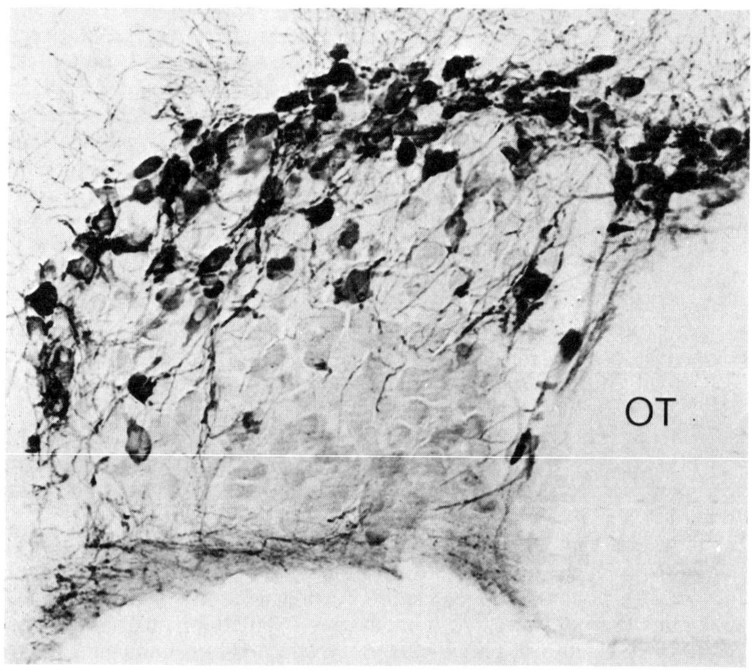

FIGURE 1. In this coronal section of the SON of the Brattleboro rat, immunohistochemical staining reveals the location of oxytocin-containing neurons following staining for rat neurophysin. They predominate in dorsal portions of the nucleus at this neuroanatomical level. Vasopressin-deficient neurons appear negatively stained and predominate in the ventral portion of the nucleus. OT, optic tract ×200.

VP-deficient neurons was characterized by a marked decrease in CA fluorescence; moreover varicosities were observed only occasionally in apposition to VP-deficient neurons (FIGURE 2b). However, more dorsally an increased number of varicosities was seen in comparison to normal, Long-Evans rats. In this OX-rich portion of the nucleus, many neurons were seen apposed by solid rims of varicosities (FIGURE 2b). This apparent shift in the noradrenergic terminal distribution within the SON of the Brattleboro rat was verified utilizing the quantitative procedure outlined above. This analysis revealed a 12% decrease in the number of varicosities in apposition to VP-deficient perikarya with a

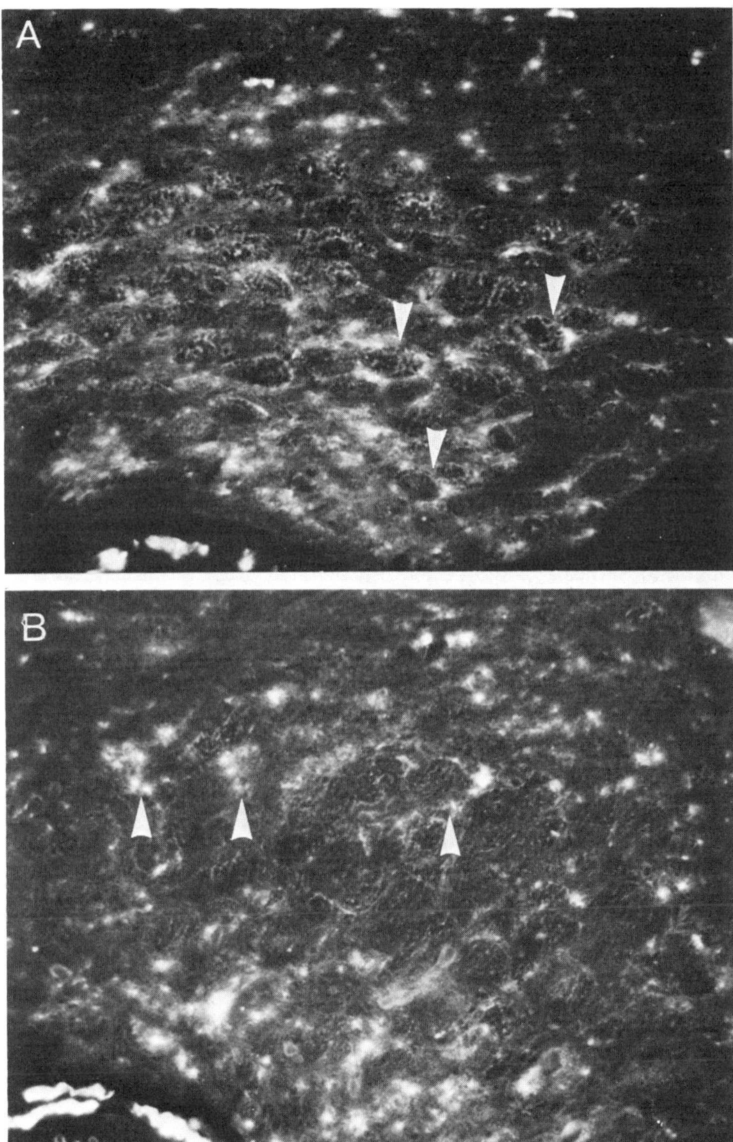

FIGURE 2. Catecholamine-containing varicosities as demonstrated with formaldehyde-induced histofluorescence reveal the innervation patterns in the SON of the normal Long Evans (A) and Brattleboro rat (B). In the normal animal, these varicosities are abundant ventrally and appear in apposition to magnocellular neurons (►) which have been demonstrated to contain vasopressin. This pattern is deficient in the Brattleboro rat, however numerous varicosities are seen in apposition to dorsally placed neurons (►), which are known to contain oxytocin. ×280.

concomitant 14% increase in the number of varicosities apposing NP-OX-producing perikarya in the Brattleboro rat when compared to controls.

Animals processed with the FAGLU technique and embedded in polyethylene glycol exhibited excellent fluorescence of varicosities and intervaricose segments in the SON. Due to the thinness of the sections, varicosities appeared more punctate than in freeze-dried tissue and could be identified clearly in apparent apposition to magnocellular perikarya. Although the injection of pargyline in control animals should theoretically increase terminal fluorescence, it was not possible to distinguish between normal Long-Evans rats and those which had received pargyline due to the dense innervation of the nucleus (FIGURES 3 & 5). However, the Brattleboro rats that received pargyline had significantly more fluorescent varicosities distributed homogeneously within the SON when compared to their saline-injected controls (FIGURES 4 & 6). Although the fluorescence was increased as a result of the treatment, the innervation was not as dense as in either the normal or pargyline-treated Long-Evans rats.

DISCUSSION

The Brattleboro rat provides a unique model for examining the maintenance of innervation patterns of target neurons devoid of their peptide content. The VP-deficient neurons in the ventral portions of the SON were less frequently apposed by varicosities whereas more dorsally, the OX neurons appeared more heavily contacted when compared to their non-mutant controls. This implies a hyperinnervation of the OX neurons. The altered innervation pattern reported here is in contrast to results recently reported by Swanson *et al.*[17] Employing dopamine-β-hydroxylase immunohistochemistry, on 10 μm frozen sections, they observed no qualitative differences in the distribution of terminals within the SON of Brattleboro rats. If an altered state of neuronal activity exists in the Brattleboro rat it might not be reflected by visible changes in synthetic enzyme levels, which could account for this apparent discrepancy.

In a study by Versteeg *et al.*,[18] catecholamine concentrations and turnover rates were measured in discrete regions of the brain of Brattleboro rats. They found increased norepinephrine concentrations in the Brattleboro rats along with an increased turnover rate. The increased turnover rate in the SON could explain the fewer varicosities in the ventral region of the nucleus, whereas absolute concentrations of NE, in light of this morphometric data, might be accounted for by the hyperinnervation of the dorsal OX-containing portion of the nucleus.

Several hypotheses can be set forth to explain the altered innervation pattern of the Brattleboro rat SON. The observed decreased innervation of the VP-deficient population of neurons in the SON of the Brattleboro rat could account for the reported increased firing rate of these neurons[19] especially since the role of norepinephrine appears to inhibitory.[20, 21] The observed alteration does not appear to be entirely morphological, however, since the blocking of monoamine oxidase activity, the enzyme responsible for transmitter breakdown both within terminals and extracellularly, caused an increase in the number of varicosities. However, since the innervation, even after this treatment, did not compare to the control animals, it is likely that a morphological change also has taken place as a result of the hormone deficiency. Since the VP-deficient

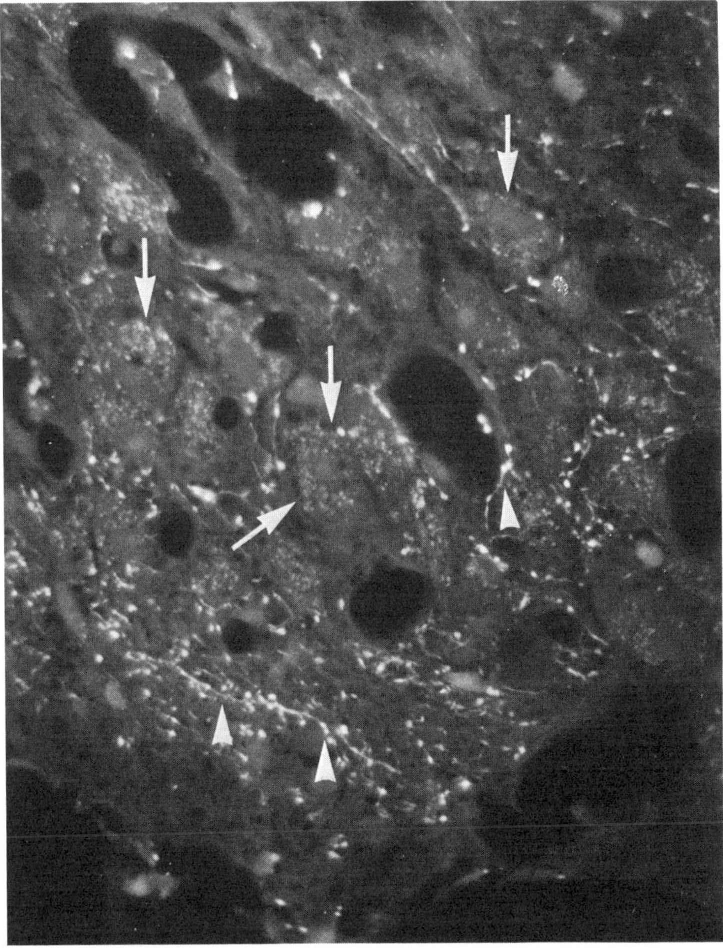

FIGURES 3–6. Aqueous aldehyde histofluorescence (FAGLU) in 2 μm thin sections at comparable levels of the SON reveals the noradrenergic innervation patterns in normal Long-Evans and homozygous Brattleboro rats.

FIGURE 3. The innervation pattern as seen in the normal Long-Evans rat following application of the FAGLU histofluorescence technique reveals fine-sized catecholamine profiles (►) which occur most abundantly in the ventral portion of the SON. In these 2 μm thin sections, magnocellular perikarya are identified in part by their granular pigment fluorescence (→). This richly vascularized nucleus is characterized by numerous capillaries. $\times$600.

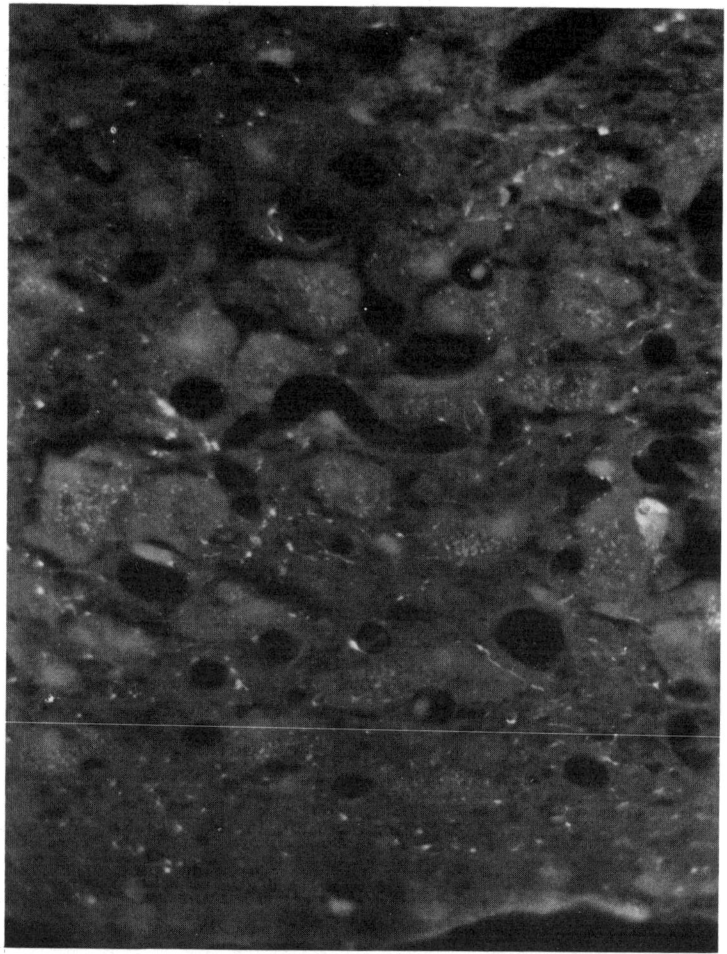

FIGURE 4. The FAGLU technique confirms the decrease in innervation pattern in the Brattleboro rat. Even with this more sensitive histofluorescence technique, the altered pattern is apparent. ×600.

population of neurons is less heavily contacted by fluorescent varicosities in the Brattleboro rat and because there is an overall increase in NE within the nucleus,[18] it is likely that a shift in the noradrenergic reinnervation has taken place that favors the OX population of neurons. Although histofluorescence as employed in the present study is not able to differentiate between the various catecholamine transmitters, it is not likely that dopamine or epinephrine interfered with the interpretations of this study since their content in the SON of both normal and Brattleboro rats is almost negligible.[18]

In addition to VP and OX, magnocellular neurons of the SON may contain additional hormones. Vanderhaeghen and Costra [22] recently reported that the peptides of the gastrin family co-exist frequently within neurons containing OX in the SON. In addition, the opiate peptide dynorphin, found in high concentrations in the Brattleboro rat[23] appears localized to the VP-deficient neurons.[24] Since NE may play a role in the modulation of a host of peptides, it is possible that a decreased number of noradrenergic fibers is maintained for these and perhaps other substances present within the SON. However, it is clear that the abnormal innervation is correlated with the absence of a specific peptide and that perhaps VP exerts an important trophic influence for the maintenance of noradrenergic interactions.

FIGURE 5. Following monoamine oxidase inhibition in the normal rat, the catecholamine innervation pattern appeared comparable to that seen in the non-pharmacologically treated animal, (FIGURE 3). Presumably because the FAGLU technique is capable of demonstrating intervaricose fluorescence the monoamine oxidase inhibition did not result in a visual increase in the histofluorescence pattern. ×600.

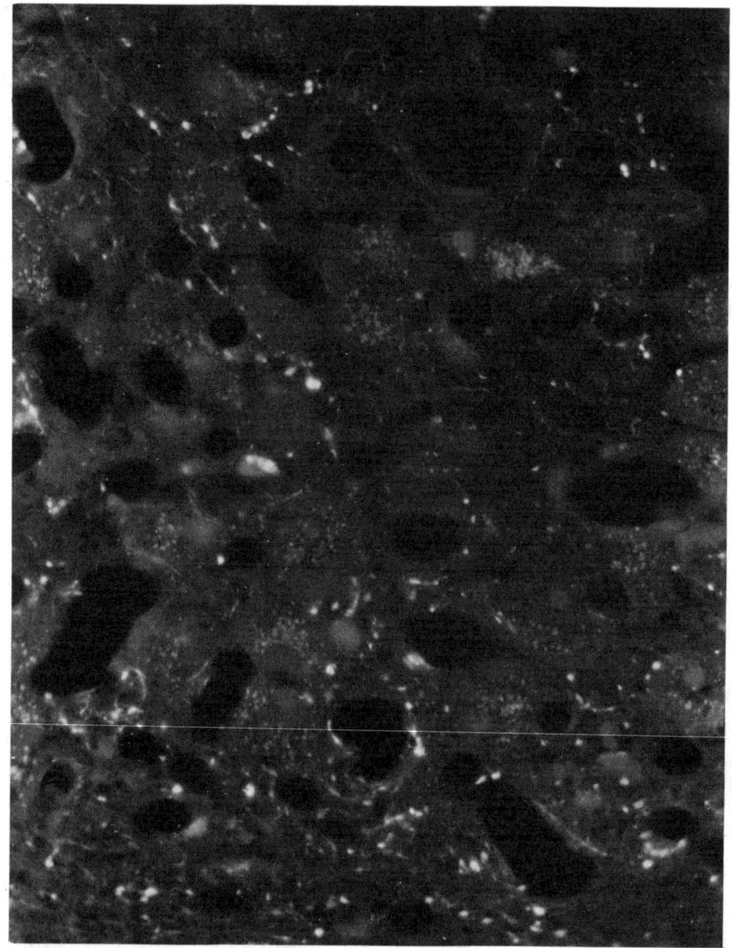

FIGURE 6. Following monoamine oxidase inhibition in the Brattleboro rat, an increase in catecholamine fluorescence is seen somewhat homogeneously throughout both ventral and dorsal regions of the nucleus. In spite of this increase, the innervation pattern is still less dense than that seen in normal animals (FIGURE 3). ×600.

REFERENCES

1. DEFENDINI, R. & E. A. ZIMMERMAN. 1978. The magnocellular neurosecretory system of the mammalian hypothalamus. *In* The Hypothalamus. S. Reichlin, R. J. Baldessarini & J. B. Martin, Eds.: 137–152. Raven Press. New York, N.Y.
2. SWANSON, L. W. 1977. Immunohistochemical evidence for a neurophysin-containing autonomic pathway arising in the paraventricular nucleus of the hypothalamus. Brain Res. **128:** 356–363.
3. NILAVER, G., E. A. ZIMMERMAN, J. WILKINS, J. MICHAEL, D. HOFFMAN & A. J. SILVERMAN. 1980. Magnocellular hypothalamic projections to the lower brain stem and spinal cord of the rat. Neuroendocrinology **30:** 150–158.

4. WARD, D. G. & D. S. GANN. 1976. Inhibitory and facilitatory areas of the dorsal medulla mediating ACTH release in the cat. Endocrinology **99:** 1213–1219.
5. KRIEGER, D. T. & E. A. ZIMMERMAN. 1977. The nature of CRF and its relationship to vasopressin. *In* Clinical Neuroendocrinology. G. M. Besser & L. Martini, Eds.: 363–391. Academic Press. New York, N.Y.
6. SIRETT, N. E. & H. G. PURVES. 1973. *In* Brain-Pituitary-Adrenal Interrelationships. A. Brodish & E. S. Redgate, Eds.: 79–98. Karger. Basel, Switzerland.
7. FUXE, K. 1965. The distribution of monoamine terminals in the central nervous system. Acta Physiol. Scand. **64** (Suppl. 247): 37–85.
8. UNGERSTEDT, U. 1971. Stereotaxic mapping of the monoamine pathways in the rat brain. Acta Physiol. Scand. (Suppl. 367): 1–48.
9. LINDVALL, O. & A. BJÖRKLUND. 1974. The organization of the ascending catecholamine neuron systems in the rat brain as revealed by the glyoxylic acid fluorescence method. Acta Physiol. Scand. (Suppl. 412): 1–48.
10. MCNEILL, T. H. & J. R. SLADEK, JR. 1980. Simultaneous monoamine histofluorescence and neurophysin immunocytochemistry. II. Correlative distribution of catecholamine varicosities and magnocellular neurosecretory neurons in the rat supraoptic and paraventricular nuclei. J. Comp. Neurol. **193:** 1023–1033.
11. KHACHATURIAN, H. & J. R. SLADEK, JR. 1980. Simultaneous monoamine histofluorescence and neuropeptide immunocytochemistry: III. Ontogeny of catecholamine varicosities and neurophysin neurons in the rat supraoptic and paraventricular nuclei. Peptides **1:** 77–95.
12. SCHÖLER, J. & J. R. SLADEK, JR. 1981. The supraoptic nucleus of the Brattleboro rat has an altered afferent noradrenergic input. Science **214** (4518): 347–349.
13. HOFFMAN, G. E. & J. R. SLADEK, JR. 1980. Age related changes in dopamine, LHRH, and somatostatin in the rat hypothalamus. Neurobiol. Aging **1:** 27–37.
14. STERNBERGER, L. A. 1974. Immunocytochemistry. Prentice-Hall. Englewood Cliffs, N.J.
15. FURNESS, J. B., M. COSTA & W. W. BLESSING. 1977. Simultaneous fixation and production of catecholamine fluorescence in central nervous tissue by perfusion with aldehydes. Histochemistry J. **9:** 745–750.
16. SCHÖLER, J. & W. E. ARMSTRONG. 1982. Aqueous aldehyde (Faglu) histofluorescence of catecholamines in 2 μm sections using polyethylene glycol embedding. Brain Res. Bull. (In press.)
17. SWANSON, L. W., P. E. SAWCHENKO, A. BÉROD, B. K. HARTMAN, K. B. HELLE & D. E. VANORDEN. 1981. An immunohistochemical study of the organization of catecholaminergic cells and terminal fields in the paraventricular and supraoptic nuclei of the hypothalamus. J. Comp. Neurol. **196:** 271–285.
18. VERSTEEG, D. H. G., M. TANAKA & E. R. DEKLOET. 1978. Catecholamine concentration and turnover rate in discrete regions of the brain of the homozygous Brattleboro rat deficient in vasopressin. Endocrinology **103**(5): 1654–1661.
19. DYBALL, R. E. J. 1974. Single unit activity in the hypothalamoneurohypophysial system of Brattleboro rats. J. Endocr. **60:** 135–143.
20. SAKAI, K. K., B. H. MARKS, J. M. GEORGE & A. KOESTNER. 1974. The isolated organ cultured supraoptic nucleus as a neuropharmacological test system. J. Pharmacol. Exp. Therap. **190:** 482–491.
21. BARKER, J. L., J. W. CRAYTON & R. A. NICOLL. 1971. Noradrenaline and acetylcholine responses of supraoptic neurosecretory cells. J. Physiol. **218:** 19–32.
22. VANDERHAEGHEN, J. J. & F. NOSTRA. 1980. Coexistence of gastrin family peptide with ACTH, α-MSH, oxytocin or dopamine in the same cell. A study in the CNS, hypophysis and gut of rat, ox, dog, monkey and human. Abstract 233–8, Society for Neuroscience.

23. Cox, B. M., V. E. Ghazarossian & A. Goldstein. 1980. Levels of immunoreactive dynorphin in brain and pituitary of Brattleboro rats. Neurosci. Lett. **20:** 85–88.

24. Watson, S. J., H. Akil, V. E. Ghazarossian & A. Goldstein. 1981. Dynorphin immunocytochemical localization in brain and peripheral nervous system: Preliminary studies. Proc. Natl. Acad. Sci. USA **78**(2): 1260–1263.

Discussion of the Three Preceding Papers

R. Bodnar (*Queens College, CUNY, New York, N.Y.*): Has the abnormal innervation in Brattleboro rats been confirmed by radioimmunoassay?

Sladek: We have not done a catecholamine-punch assay of the supraoptic nucleus. There is a report of interest in the literature. Versteeg, Tanaka, and deKloet in 1977, found an increased norepinephrine content in that particular brain area, and an enhanced turnover of norepinephrine. This could be consistent with the present finding, if indeed the apparent hyperinnervation in dorsal portions of the SON is significant enough to account for a total overall increase in norepinephrine.

G. Nilaver (*Columbia University, New York, N.Y.*): Dr. Dreifuss, in view of the observed opposite effects of oxytocin and vasopressin on memory and behavior, have you looked for any inhibitory effects of oxytocin on neuronal firing in your system?

J. J. Dreifuss (*University of Geneva Medical School, Geneva, Switzerland*): We have looked at oxytocin (in hippocampal slices) and it has the same effect as vasopressin. I alluded to the fact that we do not know whether we are dealing with an oxytocin receptor or a vasopressin receptor; if it is a vasopressin receptor, we have yet to know exactly what kind of receptor it is. It's unlikely to be the renal receptor because of the antagonist data and dDAVP results we also have, but it could well be a different receptor and we have yet to see to what it responds. But, phenomenologically oxytocin has not an opposing effect to vasopressin, but rather a similar effect. However, it is a large leap to go from our modest results to the question of memory consolidation.

G. Brito (*University of Rochester, Rochester, N.Y.*): Dr. Boer, since Brattleboro rats appear to have impaired thyroid function, their abnormalities in brain development might very well be related to that endocrinopathy.

G. J. Boer (*Netherlands Institute for Brain Research, Amsterdam, the Netherlands*): We have described the differences between homozygous and heterozygous Brattleboro rats and we are now trying to reverse or prevent that deficit by treatment with vasopressin. So far, neonatal supplementation by repeated injections has failed. If prenatal and continuous postnatal treatment with vasopressin also fails to restore the deficits in the brain in homozygous DI rats, then we must consider that other factors, like thyroxine, may be involved. However, the etiology of the present impairments has been shown to be different from anomalies in brain growth due to changes in thyroid function (Boer *et al.*, *Progr. Brain Res.* **53:**207, 1980). So, this system is not likely to be involved. Additional support can actually be distilled from the presentation of Dr. G. A. Hedge, who, although showing a clear difference in the thyroid function between homozygous DI rats and Long-Evans rats, was not able to see a similar difference between homozygous and heterozygous Brattleboro rats.

VASOPRESSIN PROJECTIONS AND CENTRAL CONTROL OF CARDIOVASCULAR FUNCTION *

K. H. Berecek,† R. L. Webb, K. W. Barron, and M. J. Brody ‡

The Cardiovascular Center and
Department of Pharmacology
University of Iowa
Iowa City, Iowa 52242

Numerous neuroanatomical studies have provided evidence that the vasopressin (VP)-synthesizing nuclei, the paraventricular (PVN), supraoptic (SON), and suprachiasmatic (SCN) nuclei, send VP-containing projections to a number of neural target areas including the nucleus tractus solitarii (NTS), dorsal vagal complex, anteroventral third ventricle region (AV3V), and central grey, all of which are involved in cardiovascular regulation.[1-3] Little is known about the functional significance of these projections. The purpose of the present studies was to examine the possible functional participation of vasopressinergic nuclei and projections in cardiovascular regulation. Specifically, we examined (1) the cardiovascular effects of electrical stimulation of the PVN and (2) compared cardiovascular responses to electrical stimulation of the AV3V region in VP-deficient Brattleboro rats (DI rats) to responses obtained in normal Long-Evans (LE) rats. The AV3V region receives vasopressinergic projections[4] and is involved in regulation of VP secretion.[5,6] In view of the known neuroanatomical connections among the VP synthesizing nuclei, the AV3V region and brain areas subserving the baroreceptor reflex,[1-4] stimulation of the PVN and AV3V regions was also performed before and after sinoaortic deafferentation (SAD).

MATERIALS AND METHODS

Bipolar stainless steel electrodes were stereotaxically placed in the PVN or AV3V region 1–4 days prior to acute experimentation. The skull was leveled between bregma and lambda and the following coordinates were used: PVN, —1.8 mm posterior to bregma, ±0.25 mm lateral to midline, and 6.9 mm below dura; AV3V, —0.3 mm posterior to bregma, on the midline, and 7.5 mm below dura. Stimulation experiments were performed in dial-urethane (0.06 ml/100 g, CIBA) anesthetized rats implanted with an arterial catheter for monitoring mean arterial pressure (MAP) and a venous catheter for injection of drugs. Miniaturized pulsed-Doppler flow probes were used to monitor simultaneously renal, mesenteric, and hindquarter blood flows.[7] Electrical stimulation (14 volt, 5–60 Hz, 30 sec pulse duration) of the PVN and AV3V

* Supported in part by U.S. Public Health Service grants HLB–14388 and HLB–07121.

† Present address: Cardiovascular Research and Training Center, University of Alabama in Birmingham, 1012 Zeigler Building, Birmingham, Al. 35294.

‡ Address all correspondence M.J.B. at the Department of Pharmacology.

0077–8923/82/0394–0729 $1.75/0 © 1982, NYAS

region were performed before and after SAD, carried out using the method described by Krieger.[8] Completeness of SAD was confirmed by absence of reflex bradycardia in response to a phenylephrine-induced increase in MAP.

Results

PVN Stimulation

Stimulation of PVN produced a frequency-dependent decrease in MAP, bradycardia, an increase in renal and mesenteric vascular resistances, and a decrease in hindquarter resistance (Figure 1, Table 1). Since these responses began immediately with onset of stimulation, they were primarily neural in origin. Following SAD the resistance changes were enhanced in the renal and

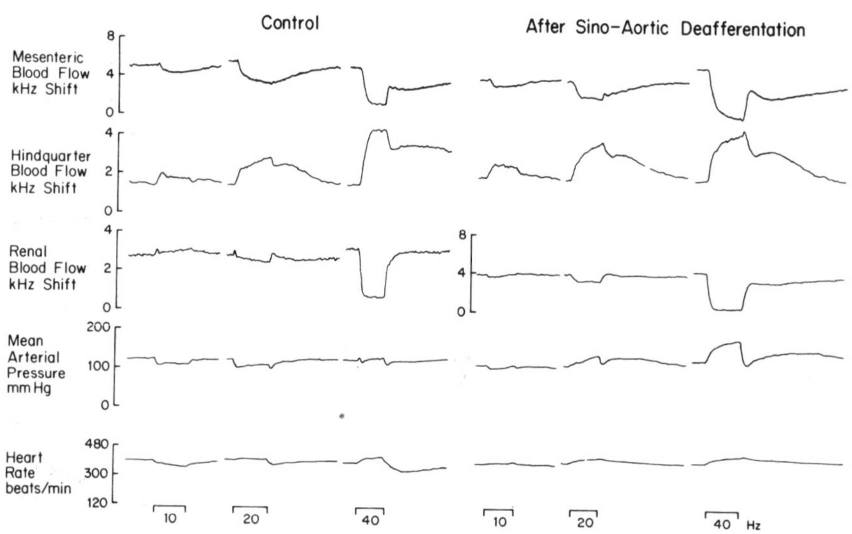

Figure 1. Tracing of the hemodynamic effects of electrical stimulation of the paraventricular nucleus before and after sino-aortic deafferentiation.

mesenteric vascular beds and decreased in the hindquarter bed. Moreover, the decrease in MAP was attenuated at low frequencies of stimulation and converted to a substantial pressor response at high frequencies. Tachycardia also occurred after SAD.

AV3V Stimulation in DI and LE Rats

AV3V stimulation produced in LE rats cardiovascular effects similar both to those seen with PVN stimulation (Figure 2) and to those observed with AV3V stimulation in Sprague-Dawley rats.[9] DI rats, however, demonstrated a significantly greater decrease in MAP and bradycardia, and lesser changes

TABLE 1

CARDIOVASCULAR EFFECTS OF PARAVENTRICULAR NUCLEUS STIMULATION

	Control (Hz)		After SAD * (Hz)	
	20	40	20	40
% Δ mean arterial pressure	−16 ± 3	−5 ± 4	27 ± 6 §	53 ± 5 §
% Δ heart rate	−2 ± 1	0.1 ± 4	9 ± 2 §	20 ± 2 §
% Δ renal vascular resistance	9 ± 5	178 ± 40	104 ± 26 §	404 ± 89 †
% Δ mesenteric vascular resistance	26 ± 3	220 ± 42	140 ± 30 §	450 ± 157 ‡
% Δ hindquarter vascular resistance	49 ± 3	60 ± 3	37 ± 5	31 ± 5 §

Responses expressed as mean ± standard error (SE). $N = 9$ rats.
* SAD = sinoaortic deafferentation.
Significant difference from control: † $p < .05$, ‡ $p < .01$, and § $p < .005$ determined by one-way analysis of variance.

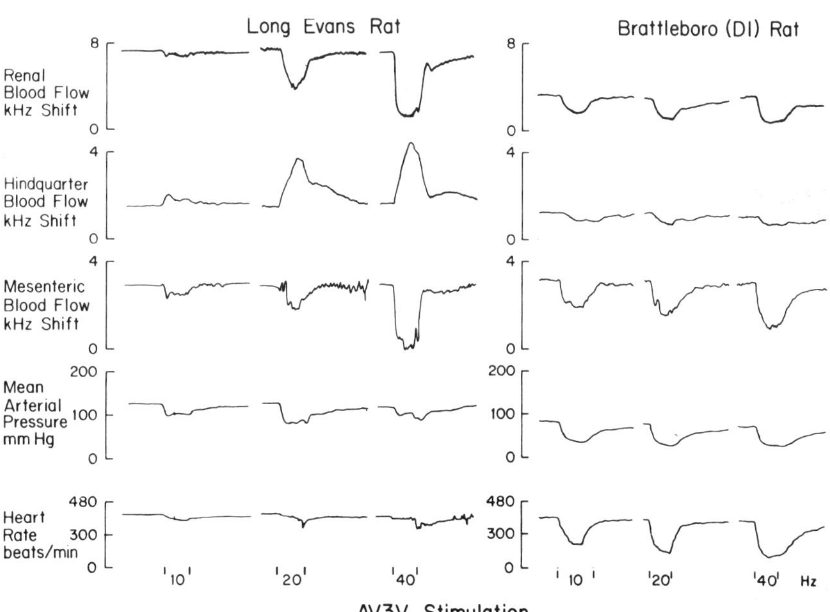

FIGURE 2. Tracing of the hemodynamic effects of AV3V stimulation in an LE and DI rat.

in renal, mesenteric, and hindquarter vascular resistances in comparison to the responses in LE rats (FIGURE 2, TABLE 2). SAD produced the expected acute rise in MAP in LE rats (control 105 ± 5 mm Hg, after SAD 146 ± 6 mm Hg); whereas, no change was found in DI rats (control 90 ± 6 mm Hg, after SAD 80 ± 5 mm Hg). SAD did not have a consistent effect on CV responses to AV3V stimulation with the exception that the bradycardia was significantly attenuated at all frequencies in the DI rat (TABLE 2).

TABLE 2

CARDIOVASCULAR EFFECTS OF ELECTRICAL STIMULATION OF THE AV3V REGION IN LE AND DI RATS

	Control (Hz)		After SAD (Hz)	
	20	40	20	40
% Δ mean arterial pressure				
LE rats (N=8)	−22 ± 5	−23 ± 4	−28 ± 6	−32 ± 5
DI rats (N=8)	−40 ± 4 †§	42 ± 3 §†	−24 ± 5	−26 ± 5
% Δ heart rate				
LE rats	−13 ± 2	−16 ± 2 §	−1 ± 3	−3 ± 3
DI rats	−33 ± 5 †§	−43 ± 4 §¶	−11 ± 6	−10 ± 4
% Δ renal vascular resistance				
LE rats	71 ± 20 §	156 ± 40	31 ± 14	96 ± 48
DI rats	7 ± 19 †	7 ± 19 †	−6 ± 13	18 ± 16
% Δ mesenteric vascular resistance				
LE rats	68 ± 14	138 ± 18	51 ± 24	221 ± 66
DI rats	4 ± 6 †	28 ± 11 ‡	13 ± 3	40 ± 10
% Δ hindquarter vascular resistance				
LE rats	51 ± 4	71 ± 2	54 ± 7	62 ± 5
DI rats	32 ± 4 ‡¶	40 ± 3 ‡¶	48 ± 4	58 ± 4

* SAD = sinoaortic deafferentation, responses are mean ± standard error.
Significantly different from LE rats: † $p < .05$, ‡ $p < .01$.
Significantly different from response after SAD: § $p < .05$, ¶ $p < .01$, determined by one-way analysis of variance and the Newman-Keuls ranking test.

DISCUSSION

Our data provide evidence that stimulation of the paraventricular nucleus produces an integrated pattern of cardiovascular effects through neural interactions and that the effects are greatly influenced by the level of activity of the baroreceptor reflex. In view of neuroanatomical evidence for pathways, identified as vasopressinergic by immunocytochemistry, that originate from the VP-producing nuclei and project to neural target areas involved in cardiovascular regulation, there is potential for these nuclei to participate in cardiovascular control.[2, 3] There have been few studies describing the functional significance of vasopressinergic projections. Morphological studies have demonstrated that these projections make axosomatic and axodendritic connections with the aforementioned neural target areas, suggesting that VP may be able to modulate the function of specific neuronal systems.[10] Evidence in support of this hypothesis has come from studies of Ciriello and Calaresu [11] showing that stimulation of the PVN and SON in cats attenuated the bradycardia produced by carotid sinus stimulation, whereas ablation of these areas increased reflex bradycardia. These data suggest that the PVN and SON, via neural projections, may exert an inhibitory influence on the baroreceptor reflex. Preliminary data in DI rats showed little or no cardiovascular response to PVN stimulation, suggesting that VP projections may mediate these effects.

The attenuation of vasodilation in the hindquarters, following SAD, was unexpected since both vasodilator and vasoconstrictor components of PVN stimulation would be expected to be exaggerated in the absence of buffer reflexes. This unusual interaction between baroreceptor mechanisms and the vasodilator response evoked from PVN needs further study.

We also studied the effects of electrical activation of the AV3V region, an area that receives VP-containing projections, in rats genetically devoid of VP (DI rats). The marked differences in the cardiovascular responses of DI rats suggests that the responses may depend, in part, upon the integrity of VP-containing neurons. Interestingly, we also found that removal of baroreceptor inhibition in DI rats did not produce hypertension. While this preliminary finding suggests that sympatho-excitation responsible for the rise in arterial pressure following SAD may be dependent in part upon central vasopressinergic mechanisms, additional studies are needed to determine whether the lower arterial pressure of DI rats influences their response to removal of baroreceptor afferents.

References

1. Saper, C. B., A. D. Loewy, L. W. Swanson & W. H. Cowan. 1976. Direct hypothalamoautonomic connections. Brain Res. **117:** 305–312.
2. Weindl, A. & M. V. Sofroniew. 1980. Immunohistochemical localization of hypothalamic peptide hormones in neural target areas. *In* Brain and Pituitary Peptides. W. Wuttke, A. Weindl, K. H. Voigt & R. R. Dries, Eds.: 97–109. S. Karger. Basel, Switzerland.
3. Buijs, R. M., D. F. Swaab, J. Dogterom & F. W. Van Leeuwen. 1978. Intra- and extrahypothalamic vasopressin and oxytocin pathways in the rat. Cell. Tiss. Res. **186:** 423–433.
4. Miselis, R. R., R. E. Shapir & P. J. Hand. 1979. Subfornical organ efferent to neural systems for control of body water. Science **205:** 1022–1025.
5. Johnson, A. K., J. Schoun, J. R. McNeill & J. Möhring. 1980. Periventricular tissue of the anteroventral third ventricle (AV3V): A role for control of vasopressin (VP) release. Fed. Proc. **39:** 986.
6. Johnson, A. K., W. E. Hoffman & J. Buggy. 1978. Attenuated pressor responses to intracranially injected stimuli and altered antidiuretic activity following preoptic-hypothalamic periventricular ablation. Brain Res. **157:** 161–166.
7. Haywood, J. R., R. A. Shaffer, C. Fastenow, G. D. Fink & M. J. Brody. 1981. Regional blood flow measurement in the conscious rat with a pulsed-Doppler flow meter. Am. J. Physiol. **241:** H273–H278.

8. KREIGER, E. M. 1964. Neurogenic hypertension in the rat. Circ. Res. **15:** 511–521.

9. BRODY, M. J. & A. K. JOHNSON. 1980. Role of the anteroventral third ventricle region in fluid and electrolyte balance, arterial pressure regulation and hypertension. *In* Frontiers in Neuroendocrinology. L. Martini & W. F. Ganong, Eds.: 249–292. Raven Press. New York, N.Y.

10. SOFRONIEW, M. N. 1980. Projections from vasopressin oxytocin and neurophysin neurons to neural target areas in the rat and human. J. Histochem. Cytochem. **28:** 475–478.

11. CIRIELLO, J. & F. R. CALARESU. 1980. Role of paraventricular and supraoptic nuclei in central cardiovascular regulation in the cat. Am. J. Physiol. **239:** R137–R142.

MODULATION OF NOCICEPTIVE THRESHOLDS BY VASOPRESSIN IN THE BRATTLEBORO AND NORMAL RAT *

R. J. Bodnar,† M. M. Wallace,† J. H. Kordower,† G. Nilaver,‡ J. Cort,§ and E. A. Zimmerman ‡

† Department of Psychology
Queens College
City University of New York
Flushing, New York 11367

‡ Department of Neurology
College of Physicians and Surgeons
Columbia University
New York, New York 10032

§ Department of Pharmacology
University of Arizona Health Science Center
Phoenix, Arizona 85724

Vasopressin, a neuropeptide of the magnocellular neurosecretory system, is present in neurons that project to the posterior lobe of the pituitary gland as well as to the hypophyseal portal system in the zona externa of the median eminence.[1] Following autoradiographic analysis of magnocellular neuronal projections to extrahypothalamic forebrain and brainstem structures,[2] Swanson [3] used antisera to the neurophysin carrier protein in immunocytochemical studies to demonstrate similar distributions for neurophysin. Further refinements revealed that vasopressin is present in neurons projecting to the dorsomedial region of the medulla (X cranial nerve), pons (V cranial nerve), and the substantia gelatinosa of the spinal cord,[4, 5] areas intimately involved in the perception and modulation of noxious input. That vasopressin is involved in antinociceptive processes was supported by the observation of Berntson and Berson [6] that both systemic and intracerebroventricular injections of lysine vasopressin at doses of 16–100 μg increased rat tail-flick latencies. Moreover, our laboratory [7] showed that Brattleboro rats, deficient in vasopressin, were hyperalgesic and displayed selective analgesic deficits as compared to controls matched for age and weight. The present study will (1) review the role of vasopressin and ACTH in the hyperalgesia and analgesic deficits of Brattleboro rats; (2) examine the effects of centrally administered antisera to vasopressin upon normal rat nociception; and (3) examine the antinociceptive properties of vasopressin and structural analogues at physiological nanomolar doses.

Pain Thresholds and the Brattleboro Rat

Method

Twenty homozygous Brattleboro female rats with diabetes insipidus and ten normal female Long-Evans rats were tested for flinch-jump pain thresholds [7]

* Supported by PSC/CUNY grant 13493.

0077–8923/82/0394–0735 $1.75/0

under baseline conditions as well as 30 min following either a forced cold-water swim at 2° C for 3.5 min or a subcutaneous injection of 5 mg/kg of morphine. In a second phase of the experiment, ten Brattleboro and ten control rats received systemic injections of 320 ng of arginine vasopressin (AVP) twice a day through the same protocol; the remaining ten Brattleboro rats received a 160 ng dose. In a third phase of the experiment, the protocols were identical except that systemic desamino-D-arginine vasopressin (DDAVP) was administered. In a final study, Brattleboro and normal rats received single intracerebroventricular injections of either AVP (5 μg) or ACTH (0.5 μg) under baseline, swim, and morphine conditions.

Results

FIGURE 1 displays the significantly lower jump thresholds of Brattleboro rats as compared to age and weight-matched controls. While both systemic doses of AVP and DDAVP increased the jump thresholds of Brattleboro rats, only systemic injections of 320 ng of AVP increased normal rat jump thresholds. As seen in FIGURE 2, Brattleboro rats exhibited a significantly impaired analgesic response following cold-water swims, an effect that was partially reinstated by systemic injections of 160 ng of AVP. By contrast, Brattleboro and normal rats were similar in their analgesic response to morphine. Finally, the central injections of ACTH and AVP significantly potentiated the impaired analgesic response following cold-water swims in Brattleboro, but not normal rats.

PAIN THRESHOLDS AND ANTISERA TO AVP

Method

Thirty-six male albino Sprague-Dawley rats were subdivided into three groups with matched tail-flick latencies. The first group was tested over three levels of radiant heat 15, 30, 60, and 120 min following intracerebroventricular injection of 5 μl of undiluted antisera to AVP or vehicle. The second and third groups were tested identically except they received 50% and 10% dilutions of the antisera respectively.

Results

TABLE 1 summarizes the alterations in tail-flick latencies following antisera to AVP as a function of concentration and level of radiant heat. Undiluted antisera to AVP elicited significant biphasic effects upon tail-flick latencies with hyperalgesia occurring 15 min after injection at the intense radiant heat level and analgesia occurring for up to 60 min at the moderate radiant heat level. Dilution of the antisera to 50% and 10% concentration eliminated these alterations. Furthermore, these effects appeared to be specific to the antisera since injections of normal rabbit serum and preimmune serum were without effect across thermal intensities and time course. Though antisera to AVP induced significant hypothermia 60 and 90 min after injection, the time course of this effect differed from changes in nociceptive reactivity.

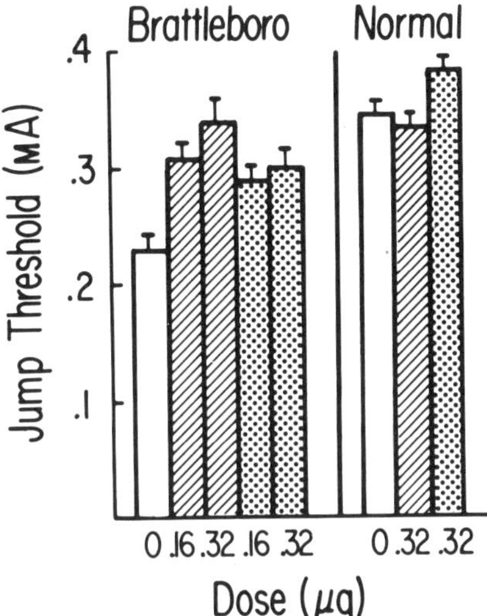

FIGURE 1. Baseline jump thresholds (± SEM) for Brattleboro and normal rats 60 min following placebo (clear), DDAVP (striped), or AVP (dotted) injections.

PAIN THRESHOLDS AND CENTRAL VASOPRESSIN INJECTIONS

Method

Two groups of eight rats were tested for tail-flick latencies under baseline conditions and 5, 10, and 15 min following lysine vasopressin (LVP) administered intracerebroventricularly (0, 150, 500 ng/5 μl) or subcutaneously (0, 150, 500, 1500 ng/ml). Three further groups received the identical intracerebroventricular regimen except they received a prolonged vasopressin an-

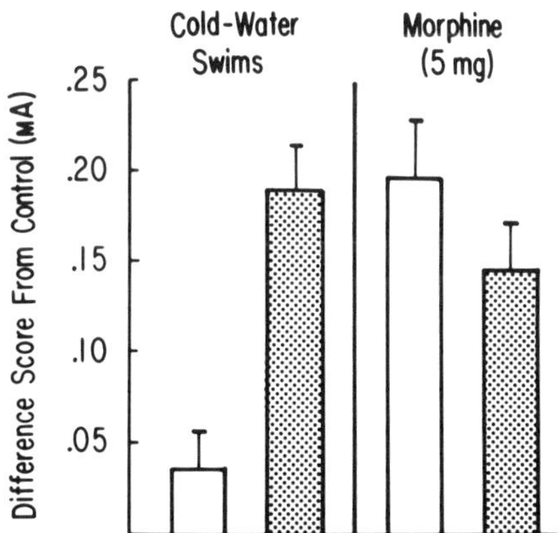

FIGURE 2. Alterations in jump thresholds (± SEM) 30 min following cold-water swims or morphine injections for Brattleboro (clear) and normal (dotted) rats.

alogue, an acetyl-inactivated analogue, and a carboxyl-inactivated analogue, respectively.

Results

TABLE 2 shows that while only the 1,500 ng dose of systemic LVP increased tail-flick latencies 15 min after injection, centrally administered LVP produced analgesia 5, 10, and 15 min after the 500 ng dose. Further, the prolonged analogue of LVP produced analgesia 5, 10, and 15 min after central injection of both the 150 and 500 ng doses. Interestingly, while a LVP analogue that inactivated the carboxyl end of the molecule could only elicit an analgesic response 5 min after the 500 ng dose, the acetyl-inactivated analogue failed to elicit significant increases in tail-flick latencies. These latter results are preliminary given the small (5) test group.

Discussion

The anatomical distribution and behavioral effects of vasopressin argue strongly for a modulatory role in the processing of nociceptive information. It is apparent that Brattleboro rats deficient in vasopressin exhibit hyperalgesia and impaired analgesic responses to stress.[7] From the present data, it appears that these deficits can be reinstated either fully or in part by vasopressin, suggesting that it is the vasopressin deficiency that causes these nociceptive impairments. Moreover, vasopressin is important in normal rodent pain modulation. Selective inactivation of vasopressin through antisera administration produces hyperalgesia under nociceptive conditions where nanomolar amounts of vasopressin increase pain thresholds. Preliminary evidence[6] suggests that

TABLE 1

ALTERATIONS IN TAIL-FLICK LATENCIES FOLLOWING INTRACEREBROVENTRICULAR ADMINISTRATION OF ANTI-SERA TO AVP OR VEHICLE

Radiant Heat Level	Post-Injection Time (min)	Mean Latency (sec) Concentration 100% VEH.	100% AAVP	50% VEH.	50% AAVP	10% VEH.	10% AAVP
High (50° C in 3 sec)	15	3.90	3.26*	3.90	3.82	4.03	4.06
	30	3.49	3.36	3.82	3.52	3.94	3.76
	60	3.62	3.35	3.53	3.54	3.53	3.67
	90	3.42	3.32	3.43	3.68	3.50	3.60
Moderate (46° C in 4 sec)	15	3.40	4.01*	5.33	4.82	5.40	5.21
	30	3.27	3.98*	4.63	5.76	4.80	5.17
	60	3.52	4.04*	5.05	4.98	5.06	4.75
	90	3.64	3.68	4.93	4.86	4.90	4.60
Low (45° C in 7 sec)	15	6.74	6.40	6.53	6.47	6.32	6.74
	30	6.29	6.02	6.06	6.67	5.80	5.87
	60	5.95	5.88	5.61	5.89	5.44	6.18
	90	6.02	6.28	5.59	6.12	5.66	5.83

TABLE 2

ALTERATIONS IN TAIL-FLICK LATENCIES FOLLOWING CENTRAL AND SYSTEMIC ADMINISTRATION OF LYSINE VASOPRESSIN (LVP) AND SELECTED ANALOGUES

			Tail-Flick Latencies (sec) Relative to Injection (min)			
Group	Route	Dose (ng)	Pre-	Post 5	Post 10	Post 15
LVP	sc	0	2.86	3.02	2.91	2.76
		150	2.98	2.60	2.53	3.16
		500	2.64	2.52	2.96	3.10
		1500	2.54	2.79	3.05	3.29*
LVP	icv	0	2.60	2.48	2.49	2.64
		150	2.72	2.99	3.22	3.38
		500	2.64	3.62*	4.16*	3.77*
Prolonged LVP	icv	0	2.70	3.01	3.03	2.69
		150	2.95	3.99*	4.25*	4.53*
		500	2.90	4.34*	4.11*	4.73*
Analogues	icv	0	3.18	2.97	3.38	3.28
Acetyl		500	3.33	3.92	4.02	4.26
Carboxyl		500	3.32	4.55*	4.13	3.96

these effects are not reversed by the opiate antagonist naloxone, suggesting the possibility that vasopressin is a suitable alternative candidate for the endogenous substance(s) that may mediate non-opioid analgesia.

REFERENCES

1. ZIMMERMAN, E. A. 1981. The organization of oxytocin and vasopressin pathways. *In* Neurosecretion and Brain Peptides. J. B. Martin, S. Reichlin & K. L. Bick, Eds.: 63–75. Raven Press. New York, N.Y.
2. CONRAD, L. C. A. & D. W. PFAFF. 1976. Efferents from medial basal forebrain and hypothalamus in the rat. II. An autoradiographic study of the anterior hypothalamus. J. Comp. Neurol. **169:** 221–262.
3. SWANSON, L. W. 1977. Immunohistochemical evidence for a neurophysin-containing autonomic pathway arising in the paraventricular nucleus of the hypothalamus. Brain Res. **128:** 346–353.
4. NILAVER, G., E. A. ZIMMERMAN, J. WILKINS, J. MICHAELS, D. HOFFMAN & A. J. SILVERMAN. 1980. Magnocellular hypothalamic projections to the lower brainstem and spinal cord of the rat: immunocytochemical evidence for predominance of the oxytocin-neurophysin system compared to the vasopressin-neurophysin system. Neuroendocrinology **30:** 150–158.
5. SWANSON, L. W., P. E. SAWCHENKO, S. J. WIEGAND & J. L. PRICE. 1980. Separate neurons in the paraventricular nucleus project to the median eminence and to the medulla or spinal cord. Brain Res. **198:** 190–195.
6. BERNTSON, G. G. & B. S. BERSON. 1980. Antinociceptive effects of intraventricular or systemic administration of vasopressin in the rat. Life Sci. **26:** 455–459.
7. BODNAR, R. J., E. A. ZIMMERMAN, G. NILAVER, A. MANSOUR, L. W. THOMAS, D. D. KELLY & M. GLUSMAN. 1980. Dissociation of cold-water swim and morphine analgesia in Brattleboro rats with diabetes insipidus. Life Sci. **26:** 1581–1590.

THE BEHAVIOR OF BRATTLEBORO RATS *

Gilberto N. O. Brito,† Garth J. Thomas,† Don M. Gash,†‡ and John H. Kitchen ‡

† Center for Brain Research and ‡ Department of Anatomy University of Rochester Medical Center Rochester, New York 14642

Introduction

Research reports suggested that vasopressin-deficient rats (Brattleboro strain—HODI) are impaired in memory,[1] and that administration of vasopressin ameliorates their memory impairment.[2] However, other investigators have not been able to demonstrate that normal (NORM) rats show better memory than HODI rats for shock-motivated behavior.[3–5]

It should be pointed out that in the studies discussed above, only shock-motivated behavioral tasks were used, and we believe that such an approach may not provide definitive evidence regarding neuropsychological assessment. Therefore, we have used a "neuropsychological-test battery" to evaluate behavioral characteristics of HODI rats, and we found that HODI rats do not have a general impairment in memory.[6] In the present studies we have expanded the neuropsychological battery to include food- and shock-motivated behavioral tasks involving different hypothetical mechanisms such as reference and working memory,[7] species-specific behavior,[11] and stress-induced interference.[8]

Data from the present study confirmed our previous observation that HODI rats have altered temperamental dispositions as compared with NORM rats.[6] However, this study did not provide evidence to support our previous suggestion that HODI rats have impaired reference-memory mechanisms.[6]

Methods

Subjects

Seven young-adult Long-Evans male rats homozygous for diabetes insipidus (HODI) and seven NORM male Long-Evans rats were obtained commercially (Blue Spruce Farms, Altamont, N.Y.). Because of growth deficiencies in HODI rats,[1] NORM animals were four weeks younger than HODIs. The younger NORM rats weighed between 175 and 191 g and the weights of HODI rats ranged from 161 to 208 g at the beginning of experimentation. The median 24-h water intake of HODI rats was 92.5 ml/100 g BW (60.0–113.5) and that of NORM rats was 15.5 ml/100 g BW (13.0–19.5). Two

* These studies were supported by National Institutes of Health grants NS–15901 and NS–17543 and a Biomedical Research Support Grant from the University of Rochester School of Medicine and Dentistry.

Address correspondence to G. N. O. Brito, Departamento de Ciencias Fisiologicas, Instituto de Biologia, Universidade do Estado do Rio de Janiero, Avenida 28 de Setembro, 87-Fundos, Rio de Janiero-RJ 20551, Brasil.

0077–8923/82/0394–0740 $1.75/0 © 1982, NYAS

HODIs and one NORM rat died toward the end of the study so data for only five HODI and six NORM rats are presented on the last behavioral task.

Housing and general adaptation procedures were as described previously.[6]

Apparatus

In addition to the T-maze, straight runway, and the open field used in a previous report,[6] the following apparatuses were used: (1) wooden and cardboard boxes for testing time-spent-eating in novel environments; (2) a step-through passive-avoidance device;[9] (3) a wooden box with Plexiglas® front for the prod-burying test; and (4) a shock chamber and a shuttlebox to test for stress-induced interference.

Behavioral Testing

The neuropsychological-test battery consisted of the tasks listed below in the order administered. Details of procedures can be found in appropriate references: (1) Time-spent-eating;[10] (2) Time-to-emerge;[6] (3) T-maze adaptation, T-maze alternation, Olfactory discrimination, and Visual discrimination;[6] (4) Runway learning and Approach-avoidance conflict;[6] (5) Step-through passive avoidance;[2, 9] (6) Prod burying[11] (Time-spent-burying prod and height of highest mound of bedding material were recorded); and (7) Stress-induced interference.[8]

Statistics

The nonparametric Mann-Whitney U statistical test[12] was used throughout these studies. All statistical tests were two-tailed, with the exception of the stress-induced interference task, which was one-tailed, because interference effects could only be one way.

Results

Time-Spent-Eating

The total time-spent-eating by HODI rats in a wooden box across nine sessions was significantly less than that of NORM rats ($U = 5$, $p < 0.01$). However, as shown in Figure 1 (panel A), HODI rats, by the seventh session, were well adapted to eating in the wooden box. When subsequently placed in a cardboard box, HODI rats spent as much time eating as NORM rats.

Time-to-Emerge

That HODI rats were significantly slower to emerge into an open field than NORM rats is indicated in Figure 1 (panel B). When total time-to-emerge across five trials for HODI rats is compared with that for NORM rats, the difference is statistically significant ($U = 7$, $p < 0.02$). By the last trial, only three HODI rats had emerged whereas six NORM rats had done so.

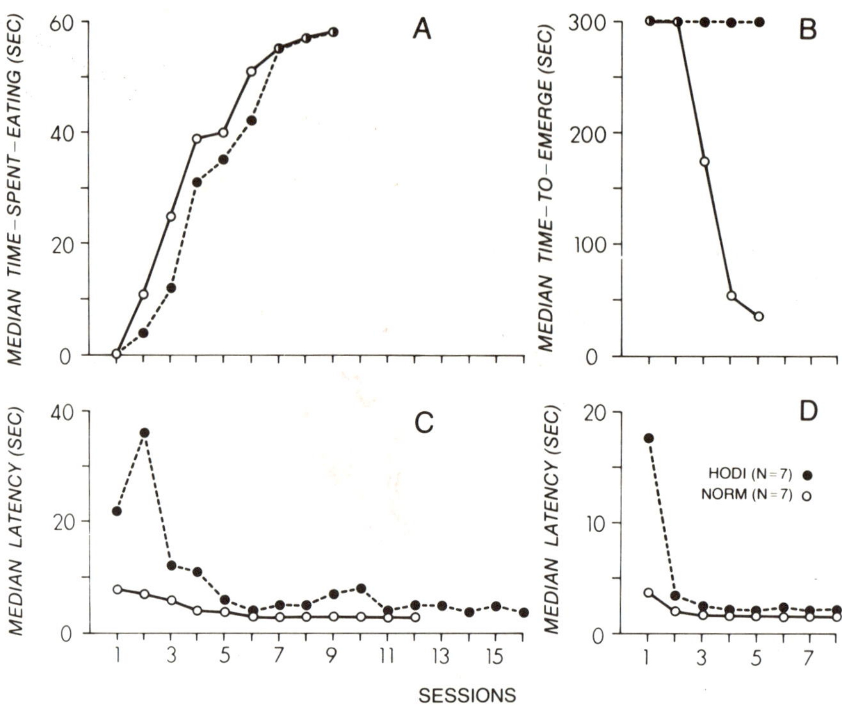

FIGURE 1. Performance of HODI and NORM rats on four food-motivated behavioral tasks. Panel A shows median time-spent-eating as a function of sessions in the first adaptation box. Panel B indicates median time-to-emerge from home cage into an open field across sessions. Panel C represents median latency to enter goal box of T-maze as a function of adaptation sessions. Panel D shows median latency to enter goal box of straight runway across sessions.

T-Maze Adaptation

FIGURE 1 (panel C) shows that HODI rats adapted to the T-maze more slowly than NORM rats. Total latency to enter the goal box of the maze across 12 sessions was significantly greater for the HODI rats compared with NORM rats ($U = 0$, $p < 0.002$). (Note: U of 0 indicates no overlap between the two distributions.) By the sixth session, NORM rats were performing at asymptote and continued to do so through their last (12th) session. However, HODI rats, even after four additional sessions, were still not performing at the level of NORM rats.

T-Maze Alternation, Olfactory Discrimination, and Visual Discrimination

HODI rats performed T-maze alternation (a working-memory task) as well as NORM rats. In addition, HODI rats performed the reference-memory tasks (olfactory and visual discrimination) as well as NORM rats (TABLE 1).

Runway Learning

As shown in FIGURE 1 (panel D), it took HODI rats significantly longer than NORM rats to reach the goal box of the straight runway on the first session ($U = 4$, $p < 0.01$). On the last session, the performance of HODI rats was much improved, and it approached that of NORM rats, although, statistically, HODI rats were still performing worse than NORM rats ($U \doteq 4.5$, $p < 0.01$).

Approach-Avoidance Conflict

FIGURE 2 (top panel) indicates that the performance of HODI rats on trials before the introduction of shock was similar to that of NORM rats, even though the difference between the two groups was statistically significant (there was little overlap between the groups). However, on the post-shock sessions, HODI rats showed memory as good as that of NORM rats. (Although the graph indicates a difference between the two groups on the last three post-shock sessions with the HODIs showing better "memory" than NORM rats, this difference was only marginally significant, $p < 0.10$).

Passive Avoidance

As shown in FIGURE 2 (bottom panel), HODI rats entered the dark box from an elevated platform on the pre-shock sessions significantly more slowly than NORM rats ($U = 6$, $p < 0.02$). On the post-shock sessions, HODI rats remembered their experience with shock in the dark box at least as well as NORM rats.

Prod Burying

There were no significant differences between HODI and NORM rats on the memorial aspects of this task (TABLE 1).

Interference Effects

Both HODI and NORM rats showed impairment of one-way active-avoidance performance following inescapable-shock stress, as shown in FIGURE 3. However, HODI rats that had not been exposed to inescapable shock performed better than their NORM counterparts on the one-way active avoidance task.

DISCUSSION

All results of the present studies are summarized in TABLE 1; the main findings were: (1) Rats lacking vasopressin (HODI) were impaired, compared with NORM rats, on temperament-related tasks, i.e., time-spent-eating, time-to-

TABLE 1

MEDIANS AND RANGES OF VASOPRESSIN-DEFICIENT (HODI) AND NORMAL (NORM) RATS IN A NEUROPSYCHOLOGICAL TEST BATTERY

	HODI ($N = 7$)	NORM ($N = 7$)	Statistics (Mann-Whitney U Test)
* Water Intake (ml/100 g BW)	92.5(60.0–113.5)	15.5(13.0–19.5)	$U = 0$, $p < 0.002$
* Time-Spent-Eating 1 (9 trials, sec)	274(174–329)	346(291–358)	$U = 5$, $p < 0.01$
Time-Spenting-Eating 2 (5 trials, sec)	259(234–287)	275(252–281)	$U = 19$, N.S.
* Time-to-Emerge (5 trials, sec)	1500(482–1500)	849(65–1430)	$U = 7$, $p < 0.02$
* T-Maze Adaptation (144 trials, sec)	146(138–537)	50(48–55)	$U = 0$, $p < 0.002$
T-Maze Alternation (72 trials, % corr)	83(69–86)	78(64–86)	$U = 15$, N.S.
Olfactory Discrimination (168 trials, % corr)			
1st half	54(52–69)	54(49–71)	$U = 20.5$, N.S.
2nd half	81(76–94)	76(56–89)	$U = 15.5$, N.S.
Visual Discrimination (240 trials, % corr)			
1st half	49(46–61)	48(44–68)	$U = 21$, N.S.
2nd half	77(53–85)	82(51–91)	$U = 16.5$, N.S.

* Runway Learning (96 trials, sec)	34.0(16.1–366.4)	15.3(13.3–20.5)	$U = 2$, $p < 0.002$
Conflict Ret.			
* Preshock (5 trials, sec)	2.0(1.6–4.5)	1.6(1.4–1.7)	$U = 4$, $p < 0.01$
Postshock (37 trials, sec)	348.0(122.1–360)	172.7(70.7–360)	$U = 18$, N.S.
Passive Avoidance Ret.			
* Preshock (4 trials, sec)	571(250–920)	79(34–452)	$U = 6$, $p < 0.02$
Postshock (5 trials, sec)	1500(930–1500)	991(774–1500)	$U = 12$, N.S.
Prod Burying			
Time (sec)	19(0–247)	158(0–574)	$U = 14.5$, N.S.
Height (cm)	7.9(3.8–10.8)	7.5(2.0–13.7)	$U = 16.5$, N.S.
Stress-Induced Interference (25 trials, sec)			
* No preshock	62.1(56.5–70.8) $N = 3$	76.5(75.6–114.2) $N = 3$	$U = 0$, $p < 0.02$
Preshock	121.8(98.4–145.2) $N = 2$	131.7(104.2–152.5) $N = 3$	$U = 2$, N.S.

* Statistically significant differences.

emerge, and adaptation to the T-maze and runway; (2) HODI rats performed as well as NORM rats on a task involving working memory, i.e., T-maze alternation; (3) HODI rats performed as well as NORM rats on tasks involving reference memory, i.e., olfactory and visual discriminations; (4) HODI rats showed memory as good as that of NORM rats on shock-motivated tests, i.e., approach-avoidance conflict and step-through passive avoidance; (5) HODI rats performed as well as NORM rats on a task involving memory for a species-specific behavioral response, i.e., prod burying; and (6) HODI rats were as susceptible as NORM rats to the deleterious effects of inescapable shock on

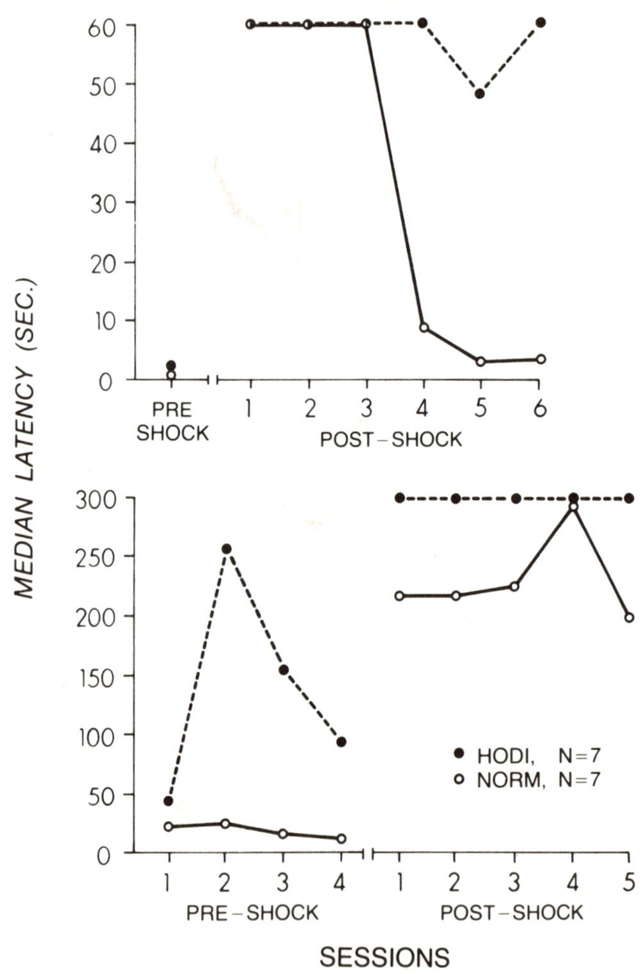

FIGURE 2. Performance of HODI and NORM rats on two shock-motivated tasks. The upper panel shows median latency as a function of sessions on the approach-avoidance conflict test. The points on the extreme left indicate preshock performance. The lower panel represents median latency as a function of pre- and postshock sessions on the step-through passive avoidance test.

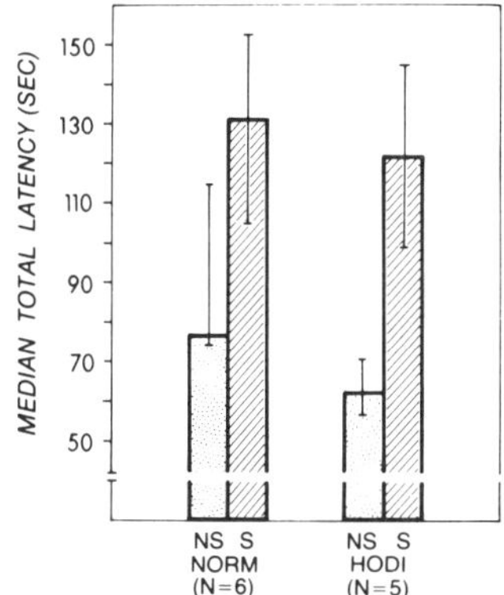

FIGURE 3. Median total latency across 25 trials on the one-way active avoidance task for HODI and NORM rats that had (S), and had not (NS), been exposed to a previous session of inescapable stress.

subsequent performance of a one-way active-avoidance task. Moreover, HODI rats learned a one-way active-avoidance task at least as well as NORM rats.

It should be emphasized that the same NORM and HODI rats participated in the sequence of behavioral tests of the battery in the order discussed, so that there was increasing opportunity for habituation to effects of handling and novelty of task situation.

We propose that HODI rats may not be impaired in memory, but they have altered temperamental dispositions compared with NORM rats, e.g., enhanced neophobia, timidity, etc. These altered temperamental dispositions can interact in misleading ways with "performance aspects" of cognitive behaviors supposedly depending on memory.

ACKNOWLEDGMENTS

The authors are grateful to Susan D. Connor for technical assistance, Louise Behrens for typing the manuscript, and Dr. Mark E. Stanton for valuable comments on the manuscript.

REFERENCES

1. BOHUS, B., TJ. B. VAN WIMERSMA GREIDANUS & D. DE WIED. 1975. Behavioral and endocrine responses of rats with hereditary hypothalamic diabetes insipidus (Brattleboro strain). Physiol. Behav. **14:** 609–615.
2. DE WIED, D., B. BOHUS & TJ. B. VAN WIMERSMA GREIDANUS. 1975. Memory deficit in rats with hereditary diabetes insipidus. Brain Res. **85:** 152–156.
3. CELESTIAN, J. F., R. J. CAREY & M. MILLER. 1975. Unimpaired maintenance of a conditioned avoidance response in the rat with diabetes insipidus. Physiol. Behav. **15:** 707–711.
4. MILLER, M., E. G. BARRANDA, M. C. DEAN & F. R. BRUSH. 1976. Does the rat with hereditary hypothalamic diabetes insipidus have impaired avoidance learning and/or performance? Pharmac. Biochem. Behav. **5** (Suppl. 1): 35–40.

5. BAILEY, W. H. & J. M. WEISS. 1979. Evaluation of a "memory deficit" in vasopressin-deficient rats. Brain Res. **169:** 174–178.
6. BRITO, G. N. O., G. J. THOMAS, S. I. GINGOLD & D. M. GASH. 1981. Behavioral characteristics of vasopressin-deficient rats (Brattleboro strain). Brain Res. Bull. **6:** 71–75.
7. HONIG, W. K. 1978. Studies in working memory in the pigeon. *In* Cognitive Processes in Animal Behavior. S. H. Hulse, H. Fowler & W. K. Honig, Eds.: 211–248. Lawrence Erlbaum. Hillsdale, N.J.
8. GLAZER, H. I. & J. M. WEISS. 1976. Long-term and transitory interference effects. J. Exp. Psychol. (Anim. Behav. Processes) **2:** 191–201.
9. ADER, R. & D. DE WIED. 1972. Effects of lysine vasopressin on passive avoidance learning. Psychon. Sci. **29:** 46–48.
10. THOMAS, G. J. & G. N. O. BRITO. 1980. Recovery of delayed alternation in rats after lesions in medial frontal cortex and septum. J. Comp. Physiol. Psychol. **94:** 808–818.
11. PINEL, J. P. J. & D. TREIT. 1978. Burying as a defensive response in rats. J. Comp. Physiol. Psychol. **92:** 708–712.
12. CONOVER, W. J. 1971. Practical Nonparametric Statistics. Wiley & Sons, Inc. New York, N.Y.

ENVIRONMENTAL ENRICHMENT IN BRATTLEBORO RATS: BRAIN MORPHOLOGY

E. Rosalie Greer, Marian C. Diamond, and Julie M. W. Tang

Department of Physiology-Anatomy
University of California
Berkeley, California 94720

Environmental enrichment is known to change certain brain parameters in normal rats,[1] and these changes may correlate with memory storage. The arousal response has been suggested as a fundamental mechanism mediating the anatomical and biochemical brain changes induced by enriched environments.[2] Because Brattleboro rats have abnormalities in learning and memory and in open-field activity indicating prolonged arousal,[3] we assessed the effects of enrichment on brain dimensions in male heterozygous (HZ) and homozygous (DI) Brattleboro rats.

Measurements were made in neocortex, subcortical telencephalon, caudal diencephalon, and hippocampus in three groups of HZ and DI rats. One group was sacrificed at 60 days of age from standard environmental conditions (60-day SC), another at 90 days of age from standard conditions (90-day SC), and a third group from enriched environmental conditions at 90 days of age after 30 days of enrichment (90-day EC). Alternate coronal sections obtained from frontal, parietal, and occipital regions were Golgi-Cox stained (for a later study of dendritic branching) and counterstained with thionine for measurement of brain dimensions. For evaluation of enrichment effects, comparisons were made: between heterozygous and homozygous rats within each condition group; between 60-day SC and 90-day SC groups to study age-related changes; and between the 90-day SC and 90-day EC groups to evaluate enrichment-related effects.

Environmental enrichment resulted in increases in dimensions in neocortex, subcortical telencephalon, caudal diencephalon, and hippocampus. With only two exceptions, the magnitude of the response of DI rat brain dimensions was greater than that of the HZ rats. Statistically significant responses were also more frequent, and they attained greater significance than in the HZ rats (TABLE 1).

Brain dimensions of HZ rats were greater than those of DI rats (as expected because of their significantly greater body weight) with only three exceptions in 24 comparisons (TABLE 2). However, the greater HZ brain dimensions reached significance in only five comparisons. These differences in brain measures between the two types of Brattleboro rats, though seldom significant, tended to increase with age (see Avg. % difference in TABLE 2). Enrichment appeared to prevent this age-related increase in the difference between HZ and DI brain dimensions. Body weight differences with age and with enrichment followed the same pattern. These trends were reflected in the greater magnitude of response to environmental enrichment in DI rats than in HZ rats.

Details of the results of cortical thickness studies[4] were in general agreement with previously reported developmental and enrichment studies in normal rats.[1] However, there were important differences in that Brattleboro rats showed greater and more generalized increases in cortical thickness in response to

0077-8923/82/0394-0749 $1.75/0 © 1982, NYAS

TABLE 1

PERCENTAGE DIFFERENCE IN BRAIN DIMENSIONS BETWEEN 90-DAY STANDARD AND ENRICHED CONDITION GROUPS OF HETEROZYGOUS (HZ) AND HOMOZYGOUS (DI) BRATTLEBORO RATS

Brain Region	Percentage Difference *			
	Heterozygous †		Homozygous ‡	
Frontal cortex	+4.5	NS §	+6.7	$p < 0.025$
Parietal cortex	+5.3	$p < 0.05$NK ¶	+5.1	$p < 0.05$
Occipital cortex	+6.3	$p < 0.01$	+8.1	$p < 0.001$
Telencephalic height	+7.0	$p < 0.05$NK	+4.7	NS
Telencephalic width	+2.4	NS	+4.7	$p < 0.05$
Diencephalic height	+3.4	NS	+7.7	$p < 0.05$NK
Diencephalic width	+3.6	$p < 0.05$	+6.3	$p < 0.01$
Hippocampus	+8.1	$p < 0.001$	+10.2	$p < 0.001$

* Percentage difference = (90-day EC/90-day SC × 100) − 100.

† Number of HZ rats: 90-day SC = 5; 90-day EC = 12.

‡ Number of DI rats: 90-day SC = 5; 90-day EC = 12.

§ NS = not significant.

¶ NK = Neuman-Keuls correction; all other p values are Scheffé correction.

TABLE 2

COMPARISONS OF BRAIN DIMENSIONS AND BODY WEIGHT BETWEEN HETEROZYGOUS (HZ) AND HOMOZYGOUS (DI) BRATTLEBORO RATS

Brain Region	Percentage Difference *					
	60-day SC †		90-day SC		90-day EC	
Frontal cortex	1.4	NS ‡	3.8	$p < 0.025$	1.6	NS
Parietal cortex	−0.3 §	NS	2.7	NS	2.9	$p < 0.05$
Occipital cortex	1.9	NS	2.4	NS	0.8	NS
Telencephalic height	1.4	NS	−1.1	NS	1.1	NS
Telencephalic width	1.7	NS	5.3	$p < 0.01$	3.0	$p < 0.01$
Diencephalic height	−1.4	NS	5.1	NS	0.9	NS
Diencephalic width	3.7	$p < 0.05$	3.3	NS	0.6	NS
Hippocampus	2.6	NS	3.4	NS	1.4	NS
Avg. % difference	1.4		3.1		1.5	
Body weight	15	$p < 0.001$	24	$p < 0.001$	14	$p < 0.001$

* Percentage difference = (HE/DI × 100) − 100.

† Number of rats in each group: 60-day SC, HZ = 8, DI = 6; 90-day SC, HZ = 5, DI = 5; 90-day EC, HZ = 12, DI = 12.

‡ NS = not significant. p values obtained by analysis of variance.

§ Negative values of percentage difference indicate that DI rats' brain dimensions were greater than those of HZ rats. Note that HZ measurements exceeded those of DI rats in 21 of the 24 comparisons.

enrichment than normal (non-vasopressin-deficient) rats, as well as a different pattern of response.[4] The significant increases in subcortical dimensions in both HZ and DI rats also suggest a greater responsiveness to enrichment than has been found in normal rats.

Increased responsiveness to enrichment in Brattleboro rats is consistent with a role of arousal in enrichment,[2] i.e., both HZ and DI rats have shown prolonged open-field activity [3] indicating a heightened arousal response. Excessive arousal may be related to abnormalities in norepinephrine (NE) concentration and/or turnover associated with vasopressin deficiency.[5] We suspect that NE may act as a neurohormone [6] and thus increase anabolic activity in certain brain regions in both Brattleboro rats. Results of a pilot study in our laboratory are consistent with greater than normal cortical NE concentration in adult Brattleboro rats, as well as with increased NE release during enrichment.

To explain the greater magnitude of response to enrichment in DI rats as compared with HZ rats, we propose that a diminishing growth rate in the DI rat in standard environmental conditions is nullified by metabolic events during enrichment. Thus, under standard conditions, the chronic dehydration that accompanies diabetes insipidus may give rise to increased sympathetic outflow with prolonged systemic stress and increased catabolism. The cumulative effect of this metabolic state would account for the increased percentage difference between HZ and DI rats' brain dimensions (and body weights) in the 90-day SC groups as compared with the 60-day SC groups.

During environmental enrichment, nullification of a diminishing growth rate could result from two responses of DI rats to an increased NE turnover elicited by enrichment-mediated arousal. First, evidence that increased central adrenergic activity reduces sympathetic outflow [7] is consistent with reduction of systemic stress and catabolic activity in DI rats during enrichment. Second, evidence that vasopressin action on the release of growth hormone (GH) is blocked by anti-adrenergic agents [8] suggests that increased adrenergic activity could facilitate GH release in DI rats. Thus both reduced catabolic activity in certain brain regions and also increased GH-induced anabolic activity may contribute to the DI response to enrichment.

In summary, while both Brattleboro rats responded to environmental enrichment with significant increases in certain brain dimensions, the magnitude of the DI response was greater. We hypothesize that prolonged arousal in Brattleboro rats during enrichment is associated with an increased central adrenergic activity with NE acting as an neurohormone to promote anabolic effects. In addition, in the DI rat, the increased adrenergic activity may reduce a catabolic effect of stress due to chronic dehydration, and may also promote systemic growth by increasing GH release.

References

1. Diamond, M. C. 1976. Anatomical brain changes induced by environment. *In* Festschrift: Knowing, Thinking, and Believing. J. McGaugh & L. Petrinovich, Eds.: 215–241. Plenum Press. New York, N.Y.
2. Walsh, R. N. & R. A. Cummins. 1975. Mechanisms mediating the production of environmentally induced brain changes. Psych. Bull. **82:** 986–1000.
3. Bohus, B., Tj. B. van Wimersma Greidanus & D. de Wied. 1975. Behavioral and endocrine responses of rats with hereditary hypothalmic diabetes insipidus (Brattleboro strain). Physiol. Behav. **14:** 609–615.

4. GREER, E. R., M. C. DIAMOND & J. M. W. TANG. 1981. Increase in thickness of cerebral cortex in response to environmental enrichment in Brattleboro rats deficient in vasopressin. Exp. Neurol. **72:** 366–378.
5. VERSTEEG, D. H. G., M. TANAKA & E. R. DE KLOET. 1978. Catecholamine concentration and turnover in discrete regions of the brain of the homozygous Brattleboro rat deficient in vasopressin. Endocrinology **103:** 1654–1661.
6. VARON, S. S. & G. G. SOMJEN. 1979. Neurotransmitters and neuroglia. Neurosci. Res. Prog. Bull. **17:** 131–146.
7. COOTE, J. H. & V. H. MACLEOD. 1974. The influence of bulbospinal monoaminergic pathways on sympathetic nerve activity. J. Physiol. (London) **241:** 453–475.
8. HEIDINGSFELDER, S. A. & W. G. BLACKARD. 1968. Adrenergic control mechanism for vasopressin-induced plasma growth hormone response. Metabolism **17:** 1019–1024.

ON THE ROLE OF ARGININE VASOPRESSIN IN CIRCUMVENTRICULAR ORGANS

A. Negro-Vilar and W. K. Samson

Department of Physiology
University of Texas
Health Science Center at Dallas
Dallas, Texas 75235

Arginine vasopressin plays a key role in osmoregulation and the control of body fluid volume. Physiological stimuli such as dehydration, acute changes in plasma osmolarity, and blood volume loss can activate the hypothalamo-neurohypophyseal system (HNS) to elicit AVP release. The peptide is synthesized in neurons located primarily in the supraoptic (SON) and paraventricular (PVN) nuclei, although AVP/neurophysin-containing perikarya have also been identified in the suprachiasmatic nucleus (SCN) and in smaller areas between these three nuclei.[1, 2] AVP fibers arising from the SCN cells are of a fine caliber and project to the lateral septum, habenula, and thalamus, among other structures.[3] In recent years, evidence has accumulated to indicate that AVP is distributed beyond the classical neurohypophyseal system, with projections to the median eminence, circumventricular organs (CVO), and many extrahypothalamic areas.[2] These anatomical findings have lent further support to the concept that AVP may subserve brain functions other than those related to its peripheral role as the antidiuretic hormone.

AVP, and its neurophysin, have been localized to different CVO's by immunohistochemistry.[2] Using a specific radioimmunoassay (RIA) for AVP, we have recently reported the presence of significant amounts of the peptide in the organum vasculosum of the lamina terminalis (OVLT),[4-6] subfornical organ (SFO),[5, 6] subcommissural organ (SCO),[6] and in the pineal gland.[4, 6, 7] These structures lack a blood-brain barrier and they purportedly have a role in mechanisms regulating water intake, thirst, and osmoregulation.

Role of Blood Osmolarity and Volume in the Regulation of AVP Release

AVP is regulated by an exquisitely sensitive osmoregulatory system, since changes in plasma osmolarity of 2% or less increase plasma AVP levels by about tenfold, this effect being independent of changes in blood volume.[8] Isotonic hypovolemia, on the other hand, can also increase plasma AVP, but in this case a reduction of 8% or more in blood volume is required to obtain a significant change in AVP levels. Moreover, in situations of hypovolemia, the osmoreceptors have a lower threshold, and there is increased responsiveness to an osmotic challenge.[8] The osmoreceptor cells are located not only in the vicinity of the supraoptic nucleus but also throughout the anterior hypothalamus[9] and these cells are not AVP-producing magnocellular neurons, as determined by electrophysiological recording techniques. Evidence has been presented to indicate that the osmoreceptors may in fact be sodium receptors

0077-8923/82/0394-0753 $1.75/0 © 1982, NYAS

rather than being stimulated by osmotically stimulated water shifts (see Reference 9 for references).

Although the connection between CVO structures and osmoreceptors is not well understood, the fact that many of these organs have direct contact with peripheral circulation as well as with subarachnoid and ventricular cerebrospinal fluid (CSF) routes, stimulates the conjecture that an active interchange between central (AVP) and peripheral (renin-angiotensin) salt-water regulatory mechanism occurs at this level.

Changes in AVP Levels in Blood and Central Sites Including the CVO's During Progressive Dehydration

We recently reported changes in AVP levels in certain specific hypothalamic areas in response to progressive dehydration.[4] AVP levels were examined at 0, 3, and 7 days after dehydration in three CNS regions containing AVP nerve terminals (ME, posterior pituitary, and OVLT), another with mostly AVP fibers (arcuate nucleus), and in the two major nuclei containing the perikarya of the magnocellular neurons (SON and PVN). This particular design allowed us to look at changes in AVP levels that might be occurring at different levels of the neuron, and also to study areas not directly related to water metabolism. The results indicate that AVP levels fall drastically in the posterior pituitary, more slowly in the ME, ARC, and PVN; and do not change significantly in the SON. Surprisingly, AVP levels in the OVLT rose significantly three days after water deprivation and returned to basal levels by seven days. This preliminary study indicated to us that different areas within (or outside) the HNS react differently to dehydration and also suggested that this might be a useful approach to determine the involvement of different areas within the AVP neuronal system with increased plasma osmolarity after prolonged dehydration. We therefore initiated a more comprehensive study in which levels of AVP in hypothalamic and circumventricular structures were correlated to the peripheral changes in serum AVP at 1, 2, 3, 5, and 7 days after water deprivation. An initial analysis of hypothalamic structures and the neural lobe (Figure 1) confirmed and extended our previous findings. A decline of AVP levels in ME, PVN (not shown), and the suprachiasmatic nucleus (SCN) at five and seven days after water deprivation (WD) was seen. The supraoptic nucleus, however, showed no change in AVP levels at any of the times studied. In agreement with previous reports,[10] this observation suggests that AVP synthesis in the SON increases after dehydration and keeps pace with release even in a situation of accelerated axonal transport and release of the peptide. AVP synthesis in the PVN and SCN is probably declining after three days, in view of the lowering of the tissue stores at five and seven days. Plasma levels of AVP (Figure 2) showed the expected progressive rise at 1, 2, and 3 days after WD, with values peaking on day 3 and then surprisingly declining at 5 and 7 days. These changes in serum AVP levels were mirrored by changes in AVP content of the OVLT and SFO (Figure 3) and also in the pineal gland (not shown). In these circumventricular structures, the changes were specific for AVP, since levels of another peptide, somatostatin, were not affected at any time. These studies clearly indicated a close relationship between levels of AVP in serum and in brain structures associated with the cerebrospinal fluid. However, the question still remains whether the simultaneous changes in AVP in plasma and CVO are

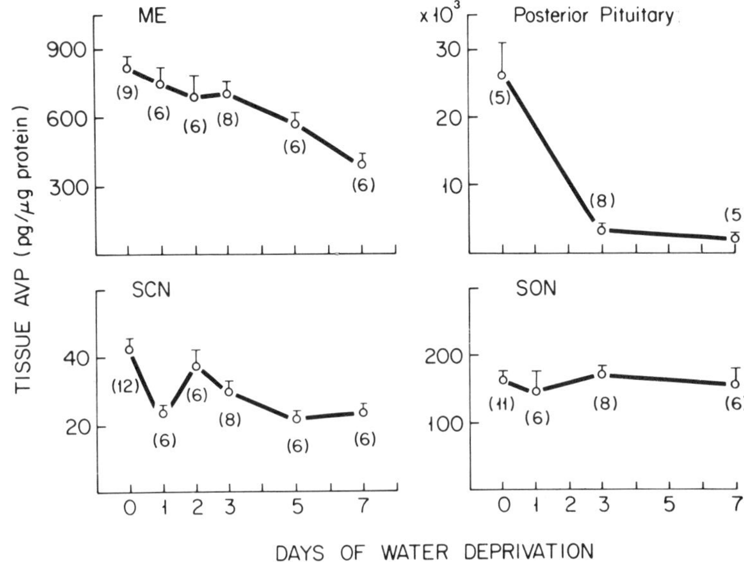

FIGURE 1. Changes in tissue AVP levels at different intervals after water deprivation. In this and subsequent figures values are mean ± SEM, and number between parentheses indicate number of animals per group. ME = median eminence. SCN = suprachiasmatic nucleus. SON = supraoptic nucleus.

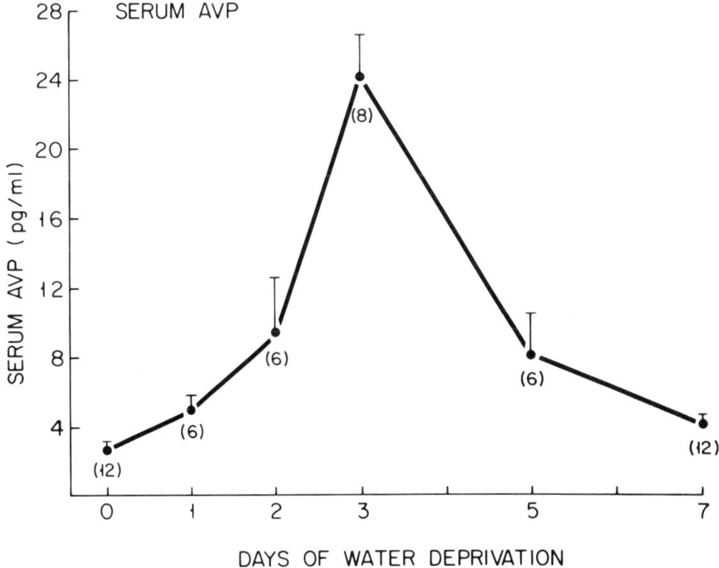

FIGURE 2. Changes in serum AVP levels at different intervals after water deprivation.

indicative of an uptake of peripheral circulating AVP into the CVO or a delivery of AVP to the CVO's for release of the peptide into the general circulation and the CSF.

Studies to Determine Uptake of AVP into Brain Areas from Peripheral Blood

To answer the above question, experiments were performed in homozygous Brattleboro rats, in which an i.v. infusion of AVP was given to determine if circulating AVP could be taken up by the different CVO. Since these animals lack AVP, any amount of the peptide present in brain regions should originate in exogenously infused AVP, which was removed from the peripheral circulation. The results of this experiment, shown in Table 1, indicate that despite attaining huge levels of AVP in peripheral plasma, no trace of the peptide could be found in any of the CVO structures analyzed. This finding suggests that AVP present in the CVO's of normal animals is not normally derived from the general circulation but instead is probably transported neuronally to these

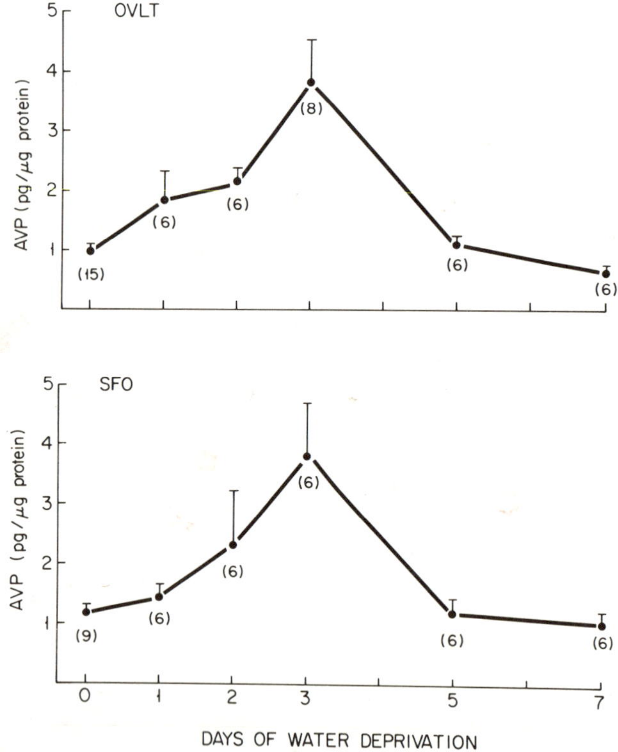

Figure 3. Changes in tissue AVP levels in organum vasculosum lamina terminalis (OVLT, upper panel) and subfornical organ (SFO, lower panel) after water deprivation.

TABLE 1

EFFECTS OF INTRAVENOUS INFUSION * OF AVP ON AVP LEVELS IN DIFFERENT BRAIN REGIONS IN HOMOZYGOUS BRATTLEBORO RATS

Serum AVP (pg/ml)		Areas †					
Before	After ‡	ME	OVLT	SFO	SCO	PI	NL
ND	928.9 ± 117	ND	ND	ND	ND	ND	ND

* Synthetic AVP (Bachem) was infused in saline, at a rate of 0.5 ml/min in rats anesthetized with tribromoethanol, over a 30 min period.

† Mean AVP levels ± SEM at the end of a 30 min infusion period. N = 5, ND = not detectable, less than 2.0 pg/area.

‡ Bilateral structures were punched out, pooled, and homogenized in 0.1 M acetic acid. AVP RIA was performed in diluted or undiluted aliquots of supernatant. ME = median eminence. OVLT = organum vasculosum lamina terminalis. SFO = subfornical organ. SCO = subcommisural organ. PI = pineal gland. NL = neural lobe.

structures from the cell bodies in the HNS. The fact that AVP concentrations in those areas paralleled the changes in peripheral levels of the peptide, suggests that the hormone is released concomitantly to the systemic circulation and to the CSF, where it has been detected under a variety of conditions.[11] It is also possible that the AVP released into the CSF may play a role in local water homeostasis perhaps by regulating water metabolism at the level of the choroid plexus.

REFERENCES

1. SOKOL, H. W., E. A. ZIMMERMAN, W. H. SAWYER & A. G. ROBINSON. 1976. The hypothalamic-neurohypophyseal system of the rat: Localization and quantification of neurophysin by light microscopic immunocytochemistry in normal rats and in Brattleboro rats deficient in vasopressin and a neurophysin. Endocrinology **98:** 1176–1188.
2. DEFENDINI, R. & E. A. ZIMMERMAN. 1978. The magnocellular neurosecretory system of the mammalian hypothalamus. *In* The Hypothalamus. S. Reichlin, R. J. Baldessarini & J. B. Martin, Eds.: 137–152. Raven Press. New York, N.Y.
3. SOFRONIEW, M. W. & A. WEINDL. 1978. Projections from the parvocellular vasopressin and neurophysin-containing neurons of the suprachiasmatic nucleus. Am. J. Anat. **153:** 391–429.
4. NEGRO-VILAR, A. & W. K. SAMSON. 1979. Dehydration-induced changes in immunoreactive vasopressin levels in specific hypothalamic structures. Brain Res. **169:** 585–589.
5. NEGRO-VILAR, A. & J. M. SAAVEDRA. 1980. Effects of stress on the vasopressin and somatostatin content of discrete brain nuclei of spontaneously hypertensive rats. Brain Res. Bull. **5:** 353–358.
6. NEGRO-VILAR, A. & W. K. SAMSON. 1979. Differential changes in vasopressin in specific hypothalamic nuclei and circumventricular organs after progressive dehydration. 9th Mtg. Soc. Neurosci. Atlanta, Georgia. p. 230.
7. NEGRO-VILAR, A., F. SANCHEZ-FRANCO, M. KWIATKOWSKI & W. K. SAMSON. 1979. Failure to detect radioimmunoassayable arginine vasotocin in mammalian pineals. Brain Res. Bull. **4:** 789–792.

8. Dunn, F. L., T. J. Brennan, A. E. Nelson & G. L. Robertson. 1973. The role of blood osmolarity and volume in regulating vasopressin secretion in the rat. J. Clin. Invest. **52:** 3212–3219.
9. Kleeman, C. R. & T. Berl. 1979. The neurohypophyseal hormones: Vasopressin. *In* Endocrinology. L. J. DeGroot, Ed. **1:** 253–275. Grunne & Stratton. New York, N.Y.
10. Brownstein, M. J., J. T. Russell & H. Gainer. 1980. Synthesis, transport and release of posterior pituitary hormones. Science **207:** 373–378.
11. Dogterom, J., T. J. B. van Wimersma Greidanus & D. F. Swaab. 1977. Evidence for the release of vasopressin and oxytocin into cerebrospinal fluid: Measurements in plasma and CSF of intact and hypophysectomized rats. Neuroendocrinology **24:** 108–118.

EXTRAHYPOTHALAMIC NEUROPHYSIN PROJECTIONS IN THE BRAIN STEM AND SPINAL CORD OF NORMAL AND HOMOZYGOUS BRATTLEBORO RATS

G. Nilaver, J. Mulhern, and E. A. Zimmerman

Department of Neurology
College of Physicians and Surgeons
Columbia University
New York, New York 10032

Introduction

The oxytocin (OT)- and vasopressin (VP)-containing neurons of the rat paraventricular nucleus have been shown to project to a variety of extrahypothalamic sites including forebrain,[1] brainstem, and spinal cord.[2,3] Most of these studies have been performed employing immunocytochemical techniques on paraffin-embedded tissue sections with antisera to neurophysins (NPS), the carrier proteins for OT and VP. Although the course of OT and VP fibers in these regions seem to parallel the neurophysin-containing fibers, their visualization in paraffin-embedded sections has been consistently less intense. The relative paucity of OT and VP in paraffin sections of these regions probably represents a technical artifact. It is most likely caused by greater tissue losses of the smaller hormones, than the associated NPS, possibly from elution by the organic solvents used in the dehydration procedures of the paraffin-embedding technique. Such losses, as have been demonstrated for LHRH in tissue prepared for immunocytochemistry,[4] become critical factors when this technique is employed to study the relative content of a peptide within different brain regions, or when comparing its distribution in different species.

Employing an antiserum directed to both rat NPS,[5] we had previously reported on the NP-containing fiber projections to the lower brain stem and spinal cord of normal and homozygous Brattleboro (DI) rats with diabetes insipidus.[3] Using paraffin-embedded sections, we demonstrated a relative predominance of oxytocin-neurophysin (OT-NP) over vasopressin-neurophysin (VP-NP) in these structures in normal rat. Losses of antigen in tissue from paraffin-embedding procedures are generally considered to affect only small peptide molecules, such as OT and VP. No data are available on the effects of organic solvents on the larger NPS. In order to account for possible losses of tissue NPS in our previous study with the paraffin-embedded sections, and to obtain a more accurate estimate of the NP content of these projections, we have extended our analysis of these regions, using the pre-embedding method of staining, a technique that avoids the dehydrating and paraffin-embedding procedures.

Materials and Methods

Five adult Long-Evans rats and five homozygous Brattleboro rats with diabetes insipidus (Blue Spruce Farms, Altamont, New York) of both sexes

0077–8923/82/0394–0759 $1.75/0 © 1982, NYAS

weighing 250–300 g were used. Animals were sacrificed by cardiac perfusion with 0.9% $NaCl_2$ under pentobarbital anesthesia, followed by 10% buffered formalin. The brainstem and spinal cord were removed, blocked, and post-fixed by immersion in the same fixative overnight. Tissue blocks were embedded in 7% agar, and coronal (brainstem and spinal cord) and longitudinal (spinal cord) sections were cut at 50 μm thickness, using a Vibratome (Oxford Instruments). The sections were incubated in 1% H_2O_2 in Tris buffer (0.1 M, pH 7.4) for 30 minutes at room temperature (to destroy endogenous peroxidase activity), rinsed extensively in the same buffer, followed by 0.1 M Tris containing 0.1% Triton X-100 (Tris-A) and 0.1 M Tris containing 0.1% Triton and 1% normal sheep serum (Tris-B) (both at pH 7.4), prior to their application in immunocytochemical procedures. All antisera dilutions were made in Tris-B, and the sections washed for 15 min in Tris-A between each successive step of the immunocytochemical procedure.

Pre-embedding immunocytochemistry was performed using modifications[6] of the peroxidase-antiperoxidase (PAP) method of Sternberger.[7] Sections were incubated with rabbit antiserum to rat neurophysins (Robinson #4, 1:4000, at 4° C for 12 hours) followed by goat anti-rabbit globulin (1:100, 27° C, 30 min) and rabbit PAP (1:400, 27° C, 1 hour). Reaction products were formed with 3,3′-diaminobenzidine tetrahydrochloride. The sections were then mounted on glass slides, dehydrated, and permanently mounted.

Results

A greater density of fiber staining was seen in all the examined regions of the normal rat, compared to the DI. This was particularly evident in the substantia gelatinosa, at all levels of the spinal cord, and in the nucleus tractus solitarius and dorsal motor vagal nucleus of the medulla. In the spinal cord, the density of innervation of the substantia gelatinosa was more evident in the longitudinal sections. In these sections, in addition to the major caudal descending pathway in the white matter of the dorso-lateral funiculus, a dense bundle of rostro-caudally oriented fibers was also observed in the region of the central canal (Rexed's lamina X) extending throughout the length of the cord. Fibers from this projection were often frequently traced perpendicularly into the subjacent regions of the grey matter. Although present in both rat strains, the greater density of fiber staining in the normal rat was more evident in this region, than all other areas examined.

Discussion

The greater density of NP fibers in the normal rat, compared to the DI, in the various regions studied suggests the contribution of both OT-NP and VP-NP to these projections in the normal rat. Such differences in fiber density were not readily apparent in our previous study with the paraffin-embedding technique.[3] Indeed, in that study we had reported a similar number of fibers in the caudal extrahypothalamic regions in both rat strains. While the inconsistency of these findings could be explained by a selective loss of VP-NP by the paraffin-embedding procedures, a more likely interpretation would be that pre-embedding staining, by employing thicker sections (with greater depth of field) and better antibody penetration (with Triton X-100), results in greater immunoreactivity, allowing even subtle differences in fiber density to be more readily discernible.

The distribution of NP fibers in the caudal extrahypothalamic regions in this study, agrees well with that reported previously.[2, 8, 9] Each of these regions has been implicated, either directly or indirectly, in the central control of the autonomic nervous system.[10] The NP fibers in Rexed's lamina X were previously described to arise from the projection in the dorso-lateral funiculus. In

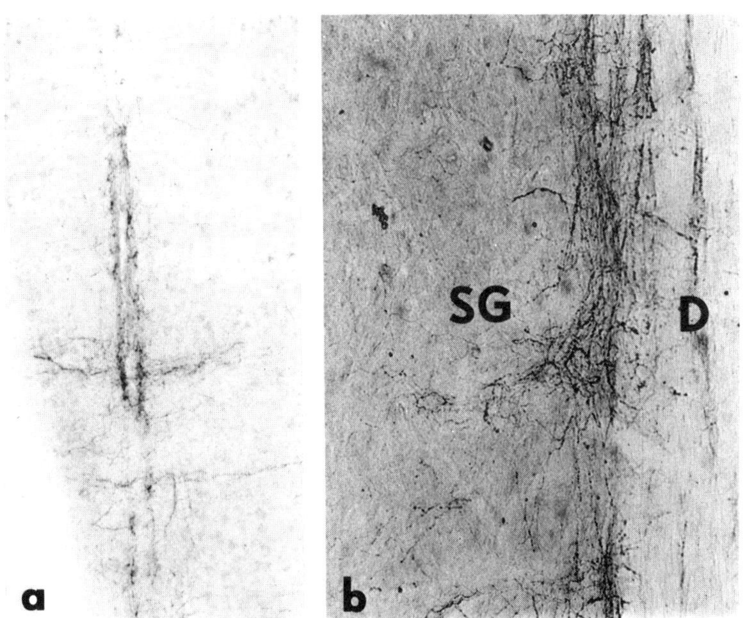

FIGURE 1. Longitudinal 100-μm thick sections of normal Long-Evans rat spinal cord demonstrating the rostro-caudal orientation of the neurophysin immunoreactive fibers in (a) the region of the central canal (Rexed's lamina X), and (b) the white matter of the dorso-lateral funiculus. In both (a) and (b) individual NP fibers can be traced to enter the adjacent regions of the cord. SG: substantia gelatinosa; D: dorso-lateral funiculus. The rostral regions of the cord are at the top of the photomicrographs (a: $\times$ 56, b: $\times$ 75).

this study, using unembedded, thick longitudinal sections, we demonstrate a distinct descending caudal projection in this region of the cord, confirming earlier findings in the rat.[9] The marked differences in fiber density of this region in the two rat strains further indicate that while an overall predominance of OT-NP exists over VP-NP in the caudal extrahypothalamic pathways,[3] VP-NP forms a major component of this lamina X projection. While the origin of this projection from the paraventricular nucleus is presumptive, its precise role in spinal cord function remains to be elucidated.

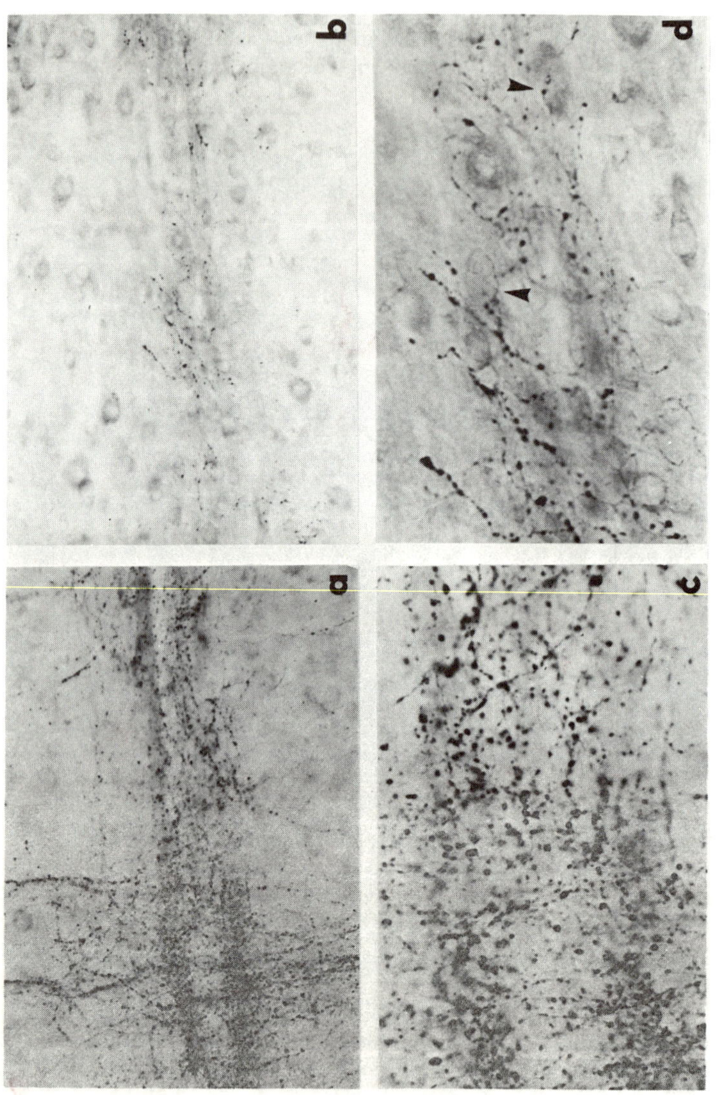

FIGURE 2. Longitudinal sections of the spinal cord through the region of the central canal (Rexed's lamina X) comparing the patterns of NP immunoreactivity in the normal Long-Evans rat (a and c) and the homozygous Brattleboro rat (b and d). The significant difference in the density of fibers in this region of the two rat strains is evident in the low power field (a and b) (× 71). At higher magnifications (c and d), individual fibers appear to contact the neuronal elements of this region (arrows) (× 228). The rostral regions of the cord are to the left of the micrographs.

References

1. Weindl, A. & M. V. Sofroniew. 1976. Demonstration of extrahypothalamic peptide secreting neurons. A morphological contribution to the investigation of psychotropic effects of neurohormones. Pharmakopsych. **9:** 226–234.
2. Swanson, L. W. 1977. Immunohistochemical evidence for a neurophysin-containing autonomic pathway arising in the paraventricular nucleus of the hypothalamus. Brain Res. **128:** 346–353.
3. Nilaver, G., E. A. Zimmerman, J. Wilkins, J. Michales, D. Hoffman & A.-J. Silverman. 1980. Magnocellular hypothalamic projections to the lower brainstem and spinal cord of rat: immunocytochemical evidence for predominance of the oxytocin-neurophysin system compared to the vasopressin-neurophysin system. Neuroendocrin. **30:** 150–158.
4. Goldsmith, P. C. & W. F. Ganong. 1975. Ultrastructural localization of luteinizing hormone-releasing hormone in the median eminence of the rat. Brain Res. **97:** 181–197.
5. Sokol, H. W., E. A. Zimmerman, W. H. Sawyer & A. G. Robinson. 1976. The hypothalamo-neurohypophysial system of the rat: localization and quantification of neurophysin by light microscopic immunocytochemistry in normal rat and in Brattleboro rats deficient in vasopressin and neurophysin. Endocrinology **98:** 1176–1188.
6. Kozlowski, G. P., M. S. Brownfield & G. Hostetter. 1978. Neurophysin and gonadotrophin-releasing hormone (Gn-RH) containing hypothalamofugal neurosecretory systems. *In* Current Studies of Hypothalamic Function. K. Lederis & W. L. Veale, Eds. Vol. 1: 15–26. Karger. Basel, Switz.
7. Sternberger, L. A. 1974. Immunocytochemistry. Prentice Hall. Englewood Cliffs, N.J.
8. Swanson, L. W. & S. McKellar. 1979. The distribution of oxytocin- and neurophysin-stained fibers in the spinal cord of the rat and monkey. J. Comp. Neurol. **188:** 87–106.
9. Sofroniew, M. V. 1980. Projections from vasopressin, oxytocin, and neurophysin neurons to neural targets in rat and human. J. Histochem. Cytochem. **28:** 475–478.
10. Swanson, L. W. & B. K. Hartman. 1980. Biochemical specificity in central pathways related to peripheral and intracerebral homeostatic functions. Neurosci. Lett. **16:** 55–60.

DEFICITS IN TOLERANCE TO ETHANOL IN BRATTLEBORO RATS *

Quentin J. Pittman,† Joseph Rogers,‡ and Floyd E. Bloom ‡

† *Department of Pharmacology and Therapeutics*
University of Calgary
Calgary, Alberta
Canada

‡ *Alcohol Research Center*
The Salk Institute
San Diego, California

Subjects exposed continuously to ethanol become tolerant to its physiological and behavioral effects. Administration of arginine vasopressin (AVP) or related analogues can influence both the acquisition of ethanol drinking behavior [1, 3] and the development of tolerance in mice to the hypothermic and sedative effects of alcohol.[2, 4] These observations raise the possibility that endogenous AVP might play a role in the development of tolerance to ethanol. To test this hypothesis, tolerance to ethanol was evaluated in Brattleboro rats, which in the homozygous state lack AVP and in the heterozygous condition exhibit reduced stores of AVP.[6] We report here that Brattleboro rats, both homozygous or heterozygous for the presence of AVP, failed to develop tolerance to hypothermic effects of ethanol after prolonged chronic exposure.

Heterozygous or homozygous naive male Brattleboro (Blue Spruce Farms, Altamont, New York) and Sprague-Dawley (Charles River Farms, Wilmington, Delaware) rats were employed in these studies. Strains were age-matched to 100–120 days old. Within each strain, rats were randomly assigned to either a chronic ethanol treatment or control condition. The ethanol inhalation paradigm of Rogers *et al.*[5] was used to provide uniform, constant, high blood alcohol levels in rats for a period of 21 days. Blood alcohol levels were checked every day by tail vein blood sample using the Sigma NAD-ADH blood alcohol test (kit 331–10). Using this exposure paradigm, blood alcohol levels in all experimental rats averaged around 200 mg% and no significant differences were seen in blood alcohol levels among the different strains. Control animals were similarly housed and tail-vein blood was sampled; however, they were not exposed to ethanol vapor. After 21 days of ethanol inhalation, subjects were withdrawn from ethanol for 48 hours, then injected intraperitoneally with a 3 g/kg dose of 20% w/vol ethanol to evaluate tolerance to its poikilothermic effects. Rectal temperatures were recorded 5 min before and 30, 60, 90, 120, 150, and 180 min after injection. These tests were conducted at room temperature (22° C) in a novel environment.

Acute ethanol injection caused time-dependent changes in body temperature in all subjects. There were no statistically significant differences in the temperature reductions attained by homozygous or heterozygous Brattleboro subjects and therefore their results have been pooled within the experimental or control

* Supported by the Alcohol Research Center (Grant AA03504) and the Medical Research Council of Canada.

0077–8923/82/0394–0764 $1.75/0

groups. Of particular interest, however, is the finding that there is a significant difference in the hypothermic response to ethanol of naive Sprague-Dawley and Brattleboro subjects ($F = 164.08$, $p < .001$, three-way ANOVA), with the latter exhibiting consistently lower temperatures at all time points after ethanol administration (FIGURE 1). Unpublished data from our laboratory indicate that Brattleboro rats do *not* show higher blood alcohol levels than do Sprague-Dawley subjects given this dose of ethanol, nor do they show reduced amounts of liver alcohol dehydrogenase activity.

A three-way ANOVA also reveals a significant difference ($F = 3.87$, $p < .05$) between the two rat strains in the way ethanol interacts with chronically exposed versus ethanol-naive subjects. For Sprague-Dawley rats, previous chronic ethanol inhalation results in significant hypothermic tolerance ($F = 7.07$, $p < .01$) (FIGURE 1). In Brattleboro rats, however, the previous ethanol inhalation fails to produce tolerance to the hypothermic effects of ethanol; indeed, after acute intraperitoneal ethanol, the temperature drop of chronic ethanol and ethanol-naive Brattleboro subjects is virtually identical.

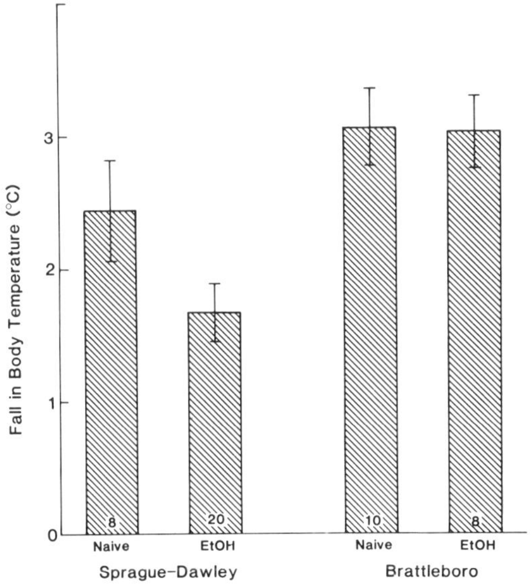

FIGURE 1. Maximum fall in body temperature 180 min after intraperitoneal injection of 3 g/kg ethanol to Sprague-Dawley or Brattleboro rats that were naive with respect to previous ethanol exposure (naive) or had been exposed to ethanol vapor continuously for 21 days (ETOH).

Our results indicate that both homozygous and heterozygous Brattleboro rats show a reduced ability to develop tolerance to the hypothermic effects of ethanol, when compared to rats of the Sprague-Dawley strain. These findings may relate to the growing body of evidence relating AVP to various tolerance phenomena.[4] In particular, administration of AVP or fragments of this molecule to mice enhances and prolongs tolerance to alcohol.[2,4] Our data support the reasoning that if AVP excesses prolong tolerance, then AVP deficits should diminish or prevent tolerance. These data, therefore, strongly suggest a functional role for endogenous AVP in the development of tolerance to hypothermic effects of ethanol. It is of interest that our data indicate that even a *relative* deficiency in AVP levels, such as are found in heterozygous Brattleboro rats,[6] may be sufficient to induce clear deficits in acquisition of tolerance phenomena. Of course, Brattleboro rats also exhibit a number of other abnormalities [6] and it is possible that their failure to develop tolerance could have more to do with these alterations than the deficit in AVP.

Brattleboro rats in our experiments not only failed to develop tolerance to ethanol, but also exhibited an increased sensitivity to the hypothermic effects of acute ethanol, when compared to Sprague-Dawley subjects (FIGURE 1). This is not explicable by higher blood alcohol levels in the Brattleboro rats; on the contrary, Brattleboro rats have lower blood alcohol levels than age-matched Sprague-Dawley controls given the same gram per kilogram ethanol dose (unpublished data). It is possible that AVP, or some undefined factor lacking or reduced in Brattleboro rats may be required for defense of body temperature against the hypothermic effects of ethanol.

The mechanisms by which AVP might enhance tolerance development are yet obscure. AVP may participate in the pharmacological changes involved in tolerance development or it may be related to certain mechanisms of conditioning that have been implicated in tolerance development. Studies in other laboratories (see References 2 and 4 for review) indicate that an action of AVP on intact noradrenergic systems may be responsible, in part, for its effects on the development of tolerance.

ACKNOWLEDGMENTS

We acknowledge with thanks the technical assistance of S. Madamba and the secretarial assistance of Veronica J. Trevor.

REFERENCES

1. FINKELBERG, F., H. KALANT & A. LE BLANC. 1978. Effect of vasopressin-like peptides on consumption of ethanol by the rat. Pharmacol. Biochem. Behav. **9:** 453–458.
2. HOFFMAN, P. L., B. TABAKOFF & R. F. RITZMANN. 1979. The influence of arginine, vasopressin and oxytocin on ethanol dependance and tolerance. *In* Currents in Alcoholism. M. Galanter, Ed.: 5–16. Grune and Stratton, Inc. New York, N.Y.
3. MUCHA, R. F. & H. KALANT. 1979. Effects of des-glycinamide[9]-lysine[8]-vasopressin and prolyl-leucyl-glycinamide on oral ethanol intake in the rat. Pharmacol. Biochem. Behav. **10:** 229–234.
4. RIGTER, H. & J. C. CRABBE. 1980. Alcohol: modulation of tolerance by neuropeptides. *In* The Psychopharmacology of Alcohol. 179–189.
5. ROGERS, J., S. C. WIENER & F. E. BLOOM. 1979. Long-term ethanol administration methods for rats: advantages of inhalation over liquid diets. Behavioural Neural Biol. **27:** 466–486.
6. VALTIN, H., W. H. SAWYER & H. W. SOKOL. 1965. Neurohypophysial principles in rats homozygous and heterozygous for hypothalamic diabetes insipidus (Brattleboro strain). Endocrinology **77:** 701–706.

CYTOPLASMIC PEPTIDE CONTENT AND NUCLEAR ESTROGEN BINDING OF MAGNOCELLULAR NEURONS IN THE HYPOTHALAMUS OF LONG-EVANS AND BRATTLEBORO RATS*

C. H. Rhodes, J. I. Morrell, and D. W. Pfaff

Department of Neurobiology and Behavior
Rockefeller University
New York, New York 10021

The rat hypothalamus contains several nuclei composed of large cells that produce the peptide hormones oxytocin and vasopressin and their associated neurophysins. This report describes several recent studies using immunocytochemical and steroid autoradiographic methods to identify subpopulations of magnocellular neurons that differ in their cytoplasmic peptide content and estrogen-binding ability.

Materials and Methods

Female rats (Long-Evans, Charles River Breeding Laboratory, Massachusetts; Brattleboro, Blue Spruce Farms, New York) were ovariectomized one month prior to use to maximize the changes seen in the quantitative immunocytochemical experiments with estrogen treatment, and to reduce the levels of endogenous estrogen that would compete with the [^{3}H]estradiol in the autoradiographic experiments. Estrogen-treated animals had 6-mm estradiol-filled Silastic® capsules implanted subcutaneously. Water-deprived animals were kept without water for nine days.[9]

The immunocytochemical technique was basically that of Sternberger[13] and has been described.[8] Antisera directed against oxytocin, vasopressin, and bovine neurophysin I were used as primary antisera. The anti-bovine neurophysin I cross-reacts with both of the rat neurophysins. The procedure used to estimate the total number of cells in each cell group has been described.[8]

The quantitative immunocytochemical experiments were done with anti-oxytocin antiserum (gift of Dr. Czernichow). The optical density of the final immunocytochemical reaction product in a 3.2-micron diameter spot in the darkest part of the cytoplasm was measured in each stained cell in a series of representative sections using equipment described elsewhere.[9]

The technique combining steroid autoradiography and immunocytochemistry has been described.[10] Frozen sections (6 μm) from hypothalami of animals injected intraperitoneally with tritiated estradiol were mounted on emulsion-coated slides and exposed for up to 18 months. They were then fixed with paraformaldehyde, photodeveloped and photofixed using standard techniques,[6] and stained using the immunocytochemical procedure and the anti-neurophysin antiserum described above.

* Supported in part by the National Institutes of Health (HD 05751 and HD 10655). C.H.R. is the recipient of National Research Service Award 5T32GM07739.

0077–8923/82/0394–0767 $1.75/0 © 1982, NYAS

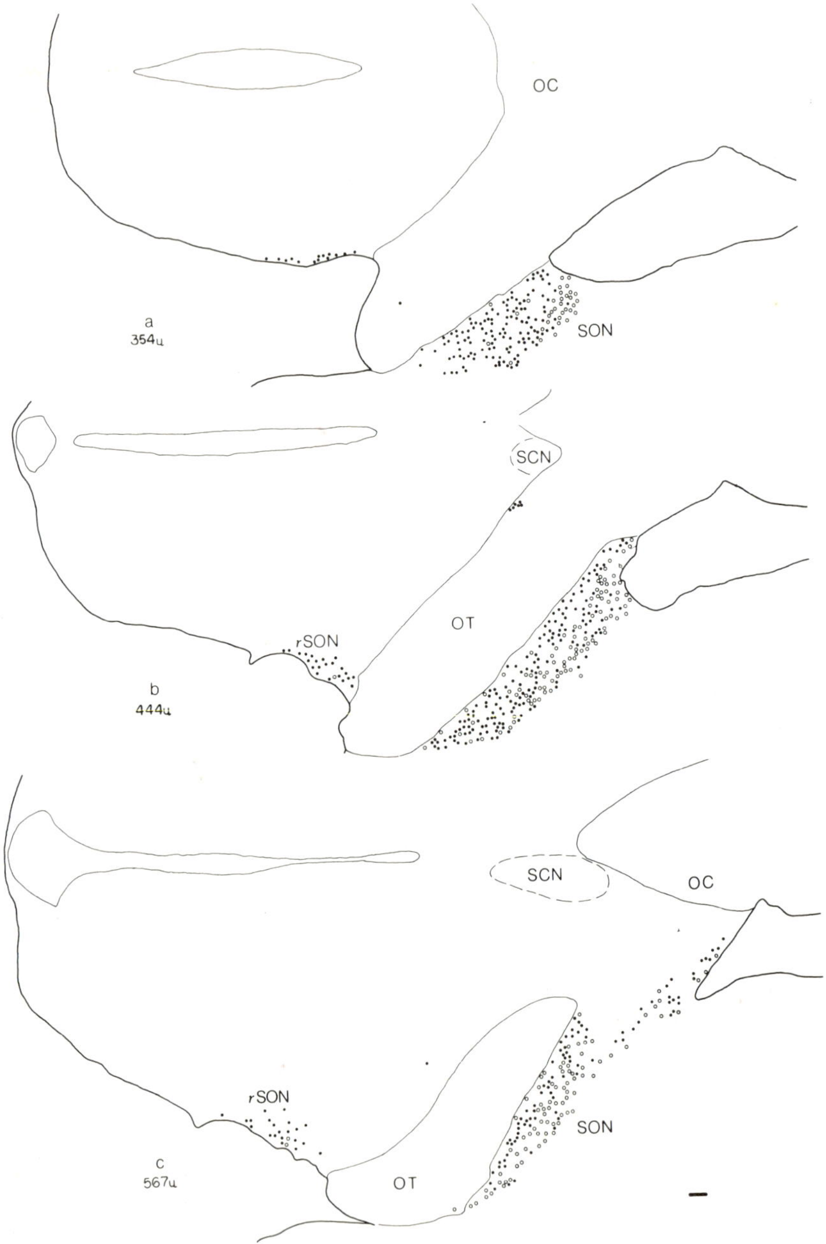

FIGURE 1. Horizontal sections through the hypothalamus of a representative Brattleboro rat. This figure was made from tracings of tissue stained for neurophysin and counterstained with cresyl violet. The open circles represent oxytocin cells and the solid dots are neurophysin-negative magnocellular cells, which would be vasopressin-containing in the Long-Evans animal. The number in the lower left of each figure is

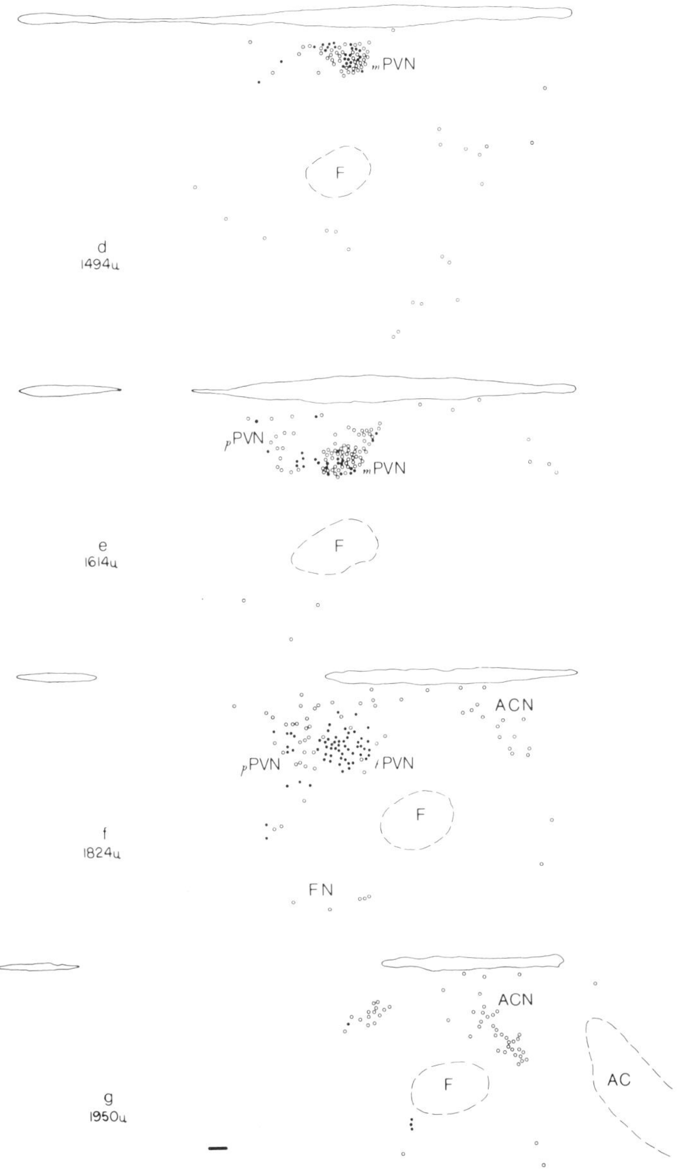

the distance in μm from the base of the brain. The bar in the lower right is 100 μm. Abbreviations used: AC–anterior commissure, ACN–anterior commissural nucleus, F–fornix, FN–fornical nucleus, mPVN–medial subnucleus of PVN, lPVN–lateral subnucleus of PVN, pPVN–posterior subnucleus of PVN, OC–optic chiasm, OT–optic tract, SCN–suprachiasmatic nucleus, rSON–retrochiasmatic SON. (From Rhodes *et al.*[8] With permission from *Journal of Comparative Neurology*.)

Results and Discussion

Anatomical Distribution and Numbers of Magnocellular Cells

The number of oxytocin, vasopressin, and neurophysin cells in each of the magnocellular nuclei was determined.[8] The cells that contain vasopressin in the Long-Evans rat were identified in the Brattleboro rat as large, neurophysin-negative cells. FIGURE 1 is a series of charts of horizontal sections through the

TABLE 1

NUMBERS OF OXYTOCIN- AND VASOPRESSIN *-CONTAINING CELLS IN SIX BRATTLEBORO STRAIN RATS

Animal	ACN	PVN	SON	rSON	AHA
Oxytocin Cells					
Males					
D2	513	1595	2160	173	974
D4	749	1161	2111	205	1258
D6	647	1042	2120	137	1279
Mean	636	1266	2130	172	1170
Females					
D1	414	1096	1075	114	975
D5	540	1431	2209	177	1175
D7	453	1555	1520	34	894
Mean	469	1360	1602	109	1015
Vasopressin * Cells					
Males					
D2	0	959	2577	495	360
D4	18	729	2301	618	517
D6	0	1054	2796	657	422
Mean	6	914	2558	590	433
Females					
D1	22	548	2169	543	566
D5	0	1077	2416	335	459
D7	0	1116	2735	549	291
Mean	7	913	2440	475	438

* The cells that would have contained vasopressin in the Long-Evans strain rat were identified in the Brattleboro animals as large neurophysin-negative cells in the appropriate regions. (From Rhodes *et al.*[8] With permission from *Journal of Comparative Neurology.*)

hypothalamus of a Brattleboro rat showing the anatomical distribution of these cells. TABLE 1 summarizes such data from six Brattleboro animals. The number and anatomical distribution of both the oxytocin-containing and vasopressin-containing (or neurophysin-negative) magnocellular cells were very similar in the Brattleboro and Long-Evans strains.

Based on the distribution of oxytocin- and vasopressin-containing cells the paraventricular nucleus (PVN) was divided into three subnuclei, medial PVN (mPVN), lateral PVN (lPVN), and posterior PVN ((pPVN). The mPVN contained primarily oxytocin-producing cells, while the lPVN consisted of a core of vasopressin-producing cells with a few oxytocin cells forming a rim around the vasopressin cells. The pPVN was composed of a mixture of oxytocin- and vasopressin-containing cells that were well separated from each other and that were spindle-shaped with their long axes running in a medio-lateral direction. This was in sharp contrast to the cells of the mPVN and lPVN, which were closely packed together and nearly round. Within the supraoptic nucleus (SON) oxytocin cells tended to be found rostrally and on the dorsal and lateral margins of the nuclei while vasopressin cells were more common caudally and ventrally.

Magnocellular cells were found outside of SON and PVN as isolated individual cells and in two nuclei, the anterior commissural nucleus (ACN) and the fornical nucleus. A small cluster of cells roughly midway between SON and PVN, nucleus circularis, was also observed. While PVN and SON account for 90% of the vasopressin-containing cells (or neurophysin-negative magnocellular cells in the Brattleboro), they include only about 67% of the oxytocin cells. ACN was found to have almost 50% as many oxytocin cells as PVN and is, therefore, likely to be a cell group of some importance.

Quantitative Immunocytochemical Studies

Oxytocin is released from the pituitary in response to stimuli associated with nursing and vaginal distension [7] and in response to estrogen treatment.[14] It is also released in response to intracarotid saline, water deprivation, or the substitution of 2% saline for drinking water.[2-5] The quantitative immunocytochemical technique was used with an anti-oxytocin antiserum to examine the possibility that two very different stimuli capable of causing oxytocin release—estrogen treatment and water deprivation—might produce changes in oxytocin content in different groups of cells.[9]

FIGURE 2 shows the data from PVN of representative control and water-deprived animals, and photomicrographs of tissue from those animals. The fact that reproducible changes in optical density were observed in some nuclei indicates that (1) these nuclei are affected by these treatments, and (2) this technique is sufficiently sensitive to detect these changes. However, because of the limited fraction of the total cytoplasmic oxytocin sampled by this technique and because the cytoplasmic oxytocin content represents a balance between the rates of synthesis and transport of the hormone, the lack of change in some areas does not show that these regions do not participate in the response to water deprivation or estrogen treatment. These results (TABLE 2) suggest that the same groups of cells are involved in the release of oxytocin in response to estrogen treatment and water deprivation.

Combined Immunocytochemistry and Steroid Autoradiography

Estrogen treatment produces the release from the pituitary of not only oxytocin [14] but also vasopressin.[12] Steroid autoradiographic and immunocytologic techniques were used to identify cells that contained neurophysin in their cytoplasm and concentrated estradiol in their nuclei.[10, 11]

In the Long-Evans animal estrogen-concentrating, neurophysin-containing

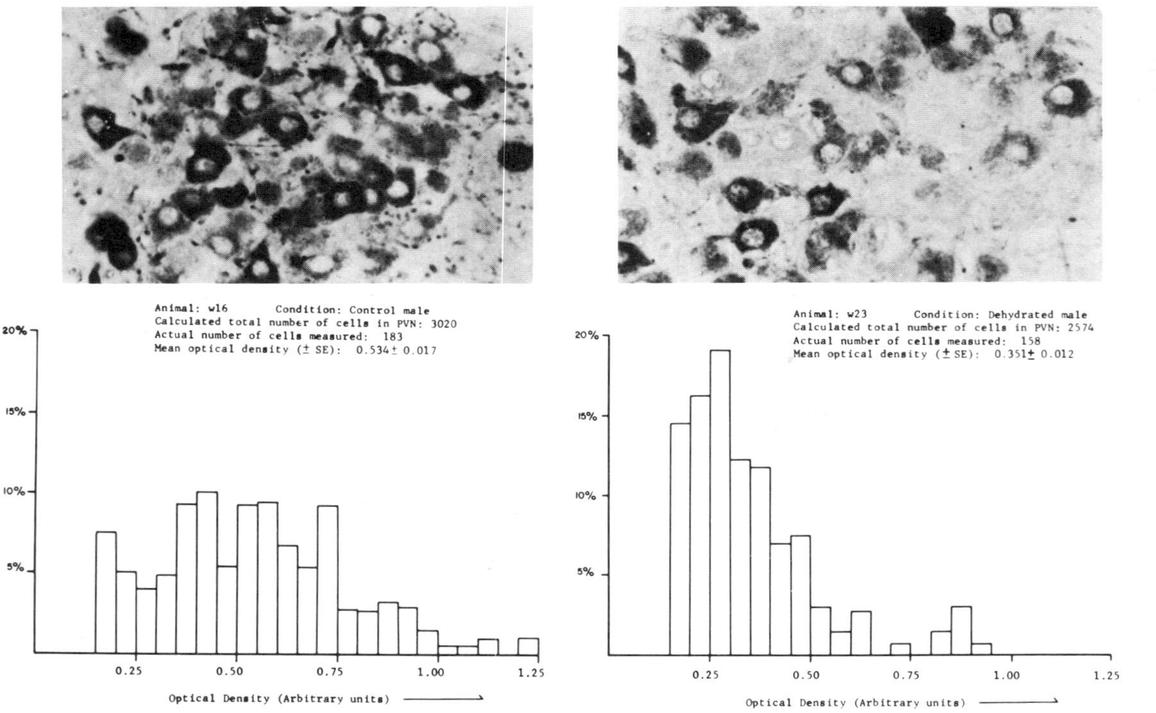

FIGURE 2. Photomicrographs of typical sections through PVN of Long-Evans control (left) and water-deprived (right) male animals stained with anti-oxytocin antiserum. × 290. The figures below the photographs are histograms of the optical densities of all the stained cells measured in PVN in those animals. The units for the ordinates of the histograms are percentage of the total cells measured. The calibration bar in the photographs is 25 μm. "Calculated total number of cells in PVN" in the figure is the average number of stained cells in the sections measured multiplied by the number of sections required to span the nucleus. (From Rhodes *et al.*[9] With permission from *Cell and Tissue Research*.)

cells were found almost exclusively in PVN, ventral and medial to lPVN, and within pPVN. Many estrogen-concentrating parvocellular cells were found near ACN, but only very few estrogen-concentrating, neurophysin-containing cells were found in ACN, mPVN, or lPVN. SON and the fornical nucleus had almost none (an average of <1 cell/section) of these cells.

These experiments were done with both Long-Evans and Brattleboro rats so that the peptide produced by the estrogen-concentrating magnocellular cells could be identified by comparing the two sets of results. The overall distribution of estrogen-concentrating magnocellular cells was found to be similar in the two strains of rat. Both neurophysin-negative and neurophysin-positive estrogen-concentrating magnocellular cells were found in the Brattleboro rat (FIGURE 3) suggesting that in the Long-Evans rat there are both oxytocin- and vasopressin-containing neurons that concentrate estradiol.

TABLE 2

MEAN OPTICAL DENSITIES OF ANTI-OXYTOCIN-STAINED CELLS IN CONTROL, WATER-DEPRIVED, AND ESTROGEN-TREATED ANIMALS

	Control	Water-deprived	Estrogen-treated
Anterior Commisural Nucleus			
Female	0.517±0.049 (5)	0.338±0.082 * (6)	0.415±0.036 * (5)
Male	0.425±0.071 (5)	0.321±0.038 * (5)	—
Paraventricular Nucleus			
Female	0.388±0.056 (5)	0.386±0.035 (6)	0.411±0.038 (5)
Male	0.459±0.061 (5)	0.317±0.028 * (5)	—
Supraoptic Nucleus			
Female	0.361±0.049 (4)	0.367±0.063 (5)	0.361±0.027 (3)
Male	0.348±0.063 (3)	0.329±0.037 (4)	—

* $p < 0.001$.

The means are given ± the SEM. p Values refer to the significance of the difference in staining between the water-deprived and corresponding control animals as determined using a two-tailed Student's *t*-test. The difference between the male and female control animals in ACN and PVN is significant at the 0.005 level. The numbers in parentheses are the number of animals in each group. (From Rhodes *et al.*[9] With permission from *Cell and Tissue Research.*)

ACKNOWLEDGMENTS

During the course of this work we used antisera from the following investigators and companies whose assistance is gratefully acknowledged. Anti-oxytocin: Dr. Greidanus, Rudolf Magnus Instituut voor Farmacologie (Utrecht) and Dr. Czernichow, Laboratoire Archambualt, Hôpital des Enfants Malades (Paris). Anti-vasopressin: Dr. Sofroniew, Universität München (Munich) and Dr. Greidanus, Rudolf Magnus Instituut voor Farmacologie (Utrecht). Anti-bovine neurophysin I: Dr. Zimmerman, Columbia University College of Physicians and Surgeons (New York, N.Y.).

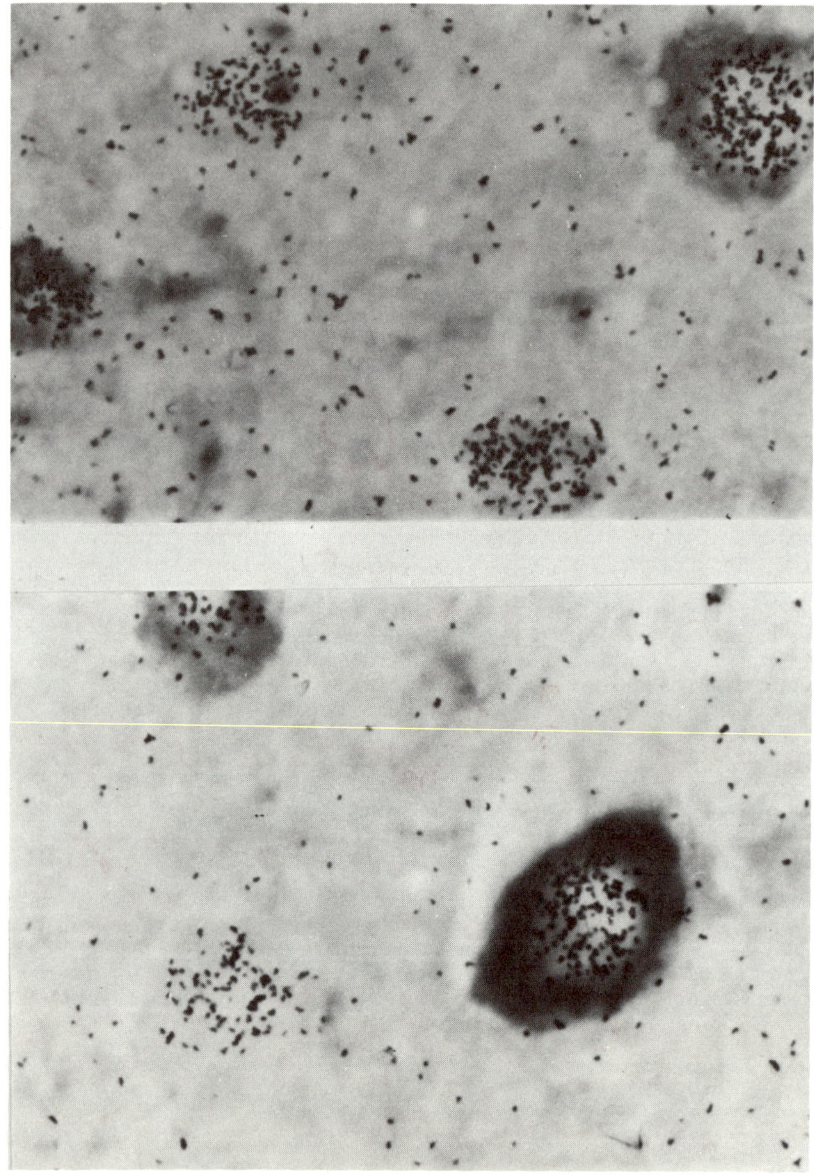

FIGURE 3. Microphotographs illustrating estrogen-concentrating cells in the posterior subnucleus of PVN of a Brattleboro animal. Silver grains indicate the autoradiographic localization of [^{3}H]estradiol; diaminobenzidine reaction product indicates the immunocytochemical localization of neurophysin. Note that some of the estrogen-binding cells contain neurophysin in their cytoplasm (for example, the cell in the lower right of the bottom panel) while others do not (for example, the cell in the lower left of the bottom panel). (×1,300).

References

1. Cajal, R. 1911. Histologie du Système Nerveux. vol. 2 Translated into French by Dr. L. Azovlay. Consejo Superior de Investigaciones Scientificas. Instituto Ramon y Cajal. Madrid, 1955.
2. Dyball, R. E. J. 1971. Oxytocin and ADH secretion in relation to electrical activity in antidromically identified supraoptic and paraventricular units. J. Physiol. **214:** 245–256.
3. George, J. M. 1976. Vasopressin and oxytocin are depleted from rat hypothalamic nuclei after oral hypertonic saline. Science **193:** 146–148.
4. Legros, J. J. & J. J. Dreifuss. 1975. Water deprivation in rats: Elevated plasma neurophysin levels. Experientia **31:** 603–605.
5. Jones, C. W. & B. T. Pickering. 1969. Comparison of the effects of water deprivation and sodium chloride imbibition on the hormone content of the neurohypophysis of the rat. J. Physiol. **203:** 449–458.
6. Morrell, J. I. & D. W. Pfaff. 1981. Autoradiographic technique for steroid hormone localization: Application to the vertebrate brain. *In* Neuroendocrinology of Reproduction. N. T. Adler, Ed.: 519–531. Plenum Publishing Corporation. New York, N.Y.
7. Renaud, L. P. 1978. Neurophysiological organization of the endocrine hypothalamus. *In* The Hypothalamus. S. Reichlin, R. J. Baldessarini & J. B. Martin, Eds.: 269–300. Raven Press. New York, N.Y.
8. Rhodes, C. H., J. I. Morrell & D. W. Pfaff. 1981. Immunohistochemical analysis of magnocellular elements in rat hypothalamus: Distribution and numbers of cells containing neurophysin, oxytocin, and vasopressin. J. Comp. Neurol. **198:** 45–64.
9. Rhodes, C. H., J. I. Morrell & D. W. Pfaff. 1981. Changes in oxytocin content in the magnocellular neurons of the rat hypothalamus following water deprivation or estrogen treatment. Cell Tiss. Res. **216:** 47–55.
10. Rhodes, C. H., J. I. Morrell & D. W. Pfaff. 1981. Distribution of estrogen-concentrating magnocellular neurons in the rat hypothalamus as demonstrated by a technique combining steroid autoradiography and immunohistology in the same tissue. Neuroendocrin. **33:** 18–23.
11. Rhodes, C. H., J. I. Morrell & D. W. Pfaff. 1982. Distribution of estrogen-concentrating, neurophysin containing magnocellular neurons in the hypothalamus of the Brattleboro rat. (Submitted to J. Neurosci.)
12. Skowsky, W. R., L. Swan & P. Smith. 1979. Effects of sex steroid hormones on arginine vasopressin in intact and castrated male and female rats. Endocrinology **104:** 105–108.
13. Sternberger, L. A. 1974. Immunocytochemistry. Prentice-Hall. Englewood Cliffs, N.J.
14. Yamaguchi, K., T. Akaishi & H. Negoro. 1979. Effect of estrogen treatment on plasma oxytocin and vasopressin in ovariectomized rats. Endo. Jap. **26:** 197–205.

ABNORMALITY OF THE FEBRILE RESPONSE OF THE BRATTLEBORO RAT *

W. L. Veale, P. C. Eagan, and K. E. Cooper

Department of Medical Physiology
University of Calgary
Calgary, Alberta
Canada T2N 4N1

Introduction

Fever is a natural response to the pyrogenic substance bacterial endotoxin in mammals, lower vertebrates, and some invertebrates as well.[1] This is not the case for the Brattleboro (DI) rat. The DI rat possesses the necessary physiological mechanisms to control body temperature and produce fever, as indicated by the increased temperatures observed when prostaglandin E_2 (a pyretogenic substance) is given into the lateral cerebral ventricles. However, if bacterial endotoxin is given to the DI rat, no fever results.[2]

The DI rat is incapable of producing arginine vasopressin (AVP) due to a genetic defect that prevents its synthesis.[3, 4] The experiments reported here were undertaken to ascertain whether this hormone deficiency is related to the lack of a febrile response to endotoxin in the DI rat.

Methods

Male Brattleboro rats (250–350 g) homozygous recessive for the genetic defect were used in these experiments. At least one week before the start of the experiments a 23 gauge stainless steel guide cannula was implanted stereotaxically into each rat to enable subsequent microinjection into a lateral cerebral ventricle. The guide cannula was anchored to the skull with stainless steel jeweller's screws and dental cement. Body temperature was measured in each animal with a thermistor probe (Yellow Springs Instruments) inserted 9 cm past the anus and taped to the base of the tail. Colonic temperature was recorded on a Digitech temperature recording system.

Bacterial endotoxin derived from *Salmonella abortus equi* (SAE) was dissolved in sterile, pyrogen-free physiological saline and stored at 4° C. Arginine vasopressin, AVP, (Pitressin) was used in two forms, one in an aqueous solution of sterile saline; the other in a suspension of sterile peanut oil.

Two sets of experiments were carried out. In the first, the DI rats received a subcutaneous (s.c.) injection of 1.0 Unit of Pitressin dissolved in 0.5 ml of sterile saline. One hour following the injection, the animals received either 1.0 μg of SAE in 10.0 μl of sterile saline into a lateral cerebral ventricle (i.c.v.), or the vehicle alone. Colonic temperature was monitored for one hour prior to AVP administration and for five hours following.

The second set of experiments consisted of daily injection of SAE (1.0 μg/10.0 μl of sterile saline) i.c.v. for seven days. On day 1, the animals received

* Supported by the Medical Research Council of Canada.

0077–8923/82/0394–0776 $1.75/0

SAE alone, on days 2, 3, and 4 AVP (1.0 Unit of Pitressin in 0.5 ml oil suspension) s.c. and SAE one hour later, and on days 5, 6, and 7, 0.5 ml of sterile peanut oil s.c. and SAE one hour following. Colonic temperature was monitored for two hours preceding SAE administration and for three hours following injection.

All appropriate data were analyzed using Student's *t*-test (two-tailed) and values for p less than or equal to 0.05 were considered significant.

Results

The changes in colonic temperature in response to s.c. administration of AVP (aqueous) and SAE or the vehicle i.c.v. are shown in Figure 1. Subcutaneous injection of aqueous AVP resulted in a rapid drop in body temperature that was significantly below baseline temperature after one hour ($p \leq 0.01$). This drop in body temperature was unaffected by i.c.v. injection of the control solution, 10.0 μl of sterile saline. However, when SAE was given i.c.v., the body temperature returned to baseline values and then continued on to plateau at $1.20 \pm 0.19°$ C above baseline three hours following injection. This increase in body temperature was achieved by vasoconstriction of peripheral blood vessels (as indicated by blanching of the skin of the ears and tail) as well as a piloerection and the assumption of a heat-conserving posture.

Figure 2 illustrates the effect of the gradual build-up and decline of AVP levels on the response of the animals to SAE. On day 1, when only SAE was given, there was no significant change in body temperature after three hours, the time at which the fever should be at its height. When AVP in an oil suspension was given s.c. (which results in a slow release of the peptide) on days 2, 3, and 4, there is a gradual development in the febrile response so that by day 4 a fever of $1.5 \pm 0.20°$ C above baseline is achieved ($p \leq 0.01$). After day 4, the vehicle (0.5 ml of sterile peanut oil) is given subcutaneously. The result is a gradual diminishment of the febrile response to endotoxin so that by day 6 and 7 there are only small changes from baseline values that are not significant ($p > 0.05$).

Discussion

These experiments were carried out in order to examine the possibility that the AVP deficiency in the DI rat is related to its lack of a febrile response to bacterial endotoxin. It appears that such a relationship might exist. When AVP is administered exogenously to these animals and they are then subsequently challenged with endotoxin, the DI rats will develop a fever. In fact, they will develop a fever that overrides the hypothermic effects of aqueous AVP s.c. The apparent necessity for AVP in the blood stream for fever production is further suggested by the second set of experiments. On day 1, when no AVP was present, there was no febrile response to endotoxin. On days 2, 3, and 4, when AVP was given in an oil suspension so that there is a slow increase of AVP levels in the plasma, there is a gradual increase in the febrile response. Once AVP injections are stopped, a gradual decrease in the febrile response is observed. Thus, when AVP is present, fever occurs and when it is absent it does not.

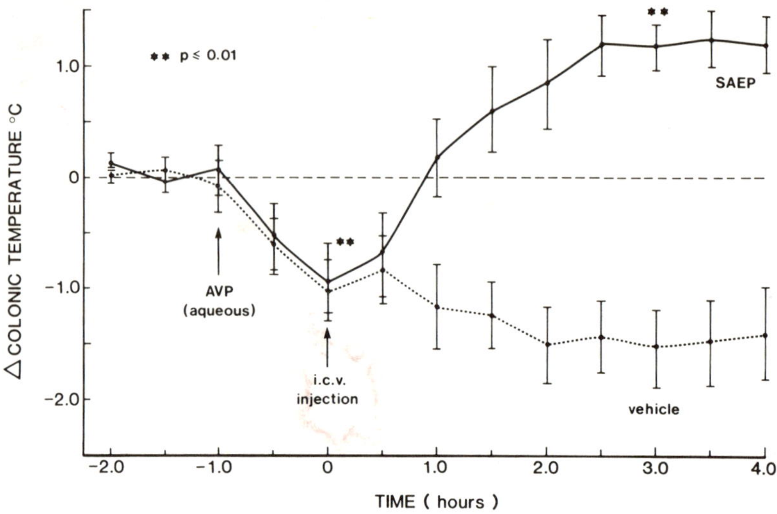

FIGURE 1. Mean changes in colonic temperature ± SEM are shown for eight DI rats. The animals received AVP (1.0 Units Pitressin, aqueous) followed one hour later by 1.0 μg of SAE (upper trace) or vehicle (lower trace) i.c.v.

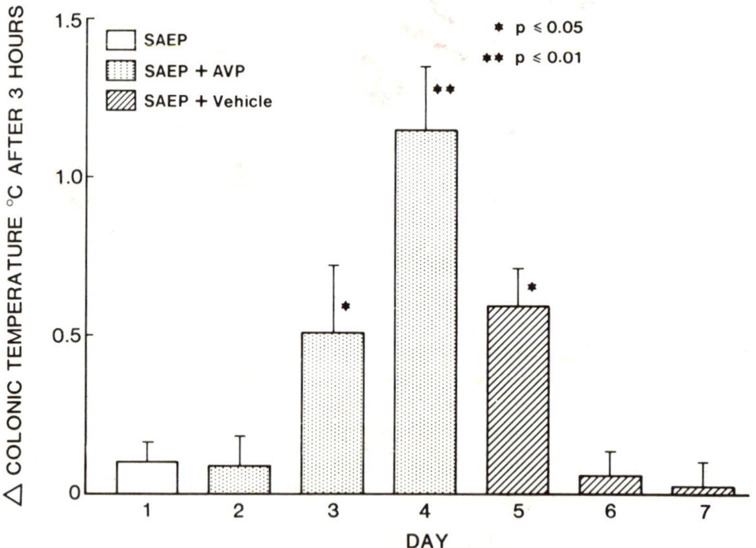

FIGURE 2. Mean change in colonic temperature ± SEM after three hours for eight DI rats. AVP concentrations were 1.0 Unit Pitressin, in oil. SAE was given i.c.v. (1.0 μg/10.0 μl sterile saline).

One possible mechanism for the role of AVP in the production of fever is its importance in the activation of the reticuloendothelial system. Altura[5] has shown that phagocytic activity of the reticuloendothelial system (RES) is significantly reduced in the DI rat following mild forms of shock. This can be corrected by giving the animals exogenous AVP (Altura, personal communication). The RES is important in the fever process, since interactions of the RES with bacterial endotoxin result in the production of a heat-labile protein called endogenous pyrogen, which acts on the brain to produce fever.[6] It is possible that AVP must be present in the plasma to cause production of endogenous pyrogen by RES. The lack of AVP in the DI rat may result in inadequate production of endogenous pyrogen by the RES in response to endotoxin and thus the observed afebrile response. Experiments are currently underway to study this possibility.

Summary

The febrile responses of the Brattleboro rat to bacterial endotoxin were studied during and following the subcutaneous injection of arginine vasopressin. When arginine vasopressin was given in an aqueous solution, endotoxin reversed the normally observed drop in body temperature and caused a fever of 1.20° C. When arginine vasopressin was administered in an oil suspension, which results in a slower release of the peptide, there is a gradual development of a fever to endotoxin that is lost gradually following the withdrawal of the vasopressin. These results suggest that arginine vasopressin is a necessary requirement in the production of fever to bacterial endotoxin in the Brattleboro rat.

Acknowledgments

P. C. Eagan is a predoctoral student of the Alberta Heritage Foundation for Medical Research. We acknowledge the secretarial assistance of Mrs. G. Olmstead in preparing the manuscript.

References

1. Kluger, M. J. 1979. Phylogeny of fever. Fed. Proc. **38:** 39–34.
2. Eagan, P. C., N. W. Kasting, W. L. Veale & K. E. Cooper. 1982. The absence of endotoxin fever but not prostaglandin E_2 fever in the Brattleboro rat. Am. J. Physiol. **242:** R116–R120.
3. Vatlin, M., W. H. Sawyer & H. W. Sokol. 1965. Neurohypophysial principles in rats homozygous and heterozygous for hypothalamic diabetes insipidus (Brattleboro strain). Endocrinology **77:** 701–706.
4. Saul, G. B., E. B. Garrity, K. Benirschke & H. Valtin. 1968. Inherited hypothalamic diabetes insipidus in the Brattleboro strain of rats. J. Hered. **59:** 113–117.
5. Altura, B. M. 1980. Evidence that endogenous vasopressin plays a protective role in circulatory shock. Role for reticuloendothelial system using Brattleboro rats. Experientia **36:** 1080–1082.
6. Cooper, K. E., W. I. Cranston & A. J. Honour. 1967. Observations on the site and mode of actions of pyrogens in the rabbit brain. J. Physiol. **191:** 325–337.

BIBLIOGRAPHY

Early in the planning stages of this symposium, the organizing committee decided that it would be useful if the proceedings of the symposium were to include a "complete" listing of those published papers that dealt, in some measure, with the Brattleboro rat. The results of our efforts are given on the following pages. So that the reader may use this bibliography advisedly and be aware of its limitations, we will here describe the strategy used in its preparation.

Two databases were searched—*Medline,* based on Index Medicus and operated by the National Library of Medicine (Washington, D.C.), and *Biosis,* based on Biological Abstracts and operated by Bibliographic Retrieval Services (Scotia, N.Y.). The search covered the period from 1961 through January 28, 1982. Publications that contained both the key words "rat" and "diabetes insipidus" were retrieved.

The computer printouts generated from this search were then distributed among the members of the organizing committee who were responsible for deleting papers unrelated to the Brattleboro rat, citations that were only abstracts, and duplications. While attempts were made to verify the accuracy of most of the citations, the limitation of time precluded as thorough a check as we would have wished. Moreover, in some instances, our library did not carry the cited journal in its collection.

The "completeness" of any bibliography generated in this way is limited by several factors: first, the number and types of journals contained in each database; second, the appropriateness of the key words used for the search; and third, the adequacy of the key words assigned to published articles by the author(s) and/or indexing services. Although the two databases used are comprehensive, they are not all inclusive. Notable exclusions are books, chapters in books, and the proceedings of many meetings and symposia.

Additional sources of information were consulted in compiling the final bibliography. For example, several participants had included with their advance registration a list of their own publications on the Brattleboro rat. In addition, Drs. Lewis Kinter, Heinz Valtin, and Josef Zicha kindly provided us with copies of their own lists of articles that have been published on this animal.

Beyond the people already mentioned, several individuals provided invaluable help to us in assembling this bibliography and deserve our very grateful thanks. Ms. Elaine A. Bent of Dana Biomedical Library at Dartmouth performed the computer searches for us. Mr. Kurt V. Knuth wrote the computer program that we used to assemble, edit, and alphabetize the bibliography. (The bibliography is alphabetized by first author only, and then by date of publication.) Ms. Terry Hall performed the onerous task of typing into the computer all the citations in the appropriate format, as well as handling the innumerable changes, additions, and deletions that we asked of her so frequently. Her patience with us seemed inexhaustible.

With a project of this nature, and despite the best of intentions, we would be surprised if the following bibliography were totally free of errors and omissions. Therefore, we should like to end by apologizing to any author whose work may be incorrectly cited, or, even more regrettably, may have been inadvertently overlooked. We ask these people, as well as the readership in general,

0077–8923/82/0394–0780 $1.75/0 © 1982, NYAS

to please notify us of any such errors. If there are to be other symposia on the Brattleboro rat in the future, and if this bibliography is regarded as a useful component, it would help future organizers of such symposia if the current bibliography were appropriately corrected, amended, and updated. We also ask that authors keep us informed of all their future publications concerning the Brattleboro rat.

Brian R. Edwards
Larry A. Walker

Department of Physiology
Dartmouth Medical School
Hanover, New Hampshire 03755

1. Abelson, I. O. 1976. Effect of fat on renal water excretion under conditions of disorders in the secretion of antidiuretic hormone. Fiziol. AH SSSR **62:** 1863–1869.
2. Adler, R. A., S. Dolphin, M. Szefler & H. W. Sokol. 1979. The effects of elevated circulating prolactin in rats with hereditary hypothalamic diabetes insipidus (Brattleboro strain). Endocrinology **105:** 1001–1006.
3. Aisenbrey, G. A., W. A. Handelman, P. Arnold, M. Manning & R. W. Schrier. 1981. Vascular effects of arginine vasopressin during fluid deprivation in the rat. J. Clin. Invest. **67:** 961–968.
4. Alexander, C. S., W. R. Swaim & M. C. Garcia. 1975. Urine concentration and dilution: effect on red cell survival. Proc. Soc. Exp. Biol. Med. **150:** 295–298.
5. Alonso, G., B. Bloch, B. Lutz-Bucher, G. Bugnon & I. Assenmacher. 1981. Light- and electron-microscope differentiation of axons containing vasopressin and oxytocin in the neurohypophyseal lobe of the rat. Neurosci. Lett. **25:** 113–118.
6. Altura, B. M. 1980. Evidence that endogenous vasopressin plays a protective role in circulatory shock. Role for reticuloendothelial system using Brattleboro rats. Experimentia **36:** 1080–1082.
7. Arimura, A., T. Saito, C. Y. Bowers & A. V. Schally. 1967. Pituitary-adrenal activation in rats with hereditary hypothalamic diabetes insipidus. Acta Endocrinol. **54:** 155–165.
8. Arimura, A., S. Sawano, T. W. Redding & A. V. Schally. 1968. Studies on retarded growth of rats with hereditary hypothalamic diabetes insipidus. Neuroendocrinology **3:** 187–192.
9. Auriac, A., J. Azam-Traves, J. C. Dumas, J. Garigue & G. Roux. 1978. Research on the influence of sympathomimetics (isoprenaline and noradrenaline) on water metabolism in the rat. Arch. Farmacol. Toxicol. **4:** 31–34.
10. Auriac, A., J. Azam, J. C. Dumas, G. Roux & J. L. Montastruc. 1981. [Effects of clonidine on diuresis and water intake in normal and Brattleboro rats]. J. Pharmacol. **12:** 277–288.
11. Baertschi, A. J., P. Vallet, J. B. Baumann & J. Girard. 1980. Neural lobe of pituitary modulates corticotropin release in the rat. Endocrinology **106:** 878–882.
12. Bailey, W. H. & J. M. Weiss. 1978. Effect of ACTH 4–10 on passive avoidance of rats lacking vasopressin (Brattleboro strain). Horm. Behav. **10:** 22–29.
13. Bailey, W. H. & J. M. Weiss. 1979. Evaluation of a “memory deficit” in vasopressin-deficient rats. Brain Res. **162:** 174–178.

14. Baisset, A., J. Cotonat, J. C. Dumas & P. Montastruc. 1973. Recherches sur l'action antidiurétique du clofibrate. Thérapie **28:** 651–661.

15. Baisset, A., J. Cotonat, J. C. Dumas & P. Montastruc. 1973. Recherches sur l'action antidiurétique de la carbamazepine. Thérapie **28:** 663–669.

16. Baker, B. I. 1973. The separation of different forms of melanocyte-stimulating hormone from the rat neurointermediate lobe by polyacrylamide gel electrophoresis, with a note on rat neurophysins. J. Endocrinol. **57:** 393–404.

17. Balment, R. J., I. W. Henderson & J. A. Oliver. 1975. The effects of vasopressin on pituitary oxytocin content and plasma renin activity in rats with hypothalamic diabetes insipidus (Brattleboro Strain). Gen. Comp. Endocrinol. **26:** 468–477.

18. Balment, R. J., I. Chester Jones, I. W. Henderson & J. A. Oliver. 1976. Effects of adrenalectomy and hypophysectomy on water and electrolyte metabolism in male and female rats with inherited hypothalamic diabetes insipidus (Brattleboro strain). J. Endocrinol. **71:** 193–217.

19. Balment, R. J., I. W. Henderson, I. Chester Jones & W. Mosley. 1977. Water and electrolyte balance in adrenalectomized rats with diabetes insipidus (Brattleboro strain) given antidiuretic hormone. Gen. Comp. Endocrinol. **33:** 428–433.

20. Balment, R. J., I. Chester Jones & I. W. Henderson. 1977. Time course of lithium-induced alterations in renal and endocrine function in normal and Brattleboro rats with hypothalamic diabetes insipidus. Br. J. Pharmacol. **59:** 627–643.

21. Balment, R. J., M. J. Brimble & M. L. Forsling. 1980. Release of oxytocin induced by salt loading and its influence on renal excretion in the male rat. J. Physiol. **308:** 439–449.

22. Bankir, L., M. M. Trinh-Trang-Tan, M. P. Nivez, J. Sraer & R. Ardaillou. 1980. Altered PGE_2 production by glomeruli and papilla of rats with hereditary diabetes insipidus. Prostaglandins **20:** 349–366.

23. Banks, R. O. & J. Di Salvo. 1976. Influence of antidiuretic hormone on intrarenal blood flow distribution in diabetes insipidus dogs and rats. Proc. Soc. Exp. Biol. Med. **151:** 547–551.

24. Barnett, J. L., P. Cheeseman, J. Cheeseman, J. M. Douglas & J. G. Phillips. 1974. Adrenal responsiveness in ageing Brattleboro rats with hereditary diabetes insipidus. Age Ageing **3:** 189–195.

25. Barnett, J. L., P. Cheeseman, J. Cheeseman, J. M. Douglas & J. G. Phillips. 1974. Changes in organ weights and blood parameters in ageing Brattleboro rats with hereditary diabetes insipidus. Age Ageing **3:** 229–239.

26. Barr, J. G. & M. L. Kauker. 1979. Renal tubular site and mechanism of clonidine-induced diuresis in rats: clearance and micropuncture studies. J. Pharmacol. Exp. Ther. **209:** 389–395.

27. Bauman, Jr., J. W., C. Van Wegen & J. Weil-Malherbe. 1969. Water metabolism in the hypophysectomized Brattleboro (DI) rat. Acta Endocrinol. **61:** 720–728.

28. Bauman, Jr., J. W., R. Manshardt & G. Levisay. 1969. Antidiuretic effect of triiodothyronine in the hypophysectomized rat. Amer. J. Physiol. **216:** 139–142.

29. Bény, J.-L. & A. J. Baertschi. 1980. Oxytocin: major corticotropin-releasing factor secreted from diabetes insipidus rat posterior pituitary in vitro. Neuroendocrinology **31:** 261–264.

30. Berndt, W. O., M. Miller, W. M. Kettyle & H. Valtin. 1970. Potentiation of the antidiuretic effect of vasopressin by chlorpropamide. Endocrinology **86:** 1028–1032.

31. Better, O. S., G. A. Aisenbrey, T. Berl, R. J. Anderson, W. A. Handelman, S. L. Linas, S. J. Guggenheim & R. W. Schrier. 1980. Role of

antidiuretic hormone in impaired urinary dilution associated with chronic bile-duct ligation. Clin. Sci. **58:** 493–500.

32. BIA, M. J., S. DEWITT & J. N. FORREST, JR. 1979. Dissociation between plasma, urine, and renal papillary cyclic AMP content following vasopressin and DDAVP. Am. J. Physiol. **237:** F218–F225.

33. BODNAR, R. J., E. A. ZIMMERMAN, G. NILAVER, A. MANSOUR, L. W. THOMAS, D. D. KELLY & M. GLUSMAN. 1980. Dissociation of cold-water swim and morphine analgesia in Brattleboro rats with diabetes insipidus. Life Sci. **26:** 1581–1590.

34. BOER, G. J., F. W. VAN LEEUWEN, D. F. SWAAB & J. L. NOLTEN. 1976. Acid phosphatase activity in the rat neurohypophysis during increased levels of gonadotropic hormones, in diabetes insipidus (Brattleboro strain), and after water loading. Acta Endocrinol. **81:** 697–706.

35. BOER, G. J., H. B. M. UYLINGS, C. M. F. VAN RHENEN-VERBERG & B. FISSER. 1978. Postnatal brain development in rats with hereditary diabetes insipidus (Brattleboro strain). *In* Hormones and Brain Development. G. Dorner & M. Kawakami, Eds.: 253–258. Elsevier. Amsterdam.

36. BOER, G. J., D. G. SWAAB, H. B. M. UYLINGS, K. BOER, R. M. BUIJS & D. N. VELIS. 1980. Neuropeptides in rat brain development. *In* Progress in Brain Research. "Adaptive capabilities of the nervous system." P. McConell, G. J. Boer, N. E. van den Poll, H. J. Romijn & M. A. Corner, Eds. Vol. 53: 207–227. Elsevier-North Holland Biomedical Press. Amsterdam.

37. BOER, G. J., R. M. BUIJS, D. F. SWAAB & G. J. DE VRIES. 1980. Vasopressin and the developing rat brain. Peptides **1** (Suppl. 1): 203–209.

38. BOER, K., G. J. BOER & D. F. SWAAB. 1974. Does the hypothalamo-neurohypophyseal system play a role in gestation length or the course of parturition? Prog. Brain Res. **41:** 307–319.

39. BOER, K., G. J. BOER & D. F. SWAAB. 1981. Reproduction in Brattleboro rats with diabetes insipidus. J. Reprod. Fertil. **61:** 273–280.

40. BOHUS, B., T. B. VAN WIMERSMA GREIDANUS & D. DE WIED. 1975. Behavioral and endocrine responses of rats with hereditary hypothalamic diabetes insipidus (Brattleboro strain). Physiol. Behav. **14:** 609–615.

41. BÖNNER, G., W. RASCHER, G. SPECK, M. MARIN-GREZ & F. GROSS. 1981. The renal kallikrein-kinin system in Brattleboro rats with hereditary hypothalamic diabetes insipidus. Missing relationship between antidiuretic hormone and the renal kallikrein-kinin system. Acta Endocrinol. **98:** 36–42.

42. BRIAUD, B., B. KOCH, B. LUTZ-BUCHER & C. MIALHE. 1978. In vitro regulation of ACTH release from neurointermediate lobe of rat hypophysis. I. Effect of crude hypothalamic extracts. Neuroendocrinology **25:** 47–63.

43. BRITO, G. N., G. J. THOMAS, S. I. GINGOLD & D. M. GASH. 1981. Behavioral characteristics of vasopressin-deficient rats. Brain Res. Bull. **6:** 71–75.

44. BROWN, D. R. & S. G. HOLTZMAN. 1981. Suppression of drinking by naloxone in rats homo- and heterozygous for diabetes insipidus. Pharmacol. Biochem. Behav. **15:** 109–114.

45. BROWNSTEIN, M. J. & H. GAINER. 1977. Neurophysin biosynthesis in normal rats and in rats with hereditary diabetes insipidus. Proc. Natl. Acad. Sci. **74:** 4046–4049.

46. BUCKINGHAM, J. C. & J. H. LEACH. 1980. Hypothalamo-pituitary-adrenocortical function in rats with inherited diabetes insipidus. J. Physiol. **305:** 397–404.

47. BUCKINGHAM, J. C. 1981. The influence of vasopressin on hypothalamic corticotrophin releasing activity in rats with inherited diabetes insipidus. J. Physiol. (Lond) **312:** 9–16.

48. BUIJS, R. M. 1978. The specific localization of vasopressin and oxytocin pathways in the rat central nervous system. *In* Hormones and Brain Develop-

ment. G. Dörner & M. Kawakami, Eds.: 223–228. Elsevier-North Holland Biomedical Press. Amsterdam.

49. Burford, G. D., C. W. Jones & B. T. Pickering. 1971. Tentative identification of a vasopressin-neurophysin and an oxytocin-neurophysin in the rat. Biochem. J. **124:** 809–813.

50. Burlet, A. & J. Marchetti. 1976. Contributions of immunoenzymatic technique to the study of diencephalic localization of vasopressin. *In* Immunoenzymatic Techniques. G. Feldmann *et al.*, Eds.: 333–343. North-Holland. Amsterdam.

51. Burlet, A., M. Chateau & P. Czernichow. 1979. Infundibular localization of vasopressin, oxytocin and neurophysins in the rat; its relationships with corticotrope function. Brain Res. **168:** 275–286.

52. Cantalamessa, F., G. de Caro & M. Perfumi. 1980. Water intake inhibition and vasopressin release induced by eledoisin and leuenkephalin in rats of the Brattleboro and Wistar strain. Pharmacol. Res. Commun. **12:** 275–278.

53. Cantalamessa, F., G. De Caro & M. Perfumi. 1981. Water intake inhibition and vasopressin release following intracerebroventricular administration of tachykinins to rats of the Wistar and Brattleboro strain. Pharmacol. Res. Commun. **13:** 641–655.

54. Carraway, R. E., L. M. Demers & S. E. Leeman. 1976. Hyperglycemic effect of neurotensin, a hypothalamic peptide. Endocrinology **99:** 1452–1462.

55. Celestian, J. F., R. J. Carey & M. Miller. 1975. Unimpaired maintenance of a conditioned avoidance response in the rat with diabetes insipidus. Physiol. Behav. **15:** 707–711.

56. Chateau, M., J. Marchetti, A. Burlet & M. Boulange. 1979. Evidence of vasopressin in adenohypophysis: research into its role in corticotrope activity. Neuroendocrinology **28:** 25–35.

57. Cheng, K. W., H. G. Friesen & J. B. Martin. 1972. Neurophysin in rats with hereditary hypothalamic diabetes insipidus (Brattleboro strain). Endocrinology **90:** 1055–1063.

58. Choy, V. J. & W. B. Watkins. 1977. Immunocytochemical study of the hypothalamo-neurohypophysial system. II. Distribution of neurophysin, vasopressin and oxytocin in the normal and osmotically stimulated rat. Cell Tissue Res. **180:** 467–490.

59. Coculescu, M., M. Oprescu & L. Zagrean. 1977. Angiotensin I-like immunoreactive substance in pineal gland of normal and Brattleboro rats with hereditary diabetes insipidus. Rev. Roum. Morphol. Embryol. Physiol. **14:** 47–52.

60. Coculescu, M., V. Matulevicius, R. Goldstein & S. Pavel. 1978. Presence and synthesis of vasotocin in the pineal gland of Brattleboro rats. J. Endocrinol. **77:** 145–146.

61. Conger, J. D. & S. A. Falk. 1977. Intrarenal dynamics in the pathogenesis and prevention of acute urate nephropathy. J. Clin. Invest. **59:** 786–793.

62. Cox, B. M., V. E. Ghazarossian & A. Goldstein. 1980. Levels of immunoreactive dynorphin in brain and pituitary of Brattleboro rats. Neurosci. Lett. **20:** 85–88.

63. Crofton, J. T., L. Share, R. E. Shade, W. J. Lee-Kwon, M. Manning & W. H. Sawyer. 1979. The importance of vasopressin in the development and maintenance of DOC-salt hypertension in the rat. Hypertension **1:** 31–38.

64. Davis, J. M. & J. Schnermann. 1971. The effect of antidiuretic hormone on the distribution of nephron filtration rates in rats with hereditary diabetes insipidus. Pfluegers Arch. **330:** 323–334.

65. De Bermudez, L. & E. E. Windhager. 1975. Osmotically induced changes in electrical resistance of distal tubules of rat kidney. Am. J. Physiol. **229:** 1536–1546.

66. De Groot, C. A. & A. M. Tijssen. 1979. The effects of long-term oral treatment with chlorothiazide or furosemide on hereditary diabetes insipidus in rats. Arch. Int. Pharmacodyn. Ther. **237:** 75–87.

67. De Groot, C. A. & A. M. I. Tijssen. 1981. Dissociation of antipolyuric action and increase in plasma renin activity caused by chlorothiazide in Brattleboro rats. Arch. Int. Pharmacodyn. Ther. **253:** 137–147.

68. De Kloet, E. R. & H. D. Veldhuis. 1980. The hippocampal corticosterone receptor system of the homozygous diabetes insipidus Brattleboro rat. Neurosci. Lett. **16:** 187–192.

69. De Sousa, R. C. 1974. Diabetes insipidus—various recent aspects. Schweiz. Med. Wochenschr. **104:** 1045–1053.

70. De Wied, D., T. B. Van Wimersma Greidanus & B. Bohus. 1974. The rat supraoptic-neurohypophyseal system and behavior: role of vasopressin in memory processes. Probl. Actuels. Endocrinol. Nutr. **18:** 323–328.

71. De Wied, D., B. Bohus & T. B. Van Wimersma Greidanus. 1975. Memory deficit in rats with hereditary diabetes insipidus. Brain Res. **85:** 152–156.

72. De Wied, D. & W. H. Gispen. 1976. Impaired development of tolerance to morphine analgesia in rats with hereditary diabetes insipidus. Psychopharmacologia **46:** 27–29.

73. De Wied, D., T. B. Van Wimersma Greidanus, B. Bohus, I. Urban & W. H. Gispen. 1976. Vasopressin and memory consolidation. *In* Progress in Brain Research. "Perspectives in brain research." M. A. Corner & D. F. Swaab, Eds. Vol. **45:** 181–194. Elsevier Scientific Publishing Company. Amsterdam.

74. De Wied, D., B. Bohus, W. H. Gispen, I. Urban & T. B. Van Wimersma Greidanus. 1976. Hormonal influences on motivational, learning, and memory processes. *In* Hormones, Behavior, and Psychopathology. E. F. Sachar, Ed.: 1–14. Raven Press. New York, N.Y.

75. De Wied, D., B. Bohus & T. B. Van Wimersma Greidanus. 1976. The significance of vasopressin for pituitary ACTH release in conditioned emotional situation. *In* Cellular and Molecular Bases of Neuroendocrine Processes. E. Endröczi, Ed.: 547–553. (Symp. Int. Soc. Psychoneuroendocrinology, 1975, Visegrad.) Akadémiai Kiádó. Budapest.

76. De Wied, D., B. Bohus, J. M. Van Ree, I. Urban & T. B. Van Wimersma Greidanus. 1977. Neurohypophyseal hormones and behavior. *In* Neurohypophysis. A. M. Moses & L. Share, Eds.: 201–210. Karger. Basel.

77. De Wied, D. & B. Bohus. 1979. Modulation of memory processes by neuropeptides of hypothalamic neurohypophyseal origin. *In* Brain Mechanisms in Memory and Learning: From the Single Neuron to Man. (Int. Brain Res. Org. Monograph Series, Vol. 4). M. Brazier, Ed.: 139–149. Raven Press. New York, N.Y.

78. De Wied, D. 1980. Behavioural actions of neurohypophysial peptides. Proc. Roy. Soc. London Ser. B **210:** 183–195.

79. Dieppa, R. A. & R. F. Gittes. 1977. Hydronephrosis and megacystis in rats with diabetes insipidus: radiographic and ureteral pressure studies. Surg. Forum. **28:** 548–550.

80. Dierickx, K., F. Vandesande & J. De Mey. 1976. Identification, in the external region of the rat median eminence, of separate neurophysin-vasopressin and neurophysin-oxytocin containing nerve fibres. Cell Tissue Res. **168:** 141–151.

81. Dlouga, G., S. P. Gambarian & L. G. Kniaz'kova. 1981. [Glomerular filtration rate in distinct nephron populations of rodents with different ecological specializations]. Zh. Evol. Biokhim. Fiziol. **17:** 534–536.

82. Dlouhá, H., J. Křeček & J. Zicha. 1976. The renal concentrating ability of newly born Brattleboro rats (hereditary diabetes insipidus). Experientia **32:** 59–61.

83. DLOUHÁ, H., J. KŘEČEK & J. ZICHA. 1977. Hypertension in rats with hereditary diabetes insipidus: the role of age. Pfluegers Arch. Eur. J. Physiol. **369:** 177–182.

84. DLOUHÁ, H., J. KŘEČEK & J. ZICHA. 1977. Blood pressure and water and electrolyte intake and excretion in rats (Brattleboro strain) after unilateral nephrectomy. Physiol. Bohemoslov. **26:** 543–556.

85. DLOUHÁ, H., J. KŘEČEK & J. ZICHA. 1977. Growth and urine osmolarity in young Brattleboro rats. J. Endocrinol. **75:** 329–330.

86. DLOUHÁ, H., J. KŘEČEK & J. ZICHA. 1979. Effect of age on hypertensive stimuli and the development of hypertension in Brattleboro rats. Clin. Sci. **57:** 273–275.

87. DLOUHÁ, H., J. KŘEČEK & J. ZICHA. 1979. The development of osmoregulation in rats with hereditary diabetes insipidus (Brattleboro strain). *In* Hormones and Development. L. Macho & V. Štrbák, Eds.: 503–514. (Proc. Int. Symp. on Hormones and Development, 1976, Bratislava.) Veda. Bratislava, Czech.

88. DLOUHÁ, H. & J. KŘEČEK. 1980. Kidney and vasopressin in the postnatal development. Zbl. Pharm. **119:** 1070–1074.

89. DLOUHÁ, H., J. KŘEČEK & J. ZICHA. 1980. Unilateral nephrectomy and intake of 0,6% NACL solution as a complex hypertensogenic stimulus in Brattleboro rats: its age-dissociation. Physiol. Bohemoslov. **29:** 97–106.

90. DLOUHÁ, H., J. KŘEČEK & J. ZICHA. 1980. Sodium metabolism in rats with hereditary defect of vasopressin synthesis. *In* Humoral Regulation of Sodium Excretion. B. Lichardus, R. W. Schrier & J. Ponec, Eds.: 129–134. Elsevier-North Holland Biomedical Press. Amsterdam.

91. DLOUHÁ, H., J. KŘEČEK & J. ZICHA. 1981. The developing kidney and the age-dependent adaptation to salt intake. *In* Kidney and Body Fluids. L. Takács, Ed. Vol. II: 111–118. (Adv. Physiol. Sci.) Akadémiai Kiádó. Budapest.

92. DLOUHÁ, H., J. KŘEČEK & J. ZICHA. 1981. Hypertension in Brattleboro rats and the injurious influence of salt in youth. Physiol. Bohemoslov. **30:** 531–538.

93. DOGTEROM, J., T. B. VAN WIMERSMA GREIDANUS & D. F. SWAAB. 1977. Evidence for the release of vasopressin and oxytocin into cerebrospinal fluid: measurements in plasma and CSF of intact and hypophysectomized rats. Neuroendocrinology **24:** 108–118.

94. DOGTEROM, J., T. B. VAN WIMERSMA GREIDANUS & D. DE WIED. 1978. Vasopressin in cerebrospinal fluid and plasma of man, dog, and rat. Am. J. Physiol. **234:** E463–E467.

95. DOGTEROM, J., F. G. M. SNIJDEWINT & R. M. BUIJS. 1978. The distribution of vasopressin and oxytocin in the rat brain. Neurosci. Lett. **9:** 341–346.

96. DOUSA, T. P., Y. S. HUI & L. D. BARNES. 1975. Renal medullary adenylate cyclase in rats with hypothalamic diabetes insipidus. Endocrinology **97:** 802–807.

97. DUMAS, J. C., J. GARIGUE & G. ROUX. 1972. Changes in water metabolism of rats with diabetes insipidus under treatment with clofibrate, with or without ADH. C. R. Soc. Biol. **166:** 460–464.

98. DUMAS, J. C., J. TRAVES, A. AURIAC & G. ROUX. 1973. Effect of carbamoyl-dibenzo-azepine on fluid balance in rats with diabetes insipidus. C. R. Soc. Biol. **167:** 161–164.

99. DUNN, M. J., H. P. GREELY, H. VALTIN, L. B. KINTER & R. BEEUWKES III. 1978. Renal excretion of prostaglandins E_2 and $F_{2\alpha}$ in diabetes insipidus rats. Am. J. Physiol. **235:** E624–E627.

100. DUNN, M. J., L. B. KINTER, R. BEEUWKES III, H. P. GREELEY & H. VALTIN. 1980. Interaction of vasopressin and renal prostaglandins in the homozygous diabetes insipidus rat. Adv. Prostaglandin Thromboxane Res. **7:** 1009–1015.

101. DURR, J. A., B. STAMOUTSOS & M. D. LINDHEIMER. 1981. Osmoregulation during pregnancy in the rat. Evidence for resetting of the threshold for vasopressin secretion during gestation. J. Clin. Invest. **68:** 337–346.

102. DYBALL, R. E. 1974. Single unit activity in the hypothalamo-neurohypophysial system of Brattleboro rats. J. Endocrinol. **60:** 135–143.

103. EDWARDS, B. R., M. GELLAI & H. VALTIN. 1980. Concentration of urine in the absence of ADH with minimal or no decrease in GFR. Am. J. Physiol. **239:** F84–F91.

104. EMMANOUEL, D. S., M. D. LINDHEIMER & A. I. KATZ. 1974. Mechanism of impaired water excretion in hypothyroid rats. J. Clin. Invest. **54:** 926–934.

105. EMMANOUEL, D. S., M. D. LINDHEIMER & A. I. KATZ. 1975. Urinary concentration and dilution after unilateral nephrectomy in the rat. Clin. Sci. Mol. Med. **49:** 563–573.

106. FERNANDEZ-REPOLLET, E., M. MARTINEZ-MALDONADO & S. OPAVA-STITZER. 1980. Role of water balance in the enhanced potassium excretion and hypokalaemia of rats with diabetes insipidus. J. Physiol. **305:** 97–108.

107. FORREST, J. N., A. D. COHEN, J. TORRETTI, J. M. HIMMELHOCH & F. H. EPSTEIN. 1974. On the mechanism of lithium-induced diabetes insipidus in man and the rat. J. Clin. Invest. **53:** 1115–1123.

108. FORREST, J. N. & I. SINGER. 1977. Drug-induced interference with action of antidiuretic hormone. *In* Disturbances in Body Fluid Osmolality. T. E. Andreoli, J. J. Grantham & F. C. Rector, Eds.: 309–340. American Physiological Society. Bethesda, Md.

109. FRIEDMAN, S. M. & C. L. FRIEDMAN. 1965. Salt and water distribution in hereditary and in induced hypothalamic diabetes insipidus in the rat. Can. J. Physiol. Pharmacol. **43:** 699–705.

110. FRIEDMAN, S. M., M. NAKASHIMA & C. L. FRIEDMAN. 1967. Neurohypophysial function in desoxycorticosterone acetate-hypertension in the rat. Endocrinology **81:** 1231–1240.

111. FRIESEN, H. G. & E. B. ASTWOOD. 1967. Changes in neurohypophysial proteins induced by dehydration and ingestion of saline. Endocrinology **80:** 278–287.

112. GALTON, V. A., H. VALTIN & D. G. JOHNSON. 1966. Thyroid function in the absence of vasopressin. Endocrinology **78:** 1224–1229.

113. GANTEN, D., J. S. HUTCHINSON & P. SCHELLING. 1975. The intrinsic brain iso-renin-angiotensin system in the rat: its possible role in central mechanisms of blood pressure regulation. Clin. Sci. Mol. Med. **48:** 265S–268S.

114. GARINA, I. A., V. G. SHALYAPINA & J. KŘEČEK. 1981. Uptake of ^{3}H-cortisol by the pituitary and brain structures of Brattleboro rats with hereditary diabetes insipidus. Physiol. Bohemoslov. **30:** 381–383.

115. GARTSIDE, I. B., A. M. JONES, J. F. LAYCOCK & S. J. WALTER. 1981. The effect of vasopressin on extracellular cation concentrations and muscle resting potentials in the rat. J. Physiol. London **318:** 317–326.

116. GASH, D., J. R. SLADEK, JR. & C. D. SLADEK. 1980. Functional development of grafted vasopressin neurons. Science **210:** 1367–1369.

117. GASH, D. & D. E. SCOTT. 1980. Fetal hypothalamic transplants in the third ventricle of the adult rat brain. Correlative scanning and transmission electron microscopy. Cell Tissue Res. **211:** 191–206.

118. GASH, D. & J. R. SLADEK, JR. 1980. Vasopressin neurons grafted into Brattleboro rats: viability and activity. Peptides **1:** 11–14.

119. GELLAI, M., B. R. EDWARDS & H. VALTIN. 1979. Urinary concentrating ability during dehydration in the absence of vasopressin. Am. J. Physiol. **237:** F100–F104.

120. GELLAI, M., F. T. LAROCHELLE, JR. & H. VALTIN. 1980. Effect of oxytocin on renal function in conscious Brattleboro homozygous rats. A model for studying renal regulation by hormones. *In* Hormonal Regulation of Sodium

Excretion. B. Lichardus, R. W. Schrier & J. Ponec, Eds.: 121–128. Elsevier-North Holland Biomedical Press. Amsterdam.

121. George, J. M. & J. Forrest. 1976. Vasopressin and oxytocin content of microdissected hypothalamic areas in rats with hereditary diabetes insipidus. Neuroendocrinology **21:** 275–279.

122. Gillies, G. & P. J. Lowry. 1980. Corticotropin releasing activity in extracts of the stalk median eminence of Brattleboro rats. J. Endocrinol. **84:** 65–73.

123. Gillies, G. & P. J. Lowry. 1980. The relationship between vasopressin and corticotrophin releasing factor. *In* Interaction within the Brain-Pituitary-Adrenocortical System. M. Jones, E. Gillham & M. Dallman, Eds.: 51–62. Academic Press. London.

124. Gillies, G. & P. J. Lowry. 1980. Corticotrophin releasing factor may be modulated vasopressin. Nature **278:** 463–464.

125. Goren, H. J., R. M. Geonzon, M. D. Hollenberg, K. Lederis & D. O. Morgan. 1980. Oxytocin action: lack of correlation between receptor number and tissue responsiveness. J. Supramol. Struct. **14:** 129–138.

126. Grainger, F. & J. C. Sloper. 1974. Correlation between microtubular number and transport activity of hypothalamo-neurohypophyseal secretory neurons. Cell Tissue Res. **153:** 101–113.

127. Grainger, F. & J. C. Sloper. 1976. Microtubular number in the tractus hypophyseus of newborn normal rats and newborn rats with congenital diabetes insipidus. Cell Tissue Res. **169:** 405–414.

128. Green, H. H., A. R. Harrington & H. Valtin. 1970. On the role of antidiuretic hormone in the inhibition of acute water diuresis in adrenal insufficiency and the effects of gluco- and mineralocorticoids in reversing the inhibition. J. Clin. Invest. **49:** 1724–1736.

129. Greer, E. R., M. C. Diamond & J. M. Tang. 1981. Increase in thickness of cerebral cortex in response to environmental enrichment in Brattleboro rats deficient in vasopressin. Exp. Neurol. **72:** 366–378.

130. Groblewski, T. A., A. A. Nunez & R. M. Gold. 1981. Circadian rhythms in vasopressin deficient rats. Brain Res. Bull. **6:** 125–130.

131. Gross, F., G. Dauda, S. Kazda, J. Kyncl, J. Möhring & H. Orth. 1972. Increased fluid turnover and the activity of the renin angiotensin system under various experimental conditions. Circ. Res. (Suppl) **31:** 173–181.

132. Güllner, H. G. & F. C. Bartter. 1980. The role of urinary prostaglandin E and cyclic AMP in the polyuria of hypokalemia in rats. Prostaglandins Med. **4:** 13–19.

133. Güllner, H. G. 1980. Measurement of plasma substance P in Sprague-Dawley rats and in rats with hypothalamic diabetes insipidus. IRCS Med. Sci. Biochem. **8:** 236–237.

134. Gussin, R. Z., E. H. Stokey, M. A. Ronsberg, L. C. Malone & J. R. Cummings. 1971. The effect of guancydine on fluid and electrolyte excretion: a comparison with hydralazine and reserpine. Arch. Int. Pharmacodyn. Ther. **192:** 135–151.

135. Gutman, Y. & F. Benzakein. 1971. Effect of an increase and of lack of antidiuretic hormone on plasma renin activity in the rat. Life Sci. (Part 1) Physiol. Pharmacol. **10** (19): 1081–1085.

136. Gutman, Y. & R. Benzakein. 1974. Antidiuretic hormone and renin in rats with diabetes insipidus. Eur. J. Pharmacol. **28:** 114–118.

137. Haack, D., E. Homsey, G. Kohrs, B. Möhring, P. Oster & J. Möhring. 1975. Studies on drinking-feeding interactions in rats with hereditary hypothalamic diabetes insipidus. *In* Control Mechanisms of Drinking. G. Peters, J. T. Fitzsimons & L. Peters-Haefli, Eds.: 41–45. (Proc. Symp. on Thirst.) Springer. Berlin.

138. Haack, D. & J. Möhring. 1978. Vasopressin-mediated blood pressure re-

sponse to intraventricular injection of angiotensin II in the rat. Pfluegers Arch. **373:** 167–173.

139. Hall, D. E., S. Ayachi & O. Hall. 1973. Spontaneous hypertension in rats with hereditary hypothalamic diabetes insipidus. Tex. Rep. Biol. Med. **31:** 471–487.

140. Hamberger, A., A. Norstrom, M. Sandberg & U. Svanberg. 1979. In vitro GABA transport in the neurohypophysis from rats with hereditary diabetes and after osmotic stimulation. Brain Res. **174:** 341–344.

141. Harmanci, M. C., W. A. Kachadorian, H. Valtin & V. A. DiScala. 1978. Antidiuretic hormone-induced intramembranous alterations in mammalian collecting ducts. Am. J. Physiol. **235:** F440–F443.

142. Harmanci, M. C., P. Stern, W. A. Kachadorian, H. Valtin & V. A. DiScala. 1980. Vasopressin and collecting duct intramembranous particle clusters: a dose-response relationship. Am. J. Physiol. **239:** F560–F564.

143. Harrington, A. R. & H. Valtin. 1965. Vasopressin effect on urinary concentration in rats with hereditary hypothalamic diabetes insipidus. Proc. Soc. Exp. Biol. Med. **118:** 448–450.

144. Harrington, A. R. & H. Valtin. 1968. Impaired urinary concentration after vasopressin and its gradual correction in hypothalamic diabetes insipidus. J. Clin. Invest. **47:** 502–510.

145. Harrington, A. R. & R. E. Rieselbach. 1970. Na conservation in rats with reduced medullary osmotic gradient due to diabetes insipidus. Am. J. Physiol. **219:** 384–386.

146. Harrington, A. R. 1972. Hyponatremia due to sodium depletion in the absence of vasopressin. Am. J. Physiol. **222:** 768–774.

147. Harris, M. C. 1971. Release of an antidiuretic substance by bradykinin in the rat. J. Physiol. **219:** 403–419.

148. Harris, R. H. & W. E. Yarger. 1977. Urine-reinfusion natriuresis: evidence for potent natriuretic factors in rat urine. Kidney Int. **11:** 93–105.

149. Haulica, I., E. Hefco, V. Rosca, D. Steanu, G. H. Petrescu, A. Stratone, M. Coculescu & I. Bordea. 1977. Studies of the presence and role of renin in the pituitary gland. Rev. Med. Chir. Soc. Med. Nat. Iasi. **81:** 425–430.

150. Henderson, I. W., R. J. Balment & J. A. Oliver. 1978. Vasopressin effects on plasma renin activity in male and female rats. Clin. Sci. Mol. Med. **55:** 301–307.

151. Henderson, I. W., A. McKeever & C. J. Kenyon. 1979. Captopril (SQ–14225) depresses drinking and aldosterone in rats lacking vasopressin. Nature (Lond.) **281:** 569–570.

152. Hitt, B. A., R. I. Mazze, T. L. Cook, W. J. Beppu & J. C. Kosek. 1977. Thermoregulatory defect in rats during anesthesia. Anesth. Analg. **56:** 9–15.

153. Hochman, S. & Y. Gutman. 1974. Lithium: ADH antagonism and ADH independent action in rats with diabetes insipidus. Eur. J. Pharmacol. **28:** 100–107.

154. Hoffman, W. E., U. Ganten, P. Schelling, M. I. Phillips, P. G. Schmid & D. Ganten. 1978. The renin and isorenin-angiotensin system in rats with hereditary hypothalamic diabetes insipidus. Neuropharmacology **17:** 919–923.

155. Hoffman, W. E., U. Ganten, M. I. Phillips, P. G. Schmid, P. Schelling & D. Ganten. 1978. Inhibition of drinking in water-deprived rats by combined central angiotensin II and cholinergic receptor blockade. Am. J. Physiol. **234:** F41–F47.

156. Holzbauer, M., D. F. Sharman, U. Godden, S. P. Mann, D. Blatchford, R. Laverty & L. G. Jarvis. 1980. Observations on pituitary and cerebral catecholamines in Brattleboro rats. Medical Biology **58:** 25–32.

157. Hornychová, H., O. Hrubik & J. Křeček. 1969. On the mechanism of urine concentration in rats with hereditary hypothalamic diabetes insipidus. Acta Biol. Iugoslav. Ser. C **5:** 269–272.

158. Huidobro-Tor, J. P. 1980. Antidiuretic effect of beta-endorphin and morphine in Brattleboro rats: development of tolerance and physical dependence after chronic morphine treatment. Br. J. Pharmacol. **71:** 51–56.

159. Hunt, N. H., A. D. Perris & P. A. Sandford. 1977. Role of vasopressin in the mitotic response of rat bone marrow cells to hemorrhage. J. Endocrinol. **72:** 5–16.

160. Hutchinson, J. S., J. Möhring, P. Schelling & D. Ganten. 1975. Differential pressor responses to centrally administered angiotensin II (A II) in conscious rats, homozygous and heterozygous for hypothalamic diabetes insipidus. Naunyn. Schmiedebergs Arch. Pharmacol. **287:** R52.

161. Hutchinson, J. S., P. Schelling, J. Möhring & D. Ganten. 1976. Pressor action of centrally perfused angiotensin II in rats with hereditary hypothalamic diabetes insipidus. Endocrinology **99:** 819–823.

162. Hutchinson, J. S., P. Schelling, J. Möhring & D. Ganten. 1976. Effect of intraventricular perfusion of angiotensin II in conscious normal rats and in rats with hereditary hypothalamic diabetes insipidus. Clin. Sci. Mol. Med. **51:** 391S–394S.

163. Hutchinson, J. S., D. Ganten, P. Schelling, P. Ylitalo, J. Möhring & M. Kalina. 1976. Central pressor actions of angiotensin II. Acta Med. Acad. Sci. Hung. **33:** 101–109.

164. Imbert-Teboul, M., D. Chabardes, M. Montegut, A. Clique & F. Morel. 1978. Impaired response to vasopressin of adenylate cyclase of the thick ascending limb of Henle's Loop in Brattleboro rats with diabetes insipidus. Renal Physiol. **1:** 3–10.

165. Imbs, J. L., M. Schmidt, A. Parrenin, H. Belhadj-Mostefa & J. Schwartz. 1977. Effect of clopamide and furosemide on blood urea in rats with congenital diabetes insipidus. J. Urol. Nephrol. (Paris) **83:** 336–342.

166. Jamison, R. L., J. Buerkert & F. Lacy. 1971. A micropuncture study of collecting tubule function in rats with hereditary diabetes insipidus. J. Clin. Invest. **50:** 2444–2452.

167. Jamison, R. L. 1972. Evidence for functional intrarenal heterogeneity obtained by the micropuncture technique. Yale J. Biol. Med. **45:** 254–262.

168. Jamison, R. L. & F. B. Lacy. 1972. Evidence for urinary dilution by the collecting tubule. Am. J. Physiol. **223:** 898–902.

169. Jamison, R. L., J. Buerkert & F. Lacy. 1973. A micropuncture study of Henle's thin loop in Brattleboro rats. Am. J. Physiol. **224:** 180–185.

170. Jard, S., D. Butlen, R. Rajerison & C. Roy. 1977. The vasopressin-sensitive adenylate cyclase from the mammalian kidney. Regulation of enzyme responsiveness to hormonal stimulation. *In* Neurohypophysis. A. M. Moses & L. Share, Eds.: 211–219. Karger. Basel.

171. Jarlstedt, J. & A. Norstrom. 1975. Content of RNA in supraoptic and paraventricular neurones of rats with hereditary diabetes insipidus (Brattleboro strain). J. Endocrinol. **66:** 435–436.

172. Johnson, A. K., M. I. Phillips, J. Möhring & D. Ganten. 1977. Angiotensin induced drinking in rats with hereditary hypothalamic diabetes insipidus. Neurosci. Lett. **4:** 327–330.

173. Johnston, P. A., F. B. Lacy & R. L. Jamison. 1977. Effect of antidiuretic hormone-induced antidiuresis on water reabsorption by the superficial loop of Henle in Brattleboro rats. J. Lab. Clin. Med. **90:** 1004–1011.

174. Jones, J. J. & J. Lee. 1967. The value of rats with hereditary hypothalamic diabetes insipidus for the bioassay of vasopressin. J. Endocrinol. **37:** 335–344.

175. Jones, R. M., C. Terhaard, J. Zullo & S. M. Tenney. 1981. Mechanism of reduced water intake in rats at high altitude. Am. J. Physiol. **240:** R187–R191.

176. Jones, R. M., F. T. Larochelle, Jr. & S. M. Tenney. 1981. Role of

arginine vasopressin on fluid and electrolyte balance in rats exposed to high altitude. Am. J. Physiol. **240:** R182–R186.

177. JOPPICH, R. & P. DEETJEN. 1971. The relation between the reabsorption of urea and of water in the distal tubule of the rat kidney. Pfluegers Arch. **329:** 172–185.

178. KARTÉSZI, M., E. STARK, G. RAPPAY, F. A. LÁSZLÓ & G. B. MAKARA. 1981. Corticoliberin activity of rat neurohypophysis is distinct from vasopressin. Am. J. Physiol. **240:** E689–E693.

179. KAKOLEWSKI, J. W. & E. DEAUX. 1972. Aphagia in the presence of drinking an isosmotic NaCl solution. Physiol. Behav. **8:** 623–630.

180. KALIMO, H. & U. K. RINNE. 1972. Ultrastructural studies on the hypothalamic neurosecretory neurons of the rat. II. The hypothalamo-neurohypophysial system in rats with hereditary hypothalamic diabetes insipidus. Z. Zellforsch. Mikrosk. Anat. **134:** 205–225.

181. KAPITOLA, J., H. DLOUHÁ, J. KŘEČEK & J. ZICHA. 1977. The effect of dehydration on the neurohypophyseal blood flow in rats with hereditary diabetes insipidus. Experientia **33:** 1615–1616.

182. KASTING, N. W., W. L. VEALE, K. E. COOPER & K. LEDERIS. 1981. Vasopressin may mediate febrile convulsions. Brain Res. **213:** 327–333.

183. KEELER, R. & N. WILSON. 1976. Vasopressin contamination as a cause of some apparent renal actions of prolactin. Can. J. Physiol. Pharmacol. **54:** 887–890.

184. KENYON, C. J., G. HARGREAVES & I. W. HENDERSON. 1978. Adrenocortical function in rats with inherited hypothalamic diabetes insipidus (Brattleboro strain). J. Steroid Biochem. **9:** 345–348.

185. KENYON, C. J., W. MOSLEY, G. HARGREAVES, R. J. BALMENT & I. W. HENDERSON. 1978. The effects of dietary sodium restriction and potassium supplementation and hypophysectomy on adrenocortical function in the rat. J. Steroid Biochem. **9:** 337–344.

186. KETTYLE, W. M., M. HORSTER, K. THURAU & H. VALTIN. 1970. Cryoscopic determination of renal tissue osmolality by an ultra micro method. Pfluegers Arch. Eur. J. Physiol. **321:** 83–89.

187. KILCOYNE, M. M., D. L. HOFFMAN & E. A. ZIMMERMAN. 1980. Immunocytochemical localization of angiotensin II and vasopressin in rat hypothalamus: evidence for production in the same neuron. Clin. Sci. **6:** 57s–60s.

188. KLEEMAN, C. R., H. VORHERR & M. HOUGHOUGHI. 1967. Sensitivity to ADH of normal and diabetes insipidus rats. Endocrinology **80:** 1168–1169.

189. KLEIN, L. A. 1975. Beta-adrenergic (isoproterenol) regulation of antidiuretic hormone. Invest. Urol. **12:** 285–290.

190. KLEIN, L. A. 1975. Alpha-adrenergic (norepinephrine) effect on antidiuretic hormone activity. Invest. Urol. **13:** 159–164.

191. KNEPEL, W., H. ANHUT, D. NUTTO & G. HERTTING. 1980. Evidence that vasopressin is involved in the isoprenaline-induced beta-endorphin release. Eur. J. Pharmacol. **68:** 359–363.

192. KNEPEL, W. & D. K. MEYER. 1981. Inhibition of the isoprenaline-induced renin release by bilateral vagotomy is mediated by vasopressin. Endokrinologie **77:** 325–332.

193. KOEHL, W. 1978. Experimental studies of the subcommissural organ (SCO). Verh. Anat. Ges. **72:** 717–718.

194. KONRADS, A., K. G. HOFBAUER, K. BAUEREISS, J. MÖHRING & F. GROSS. 1979. Glycerol-induced acute renal failure in Brattleboro rats with hypothalamic diabetes insipidus. Clin. Sci. **56:** 133–138.

195. KOVACS, G. L., G. SZABÓ, L. SZONTÁGH, L. MEDVE, G. TELEGDY & F. A. LÁSZLÓ. 1980. Hereditary diabetes insipidus in rats. Altered cerebral indolamine and catecholomine metabolism. Neuroendocrinology **31:** 189–193.

196. KŘEČEK, J., H. DLOUHÁ & J. ZICHA. 1979. The role of the age factor in the

development of experimental hypertension in rats with hereditary diabetes insipidus (Brattleboro strain). *In* Hormones and Development. L. Macho & V. Strbak, Eds.: 473–487. (Int. Symp. on Hormones and Development, 1976, Bratislava.) Veda. Bratislava, Czech.

197. Krieger, D. T., A. Liotta & M. J. Brownstein. 1977. Corticotropin releasing factor distribution in normal and Brattleboro rat brain and effect of deafferentation, hypophysectomy and steroid treatment in normal animals. Endocrinology **100:** 227–237.

198. Krieger, D. T. & A. Liotta. 1977. Pituitary ACTH responsiveness in the vasopressin deficient rat. Life Sci. **20:** 327–335.

199. Kutscher, C. L. & W. A. Wright. 1977. Drinking characteristics of normal rats and rats heterozygous or homozygous for diabetes insipidus. Physiol. Behav. **18:** 833–839.

200. Kyncl, J., V. Jelinek & J. Rudinger. 1969. The use of rats with lesion-induced or hereditary hypothalamic diabetes insipidus for evaluating prolonged responses to antidiuretic hormone preparations. Acta Endocrinol. **60:** 369–379.

201. Kyncl, J. & J. Rudinger. 1970. Excretion of antidiuretic activity in the urine of cats and rats after administration of the synthetic hormonogen, N alpha-glycyl-glycyl-gylcyl-[8-lysine]-vasopressin(triglycylvasopressin). J. Endocrinol. **48:** 157–165.

202. Laczi, F., E. Nagy & F. A. László. 1978. The ADH-reserve capacity in Brattleboro rats. Acta Med. Acad. Sci. Hung. **35:** 173–179.

203. Laczi, F., F. A. László, G. Keri & I. Teplan. 1980. Study on the biological half-life and organ-distribution of tritiated lysine-vasopressin in Brattleboro rats. Acta Physiol. Acad. Sci. Hung. **55:** 129–133.

204. Laycock, J. F. & P. G. Williams. 1973. The effect of vasopressin (Pitressin) administration on sodium, potassium and urea excretion in rats with and without diabetes insipidus, with a note on the excretion of vasopressin in the DI rat. J. Endocrinol. **56:** 111–120.

205. Laycock, J. F., J. Lee & A. F. Lewis. 1974. The effect of chlorpropamide on water balance in pitressin-treated Brattleboro rats. Br. J. Pharmacol. **52:** 253–263.

206. Laycock, J. F. 1976. Review: The Brattleboro rat with hereditary hypothalamic diabetes insipidus as an ideal experimental model. Lab Anim. **10:** 261–270.

207. Laycock, J. F. 1977. The Brattleboro rat with hereditary hypothalamic diabetes insipidus. Gen. Pharmacol. **8:** 297–302.

208. Laycock, J. F. & A. F. Lewis. 1977. Potentiation of the response to vasopressin (Pitressin) by treatment with a combination of chlorpropamide and chlorothiazide in Brattleboro rats with hereditary hypothalamic diabetes insipidus. Br. J. Pharmacol. **59:** 11–16.

209. Laycock, J. F., W. Penn, D. G. Shirley & S. J. Walter. 1979. The role of vasopressin in blood pressure regulation immediately following acute haemorrhage in the rat. J. Physiol. **296:** 267–275.

210. Leclerc, R. & G. Pelletier. 1974. Electron microscope immunohistochemical localization of vasopressin in the hypothalamus and neurohypophysis of the normal and Brattleboro rat. Am. J. Anat. **140:** 583–588.

211. Leclerc, R. & G. Pelletier. 1976. Electron microscope immunohistochemical localization of neurophysin in the rat with hereditary diabetes insipidus. Virchows Archiv. Cell. Pathol. **22:** 233–243.

212. Lee, J. & P. G. Williams. 1972. The effect of vasopressin (Pitressin) administration and dehydration on the concentration of solutes in renal fluids of rats with and without hereditary hypothalamic diabetes insipidus. J. Physiol. **220:** 729–743.

213. Leng, G. & J. Wiersma. 1981. Effects of neural stalk stimulation on phasic

discharge of supraoptic neurones in Brattleboro rats devoid of vasopressin. J. Endocrinol. **90:** 211–220.

214. LEONARD, B. E., F. RAMAEKERS & H. RIGTER. 1976. Monoamines in brain and urine of rats with hereditary hypothalamic diabetes insipidus. Experientia **32:** 901–902.

215. LEVI, J., J. GRINBLAT & C. R. KLEEMAN. 1971. Effect of isoproterenol on water diuresis in rats with congenital diabetes insipidus. Am. J. Physiol. **221:** 1728–1732.

216. LINAS, S. L., T. BERL, G. L. ROBERTSON, G. A. AISENBREY, R. W. SCHRIER & R. J. ANDERSON. 1980. Role of vasopressin in the impaired water excretion of glucocorticoid deficiency. Kidney Int. **18:** 58–67.

217. LINAS, S. L., R. J. ANDERSON, S. J. GUGGENHEIM, G. L. ROBERTSON & T. BERL. 1981. Role of vasopressin in impaired water excretion in conscious rats with experimental cirrhosis. Kidney Int. **20:** 173–180.

218. LINDHEIMER, M. D., A. REINHARZ, A. GRANDCHAMP & M. B. VALLOTTON. 1980. Fate of vasopressin perfused into nephrons of Wistar and Brattleboro (diabetes insipidus) rats. Clin. Sci. **58:** 139–144.

219. LUTTERODT, A. T., G. D. BURFORD & K. LEDERIS. 1976. Production of antisera to (8-arginine)-vasopressin in homozygous Brattleboro rats. J. Endocrinol. **71:** 161–162.

220. MANN, J. F. E., W. RASCHER, A. SCHÖMIG & R. DIETZ. 1978. Inhibition of the renin-angiotensin-system in Brattleboro rats with hereditary hypothalamic diabetes insipidus. Klin. Wochenschr. **56** (Supp 1): 67–70.

221. MANN, J. F. E., K. A. JOHNSON & D. GANTEN. 1980. Plasma angiotensin II dipsogenic levels and angiotensin-generating capacity of renin. Am. J. Physiol. **238:** R372–R377.

222. MANNING, M., L. BALASPIRI, M. ACOSTA & W. H. SAWYER. 1973. Solid phase synthesis of (1-deamino, 4-valine)-8-D-arginine-vasopressin (DVDAVP), a highly potent and specific antidiuretic agent possessing protracted effects. J. Med. Chem. **16:** 975–978.

223. MANNING, M., L. BALASPIRI, J. JUDD, M. ACOSTA & W. H. SAWYER. 1974. Probing the molecular basis of antidiuretic specificity and duration of action with synthetic peptides. FEBS Lett. **44:** 229–232.

224. MANNING, M., J. LOWBRIDGE, C. T. STIER, JR., J. HALDAR & W. H. SAWYER. 1977. [1-deaminopenicillamine,4-valine]8-D-arginine-vasopressin, a highly potent inhibitor of the vasopressor response to arginine-vasopressin. J. Med. Chem. **20:** 1228–1230.

225. MARTIN, R. & K. H. VOIGHT. 1981. Enkephalins co-exist with oxytocin and vasopressin in nerve terminals of rat neurohypophysis. Nature **289:** 502–504.

226. MARTINEZ-MALDONADO, M., A. STAVROULAKI-TSAPARA & G. EKNOYAN. 1974. Renal effects of cyclic AMP in normal and congenital diabetes insipidus rats. Life Sci. **14:** 2025–2030.

227. MATA, M. M., H. GAINER & W. A. KLEE. 1977. Effect of dehydration on the endogenous opiate content of the rat neurointermediate lobe. Life Sci. **21:** 1159–1162.

228. MATSUI, K., T. KIMURA & K. YOSHINAGA. 1980. Animal models for diabetes insipidus. Nippon Rinsho **38:** 2936–2941.

229. MCAULIFFE, W. G. 1980. Histochemistry and ultrastructure of the interstitium of the renal papilla in rats with hereditary diabetes insipidus (Brattleboro strain). Am. J. Anat. **157:** 17–26.

230. MCCANN, S. M., J. ANTUNES-RODRIGUES, R. NALLAR & H. VALTIN. 1966. Pituitary-adrenal function in the absence of vasopressin. Endocrinology **79:** 1058–1064.

231. MCDONALD, K. M., K. C. KURUVILA, G. A. AISENBREY & R. W. SCHRIER. 1977. Effect of alpha and beta adrenergic stimulation on renal water excretion and

medullary tissue cyclic AMP in intact and diabetes insipidus rats. Kidney Int. **12:** 96–103.

232. MICHAEL, U. K., J. KELLEY, H. ALBERT & C. A. VAAMONDE. 1976. Role of distal delivery of filtrate in impaired renal dilution of the hypothyroid rat. Am. J. Physiol. **230:** 699–705.
233. MICHAJLOVSKI, J. N., J. PONEC, A. D'ZURBA & B. LICHARDUS. 1980. (Na-K) ATPase activity and corticopapillary osmotic gradient in the kidney of rats with hereditary hypothalamic diabetes insipidus after administration of 1-deamino-8-D-arginine vasopressin (DDAVP). Bratisl. Lek. Listy. **73:** 741–748.
234. MILLER, M. & A. M. MOSES. 1969. Radioimmunoassay of vasopressin with a comparison of immunological and biological activity in the rat posterior pituitary. Endocrinology **84:** 557–562.
235. MILLER, M. & A. M. MOSES. 1970. Potentiation of vasopressin action by chlorpropamide in vivo. Endocrinology **86:** 1024–1027.
236. MILLER, M. & A. M. MOSES. 1971. Radioimmunoassay of urinary antidiuretic hormone with application to study of the Brattleboro rat. Endocrinology **88:** 1389–1396.
237. MILLER, M. 1975. Inhibition of ADH release in the rat by narcotic antagonists. Neuroendocrinology **19:** 241–251.
238. MILLER, M. & M. VAN GEMERT. 1976. Antidiuretic action of prolactin in the rat with diabetes insipidus. Horm. Res. **7:** 319–332.
239. MILLER, M., E. G. BARRANDA, M. C. DEAN & F. R. BRUSH. 1976. Does the rat with hereditary hypothalamic diabetes insipidus have impaired avoidance learning and/or performance? Pharmacol. Biochem. Behav. **5:** 35–40.
240. MILLER, M. 1980. Clonidine-induced diuresis in the rat: evidence for a renal site of action. J. Pharmacol. Exp. Therap. **214:** 608–613.
241. MÖHRING, B., J. MÖHRING, G. DAUDA & D. HAACK. 1974. Potassium deficiency in rats with hereditary diabetes insipidus. Am. J. Physiol. **227:** 916–920.
242. MÖHRING, B. & J. MÖHRING. 1975. Plasma ADH in normal Long-Evans rats and in Long-Evans rats heterozygous and homozygous for hypothalamic diabetes insipidus. Life Sci. **17:** 1307–1314.
243. MÖHRING, J., G. DAUDA, D. HAACK, E. HOMSY, G. KOHRS & B. MÖHRING. 1972. Increased potassium intake and kaliopenic nephropathy in rats with genetic diabetes insipidus. Life Sci. **11:** 679–683.
244. MÖHRING, J., A. SCHÖMIG, H. BREKNER & B. MÖHRING. 1972. ADH-induced potassium retention in rats with genetic diabetes insipidus. Life Sci. **11:** 65–72.
245. MÖHRING, J., B. MÖHRING, A. SCHÖMIG, H. SCHÖMIG-BREKNER & D. HAACK. 1974. Acute effects of vasopressin on potassium and water balance in rats with diabetes insipidus. Am. J. Physiol. **227:** 921–926.
246. MÖHRING, J., B. MÖHRING, M. PETRI & D. HAACK. 1977. Vasopressor role of ADH in the pathogenesis of malignant DOC hypertension. Am. J. Physiol. **232:** F260–F269.
247. MÖHRING, J., G. KOHRS, B. MÖHRING, M. PETRI, E. HOMSY & D. HAACK. 1978. Effects of prolonged vasopressin treatment in Brattleboro rats with diabetes insipidus. Am. J. Physiol. **234:** F106–F111.
248. MOORE, L. C., D. J. MARSH & C. M. MARTIN. 1980. Loop of Henle during the water-to-antidiuresis transition in Brattleboro rats. Am. J. Physiol. **239:** F72–F83.
249. MOREL, F., M. IMBERT-TEBOUL & D. CHABARDÈS. 1980. Cyclic nucleotides and tubule function. Adv. Cyclic Nucleotide Res. **12:** 301–313.
250. MORRIS, J. F., H. W. SOKOL & H. VALTIN. 1977. One neuron—one hormone? Recent evidence from Brattleboro rats. *In* Neurohypophysis. A. M. Moses & L. Share, Eds.: 58–66. Karger. Basel.

251. Moses, A. M. & M. Miller. 1970. Accumulation and release of pituitary vasopressin in rats heterozygous for hypothalamic diabetes insipidus. Endocrinology **86:** 34–41.
252. Moses, A. M., J. Howanitz, M. Van Gemert & M. Miller. 1973. Clofibrate-induced antidiuresis. J. Clin. Invest. **52:** 535–542.
253. Moses, A. M., M. Van Gemert & M. Miller. 1974. Evidence that glyburide-induced diuresis is not mediated by inhibition of ADH. Horm. Res. **5:** 359–366.
254. Moses, A. M. & R. Coulson. 1980. Augmentation by chlorpropamide of 1-deamino-8-D-arginine vasopressin-induced antidiuresis and stimulation of renal medullary adenylate cyclase and accumulation of adenosine 3′,5′-monophosphate. Endocrinology **106:** 967–972.
255. Mouw, D. R., A. J. Vander, C. Landis, S. Kutschinski, N. Mathias & D. Zimmerman. 1980. Dose-response relation of CSF sodium and renal sodium excretion and its absence in homozygous Brattleboro rats. Neuroendocrinology **30:** 206–212.
256. Murray, R. D., K. W. Cho & R. L. Malvin. 1980. Deoxycorticosterone acetate-induced renin suppression in the absence of antidiuretic hormone. Proc. Soc. Exp. Biol. Med. **165:** 137–140.
257. Needleman, P., J. R. Douglas, Jr., B. A. Jakschik, A. L. Blumberg, P. C. Isakson & G. R. Marshall. 1976. Angiotensin antagonists as pharmacological tools. Fed. Proc. **35:** 2488–2493.
258. Nilaver, G., E. A. Zimmerman, J. Wilkins, J. Michaels, D. Hoffman & A. J. Silverman. 1980. Magnocellular hypothalamic projections to the lower brain stem and spinal cord of the rat. Neuroendocrinology **30:** 150–158.
259. Norstrom, A. 1973. Subcellular distribution of neurophysin in rats subjected to haemorrhage, salt-loading and lactation and in rats with hereditary diabetes insipidus (Brattleboro strain). Z. Zellforsch. Mikrosk. Anat. **140:** 413–424.
260. Norstrom, A. 1974. Biosynthesis of neurohypophysial proteins in rats with hereditary hypothalamic diabetes insipidus (Brattleboro strain). Brain Res. **68:** 309–317.
261. North, W. G., F. T. Larochelle, J. F. Morris, H. W. Sokol & H. Valtin. 1978. Biosynthetic specificity of neurons producing neurohypophysial principles. *In* Recent Studies of Hypothalamic Function. Part 1. Hormones. K. Lederis & W. L. Veale, Eds.: 62–76. Karger. Basel.
262. Olesen, O. V. & K. Thomsen. 1974. Effect of prolonged lithium ingestion on glucagon and parathyroid hormone responses in rats. Acta Pharmacol. Toxicol. **34:** 225–231.
263. Pennell, J. P., F. B. Lacy & R. L. Jamison. 1974. An in vivo study of the concentrating process in the descending limb of Henle's Loop. Kidney Int. **5:** 337–347.
264. Peter, S. & J. Möhring. 1978. The juxtaglomerular apparatus of rats with hereditary hypothalamic diabetes insipidus. Cell Tissue Res. **188:** 335–339.
265. Pickering, B. T., C. W. Jones, G. D. Burford, M. McPherson, R. W. Swann, P. F. Heap & J. F. Morris. 1975. The role of neurophysin proteins: suggestions from the study of their transport and turnover. Ann. N.Y. Acad. Sci. **248:** 15–35.
266. Ponec, J. & B. Lichardus. 1975. Water and sodium excretion during extracellular fluid volume expansion in hereditary diabetes insipidus rats. Curr. Probl. Clin. Biochem. **4:** 237–239.
267. Ponec, J. & B. Lichardus. 1980. Decreased free water excretion after indomethacin in the absence of antidiuretic hormone in saline loaded hypophysectomized Wistar and hydropaenic Brattleboro rats. Endokrinologie **75:** 67–76.

268. RAJERISON, R. M., D. BUTLEN & S. JARD. 1977. Effects of in vivo treatment with vasopressin and analogues on renal adenylate cyclase responsiveness to vasopressin stimulation in vitro. Endocrinology **101:** 1–12.

269. RAMAEKERS, F., H. RIGTER & B. E. LEONARD. 1977. Increased membrane-bound polyribosome fraction in the brains of rats with hereditary diabetes insipidus. Experientia **33:** 1326–1327.

270. RECTOR, JR., F. C., M. MARTINEZ-MALDONADO, N. A. KURTZMAN, J. C. SELLMAN, F. OERTHER & D. W. SELDIN. 1968. Demonstration of a hormonal inhibitor of proximal tubular reabsorption during expansion of extracellular volume with isotonic saline. J. Clin. Invest. **47:** 761–773.

271. RHODES, C. H., J. I. MORRELL & D. W. PFAFF. 1981. Immunohistochemical analysis of magnocellular elements in rat hypothalamus: distribution and numbers of cells containing neurophysin, oxytocin, and vasopressin. J. Comp. Neurol. **198:** 45–64.

272. RIGTER, H., R. B. MESSING, B. J. VASQUEZ, R. A. JENSEN, J. L. MARTINEZ, JR., J. C. CRABBE, JR. & J. L. MCGAUGH. 1979. Regional analysis of brain opiate receptors in rats with hereditary hypothalamic diabetes insipidus. Life Sci. **25:** 1137–1144.

273. ROBINSON, A. G. & A. G. FRANTZ. 1973. Radioimmunoassay of posterior pituitary peptides: A review. Metabolism **22:** 1047–1057.

274. ROBINSON, A. G. 1978. Neurophysins, an aid to understanding the structure and function of the neurohypophysis. *In* Frontiers in Neuroendocrinology. W. F. Ganong & L. Martini, Eds. Vol. 5: 35–59. Raven Press. New York, N.Y.

275. ROBINSON, R. R. & V. W. DENNIS. 1980. Spontaneously occurring animal models of human kidney diseases and altered renal function. Adv. Nephrol. **9:** 315–366.

276. ROLLS, B. J. 1970. Drinking by rats after irritative lesions in the hypothalamus. Physiol. Behav. **5:** 1385–1393.

277. ROSENBLOOM, A. A. & D. A. FISHER. 1975. Radioimmunoassayable AVT and AVP in adult mammalian brain tissue: comparison of normal and Brattleboro rats. Neuroendocrinology **17:** 354–361.

278. ROSIER, J., E. BATTENBERG, Q. PITTMAN, A. BAYON, L. KODA, N. MILLER, R. GUILLEMIN & F. BLOOM. 1979. Hypothalamic enkephalin neurons may regulate the neurohypophysis. Nature **277:** 653–655.

279. RUSSELL, J. T., M. J. BROWNSTEIN & H. GAINER. 1980. Biosynthesis of vasopressin, oxytocin, and neurophysins: isolation and characterization of two common precursors (propressophysin and prooxyphysin). Endocrinology **107:** 1880–1891.

280. RUSSELL, J. T., M. J. BROWNSTEIN & H. GAINER. 1980. [^{35}S]cysteine labeled peptides transported to the neurohypophyses of adrenalectomized, lactating, and Brattleboro rats. Brain Res. **201:** 227–234.

281. SAITO, T. & A. ARIMURA. 1968. Pituitary-adrenal function in rats with hereditary diabetes insipidus. Saishin. Igaku. **23:** 822–823.

282. SAITO, T., Y. YAJIMA & T. WATANABE. 1981. Involvement of vasopressin in the development and the maintenance of hypertension in rats. *In* Antidiuretic Hormone. S. Yoshida, L. Share & K. Yagi, Eds.: 215–225. University Park Press. Baltimore, Md.

283. SAUL, G. B., E. B. GARRITY, K. BENIRSCHKE & H. VALTIN. 1968. Inherited hypothalamic diabetes insipidus in the Brattleboro strain of rats. J. Hered. **59:** 113–117.

284. SAWANO, S., A. ARIMURA & S. ITO. 1969. Growth hormone secretion stimulating hormone. Clin. Endocrinol. (Tokyo) **17:** 977–984.

285. SAWYER, W. H., H. VALTIN & H. W. SOKOL. 1964. Neurohypophysial principles in rats with familial hypothalamic diabetes insipidus (Brattleboro strain). Endocrinology **74:** 153–155.

286. SAWYER, W. H. & H. VALTIN. 1965. Inhibition of vasopressin antidiuresis by extracts of pituitaries from rats with hereditary hypothalamic diabetes insipidus and oxytocin. Endocrinology **76:** 999–1001.

287. SAWYER, W. H. & H. VALTIN. 1967. Antidiuretic responses of rats with hereditary hypothalamic diabetes insipidus to vasopressin, oxytocin and nicotine. Endocrinology **80:** 207–210.

288. SAWYER, W. H., M. ACOSTA, L. BALASPIRI, J. JUDD & M. MANNING. 1974. Structural changes in the arginine vasopressin molecule that enhance antidiuretic activity and specificity. Endocrinology **94:** 1106–1115.

289. SAWYER, W. H., M. ACOSTA & M. MANNING. 1974. Structural changes in the arginine vasopressin molecule that prolong its antidiuretic action. Endocrinology **95:** 140–149.

290. SAWYER, W. H., M. ACOSTA, L. BALASPIRI, J. JUDD & M. M. MANNING. 1974. Structural changes in the arginine vasopressin molecule that enhance antidiuretic activity and specificity. Endocrinology **94:** 1106–1115.

291. SCHLEIFFER, R., B. KOCH & C. MIALHE. 1976. Depletion in ACTH content of the posterior pituitary in vasopressin deficient Brattleboro rats after adrenalectomy. Horm. Metab. Res. **8:** 495.

292. SCHLEIFFER, R., C. MIALHE, B. BRIAUD, B. LUTZ-BUCHER & B. KOCH. 1979. Effects of adrenalectomy and hypercorticism on the ACTH content of the anterior and posterior pituitary in rats with inherited diabetes insipidus (Brattleboro strain). Horm. Metab. Res. **11:** 130–135.

293. SCHNERMANN, J., M. WAHL, G. LIEBAU & H. FISCHBACH. 1968. Balance between tubular flow rate and net fluid reabsorption in the proximal convolution of the rat kidney. I. Dependency of reabsorptive net fluid flux upon proximal tubular surface area at spontaneous variations of filtration rate. Pfluegers Arch. **304:** 90–103.

294. SCHNERMANN, J., H. VALTIN, K. THURAU, W. NAGEL, M. HORSTER, H. FISCHBACH, M. WAHL & G. LIEBAU. 1969. Micropuncture studies on the influence of antidiuretic hormone on tubular fluid reabsorption in rats with hereditary hypothalamic diabetes insipidus. Pfluegers Arch. **306:** 103–118.

295. SCHÖLER, J. & J. R. SLADEK, JR. 1981. Supraoptic nucleus of the Brattleboro rat has an altered afferent noradrenergic input. Science **214:** 347–349.

296. SCHRIER, R. W., T. BERL, R. J. ANDERSON & K. M. MCDONALD. 1977. Nonosmolar control of renal water excretion. *In* Disturbances in Body Fluid Osmolality. T. E. Andreoli, J. J. Grantham & F. C. Rector, Eds.: 149–178. American Physiological Society. Bethesda, Md.

297. SCHULTZ, G., K. H. JAKOBS, E. BOHME & K. SCHULTZ. 1972. Influence of various hormones on formation of adenosine-3′:5′-monophosphate and guanosine-3′:5′-monophosphate by particulate preparations from rat kidney. Eur. J. Biochem. **24:** 520–529.

298. SCHWABEDAL, P. E., R. BOCK, W. B. WATKINS & J. MÖHRING. 1977. Influence of adrenalectomy on "Gomori positive" substances in the hypothalamo-neurohypophysial system of rats heterozygous and homozygous for hypothalamic diabetes insipidus. Anat. Embryol. **151:** 81–89.

299. SCOTT, D. E. 1968. Fine structural features of the neural lobe of the hypophysis of the rat with homozgous diabetes insipidus (Brattleboro rat strain). Neuroendocrinology **3:** 156–176.

300. SEALEY, J. E., J. D. KIRSHMAN & J. H. LARAGH. 1969. Natriuretic activity in plasma and urine of salt loaded man and sheep. J. Clin. Invest. **48:** 2210–2224.

301. SEALEY, J. E. & J. H. LARAGH. 1971. Further studies of a natriuretic substance occurring in human urine and plasma. Circ. Res. **2:** 32–43.

302. SENFT, G., K. MUNSKA, G. SCHULTZ & M. HOFFMANN. 1968. Influence of hydrochlorothiazide and other sulfamoyl diuretics on the activity of 3′,5′-AMP phosphodiesterase in rat kidney. Naunyn. Schmiedebergs Arch. Pharmacol. **259:** 344–359.

303. Shirley, D. G., S. J. Walter & J. F. Laycock. 1978. The role of sodium depletion in hydrochlorothiazide induced antidiuresis in Brattleboro rats with diabetes insipidus. Clin. Sci. Mol. Med. **54:** 209–215.

304. Shirley, D. G. & J. Skinner. 1978. Acute compensatory adaptation of renal function following contralateral kidney exclusion in Brattleboro rats with diabetes insipidus. J. Physiol. **283:** 425–438.

305. Skopkova, J., O. Schuck & J. H. Cort. 1979. Oral urine recycling in diabetes insipidus rats. The role of the gut in absorption of some urinary components. Physiol. Bohemoslov. **28:** 201–208.

306. Sladek, J. R., H. Khachaturian, G. E. Hoffman & J. Schöler. 1980. Aging of central endocrine neurons and their aminergic afferents. Peptides **1** (Suppl. 1): S141–S157.

307. Sloper, J. C. & F. Grainger. 1975. Quantitation of microtubules in secretory neurons. *In* Microtubules and Microtubule Inhibitors. M. Borgers & M. De Brabander, Eds.: 281–287. (Proc. Int. Symp. 1975, Beerse, Belgium). North-Holland Publishing Co. Amsterdam.

308. Sofroniew, M. V. & A. Weindl. 1978. Projections of the parvocellular vasopressin- and neurophysin-containing neurons of the suprachiasmatic nucleus. Amer. J. Pathol. **153:** 391–430.

309. Sokol, H. W. & H. Valtin. 1965. Morphology of the neurosecretory system in rats homozygous and heterozygous for hypothalamic diabetes insipidus (Brattleboro Strain). Endocrinology **77:** 692–700.

310. Sokol, H. W. & H. Valtin. 1967. Evidence for the synthesis of oxytocin and vasopressin in separate neurons. Nature **214:** 314–316.

311. Sokol, H. W. 1970. Evidence of oxytocin synthesis after electrolytic destruction of the paraventricular nucleus in rats with hereditary hypothalamic diabetes insipidus. Neuroendocrinology **6:** 90–97.

312. Sokol, H. W. & J. Sise. 1973. The effect of exogenous vasopressin and growth hormone on the growth of rats with hereditary hypothalamic diabetes insipidus. Growth **37:** 127–142.

313. Sokol, H. W., E. A. Zimmerman, W. H. Sawyer & A. G. Robinson. 1976. The hypothalamic-neurohypophysial system of the rat: Localization and quantitation of neurophysin by light microscopic immunocytochemistry in normal rats and in Brattleboro rats deficient in vasopressin and a neurophysin. Endocrinology **98:** 1176–1188.

314. Sokol, H. W. 1979. Diabetes insipidus. *In* Spontaneous Animal Models of Human Disease. B. C. Ward & N. H. Altman, Eds. Vol. **2:** 293–295. Academic Press. New York, N.Y.

315. Stamoutsos, B. A., R. G. Carpenter & S. P. Grossman. 1981. Role of angiotensin-II in the polydipsia of diabetes insipidus in the Brattleboro rat. Physiol. Behav. **26:** 691–693.

316. Stavroulaki-Tsapara, A., D. Haley, G. Eknoyan & M. Martinez-Maldonado. 1974. Changes in electrolyte excretion following intraperitoneal injection of dibutyryl cyclic AMP in the Brattleboro rat. Life Sci. **14:** 2031–2036.

317. Stephan, F. K. & I. Zucker. 1974. Endocrine and neural mediation of the effects of constant light on water intake of rats. Neuroendocrinology **14:** 44–60.

318. Stern, P. & H. Valtin. 1978. Lack of clear-cut antidiuretic effect of 8-*p*-chlorophenylthio-cyclic AMP. Min. Elect. Metab. **1:** 330–335.

319. Stillman, M. A., L. D. Recht, S. L. Rosario, S. M. Seif, A. G. Robinson & E. A. Zimmerman. 1977. The effects of adrenalectomy and glucocorticoid replacement on vasopressin and vasopressin-neurophysin in the zona externa of the median eminence of the rat. Endocrinology **101:** 42–49.

320. Stoff, J. S., R. M. Rosa, P. Silva & F. H. Epstein. 1981. Indomethacin impairs water diuresis in the DI rat: role of prostaglandins independent of ADH. Am. J. Physiol. **241:** F231–F237.

321. SUESS, U. & V. PLIŠKA. 1981. Identification of the pituicytes as astroglial cells by indirect immunofluorescence-staining for the glial fibrillary acidic protein. Brain Res. **221:** 27–33.
322. SUN, C. N., H. J. WHITE & E. J. TOWBIN. 1972. Histochemistry and electron microscopy of the renal papilla in a genetic strain of rats with diabetes insipidus. Nephron **9:** 308–317.
323. SUN, C. N., H. J. WHITE & E. J. TOWBIN. 1972. Crystalloid inclusions in mitochondria of renal tubules. Exp. Pathol. **7:** 166–171.
324. SUNDE, D. A. & H. W. SOKOL. 1975. Quantification of rat neurophysins by polyacrylamide gel electrophoresis (PAGE): Application to the rat with hereditary hypothalamic diabetes insipidus. Ann. N.Y. Acad. Sci. **248:** 345–364.
325. SWAAB, D. F. & C. W. POOL. 1975. Specificity of oxytocin and vasopressin immunofluorescence. J. Endocrinol. **66:** 263–272.
326. SWAAB, D. F., C. W. POOL & F. NIJVELDT. 1975. Immunofluorescence of vasopressin and oxytocin in the rat hypothalamo-neurohypophyseal system. J. Neural. Transm. **36:** 195–215.
327. SWAAB, D. F., K. BOER & W. J. HONNEBIER. 1977. The influence of the fetal hypothalamus and pituitary on the onset and course of parturition. CIBA Found. Symp. **47:** 379–400.
328. SWAAB, D. F., C. W. POOL & F. W. VAN LEEUWEN. 1977. Can specificity ever be proved in immunocytochemical staining? J. Histochem. Cytochem. **25:** 388–390.
329. SWANSON, L. W., P. E. SWACHENKO, A. BEROD, B. K. HARTMAN, K. B. HELLE & D. E. VANORDEN. 1981. An immunohistochemical study of the organization of catecholaminergic cells and terminal fields in the paraventricular and supraoptic nuclei of the hypothalamus. J. Comp. Neurol. **196:** 271–285.
330. TANKOSIC, P., M. CHATEAU & A. BURLET. 1980. Postnatal changes in the content of neurohypophyseal enzymes in the Brattleboro rat. C. R. Soc. Biol. **174:** 21–27.
331. TASSO, F., D. PICARD & J. L. DREIFUSS. 1976. Ultrastructural identification of granules containing oxytocin and vasopressin. Nature **260:** 621–622.
332. TASSO, F., S. RUA & D. PICARD. 1977. Cytochemical duality of neurosecretory material in the hypothalamo-posthypophysial system of the rat as related to hormonal content. Cell Tissue Res. **180:** 11–29.
333. TASSO, F. & S. RUA. 1978. Ultrastructural observations on the hypothalamo-posthypophysial complex of the Brattleboro rat. Cell Tissue Res. **191:** 267–286.
334. TELEGDY, G. & G. L. KOVÁCS. 1979. Role of monoamines in mediating the action of ACTH, vasopressin and oxytocin. *In* Central Nervous System Effects of Hypothalamic Hormones and Other Peptides. R. Collu, A. Barbeau, J. R. Ducharme & J. G. Rochefort, Eds.: 189–205. Raven Press. New York, N.Y.
335. THOMSEN, K. & M. SCHOU. 1973. The effect of prolonged administration of hydrochlorothiazide on the renal lithium clearance and the urine flow of ordinary rats and rats with diabetes insipidus. Pharmakopsychiatr. Neurosynchopharmakol. **6:** 264–269.
336. THOMSEN, K. & O. V. OLESEN. 1974. Long-term lithium administration to rats. Lithium and sodium dosage and administration, avoidance of intoxication in polyuric and control rats. Int. Pharmacopsychiatry **9:** 118–124.
337. THOMSEN, K. 1977. The renal handling of lithium: relation between lithium clearance, sodium clearance and urine flow in rats with diabetes insipidus. Acta Pharmacol. Toxicol. **40:** 491–496.
338. TISHER, C. C., R. E. BULGER & H. VALTIN. 1971. Morphology of renal medulla in water diuresis and vasopressin-induced antidiuresis. Am. J. Physiol. **220:** 87–94.

339. Tisher, C. C. & W. E. Yarger. 1975. Lanthanum permeability of tight junctions along the collecting duct of the rat. Kidney Int. **7:** 35–44.
340. Trinh-Trang-Tan, M. M., M. Diaz, J. P. Grünfeld & L. Bankir. 1981. ADH-dependent nephron heterogeneity in rats with hereditary diabetes insipidus. Am. J. Physiol. **240:** F373–F380.
341. Trinh-Trang-Tan, M. M., L. Bankir, J. P. Grünfeld, M. Diaz & J. L. Funck-Brentano. 1980. Glomerular extraction of antiglomerular basement membrane antibody in normal Wistar and in Brattleboro rats with hereditary diabetes insipidus. *In* Radionuclides in Nephrology. N. K. Hollenberg & S. Lange, Eds.: 129–133. Thieme. Stuttgart.
342. Uhlich, E., J. Eigler, G. Tryzna, K. Finke, W. Winkelmann & E. Buchborn. 1971. The mode of action of chlorpropamide in diabetes insipidus. Klin. Wochenschr. **49:** 314–322.
343. Uhlich, E., K. Loeschke & J. Eigler. 1972. Antidiuretic effect of carbamazepine in diabetes insipidus. Klin. Wochenschr. **50:** 1127–1133.
344. Urban, I. & D. De Wied. 1975. Inferior quality of rhythmic slow activity during paradoxical sleep in rats with hereditary diabetes insipidus. Brain Res. **97:** 362–366.
345. Urban, I. & D. De Wied. 1978. Neuropeptides: effects on paradoxical sleep and theta rhythm in rats. Pharmacol. Biochem. Behav. **8:** 51–59.
346. Valtin, H., H. A. Schroeder, K. Benirschke & H. W. Sokol. 1962. Familial hypothalamic diabetes insipidus in rats. Nature **196:** 1109–1110.
347. Valtin, H. & H. A. Schroeder. 1964. Familial hypothalamic diabetes insipidus in rats (Brattleboro strain). Am. J. Physiol. **206:** 425–430.
348. Valtin, H., W. H. Sawyer & H. W. Sokol. 1965. Neurohypophysial principles in rats homozygous and heterozygous for hypothalamic diabetes insipidus (Brattleboro Strain). Endocrinology **77:** 701–706.
349. Valtin, H. 1966. Sequestration of urea and nonurea solutes in renal tissues of rats with hereditary hypothalamic diabetes insipidus: effect of vasopressin and dehydration on the countercurrent mechanism. J. Clin. Invest. **45:** 337–345.
350. Valtin, H. 1967. Hereditary hypothalamic diabetes insipidus in rats (Brattleboro strain). A useful experimental model. Am. J. Med. **42:** 814–827.
351. Valtin, H. 1969. Hereditary diabetes insipidus. Lessons learned from animal models. *In* Progress in Endocrinology. C. Gual, Ed.: 321–327. Excerpta Medica Foundation. Amsterdam.
352. Valtin, H. & A. R. Harrington. 1970. Accumulation of medullary urea during prolonged treatment of hypothalamic diabetes insipidus with vasopressin (ADH). *In* Urea and the Kidney. B. Schmidt-Nielsen, Ed.: 430–442. (Int. Congr. Ser. Vol. 195.) Excerpta Medica Foundation. Amsterdam.
353. Valtin, H. 1974. Genetic models in biomedical investigation. N. Engl. J. Med. **290:** 670–675.
354. Valtin, H., J. Stewart & H. W. Sokol. 1974. Genetic control of the production of posterior pituitary principles. *In* Endocrinology: The Pituitary Gland and Its Neuroendocrine Control. Handbook of Physiology. E. Knobil & W. H. Sawyer, Eds. Section 7, Vol. 4, Part 1, pp. 131–172. Washington, D.C.
355. Valtin, H., H. W. Sokol & D. Sunde. 1975. Genetic approaches to the study of the regulation and actions of vasopressin. Rec. Prog. Horm. Res. **31:** 447–486.
356. Valtin, H. 1976. Animal model of human disease: hereditary hypothalamic diabetes insipidus. Am. J. Pathol. **83:** 633–636.
357. Valtin, H. 1977. Genetic models for hypothalamic and nephrogenic diabetes insipidus. *In* Disturbances in Body Fluid Osmolality. T. E. Andreoli, J. J. Grantham & F. C. Rector, Jr., Eds.: 197–215. American Physiological Society. Bethesda, Md.

358. VALTIN, H., M. GELLAI & B. R. EDWARDS. 1981. Concentration of urine in the absence of vasopressin: Effect of decreased renal perfusion pressure in conscious Brattleboro homozygotes. *In* Kidney and Body Fluids. L. Takács, Ed. Vol. **2:** 575–579. Adv. Physiol. Sci. Akadémiai Kiádó. Budapest.

359. VAN LEEUWEN, F. 1980. Immunocytochemical specificity for peptides with special reference to arginine-vasopressin and oxytocin. J. Histochem. Cytochem. **28:** 479–482.

360. VAN LEEUWEN, F. W., D. F. SWAAB & H. J. ROMIJN. 1976. Light and electron microscopic immunolocalization of oxytocin and vasopressin in rats. *In* Immunoenzymatic Techniques. G. Feldmann *et al.*, Eds.: 345–353. North-Holland. Amsterdam.

361. VAN LEEUWEN, F. W. & D. F. SWAAB. 1977. Specific immunoelectronmicroscopic localization of vasopressin and oxytocin in the neurohypophysis of the rat. Cell Tissue Res. **177:** 493–501.

362. VAN LEEUWEN, F. W., C. DE RAAY, D. F. SWAAB & B. FISSER. 1979. The localization of oxytocin, vasopressin, somatostatin and luteinizing hormone releasing hormone in the rat neurohypophysis. Cell Tissue Res. **202:** 189–201.

363. VAN WIMERSMA GREIDANUS, T. B., R. M. BUYS, H. J. HOLLEMANS & W. DE JONG. 1974. A radioimmunoassay of vasopressin. A note on pituitary vasopressin content in Brattleboro rats. Experientia **30:** 1217–1218.

364. VAN WIMERSMA GREIDANUS, T. B., B. BOHUS & D. DE WIED. 1975. The role of vasopressin in memory processes. *In* Hormones, Homeostasis and Brain. W. H. Gispen, T. B. Van Wimersma Greidanus, B. Bohus & D. De Wied, Eds. Prog. Brain Res. **42:** 135–141. Elsevier. Amsterdam.

365. VAN WIMERSMA GREIDANUS, T. B. & D. DE WIED. 1980. Physiological significance of neurohypophyseal hormones in memory processes. Acta Psychiatr. Belg. **80:** 721–727.

366. VANDESANDE, F., K. DIERICKX & J. DE MEY. 1975. Immunohistochemical demonstration of the oxytocin- and the vasopressin-producing neurons in the magnocellular hypothalamic neurosecretory system of the cow, the rat and the Brattleboro rat. Ann. Endocrinol. **36:** 379–380.

367. VANDESANDE, F. & K. DIERICKX. 1976. Immuno-cytochemical demonstration of the inability of the homozygous Brattleboro rat to synthesize vasopressin and vasopressin-associated neurophysin. Cell Tissue Res. **165:** 307–316.

368. VERSTEEG, D. H., M. TANAKA & E. R. DE KLOET. 1978. Catecholamine concentration and turnover in discrete regions of the brain of the homozygous Brattleboro rat deficient in vasopressin. Endocrinology **103:** 1654–1661.

369. VIERLING, A. F., J. B. LITTLE & E. P. RADFORD, JR. 1967. Antidiuretic hormone bio-assay in rats with hereditary hypothalamic diabetes insipidus (Brattleboro strain). Endocrinology **80:** 211–214.

370. VINSON, G. P., C. GODDARD & B. J. WHITEHOUSE. 1978. Corticosteroid production in vitro by adrenal tissue from rats with inherited hypothalamic diabetes insipidus (Brattleboro Strain). J. Steroid Biochem. **9:** 657–665.

371. WAHL, M., G. LIEBAU, H. FISCHBACH & J. SCHNERMANN. 1968. Balance between tubular flow rate and net reabsorption in the proximal convolution of the rat kidney. II. Reabsorptive characteristics during constriction of the renal artery. Pfluegers Arch. **304:** 297–314.

372. WALKER, L. & J. C. FRÖHLICH. 1980. Differential effects of deamino-8-D-arginine vasopressin on urinary prostaglandins E_2 and $F_{2\alpha}$ excretion. *In* Advances in Prostaglandin and Thromboxane Research. B. Samuelsson, P. W. Ramwell & R. Paoletti, Eds. Vol. **7:** 1107–1110. Raven Press. New York, N.Y.

373. WALKER, L. A., A. R. WHORTON, M. SMIGEL, R. FRANCE & J. C. FRÖLICH. 1978. Antidiuretic hormone increases renal prostaglandin synthesis in vivo. Am. J. Physiol. **235:** F180–F185.

374. Walker, L. A. & J. C. Frölich. 1981. Dose-dependent stimulation of renal prostaglandin synthesis by deamino-8-D-arginine vasopressin in rats with hereditary diabetes insipidus. J. Pharmacol. Exp. Ther. **217:** 87–91.

375. Wallin, J. D. & P. A. Lee. 1976. Effect of prolactin on diluting and concentrating ability in the rat. Am. J. Physiol. **230:** 1524–1530.

376. Wallin, J. D. & R. A. Kaplan. 1977. Effect of sodium fluoride on concentrating and diluting ability in the rat. Am. J. Physiol. **232:** F335–F340.

377. Walter, S. J., J. F. Laycock & D. G. Shirley. 1979. A micropuncture study of proximal tubular function after acute hydrochlorothiazide administration to Brattleboro rats with diabetes insipidus. Clin. Sci. **57:** 427–434.

378. Watkins, W. B. 1975. Presence of neurophysin and vasopressin in the hypothalamic magnocellular nuclei of rats homozygous and heterozygous for diabetes insipidus (Brattleboro strain) as revealed by immunoperoxidase histology. Cell Tissue Res. **157:** 101–113.

379. Watkins, W. B., S. S. Yen & R. Y. Moore. 1981. Presence of beta-endorphin-like immunoreactivity in the anterior pituitary gland of rat and man and evidence of the differential localization with ACTH. Cell Tissue Res. **215:** 577–589.

380. Wiley, M. K., A. F. Pearlmutter & R. E. Miller. 1974. Decreased adrenal sensitivity to ACTH in the vasopressin-deficient (Brattleboro) rat. Neuroendocrinology **14:** 257–270.

381. Wilson, D. R., G. Thiel, M. L. Arce & D. E. Oken. 1969. The role of the concentration mechanism in the development of acute renal failure: micropuncture studies using diabetes insipidus rats. Nephron **6:** 128–139.

382. Windhager, E. E. & L. De Bermudez. 1976. Effect of osmotic gradients on distal tubular electrical resistance. F. Giovannetti, *et al.,* Eds.: 113–117. *In* Sixth International Congress of Nephrology. Karger. Basel.

383. Woodhall, P. B. & C. C. Tisher. 1973. Response of the distal tubule and cortical collecting duct to vasopressin in the rat. J. Clin. Invest. **52:** 3095–3108.

384. Wooten, G., T. Hanson & F. Lamprecht. 1975. Elevated serum dopamine-beta-hydroxylase activity in rats with inherited diabetes insipidus. J. Neural Transm. **36:** 107–112.

385. Wright, W. A. & C. L. Kutscher. 1977. Vasopressin administration in the first month of life: effects on growth and water metabolism in hypothalamic diabetes insipidus rats. Pharmacol. Biochem. Behav. **6:** 505–509.

386. Wu, W. H. & V. Zbuzkova. 1979. Vasopressin. S. B. Day, Ed.: A companion to the life sciences, Vol. 1: 431–434. Van Nostrand Reinhold Company. New York, N.Y.

387. Yates, F. E., S. M. Russell, M. F. Dallman, G. A. Hedge, S. M. McCann & A. S. Dhariwal. 1971. Potentiation by vasopressin of corticotropin release induced by corticotropin-releasing factor. Endocrinology **88:** 3–15.

388. Zicha, J., H. Dlouhá & J. Křeček. 1979. Blood pressure, water and electrolyte intake and excretion in Brattleboro rats and the effect of unilateral nephrectomy. *In* Hormones and Development. L. Macho & V. Štrbák, Eds.: 489–501 (Proc. Int. Symp. on Hormones and Development, 1976, Bratislava.) Veda. Bratislava, Czech.

389. Zimmerman, E. A., R. Defendini, H. W. Sokol & A. G. Robinson. 1975. The distribution of neurophysin-secreting pathways in the mammalian brain. Light microscopic studies using the immunoperoxidase technique. Ann. N.Y. Acad. Sci. **248:** 92–111.

390. Zimmerman, E. A. 1976. Distribution of the neurophysin vasopressin system in the mammalian hypothalamus. J. Med. Sci. **12:** 994–1003.

Index of Contributors

(Italic page numbers refer to comments made in discussion.)

Subject Index